Thanks for
Honeywell Partnership
with PRC

5/99

Microelectronics Packaging Handbook

Microelectronics Packaging Handbook

Semiconductor Packaging

Part II

Second Edition

Edited by
Rao R. Tummala
Georgia Institute of Technology
Eugene J. Rymaszewski
Rensselaer Polytechnic Institute
Alan G. Klopfenstein
AGK Enterprises

KLUWER ACADEMIC PUBLISHERS
BOSTON/DORDRECHT/LONDON

Distributors for North, Central and South America:
Kluwer Academic Publishers
101 Philip Drive
Assinippi Park
Norwell, Massachusetts 02061 USA
Telephone (781) 871-6600
Fax (781) 871-6528
E-Mail <kluwer@wkap.com>

Distributors for all other countries:
Kluwer Academic Publishers Group
Distribution Centre
Post Office Box 322
3300 AH Dordrecht, THE NETHERLANDS
Telephone 31 78 6392 392
Fax 31 78 6546 474
E-Mail <orderdept@wkap.nl>

Electronic Services <http://www.wkap.nl>

Library of Congress Cataloging-in-Publication

Microelectronics packaging handbook / edited by Rao R. Tummala,
Eugene J. Rymaszewski, Alan G. Klopfenstein.
p. cm.
Includes bibliographical references and index. Contents: pt. 1. Technology drivers -- pt. 2. Semiconductor drivers -- pt. 3. Subsystem packaging.
ISBN 0-412-08431-7 (pt. 1) -- 0-412-08441-4 (pt. 2) -- 0-412-08451-1 (pt. 3)
1. Microelectronic packaging--Handbooks, manuals, etc. 1. Tummala, Rao R., 1942 --.
TK7874.M485 1996 96-37907
621.381'046--dc 21 CIP

Printed on acid-free paper.

Printed in the United States of America

FOREWORD

Electronics has become the largest industry, surpassing agriculture, auto, and heavy metal industries. It has become the industry of choice for a country to prosper, already having given rise to the phenomenal prosperity of Japan, Korea, Singapore, Hong Kong, and Ireland among others. At the current growth rate, total worldwide semiconductor sales will reach $300B by the year 2000.

The key electronic technologies responsible for the growth of the industry include semiconductors, the packaging of semiconductors for systems use in auto, telecom, computer, consumer, aerospace, and medical industries, displays, magnetic, and optical storage as well as software and system technologies. There has been a paradigm shift, however, in these technologies, from mainframe and supercomputer applications at any cost, to consumer applications at approximately one-tenth the cost and size. Personal computers are a good example, going from $500/MIP when products were first introduced in 1981, to a projected $1/MIP within 10 years. Thin, light portable, user friendly and very low-cost are, therefore, the attributes of tomorrow's computing and communications systems.

Electronic packaging is defined as interconnection, powering, cooling, and protecting semiconductor chips for reliable systems. It is a key enabling technology achieving the requirements for reducing the size and cost at the system and product level.

The recent paradigm shifts in packaging processes such as direct flip-chip attach to organic board, the ability to increase wiring in the organic board away from expensive drilled technology by the direct deposition of thin organic and metal films, the achievement of reliability without hermeticity previously achieved only in ceramic packaging, and the potential integration of all passive components within the interconnect board are expected to lead to revolutionary products in all segments of electronics.

To describe the status and future developments in these technologies, editors Profs. Tummala and Rymaszewski and Mr. Klopfenstein have assembled an outstanding team of 74 packaging practitioners from across the globe. Together, they produced in three parts the *Microelectronics*

Packaging Handbook, an unmatched book needed by industry and universities. It is equally appropriate both as an introduction to those just entering the field, and as an up-to-date reference for those already engaged in packaging design, first and second level packages and their interconnections, test, assembly, thermal management, optoelectronics, reliability, and manufacturing.

I applaud the authors and editors for their great contributions and hope that this book will help focus the attention of outstanding faculty, students, and industry researchers on the complex packaging challenges that they face in the 21st century.

Bertrand Cambou
Motorola Senior Vice-President
Sector Technology
Semiconductor Productor Sector

PREFACE

"This book reflects a need based on an ever-increasing realization by the microelectronics community that while the semiconductors continue to be improved upon relentlessly for performance, cost, and reliability, it is the packaging of these microelectronic devices that may limit the systems performance. In response to this need, the academic community began to ask its industrial counterparts what packaging is all about, and what scientific and technological areas should be pursued in collaboration with industry. As a result of these discussions, a number of universities in the United States and throughout the world already have research and/or teaching programs in various aspects of packaging, ranging from materials, assembly, electrical, and thermal modeling, thin films, and many others. The multidisciplinary nature of packaging technology clearly poses challenges not only to this community, but also to industrial colleagues who will have to use scientific fundamentals from a cross section of disciplines to bring about advanced products." That's the way we began the preface of the 1st edition of the *Microelectronics Packaging Handbook* in 1989. The first sentence is true today; however, two additional factors have become more important: cost and size. This is readily apparent when one considers consumer electronics. The use of packaged microelectronics has increased tremendously in a great variety of applications—television, computers, communications, navigation, automotive, avionics, medical, and so forth. Perhaps the biggest change is the degree to which universities have embraced the field of microelectronics and established major research and teaching programs. Also, the growth in coverage of microelectronics packaging by technical societies—both in technical symposia and in tutorial sessions—has been dramatic.

While the book is not intended to be a classic textbook for any course in packaging, it does attempt to define what microelectronics packaging is all about, including the current state of the technology across all engineering and scientific disciplines, and the fundamental areas that could impact the industrial needs. It also provides a comprehensive source of information on all aspects of microelectronics packaging.

Packaging of electronic circuits is the science and art of establishing interconnections and a suitable operating environment for primarily electronic circuits to process and store information. Accordingly, microelectronics packaging in this book is assumed to mean those designs and interconnection technologies necessary to support electrically, optically, thermally, mechanically, and chemically those semiconductor devices with micron and submicron dimensions that are often referred to as integrated circuits. Electronic components and their packaging are the building blocks for a vast variety of equipment. Customer demand and competition among equipment suppliers result in continuing enhancements and evolution of these building blocks, particularly in terms of cost, performance, quality, and reliability.

Since publishing the first edition of this book, a number of paradigm shifts have taken place in the electronics industry. For one, the emphasis on systems has changed from mainframe computers to personal computers and portable, wireless systems requiring ultra low-cost, thin, light, and portable packages. Second, the semiconductor technologies have shifted emphasis from bipolar to CMOS. A number of paradigm shifts have taken place in packaging as well. Direct chip attach was possible in 1989 only to such inorganic substrates as alumina and silicon. Today, chips can be bonded and used directly on printed wiring boards. The board itself is undergoing major change, allowing very high density wiring by photolithograpic processes in contrast to mechanical drilling, which became expensive and obsolete. The very high reliability that was achieved previously only with ceramic packages is now beginning to be achieved with organic packages and boards without hermeticity. This book is written with these changes in mind and consistent with the needs of the industry.

This collection of books is organized into three parts. Part 1 includes chapters 1 through 6 and covers the driving forces of microelectronics packaging—electrical, thermal, and reliability. In addition, Part 1 introduces the technology developer to aspects of manufacturing that must be considered during product development. Part 2, chapters 7 through 14, covers the interconnection of the IC chip to the first level of packaging and all first level packages. Part 2 also includes electrical test as well as sealing and encapsulation technologies. Part 3, chapters 15 through 20, covers board level packaging as well as connectors, cables, and optical packaging. The general overview of packaging is repeated in each part as chapters 1, 7, and 15.

The main problem of creating a handbook in any field is obvious: by the time the book is completed, much of the information in it may require updating. This is particularly true with microelectronics packaging, which is one of the fastest, if not the fastest, growing of all technologies. The first edition was a beginning and this 2nd edition is long overdue. Further development and refinement, based on the comments and suggestions of the users, can appear in future editions. The book in its present state reflects what we believe the community wants currently. We are exploring new ways of providing the information on a more current basis including electronic publishing and interchange.

The first edition was written entirely by IBMers. We tried to include as many non-IBM technologists as possible and to minimize the use of IBM jargon. This 2nd edition includes 74 authors, each an expert in his or her own field, from many companies, universities, and countries. We believe that this second edition provides a very representative and comprehensive look at the field of microelectronics packaging.

Any handbook requires the dedication of a number of individuals involved in writing, typing, graphics preparation, manuscript reviewing, copyediting, and publishing, and managing all these operations in such a way that the final book is available in a timely manner. Above all, a free and stimulating attitude on the part of all the participants is necessary. In addition to the chapter authors, we would like to acknowledge the work of Debra Kelley in helping to keep us on track and for her efforts in preparing some of the manuscripts. It should be pointed out that extensive use of the Internet permitted us to work together more easily and cost effectively. Our thanks to Jim Geronimo and Barbara Tompkins for the preparation of numerous drafts, extensive copyediting, and their willingness to be sure that all appropriate authors had timely copies to review. Also to Kristi Bockting and the staff at WorldComp for the monumental job of incorporating all author comments into the final "camera-ready" manuscript. Our greatest thanks go to our wives, Anne Tummala and Jean Rymaszewski, and MaryAnn Klopfenstein for their patience and full support. We thank Bertrand Cambou, Motorola Senior Vice-President for the insightful Foreword.

Rao R. Tummala
Eugene J. Rymaszewski
Alan G. Klopfenstein

Part 2

TABLE OF CONTENTS*

*See Summary of Contents for 3 part set, p. xxxi.

CHAPTER 11. POLYMERS IN PACKAGING II-509

CHAPTER 13. PACKAGE ELECTRICAL TESTING II-814

CHAPTER 14. PACKAGE SEALING AND ENCAPSULATION II-873

CONVERSION FACTORS

Length

$1\ m = 10^{10}$ Å	1 Å $= 10^{-10}$ m
$1\ m = 10^{9}$ nm	$1\ nm = 10^{-9}$ m
$1\ m = 10^{6}\ \mu m$	$1\ \mu m = 10^{-6}$ m
$1\ m = 10^{3}$ mm	$1\ mm = 10^{-3}$ m
$1\ m = 10^{2}$ cm	$1\ cm = 10^{-2}$ m
1 mm = 0.0394 in.	1 in. = 25.4 mm
1 cm = 0.394 in.	1 in. = 2.54 cm
1 m = 3.28 ft	1 ft = 0.3048 m

Area

$1\ m^2 = 10^4\ cm^2$	$1\ cm^2 = 10^{-4}\ m^2$
$1\ mm^2 = 10^{-2}\ cm^2$	$1\ cm^2 = 10^2\ mm^2$
$1\ m^2 = 10.76\ ft^2$	$1\ ft^2 = 0.093\ m^2$
$1\ cm^2 = 0.1550\ in.^2$	$1\ in.^2 = 6.452\ cm^2$

Volume

$1\ m^3 = 10^6\ cm^3$	$1\ cm^3 = 10^{-6}\ m^3$
$1\ mm^3 = 10^{-3}\ cm^3$	$1\ cm^3 = 10^3\ mm^3$
$1\ m^3 = 35.32\ ft^3$	$1\ ft^3 = 0.0283\ m^3$
$1\ cm^3 = 0.0610\ in.^3$	$1\ in.^3 = 16.39\ cm^3$

Mass

$1\ Mg = 10^3$ kg	$1\ kg = 10^{-3}$ Mg
$1\ kg = 10^3$ g	$1\ g = 10^{-3}$ kg
$1\ kg = 2.205\ lb_m$	$1\ lb_m = 0.4536$ kg
$1\ g = 2.205 \times 10^{-3}\ lb_m$	$1\ lb_m = 453.6$ g

Density

$1\ kg/m^3 = 10^{-3}\ g/cm^3$	$1\ g/cm^3 = 10^3\ kg/m^3$
$1\ Mg/m^3 = 1\ g/cm^3$	$1\ g/cm^3 = 1\ Mg/m^3$
$1\ kg/m^3 = 0.0624\ lb_m/ft^3$	$1\ lb_m/ft^3 = 16.02\ kg/m^3$
$1\ g/cm^3 = 62.4\ lb_m/ft^3$	$1\ lb_m/ft^3 = 1.602 \times 10^{-2}\ g/cm^3$
$1\ g/cm^3 = 0.0361\ lb_m/in.^3$	$1\ lb_m/in.^3 = 27.7\ g/cm^3$

Unit conversion factors begin on previous page.

Force

$1\ \text{N} = 10^5$ dynes 1 dyne $= 10^{-5}$ N

$1\ \text{N} = 0.2248\ \text{lb}_f$ $1\ \text{lb}_f = 4.448$ N

Stress

1 MPa = 145 psi $1\ \text{psi} = 6.90 \times 10^{-3}$ MPa

$1\ \text{MPa} = 0.102\ \text{kg/mm}^2$ $1\ \text{kg/mm}^2 = 9.806$ MPa

$1\ \text{Pa} = 10\ \text{dynes/cm}^2$ $1\ \text{dyne/cm}^2 = 0.10$ Pa

$1\ \text{kg/mm}^2 = 1422$ psi $1\ \text{psi} = 7.03 \times 10^{-4}\ \text{kg/mm}^2$

Fracture Toughness

$1\ \text{psi}\sqrt{\text{in.}} = 1.099 \times 10^{-3}\ \text{MPa}\sqrt{\text{m}}$ $1\ \text{MPa}\sqrt{\text{m}} = 910\ \text{psi}\sqrt{\text{in.}}$

Energy

$1\ \text{J} = 10^7$ ergs $1\ \text{erg} = 10^{-7}$ J

$1\ \text{J} = 6.24 \times 10^{18}$ eV $1\ \text{eV} = 1.602 \times 10^{-19}$ J

1 J = 0.239 cal 1 cal = 4.184 J

$1\ \text{J} = 9.48 \times 10^{-4}$ Btu 1 Btu = 1054 J

$1\ \text{J} = 0.738\ \text{ft-lb}_f$ $1\ \text{ft-lb}_f = 1.356$ J

$1\ \text{eV} = 3.83 \times 10^{-20}$ cal $1\ \text{cal} = 2.61 \times 10^{19}$ eV

$1\ \text{cal} = 3.97 \times 10^{-3}$ Btu 1 Btu = 252.0 cal

Power

1 W = 0.239 cal/s 1 cal/s = 4.184 W

1 W = 3.414 Btu/h 1 Btu/h = 0.293 W

1 cal/s = 14.29 Btu/h 1 Btu/h = 0.070 cal/s

Viscosity

1 Pa-s = 10 P 1 P = 0.1 Pa-s

Temperature, *T*

$T(\text{K}) = 273 + T(°\text{C})$ $T(°\text{C}) = T(\text{K}) - 273$

$T(\text{K}) = \frac{5}{9}[T(°\text{F}) - 32] + 273$ $T(°\text{F}) = \frac{9}{5}[T(\text{K}) - 273] + 32$

$T(°\text{C}) = \frac{5}{9}[T(°\text{F}) - 32]$ $T(°\text{F}) = \frac{9}{5}[T(°\text{C})] + 32$

Specific Heat

$1\ \text{J/kg-K} = 2.39 \times 10^{-4}$ cal/g-K 1 cal/g-°C = 4184 J/kg-K

$1\ \text{J/kg-K} = 2.39 \times 10^{-4}\ \text{Btu/lb}_m\text{-°F}$ $1\ \text{Btu/lb}_m\text{-°F} = 4184$ J/kg-K

$1\ \text{cal/g-°C} = 1.0\ \text{Btu/lb}_m\text{-°F}$ $1\ \text{Btu/lb}_m\text{-°F} = 1.0$ cal/g-K

Thermal Conductivity

$1\ \text{W/m-K} = 2.39 \times 10^{-3}$ cal/cm-s-K 1 cal/cm-s-K = 418.4 W/m-K

1 W/m-K = 0.578 Btu/ft-h-°F 1 Btu/ft-h-°F = 1.730 W/m-K

1 cal/cm-s-K = 241.8 Btu/ft-h-°F $1\ \text{Btu/ft-h-°F} = 4.136 \times 10^{-3}$ cal/cm-s-K

Unit Abbreviations

A = ampere	in. = inch	N = newton
Å = angstrom	J = joule	nm = nanometer
Btu = British thermal unit	K = degrees Kelvin	P = poise
C = Coulomb	kg = kilogram	Pa = pascal
°C = degrees Celsius	lb_f = pound force	s = second
cal = calorie (gram)	lb_m = pound mass	T = temperature
cm = centimeter	m = meter	μm = micrometer (micron)
eV = electron volt	Mg = megagram	W = watt
°F = degrees Fahrenheit	mm = millimeter	psi = pounds per square inch
ft = foot	mol = mole	
g = gram	MPa = megapascal	

SI Multiple and Submultiple Prefixes

Factor by Which Multiplied	*Prefix*	*Symbol*
10^9	giga	G
10^6	mega	M
10^3	kilo	k
10^{-2}	centi[a]	c
10^{-3}	milli	m
10^{-6}	micro	μ
10^{-9}	nano	n
10^{-12}	pico	p

[a] Avoided when possible.

Parts 1, 2, and 3

SUMMARY OF CONTENTS

PART 1. MICROELECTRONICS PACKAGING HANDBOOK: TECHNOLOGY DRIVERS

Part 2

MICROELECTRONICS PACKAGING HANDBOOK: SEMICONDUCTOR PACKAGING

RAO R. TUMMALA—GEORGIA TECH
EUGENE J. RYMASZEWSKI—RENSSELAER
ALAN G. KLOPFENSTEIN—AGK ENTERPRISES

7

MICROELECTRONICS PACKAGING—AN OVERVIEW

EUGENE J. RYMASZEWSKI—*Rensselaer*
RAO R. TUMMALA—*Georgia Tech*
TOSHIHIKO WATARI—*NEC*

7.1 INTRODUCTION

Packaged electronics is the embodiment of all electronic equipment—calculators, personal computers (PCs), mainframe computers, telephones, television, and so forth. It consists of the active components, such as integrated circuit (IC) chips, flat-panel and cathode-ray-tube displays, loudspeakers, and so on. The active components are interconnected, supplied with electric power and housed in packaging. Success in a very competitive marketplace hinges on superior performance and price.

Performance is a broad term which encompasses the ability to perform a multitude of functions—word processing with embedded graphics or spreadsheet calculations, airline reservations, banking transactions, and so on—within a certain time, preferably without imposing an unacceptable

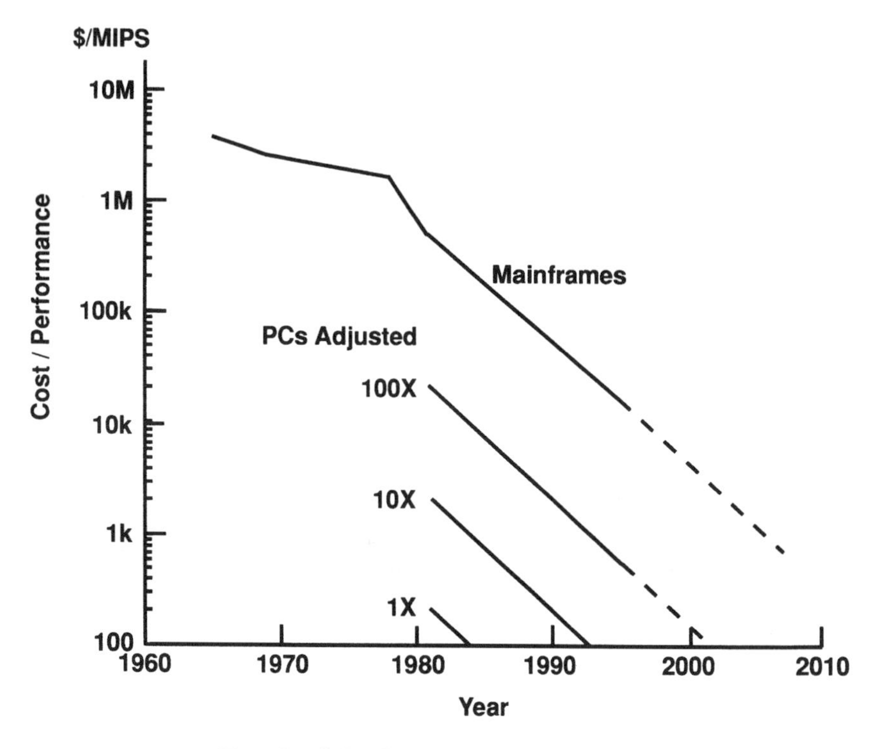

Figure 7-1. Trends of the System Price-to-Performance Ratio

wait on the users. It is task dependent, so only the general trends will be discussed throughout this book. In the last several decades, the price (in $) for performance (measured in millions instructions per second—MIPS) decreased dramatically as shown in Figure 7-1 which is derived from Ref. 1.

This gain with PC was in improving the performance at relatively constant cost and price, thus increasing their value every 2–3 years. These humble PCs, priced at a (very) few thousand dollars, easily outperform (with certain tasks) the yesteryear's mainframes which had been priced in millions of dollars. And they are much easier to use.

The mainframe improvements now (the second half of 1990s) mainly result from the cost reductions enabled by higher levels of integration on the ICs and in the packaging, thus reducing the size and number of parts, and from simplifying the packaging complexity (e.g., switch from the water to air cooling).

Figure 7-2 illustrates the interdependency between processor performance and its clock frequency. Another strong variable is the number of cycles per instruction [2] which depends on the computer design in general

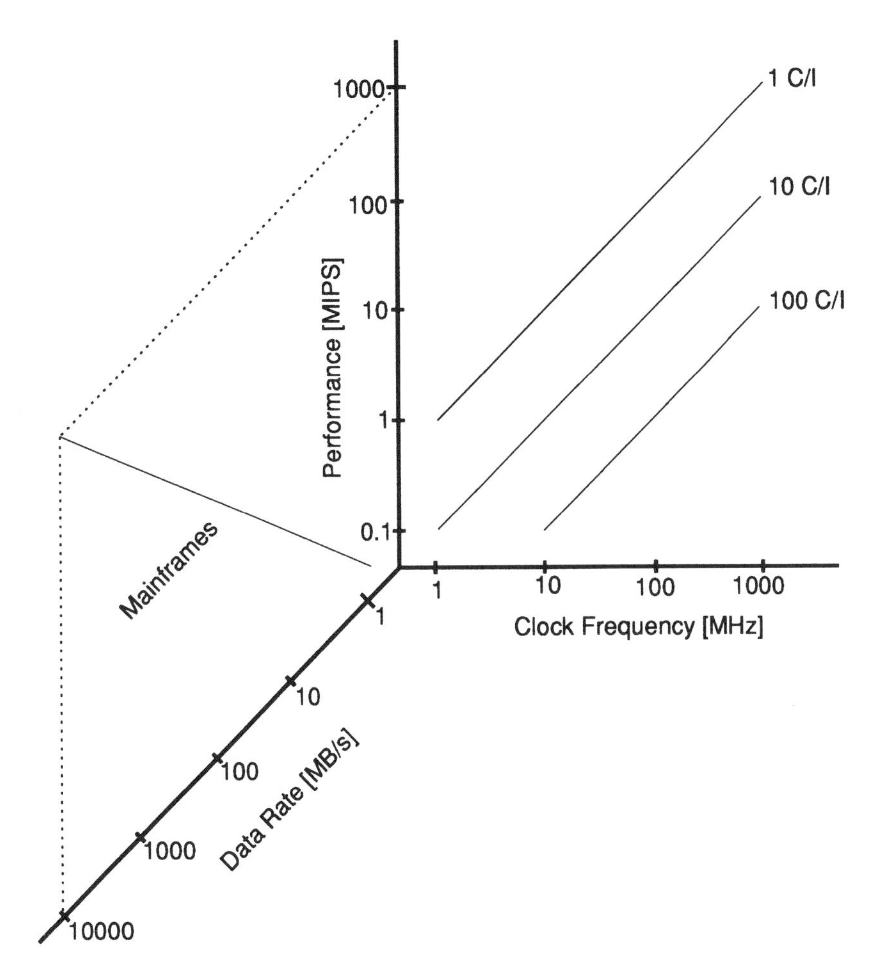

Figure 7-2. System Performance Dependency on Cycle Time and Cycles Per Instruction, and the Input–Output Data Rate

and on the number of its circuits in particular, as is discussed in Section 7.2.11. The high performance (number of operations per unit of time, usually 1 s) requires processing and storing of data within short time intervals, on the order of nanoseconds (ns or 10^{-9} s). Such high-performance equipment must be of small physical size because of the finite time a signal takes to travel through an interconnection—1 ns through an interconnection 10–30 cm (4–12 in.) long. The evolved, and still evolving, semiconductor and packaging technologies have enabled this progress to occur and to continue.

The above scenario is grossly incomplete without highlighting another, perhaps even more significant, paradigm shift—the emergence of technologies for the *portable* systems (Fig. 7-3) with wired (via modem and telephone lines) and wireless (like cellular telephones) connections to a rapidly expanding network of interconnected computers, already providing near-unlimited amounts of information and vast opportunities for person-to-person as well as group communications. The *World Wide Web* name is not an exaggeration.

As already indicated in Figure 7-2 and discussed in more detail in Section 7.2.11, the communications between the computers, and even within a powerful single computer, demand very high data rates, on the order of 10 megabytes (MB) per second (1 byte = 8 bits—binary units) transmitted in and out of the processor per 1 MIPS; see Figure 7-18. The past, present, and future mainframes maintain this relationship to balance their performance and communications capabilities because of the heavy communications demand of all tasks they perform. The seemingly imbalanced ever-increasing performance of PCs without a correspondingly high communication bandwidth constitutes the differentiating property between the two. The PC performance is used to continually enhance the user friendliness and to perform computation-intensive, but not communication-intensive, tasks such as pattern manipulation (e.g., maps, images, voice recognition, and so forth).

Customer expectations are also very high indeed in terms of quality *(work perfectly out of the box),* reliability *(no failures during useful life),* and rapid and inexpensive service in the unlikely event it is needed. Two other main branches of electronics, consumer and communications, have almost completed their transitions from analog to digital circuits wherever practical. Thus, the long-predicted convergence of these three main branches of electronics (computer, consumer, communications) is well underway. The packaging continues to provide strong competitive leverages as it has in the past, in the context of totally different IC chips and much broader and still widening applications.

Thus, the second edition of this book endeavors to preserve and build on the *timeless* fundamentals presented in the first edition and to update its *timely* content by splicing in the various advances of the state of the art. It must be noted, however, that designs for various applications undergo much more rapid changes and suffer obsolescence much sooner than at the time of work on the first edition. Consequently, the applications chapters are no longer included. The interested readers are referred to many good periodicals, conferences, and workshops. The changing world of electronics required a different approach to this second edition. First, its chapter authors represent the broad international professional community. Second, the book is now being divided into three parts to make it more complete and comprehensive without approaching a cubical form factor and to enhance its use as a study book.

The three parts are basically the original book, but now they concen-

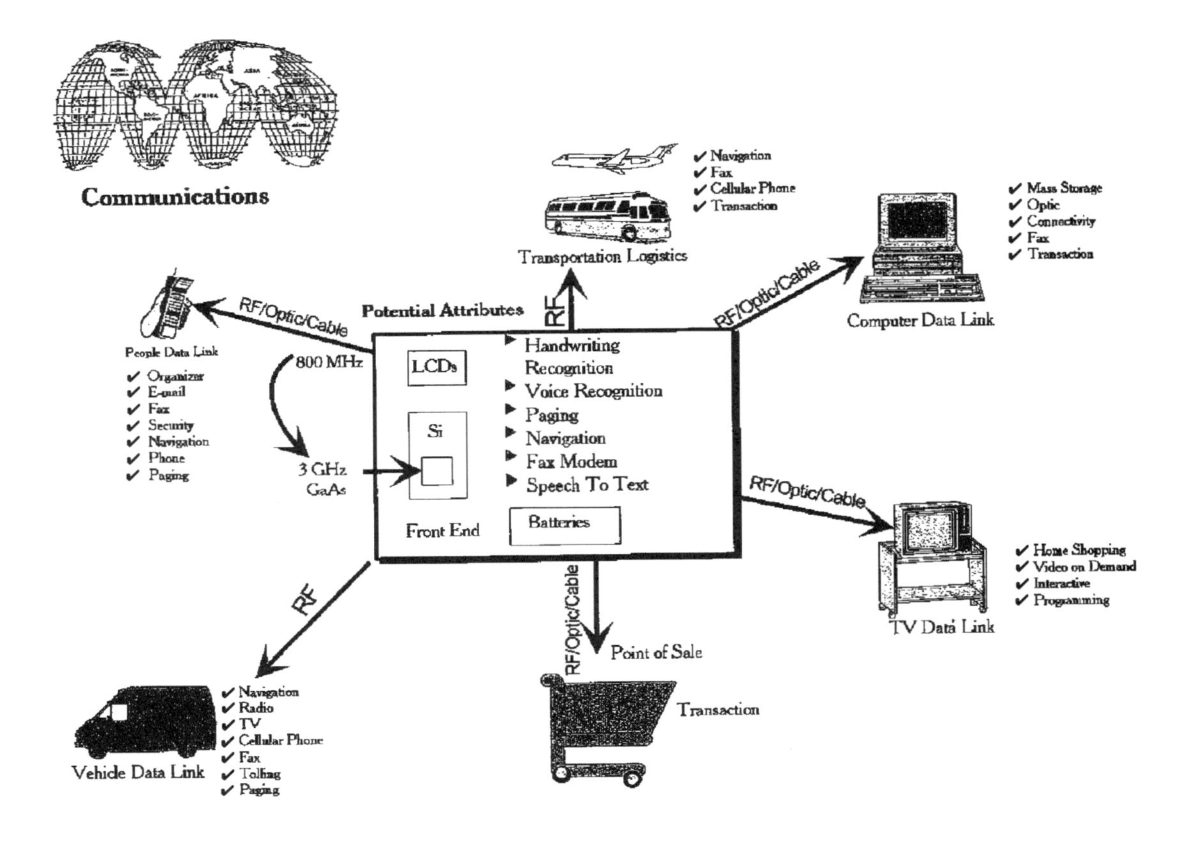

Figure 7-3. Portable Electronics (Courtesy of NEMI)

trate on technology drivers and technologies. The first part, "Technology Drivers," begins with an overview followed by an updated in-depth treatment of the key packaging design aspects—topological, electrical, thermal, reliability, and manufacturability. The second part, the "Semiconductor Packaging," focuses on the extremely important and very diverse technology options (materials and processes) to package semiconductor chips in single-chip (SCM) and multichip (MCM) modules. The third part, "Subsystem Packaging," completes the coverage of the technology options available to the designer to form a system or a major subsystem. New chapters were added to address manufacturing considerations, electrical testing, polymers used in packaging, and optical packaging.

Another aspect of packaging is microelectronics versus general electronics, in which a great deal of components are heavy and/or bulky, such as power-supply transformers, filter capacitors, cathode-ray-display tubes, and flat-panel displays. This book, as its title indicates, concerns itself mainly with packaging of the wide range of microelectronic components.

The next subsection of this chapter, "Packaged Electronics," highlights the two pivotal roles of the electronic packaging—the ability to support and preserve the on-chip performance, and the main limiting factors to such an ability. It also raises many questions triggered by these initial highlights to introduce the subsequent section on packaging definition, to describe its functions and hierarchy, and to give a brief overview of these books. The detailed chapter-by-chapter description is given in Section 7.10, "Book Organization and Scope."

7.1.1 Packaged Electronics

As already mentioned in Section 7.1, the electronics packaging must support the original design objective/intent of its IC chips, yet it may fail to do so to a greater or lesser degree, either by limiting the realization of the intended IC performance potential or by "diluting" the on-chip performance, or by both. To illustrate these points for digital applications, let us consider an assembly (e.g., in a workstation) of 750,000 computing elements, all of which having the same value of the "speed-power product" (actually energy), assumed to be 0.1 pJ (0.1×10^{-12} J).

Let us select three operating points (signal delay at an average power dissipation) roughly representative of *performance* (but not the power) of yesterday's, today's, and tomorrow's technologies: 10 ns at 10 μW, 1 ns at 100 μW, 0.1 ns at 1 mW, respectively. During each system cycle, the signal being processed has to traverse 20 of these computing elements. Furthermore, there are three packaging options, each assumed to be a square with a 1-cm edge typical of an IC chip, 6-cm edge typical of a multichip module, and a 36-cm edge typical of a printed wiring board (all these terms will soon be discussed). Insulating material of the signal

Table 7-1. Attributes of Nine Workstation Implementations

		IC Chip	Packaged Chips	
			Multichip Module	Printed Wiring Board
Edge dimension		**1 cm**	**6 cm**	**36 cm**
Manhattan t_d		0.13 ns	0.8 ns	4.8 ns
Area		1 cm^2	36 cm^2	1296 cm^2
"Yesterday's designs," 10 ns				
	Wire delay adder	~0%	0.4%	2.5%
10 μW	Clock	~5 MHz	4.98 MHz	4.88 MHz
7.5 W	Power density	7.5 W/cm^2	0.21 W/cm^2	0.006 W/cm^2
"Today's designs," 1 ns				
	Wire delay adder	0.5%	3.8%	19.3%
100 μW	Clock	49.7 MHz	48.1 MHz	40.3 MHz
75 W	Power density	75 W/cm^2	2.1 W/cm^2	0.057 W/cm^2
"Tomorrow's designs," 0.1 ns				
	Wire delay adder	6.1%	28%	67%
1 mW	Clock	463 MHz	357 MHz	147 MHz
750 W	Power density	750W/cm^2	20.8 W/cm^2	0.58 W/cm^2

interconnections has a dielectric constant of 4, an approximate value of many actual materials.

In a fairly representative situation, the signal will originate by the circuits located in one corner of the square package and will have to travel ultimately to the diagonally opposite corner via paths mostly parallel to the edges, commonly called "Manhattan delay" (from the streets and avenues of the borough of Manhattan of New York City). In such an intentionally oversimplified design, the cycle time is determined by the sum total of the two signal propagation times:

Through the 20 computing elements, connected in series (on-chip delay)

Manhattan (along two packaging edges) wire delay for $\varepsilon_r = 4$, which is $1/15$ ns/cm.

Thus, there are nine technology design options presented in detail in Table 7-1. The three columns correspond to the designs with each of the three edge dimensions (a single IC chip and two packaging structures). The three rows reflect the designs with each of the three performance levels representative of the three time frames. The individual cells of Table 7-1 list the resulting clock frequency and power density, and the % adder of the wire delay. The individual cells of Table 7-1 present a more complete picture by listing the total signal delay of the on-chip computing elements (mostly logic circuits), the cycle path delay (and the % adder of the wire delay), the corresponding clock frequency, and the power density.

Contemplating the key system attributes of this table, we note minuscule effects of the wire delay on the cycle time for all 10-ns ("yesterday's") designs, regardless of the edge dimension. The added delay is 2.5% or less of the total, and the power densities are easily managed, as will become more apparent in Section 7.2.4 (note that this subject is fully treated in Chapter 4). Section 7.2.3 will show another aspect of minimal packaging effects.

The 1-ns ("today's") designs begin to show noticeable Manhattan wire delay adder (~20%) for the 36-cm-edge case and a rather challenging power density (75 W/cm^2) for the 1-cm edge. Again, Section 7.2.3 will show similar situation.

The 0.1-ns ("tomorrow's") designs are most severely affected by the packaging: up to 67% added wire delay (3.4× performance degradation) on the 36-cm-edge package and, on the 1-cm-edge package, a 750-W/cm^2 power density—on the order of the power density of a nuclear blast 1 mile from its center, as shown in Figure 7-12. Such a power density cannot be supported by any practical packaging design. Either a circuit power reduction (and correspondingly greater signal delay) or an increased edge dimension (greater Manhattan delay), or both, are required. Consequently, they preclude any practical full utilization of the theoretically possible 2-ns cycle time, or 500-MHz clock, and leave one to hope that the future ICs will achieve the same 0.1-ns performance at the circuit power levels lower than 1 mW (i.e., will have a speed–power product less than 0.1 pJ). The power distribution for these tomorrow's ICs will also need significant design innovations.

In these examples, only the power density and wire delay were considered, which are, in fact, the primary variables. Further considerations raise some additional intriguing questions.

The 1-cm edge is most likely to be that of an IC chip itself. The larger edges imply a somewhat uniform distribution of the circuits over the larger area, which is possible only with *partitioning* of the total circuit content into several IC chips. How many of them? Just a few or quite a few (a high number)? What are the consequences of such partitioning in terms of chip-to-package interface on the number and arrangements of the signal, ground and power terminals, the on-packaging interconnections between them, and the thermal interfaces/paths to the equipment exterior?

What are the topological aspects, the physical dimensions, and the electromagnetic characteristics of the interconnections? How does one deal with the chip-to-package interface regions? What are the materials, in what shapes and dimensions, and how to process them during manufacturing?

What are the values of stray electric and magnetic fields, and the resultant capacitances and inductances, and what impact do they have on the system performance and/or reliability of its operation (intermittent failures—the most onerous kind)? How well can it all be manufactured and what reliability (hard failure rates and lifetimes) to expect?

How to integrate this portion of a larger data processing system with the rest of it in a competitively advantageous manner? What performance leverage, or penalty, results from employing either less or more than the 750,000 computing elements exemplified above?

This book endeavors to deal with all of these questions, and raise and answer a few more. This chapter sets the stage for the entire book (all three parts) by introducing and highlighting the topics of all subsequent chapters. This concludes with the packaging functions and structural hierarchy, and presents the evolving trends of semiconductors and packaging.

The next major section, Section 7.2, deals with the two sets of driving forces—the design considerations (topological, electrical, thermal, reliability, manufacturability, and testability) and the applications (memories, PCs and workstations, portable electronics, mainframes, optical interconnects) which optimally integrate the various technologies. Section 7.2.13 concludes the description of driving forces, gives a bird's-eye view of evolution of the entire range of digital-electronics-based products, and sets the stage for the next six sections to deal with the various packaging technologies (structures, materials, and processing) and their trends and applications.

Section 7.3 introduces the packaging technology and describes its trends. Section 7.4 describes the various technologies for chip-to-package connections. Section 7.5 focuses on the single-chip and multichip modules—the first-level packages. Section 7.6 covers the package-to-board interconnections. Section 7.7 deals with the second-level packaging technologies. The packaging cooling is the subject of Section 7.8. Section 7.9 handles the packaging sealing and encapsulation.

Section 7.10 summarizes this book with a schematic overview and a detailed chapter-by-chapter description of all three parts. It is followed by listing of references.

7.1.2 Packaging Functions and Hierarchy

Electronic packages contain many electrical circuit components—up to several millions or even tens of millions—mainly transistors assembled in integrated circuit (IC) chips, but also resistors, diodes, capacitors, and other components. To form circuits, these components need interconnections. Individual circuits must also be connected with each other to form functional entities. Many of the intercircuit connections are steadily migrating into the IC chips as the levels of integration continue to increase. Mechanical support and environmental protection are required for the ICs and their interconnections. To function, electrical circuits must be supplied with the electrical energy which is consumed and converted into the thermal energy (heat). Because all circuits operate best within a limited temperature range, packaging must provide an adequate means for removal of heat.

Thus, the packaging has four major functions:

- Signal distribution, involving mainly topological and electromagnetic consideration
- Power distribution, involving electromagnetic, structural, and materials aspects
- Heat dissipation (cooling), involving structural and materials consideration
- Protection (mechanical, chemical, electromagnetic) of components and interconnections

as illustrated in Figure 7-4. They are discussed throughout this series of books.

Arguably, the thin-film wires (interconnections) within semiconductor IC chips are an important packaging component. Increased integration had migrated onto the IC chip at a steadily increasing percentage of the wires previously provided by the packaging structures outside the chips.

Packaged components must be manufactured and assembled at affordable costs. Repairs must be few, quick, and inexpensive. In many applications, it is also desirable/mandatory to enable the addition or rearrangement of functional features (e.g., expansion of storage capacity, addition of a coprocessor or a modem, or even a processor upgrade).

All interconnections on an IC chip (even though some of them might have been part of packaging in previous—lower level of integration—technology) or all internal structures of any other component (e.g., a crystal or those of discrete capacitors and inductors) are not considered packaging in the common parlance. Thus, the packaging begins at the interfaces to these ICs and other components: mechanical, to provide support and, if needed, an enhanced thermal path to facilitate heat transfer; and electrical, for power and signal connections, whose electromagnetic properties also play an important, but not always desirable, role.

These interfaces may be the exterior surfaces of an IC chip (die) itself—the common term being a bare chip—often in the context of a "direct chip attach." Or, as it is still often the case, the IC chip has an enclosure, commonly referred to as a single-chip module (SCM), or a chip carrier, or a header. In either case, it is considered a **first-level package.** This is an extremely important and very diversified level of packaging which fills the entire Part 2, Semiconductor Packaging. This book also deals with a direct attach of several unencapsulated (bare) chips to a **substrate** of various materials produces a multichip module (MCM)

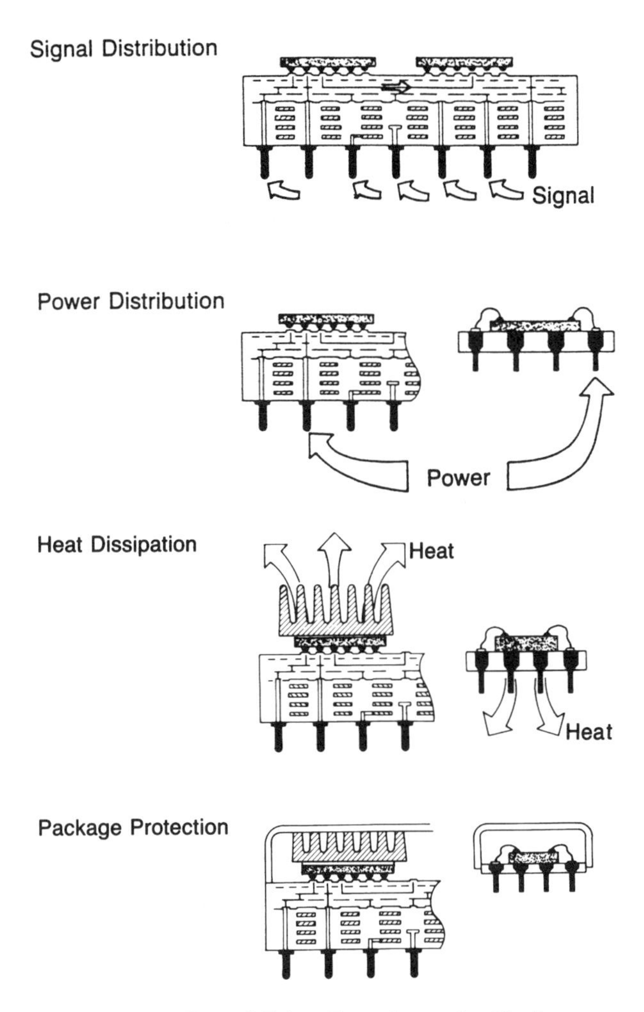

Figure 7-4. Four Major Functions of a Package

or multichip packaging (MCP). MCMs are often assembled on PWB along with SCMs and other discrete components.

The groups of chips (packaged in SCMs), along with other components such as capacitors, resistors, inductors, filters, switches, and optical and RF components, are assembled on a **second-level packaging**—covered in Part 3—usually copper-clad sheets of epoxy–glass laminates. After patterning of the metallic sheet followed by mechanical drilling and lamination to form interconnections, such structures are commonly called "printed wiring boards" (PWB) or card. See Figure 7-5.

The next level packaging **(3rd or higher)** may be the outer shell of a small piece of equipment (e.g., a hand-held calculator, a portable phone, or a pluggable unit to supplement or to enhance functions of a

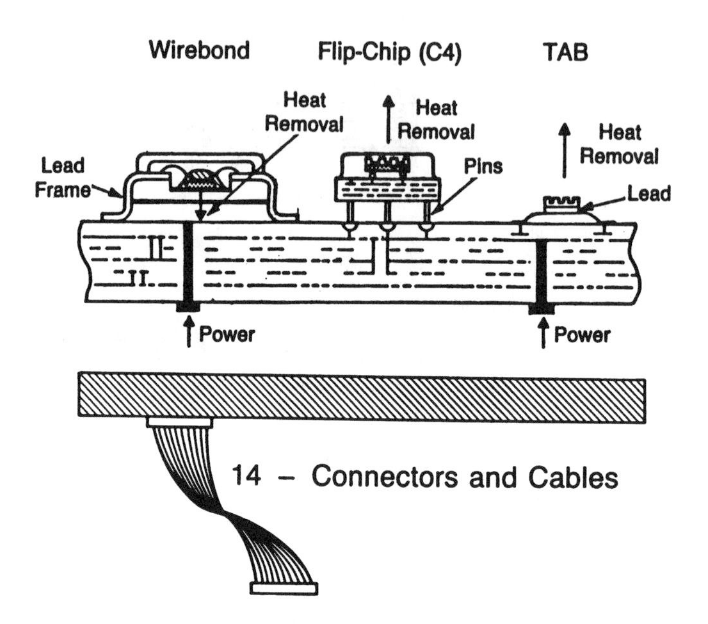

Figure 7-5. Packaging Levels, Interconnections, and Technologies

larger equipment—a pluggable PC card with a hard disk or with a wireless modem). In a larger equipment (desktop or floor standing), several cards are plugged into a PWB. Such cards are sometimes referred to a "daughter cards" and the PWB as "mother boards." There may be several boards within a box/frame of large equipment such as a workstation or, especially, a mainframe. As already mentioned, the continuing increases in the number of circuits and/or bits per IC chip tend to simplify the packaging hierarchy. In addition to the relatively small and light IC chips, many electronic applications require bulkier and heavier components such as power transformers, large electrolytic capacitors, disk drives, display tubes, and so on. They are mounted separately and are electrically connected to the rest of the electronics via various cables. Cables also are used to connect boards either to each other or to the rest of the equipment, including its external terminals (e.g., a PC to its monitor, keyboard, mouse, printer, and, via modem, to the house telephone line).

7.1.3 Evolving Trends

7.1.3.1 Product Applications

The main driver is the price for performance shown in Fig. 7-1 for the digital data processing. The general trends of the attributes of packaged electronics are shown in Figure 7-6. Note how lower prices enhance production volume.

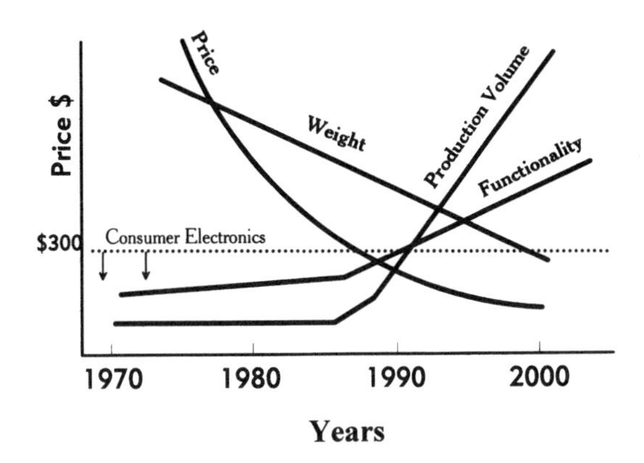

Figure 7-6. Attributes of Next-Generation Electronics

The mainframe improvements were progressing at a slow rate until emergence of the microprocessors based on the complementary-metal-on-silicon (CMOS) technology. This technology was initially too slow to challenge the bipolar-based emitter (driver output) complementary logic (ECL), as shown in Figure 7-7. The advent of better (smaller dimensions) photolithography greatly benefited the CMOS-based logic-circuit performance, at a better speed–power product. It had also enabled higher circuit density leading to lower circuit costs. These higher levels of integration on the ICs lead to higher packaged circuit density, thus reducing the size and number of parts. Lower power densities have simplified the packaging complexity, mainly by the switch from water cooling to air cooling.

The clock frequencies of the CMOS-based microprocessors (used

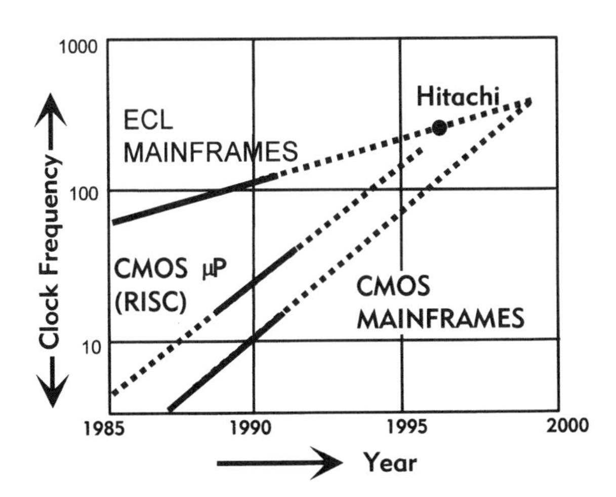

Figure 7-7. Performance Trends of CMOS and Bipolar Logic Technologies

mainly in PCs) have caught up with the mainframes and are chasing the supercomputers (in the range of 200–350 MHz). The CMOS-based mainframes are expected to do so before the year 2000. It must be noted, however, that it is extremely difficult to compare PC performance (essentially a single-user system) with the mainframes, which have a totally different instruction set and serve a vast number of users. In Figure 7-1, 1 mainframe MIPS was assumed to equal approximately 10–100 PC MIPS. The main differences between the two are in the instruction sets (tasks) and in the communications bandwidth—the number of bits per second transmitted into and out of the processor for each MIPS, as will be presented in detail in Figure 7-18 of Section 7.2.11.

The zero power consumption of the CMOS circuits when in the standby mode (no data processing) contrasts sharply with the near-constant power consumption of the ECL technology. It has enabled portable, battery-powered, equipment (e.g., PCs, cellular phones) with reasonably long operating times. The trend here is toward smaller dimensions, lower weight, more functionality, and very attractive prices, leading to high volumes.

Low technology costs and the better understanding and control of the failure mechanisms resulted in highly reliable products (see, for example, Fig. 7-13), with the new applications, such as automotive electronics.

7.1.3.2 Semiconductors

Until the mid-fifties, most of the electronic circuits were produced from individual discrete components: vacuum tubes, resistors, capacitors, and, notably, discrete wires. The first step toward integration occurred with the emergence of the printed wiring board (PWB). An insulating layer, a fraction of 1 in. thick, was clad with metal layers on one or both sides. Interconnections were photolithographically patterned. Wire terminals of the still discrete components were inserted into prepunched holes in the PWB and soldered, all at once.

It all began to change rapidly in late sixties, with the advent of integrated circuits (IC) which contained in a little semiconductor (mostly silicon) "chip" an ever-increasing number of planar devices (mainly, transistors, field-effect—FET, or bipolars) isolated from each other, with their terminals on the surface of the chip. These devices were interconnected by "wires" photolithographically formed from one, and later more, layers of metallic thin film, predominantly aluminum. Enormous density improvements (Fig. 7-8) resulted from advances in the photolithography, for instance more than a fourfold increase in storage cells per chip every 3 years or approximately a 100-fold increase every decade. This mass production of electronic circuits dramatically reduced (and continues to reduce) their cost. Table 7-2 highlights the evolution of the key IC parame-

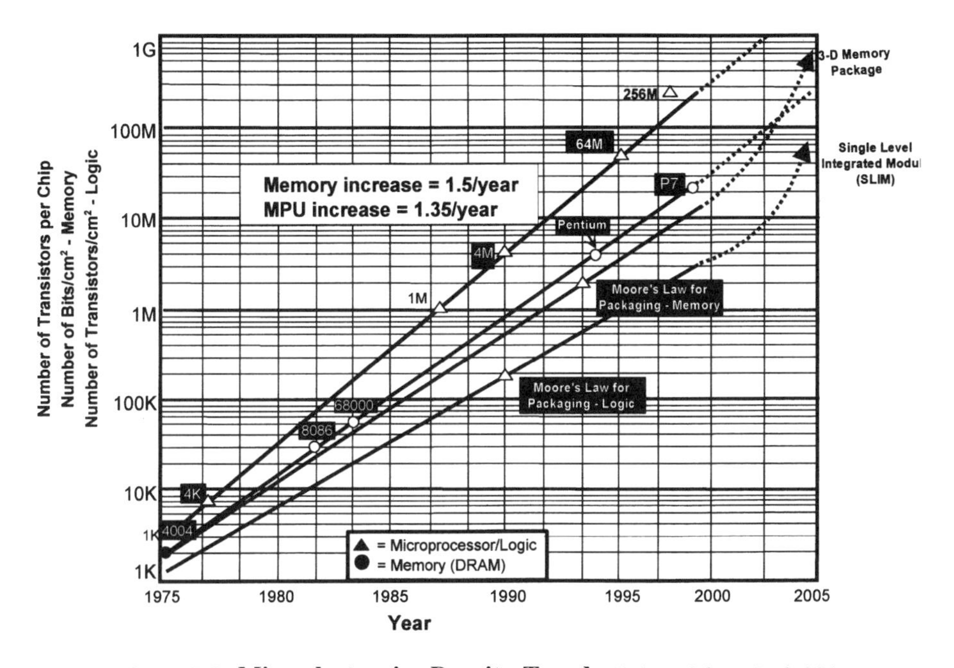

Figure 7-8. Microelectronics Density Trends (Adapted from Ref. 22.)

ters and attributes from 1992 to 2007. Figure 7-9 offers comparisons of the major ICs—bipolars (emitter-current logic, ECL), complementary metal on silicon (CMOS), BICMOS, GaAs—in terms of their major attributes.

These semiconductor advancements have put, and are still putting, an ever-increasing burden on the packaging—in terms of interconnecting, powering, cooling (Fig. 7-10), and protection—to keep pace with and provide sufficient support to the ICs as it is highlighted in Table 7-1 and will be discussed in more detail in Section 7.2. An added demand to reduce cost is becoming stronger and more pervasive since the shift in the application thrusts from the large (and still expensive, but getting less so) mainframes and supercomputers to desktop PCs and workstations (shown in Fig. 7-1) and portable electronics. Portable electronics also demand very energy-efficient designs to prolong operating time on one battery charge.

7.1.3.3 Future Packaged-Electronics Considerations

The quest for the further cost reduction opportunities now focuses on those attributes/elements of technology whose contributions to the total cost are becoming disproportionaly large (e.g., pins, individual chip carriers, individual discrete components, extensive rework to replace faulty

Table 7-2. Semiconductor Evolution

	1992	1995	1998	2001	2004	2007
Feature size (μm)	0.5	0.35	0.25	0.18	0.12	0.10
Gates/chip	300K	800K	2M	5M	10M	20M
Bits/chip						
DRAM	16M	64M	256M	1G	4G	16G
SRAM	4M	16M	64M	256M	1G	4G
Wafer processing cost ($/cm^2)	$4.00	$3.90	$3.80	$3.70	$3.60	$3.50
Chip size (mm^2)						
Logic/unprocessor	250	400	600	800	1000	1250
DRAM	132	200	320	500	700	1000
Wafer diameter (mm)	200	200	200–400	200–400	200–400	200–400
Defect density (defects/cm^2)	0.1	0.05	0.03	0.01	0.004	0.002
No. of interconnect levels—logic	3	4–5	5	5–6	6	6–7
Maximum power (W/die)						
High performance	10	15	30	40	40–120	40–200
Portable	3	4	4	4	4	4
Power supply voltage (V)						
Desktop	5	3.3	2.2	2.2	1.5	1.5
Portable	3.3	2.2	2.2	1.5	1.5	1.5
No. I/Os	500	750	1500	2000	3500	5000
Performance (MHz)						
Off chip	60	100	175	250	350	500
On chip	120	200	350	500	700	1000

components). These cost-reducing efforts may disrupt the continuity of the technology evolution; they may even traumatize the established technology-flow and business-flow patterns; for example, the replacement of the surface-mount technology (SMT) for discrete components (mainly resistors and capacitors) by embedding them into the substrate to enable higher density, better performance, and lower costs.

Equally traumatic are transitions from one technological approach to another which is vastly different. Examples from the past are replacements of the ferrite memory-core arrays with the integrated-circuit (IC) dynamic and static random-access memories (DRAMs and SRAMs), or advent of the microprocessors and application-specific integrated circuits (ASICs). In the examples just cited, the cost and performance improvements went approximately hand in hand.

Introduction of the dense liquid-cooled multichip modules (MCM), with well over 100 bipolar IC chips directly attached (without benefit of individual packages—SCMs) to multilayer ceramic substrates, provided very significant performance leverages, but it was not a low-cost approach. Yet, it lasted for several generations of IBM mainframes and was emulated by mainframe and supercomputer producing competitors; see Figures 7-32

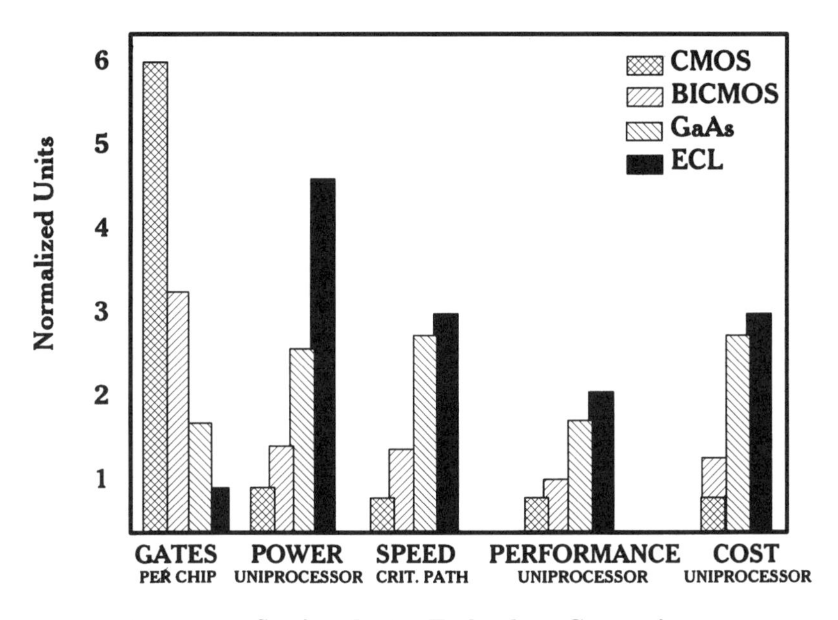

Figure 7-9. Semiconductor Technology Comparisons

and 7-33. The direct attach of bare chips to MCM substrates is still resisted by some suppliers of electronics technology, at least in part due to the business separation between the makers of the ICs, the makers of the PWBs, and the makers of the end products or subproducts.

The traditional single-chip-module approach, when extended to giga-scale integration (GSI, with large IC edge dimensions—25 mm and beyond) and, especially, combined with decreasing cross-sectional geometries of interconnections (down to 0.1 μm wide and thick), is likely to be competitively inferior to the alternative implementation in MCMs or MCPs on two most significant counts: cost and performance. The following is a generalized qualitative examination of both aspects.

Manufacturing costs: Inscribing *large* chips within a circular wafer produces a disproportionately smaller number of them than a smaller chip-size alternative. Given identical defect density on wafers, a higher percentage of larger chips will be defective. Consequently, the cost of good chips will suffer disproportionately. The chip yield problems will be further aggravated by the need for additional metal levels. *The cost differential is likely to be large enough to offset the additional MCM/MCP costs.*

Performance: Smaller device geometries necessitate smaller contact areas and narrower interconnections. To avoid "degenerate" aspect ratios, interconnections also have to become thinner. These two factors will reduce the cross-sectional area at a faster rate than

> the device size and interconnection lengths reductions. The smaller area increases the resistance and, therefore, reduces bandwidth, unless the interconnection length is also reduced by the same factor. In fact, the interconnection lengths become greater with the increasing IC chip size, further aggravating the performance degradation. An attempt to widen the interconnections will lead to an even higher number of metal layers to render the chip wireable—hardly a desired by-product because of the added complexity and increased number of defects.

Note that many GSI chips consist of "functional islands," and busses: 16-bit or 32-bit (or more) sets of interconnections between them. The wiring within functional islands is at least partially random. A properly partitioned GSI chip may be readily subdivided into smaller chips. The interconnections between the subchips will have increased performance when their cross-sectional geometries are increased: an easy task, if done on a physically separate chip carrier (MCM or MCP). The chip-to-carrier interface must present minimum electromagnetic loading. Indeed, it can be done with the advanced flip-chip technologies. *Performance of such nets will be vastly superior to those implemented on a single chip of reasonably complexity—with a limited number of thin-film metal layers.*

As most of the web-surfers already know, the wait inflicted by the interconnection's bandwidth is usually much longer and, therefore, less tolerable than that caused by the signal round-trip delay. This bottleneck is very typical for distributed computing (actually data processing). Ultimately, it is also gated by the affordable price/cost. To continue with this example, the service provider accepts a regular telephone line with its severe data-rate limitation, but will be happy to provide a data link of (much) greater bandwidth at a correspondingly higher price.

In high-speed data processing, both the round-trip delay and the bandwidth are important. The shorter round-trip delay is achieved by denser (usually a smaller area for the same electronic content) technology. Denser interconnections have smaller cross-sectional areas and, consequently, reduced bandwidth simply because the cross-sectional area scales down with the square of the scaling factor while the distance reduces linearly. Such a dimensional scaling-down process has been continuing since the early seventies, as already indicated. The bandwidth reductions were barely noticeable, if at all, until recently,when detailed design and performance simulation work on multi-GHz clock systems required extremely wide busses (on the order of 1000 lines) between the processor and its temporary (cache) memory sections; see Figure 7-24 and Refs. 3 and 4. For a single processor configuration, a multilayer planar packaging structure is manufacturable. A multiprocessor configuration will probably choke on a submarginal data-transfer rate between the processors and their caches. Optical connections are coming to the rescue, with their much greater bandwidths.

As already mentioned in Section 7.1, a great deal of digital data processing in the mid-nineties is microprocessor based. The microprocessors are often contained within a single chip (characteristically, such was the case in the seventies, eighties, and early nineties) or the higher performers are partitioned into a small group of chips, typically between two (Intel Pentium-Pro) to about nine (eight in the newer IBM workstations [5] and Fig. 7-20).

This represents two competing schools of thought—and ways of doing business: single-chip design, with chips projected to grow rather large; and multichip design, with several smaller chips tightly packaged on an MCM or MCP.

Although the MCM approach had been the technology of choice for the mainframes, its use in PCs is still emerging (in 1996). A study reported in Ref. 6 substantiates the above conclusions with a quantitative analysis. Such studies confirm the general applicability of the insights gained from R&D of the high-end mainframe and supercomputer packaged electronics, as discussed in detail in Section 7.2.13.

A multichip approach seems the only practical choice when heterogeneous semiconductors are a necessity. The most obvious case is with wireless communications, with the silicon-based ICs handling the digital signal processing and the compound semiconductors-based (e.g., gallium arsenide) devices handling the radio-frequency (RF) transmitting and receiving tasks [7]. The SCM carriers on a PWB for such ICs are, in fact, inferior solutions to the MCMs in terms of higher packaging volume (premium for portable equipment) and poorer performance. Thus, there is a strong drive for dense, low-cost interconnections capable of providing high degree of electromagnetic separation between the digital and RF signal and power interconnections. The detailed packaging technology trends are given in the introduction to Section 7.3.

Yet another set of challenges stems from distributed computing (actually data processing) with its need for rapid and faultless transmission of enormously large data/information volumes. Each challenge involves several key parameters or characteristics which often counteract each other—improvements in one often cause deterioration of another. Furthermore, the current technologies are approaching numerous fundamental limits set by the properties of materials and by the structural geometries (form factors and dimensions) of packaging.

7.2 TECHNOLOGY DRIVERS

A successful package design will satisfy all given application requirements at an acceptable design, manufacturing, and operating expense. As a rule, application requirements prescribe the number of logic circuits and bits of storage (possibly along with other components, such as switches, lights, displays, etc.) which must be packaged—interconnected, supplied with appropriate electric power, kept within proper temperature range, mechanically supported, and protected against the environment. Ever-increasing integration at the chip level toughens demands on the number and density of signal terminals and interconnections. The resulting smaller physical dimensions tax both the properties of materials, and the production assembly and test techniques. These challenges are further compounded by the need to preserve or even to enhance the thermal and electromagnetic characteristics while maintaining or improving the quality and reliability.

There are two major sets of technology drivers—one governs the technology selection for a particular application (e.g., mainframe processor, its main memory, PC), the other set is derived from the key design considerations: topological (wiring, number of terminals), electrical (signal and power interconnections), thermal, reliability, manufacturability, and testing. The technology selection for a particular application integrates and optimizes these design considerations and specific parameter values—power density, electrical characteristics, number and density of terminals and interconnections, and so forth. In this section, the design considerations are reviewed first, following the subject of Chapters 2–6 (the first part). Test design consideration are covered in Part 2, Chapter 13 and design considerations for optical are covered in Part 3, Chapter 20. Sections 7.2.8–7.2.12 highlight the various generic application considerations—memories, PCs and workstations, portable equipment, mainframes, and optical interconnects. Section 7.2.13 gives a bird's-eye view of the evolution of the entire range of products with regard to several key parameters.

7.2.1 Wireability, Number of Terminals, and Rent's Rule

The total number of terminals and wires at packaging interfaces (module pins, card connectors) is a major cost factor. As will be seen later, the electrical characteristics of the interconnections are also the key performance determinators, along with the limits in the ability to supply power and remove heat. They can also pose serious reliability challenges.

Essentially all of the electronic equipment components and subassemblies have two distinct sets of terminals (individual electrical connection)—signals and power. Common to both is usually the "ground" terminal(s), a carryover from the old radios which had to be connected to an areal (antenna) and to the ground, literally a spike driven into the ground or other suitable artifact such as a water pipe. House electrical wiring, telephones, and so forth still have such a connection to prevent an electrical shock to a person in contact with an electrically conducting part of an appliance. However, in electronic applications, the "ground" designates a neutral reference voltage to be considered with a much greater care than frequently done (for more details see Section 7.2.3).

Signal interconnections and terminals constitute the majority of conducting elements, especially in low-cost packaging. Other conductors supply power and provide ground or other reference voltages. Packaging designed for the high-performance circuits must contain more elaborate power and ground connections to ensure acceptably low levels of electrical interference (noise). This significantly complicates the design and, therefore, raises the cost.

The number of terminals supporting a group of circuits is strongly dependent on the function of this group. The smallest number goes with the memories (storage of binary coded information) because the stream of information can be as limited as a single bit, requiring only one data line. The address data depend on the storage capacity. For example, a 4K bit ($4096 = 2^{12}$) random access memory (RAM) will require 12 address lines (terminals) and 16M RAM bits will be satisfied with 24 address terminals. Controls (e.g., read, write, clock) require a few more lines. Thus, the packaging of memories is usually simple and inexpensive.

Exactly the opposite is the case with the groups of logical circuits which result from random partitioning of a computer. Repeated studies of the relationship between the number of circuits within an enclosure (module, card, board, gate of Figure 7-10) and the number of terminals required have reconfirmed findings of Rent [8,9,12], as discussed in depth in Chapter 2. The relationship is known as Rent's rule [8] which can be generally expressed as[1]:

$$N = K\,M^{p} \qquad [7\text{-}1]$$

where N is the number of required input–output terminals, K is the average number of terminals used for an individual logic circuit, M is the total number of circuits, and p ($0 \le p \le 1$) is a constant, called **Rent's constant**. A detailed discussion is given in Section 2.2.2.

Of main concern here is the number of signal terminals in digital applications, especially digital data processing as opposed to data storage. The more circuits (gates) the data processing IC (or other packages) have, the more terminals will be required at their interface. Although the general dependency follows the same rule, different classes of applications demand somewhat smaller or higher number of terminals for the same number

---

1 This expression stresses the fact that the average number of wired circuits depends on the pinout provided by the package. Maximum available circuits must exceed the average by approximately a factor of 2. Other limitations are wiring channels, power supply, and cooling.

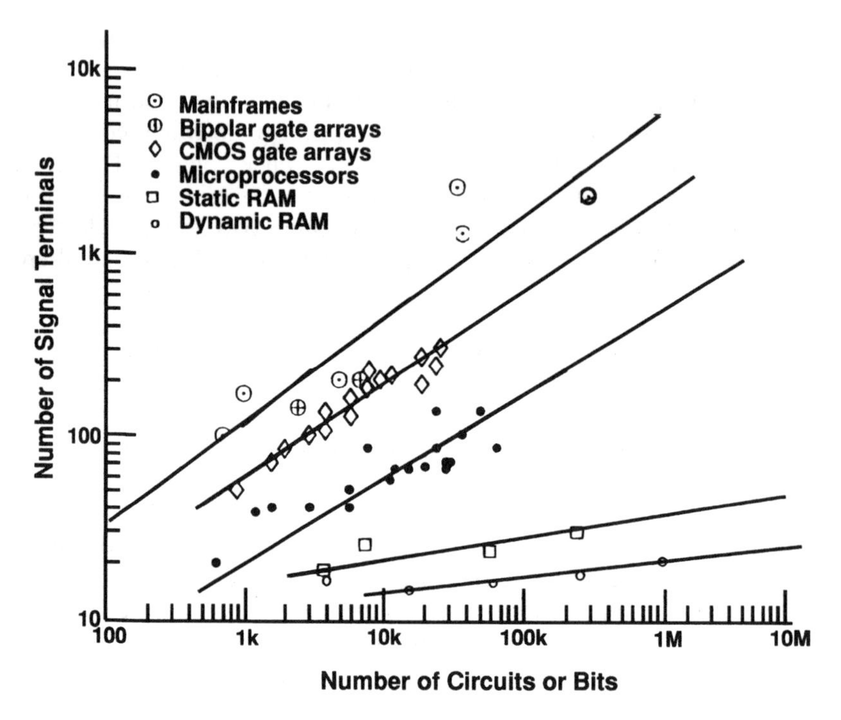

Figure 7-10. Number of Terminals and Number of Circuits (Adapted from Ref. 8.)

of circuits (Fig. 7-10). The highest number of terminals with randomly partitioned logic is needed for the highest-performance applications such as supercomputers or mainframes.

Generally, the "dictate" of Rent's rule applies to information being conveyed to the assembly of circuits, not necessarily to the number of actual lines or terminals. Information can be coded on several lines, somewhat analogous to memory address, to trade the total number of terminals against the need to code–decode and to buffer data, with the resultant performance penalty. Such a penalty is often small when many logic (and memory) blocks are encountered by the data between entering and exiting the enclosure (chip, card, board), as it occurs in a microprocessor. Therefore, significant pinout reduction can be utilized with an acceptable performance degradation. In the case of high-performance logic, only very few logic blocks are present between the input and output terminals. Consequently, to utilize most circuits and to avoid severe performance degradation, Rent's rule of Equation [7-1] has to be followed.

7.2.2 Electrical Design Considerations

Objectives of the electrical package design are to assure reliable and predictable signal transmission through the on-chip and off-chip interconnections. It is accomplished by numerous design considerations, all of which have to be met as follows.

- Assure reliable functioning of the *interchip signal interconnections* (nets):
  - By preservation of the transmitted signal (within acceptable tolerances): DC levels, length of signal transitions, and distortions caused by reflections
  - By containment of the noises (unwanted signals) injected through various paths, such as interline coupling (controlled by the lines' cross sections and shared lengths) and the common path impedances
- Assure *reliable on-chip nets performance:*
  - By ascertaining integrity of power distribution
  - By providing integrity of the chip-to-package interface regions: self-inductances and mutual inductances and capacitances
- Optimize the packaging cross-sectional geometries and, if possible, materials:
  - By avoiding insufficient/excessive interline separation and line cross sections
  - By providing adequate ground and power planes
- Devise and verify "wiring rules":
  - For the "legal" signal nets configurations—and to grant expections
  - To facilitate implementation of the noise containment constraints
- Enable performance prediction, or at least assessment

These objectives were pioneered by the developers of packaged electronics for the mainframes and supercomputers; now they are becoming increasingly pervasive for the entire range of digital processors, because their clock frequencies are going well beyond the 100-MHz mark. As was mentioned in Section 7.1 and will be shown in Section 7.2.11, the demand for higher performance had been driving down the overall dimensions of a computer while demanding more circuits and more bits of storage. The higher level of IC integration enables such size reductions. In addition to the impact on physical dimensions, two other key considerations need further discussion—the electrical considerations, to be dealt with in this section, and the thermal considerations, which will be covered in the Section 7.2.4. The in-depth coverage of these topics will be given in Chapters 3 and 4, respectively.

7.2.2.1 General

Digital information processing is based on assigning a discrete (binary) value to a small range of an electrical signal and another value to another range. This value can the be transmitted through a wire to another location, stored, or applied to a logic circuit as its input signal and, as a rule, combined with the other input signal(s) to form an output signal, the value of which is governed by the desirable logical function (e.g., INVERT, OR, AND, etc.). New information is processed by changing the input signals and using the output signal as an input to the subsequent operation(s).

Changing information then requires a change in the signal level from one value to the other. The transition to a new signal value cannot occur instantaneously. The time it takes is called the signal transition time. Short transmission times facilitate better performances but also cause a variety of problems.

Unlike the liquid or heat flow, or the propagation of light and other electromagnetic radiation, transmission of electrical signals requires the use of two conductors, such as the wires to a light bulb or to a loudspeaker. In the majority of data-processing applications, the one conductor is well defined, frequently used terms being a **wire,** a **connection,** or a **signal line,** whereas the other conductor is shared by many other signal and power circuits. From the early days of electronics, that second conductor has been usually provided by the metallic packaging structures, such as the chassis of an amplifier or a TV set. The presence of the second conductor is often only implied in the schematic diagrams, based on the assumption that no special attention is required to its role in conveying signals.

Similarly, properties of the connecting conductor are also frequently assumed to be ideal, and of no particular consequence to the task of connecting two circuit elements or two complete circuits. Again, no further attention is paid to any of its properties. Such lack of attention may, and in the case of high-performance electronic circuits definitely will, cause the appearance of several phenomena that are undesirable and often harmful to the intended function.

A particularly lucid treatment of this subject—in fact, an excellent supplement to Chapter 3—is accomplished by Ron K. Poon in Section 8 of his very comprehensive book published in late 1995 [10].

7.2.2.2 Electromagnetic Properties of Signal Wire

Three electrical parameters—resistance, capacitance, inductance—are always present[2] and cause signal delays and many forms of signal distortions, including the unwanted appearance of signals on supposedly quiescent (not switched) lines. It is these unwanted noise signals that are

---

2 Except zero DC resistance in a very special case of superconductivity.

the biggest troublemakers, because they are hard to eliminate from a finished design; they must be addressed during the initial design phase.

All signal lines and also the input electrodes of switching devices (e.g., gate, base) have a certain capacitance; they store an electric charge when a voltage is applied. To change the voltage, the amount of charge must be changed by the flow of electric current. This flow causes the buildup of a magnetic field that induces voltage into the current-carrying line (self-inductance), resisting the current flow, and into the other lines in the proximity of the line being switched (mutual inductance), causing a current flow there. Furthermore, the source resistance of the switching circuit and the signal line resistance also limit the amount of current flow and, thus, control the time during which the signal transits from its old value to the new. The transition time increases with the increase in capacitance.

The design of faster devices and circuits concerns itself, among other things, with the faster switching speeds of devices, with the reduction of the device/circuit input capacitance, and with the optimization (reduction) of the driving (source) impedance. The packaging design (geometries and materials) can control the signal line capacitance to the reference plane, the line resistance, its self-inductance, and interline capacitances and inductances. Consequently, it can control the signal distortions and the appearance of unwanted interferences. In spite of best efforts, it may even limit performance of the high-speed computers.

The signal line capacitance is directly proportional to the line length, whereas the capacitance per unit length is a function of cross-sectional geometry and the value of the dielectric constant of chosen materials. Note that the scaling of geometry, preserving all ratios of dimensions, preserves the capacitance per unit length.

7.2.2.3 Signal Degradation

There are two internal causes of degradation (omitting externally induced interferences) as the signal propagates through the package—reflections and the effects of line resistance. Resistance mainly causes voltage drops and increases in the signal transition time. Reflections, in addition to causing increases in the transition time, may split the signal into two or more pulses, with the potential of causing erroneous switching in the subsequent circuits and, thus, malfunction of the system. As will be shown in detail in Chapter 3, the reduction in signal level due to the resistive voltage drop will decrease immunity to the signal noise which has several causes; mainly cross-coupling noise, switching (ΔI) noise, and signal reflections.

7.2.2.4 Cross-Coupling Noise

The cause of cross-coupling noise is illustrated in Figure 7-11, along with the causes of switching and reflected noises. The structure shown

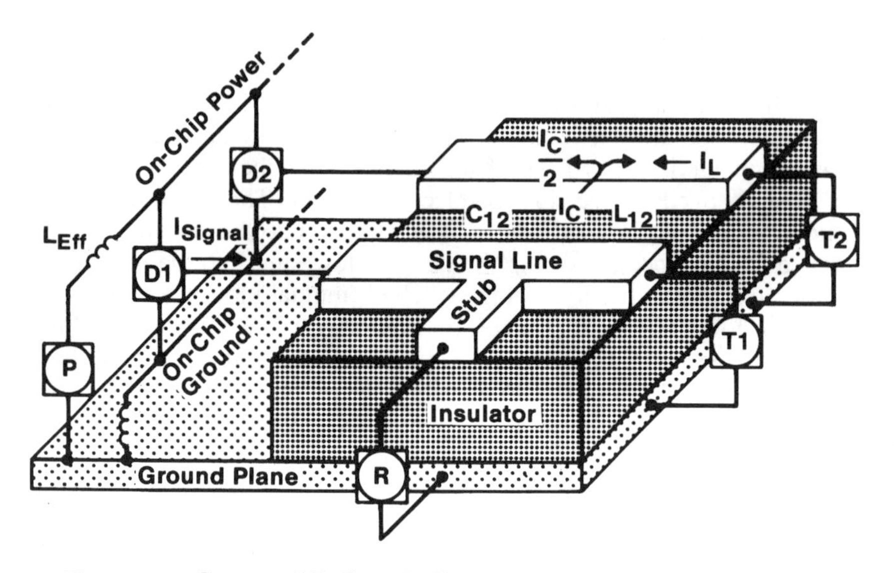

Figure 7-11. Causes of Reflected, Coupled, and Switching (ΔI) Noises

represents two short sections of typical wiring. Each signal line, in addition to its own resistance, capacitance to the reference (ground) plane, and self-inductance, has mutual inductance L_{12} and interline capacitance C_{12} to its neighbor.

When driver D1 sends a changing signal voltage and current through the first signal line toward its terminating resistor T1, line 2 between D2 and T2 is supposed to be in a steady state, not switched. Voltage change on line 1 induces current I_c through C_{12} which splits and flows toward both ends of line 2. The current change on line 1 induces a voltage on line 2, which, in turn, causes the current I_L to flow in the direction opposite to the signal current on line 1.

In many cases, induced currents $I_c/2$ and I_L, flowing toward the far end of line 2, cancel each other. By contrast, the currents flowing toward D2 (near end) add up, are reflected by the low impedance of D2, propagate toward T2, and contribute to the signal noise.

7.2.2.5 Switching (ΔI) Noise

Another serious, and potentially the worst, noise originates by the driver switching current drawn from the power distribution system P, through the inductance L_{eff} common to some, many, or all drivers of the same chip. A voltage spike across L_{eff} will usually propagate through all drivers (even the quiescent D2 in this example) and, because of the additive effect of simultaneously switching drivers, can cause severe problems. The remedy may be either in reducing the value of L_{eff} by changing packaging geometries or in restricting the total switched current by either slowing down the driver speed, by limiting the number of drivers allowed to switch at the same time, or by incorporating decoupling capacitors, which shunt some of the switching current, effectively reducing ΔI through L_{eff} and the noise generated by it.

7.2.2.6 Signal Reflections

A potentially very serious signal degradation occurs at the **junction** between the signal line and the **stub** of Figure 7-11, providing connection to the input of a circuit. If the stub delay is less than one-fourth of the signal transition time, then it acts as a capacitive load, causing a dip in the signal transition waveform and sending a negative reflection toward the driver. This by itself is undesirable but unavoidable. It may be intolerable if the stub capacitance in combination with receiver R1 input capacitance generates a large reflection that can be magnified by several other closely spaced (less than one-fourth the transition time) stubs and receivers. The resultant degradation of the signal transition waveforms may also be unacceptable.

These problems are further magnified by the long(er) or multiple stubs. When the signal transition arrives at the stub location, it splits. One part, with the reduced amplitude, travels along the signal line and the remainder travels along the stub toward the receiver input. A negative (reflected) signal is sent on the signal line toward the driver. After charging the receiver input capacitance, it is reflected back, traveling on the stub toward the signal line. Upon arrival there, it restores the signal to its original value. In the meantime, the negative reflection may affect receivers located between the driver and this location. Other receivers, located between the stub and the line terminator, may not switch until the restored signal amplitude arrives at their respective locations.

Shorter signal transition times improve performance. They also cause increased noise and restrict the maximum stub length and number of stubs. Their generation usually requires higher power, just as the achievement of higher circuit performance does.

7.2.3 Electrical Performance—Power Distribution

- Leverages of a "quiet" power distribution:
  - Reduced noise due to circuits switching activity
  - Enables either wider processing tolerances or *better performance*
  - Permits lower power and signal voltages ⇔ better speed–power product
- How to achieve a "quiet" power distribution:
  - Use high-speed voltage regulator and enough decoupling capacitors

"Hierarchical" decoupling ⇔ control of resonance peaks, as shown in Figures 3-12 and 3-13
Need to contain deleterious effects of interconnection inductances
Employ distributed decoupling *with low inductance*

7.2.3.1 Rigorous Analysis

Following is the rigorous analysis of the noise generated in any power distribution system, derived from Ref. 11. Let us designate:

V_s = signal voltage, as a function of time
V_p = voltage variation at the power terminals of a circuit or a circuit group
L = inductance of the decoupling capacitor and its connections to the circuit(s)
i_p = current drawn by *ckt*(s) from the power distribution
C_p = decoupling (power distribution) capacitance
R_c = intrinsic resistance of the decoupling capacitor
V_{ci} = voltage change across the decoupling capacitor caused by the dielectric rolloff

The rigorous equation for V_p is

$$V_p(t) = L\frac{di_p}{dt} + \frac{1}{C_p}\int i_p dt + R_c i_p + V_{ci} \qquad [7\text{-}2]$$

For unterminated lines, with pure capacitive load C_s,

$$i_p = C_s \frac{dV_s}{dt} \qquad [7\text{-}3]$$

Sustitution yields

$$V_p(t) = LC_s\frac{d^2V_s}{dt^2} + \frac{C_s}{C_p}V_s(t) + R_cC_s\frac{dV_s}{dt} + V_{ci} \qquad [7\text{-}4]$$

The ratio of noise-to-signal voltage is

$$\frac{V_p(t)}{V_s(t)} = LC_s\frac{1}{V_s(t)}\frac{d^2V_s}{dt^2} + \frac{C_s}{C_p} + R_cC_s\frac{1}{V_s(t)}\frac{dV_s}{dt} + \frac{V_{ci}}{V_s(t)} \qquad [7\text{-}5]$$

This is a precise but rather complex depiction. Note the significance of the four terms:

LC_s indicates the importance of keeping small the value of L—the inductance of the "last" power-distribution link between the decoupling capacitor and the circuits being decoupled—for high-performance applications with high values of dV_s/dt and d^2V_s/dt^2. It will be rather crudely illustrated on the following pages in which d^2V_s/dt^2 is replaced with $(dV_s/dt)^2$.

C_s/C_p indicates the fraction of the charge the decoupling capacitor(s) are to lose when supplying current to (re)charge the signal lines through the driver circuits. Thus, it determines the minimum value of C_p for a given C_s.

R_cC_s is the intrinsic time constant of the decoupling capacitor(s), assuming no dielectric rolloff. It indicates the need for high-quality capacitors (low R_s) for high-performance (high dV_s/dt) applications.

$V_{ci}/V_s(t)$ indicates the additional voltage drop caused by the dielectric constant rolloff at high frequencies, a common occurrence with the high-dielectric-constant materials which can negate the high values of ε_r occurring at low frequencies—typical for very high values of ε_r. As a rule, the dielectrics with the higher values of ε_r begin to drop it at lower frequencies and rolloff more than those with the lower values. The breakdown field (V/cm) is also lower for the dielectric materials with higher values of ε_r.

7.2.3.2 Very Rough Estimate of Acceptable Lead Inductance $L = L_{max}$

Note that rigorous calculations involve second-order nonlinear differential equations, as shown above.

One circuit is driving a capacitive load $C_{signal} = C_s$ with the signal amplitude V_s. During the signal transition time Δt, there will be a surge of current ΔI through the signal and ground/power leads.

$$\Delta I = C_s(V_s / \Delta t) \qquad [7\text{-}6]$$

(With more outputs n, simultaneously switching, the total current surge is $\Delta I = n\Delta I_{single}$). This current surge ΔI during Δt_s will produce a voltage spike ΔV at the corresponding parasitic inductances L of signal, ground, and power leads.

$$\Delta V = L\Delta I / \Delta t_v \qquad [7\text{-}7]$$

Substituting equation [7-6] for ΔI in equation [7-7]:

$$\Delta V = V_sLC_s[1 / (\Delta t\Delta t_v)] \qquad [7\text{-}8]$$

Table 7-3. Values of Maximum Power Lead Inductances L_{max}

For $\Delta V/V_{signal} = 0.1$ and $n = 1$	$C_S = 10$ pF	$C_S = 1$ pF	$C_S = 0.1$ PF
$\Delta t = 10$ ns $= 10{,}000$ ps $(\Delta t)^2 = 10^8$ **"Yesterday's designs"**	$L_{max} = 1\ \mu H$	$L_{max} = 10\ \mu H$	$L_{max} = 100\ \mu H$
$\Delta t = 1$ ns $= 1000$ ps $(\Delta t)^2 = 10^6$ **"Today's designs"**	$L_{max} = 10$ nH	$L_{max} = 100$ nH	$L_{max} = 1\ \mu H$
$\Delta t = 0.1$ ns $= 100$ ps $(\Delta t)^2 = 10^4$ **"Tomorrow's designs"**	$L_{max} = 100$ pH	$L_{max} = 1$ nH	$L_{max} = 10$ nH
$\Delta t = 10$ ps $(\Delta t)^2 = 100$ **"Future designs"**	$L_{max} = 1$ pH	$L_{max} = 10$ pH	$L_{max} = 100$ pH

By rearranging,

$$\Delta V / V_{signal} = LC_s 1 / (\Delta t \Delta t_v) \approx LC_s (1 / \Delta t)^2 \; for \; \Delta t_v \approx \Delta t \qquad [7\text{-}9]$$

Note that if L is in pH, C_s is in pF, and Δt is in ps, no multiplier is required.

A typical interconnecting wire in a PWB is a transmission line with the characteristic impedance $Z_0 \approx 80\ \Omega$ and $\varepsilon_r \approx 4.5$ [12]. Such lines have $L = 550$ pH/mm = 5.5 nH/cm and $C = 85$ fF/mm = 0.085 pF/mm = 0.85 pF/cm.

Contemplating Table 7-3, we notice that the typical PWB wire lengths of "yesterday's designs" were on the order of 1–50 cm. The corresponding C_s were then in the 1–50-pF range, roughly represented in the first two columns of Table 7-3. The long signal rise times permitted the power interconnectin inductances in the 0.2–10-μH range, or > 40 cm—an easily manageable length of a signal wire serving as a power-distribution connection.

By contract, the "today's designs," which may well have become those of yesterday when you read this, have similar wire lengths, but the signal transition times shrunk to about 1 ns, forcing the power distribution wires into the several-centimeter range—still an easy to meet requirement.

The today's (ca. 1996) mainframe designs, with ~0.1 ns signal transition times, will be almost all designs around the year 2000. The signal transmission lines become shorter, with capacitances in the $0.1–1^+$ pF. These designs can tolerate the power-distribution interconnection lengths of only a few millimeters as, for example, was done for the IBM ES/9000 mainframe MCM packaging [11,13].

Very advanced designs, still in the research phase in the late nineties, aim at clock frequencies of 1 GHz and higher. This drives the signal transition times into the 10-ps range. The signal-line capacitances are still in the 0.1–1-pF range. Consequently, the affordable power-connection inductances must be in the few pH range, requiring the interconnection lengths of a fraction of 1mm! This is a significant paradigm shift, which amounts to replacement of discrete decoupling capacitors, such as in Refs. 11 and 13, with the low-impedance power-distribution systems, such as demonstrated with thin high-dielectric films sandwiched between the power and ground planes [14,15].

7.2.4 Thermal Design Considerations

A somewhat humorous way to describe electronic data processing might be by saying that its prime purpose is the conversion of the electrical energy into thermal—generation of heat. Some circuits, notably those employing bipolar devices, generate most of the heat even if they idle (i.e., process no information at all). Others, notably CMOS, generate heat only when they switch and do so in direct proportion to their operating (clock) frequency. High concentration of circuits within a small area of an IC, leads to high **power density,** measured in W/cm^2. The power density is actually another term for a heat flux—the number of cal/s (1 W = 0.239 cal/s) to be transported from the location of their generation to the location of their removal. Such transport is associated with a temperature difference between the two locations. **Thermal resistance** quantifies the temperature rise for a given flux and is expressed in °C/W or K/W.

The quest for higher and higher levels of integration drives technologies to produce ever-smaller devices, interconnections, and terminals. As already illustrated in Table 7-1, the demands on heat-removal capability can easily snowball. Consider, for example, Figure 7-12 which relates the power density, in W/cm^2, demands of 10-mm × 10-mm IC chips with various numbers of circuits and operating at various levels of circuit power. It also benchmarks temperatures and power densities of a light bulb and the sun surface. A nuclear blast, a mile away, fits in between. So will chips with a circuit count range around 1990 (10,000–100,000) if each were to consume 1–10 mW. But the **chip temperature** must be below 100°C to assure proper electrical performance and to contain the propensity to fail.

Figure 7-12 clearly demonstrates the need to enhance heat removal and to contain the power hunger of the logic circuits. It also indicates that the power density, and not the more commonly used power-per-chip

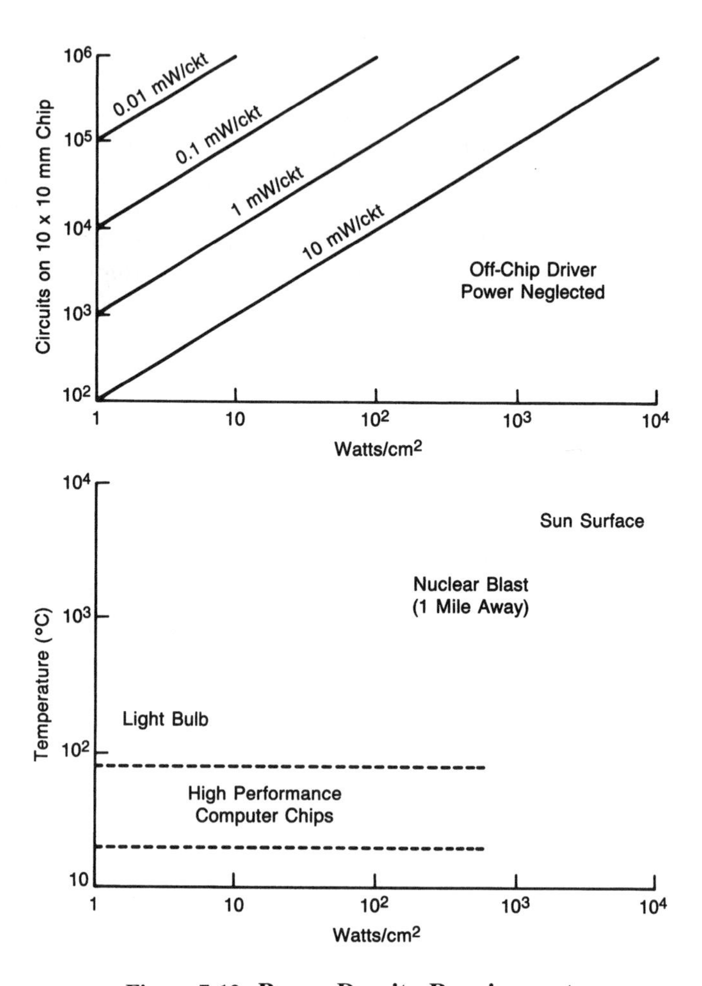

Figure 7-12. Power Density Requirements

figure, is the measure of the heat-removal capability of a package. This capability is particularly important when large assemblies of closely spaced high-density high-performance chips must be employed, as in a powerful mainframe.

Figure 7-13 presents an evolution summary of the heat-flux demands for a wide range of applications, from the hand-held equipment to high-performance mainframes and supercomputers. It also indicates the projected increases in the chip (die) dimensions. The conventional package consisting of single-chip modules and card on board is capable of about an order-of-magnitude lower power density than the state-of-the-art multilayer ceramic technology with liquid cooling. However, significant progress has been made, and continues to be made, in increasing the manageable power density, as shown in Figure 7-14.

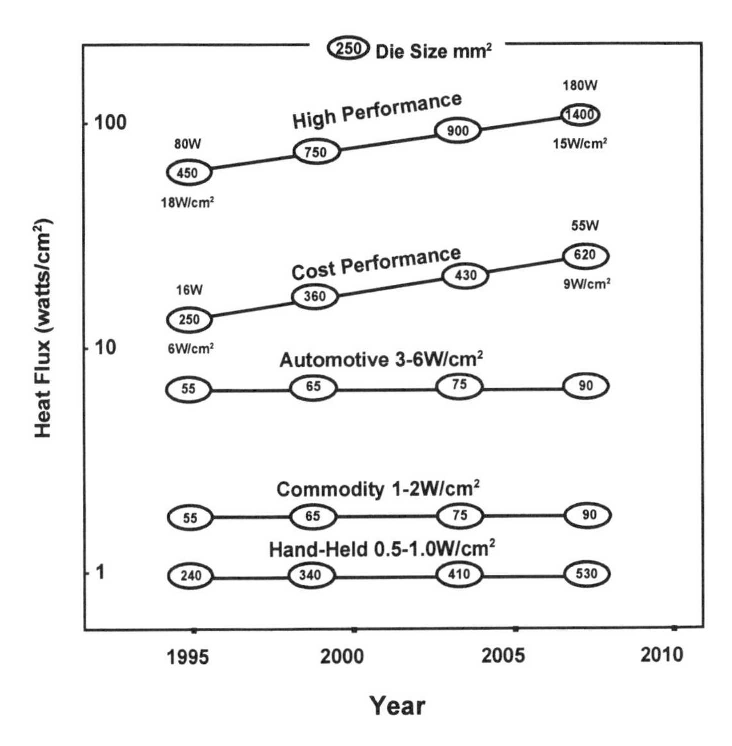

Figure 7-13. Evolution of the Thermal-Dissipation Requirements

Thermal expansion caused by heating up the packaging structure is not uniform—it varies in accordance with the temperature gradient at any point in time and with the mismatches in the thermal coefficient of expansion. Mechanical stresses result from these differences and are one of the contributors to the finite lifetime and the failure rate of any packaging structure.

7.2.5 Reliability

There are two fundamentally opposite approaches to the reliability aspect of a design. The first one is to build the hardware of a given design and either just hope for the best or subject it to some sort of stress test(s) to assess the potential for performing at acceptable failure rates within the anticipated/specified lifetime and under expected/specified operating conditions.

When the risk of failing rules out such an approach (e.g., in the case of nuclear plants, aircraft, or electronics for critical applications), then the second alternative is the only choice. It involves understanding the failure mechanisms, systematic elimination of their causes during the design phase, and, finally, passing a set of qualification tests. Such tests

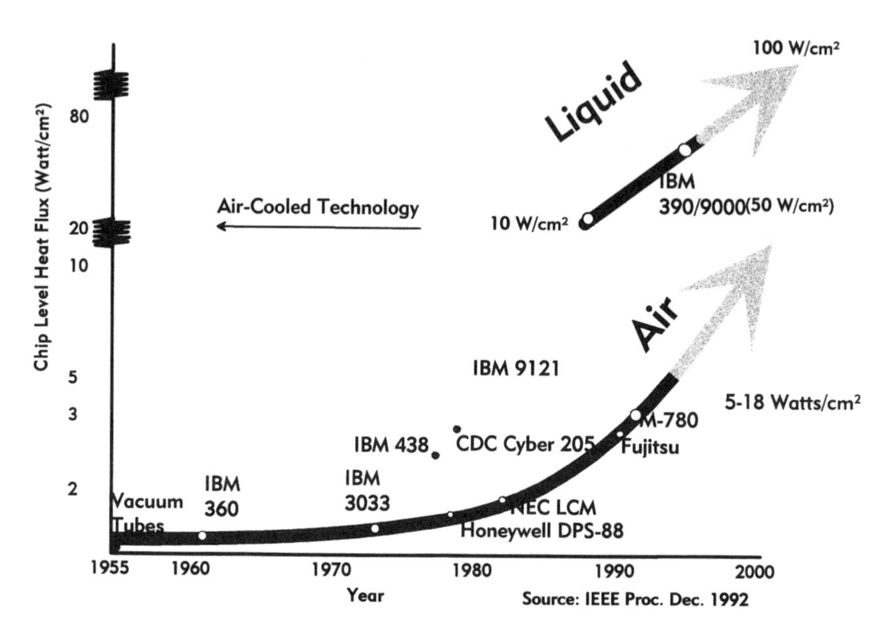

Figure 7-14. Evolution of Air/Liquid Cooling Capabilities. (From *IEEE Proc.* Dec. 1992.)

(or subsets of them) may also be repeated at prescribed intervals to ascertain continued stability of the product manufacturing process.

In the case of microelectronics, the failure to function may have two basically different causes—error signals occurring in the otherwise fully functional product and a compromise to the product integrity, such as opens and shorts in the signal lines. Containment of error signals has already been highlighted in Section 7.2.2 and is covered in great detail in Chapter 3.

Product integrity disruptions are caused either by the intrinsic wear-out mechanisms or by hostile environmental factors, such as corroding agents penetrating the package interior and attacking sensitive parts or their interfaces. The formation of cracks in the solder joints caused by mechanical stresses which result from nonuniform thermal expansions and contractions is an example of the wear-out mechanism. Corrosion of connector contacts may result either from a process residue not fully removed or from the environmental contaminants. Chapter 5 deals with the determination of reliability parameters and Chapter 10 describes technologies specifically aimed at protection against the environment. Chapters 7–13 include discussions of the reliability aspects.

The failure-rate reductions in leading-edge electronic products were quite phenomenal—about four orders of magnitude in a quarter of century, or a twofold reduction every 2 years—as presented in Figure 7-15.

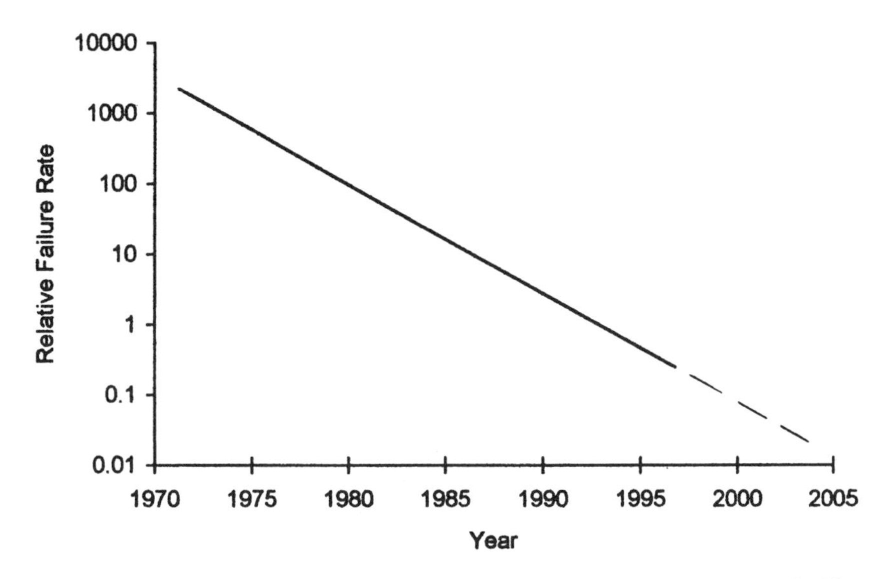

Figure 7-15. Intrinsic Failure-Rate Improvement Trends for IBM Logic-Circuit Reliability

7.2.6 Manufacturability and Quality

For the majority of the packaged electronic products, the emphasis has shifted from performance—the major preoccupation at the time of writing the first edition—to cost. Performance is still very important as, for example, manifested by the continuing new PC announcements with improved price per performance ($/MIPS) shown in Figure 7-1. But the drive for a constant cost–performance factor has been replaced by the drive to maintain or reduce the cost or raise the performance without sacrifice in quality or reliability. There are two major enabling factors to accomplish these goals:

- Establishment and maintenance of manufacturing process tolerance windows necessary and sufficient for the product to pass its tests, actually with ever decreasing incidence of fails as illustrated in Figure 7-16. Elimination of causes of "killer defects," such as sloppy handling of parts, is also an essential part of this cost-reduction effort.
- Elimination (or a drastic reduction) of buffer inventories by a balanced manufacturing throughput capability and an uninterrupted flow of product through various manufacturing steps. Increased throughput lowers manufacturing costs in proportion.

In recognition of the pivotal role these factors play in product competitiveness and success in the marketplace, a new Chapter 6, "Package

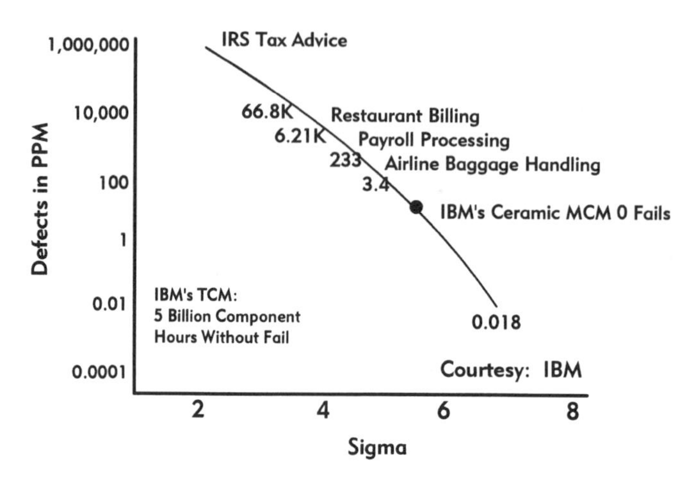

Figure 7-16. Defects Versus Statistical-Distribution Limits, Trend Is Toward ≥6σ

Manufacturing" has been added in this second edition. It reviews these five key principles:

- Quality in product and process designs
- Product qualification
- Manufacturing line modeling and simulation
- Process control
- Worldwide quality standard—ISO-9000—for quality systems certification and verification procedures. In many cases, it has already become a necessary element when participating in trade within and between certain countries.

7.2.7 Testability

Implicit in reliability considerations is the assumption of a flawless product function after its initial assembly—a zero-defect manufacturing. Even though feasible in principle, it is rarely practiced because of the high costs and possible loss of competitive edge due to conservative dimensions, tolerances, materials, and process choices. The commonly practiced alternative is to balance the above variables against loss of defective product, which must be screened out by testing.

At the dawn of electronics, discrete components were individually tested prior to assembly. Such tests were often simple; for example, the value of a resistor or capacitor. Completed cards were tested for functionality. Again, such tests involved a small number of steps necessary to ascertain freedom of assembly defects. With the introduction of statistical designs in the late fifties, additional tests were required to weed out the small percentage of circuits (cards) whose components contained extreme values of their parameters and caused at least one important circuit parameter to fall outside of specified tolerances. Such cards were commonly reworked by identifying, desoldering, removing, and replacing the likely culprit—an item still within its own tolerance band but not blending well with other components on that particular card. Additional sophistication was required in devising and performing tests and in analyzing the test results. Similar methods were employed in determining a faulty component, one which either slipped through its testing screen or was damaged during the card assembly. These techniques also worked well with cards that failed in the field, which were found, replaced, and returned for repair and reuse.

Finding a faulty card was facilitated by connector pins protruding out of the other side of the board and offering convenient test points for a voltmeter or an oscilloscope. “Swapping” two cards of the same design (part number) often could cure a problem arising from the statistical design of the off-card signal nets.

None of these techniques work with the large-scale integration (LSI) and very large-scale integration (VLSI) chips. Circuit terminals are buried in the chip wiring, and of course, nothing on the chip can be removed and replaced. Whereas at lower levels of integration, some external test points (terminals) would permit access to internal nodes, the higher level of integration and the resultant pinout demands severely limit such an option. Two different solutions have evolved, each suitable for different designs. Ultimately, a combination of these solutions, and other ideas already on the horizon, will provide an adequate level of testability.

Storage arrays lend themselves to testing through their regular terminals by writing in and reading out patterns that determine the ability of each individual cell to store ones and zeros and to be immune to the switching activity of other cells or groups of cells. By contrast, random logic requires strict design discipline, sophisticated test pattern generation, and involved analysis to verify proper functionality and to uniquely identify causes of malfunction [9].

Multichip modules (MCMs) or multichip packages (MCPs) populated with VLSI chips and/or with specialized microwave transmitter/receiver chips have become high-value products because of the expensive chips and because of the relatively complex, high-cost substrates. Avoidance of nonshippable modules is an economic must. It is accomplished by the use of so-called “known good die” (KGD)—pretested chips, including burn-in—and thorough pretesting of the substrates. Doing both minimizes the risk of manufacturing defective MCMs but does not reduce it to zero. Therefore, testing of populated MCMs is also required. This testing has two major facets. One is an extension of the design for testability methodologies and techniques developed for the KGD. The second is akin to functional testing of subsystems which, in fact, many MCMs are.

The subsystem testing is highly application dependent. It is still beyond the scope of this book. Testing of the substrates and KGD-type testing (functional and diagnostic) of populated MCMs has been added to this second edition in Chapter 13, "Electrical Testing." It covers various techniques (both, currently practiced and still in the R&D phase) for opens and shorts testing of the wiring nets, contacting the substrates with single-terminal and multiterminal probes, the boundary-scan and built-in self-tests of the populated modules/packages.

7.2.8 Memory Packaging

The requirements for the main memory packaging—as opposed to various buffers and caches which are often integral with logic—were least affected by the performance race, primarily because the speed of the dynamic random-access memory (DRAM) chips is improving at a very low rate. The cost improvements are the main driving force. This improvements were leveraged by several key factors. One, very consistent of them, is the steadily increasing number of bits per chip and the bits density on the chip, a planar—two-dimensional—density increase of approximately 2 times every 3 years, or 10 times every 10 years due to improvements in the photolithography. In the past, there had been an approximately sixfold gain in the on-chip density produced by the circuit inventions which replaced a six-transistor storage cell with a one-transistor plus capacitor cell in DRAMs. This maps into a prorated annual density improvement of 9.4% (1.094×), or 2.45× in 10 years. Thus, the on-chip advancements contribute to approximately 25-fold density improvements every 10 years.

Figure 7-17 shows an interesting long-range trend of memory density

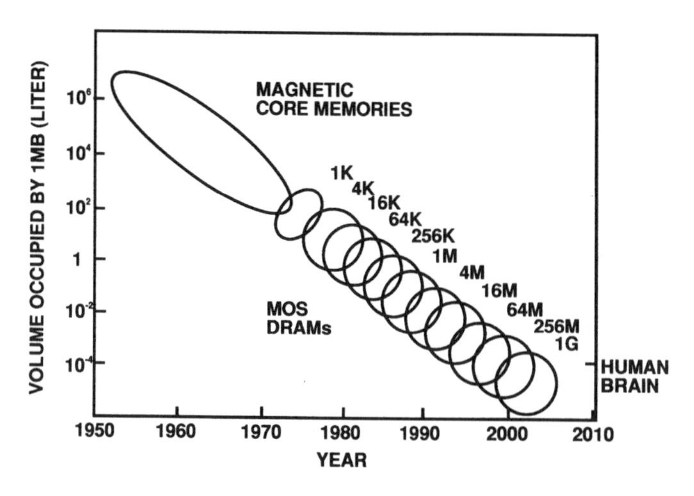

Figure 7-17. Trend in the Volume Required for 1 MB of Storage (Courtesy of SRC, reproduced with permission.)

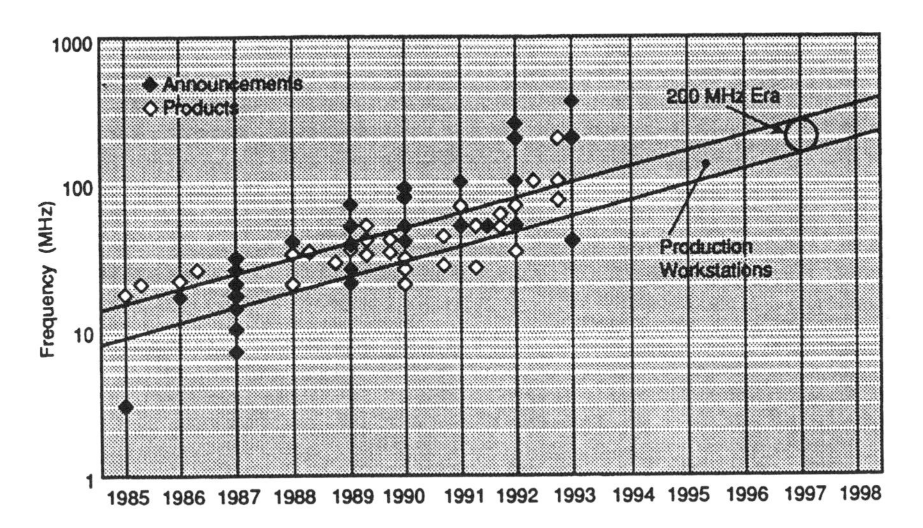

Figure 7-18. Microprocessor Clock Frequencies (Courtesy of TechSearch International; from Ref. 26.)

improvements over a 60-year span. The continuous trend is approximately 200× every 10 years, even though the ferrite cores of the 1950–1970s were replaced by the semiconductor chips. With the on-chip improvements contributing ~25×, the balance of 8× in 10 years must be, and in fact is, contributed by the packaging advancements such as increased planar and volumetric silicon efficiency.

Finally, also note that in the first decade of the next century, we anticipate the volumetric bit density of the electronic memories to match that of the human brain.

The requirements for memory packaging are somewhat different from those for the packaging of logic discussed in almost all the chapters. At the systems level, the demand for memory in terms of numbers of active elements per chip has outpaced all the other packaging requirements (Figs. 7-8 and 7-17). Today's supercomputers, for example, require as much as 2–10 gigabytes of main memory, which must be packaged in such a way that the access time to it, as well as retrieval from it, is fast, often no more than a few cycles of the machine. All of this clearly requires that the memory package be as dense as possible and either be part of the logic or be very close to it. Also see Section 7.2.11 and Figure 7-24. Memory chips generally have very low-power and I/O requirements, as pointed out earlier in the chapter, and thus lend themselves to the low-cost, high-density packaging.

The card-on-board technology is the most common technology used for packaging memory to date. However, a departure from this, particularly

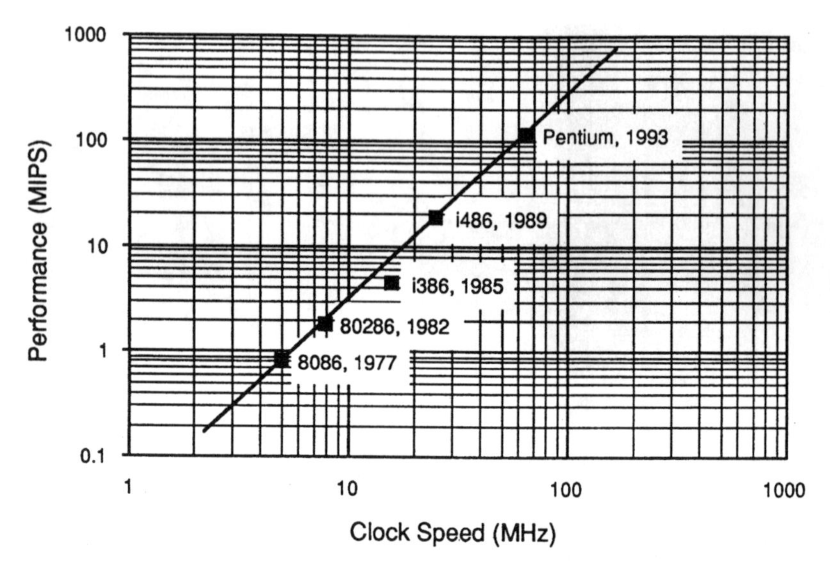

Figure 7-19. Intel Microprocessor Clock Frequency Versus Performance (From Ref. 26.)

for high-end mainframes and top-of-the-line supercomputers that can process information in a cycle of a few nanoseconds, is expected. In the early 1970s DIPs with 16 leads were capable of handling 4K dynamic random access memory (DRAM). Single inline packages (SIPs) and DIPs handled up to 256K DRAM's in the late 1970s. The megabit DRAMs that were beginning to be available in the mid-1980s were expected to be packaged in PLCC or quad packs. Each card (250 × 270 mm) is capable of holding about 160 chips, along with logic support chips, to store 1 megabyte with 64K DRAMs, 4 megabytes with 256K DRAMS, and 16 megabytes with 1-megabyte chips. Both sides mounting with six layers of interconnecting lines within the card could be used to double the memory capacity to 32 megabytes. Half a gigabyte can be achieved by edge connecting 16 of these cards in the width of an 18-layer board.

7.2.9 Personal Computers and Workstations

At first it would seem that personal computers should be among the easier products to design; that is, with the large systems driving performance technology development and the consumer products driving cost, designing a personal computer should simply mean taking the best from both disciplines. This would indeed be the case if it were not for the many contending user and business requirements for entry systems which include, but are not limited to, the following.

First and foremost, the machine must be **cost-competitive** and offer

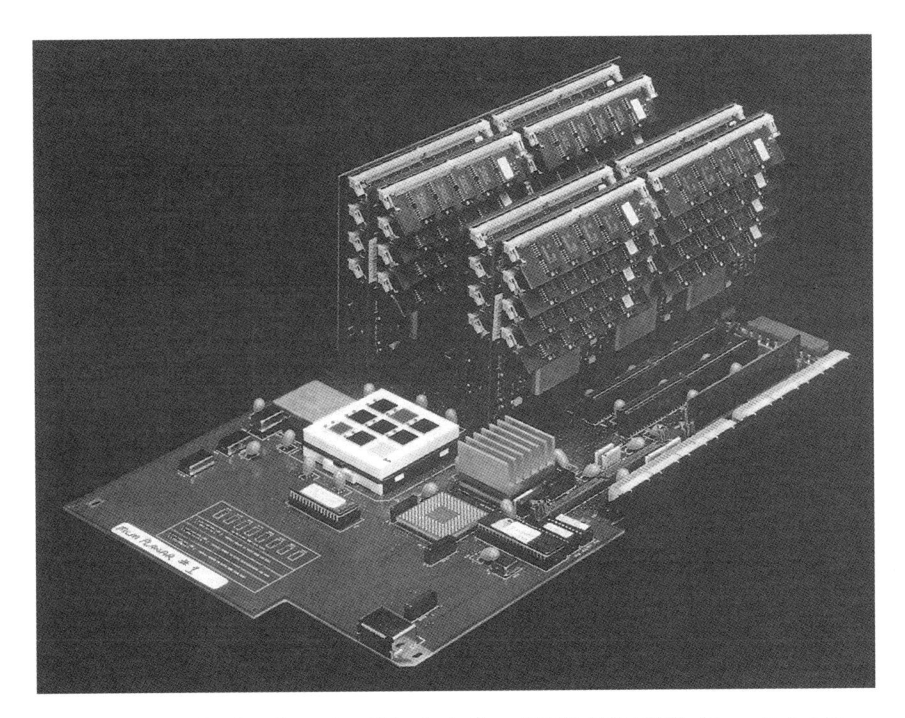

Figure 7-20. Packaging of a Workstation (IBM RS/6000) (From Ref. 26.)

on the order of a factor of 2 improvement in value in comparison to the products marketed just 2 or 3 years before. In the short history of personal computers this has been the compound rate of cost–performance improvement as already indicated in Figure 7-1. Today's machines must provide **performance** in the range of mainframe processors of just 7 years ago. The **reliability** of the machine means that it must work right out of the box just as any other modern electrical appliance, and for a comparable period of time. Also it should be easily assembled and operated by the customers who usually do not possess a high level of technical skill.

The personal computer, more than any other data-processing machine, serves a broad range of applications and therefore must offer "expansion" cards for configuration to the problems of that specific user. As personal computers are typically customer-operated systems, provisions for automatic-fault detection and location are required to enable the owner to make repairs with a reasonable level of confidence. The "footprint" of the machine should be as small as possible, so that the machine will consume the minimum of desktop work space and be an unobtrusive part of the office environment. It should be of reasonable weight for portability, whether for office rearrangement or for relocation to where the work is to be done.

The display capabilities expected by the customers are far more than the text-quality graphics provided by the TV-like screens of the first-generation personal computers. Current product generations approach CAD/CAM-like system's display capability requiring high speed and high bandwidth for the manipulation of these data. Video-processing requirements and the hardware content for this aspect can easily surpass the complexity of the "central processor" hardware of the machine. The personal computer must not only be sparing in its power requirements but also should be tolerant of electrically noisy power environments. Its electromagnetic radiation must achieve legislated standards which have been tightened worldwide in the recent past. Finally, the personal computer must "look good," be "human engineered," and also "user friendly."

The mid-range data-processing systems require a blend of performance and low cost optimized for the lowest possible cost for a given performance. The typical performance range of these is between about one MIPs (million instructions per second) or less, to almost the bottom end of mainframes, which in the mid-nineties is around 100 MIPS per uniprocessor. This range is generally achieved by a single processor, but sometimes with multiple processors, particularly in the top range of performance.

The semiconductor technology capable of providing the function necessary to achieve these performances has been primarily ECL and TTL, but CMOS is emerging as the leading candidates offering much higher integration levels with lower power requirements [5,8]. Single-card central-processing units are marketed with those silicon technologies and improved system's performance. BICMOS is a strongly emerging technology taking the best of bipolar and CMOS transistors on the same die, providing an alternative to both those technologies in many applications.

The internal CMOS circuits offer very low-power and low-current operation sufficient for the short on-chip circuit-to-circuit communications and the bipolar circuits are used primarily for high-speed off-chip drivers. In comparison, this technology provides a lower speed-power product than can be achieved by either of the semiconductor technologies.

7.2.10 Portable and Mixed-Signal Equipment

All portable equipment, from hand-held calculator, to lap-top PCs, to various communications equipment (e.g., pagers, portable telephones, geo-positioning systems, etc.) must consume very little power to prolong the operating time on one battery charge. As a by-product, the thermal management is simplified. However, the operating environment, especially automotive and marine, is considerably more hostile than a typical office environment. The ever-smaller size and mass of the equipment (Fig. 7-21)

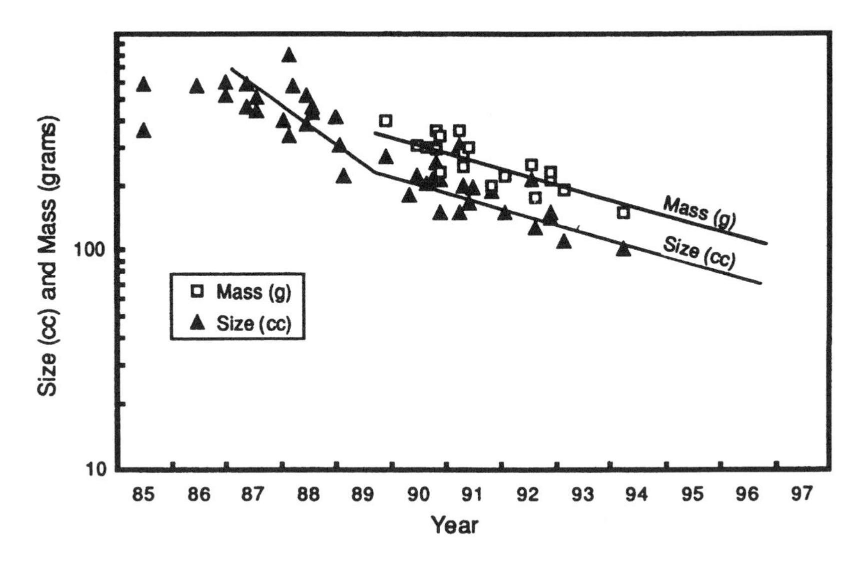

Figure 7-21. Trends in Mass and Size of Cellular Phones (From Ref. 26.)

leads to small dimensions of the components and interconnections and adds to the challenge of design for ruggedness.

The mixed-signal applications environment has its own set of design challenges: good electromagnetic isolation between the digital and the analog, usually radio-frequency (RF), circuits to avoid malfunctions due to injections (either way) of potentially strong noise signals; and well-controlled low-loss properties of interconnections which, in fact, may often constitute various circuits and circuit elements—resonance circuits and filters, impedance transformers, and even receiving and transmitting antennas [7]. Once again, the cost-competitiveness is a very important design and manufacturing parameter.

7.2.11 High-Performance Processors

The general trends and key dependencies were highlighted in Section 7.1. Figure 7-22 quantifies the performance trends as determined by the two key parameters: the cycle time (reciprocal of the clock frequency) and the average number of cycles the processor needs to execute an instructions—cycles per instruction or C/I. The short cycle times require correspondingly short interconnections—signal propagation time through the interconnections must be less than the cycle time. The scale on the right side of Figure 7-22 indicates the propagation times though the interconnections whose insulating materials have dielectric constants of 1.6—an extremely small number, not yet available—and 10—a number

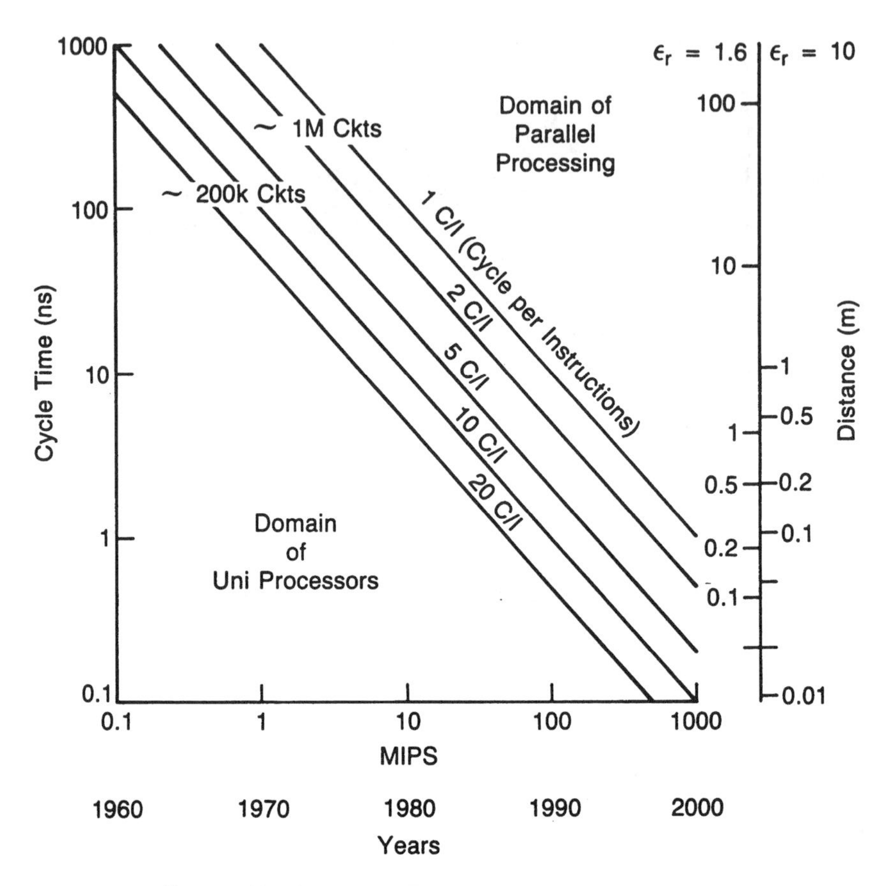

Figure 7-22. System Performance Dependency on Cycle Time and Cycles per Instruction

somewhat higher than present with alumina-based ceramics. These two numbers, therefore, bracket all practical materials.

Practically all packaging technologies have predominantly planar interconnections. Thus, the distances of Figure 7-22 apply to the maximum Manhattan length—the sum of the two packaging edges. All circuits and associated bits of storage (e.g., buffers and first-level caches) must be contained within these dimensions. Therefore, the maximum available power density and the total packaged electronics area determine the total power available for all circuits. The total number of circuits depends greatly on the intended number of cycles per instruction as shown in Figure 7-23. Obviously, the lower number of C/I may lead to reduced circuit performance because of the lower power per circuit, if the design calls for higher than available power density.

The mainframes have provisions for high data-rate transmission into

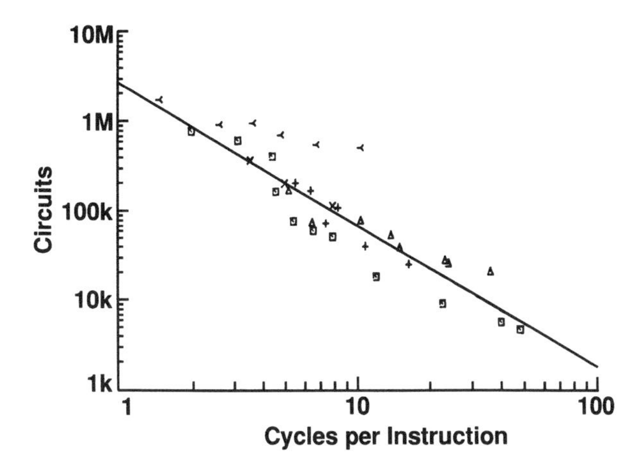

Figure 7-23. Circuit Count Dependency on the Number of Cycles per Instruction

and out of the processor. This interdependency is quantified in Figure 7-24 and amounts to ≥ 8 MB/MIPS. Note the three data points obtained by independent researchers (Refs. 3 and 4 and Chapter 20) in their work on very advanced system applications. They confirm continuation of the same numerical relationship to at least 1000 MIPS.

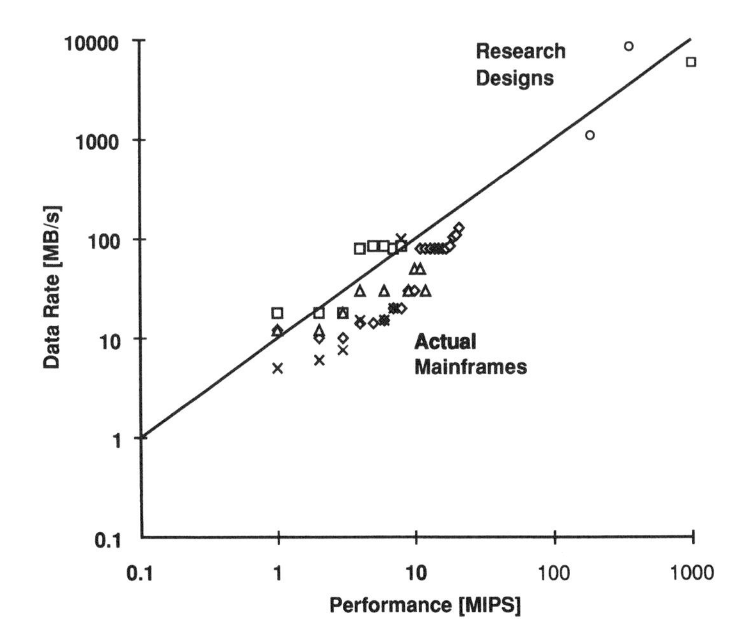

Figure 7-24. Input–Output Data-Rate Dependency on Performance

Section 7.2.13 offers an in-depth study of the simplistic examples of Section 7.1.1 and shows the evolution trajectories of various semiconductor technologies in a wide range of applications.

7.2.12 Optical Interconnects

All electrical interconnections have a limited bandwidth—the highest transmittable number of pulses per unit time, such as megabits per second. The bandwidth with a given cross-sectional geometry and materials is reciprocal to the distance (cable length). The interconnection bandwidth must keep up with the increasing performance, whereas higher densities and reduced geometries do the opposite—decrease the bandwidth per unit length. A very practical alternative for higher data rates is modulation of light, especially coherent light, and its channeling from the point of origination (transmitter) to the destination (receiver). A description of such technologies in Chapter 20, "Optical Packaging," concludes Part 3 of this book.

7.2.13 Summary of Packaging Performance Considerations and Evolution

In developing digital technologies, various parameters have been applied to circuit-performance evaluations. The **power–time product** of a gate is one of the well-known parameters. The advances in digital circuits have resulted in remarkable reduction of this equivalent energy. Energy levels of electromagnetic relays (early logic blocks) were several millijoules (mJ) and the levels of magnetic switching cores for memories had approached the microjoule (μJ) level. Silicon transistor gates had opened the way to the nanojoule (nJ) level, and large-scale integration (LSI) technology have been developed with picojoule (pJ) and sub-pJ energies.

This progress indicates that the equivalent energy driven by power is one of the fundamentals of understanding the characteristics of digital circuit design. The performance of a digital functional block is quantitatively measured by their equivalent energy values and thermal resistance to the ultimate heat sink. The hardware design concept of processors (packaged electronics) is described with both chips and packaging technologies. Energy reduction of chips and packaging and appropriate thermal resistance are essential for appropriate processor performance. The following subsection discusses the Power–Time Product Theory proposed by H. Kanai in 1967 [19].

7.2.13.1 Power–Time Product Theory

Given a digital functional block B, its equivalent energy U_B can be defined as the product of its average power P_B and signal delay time t_B of performing the function B:

$$U_B = P_B t_B \tag{7-10}$$

Another important factor in the design of the functional block B is the thermal characteristic, represented by

$$R_{\Theta B} = \Theta_B \,/\, P_B \tag{7-11}$$

where Θ_B is the temperature rise in the functional block B and $R_{\Theta B}$ is the equivalent thermal resistance defined by Equation 7-11. The failure rate λ_B of the block B can be represented by

$$\lambda_B = \lambda_0 \exp(\alpha\Theta_B) \tag{7-12}$$

where λ_0 and α are given constants.

The major target of digital system design is to provide satisfactory service to users not only in terms of high-speed data processing but also in terms of high reliability. The performance (PF) of a digital functional block B in terms of speed and reliability can be represented as

$$PF = (1 - b\lambda_B) \,/\, T_B(\text{Logical function of block B}) \tag{7-13}$$

where b is a given constant. By introducting Equations 7-10 through 7-12 into Equation 7-13,

$$PF = \frac{\Theta_B[1 - b\lambda_0 \exp(\alpha\Theta_B)]}{U_B R_{\Theta B} \,/\, \text{Logical function}} \tag{7-14}$$

In the case of digital processors, the logical function of generalized blocks correspond to instructions of a processor and T_B corresponds to instruction execution time:

$$PF = \frac{\Theta_B[1 - b\lambda_0 \exp(\alpha\Theta_B)]}{U_B R_{\Theta B} \,/\, \text{Instruction}} \tag{7-15}$$

The upper part of Equation 7-15, $\Theta_B\ [1 - b\lambda_0 \exp(\alpha\Theta_B)]$, describes the reliability of the logic-function block B. This value also indicates the long-term performance of the system because the long-term performance, MTBF (mean time between failure) can be represented by a reciprocal of the failure rate. The lower part of Equation 7-15, $U_B R_{\Theta B}$ / Instruction, shows the short-term performance—the processing time of a unit instruction. Thus, this equation summarizes the total system performance, the circuit speed, and circuit reliability.

Equation 7-14 or 7-15 gives us ways to maximize the performance of either generalized function blocks or processors:

1. Reduction of $U_B R_{\Theta B}$ / function (or instruction)
   (a) Reduction of energy U_B
   (b) Reduction of thermal resistance $R_{\Theta B}$
   (c) Minimization of $U_B R_{\Theta B}$ by optimization of packaging structural dimensions in functional blocks
2. Reduction of constants α, λ_0, and b by passivation of devices, reduction of failure rates, and high-availability design of blocks in system
3. Optimized Θ_B to maximize $\Theta_B\,[1 - b\lambda_0 \exp(\alpha\Theta_B)]$; design value of P_B given by (Optimized Θ_B) / $R_{\Theta B}$.

From Equations 7-10 and 7-12, a nominator of Equation 7-14 can be represented as

$$U_B(\text{Instruction}) = t_B \Theta_B \,/\, \text{Logic function} \qquad [7\text{-}16]$$

As the temperature rise Θ_B is obtained by maximization of the numerator of Equation 7-14, $U_B R_{\Theta B}$ is directly related to the function time t_B.

7.2.13.2 Reduction of $U_B R_{\Theta B}$ by LSI Technology

The progress in LSI technology has had a major effect on increasing the performance and reducing the cost of digital systems. The design concept of digital processors represented by Equation 7-15 and items 1–3 enables us to comprehend this effect. The remarkable reduction of logic-block energy U_B resulted from the highly integrated fine patterns of transistor circuits and their on-chip interconnections. Figure 7-25a shows a typical model of sequentially connected logic circuits (gates), called switches. Each circuit is supposed to have the same power supply voltage V_0 and resistor R_0. It stores energy in its two parts: the switching energy and the connecting energy. Therefore, to reduce the equivalent stored energy U_B represented by Equation 7-10, both parts should be reduced. Even if the stored switching energy could be reduced, the total reduction is insufficient unless the stored interconnection energy is also reduced.

If the switching energy in Figure 7-25a is proportional to the current $I_0 = V_0 / R_0$ to be switched, then the interdependence between circuit power and delay time are shown in Figure 7-25b. In this circuit model, the reduction in interconnection size is effective in decreasing the circuit energy, and the reduction switch (device) size moves the reduced energy area into a higher-speed domain.

The LSI technology is one of the best approaches available to achieve these objectives because of the microscopic dimensions of its devices and

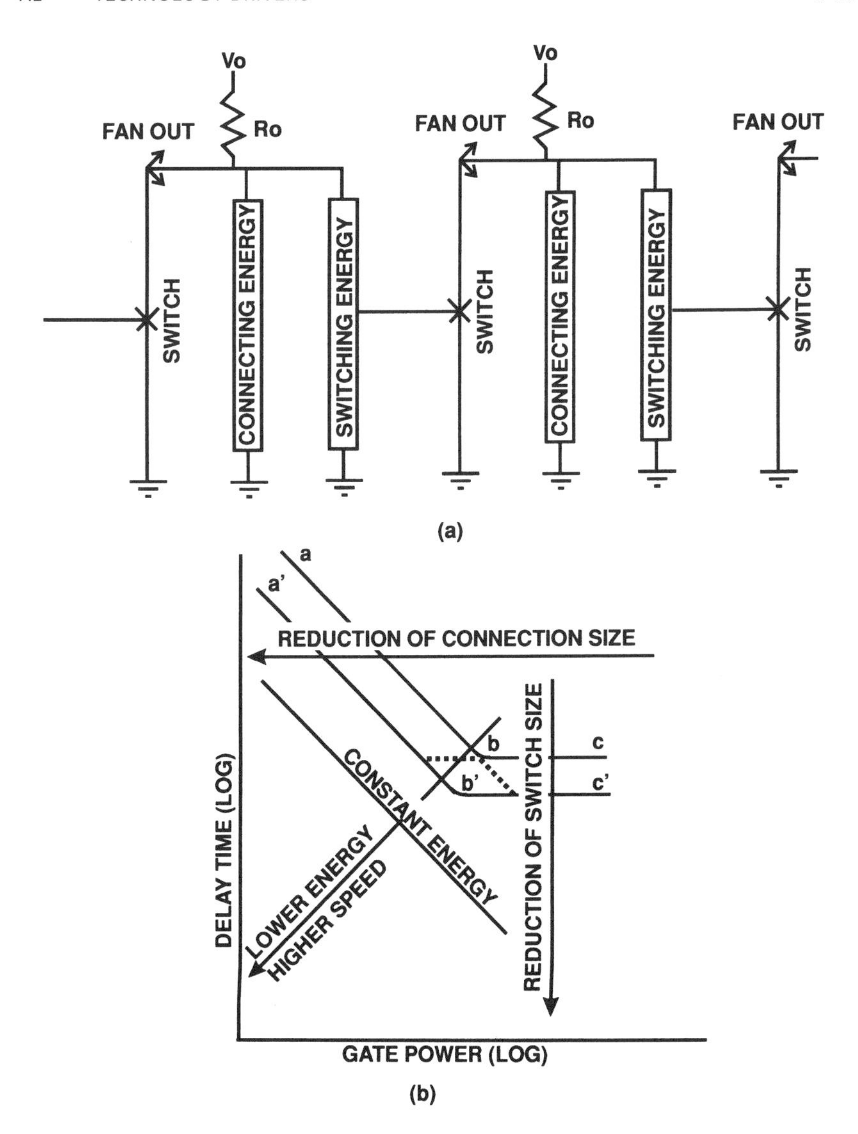

Figure 7-25. Typical Model of a Sequentially Connected Logic Circuits (a) and Its Power–Delay-Time Characteristics (b). Reduction of geometrical dimensions of connections and switches of ICs is effective in reducing energy.

interconnections arranged in a two-dimensional plane. The advances in silicon LSI technology have enabled a remarkable reduction of the energy U_B of the digital-function blocks and, thus, contributed to increases in processor performance as indicated in Equation 7-15.

In addition to the LSI advancements, the reduction in equivalent thermal resistance $R_{\Theta B}$ of LSI MCMs is pivotal in the minimization of

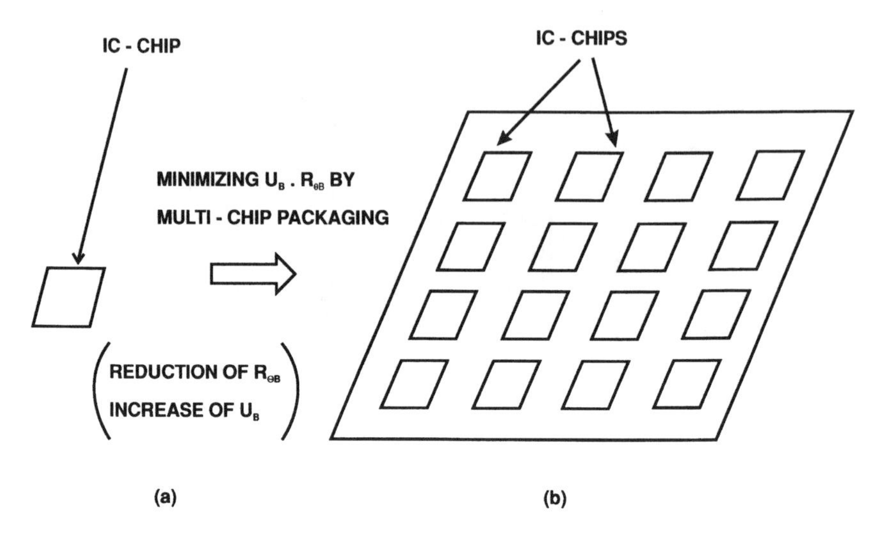

Figure 7-26. Comparison Between Single-Chip and Multichip Packages. The value of $U_BR_{\Theta B}$ of a processor can be minimized by optimizing design relative to the number of chips, chip size chip locations, and intrachip connections for multichip structure. (a) Single-chip package for small processor and (b) multichip package for large processor.

$U_BR_{\Theta B}$, essential for high-performance processors. Figure 7-26 shows a comparison of SCM and MCM structures. A processor is considered for integration on a single silicon chip in Figure 7-26. The value of $U_BR_{\Theta B}$ of a processor in an SCM is adequate for a low-performance design, because of the value of PF corresponding to a low U_B associated with the $U_BR_{\Theta B}$ of an LSI chip on an SCM.

However, the small value of $U_BR_{\Theta B}$ required for a high-performance processor cannot be realized with an SCM. Further chip improvements may sufficiently reduce U_B, but the effective approach for a small $U_BR_{\Theta B}$ is to use an MCM. The chip of Figure 7-26a can be subdivided into m chips and packaged using thin-film interchip connections as shown in Figure 7-26b. With this MCM structure, the equivalent thermal resistance $U_BR_{\Theta B}$ can approach the level of $U_BR_{\Theta B}/m$ of an SCM structure. On the other hand, the value of U_B of the processor increases slightly due to the additional energy of the chip interconnections. Therefore, the value of $U_BR_{\Theta B}$ can be minimized in MCM designs by optimization of the number of chips, chip size, chip location, and interchip connections. Thus, a sufficiently low level of $U_BR_{\Theta B}$ for high-performance processors is achievable with MCMs.

In the case of high-performance processors, large numbers of chips are to be interconnected not only on MCMs but also on cards and/or boards which support MCMs. The same design method for minimization of $U_BR_{\Theta B}$ can be employed for optimization of the processor performance. These processor design concepts based on energy and thermal resistance

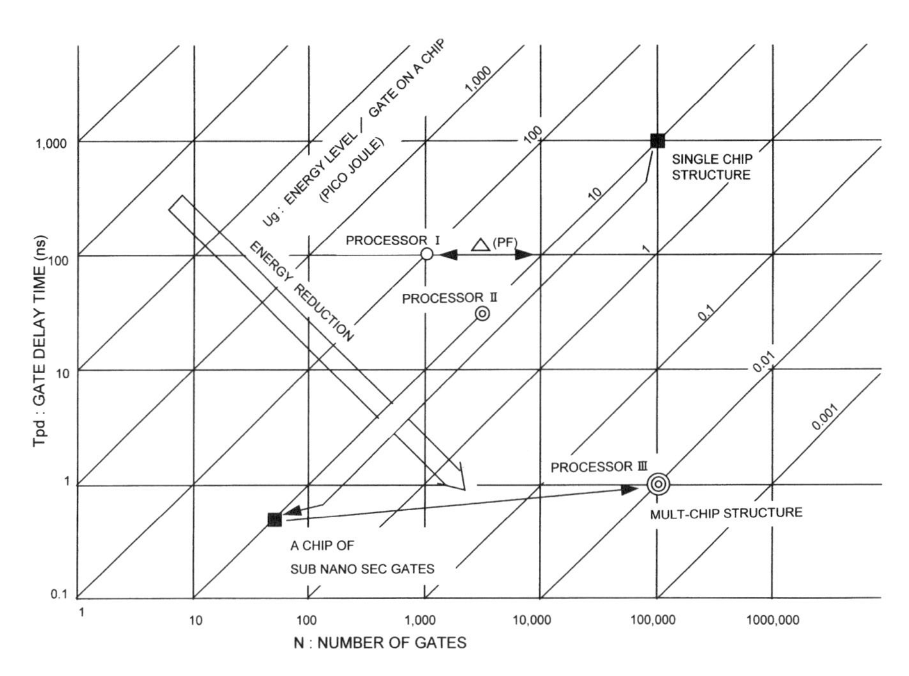

Figure 7-27. Correlation Between Energy Level of Gates on Silicon chips and Processor Design. The three coordinates are *N:* number of gates, t_{pd}: average gate delay time, and U_g: required energy level of gates on a chip when maximum power of a chip is 1 W. Design points of three types of processors, I, II, and III, are illustrated in the case of 10-pJ energy level of gates. Processor II fits in one chip structure. Processor III requires multichip packagings. Processor I can be designed with additional functions on one chip.

are illustrated in Figure 7-27. This graph has three coordinates: N is the number of circuits (gates), t_{pd} is the average circuit delay time, and U_g is the average energy level per circuit (gate) on a chip. In Figure 7-27, the thermal resistance of a semiconductor chip is assumed to be constant, and the maximum power of each chip is supposed to be 1 W to give the optimized temperature rise Θ_B for processor performance. If a circuit (gate) should be designed for a chip containing 100 circuits, each is allowed to consume 10 mW. Therefore, in the case of t_{pd} = 100 ns, the average energy level of the circuit U_g = 1000 pJ. In the case of t_{pd} = 10 ns, U_g = 100 pJ. In the case of t_{pd} = 1 ns, U_g = 10 pJ, and so forth. If the circuit should be designed for a chip of 1000 circuits, each of them can have 1 mW. Then, in case where t_{pd} = 100 ns, U_g = 100 pJ; t_{pd} = 10 ns, U_g = 10 pJ, and so forth. Therefore, the coordinate U_g can be indicated diagonally on the t_{pd}–N plane as shown in Figure 7-27.

7.2.13.2 System LSI in Chips and Packages

As is well known, digital processors enabled significant advancements in computer and communication systems consisting of hardware and software. At the dawn of the integrated circuit (IC) development, small-scale integrated (SSI) chips had limited functionality as basic logic gates, flip-flops, and so on. The SSI–MSI–LSI–VLSI–ULSI evolution expanded the LSI functional and performance capabilities due to the remarkable reduction of the circuit energy U_g. Likewise, the semiconductor and packaging improvements have contributed to significant enhancements of the processor figure of merit $U_B R_{\Theta B}$ as illustrated in Figure 7-28 which has the same coordinates N, t_{pd}, and U_g as Figure 7-27. The initial U_g is valid only for 1 W maximum chip power. In the case of the chip power being nW, the value of U_g will be $1/n$ times the value from Figure 7-28. In both figures, seven various processor domains (very large, large, medium, small, mini, micro, and calculators) are approximated by their respective t_{pd} (circuit delay time) and N (number of gates) plane.

Figure 7-28a shows the integrated structure of processors of early IC development (SSI) at microjoule energy levels. Each SSI chip had a small number of MOS, TTL, or ECL gates and was packaged in an SCM. These SCMs were soldered or socketed on printed wiring boards; and the processors had many cards plugged into boards assembled in cabinets. Figure 7-28b shows the technology evolution at its second stage. The gate energy level had dropped to approximately 100 pJ, with a MSI or even LSI level of integration. During this stage, low-performance processors could be implemented within a single MOS LSI chip. Production and marketing of the calculator LSI and four-bit microprocessors had a remarkable effect on the growth of the electronic industry. At the high level of processor hierarchy, large processors had been implemented with the high-speed bipolar MSI chips in SCMs. Packaging hierarchy of these high-performance processors was similar to the preceding SSI implementations of Figure 7-28a.

Figure 7-28c shows the packaging structures of the third-generation ICs, when the gate energy dropped to the several picojoule level and the integration levels reached the advanced LSI, the functionality of which was custom-tailored to a particular system design. For lower-level processors, the functional capability of a (MOS) LSI chip exceeded that required for Processor I, as shown in Figure 7-27. This excess capability was utilized to incorporate software elements onto the chip via inclusion of memory circuits. Thus, a single-chip (low-level) processor had emerged, capable of performing various functions such as calculations, conversions, controls, and so forth. This high functional ability also provided for great architectural design flexibility of processors on a chip. For medium-class processors, the number of chips and the total processor size have been greatly reduced, enabling more compact lower-cost higher-performing machines than the second stage of Figure 7-28b.

At the high end, large or very large processors were implementable with the third-generation LSI chips having more than 100 gates of less

(a)

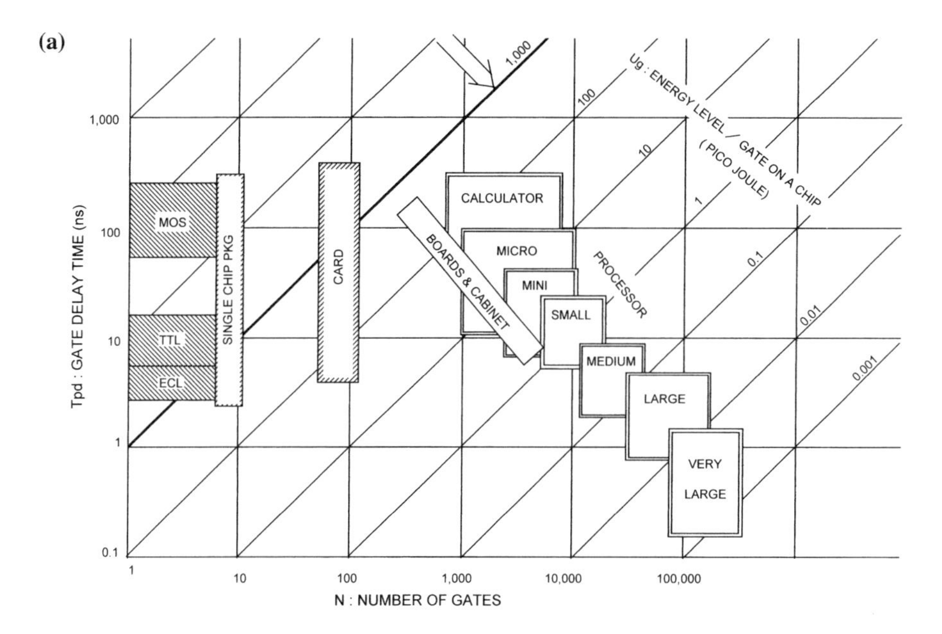

(b)

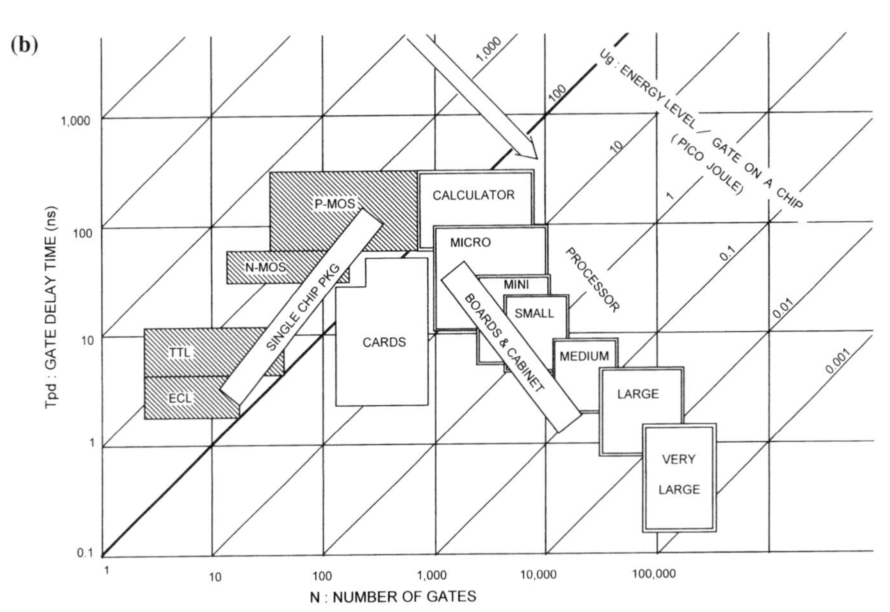

Figure 7-28. Correlation Between Energy Levels on IC Chips and LSI Design of Processors from Calculators to Very Large Mainframes. Coordinates are the same as shown in Figure 7-27. The progress of LSI technology can be viewed by energy-level reduction of gates on chips. LSI chip technology and LSI packaging technology are closely related to design processors in each level of progression.

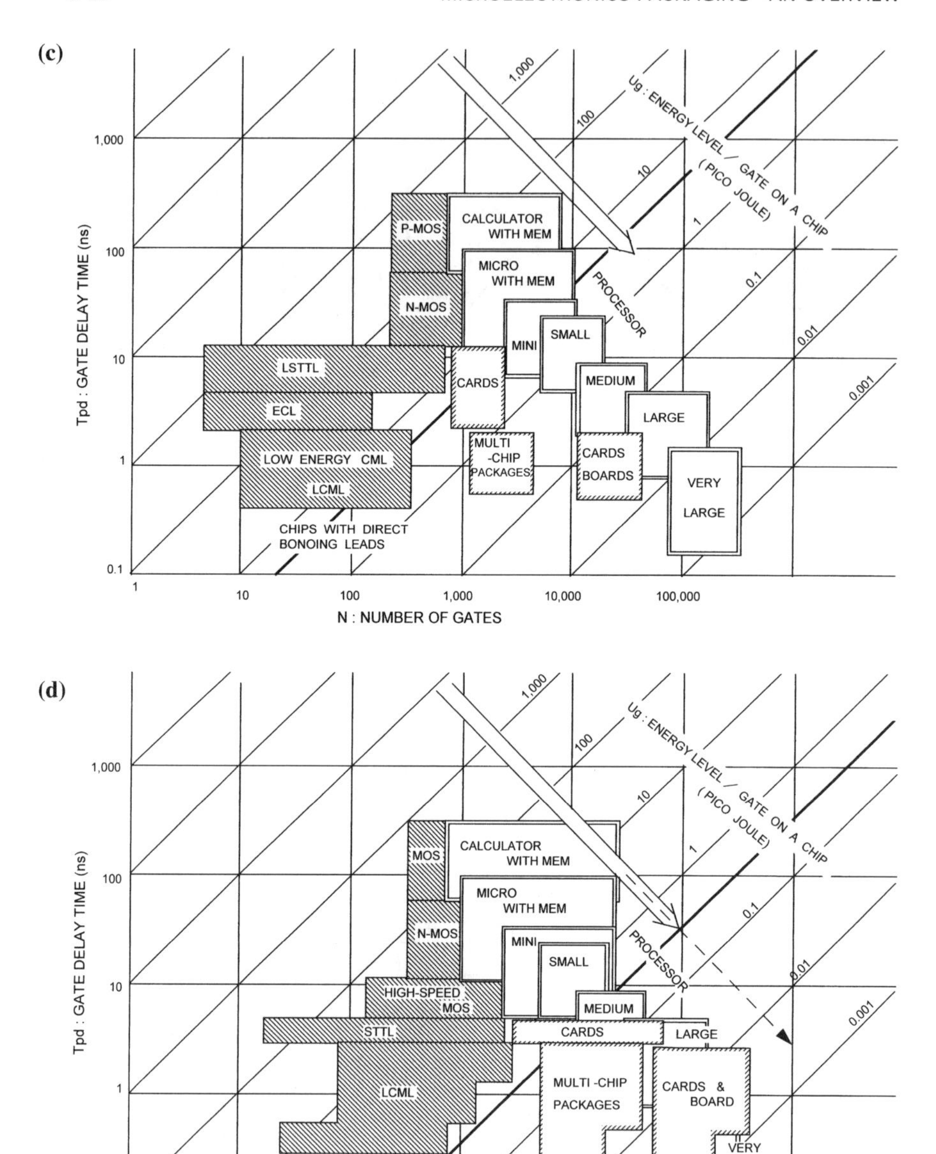

Figure 7-28. (*Continued*)

(e)

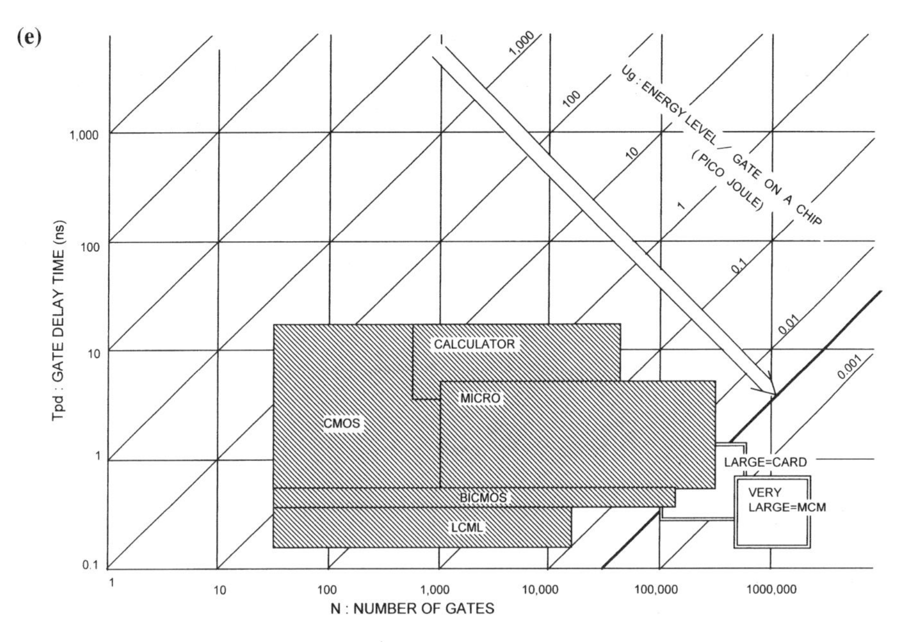

Figure 7-28. (*Continued*)

than 1 ns delay each, Figure 7-28c. To maximize performance by minimizing $U_B R_{\Theta B}$/instruction of Equation 7-15; these large processors require packaging technology as Processor III of Figure 7-27, with a special consideration toward dense chip interconnections (linewidths and pitch) and low thermal resistance between device junctions and cooling air. To achieve this, one of the best methods is to use MCMs on cards and boards, as indicated in Figure 7-28c. Without the effort to reduce the energy stored in the chip interconnections, it is difficult to reduce the equivalent energy level of large high-speed processors, even if low-energy chips can be developed with low-energy bipolar CML gates.

Figure 7-28d shows the integration technology for processors at its fourth step of evolution. The gate energies are in the sub-picojoule range. The level of integration has reached VLSI. At this stage, the energy reduction provided by the technology has more effect on the data processing system than during the third stage. For low-class processors, the utilization of the increased/excess chip functional capabilities due to further gate-energy reductions gives more intelligence to the chip by further expansion of the built-in software. This evolutionary direction will continue because an almost infinite demand for intelligence improvement exists in the data-processing market. Other applications are to extend the processing capability from 16 to 32 bits at high speed or to have

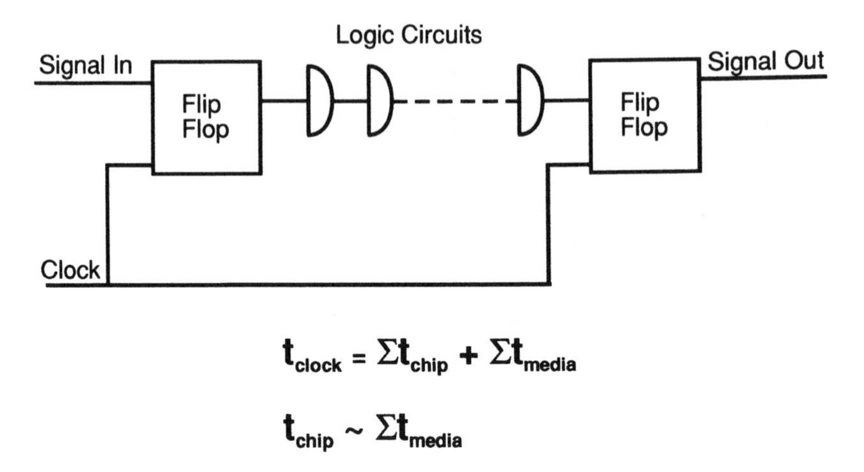

$$t_{clock} = \Sigma t_{chip} + \Sigma t_{media}$$

$$t_{chip} \sim \Sigma t_{media}$$

Figure 7-29. Basic Chain of Logic Blocks Operating in One Clock Cycle

multiprocessors on the chip. (Also consider arguments for smaller chips made in Section 7.1.3.3).

Medium to large processors can be produced with better performance and lower cost, in one chip or one-card packaging by application of bipolar or high-speed MOS VLSI chips with an integration range of more than several thousand gates per chip. Very large computers will use VLSI MCM packages, with 100–500-ps gates. These large computers also will require very fine and highly multilayered PWB technology for interconnections between MCMs.

7.2.13.3 Media Delay Factor

In digital data processing, the information to be processed is stored in a register (flip-flop) as shown in Figure 7-29. At the arrival of the first clock pulse it is sent for processing through a chain of logic gates (circuits) and has to arrive at the receiving register (flip-flop) ahead of the second clock pulse to enable its error-free storage and further subsequent processing. Therefore, the clock cycle time t_{clock} is driven by the sum-total delay of the logic circuits between the registers. It consists of the sum of delays within chips (on-chip circuits) t_{chip} and the chip interconnections delay t_{media}. The values of these two time intervals are similar. Therefore, to minimize the clock time, the reduction in the interconnections delay t_{media} is an effective as the use of higher-speed VLSI chips. Figure 7-30 shows the relationship between media delay and gate density. In an attempt to cut the media delay in half, the packaging edge L which contains chips must become $L/2$, assuming t_{media} proportionality with the packaging edge. Consequently, the gate density in the half-edge package becomes fourfold,

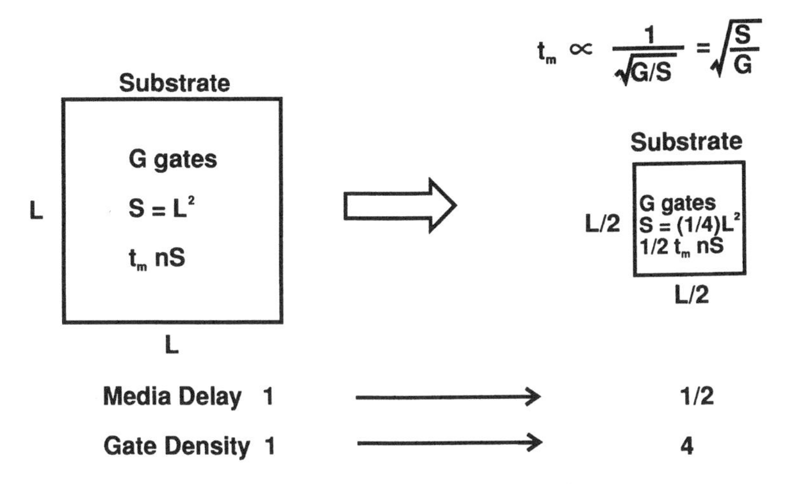

Figure 7-30. Relationship Between Media Delay and Circuit Density

because the new package area is one-fourth of the original. The media delay t_m is, therefore, reciprocal to the square root of the gate density *G/S,* where *G* is the number of gates within the package and *S* is the area.

The unit signal delay τ in a medium of dielectric constant ε is

$$\tau = \frac{\sqrt{\varepsilon_r}}{c} \qquad [7\text{-}17]$$

where c is the speed of light in vacuum. The media delay factor can, therefore, be expressed as

$$t_{\text{media}} = \tau\sqrt{\frac{S}{G}} \qquad [7\text{-}18]$$

To evaluate the packaging performance in the digital systems, the t_m value is one of the figures of merit; the smaller it is, the better. Figure 7-31 shows t_m values for various substrates. Earlier substrates, PWBs, were made from organic materials such as FR-4 and polyimide. Those PWBs, however, had wiring-density limitations because of plated-through-holes (PTH) which required mechanical drilling. Even though the dielectric constant of PWB is small, the packaging density was not high, caused by limited PTH density. By contrast, the ceramic cofired substrates, such as alumina, have much denser vias and more closely spaced chips but have a higher dielectric constant. The t_m values of the ceramic substrates, therefore, could not be dramatically improved versus the PWBs. Several improvements were made by using glass–ceramic substrates, which have

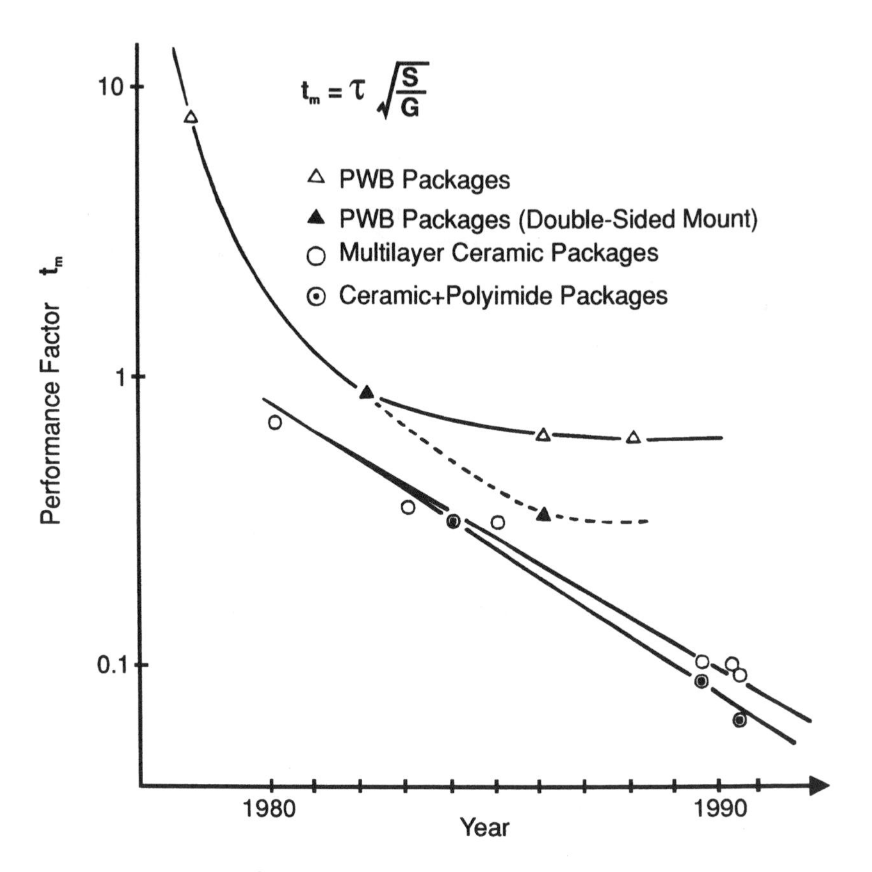

Figure 7-31. Preformance-Factor t_m Trends in Various Packages

lower dielectric constants of $\varepsilon_r = 5$ to 7 and polyimide–ceramic substrates ($\varepsilon_r = 3.5$) for the application of high-end computers and supercomputers, as shown in Figure 7-32. By using those high-speed (τ) and high-density *(G/S)* substrate techniques, the t_m value has been dramatically improved, as shown in Figure 7-31. In order to improve the gate density on the substrate, chip mounting and its cooling techniques are important as well as dense wirings. C4 technology (IBM) is the best way for highest chip mounting when the power dissipation of the LSI chips are optimized to match cooling methods. In case of up-to-date high-end computers, such as supercomputers, the LSI chip design is rather optimized for the highest-speed and high-density integration, and results in a higher power density by depending on lower-thermal-resistance chip-mounting techniques and cooling techniques. Flipped TAB carrier (FTC) (NEC, Fujitsu) and compact chip carrier (Hitachi) technologies are widely used as the high-density chip mountings for high-performance computers, as shown in Figure 7-33.

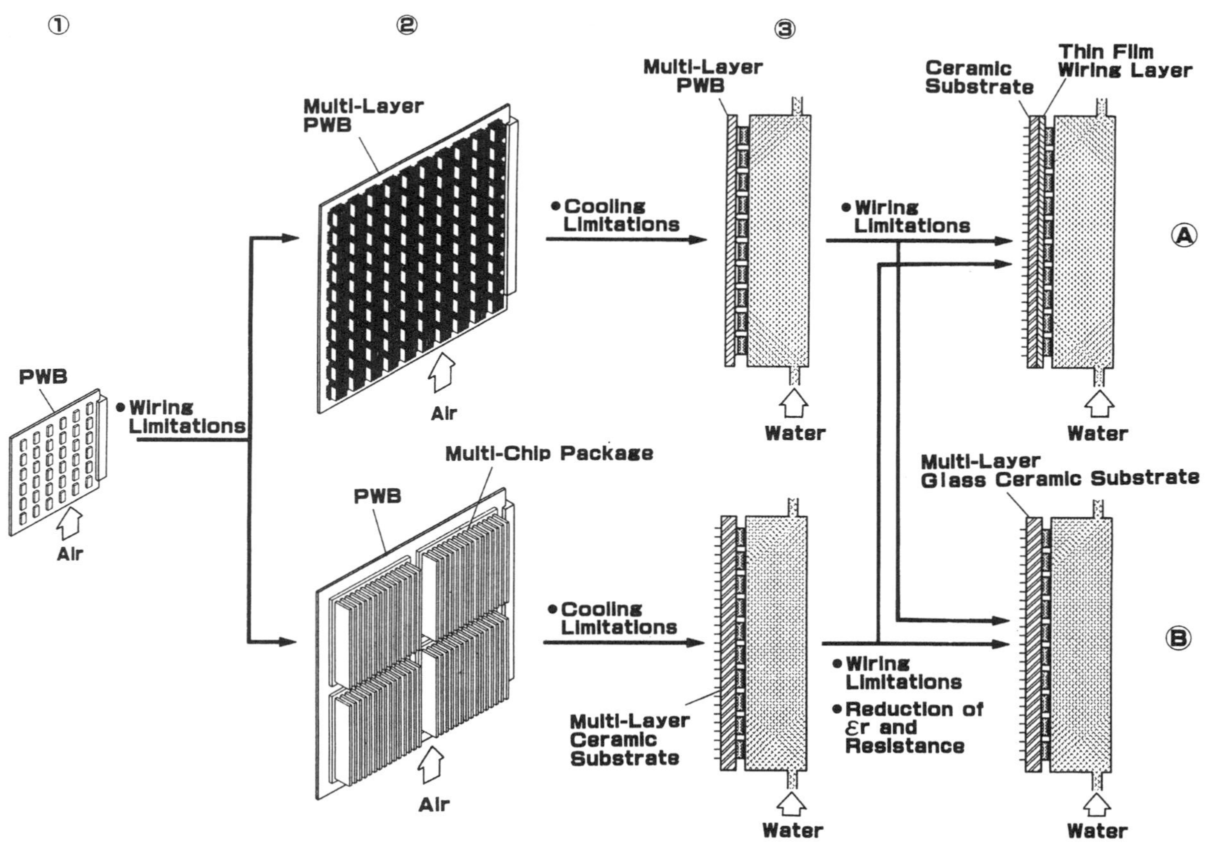

Figure 7-32. MCM Designs with Significantly Improved Values of t_m

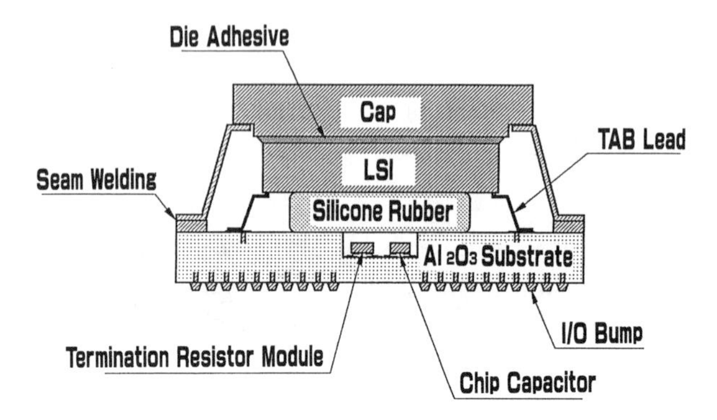

Figure 7-33. Cross-Section of SX-3 Flipped-TAB Carrier

Those chip carriers have the excellent features of high pin counts and lower thermal resistance. As shown in Figure 7-33, I/O pads are arranged in a pad-grid array on the rear surface of the substrate which includes many wiring layers inside for the interconnection between the LSI chip and the pads. The cap of the chip carrier acts as a heat spreader which can conduct heat generated on the LSI chips to the cooling mechanism effectively.

7.3 PACKAGING TECHNOLOGIES

As pointed out earlier and throughout the book dramatic changes are underway in the computer, telecommunication, automotive, and consumer electronics industries. In computers, the cost per unit of computing has come down from $4M/MIPS (millions of instructions per second) with mainframes to less than $10 with PCs and some workstations, as indicated in Figure 7-1. In addition to increased performance, computing is increasingly being performed in hand-held next-generation computers. The overall trend during the last three decades has been from centralized systems such as mainframe to distributed systems such as PCs to portable, wireless, and network systems. There is a similar trend in the telecommunication industry where the user is employing high-performance, multifunctional, portable units. In the consumer industry, multimedia products with voice, image, video, text, speech recognition, speech to text, and other functions as illustrated in Figure 7-3 in wireless applications are expected to be commonplace within the next decade. Major changes in the automotive industry, both in the amount of electronics in automobiles and the functions needed to support the intelligent vehicle highway systems (IVHS) are occurring in this decade. The common and pervasive requirements in all of these electronics are (1) ultra-low-cost, (2) thin, light, and portable, (3) very high performance, (4) diverse functions involving a variety of semiconductor chips, and (5) user friendliness. Figure 7-6 reflects this trend not only for consumer products but most of the electronics of the next century.

7.3.1 Packaging Evolution

Tables 7-4A through 7-4B project the packaging trend in consumer, automotive, and high-performance systems in I/Os, I/O density, power, and acceptable cost for board and wiring dimensions. The packaging technologies summarized in this chapter as well as the details of the rest of the technologies required to meet these needs is what the book is all about. All these technologies from chip-level connections to the first-level package and to the printed wiring board is referred to as the packaging hierarchy, as illustrated in Figure 7-5. The evolution of this hierarchy is

Table 7-4A. Packaging Trends in Consumer Products

	1996	2000	2004
I/O per chip	100	200	375
I/O density/in.^{2}	100	200	375
Area array pitch (mm)	0.4	0.25	0.20
Board lines/spaces (mil)	5	4	2
Power/chip (W/chip)	1	1	1
Board cost ($/in.^{2})	1.0	0.6	0.4

Source: Courtesy of NEMI.

Table 7-4B. Packaging Trend in Automotive Electronics

	1996	2000	2004
I/O per chip	150	200	500
I/O density/in.^{2}	30	100	150
Area array pitch (mm)	0.4	0.25	0.20
Board lines/spaces (mil)	6	3	2
Power/chip (W/chip)	6	12	15
Max. board temp. (°C)	125	155	170
Board cost ($/in.^{2}) (6 layers)	1.0	0.70	0.40

Source: Courtesy of NEMI.

Table 7-4C. Packaging Trend in High-Performance Systems

	1996	2000	2004
I/O per chip	600	1,500	3,000
I/O per MCM	3,000	6,000	10,000
I/O density/in.^{2}	150	300	400
Board lines/spaces	3	2	1
Power/chip (W/chip)	40	60	80
Board cost (\$/in.^{2}) (30 layer)	10	5	2.5

Source: Courtesy of NEMI.

depicted in Figures 7-34 and 7-35 from dual-in-line packaging (DIPS) in the 1970s to quad flat package (QFP) in the 1980s to ball-grid arrays in the 1990s and to a single-level integrated module referred to as SLIM in the next century. In contrast to all the current and previous packages which not only have a complex hierarchy involving two or more levels and the interconnections between them, but also 5–10 components such as resistors, capacitors, inductors, filters, switches, RF and optoelectronic and so forth, the SLIM package under development at the Georgia Institute of Technology is a single hierarchy-level and single component system-level package. Table 7-5 illustrates this trend in packaging hierarchy, number of levels required to complete a high-performance system, and the ultimate measure of packaging referred to as silicon efficiency, defined as the total area of all silicon chips divided by the area of the board. The silicon efficiency of 2% achieved in the 1970s is both due to the large size of the package as well as the area occupied by discrete components. The BGAs and chip-scale packages currently in development will increase

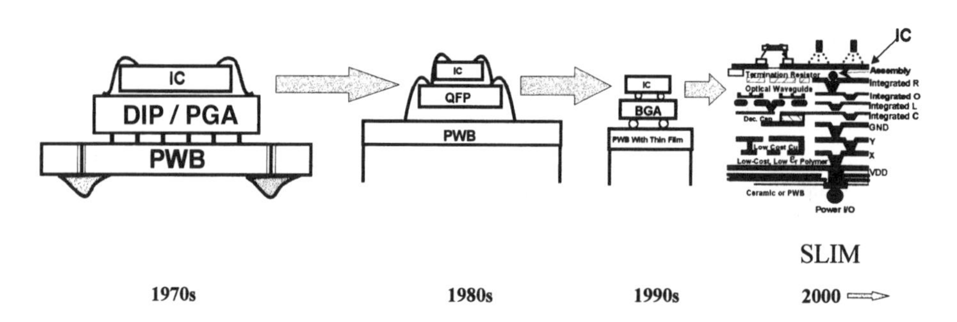

Figure 7-34. Packaging Evolution

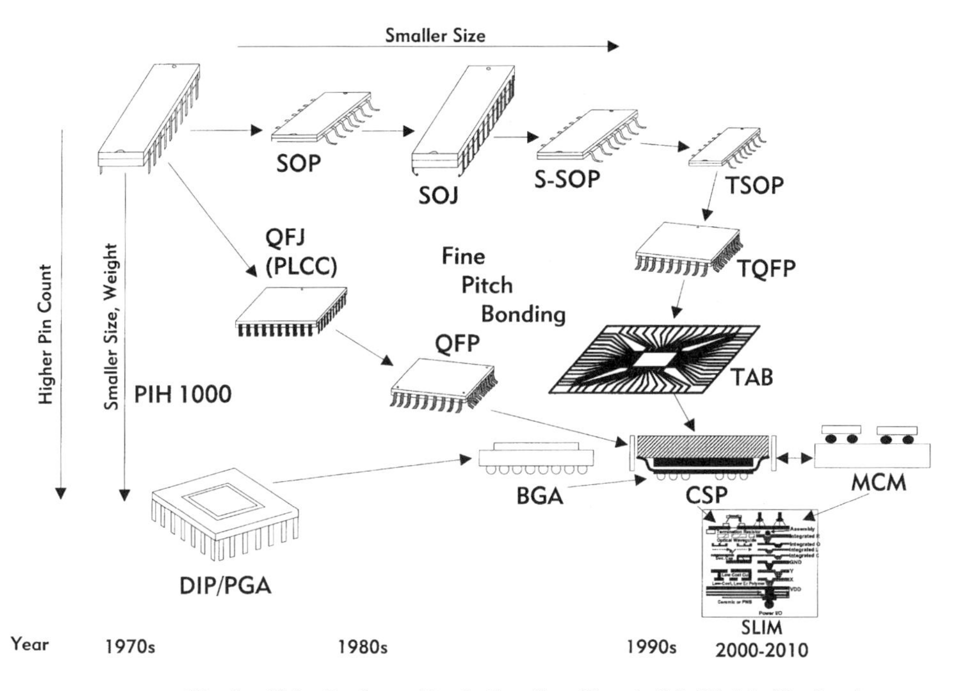

Figure 7-35. Single-Chip Package Evolution Leading to Multichip Packaging

Table 7-5. Packaging Evolution

	Past		Current		Future
	1970s	1980s	1990s	2000	2005
Chip connection	Wire bond	Wire bond	Wire bond	Flip chip	Low-cost flip chip
Package	DIP	PQFP	P/C-BGA		None
Package assembly	PTH	SMT	BGA-SMT		None
Passives	C-discretes	C-discretes	C-discretes	C-discretes	Integrated
Board	Organic	Organic	Organic	DCA to board	SLIM
No. of levels	3	3	3	1	1
No. of types of components	5–10	5–10	5–10	5–10	1
Si efficiency (%)	2	7	10	25	>75

Source: Courtesy of Georgia Tech.

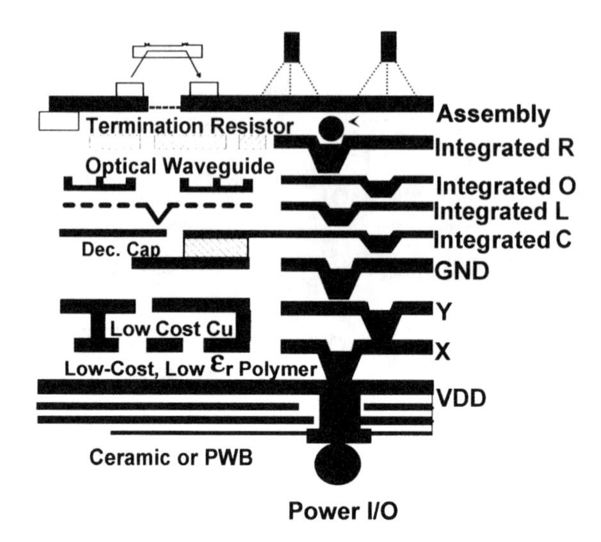

Figure 7-36. Single-Level Integrated Module (SLIM) (Courtesy of Georgia Tech.)

the efficiency to about 25% because discretes are still on the board. The SLIM package illustrated in Figure 7-36 integrates all or most of the discretes as well as RF and optical components into a single-level board to which the chip is attached in area-array fashion. This package, consequently, provides packaging efficiency in excess of 75%, perhaps close to 90%, thus maintaining almost the on-chip transistor density. This is referred to as "Moore's Law of Packaging" as illustrated in Figure 7-8 in both Si efficiency or in transistor density on board.

Table 7-6 illustrates a more detailed evolution for each of the major packaging hierarchy technologies. Interconnections at the chip level were and are being achieved by wire bonding to plastic or ceramic single-chip packages which are then bonded to printed wiring boards using surface-mount technology (SMT). The trend is toward flip chip and ball-grid array (BGA) in the short term, and to direct-chip attach to the board in the long run. The board will be fabricated using fine-line photolith via as contrast to drilled via in laminates and to photolith wiring on greensheets in ceramic boards.

There appears to be a paradigm shift from inorganic to organic packaging; however:

- Move to nonhermatic packaging
- Direct-chip attach to organic board
- BGA–SMT reliability of ceramic packages in contrast to plastic BGAs
- Cost and size requirements driving to a large organic board with thin film fabrication

Table 7-6. Packaging Technology Evolution

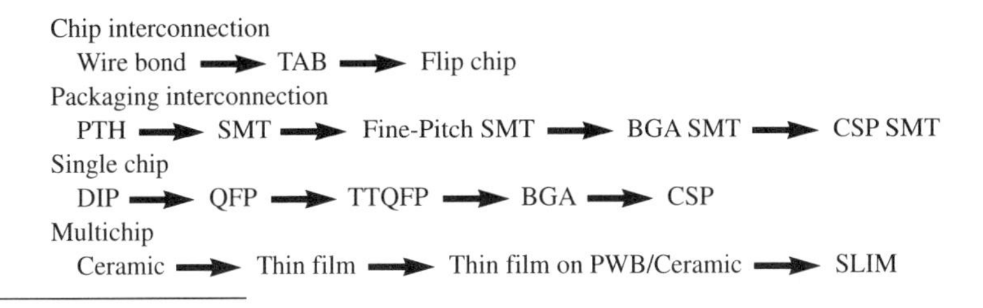

Chip interconnection
Wire bond → TAB → Flip chip
Packaging interconnection
PTH → SMT → Fine-Pitch SMT → BGA SMT → CSP SMT
Single chip
DIP → QFP → TTQFP → BGA → CSP
Multichip
Ceramic → Thin film → Thin film on PWB/Ceramic → SLIM

Source: Courtesy of Georgia Tech.

7.3.1.1 Packaging in the Past

The packaging technologies that span the microelectronics industry from consumer electronics to low-end systems to high-performance systems are very diverse, as illustrated in Table 7-7. Several important observations can be made from this table. The number of chips needed to form a system in the past increased from a few, in consumer electronics, to several thousands, in supercomputers. Given this, the packaging hierarchical technologies necessary to interconnect all these chips became complex; consumer electronics typically required a card or flexible-circuit carrier, and supercomputers required several boards, each containing several cards or multichip modules. In addition, these boards needed to be interconnected by high-speed cables and connectors. NEC systems, for example, used as many as four levels of packaging hierarchy to complete the system. They are, referring to Table 7-7, and Figs. 7-32 and 7-33, flip-TAB carrier (FTC, first level) bonded to 100-mm multilayer ceramic substrate (MLC-MCM, second level) plugged into polyimide-glass board (third level) interconnected by cables and connectors (fourth level) [20], as illustrated in Figs. 7-32 and 7-33.

The power and cooling capabilities of these systems range from no special provisions for cooling in consumer electronics to very complex cooling technologies in mainframes and supercomputers. The driving forces for the latter stem from the fact that the chips need to be placed and bonded very close to each other in order to minimize signal delay in addition to the use of high-speed high-power chips.

The main emphasis in consumer electronics is cost. Given the need for smaller functions to be packaged, chips that are wire-bonded or tape-automated-bonded (TAB) to such low-cost plastic carriers as plastic leaded chip carriers (PLCC) and plastic dual-in-line (DIPs) packages on cards with minimum internal planes generally suffice. Space limitations in these

Table 7-7. Packaging Technologies in Some Previous Electronic Products

	Chip Conn.	1st Level Package	1st to 2nd Level Conn.	2nd Level Package	2nd to 3rd Level Conn.	3rd Level Package	Chip Cooling	Max. Chips/ Systems
Sony CD	WB	PSCM	SMT/ PTH	Card	—	—	—	<10
TFTV	WB/TAB	PSCM	SMT/ PTH	Card	—	—	—	<10
Apple PC	WB	PSCM	SMT/ PTH	Card	Conn.	Board	—	10's
HP Work-station	Wire-bond	P-SCM	SMT	Board	Conn.		Air	10's
IBM 390 Mainframe	C4	C-MCM	PGA	FR-4	Conn.	Cable	Water	1000's
NEC SX-4 Supercomputer	TAB	FTC	SMT	LCM	PTH	P-G Board	Water	1000's

Key:

C-MCM: Ceramic multichip module
Conn.: Connector
FR-4 Board: Epoxy-glass board
PC: Personal computer
PGA: Pin-grid array
P-G Board: Polyimide-glass board
PSCM: Plastic single-chip module
PTH: Pin-through-hole
SMT: Surface mount technology
TAB: Tape automated bonding
TFTV: Thin-film television
WB: Wirebond

systems have brought about the need to package these components with surface-mount technology with lead spacing as low as 0.4 mm. In some circumstances, a flexible-circuit carrier is used that utilizes fully the three-dimensional space available in small systems, such as the computer terminal illustrated in Figure 7-37.

The low-end systems, such as personal computers and printers, need additional functions, thus requiring increased sophistication in package technologies. Surface-mounted as well as pinned components containing wire-bonded or TAB chips are assembled on cards which are then plugged into back panels or boards. Power requirements are generally low, not requiring special provisions for cooling of components, in addition to heat dissipation by convection and by conduction through the structural members. In applications that require higher-power dissipation, as in power supplies or motors to drive typewriter hammers, chips can either be backbonded with adhesives or metal eutectics to a thermally enhanced epoxy–glass or metal core substrates or, alternatively, to TAB chips with heat sinks on the back of the chips. As in consumer electronics, the use

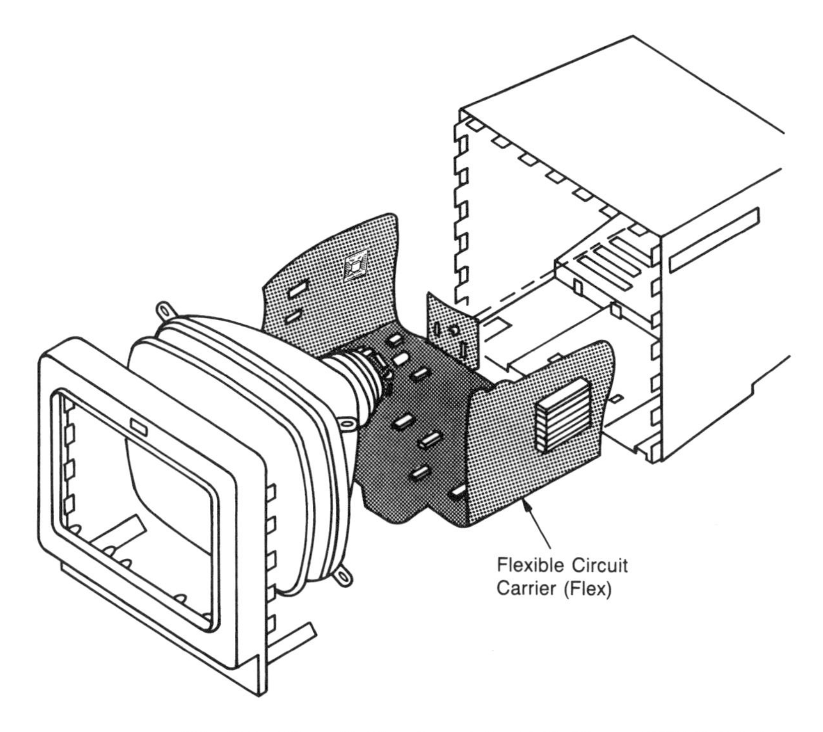

Figure 7-37. Flexible Circuit Carrier (After R.K. Hayes, IBM Raleigh.)

of a flexible-circuit carrier to provide design freedom and to minimize cables and connectors is being explored (Figure 7-37).

Midrange systems and workstations that contain a few dozen chips required single-chip or small multichip packages with I/Os in excess of 200. Surface-mounted chip carriers or pinned modules are generally mounted on cards which are then plugged into boards. The use of single-chip packages, particularly single-chip ceramic packages, with closely spaced I/O connections is often extended to high-performance systems such as mainframes and supercomputers. The large computers of Fujitsu and Hitachi, for example, use single-chip ceramic modules surface mounted to very sophisticated boards. In contrast, IBM and NEC computers use very sophisticated multichip pin-grid modules that are plugged into large multilayer boards. The single-chip and small-number multichip-modules approach generally allows air cooling to achieve the desired power dissipation (except the Fujitsu M780 which uses water), whereas the large-number multichip-module approach, with its very close chip-to-chip spacing, demands power densities beyond the capabilities provided by air cooling. The closer chip pitch, as in IBM's Thermal Conduction Technology with a center-to-center chip spacing of about 9 mm, accounts

Table 7-8. Packaging Technologies and Processes

Technology Function	Technology Options	Typical Materials	Typical Process	Typical Process Temp. (°C)
Connection to chip	Wirebond	Gold, aluminum	Wirebond	225
	Solder bond (C4)	Pb-Sn	Reflow	360
	TAB	Copper, gold, aluminum, polyimide	Thermo-compression	550
First-level package	High temperature	Al_2O_3, SiC, BeO	Sintering	1500–2000
	Low-Temp. Ceramic	Glass–ceramic	Sintering	900
	Plastic	Epoxy	Molding	200
	TAB	Cu on Kapton*	Adhesive-Bond Molding	200
First-to-second-level Connection	Surface mount	Pb-Sn	Reflow	220
	Solder	Kovar, Pb/Sn	Reflow	220
	Pin-in-hole solder pin-braze	Kovar, Au/Sn	Braze	400
Second-level package	Card	Epoxy–glass	Cure	200
	Metal carrier	Glass on steel, Invar	Fuse	1000
	Flex	Cu on Kapton	Adhesive-	200
	Inj. mold. card	Resin	Bond Molding	200
Third-level package	Board	Epoxy–glass	Cure	175
		Polyimide, glass	Cure	200
Second-to-third level connection	Connector	Polymer, BeCu	Cure	200
	Cable	Polymer, copper	Cure	200

*Kapton is a trademark of Dupont Company.

for its very high performance while handling high-power densities. Thus, power densities and performance generally go hand in hand.

The packaging technologies required to form a logic function or gate described earlier, involving three levels of package as well as an interconnection between these and the chip, are illustrated in Table 7-8. These technologies, together with thermal technology for cooling the chip, and sealing and encapsulation technologies required to protect the package and device circuitry, form the basis of packaging-materials science and technology. *The basic requirements, the status of, and the future direction for each of these technologies are summarized in this section.* Detailed discussions and additional references can be found for each technology in the appropriate chapter.

Because each packaging level is manufactured independently of the other, the selection of materials and processes, including the process temperatures for fabrication of the package, is determined for that level. However, the packaging hierarchy does impose thermal-hierarchy consideration in the assembly of the total package system. For example, if the first level requires a brazed pin in forming a pin-grid array module, the maximum temperature at which the chip attachment (to the first-level package) can take place is the lowest temperature at which the braze is dimensionally stable. Assuming the chip joining is done by the use of conventional Pb-Sn solders, thermal-assembly-hierarchy considerations require that the connection between the first and second levels is done at yet a lower temperature than the temperature at which the chip solder is stable. If a third level exists in the function, further thermal hierarchy is extended. Rework of each of the packages and chips requires similar considerations as well. Although in principle the thermal-assembly hierarchy could be set up in either direction of the packaging hierarchy, it is determined by the maximum temperature a particular level of packages can withstand. Because most conventional second- and third-level packages are made of organic cards and boards that are limited to about a 250°C joining temperature, the assembly of the package system involving these is done as the last and lowest temperature operation.

7.3.1.2 Packaging in the Future

As indicated above, the future systems for the most part are either portable, wireless, network or other systems with the same packaging attributes as these. The system-level attributes include weight, cost, and system functions as indicated in Figure 7-6.

The second important change driving packaging technology is the type of semiconductor used to form the systems of the future. In the past, bipolar was the primary engine of high-performance systems, and CMOS met all other requirements. As illustrated in Figures 7-7 and 7-9, CMOS provides the lowest cost and highest integration while it approaches the performance of a bipolar but at much lower cost at the system level [22]. As a result, CMOS becomes the most perversive technology from PCs to workstations to mainframes. The packaging in the future, therefore, must be consistent with this evolution.

7.4 CHIP-LEVEL INTERCONNECTIONS

Connections between chip and package are commonly performed by one of three technologies: wirebond, solder, or controlled collapse chip connection (C4), also called “flip chip (FC)”, or “solder bump,” and tape automated bonding (TAB), depending on the number and spacing of I/O connections on the chip and the substrate as well as permissible cost. The possible number of connections per chip with each of the

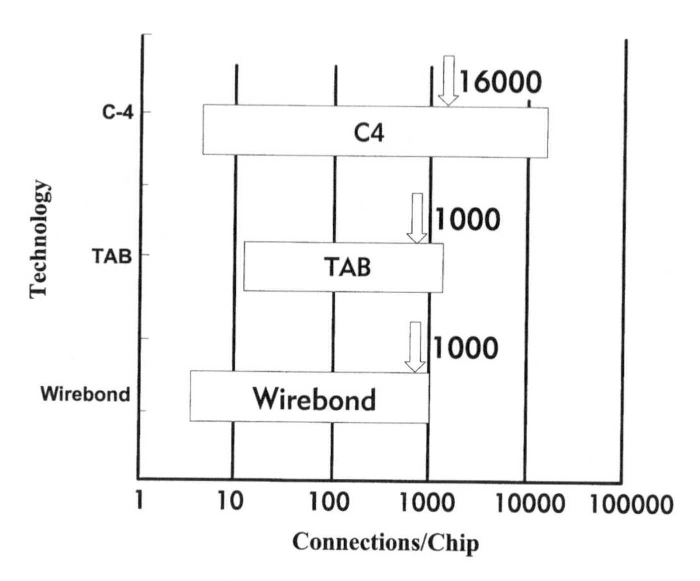

Figure 7-38. Chip-Connection Technology Status

technologies is illustrated in Figure 7-38, where the arrow indicates the current technology in manufacturing (1995) and the higher number indicates the technology feasibility.

7.4.1 Wirebond

Wirebonding is the most common chip-bonding technology, spanning the needs from consumer electronics to mainframes. Its widespread use is primarily based on the fact that the maximum number of chip connections in products in use can be accomplished with this technology in addition to providing the lowest cost per connection. It should be pointed out that wirebonding concurrently provides for thermal dissipation by backbonding or diebonding the chip to the substrate. The semiconductor trend for the next decade as indicated in Table 7-2 illustrates the leading-ledge I/Os, lithographic dimensions, chip and wafer size, and power requirements for portable and high-performance semiconductor chips. The maximum number of I/Os in the leading-edge high-performance semiconductors in Table 7-2 calls for as many as 5000 I/Os on a single chip, which is best accomplished by area-array connections such as flip chip using solder or conductive adhesives. There are several reasons for this. First, referring to Table 7-1, the connection density (I/O/chip size in mm) increases by a factor of 2 (from 750/400) to 4 (5000/1250). Thus, if leading-edge wirebonding is to meet today's requirement, it needs to be enhanced by a factor of 4 in pitch—a major challenge. Second, the peripheral pad connections on the edge of the chip, particularly in large

Table 7-9. Materials for Chip Connections

	Flip Chip	TAB	Wirebond
Metal	Solder, conductive adhesive	Copper on polymide	Gold, Al-Si
Process	Melt/reflow	Thermo and Compression	Thermo/ultrasonic
Chip metallurgy	Cr/Cu/Au	Ti/W/Au	Al, Ti/W/Au
Substrate metallurgy	Ni/Au	Pb-Sn/Au	Ni/Au

Source: From Ref. 23.

chips, do not provide optimal power distribution to all the transistors on that chip. Perhaps, the most important limitation of wirebond has to do with the economics of chip usage itself. Because wirebonding cannot be performed easily in manufacturing over the active area of chip compared to area-array solder flip-chip connections, the size of the chip [21] is larger by about one-third with the wirebonded chip. This difference will account for tens to hundreds of dollars, as it directly relates to the number of chips per wafer. In addition to the above reasons, the reliability of flip chip in Table 7-10 has been shown to be three orders of magnitude better than wirebond. Because the leading-edge products constitute a small fraction of all semiconductors, all three chip-level interconnection technologies will be used and enhanced. The basic materials and processes that are currently in use in these technologies are indicated in Table 7-9 [23], whereas the electrical and other parameters, including reliability, are illustrated in Table 7-10 [23].

Table 7-10. Chip-Level Connection Parameters

	Wirebonding		TAB	Flip Chip
Conn. Metallurgy	Al	Au	Cu	Pb/Sn
Resistance (Ω)	0.035	0.03	0.02	0.002
Inductance (nH)	0.65	0.65	2.10	0.200
Capacitance (pF)	0.006	0.006	0.04	0.001
I/O density (cm^{-2})	400	400	400	1600
Rework	Poor	Poor	Poor	Good
Failure rate (%/1000 h)	1×10^{-5}	1×10^{-5}	N/A	$<1\times10^{-3}$

Source: From Ref. 23.

Table 7-11. Chip-Level Interconnection Development Trend for Wirebond (WB), TAB, and Flip Chip (FC).

		1990	1995	2000	2005
WB	LSI chip size (mm) I/O count	15 × 15 / 700	20 × 20 / 1000	20 × 20 / 1200	25 × 25
	Wire pitch ball bond (μm)	105 (line)	70 (staggered)	60 (staggered)	50
	Wedge bond (μm)	100	65	55	50
TAB	LSI chip size (mm) I/O count	15 × 15 / 500	20 × 20 / 1000	—	
	Inner lead pitch (μm)	100	75	50	30
	Outer pitch bond (μm)	(wire bump) 100	(Plating bump) 75	50	30
FC	LSI chip size (mm) I/O count	15 × 15 / 900	15 × 15 / 1000	20 × 20	25 × 25
	Bump pitch (μm)	250	250	200	150
	Bump diameter (μm)	130	125	100	75

Table 7-11 demonstrates the chip-level developments underway in all three technologies—wirebond (WB), tape automated bonding (TAB), and flip chip (FC). The development goals for wire bond by ball-bond and wedge-bond technologies at a pitch of 50 μm and for TAB at a pitch of 75 μm are capable of providing 1200 I/Os on a 20 × 20-mm chip using a staggered pin configuration.

In addition to the thermal-hierarchy considerations discussed previously, an important factor in deciding the chip-connection technology is the lead inductance, resistance, chip rework, and failure rate as indicated in Table 7-10 that a particular technology provides; wirebond having the highest value and C4 the lowest. Detailed discussions on this and other aspects of chip-level interconnections can be found in Chapter 8, "Chip-Level Interconnections."

7.4.2 TAB

The progress and trend in TAB listed in Table 7-12 show two or more layers with an inner lead bonding pitch of 60 μm and an outer lead bonding pitch of 90 μm. On a 28 mm size chip, these leads provide in excess of 1100 I/Os. TAB has been a major accomplishment in Japan but not in the United States. Various advancements in TAB technology are being pursued by such Japanese companies as Shinko-Denshi, NEC, Fujimitso, Mitsui-Kinzoku, Oki, and Nitto Denko. One particular enhancement of TAB being pursued by Nitto Denko is illustrated in Figure

Table 7-12. TAB Trend

		1990	1995	2000
TAB Tape	Signal layer	1 layer	2 layer	2 layer
	Max. tape width	35 mm	70 mm	70 mm
	Cu thickness min. plating	35 μm	18 μm	18 μm
Min. pattern	Inner lead pitch	100 μm	60 μm	60 μm
	Outer lead pitch	200 μm	90 μm	90 μm
	Pattern pitch	100 μm	60 μm	60 μm
Max. height		1.0 mm	0.7 mm	0.6 mm
28 mm Max. Pin count		520 pins	1152 pins	1152 pins

Source: From Ref. 24; courtesy of Oki.

7-39a, comparing the new two-layer direct copper bonding process with a conventional three-layer process. The new process coats polyimide onto copper, the opposite of the 3M process in the United States that coats copper on Kapton or other polyimides by electroplating. The advantages claimed for this new process, shown in Table 7-13, include high heat

Table 7-13. Advantages of the Nitto Process

- High Heat Resistance → No Adhesive Layer
- No Curling → CTE = 16 ppm (0–300°C)

	Nitto	Kapton-H	Upilex-S	Copper
CTE (ppm)	16	20	8	16

- Low Moisture Absorption → 1% 24 hours water immersion at 23°C

	Nitto	Kapton-V	Kapton-E	Upilex-S
(%)	1.0	3.1	2.4	1.3

- Strong Peel Strength → 6.7 lbs/in (1.2 kg/cm)

	Nitto	3M	
Peel Strength	6.7	4.0	(lbs/in)
Minimum Value	1.2	0.7	(kg/cm)

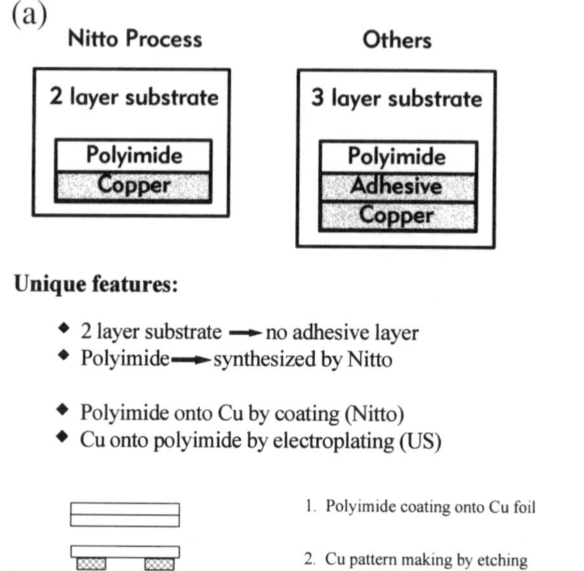

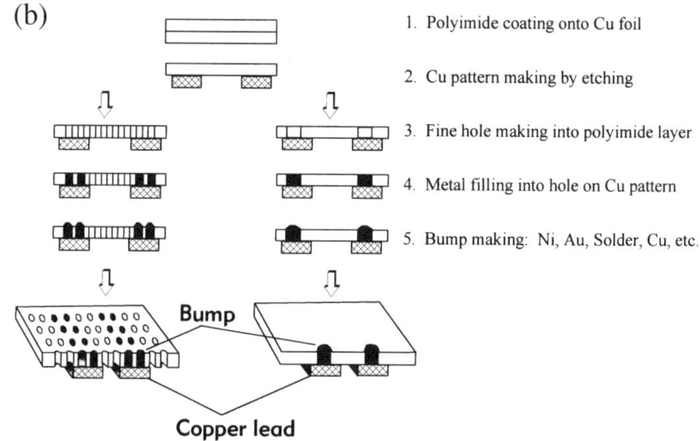

Figure 7-39. Nitto Process (a) TAB and (b) Bump (Courtesy of JTEC.)

resistance, low moisture absorption by proper selection of polyimide, and better adhesion. There are other advantages to using this new process (1) a very high aspect ratio—25-μm-diameter holes in a 50-μm-thick film; (2) large area processing (300 × 300 mm); and (3) complete wiring patterns (both vias and lines). Nitto is also applying this technology for burn-in and electrical testing, as illustrated in Figure 7-39b.

7.4.3 Flip Chip

Flip-chip technology is being extensively studied by almost all major companies and is already used in products by Hitachi, IBM, and Delco.

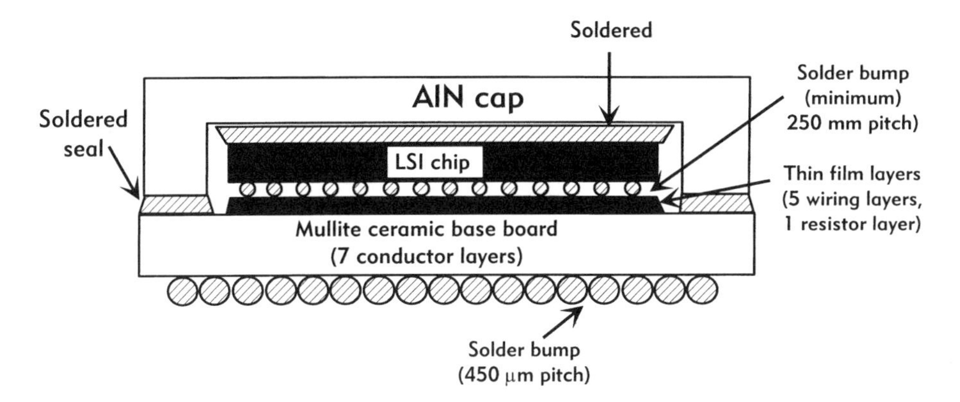

Figure 7-40. Microprocessor Carrier (BGA) for LSI (Courtesy of Hitachi.)

The Hitachi flip chip, together with microcarrier BGA assembly to next-level (mullite glass–ceramic) package, is illustrated in Figure 7-40 [25]. The microcarrier, which is only larger than the chip itself by about 2 mm, and henced can be called a chip scale package, is a single-chip carrier fabricated with seven layers of mullite ceramic and first levels of polyimide–aluminum thin-film technology. Flip-chip enhancements being pursued in the industry generally consist of one of two approaches—solder bonding (including Pb-Sn, Pb-In) and conductive adhesive bonding. Bump technology itself, like Fujitsu's bump integration technology (BIT), is generating considerable interest. One example is illustrated in Figure 7-41 using thin-film and electroplating processes. In contrast, the Germans are pursuing electroless-plate bumping, and the British are trying gold-ball bumping by wire-bonding tools. Bumping, because of its high cost has received considerable attention leading to various processes as illustrated in Figure 7-42. These include IBM's original process based on high lead solder, followed by high lead solder on chip and eutectic on board, elimination of high lead with the substitution of eutectic on chip and conductive adhesives on substrate. Low cost bumping processes, as illustrated in Figure 7-43 are also being pursued.

The biggest breakthrough in flip-chip bonding is the technology IBM (Japan) pioneered as an extension of IBM (U.S.) flip-chip technology developed three decades ago [25]. It involves direct bonding of a bumped chip to a PWB by the use of low-temperature solder that is hot-injection deposited onto PWB through a mask. The challenge here is to develop a thermally compatible encapsulant to reduce the strain on the solder joint arising from the great mismatch in thermal expansions between the PWB (17 ppm) and the chip (3 ppm). This invention together with fine-line thin-film technology is capable of revolutionizing packaging technology

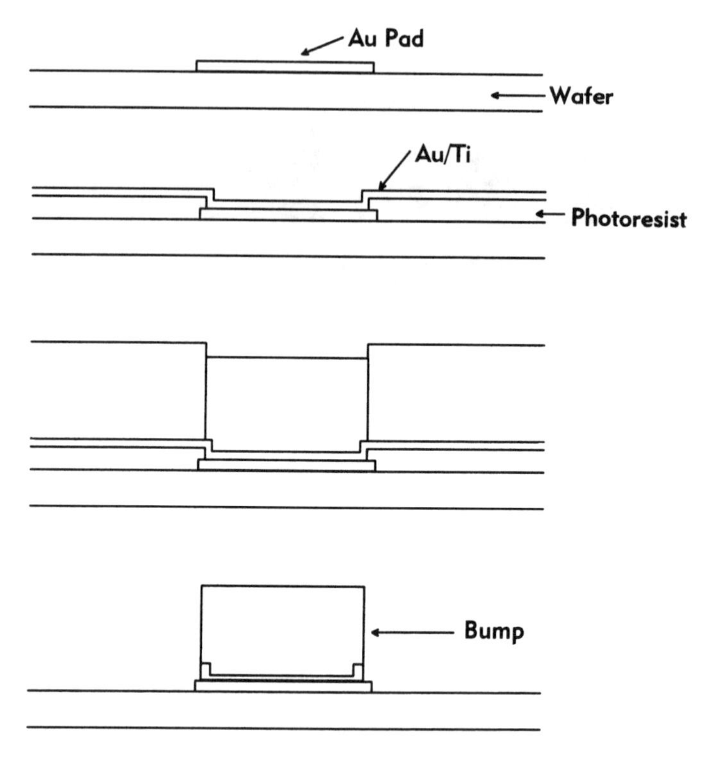

Figure 7-41. Bump-Fabrication Process (Sharp)

for minimizing the number of package levels and reducing size, weight, and cost.

Figure 7-44 illustrates a tenfold strain reduction when the encapsulant is used between the PWB and the chip. This discovery has major implications for the Japanese packaging industry, particularly for consumer electronics, as it allows Japan to continue to use its existing infrastructure for PWB. Figure 7-45 shows the eutectic solder to be more effective than high-Pb solder (95/5) in achieving the desired fatigue life.

7.5 FIRST-LEVEL PACKAGES

7.5.1 Single-Chip Packages

A summary of all first-level packages, both plastic (including TAB) and ceramic, is given in Table 7-14, wherein each package and its accepted abbreviation is described in terms of the insulator material it is made of and the maximum number of I/O connections and spacings that were available in 1995 as well as year 2000 projections. These packages as illustrated in Figure 7-46 and as illustrated in Table 7-14 fall into six

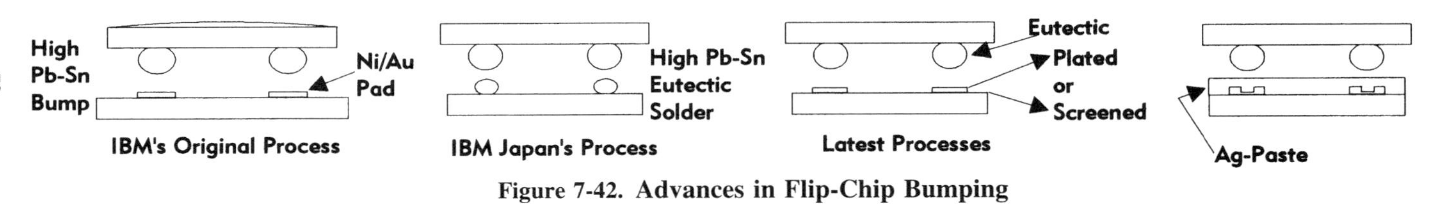

Figure 7-42. Advances in Flip-Chip Bumping

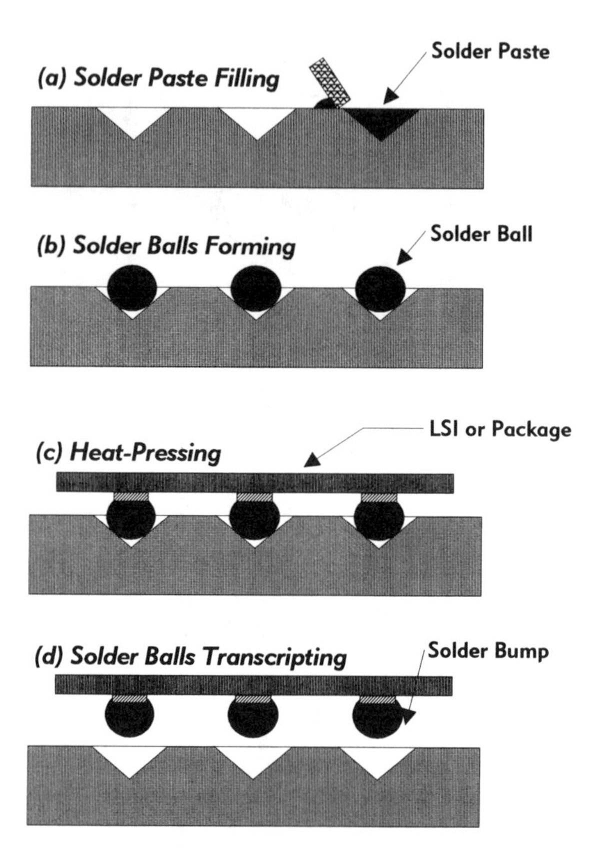

Figure 7-43. Low-Cost Bumping (Courtesy of Fujitsu.)

types: (1) those with pins often referred to as pin-grid array (PGA) for pin-through-hole or butt surface mounting, (2) those with lead-frame leads that are meant for surface mounting, (3) those with pads or balls on area-array pads that are called ball-grid arrays (BGA), (4) solder or other columns on an area array called solder column packages, (5) TAB bonded to the board typically with solder, and (6) chip-scale packages that are bonded to the board by one of the above connection technologies. The substrate materials these are made of can be plastic, ceramic, printed wiring board, or thin film. The development of all of these first-level packages over the last three decades was aimed at two categories: those that contain one chip, namely single-chip modules (SCM), and those that can support more than one, called multichip module (MCM). Multichip modules sometimes support up to and in excess of 100 chips, as discussed in Chapter 9, "Ceramic Packaging," and in Chapter 12, "Thin-Film Packaging." These latter ones are referred to as thermal-conduction modules (TCM) or liquid-cooled modules (LCM) in this book. The chronological

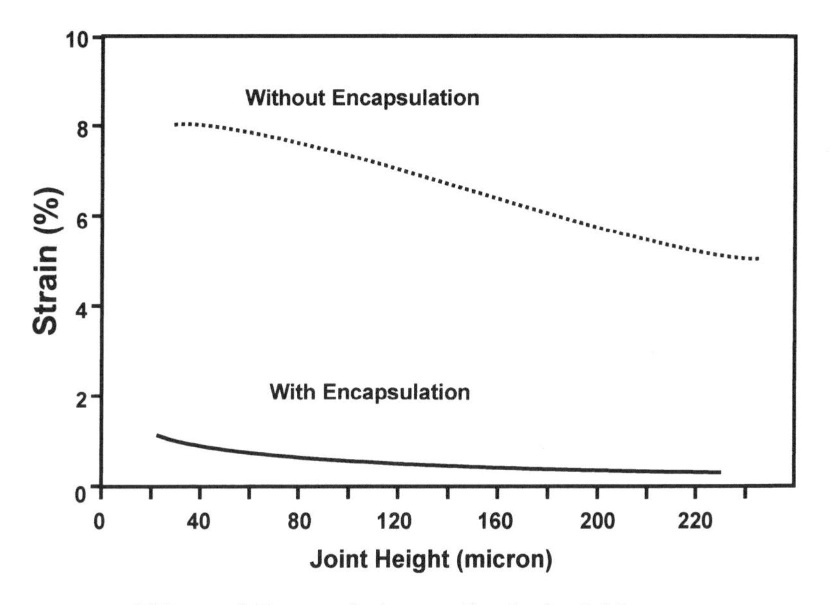

Figure 7-44. Effects of Encapsulation on Strain in Solder (Courtesy of IBM Japan.)

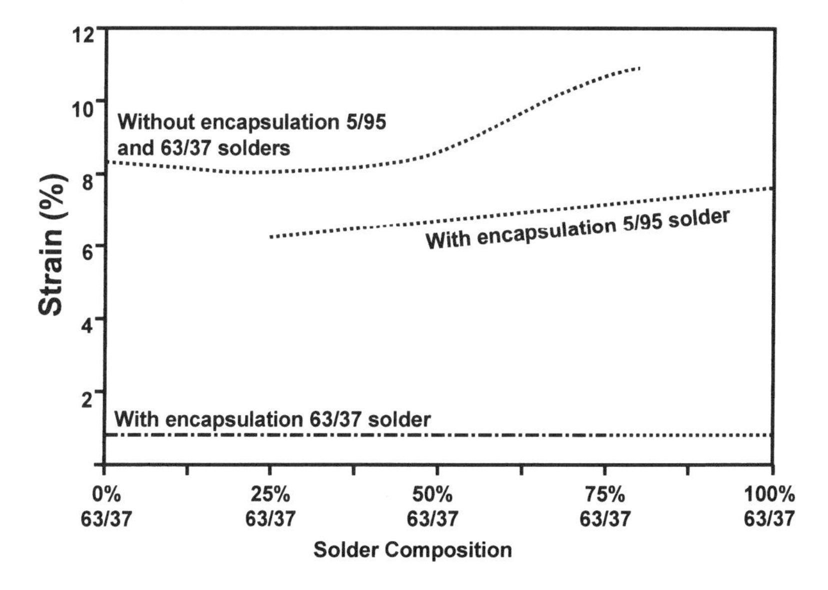

Figure 7-45. Effects of Encapsulation and Solder Composition on Strain in Solder (Courtesy of IBM Japan.)

Table 7-14. First-Level Single-Chip Packages and Their Characteristics

Package	Package Materials	Minimum No. of Interconnections Available 1995	Minimum No. of Interconnections Future 2000	I/O Spacing (mm)
Dual-in-line (DIP)	Alumina ceramic,	64		2.54
	plastic	64		2.54
Shrink DIP	Plastic	64		1.77
Skinny DIP	Plastic	64		2.54
Single-in-line (SIP)	Plastic	21		2.54
Leadless chip carrier (LLCC)	Ceramic	132		1.27
		300	400	0.63
	Plastic	180		1.00
Small outline package (SOP)	Plastic	40		1.27
Leaded chip carrier (LCC)	Plastic	84		1.27
		144		0.63
Quad flat pack (QFP)	Plastic	130		1.00
		160	500	0.65
	Ceramic	180		0.40
		200		0.63
Very small periphery array	Plastic	300	600	0.40
			500	0.40
Pin-grid array (PGA)	Alumina (single chip)	312		2.54
			>1000	1.27
	Ceramic (multichip)	2177	>5000	2.54
	Plastic	240		2.54
			>500	2.54
Tape automated bonding (TAB)	Plastic	300		0.50
			>1000	0.25
Ball grid array	Plastic	300	>500	0.50
	Ceramic	604	>1000	0.40
Chip scale packages	Thin film	300–1000	>1000	0.5
	Ceramic	300–1000	>1000	0.5
SLIM	Thin film	—	>5000	0.25

development of those packages as a function of single and multichip I/Os is illustrated in Figure 7-47.

7.5.2 Chip-Scale or Chip-Size Packages

An improvement in packaging efficiency beyond BGAs is being achieved by so-called chip-scale or chip-size packages (CSP) which are hardly bigger than the chip itself. Similar to QFP and BGA, CSP offers

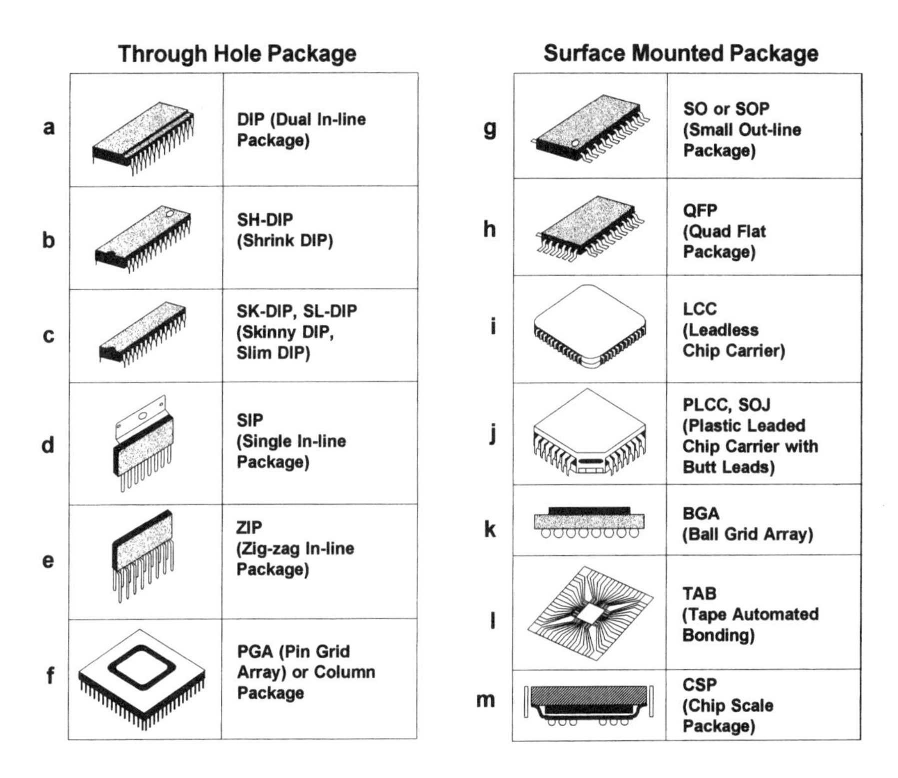

Figure 7-46. Types of First-Level Packages

burn-in and testability of ICs prior to joining to the printed-wiring board. There are a number of types of these packages (Fig. 7-48) that include such rigid substrates as ceramics with straight-through vias, flex or TAB-like connections with bumps or balls on either end and wafer-level assembly such as redistribution of peripheral I/Os to area array I/Os on chip and lead-on-chip (LOC). Two of the most popular of these packages are by Tessera and by Hitachi. The heat transfer characteristics of these packages range from 0.5 to 10°C/watt.

First-level packages, in addition to providing the required number of contacts for power and signal transmission, need to have the desired number of wiring layers discussed previously, provide thermal expansion compatibility with the chip, provide a thermal path for heat removal from the chip, and keep signal transmission delay and electrical noise to a minimum. Package sealing, to provide protection of both the package metallurgy and chip metallurgy, is yet another important function of the first-level package. Minimum electrical delay requires a low dielectric constant. Low electrical noise requires low self-inductances and low inter-line capacitances and inductances. Maximum heat removal requires high

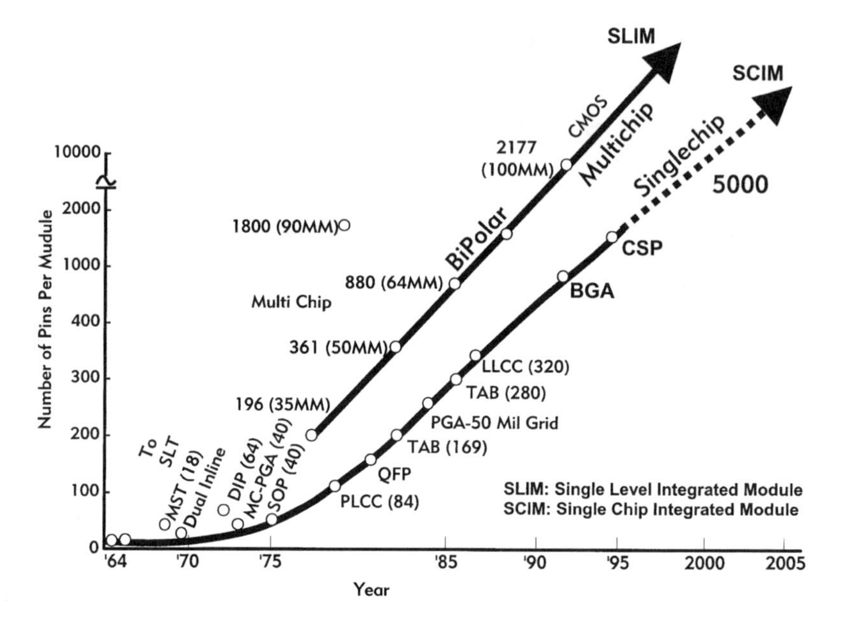

Figure 7-47. Chronological Development of Single-Chip and Multichip Modules

Example	CSP Type	Companies
	Flex circuit interposer	General Electric NEC Nitto Denko Tessera
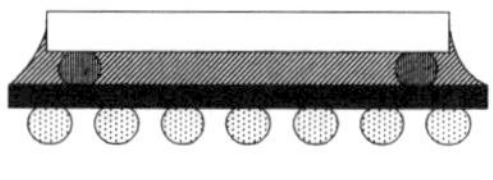	Rigid substrate interposer	IBM, Kyocera Matsushita Motorola Toshiba
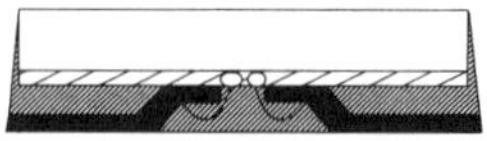	Custom lead frame	Fujitsu Hitachi LG Semicon Mitsubishi Electric Rohm
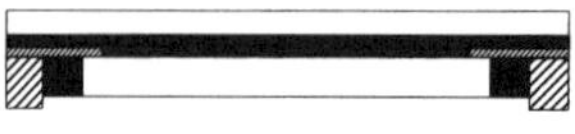	Wafer-level assembly	ChipScale ShellCase Sandia

Figure 7-48. Chip-Scale Packages

Table 7-15. Properties of Package Insulator Materials.

	Dielectric Constant	Thermal Expansion Coefficient $10^{-7}/°C$	Thermal Conductivity W/m K	Approximate Processing Temperature °C
Non-organics				
92% Alumina	9.2	60	18	1500
86% Alumina	9.4	66	20	1600
Si_3N_4	7	23	30	1600
SiC	42	37	270	2000
AlN	8.8	33	230	1900
BeO	6.8	68	240	2000
BN	6.5	37	600	>2000
Diamond				
High pressure	5.7	23	2000	>2000
Plasma CVD	3.5	23	400	≃1000
Glass-Ceramics	4–8	30–50	5.0	1000
Copper Clad Invar (10% Copper)/ (Glass Coated)	—	30	100	800
Glass Coated Steel	6	100	50	1000
Organics				
Epoxy-Kevlar (x-y) (60%)	3.6	60	0.2	200
Polyimide-Quartz (x-axis)	4.0	118	0.35	200
Fr-4(x-y plane)	4.7	158	0.2	175
Polyimide	3.5	500	0.2	350
Benzocyclobutene	2.6	300–600	0.2	240
Teflon®*	2.2	200	0.1	400

*Teflon is a trademark of Dupont Company.

thermal conductivity. Maximum power distribution requires high electrical conductivity of package metallization, and high reliability requires a close thermal-expansion match between the chip and substrate. The thermal and electrical properties of substrate materials in use in 1995 are shown in Table 7-15 for insulating materials and in Table 7-16 for conductors. The trend in the 1990s is toward glass±ceramic and organic materials with low dielectric constants. Thin-film multilevel wiring embedded in thin-film dielectrics of these materials and other low-dielectric-constant materials are important in the minimization of transmission delays that will be required in the 21st century.

Table 7-16. Properties of Package Conductor Materials

Metal	Melting Point °C	Electrical Resistivity $10^{-6}\Omega\cdot cm$	Thermal Expansion Coefficient $10^{-7}/°C$	Thermal Conductivity W /m·K
Copper	1083	1.7	170	393
Silver	960	1.6	197	418
Gold	1063	2.2	142	297
Tungsten	3415	5.5	45	200
Molybdenum	2625	5.2	50	146
Platinum	1774	10.6	90	71
Palladium	1552	10.8	110	70
Nickel	1455	6.8	133	92
Chromium	1900	20	63	66
Invar	1500	46	15	11
Kovar	1450	50	53	17
Silver-palladium	1145	20	140	150
Gold-platinum	1350	30	100	130
Aluminum	660	4.3	230	240
Au–20% Sn	280	16	159	57
Pb–5% Sn	310	19	290	63
CU–W[20%*Cu*)	1083	2.5	70	248
Cu–Mo(20%*Cu*)	1083	2.4	72	197

7.5.3 Single-Chip Package Markets

The worldwide package demand for the single chip packages discussed above is indicated in Table 7-17, which includes the application for diodes, transistors, optoelectronics, and discretes, both for memory and logic. Ceramic PGAs are used primarily for military and for high-reliability commercial applications (PCs to mainframes) that require either hermeticity or pluggability. Of the total ceramic PGA usage of about 150 million units, Intel alone uses about 50 million units for its 486, Pentium, P6 and others, and IBM uses about 30 million units for PC, workstation, midrange, and mainframe computers.

Plastic quad flat package is the most common package in use today, exceeding 10 billion units worldwide in 1995 and doubling by the year 2000. A variation of this package as indicated in Figure 7-46 is used in much smaller size with less pins [hence, on only two sides called small outline package (SOP) or small outline J-lead package (SOJ)], is actually manufactured in higher volume, almost by a factor of 3, than PQFP. Because Japan produces most of these packages, Figure 7-49 that lists the Japanese production of these packages is of great use. It should be

Table 7-17. Worldwide Package Demand by Package Type (in Millions of Units)

Package/Year	1994	1995	1996	1997	1998
CPGA*	151	166	168	181	203
PPGA	10	10	15	15	15
CQFP*	7.8	7.7	7.8	8.1	8.7
PQFP*	10,122	10,875	12,721	15,214	18,897
PBGA	14	17	32	41	80
CBGA	<1	2	4–10	4–10	4–10
TBGA	<1	1	2–8	2–8	4–10
TCP	0.5	1	1.5–2	2–4	2–4

Source: TechSearch International, Inc.
*VLSI Research, Inc.

noted that a number of these packages, particularly PGA, QFP, DIP as well as BGA, are available in ceramics from Kyocera, NTK, Sumitomo, IBM.

Ball-grid arrays (BGA) are the natural outgrowth of flip-chip technology and pin-grid-array connections, providing area-array connections on a smaller grid. The need for this type of package arose for mainframe

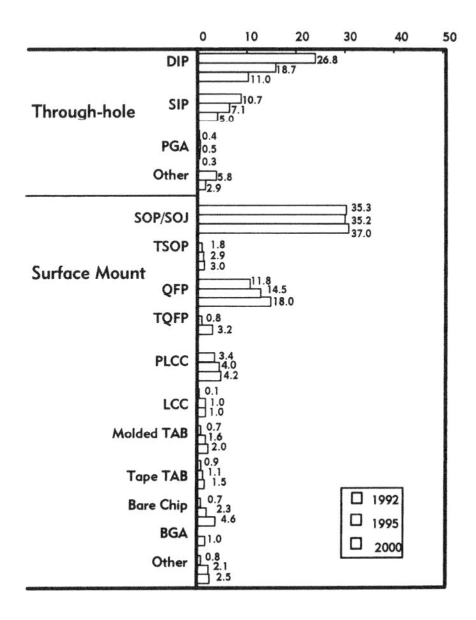

Figure 7-49. Japanese Packaging Production (Percent) (From Ref. 27.)

Table 7-18. BGA Market Trends (Packages in Millions)

BGA Type/Year	1995	1998	2000
PBGA	17	100	200
CBGA	2	4–10	10–15
TBGA	1	4–10	10–15

Source: TechSearch International, Inc.

and supercomputer applications by IBM, Hitachi, and NEC. NEC applied a modification of this package called flip TAB carrier (FTC) with solder ball connections in 1985 and Hitachi has been using, since 1990, a microchip carrier (MCC) with 7 layers of thin film, and resistor wiring, providing in excess of 600 balls or I/Os. As discussed in the trade-off section (Section 7.5.8) and as indicated in the chip-level interconnections, the BGA strategy is clearly the current wave in single-chip packages, providing cost-effective packaging solution at almost one-fourth in size from current ceramic PGAs for microprocessor PC applications. Of the dozens of companies currently evaluating the plastic or ceramic BGAs, IBM, Motorola, Sun Microsystems, COMPAQ, Ericsson, and BULL, among others, are using these in their products. Table 7-18 lists the market trend that is expected of these packages.

7.5.4 Plastic Packages

Plastic package developments currently underway include the following

1. New molding materials and processes
2. Process controls and quality assurance
3. Fine-pitch interconnections
4. New lead frame materials
5. Multilayer plastic PGA packages up to 2000 I/Os

Table 7-19 lists the historical and projected development in molding compounds by Nitto. Future developments are aimed at improving processes for moldability into thinner packages, antidelamination, higher-temperature solder resistance, and low stress.

In addition, the molding compound developments include composi-

Table 7-19. Molding Compound Development in Japan

	Discrete	MP-2000	Modability
1970s	Small DIP	MP-3000	Moisture resistance, moldability
	Middle DIP	HC-10-2	Moisture resistance, moldability
	Large DIP, SOP	MP 150SG	Low stress
1980s	PLCC	MP-180	Low stress
	QFP	MP-190	Low stress, processability
	SSOP, TSOP	MP-7000	Soldering resistance
	SQFP, TQFP	MP-7000	Soldering resistance
1990s	SOP, DIP	MP-80	Antidelamination, processability
	PLCC, QFP	MP-8000	Floor life for soldering, low stress, high adhesion
	Large package	Target 1	Soldering resistance, low stress, moldability
	Ultrathin package	Target 2	Moldability, soldering resistance
	Specific package	Target 3	Adaptability to components

Source: Nitto Denko

tion and formulation, moldability, and temperature and humidity bias reliability. These developments are summarized in Figure 7-50.

7.5.4.1 Process Controls

The increasing use of analytical techniques and microsensors to fine-tune materials and control process-induced defects falls into the category of enhanced process control and quality assurance. X-ray radiography and C-mode scanning acoustic microscopy–tomography are finding increased use as nondestructive techniques for evaluating a number of processes—plastic delamination; die metal, die attach, and bonding wire deformation; die metal and wire-bond voiding; lead frame, die passivation, die match, wire, wire bond, and case brittle fracture; and dendritic growth under bias. Mercury porosimetry has also been used successfully to find the number and size of epoxy and epoxy–metal pores in epoxy-encapsulated packages. Epoxy pores are less than 0.2 μm in diameter, epoxy–lead frame voids are about 1 μm in diameter, and surface pores range from 5 to 500 μm in diameter. Piezoresistive strain gauges integrated into test chips will continue to be used to directly measure the mechanical stress in a plastic encapsulated module (PEM)—caused by encapsulation, die bonding, and other factors, either during fabrication or under environmental stress testing. Solid-state moisture microsensors will also be used to measure the moisture content at any specific location inside a PEM.

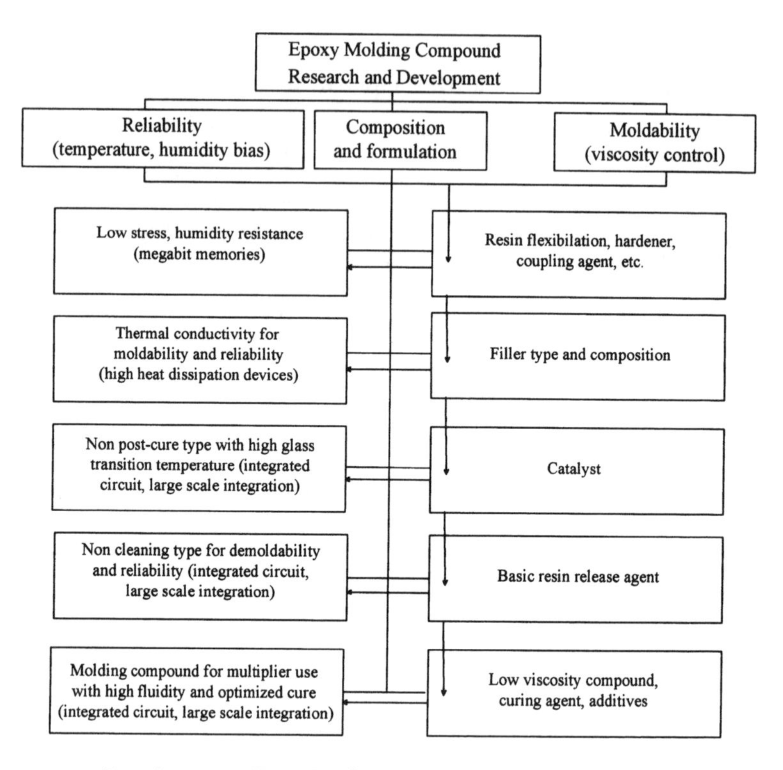

Figure 7-50. Development Trends of Epoxy-Molded Compounds (From Ref. 24.)

7.5.5 Plastic PGA Technology

Even though plastic PGA is not used as extensively as QFP or SOP, it is clearly the leading-edge technology providing very high I/Os with multilayer structures. As shown in Table 7-20, Ibiden's plan for future plastic PGA technology is expected to increase the pin count from the current 500–600 pins to 2000 pins by the year 2000. This increase in pin count is possible with multilayer plastic packaging technology that includes new materials with lower dielectric constant and high glass-transition temperature, finer-line wiring, down to 25 μm lines to 100 μm pitch, and layer counts up to 15. Ibiden expects land-grid arrays (LGAs) and ball-grid arrays (BGAs) to approach 600 pins before the end of the decade.

7.5.6 Lead-Frame Fabrication

Trends in lead-frame technology have important implications for the future of molded plastic packaging. Japanese companies are replacing copper alloys with 42Fe/58Ni (Alloy 42) as the lead-frame material,

Table 7-20. Plastic PGA Technology Trends (Ibiden) (From Ref. 24)

	1992	1995	1998
Pin count	300–400 pin	1000 pin	2000 pin
Pin pitch	2.54 mm	1.5 mm (LGA) 1.27 mm (PGA)	1.27 mm
B'g pitch	180 μm	~120 μm	~100 μm
Linewidth	100 μm	~50 μm	~25 μm
Trace thickness	18 μm	~10 μm	~10 μm
Multilayer	6 layers	10 layers	15 layers
Via hole size (diam.)	0.3 mm	0.1 mm	0.1 mm
Insulation-layer thickness	0.1 mm	0.06 mm	0.06 mm
Dielectric constant	4.4	3.5	3.5
Impedance control	±15%	±10%	±5%
TCE. (ppm)	14	10	10
Tg Point	180°C	200°C up	200°C up
Material	BT	New BT	New BT

Source: Courtesy of Ibiden.

especially for moderate to high heat-dissipating devices such as processors. For thin (≤0.15 mm) lead frames with close lead tips for very high lead-count packages, chemical etching has replaced mechanical punching.

7.5.7 Logic and Memory Applications

The technology developments discussed above are expected to lead to enhancing both memory and logic chips. A typical representative roadmap from Oki for both these are included in Tables 7-21 and 7-22. The memory roadmap indicates reduction in mounting and wire loop heights, leadframe thickness and reliability improvements as a result of new materials. The logic roadmap, in addition to indicating the role of plastic QFP, also indicates high I/O achievability with ceramic PGAs and thin-film packages.

7.5.8 Ceramic Packaging Technologies

The chronological development in ceramic packaging is indicated in Figure 7-51, which illustrates the recent materials development from alumina to AlN, mullite and glass–ceramic and future developments in composite ceramics. The properties of the substrate materials currently in use are indicated in Table 7-23. The process advances in cofiring in

Table 7-21. Logic LSI Package Roadmap

Year (CY)		1990	1993	1996	1999	2002
LSI chip trend	Gate density	200K	300K	500K	1M	2M
	Max. chip size	155 mm	15 mm	17.5 mm	20 mm	25 mm
	Max. power	1.5 W	2 W	3 W	5 W	10 W
	Max. speed	50 MHz	150 MHz	300 MHz	500 MHz	500 MHz
	Max. pin count	250	350	500	750	1000
Package technology	QFP size (mm)	40	40	28	40	—
	QFP pitch (mm)	0.65	0.5	0.4	0.3	—
	C-PGA I/O	—	401	526	750	1000
	Size (mm)	—	40	32	36	40
	Pitch (mm)	—	1.778	1.27	1.27	1.27
	Thin package					
	I/O	—	1.44	216	344	2966
	Size (mm)	—	20	24	28	24
	Height (mm)	1.2	0.5	0.4	0.3	0.3
	Cooling	Cu lead-frame	QFP with Fn	QFP with heat spreader	QFP with cool module	

Source: Courtesy of Oki.

terms of line and via dimensions and the number of layers that are cofired are indicated in Figure 7-51. The state of the art in ceramic packaging involves 50–100 μm lines and vias placed on 225–450 μm apart and the substrate cofired with dimensional control better than ±0.1% in 63 layers of metal and ceramic. It appears that these dimensions, dimensional controls, and the number of layers can be improved technically by about a factor of 2 within the next decade. Whereas ceramic packaging was used primarily for either high performance or higher-reliability applications in the past, Panasonic, Kyocera, and others have begun to apply to consumer products.

Table 7-23 shows the variety of materials being used as ceramic substrates. These materials include Al_2O_3, AlN, mullite, and a variety of glass–ceramics that include both glass added to alumina and crystalizable glasses. Whereas most of these low-temperature ceramics are metalized with Ag, Ag-Pd, or Au as fired in air, a few firms are beginning to cofire with copper using special binders, or special atmosphere cycles, to remove organics from greensheets. Table 7-23 also illustrates the properties of some of the glass–ceramics being pursued by Kyocera, Panasonic, Oki, Fujitsu, NEC, and NTK.

Table 7-22. Memory Package (TSOP) Technology Roadmap

		1990	1995	2001
Device Configuration	Density	4M	64M	1G
	Die size (mm)	90	190	400
Package mounting height (mm)		1.2	1.2	0.8
Customer usage condition	Storage	1 day usage		
	Reflow Condition	IR (220°C)	IR (240°C)	
Ratio of die size and package size (%)		64	79	80
Spacing between package edge and die edge (mm)		1.0	0.4	
Packaging technology	Minimization	Conventional	LOC	
	Low profile wire loop height	200 μm max.	120 μm max	
	Lead-frame thickness	0.15 μm	0.15 μm	
Reliability (reflow resistance) molding epoxy resin		Conventional (EOCN)	Biphenyl	Imid modified epoxy (high Tg)

Source: Courtesy of Oki.

7.5.9 BGA Packages

There are several BGAs including micro-BGAs (μ-BGA) that are in the horizon, as illustrated in Figure 7-52. These are packages with reduced I/O pitches but still compatible with the leading-edge printed wiring board technologies. These technologies and their I/O spacings are

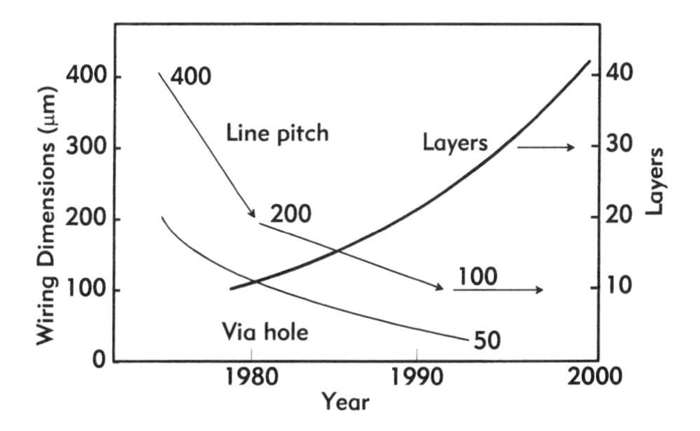

Figure 7-51. Ceramic Packaging Trends (Courtesy of Kyocera.)

Table 7-23. Ceramic Substrate Materials

	Alumina	Mullite	AIN (Ibiden)	Glass-Ceramic Kyocera	Pansonic	IBM	Fujitsu	NEC	SiO_2 + B_2O_3
Thermal conductivity (W/mk, RT)	18	5	180	2	2	2	2	2	2
TCE. (10^{-6}/°C) 40–400°C	7.0	4.4	4.5	4.0	—	3.0	4.5	3.5	2.5
Dielectric constant (1 MHz, RT)	10.0	6.8	8.9	5.0	7.4	5	5.6	4.4	4.0
Tan δ (1×10^{-4})	24	10	1	20	—	—	—	—	—
Conductor metal	W, Mo	W, Mo	W	Cu	Ag, Cu	Cu	Cu	Au	Ag-Pd
Sheet resistance (mΩ/sq)	10	10	15	3	2–5	2	3	3	2–5

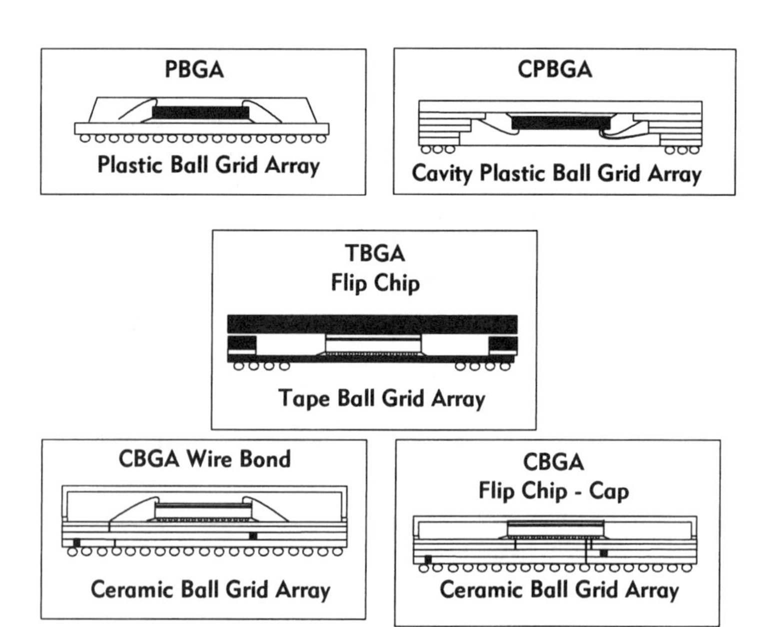

Figure 7-52. Various Types of BGA Packages

Table 7-24. BGA Packages

Technology (Company)	I/O Configuration	I/O Pitch (μm)
Micro SMT (Mplus Microwave)	Peripheral	100–150
m-BGA (Sandia)	Part area array	500–750
μBGA (Tessera)	Peripheral	500–1000
DCAM (IBM)	Part area array	200–500
Slicc (Motorola)	Area array	825–1500
MC2M (Valtronic USA)	Peripheral	1000–1500

indicated in Table 7-24. Some of these packages are also referred to as chip scale packages, particularly in harsh environments. While the BGA trend continues, there are certain reliability issues to which BGAs are prone. Quad flat pack, although low in cost, is too large in size and too low in thermal performance. VSPA (very small peripheral array) illustrated in Figure 7-53 seems to offer the best of both worlds of QFP and BGA [28].

The trend in the past has been from pin-through-hole to SMT as electronic products shifted from mainframes to desktop. The most recent trend, however, is from SMT to BGA, consistent with the product trend from desktop to laptop to notebooks.

QFP 304 leads

36 mm X 36 mm
Plastic QFP
0.5 mm pitch
Easily Reworkable
4-6 layers required in PCB
Visual Inspection OK
Normal SMT

Back side card OK
$0.011/pin

BGA 313 leads

35 mm X 35 mm
Peripheral BGA
1.2 mm pitch
Reworkable
6-8 layers required in PCB
No Visual Inspection
Not "normal" SMT
Thermal via array
Back side NOT OK
$0.02/pin

VSPA™ 320 leads

27 X 27 mm (including leads)
Cavity Down VSPA
.5 mm/.8 mm pitch (50/50%)
Reworkable
4-6 layers required in PCB
Visual Inspection OK
Normal SMT

Back side card OK
$0.01/pin

Figure 7-53. QFP, BGA, and VSPA™ Package Comparison (From Ref. 28.)

The advantages of BGA are as follows:

- Assembled with same SMT equipment as QFP
- Smaller area board and yet, larger I/O Pitch
- Self alignment during soldering process as with flip chip
- Low assembly cost
- Repairable
- Component cost reduction
- Inspection unnecessary compared to QFP

7.5.10 Package Trade-offs

Figure 7-54 plots the cost of various single-chip packages as a function of I/Os. Given the cost-sensitive nature of electronic packaging, plastic quad flat pack and single-layer TAB are the leading candidates, compared to ceramic and plastic PGAs. This is already reflected in the huge volume of these packages as described in Section 7.5.1, "Single-Chip Packages." As pointed out earlier, the QFP is being improved upon for both increased I/Os by improving on lead-frame pitch from the current 0.4 mm to 0.15 mm and for thinness by reduction of molded or ground thickness. In spite of these improvements, the problems with QFP are its huge size and poor thermal and electrical performance. The popularity of BGAs and their exponential growth discussed above can be attributed to this deficiency. In addition, a new revolutionary, simple, and elegant package known as very small peripheral array (VSPA) is beginning to emerge as the best overall medium I/O package from the Panda Project

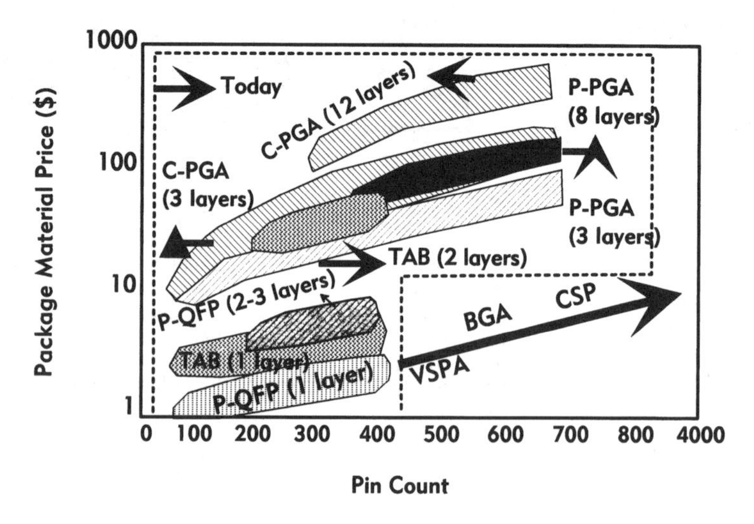

Figure 7-54. Single-Chip Packaging Cost (Courtesy of Otsuka.)

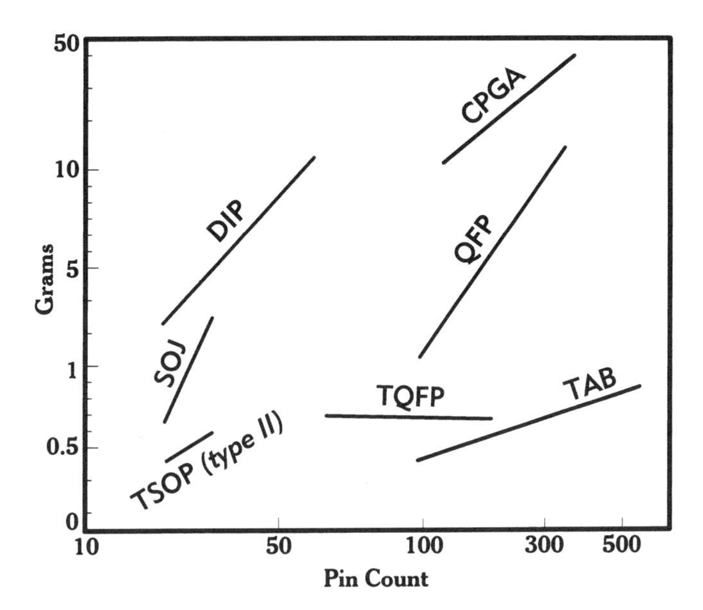

Figure 7-55. Package Weight Versus Pin Count (Courtesy of JTEC.)

[18]. As compared in Figure 7-53, this package is smaller in size than QFP and BGA and yet provides far superior thermal performance, due to its metallic heat sink to which the die is bonded. Because the heat sink can be copper, thermal conductivity improvement of the order of 20 over ceramic and 500–1000 over plastic packages can be achieved. The cost per I/O is expected to be the same as QFP. Because of these advantages and better electrical performance, this package may emerge as the most competitive package to QFP.

Figure 7-55 and 7-56 compare the package weight and mounting height on PWB for a number of single-chip packages as a function of pin count. Only TAB and TSOP, besides QFP, BGA and VSPA, can meet future competitive requirements for thinness and lightweight in packages.

As shown in Figure 7-57, a number of companies such as Hitachi has selected QFP and TAB, as their front up packages for low to medium I/Os. The only exception seems to be for very high I/Os, where PGA and BGA will be their strategic packages. There is general agreement that the best alternative to QFP in low-pin-count consumer products and high-pin-count computer products will be BGA or surface-mount PGA, the new VSPA package. BGA is seen as a high-speed, high-pin-count package that also provides a compact solution. Figure 7-58 shows how BGA provides a smaller footprint at a 1-mm pitch than the ultimate 0.15-mm-pitch QFP beyond 600 I/Os. This is one of the primary reasons for the emergence of BGA.

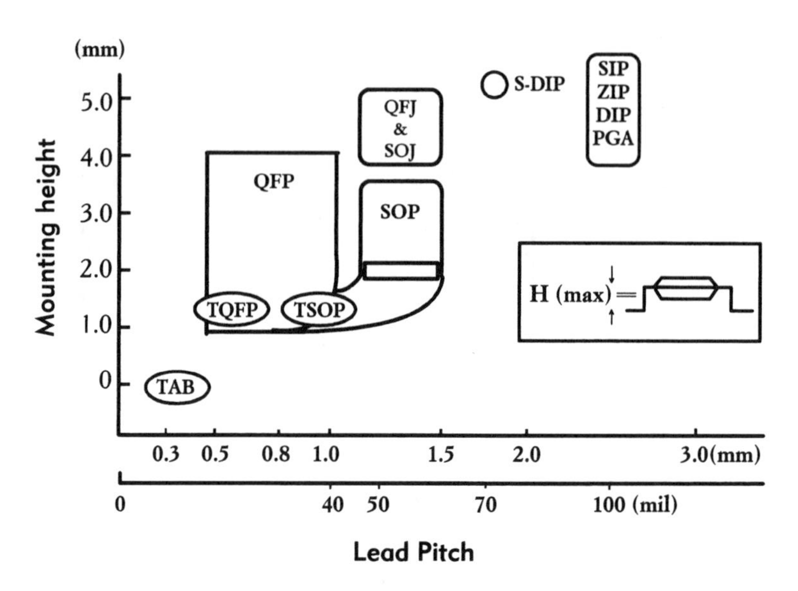

Figure 7-56. Lead Pitch and Mounting Height

In contrast to the inspection needs of fine-pitch QFP, manufacturers like Hitachi and IBM, do not believe there is a need to inspect the BGA joint, even though X-ray inspection may be possible. This is so because of the large pitch that BGA provides. Hitachi also claims great flexibility in circuit design using BGAs, allowing V_{cc} and V_{dd} connection everywhere,

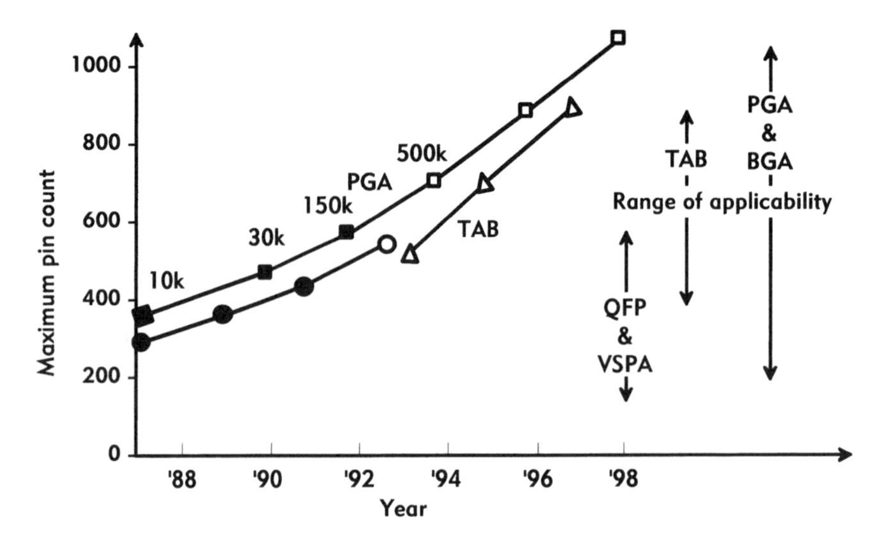

Figure 7-57. High-Pin-Count Packages (Courtesy of Hitachi.)

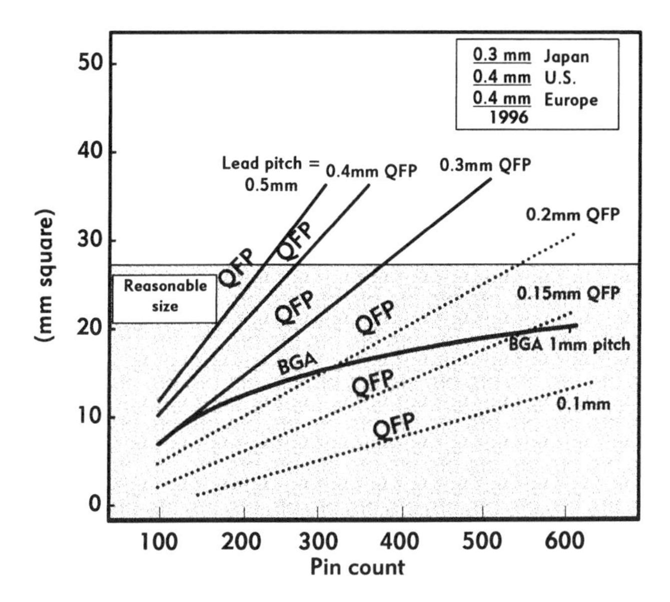

Figure 7-58. Relative Package Areas: BGA Versus QFP, Illustrating Superiority of BGA

in addition to providing power and ground for each group of output buffers, reducing the simultaneous switching noise. The QFP approach does not provide this flexibility.

In parallel to surface-mount options, the electronics industry is pursuing PGA options in limited fashion both in ceramics and plastics. Toshiba, for example, is already pursuing an 820-pin ceramic PGA on a 60-mm square ceramic substrate. Kyocera supplies a 1000-pin surface-mount PGA in ceramics with its improved PGA that compares favorably with QFP, as illustrated in Figure 7-59. The pin-grid pitch in these packages is 1.27 mm (50 mils). Toshiba chose TAB connection to the 20-mm chip using gang inner lead bonding and single-point outer lead bonding. The plastic PGA trend providing in excess of 1000 I/Os was discussed earlier (Section 7.5.3).

7.5.11 First-Level Multichip Packaging

Multichip packaging is one of the fastest growing segments of all the packaging technologies. It came into existence for mainframes and supercomputer applications that required the most number of circuits to be packaged in the least amount of space. These attributes are now expected to be expanded into workstations, consumer electronics, medical, aerospace, automotive and telecommunication technology. This subject is

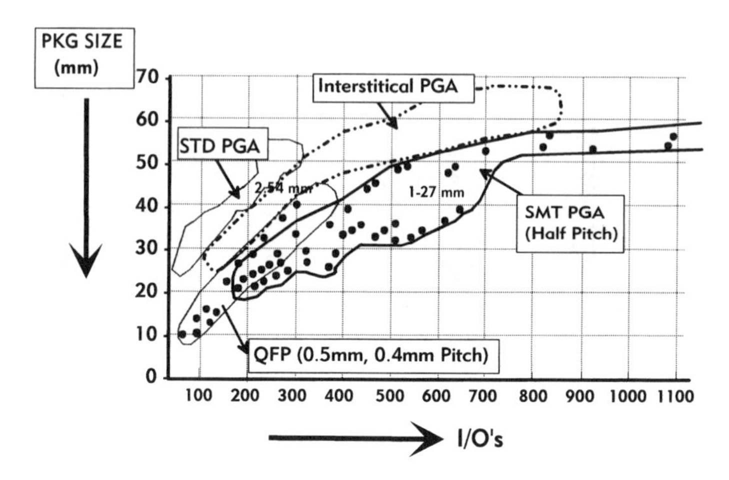

Figure 7-59. Package Body Size Versus Terminal Count; SMT PGA Same Size as QFP

discussed in more detail in Part 2, Chapter 9, Ceramic Packaging, Part 2, Chapter 12, Thin Film Packaging, and elsewhere [29].

7.5.11.1 Definition

Multichip packaging requires the definitions of *multichip* and *packaging*. Packaging has been defined elsewhere as interconnecting, powering, cooling, and protecting the semiconductor chips. Because single-chip package is defined as interconnecting bare chips onto a single plastic, ceramic, or some other carrier, the multichip package has been defined as providing the same function with more than one bare chip on to a single carrier or substrate. With this definition, however, there are very few current manufacturers, such as IBM, that qualify as the multichip fabricators, as there are very few companies that bond the bare chips directly to the multichip substrate.

The benefits of multichip packaging, however, can be realized, although not as fully as in direct bare-chip attachment, by bonding prepackaged chips in their single-chip carriers to the multichip substrate. This is the most common multichip packaging that is currently in use. Both of the above types are illustrated in Figure 7-60. The multichip carrier on which the chips are bonded can be ceramic, printed wiring board, or thin films deposited on any substrate, including metals like silicon, copper, and aluminum, and ceramics like AlN, diamond, and alumina, and organics like polyimide–glass, composite organics–inorganics.

7.5.11.2 Functions of Multichip Package

Consistent with the above definitions of multichip and package, the multichip package must perform the following functions:

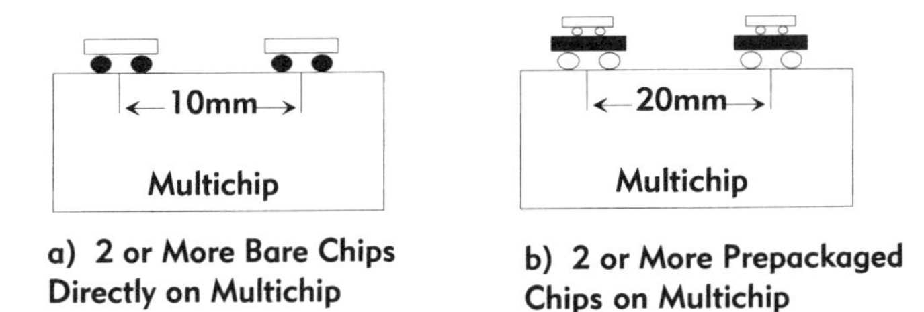

Figure 7-60. Multichip Module Defined as ≥ 40% Silicon Efficiency

1. Remove heat effectively from the chips
2. Provide interconnections between all the chips with as many circuits and as high a circuit performance as available to the multichip substrate
3. Provide wiring density with high-conductivity metal to interconnect all the chips with minimum chip-to-chip spacing
4. Provide multichip substrate connections for signal and power distributions
5. Provide protection to all the chips and the multichip substrate itself.

The schematic in Figure 7-61 illustrates these functions.

7.5.11.3 Packaging Efficiency

High packaging efficiency results as a result of placing a large number of chips closely spaced on an individual multichip package to meet a certain function (e.g., processor on a module).

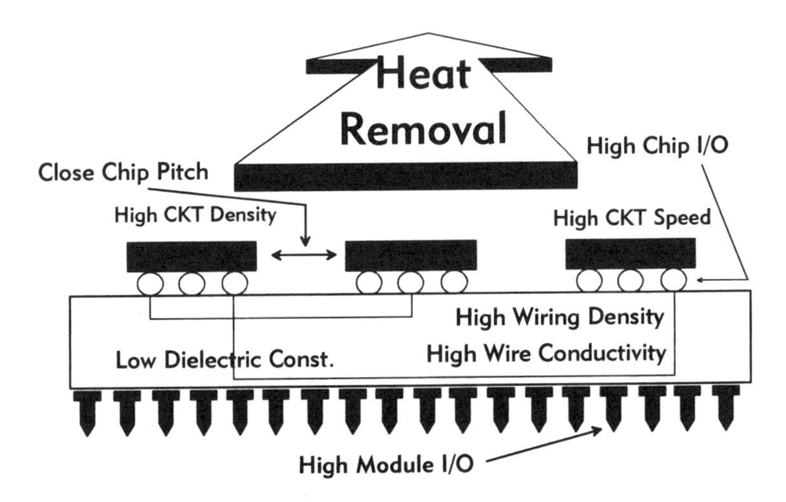

Figure 7-61. Functions of Multichip Packages

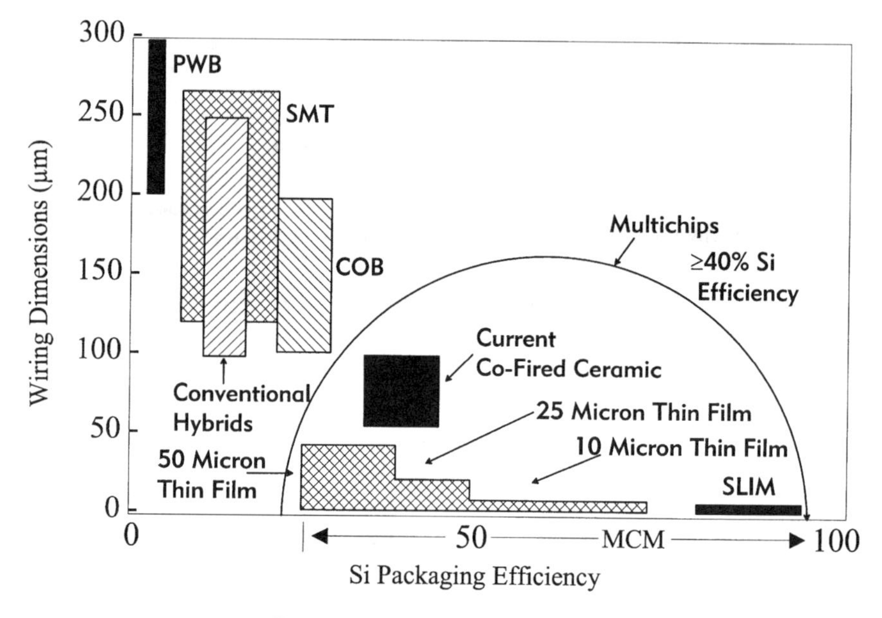

Figure 7-62. Packaging Efficiency of Various Packages

Packaging efficiency (P_{eff}) is defined by

$$P_{eff} = \frac{\text{Total active silicon area}}{\text{Total active multichip packaging area}} \qquad [7\text{-}19]$$

Figure 7-62 illustrates packaging efficiency of various packages including multichip with various line widths that are available in various packages, illustrating clearly the leverage of multichip module in packing more chips in least space than any other packaging including wafer scale integration (WSI). This advantage of multichip is expected to provide a wide array of applications in aerospace, medical, consumer, telecommunication, and computer industries.

7.5.11.4 Electrical Performance

Electrical performance is the primary reason for the use of multichip packages in computers. There are two parameters that contribute to the system-level performance, which is often quoted by a throughput parameter, MIPS (millions of instructions per second). These are cycle time and cycles per instruction and are related to the MIPS by the following equation, as discussed earlier in this chapter:

$$\text{Number of MIPS} = \frac{10^3}{\text{(Cycle time)(Cycles per instruction)}} \qquad [7\text{-}20]$$

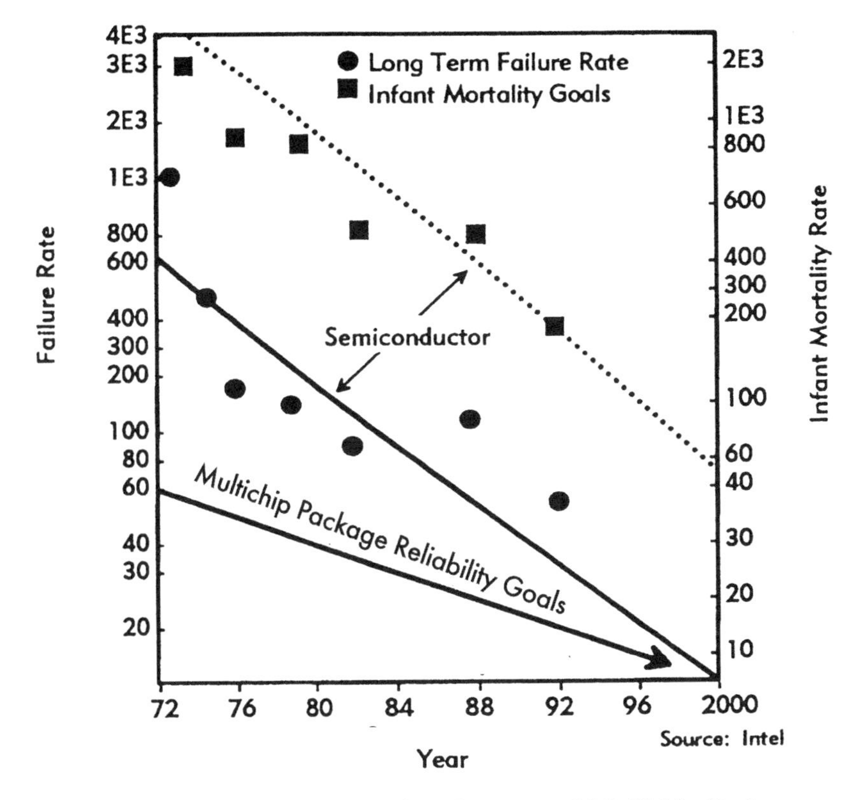

Figure 7-63. Reliability of Semiconductors and Multichip Packages

Leading-edge computer performance in MIPS requires leading-edge cycle time and leading-edge cycle per instruction. The latter parameter is often ignored, and as can be calculated from Equation 7-20, it will have a great effect on the MIPS. The multichip package influences both of these parameters, as both of these are influenced by packaging efficiency of circuits.

7.5.11.5 Reliability

Reliability enhancement is an essential function of any multichip *package,* particularly when it carries so many leading-edge and expensive chips, making it prohibitively expensive to discard due to any field failure. Any multichip *package* must inherently be more reliable than the semiconductors it packages, the reliability of which continues to be improved upon according to Figure 7-63, which plots both the infant mortality and long-term failure rate in parts per million as a function of time [30].

Fortunately, multichip packaging provides the fundamental basis for improvement of reliability, as illustrated in Figure 7-45, wherein the total number of interconnections and wiring lengths in forming an equivalent processor involving chips and various packaging levels is illustrated. Reliability of multichip packages must be treated at three different levels to ensure its success:

1. Design for reliability upfront
2. Build-in reliability by process controls during manufacturing
3. Accelerate test of final product to guarantee the design and the process

7.5.11.6 Leverages

There are four major leverages of multichip packaging:

1. Packaging efficiency
2. Electrical performance
3. Reliability
4. Cost

7.5.11.7 Types of Multichips

Consistent with the definition of multichip packaging technology in this chapter, one can visualize three different multichip technologies (Fig. 7-64): (1) those formed with laminated, organic boardlike technology referred to as MCM-L, to which chips can be bonded by TAB or wire bond, (2) those formed with screen-printed ceramic thick film, either sequentially formed by print and fire process, or parallel formed by lamination of screened greensheets and subsequently cosintered, referred to as MCM-C, to which chips are attached directly by flip-chip or indirectly in prepackaged carriers, and (3) those formed with deposited thin-film technology using thin films of polymers and metals, referred to as MCM-D, to which the chips are attached. Variations of these can also be considered multichips. For example, a variation of laminated MCM using plastic packages, involving molding compounds and lead frames, so-called multichip plastic quad packs (MCM-P), containing four chips became available.

7.5.11.8 Cost

The cost of multichip modules varies with the type of multichip and wiring density it provides.

Within various multichip options, it is controversial as to which type is cheaper. Figure 7-65 illustrates the relative costs of various multichip technologies as a function of wiring density. It is generally agreed that

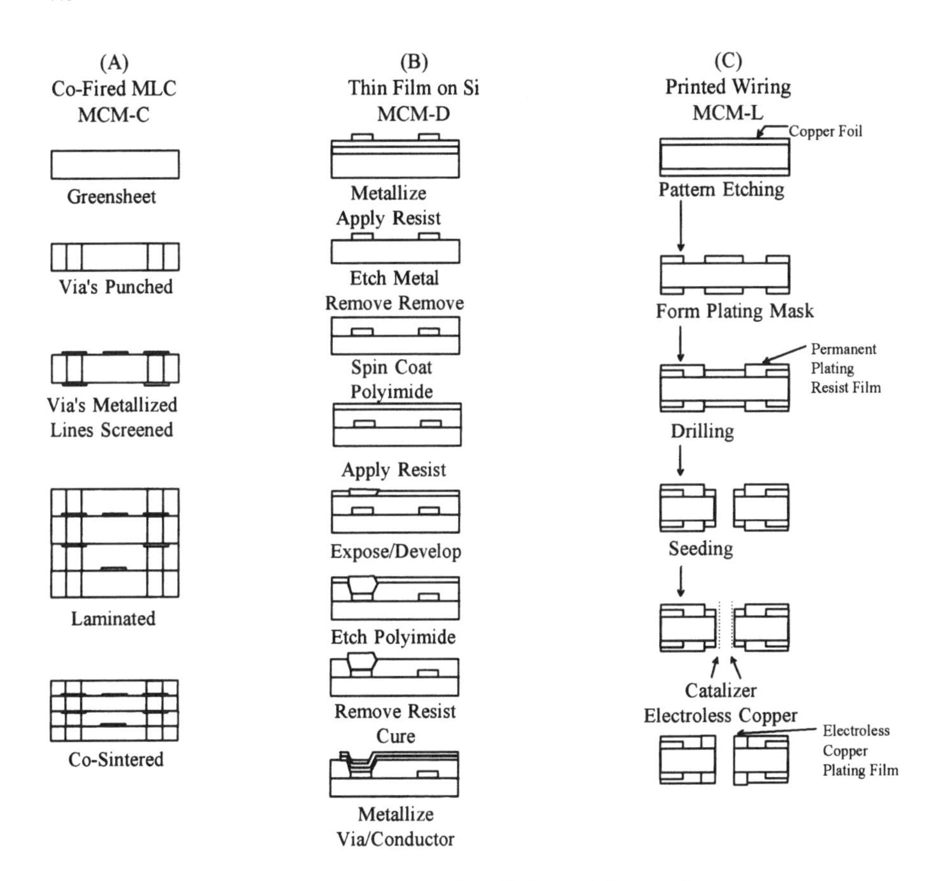

Figure 7-64. Three Types of MCMs and Their Processes

the laminated technology, although it is cheapest at low wiring densities, becomes very expensive at medium to high wiring densities. The controversy is mainly in the multilayer ceramic versus thin-film multichip packages. Whereas the former is based on a parallel process involving screened greensheets and a final high-yield cosintering process that assembles all the layers, the latter is a sequential and an expensive ultraclean-room process. Thus, even if all operations have better than 95% yield in the sequential thin-film process, the final yield can be significantly lower than desired. IBM's experience with multilayer ceramic during the parallel process approach has given yields in excess of 90%. Because of these and other reasons, the cost of multichip technologies, as illustrated in Figure 7-65 has been determined [29].

There is a need yet to have another category, based on deposition of thin films (MCM-D) on ceramic multichip modules. The wiring in this type of multichip is shared in cofired ceramic thick film and polymer–metal

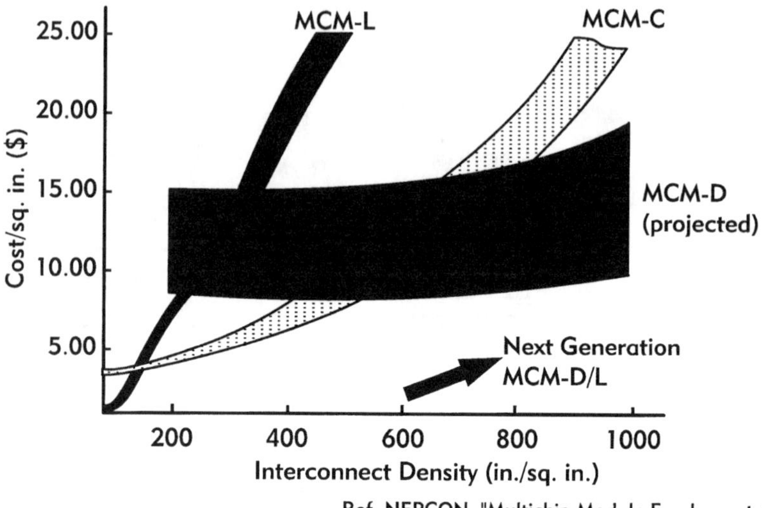

Figure 7-65. Interconnect Density Versus Cost for Various MCMs

thin film, offering the best of both technologies in cost and function. This type of multichip is expected to be the most predominant of all multichips and has already formed the basis of two multichips at two major companies, NEC [31] and IBM [32].

The multichip package that is expected to be dominant, however, in the next decade is thin film on laminate MCM-D on MCM-L. A good example of this is the IBM SLC surface laminar circuitry discussed in Section 7.7, "Second-Level Packages."

Table 7-25 illustrates typical package parameters for each of the three multichip types. The materials and their properties are illustrated in Table 7-26 and in Figure 7-66.

7.5.11.9 Applications

Because of high electrical performance, reduced space and weight, and improved reliability over single-chip modules, as discussed above, to form the required system-level function, multichip packaging is expected to have a broad spectrum of applications. Some of these are shown in Table 7-27.

7.6 PACKAGE-TO-BOARD INTERCONNECTIONS

As pointed out earlier, the system needs such as cellular phones or disk drives, as illustrated in Figure 7-67, drives the technologies toward small sizes and low weight but with high functionality and at low consumer

Table 7-25. Typical Package Parameters of the Three Multichip Module Types

Characteristic	MCM-L	MCM-C	MCM-D
Description	High-density laminated printed circuit board	Cofired low-dielectric-constant ceramic substrate	Thin film on silicon
Max. wiring density (cm/cm^2)	300	800	250–750
Min. linewidth (μm)	60–100	75–100	8–25
Line space (μm)	625–2250	125–450	25–75
Max. substrate size (mm)	700	245	50–225
Dielectric constant	3.7–4.5	5–5.9	3.5
Pinout grid (mm)	Array 2.54	Array 2.54 (staggered)	Peripheral 0.63 mm
Max. No. of wiring layers	46	63	4
Via grid (μm)	1250	225–450	25–75
Via diameter (μm)	300–500	100	8–25

prices. These requirements translate directly into component density—number of components per unit area of board. The components include active plastic or ceramic packages containing logic and memory chips as well as nonactive components like capacitors, resistors, inductors, and so forth. Figure 7-68 illustrates the trend in component density which is currently at about 20–25/cm^2 and which is expected to be 50/cm^2 by the years 2000–2005. This figure also illustrates, to some extent, how this component density will be achieved by reducing the size of each component, both active and passive while providing more pins or connections. The active component reduction is achieved for most part by reducing the I/O pitch or by area-array connections or both. The passive component improvements are made by reducing the sizes, as also illustrated in Figure 7-68 from 3.2×1.6 mm to 0.8×0.4 mm.

There are three basic types of connections between the first- and second-level packages: those with pins, requiring pin-through-holes (PTH), and others with lead frames meant for surface-mounting the device (SMD) by the use of surface-mounting technology (SMT). The third category is BGA, using balls in area-array fashion in contrast to peripheral lead frames. Package assembly includes, in addition to, an active plastic or ceramic components containing logic and memory chips, nonactive components like capacitors, resistors, and inductors. All the available

Table 7-26. Properties of Multichip Substrate Materials

MCM Material	Dielectric Constant (1 MHz)	Thermal Expansion Coefficient ($10^{-7}/°C$)	Reuse/ Fabr./ Temp. (°C)	Thermal Conductivity (W/mk)
Alumina	9.4	66	1600	20
Aluminum nitride	8.8	35	1600	20
Mullite + glass	5.9	35	1200	6
Borosilicate + alumina	5.6	40	900	6
Lead borosilicate + alumina	7.8	42	900	5
Glass–ceramic	5.0	30	950	7
Composite SiO_2 + glass	3.9	19	900	4
Epoxy Resin	4.7	1400	200	0.1
Polyimide resin	4.7	750	300	0.1
MS resin	3.7	680	250	0.1
Polyimide				
PMDA-ODA	3.5	250–400	400	0.2
BPDA-ODA	3.0	20–60	400	0.2
Benzocyclobutene	2.7	650	350	0.2
Polyphenyl quinoxaline	2.7	350	450	0.2

packages were previously classified in Figure 7-64 into one of these categories.

The surface-mount technology was the fastest growing technology in small systems and consumer electronics during the 1980s. It presents a number of advantages. The packages for SMT can be developed with pads or connections out of the package at the tightest spacings, thus allowing more I/Os per unit area of the package. Second, the technology can be used to increase the packaging density by requiring less board area and, thus, lowering the cost of packaging the total system. As a result, surface mounting, which accounted for roughly 20% of all consumer and low-end packages in the late 1980s, is already at about 80%. This technology will, however, pave the way for BGAs in the late 1990s.

The overall trend in package-to-board assembly is illustrated in Figure 7-69 showing the expected trend to fine-pitch quad flat pack (from

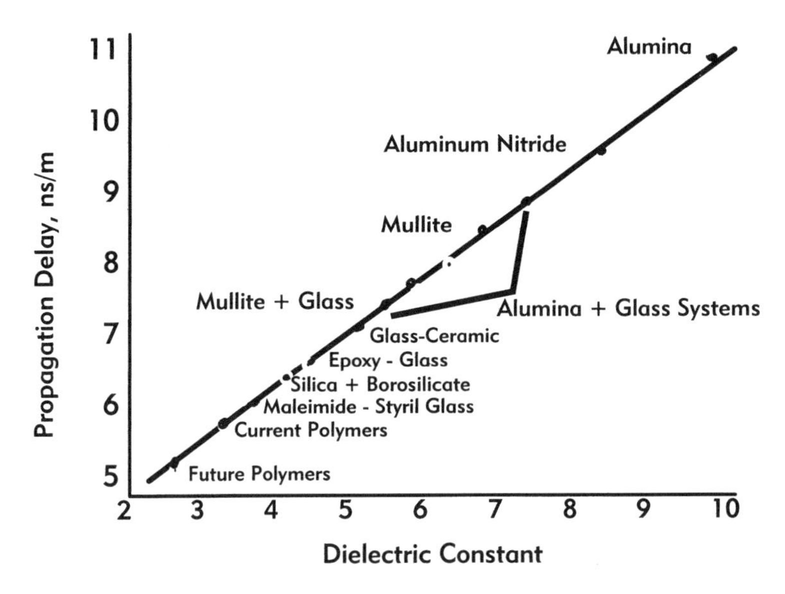

Figure 7-66. MCM Substrate Materials and Their Dielectric Constants

Table 7-27. Potential Multichip Package Applications and Their Characteristics

Application	Characteristics
Computers	High electrical performance Less space, high reliability
Military	Less space and weight
Telecommunications	High bandwidth performance
Consumers electronics	Less space and weight
Medical	Less space
Automotive	Less space

0.4 mm pitch in 1995 to 0.15 mm in 2000) as well as to ball-grid array, chip scale package, and direct flip-chip attach to organic board. The QFP to BGA change appears in Figure 7-70 around 400 I/Os, but as pointed out earlier, a number of companies, particularly Japanese with massive investments in peripheral SMT, will continue to push beyond 400 I/Os. The lead-pitch improvement consistent with this trend is indicated in Figure 7-71 for a number of actual Japanese packages.

Consumer products require thin and lightweight packaging. Plastic packages such as QFP that are surface-mounted onto PWB have effectively

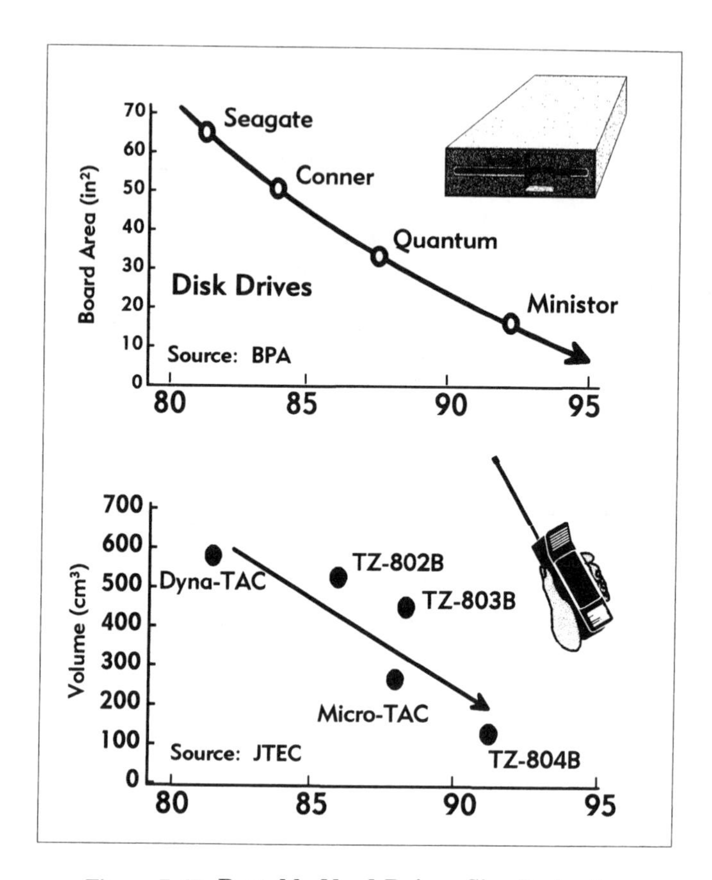

Figure 7-67. Portable Need Drives Size Reduction

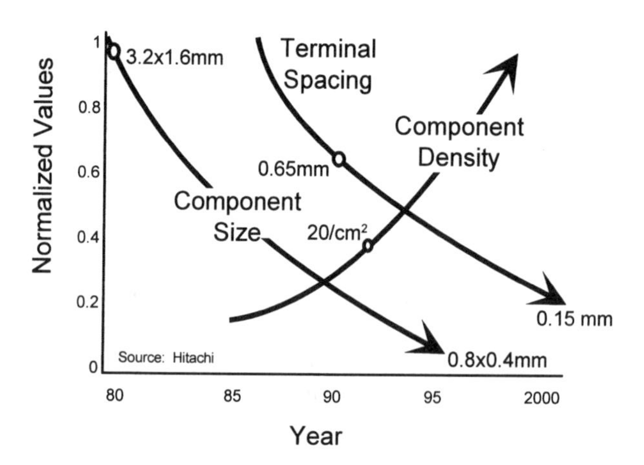

Figure 7-68. Portable Component Density (Courtesy of JTEC.)

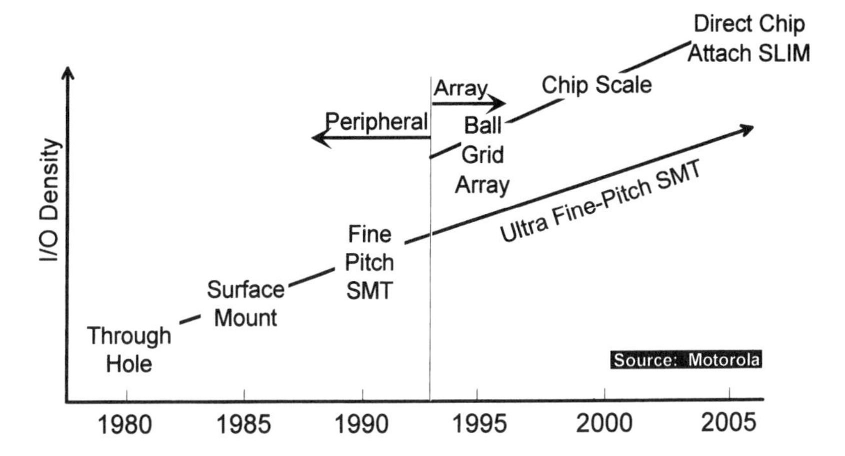

Figure 7-69. Package-to-Board Interconnection Density Trends

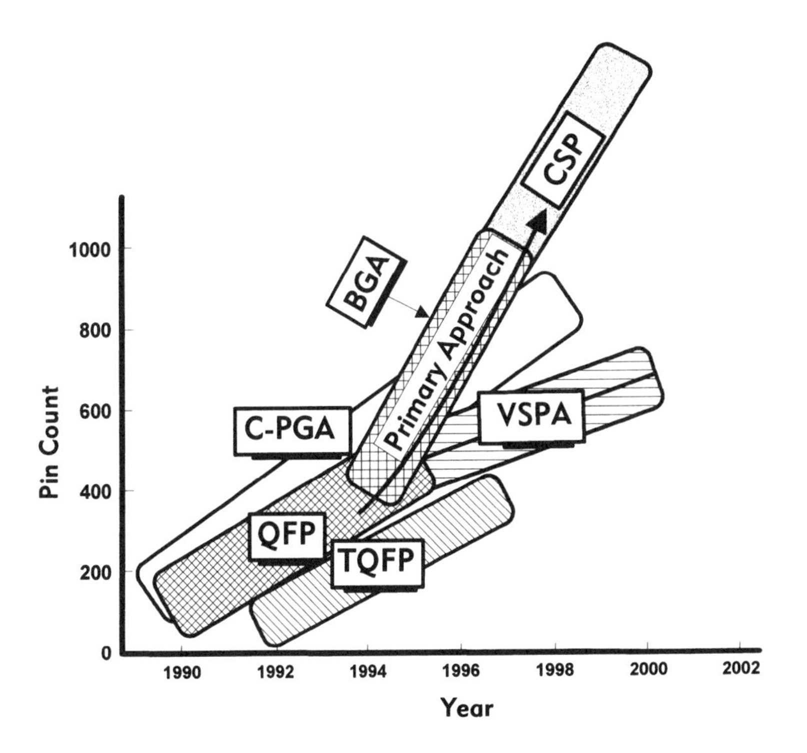

Figure 7-70. Pin-Count Evolution

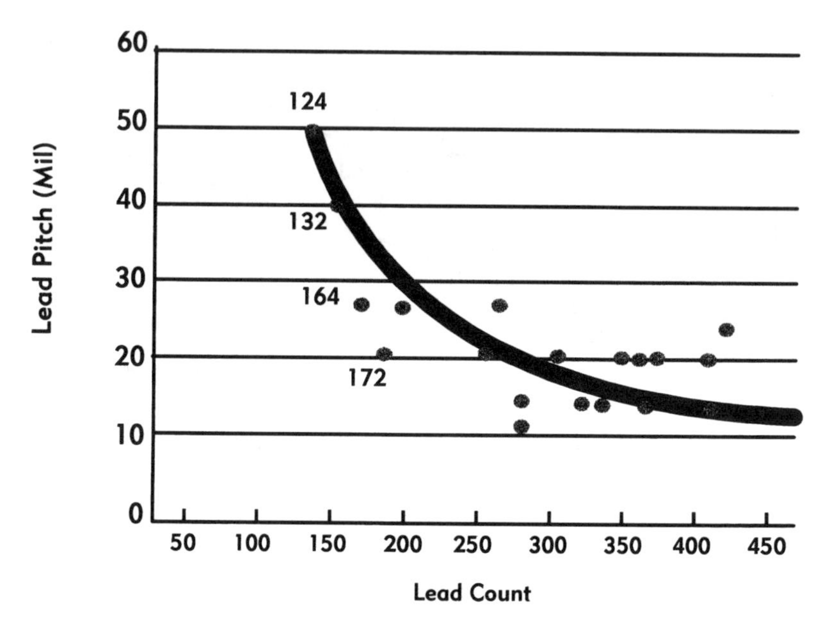

Figure 7-71. QFP Lead Pitch Versus Lead Count

met consumer product requirements. The industry vision of next-generation products requires packages that are smaller and cheaper than in the past, roughly 50% smaller for each new generation. Given industry's past investments in PWB and SMT technologies, and given increased global cost-competitive pressures, the industry is expected to pursue the use of plastic packages to the ultimate limit. The ultimate limit accepted by Japan currently is 0.15 mm lead-frame pitch, giving rise to 800 pins in 30-mm sizes and 1000 pins in 38-mm sizes. The United States seems to follow the path of BGA.

It should be noted that while the QFP and BGA improvements as discussed above are taking place, the pin-grid-array technology is also capable of being enhanced for those limited applications that require insertable components for upgradeability, for example. Both ceramic and plastic PGAs are capable of being enhanced to 1000 I/Os at a pin pitch of 1.27 mm, or 50 mils. It should also be noted that these enhanced PGAs can be surface-mountable as well using standard or modified SMT equipment. Figure 7-59 illustrated the advantage of this approach—that the size of this new PGA is no bigger than the current QFP at 0.4 mm and that a 1000 I/O PGA is readily available.

The continued use of P-QFP beyond the current 0.4-mm pitch toward the 0.15-mm pitch, however, requires major enhancements in SMT pick and placement tools, solder deposition technologies, reflow tools and technologies, inspection, solder repair for opens and shorts, and electromigration resistance of both the plastic package and the printed wiring board. Contrary to what one might expect, the industry will incrementally enhance each of these to a level that will guarantee high yield and high reliability. This conclusion is supported by (1) Sony's advancements in factory automation and (2) Oki's single-ppm-defect soldering systems. These systems and processes have lowered assembly defects to less than 20 ppm, as shown in Figure 7-72. Sony's precision robots have improved placement repeatability to 0.01 mm from 0.05 mm during the last 6 years. Matsushita's new SMT machine has 11 placement heads with 0.01 mm repeatability. Toshiba's advanced TAB equipment can place 0.2-mm-pitch parts using CCD vision, because pitch size has reached the limits of human vision.

With increased miniaturization, soldering technologies continue to evolve. For example, Oki's single-ppm-defect technology includes developments in the following:

- New wave soldering machine for zero defects
- Nitrogen flow soldering process technology
- Rheology and printability of solder paste
- Inspection technique for solder paste printability and printing parameter optimization
- Development of an automatic solder-joint inspection system

Figure 7-73 illustrates the general trend in soldering techniques that the Japanese microelectronics industry is expected to follow, shifting from reflow to local soldering techniques in order to meet ultrafine pitch assembly requirements.

Table 7-28 summarizes the overall packaging that can be expected in terms of chip and package connections, board substrate wiring, chip

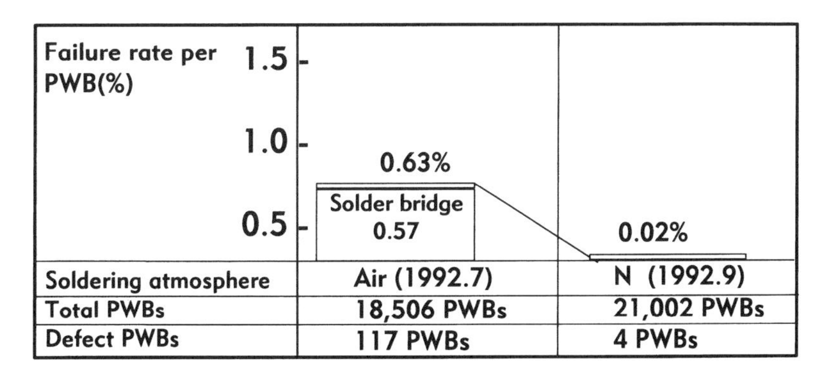

Figure 7-72. Soldering Defects Improvement Achieved at Oki (From Ref. 24.)

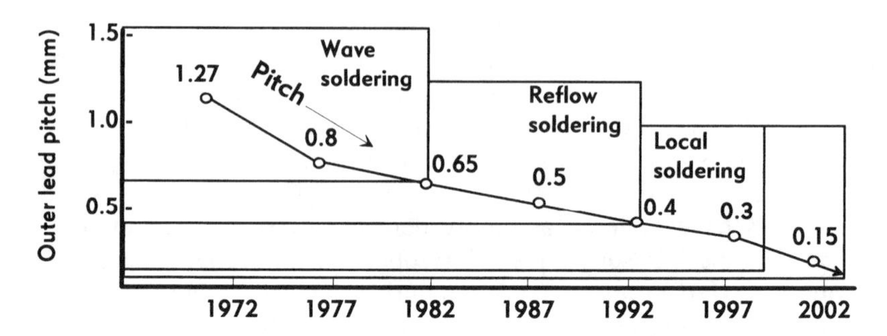

Figure 7-73. Soldering Technology Trend in Japan (Courtesy of Hitachi.)

placement accuracy and cost for leading-edge consumer products. Similar parameters were reviewed earlier in Tables 7-4A, 7-4B, and 7-4C for consumer, automotive, and high-performance systems, respectively.

7.7 SECOND-LEVEL PACKAGES

Four technologies capable of meeting the requirements of the second-level package are as follows:

Table 7-28. Package Assembly Trend for Consumer Electronics

	1995	1998	2000
Cost			
Component assembly (Conversion) cost (¢ per I/O)	2	1	0.4
Substrate cost (6-layer H.D.→ 4 layer) ($ per in.^{2})	1	0.5	0.25
IC package cost (¢ per I/O)	1.5	0.4	0.2
Package connections			
Component I/O density (I/O per in. ^{2})	100	400	600
Component I/O package pitch—peripheral (mm)	0.4	0.25	0.15
Component I/O Package pitch—area array (mm)	0.4	0.25	0.25
Chip connections			
Component I/O perimeter flip-chip pitch (mm)	0.15	0.05	0.05
Component I/O area chip pitch (mm)	0.30	0.20	0.15
Substrate wiring			
Substrate lines and spaces (mils)	5	2	1
Substrate pad diameter (mils)	20	2	1
Equipment			
Chip placement accuracy (mils)	2	0.5	0.25

Source: Courtesy of NEMI.

1. Organic cards and boards
2. Polymer- or glass-coated metal carrier
3. Flex film carrier
4. Injection-molded card
5. Ceramic boards

These are schematically illustrated in Figure 7-74 and their characteristics described in Table 7-29. Additional details can be found in Chapter 17, "Printed-Circuit Board Packaging," and Chapter 18, "Coated-Metal Packaging." Additionally, ceramic boards, described in Chapter 9, "Ceramic Packaging," can be considered second level as well.

Card and board technologies based on epoxy-glass, and enhancements in these by the use of such materials as polyimide, Telflon, benzocyclobutene organics, and quartz glass to provide better electrical characteristics and packaging densities, form the most common second-level package. The technology development is summarized in Figure 7-75 in terms of such package parameters as linewidths, intervia diameter, plated-through-hole diameter, aspect ratio, and number of layers. Printed-circuit boards with up to 42 layers capable of supporting as many as 336 chips on an individual board of 600 × 700 mm have been reported [33]. Very

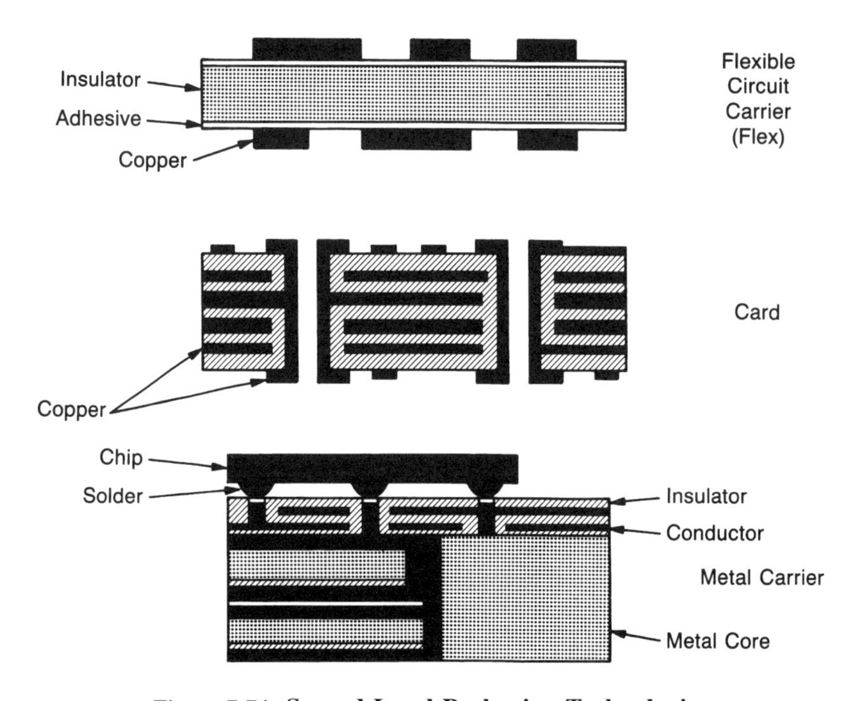

Figure 7-74. Second-Level Packaging Technologies

Table 7-29. Second-Level Package Characteristics

	Organic Card	Inorganic Board	Inorganic Board	Coated Metal	Flexible Carrier	Inj.-Molded Card
Insulator material	Epoxy-glass	Glass-ceramic	alumina	Glass or polyimide	Polyimide	Resin
Conductor materials	Cu	Cu	Mo or W	Ag-Pd or Cu	Cu	Cu
Number of layers	8	63	>33	4	2	2
Coefficient of thermal expansion (RT*–100°C) $\times 10^{-7}/°C$	150	30	66	30–70	200	300
Dielectric constant	4.0	5.0	9.5	3.5–6.0	3.5	4
Thermal conductivity (W/m K)	0.2	4	20	150	5	0.1

*Room temperature.

high-performance boards with materials of dielectric constant around 2.2 with five or more lines per channel are under development for potential applications in the next decade. Injection-molded cards (IMC), as the name implies, involve molding plastics into two- and three-dimensional shapes, and subsequently metallizing with seeding and plating technologies. Only surface wiring, with no internal planes (NIP), has been accomplished to date. A 50% cost savings, primarily associated with the reduced number of wet process steps, compared with the epoxy–glass card process, have been reported by a number of companies. Accordingly, injection-molding card technology may find widespread use in low-end systems.

The flexible carrier, which consists of two surface layers of thin-film copper wiring on each side of polyimide or other polymeric film on which protective epoxy could be masked and screened to form terminals for bonding the first-level package is a growing second-level package. This kind of second level, which is most commonly used in consumer electronics, can eliminate the need for a number of cables and connectors that are otherwise required to interconnect packages and can thus provide significant cost savings. Additionally, this type of package can be considered a three-dimensional package, as the flex film with mounted components could be bent in any dimension, as conceptually illustrated earlier in Figure 7-37 for display applications.

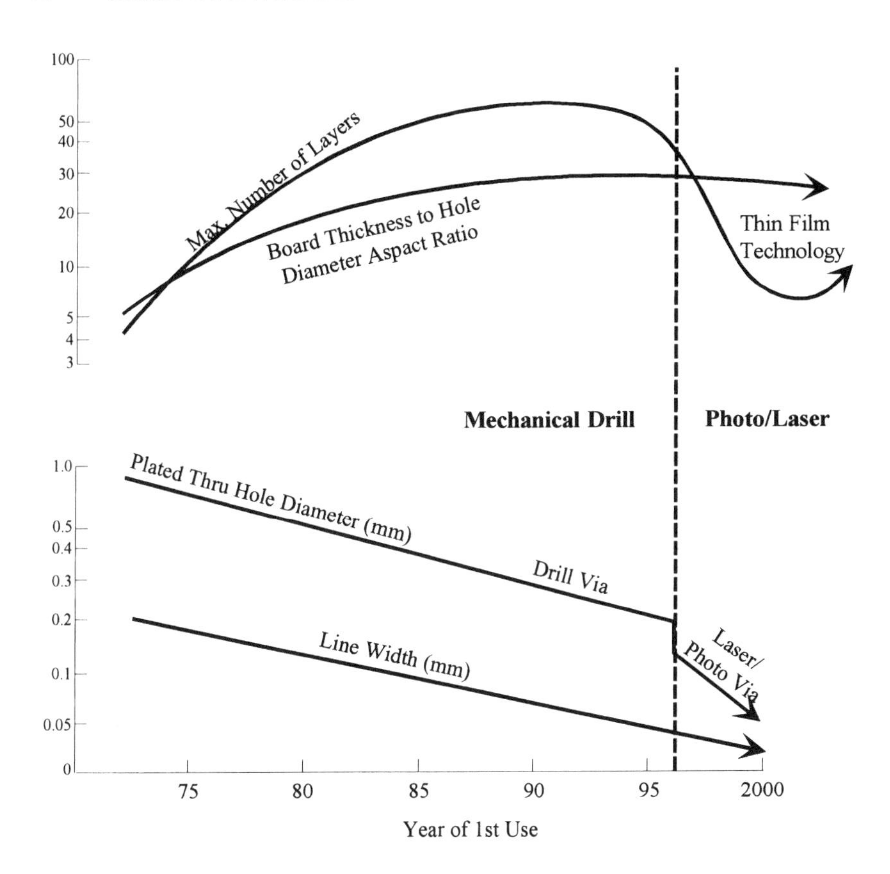

Figure 7-75. Evolution of Second-Level Packaging Parameters

The coated metal (sometimes referred to as metal carrier or metal substrate), with a coating of glass, glass-ceramic, or polymer on such metal carriers as low-carbon steel (as in porcelain enamels for dishwashers, sinks, and so forth, or aluminum, stainless steel, or copper–invar–copper) can form excellent second-level packages with a built-in ground plane, mechanical stiffness, and thermal-expansion tailorability. This technology, discussed in Chapter 18, "Coated-Metal Packaging," can be applied for removal of heat without heat sinks on the back of the chip by the direct attachment of the chip backside to the metal carrier. Alternatively, solder bonding of chips (C4) can be accomplished with copper–invar–copper substrates, the thermal expansion of which can be made to approximate that of silicon.

A revolutionary approach to organic board enhancements has been advanced by IBM Japan [26] that preserve low-cost attributes, such as large-area processing, low-cost wet processing, but enhance the wiring

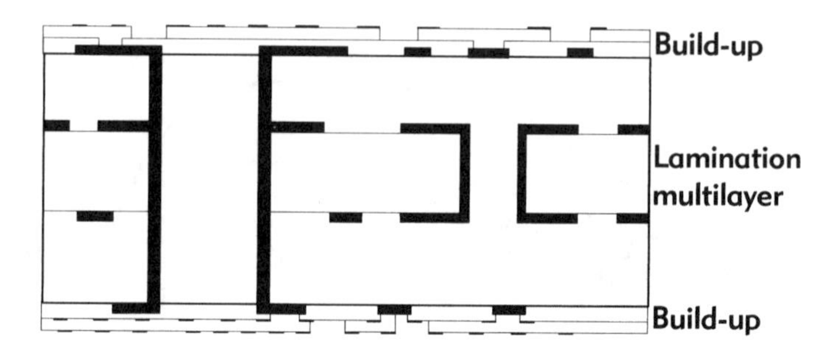

Figure 7-76. Additive Process Enhancement (Courtesy of IBM Japan, Ibiden.)

density by photolithographic via and line processes in contrast to drilled via and laminate technologies. This is referred to as thin film on printed wiring laminates and also as surface laminar circuitry (SLC). A typical structure is illustrated in Figure 7-76. These enhancements fall into several categories:

- Thin-line and fine-line conventional subtractive etching process.
- Low-cost, fine-line, thin-film additive process. Shown in Figure 7-77 is a new sequential process involving deposition of a photosensitive polymer or epoxy, formation of via holes by large-area photo exposure, and subsequent chemical etching and metalization by catalytic chemical seeding and electroless plating. The advantages of this process are many, including pattern shape, pattern thickness control, pattern width control, mounting reliability, and, most importantly, small via size. Ibiden compares this additive process with subtractive processes in Figure 7-78 [24].
- Minimization of solder bridge by using dry film.
- New materials. Examples include aramid-based laminates with low thermal expansion, good electromigration resistance, high glass-transition temperature, and excellent processability. Another material is ceracom, a combination of porous ceramic laminated with glass and epoxy resin to form very low-TCE boards suitable for direct chip bonding. Chapter 9, "Ceramic Packaging."
- Direct bonding of chip. This may be by COB (wire bonding), tape on board (TAB), and flip chip on board to the printed wiring board with appropriate low-stress encapsulants.

Japan has invested in a variety of PWB materials that include FR-4, polyimide–glass, maleimide styryl, BT resin, and a new aramid-based laminate consisting of aramid-based paper as a reinforcement in a matrix of a new epoxy resin by Teijin Limited. The superior properties of this

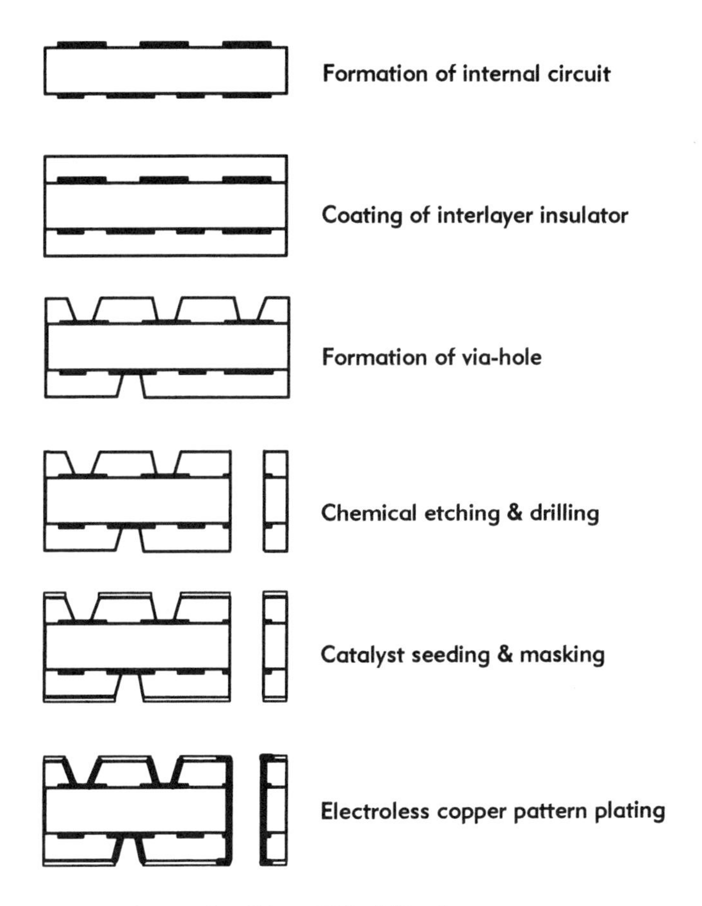

Figure 7-77. Low-Cost, Fine-Line, Thin-Film Process (Courtesy of IBM Japan, Ibiden.)

aramid-based board for potential MCM applications include low TCE (6–16 ppm/°C), very high electrical resistance, very low impurities in the aramid fiber, and processability with fine via holes.

The shape and accuracy of conductor patterns, as well as the mounting reliability of solder bridging, as illustrated in Figure 7-78, compares with the standard substractive process. Table 7-30 indicates the dielectric and metal ground rules, as well as the drilled and photolith dimensions resulting in fine-line, thin-film structures on both sides of a PWB. The structure of the additive process as practiced by IBM (Japan) and Ibiden is illustrated in Figure 7-76, using two layers on each side of the PWB.

7.8 PACKAGING COOLING

Removal of heat from today's highly integrated ICs remains a major bottleneck. The dissipative needs of microprocessors are in the range of

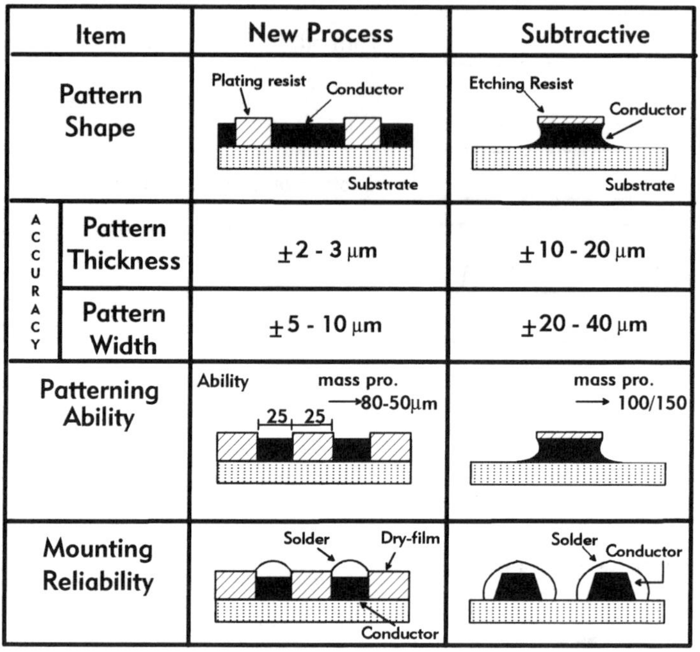

Item		New Process	Subtractive
Pattern Shape		Plating resist, Conductor, Substrate	Etching Resist, Conductor, Substrate
ACCURACY	Pattern Thickness	$\pm 2-3\ \mu m$	$\pm 10-20\ \mu m$
	Pattern Width	$\pm 5-10\ \mu m$	$\pm 20-40\ \mu m$
Patterning Ability		Ability; mass pro. → 80-50μm; 25, 25	mass pro. → 100/150
Mounting Reliability		Solder, Dry-film, Conductor	Solder, Conductor

Figure 7-78. Shape and Accuracy of Conductor Pattern by Additive Process (Courtesy of Ibiden.)

Table 7-30. Characteristics of Additive-Plated PWB

Number of layers	6	
Thickness	0.8 mm	
Insular thickness	50 μm	
Conductor thickness	15 μm	
Minimum wiring width/space/pitch	50 μm/50 μm/10 μm	
Minimum via φ	100 μm	
	Drill φ	Land φ
1. Inner via (drill) (mm)	φ 0.2	φ 0.4
2. Blind TH (photo) (mm)	φ 0.3	φ 0.5
3. Through hole (drill) (mm)	φ 0.5	φ 0.7

Source: From Ref. 14.

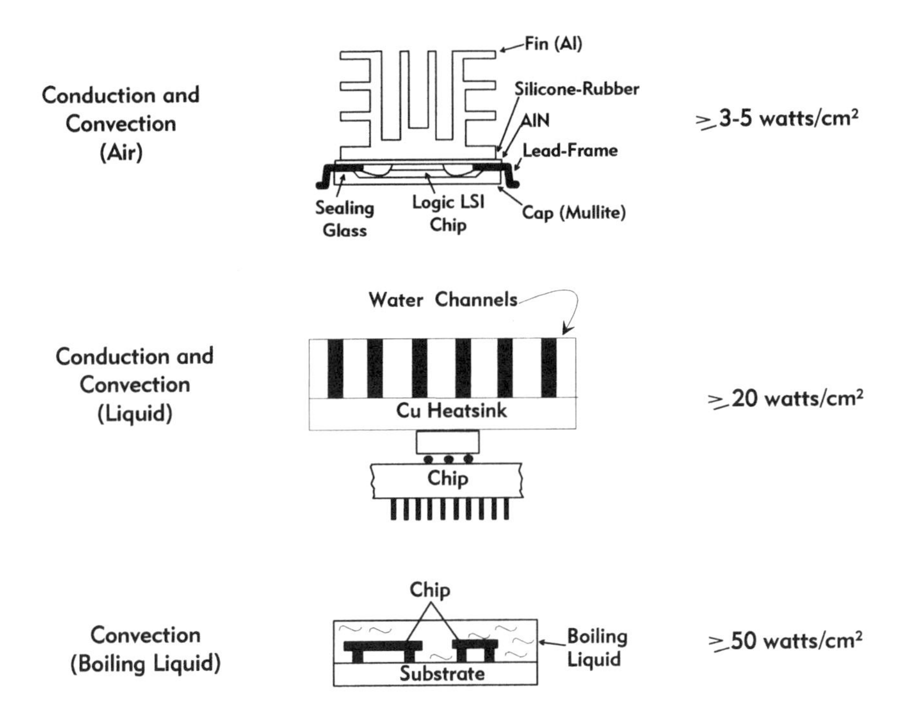

Figure 7-79. Cooling Technology Options

5–20 W/chip, whereas those in leading-edge ECL and CMOS technologies are about 20–100 W/chip.

Even though the power per circuit has been steadily decreasing with time because of improvements in the device and circuit technologies, the growing total number of circuits per chip and the growing total number of circuits per system are continually increasing the power and power density requirements of logic functions. Figure 7-13 illustrated the package power cooling requirements in watts per chip area.

The low-end systems have used and will continue to use forced-air cooling of single-chip modules, surface mounted onto cards or boards. The future use of chips with very high integration levels is still expected to use this kind of cooling technology enhanced by heat sinks mounted on the backside of chips or, alternatively, chips back-bonded to either thermal cards based on coated-metal substrates or to high-conductivity ceramics (AlN) or metallic sinks (Cu-W) joined to the ceramic packages by solder glasses or braze alloys, as discussed in Chapter 9, "Ceramic Packaging."

Cooling technologies used to provide these capabilities generally fall into three categories, as shown in Figure 7-79, and are further discussed in Chapter 4, "Heat Transfer in Electronic Packages."

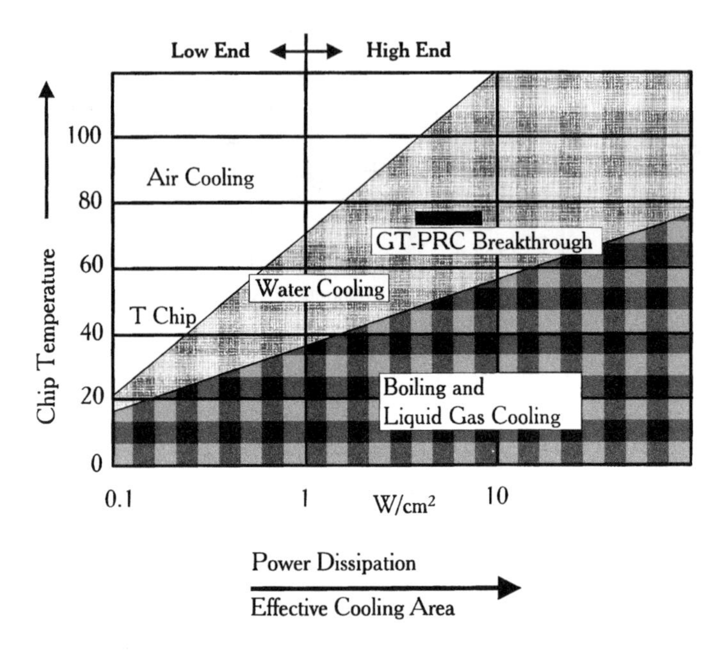

Figure 7-80. Cooling Regimes by Air, Liquid, and Boiling Liquid

1. Those that take advantage of the higher thermal conductivity of the substrate, such as aluminum nitride ceramics, with thermal conductivity approximately 10 times better than alumina
2. Those that do not require the high thermal conductivity of the first-level package, and thus depend on external cooling by means of heat sinks bonded on the backside of the chip or package
3. Those that involve direct immersion of the chip or package in inert liquids such as fluorocarbons

Figure 7-80 illustrates the cooling capabilities provided by air, water, and boiling liquids as a function of both chip temperature and power dissipation density. This figure clearly indicates the cooling challenges being faced by PCs and workstations that require high heat flux removal, which must be accomplished by air cooling.

In most of the systems shipped in the eighties and early nineties, a combination of conduction, convection, and sometimes radiation has been used with liquid or air. In the future, air cooling is the predominant choice, and only if it cannot be achieved, liquid cooling will be pursued. A major breakthrough in air cooling as indicated in Figure 7-80 appears to be the microjet technology being pursued by the Packaging Research Center at Georgia Tech [34] that has the potential to enhance the air-cooling regime to in excess of 10 W/cm^2 from about 3 W/cm^2 currently. It involves

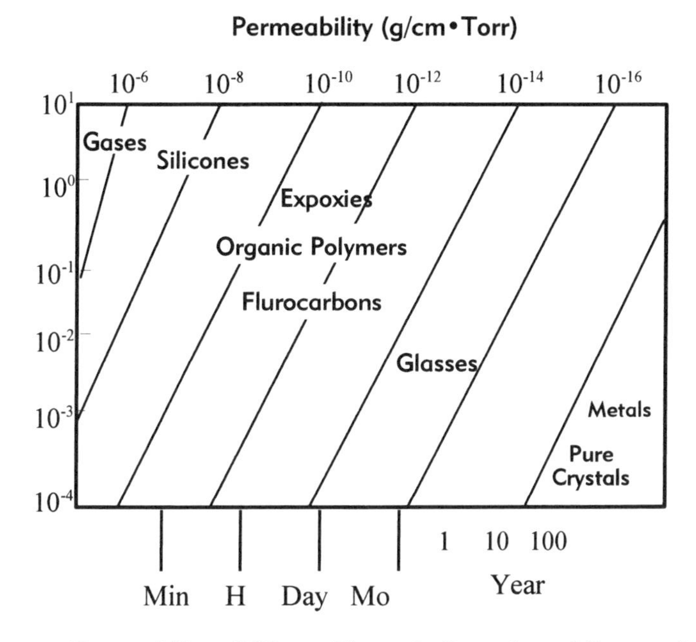

Figure 7-81. Permeability of Water Through Organic and Inorganic Materials

microjet air cooling. It focuses on the development of zero net mass flux synthetic jet technology for integrated cooling of single-chip and multichip modules in open and closed flow systems. An experimental technology capability of about 800 W/cm^2 has been demonstrated [35] by direct liquid cooling of silicon with fine groves directly etched in silicon as discussed in Chapter 4, "Heat Transfer in Electronic Packages."

7.9 PACKAGE SEALING AND ENCAPSULATION

Package sealing is performed to protect the device and package metallization from corroding environments and from mechanical damage due to handling. Moisture is one of the major sources of corrosion of the semiconductor devices. Electro-oxidation and metal migration are associated with the presence of moisture. The diffusion rate of moisture also depends on the encapsulant material, as shown in Figure 7-81, which compares the permeability of water for various organic and inorganic materials. In spite of the poor permeability of organic materials in the figure, a number of encapsulants, particularly silicones, certain epoxies, polyimides, and imide-siloxanes, have been shown to have great promise [36]. The characteristics of these materials are listed in Table 7-31. In plastic packages for consumer electronics and low-end systems, as discussed in greater detail in Chapter 8, "Plastic Packaging," a variety of plastic encapsulants have been developed to minimize the diffusion of

Table 7-31. Characteristics of Potential Encapsulants

Materials	Application	Advantages	Disadvantages
(1) Epoxies	• Normal dispensing or molding	• Good solvent resistance • Excellent mechanical strength	• Nonrepairable • High stress • Marginal electrical performance
(2) Polyimides	• Normal dispensing (spin-coat)	• Good solvent resistance (~500°C)	• High-temperature cure • High stress
(3) Polyxylylene (Parylene® Trademark from Union Carbide)	• Thermal deposition (reactor)	• Good solvent resistance • Conformal coating	• Thin film only
(4) Silicone-polyimides	• Normal dispensing	• Less stress vs. polyimide • Better solvent resistance vs. polyimide resistance vs. silicone	• Higher thermal expansion • Thin-film only
(5) Silicons (RTV, Gel)	• Normal dispensing	• Good temperature cycling • Good electrical properties • Very low modulus	• Weak solvent resistance • Low mechanical strength

Source: After Ref. 38.

water. This results in significant improvements in reliability of plastic-encapsulated packages. The diffusion coefficients of these materials range from 1×10^{-8} cm^2/s for epoxy resins, 12×10^{-8} cm^2/s for typical silicone 305, to 0.1×10^{-8} cm^2/s for NYSOL transfer molding material. The developments with silicone rubber have recently led to further improvements with silicone gel [37], which is soft and can be deposited to thinner dimensions with minimal stress problems.

High-performance systems, however, have depended on hermetic packages. Hermeticity of the package requires a He leak rate of better than 1×10^{-8} cm^2/s. The usual methods used in hermetic sealing are as follows:

- Welding
- Brazing
- Soldering
- Glass sealing

Three basic welding techniques are used: resistance welding, seam welding, and cold welding. Brazing is performed by melting metal or an alloy to join package metals of higher melting points. The common brazes are gold-tin eutectic with a melting point of 280°C, In-Cu-Sil with a melting point of 680°C, and Cu-Sil with a melting point of 750°C. Solder sealing is accomplished in the same way, except by the use of solders, the most common being Pb-Sn eutectic at 163°C. Both soldering and brazing require that the packages and lids (covers) to be sealed be metallized by thick-film, thin-film, or plating technologies. Glass sealing, which requires an excellent thermal expansion match between the composite package to be sealed and the glass used for sealing, requires glass of a slightly lower thermal expansion coefficient so as to put the seal in slight compression. A number of solder glasses have been developed in the PbO-ZnO-Al_2O_3-B_2O_3-SiO_2 systems to meet a variety of sealing requirements [29].

7.10 BOOK ORGANIZATION AND SCOPE

For the readers convenience, an extensive glossary of terms and symbols are included at the end of each part and are available at our publishers—Chapman and Hall—internet address: ftp://ftp.thomson.com/chapman&hall/tummala/glossary. As mentioned in early sections of this chapter, this second edition is partitioned into three parts. Figure 7-82 depicts how the 20 chapters are organized. The introductory chapters of each part—1, 7, 15—are identical. This was done intentionally for those that do not have access to the complete set.

Part 1, "Technology Drivers", deals with several driving forces, design considerations and technology options available. Chapter 2, "Packaging Wiring and Terminals," appropriately begins this part by addressing the topological (layout, placements, and wiring) considerations. The electromagnetic design considerations are followed in Chapter 3, "Package Electrical Design." The thermal design considerations, driven to a great degree by the electrical design, are presented in Chapter 4, "Heat Transfer in Electronic Packages." The reliability considerations are in Chapter 5, "Package Reliability." A new Chapter 6, "Package Manufacturing," has been added in this edition for the competitively pivotal manufacturability considerations.

Previously, it was noted that all interconnections on an IC chip or all internal structures of any other component (e.g., a crystal or those of discrete capacitors and inductors) are not considered packaging in the common parlance. Thus, the packaging begins at the interfaces to interconnect, power, cool, and protect all ICs.

Parts 2 and 3 deal with packaging technologies; Part 2 relating to so-called semiconductor packaging and Part 3 relating to subsystem or board-level packaging. Semiconductor packaging relates all technologies involved in the packaging of semiconductors ready to be assembled onto

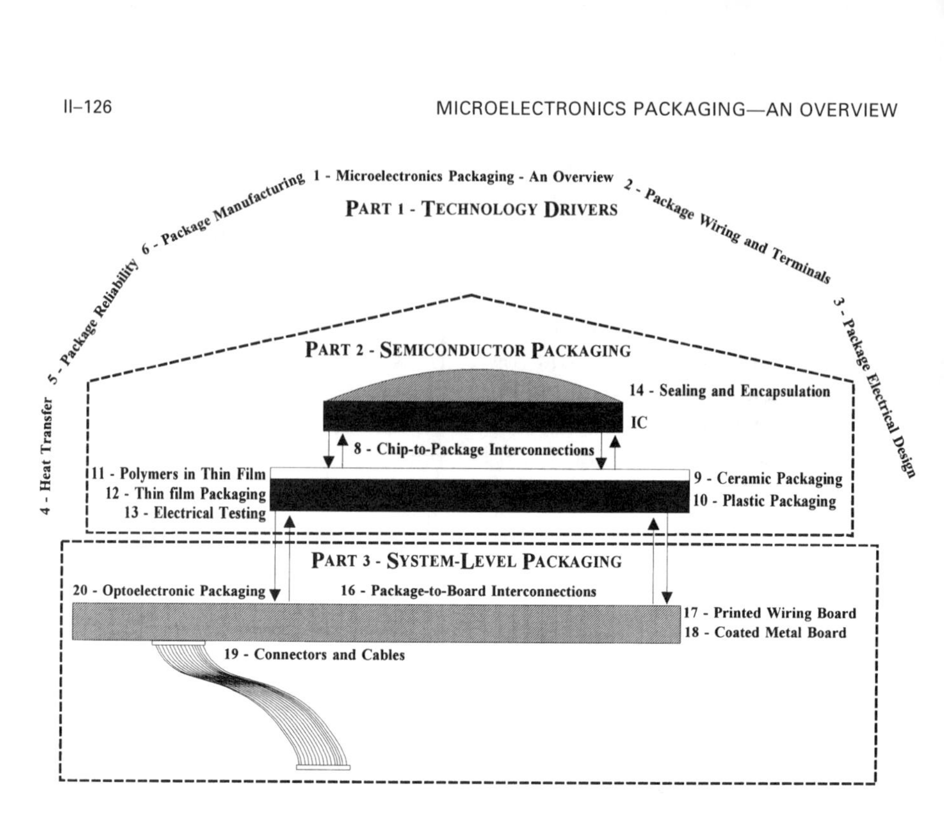

Figure 7-82. Book Organization. Numbers refer to chapter numbers in this three part book. Chapter One is repeated in each part as Chapters 7 and 15.

board. Accordingly, it includes all chip-to-package connections such as TAB, flip chip, and wire bond in Chapter 8 and all first-level packages such as ceramic and plastic as in Chapters 9 and 10. Because some of these packages, particularly the most advanced ones, contain thin-films of polymers and metals, they are covered in Chapters 11 and 12. All those packages can be single-chip or multichip and these chapters therefore review both of these. Some of these are very complex and involve I/Os in excess of 1000 per IC and wiring density in excess of 500 cm/cm^2. To assure "known-good substrate" before assembled onto board, electrical testing is performed. This technology is covered in Chapter 13. The semiconductor packages is ready to be shipped for assembly provided it is sealed or encapsulated to guarantee reliability with or without hermeticity. This final technology operation is covered in Chapter 14.

Part 3 deals with all technologies required to assemble into subsystems or total systems depending on the need for systems from consumer electronics to mainframes. The first-level packages described in Part 2 are typically assembled as pin-through-hole (PTH) in the board, surface-mount (SMT) onto the board and more recently ball-grid array or chip-scale package connection to the board. These package-to-board connections are reviewed in Chapter 16. The board technology itself is undergoing major changes to accommodate the high I/O IC demand. Some of these changes include alternatives to drilling, alternatives to epoxy–glass dielectric and photolithography. All these printed wiring board technologies are covered in Chapter 17. There are alternative technologies to printed wiring board that include metal or ceramic boards. The ceramic boards are covered in the chapter "Ceramic Packaging" in Part 2 and, therefore, it is not repeated here. The metal board, however, referred to as coated-metal that involves polymer, glass and ceramic coatings is covered in Chapter 18. For very high performance applications such as mainframe and supercomputers, several boards are required to meet their semiconductor needs. The interconnections between these are referred as connector and cable packaging. Optoelectronic packaging is a fast-moving technology because of its advantages both for long-distance communications replacing copper and for short interconnections because of its bandwidth and data rate. This technology is covered in Chapter 20.

7.11 REFERENCES

1. S. Sherman. "The New Computer Revolution," *Fortune,* 14 June 1993, pp. 56–80.
2. J. L. Hennesy and D. A. Patterson. *Computer Architecture: A Quantitative Approach,* 2nd ed. Morgan Kaufmann Publishers, San Matao, CA, 1966, pp. 1–59.
3. H. J. Greub, J. F. McDonald, and T. Creedon. "FRISC-E: a 250-MIPS Hybrid Microprocessor," *Circuits & Devices,* 6(3): pp. 16–25, 1990.
4. R. Philhower. "Spartan RISC Architecture for Yield-Limited Technology," Ph.D. Dissertation, Rensselaer Polytechnic Institute, 1993.
5. R. F. Sechler and G. F. Grohoski. "Design at the System level with VLSI and CMOS," *IBM J. Res. Devel.,* 39(1/2): pp. 5–22, 1995.
6. P. D. Franzon, A. Stanaski, Y. Tekmen, and S. Banerjia. "System Design Optimization for MCM-D/Flip-Chip," *IEEE Trans. Components, Packaging, and Manuf. Technol., Part B: Advanced Packaging,* CPMT-18(4), 1995.
7. Mixed Signal Design, Short Course, *GaAs IC Symposium co-sponsored by IEEE Electron Devices Society and IEEE Microwave Theory and Techniques Society,* 16 October 1994.
8. H. B. Bakoglu, *Circuits, Interconnections and Packaging for VLSI,* Addison-Wesley, Reading, MA, 1990, Fig. 9.12.
9. See B. J. Landman and R. L. Russo. "Pin vs. Block Relationship for Partitions of Logic Graphs," *IEEE Trans. Computers,* C-20(12): pp. 1469–1479, 1971.
10. R. K. Poon. *Computer Circuits Electrical Design,* Prentice-Hall, Englewood Cliffs, NJ., 1995.
11. J. N. Humenik, J. M. Oberschmidt, L. L. Wu, and S. G. Paul. "Low-Inductance Decoupling Capacitor for the Thermal Conduction Modules of the IBM Enterprise System/9000 Processors," *IBM J. Res. Devel.,* 36(5): pp. 935–942, 1992.
12. D. P. Seraphim, R. Lasky, and C.-Y. Li (eds.). *Principles of Electronic Packaging,* McGraw-Hill, New York, 1989.
13. E. E. Davidson, P. W. Hardin, G. A. Katopis, M. G. Nealon, and L. L. Wu. "Physical and Electrical Features of the IBM Enterprise System/9000 Circuit Module," *IBM J. Res. Devel.,* 36(5): pp. 877–877, 1992.
14. W.-T. Liu, S. Cochrane, S. T. Lakshmikumar, D. B. Knorr, E. J. Rymaszewski, J. M. Borrego, and T.-M. Lu. "Low Temperature Fabrication of Amorphous $BaTiO^3$ Thin Film Bypass Capacitors," *IEEE Electron. Device Lett.* 14(7): p. 320, 1993.

15. P. K. Singh, R. S. Cochrane, J. M. Borrego, E. J. Rymaszewski, T.-M. Lu, and K. Chen. "High Frequency Measurement of Dielectric Thin Films," *Proceedings of Microwave Theory and Techniques Conference,* 1994.
16. R. Crowley, E. J. Vardaman, and I. Yee. "Worldwide Multichip Module Market Analysis," A Multi-Client Study, TechSearch International, Austin, TX, July 1993.
17. D. A. Doane and P. D. Franzon (eds.). *Multichip Module Technologies and Alternatives: The Basics,* Van Nostrand Reinhold, New York, 1993.
18. E. L. Boyd, Guest Lecture Notes for Electronic Packaging course, Rensselaer Polytechnic Institute, 1993.
19. Hisao Kanai, "Low Energy LSI and Packaging for System Performance," IEEE Trans. on CHMT, CHMT-4, No. 2, pp. 173–180, June 1981.
20. T. Watari and H. Murano. "Packaging Technology for the NEC SX Supercomputer," *Proceedings of Electronic Components Conference,* 1985, pp. 192–198.
21. *The National Technology Roadmap for Semiconductors,* Semiconductor Industry Association, San Jose, CA, 1994.
22. R. Hillemann. "Bipolar-based Mainframes vs. CMOS-based Mainframes," *Computer Packaging Workshop,* 1992.
23. R. R. Tummala. "Next Generation of Packaging beyond BGA, MCM and Flipchip," *Proceedings of IMC,* 1996.
24. R. R. Tummala and M. Pechi. "Japan's Electronic Packaging Technologies," JTEC Panel Report, pp. 59–96, 1995.
25. F. Kobayashi, Y. Watanabe, M. Yamamotu, A. Anzai, A. Takahashi, T. Daikoku, and T. Fujita. "Hardware Technology for Hitachi M-880 Processor Group," *Proceedings of 42nd IEEE–ECTC,* pp. 693–703, 1992.
26. Y. Tsukada, "Surface Laminar Circuit Packaging," *Proceedings of 42nd IEEE–CHMT,* pp. 22–27, 1992.
27. J. Vardaman, Tech Search International, Austin, Texas, private communication.
28. R. R. Tummala. *Proceedings of Emerging Microelectric and Information Technologies,* pp. 1–8, 1996.
29. R. R. Tummala. "Multichip Packaging—A Tutorial," *Proceedings of IEEE,* Vol. 80, No. 12, pp. 1–17, 1992.
30. D. L. Crook. "Evolution of VLSI Reliability Engineering Meeting," *Proceedings 2nd European Symposium on Reliability of Electronic Devices,* pp. 293–312, 1991.
31. T. Watari. "Computer Packaging Technology for System Performance," *NEC Res. Devel.,* No. 98, pp. 45–59, 1990.
32. R. R. Tummala and B. T. Clark. "Multichip Packaging Technologies in IBM," *Proceedings of 42nd IEEE–CHMT,* pp. 1–9, 1992.
33. M. Nishihaka, K. Kurosawa, and T. Sakamura. "Facom M-780," *Fujitsu J.,* 32(2): pp. 2–8, 1986.
34. A. Glezer and M. Thompson. "High Heat Flux Air Cooling Using Synthetic Microjets," *Semicon West,* 1995.
35. Y. Tsukada. "Surface Laminar Circuit Packaging," *Proceedings of 42nd IEEE–CHMT,* pp. 22–27, 1992.
36. J. Vardaman, Tech Search International, Austin, Texas, private communication.
37. R. R. Tummala, *Proceedings of Emerging Microelectronic and Information Technologies,* pp. 1–8, 1996.
38. C. P. Wong. "High Performance Silicone Gel as IC Device Chip Protection," *Mat. Res. Symp.* (108): pp. 175–187, 1988.

8

CHIP-TO-PACKAGE INTERCONNECTIONS

PAUL A. TOTTA—*IBM*
SUBASH KHADPE—*Semiconductor Technology Center*
NICHOLAS G. KOOPMAN—*Microelectronic Center of NC*
TIMOTHY C. REILEY—*IBM*
MICHAEL J. SHEAFFER—*Kulicke & Soffa*

8.1 INTRODUCTION

Integration of circuits to semiconductor devices, driving the need for improvements in packaging, has been discussed in Chapter 7, "Microelectronics Packaging—An Overview." This is further illustrated in Figure 8-1, wherein the cost of interconnecting on silicon is compared with interconnecting on ceramic substrates and on organic boards, clearly showing the lower cost of interconnecting on silicon [1]. Although the trend is toward total integration on Si there is, however, a practical, growing limit to the number of circuits which can be made on a single piece of silicon, which is currently at about 1.6 million circuits for CMOS logic, 40,000 circuits for bipolar logic, and 64 megabits for memory. The highest integrated transistor counts are approximately 5 million on advanced microprocessors. Therefore, because most current information systems

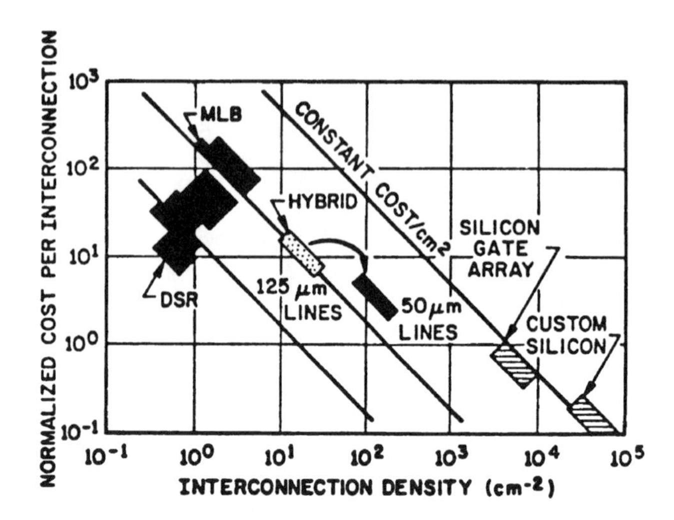

Figure 8-1. Cost of Interconnections. The relative cost of interconnection for DSR (double-sided rigid printed wiring board), MLB (multilayer printed wiring board), ceramic hybrids, gate array, and custom silicon. The curves nearly show the advantage of silicon integration in decreasing interconnection costs. (From Ref. 1.)

require a greater number of circuits and interconnections, a number of chips still need to be interconnected on organic or ceramic first-level packages. The electrical connections between the chip and the package, referred to here as chip-level interconnections, are the subject matter of this chapter. Because, for the systems considered here, no first-level package can usually accommodate all the required chips, a second-level package interconnecting the first levels is often required. These interconnections, referred to as package-to-board interconnections, are reviewed in Chapter 16. A recent deviation from this packaging pattern, also to be discussed, is the direct surface mounting of flip chips on FR4 cards or flexible circuits which is referred to as direct chip attach (DCA) or chip-on-board (COB).

An abstract of the SIA national roadmap for semiconductors and packages for various applications is presented in Table 8-1 [2]. Very clearly the trend for input–output (I/O) on chips is upward with advanced application-specific integrated circuits (ASICs) requiring in excess of 1000 I/O by the turn of the century. For other applications, the growth of I/O is not as rapid, with automotive and commodity applications requiring only modest increases.

There are three principal chip-level interconnection technologies currently in use:

Table 8-1. SIA Roadmap for chip interconnections. Extraction of estimated I/O Requirements on chips in different applications for each step in feature size reduction.

Years	1995 1998	1988 2000	2001 2003	2004 2006	2007 2009	2010 2012
Smallest Feature	0.35 μm	0.25 μm	0.18 μm	0.13 μm	0.10 μm	0.07 μm
	Chip Pad Count: (chip-to-package)					
Commodity	208	256	324	420	550	700
Hand held	300	450	675	880	1140	1500
Cost/performance	540	810	1200	1600	2000	2600
High-performance	900	1350	2000	2600	3600	4800
Automotive	132	200	300	400	500	700

Pad/pin-count ratios: mem, comm, port, and auto = 1 : 1; performance classes = 1 : 1.2
Source: From Ref. 2.

- Wirebonding (WB)
- Tape automated bonding (TAB)
- Flip-chip solder connection (C4)

Less common, but also important, are some chip-to-package interconnections such as conductive adhesives, elastomeric connections, and combinations of these. The latter connections dominate the flat panel, liquid crystal display applications.

This chapter reviews primarily WB, TAB, and C4, indicating the advances made over the last two to three decades in materials, design factors, fabrication processes, tools, assembly, rework, and reliability. Future trends in this very important area are also discussed.

The most basic function of the chip-level interconnections is to provide electrical paths to and from the substrate for power and signal distribution. Electrical parameters such as resistance, inductance, and capacitance need to be quantified for each interconnection design, as each of these will affect the total system's performance. In addition to the electrical functions, the C4 and TAB designs provide mechanical support for the chip, which is augmented by the encapsulants used to seal the chip metallization or to modify strain in the interconnections. Because almost any electrical conductor is also a good thermal conductor, some interconnections together with package materials are designed for removal of heat from the chip. Thus, each interconnection provides electrical, mechanical, and thermal functions which are discussed in the chapter.

8.2 CHIP-LEVEL INTERCONNECTION EVOLUTION

The silicon integrated-circuit chip was invented in the late 1950s and was in manufacturing by the mid-1960s. Prior to its existence, chips or die of silicon or germanium were single, discrete transistors or diodes for a number of years. These devices were initially packaged in hermetically sealed metal cans and interconnected with passive components such as resistors and capacitors on printed-circuit boards. The first transistorized generation of computers and entertainment systems were built this way.

During the early, discrete transistor era (1955–1960) there was only one kind of chip-level interconnection, the thermocompression wirebond. This technology came from Bell Laboratories simultaneously with the germanium and silicon semiconductor technology [3]. The implementation of thermocompression (TC) wirebonding at that time led to many problems with reliability, manufacturability, and cost. Simple TC bonding of gold wires on aluminum chip metallization was a hot process, typically in the 350–400°C region, which resulted in excessive intermetallic formation between Au and Al (observed as "purple plague") and caused bonding weakness and field-failure problems. The manual nature of the bonding also resulted in great bond variability and operator dependence. The cost was therefore very high and the supply of transistors limited. For all these reasons there was a quest in industry for a better way.

At Bell Laboratories and at IBM, the chip was taken out of its hermetic enclosure and made into a smaller, less vulnerable component. The approach at Bell Laboratories was to use a combination of silicon nitride on the chip to protect junctions, and surface gold wiring with "beam leads" to keep the interconnections free of corrosion [4]. At IBM, a thin layer of glass passivation sealed the chip surface and its aluminum-based wiring [5,6]. Solder–bump interconnections joined the chip to its package [7,8]. Both the Bell and the IBM chips were usually mounted face down in the "flip-chip" configuration on ceramic substrates that had thin-film or thick-film wiring and passive components. The chips were then encapsulated in silicone gel, which prevented the formation of continuous water films and electrolytic metal migration between exposed electrodes. Thus began the nonhermetic, hybrid packaging era, which was to accompany second-generation transistors in computer and telecommunications applications.

The first integrated circuits became available in manufacturing quantities in the mid-1960s. As with their hybrid module predecessors, there were few circuits on each chip of silicon. Aluminum-based or gold-based thin-film wires were used to "integrate" the active and passive devices embedded in the silicon. The exponential growth of the circuit count per die during the following decades has been phenomenal and was faster than expected. At the start there were only one to five bipolar logic circuits on a chip, and the first bipolar memory chips had a modest 16 bits in a scratch pad memory application. In the early 1970s, the bipolar logic count grew to about 100 circuits and monolithic memory to 128 bits for the first commercial, bipolar main memory replacing ferrite cores, as, for example, in the IBM 370 system. Today, the number of logic circuits has grown to about 40,000 per chip (bipolar) or 1.6 million circuits (CMOS) and up to 64-megabit memory arrays with single FET (field-effect transistor) cells totally replacing multi-transistor main memory bits. Microprocessors, essentially computers-on-a-chip, are commonplace in personal computers, workstations, and highly parallel supercomputers. Very-large-scale integration (VLSI) has turned into ultralarge-scale integration (ULSI).

The integration and densification process in integrated circuits has caused the continuous migration of intercircuit wiring and connections from boards, cards, and modules to the chip itself. The surface of the chip, with its multilayer wiring, has become a microcosm of the conductor and insulator configurations that were common on previous multilayer printed-circuit boards and on multilayer ceramic packages (Figs. 8-2 and 8-3). As many as four to six levels of wiring have been created on the chip. Advances in interlevel vias (etched, tapered holes) to vertical wires or studs and significant planarization of wiring layers has led to the ability to have 1.6 million circuits on a 15–20-mm chip with five levels of wiring. Even with this progress, the wiring capability on the chip continues to lag the potential density of silicon devices. Therefore, wiring pitches are rapidly dropping into the low-submicron dimensions. It is anticipated that

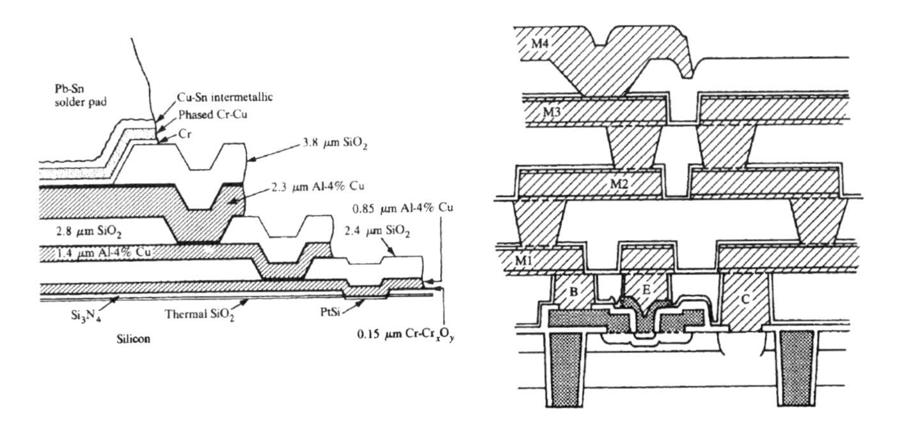

Figure 8-2. Chip Multilayer Wiring. Contrast of conventional etched via, three-layer wiring with excessive topography (left) with studded, planarized wiring for four-level structures (right). (From Refs. 9 and 10.)

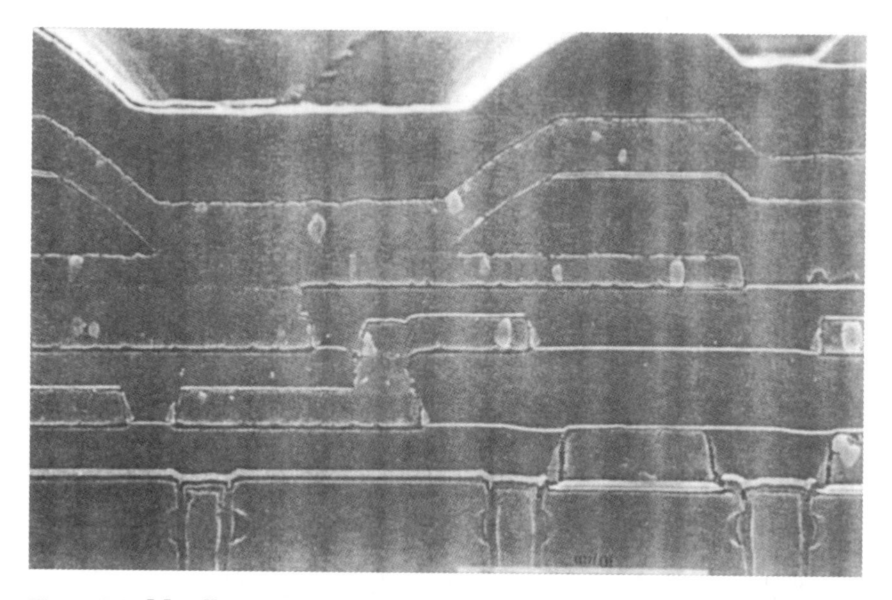

Figure 8-3. Metallographic Cross Section of Four-Level Wiring on IBM Bipolar Hip, Illustrating Damascene Studs and Chem-mech Polishing Planarization. (From Ref. 10.)

0.25 μm wiring will be commonplace by the end of the decade and chip sizes up to 30 mm are envisioned for the logic circuits or microprocessors of the future.

The technology of chip surface wiring is truly a part of the packaging technology, but it is too complex and different from typical board and substrate technology to be treated adequately in this chapter. It is a fact, however, that all the physics and engineering of high-speed transmission line theory applies in the chip-level wiring as well. The role of multilevel wiring in reducing on-chip delay is shown in Figure 8-4. Compared are the distributions and magnitudes of on-chip delays attributable to wiring versus Si for three generations of IBM logic chips for mainframe computers. The importance of effective wiring is illustrated. In the oldest, first family shown, only 10% of the delay was attributed to the three-level chip wiring. This grew to 27% of the delay in the third family with four-level wiring and would have been 45% if the progress from three- to four-level wiring had not been made. The gains in circuit performance for five and six levels are not illustrated but might be anticipated by projection.

The progress in integrated circuits has led not only to enormous densification of circuits on a chip but also to the total integration of a computer on a chip. Modern microprocessor chips in hand-held calculators have the computing power of second-generation large-scale computers of the mid-sixties. Powerful personal computers or workstations in parallel

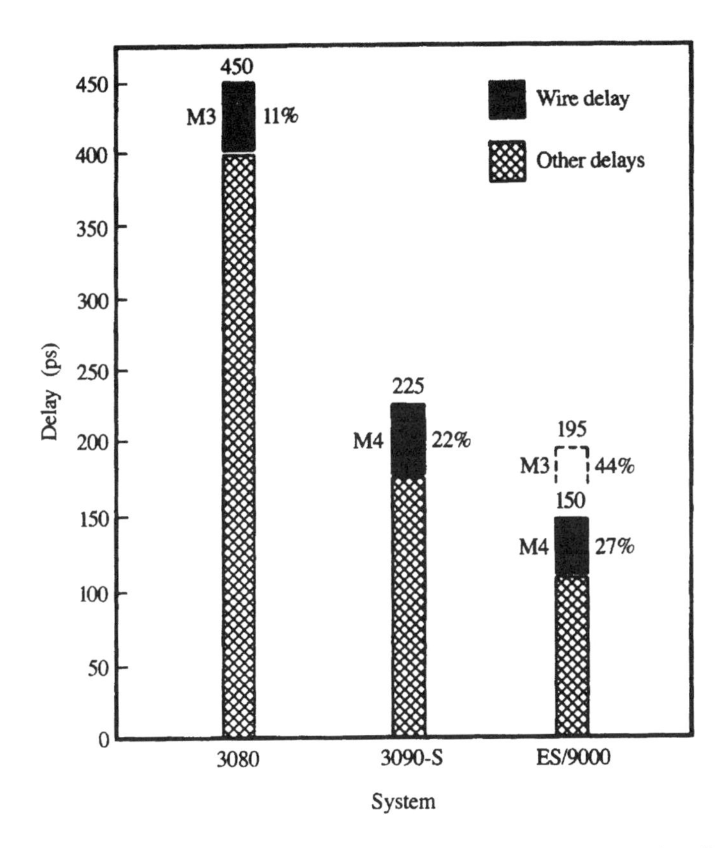

Figure 8-4. Relative Contributions of Wiring and Silicon to the On-Chip Propagation Delays in Three Generations of Bipolar Switched Computers. The third bar contrasts difference in delays between the three and four wiring levels for the same chip. (From Ref. 10.)

are so capable that they are beginning to consume some of the market share that was the domain of supercomputers and mainframes.

The advanced VLSI era has put great demands on the functionality and reliability of ever-increasing numbers of input–output (I/O) connections. An empirical relationship between I/O and the number of circuits to be wired (Rent's Rule) appears to be holding well for mid-to-large systems. This means that the logic I/O demand will expand from about 500 today to 5000 I/Os by the year 2010 [2].

Heretofore, serial wirebonding of one or two rows of I/Os around the perimeter of the chip has satisfied the needs of ceramic or plastic packages for logic packaging. Automated wirebonding today is very fast, efficient, and reliable compared to the manual bonding of the early sixties. Wirebonding has been displaced with TAB bonding in instances where the perimeter pitch must be finer to keep the chip size smaller, yet accommodate 500–600 connections. On the other hand, the solder–bump peripheral counterpart evolved differently into an area-array C4 configuration in which the entire surface of the chip is covered with C4s for the highest possible I/O counts, as high as 2000–3000 pads. Unlike wirebonding, C4 and usually TAB demand bump formation on the surface of the chip when the chip is in wafer form. Bumping the chip has been and is a constraint to the widespread adoption of C4 or TAB in commercial devices. It is an added expense and a commitment to fixed bonding patterns, but, in the VLSI era, perhaps a necessary one. Typically, for bumping a wafer, a layer of silicon oxide, silicon nitride, or polyimide passivation must be used on the final wiring level of the chip. But this has become a commonplace precaution anyway to protect the fine chip wiring from corrosion and mechanical damage, even in advanced wirebonded chips. The details of the required bump fabrication will be discussed later in this chapter.

The die then has evolved to become a total "minipackage" of its own. In a sense, it is the world's smallest electronic package. The role of future packaging will be that of protecting the chip, getting power in and heat and signal out, which will, of course, become more challenging as microprocessors grow larger and more powerful. The chip-level connections toward achieving this will be discussed in the following sections.

8.3 FLIP-CHIP SOLDER-BUMP CONNECTIONS

The solder-bump flip-chip interconnection was initiated in the early 1960s to eliminate the expense, unreliability, and low productivity of manual wirebonding. Whereas the initial, low-complexity chips typically had peripheral contacts similar to wirebonded chips, this technology has allowed considerable extendibility in I/O density as it progressed to full-population area arrays. The so-called controlled-collapse-chip connection (C^4 or C4) utilizes solder bumps deposited on wettable metal terminals on the chip and joined to a matching footprint of solder wettable terminals on the substrate [11]. The upside-down chip (flip chip) is aligned to the substrate, and all joints are made simultaneously by reflowing the solder (Fig. 8-5).

The C4 joining process has been described in the literature with numerous acronyms. In the industry, it has also been called CCB—controlled-collapse bonding—and flip-chip joining, recognizing the fact that it is opposite to the traditional backside-down method of bonding, in which the active side of the chip, facing up, is wirebonded. The terms C4 or flip chip are used interchangeably in this book.

Two other acronyms are also used in this section: BLM and TSM. These refer to the terminal-connecting metallurgies at the chip and substrate, respectively. BLM stands for ball-limiting metallurgy and refers

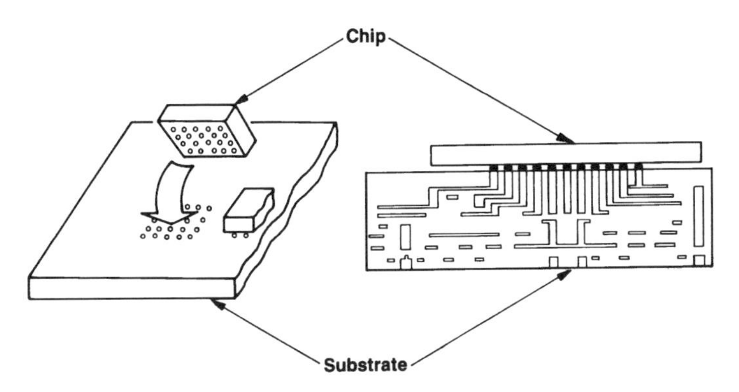

Figure 8-5. Controlled Collapse Chip Connection (C4). The upside-down chip (flip chip) is aligned to the substrate and all joints are made simultaneously by reflowing the solder. (From Refs 9 and 12.)

to the region of terminal metallurgy on the top surface of the chip that is wettable by the solder. Alternate references to BLM are PLM (pad-limiting metallurgy) and UBM (under-bump metallurgy). TSM stands for top-surface metallurgy and refers to the terminal metallurgy on the substrate to which the chip and its associated solder balls are joined.

8.3.1 C4 History

The solder-bump interconnection of flip chips, the face-down soldering of silicon devices to alumina substrates, has been practiced for approximately 30 years. First introduced in 1964 with the solid logic technology (SLT) hybrid modules of IBM's System/360, it was part of a design to eliminate the expense, unreliability, and low productivity of the early manual wirebonding [5]. The solder bump was also an integral part of a chip-level hermetic sealing system created by the glass passivation film on the wafer [6]. Most semiconductor devices of that era were, in contrast, protected by expensive, hermetically sealed can enclosures. The terminal bump design was created to hermetically reseal the access or "via" hole through the glass protection layer as well as to provide a means for testing and joining the chip (Fig. 8-6).

Initially, for the discrete transistors or diodes of the hybrid SLT, copper ball, positive standoffs, embedded in the solder bumps, were used to keep the unpassivated silicon edges of the chips from electrically shorting to solder-coated thick-film lands [7]. Later, in the early integrated-circuit era, the controlled collapse chip connection was devised. In this technique, a pure solder bump was restrained (controlled) from collapsing or flowing out over the electrode land by using thick-film glass dams, or stop-offs [11], which limited device solder-bump flow to the tip of the substrate metallization (Fig. 8-7).

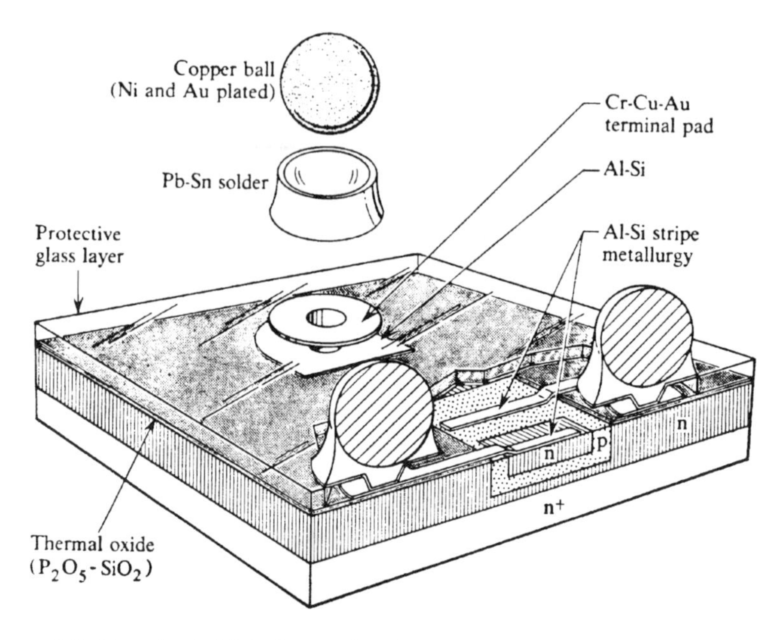

Figure 8-6. Terminal Metallurgy Design. The original SLT flip chip (0.69 mm square) with glass passivation, BLM sealing of via holes, and Cu ball solder bumps. (From Ref. 7.)

As with SLT, the solder flow on the chip surface is limited by a ball-limiting metallurgy (BLM) pad, which is a circular pad of sequentially evaporated thin-film metals. Chromium, copper, and gold, in that order, provide the sealing of the via as well as a solderable, conductive base for the solder bump. A very thick deposit (100–125 μm) of evaporated 97 Pb/3 Sn solder acts as the primary conduction and joining material between chip and substrate [7,8].

The earliest integrated-circuit chips typically had peripheral C4 I/O pads like their wirebonded counterparts. The pads were 125–150 μm in diameter, on 300–375-μm centers. The spacing (pitch) of connections was compatible with the screening resolution and pitch capability of thick-film—AuPt or AgPd—electrodes on the ceramic substrates.

Occasionally it was convenient to have an "in-board" power pad or two in the thick-film technology, but larger numbers of inside I/O pads could not be used until thin film metallized ceramic (MC) technology became available in the mid-1970s, as discussed in Chapter 9, "Ceramic Packaging," and Chapter 12, "Thin-Film Packaging." The narrower lines and spaces made possible with etched thin-film Cr–Cu–Cr on ceramic, as fine as 20-μm lines with 60-μm pitches, allowed wiring escape for double rows of I/O pads and many internal connections [13]. Later, a "depopulated" grid of bumps allowed the interconnection of 700 circuit

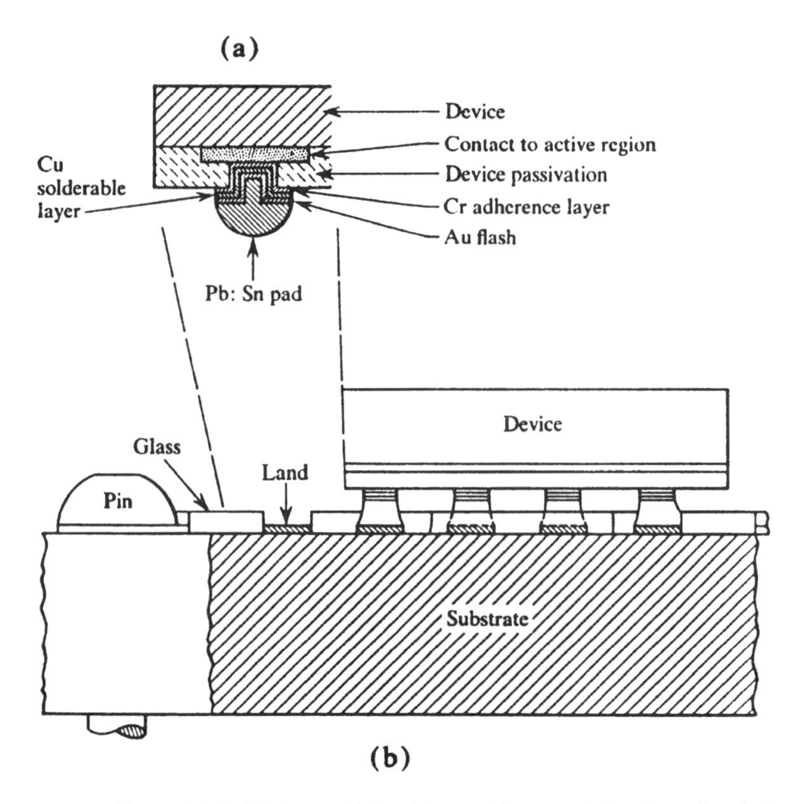

Figure 8-7. Controlled Collapse Chip Connection. (a) Side view of a device or chip; (b) Side view of a device or chip on a substrate (dam method). (From Ref. 11.)

logic chips. The fully populated area-grid array, in which every grid point is occupied by a solder bump, required the complexity of multilayered, cofired ceramic packages. In these packages, the distribution of I/O wiring could be accommodated by via "microsockets" and multiple buried redistribution layers of wiring as opposed to single-level wiring, where the "escape" of wires is geometrically restricted by the maximum number of lines per channel between I/O connections [14]. The progression of C4 "footprint" geometry on chips is shown in Figure 8-8.

An example of an early full area-array C4 configuration is shown in Figure 8-9. The I/O count was only 120 in an efficient square grid array, which is 11 C4 pads long by 11 pads wide on 250-μm (10-mil) centers. A 125-μm (5-mil) solder bump is located at every intersection in the grid except one, which is displaced for orientation purposes [16]. Some packages, such as the cofired alumina multilayer ceramic (MLC) (Figure 8-10) used 9–133 area-array chip sites per package to attain high bipolar circuit densities in IBM's 4300 and 3081 series computers. Logic

Figure 8-8. Progression of C4 area array from (a) essentially peripheral I/O to (b) staggered double row to (c) depopulated array to (d) full-area arrays. (From Ref. 15; reprinted with permission of *Solid State Technology*.)

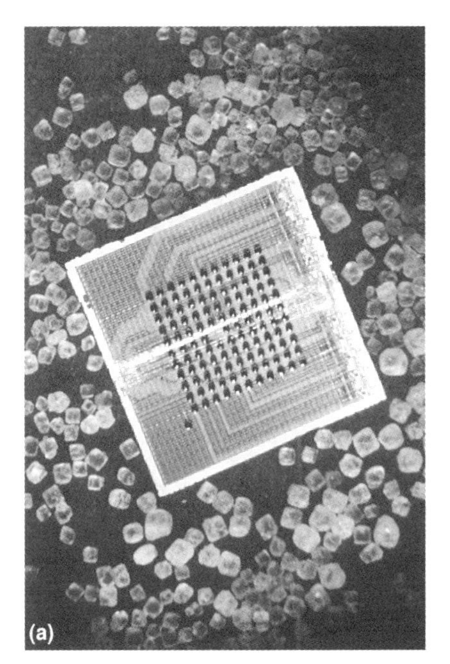

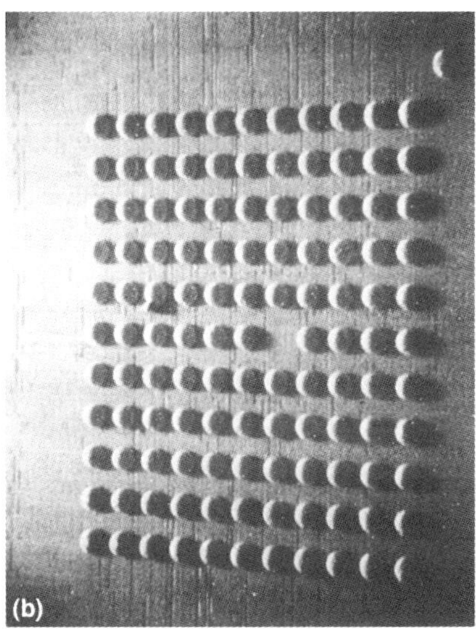

Figure 8-9. Area Array C4 Configuration. (a) An 11 × 11 full area array of solder bumps on a 700-circuit logic chip for use with multilayered ceramic. (b) SEM view of solder bumps. After Goldmann, Ref. [6], 1983, reprinted with permission of *Solid State Technology*.

and memory chips were mixed as required. As many as 25,000 logic circuits or 300,000 memory bits were packaged on a single TCM substrate in this technology [14,17–20]. By adding thin films to a glass ceramic MLC base, IBM was able to achieve 70,000 C4's on a single module for its 9000 series computers [21]. These modules hold 121 integrated-circuit (IC) chips with 648 pads each, and 144 special low-inductance C4 MLC capacitors with 16 pads each (Fig. 8-11). A highly populated logic flip chip of earlier times was a "computer-on-a-chip" which had 762 C4 solder bumps in a 29 × 29 area array (Fig. 8-12) [24]. Four levels of metal wiring in this chip were used, compared to the two or three levels used previously.

More recently, the tremendous improvement in lithography with minimum dimensions in fabs at 0.5 μm (0.35 just starting) and advanced chip wiring rapidly advancing with chemical vapor deposition stud formation and chemical-mechanical polishing to planarize wiring (five and six levels), great strides have occurred in device density, chip size, and I/O count. Leading-edge ASICs have 40,000 bipolar circuits (Hitachi Skyline machine) with approximately 2250 I/Os which is well ahead of the SIA roadmap. Power PC microprocessor flip chips from IBM and Motorola have area arrays of approximately 650 I/O.

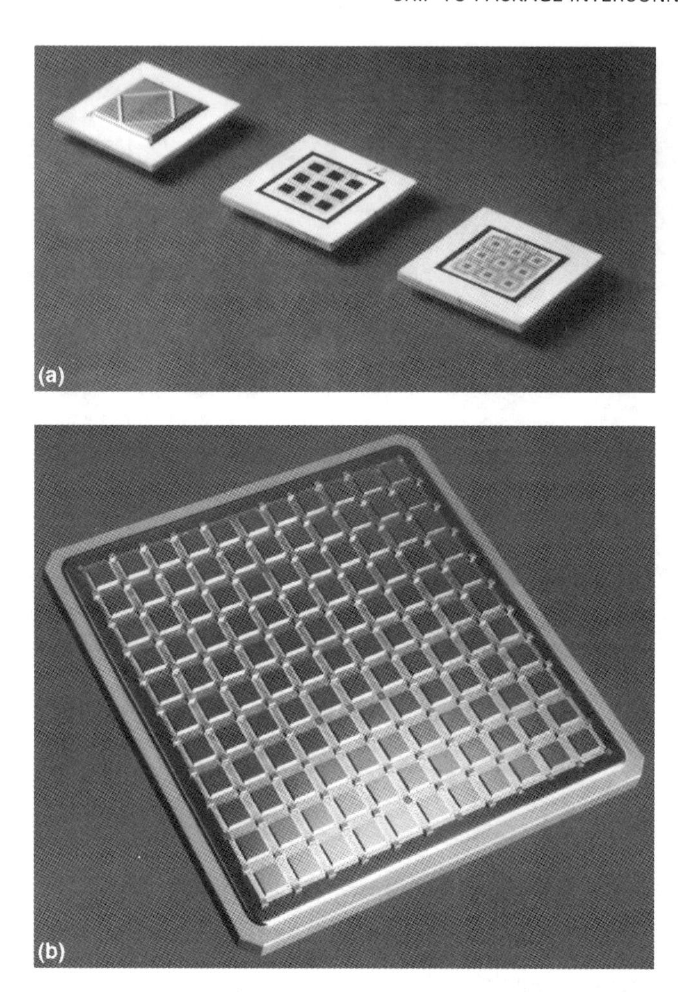

Figure 8-10. Cofired Ceramic Modules. Multilayered ceramic packages: The 9-chip multichip module (MCM) in alumina/molybdenum technology (a), and the 121-chip thermal-conduction module (TCM) in glass–ceramic/Cu technology (b).

Over time, the C4 technology has been extended to other applications. C4s are used on thin-film resistor and capacitor chips in hybrid module applications [25]. Solder pads for some applications can be very large—750 μm in diameter. At the other extreme, Schmid and Melchoir [26] used C4's for precision registration and alignment in the joining of a GaAs waveguide. The C4's in this case were only 25 μm wide and high. The most dense area-array application reported has been a 128 × 128 array of 25-μm bumps on 60-μm centers, resulting in 16,000 pads [27]. The photolith process for forming these is discussed later. C4's or solder connections on substrates similar to C4's and referred to as ball-grid

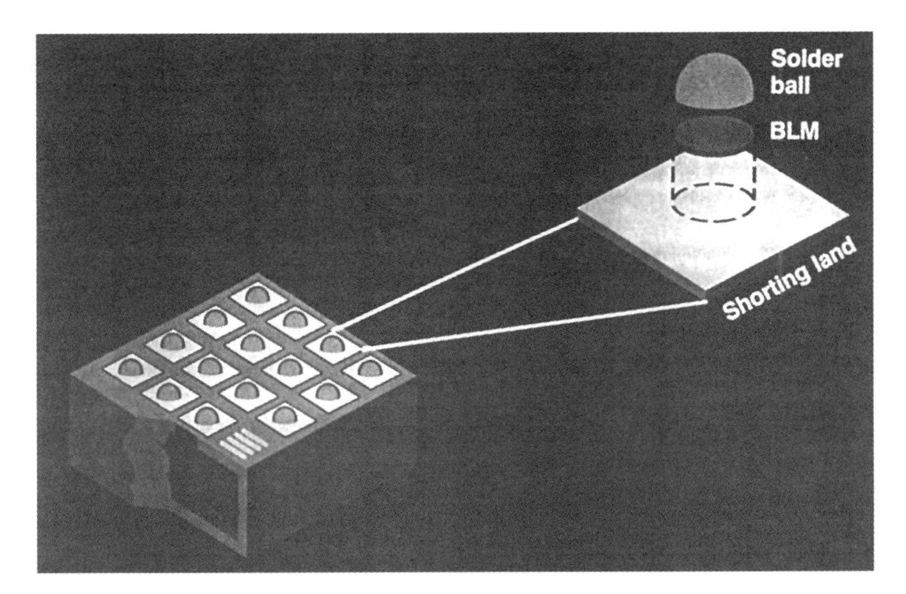

Figure 8-11. Flip-Chip Decoupling Capacitor Made with Pt Metallized $BaTiO_3$ Platelets Terminated with 16 C4's. (From Ref. 22.)

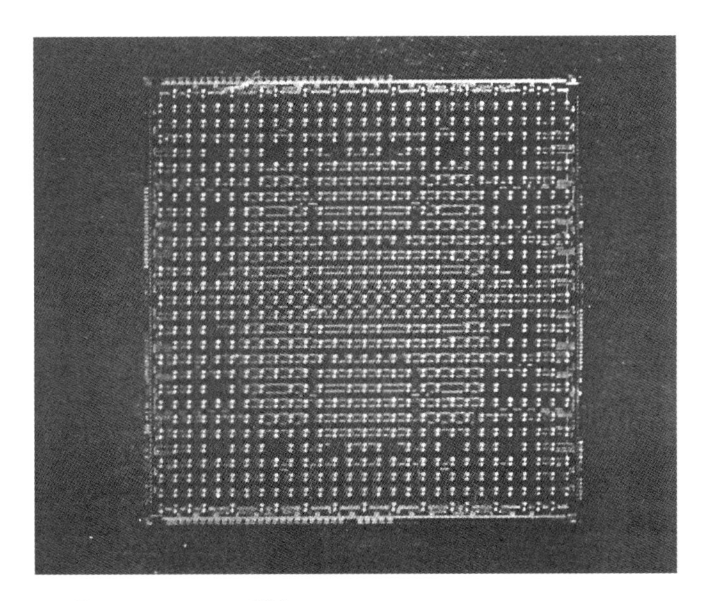

Figure 8-12. Computer-on-a-Chip. An 11,000-circuit "computer-on-a-chip" with 762 solder bumps in a 29 × 29 area array. (From Ref. 23.)

arrays (BGA's) are used for attaching chip carriers to boards and have become part of the surface-mount revolution discussed in Chapter 7, "Microelectronics Packaging—An Overview." New applications of this technology are being explored continually.

8.3.2 Materials

Melting point has been a prime consideration in the choice of solder alloys for C4's. High-lead solders, especially 95 Pb/5 Sn and 97 Pb 3 Sn, have been most widely used with alumina ceramic substrates because of their high melting point, approximately 315°C. Their use for the chip connection allows other, lower-melting-point solders to be used at the module-to-card or card-to-board packaging level without remelting the chip C4's.

A reverse order of assembly (e.g., modules-to-board, then chips-to-module) would require a reverse order of melting point. Josephson superconducting devices have been joined in such a fashion, using an alloy of 51 In/32.5 Bi/16.5 Sn (having a melting point of 60°C) for the chip C4's, whereas a higher-melting-point eutectic alloy, 52 In/48 Sn (having a melting point of 117°C), was used for pins and for orthogonal connections to the chip carrier [28–30].

Joining to advanced organic carriers such as polyimide–Kevlar®[1] [16] or ordinary FR-4 printed-circuit boards [31–33] also requires lower processing temperatures. Here, intermediate-melting-point solders, such as eutectic 63 Sn/37 Pb (melting point 183°C), and PbIn alloys, such as 50 Pb/50 In (melting point of approximately 220°C), have been used. Joining with mixed solders has also been demonstrated. IBM has joined chips with high-melting-point (315°C) PbSn solder balls to printed-circuit boards with low-melting-point (183°C) eutectic PbSn solder [34]. In this case, the processing temperature is intermediate (250°C) and the high-lead solder ball does not melt but is wetted by the low-melt solder on the board. A listing of solder alloy compositions and melting points is shown in Table 8-2. Some phase diagrams relevant to C4 solder joints are shown in Figure 8-13.

The choice of terminal metals, which is described in detail later, will depend on the choice of solder. For example, silver and gold are poor terminal metals to be used with the SnPb alloy. In only a few seconds, gold completely dissolves into the liquid solder. In these cases, another solder alloy could be used, such as indium, which has a much lower solubility for gold; or one of the other lower-solubility metals could be used for the terminal. Thus, Cu, Pd, Pt, and Ni are very commonly used

---

1 Trademark of Dupont Corp.

Table 8-2. Selection of Low-Melting Solder Alloys.

Melting Point	Composition in Mass Percent					Other
(°C)	Sn	Pb	Bi	In	Cd	Elements
16				24		76 Ga
20	8					92 Ga
25						95 Ga; 5 Zn
29.8						100 Ga
46.5	10.8	22.4	40.6	18	8.2	
47.2	8.3	22.6	44.7	19.1	5.3	
58	12	18	49	21		
61	16		33	51		
70	13.1	27.3	49.5		10.1	
70–74	12.5	25	50		12.5	
72.4			34	66		
79	17		57	26		
91.5		40.2	51.7		8.1	
93	42			44	14	
95	18.7	31.3	50			
96	16	32	52			
96–98	25	25	50			
103.0	26		53.5		20.5	
96–110	22	28	50			
117	48			52		
125		43.5	56.5			
127.7				75	25	
139	43		57			
144			62		38	
145	49.8	32			18.2	
156.4				100		
170	57					43T1
176	67				33	
178	62.5	36				1.5 Ag
180	63	34	3			
183	61.9	38.1				
183	62	38				
198	91					9 Zn
215		85				15 Au
221	96.5					3.5 Ag
232	100					
248		82.6			17.4	
251		89				11 Sb
266					82.6	17.4 Zn
271			100			
280	20					80 Au
288		97.2				2.8 As
304		97.5				2.5 Ag
304–312	5	95				
318		99.5				0.5 Zn
321					100	
327		100				
356						88 Au; 12 Ge
370						97 Au; 3 Si
420						100 Zn

Source: From Ref. 35; reprinted with permission of Electrochemical Publications Ltd., Ayr, Scotland.

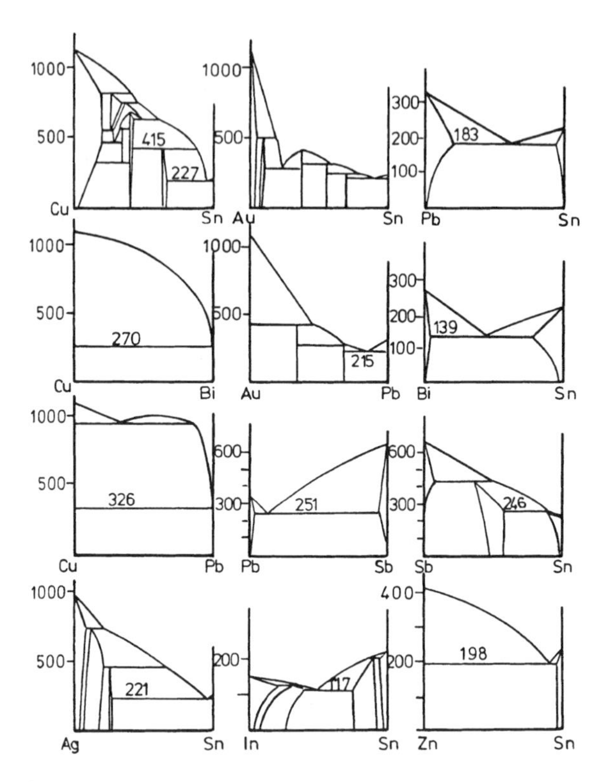

Figure 8-13. Some Phase Diagrams Relevant to Soldering. Temperature, along the vertical axes, is given in °C; concentration, along the horizontal axes, is given in mass percent (Hansen and Smithells give concentrations in atomic percentages). (From Ref. 35; reprinted with permission of Electrochemical Publications Ltd., Ayr, Scotland.)

for both BLM and TSM. All of these metals form intermetallics with Sn, which limits the reaction rates with PbSn solders. On the chip side, this terminal metal is normally sandwiched between an adhesion metal layer of Cr or Ti, and a passivation metal layer, usually of thin gold. The copper, palladium, or nickel thin films on the substrate are typically protected with gold [17–20] or are tinned with solder. In the latter case, some of the solder for the C4 joint is supplied by the substrate [13,36]. MLC substrates usually use a flash of gold on nickel [17–20,25,37–39]. Thick-film substrates have the palladium or platinum alloyed with gold or silver and are dip-soldered prior to the joining operation. AuPt, AgPd, AgPdAu, and AgPt have been reported [8,11,40–44] as thick-film TSM pads.

8.3.3 Design Factors

Some of the factors affecting the material choices for the terminal and solder have already been discussed, but other variables must also be considered in C4 design. The joints must be high enough to compensate for substrate nonplanarity, especially for the older version of thick-film substrates. Because solder surface tension "holds up the chip," a sufficient number of pads must be provided to support the weight of the chip. Typically, this only becomes a cause for concern with very low I/O devices, such as memory chips or chip carriers, which are very bulky. Numerous studies have been published showing the interrelationship between BLM and TSM size, solder volume, chip weight, and C4 height. Figure 8-14 shows these relationships over a very wide range [25].

Extra "dummy" pads, added to supplement those needed for simple electrical connection, have often been used to enhance the mechanical behavior, reliability, or thermal performance of the assembly [45]. Pad location is important from the standpoints of both electrical design and reliability. The effect of distance to neutral point (dnp) is discussed later in this chapter as it relates to thermal-cycle fatigue.

The original solder bumps were placed over inactive Si in SLT, or even over diode-isolated regions in early integrated circuits, just as wire-bond or TAB pads are placed over inactive Si today. But it was soon discovered in the early 1970s that C4 solder bumps could be safely placed

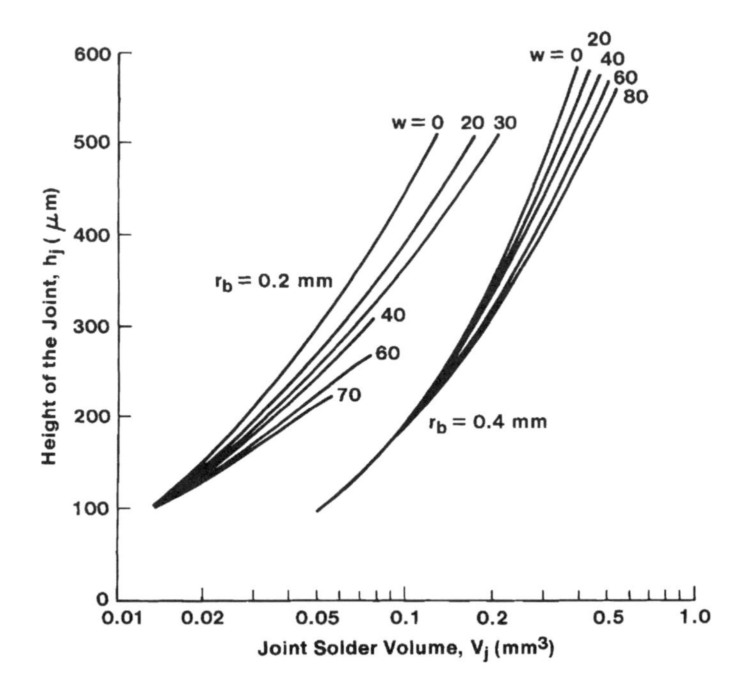

Figure 8-14. Interrelationship Between BLM and TSM Size. Height of the joint (h_j) vs. solder volume of the joint (V_j) as a function of chip weight (W in grams); r_b is radius of joint. (From Ref. 25.)

over active silicon devices and multilayer wiring. This unique capability has made area-array bumping a very powerful packaging attribute and has provided the designer with much freedom in wiring complex ULSI devices. Automatically, area-array flip chips are smaller than their peripheral counterparts, meaning more chips per wafer and lower chip cost.

As VLSI logic chips become more and more dense, higher I/O counts will drive full area arrays of terminals. In this case, the pad size and location are fixed by the chip size and available real estate per pad allocated by a fully populated area array.

The number of C4 pads as a function of chip size and pad geometries is shown in Table 8-3 wherein the opportunity for 155,000 pads on a 20-mm chip is indicated. Figure 8-15 shows the pronounced density advantage of area array versus a single-perimeter row, as pad sizes and spacings decrease.

8.3.4 Bump Fabrication Processes and Tools

Evaporation through a metal stencil mask [7,8,15,25,29,41,42,46,47] is still the most widely used technology for C4 terminal fabrication. BLM and solder are both evaporated through holes in the metal mask and deposited as an array of pads onto the wafer surface. This has proven to be a practical and reliable batch process, simultaneously processing many chips per wafer and many wafers per evaporation. However, it is not the lowest-cost process nor will it accommodate finer bumps and pitches, larger wafers, more precise overlays, or low-melting solders rich in Sn which has a lower vapor pressure than Pb. Therefore, a number of innovative and alternative fabrication technologies for bumping wafers have begun to appear. Photolithographic mask fabrication with thin or thick polymer films, screening of solder creams, electroplating of solders through a photomask where the seed layer for plating becomes the BLM, deposition of blanket sputtered BLM films followed by subtractive etching, or combinations of these process designs are all beginning to appear to bump wafers for various applications.

8.3.4.1 Metal Mask Evaporation Technology

IBM's C4 evaporated bump technology has been practiced for about 30 years and would be typical of this methodology. A passivated wafer with via holes opened is aligned and assembled under a thin Mo stencil mask (Fig. 8-16). The spring-loaded mask assembly is clamped tightly against the wafer to avoid shifting during handling and deposition.

Argon ion sputter cleaning or etching of the via hole and the passivation surface around it is routinely used to remove aluminum oxide from the last metal film on the chip and to remove any photoresist residues from the passivation. This guarantees low contact resistance (typically

Table 8-3. Pad Density Comparisons.

Chip Size (mm)	Pad Diam. (μm)	Pad Pitch (μm)	# Pads Perimeter	# Pads Array
5	762	1524	8	9
	635	1270	12	16
	508	1016	16	25
	381	762	24	49
	305	610	28	64
	254	508	36	100
	203	406	44	144
	152	305	60	256
	127	254	76	400
	102	203	96	625
	76	152	128	1,089
	51	102	192	2,401
	25	51	388	9,604
10	762	1524	24	49
	635	1270	28	64
	508	1016	36	100
	381	762	48	169
	305	610	60	256
	254	508	76	400
	203	406	96	625
	152	305	128	1,089
	127	254	152	1,521
	102	203	192	2,401
	76	152	260	4,356
	51	102	388	9,604
	25	51	784	38,809
20	762	1524	48	169
	635	1270	60	256
	508	1016	76	400
	381	762	100	676
	305	610	128	1,089
	254	508	152	1,521
	203	406	192	2,401
	152	305	260	4,356
	127	254	312	6,241
	102	203	388	9,604
	76	152	520	17,161
	51	102	784	38,809
	25	51	1,572	155,236

Source: H. Nye, personal communication, 1986.

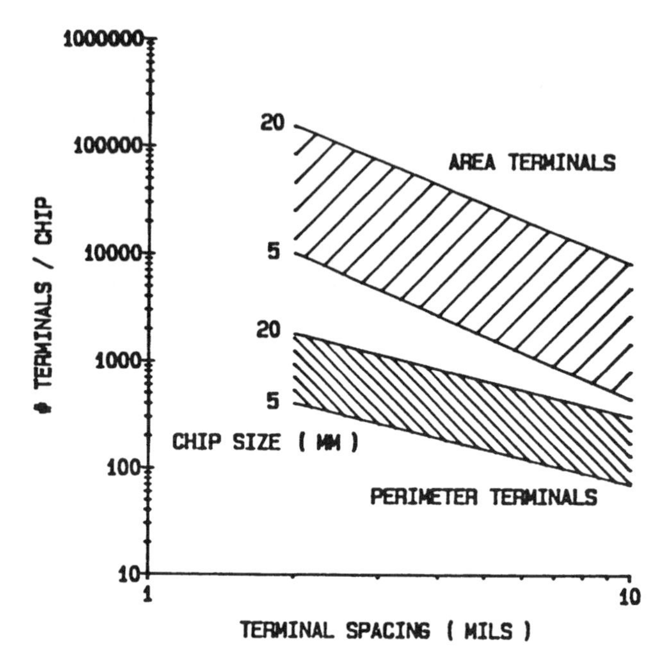

Figure 8-15. Number of Pads Versus Pad Separation for Different Chip Sizes. Input/output terminal trends. (Courtesy of H. Nye, 1986.)

less than 10 mΩ) and good adhesion to the SiO_2 or polyimide at this interface. Originally, the plasma was created by a high DC voltage but has migrated to an RF/AC process due to charge accumulations at unconnected mechanical or thermal pads and because FET devices are more sensitive to damage or parameter shifts.

The multilayered metals of the BLM are then deposited in the following way. A typical evaporator would have numerous metal sources with thermal energy supplied by either resistance, induction, or electron beam (e-gun). Cr is sublimated first to provide adhesion to the passivation layer, as well as to form a solder reaction barrier to the aluminum. A "phased" or gradually mixed layer of Cr and Cu is coevaporated next, to provide resistance to separation during multiple reflows. This is followed by a pure Cu layer to form the basic solderable metallurgy. Finally, a flash of gold is provided as an antioxidation protection layer and to promote wettability. This is desirable because the wafers are normally exposed to air before going on to the next step which is the solder evaporation. Solder deposition requires a "thick" (on the order of 100 μm) stencil mask. Although lead and tin are usually in the same charge (single alloy, molten pool), the higher vapor pressure component, Pb, evaporates and deposits

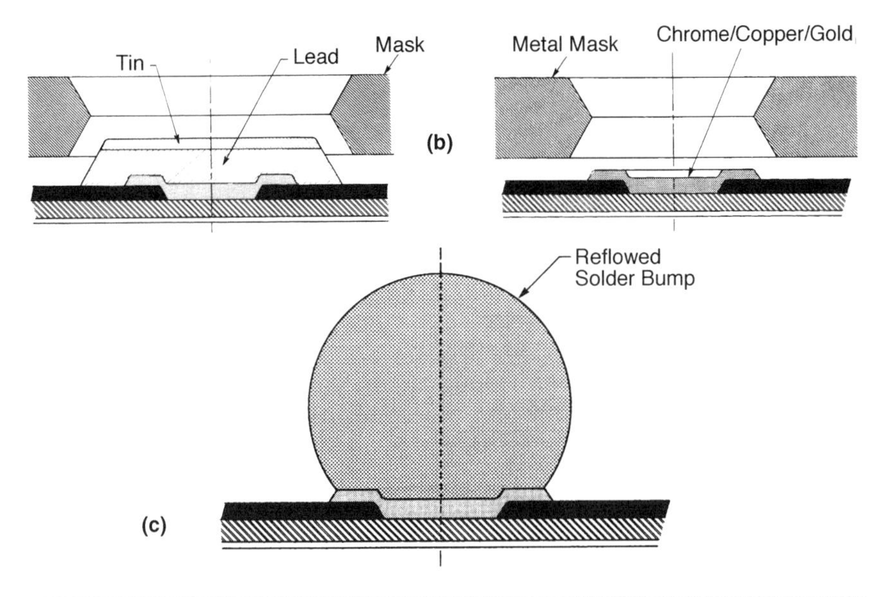

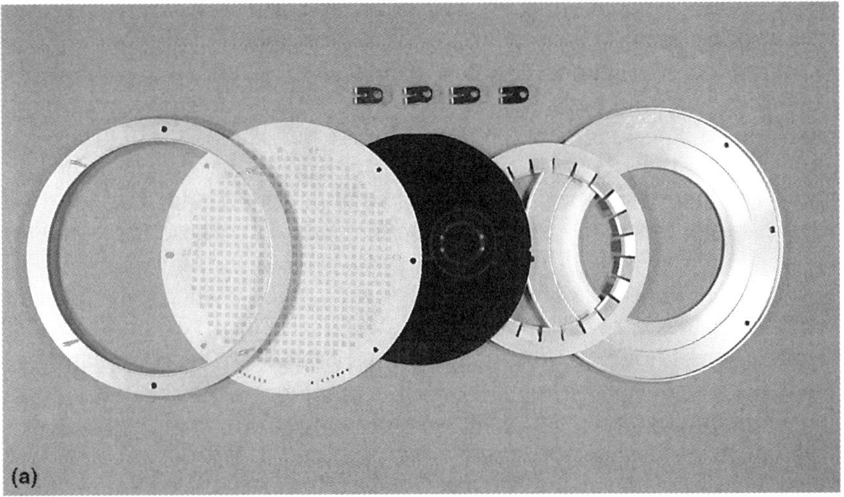

Figure 8-16. Metal Mask Technology. (a) Tooling for alignment of mask to wafer; (b) masking and evaporation of chromium/copper/gold; (c) masking and evaporation of lead/tin; (d) reflowed solder bump. (From Ref. 42; reprinted with permission of ISHM.)

first, followed by tin on top of the lead. Reflow in a subsequent hydrogen ambient furnace at about 350°C melts and homogenizes the two-layer solder pad and brings it to a spherical shape. In addition, H_2 promotes some reduction of oxides of the Pb and avoids excessive oxidation of the Sn.

Photolithographic processes and combinations of photolith and metal mask are becoming more and more popular for fabricating terminals. Most common is a sequence of blanket deposition of the BLM, application of photoresist, development of a pattern in the resist followed by electrodeposition of the solder; then removal of the resist and subetching of the BLM, using the plated solder bumps as a "mask" [37,48–51]. An alternative sequence is to blanket-deposit BLM by sputtering, apply photoresist, subetch the BLM through the resist, then deposit solder by a variety of techniques including solder dip [52], solder ball placement [53], or stencil mask evaporation [54]. The single-masking processes ("unimask") are significantly simpler and cheaper than multiple-mask techniques, although they are not as flexible in providing varying amounts of solder. In some applications, higher volumes of solder are able to decrease strain, as discussed in Section 8.3.6, "C4 Reliability."

8.3.4.2 Electroplated Bumps

It has been pointed out that extendability issues may place restrictions on metal mask evaporation. As semiconductor wafers have increased in size, it is becoming more and more difficult to hold height and volume uniformity of the bumps across a wafer. Very small bumps, typically desired for optoelectronic applications, are difficult to make with thick metal masks, and alignment tolerances limit extendability for high-density arrays. In addition, the drive to direct chip attach (DCA) on organic boards is driving more users to demand lower-melting-point solders to be used for the bump, especially SnPb eutectic with 63% Sn. These very high tin contents are extremely difficult to evaporate because the tin vapor pressure is so low as to require hard driving of the sources which can result in meltback of the evaporated solder on the wafers during the evaporation. For these reasons, photolithographic patterning, coupled with electroplating, is gradually displacing the metal mask evaporation.

Photolithographic processes use photoresist layers directly on the wafer for the terminal or solder bump definition and do not have the problems of extendability that metal masks have. Smaller bumps are in fact easier to make with resist. Electroplating various compositions can be accomplished by adjusting the plating bath composition or by the sequential plating of the individual components. Various combinations of these methodologies have been reported [51,55,56]. Most common is a sequence of blanket deposition of the "seed layer" (to be the BLM), application of photoresist, development of a pattern in the resist, followed by electrodeposition of the solder, then removal of the resist and subetching of the BLM using the plated solder bumps as a "mask." A typical flowsheet for the MCNC electroplating process is shown in Figure 8-17.

Hitachi [48] and Honeywell-Bull [57] have also created solder bumps

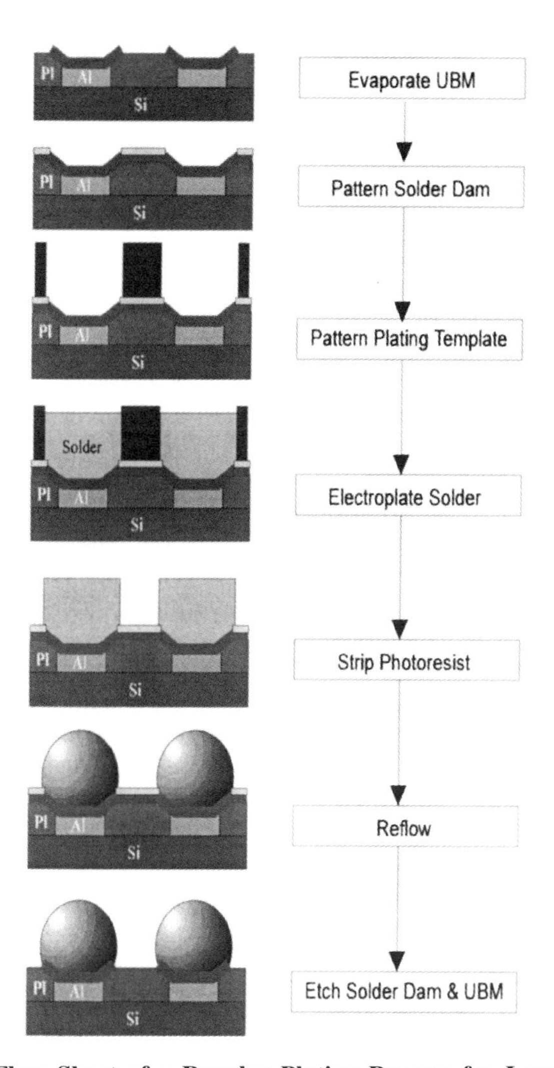

Figure 8-17. Flow Sheet of a Regular Plating Process for Low-Cost Solder Bumps at Microelectronic Center of North Carolina (MCNC). (From Ref. 51.)

in the past by electroplating solder on a seed layer. Hitachi used its bumped chips for early hybrid circuits [25]. Honeywell-Bull replaced conventional TAB Au bumps with solder bumps to gang bond inner lead bonds by soldering instead of using AuSn eutectic attachment. The similar Hitachi and Honeywell-Bull structures are shown in Figure 8-18.

A unique extension of the single-mask concept has been demonstrated at Microelectronic Center of North Carolina (MCNC), where not only the solder and terminal are patterned simultaneously but also the redistribution layer [B]. When a wafer has been designed for wirebonding

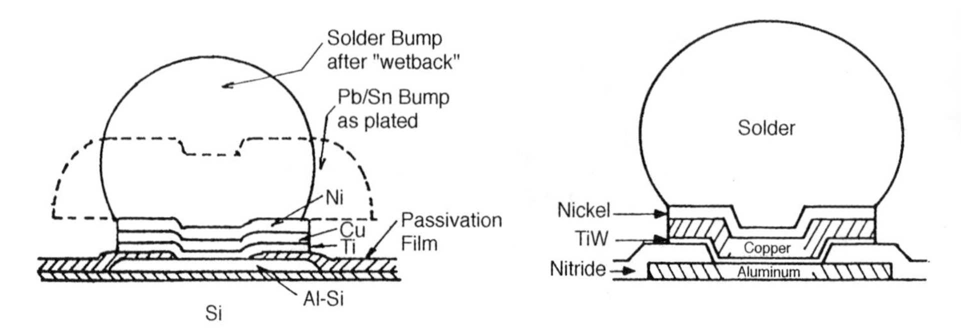

Figure 8-18. Electroplated Bump Structures from Hitachi and Honeywell-Bull (Left and Right). (From Refs. 48 and 57.)

and one wishes to convert it to solder-bump flip-chip without extensive redesign of chip surface wiring, this system combines inward redistribution of the wiring simultaneous with bump formation to have a cost-effective, dual solution. The process is referred to as single-mask redistribution (SMR). The uniformly plated solder flows during wafer reflow to distribute the solder primarily to the bump region as shown in Figure 8-19. The resultant structure contains a very high-conductivity redistribution line lightly coated with solder, connected to a larger-volume solder bump, all fabricated from the same metallization/patterning steps.

Examples of MCNC electroplated bumps are shown in Figure 8-20. A high-density array of as-plated bumps is shown in Figure 8-20(a). The uniformity of solder deposit on this chip with 1679 bumps is evident. After reflow, the bumps appear spherical as shown in Figure 8-20(b). The process has been exercised on bumps as small as 40 μm and as large as 600 μm in diameter.

The final operation before dicing of the wafer into individual chips is the electrical testing of each chip. Mechanical probes are used to contact the soft solder bumps fabricated earlier. Area-array bumps are tested with special buckling beam assemblies of wires [58].

8.3.5 Assembly/Rework

The formation of wettable-surface contacts on the substrate (providing a mirror image or "footprint" to the chip contacts) is achieved by thick- or thin-film technologies. Thin-film contact technology is similar to the BLM described previously, but thick-film technology involves the development of wettable surfaces by plating nickel and gold over generally nonwettable surfaces such as Mo or W (conductors usually used within the ceramic substrate). Solder flow may be restricted by the use of glass or chromium dams where necessary. Various thin- and thick-film processes which are typically used are illustrated in Figure 8-21.

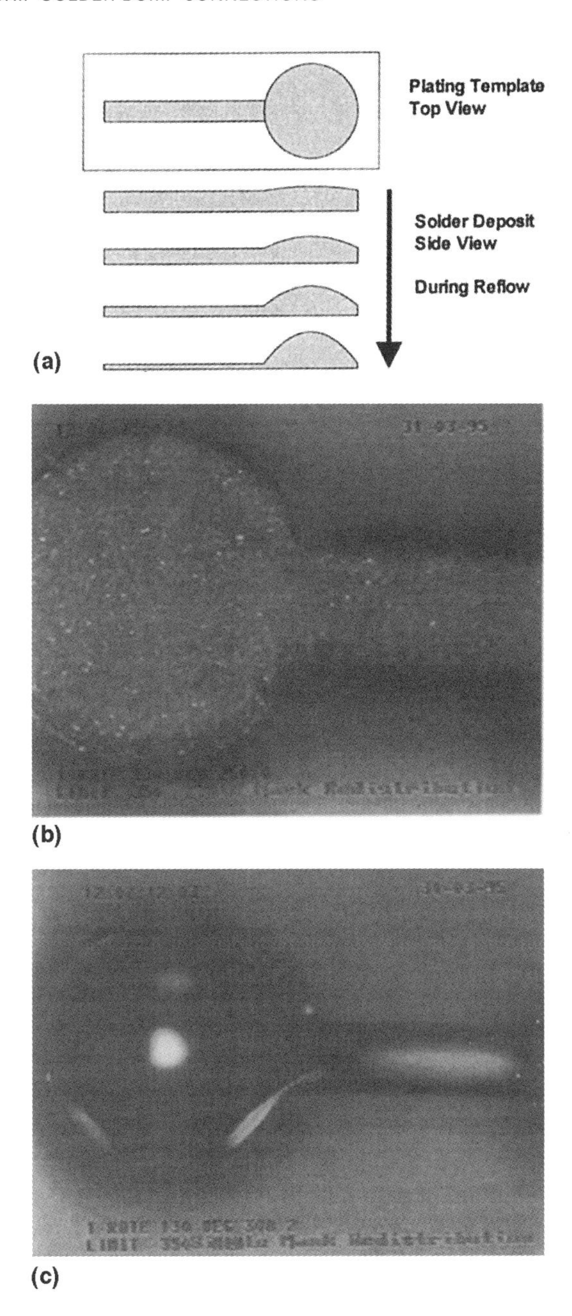

Figure 8-19. Flow Diagram for SMR Template Formation [B]; (b) Plated SMR Template [B]; (c) Completed SMR Structure, Including the Redistribution Trace and Solder Bump [B]. (From Ref. 56.)

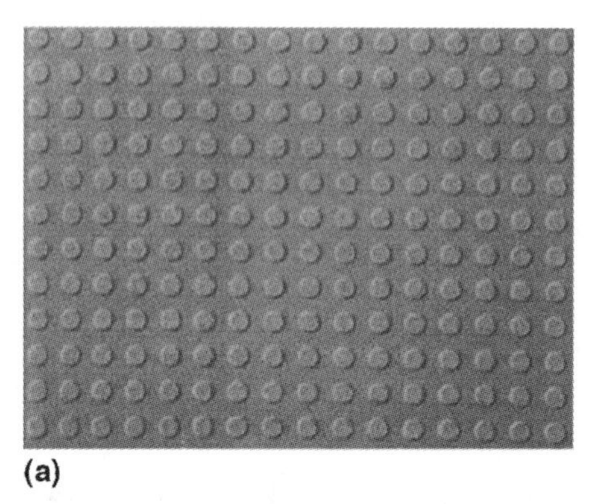

(a)

(b)

Figure 8-20. Electroplated C4 Bumps in the as Deposited (a) and Reflowed Conditions (b). (From Ref. 56.)

8.3.5.1 Self-Alignment

One of the finest attributes of the C4 process is its self-alignment capability arising from the high-surface-tension forces of solders [47]. Chip pads and their counterparts on the substrate may be separated by as much as three times the average bump radius, but if the mating surfaces touch and are reasonably wettable, self-alignment will occur (Fig. 8-22). Optoelectronic package assemblies use this phenomenon extensively [60–64] and allows precision ± 1 μm self-alignment of optical devices to waveguides and optical fibers. The productivity of the reflow joining process when done in an oven is extremely high. As many as a million C4 bonds can be made in 1 h using automated tools.

The self-alignment dynamics have been modeled as in Ref. 65. Figure 8-23a shows the restoring force versus misalignment for a chip

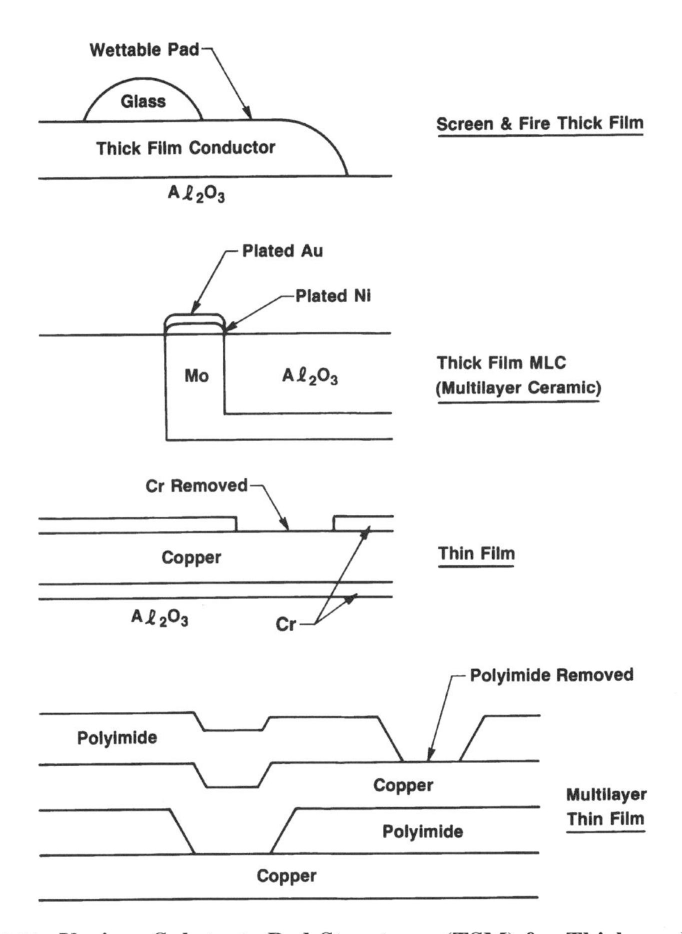

Figure 8-21. Various Substrate Pad Structures (TSM) for Thick- and Thin-Film Substrates.

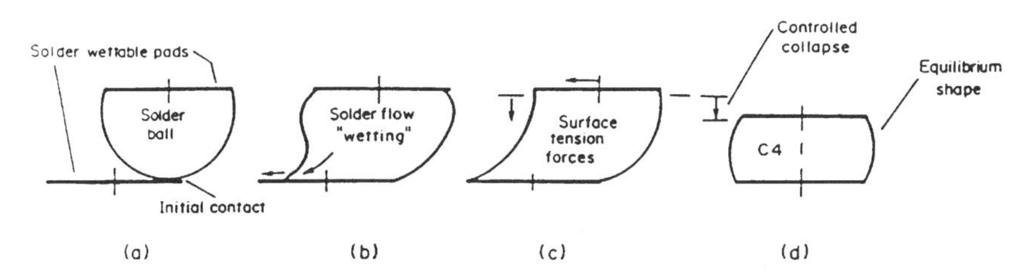

Figure 8-22. Stages of Joining Process for C4: (a) Rough Alignment of Components in Initial Placement; (b) Joining; (c) Self-Alignment Begins; and (d) Bond Complete, Components Aligned. (From Ref. 59.)

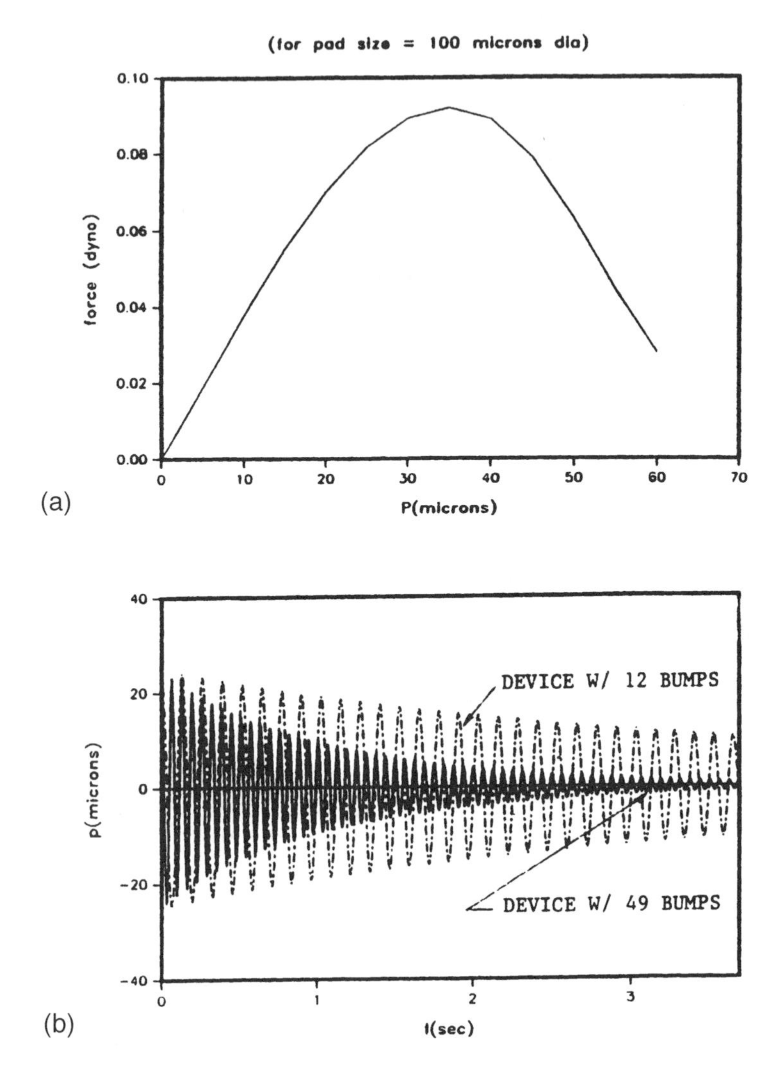

Figure 8-23. Self-Alignment Restoring Forces and Oscillations During Dynamics of Solder Wetting. (From Ref. 65.)

with 49 bumps 100-μm in diameter. During joining, a dynamic oscillation is predicted which decreases in amplitude due to the viscous damping forces of the liquid solder. This is illustrated in Figure 8-23(b). The oscillation frequency is very high, about 10 times per second. The viscous damping gradually reduces the amplitude of the oscillation, and reaches the 1-μm range after 3 s. Another chip with only 12 bumps has restoring and damping forces only one-fourth of those in the first case. These levels are small compared to the chip inertia and result in the damping being not as effective, reaching 10 μm amplitude after 3 s. Thus, a minimum time is required for the solder to be in the liquid state to allow the full damping to occur before the solder freezes. Bache (Fig. 8-24) has used two solder bump sizes on the same chip to allow the self-alignment to occur in stages. The larger bumps "catch" first and in self-aligning allow the smaller bumps to "catch" and finish the precision alignment. A better than 1-μm accuracy is attained with the process.

Inspection of the joined chips to assess the degree of self-alignment is difficult due to the joint being "hidden." Optical techniques can be used only if the devices are transparent to the optical wavelengths—such as infrared for silicon devices. Bache [62] has used a vernier made of high-density solder bars on both chip and substrate coupled with x-ray-radiography to measure the degree of alignment as shown in Figure 8-25. The vernier structures consist of a series of aligned bars on each of the two components. These bars have slightly different spacings on the two components such that any misalignment will result in the bars of the two halves of the vernier coinciding away from the central bar. The actual misalignment of the assembly can then be measured by counting the number of bars between the central bar and that bar where the registration occurs, and multiplying this by the difference in spacing of the two sets of bars. The *z* dimension (chip–substrate gap) is also extremely important for certain optical applications. This has been measured [66] by using Fourier transform infrared spectroscopy (FTIR) and measuring the spacing

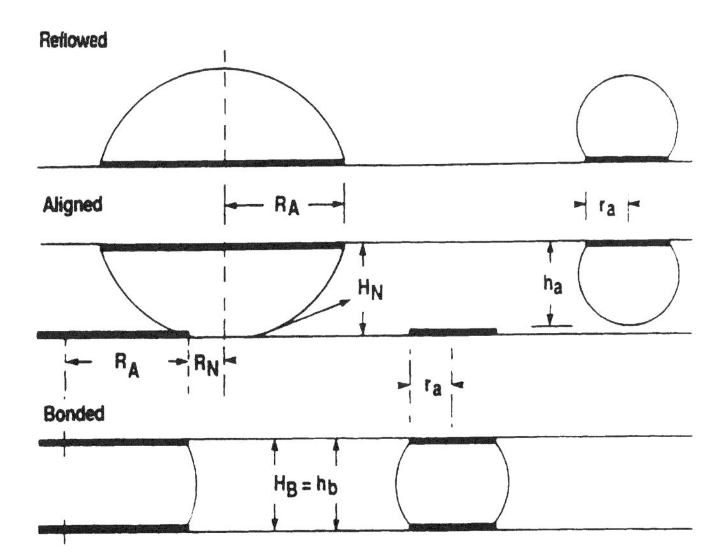

Figure 8-24. Alignment of Fine Bumps with the Use of Coarser Bumps. (From Ref. 62.)

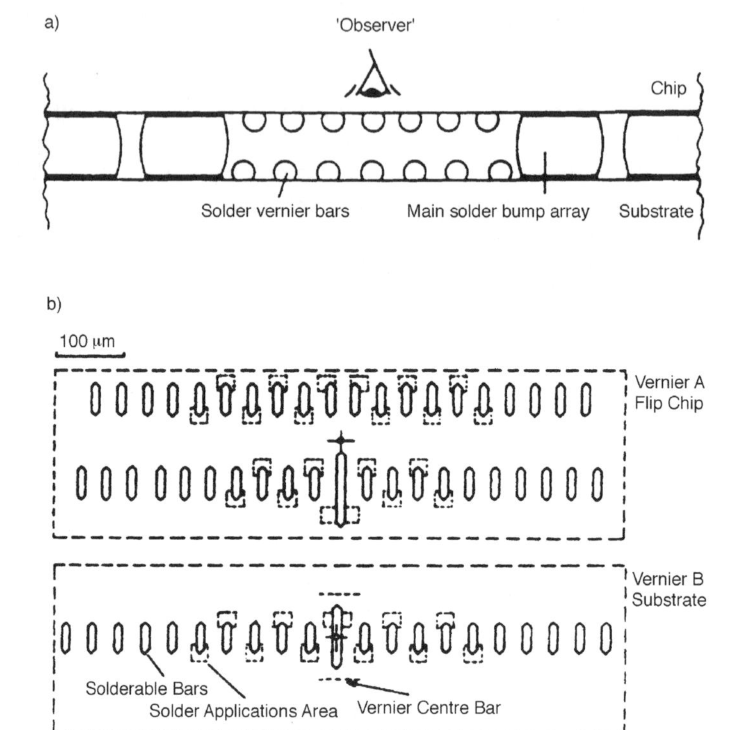

Figure 8-25. Solder Vernier Structure for Monitoring Flip-Chip Component Alignment. (a) Cross Section, Schematic; (b) Plan View of 2-µm Resolution Vernier. (From Ref. 62.)

between the interference peaks (Fig. 8-26). Bond heights to ± 0.5 µm was achieved with nominal 16-µm-high joints.

8.3.5.2 Conventional Flux-Assisted Joining

Once the BLM, TSM, and solder are in place, as described previously, the joining of chips to the substrate using C4 technology is straightforward. The chip needs to be aligned and placed upside-down on the substrate. Commercial tools are available to give ± 2-µm placement accuracy in the laboratory, and ± 8-µm in manufacturing [67]. Flux, either water-white rosin [50,68–70] for high-lead solders or water-soluble flux [28,29,71] for low-lead and other low-melting solders, is normally placed on the substrate as a temporary adhesive to hold the chips in place. Fluxless and no-clean processes are being explored as well and will be described in the next section.

Once the chip-joining operation is complete, cleaning of flux residues

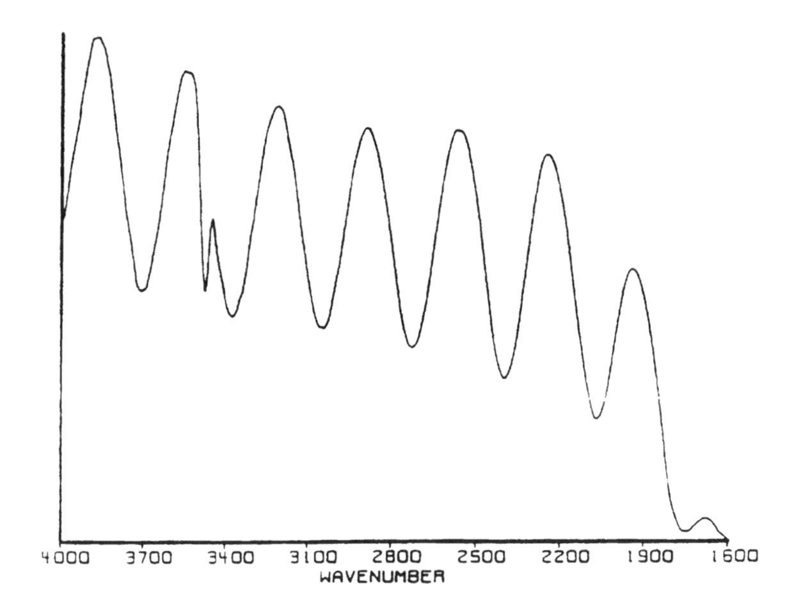

Figure 8-26. Interference Fringes Obtained by FTIR Spectroscopy for the Measurement of Solder Bond Height. The height in this example in 15.5 ± 0.2 μm. (From Ref. 66.)

is accomplished with such solvents as chlorinated solvents or xylene for rosin flux and water for water-soluble flux. The assembly is then electrically tested.

8.3.5.3 Fluxless Joining with Plasma-Assisted Dry Soldering

For fluxless plasma-assisted dry soldering (PADS) processing, flux is replaced with a dry pretreatment performed using a commercial plasma pretreatment tool (see Fig. 8-27). During this pretreatment, the oxide which inhibits joining is converted to oxyfluoride. This conversion film passivates the solder and has the unique property of breaking up when the solder melts, exposing a free solder surface and allowing reflow and joining to occur in the absence of a liquid flux or reducing gas.

The chip-joining task involves first flipping the chip over and placing it precisely on the matching pads of the substrate or carrier. Both manual [73] and automated tools [67] are available. Manual tools use an optical head incorporating a microscope and vertical illumination system that enables the operator to view both the chip and substrate simultaneously. Coupled with a joy-stick-controlled sample motion with 6 degrees of freedom, the chip and substrate can be aligned within 2 μm [73]. For rough surfaces, the optical probe have been modified by adding two illuminators and fiber-optic light pipes which introduce the light to the

Figure 8-27. Fluxless Pretreatment PADS Tool. (From Ref. 72.)

substrate at an angle. This arrangement allows any type of substrate to be used. An example is shown in Figure 8-28, where silicon chips with high-lead (95 Pb Sn) solder bumps were fluxlessly joined to a eutectic solder-dipped FR-4 organic substrate made from printed circuit board stock. The rough surface of the FR-4 is visible in the bottom part of the lower cross section (Fig. 8-28b). It is also visible on the virgin surfaces before joining in Figure 8-28c. Automated tools use pattern recognition to place chips very rapidly on the substrate with accuracies in the ± 5–8-μm.

In flux joining, flux acts as an adhesive to hold the chip in place during the transfer and handling involved in mass reflow operations in belt furnaces. For fluxless joining there is no "glue." Instead, a tacking regimen has been employed which utilizes various combinations of pressure, time, and/or temperature to temporarily bond the chip to the substrate in the aligner bonder during the chip placement operation [73]. Both upper and lower chucks could be heated, and the chip placement control computer set to apply the required loads/times/temperatures needed for the application. Once tacked, the assemblies are transported to the nitrogen reflow oven where the permanent chip joining is accomplished and self-alignment occurs.

The accuracy of the chip placement is most important for small bumps. In addition, the position of the chip relative to the substrate in the final assembly is critical for optoelectronic applications [62–64,66,74–76]. Self-alignment takes place when the solder reflows [47] and allows the final chip to be located ± 1-μm with precision [73]. The relative positions of the chip and substrate after placement or after reflow can be observed

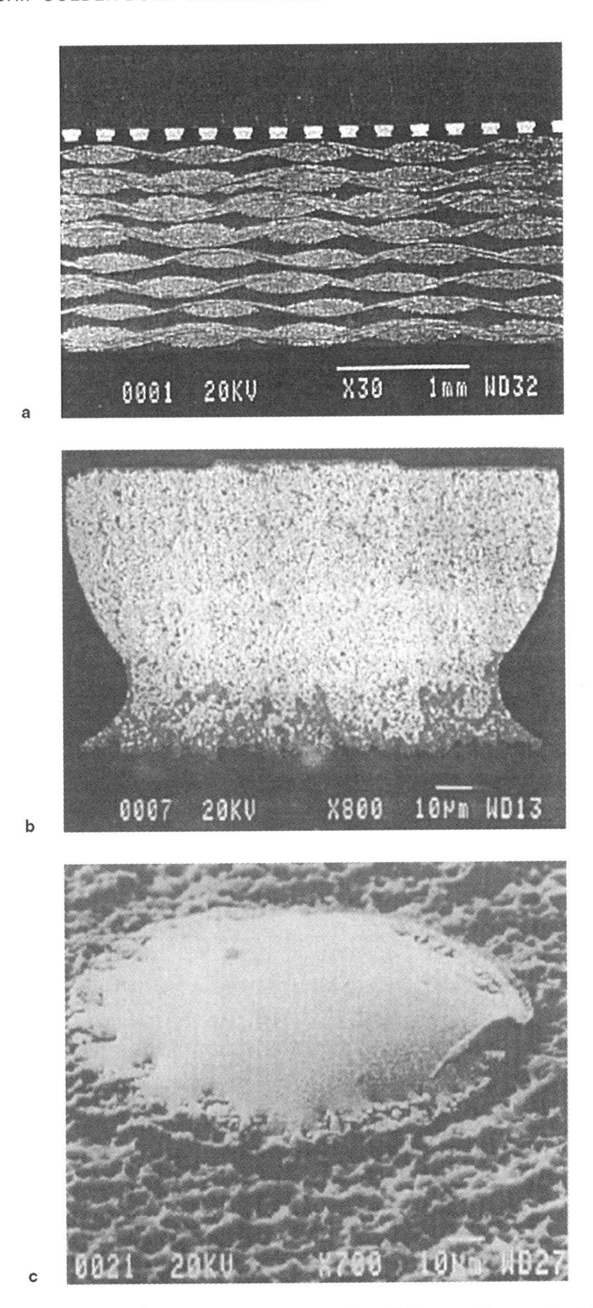

Figure 8-28. Fluxless Joining of 95 Pb 5 Sn C4's to Eutectic Tinned FR4 Terminations by PADS Fluxless Joining. (a) Cross section of chip-on-board; (b) cross section showing 95 Pb 5 Sn bump soldered to card with 63 Sn 37 Pb eutectic, and (c) eutectic tinned FR4 terminals and rough FR4 surface. (From Ref. 72.)

and measured optically with white light for glass substrates, or with infrared (IR) for silicon which is IR transparent. Separate infrared microscopes are available which, when used in conjunction with thin-film orientation patterns and verniers built into the components, allow the relative chip-to-substrate positions to be measured very accurately. Several thin-film patterns are shown in Figure 8-29a, whereas Figure 8-29b shows the vernier patterns imaged with IR [18]. The images are slightly out of focus because both the chip and substrate surfaces are being viewed simultaneously with the focal plane halfway between the two surfaces. They are sufficiently clear to be able to demonstrate better than 1-μm self-alignment accuracy using photolithographically defined solder-bump structures with 40-μm diameters.

The final chip joining is usually performed in a belt furnace with inert nitrogen. The same furnace can be used with either flux or fluxless joining. Belt furnaces have extremely high productivities—as many as a million flip-chip solder bonds can be made in 1 h. For laboratory use, local heat sources have been employed to bond one chip at a time [51]. An infrared belt furnace was used to produce the fluxlessly joined and

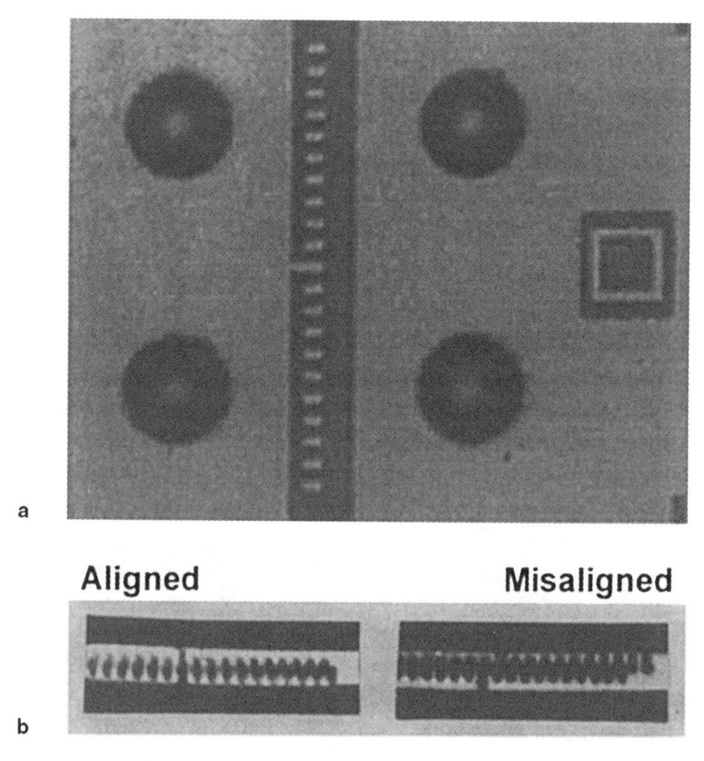

Figure 8-29. (a) Positional Test Patterns (Thin Film); (b) IR Images.

self-aligned eutectic solder bump shown in cross section in Figure 8-30. The substrate used had bare copper terminals. No post-join-cleaning is required for the fluxless process. However, cleaning of the flux residues is normally required for flux-based processes. Chlorinated solvents had been widely used for many years but have been banned due to their propensity to deplete the ozone layer. Other solvents such as xylene are now used for rosin fluxes, and water-based formulations are used for water-soluble fluxes. Once the chip-join operation is complete, the assembly is electrically tested.

8.3.5.4 Rework of Multichip Assemblies

One of the most important issues in high-end multichip module products is that of rework. Whereas the chip-join process is an extremely high-yield process, and defective joints are rarely found in manufacturing, defective chips are occasionally encountered which require replacement. Even when "known good die" are used, it is sometimes necessary to replace a die in a multichip assembly for engineering changes. The rework process for a flip chip usually requires three steps: (1) chip removal, (2) site dress, and (3) placement and joining of a new chip.

The chip removal is accomplished mechanically by ultrasonic torquing of the chip (Fig. 8-31) or (Fig. 8-32). This minimizes any damage that can occur on the substrate or chip (as can happen, for example, when a cold mechanical pull, shear, or torque is used). Local heat can be supplied by IR [69,77] or by conduction through upper and lower chuck heaters in an aligner/bonder [55]. In order not to melt all the solder joints on passenger chips—those chips on the assembly not being reworked—the

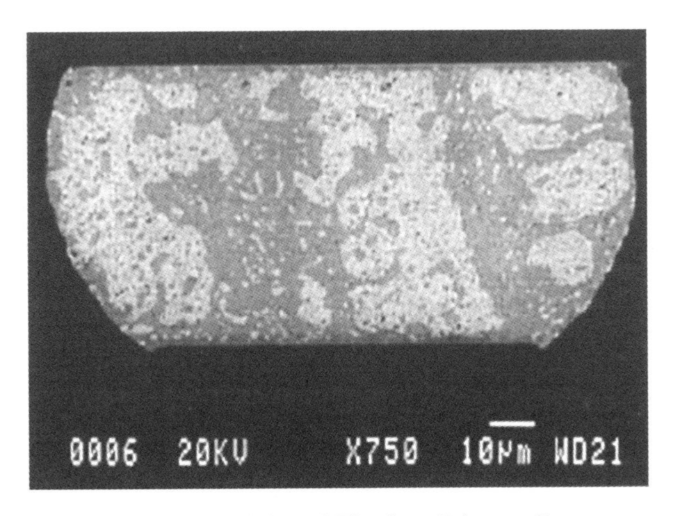

Figure 8-30. Self-Aligned Fluxless Joint to Copper.

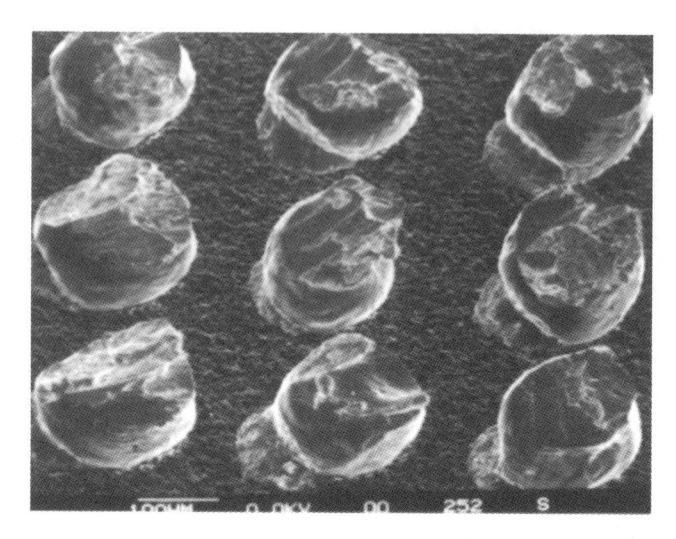

Figure 8-31. Excess Solder. SEM photograph of residual solder left on a typical substrate microsocket after mechanical removal of the device. (From Ref. 69; reprinted with permission from *Journal of Electronic Materials.*)

chuck holding the substrate is heated to a temperature slightly below the melting point of the solder while the upper chuck is heated to a temperature above the melting point of the solder. The pedestal is aligned to the die and the lower chuck raised until the chip is secured to the upper chuck by means of the vacuum contact on the pedestal. The chip stays on the pedestal when the bottom chuck is lowered.

An example of a substrate surface after hot chip removal is shown in Figure 8-32. The solder is 95/5 Pb/Sn in a 41 × 41 array of 125-mm-diameter bumps on 250-mm centers. This was performed in air without flux. To accommodate the solder melting temperature of approximately 315°C, the upper chuck was heated to 350°C while the lower chuck was heated to 300°C.

After chip removal, the site usually needs to be dressed. This entails removing most of the solder so the residual is very uniform and does not interfere with the joining yield of the new device or the reliability goals of the assembly. Hot-gas tools [69] have been used in the past (Fig. 8-33), but currently the most common practice is to use a solder wick made out of a powder metallurgy block of copper [78] (a proprietary development of IBM corporation). The block is placed on the substrate site either manually (alignment is not critical) or with the aligner bonder as if it were a chip to be joined. The assembly is then given a reflow during which the excess solder is absorbed into the copper block by surface tension. It can be picked off the site which is now ready for the new chip to be attached. A dressed site is shown in Figure 8-34.

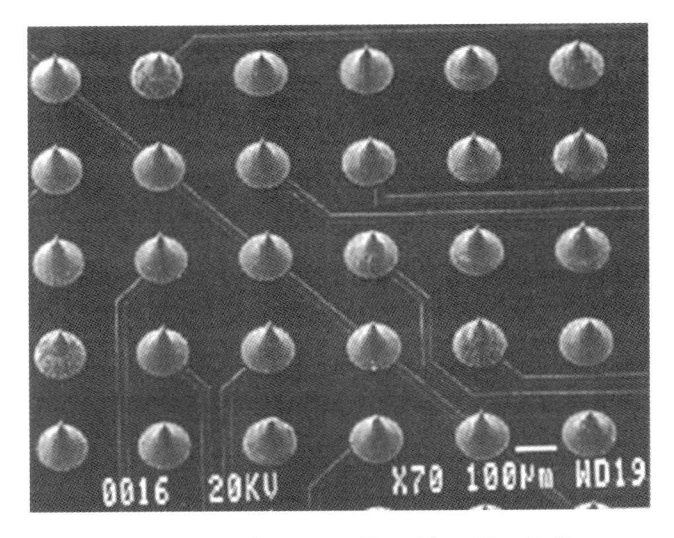

Figure 8-32. **Substrate Site After Hot Pull**

The new chip is joined to the substrate in exactly the same fashion as the original chips, using a chip placement tool in its normal fashion to align and place the new chip. Joining reflow (and clean if necessary) is accomplished also in the normal fashion as with the original chips.

Besides the electrical testing, numerous techniques are available to characterize the joints. The concern for quality of the joints is often raised

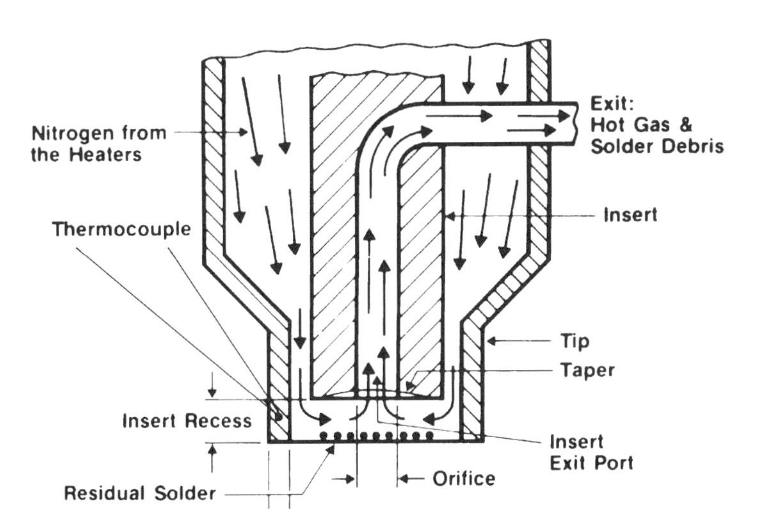

Figure 8-33. Hot-Gas Tool. Side-view sketch showing the direction of gas flow within the probe tip of the hot-gas dress tool. (From Ref. 69; reprinted with permission from *Journal of Electronic Materials.*).

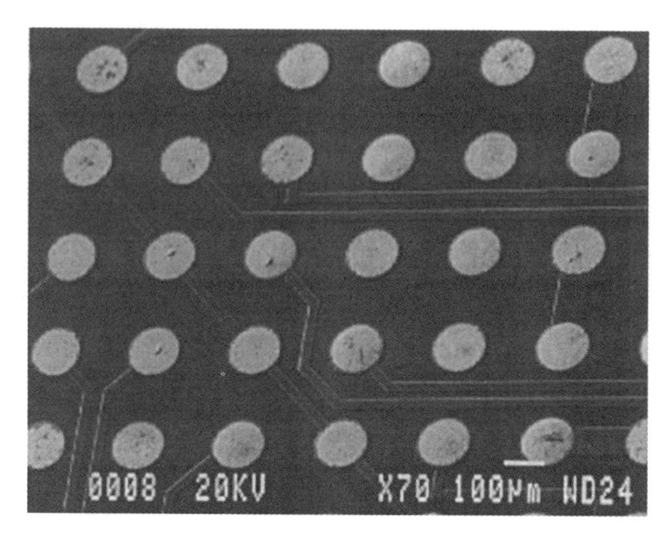

Figure 8-34. Substrate Site After Copper Block Dress

as an issue for new users or developing processes because the joints are hidden and not visible with standard inspection tools. X-radiography [75,79–81] has been used to supplement the optical techniques described earlier. The x-rays can detect misshaped joints, voiding, and low-volume solder pad defects. Acoustic imaging has also been used to detect interfacial debonding of solder bumps to pads [80].

The solder-joint height, usually the chip-to-substrate gap, is also an important factor for thermal-cycle fatigue reliability and optoelectronic device alignment (z axis). This has been measured by Fourier transform infrared spectroscopy [75]. Reflections at the chip and substrate give rise to interference peaks whose separation is a function of the chip-to-substrate spacing.

Several application examples of solder flip-chip assemblies are shown below to illustrate the technology. Chips joined in flip-chip fashion to FR4 are shown in Figure 8-35.

A cross section of these joints was shown earlier in Figure 8-28. The chips used in this demonstration were fabricated test vehicles with 1679 C4 bumps. Each chip contains many unique structures, such as temperature-sensing diodes, Kelvin connections for four-point contact resistance measurements, daisy chains for yield determinations, piezoresistive strain gauges, and reliability test patterns. These structures allow one to fully characterize the joining and rework processes as well as the chip and substrate fabrication processes. A MEMS device (microelectromechanical structure) device assembly is shown in Figure 8-36. The fine movable structures of the thin films are not disturbed by joining process because

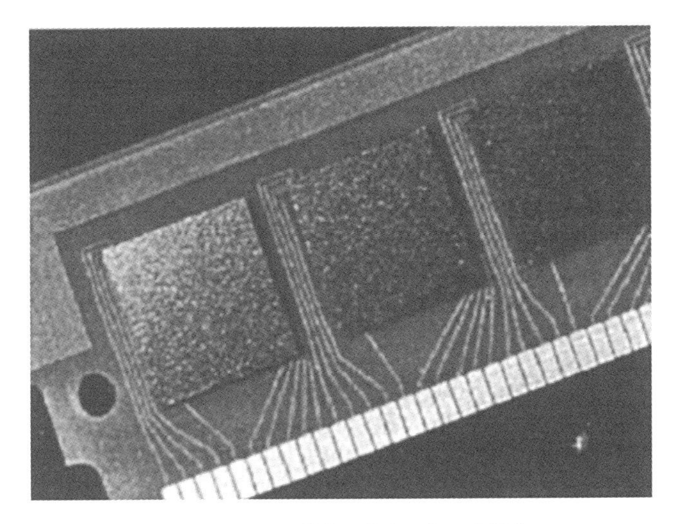

Figure 8-35. Chips Joined to FR4

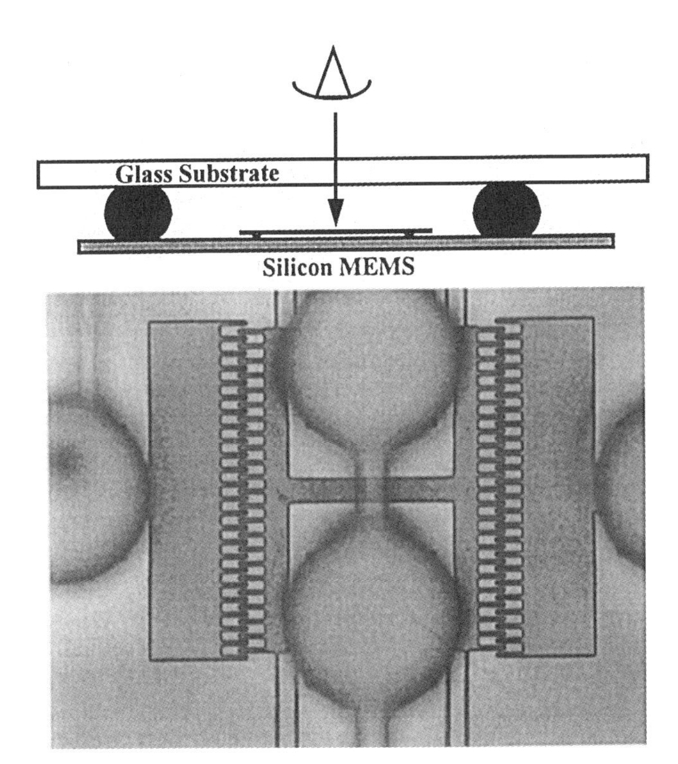

Figure 8-36. Fluxless Flip-Chip Joined MEMS Device

no flux was used in the assembly. A cross section of one of the joints was shown in Figure 8-30 and illustrated the compatibility of the fluxless process with bare copper metallurgy.

An application of flip chip to flexible circuit attachment is shown in Figure 8-37. Here, as well as in the case with FR4 organic substrates, low process temperatures are required to be compatible with the organic materials involved and low-melting eutectic lead tin alloys are preferred.

The joints in Figure 8-37 are under the flex circuit copper leads, therefore not visible. However, they are stronger than the leads which tore from the flex circuit in a pull test evaluation.

The fluxless joining of high-lead alloys in the range 90/10 PbSn to 97/3 PbSn is done at higher temperatures, typically 350°C. An example of a 97/3 PbSn flip chip joined to MoNiAu microsockets on a multilayer ceramic substrate is shown in Figure 8-38. These are fracture surfaces after pull testing indicating ductile taffy pulls which indicate excellent wetting to the MLC microsockets occurred.

8.3.6 C4 Reliability

This subject is covered in detail in Chapter 5, "Package Reliability," but because one of the major limitations of this technology is how large a chip area can be bonded and still remain reliable, a brief discussion is included here.

A question often raised regarding flip-chip bonding is the ability of the joint to maintain structural integrity and electrical continuity over a lifetime of module thermal cycling. A thermal expansivity mismatch between chip and substrate will cause a shear displacement to be applied on each terminal. Over the lifetime of a module, this may lead to an accumulated plastic deformation exceeding 1000% [83]. A quasi-empirical

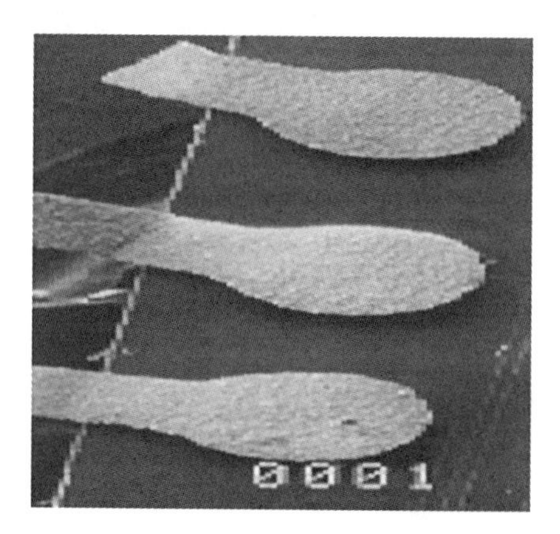

Figure 8-37. Flip Chip on Flex

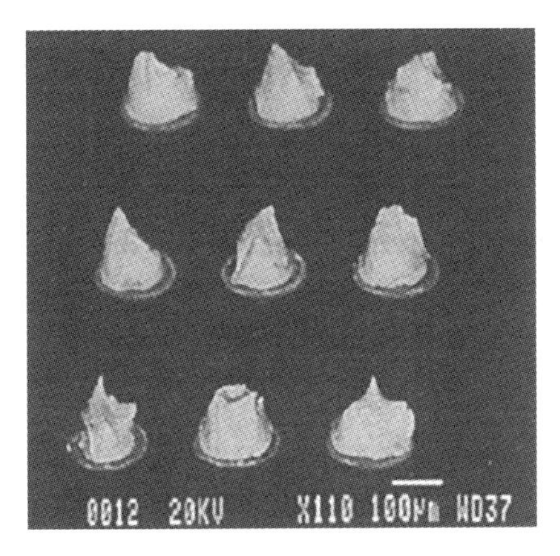

Figure 8-38. 97/3 PbSn Fluxless Flip-Chip Joined to MLC. (From Ref. 82.)

model was developed by Norris and Landzberg [40] that relates the cyclic lifetime to cyclic deformation parameters. The modified Coffin-Manson relationship after Norris and Landzberg is $N_f = \left(\frac{A}{\varepsilon_p}\right)^{1.9} f^{0.3}\, e^{-(0.123/kT_{max})}$. It is based on the Coffin–Manson relationship [84] between fatigue life and plastic strain amplitude but with two terms added to account for time-dependent behavior: a frequency term, where lifetime increases with frequency to a lower power, and a maximum-temperature term, where lifetime decreases with maximum temperature. With the assumption of a lognormal failure distribution, a product sample may be tested in an accelerated thermal cycle; then, based on the statistical lifetime to electrical failure, the projected field lifetime may be extrapolated. Using this technique, an interconnection failure rate projection was made for logic chips in System/370 [40] of no more than 10^{-7}%/1000 h per bond at the end of life. By the end of 1975, 540 billion MST interconnection hours had been accumulated with no wear-out failures reported [85], yielding a 50% confidence-level-estimated failure rate of 1.3×10^{-7}%/1000 h, in good agreement with the earlier projection. In 1991, IBM reported that there had been zero failures due to thermal-cycle fatigue wear-out on the MLC C4 technology over a 12-year period since its introduction in 1980 [86]. This is more than 10^{12} Power-on-Hours.

These results demonstrate the inherent capability of solder interconnections to withstand high-strain accumulations; the results also demonstrate the approximate validity of the projected failure rate. However, the early MST chip was only about 1 × 1 mm in size and had only 12 peripheral bumps. As discussed in Chapter 5, "Package Reliability," the mismatch shear deformation is proportional to the distance between a given pad and the neutral point referred to as dnp (the point on the chip that remains stationary relative to the substrate during a thermal excursion). Because the neutral point is near the chip center, the maximum value of dnp is roughly proportional to chip size. Moreover, the Norris–Landzberg model and most subsequent experiments show that lifetime is inversely proportional to shear deformation raised to a power that approaches 2. Thus, with the evolution from MST to much larger and denser C4 footprints, thermal wear-out was considered with renewed interest.

The wear-out model subsequently has come under close scrutiny. It has been suggested, for instance, that a dwell time factor be added [87]. Also recommended is a complete reformulation starting with the constitutive equation for solder that incorporates crack growth and creep deformation [86]. It has been further suggested that several competing mechanisms come into play in thermal wear-out, including cavitation. Recent experimentation [88] also suggests that wear-out is much more complex than implied by the simple equation discussed above. Scanning electron microscopy (SEM) of joints at various stages of thermal cycling show that mismatch between the solder and chip plays a role in the damage. Also, low and high volumes of solder joints fail by different mechanisms, thus emphasizing the effect of C4 shape.

A further complicating factor in modeling wear-out is that thermally induced strains are not uniform within the joint. An initial attempt [89] was made to incorporate joint shape into failure-rate projections. Despite the simplicity of the model, reasonable agreement has been obtained in experiments where joining geometry has been intentionally varied and the joints mechanically tested by single-cycle or cyclic torquing of the chip [25,89]. More sophisticated techniques, taking into account time-dependent and temperature-dependent solder properties, will be required to understand geometric effects fully. Chip bending also must be included in a complete analysis [90,91].

In summary, simple models that accompanied the introduction of C4 joining have, until now, proven adequate to estimate field behavior and as an aid in product design. This can be attributed not only to their simplicity and their qualitative rationality but also primarily to the fact that existing products have experienced wear-out failure rates too low to be of concern. As chips grow larger and pad counts in the hundreds become common, new or revised models will be required that reflect a greater understanding of the wear-out mechanism and better precision in failure projection.

8.3.7 Thermal Mismatch Reliability: Extensions

Together with a greater understanding must come an extension of the C4 in its ability to accommodate larger and denser chips without affecting system reliability. Existing modeling and testing techniques, however imperfect, have been used to evaluate various extension schemes, several of which show promise. They may be subdivided into strain reduction, geometry or shape improvement, alternate solders, and underfill encapsulant effects.

8.3.7.1 Design and Geometry

The thermal mismatch displacement across the pad can be kept to a minimum by arranging the footprint to minimize the dnp: for example, by deleting corner pads or, in an extreme case, by a quasi-circular array [15].

Joint geometry is dictated by the wetting areas on chip and substrate, solder volume, and the weight of the chip. Unless the chip is very heavy, the joint has the shape of a "truncated" sphere [25,89,92], and its height is uniquely determined by interface radii and volume. The previously mentioned geometric model [89] claims that wear-out depends on shape and that a joint can be geometrically optimized to extend lifetime. Although verifications have been primarily by mechanical testing, with its inherent limitations, the following optimization philosophy and sequence are probably valid:

1. Consistent with other design and process constraints, the interface areas should generally be as large as possible.
2. There is an optimum ratio of substrate-wetting area to chip-wetting area, which must be determined experimentally for each particular material set. For thin-film copper lands on ceramic (discussed in Chapter 9, "Ceramic Packaging"), the ratio is about 1.2, with the substrate pad larger. The optimum ratio is characterized by a roughly even distribution of thermal-cycle fails between chip and substrate (solder crack near BLM or TSM intermetallics).
3. For fixed interface radii, there is an optimum solder volume. A model exists, but because verification has been by mechanical testing only [25], thermal-cycle testing for the optimum is recommended.

The preceding discussion pertains to joints whose shapes are dominated by solder surface tension. Under these conditions, the joint takes the shape of a doubly "truncated" sphere, truncated at each end by the contact metallurgies. Normally, optimization of the spherical segment joint is of limited extendibility value, improving lifetime by less than 50%. Enforced changes in shape away from a spherical segment, on the other hand, can produce very large effects. Heavy chips that depress the joint, causing deep notches [25], severely reduce lifetime, whereas stretched or elongated joints substantially extend lifetime. Mechanical testing has shown an order-of-magnitude difference in fatigue life between an hourglass and a barrel-shaped joint, with the fracture location of the hourglass joint shifted to the center of the joint [93]. Stretched pads have been fabricated by a number of techniques including using two different solders on the same chip [93]. Called SST (self-stretching soldering technology), this technique makes use of the surface-tension forces of larger bumps of one solder to stretch the lower-volume functional bumps (Fig. 8-39). Two additional concepts being pursued in Japan are illustrated in Figure 8-40, wherein solder columns are being stacked to achieve improved fatigue life.

In addition, solder has been cast into helical copper coils to form very-high-aspect-ratio solder columns. It has been applied to joining leadless ceramic chip carriers (LCCC) to glass–epoxy printed-circuit boards (Fig. 8-41). This structure has not been scaled down to sufficiently fine dimensions to be applicable to integrated-circuit chip interconnections. Free-standing cast-solder pillars (without the copper helix) have been developed for package-to-board interconnections [33]. See Chapter 16, "Package-to-Board Interconnections."

8.3.7.2 Solder Composition

Among solders, which have been evaluated as alternatives to 95 Pb/5 Sn, the Pb/In system has shown the most fatigue enhancement. Thermal-cycle lifetime has been shown to be quite sensitive to composition with a minimum at 15–20% In [68,97]. There is a 2 times improvement over 95 Pb/5 Sn at 5% In, 3 times at 50%, and 20 times at 100% In [97]. Pure In is being used for optoelectronic device joining [26,27]. Early work on integrated circuits emphasized 50 Pb/50 In as a compromise between ultrahigh thermal-cycle reliability and processing constraints. Implementation of this alloy was limited by two factors: increased corrosion susceptibility in nonhermetic packages [68,97,98], and a substantially accelerated thermomigration rate over Pb/Sn alloy [99]. By the latter mechanism, the thermal gradient between chip and substrate causes a condensation of vacancies at the chip BLM region, leading to premature high resistance or mechanical wear-out. Later work emphasized low (3–5%) In alloys [68], which provided less fatigue enhancement but were

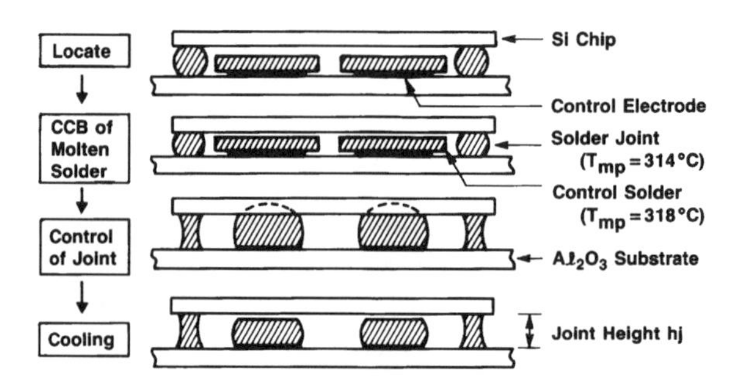

Figure 8-39. Self-Stretching Soldering Technology. (From Ref. 93.)

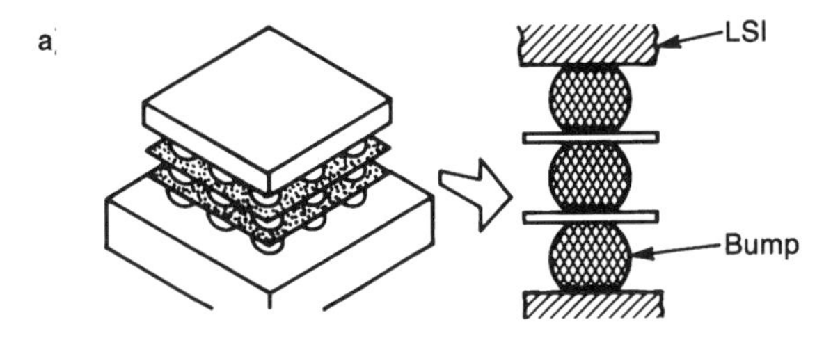

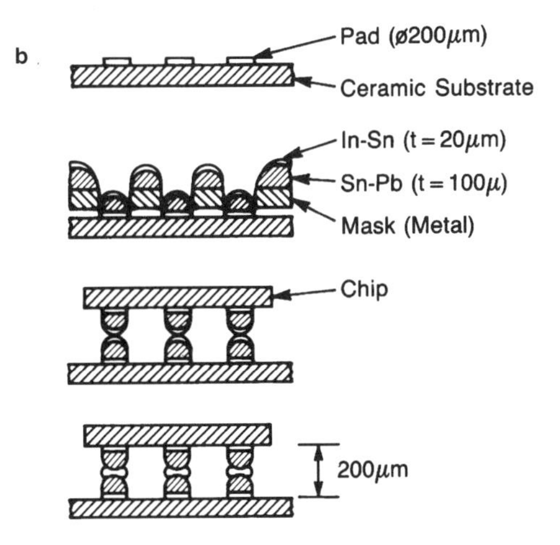

Figure 8-40. Some Methods Being Pursued to Extend C4 Life. (a) Stacked solder bumps using polymide (from Ref. 94); (b) stacked solder bumps using multiple solders (from Ref. 95).

not as susceptible to the corrosion and thermomigration difficulties of 50 Pb/50 In. Testing 95 Pb/5 In at several thermal-cycle frequencies showed a similar relationship to that found by Norris and Landzberg for 95 Pb/5 Sn.

Subsequent optimization of the 95 Pb/5 Sn system showed that a

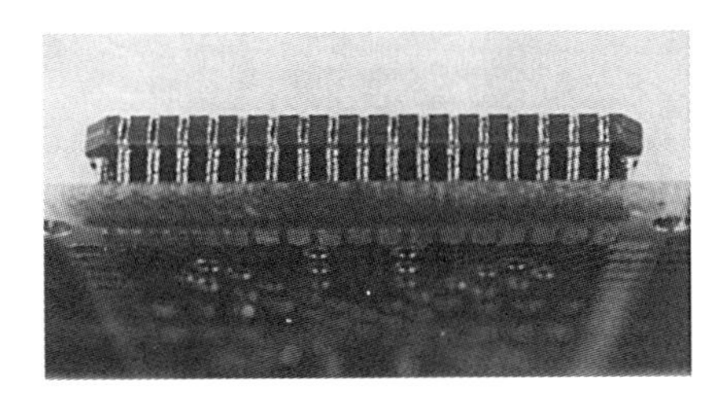

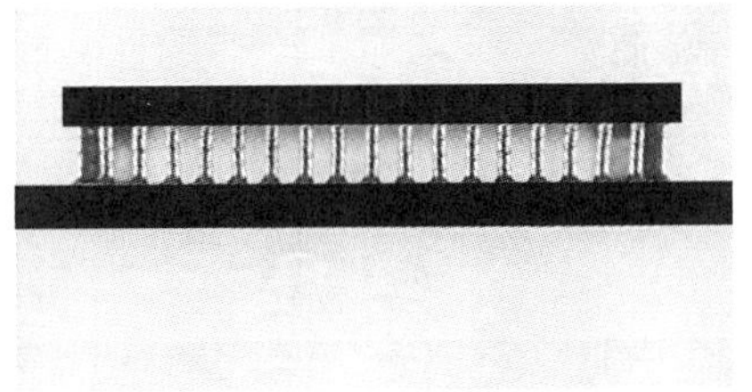

Figure 8-41. Solder Columns. Closeup view of the solder columns connecting a LCCC to a glass–epoxy printed-circuit board. (From Ref. 96.)

97 Pb/3 Sn composition also brought nearly a 2× improvement in fatigue life over 95 Pb/5 Sn without the thermomigration and corrosion concerns of the Pb/In alloy. Therefore, when IBM System 3080 was introduced, the solder composition was altered to realize the increased reliability limits. At about the same time, it was appreciated that tinning metallized ceramic Cr–Cu–Cr C4 pad tips was deleterious to fatigue life because the solder used had 10% Sn—the wrong Sn direction for long fatigue life. A process was therefore invented for putting solder balls on pin heads only, leaving the C4 sites bare [79]. All the C4 solder then came from the chip only, and both fatigue life and chip joining yields were simultaneously improved.

8.3.7.3 Substrate Materials

By far, the largest effort to reduce the strain drastically has been by matching the substrate thermal expansion to that of silicon, as demonstrated in Figure 8-42.

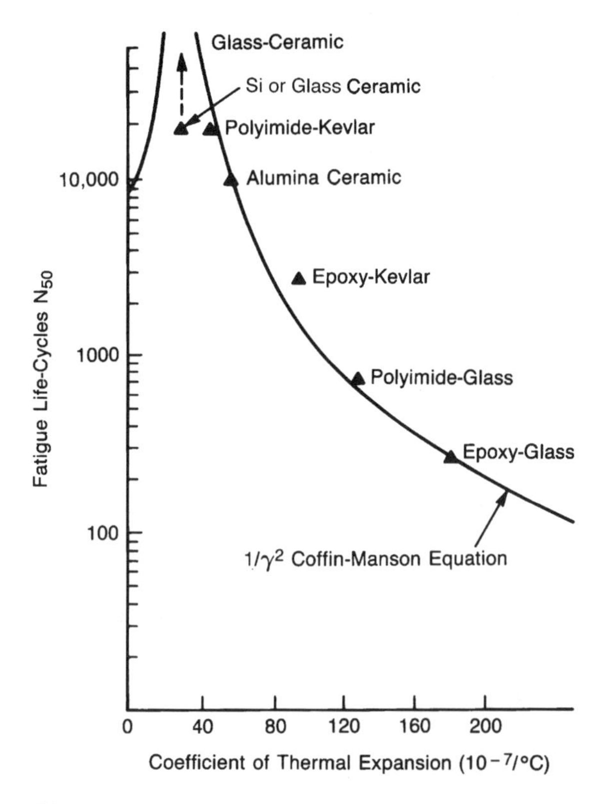

Figure 8-42. Effect of Thermal Expansion Coefficient of Substrate on C4 Fatigue Life. (Modification of Figure from Ref. 16.)

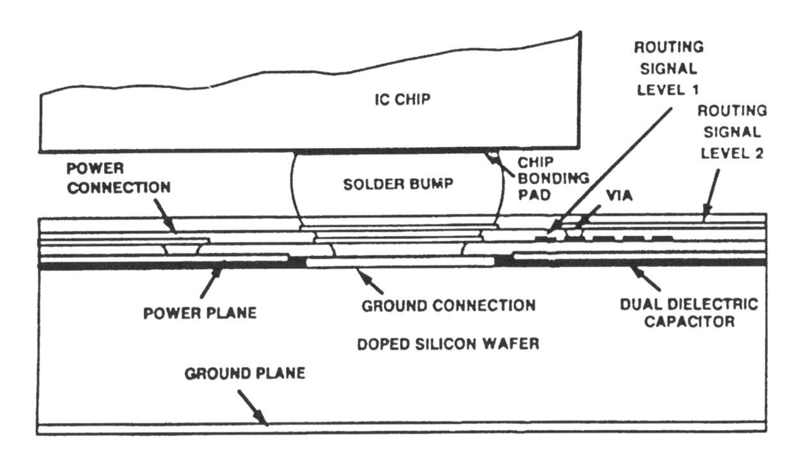

Figure 8-43. Silicon Flip-Chip Joined to Silicon Carrier with Thin-Film Interconnections. (From Ref. 100.)

Pure Si substrates with multilayer Al and polyimide or SiO_2 insulators have been fabricated and demonstrated. Figure 8-43 illustrates a typical Si carrier system [100]. An application of such a system is shown in Figure 8-44 which is essentially the core of an IBM RISC 6000 workstation on a single piece of silicon. There are two weaknesses in such a system: (1) The lack of vias through the silicon means that all wiring must lead to the perimeter for wirebonding or TAB bonding, and (2) thin-film Al

Figure 8-44. Workstation (IBM RS 6000) on Silicon Carrier. (Courtesy of IBM).

wires are more resistive and longer than multilevel Cu wiring on ball-grid-array-type substrates. Therefore, this approach typically falls short in optimum performance.

Glass–ceramic substrates in which crystallizable glass, cordeirite-like ceramic has been used as the carrier also matches silicon CTE near perfectly. The multilayer structure with punched and filled vias and Cu wiring allows short paths in good conductors to reach a pin-grid-array configuration on the bottom [21]. This approach works well but is probably too costly for many applications. Also, the mismatch in the TCE problem is now transferred to the package-to-board interface and would have to be solved to avoid premature failure of those joints (e.g. soldered BGA joints).

An early effort using a polyimide–Kevlar® organic substrate [16] with near-matching TCE has been followed by a number of packaging efforts that take advantage of the improvement in thermal fatigue lifetime of the C4 connections. The use of AlN and SiC as first-level packages (as discussed in Chapter 18, "Coated-Metal Packaging") were actively pursued for direct chip-attach applications. Gallium arsenide has also been matched using sapphire [26] or Al_2O_3, which matches TCE's well.

Power cycling is complementing thermal cycling to evaluate these material combinations [91,101,102] because it is more realistic in simulating temperature differences between the chip and substrate. Process constraints, wireability, dielectric constant, and heat dissipation must, of course, be among other factors considered in selecting an alternative substrate material.

8.3.7.3.1 Direct Chip Attach (DCA) to PWB with Underfill

Perhaps one of the most innovative developments recently has been the discovery that certain polymers such as epoxies heavily filled with particles of SiO_2, to match the solder TCE and bond chip to substrate, will increase the fatigue life of the solder by a factor of 10–100 [31,103]. Nakano (Fig. 8-45) showed this in 1987 [103], and the work was applied by Tsukada [34] in attaching silicon chips directly to thin films on FR4 in a product called surface laminar circuitry (SLC). For the first time, Si chips with low expansivity could be made reliable in direct flip-chip attachment to high-expansivity FR4 PWB. Some of the first products to use such technology commercially are IBM PCMCIA modules for portable computers (Fig. 8-46) and Motorola pen-sized pagers (Fig. 8-47).

The technology appears to work well because the encapsulant shares and reduces the all-important solder strain level to 0.10–0.25 of the strain in joints which are not encapsulated. This has been shown by finite element analysis calculation (Fig. 8-48) [34] and by Micro-Moire experimental analysis (Fig. 8-49) [104]. The latter shows the strain to be 10% relative to the solder of the unencapsulated solder and indistinguishable in solder or filler. Both ceramic and polymeric flip-chip substrates benefit from this underfill enhancement.

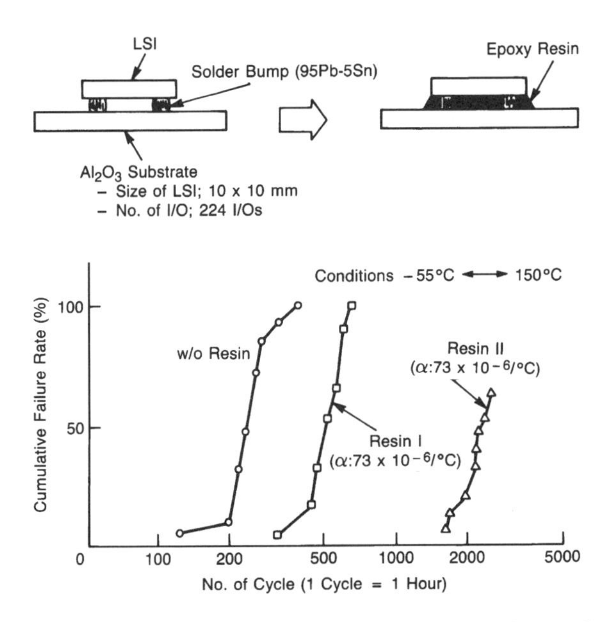

Figure 8-45. C4 Life Extension by the Use of Thermal Expansion Matched (to Solder) Resins. Resin II is assumed to include filler to match solder CTE of 27 ppm. (From Ref. 103.)

Perhaps the only negatives with the underfill process are the difficulty with reworkability and the slowness of fill and cure. All of these things are being investigated and developed. Cleavable epoxides have been demonstrated. Thermal degradable encapsulants show promise. Until these materials become qualified, the option remains practical to apply the encapsulant only at the end of processing, after test, when it is most probable that the chip is good.

8.3.7.4 Other Reliability Issues

Among other reliability concerns that have been reported in the literature, thermomigration has been shown to be a major concern for applications where high-temperature gradients are coupled with high-diffusivity, low-melting-point solders [99]. Corrosion has been encountered in high-humidity testing of PbIn [98]. Solder void defects are addressed in Ref. 85. Palladium depletion of AgPd thick films by PbSn can allow Ag corrosion and prompted a change to AgPdAu ternary alloy for the substrate electrode [105,106].

8.3.7.5 Alpha-Particle Emission

A more fundamental problem receiving increased attention is that of soft errors in devices caused by alpha-particle emission of trace quanti-

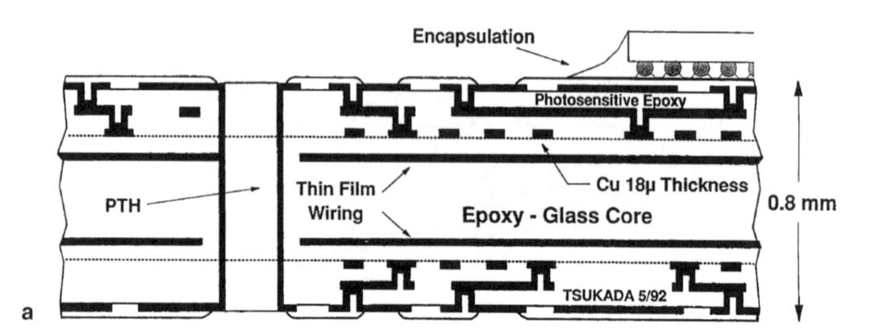

Figure 8-46. **(a) SLC Cross Section Showing Thin-Film Surface Layer Wiring on FR4 Base and Underfilled Flip Chip. (b) SLC Token Ring Adapter Card Compared to Regular Surface Mount Technology.** (From Ref. 34.)

Figure 8-47. Motorola Pen-Sized Pager

ties of radioactive materials in the packaged assembly [36,107,108]. In the case of C4, high-lead alloys are mostly used for computer applications, and lead almost invariably brings with it trace amounts of uranium and thorium. This problem will demand more attention as VLSI devices become more dense and as the critical charge levels in the device become smaller. U and Th contaminants have been successfully removed from

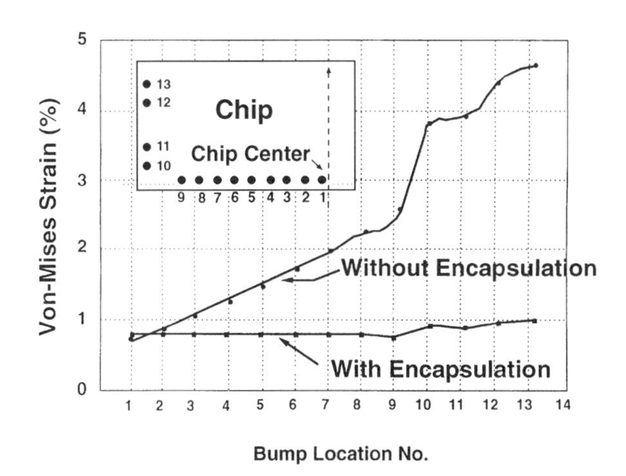

Figure 8-48. Finite Element Analysis of Strain on Chip C4's for SLC. (From Ref. 34.)

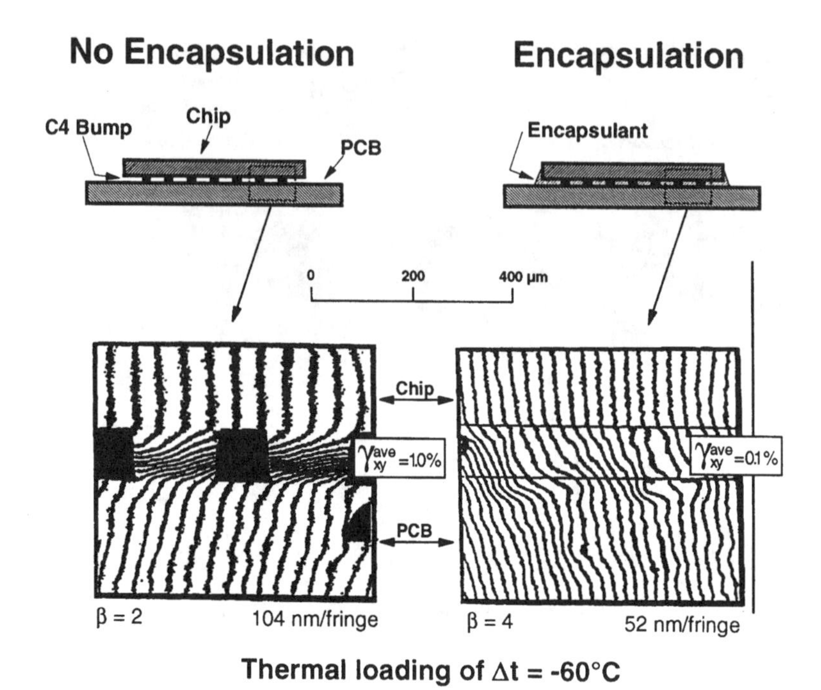

Figure 8-49. Micro-Moire Interferometry of Underfilled Chip's Showing Stain Reduction and Distribution. (From Ref. 104.)

Pb. There is even less of a problem with Sn. However, there are isotopes of Pb, ^{214}Pb and ^{210}Pb which are a part of the ^{238}U radioactive decay chain which themselves decay to Bi and Po on the way to ^{206}Pb, the ultimate, stable form of Pb. It has been discovered that some naturally occurring deposits of Pb are very low in the alpha-producing isotopes and therefore have several orders of magnitude lower alpha radiation [109]. Isotope separation of other deposits or secondary Pb were thought to be prohibitively expensive, but are still under active consideration [110].

8.3.7.6 Burn-In/Known Good Die for Flip Chip

C4 chips can be given a burn-in cycle on a temporary substrate to improve chip defect-related reliability [111]. To facilitate the removal of the chip from the temporary substrate after the burn-in, IBM has used a reduced radius removal (R3) process. As shown in Figure 8-50, the substrate pad is reduced in size to approximately one-fifth of the area of a functional terminal. After joining and burn-in, the chip is mechanically torqued off the temporary substrate, with the fracture being in the solder adjacent to the small-sized terminal. Because most of the solder remains with the chip, there is no need to dress or rework the chip except for a

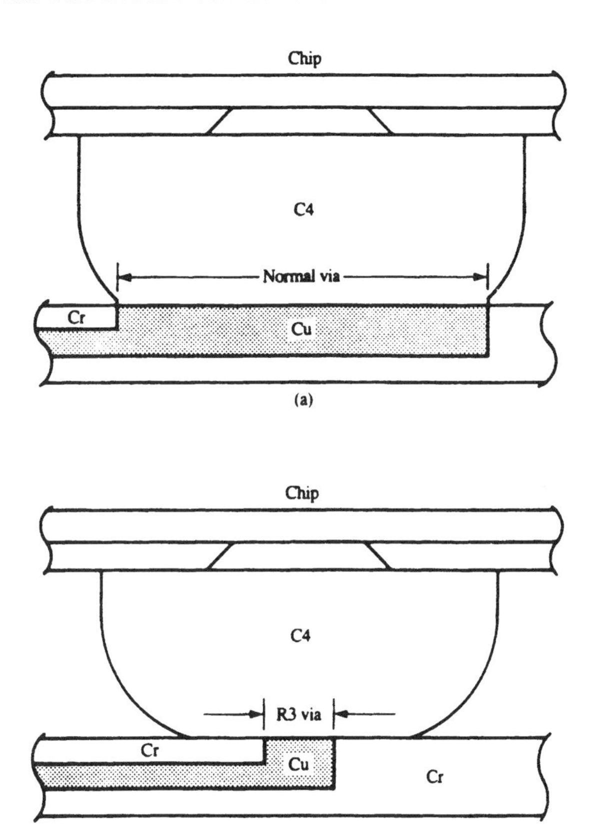

Figure 8-50. R3 Burn-In Methodology for C4 Flip Chips Featuring Solder Attach to Greatly Reduced Substrate Pads for Burn-In. (From Ref. 111.)

reballing in a dry H_2 reflow furnace. The process has been verified on 10,000 CMOS memory chips in the field, with early life-defect reliability being improved by more than a factor of 30. It should be understood that the defects referred to are not C4 defects, but internal defects in the IC chip induced to fail during the burn-in cycle instead of being allowed to fail in the field application. The temporary substrate is reusable—a sample substrate was used repeatedly for over 20 cycles.

8.3.7.7 Heat Dissipation with C4

Heat dissipation is covered in great detail in Chapter 4, "Heat Transfer in Electronic Packages," and only a brief discussion is included here. Due to the very short length and large area of contact of a C4 joint, the thermal path through the interconnection itself is very low and far superior than for TAB or wirebond. Typical dimensions are a 125-μm diameter and a 60-μm "length" for C4 as compared to a 25-μm diameter and a 2500-

μm length for wirebond, and a 50-μm square and a 1750-μm length for TAB. The chip-to-substrate resistance is so low for C4's that for many years it was not necessary to supplement it at all. Early chips with 10–20 peripheral C4 bumps bonded to an alumina substrate could dissipate approximately 0.5 W [112]. An air-cooled module with six 4.5-mm chips, each having an 11 × 11 array of solder joints, can dissipate approximately 1.5 W per chip [8]. Somewhat higher power levels have been achieved with new, high-thermal-conductivity ceramics such as AlN [113] and SiC [91]. Numerical analysis techniques are required for reasonable performance projections because the thermal path for flip chips depends on device location, size, metallization, number of terminals, and the thermal resistance of the substrate [112,114].

Today's high-power-level devices, however, require a supplemental heat path through the back of the chip. Die bonding accomplishes this for wire or TAB-bonded chips. In the C4 case, because the back of the chip is free of mechanically or electrically delicate surface features, it is amenable to direct contact by a wide variety of heat sinks, thermal greases, or solders, whose thermal conductivity is often better than the plastic or ceramic package to which they are backbonded. An example is the IBM multichip module [114,115] where spring-loaded pistons transfer heat from the back of each chip to a water-cooled plate, augmenting the traditional solder-joint thermal path. Four watts per chip or 300–400 W for a 100-chip module can be dissipated. This has been enhanced in the S/390/ES9000 series computers to a 16.7-W/cm^2 capability [115].

Full bonding to the back of the chip entails more risk for C4 mounted chips because the solder joints, being very short, are not as compliant as wires and the designer must be concerned with possible solder-joint fatigue. Bonding to the back of a C4 mounted chip may impose additional forces on the C4's and lower the fatigue life (see Chapter 5). Designs have been proposed which combine a solder backside die bond with C4's to give extremely low thermal resistances—on the order of 0.4°C/W for chips joined directly to a water-cooled plate in a MCM [41,42,44]. This would give a power capability of over 100 W per chip. Such power levels have been projected for the year 2000 [41]. Hitachi has implemented a slightly less efficient single-chip module version of this design (Fig. 8-51) in their M880 Processor Group computers [116]. Chip to water thermal resistance is 2°C/W, not quite as good as the multichip design because more interfaces and layers are between the chip and the flowing water. Various conduction cooling designs for high-power C4 mounted chips have been compared (see review by Darveau [117]).

Finally, liquid immersion cooling [117] and cryogenic applications [28,29] have also been demonstrated for C4 interconnected structures. The latter application, for Josephson devices that operate at 4.2 K, was especially noteworthy, because all of the materials were subjected to a

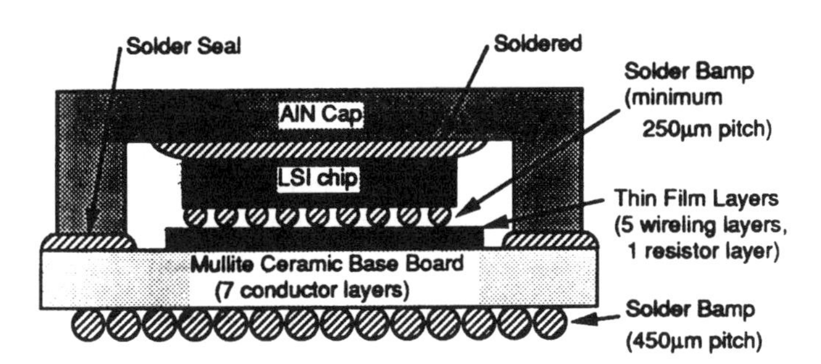

Figure 8-51. Micro Chip Carrier in Hitachi M880 Mainframe; Enhanced Heat Dissipation Through Back-Side Soldering and High-Density C4 Area Arrays on Chip and Package. (From Ref. 116.)

very large ΔT between room temperature and operating temperature and because the materials' intrinsic properties, such as resistance to cracking, are quite degraded at low temperatures. Orthogonal solder connections are also used to join silicon slices at right angles to each other. For such a joint, matched expansion materials are required [28–30]. Compatibility of standard 95/5 Pb/Sn C4's in liquid-nitrogen environments for CMOS applications has been demonstrated on alumina ceramic packages [118]. Parylene coatings have been shown to be effective in lowering the thermal shock during transfer into and out of the liquid nitrogen [119].

8.3.8 C4—Future Trends

As VLSI proceeds to denser chips, C4 densification will follow. Area-array structures will dominate over traditional peripheral devices. Bumps of 25 μm, resulting in over 10,000 pads per chip in experimental devices, have already been fabricated [27], ahead of the predictions of the SIA roadmap. Photolithographic processing will be preferred for such devices, and with that, area arrays will become more and more common. Significant activity in matched-expansion substrates, or alternatives that alleviate the fatigue limitations such as under chip fillers are being actively applied. High-thermal-conductivity packaging materials and innovative configurations to improve heat dissipation of the devices are also being used and extended. More attention will be given to defects, for both yield and reliability needs. Efforts to purify all packaging materials for low alpha emission or substitute new ones which are inherently better will continue.

The range of applications for C4-like structures is becoming much wider, given the more diverse materials used in new electronic devices and substrates (for example, In C4's on GaAs chips on sapphire substrates for optoelectronics) [26], chips C4 bonded to flexible circuits [120] and printed-circuit boards [121]. The latter developments suggest a future, strong presence in the commodity world as well as high-end computers.

8.4 WIREBONDING

Wirebonding has totally redeemed itself in reliability, manufacturability, and cost from where it was in the earliest era of semiconductor devices. It is still the dominant chip-connection technology in the semiconductor industry. Virtually all dynamic random access memory (DRAM) chips and most commodity chips are still in plastic packages with wirebonds. Therefore, because the vast majority of all integrated-circuit (IC) packages are assembled by wirebonding [122], it is important to understand this technology. This section explains the basic wirebonding process, emphasizing the metallurgy, materials, and typical applications. Guidelines are provided for selecting the appropriate methods for various package types and optimizing these methods.

Wirebonding differs from C4 and TAB techniques in that electrical package connections are created at this assembly stage by attaching a fine wire between each device I/O and its associated package pin (one at a time), as opposed to attaching prefabricated patterns of interconnecting materials. This may not seem advantageous for wirebonding, but it is actually the basis for one of its biggest strengths: flexibility.

Wiring changes are easily accomplished without the need for expensive tooling and material changes. A new "bond program" can be taught/saved in a matter of minutes, followed by immediate production of the new devices, saving time and money.

The speed and cost-effectiveness of changes also applies to the initial package design. Chip preparation cost is significantly lower than C4 and TAB devices because the wirebond welds are made directly to the chip (wafer) metallization, saving substantial tooling and production costs involved in the bumping process. Further tooling savings can be realized through the use of a common package or substrate which can be used for a number of devices using various wire configurations/lengths. Therefore, flexibility and cost are two of the primary reasons for selecting wirebonding.

A third reason for the prominence of the wirebonding process is much improved reliability. In the early 1990s, 1.2–1.4 trillion wire interconnections were produced annually (roughly 2.6 trillion welds). Manufacturing losses and test failures are extremely low (40–1000 ppm) and trending downward each year. High yields and reliability rates continue to be achieved around the world in a large variety of facilities, from completely automated, mass production factories making millions of packages per week, to single-room facilities producing low-volume, custom parts.

Today, a global network of manufacturing people, vendors/suppliers of equipment, and materials comprise a large wirebonding knowledge base. The infrastructure is vast and tends to keep its user base unless there are extremely strong reasons to deviate to a different technology.

Wirebonding is popular as a result of continuous process improvement achieved through the development of sophisticated, automated equipment. Today's production lines can assemble the latest generation of packages with the evolutionary derivative of the welding technique used to manufacture the first transistor. It is the fundamental Au to Al interconnection technology that has continued to grow with the semiconductor industry. Wirebonding appears to be the method for connecting the majority of packages in the near future, with C4 and TAB providing the techniques needed for specialty packages and high-end, high-pin-count devices.

Today, hundreds of diverse plastic and ceramic packages utilize wirebonding. Despite this diversity, commonalities do exist. First, the chip is normally mounted with the back side attached to a metal frame or substrate, although there is at least one noteworthy exception where the metal frame/leads are attached to the top side of the chip, called lead-on-chip (LOC) [123]. This mounting process, called die bonding or die attach, is achieved by a variety of conductive epoxies or solders. The chip and substrate assembly are then presented to the wirebonder. The wires are attached to the chip and package substrate, one at a time, using either a thermosonic or ultrasonic welding process. (A third method, thermocompression welding, may also be used, but it is not commonly used.) State-of-the-art bonding technology produces two welds and a precisely shaped/routed wire loop in 100–125 ms, depending on the process used.

In factories throughout the world, thousands of automatic machines produce wirebonded packages with little, or no human intervention. In each machine, die attached parts are delicately transported and presented to the bonding area. Key features of the chip and package carrier are located by a video camera and recognition system using artificial intelligence to identify individual targets for the intended wires. As the welds and interconnecting loops are made, the machine uses closed-loop servo systems to ensure that each step of the process is being performed as intended. After all the connections are made, devices are then transported to an output station where they are staged for the next operation (usually plastic encapsulation or application of a lid to the hermetic devices).

8.4.1 Joining Technology

Most wirebonding processes combine either thermosonic (T/S) or ultrasonic (U/S) welding methods, with two different bonding methods for applying the wire to the chip and to the package surfaces. The two basic wirebonding techniques are ball bonding and wedge bonding. Approximately 93% of all semiconductor packages are manufactured using ball bonding method, while wedge bonding is used to produce about 5% of all assembled packages. Both ball bonding and wedge bonding were once pure thermocompression (T/C) bonding, but because of the high bonding temperatures required (approximately 350°C) have changed to lower-temperature (150–200°C) T/S Au ball bonding and room-temperature Al wedge bonding.

Ball bonding. In this technique (Fig. 8-52), wire is fed vertically through a tool called a capillary. The wire is heated to a liquid state with an electronic spark discharge called an electronic flame-off (EFO). The surface tension of the molten metal forms a spherical shape, or ball, as the wire material solidifies, hence the process name "ball bonding."

The capillary descends and positions the ball on a bond pad of the chip using the bonder's video and servo systems. Thermosonic welding is then performed. Once a weld is completed, the capillary rises and follows a prescribed trajectory calculated for each wire. A precisely shaped wire connection called a wire loop is created as the capillary descends to a target position for the second bond. A second weld, a crescent bond, is now formed, with the wire extending through the center of the capillary, under its face, and out to the newly formed loop. The finished weld has a crescent or fishtail shape made by the imprint of the capillary's outer geometry. The capillary ascends once again, breaking the wire at its thinnest point near the bond and, simultaneously, feeding an exact wire length needed for the EFO to form a new ball to begin bonding the next wire. Figure 8-53 shows typical ball and crescent bonds, as well as loop shapes.

Wedge bonding. Named for the shape of the bonding tool used in the process, this technique feeds wire through a hole in the back of a bonding wedge at an angle usually 30°–60° from the horizontal chip surface (Fig. 8-54). As the wedge descends onto the IC bond pad, it pins the wire against the surface and performs either an U/S or T/S weld. Next, the wedge rises and executes a series of calculated motions, creating a desired loop shape.

In addition to vertical and horizontal motions needed for ball bonding, a rotational axis (called θ) is required to align the package so that the wire about to be bonded is on the same axis as the bonding wedge feed hole. This allows the wire to feed freely through the hole in the wedge during the loop formation. At the second bond location, the wedge descends to make the second weld. A mechanical clamp is closed behind the wedge and the clamp is moved to the rear to tear the wire. As the wedge ascends, the clamped wire is fed under it to begin bonding the next wire. Figure 8-55 shows examples of first and second wedge bonds.

8.4.2 Wirebonding Joining Mechanism

Like other technologies that join together two materials, some of the exact metallurgical mechanisms of joining in wirebonding are still

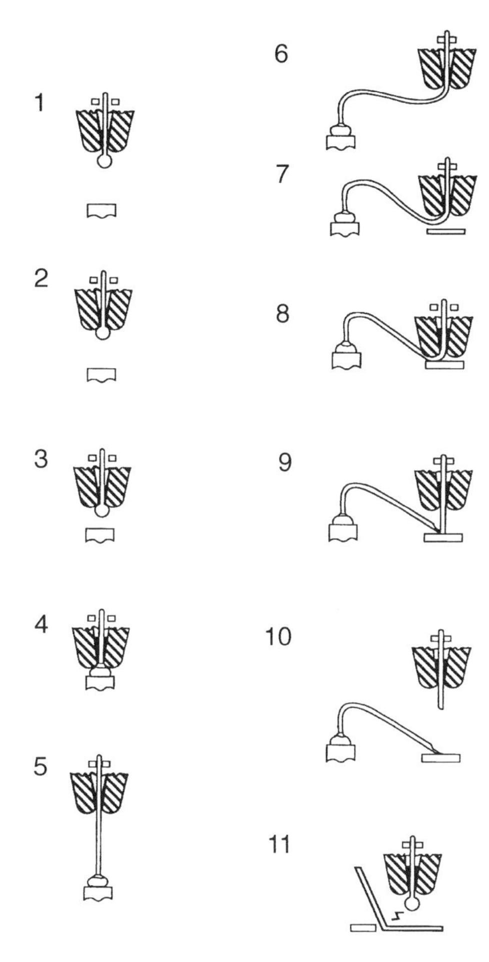

Figure 8-52. Ball-Bonding Steps to Complete One Cycle. 1. Ball is formed below capillary (note that it should not contact the capillary during the formation process). 2. Capillary descends, capturing/centering the ball so that it is seated in the capillary's inner chamfer. 3. High velocity is reduced to contact velocity at a programmed distance above the chip. The point where this velocity transition occurs is called the tool inflection point (TIP). 4. Bond force and ultrasonic energy is applied to form the ball bond. 5. The capillary ascends vertically to pay out sufficient wire to form the loop of wire between bonds. The wire clamps are typically closed at the highest point, prior to movements toward second bond. 6. High-velocity motions form the wire shape as the tool moves over the second bond site. 7. Vertical motions are slowed to contact velocity at the TIP above the lead. 8. The crescent bond is formed through the application of force and ultrasonic energy. 9. The capillary ascends a prescribed distance to the pay-out wire (called a tail). The material contained in this wire length will be used to form the correct ball size. The wire clamps are closed again when this motion is completed. 10. The capillary moves vertically to break the bottom of the tail (tack welded to the lead). 11. The capillary and the EFO wand/electrode are brought in close proximity while a spark discharge provides heat to form a molten ball.

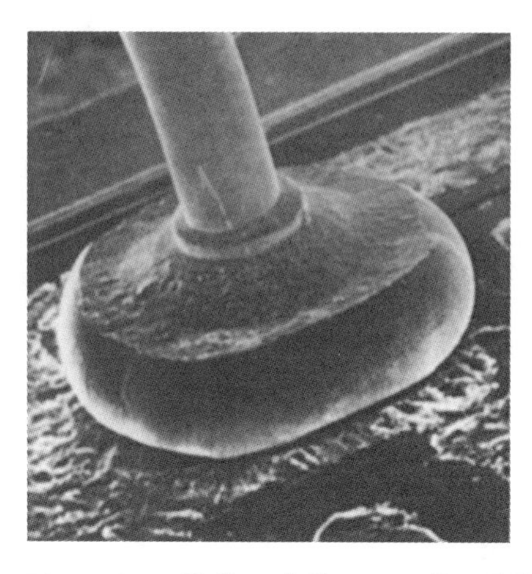

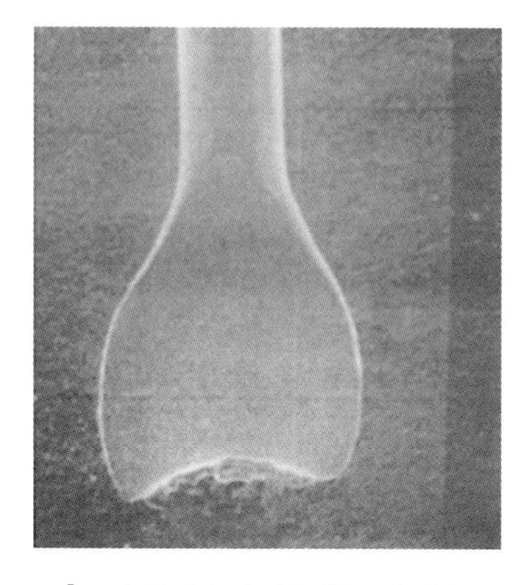

Figure 8-53. Ball and Crescent Bond Examples. *Left:* A typical ball bond without fine pitch restrictions (25-μm wire with 80-μm bond squash). Note the small amount of material around the wire that was extruded into the capillary bore. Examination of the top surface reveals the shape of the tool's inner chamfer and face. Surface roughness is typical of "secondary bonding" that causes the ball to adhere to the capillary tip while the primary bond is being formed between the ball and chip. *Right:* A typical crescent bond with the familiar fishtail shape. Small irregularities at the bottom edge of the bond are remnants of the tack weld that temporarily attached the wire tail before it was pulled away. (Courtesy of K&S.)

subject to discussion. Methods for attaching metals include a range of processes from adhesive metal to metal bonds to metallurgical mechanisms (diffusion bonds with intermetallic compound formation) and liquid metal brazing. Often there is a combination of mechanisms which results in "bonding." The following section will summarize the most accepted views about U/S and T/S wirebonding which have emerged.

8.4.2.1 Ultrasonic Welding

This method of welding is most commonly used to join aluminum wire to a variety of package materials, including the aluminum metallization found on the chip, as well as thick- and thin-film package metals consisting of gold, silver, nickel, or palladium. The bond is made at room temperature without an applied heat source. It is accomplished by combining two primary elements: downward force and horizontal oscillatory motions.

The frequency of these oscillations range from 60 to 120 kHz, hence the name "ultrasonic." The downward clamping force is continually applied to the bonding wedge to assure that the ultrasonic motion of the tool remains in contact with the wire as it plastically deforms/yields to the applied energy. A weld results from this intimate contact between the

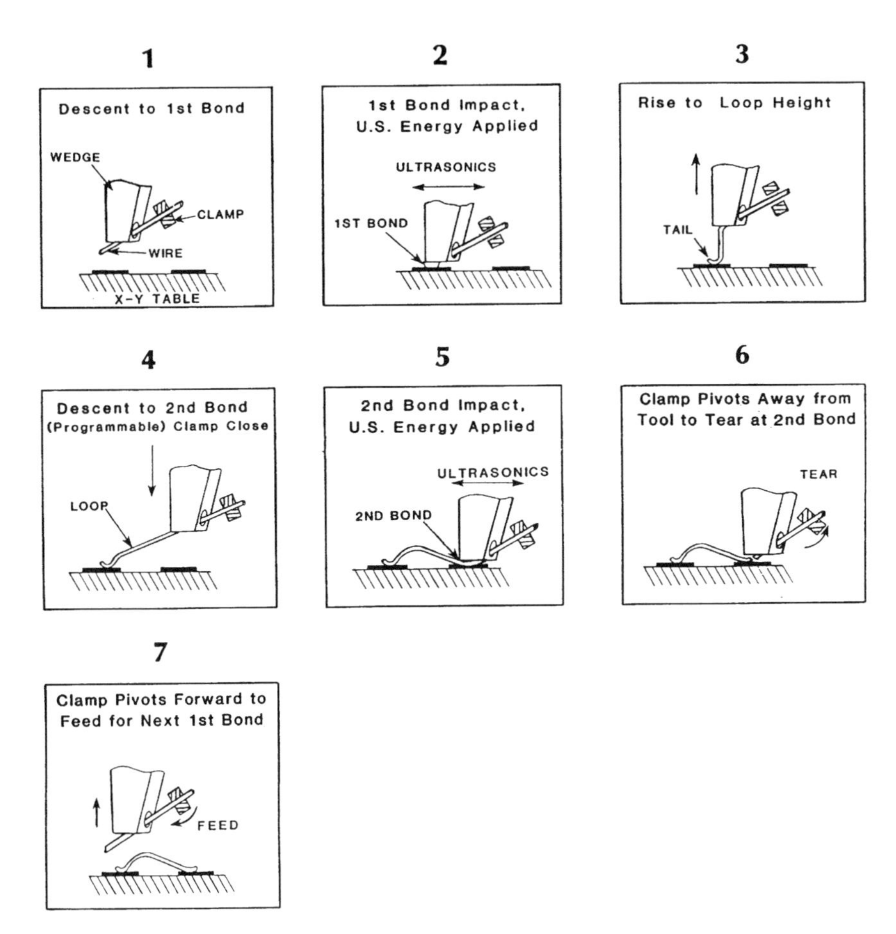

Figure 8-54. Wedge-Bonding Steps to Complete One Cycle. 1. The wedge descends at high velocity (not shown) and makes transition to contact velocity at TIP. The wire clamps are closed to assure the wire does not move under the tool. 2. The first bond is formed by application of bond force and ultrasonic energy. 3. The clamps are opened and the wedge ascends vertically, then it moves horizontally to a position over the second bond site. 4. The clamps are closed again and high-speed motions move the wedge downward until it reaches the TIP distance above the package. 5. The second bond is formed. 6. Articulated clamps move away from the wedge to break the wire at the back radius of the tool. 7. The wedge ascends from second bond (not shown) while the clamps push the wire through the wedge feed hole to provide the wire for the next bond cycle.

wire and the surface in less than 25 ms. The ultrasonic energy aids in the wire deformation and the breakup of the hard aluminum oxide on the surface of the bond pad.

Although other wires can be used, aluminum appears best suited for the U/S technique due to the thin, naturally occurring aluminum oxide layer on its surface. The ultrasonic movement of this abrasive material

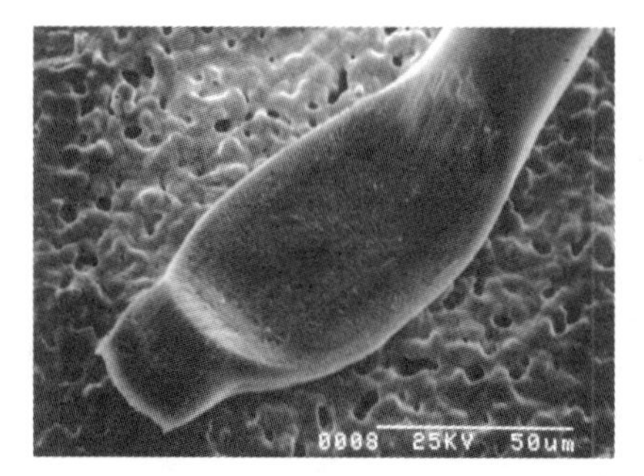

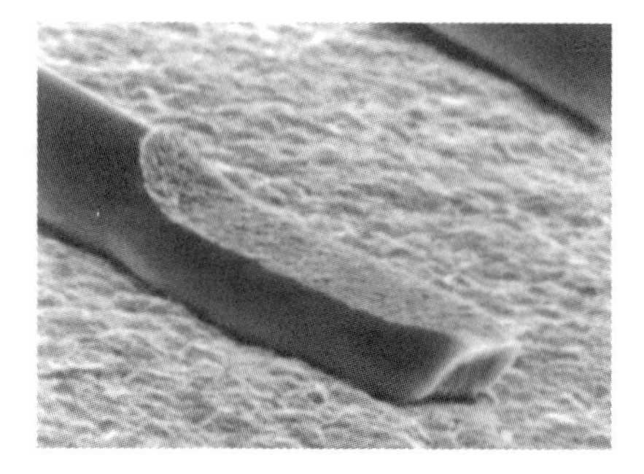

Figure 8-55. Aluminum Wedge Bonds. *Left:* Typical first bond with tail; *right:* Typical second bond.

removes surface contaminants/oxides on the mating surfaces to expose pure metal for the joining process.

In 1978, Winchell and Berg published their findings on ultrasonic bond development [124] that is still a definitive paper for ultrasonic wirebonding. Briefly summarized, they conclude that an ultrasonic bond occurs from the material flow associated with several static and dynamic forces applied by the wedge-bonding machine.

First, the aluminum wire is softened and plastic deformation occurs when ultrasonic energy is applied horizontally by the bonding wedge. It is interesting to note that ultrasonic energy needed for the elongation of aluminum is about 100 times less than the heat energy required for the same result [125]. In fact, no external heat is required, nor is a large amount generated internally by the movement of plane dislocations.

Experimental studies show that the temperature at the bond interface only reaches 70–80°C, which is too low for the bond to support a thermal-diffusion process. The softened wire is then deformed by the downward force of the bonding wedge. Although some surface oxides may be removed by this initial plastic flow of the wire over the substrate surface, intimate contact (and the resultant bonding) do not occur. The amount of deformation, or bond squash, is a result of this first stage.

As the bonder's vertical force and stress is distributed over the expanding wire surface, it decreases with time and distance from the center of the wire (Fig. 8-56). The vertical stress in the center is higher than the horizontal stress applied by the ultrasonic energy. Therefore, no wave formation is seen in this region.

Near the periphery of the wire, the horizontal stress applied by the wedge is greater than the downward stress, so wave motions are observed in this area (Fig. 8-57). This explanation coincides with the actual ultrasonic bond region observed as an ellipse with a central, unbonded area (Fig. 8-58) and refutes the theory of bonding by friction or a sliding

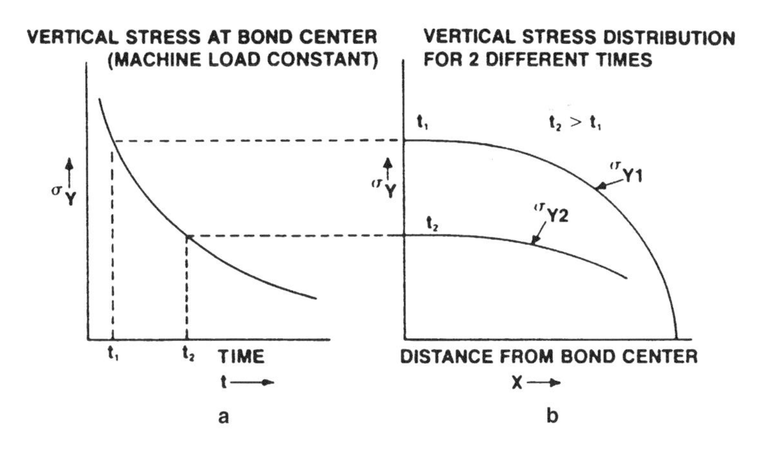

Figure 8-56. Vertical Stresses During Wedge Bond Formation. (a) For constant machine load; vertical stress decreases with time because of increased contact area; (b) magnitude of vertical stress also decreases with distance from bond center. (From Ref. 124.)

model. (Both of these cases predict preferential bonding at the center of the wire.)

The peripheral material that is subjected to the wave motion generates a strong cleansing action and pressure at the substrate. If aluminum wire is bonded directly onto a hard material like silicon, the wave motion is great enough to permanently deform the surface into ridges and troughs that are perpendicular to the direction of the applied energy of the wedge. The depth and distance between these features is directly related to the amount of power applied to the wedge. Along the sides of these valleys, two atomically clean surfaces are brought into intimate contact. Both the wire and substrate have surface atoms that are free to interact. Therefore, high-strength atomic bonding takes place to satisfy the atoms of both surfaces. The result is a continuous, metal lattice structure in these areas, which is equivalent to a single homogeneous material.

8.4.2.2 Thermosonic Welding

This method of welding is most often used with gold or copper wire. Although the technique uses the same ultrasonic energy application, there are two notable differences between this process and the U/S process—the need for an external heat source, and wire material without an abrasive oxide surface layer. Because gold and copper will not form an acceptable weld at room temperature in a reasonable amount of time, heat is applied to provide the activation energy levels to the materials so there is an effective joining and intermetallic diffusion. Typical bonding temperatures range from 150 to 200°C, with bonding times from 5 to 20 ms.

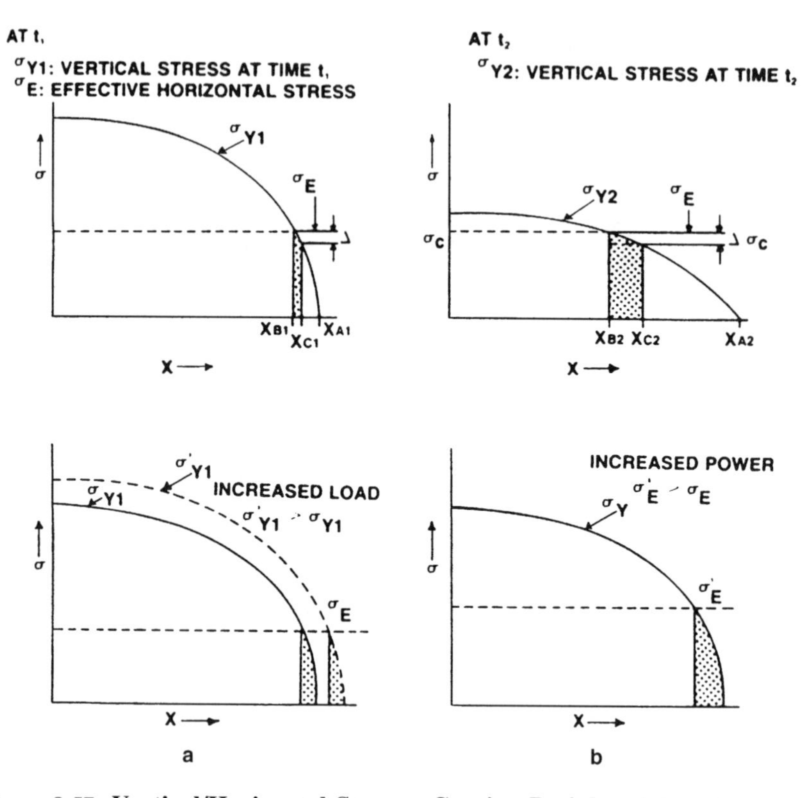

Figure 8-57. Vertical/Horizontal Stresses Causing Peripheral Bonding. Horizontal stress, S_E which produces wave form deformation is imposed upon S_Y. Bonding occurs from X_B to X_C with obvious grooving between X_C and X_A. With time, bond zone moves toward center, and previously bonded material is destroyed. Note that the S_E curve is drawn to conceptually illustrate points of model. (a) Greater machine load results in larger undeformed central area; (b) increasing power widens wave affected region. (From Ref. 124.)

Downward programmed forces for ball bonds are well under 100 g. This low force means that the capillary tool's vertical motion will "stall" and the bond squash will reach an equilibrium state before it is totally flattened. The partially deformed ball is now ready for the welding process, because it is physically wetted to the aluminum bond pad.

Several variables control the actual diffusion of two metals into each other. The diffusion rate is affected by the amount of energy applied through heat, ultrasonic energy, and time. It is also affected by the purity (concentration) of the two materials being welded. Therefore, the bond strength/depth is the sum of the combined energies applied for a given time. This means that a large number of parameter combinations can create an equally strong bond. Ultrasonic energy is most critical because

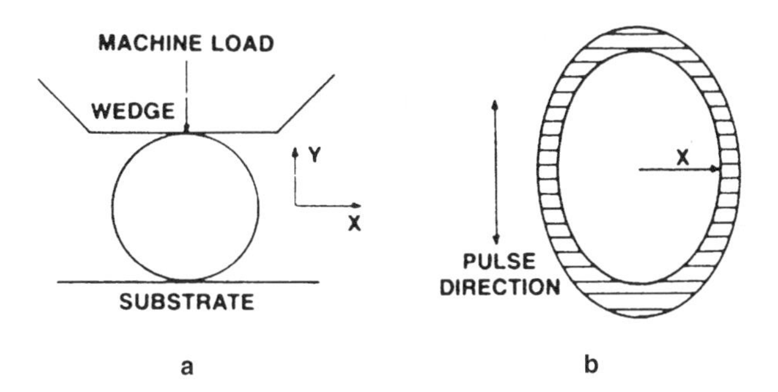

Figure 8-58. Peripheral Wedge Bonding Pattern. (a) Bond configuration for applied two-force model; ultrasonic energy applied to the wedge produces horizontal (vibrational) stress along the wire while the machine load acts vertically. (b) Bond pattern on substrate is characterized by rough peripheral region at distance *X* from the bond center. (From Ref. 124.)

it reduces the gold's yield strength [126] by increasing the mobility and density of lattice dislocations. This plastic flow begins when ultrasonic energy is applied to sweep away the brittle oxide layer on the aluminum bond pad to expose a clean metal surface (Fig. 8-59).

With incomplete lattice structures at the newly exposed surface of each metal, a migration of atoms begins from one material to the other. These diffused atoms form bonds with their neighbors' shared outer shells. As temperature and ultrasonic energy are applied over time, additional interdiffusion results.

8.4.2.3 Thermocompression Welding

Although this technique was one of the original methods for the manufacture of semiconductor packages, it is only used for a few specialty

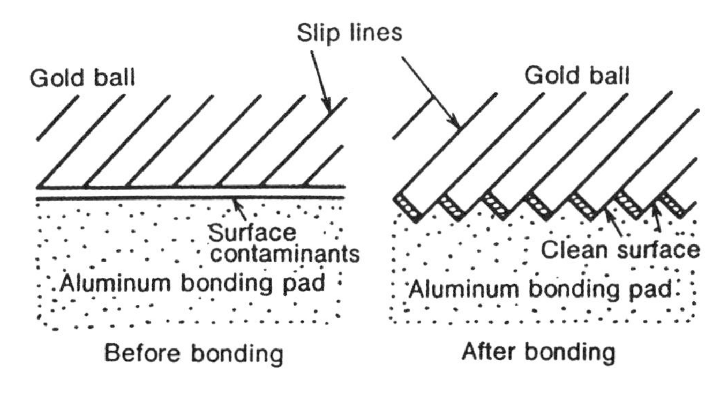

Figure 8-59. Gold Ball Slip Planes Form Clean Surfaces. Ultrasonic energy applied to the gold material causes slip-plane mobility which exposes clean Au and Al material that are in intimate contact with each other. (From Ref. 127.)

applications today. It utilizes only two energy sources: heat and a high downward force (compression) to form the weld. Today the process generally requires temperatures ranging from 280° to 380°C and is much slower than T/S bonding. The time required to make a T/C bond can approach 1 s. In most cases, the extremes of heat and time for this process make it less attractive to manufacturing engineers.

8.4.3 Wirebonding History

Wirebonding was the earliest technique for device assembly. Bell Laboratories published their first wirebonding results in 1957 on a technique called "thermocompression bonding" and began contacting equipment manufacturers to provide production machines. After developing several types of laboratory apparatus, Kulicke and Soffa Industries produced the first commercial bonding machines in 1963 for the infant semiconductor industry. With machine availability, the wirebonding technique became generally applied. By 1970, U/S bonding and T/S bonding techniques were introduced which reduced bonding temperatures.

The early machines were extremely labor-intensive, with all the positioning and motions applied manually. The machine operator controlled the entire process by moving levers that scaled down gross hand motions through micromanipulators. This made bond placement on the small bond pads a learned skill that varied greatly among operators. Loop shapes were also completely dependent on the dexterity of the operator.

The T/C bonding process became the preferred method for assembling commercial-grade parts. Aluminum wedge bonding provided a "monometal" connection on the bond pads of military-grade devices. As cam-driven, semiautomatic equipment was introduced to the marketplace, the T/C gold-ball bonder became the favored machine for two reasons: (1) the process was faster, as it did not require device rotation like the wedge bonder and (2) the gold wire could be encapsulated in nonhermetic, plastic encapsulants and not be susceptible to moisture corrosion like its aluminum counterpart.

However, the early bonding process was not without problems. Prior to the invention of the electronic flame-off, ball formation was accomplished by heat application with a hydrogen torch. This method did not provide sufficient ball size consistency and often caused damage to the wire just above the ball. In fact, a partial melting of the area above the ball was considered normal. It was called the "necked down" or simply "neck" area (a term used by many assembly engineers today, even though the "neck" phenomenon has long since disappeared from the normal process). Using today's highly controlled EFOs (electronic flame-off) it is possible to form consistent balls without visibly distorting this region.

The most serious problem associated with the early bonding process was metallurgical weld failure that led to device reliability problems. In the early 1960s, it was common practice to wirebond at 350°C because the high temperature provided the activation energy for interdiffusion gold wire and aluminum bond pads. When this occurs, five intermetallic alloys form between the pure gold and aluminum areas [126]. Some of them are quite brittle and prone to failure during temperature cycling. One of these brittle compounds, Au–Al_2 is purple in color. Its distinctive presence at failed bond locations soon led to the term "purple plague."

If the formation of the purple Au–Al_2 was not bad enough, a second failure mechanism was found to be associated with these compounds. At the high temperatures used for epoxy curing, ceramic lid sealing, and stabilization baking, the Au continues to rapidly diffuse into the Au-rich compounds like Au_5Al_2, leaving voids at the boundary between these two regions, known as Kirkendall voids [126]. The same condition occurs where the Al rapidly diffuses into the Al-rich $AuAl_2$, leaving Kirkendall voids at their shared boundary. In extreme cases, the voids grew/coalesced into planes of weakness which could easily fracture, resulting in open circuits or lifted ball bonds.

These reliability problems were so serious that minor process changes could not be considered. To solve these problems, engineers considered alternative connection methods such as flip-chip solder bumps, beam leads, and TAB, and a new wirebonding process. In 1970, U/S energy was introduced, which reduced the heat input to 200°C. This enabled interdiffusion without a high rate of brittle phase formations. Since the advent of T/S bonding equipment, the earlier problems are easily avoided. In fact, most packaging engineers have never actually seen purple intermetallic or Kirkendall voiding outside of a historical reference. It is perhaps ironic that the wirebonding process began with such basic problems and has evolved to such excellent yield and reliability levels.

Today, most reliability problems are eliminated with properly controlled and much improved tools and processes. Several generations of servo controllers have long since replaced hand-manipulated mechanisms to apply the bonding force. Free-running oscillators for the ultrasonic energy changed to phase-locked-loop circuitry and, later, microprocessor-controlled frequency, voltage, current, and time.

As mentioned previously, EFO circuitry now produces a completely repeatable ball formation process. This precise control of the bonding parameters has also been matched by the purity of materials for wirebonding. Wire and plating alloys are now specified accurately. Wire mechanical properties are now highly reproducible between lots through automated manufacturing and tight process control of the wire chemistry and heat treatment/annealing. The same servo control technology utilized in the wirebonding equipment is also used in the final grinding/polishing operations for the manufacturing of bonding tools (capillaries and wedges), which were manually produced in the past.

Today's packaging engineers can select application-specific tools and wire to meet the requirements unique to their package design. These products can be repeatably manufactured as a result of three decades of controlling and understanding each manufacturing process.

8.4.4 Wire-Bond Applications

It is beyond the scope of this section to list all current wirebonded packages. An acronym dictionary, with annual updates, is required to keep abreast of the package explosion.

Since the mid-eighties, the "end of wirebonded packages" has been predicted to be imminent. Each new package type has presented challenges that initially appeared too difficult for the current wirebonding capabilities. Yet equipment advances meet these challenges and provide the required capabilities. The semiconductor industry is strongly motivated to continue investing time and energy into an ever-evolving wirebonding technology because of its much improved reliability and historically successful track record.

One such application challenge was reducing package thickness. Portable, hand-held electronics require packages less than 1 mm thick. This translates to loop heights less than 150 μm above the chip (Fig. 8-60). Low, consistent loops (Fig. 8-61) with less than a 40-μm, 6-sigma

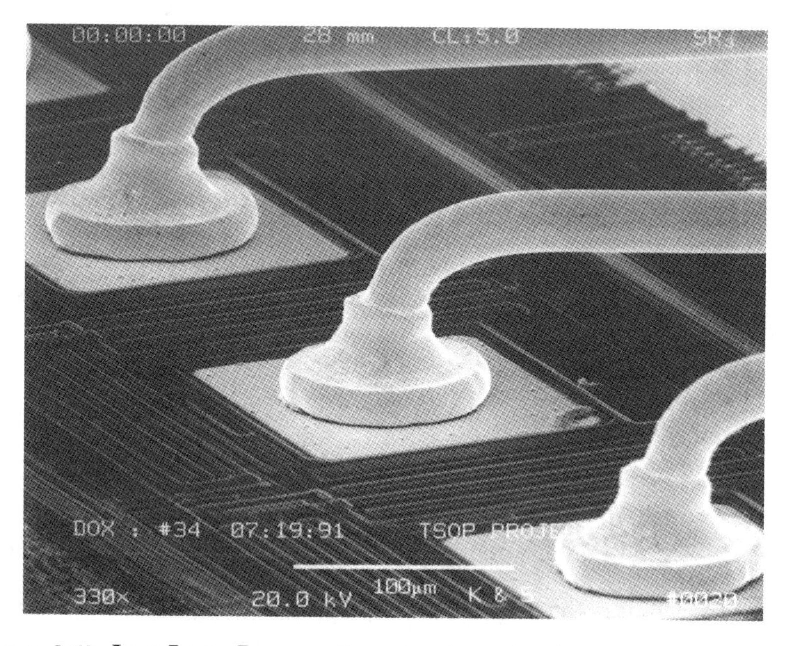

Figure 8-60. Low-Loop Process. Example of very low loops without evidence of stress damage to the wire above the ball bonds. Loop heights shown are 96 μm (top of wire to the chip surface). (Courtesy of K&S.)

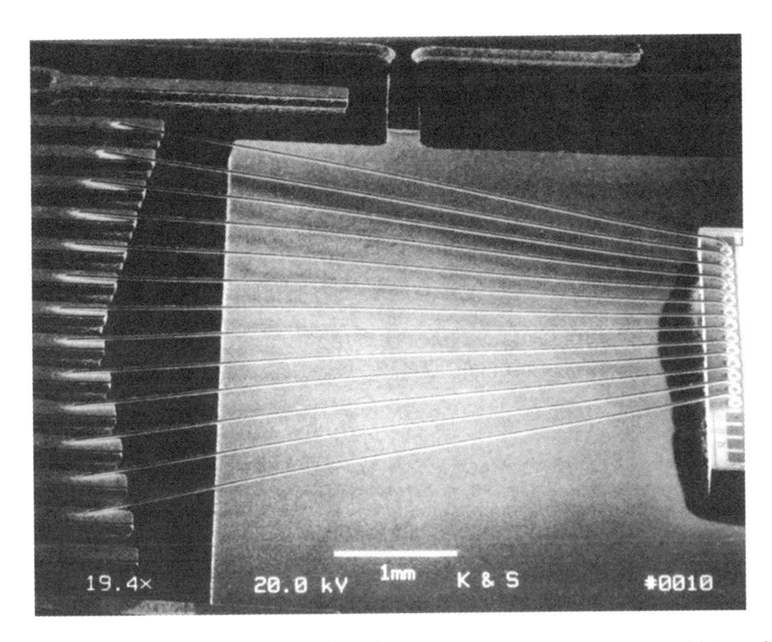

Figure 8-61. Low-Loop Process Consistency. These low loops are 114 μm above the chip surface and illustrate 5-mm lengths. (Courtesy of K&S.)

range, are achieved using high-resolution, coordinated servo systems that control the capillary's three-dimensional trajectory from first to second bond. Because an error of a few microns in the actual tool path could either buckle or break wires (Fig. 8-62), real-time microprocessor control for each servo loop is critical.

In addition to reducing thickness, the market demanded smaller package lengths and widths, made possible by advances in wafer processing, through continuous shrinking of device features and lead-frame etching improvements. The term "fine pitch" has become an ever-decreasing, moving target. Even with machine accuracy/repeatability made possible by new vision systems and advanced servo systems that deliver the wire to its intended position, two other factors needed improvement: process control and bonding tool design. Each step of the bond formation process that was responsible for the finished bond size needed to be improved. These included optimizing tail length, EFO voltage, current, time, U/S power, and bonding force for more consistent bond squash.

However, the largest factor in achieving fine-pitch goals was the accurate manufacturing of reduced-diameter bonding tools. Bottleneck capillaries and side-relieved wedges were needed to ensure bonding that did not touch the previous wire loop (Fig. 8-63). To illustrate the progress of fine-pitch bonding from 1990 to 1995, the possible bonding pitch (measured from center to center) moved from 150 μm to 90 μm for

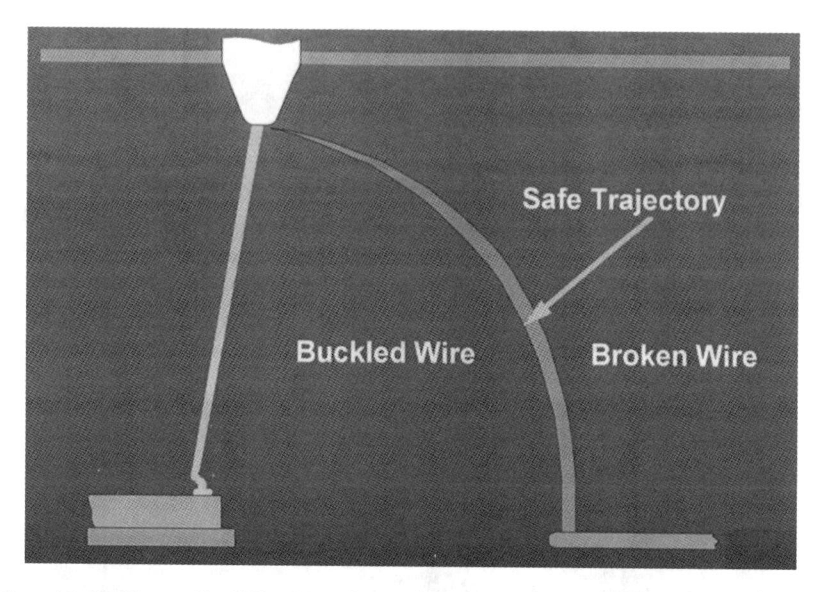

Figure 8-62. Prescribed Tool Trajectory for Loop Consistency. Looping inconsistencies may be caused by small servo lead/lag errors which are translated into irregular wire shapes. Simultaneous coordination of three axes with real-time feedback is critical.

ball bonding and 75 μm for wedge bonding. Several chip manufacturers designed staggered bond pads in two rows with staggered lead-frame fingers or two-tier ceramic packages, which produced effective pitches of 45 and 37 μm microns for ball and wedge bonds, respectively (Fig. 8-64). Both process development and refined tool designs continue to provide engineering solutions for fine-pitch applications (Fig. 8-65).

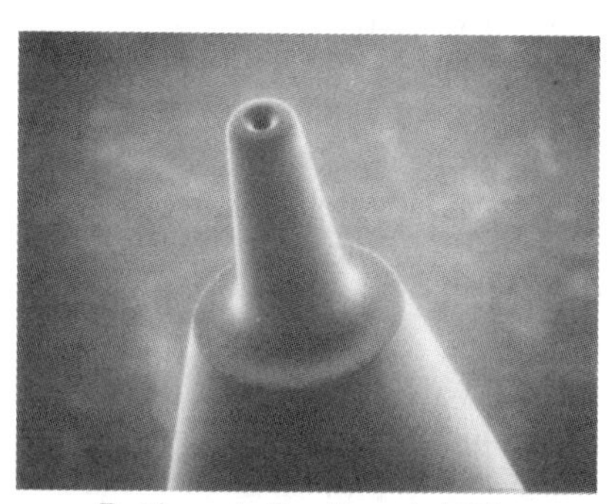

Bottleneck Capillary (x100)

Side Relief Wedge (x150)

Figure 8-63. Bottleneck Capillary and Side Relief Wedge. Examples of fine-pitch tools that allow close proximity bonding to a previously bonded wire without interference. The capillary shown is called a bottleneck. It has 360 degrees of relief because ball bonding may be omnidirectional. Only the sides of the wedge are cut away for its unidirectional process. (Courtesy of Micro-Swiss.)

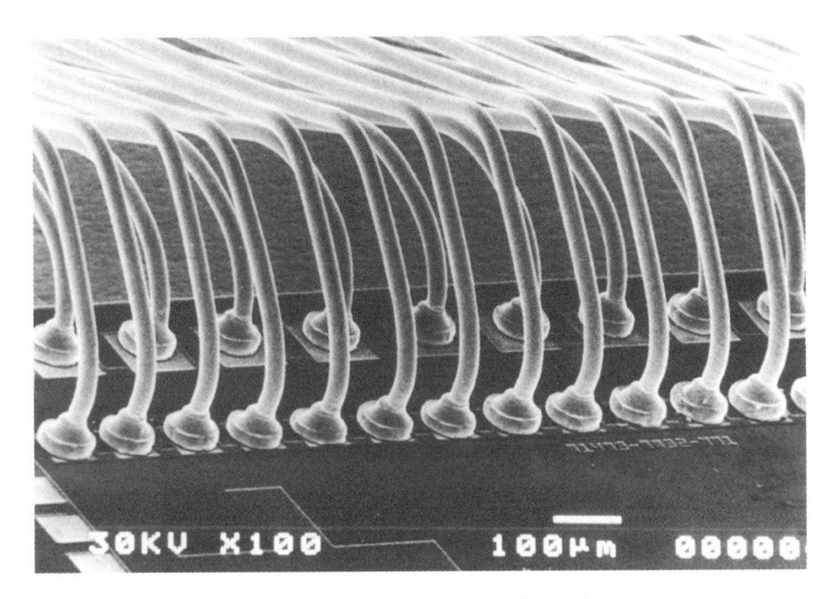

Figure 8-64. Staggered Bonds to Reduce Effective Pitch. Shown is a test chip with staggered ball bonds and two levels of loops to maintain a safe distance around each wire. The effective pitch is one-half the pitch bonded on each row. (Courtesy of K&S.)

Although wirebonded packages traditionally have bond pads located at the chip periphery, there are some applications that fall outside this design guideline. Some loops are needed to traverse the die surface, remaining parallel, then forming a bend and descending to the lead frame/ substrate. This bend near the die edge is particularly needed in thin packages to avoid clearance problems between the edge of the die and

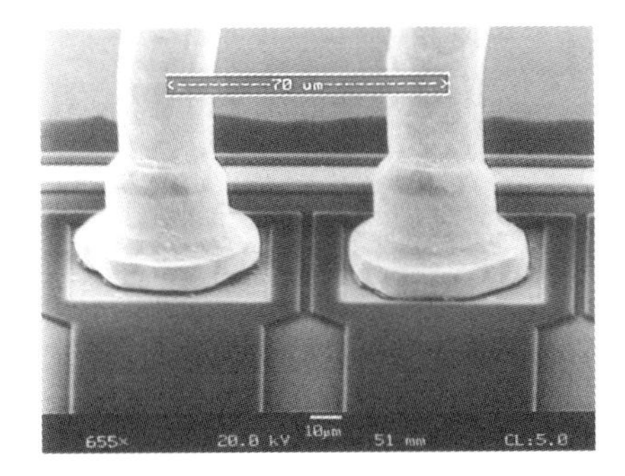

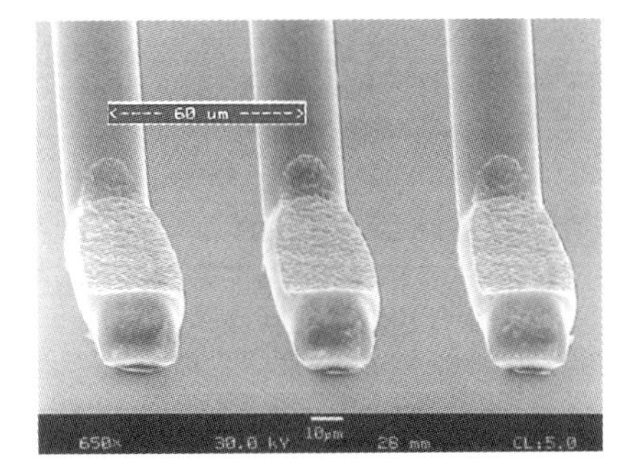

Figure 8-65. Fine-Pitch Ball and Wedge Bonding. *Left*: Shown is an example of 70-µm ball bonding (1998 SIA benchmark). Note the amount of bond squash compared to the example shown in Figure 8-53 without pitch constraints. *Right*: This is an example of 60-µm wedge bonding. High-frequency ultrasonics produces these bonds with minimal squash out. An added benefit is the increased thickness which contributes to a larger cross-sectional area of material at the heel of the bonds. (Courtesy of K&S.)

the wire. Special tool trajectories have been developed to place a calculated angle bend into the wire and lower the preformed wire to its final position without distortion (Fig. 8-66).

Although extra motions require additional time to form this type of loop shape, the benefits of bonding interior bond pads and providing a stiff wire that resists mold sweep (wire movement when viscous liquid plastic encapsulant flows over it) far outweigh the extra time penalty of a few milliseconds per wire.

Another application for the wire loop with a second bend at the die edge is close proximity "down bonding" (Fig. 8-67). This connection is used by MCM package designs as well as ICs with connections from the die to the ground plane to which it is mounted. Once again, this application requires proper clearance between the wire and die edge to eliminate the possibility of an electrical short circuit.

Most ball-bonded applications place the second bond at (or below) the first bond level for the best loop shape. Normally, second bonds located in a plane higher than the first bond are reserved for wedge bonding. However, there is one notable exception called lead-on-chip (LOC). In this application, the lead frame is mounted to the top of the

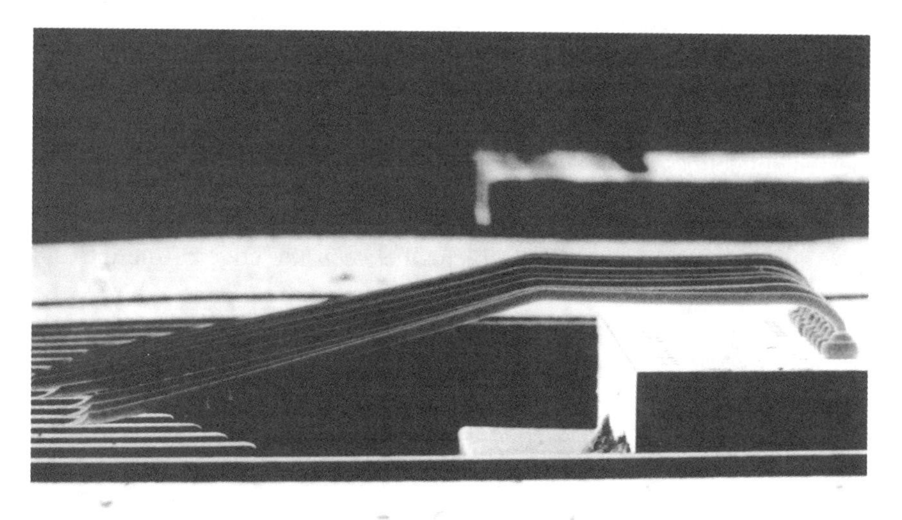

Figure 8-66. Low Loops Preformed Before Attachment. Low loops shown in this photo are 150 μm (from the top of the wire to the chip surface). The bends above the ball and beyond the die edge provide proper wire clearance and the second bend adds stiffness to resist wire movement during encapsulation. Bend angles are calculated for each wire and preformed before attachment of the crescent bond.

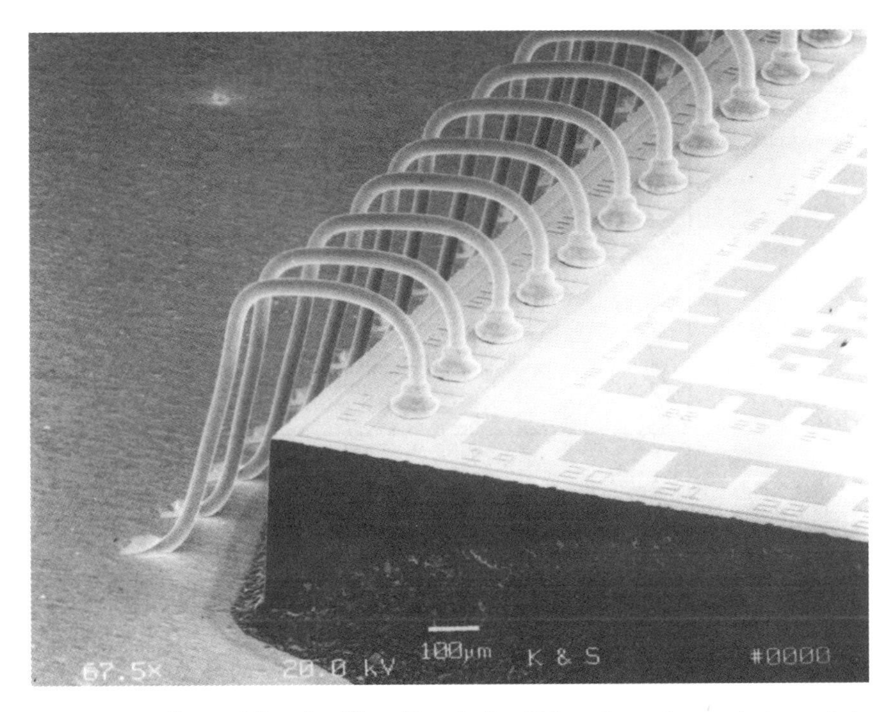

Figure 8-67. Ground Bonds: Close Proximity. Using wire preform techniques similar to loops shown in Figure 8-66, these wires have second bonds placed 330 µm from the base of the chip. (Courtesy of K&S.)

die using double-sided adhesive polyimide tape to electrically isolate the metal frame from the die. In addition to a unique looping situation, the U/S energy can be attenuated/absorbed by the tape, forming a challenge to achieve a reliable second bond. This application could be accomplished by understanding the previously discussed bonding dynamics and the variables responsible for a reliable weld. Therefore, U/S power was de-emphasized while providing activation energy in the form of kinetic force.

8.4.5 Materials

8.4.5.1 Bonding Wires

Wirebonding materials have been reviewed by Gehman [128]. It is customary to specify the mechanical properties of bonding wire by setting acceptable ranges for the break strength (BS) and elongation (EL). Tensile properties are determined from a standard stress-strain curve (Fig. 8-68).

8.4.5.2 Aluminum Wire with 1% Silicon

During the early attempts to manufacture fine aluminum wire (less than 50 µm) it was seen that pure aluminum was difficult to draw through

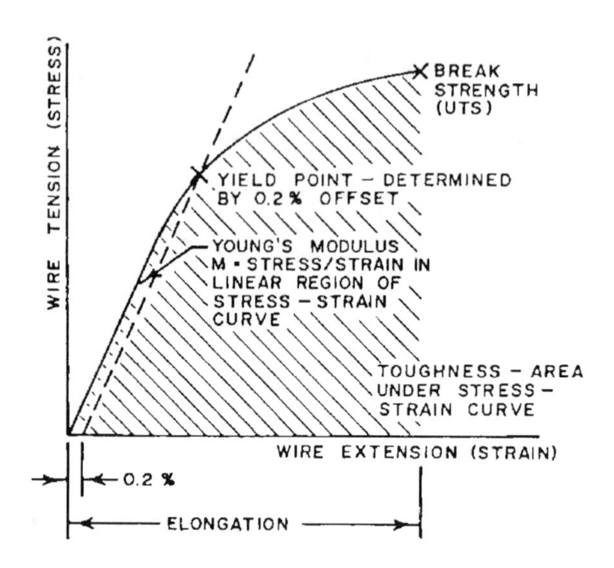

Figure 8-68. Stress-Strain Curve. Tensile test tension versus extension (stress-strain) curve. (From Ref. 128.)

the decreasing die sets that produced the wire. It was equally difficult to handle the wire in a bonding machine, so alloying was required to produce a tougher wire material. Combining aluminum with silicon was a "safe" choice because it would not introduce foreign elements into the package, and the AlSi alloy was well established in commercial practice. Today's wire alloy (AlSi 1%) became the standard composition in the early sixties and remains today.

In retrospect, from a metallurgical standpoint, 1% Si as a solute in Al for U/S bonding wire was an unfortunate choice. The equilibrium solid-state solubility of Si in Al at 20° is of the order of 0.02% by weight. Only at temperatures above 500°C is Si at 1% in equilibrium solid solution. Thus, at ordinary bonding temperatures, there is always a tendency for Si to precipitate, forming a silicon second phase. When uncontrolled, excessive Si segregation may degrade wirebondability or bond integrity.

8.4.5.3 Aluminum–Magnesium Wire

Aluminum–magnesium alloyed with 0.5% to 1% Mg can also be drawn into fine wire. The finished product exhibits breaking strength and elongation similar to that obtained in AlSi 1%. This alloy wirebonds satisfactorily and is superior to AlSi 1% in resistance to fatigue failure at first-bond heels and to break-strength degradation after exposure to elevated temperatures. An early hypothesis that device field failures were due to the presence of Mg in the wire alloy has been conclusively disproven. In comparable production volumes, AlMg wire should be no more expensive than AlSi 1%.

The advantage of Mg over Si as a solute in Al wire alloys is that the equilibrium solid solubility of Mg in Al is about 2% by weight. This compares, as noted previously, with about 0.02% Si equilibrium solid solubility in Al at 20°C. At 0.5% to 1.0% Mg concentration, there is no tendency toward second-phase segregation, as is the case with AlSi 1%.

8.4.5.4 Gold Bonding Wire

There are several parameters that must be precisely controlled to provide acceptable gold wire for high-speed wirebonding. First, the surface cleanliness must be highly controlled. This is a difficult task, as the entire surface must be well lubricated during the drawing operation in order to maintain a good surface finish and draw through the final (25-μm-diameter) diamond die set. If the lubricant is not removed, it contaminates the capillary tip during the EFO firing by deposition of the ionized material and it can also clog the feed hole as it is fed through at high velocity. Second, it reduces the area of exposed clean metal during bonding, as this thin film prohibits the joining of the two clean materials from joining/welding.

Once the wire manufacturer has produced clean wire, it must be wound on a spool with precise tension: too little causes bunching/overlapping and despooling problems, and too much results in the gold bonding to itself, making despooling difficult.

Mechanical wire properties are determined by two parameters: wire chemistry and heat treatment. Like aluminum, very pure gold (99.99%, called four 9's) is very soft and difficult to draw into fine diameters. Small concentrations of impurity atoms (dopants) make the gold workable.

Several foreign elements may be added to the pure gold to toughen it; Be and Ca are commonly used, with the latter used exclusively by Japan. The levels of the dopants varies from one manufacturer to another, but typical levels of Be are 5–10 ppm by weight.

Other dopants are used in the manufacture of specialty wires for certain applications. For example, Pd is often used in "bumping wire," which forms a gold ball on the chip pad (Fig. 8-69). The wire is then intentionally broken above the ball to provide a gold bump for the bonding of a TAB lead. The Pd dopant raises the recrystallization temperature of the material so that the fully annealed area above the ball resulting from the EFO heat applied to that area, called the heat-affected zone (HAZ), is extremely short (Fig. 8-70). Conversely, other dopants, such as La, lower the recrystallization temperature, creating a very long HAZ. This wire can be used for specific loop-height applications by providing a naturally bending area without wirebonder motions creating kinks in the wire to achieve the required loop height/shape.

Figure 8-69. Wire-Bonded Wafer Bumps. *Right*: Ball bonds exhibiting short tail broken within the small HAZ created with Pd dopant. *Left*: Bonds that are coined, using a secondary operation to flatten the tail feature. (Courtesy of K&S.)

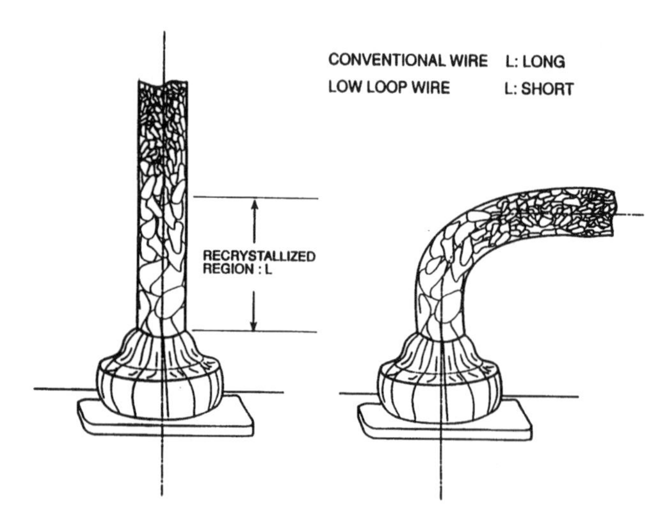

Figure 8-70. Heat-Affected Zone. Heat energy from the ball formation process is conducted along the wire length which recrystallizes/anneals the affected region. The recrystallization temperature of the material is controlled by the type of dopants added, which determines the length of this "zone."

Finally, heat treatment of the cold-worked wire is necessary to adjust the temper because the as-drawn properties are too stiff/brittle to withstand the movements of an automatic wirebonder and the rigors of the molding/encapsulation process. Annealing the material, combined with the proper dopants, produces fine gold bonding wire that is soft enough to be shaped into loops, yet stiff enough to hold its shape after forming.

8.4.5.5 Gold Wire Substitutes

Reducing packaging costs has often led designers to investigate replacing expensive gold material with less expensive ones. Many materials were able to be drawn into fine wire diameters and balls were successfully produced but only two (aluminum and copper) generated real interest over the last two decades. While both materials required changes to the flame retardant contents of the plastic encapsulant, aluminum was soon discarded because of ball porosity. The amount of this porosity could not be reliably predicted. Copper remains the major viable alternative to gold for certain applications such as power devices, but it has been slow to gain popularity due to reliability fears associated with oxides trapped at second bond and corrosion issues in a non-hermetic package, (even though its electrical/thermal conductivity, tensile strength, and price are superior to gold).

8.4.5.6 Die Attach Materials

Shukla [129] has reviewed die-bonding materials. The die-bonding process includes a large choice of materials, which may be broadly classified into the three categories described below.

8.4.5.6.1 Solders

These include single, binary, and ternary metallic compositions and may be further classified into two subgroups: hard and soft solders. Hard solders have rather high flow stresses (onset of plastic flow), offering excellent fatigue and creep resistance. The disadvantage of using hard solders stems primarily from their lack of plastic flow, which leads to high stresses in the silicon chip because of the thermal-expansion mismatch between the die and the substrate. Soft solders, on the other hand, are low-melting binaries and ternaries which have a high degree of plastic strain capability.

8.4.5.6.2 Organic Adhesives

Epoxies and polyimides filled with precious metals have found widespread acceptance as die bond materials in low-cost packaging (plastic packages). They offer the advantage of lowest processing temperatures and lower cost compared to the gold-based hard solders.

Numerous articles have claimed that organic adhesives lower thermal stresses in silicon. These adhesives are typically filled with a metal to provide the thermal and electrical conductivity required in most applications (silver being the most common filler). However, the use of organic adhesives in high-reliability packages (hermetically sealed) has been slow in acceptance because of their poor thermal stability, especially when filled with silver. Poor thermal stability, coupled with the outgassing of entrapped solvents and other gaseous species, remains a challenge in the application of these materials to high-reliability VLSI applications.

Curing epoxy requires oven baking for several hours, adding significant processing time to the assembly cycle rate. If copper leadframes are used, the oven must be filled with oxygen-free gas and coverage must continue until the oven is cooled to ambient temperature, lengthing the process even further. A considerable amount of research in epoxy chemistry resulted in fast-cure or snap-cure epoxies, capable of curing in a few minutes or less. Curing may be performed in a flow-through oven connected to the wirebonder, greatly reducing total product cycle time.

8.4.5.6.3 Glass Adhesives

Silver-filled specialty glass materials have emerged in the past few years as an alternative for VLSI die bond applications. These materials offer the possibility of a void-free, die bond interface with excellent thermal stability. However, the currently available materials require high processing temperatures (400°C) and oxidizing ambients for optimum adhesion, creating some special processing concerns. In addition, the glass die bond materials also need solvents and binders for processing purposes, leading to the problem of complete solvent removal (similar to the case of organic adhesives).

8.4.5.7 Bonding Tools: Capillaries and Wedges

The microscopic features of these replaceable tools are responsible for a significant part of the bonding process. The final shape of the ball/crescent and wedge bonds are a direct imprint of the tool itself. Consistent loop shapes cannot be attained without correct feed hole dimensions, leading-edge funnel shape, surface finish, and cleanliness.

A tool expert should always be consulted to select the correct combination of features for a particular application. Even then, it may be necessary to create a designed experiment to select a final choice from several candidates.

When tools were made using manual grinding techniques, there was a significant variation between tools. A bonding process could run out of its control limits simply by exchanging a worn tool for a fresh one. The advent of computer-controlled, micromachining centers now produces tools that can be assumed to have dimensions without variance from one to another.

The most common materials for capillaries and wedges are alumina ceramic and tungsten carbide, respectively, with both materials originating as a sintered powder. Alternative materials such as glass, ruby, and titanium carbide are sometimes used for unique applications. Research continues to find better material for fine-pitch requirements.

Defining tool life is usually at the discretion of each semiconductor manufacturer. Depending on the number of wires bonded, some high-production facilities change tools at the end of every shift, whereas others are able to last four to five working shifts using slower bonding equipment. All facilities generally agree that the tool is considered a consumable and should be replaced on a regularly scheduled basis before marginal bonds are created, or the tool breaks.

8.4.6 Process Automation

Since the mid-seventies, process engineers and production management for packaging have continually demanded higher levels of automation and process control for wirebonding. It is desirable to remove human handling of packages to improve both quality and throughput. Observations of factory production activities reveal that wirebonding defects are not randomly distributed throughout the production shift but occur most often in "clusters" surrounding operator intervention that interrupts the machine's automatic operation. This machine assist is typically required when defective/contaminated material prevents formation of a proper bond or causes the transfer handling system to jam.

Although wirebonder operators have always been restrained from changing bonding parameters beyond specified limits, some parameter adjustment has always been tolerated to adapt bonding energy to slight material differences from lot to lot. Machine designers have continually upgraded hardware and software in the areas of bonding ultrasonics and force application, as well as the wire feed system to reduce/eliminate the amount of human adjustment by closing system control loops to create self-monitoring and adjusting systems.

Host computer control for the entire wirebond manufacturing cell has been possible for a number of years, but it was not feasible until the mid-nineties because there was too much variation between machines. This prohibited downloading of a common parameter set and achieving identical bonding results. Individual machine calibration techniques and closed-loop systems now make downloading of host parameters and uploading of system states a reality that actually provides new levels of bonding/looping quality.

Further process automation includes the linking of pre/postwirebonding processes to eliminate human handling between assembly steps and to further reduce assembly cycle time. This has been achieved by physically linking epoxy cure ovens and die bonders (Fig. 8-71) to the wirebonder for some chip manufacturers, whereas others have provided a variety of robots to shuttle material on–off machine docking stations.

8.4.7 Process Choice Decisions/Guidelines

This section will provide basic guidelines for selecting a wirebonding process for various packages. There are many devices that do not fit into standard categories, so packaging engineers/designers should consult with a wirebonder manufacturer to avoid designing a package that does not allow a robust assembly process to be used.

8.4.7.1 Wire Type

The advantages of aluminum wire are that it provides the highest reliability bond to the aluminum metallization on the chip, and it forms a monometallic bond at room temperature. It has two disadvantages: low resistance to corrosion in the presence of moisture and it does not form consistent, nonporous balls when melted. This means aluminum wire is only suitable for wedge bonding (normally in high-end hermetic packages). It cannot be used in molded plastic packages because of its susceptibility to corrosion, which will cause reliability problems.

Figure 8-71. Flex-Line System. Die attach (not shown), cure oven, and wirebonders linked in-line reduce floorspace, work-in-process, and handling induced yield issues. (Courtesy of K&S.)

Although gold wire is more expensive than aluminum, it remains unharmed by moisture, making it the best choice for a plastic package. Its two biggest disadvantages are cost and 200°C bonding temperature. Gold is equally suited for ball- or wedge-bonding processes.

8.4.7.2 Ball-Bonding Guidelines

Because bonding equipment, capillaries, wire, devices, die attach, and lead frames are all changing constantly, it is impossible to set down exact guidelines for any process. However, a few basics about ball bonding are worth noting.

Bond pad size on the chip is dictated by the area needed for the bond itself. The smallest bond size is a function of the unbonded ball size, the diameter of the wire, and the bore size of the capillary. A ball must be formed by the EFO that is large enough so that it is not extruded up the bore as bonding pressure is applied. Therefore, these three factors must be optimized before a minimum bond size can be defined. The amount of pad real estate surrounding the bond allotted to accuracy/repeatability of the bonder will continue to shrink as new generations of machines are developed. The equipment supplier should be contacted for the latest performance specifications.

The main factor dictating bond pad pitch is the outside dimensions of the capillary, with the system accuracy/repeatability being statistical error terms that must be considered when calculating the minimum space between adjacent pads. This space is required to prevent a collision between the capillary and the wires previously bonded. Actual minimum pitches that can be achieved change annually and generally follow the SIA roadmap. Detailed models are available to perform the actual calculations if in doubt about a particular application.

It is somewhat ironic that silicon shrinkage and fine pitch at the first bond has significantly increased the wire lengths within the package. The reason is lead-frame pitch. The technology for stamping and etching lead frames cannot match the fine geometry possible on the chip. An increasing number of package I/Os means a larger radius around the die. Wire lengths have increased from 1 mm to 5 mm with foreseeable need for 7-mm lengths. Figure 8-72 shows the wire lengths that result from various pad and lead-frame pitches.

Wire lengths greater than 4 mm present a number of assembly challenges. The bonder must suspend the length of wire between the die and lead frame without vertical sagging or horizontal swaying. This is particularly important as device signal speeds are increased and electrical performance of the device is dependent on the shape of the loop and distance between wires.

The lead frame should be designed so that the wire loop is routed

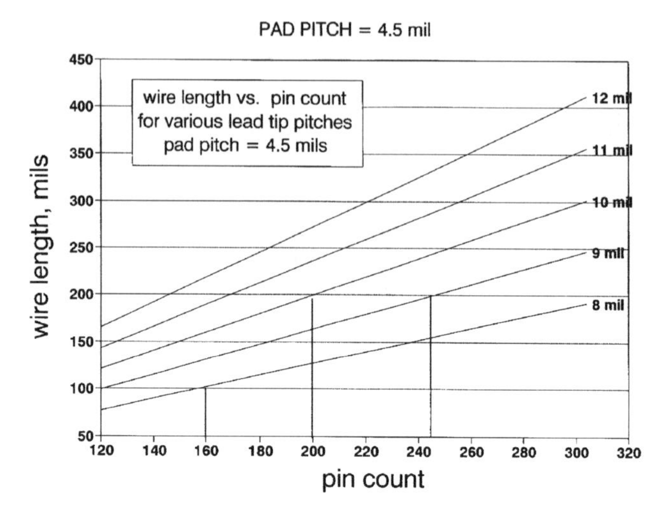

Figure 8-72. Package Wire Lengths. Example of resultant wire lengths for a radially wired package with a 4.5-mil bond pad pitch and various second-bond pitches. (From Ref. 130.)

over the tip of each lead finger. If the wire path is over the side of the finger, it will have a tendency to "roll" further to the side, inducing sway in the loop. This is more likely to happen if the angle between the wire and the finger is greater than 20 degrees.

The top surface of the lead finger is often wider than the bottom side. If the capillary must be targeted off the exact center of the finger, to keep the loop within its correct path/corridor, the delicate finger can often twist or roll. If this happens, neither the bond force or ultrasonic power will be transmitted correctly and a substandard weld can result.

The size and shape of the crescent bond are directly related to the diameter of the wire and the geometry of the capillary. This is a very important consideration when reducing capillary dimensions for fine-pitch considerations. Although it may be tempting to use a small capillary for first-bond pitch, adequate face dimensions must be provided in order to produce a crescent bond with minimum welded area.

8.4.7.3 Wedge-Bonding Guidelines

Advances in ultrasonic systems have made it possible to produce a high-strength wedge bond that is only 2–3 μm wider than the wire diameter. Therefore, the minimum width of the chip bond pads is largely dictated by the wire diameter (plus the necessary pad real estate for system inaccuracies). However, the pad layout is not nearly as straightforward as ball bonding. First, the pad length must support the long dimension of the bond (usually two to three times the wire diameter) as well as the tail protruding from the first bond. Next, the pad's long axis should be oriented along the intended wire path.

Generally, packages can be designed with either orthogonal wires or radial wires. Pad pitch for orthogonal wires is mainly governed by the wedge's outer geometry, with consideration given to the bonding system's statistical error. Radial wiring will have the same pitch as orthogonal wiring at the center of each side of the chip. As the pad locations approach the corners of the chip, the bond pads must be rotated at an angle so that it is aligned with the wire path. Furthermore, the pitch must be increased to maintain consistent distances between wires.

Up-bonding, where the second-bond location is located above the chip plane, is easily achieved with the wedge bonder while it is more difficult with ball bonding. The bonding tool movement through three-dimensional space will actually contact and "curl" the bottom side of the wire to provide extremely consistent loop heights with this process, even at wire lengths of 4 mm.

Second bonds are typically located on a ceramic substrate with either embedded lead-frame fingers or printed/fired metal paste. Bonds can be easily located in the center of these locations, but specialized vision/lighting systems must be developed to accommodate certain low-contrast packages. The most important application issue with second-bond location is the position of the bonder's wire clamps to the rear of the wedge. Many packages have a lip or wall surrounding the second-bond locations for the attachment of the hermetic lid. Horizontal and vertical clearance must be provided to prevent interference with the wire clamps' operation during second-bond formation. Consult the equipment specifications for the minimum distances required for various machine/wire path configurations.

8.4.8 Evaluating/Optimizing the Process

This section divides wirebonding evaluation into two groups: engineering methods and production tools. Some engineering methods are sometimes used for production purposes, depending on a need/situation that may arise from previous experiences with a particular package or device. Additionally, each group may be further divided into visual and mechanical techniques for wirebond evaluation.

8.4.8.1 Engineering Evaluation Methods (Visual)

1. A toolmaker's microscope equipped with 0.1-μm resolution micrometers is used to accurately measure two-dimensional (*X* and *Y*) bond features while a third micrometer mounted to the focus adjustment is used for height (*Z*) measurements. When this instrument is connected to a data-collection PC, it allows the engineer to measure statistically significant

sample sizes quickly and is considered the standard visual tool for measuring bond squash, height, concentricity, loop heights, and straightness. It is also valuable for evaluating qualitative attributes such as cracks, scratches, or surface finish while setting up or troubleshooting a wirebonder.

2. A scanning electron microscope (SEM) may also be used for optimizing a process because it provides an infinite depth of field to the observer so that features are seen clearly throughout the field-of-view. Optional features include measurement at very high magnification, providing indisputable data for optimization experiments. Auger and beta-backscatter probes may be added to the SEM to answer questions about contamination or elemental analysis.

3. The SEM also provides qualitative results for bond peel testing. Tweezers are used to gently peel a crescent bond or wedge bond away from the substrate to reveal the amount of wire material still bonded to the surface (Fig. 8-73).

4. Bonds may also be chemically removed from the chip to gain visual access to the underlying layers to look for cracks caused by the bonding process. A toolmaker's microscope fitted with Nomarski filters is ideal for finding slip-plane dislocations or cracks. Chemicals such as sodium hydroxide and aqua-regia are used to etch away the aluminum and gold, respectively. Both reagents can be harmful and should be used by an experienced technician.

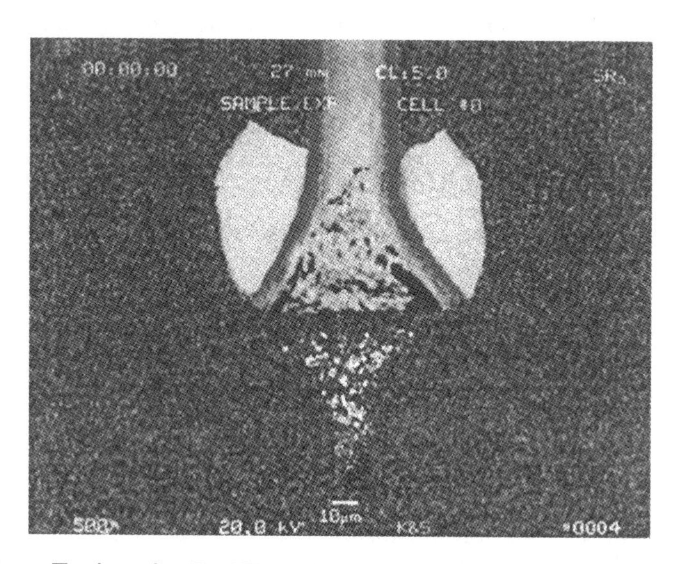

Figure 8-73. Engineering Peel Test: Crescent Bond. Crescent bond shown partially peeled away from the lead-frame finger. The beta-backscatter mode on the SEM may be used to evaluate location of Au material still bonded to a Ag substrate.

5. Perhaps the most tedious visual method is a cross-section inspection. The device is first encapsulated and placed on grinding/polishing wheels to remove one half of the bond. A light etch following the polishing reveals the amount of thickness of intermetallic between the bond and pad.

8.4.8.2 Engineering Evaluation Methods (Mechanical)

1. Mid-span pull testing is a traditional test for bond evaluation. Unfortunately, the results are more dependent on the wire length and height than on the strength of the bonding. Normally, the wire breaks near first bond, leaving both first and second bonds intact. However, spreadsheet calculations using the traditional formulas, with some refinements can now be used to select unique pull test target strengths for each application [131]. This makes pull testing much more valuable than previous specifications, where one target was used for all packages. Although the test is relatively easy to perform, it provides the least amount of usable data for the process engineer.

2. Shear testing ball and wedge bonds provides a very reasonable approximation of the weld area between the wire and substrate. Harman describes [132] this correlation as well as analyzing failure modes associated with the shear test, which is extremely important, yet rarely used.

3. A variation of the standard pull test involves placing the hook as close to second bond as practical, for a particular device. This second-bond pull test is quite useful to optimize bonding parameters for second bonds. In both ball and wedge processes, the wire above the second bond is not weakened, so pull strengths give an indication of second-bond weld strength because the bonds often lift before the wire breaks. (*Note:* In ball bonding, the EFO/ball formation recrystallizes and weakens the wire above first bond. In wedge bonding, the heel of the first bond is weakened by the smaller, back radius of the wedge, which is used to promote consistent tail lengths when the wire is pulled/severed at second bond.)

8.4.8.3 Production Evaluation Methods

1. The 30–60× optical visual inspection method is the most common technique used around the world. It is either performed on the wirebonder itself, or after bonding at an station called "Third Op." Although an inspector can scan an entire package quickly, looking for missing bonds/wires, bond placement, wire straightness, chip defects, and contamination, it becomes tedious and inefficient, like all visual inspection methods. It is also performed too late to provide real-time feedback to the wirebonder.

2. Wirebonder machine vision systems can now replace this manual technique, performing the inspection while the device is still located in the bonder itself, so real process feedback is possible. Using a pattern recognition system and high-resolution camera/optics, this technique can

perform sampling inspection of most visual attributes for the bonds and wire loop, including actual variables measurements that are not part of a typical third op visual.

3. The pull test has remained a mainstay of production testing, even with the shortcomings listed above. Its value to the production line is to highlight an out-of-control situation that may not be detected by the visual inspection. Because of the destructive nature of the test, it is only performed on a daily sampling basis for each machine.

4. Although shear testing is also destructive, it has a much higher value for production. This test can provide meaningful trend data for each bonder's true weld strength.

5. Open/short and electrical testing after encapsulation provide the final test method for wirebond yield, although traceability to the wirebonder must be maintained to make the data completely meaningful.

8.4.9 Wirebonding 2000: The Future

The Semiconductor Industry Association (SIA) has produced a national technology roadmap, updated every 2 years. Although it avoids specifying any preferred technology solutions or specific agendas, it does provide an accurate description of semiconductor advancements in the technology requirements and performance of ICs. It is the marketing and R&D departments of the equipment and consumables suppliers that must keep their products aligned with this roadmap.

Bond pitch will continue to be driven smaller, with the goal stated at 50 μm for the years 2000–2010. This will be partially achieved through continued capillary development, enhanced wire alloys, and "no wire sweep" molding compounds. In addition, machine developments that eliminate operator "teach" error, inspection systems with high magnification and positional feedback, and high-resolution servo systems will meet/exceed the fine-pitch goals.

Ball-grid-array (BGA) packages will have continued growth and the roadmap itemizes several wirebonding technology needs. The bond pad pitch is somewhat less aggressive than other package types, but wire interconnects will be greater than 600 per package. Faster bonding cycle times and low-temperature gold wedge bonding (<170°C) must be developed for this package. If a strip format for BGAs is adopted for wirebonding, a machine with a rotating head will be required. Otherwise, singulated packages can still be handled by rotating the individual part and placing into a boat/carrier.

Although many of the packaging needs listed for the future of wirebonding are difficult to achieve, perhaps the greatest challenge is the sheer number of bonds that will need to be produced per square meter of production floor space. The chip pad count (chip to package) increases more than any other package feature in the roadmap (Table 8-1). Computer-aided design/bond parameter files will need to be downloaded into the bonder to completely eliminate the huge programming functions associated with teaching more than 300 wires (600 bond sites). Servo computer systems in the bonder will have to execute calculations and drive each electromechanical axis without interrupts/delays. Low-mass bond heads are certainly required to achieve higher-velocity, accurate movements to create the bonding and looping motions associated with large numbers of wires per minute. It will ultimately be the top end speed of the bonder of the future that will determine the life of this interconnect technology.

8.5 TAPE AUTOMATED BONDING

8.5.1 Introduction

The tape automated bonding (TAB) concept originated in the 1960s as the "miniMod" project at General Electric Research Laboratories in New York. The name is derived from the French term "Transfert Automatique sur Bande" coined in 1971 by Gerard Dehaine at Bull-General Electric in France (later known as CII Honeywell-Bull, and now as Bull S.A.). In Japan, TAB is known as tape carrier packaging and a TAB package is called a tape carrier package (TCP).

TAB was originally developed as a labor-saving, highly automatable reel-to-reel "gang bonding" technique for packaging high-volume, low-I/O devices at a much lower cost than wirebonding [133]. It is worth repeating that in the 1960s and early 1970s, wirebonding equipment was slow and operator dependent. The process itself was beset by reliability problems, mostly related to operator inconsistencies, and metallurgical failures due to brittle gold–aluminum intermetallics and corrosion from ionic impurities in the molding compounds. So some companies looked for better alternatives. The result was that IBM developed the "flip-chip" process, Bell Telephone Laboratories (now known as Lucent Technologies) developed the "beam lead" process, and General Electric developed the "miniMod" process. All three processes allowed the electrodes (bond pads) on the chip to be connected simultaneously in one operation to the next interconnection level. The beam lead process featured chips with cantilevered gold beams plated onto the bonding pads at the wafer level [4]. After chip separation from the wafer, all the beams on the chip were gang bonded simultaneously to gold pads on the package or substrate using heat and pressure. The miniMod process used a three-layer, etched, 35-mm tin-plated copper tape laminated to a sprocketed polyimide film (similar to movie film). The inner bond "fingers" were simultaneously gang bonded to gold bumps on the semiconductor chip using a reel-to-reel assembly machine called the "inner lead bonder" (ILB). After inner lead bonding, the chip was tested. Good chips were then excised (cut) from the tape, and the outer tips were "outer lead bonded" (OLB) to electrodes on second-level substrates or cards.

Japan was quick to recognize the high-productivity potential of this new technology and adopted the GE process in the 1970s for low-cost consumer devices such as calculators and watches. Sharp Corporation and Shindo Company Ltd. were the pioneers. Today, Sharp is the largest user of TAB in Japan, and Shindo Company is the largest maker of TAB tape. In Europe, Bull was the first to adopt the miniMod process for computers in the seventies [134]. In the United States, most of the TAB activities in the 1970s and early 1980s were focused on assembling bipolar transistor–transistor logic (TTL) devices using single-layer or two-layer tape by leading semiconductor makers such as Fairchild Semiconductor (which merged with National Semiconductor in 1987), Motorola, National Semiconductor, RCA, and Texas Instruments. 3M became the leading supplier of two-layer tape in the seventies and eighties and still holds that position today. Over the years, TAB activities in the United States have been sporadic. At one time in the eighties, National Semiconductor was the largest producer of TTL TAB devices in the world, with an annual run rate of more than 400 million units. However, National and others abandoned TTL TAB assembly in the late 1980s in favor of wirebonding for cost reasons. Hewlett-Packard (HP) built calculators with TAB for a few years in the eighties before switching to wirebonded packages. HP also began using TAB for its inkjet printers in the eighties and is now the largest volume user of TAB in the United States. ETA, a subsidiary of Control Data Corporation used a unique TAB process with plated solder bumps in its "all-TAB" supercomputer in the eighties. The TAB parts were made by Honeywell. Around 1986, National developed a TAB package called the "TapePak®" which used a single-layer bumped tape and a unique molded ring with test pads and licensed it to Delco Electronics which continues to make TapePaks today for automotive applications. Digital Equipment Corporation (DEC) started a large TAB project for its VAX® 9000 computer line in the eighties which led to a flurry of activity by 3M, Olin Mesa, Rogers Corporation, and others to develop a two-conductor tape. At about the same time, IBM licensed the 3M two-layer tape process and set up a production line in Endicott, New York. After DEC stopped using TAB in its VAX 9000 line a few years later, both Olin Mesa and Rogers abandoned the TAB tape business. In the equipment arena, Jade Corporation of the United States was an early leader of ILB and OLB gang bonding equipment in the seventies and eighties but is not active today. The same is true of Farco, a Swiss equipment company which was prominent in the late seventies and eighties but left the business in the early nineties.

As the I/O, speed, and performance requirements of ICs increased in the eighties and nineties, TAB became more suited to applications in liquid crystal displays (LCDs), printheads, very-high-speed integrated circuits (VHSIC), workstations, and high-end computers (see Fig. 8-74) [135–137]. The users included Bull, Casio, Cray Research, Digital Equipment Corporation (DEC), Fujitsu, Hitachi, HP, Matsushita, NEC, IBM-Japan, Seiko-Epson, Siemens, Sharp, Sun Microsystems, TI-Japan, and Toshiba.

Figure 8-75 shows a typical TAB process using bumped chips and planar tape. After wafer bumping, the wafer is mounted on a dicing tape and diced. The next step is inner lead bonding (ILB) in which the metallized

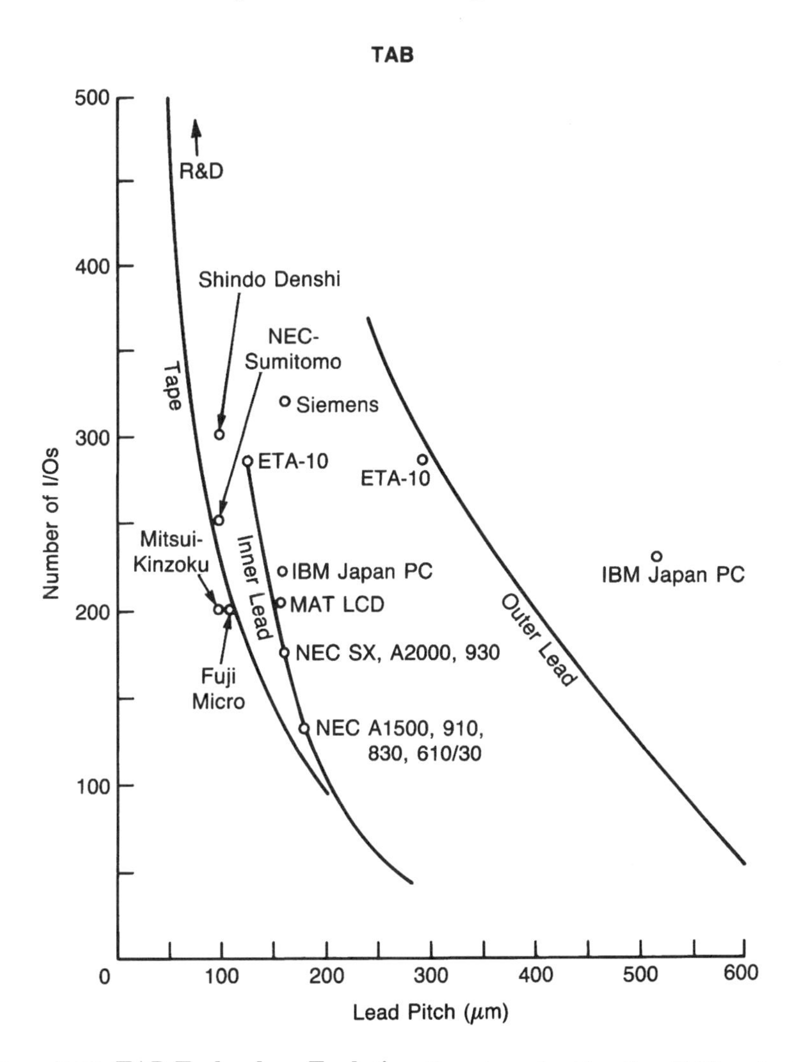

Figure 8-74. TAB Technology Evolution. Tape, inner lead bonding (ILB), and outer lead bonding (OLB).

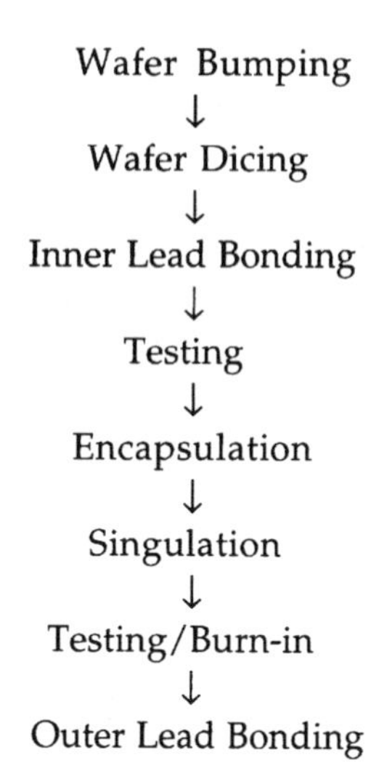

Figure 8-75. Typical TAB Fabrication Process

tape fingers are bonded to the bumps on the chip pads (see Fig. 8-76). After ILB, the chip can be tested and/or encapsulated on one or both sides. Figure 8-77 shows a schematic of the encapsulation step. Next, the device can be singulated and placed in a "slide carrier" for testing and/or burn-in prior to outer lead bonding (OLB), or it can be processed in strips or reel. Prior to OLB, each device is excised from the tape, the leads are "formed" if necessary, and aligned with the bonding pads on the substrate or card. The OLB is usually a solder bond (see Fig. 8-78). Board-mounting options include chip-up or chip-down (called "flip TAB") configurations.

There are many variations of TAB—both as a first-level interconnect

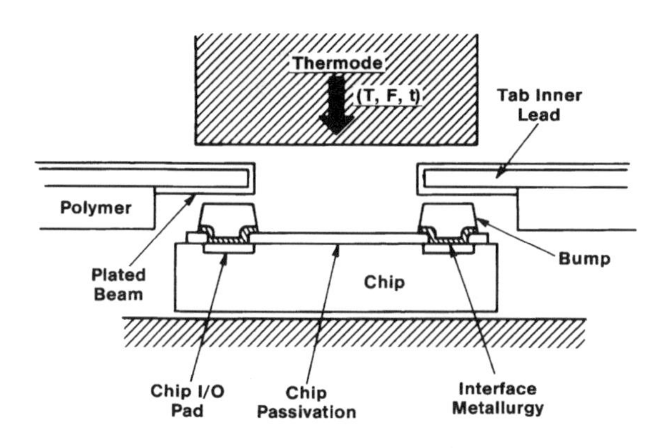

Figure 8-76. TAB Inner Lead Bonding

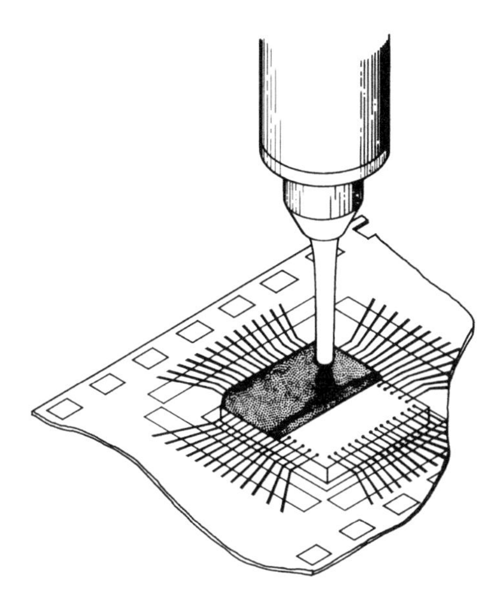

Figure 8-77. TAB Encapsulation. A single-orifice nozzle dispenses encapsulant on the bonded chip.

and as a second-level package. These include: bumped tape (BTAB), transferred bump tape (TBTAB), demountable TAB (DTAB), TapePak®, Tape Quad Flat Pack, Mikropack, and Tape Ball-Grid Array (TBGA).

Sections 8.5.2 through 8.5.7 cover the various tape structures, bumping methods, ILB and OLB processes, package options, and applications. Electrical design, thermal design and reliability issues are covered in

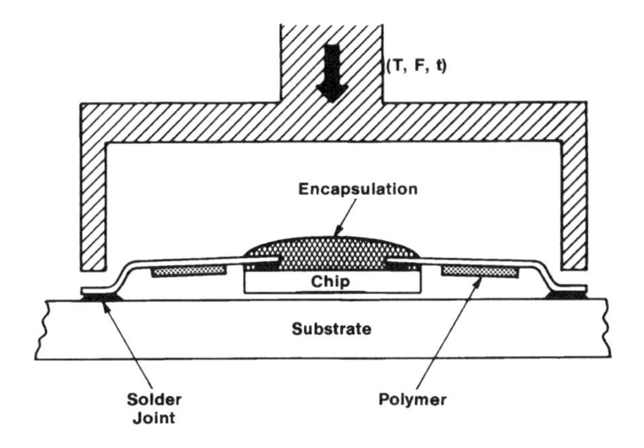

Figure 8-78. TAB Outer Lead Bonding. After the TAB package is excised from the tape and placed on the substrate or board, it is normally solder bonded.

Chapters 3, 4, and 5, respectively, and are not discussed separately in this section.

8.5.2 TAB Tape

The TAB tape (also called "tape carrier" or "film carrier") comes in a variety of shapes, sizes, base materials, and surface finishes [138–147], as shown in Table 8-4. It has three basic formats: one layer, two layer, and three layer (see Fig. 8-79). Table 8-5 shows the process steps for the fabrication of the three basic tape formats.

8.5.2.1 One-Layer Tape

One-layer tape is made of metal, typically copper, 35–70 μm thick. It can be etched or stamped and has flat (planar) lead tips or "bumped" lead tips. Both plated and unplated single-layer tapes have been used in the past. In addition to copper, aluminum, steel, alloy 42, and other conductive materials can be used for the base metal.

In general, a device bonded to single-layer tape is not testable on tape because all the leads are shorted together (except in the case of the TapePak® design which has a special test ring). One-layer etched tape does not need special punching tools for punching the sprocket holes and personality window (also called "device hole") because these are etched at the same time as the leads. This makes one-layer tape the least expensive of the three formats. It is therefore widely used in cost-sensitive, low-lead-count applications such as watches, security tags, IC cards, blood pressure monitors, and car radios.

8.5.2.2 Two-Layer Tape

Two-layer tape typically consists of a copper layer directly bonded to polyimide dielectric film without an organic adhesive. The fabrication

Table 8-4. TAB Tape Options

1. Format	One, two, or three layers
2. Conductors	One, two, or more
3. Copper foil	Rolled (as rolled or annealed); electrodeposited
4. Dielectric	Polyimide, epoxy–glass, polyester, BT
5. Adhesive	Modified epoxy, phenolic butyral, polyester, polyimide
6. Width (mm)	Super 8, 8, 11, 16, 19, 24, 35, super 35, wide 35, 48, super 48, wide 48, 70, super 70, wide 70, 120, 140, 158
7. Surface structure	Planar; bumped
8. Surface finish	Tin, gold, nickel–gold, solder (tin–lead); selective plating

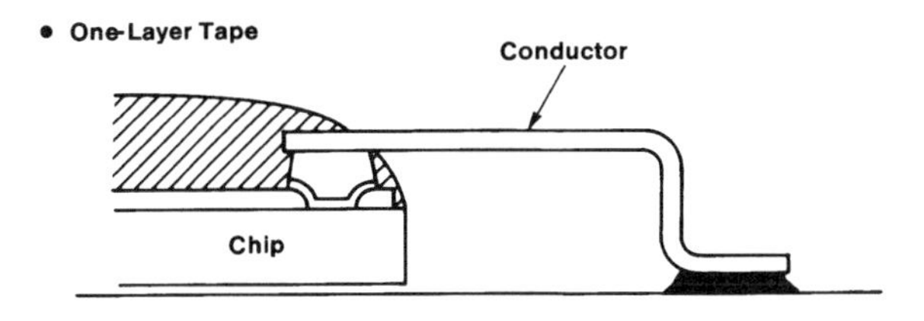

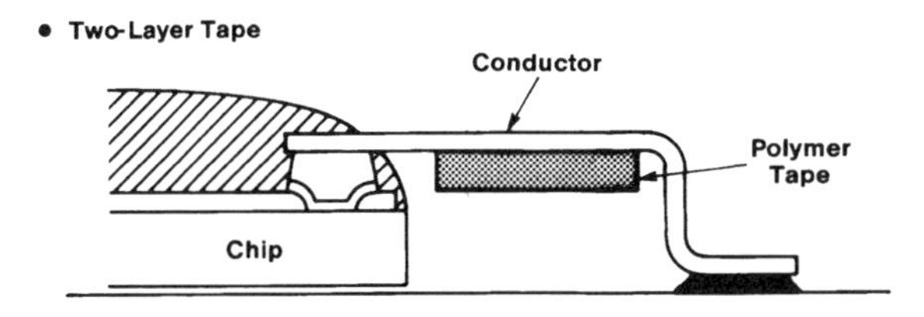

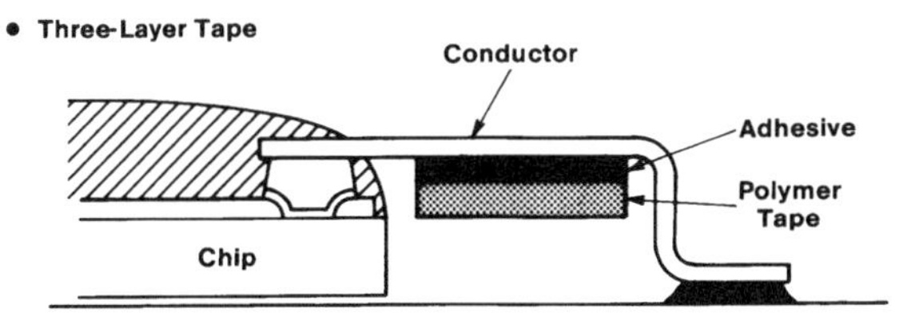

Figure 8-79. Generic Types of TAB Tape

Table 8-5. TAB Tape Fabrication Steps

1-Layer All Metal	2-Layer Additive Process	3-Layer Subtractive Process
1. Slit metal foil	1. Deposit metal adhesion layer/common electrode	1. Punch adhesive-coated polymer
2. Apply photoresist	2. Apply photoresist (2 sides)	2. Laminate metal foil
3. Expose	3. Expose (2 sides)	3. Cure adhesive
4. Develop	4. Develop	4. Apply photoresist
5. Etch metal pattern	5. Pattern plate	5. Expose
6. Strip photoresist	6. Etch polymer	6. Develop
7. Clean	7. Strip photoresist	7. Coat back side for lead protection
8. Surface plate	8. Etch common electrode	8. Etch metal
	9. Clean	9. Strip photoresist
	10. Surface plate	10. Clean
		11. Surface plate

process can be either subtractive (called "two-layer thin" by 3M, the leading two-layer TAB tape supplier in the world) or additive (called "two-layer thick" by 3M). In the subtractive process, a thin (typically 12 μm) liquid polyimide film is cast (coated) on 35-μm-thick copper, and the circuit pattern is obtained by etching the copper and polyimide. The process has two inherent disadvantages: The polyimide shrinks after cure, causing dimensional and adhesion problems, and second, the polyimide cure process anneals the copper foil. The subtractive process is not used much today.

In the additive fabrication process for two-layer thick tape, thin layers of chromium and copper are sequentially sputtered onto the polyimide film which is typically 50–75 μm thick. Chromium promotes adhesion between the polyimide and the copper, the latter acting as the seed layer for the final pattern plating through a photoresist.

Two-layer tape allows testing and burn-in because of the electrical isolation between leads provided by the polyimide. Also, it offers better handling and lead stability than one-layer tape. Two-layer tape does not require hard tooling and has a faster turnaround time than three-layer tape.

8.5.2.3 Three-Layer Tape

Three-layer tape consists of a metal foil, usually copper, laminated with an organic adhesive to a base film such as polyimide. The polyimide film, typically 75–125 μm thick, and the adhesive layer are mechanically punched with steel dies to form sprocket holes and inner lead and outer lead windows. The copper foil is either electrodeposited (ED) or rolled (wrought). Each has advantages and disadvantages depending on the application. ED foil is more popular in Japan and Europe. Rolled foil is more common in the United States. Copper foil thickness is usually specified in "ounces per square foot" ("ounces"). "One-ounce" copper is 35 μm thick and weighs 1 oz/ft^2 or 305 g/m^2. A two-ounce copper foil is 70 μm thick, and a half-ounce foil is 18 μm thick. Copper foil properties such as crystal structure, grain size, hardness, tensile strength, surface roughness, yield strength, ductility, adhesion to dielectrics, and so forth are controlled by a variety of proprietary processes and treatments. ED foil has a columnar grain structure which can be modified by annealing to produce equiaxial grains. Because of the fabrication process, one side of the ED foil is much smoother than the other side. Rolled foil has a laminar crystal structure, with the grains aligned along the length of the material due to the rolling process. This results in anisotropy—the material behaves differently in the longitudinal direction (also called "machine direction") than in the transverse direction. Like ED foil, as-rolled (AR) foil can be annealed and treated to improve its properties for a particular application.

The most common dielectric film materials used in three-layer tape are polyimide, polyester, glass–epoxy, and bismaleimide triazine (BT). Polyimides such as Kapton® and Upilex®, are preferred in high-performance applications. Lower-cost glass–epoxy and polyester base materials are used in consumer products such as calculators and smart cards. BT is used in applications requiring a higher glass-transition temperature than glass–epoxy or polyester.

The adhesives used in three-layer tape are mostly proprietary formulations. The formulations include modified epoxy, polyimide, phenolic butyrals, polyesters, and acrylics. Adhesive characteristics are extremely important for high-I/O, high-performance applications where high thermal stability, high adhesion, low moisture absorption, and low ionic impurities are critical.

Figure 8-80 shows a three-layer TAB tape layout and cross section listing the terms used to describe the tape parameters. Figures 8-81 and 8-82 show two tape designs: a 200-pin EIAJ 35-mm design, and a "super slim TAB" (SST) LCD driver design, respectively. Figures 8-83 and 8-84 show photographs of low- and high-pin-count tapes, respectively. Figure 8-85 shows a variety of tape formats.

Because the three-layer process uses subtractive methods to etch copper foil, the thickness of the copper foil determines the lead pitch (width and spacing) that can be etched. Finer lead pitches therefore require thinner copper foils [148].

Plating plays a key role in the success of inner lead bonding (ILB) and outer lead bonding (OLB). It also affects other parameters such as storage life, burn-in capabilities, long-term reliability, and cost. Plating options include tin plating, gold plating, solder (tin–lead) plating, and "selective plating" where inner leads and outer leads have different materials or thicknesses.

The most common selective plating is tin for ILB and solder for OLB. Other choices include gold for ILB and solder for OLB, tin for ILB and gold for OLB, and thin solder for ILB and thick solder for OLB.

The plating thickness depends on the application. Typical ranges are 0.3–0.6 μm for tin and solder, 0.3–4.0 μm for gold, and 0.3–1.0 μm for a nickel underlayer for gold plating [149].

Tin plating has not been popular in the United States because of tin's tendency to form "whiskers"—single crystals of tin which grow under certain conditions after tin plating [150–152]. They can be as long as a few millimeters and cause electrical shorts. Whisker growth is associated with built-in stresses during plating. It can be minimized or suppressed by heat treatment (annealing), alloying the tin with lead, nickel, or other materials, reflowing the tin after plating, or using a solder mask to cover unbonded areas. Tin also has a tendency to oxidize and has to be stored and shipped in a nitrogen ambient.

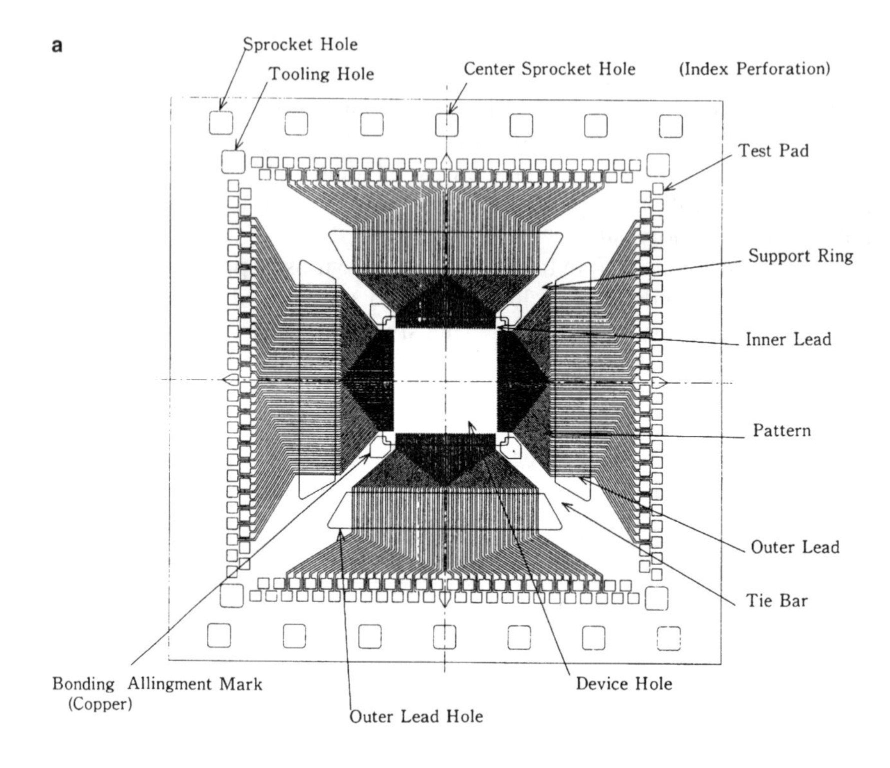

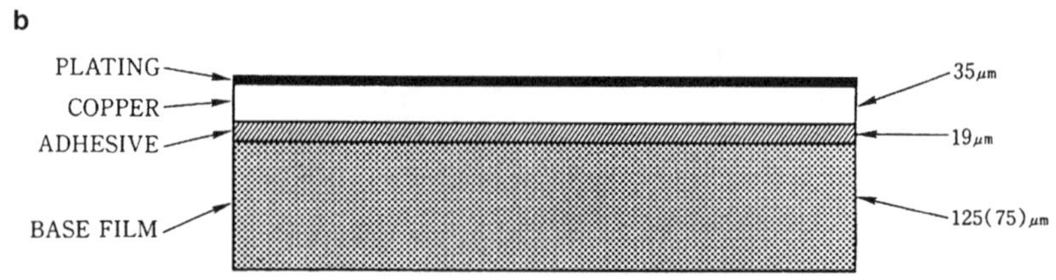

Figure 8-80. Tape Layout (a) and Cross Section (b) of Three-Layer Tape. (Courtesy of Shindo Co. Ltd.)

8.5.2.4 Bumped Tape

Most TAB applications use bumped chips and planar (unbumped) tape. However, there have been some applications over the years where bumped tape has been preferred. One of these is the TapePak which is described later in the chapter. Figure 8-86 shows the basic structure of (a) a bumped chip and (b) a bumped tape. A big advantage of bumped tape is that it can be used with standard wafers (chips) just as in wirebonding. Single-layer bumped tape is the easiest to manufacture and has been used in watches, sensors, and car radios [153]. An example of a single-

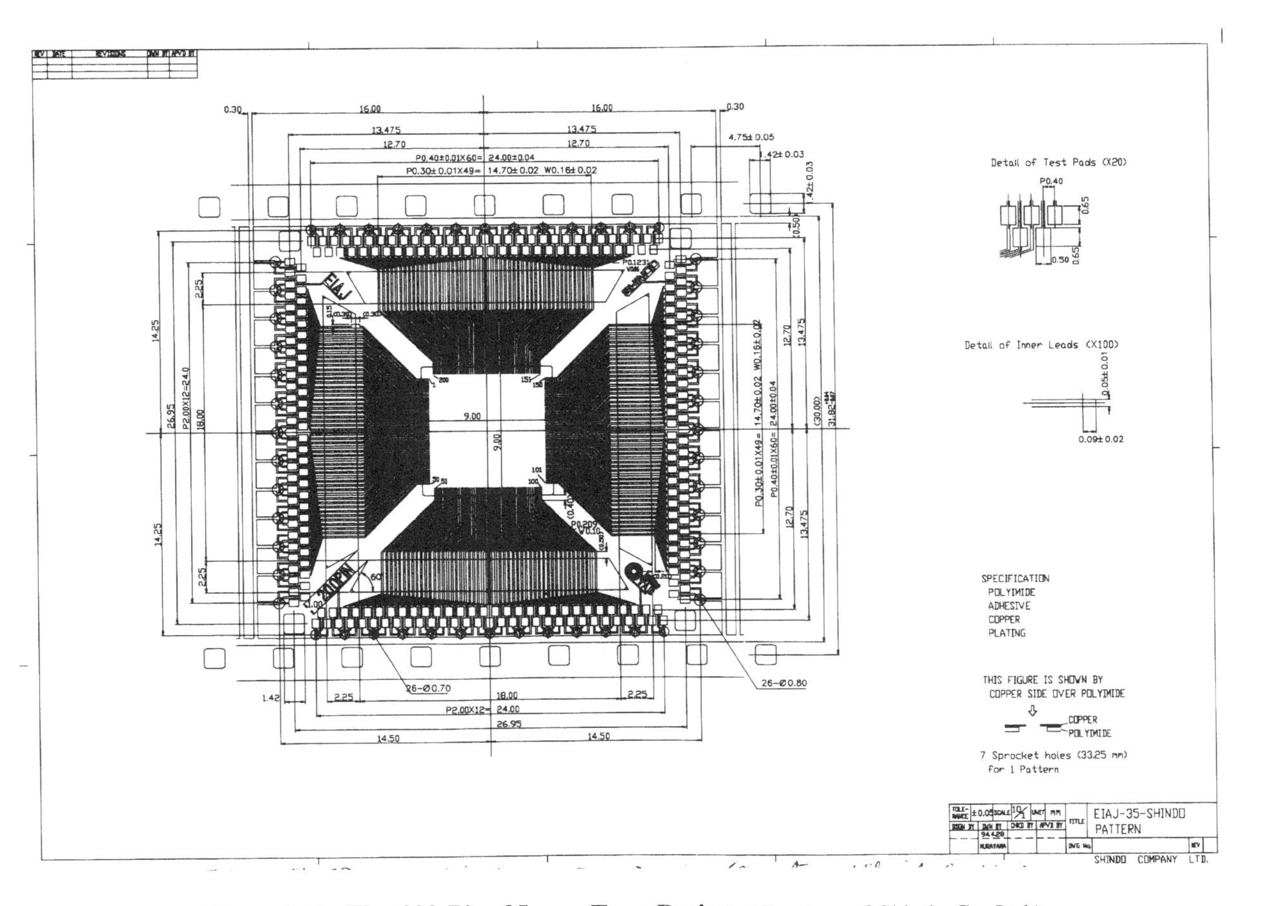

Figure 8-81. The 200-Pin, 35-mm Tape Design. (Courtesy of Shindo Co. Ltd.)

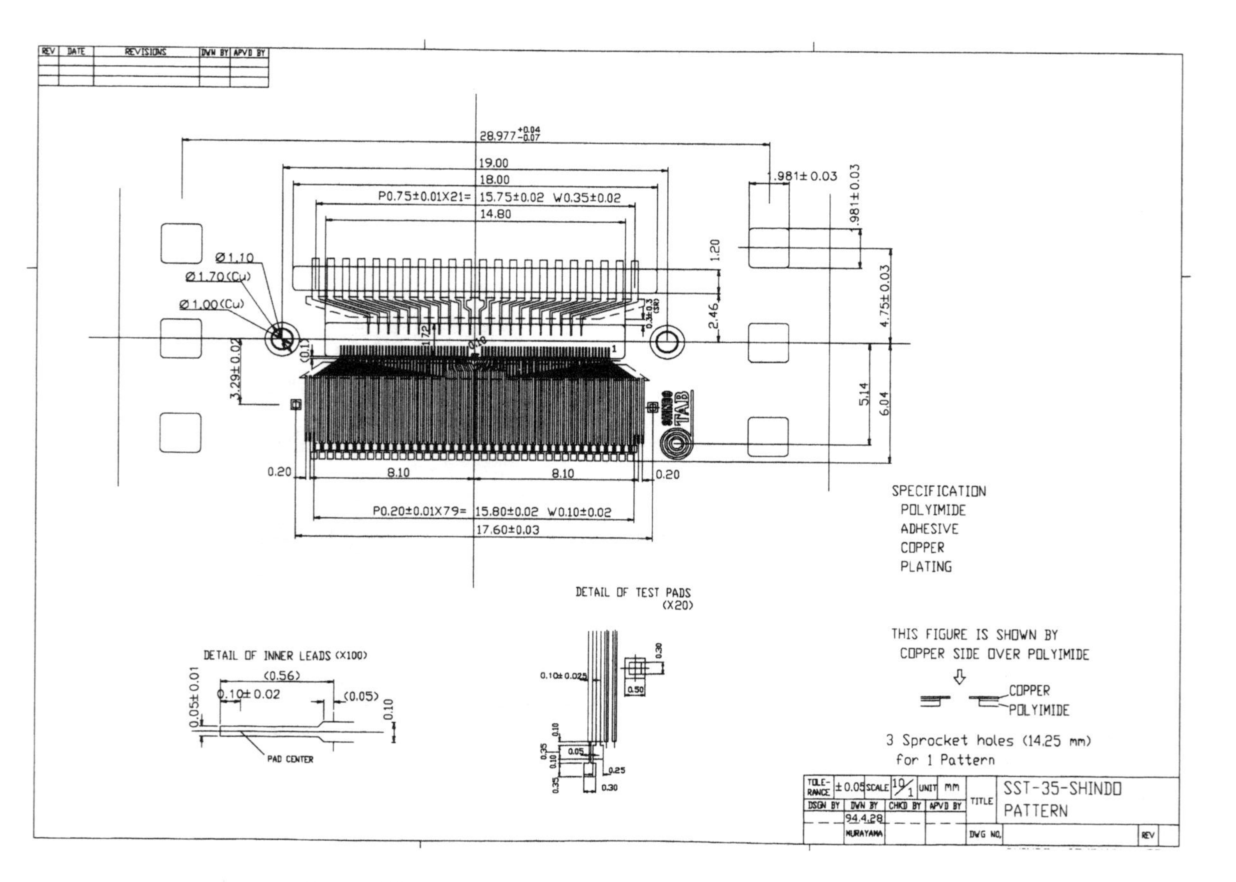

Figure 8-82. Super Slim TAB LCD Tape Design. (Courtesy of Shindo Co. Ltd.)

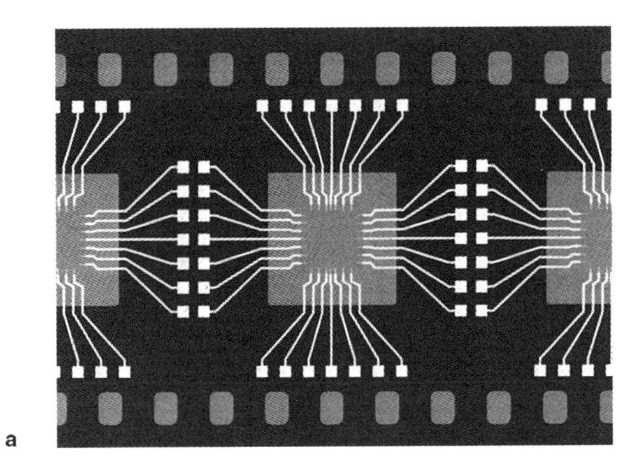

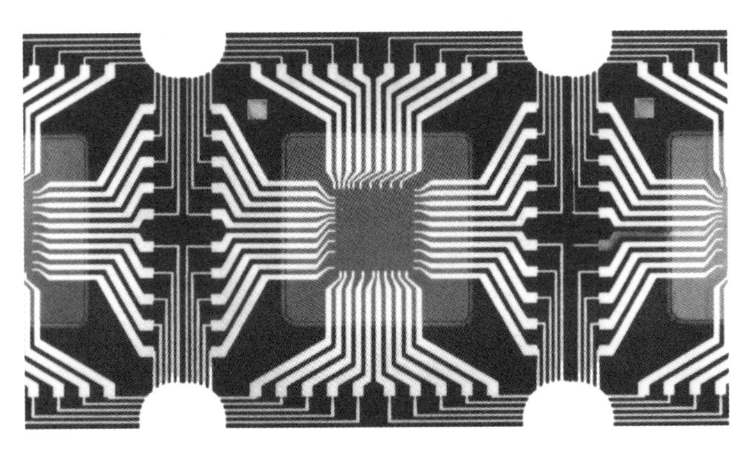

Figure 8-83. Low-I/O Tab. (a) A 28-lead three-layer tape (courtesy of Shindo Denshi Ltd.); (b) a 16-lead one-layer tape (courtesy of 3M Electronic Products Division); (c) a 40-lead three-layer tape (courtesy of Mesa Technology).

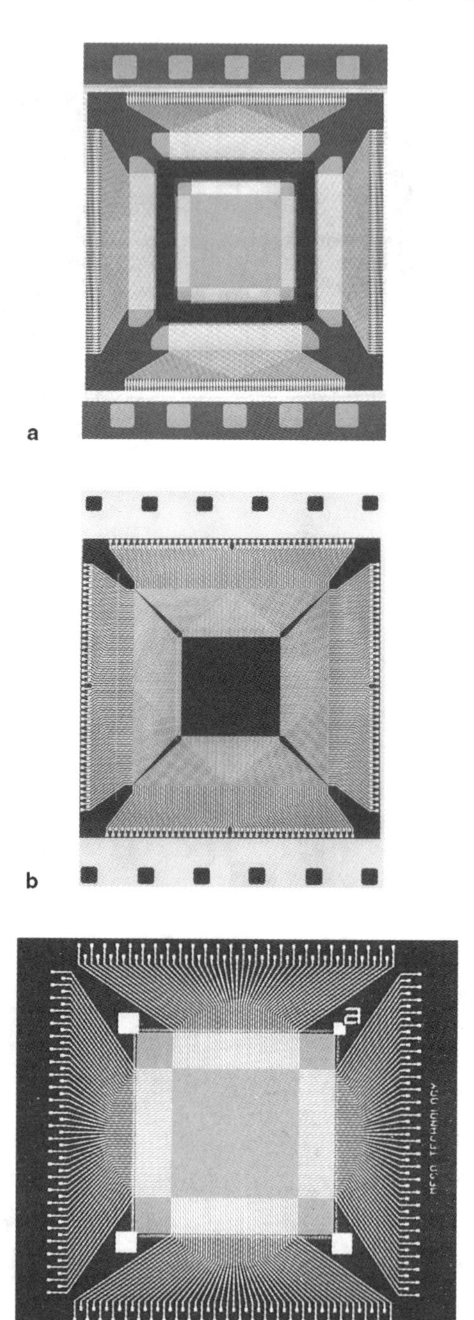

Figure 8-84. High-I/O TAB. (a) A 308-lead three-layer tape (courtesy of Shindo Denshi, Ltd.); (b) a 328-lead two-layer tape (courtesy of 3M Electronic Products Division); (c) a 204-lead three-layer tape (courtesy of Mesa Technology).

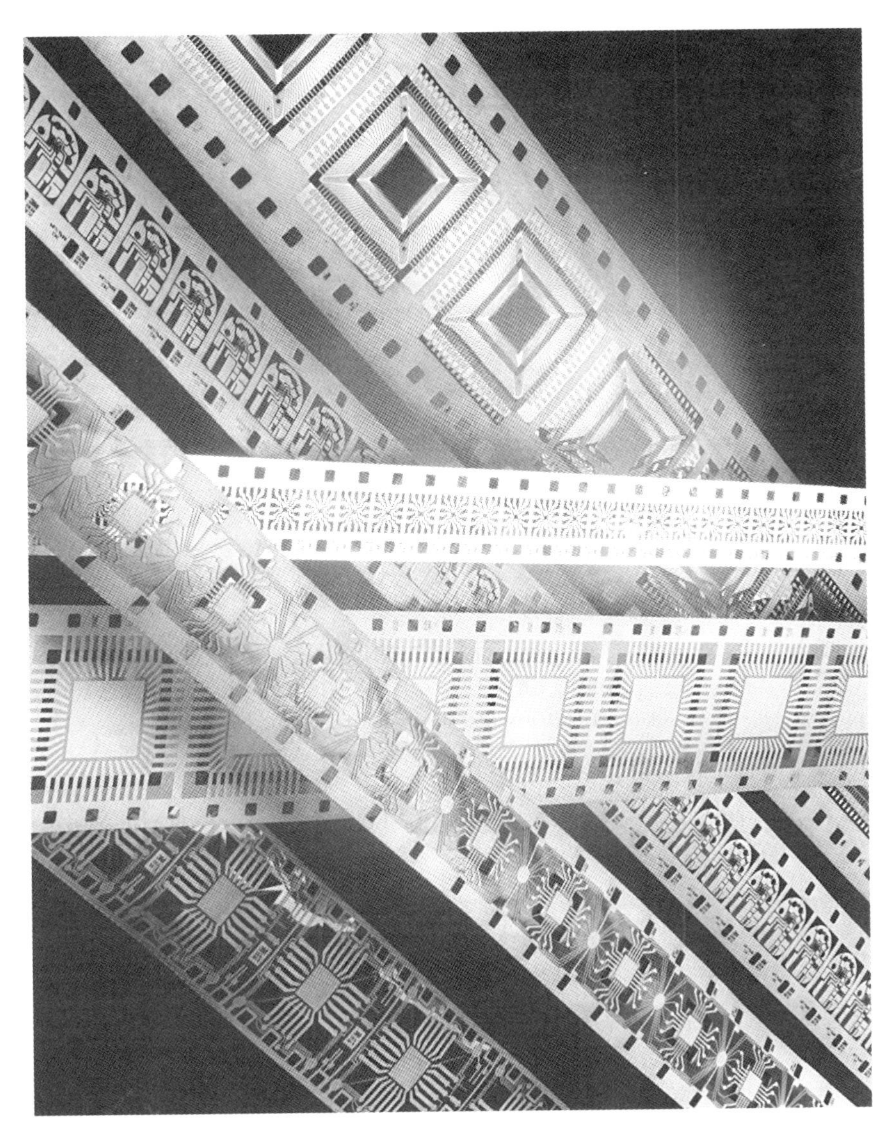

Figure 8-85. Examples of TAB Tape Formats. (Courtesy of 3M Electronic Products Division and Shindo Denshi, Ltd.)

layer bumped tape is shown in Figure 8-87. Attempts to use three-layer bumped tapes have not generally been commercially successful except for Matsushita's "transferred bumped tape" process which is discussed in the next section. One reason was that it was difficult to manufacture a three-layer BTAB polyimide-up configuration, which severely limited surface-mount options. Masuda et al. have solved this problem and devel-

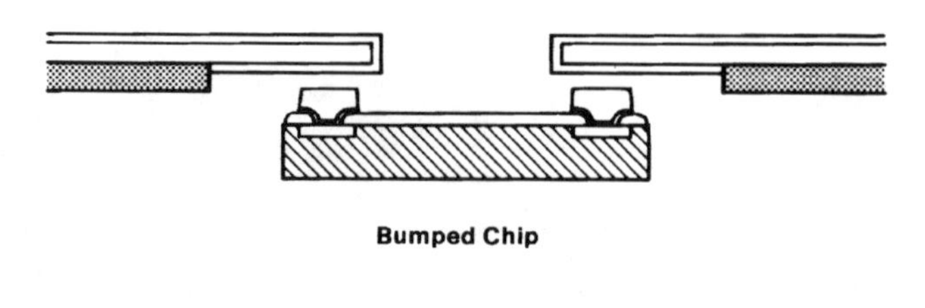

Bumped Chip

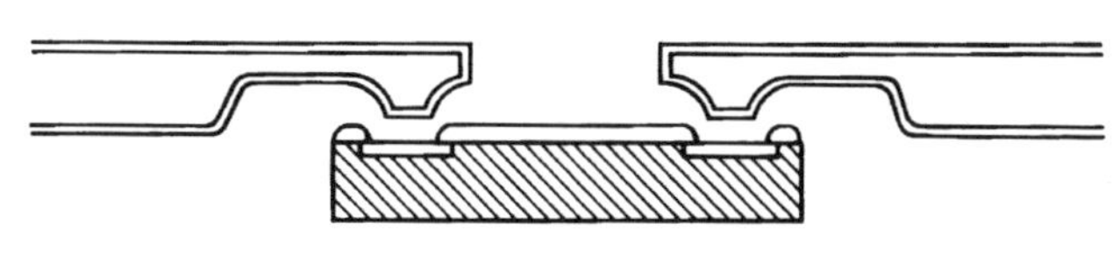

Bumped Tape

Figure 8-86. Basic Structure of (a) Bumped Chip and (b) Bumped Tape

oped a quad tape carrier package (QTCP) for a CPU card [154]. The various mounting options with the new BTAB tape with polyimide-up or polyimide-down configurations are shown in Figure 8-88.

8.5.2.5 Transferred Bump TAB

A novel approach to making bumped tape was developed by Matsushita Electric in the mid-eighties [155,156]. Instead of etching the bump

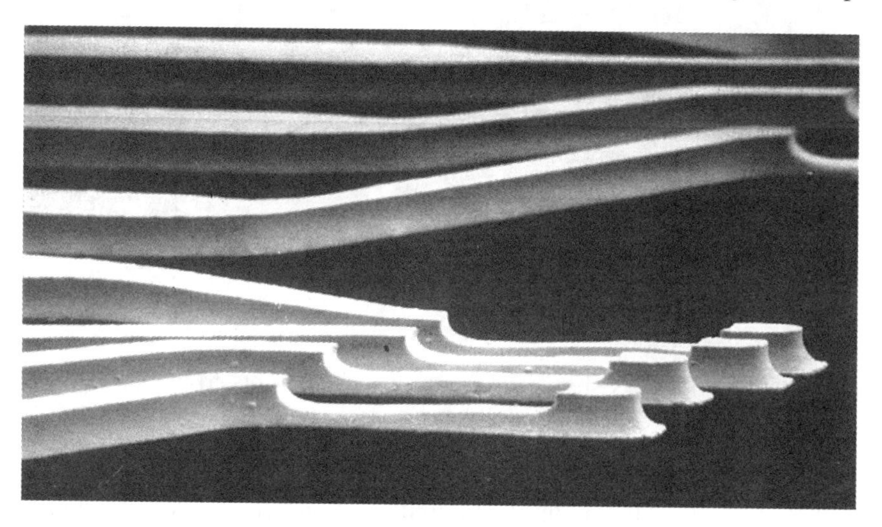

Figure 8-87. Bumped One-Layer TAB Tape. (Courtesy of Mesa Technology.)

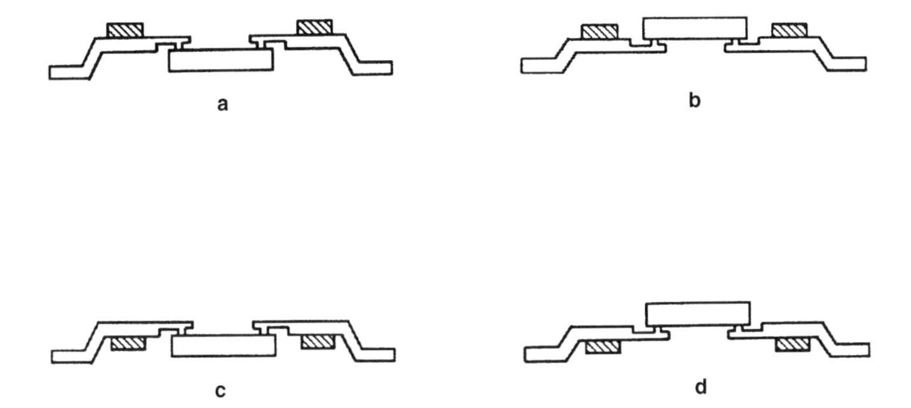

Figure 8-88. Surface Mount Structure of TCP. (a,b) Polyimide up; (c,d) polyimide down. (From Ref. 154.)

using double-sided photolithography, the bumps are first formed on a glass substrate by selective masking and gold plating over a conductive layer. The bumps are then transferred to the tape ILB fingers using heat and pressure. This is called "first bonding." Next, the bumped tape is aligned over the die pads and inner lead bonded to the die. This is called "second bonding" (See Fig. 8-89).

This process was well suited for consumer products such as hearing aids and LCD TVs. For high-I/O, high-performance application-specific integrated-circuit (ASIC) devices which require a finer pad pitch and smaller volumes, Matsushita then developed an improved process [157] using a 127-mm square substrate with 250,000–500,000 bumps, and a single-point bonding process for both the first bond (bump transfer to the leads) and the second bond (ILB to the chip). Figure 8-90 shows the new bonding technique. Toray, a Matsushita licensee, further improved the process by developing a straight-wall bumping technique for 80-μm-pitch bumps, and a method to transfer them to either the "film side" or "copper side" of the bonding fingers [158].

Tatsumi et al. have developed a variation of the transferred bump process by using a wirebonder to make gold balls which are then transferred to TAB tape [159]. The technology is said to be capable of bonding to lead pitches of 80 μm or less.

8.5.2.6 Two-Conductor Tape

The electrical performance of a TAB device is closely tied to the tape design. For high-lead-count, high-performance applications, a second metal layer can be added to be used as a ground plane or to distribute both power and ground to the chip [160]. This type of tape is called

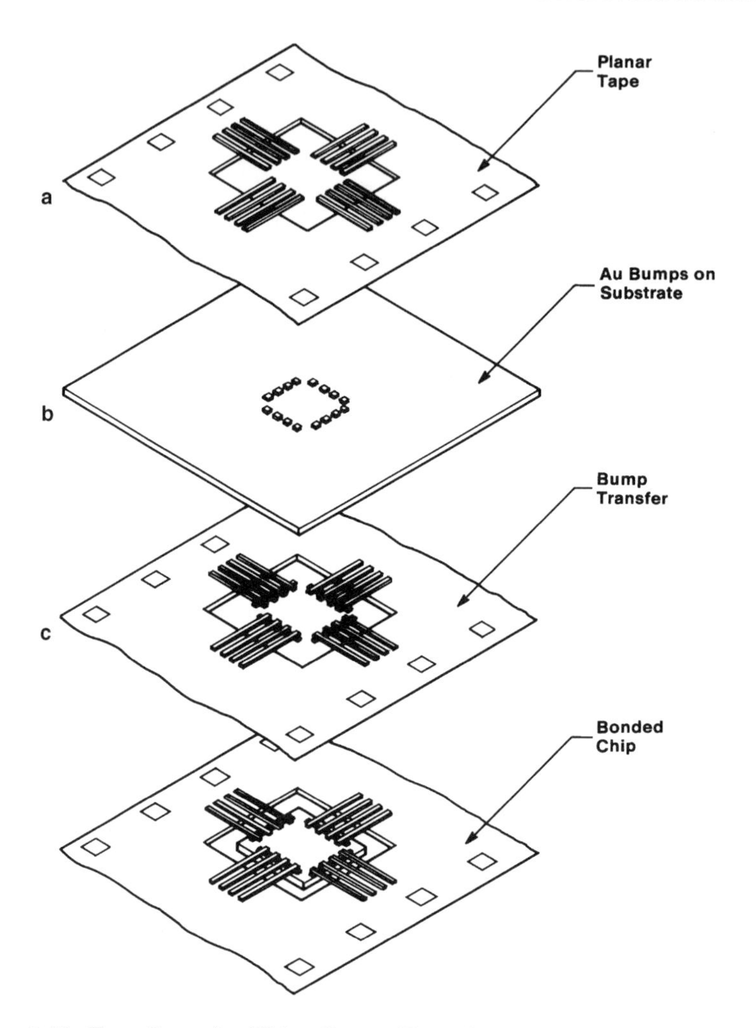

Figure 8-89. Tape Bumping Using Bump-Transfer Process. In this process, planar tape is modified by transferring gold bumps to the tape. (From Ref. 155.)

"ground-plane TAB tape" (GTAB), "two-metal-layer tape," "two-conductor tape," "double-level metal tape," or "double-metal-layer tape" [148,161–165]. It reduces the inductance of the power and ground lines, provides better impedance matching, and also reduces cross-talk. Figure 8-91 compares the noise performance of a single-level metal TAB, a double-level metal TAB, and a double-level metal area-array TAB for a given set of constant conditions. The area-array tape had the best noise reduction. The standard two-metal-layer tape offered at least a 50% noise reduction compared to the single-metal tape [166].

Motorola was the first volume producer of two-metal-layer 360-lead

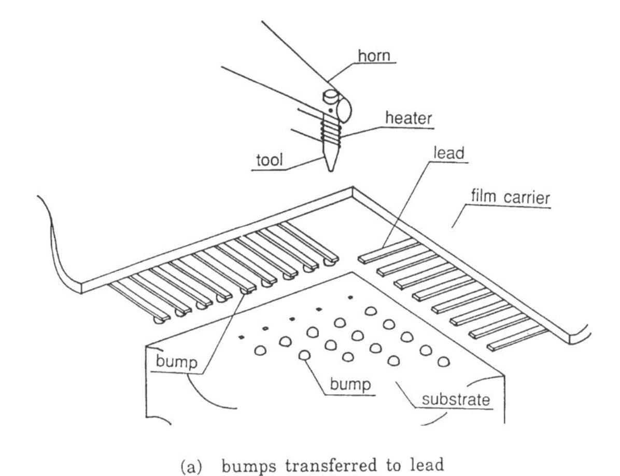

(a) bumps transferred to lead

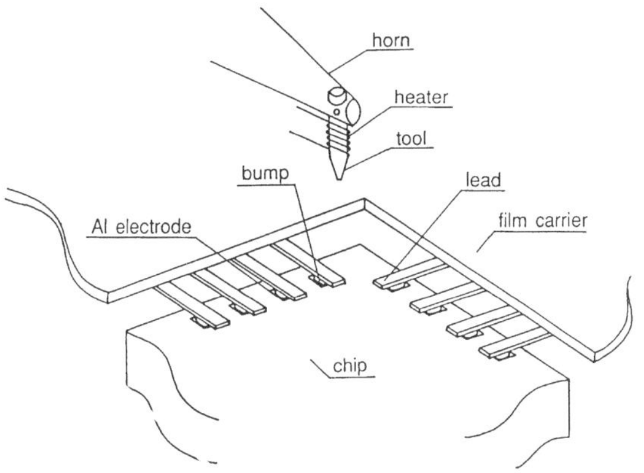

(b) bumps bonded to chip pads

Figure 8-90. Principle of New TBTAB Bonding Technique. (From Ref. 156.)

emitter coupled logic (ECL) TAB gate arrays for DEC's VAX 9000 computer [167]. Other companies in the United States and Japan are also using two-conductor tape for special applications.

The starting material for two-conductor tape can be either the three-layer tape with an additional copper foil laminated to the polyimide with an adhesive (making it a five-layer starting material) or it can be a two-layer adhesiveless process starting material with an additional copper plane.

For the five-layer starting material, a typical manufacturing process would be to punch the sprocket holes, device holes, alignment holes, debuss holes, and so forth, chemically etch the copper circuit pattern, form the vias with a laser, interconnect the vias by an electroless prime

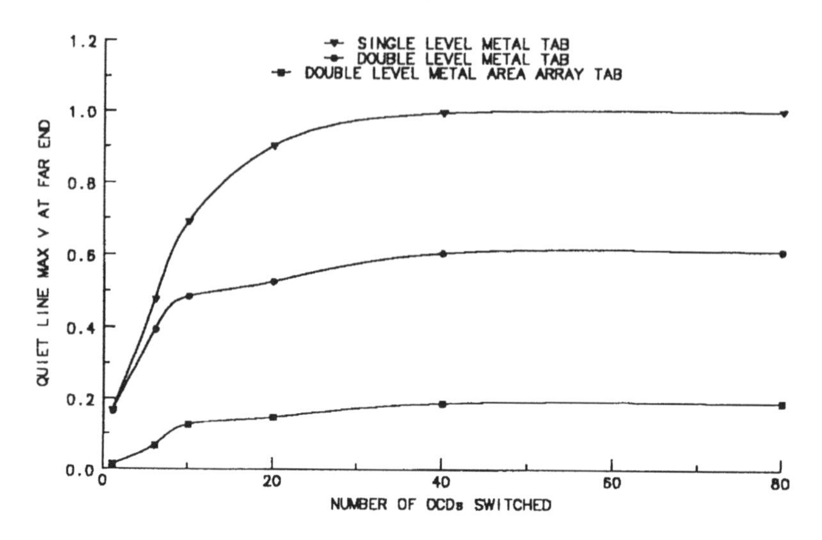

Figure 8-91. Noise Versus Simultaneous Switch for Three TAB Configurations. (From Ref. 166.)

followed by electroplated copper, and, finally, cover plate with gold or tin. Figure 8-92 shows a cross section of a two-conductor tape.

A significant benefit of two-conductor tape is that it allows "area-array" TAB (ATAB) design (see Fig. 8-93). ATAB improves I/O density by overcoming the limitations of peripheral-only interconnections. The IBM tape ball-grid array, discussed in Section 8.5.6, uses this approach.

8.5.2.7 Electrostatic Discharge Protection

In reel-to-reel high-volume production of TAB devices, the tape may be 20–100 m long. The devices are shipped to the customer in

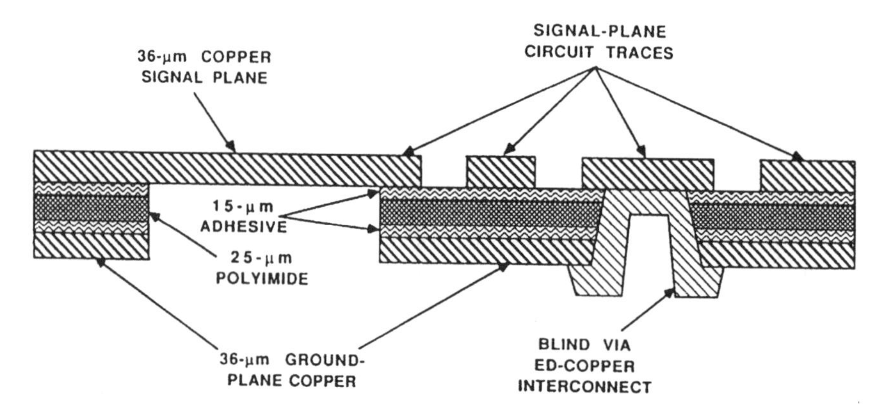

Figure 8-92. Cross-Section of Two Conductor Tape (After Hoffman, Ref. SK27, © STC, 1990)

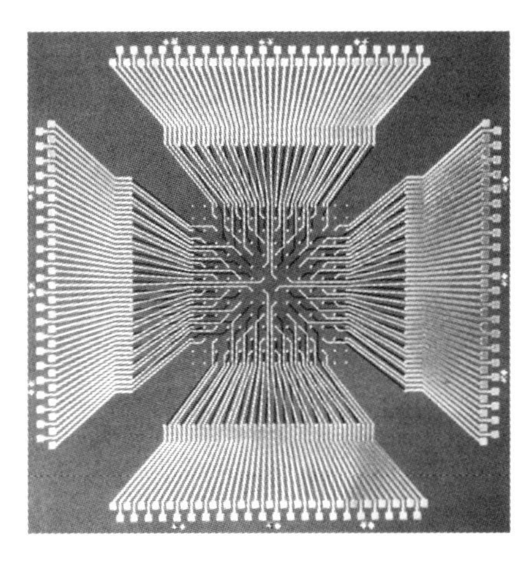

Figure 8-93. Area TAB. A 180-lead tape format is shown arranged for area bonding to the chip. (Courtesy of 3M Electronic Products Division.)

reel form after inner lead bonding with a "separator" (typically made of polyester film) which is taken up along with the tape as each device is processed on the production line. Researchers at Sharp Corporation [168] found this process capable of generating electrostatic discharge (ESD) failures due to (1) the tape producing exfoliative static charges when separated from the separator on the supply reel, (2) very large device surface areas contributing to the generation of frictional charges, and (3) high probability that the tape will be subjected to discharge from operators or equipment. Typical failures included dielectric breakdown of field-effect transistor (FET) gate-oxide layer, molten diffusion resistance in the vicinity of I/O pins, and field-oxide layer breakdown.

In spite of taking the normal anti-ESD measures, Sharp researchers found that one device mode ESD damage persisted at a low but constant rate. This was the "charged device model" (CDM), where ESD damage occurs when static charges built up in a device discharge through the equipment or human body. There were two solutions: (1) increase the material's relative dielectric constant and (2) optimize patterning. The first approach is not desirable for high-speed devices because a higher dielectric constant means reduced speed. So Sharp used the second approach to solve the ESD problem. Figure 8-94a shows four TAB patterns: pattern A has no guard ring; pattern B has the devices surrounded by a common guard ring; pattern C has each device surrounded with a separate guard ring which is connected to the chip substrate; and pattern D is a version of pattern B in which the guard ring is connected to the chip substrate. Figure 8-94b shows the potential distributions of patterns A and D. Pattern A had 23% failures; pattern B had 13.5% failures; pattern C had 19.5% failures; and pattern D had 3.9% failures.

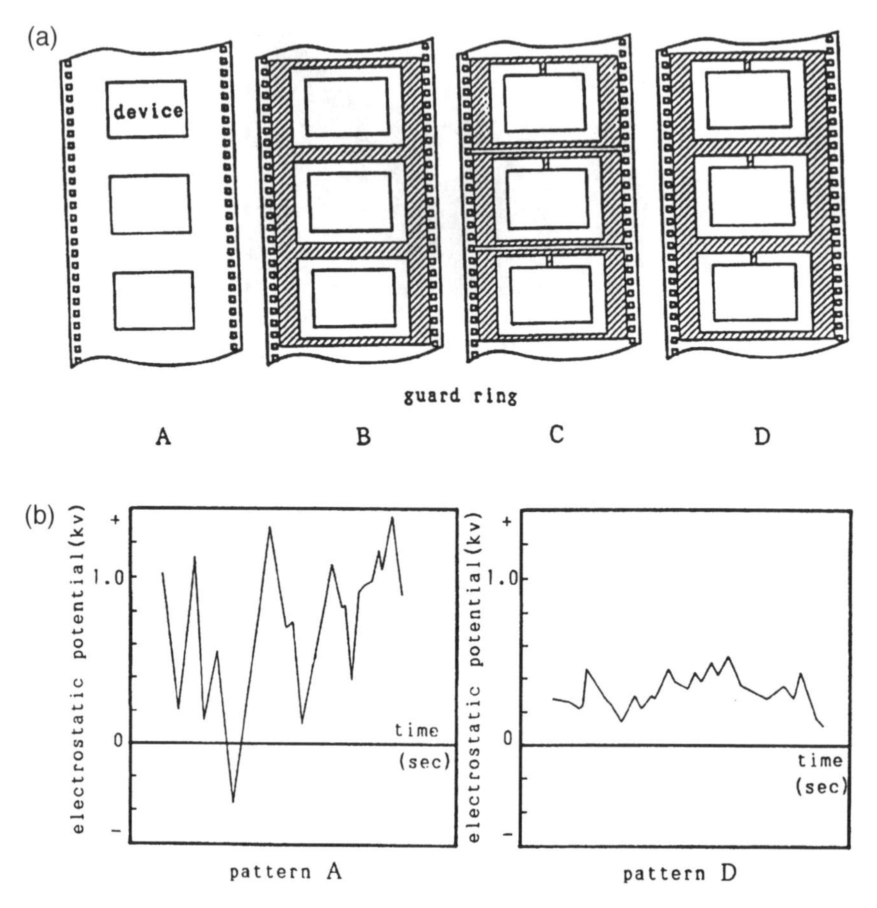

Figure 8-94. (a) TAB Patterns for ESD Experiment; (b) Potential Distribution of Patterns A and D. (After Tajima et al, Ref. SK37, © STC 1990)

8.5.3 Wafer Bumping

Wafer bumping is the process of forming "bumps" over the bonding pads at the wafer level or at the chip level. Typical bump materials include gold, copper, and solder (Sn/Pb) singly or in combination. Other conductive metals such as aluminum, indium, nickel, and silver, and various alloys have also been used in different applications. The bumps serve several purposes: They provide the right metallurgy for a variety of chip-joining processes; they provide "standoffs" which prevent shorting of the lead fingers to the chip edge; they protect the bond pad metallization [usually Al or Al–alloy (Si/Cu)] from corrosion and contamination (see Fig. 8-95); and they act as a deformable (ductile) buffer during gang inner lead bonding.

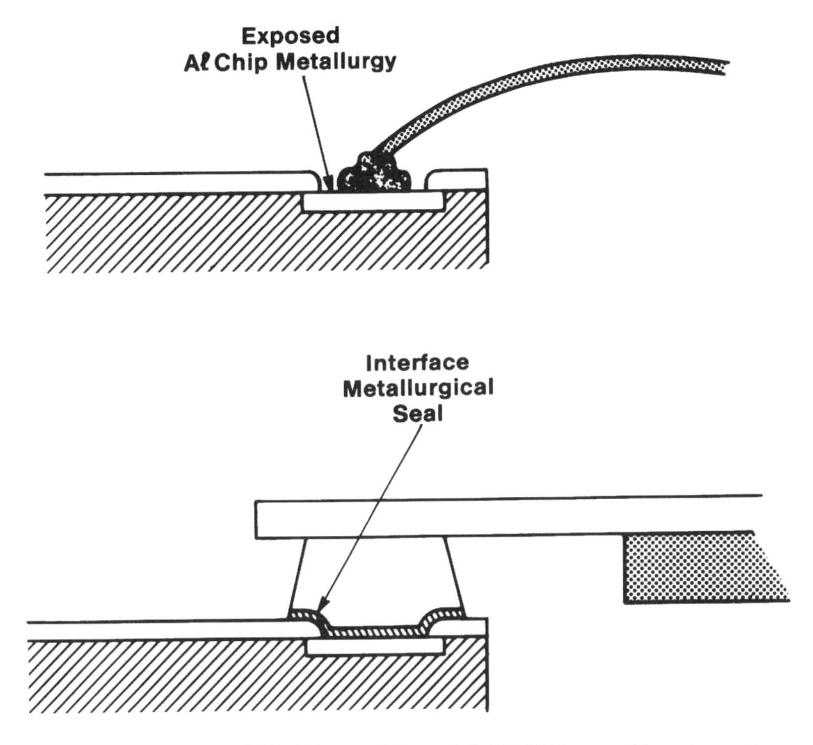

Figure 8-95. Improved Chip Hermeticity with TAB Bump Structure. A potential corrosion-resistance advantage may accompany the full coverage of the aluminum bonding pad.

Gold bumps are the most common. A typical gold bumping process (see Fig. 8-96) [169] starts with sputter-cleaning of the Al–alloy bond pads on a passivated silicon wafer in a vacuum system. A thin layer (about 200 nm) of Ti–W alloy is sputtered over the wafer. Alternately, Ti or Cr is used as the bottom "adhesive" layer, and W is used as a "barrier" layer to prevent interdiffusion between Au and Al. Other diffusion barrier metals include Cu, Mo, Ni, Pd, and Pt. Next, a thin layer of Au is sputtered on top of the barrier layer. It protects the barrier layer from oxidation and also serves as the plating base for the electroplated bump. Prior to electroplating, a liquid or dry film photoresist is applied to the wafer and serves as the plating mask for the selective gold plating of the bonding pads. The photoresist is baked, exposed, and developed. Plasma cleaning is used to remove any organic residues prior to electroplating to the desired gold thickness (typically 20–25 μm). The final step is to "anneal" the bumps in a nitrogen atmosphere to reduce the as-plated hardness to a level suitable for inner lead bonding.

For low-cost TAB applications, a modified wirebonder can be used to

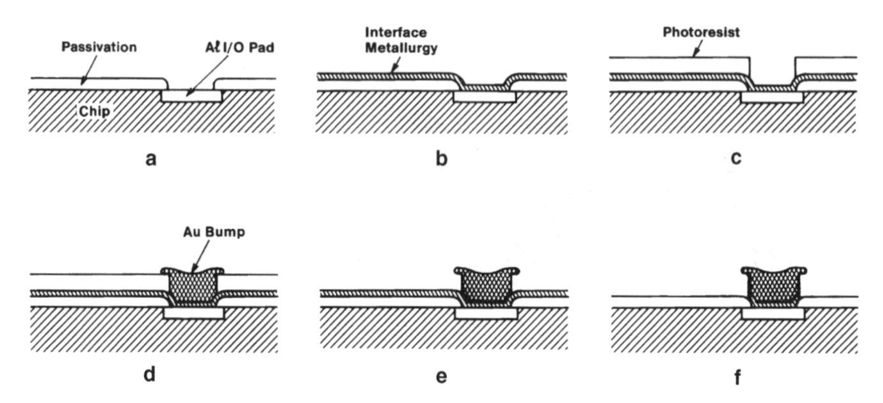

Figure 8-96. Gold-Bump Processing. (From Ref. 169; reprinted with permission of *Solid State Technology.*)

put down gold balls directly over the aluminum pads using a conventional thermosonic or thermocompression ball bonder [170]. A separate coining operation with a flat tool may be used to flatten the tail above the ball and provide a planar surface for ILB (see Fig. 8-69). NEC has implemented an ILB method using this technology for volume production of application-specific packages [171].

Solder bumps are becoming more important because they can be used both for TAB and flip-chip bonding. They are formed by evaporation of Sn/Pb through a metal mask over a suitable under bump metallurgy (UBM) such as Cr–Cu–Au, or by electroplating [172,173]. Researchers at Toshiba Corporation have developed a method of forming solder bumps by electroless plating and ultrasonic soldering [174]. Tanaka Denshi Kogyo Co. has demonstrated a method of forming solder balls over Al pads (with Cr–Cu–Au metallization) at the chip level using a wirebonder and a specially formulated 2 Sn/98 Pb solder wire [175].

8.5.4 Inner Lead Bonding

TAB inner lead bonding (ILB) was originally conceived as a gang bonding step in which all the TAB leads are simultaneously bonded to the chip electrodes using heat and pressure [thermocompression (T/C) bonding]. The objective is to provide a strong metallurgical bond with high strength. Another ILB method used predominantly by U.S. companies is called single-point bonding, where each connection between the chip pad and the tape finger is made sequentially, one at a time. In addition to thermocompression and thermosonic bonding, other options include the following:

- Eutectic/solder/hot gas
- Laser
- Laser sonic

8.5.4.1 Thermocompression Bonding

For T/C bonding, the main bonding parameters are temperature, pressure, and bonding (dwell) time. They are selected on the basis of bond reliability and manufacturing throughput trade-offs [176]. T/C gang bonding is usually done with a constant heat thermode or a pulse heat thermode. The thermode can be a solid tool or a bladed tool (Fig. 8-97). Constant heat thermode shanks are typically made of low-thermal-expan-

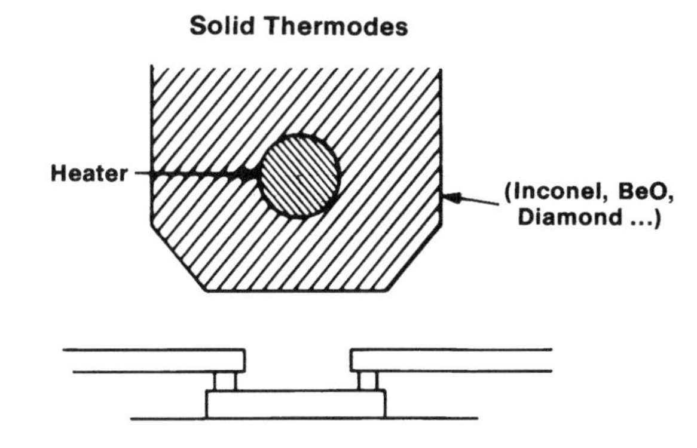

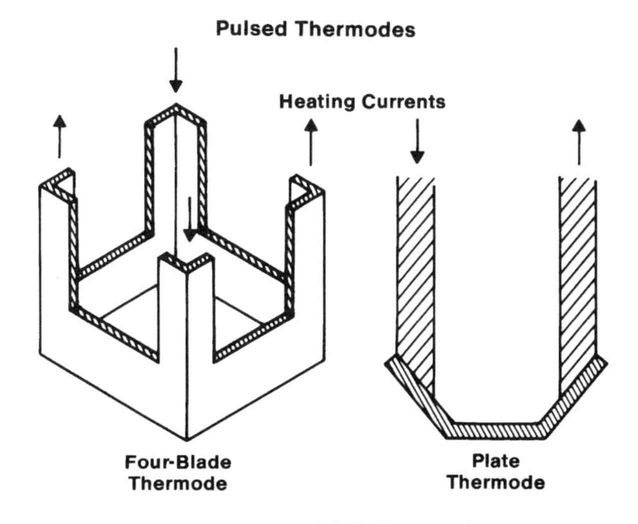

Figure 8-97. TAB Thermodes

sion alloys such as iron–nickel–cobalt ("Inconel"), or stainless steel/tungsten alloy with tips of natural or synthetic diamond (single crystal or polycrystalline), or cubic boron nitride [177]. Pulse bonding thermodes are usually made of molybdenum or titanium blades. Chemically vapor deposited (CVD) diamond is finding increasing use in TAB gang bonding because of its thermal and wear characteristics. It is a binderless polycrystalline material deposited on a special ceramic base [178].

The thermode has to be carefully selected for a particular application based on cost, bonding materials, processing temperature, leadcount, die size, and so forth. The key attributes should include excellent thermal conductivity, stability and wear resistance, a smooth surface (0.1–0.2 μm), excellent adhesion resistance to metal oxides, no bending at high temperature, and ease of cleaning by lapping with an alumina plate.

8.5.4.2 Gang Bonding Process Flow

A typical reel-to-reel gang bonding process flow is as follows [179]:

1. The TAB tape from the input reel is separated from the spacer tape and indexed to the bonding station.
2. The sawed (diced) wafer from the wafer cassette (magazine) is automatically fed to the wafer holder for die pickup.
3. The wafer is positioned and expanded.
4. The pattern recognition system locates the good die on the wafer.
5. Each good die is plunged up and transferred to the bonding stage.
6. The die is aligned in X, Y, and ϕ.
7. The bonding stage with the die moves to the bonding station.
8. The tape inner leads are inspected by the lead locator and aligned in X, Y, and ϕ. If the site is defective, it is skipped.
9. The bonding head makes the bond with a preprogrammed bond force, dwell time, and impact force.
10. The bonded site is taken up on the output reel with a protective spacer tape.
11. The bonding cycle is repeated with the next good die and the next good tape site.
12. After a certain number of bonds (typically 100), the thermode is automatically lapped (cleaned) according to preprogrammed instructions (for x–y motion, lapping interval, and duration).

8.5.4.3 Tape and Die Presentation Options

Depending on the equipment design, there are a number of options available for presenting the TAB tape and the chips to the inner lead bonder. For high-yield, high-volume runs, the chips are usually presented in wafer form on a sawed film frame as described earlier. In some cases, the good chips are presorted from the sawed frame and put in "waffle packs." Most bonders are designed to accept chips in waffle packs at the input stage. Each chip is then picked up and transferred to the bonding station. A third option is to sort the good chips and put them in an "embossed tape" (also called "pocket tape") similar to that used for surface-mount components such as capacitors, resistors, and molded semiconductor devices.

In addition to the reel-to-reel option described earlier, the TAB tape can be supplied to the inner lead bonder in strips (similar to conventional lead frames) or as individual sites in "slide carriers." The carrier, which allows testing and burn-in, is typically made of an antistatic material and has a high-temperature burn-in capability up to 180°C. The slide carriers are usually handled in magazines (coin-stack tubes).

8.5.4.4 Single-Point Bonding

Single-point bonding is the process of bonding one lead at a time just like wirebonding (wirebonding is discussed in Section 8.4). It can be used for both inner lead bonding and outer lead bonding. Single-point bonding offers many advantages over gang bonding [175,180–185].

1. The bond force is much lower—passivation cracking, silicon cratering and other problems are reduced or eliminated.
2. There are more process choices: thermocompression, thermosonic, ultrasonic, laser, or laser ultrasonic.
3. Because only one lead is bonded at a time, the height of each bond is sensed and the bond force is controlled individually, eliminating planarity problems associated with the bonding thermode, tape, bumps, and substrates. This becomes more critical with larger die size and higher I/Os.
4. The bonding tool is easier to install and change, and costs much less than a gang bond thermode.
5. The process is more flexible for small lots. The same bonding tool can be used for bonding different devices. Setup and changeover times are much less.
6. Defective leads can be repaired one at a time.
7. The same bonder can be used for ILB, OLB and repair.
8. Thermosonic and ultrasonic single-point TAB bonding is similar to wirebonding with a mature infrastructure and an extensive reliability database.

9. Most bonders are programmable and allow individual bond parameter control. Programs can be ported from one machine to another, assuring repeatability.
10. Single-point bonding is more effective with substrates which are not flat, such as ceramic and printed-circuit boards.

The bonding tools (thermodes) are usually made of ceramic, tungsten carbide, titanium carbide, and diamond-tipped ceramic. The thermode selection is based on cost, throughput, yield requirements, and assembly process compatibility. The thermode should have high thermal conductivity, high resistance to wear (hardness, density, and compressive strength), and good acoustic properties for transmitting ultrasonic energy (for thermosonic or ultrasonic bonding).

8.5.4.5 Laser Bonding

Laser bonding overcomes the two most serious limitations of thermocompression bonding, namely the need for high temperature and high pressure, which are known to cause damage to the device and affect its long-term reliability [186]. Because laser bonding is a noncontact method of providing heat energy through a finely focused beam, it can be used both on the periphery as well as in an area-array design over the surface of the chip. Laser bonding works well with materials which have high absorptivity.

Hayward [187] studied the ILB process using a Q-switched Nd : YAG laser operating at 1.064 μm developed by the Microelectronics and Computer Technology Corporation (MCC) and built by Electro Scientific Industries (ESI). Figure 8-98 shows the cross section of the fixture holding the device and tape during bonding. The device and tape are held in place by vacuum. The device had 152 gold bumps, 100 μm square and 25 μm high on a 150–200-μm pitch. The tape was a three-layer, tin plated tape with 0.6 μm of tin which had been reflowed after plating. The inner

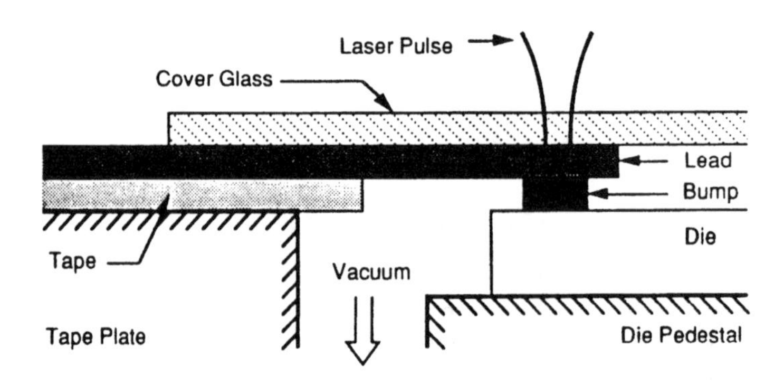

Figure 8-98. Laser Bonder ILB Tape/Die Fixture

leads were 76 μm wide and 35 μm thick. The three principal parameters studied were laser power, pulse time, and bond height (distance the die pedestal was driven above the point of first contact between the bump and tape). The baseline parameters were set at 37 W (power), 5 ms (pulse time), and 25 μm (height). A comparison with T/C gang bonding was made using an identical lot of 300 devices bonded with a Shinkawa IL-20 bonder set at an anvil temperature of 250°C, a thermode temperature of 500°C, and a force of 16 kgf. Ten units from each lot were pull tested in the as-bonded condition. The remaining were split between temperature-cycle testing (condition C, 1000 cycles) and high-temperature storage (150°C, 1000 h). Table 8-6 shows the results. The laser bonded units have lower average bond strength but a tighter distribution than the T/C bonded units and show a slower rate of degradation over time. Figure 8-99 shows SEM photographs of laser bonded leads.

Spletter [188] studied Au-to-Au laser ILB with a 328-lead, 3-layer TAB tape with 45-μm-wide, 35-μm-thick Cu leads on a 100-μm pitch. The leads were plated with 50 μm of gold and were bonded to a 0.63-cm^2 Si chip with 50-μm^2 gold bumps, 22 μm high. A frequency-doubled pulsed Nd : YAG laser was used to increase the absorptivity from 2–5% at 1.064 μm to 25–50% at 0.533 μm. The ILB setup was similar to that shown in Figure 8-98. Thirty devices were bonded and subjected to high-temperature storage (150°C) for 1000 h, temperature cycling (–65°C to 150°C) for 1000 cycles, and liquid-to-liquid thermal shock (–55°C to 125°C) for 1000 cycles. The average pull strength was 35 g with a standard deviation of 2.5 g. There was no degradation of good bonds after environmental testing.

Zakel et al. studied laser ILB of Sn, Ni–Sn, and Au tape metalliz-

Table 8-6. Comparison of Laser- and Thermocompression (T/C)-Bonding Test Results

Bond Type	Mean Value (g)	Standard Deviation (g)	Lift Failure (%)
As-bonded			
Laser	33.6	3.9	0.0
T/C	50.3	7.3	0.9
1000 Temperature Cycles			
Laser	25.8	3.8	1.6
T/C	38.8	9.2	8.9
1000 h Temperature Storage			
Laser	20.1	3.4	1.9
T/C	39.8	7.0	10.9

Figure 8-99. Laser Bonded Inner Leads

ations to gold and gold–tin bumps [189]. Laser soldering metallurgy was found to be different and more critical to long-term reliability than that of T/C gang ILB even if identical tape and bump materials are used. Accumulation of eutectic Au–Sn (80%/20%) in the bonded interface results in strong degradation due to Kirkendall pore formation in the ternary Cu–Sn–Au system. The Ni barrier inhibits this effect. However, thermal aging formed brittle intermetallics of Ni, Sn, and Au. Laser ILB of gold plated tape to Au–Sn solder bumps showed minimal degradation after thermal aging due to the formation of an intermetallic compound with high stability.

Azdasht et al. [190] have developed a laser bonder with a glass fiber for making a windowless flip–TAB connection on a flexible substrate. The tip of the fiber serves to press down on the part.

8.5.4.6 Factors Affecting ILB

There are many variables which can affect ILB bond quality:

- Tape material hardness
- Tape plating thickness and uniformity
- Tape surface cleanliness, roughness, and oxidation level
- Bump hardness
- Bump flatness
- Bump design (size with respect to pad structure)
- Under bump metallurgy
- Bump adhesion and barrier layer integrity
- Thermode design
- Thermode mass

- Thermode flatness at room temperature and bond temperature
- Thermode thermal conductivity and temperature uniformity
- Thermode wear resistance and oxidation resistance
- Thermode adhesion to plating materials (especially tin)
- Tape and die alignment
- Bond interface planarity
- Interface temperature, pressure, ultrasonic/laser energy (if used), and dwell time control
- Stage temperature
- Descent velocity and profile

Thermocompression bonding is a complex process that requires an understanding of a wide variety of mechanical and thermal interactions [191]. Several studies have been made to understand these interactions [151,192–195]. Figures 8-100 and 8-101 show the dependence of ILB bond strength on temperature and dwell time based on experimental work by Kawanobe et al. [151] and Spenser [194], respectively. Surface contaminants also

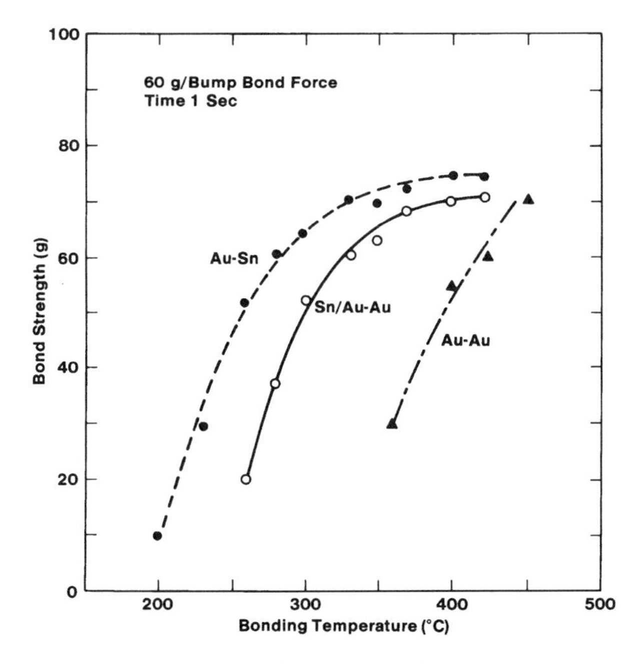

Figure 8-100. Dependency of Bond Strength on ILB Temperature. Bond strength versus bond temperature for several metallurgies. (From Ref. 151.)

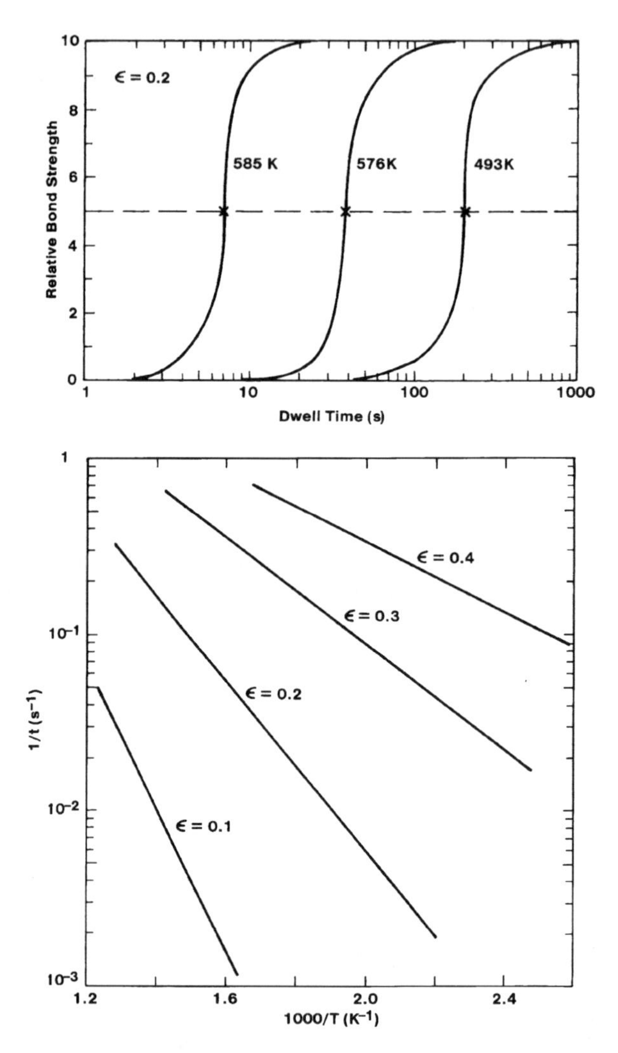

Figure 8-101. Interdependence of Dwell Time, Temperature, and Deformation on Bond Strength. Arhenius relationship between the dwell time of the bond and the interface temperature at several deformation levels. Data from plots of bond strength versus dwell time, at constant strain as shown in (a), were used to study the thermally activated process responsible for the bonding process (b). Specimens of gold plated copper were tested in shear. (From Ref. 194.)

play a large part in Au-to-Au T/C bonding as shown in Figure 8-102 from the work of Jellison [196].

The success of the ILB process also depends on the proper choice of interface metallurgy (IFM). A wide variety of IFM structures have been used in the past, as shown in Table 8-7. Today, Ti–W–Au is widely

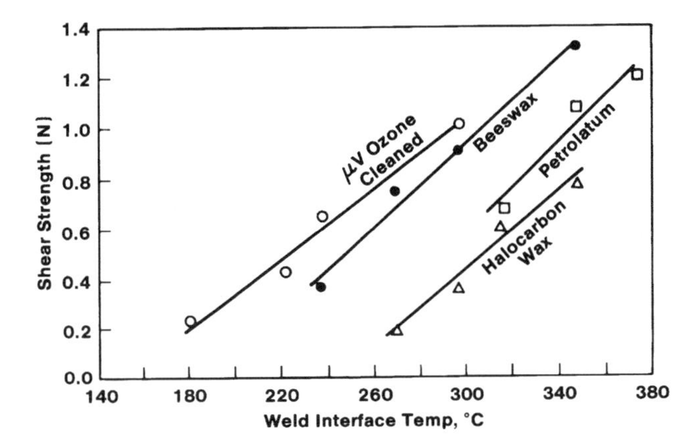

Figure 8-102. Effect of Contamination on Thermocompression Bonding. Effect of organic contamination on gold-to-gold bond strength. (From Ref. 196.)

used as the under bump metallization (UBM), along with gold–solder bumps and tin–gold plated tape.

It is well known that gang T/C bonding of high-lead-count devices at high temperatures (>400°C) and pressures can cause mechanical and thermal stresses to the UBM and affect the long-term reliability of the device [151,192,197–199].

Table 8-7. Metallurgies Used for TAB-ILB

	Bump Structure			Tape Structure	
Company	Adhesion Metal	Barrier Metal	Bump Metal	Plating Metal	Lead
GE	Cr	Cu	Au	Sn	Cu
CII Honeywell-Bull	—	—	Au	Sn	Cu
Sharp	Cr	Cu	Au	Sn	Cu
Honeywell	Ti	Pd	Au	Sn	Cu
Fairchild	—	—	Au	Sn	Cu
Signetics	Ti	Pt	Au	Sn	Cu
Nippon Electric	Ti	Pt	Au	Au	Cu
National Semiconductor	Cr	Cu	Au–Cu	Au (Thin)	Cu
BTL	Ti	Pt	Au	Au	Cu
Philips	Al	Ni	Solder–Cu	Au–Ni	Cu
Siemens	—	—	Au	Sn	Cu
RCA	Ti	Pt	Au	Au	Cu
Motorola	—	—	Au	Sn	Cu
TI	—	—	Au	Sn	Cu

Source: From Ref. 151.

8.5.4.7 ILB Evaluation

As in wirebonding, bond evaluation methods for ILB can be classified as destructive or nondestructive. The two most common are visual inspection (optical/SEM) and the pull test. For bumped-chip bonding, the destructive pull test is preferred for evaluating bond strength and quality. The die is held down by vacuum; a wire hook is placed under the inner lead (to be tested) midway between the die edge and the polyimide opening [200]. The hook is raised at a controlled rate and the force necessary to break the bond or the inner lead is recorded. The bump is then removed and each bond is inspected for passivation or barrier layer cracks and other damage such as silicon cratering [196]. Typical failure modes for Au-to-Au T/C bonding are as follows:

- Lead break—at the bump or in mid-span
- Lead lift
- Bump lift
- Bump fracture
- Silicon cratering

In addition to optical/SEM microscopy, other nondestructive methods commonly used to check the quality of bonded parts include scanning acoustic microscopy, scanning beam x-ray laminography, conventional x-ray inspection, acoustic emission detection, and laser/infrared inspection [201,202].

8.5.5 Outer Lead Bonding

Outer lead bonding (OLB) is the process of connecting the outer leads of the tape to the package, lead frame, substrate, or card. Like ILB, OLB can be a gang or a single-point process. There are many OLB processes available. The choice depends on the application and includes the following:

- Thermocompression
- Thermosonic
- Hot bar
- Laser
- Ultrasonic
- Lasersonic
- Infrared
- Hot gas
- Vapor Phase
- Anisotropic conductive film
- Mechanical (DTAB, connector, socket, elastomeric clamp, etc.)

After ILB, the chips on tape usually undergo additional processing such as encapsulation, test, and burn-in. These steps can be performed in reel form, strips, or individual sites mounted in slide carriers [201,203,204]. The encapsulation process depends on the end-use application and can include chip-size coating, potting, or transfer molding [205–207] (see Fig. 8-103). For transfer molding, the ILB device is usually bonded to a lead frame prior to molding (Fig. 8-103c). Current encapsulation processes and materials are covered in Chapter 10, "Plastic Packaging," Chapter 14, "Package Sealing and Encapsulation," and Chapter 17, "Printed-Circuit Board Packaging."

Much of the previous discussion regarding ILB processes (Section 8.5.4) is also applicable to OLB processes with some modifications. Chapter 16 ("Package-to-Board Interconnections") includes details of surface-mount joining processes such as vapor phase, infrared reflow, laser reflow, and pressure connections. This section provides a brief overview of hot bar OLB, anisotropic conductive film OLB, demountable TAB, and some recent experimental results on T/C–Laser OLB.

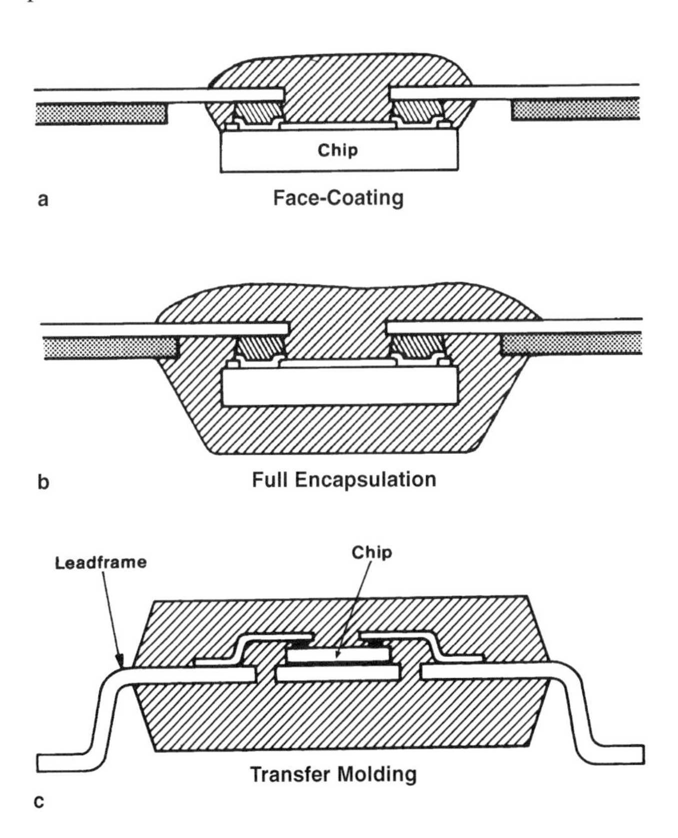

Figure 8-103. TAB Encapsulation. (a) face coating; (b) full encapsulation; (c) transfer molding.

8.5.5.1 Hot Bar OLB

Pulsed hot bar soldering has been in use for over 20 years [208]. In the OLB process, the hot bar, or thermode, mechanically presses the leads onto the bonding pads. The thermode blades are resistively heated to provide a preprogrammed temperature profile optimized for a specific application (see Fig. 8-104) [209]. The thermal cycle begins with the blades in contact with the leads. The programmable parameters include thermode idle temperature, ramp rates, flux activation time, flux activation temperature, bonding time, bonding temperature, tool-up temperature, and spindle-up time.

The hot bar process sequence depends on the application but can include the following five steps [210]:

1. Excise and form
2. Die attach
3. Fluxing
4. Placement and alignment
5. Solder reflow

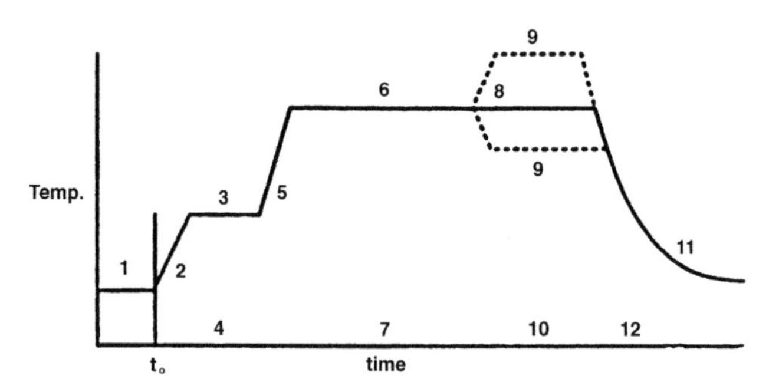

User programmable parameters for the bonding cycle:

1. Idle temperature
2. First temperature slope
3. First temperature level
4. First temperature time
5. Second temperature slope
6. Second temperature level
7. Second temperature time
8. Third temperature slope (optional)
9. Third temperature level (optional)
10. Third temperature time (optional)
11. Thermode lift temperature
12. Spindle lift time

Figure 8-104. Bonding Profile and Parameters for Hot Bar OLB. (From Ref. 209.)

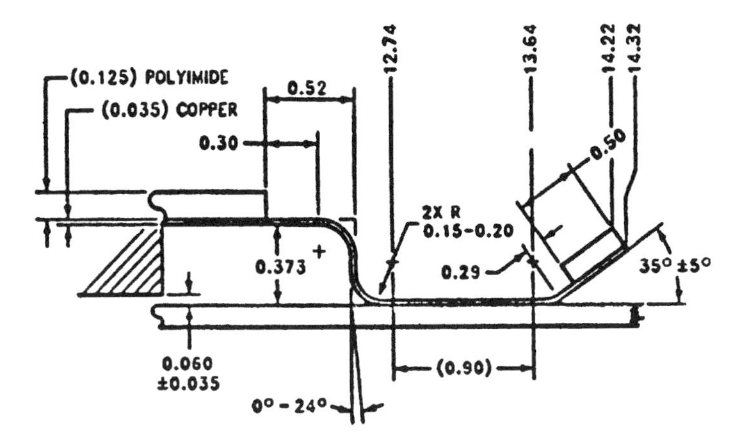

Figure 8-105. Sample Lead Form Configuration. (Courtesy of Intel Corporation.)

The excise and form operation includes cutting the leads from the tape support and forming them into a gullwing or modified gullwing shape according to specifications. Excising (cutting out) the component occurs just before fluxing and placement. A "keeper bar" helps in maintaining lead coplanarity and spacing during subsequent operations. (The keeper bar is a narrow strip of polyimide which remains in place on each row of leads after excise). Figure 8-105 shows a recommended lead form configuration for Intel's Pentium© processor, and Table 8-8 lists the important lead form dimensions.

Kleiner [211] has studied the effects of various parameters such as

Table 8-8. Key Lead Form Dimensions

Controlled Dimension	Recommended Range
"Standoff" or "set back" of die above die pad	0.025–0.095 mm
Lead foot length minimum	0.90 ± 0.05 mm
Lead foot angle	0°
Keeper bar toe angle (minimum)	20°–35°
Keeper bar width	0.5 ± 0.1 mm
Toe radius	0.2 mm (min)
Heel radius	0.2 mm (min)
Lead shoulder length	0.3 ± 0.05 mm
Lead shoulder radius	0.2 mm (min)

Source: From Intel Corporation.

bend radius, material type and thickness, plating type and thickness, draft angle, punch angle, force and speed on various lead form geometries, and developed tooling principles and requirements for formed leads which are parallel to the chip surface, coplanar and free of cracks at the bend radii.

If a die attach is needed, the lead form design should allow sufficient clearance between the bottom of the silicon die (chip) and the die attach pad. The typical die attach material is a silver-filled thermoset polymer.

The choice of flux and its method of application depend on the interface materials. For the Pentium TCP, Intel recommends immersing the leads in a rosin, mildly activated, halide-free, no-residue flux. The entire surface of the foot, top and bottom, up to the top of the heel radius should be immersed. The specific gravity of the flux, its solids content, and the surface insulation resistance between adjacent leads should be closely controlled to 0.80–0.81, 1–3%, and $>10^9$ Ω, respectively. Solder reflow should be completed within 30 min of fluxing. A maximum thermode temperature of 275°C ± 2°C with a dwell time of 5 s was shown to yield acceptable fillets. Force should be kept at ≤1.36 kg (3 lb) per blade.

The placement accuracy requirements for TAB components are much more stringent than for conventional surface-mount devices. The factors which affect accuracy are locational tolerances of the leads, the lands, and fiducials on the substrates [209]. A good pick and place system with a pattern recognition system (PRS) should be capable of ± 0.01 mm repeatability (lead to land) in x and y and 3° of rotation [210].

The reflow process can be affected by a number of parameters such as

- Blade design
- Thermode compliancy
- Nonuniform temperature and flatness along the length of the blades
- Thermode expansion
- Substrate warpage
- Substrate support
- Bonding force
- Dwell time
- Thermode temperature

Equipment manufacturers are meeting the new challenges of sub-0.3-mm pitch TAB by developing “self-planarizing” thermodes with spherically compliant suspensions and better alignment and vision capabilities [209,212–215]. Figure 8-106 shows a schematic of a self-planarizing thermode, a thermode assembly, a titanium thermode blade, and a cleaning station [216].

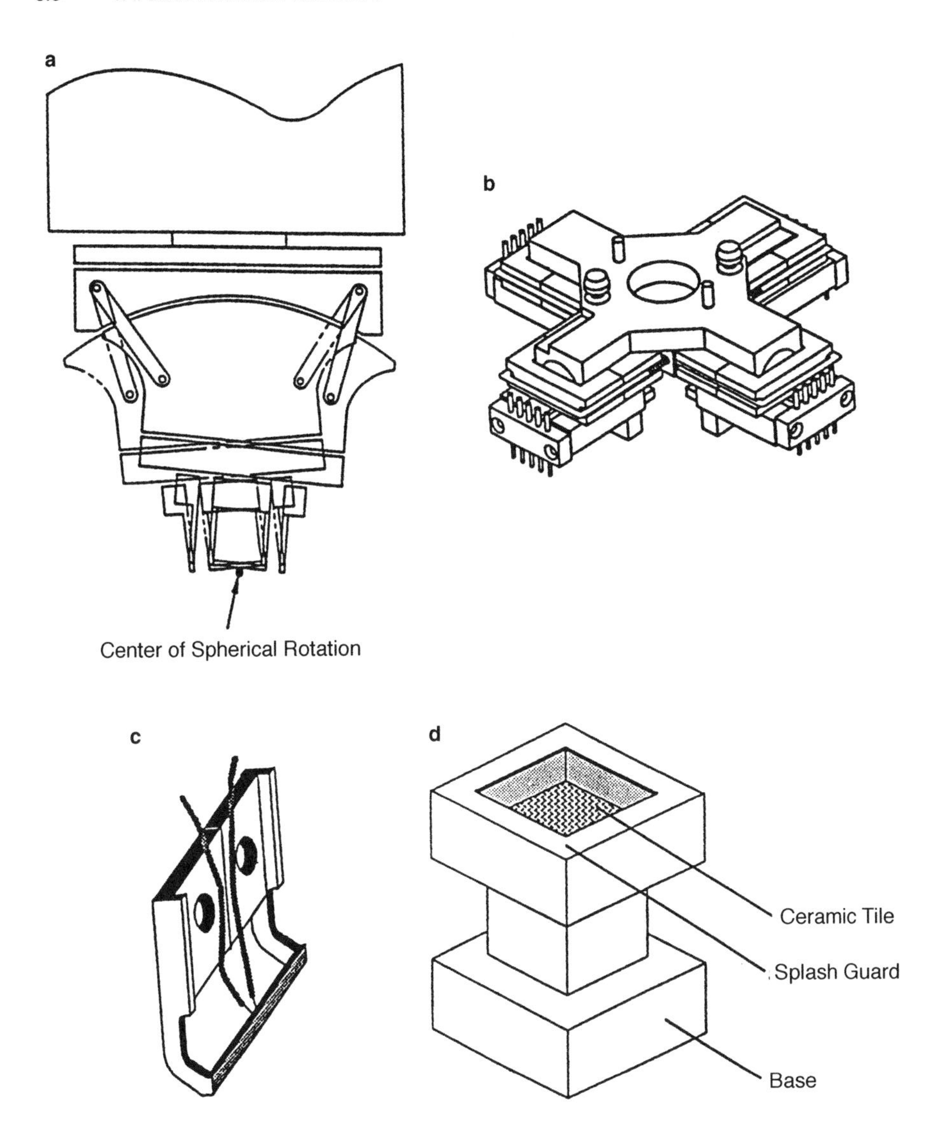

Figure 8-106. Hot Bar Process Components. (a) Self-planarizing thermode; (b) thermode assembly; (c) titanium blade, and (d) cleaning station. (Courtesy of Universal Instruments Corp.)

8.5.5.2 Anisotropic Conductive Film

In some applications such as LCDs, soldering is difficult due to the temperature limitations of piece parts to be bonded. Anisotropic conductive film (ACF) is an alternate method used in LCDs to connect the TAB outer leads and the ITO (indium–tin–oxide) electrodes on the glass substrate. The ACF consists of fine conductive particles uniformly scattered in an epoxy binder. ACFs can also be made of plastic cores plated with conductive materials. They conduct only in the vertical direction under pressure. ACFs are also known as "Z-axis conductive adhesives." Typical ACF conductors include nickel, gold, and carbon. Particle size and distribution in the film are critical to prevent shorting between particles in a horizontal direction. For TAB OLB of LCDs, heat and pressure are used to bond the outer leads to the ITO electrodes.

Casio Computer Company has developed a "microconnector" (MC) using ACF which allows OLB of 80-μm-pitch LCDs [217]. Conductive particles with plastic cores are plated with Ni–Au, and smaller particles coated with very thin (1000 Å or less) insulating film are applied to the conductive particles. The latter are then combined with an epoxy binder and T/C bonded in the same way as conventional ACF. During bonding, the insulating film in the direction of pressure is broken to achieve conductivity. The binder can be either thermoplastic (for rework) or thermoset. Figure 8-107 shows a schematic of the conductive particles and the microconnector OLB method for 80-μm pitch.

8.5.5.3 ILB and OLB Pitch and Lead-Count Trends

Figures 8-108 and 8-109 show the decreasing trend in ILB and OLB pitches for the last 5 years, and the increase in lead count since 1986,

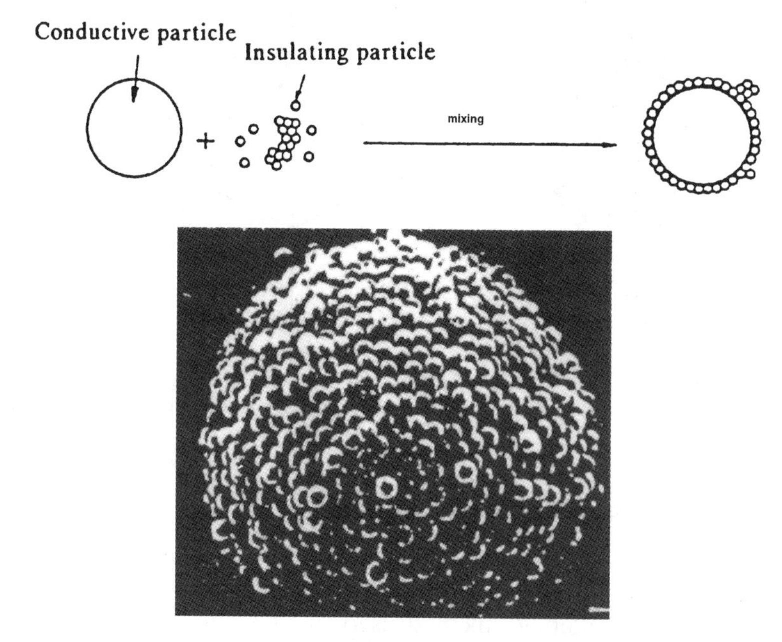

Figure 8-107. (a) MC Conductive Particles; (b) 80-μm-pitch OLB

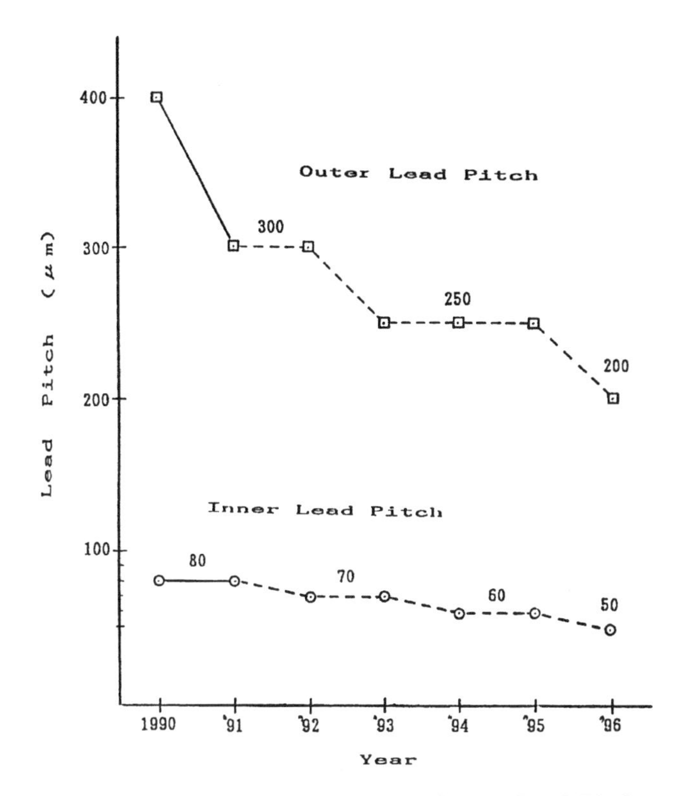

Figure 8-108. Decreasing Trends of TAB Lead Pitch

respectively [136,142]. The minimum ILB pitch declined from 80 μm in 1990–1991 to 70 μm in 1992–1993 and 60 μm in 1994–1995. It is expected to reach 50 μm in 1996. The minimum OLB pitch has dropped from 400 μm in 1990 to 300 μm in 1991–1992 and 250 μm in 1993–1995. It is expected to reach 200 μm in 1996. The maximum lead count has increased from about 250 leads in 1986 to over 700 leads in 1995.

8.5.5.4 Effect of High I/O and Pitch on ILB and OLB

When designing a TAB package, the outer lead pitch affects the package size and ease of assembly. The smaller the OLB pitch, the more difficult it is to assemble the device to the next level. The standard JEDEC OLB pitches are 0.65, 0.5, 0.4, and 0.3 mm [166].

For tin plated tape, as the pitch becomes smaller and smaller, it is important to reduce the tin thickness to ensure that excess tin does not produce shorts between adjacent leads. At the same time, tin thickness plays a major role in OLB. If there is not enough tin, the wettability of

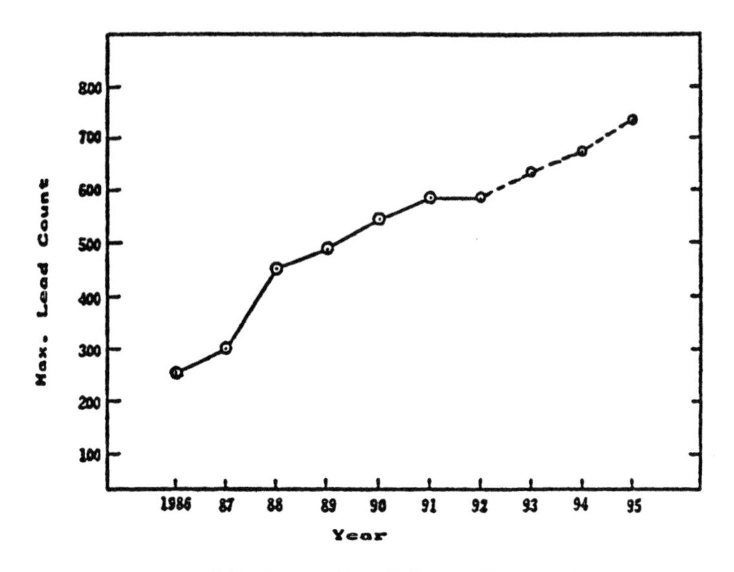

Figure 8-109. Maximum Lead-Count Trends for TCPs

the outer leads will be degraded. This will lead to weaker outer lead bonds [218].

Saito [149] studied the relationship between the solder plating thickness and bondability of outer leads using pulsed hot bar OLB. The acceptable thickness range for 0.3-mm OLB pitch was 3–10 μm of solder, with 5–10 μm giving the best results. For smaller pitches (0.25 and 0.15 mm), the acceptable range was 3–7 μm with 5–7 μm being the optimum.

8.5.6 TAB Package Applications

8.5.6.1 Tape Ball-Grid Array

The tape ball-grid array (TBGA) is a ball-grid array TAB package developed by IBM which overcomes the OLB limitations of conventional peripheral TAB packages [219]. Figure 8-110 shows the TBGA schematic. It uses area-array TAB (ATAB) with a standard ground plane and has the following performance advantages over conventional TAB and pin-in-hole packages at the card level:

- Lower lead inductance
- Lower power-supply inductance
- Lower line-to-line capacitance
- Lower average signal delay

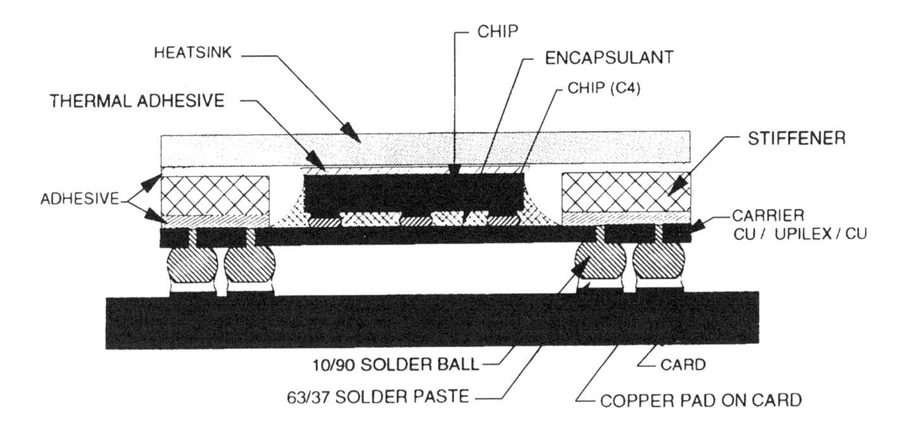

Figure 8-110. TBGA Cross Section. (From Ref. 219.)

The TBGA is constructed using an adhesiveless two-metal-layer tape. One side is used for signals and the other for power and ground interconnections. Solder balls joined to the tape interconnect the completed package to the next assembly level (PCB/card). A "stiffener plate" maintains planarity of both the solder bumps and the tape and serves to minimize thermal stresses during temperature cycling because its TCE matches that of the carrier/card.

The ILB process can be a conventional peripheral one or a modified C4 flip-chip process known as solder attach tape technology (SATT). SATT allows bonding over active circuitry and uses less bond pad area than T/C ILB and wirebonding, thus reducing the die size and improving the wafer productivity.

After ILB, an epoxy encapsulant is used to provide mechanical protection and enhance the fatigue life performance of the inner lead bonds.

A standard infrared reflow process using 63/37 Sn/Pb solder paste can be used for card assembly. The TBGA height is 1.4 mm. It weighs less than 5 g. The TBGA is registered as a JEDEC standard S-XXGA-N/TBGA. Figure 8-111 shows a photograph of the TBGA.

8.5.6.2 TapePak®

The TapePak (Fig. 8-112) was developed by National Semiconductor around 1986. It is a testable, plastic molded quad surface mount package with an OLB pitch of 0.51 mm for a lead-count range of 20–124, and 0.38 mm for higher lead counts. It uses a single-layer bumped copper tape, 70 μm thick. The bond pads are "capped" with a thin-film metallization at the wafer level. Thermocompression bonding is used for ILB. After ILB, the tape is cut into strips and the devices are molded with epoxy novolac. Next, the dam bars are removed, the devices are singulated and solder

Figure 8-111. A 736-Lead TBGA. (Courtesy of IBM Microelectronics.)

plated. The package is designed with molded-in test pads to allow test and burn-in. A trim-and-form operation is done at the pick-and-place station prior to board assembly.

Because of its compact size, the TapePak offers significant improvement in electrical characteristics. For example, a 40-lead, 7.6-mm square TapePak has a worst-case lead length of 2.54 mm, a lead resistance of 2.4 mΩ, inductance of 1.2 nH, and lead-to-lead capacitance of 0.2 pF [220].

8.5.6.3 Pentium® TCP

The increasing importance of "small form-factor" packages for what is called the "mobile revolution" has led to TAB's resurgence in the United States. The world's largest semiconductor company, Intel, has developed a TCP version of the Pentium® for mobile applications such as laptop, notebook and palm top computers, and related products [210]. Figure 8-113 shows the topside and bottomside of the Pentium in a slide carrier. The chip is inner lead bonded to a JEDEC-style UO-018, 48-mm three-layer one-ounce copper (35.56-μm-thick) tape. The inner lead and outer lead bonding areas are gold plated over a nickel flash. The inner lead bonding (ILB) is performed with a thermosonic single point bonder.

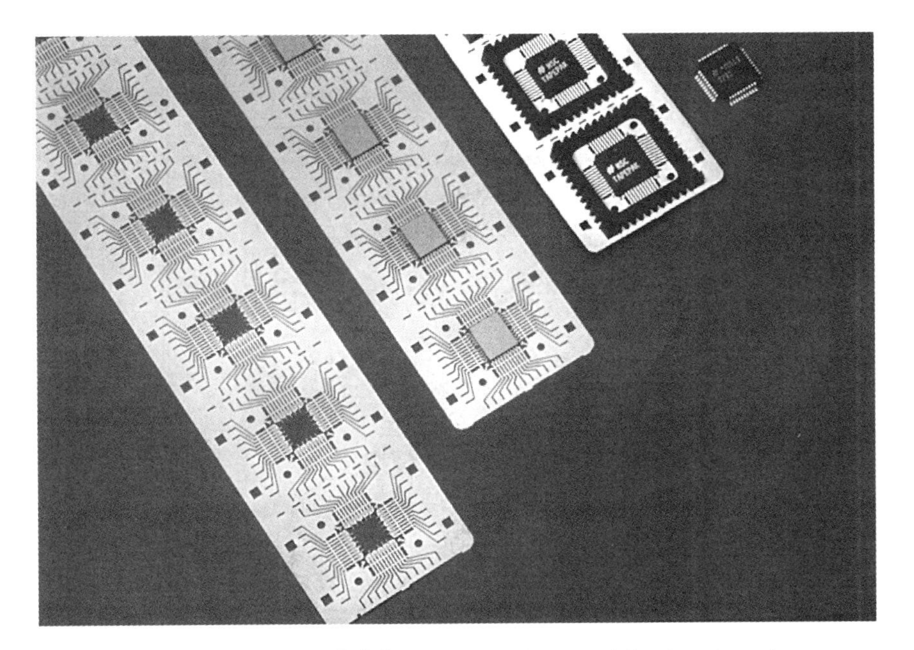

Figure 8-112. TapePak®. (Courtesy of National Semiconductor.)

After ILB, the chip and ILB area is encapsulated with a high-temperature thermoset polymer coating. It covers the top surface of the device, the sides, and the ILB area to the polyimide carrier ring (see Fig. 8-114). The backside of the chip is left bare for backside bias to the printed-circuit board (PCB).

The outer lead pitch is either 0.25 mm or 0.2 mm, depending on the device. The OLB lead width is 0.10 mm. The test pads are 0.5 mm square on 0.40-mm pitch outboard of the OLB window. The devices are shipped in individual plastic slide carriers packed in coin-stack tubes. Each component is 0.615 mm thick. After excising, lead forming, and mounting to the PCB, the total height of the component above the PCB is less than 0.75 mm. The package body is either 24 mm or 20 mm square after excise, depending on the component. The 320-lead 0.25-mm component in a 24-mm body size weighs a maximum of 0.5 g. In comparison, a 296-lead multilayer PQFP weighs 9.45 g.

For board mounting, Intel recommends either a hot bar, hot gas, or laser reflow process after all other board components have been mounted and cleaned. Figure 8-115 shows the recommended land pattern design.

The Pentium requires backside electrical and thermal contact. So it is necessary to provide a 3.75 mm ± 0.025 mm die attach pad. Intel recommends a silver-filled thermoset polymer with a cure profile of 6 min above 130°C.

The thermal resistance of the TCP package (Θ_{jc}) is 0.8°C/W to

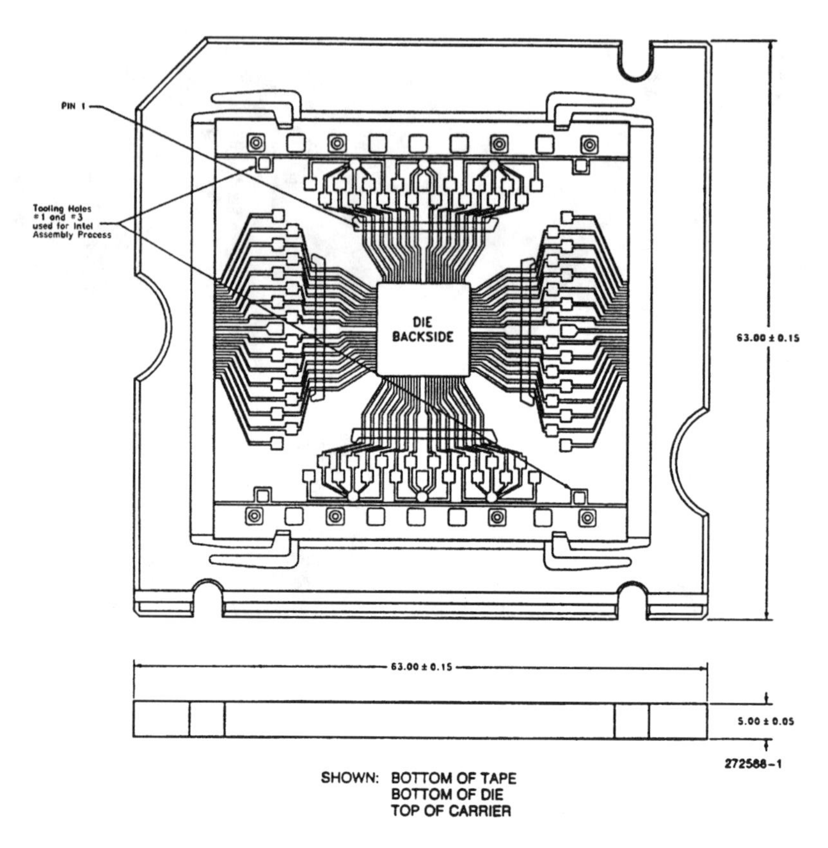

Figure 8-113. TCP Site in Carrier. (a) Bottom view of die; (b) top view of die. (Courtesy of Intel Corporation.)

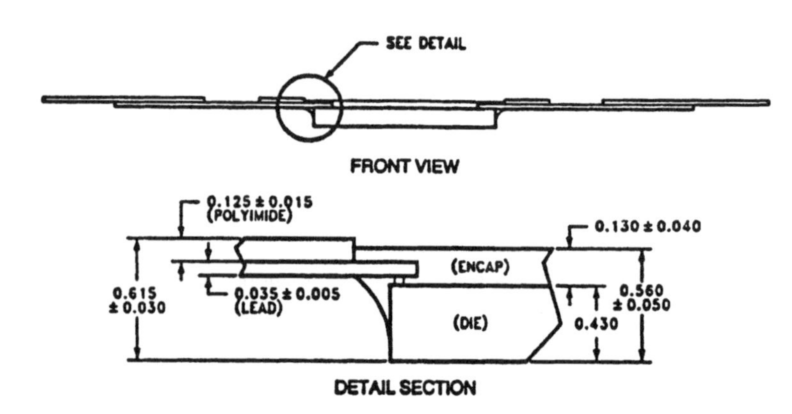

Figure 8-114. TCP Site (Cross-Sectional Detail). (Courtesy of Intel Corporation.)

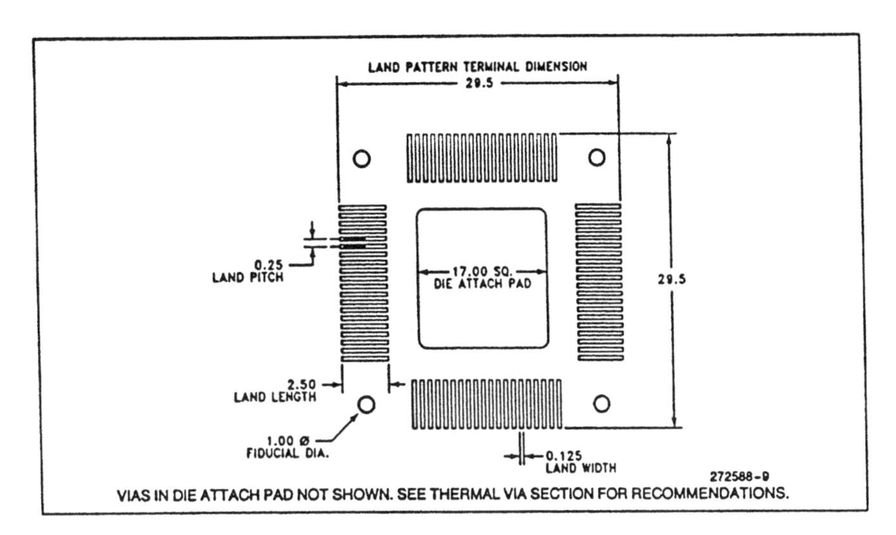

Figure 8-115. Land Pattern for Intel 24-mm Body Size TCP. Hot Bar Reflow Process. (Courtesy of Intel Corporation.)

2°C/W, depending on the product. PCB enhancements such as thermal vias with or without low-profile heat sinks brings the thermal performance in line with mobile computing requirements which do not have forced convection cooling. Figure 8-116 shows a schematic of heat transfer through the PCB.

8.5.6.4 ETA Supercomputer

It is instructive to review ETA's use of TAB for its supercomputer. ETA Systems, Inc. (which is no longer in business) was a subsidiary of

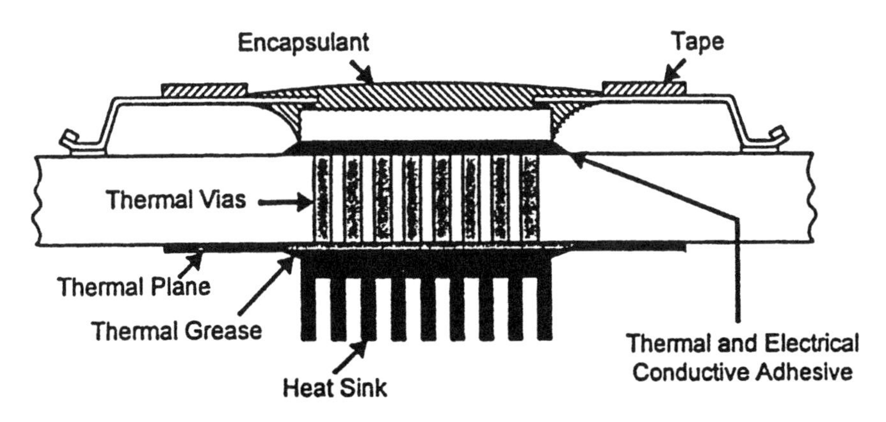

Figure 8-116. Heat Transfer Through the PCB. (Courtesy of Intel Corporation.)

Control Data Corporation in the 1980s. Its mission was to build "the world's fastest supercomputer" [221].

ETA chose to go with TAB devices mounted on a single central processing board. Each device was a 284-pin TAB ceramic quad flat pack, with 20,000 CMOS gate arrays. The TAB devices were supplied by Honeywell. The chips were solder bumped with 95% Pb–5% Sn. Each solder bump was 100 μm in diameter. The underbump metallurgy (UBM) was TiW–Cu–Ni. The bond pads were at a 254-μm pitch, arranged in two rows, at an effective 127-μm ILB pitch (see Fig. 8-117). A polyimide "through-hole" design allowed each finger to be suspended over a hole in the polyimide base corresponding to each individual solder bump. After melting, the solder was constricted within the hole and reflowed to the precise area on the lead [57].

Honeywell's solder bump reflow TAB outperformed its gold bump thermocompression bonding process in three areas: (1) lower bonding force (less than 1 kg for a 300-bump device versus about 11 kg for gold); (2) inherent planarity compensation—a 100-μm-high solder bump typically was 50 μm after bonding, thus allowing for a 50-μm variation in planarity versus about 10 μm for gold, and (3) area-array design capability—the solder bumps could be put over the whole chip surface, unlike gold bumps which were limited to the periphery.

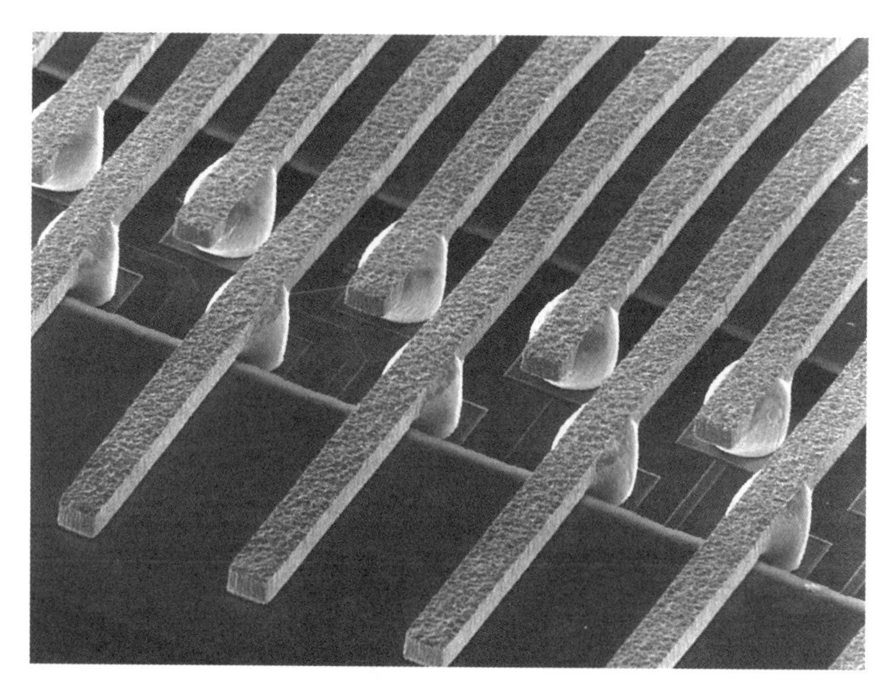

Figure 8-117. Double-Row Solder Bump TAB. (Courtesy of Honeywell.)

8.5.7 Future Trends

Tape carrier packages (TCPs) will continue to be the dominant interconnect medium for the LCD industry for the immediate future. LCDs accounted for the largest share of tape carriers in 1994 [222]. The next largest application was smart cards, mostly in Europe, where the tape is of the three-layer variety, and the chips are wirebonded. The largest application in the United States today is for inkjet printers, where the tape is of the two-layer type. The connection is by gang bonding or single-point bonding. All three of these applications will continue to thrive as the demand for notebook computers, smart cards, and printers grows every year.

Intel's decision to provide Pentiums in TCPs may serve as the catalyst for a significant broadening of the worldwide TCP infrastructure. The next generation of microprocessors will require more I/Os and finer ILB and OLB pitches. Pin counts will continue to grow with microprocessor bit size. ASICs will also act as TCP drivers with multimillion-gate sub-0.4-μm CMOS five-level metal processes becoming common and forcing the need for 1200+ I/Os per chip, which can be met only by TCP and/or flip-chip designs.

TAB's inherent advantage of being light, thin, short, and small (the Japanese call it "Kei-haku-tan-sho"), and its flexibility will enable it to maintain its edge in a wide variety of applications such as watches, calculators, cameras, sensors, and hearing aids.

The proliferation of lower-cost bumping processes and materials can expand TAB's application base into new products such as PC cards, smart cards, and multichip modules. This could lead to a strong TAB resurgence in the United States.

8.6 PRESSURE CONNECTS

Pressure connects are not actually bonded connections but maintain contact only by an external, continuously applied force (such as a metal spring or elastomer retainer). Several types have been described. Citizen Watch [223], for example, joins Au bumped chips onto Au bumped substrates using conductive rubber contacts embedded in a polyimide carrier (Fig. 8-118).

Typical loading is 2 g per pad to provide adequate deformation of the rubber for needed contact on all pads; thus, a 300-pad chip requires 600 g continuous loading. The conductive rubber is not a "good" conductor: A 200 μm × 200 μm pad has 35 mΩ resistance. Thus, it has been applied in special applications (such as liquid crystal display panels) that can tolerate the high resistivity.

Significantly lower resistances per contact, of the order of 10 mΩ, have been achieved by using an all-metal system. However, this requires

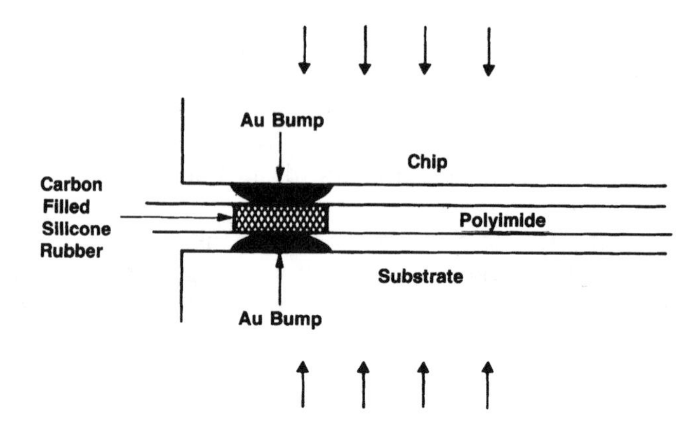

Figure 8-118. Pressure Connect (Citizen Watch). (From Ref. 223.)

a “flexible” substrate to which the chip is joined. Tektronics [224] has joined GaAs chips to a flexible circuit, as shown in Figure 8-119. The bumps are plated on the thin-film flexible circuit. Two hundred pads per chip have been fabricated with 50-μm pads on 100-μm centers. Silicon Connection has a very similar approach called “SICONS” [223]. Applications include ROMS in ATARI computer games.

Although such applications are limited at present, there are several reasons for the increased research in these areas. This type of contact eliminates the chip-size restrictions of the C4 type of full bonded connection; there is no fatigue mechanism. However, this is replaced by the concern of a sliding or point contact developing high resistance by debris or oxidation/corrosion. The second reason for research activity is the ease of assembly and reworking—no connections need to be “broken” or unbonded in order to take the package part.

8.7 ADHESIVE BONDING

A number of companies are using adhesives in novel ways to provide flip-chip options other than solder reflow and pressure connects. The driving forces are typically cost reduction, lower process temperatures, or fatigue elimination. Although still in their infancy, the following three approaches are worthy of note. All require precision placement of the chip, as no self-alignment is involved.

8.7.1 Bumps and Nonconducting Resins

Matsushita [225] has reported on a gold bump bonding process for flip chips. Chip-to-board connection is made by using pressure to force gold bumps on the chip into direct contact with gold pads on the circuit

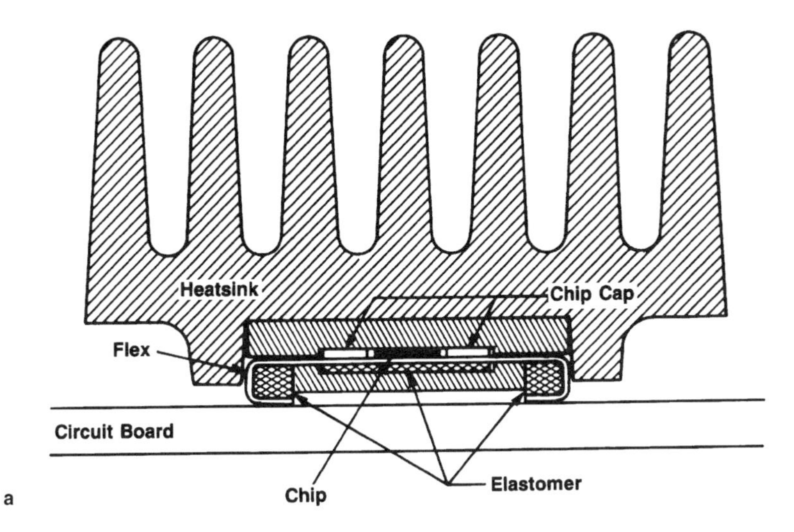

Figure 8-119. GaAs Chips Joined to a Flexible Circuit. (a) IC flex circuit elastomer pressure pad sandwich cross section; (b) I/O pressure bumps establish firm electrical contact. (From Ref. 224.)

board through a layer of acrylic resin. Using light to cure the polymer, an adhesive bond is made between the chip and board which keeps the mechanical connection under compression. Experimental assemblies have been fabricated down to a 10-μm pitch. A pitch of 100 μm has been used on thermal print heads, LED array sensor, and a memory card.

A modification of this approach has been described by TOSHIBA [226] for chip-on-glass applications in a video camera. The modification involves having tin–lead lands on the substrate, gold bumps on the chip, and supplying thermal energy below the eutectic temperature of 183°C to allow diffusion between the two metal contacts. Again, acrylic resin is used in the gap between the chip and substrate to strengthen the assembly.

8.7.2 Anisotropic Conductive Adhesives

Anisotropic conductive materials display unidirectional conductivity [227]. For flip-chip bonding, conductive particles are suspended in a dielectric medium so that they do not come into physical contact. Conductivity, in the z direction perpendicular to the mating surfaces, only occurs when protruding conductor pads on the chip or substrate (or both) are squeezed against the z-axis film. The electrically isolated particles simply make contact with the opposing conductor pads to provide electrical conduits. Seiko Epson [228,229] has used NiAu-coated resin balls, 7.5 μm in diameter, embedded in adhesive. The process temperature used to cure the adhesive is 150–200°C with a load of approximately 5 kg/cm^2. The application was chip-on-glass for high-resolution liquid crystal displays.

One unsuccessful implementation [230] demonstrated that clusters of small particles gave more reproducible electrical contact than did larger discrete particles. Many different particle materials have been explored including solid metals (Ni, solder), carbon, and metal-coated plastic particles. Thermoplastic, thermosetting, and light setting polymers have been tried as adhesives with varying degrees of success. Coupling very small particles has proven to be difficult, with intermittents showing up in thermal-cycle or humidity-cycling tests. Improvements in reliability are expected as materials are optimized.

8.8 ELECTRICAL PARAMETERS OF INTERCONNECTIONS

The electrical performance of the packaged chip may well be determined by the mode of chip interconnection. The electrical parasitics are very different for the three major interconnection techniques, with the flip chip offering the best properties due to its small size—nearly a point contact (see Table 8-8). Package geometry, however, also plays a significant role. In fact, it can reduce the intrinsic differences between the three interconnection approaches. For example, a wirebonded silicon device is normally associated with a lead frame, which, for fabrication reasons or board-level interconnection geometry requirements, may require rather long leads, well separated from a ground or reference plane. (Alternative wirebonded structures, such as pin-grid arrays or direct-chip attach to a substrate, may provide a much different electrical signal environment). High-end computer packaging designers are aware of the electrical performance issues associated with packaging, usually for their own customized applications (see Chapter 1). For this reason, as well as because low-performance systems are not so often package limited, focus on mid-range structures are frequently gated by package configuration and performance. To be considered are chips having 84 and 180 I/Os packaged using each of the three primary chip interconnection approaches addressed in this chapter: C4 solder-joined, wirebonded and TAB-bonded packages.

The packages to be compared are (1) a C4-joined, metallized ceramic single-chip carrier having swaged pins, (2) a C4-joined, multilayer ceramic, single-chip carrier having brazed pins, (3) a wirebonded pin-grid array, (4) a wirebonded plastic-leaded chip carrier (PLCC) (84 I/Os only), and (5) a TAB-packaged device. The pin grid for packages 1–3 is fixed at 2.54 mm; the PLCC is fixed at 1.27-mm lead pitch; the TAB package is fixed at 0.5-mm lead pitch. (Note, package 2 is a somewhat artificial configuration, given that usually more than one C4-joined chip is joined to an MLC module.) The objective here is to give a rough basis for comparing package performance, using the lead geometry of the packages. The basis for comparison is the calculated range in the value of lead inductance. This provides a starting point for estimating the effective package inductance, L_{eff}. This quantity is used to predict the inductance-based voltage swing (or ΔI noise), $\Delta V = L_{eff}\, n(dI/dt)$, where n is the number of simultaneously switching drivers, having a given current surge dI/dt (as discussed in more detail in Chapter 3). Shown in Table 8-9 is the lead inductance calculated for various packages using a three-dimensional inductance program [231]. Lead inductance may be used to estimate the effective package inductance, given a detailed knowledge of chip configuration and the number and location of power and ground leads (all of which should be optimized for the package and chip of interest [232]). The relation between lead inductance and L_{eff} is quite geometry-specific and will not be expanded upon further. Signal coupled noise, which is equally dependent on a detailed package description, will also not be considered here. Referring to the table of package inductances, it should be noted that for most of the package geometries chosen, a broad range of lead inductance is observed, indicating the need for personalization of signal and power/ground assignments.

8.9 DENSITY OF CONNECTIONS

A dramatic difference in attainable connection density between area arrays and peripheral pads was shown previously in Table 8-3 and Figure 8-15. With the advent of area TAB and multiple wirebond layers, improvements in density are being made (Fig. 8-120). High performance and I/O technologies in both leading edge CMOS and bipolar chips will favor area-array technologies over peripheral. The present area of applicability of wirebond, TAB, and C4 is shown in Figure 7-16 in Chapter 7, "Microelectronics Packaging—An Overview."

A typical process comparison of C4, wirebond, and TAB for a

Table 8-9. Calculated Lead Inductance for Single-Chip Modules

	C4 Bonded, MC Pin Grid Array	C4 Bonded, MLC Pin Grid Array	Wirebonded Pin Grid Array	Wirebonded Plastic Leaded Chip Carrier	TAB Package
84 I/O Module					
Substrate size (mm)	24	24	24	30	18
Lead length (mm)					
(min.)	2	1	2	12.5	6.5
(max.)	17	13.5	17	17.7	9.0
Lead inductance (nH)					
(min.)	3.5	2.5	7.0	16.0	5.0
(max.)	13.3	6.3	19.0	23.0	7.2
180 I/O Module					
Substrate size (mm)	36	36	36	—	30
Lead length (mm)					
(min.)	2	1	2	—	11.3
(max.)	23	20	23	—	16.3
Lead inductance (nH)					
(min.)	3.5	2.5	7.0	—	8.5
(max.)	17	8.2	24	—	12.5

Source: From Ref. 232.

multichip module application is shown in Table 8-10. Although process costs and density are major influencing factors, reliability, component and tooling costs, thermal performance, hermeticity requirements, need for reworking, turnaround time, and electrical parameters have all been shown to play major roles in the selection process. Although reworking is easy for C4 interconnections, inspectability of the joints is not. Although wirebonding and TAB do not restrict chip sizes and thermal-expansion coefficients of the substrates, these restrictions are inconsequential for C4 when matched expansion substrates are used. High thermal capability options have been shown for all of the technologies.

8.10 SUMMARY

It has been shown that wirebond, TAB, and C4 have all evolved and improved over the past 25–40 years. All have found niches and

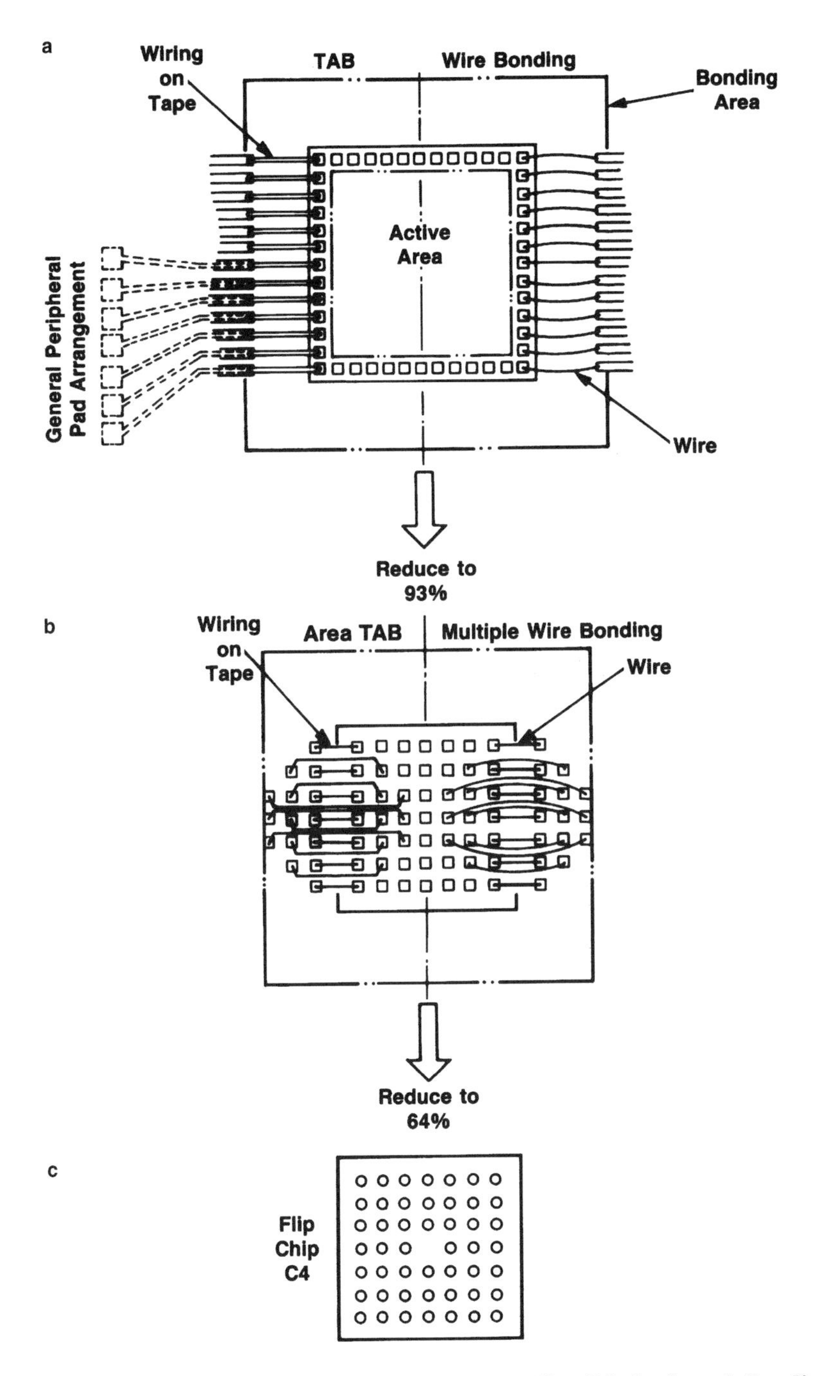

Figure 8-120. Relationships Among Various Bonding Methods and Bonding Space. (From Ref. 233.)

Table 8-10. Chip Interconnection Assembly Comparison for Multichip Module

C4	Wire Bond	TAB
1. Flux site	1. Solder or epoxy preform place	1. Reel tape into bond position
2. Align and place chip(s)	2. Align and place chip	2. Align and place chip
3. Reflow to bond all pads (on all chips)	3. Die bond	3. Inner lead bond (one chip at a time)
4. Clean flux	4. Wirebond each wire on chip and S/S (one wire at a time)	4. Encapsulate
5. Test	5. Test	5. Test/burn-in (optional)
6. Encapsulate/finish module assembly	6. Encapsulate/finish module assembly	6. Excise chip
		7. Align and place assembly on S/S
		8. Outer lead bond
		9. Test

applications which make them valuable technologies; but, as of today, wirebonding remains the incumbent king of the broad market and is the chip interconnect of choice for about 95% of the packages. The reasons are many. First, the process has transformed itself from where it was in 1960 to a highly reliable, inexpensive, very manufacturable process with greatly appreciated flexibility and versatility which does not constrain or commit the manufacturer. Au ball-bond wires coupled with greatly improved plastic molding processes and materials have made plastic dips and quad flat packs the low-cost, highly reliable package for the majority. Virtually all memory and all commodity electronics utilize these packages. The wirebonding infrastructure is so huge worldwide that there would have to be extremely strong and compelling reasons not to use today's wirebonding.

TAB was invented to replace wirebonding in the early 1970s. Its fine-pitch, peripheral nature was intended to extend wirebonding to significantly higher I/Os on chips with greatly improved reliability and manufacturability. Some referred to it as "poor man's beam lead," alluding to Bell's own replacement technology for the T/C wirebond. The technology is good. It did what it was designed to do, but the rate of improvement and cost of wirebonding have made it extremely difficult for TAB to displace WB. The Japanese manufacturers have had much more success with TAB than American or European manufacturers. The biggest use of TAB there is in the niche of LCD peripheral connections. Sharp has also been very successful with commodity items because of the inherently manufacturable nature of the process, combined with the lightness and low profile.

The overriding weakness of WB and TAB has been the peripheral nature of the interconnect, as opposed to C4, which has its greatest strength in area-array connectability. The rapid rate of integration of all devices in the ULSI era has driven the need for I/O, particularly in logic ASICs and microprocessors. The ease with which C4's satisfy the 1000–5000 I/O needs of current and future semiconductors has made its choice easy. The coupling of multilayer chip wiring and multilayered substrate wiring has also brought the best performance for high-speed computers and telecommunications devices. Also of value is the fact that the same chip functionality can be realized in a much smaller chip with the resulting economies of more chips per wafer.

The development of chip underfills has made possible direct chip attach to FR-4 cards in a reliable way for the first time, making the chip the ultimate "smallest package." Miniature pagers and telephones are beginning to appear, and low-cost bumping consortia have the support of government and industrial groups to make low-cost chip-on-board a growing interest. For the first time, flip chip is a serious consideration for low-cost commodity electronics as well as the up-scale applications.

It has been gratifying to watch these three interconnect technologies grow and develop, particularly the sharing of ideas. Features of one system are used freely in another. For example, single-point thermosonic wirebonders are excellent for TAB inner lead bonds, Au ball-bond bumping on flip chips provides bumping without major reinvestment. Flip-on-flex has a lot of commonality with TAB.

All three interconnect technologies will continue to coexist for a long time. Wirebonding will continue to dominate memory and commodity packaging requiring relatively small I/O counts. Flip chip will dominate high-end packages with greater than 500 I/O, and perhaps some commodity applications; TAB will find the special niches featuring thinness and lightness.

8.11 REFERENCES

1. C. T. Goddard. "The Role of Hybrids in LSI Systems," *IEEE TR*, TR-2: 367, 1979.
2. *The National Technology Roadmap for Semiconductors.* Semiconductor Industry Assn., 1994.
3. O. L. Anderson. *Bell Lab. Rec.*, November 1957.
4. M. P. Lepselter. "Beam Lead Technology," *Bell Sys. Tech. J.*, 45: pp. 233–253, 1966.
5. E. M. Davis, W. E. Harding, R. S. Schwartz, and J. J. Corning. "Solid Logic Technology: Versatile High Performance. Microelectronics," *IBM J. Res. Devel.*, 8: p. 102, 1964.
6. J. A. Perri, H. S. Lehman, W. A. Pliskin, and J. Riseman. "Surface Protection of Silicon Devices with Glass Films," *Electrochemical Society Meeting*, p. 102, 1961.

7. P. A. Totta and R. P. Sopher. "SLT Device Metallurgy and Its Monolithic Extension," *IBM J. Res. Devel.,* 5: pp. 226–238, 1969.
8. P. A. Totta. "Flip-Chip Solder Terminals," *21st Electronics Components Conference,* p. 275, 1971.
9. L. J. Fried, J. Havas, J. S. Lechaton, J. S. Logan, G. Paal, and P. Totta. "A VLSI Bipolar Metallization Design with Three-Level Wiring and Area Array Solder Connections," *IBM J. Res. Devel.,* 26: pp. 362–371, 1982.
10. K. H. Brown, D. A. Grose, R. C. Lange, T. H. Ning, and P. A. Totta. "Advancing the State of the Art in High-Performance Logic and Array Technology," *IBM J. Res. Devel.,* 36: p. 821, 1992.
11. L. F. Miller. "Controlled Collapse Reflow Chip Joining," *IBM J. Res. Devel.,* 13: pp. 239–250, 1969.
12. R. Darveaux, I. Turlik, L. Hwang, and A. Reisman. "Thermal Analysis of a Multichip Package Design," *IEEE Trans. Components Hybrids Manuf. Technol.* CHMT-12(4): pp. 663–672, 1989; *J. Electron. Mater.,* 18(2): pp. 267–274, 1989.
13. D. J. Bendz, R. W. Gedney, and J. Rasile. "Cost/Performance Single Chip Module," *IBM J. Res. Devel.,* 26: pp. 278–285, 1982.
14. A. J. Blodgett. "A Multilayer Ceramic Multichip Module," *IEEE Components Hybrids Manuf. Technol.* CHMT-3: p. 634, 1980.
15. L. S. Goldmann and P. A. Totta. "Area Array Solder Interconnections for VLSI," *Solid State Technol.,* 1983.
16. S. E. Greer. "Low Expansivity Organic Substrate for Flip-Chip Bonding," *28th Electronic Components Conference Proceedings*, pp. 166–171, 1978.
17. T. Inoue, H. Matsuyama, E. Matsuzaki, Y. Narazuka, M. Ishino, M. Tanaka, and T. Takenaka. "Micro Carrier for LSI Chip Used in the HITAC M-880 Processor Group," *Proc. ECTC,* pp. 704–711, 1991.
18. B. T. Clark. "Design of the IBM Thermal-Conduction Module," *IEEE Components Hybrids Manuf. Technol.* CHMT-4: 1981.
19. A. J. Blodgett and D. R. Barbour. "Thermal Conduction Module: A High-Performance Multilayer Ceramic Package," *IBM J. Res. Devel.,* 26: pp. 30–36, 1982.
20. D. R. Barbour, S. Oktay, and R. A. Rinne. "Thermal-Conduction Module Cradles and Coops up to 133. LSI Chips," *Electronics,* June 1982.
21. S. Ahmed, R. Tummala, and H. Potts. "Packaging Technology for IBMs Latest Mainframe Computers (S1390/ES 9000)," *Proc. ECTC,* pp. 682–688, 1991.
22. J. Oberschmidt and J. Humenik. "A Low Inductance Capacitor Technology," *Proc. ECTC,* pp. 284–288, 1990.
23. N. G. Koopman and P. A. Totta. "Controlled Collapse Chip Connections (C4)," *ASM World Material Congress,* 1988.
24. J. P. Dietz. "East Fishkill Advanced Technology in New 9370 Series," IBM Press Release, October 1986.
25. T. Kamei and M. Nakamura. "Hybrid IC Structures Using Solder Reflow Technology," *28th Electronic Components Conference Proceedings,* pp. 172–182, 1978.
26. P. Schmid and H. Melchior. "Coplanar Flip-Chip Mounting Techniques for Picosecond Devices," *Rev. Sci Instrum.,* 55(11): p. 1854, 1984.
27. W. Weston. "High Density 128 × 128 Area Arrays of Vertical Electrical Interconnections," *4th Annual Microelectronic Interconnect Conference,* 1985.
28. S. K. Lahiri, H. R. Bickford, P. Geldermans, K. R. Grebe, and P. A. Moskowitz. "Packaging Technology for Josephson Integrated Circuits," *IEEE Components Hybrids Manuf. Technol.,* CHMT-5(2): p. 271, 1982.
29. C. Y. Ting, K. Grebe, and D. Waldman. "Controlled Collapse Reflow for Josephson Chip Bonding," *J. Electrochem. Soc.,* 129(4): 859–864, 1982.

30. K. R. Grebe. "Orthogonal Solder Interconnects for Josephson Packaging," *J. Electrochem. Soc.,* 127: 1980.
31. I. Lang. "Surface Mounting of Leadless Chip Carriers on Various Printed Circuit Board Type Substrates," *Electron. Sci. Technol.,* 9: p. 296, 1982.
32. J. D. Leibowitz and W. E. Winters. "Shirline A Coefficient of Thermal Expansion (CTE) Controllable Family of Materials," *IEPC,* 1982.
33. J. Fisher. "Cast Leads for Surface Attachment," *ECC,* p. 219, 1984.
34. Y. Tsukada. "Surface Lamilar Circuit and Flip Chip Attach Packaging," *ECTC,* 1992.
35. R. K. Wassink. *Soldering in Electronics,* Electrochemical Publications Limited, Ayr, Scotland, 1984.
36. J. Gow III and A. R. Leach. "Memory Packaging at IBM," *Circuits Manuf.,* p. 72, May 1983.
37. M. Ohshima, A. Kenmotsu, and I. Ishi. "Optimization of Micro Solder Reflow Bonding for the LSI Flip Chip," *The Second International Electronics Packaging Conference,* 1982.
38. T. Takenaka, F. Kobayashi, and T. Netsu. "Reliability of Flip-Chip Interconnections," *1984 ISHM Conference Proceedings,* p. 419, 1984.
39. F. Kobayashi, T. Takenaka, M. Ishida, and T. Netsu. "Fabrication and Reliability Evaluation of the Multi-Chip Ram Module," *Proc. 1982 IMC,* p. 425, 1982.
40. K. C. Norris, and A. H. Landzberg. "Reliability of Controlled Collapse Interconnections," *IBM J. Res. Devel.,* 13(3): pp. 266–271, 1969.
41. V. K. Nagesh. "Reliability of Flip Chip Solder Bump Joints," *Reliability Physics Symposium,* 1982.
42. D. L. Brownell and G. C. Waite. "Solder Bump Flip-Chip Fabrication Using Standard Chip and Wire Integrated Circuit Layout," *1974 ISHM Conference Proceedings,* p. 77, September 1974.
43. W. Roush. "Effect of Gold Additions on High Strain-Low Cycle Fatigue Behavior of 95/5 Pb/Sn Solder Connections," *Proc. 2nd International Conference on Behavior of Materials,* p. 780, 1976.
44. G. Clatterbaugh and H. Charles. "Design Optimation and Reliability Testing of Surface Mounted Solder Joints," *ISHM,* p. 31, 1985.
45. T. Yamada, K. Otsutani, K. Otsukani, K. Sahara, and K. Otsuka. "Low Stress Design of Flip Chip Technology for "Si on Si Multichip Modules," *IEPS,* pp. 551–557, 1985.
46. H. Curtis. "Integrated Circuit Design Production and Packaging for System 38," *IBM System 38 Technical Development,* December 1978.
47. L. S. Goldmann. "Self-Alignment Capability of Controlled-Collapse Chip Joining," *Proceedings 22nd Electronic Components Conference,* p. 332, 1972.
48. T. Kawanobe, K. Miyamoto, and Y. Inaba. "Solder Bump Fabrication by Electrochemical Method for Flip-Chip Interconnections," *31st Electronic Components Conference Proceedings,* p. 149, 1981.
49. J. Marshall and F. Rode. "Solder Bump Interconnected Multiple Chip Thick Film Hybrid for a 40-Character Alphanumeric LCD Application," *Solid State Technol.,* p. 87, 1979.
50. E. Rasmanis. "Fabrication of Semiconductor Devices with Solder Terminals," *1973 ISHM Conference Proceedings,* pp. 3B-3–1, 1973.
51. A. VanDer Drift, W. G. Gelling, and A. Rademakers. "Integrated Circuits with Leads on Flexible Tape," *Philips Tech. Rev.,* 34(4): p. 85, 1974.
52. T. R. Myers. "Flip-Chip Microcircuit Bonding System," *Proc. ECC,* p. 131, 1969.
53. N. Tsujimoto. "Flip Chip and Micropackaged Chips," *Electron. Parts,* 19(5): p. 51, 1980.

54. N. Nuzaki, K. Nakamura, and I. Taubokawa. "A Thick Film Active Filter for PCM Communication Systems," *28th Electronic Components Conference Proceedings,* p. 183, 1977.
55. G. Adema, C. Berry, N. Koopman, G. Rinne, E. Yung, and I. Turlik. "Flip Chip Technology: A Method for Providing Known Good Die With High Density Interconnections", *3rd International Conference and Exhibition on Multichip Modules,* 1994.
56. P. Magill, P. Deane, D. Mis, and G. Rinne. "Flip Chip Overview," *IEEE Multichip Module Conference,* 1996.
57. C. J. Speerschneider and J. M. Lee. "Solder Bump Reflow Tape Automated Bonding," *Proceedings 2nd ASM International Electronic Materials and Processing Congress,* pp. 7–12, 1989.
58. D. J. Genin and M. M. Wurster. "Probing Considerations in C-4 Testing of Wafers," *Microcircuits Electron. Packaging,* 15-4: p. 229, 1992.
59. N. Koopman. "Flip Chip Interconnections," in *Concise Encyclopedia of Semiconducting Materials and Related Technologies,* pp. 184–187, Pergamon Press, London, 1992.
60. V. Iinuma, T. Hirohara, K. Inove, "Liquid Crystal Color Television," *ISHM Proc* 1987, pp. 635–640.
61. Y. Kondoh, M. Saito, "A New CCD Module Using the Chip–On–Glass (COG) Technique, *ISHM Proc,* 1990, pp. 487–494.
62. R. Bache, P. Burdett, K. Pickering, A. Parsons, D. Pedder. "Bond Design and Alignment in Flip-Chip Solder Bonding, *Proc. IEPC,* pp. 830–841, 1988.
63. M. Goodwin, C. Kirkby, A. Parsons, I. Bennion, and W. Stewart. "8 × 8 Element Hybridized PLZT/Silicon Spacial Light Modulator Array," *Electron. Lett.,* 25(18): pp. 1260–1262, 1989.
64. P. Burdett, K. Lodge, and D. Pedder. "Techniques for the Inspection of Flip Chip Solder Bonded Devices," *Proc IEPS,* pp. xxx-xxx, 1988, reprinted in *Hybrid Circuits,* 19: pp. 44–48, 1989.
65. S. Patra and Y. Lee. "Modelling of Self Alignment in Flip Chip Soldering—Part 2: Multichip Solder Joints," *Proc. ECTC,* pp. 783–788, 1991.
66. M. Wale. "Self-Aligned, Flip Chip Assembly of Photonic Devices with Electrical and Optical Connections," *Proc. ECTC,* pp. 470–476, 1990.
67. A. Burkhart. "Recent Developments in Flip Chip Technology," *Surface Mount Technol.,* pp. 41–45, 1991.
68. R. T. Howard. "Optimization of Indium-Lead Alloys for Controlled Collapse Chip Connection Application," *IBM J. Res. Devel.,* 26(3): pp. 372–389, 1982.
69. K. J. Puttlitz. "Flip-Chip Replacement Within the Constraints Imposed by Multilayer Ceramic (MLC) Modules," *J. Electron. Mater.,* 13(1): pp. 29–46, 1984.
70. S. Teed and V. Marcotte. "A DSC Technique to Measure the Amount of Indium or Tin Leached from Lead Base Solders by Rosin Fluxes," *NATAS,* p. 77, 1981.
71. J. Temmyo, K. Aoki, H. Yoshikiyo, S. Tsurumi, and Y. Takeuchi. "Solder Bump Height Dependence of Josephson Chip to Card Interconnection Inductance Using Flip-Chip Bonding Technique," *J. Appl. Phys.,* 54(9): p. 5282, 1983.
72. N. Koopman and S. Nangalia. "Fluxless Flip Chip Solder Joining," *NEPCON WEST,* 1995.
73. N. Koopman, G. Adema, S. Nangalia, M. Schneider, and V. Saba. "Flip Chip Process Development Technique Using a Modified Laboratory Aligner Bonder," *IEMT Symposium,* 1995.
74. H. Markstein, "Multichip Modules Pursue Water Scale Interaction,", *EP&P,* October, 1991.
75. K. Lodge, D. Pedder. "The Impact of Packaging on the Reliability of Flip Chip Solder Bonded Devices," *Proc. ECTC,* pp. 470–476, 1990.

76. A. Munns. "Flip-Chip Solder Bonding in Microelectronic Applications," *Metals Mater.*, pp. 22–25, 1989.
77. K. Puttlitz. "Preparation, Structure, and Fracture Modes of Pb–In Terminated Flip Chips Attached to Gold-Capped Microsockets," *Proc. ECTC,* pp. 360–367, 1990.
78. B. Le Pape, "Use of a Tinned Copper Slug for Module Reworking." *IBM Tech. Disclosure Bull.,* 24(7A): pp. 3481, 1981.
79. A. Ingraham, J. McCreary, and J. Varcoe. "Flip-Chip Soldering to Bare Copper Circuits," *Proc. ECTC,* pp. 333–337, 1990.
80. T. Adams. "Acoustic Micro-Imaging of Multichip Modules," *Surface Mount Technol.,* June 1991.
81. B. Gilbert, W. McNeff, P. Zabinski, W. Kornrumpf, S. Tead, R. Maki, K. Carlson, and J. Gerber. "Development of Deposited Multichip Modules with Unique Features for Application in GaAs Signal Processors Operating Above 1 GHz Clock Rates," *Proc. IEPS,* pp. 526–541, 1992.
82. N. Koopman, G. Rinne, P. Magill, S. Nangalia, C. Berry, D. Mis, V. Rogers, G. Adema, and M. Berry. "Solder Flip Chip Developments at MCNC," *1996 International Flip Chip, Ball Grid Array, TAB and Advanced Packaging Symposium—ITAP,* 1996.
83. D. A. Jeannotte. "Solder as a Structural Member for Chip Joining," *Electronic Components Conference Proceedings,* p. 334, 1969.
84. L. F. Coffin. "Low Cycle Fatigue: A Review," *Appl. Mech. Res.,* 1(3): pp. 129–141, 1962.
85. P. A. Tobias, N. A. Sinclair, and A. S. Van. "The Reliability of Controlled-Collapse Solder LSI Interconnections," *ISHM Proc.,* p. 360, 1976.
86. N. A. Sinclair. "Thermal Cycle Fatigue Life of LSI Solder Interconnections," *International Electrical Proceedings,* p. 56, 1982.
87. H. J. Shah and J. H. Kelly. "Effect of Dwell Time on Thermal Cycling of the Flip-Chip Joint," *1970 ISHM Proc.,* paper 3.4, 1970.
88. E. Levine and J. Ordonez. "Analysis of Thermal Cycle Fatigue Damage in Microsocket Solder Joints," *31st Electronic Components Conference Proceedings,* pp. 515–519, 1981.
89. L. S. Goldmann. "Geometric Optimization of Controlled Collapse Interconnections," *IBM J. Res. Devel.,* 3: pp. 251–265, 1969.
90. E. Anderson. "Flip-Chip Bonding: The Effect of Increased Solder Bump Spacing," *Microelectronic Interconnect Conference,* 1984.
91. R. T. Howard, S. W. Sobeck, and C. Sanetra. "A New Package-Related Failure Mechanism for Leadless Ceramic Chip Carriers (LC-3s) Solder-Attached to Alumina Substrates," *Solid State Technol.,* 26(2): pp. 115–122, 1983.
92. P. M. Hall. "Solder Post Attachment of Ceramic Chip Carriers to Ceramic Film Integrated Circuits," *31st Electronic Components Conference Proceedings,* pp. 172–180, 1981.
93. R. Satoh, M. Ohshima, H. Komura, I. Ishi, and K. Serizawa. "Development of a New Micro-Solder Bonding Method for VLSI," *IEPS,* p. 455, 1983.
94. N. Matsui, S. Sasaki, and T. Ohsaki. "VLSI Chip Interconnection Technology Using Stacked Solder Bumps," *Proceedings IEEE 37th Electronic Components Conference,* pp. 573–578, 1987.
95. E. Horikoshi, K. Hoshimoto, T. Sato, and H. Nakajima. "Published unexamined patent application," Japanese Patent No. 62-117346, 1987.
96. G. Cherian. "Use of Discrete Solder Columns to Mount LCC's on Glass/Epoxy Printed Circuit Boards," *IEPS,* p. 710, 1984.
97. L. S. Goldmann, R. J. Herdizk, N. G. Koopman, and V. C. Marcotte. "Lead Indium for Controlled Collapse Chip Joining," *27th Electronic Components Conference Proceedings,* 27: p. 25, 1977.

98. R. T. Howard. "Packaging Reliability and How to Define and Measure It," *32nd Electronic Components Conference Proceedings,* pp. 367–384, 1982.
99. W. Roush and J. Jaspal. "Thermomigration in Lead-Indium Solder," *32nd Electronic Components Conference Proceedings,* p. 342, 1982.
100. C. J. Bartlett et al. "Multichip Packaging Design for VLSI-Based Systems," *Proc. ECTC,* p. 518, 1987.
101. J. H. Lau and D. W. Rice. "Solder Joint Fatigue in Surface Mount Technology State of the Art," *Solid State Technol.,* p. 91, 1985.
102. W. Engelmaier. "Effects of Power Cycling on Leadless Chip Carrier Mounting Reliability," *IEPS,* p. 15, 1982.
103. F. Nakano, T. Soga, and S. Amagi. "Resin Insertion Effect on Thermal Cycle Resistivity of Flip Chip Mounted LSI Devices," *Proceedings 1987 ISHM Conference,* pp. 536–541, 1987.
104. B. Han and Y. Guo, "Thermal Deformation Analysis of Various Electronic Packaging Products by Moire and Microscope Moire Interferometry," *J. Electron. Packaging,* 117: p. 185, 1995.
105. V. Marcotte, M. Ricker, and N. Koopman. "Metallography of Soldered Joints," *IMS/ASM/AWS Symposium,* 1985.
106. V. C. Marcotte and N. G. Koopman. "Palladium Depletion of 80Ag20Pd Thick Film Electrodes," *ECC,* pp. 157–164, 1981.
107. D. Bouldin. "The Measurement of Alpha Particle Emissions from Semiconductor Memory Materials," *J. Electron. Mater.,* 10(4): p. 747, 1981.
108. R. Howard. "Characterization of Low-Alpha-Particle Emitting Ceramics," *ECC,* p. 453, 1984.
109. Cominco/Johnson Matthry. Private Communications, 1987.
110. K. Scheibner. "Laser Isotope Purification of Lead for Use in Semiconductor Chip Interconnect", *ECTC,* 1996.
111. J. L. Chu, H. R. Torabi and F. J. Towler, "A 128 Kb CMOS Static Random—Access Memory," *IBM J. R&D* pp. 321–329, 1991.
112. S. Oktay. "Parametric Study of Temperature Profiles in Chips Joined by Controlled Collapse Technique," *IBM J. Res. Devel.,* 13: pp. 272–285, 1969.
113. V. Kurokawa, K. Utsumi, H. Takamizawa, T. Kamata, and S. Noguchi. "AlN Substrates with High Thermal Conductivity," *IEEE TR.,* Tr-2: p. 247, 1985.
114. S. Oktay and H. C. Kammerer. "A Conduction-Cooled Module for High-Performance LSI Devices," *IBM J. Res. Devel.,* 26(1): pp. 55–66, 1982.
115. R. C. Chu, U. P. Hwang, and R. E. Simons. "Conduction Cooling for an LSI Package: A One-Dimensional Approach," *IBM J. Res. Devel.,* 26: pp. 45–54, 1982.
116. F. Kobayashi, Y. Watanabe, M. Yamamoro, A. Anzai, A. Takahashi, T. Daikoku, and T. Fujita. "Hardware Technology for Hitachi M880 Processor Group," *Proc. ECTC,* pp. 693–704, 1991.
117. R. E. Simmons and R. C. Chu. "Direct Immersion Cooling Techniques for High Density Elect. Packaging and System," *ISHM,* p. 314, 1985.
118. S. Aoki and Y. Imanaka. "Multilayer Ceramic Substrate for HEMT Packaging (Liquid Nitrogen Packaging) for GaAs Devices," *Proc. IEPS,* pp. 329–341, 1991.
119. H. Tong, L. Mok, K. Grebe, and H. Yeh. "Paralyne Encapsulation of Ceramic Packages for Liquid Nitrogen Applications," *Proc. ECTC,* pp. 345–350, 1990.
120. K. Casson, B. Gibson, and K. Habeck. "Flip-on-Flex: Solder Bumped ICs Bond to a New High Temperature, Adhesiveness, Flex Material," *Surface Mount International Proceedings,* pp. 99–104, 1991.
121. K. Gilleo. "Direct Chip Interconnect Using Polymer Bonding," *Proc. 39th ECC,* pp. 37–44, 1989.

122. *The VLSI Manufacturing Outlook,* vol. I, sects. 52 and 54, VLSI Research Inc., 1992.
123. Pashby et al. International Business Machines Corporation. "Lead on Chip Semiconductor Package," U. S. Patent 4,862,245.
124. Winchell and Berg. "Enhancing Ultrasonic Bond Development," *IEEE Trans. Components Hybrids Manuf. Technol.,* CHMT-1 (3): 1978.
125. Harman and Leedy. "An Experimental Model of the Microelectronic Ultrasonic Wire Bonding Mechanism," *Proceedings 10th Annual Reliability Physics Symposium,* pp. 49–56, 1972.
126. E. Philofsky. "Design Limits When Using Gold–Aluminum Bonds," *Proceedings 9th Annual IEEE Reliability Physics Symposium,* pp. 177–185, 1970.
127. Mitsubishi Semiconductor IC Packages, *Databook,* pp. 2–14, 1988.
128. B. L. Gehman, "Bonding Wire Microelectronics Interconnections," *IEEE Trans. on Comp.*, CHMT-3(3): p. 375, 1980.
129. R. Shukla and N. Mencinger, "A Critical Review of VLSI Die-Attachment in High Reliability Applications," *Solid State Tech.*, p. 67, July 1985.
130. Presented at: IEE/CHMT International Electronics Manufacturing Technology Symposium, San Francisco, Ca., September 16–18, 1991.
131. Cindy Enman and Gil Perlberg, "Wirebond Pull Testing: Understanding the Geometric Resolution Forces," *Advanced Packaging*, pp. 21–22, 1994.
132. G. Harman, "Reliability and Yield Problems of Wire Bonding in Microelectronics," *ISHM*, pp. 19–20, 1991.
133. A. D. Aird. "Method of Manufacturing a Semiconductor Device Utilizing a Flexible Carrier," U. S. Patient 3,689,991, 1972.
134. G. Dehaine and M. Leclercq. "Tape Automated Bonding, a New Multichip Module Assembly Technique," *Electronic Components Conference Proceedings,* pp. 69–73, 1973.
135. T. Watari and H. Murano. "Packaging Technology for the NEC SX Supercomputer," *Proc. 35th ECC,* p. 192–198, 1985.
136. K. Saito. "TAB Applications in Japan," *Proceedings Fifth International TAB and Advanced Packaging Symposium,* pp. 9–19, 1993.
137. J. Lyman. "How DOD's VHSIC Is Spreading," *Electronics,* pp. 33–37, 1985.
138. R. L. Cain. "Beam Tape Carriers—a Design Guide," *Solid State Technol.,* pp. 53–58, March 1978.
139. *TAB Design Guide,* Shindo Company Ltd., 1993.
140. K. Fukuta, T. Tsuda, and T. Maeda. "Optimization of Tape Carrier Materials," *Proceedings Fourth International TAB Symposium,* pp. 283–312, 1992.
141. J. M. Smith and S. M. Stuhlbarg. "Hybrid Microcircuit Tape Chip Carrier Materials/Processing Tradeoffs," *Proc. 27th ECC,* pp. 34–47, 1977.
142. K. Saito, K. Nomoto, and Y. Nishimoto. "The Properties of Adhesive in Three-Layer TAB Tape," *Proceedings Fourth International TAB Symposium,* pp. 210–235, 1992.
143. K. Doss and S. Holzinger. "Materials Choices for Tape Automated Bonding," in *Handbook of Tape Automated Bonding,* ed. by J. Lau, pp. 99–134, Van Nostrand Reinhold, New York, 1992.
144. T. Mugishima, O. Seki, and K. Ohtani. "Improvement of Adhesion Strength in Two-Layer Tape," *Proceedings Fourth International TAB Symposium,* pp. 198–209, 1992.
145. R. S. Dodsworth and R. T. Smith. "TAB Tape Design and Manufacturing," in Handbook of Tape Automated Bonding, ed. by J. Lau, pp. 135–175, Van Nostrand Reinhold, New York, 1992.
146. P. Chen, K. Blackwell, and A. Knoll. "Curl and Residual Stress Control of Metallized Polymides," *Proceedings Fifth International TAB/Advanced Packing Symposium,* pp. 85–91, 1993.

147. H. Nakayama, Y. Satomi and T. Natsume. "Characteristics of New TAB Products using Special Copper Alloys," *Proceedings Third International TAB Symposium,* pp. 180–191, 1991.
148. K. Saito. "Fine Lead Pitch Technology in Tape Carriers and Double Metal Layer Tape Carrier," *Proceedings Second International TAB Symposium,* pp. 199–213, 1990.
149. K. Saito. "Selective Plating," *Proceedings Third International TAB Symposium,* pp. 146–156, 1991.
150. S. Britton. "Spontaneous Growth of Whiskers on Tin Coatings: 20 Years of Observation," *Trans. Inst. Metal Finishing,* pp. 95–102, 1974.
151. T. Kawanobe, K. Miyamoto and M. Hirano. "Tape Automated Bonding Process for High Lead Count LSI," *Proc. 33rd ECC,* pp. 221–226, 1983.
152. P. Hoffman, T. Kleiner, and H. Phan. "Characteristics of Tin Plated TAB Tape," *Proc. IEPS,* pp. 1307–1326, 1989.
153. A. Juan. "TAB Applications for Miniaturization," *Proceedings Fourth International TAB Symposium,* pp. 26–41, 1992.
154. M. Masuda, K. Sakurai, K. Haley, Y. Hiraki, and S. Inoue. "New Configuration Three-Layer BTAB Technology Application for CPU," *Proceedings Fifth International TAB Symposium,* pp. 112–117, 1993.
155. K. Hatada, J. Okamoto, M. Hirai, K. Matsunaga, and I. Kitahiro. "New Film Carrier Assembly Technology: Transferred Bumped TAB," *Nat. Tech. Rept. Japan,* 31: pp. 412–420, 1985.
156. K. Hatada. "Application to the Electronic Instrument by Transferred Bump-TAB Technology, *Proceedings International Symposium on Microelectronics (ISHM),* pp. 649–653, 1987.
157. K. Hatada. "New Bonding Technology Using Transferred Bumped TAB and Single Point Bonding," *Proceedings Third International TAB Symposium,* pp. 124–130, 1991.
158. K. Taguchi. "Small Bump Development on using TB-TAB for Fine Pitch Interconnect," *Proceedings Fourth International TAB Symposium,* pp. 236–241, 1993.
159. K. Tatsumi, T. Ando, Y. Ohno, M. Konda, Y. Kawakami, N. Ohikata, and T. Maruyama. "Transferred Ball Bump Technology for Tape Carrier Packages," *Proceedings International Symposium on Microelectronics,* pp. 54–59, 1994.
160. P. Hoffman. "Design, Manufacturing and Reliability Aspects of Two Metal Layer TAB Tape," *Proceedings Second International TAB Symposium,* pp. 124–146, 1990.
161. K. Saito. "Fine Lead Pitch Technology in Tape Carriers and Double Metal Layer Tape Carrier," *Proceedings Second International TAB Symposium,* pp. 199–213, 1990.
162. Y. Mashiko and S. Nishiyama. "Comparison of Characteristics and Reliability Between Two Types of Adhesiveless 2-Metal TAB Tape," *Proceedings Fifth International TAB/Advanced Packaging Symposium,* pp. 62–69, 1993.
163. D. Aeschliman. "Multiconductor TAB Tape Materials," *Proceedings Second International TAB Symposium,* pp. 250–257, 1990.
164. B. Lynch. "Two Metal Layer COT Package Enhancement," *Proceedings Fifth International TAB/Advanced Packaging Symposium,* pp. 74–84, 1993.
165. G. Cox, J. Dorfman, C. Fedetz, T. Lantzer, R. Lawton, and J. Lott. "DuPont Process for High Performance Two Conductor TAB," *Proceedings Second International TAB Symposium,* pp. 222–229, 1990.
166. J. M. Milewski and C. G. Angulas. "TAB Design for Manufacturability and Performance," *Proceedings Second International TAB Symposium,* pp. 121–132, 1992.
167. S. C. Lockard, J. M. Hansen, and G. H. Nelson. "High Leadcount Multimetal Layer TAB Circuits," *Proceedings Second International TAB Symposium,* pp. 214–221, 1990.

168. N. Tajima, Y. Chikawa, T. Tsuda, and T. Maeda. "TAB Design for ESD Protection," *Proceedings Second International TAB Symposium,* pp. 77–87, 1990.

169. R. G. Oswald and W. R. de Miranda. "Application of Tape Automated Bonding Technology to Hybrid Microcircuits," *Solid State Technol.,* pp. 33–38, 1977.

170. P. Elenius. "Au Bumped Known Good Die," *Proceedings First International Symposium on Flip Chip Technology & Sixth International TAB/Advanced Packaging Symposium,* pp. 94–97, 1994.

171. M. Bonkohara, E. Hajimoto, and K. Takekawa. "Utilization of Inner Lead Bonding Using Ball Bump Technology," *Proceedings Fourth International TAB Symposium,* pp. 86–96, 1992.

172. J. Wolf, G. Chmiel, J. Simon, and H. Reichl. "Solderbumping—A Comparison of Different Technologies," *Proceedings First International Symposium on Flip Chip Technology & Sixth International TAB/Advanced Packaging Symposium,* pp. 105–110, 1994.

173. T. Yokoyama, H. Ikeda, K. Oshige, M. Kimura, and K. Utsumi. "Bare Chip Bump Technology Using Advanced Electroplating Method," *Proceedings 8th International Microelectronics Conference,* pp. 402–407, 1994.

174. M. Inaba, K. Yamakawa, and N. Iwase. "Solder Bump Formation Using Electroless Plating and Ultrasonic Soldering," *Proceedings 5th IEEE/CHMT IEMT Symposium,* pp. 13–17, 1988.

175. S. Khadpe. *Introduction to Tape Automated Bonding,* pp. 28–29, Semiconductor Technology Center, Inc., 1992.

176. K. Atsumi, N. Kashima, and Y. Maehara. "Inner Lead Bonding Techniques for 500 Lead Dies Having a 90 μm Lead Pitch," *Proceedings 39th IEEE Electronic Components Conference,* pp. 171–176, 1989.

177. R. C. Kershner and N. T. Panousis. "Diamond-Tipped and Other New Thermodes for Device Bonding," *IEEE Trans. Components Hybrids Manuf. Technol.,* CHMT-2, pp. 283–288, 1979.

178. K. Tanaka, "The Performance of a CVD Diamond Bonding Tool," *First VLSI Packaging Workshop of Japan Abstracts,* pp. 105–107, 1992.

179. Shinkawa Inner Lead Bonder Brochure, Shinkawa Ltd., Japan, 1992.

180. L. Levine and M. Sheaffer. "Optimizing the Single Point TAB Inner Lead Bonding Process," *Proceedings Second International TAB Symposium,* pp. 16–24, 1990.

181. S. Patil. "Chip-on-Tape Single Point Bonding and Interconnect Reliability," presented at the *Second International TAB Symposium,* 1990.

182. G. Silverberg, "Single Point TAB (SPT): A Versatile Tool for TAB Bonding," *Proceedings ISHM Symposium,* pp. 449–456, 1987.

183. G. Dehaine, "Ceramic Package Leadframe Replaced by TAB Carrier,"*Proceedings Second International TAB Symposium,* pp. 54–61, 1990.

184. G. Kelly. "Automated Single Point TAB—High Quality Bonding for High-Density Chips," *Microelectronic Manufacturing and Testing,* 1989.

185. G. Dehaine. "Single Point ILB at Narrow Pitch," *Proceedings Fifth International TAB/Advanced Packaging Symposium,* pp. 153–157, 1993.

186. J. D. Hayward. "Preliminary Evaluation of Laser Bonding for TAB ILB," presented at the *Third International TAB Symposium,* 1991.

187. J. D. Hayward. "Optimization and Reliability Evaluation of a Laser Inner Lead Bonding Process," *Proceedings Fifth International TAB Symposium,* pp. 52–61, 1993.

188. P. Spletter. "Gold to Gold TAB Inner Lead Bonding with a Laser," *Proceedings Fourth International TAB Symposium,* pp. 58–71, 1992.

189. E. Zakel, G. Azdasht, and H. Reichl. "Investigations of Laser Soldered TAB Inner Lead Contacts," *Proceedings 41st ECTC,* pp. 497–506, 1991.

190. G. Azdasht, E. Zakel, and H. Reichl. "A New Chip Packaging Method Using Windowless Flip-TAB Laser Connection on Flex Substrate," *Proceedings International Flip Chip, BGA, TAB and Advanced Packaging Symposium,* pp. 237–244, 1995.
191. N. G. Koopman, T. C. Rieley, and P. A. Totta. "Chip-to-Package Interconnections," in *Microelectronics Packaging Handbook,* pp. 409–435, 1989.
192. T. A. Scharr. "TAB Bonding a 200 Lead Die," *Proceedings ISHM Symposium,* pp. 561–565, 1983.
193. M. Hayakawa, T. Maeda, M. Kumura, R. H. Holly, and T. H. Gielow. "Film Carrier Assembly Process," *Solid State Technol.,* pp. 52–55, March 1979.
194. T. H. Spenser. "Thermocompression Bond Kinetics: The Four Principal Variables," *Int. J. Hybrid Microelectron.,* pp. 404–410, 1982.
195. N. Ahmed and J. J. Svitak. "Characterization of Gold-Gold Thermocompression Bonding," *Proceedings ECC,* pp. 52–63, 1975.
196. J. L. Jellison. "Effect of Contamination on the Thermocompression Bondability of Gold," *Proceedings ECC,* pp. 271–277, 1975.
197. P. Hedemalm, L.-G. Liljestrand, and H. Bernhoff. "Quality and Reliability of TAB Inner Lead Bond," *Proceedings Second International TAB Symposium,* pp. 88–102, 1990.
198. E. Zakel, R. Leutenbauer, and H. Reichl. "Investigations of the Cu-Sn and Cu-Au Tape Metallurgy and of the Bondability of TAB Inner Lead Contacts after Thermal Aging," *Proceedings Third International TAB Symposium,* pp. 78–93, 1991.
199. D. B. Walshak, Jr. "TAB Inner Lead Bonding," in *Handbook of Tape Automated Bonding,* ed. by J. Lau, pp. 202–242, Van Nostrand Reinhold, New York, 1992.
200. B. Lynch. "Optimization of the Inner Lead Bonding Process Using Taguchi Methods," *Proceedings Fourth International TAB Symposium,* pp. 72–85, 1992.
201. T. C. Chung, D. A. Gibson, and P. B. Wesling. "TAB Testing and Burn-in," in *Handbook of Tape Automated Bonding,* ed. by J. Lau, pp. 243–304, Van Nostrand Reinhold, New York, 1992.
202. G. G. Harman. "Acoustic-Emission Monitored Tests for TAB Inner Lead Bond Quality," *Trans. IEEE Components Hybrids Manuf. Technol.,* CHMT-5: pp. 445–453, 1982.
203. R. W. Shreeve. "Automated On-Chip Burn-In System for TAB Reels," *Proceedings 39th IEEE Electronic Components Conference,* pp. 187–189, 1989.
204. J. L. Kowalski. "Individual Carriers for TAB Integrated Circuit Assembly," *Proceedings 29th ECC,* pp. 315–318, 1979.
205. K. Fujita, T. Onishi, S. Wakamoto, T. Maeda, and M. Hayakawa. "Chip-Size Plastic Encapsulation on Tape Carrier Package," *Int. J. Hybrid Microelectron.,* 8(2), pp. 9–15, 1985.
206. K. Mori, M. Ohsono, and T. Maeda. "High-Reliability TAB Encapsulation Technology," *Proceedings Third International TAB Symposium,* pp. 43–53, 1991.
207. B. Bonitz, M. DiPietro, and J. Poole. "Encapsulation of TAB Devices," *Proceedings Fourth International TAB Symposium,* pp. 133–139, 1992.
208. G. Zimmer. "Using Advanced Pulsed Hotbar Solder Technology for Reliable Positioning and Mounting of High Lead Count Flat Packs and TAB Devices," *Proceedings Second International TAB Symposium,* pp. 230–249, 1990.
209. G. Westby, R. Heitmann, and S. Haruyama. "Hot Bar Soldering for TAB Attachment," *Proceedings Fourth International TAB Symposium,* pp. 332–341, 1992.
210. *Advanced Packaging Applications: Tape Carrier Package,* Addendum to Intel's 1995 Packaging Databook, pp. 17–18, Intel Corporation, 1995.
211. T. Kleiner. "Excise & Form Tooling Principles," *Proceedings Second International TAB Symposium,* pp. 62–76, 1990.

212. E. Zakel, G. Azdasht, P. Kruppa, and H. Reichl. "Reliability Investigations of Different Tape Metallizations for TAB-Outer Lead Bonding," *Proceedings Fourth International TAB Symposium,* pp. 97–120, 1992.
213. M. C. Miller and J. Glazer. "Equipment Selection for TAB Outer Lead Bonding," in *Handbook of Tape Automated Bonding,* ed. by J. Lau, pp. 338–357, Van Nostrand Reinhold, New York, 1992.
214. M. Kasahara, T. Oyanagi, and Y. Honma. "Hot Bar Technology for Gang Bond," *Proceedings 8th International Microelectronic Conference,* pp. 399–401, 1994.
215. A. Eldred, R. Heitmann, and M. Yingling. "TAB Outer Lead Bonding at 96 Micron Pitch," presented at *Fifth International TAB/Advanced Packaging Symposium,* 1993.
216. Fine Pitch Bonder Brochure, Universal Instruments Corporation, 1991.
217. I. Kurashima. "The New Technology of Anisotropic Conductive Adhesive for Fine Pitch TAB," *Proceedings Third International TAB Symposium,* pp. 212–220, 1991.
218. E. Hosomi, C. Takubo, Y. Hiruta, T. Sudo, F. Takahashi, K. Saito, N. Nakajima, and M. Nakazono. "The Wettability and Bondability Degradation Mechanisms of Tin-Plated Copper Leads of a Fine-Pitch TCP during Thermal and Humid Aging," *Proc. IMC* 1994, pp. 128–133, 1994.
219. F. Andros and R. Hammer. "IBM's Area Array TAB Packaging Technology," *Semicond. Packaging Update,* 8(3): pp. 1–8, 1993.
220. J. Walker. "TapePak: The Semiconductor Package for the Future," *Natl. Semicond. Technol. Rev.,* 1: pp. 69–72, 1988.
221. D. Card. "How ETA Chose to Make a Megaboard for its Supercomputer," *Electron. Bus.,* pp. 50–52, July 1, 1988.
222. S. Khadpe. "Worldwide Activities in Flip Chip, BGA and TAB Technologies and Markets," *Proceedings 1995 International Flip Chip, Ball Grid Array, TAB and Advanced Packaging Symposium,"* pp. 290–293, 1995.
223. "New Mounting Methods to Directly Connect LSI Chips with a Large Number of Pins," *Nikkei Electronics Microdevices,* June 11, 1984.
224. K. Smith. "An Inexpensive High Frequency High Power VLSI Chip Carrier," *IEPS,* 1985.
225. K. Hatada, H. Fujumoto, and T. Kawakita. "Insulation Resin Bonding: Chip on Substrate Assembly Technology," *ICICE,* 1987.
226. K. Chung, R. Fleishman, D. Bendorovich, M. Yan, and N. Mescia. "Z-Axis Conductive Adhesives for Fine Pitch Interconnections," *ISHM Proceedings,* pp. 678–689, 1992.
227. Epoxy Technology, Inc. Product Literature, 1992.
228. N. Masuda, K. Sakuma, E. Satoh, Y. Yamasaki, H. Miyasaka, and J. Takeuchi. *IEEE/CHMT Japan IEMT Symposium,* pp. 55–58, 1989.
229. H. Kristiansen and A. Bjorneklett. "Fine Pitch Connection of Flexible Circuits to Rigid Substrates Using Non-Conductive Epoxy Adhesive," *IEPS,* pp. 759–773, 1991.
230. K. Sakuma, K. Nozawa, E. Sato, Y. Yamasaki, K. Hanyuda, H. Miyasaka, and J. Takeuchi. "Chip on Glass Technology with Standard Aluminized IC Chip," *ISHM,* pp. 250–256, 1990.
231. M. F. Bregman. Private communication.
232. G. A. Katopis. "Delta-I Noise Specification for a High Performance Computing. "Machine," *Proc. IEEE,* 73: pp. 1405–1415, 1985.
233. K. Ohtsuka. "Special Article 2 Diversifying LSI Packages 1," *Nikkei Electronics Microdevices 2,* June 11, 1984.

9

CERAMIC PACKAGING

RAO R. TUMMALA—*Georgia Tech*
PHIL GARROU—*Dow Chemical*
TAPAN GUPTA—*Westinghouse*
N. KURAMOTO—*Tokuyama Soda*
KOICHI NIWA—*Fujitsu*
YUZO SHIMADA—*NEC*
MASAMI TERASAWA—*Kyocera*

9.1 INTRODUCTION

Ceramics and glasses, defined as inorganic and nonmetallic materials, have been an integral part of the information-processing industry. Glasses and ceramics are used for data processing by the use of semiconductor devices interconnected onto packages. They are also used for storing information by the use of such magnetic materials as iron and chromium oxides and ferrites; for displaying information by the use of low-temperature solder glasses sealed to transparent faceplate glasses; for printing information by the use of ceramic–metal print heads and corrosion-resistant glasses; and for transferring information by the use of very-low-loss glass fibers. The materials currently used for these applications are shown in Table 9-1.

Integrated-circuit (IC) packaging is one of the most important appli-

Table 9-1. Application of Ceramics in Microelectronics

Information Processing Function	Application	Ceramic Materials
Information processing		
Package	Substrate	Al_2O_3, BeO, AlN
Device	Dielectric	$Si_3 N_4$, SiO_2
	Mask	Borosilicate glass
Information storage		
	Disk	Iron oxide, Ferrite
	Tape	Chrome oxide
	Head	Ferrites, glass
		Al_2O_3 + TiC substrate
Information display		
	Dielectric	Lead borosilicates
	Seals	Lead zinc borosilicates
	Faceplate	Soda-lime glass
Information printing		
	Ink jet	ZrO_2-containing glass
	Electro-erosion	Ceramic–metal composites
Information transfer		
	Optical fiber	SiO_2, B_2O_3–SiO_2

cations of ceramics in microelectronics. Ceramic materials in the form of dual-in-line packages, chip carriers, and pin-grid arrays are used throughout the industry, from consumer electronics to mainframe computers. Although plastic packages outnumber ceramic packages, ceramic packages constitute roughly two-thirds of the total packaging market, which is currently considered a multibillion dollar industry worldwide. The Japanese market illustrated in Figure 9-1 is considered roughly two-thirds of the total market, as it does not include the ceramic substrates manufactured by United States and European computer and ceramic manufacturers. The ceramic package market in the United States along with a list of ceramic package suppliers to this market and their market share are shown in Tables 9-2 and 9-3, respectively [1]. It should be noted, again, that these numbers do not include those markets for internal consumption of individual companies, such as IBM, which manufactures ceramics for its own high-performance applications.

9.1.1 Unique Attributes of Ceramic Packages

9.1.1.1 Unique Properties

Ceramics possess a combination of electrical, thermal, mechanical, and dimensional stability properties unmatched by any group of materials.

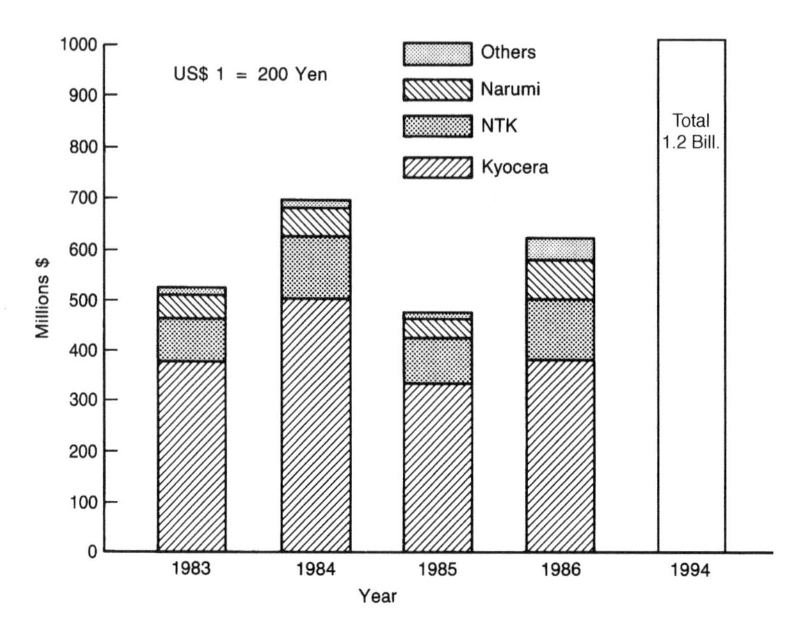

Figure 9-1. Japanese Ceramic Companies' Shipments

Table 9-2. Package Suppliers to U.S. Market. Market share based on sales volume.

Primary	Percent
Kyocera	55–70
NTK	15–25
Sumitomo	5–10
ALCOA	<4
COORS	<4
Secondary	
American Hoescht (Metceram)	<2
Cabot	<2
Raychem/Interamics	<2
General Ceramics	<2
Brush Wellman (EPI)	<2
LDC	<2
Diacon	<2
Bourns	<2
Shinko	<2
TOTAL	100

Source: Data from Ref. 1.

Table 9-3. Worldwide Ceramic and Other Packaging Materials Market
Worldwide Materials Consumption ($ millions)

	1982	1988	1993	1996	2000
Ceramic Packages, Metallized	230	1,073	1,230	1,480	1,300
CERDIP	200	363	110	80	50
Multilayer Capacitors	350	610	1,800	2,080	2,470
Leadframes	600	1,146	1,830	2,460	3,010
Encapsulation Resins	180	394	565	680	780
Bonding Wire	117	253	363	479	710
Seal Lids with Preforms	190	222	110	135	120
Die Attach Materials	55	75	100	118	140
Headers and Cans	110	105	90	85	60
Hybrid Materials	340	402	314	350	290
Other	90	175	200	230	390
	$2,462	$4,208	$4,912	$6,097	$9,320

For example, ceramics have dielectric constants from 4 to 20,000, thermal-expansion coefficients matching silicon (3×10^{-6}/°C) to matching copper (17×10^{-6}/°C), and thermal conductivities from that of insulated brick (0.1 W/m K) to better than aluminum metal (220 W/m K), which is one of the best metallic thermal conductors known. Dimensional stability as measured by either change of dimensions as a function of temperature and time or shrinkage control during substrate processing has been achieved at better than ±0.05% nominal shrinkage, allowing as many as 100 layers of ceramics and metal to be cofired into one substrate [2]. Such a substrate is illustrated in Figure 9-2 and is discussed in the glass–ceramic section of this chapter. A number of manufacturers, such as IBM, are routinely manufacturing these type of substrates but with a lower number of layers with nominal zero shrinkage and tolerance better than ±0.1%. Recent technology breakthroughs have led to the use of ceramics as superconductors, offering no electrical resistance to the flow of current at low temperatures. Figure 9-3 illustrates the development in this area over the last eight decades. As a result, a number of commercial applications in medical, electronic, and transportation industries are being developed. If the temperature could be pushed closer to room temperature, a direct and profound application for power distribution of electronic packages could be developed.

9.1.1.2 Highest Wiring Density

Ceramic substrates provide the highest wiring density of all substrate technologies. Even though the line and via dimensions as well as their

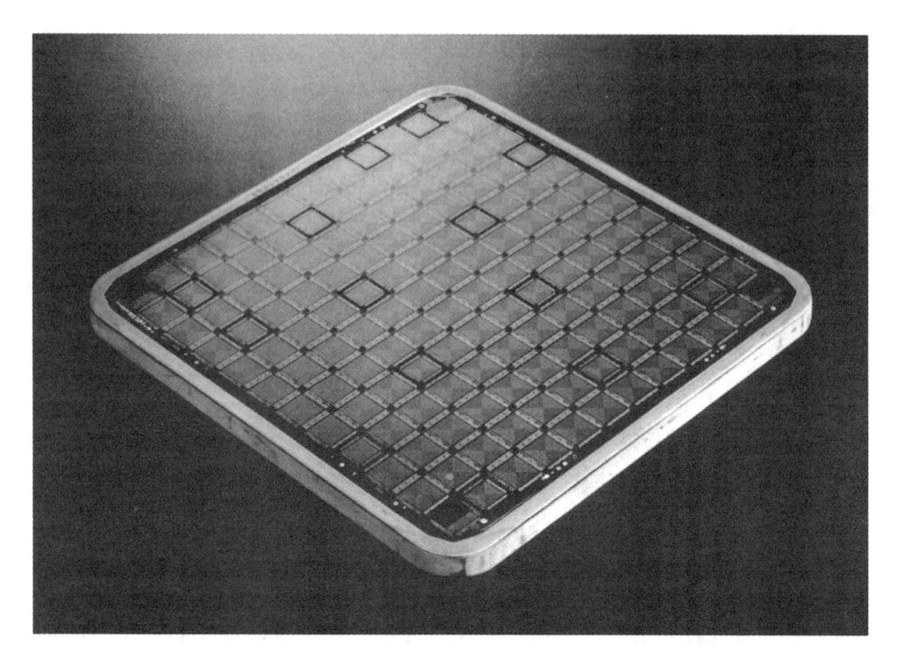

Figure 9-2. State-of-the Art 166mm Ceramic Substrate with 107 Layers. (Courtesy of IBM.)

pitch are coarser and further apart compared to thin-film technologies, as illustrated in Figure 9-4, and hence the wiring density is lower per layer, the total wiring density that has been achieved is much higher, because of cofiring of as many as 63 layers in full production and 100 layers in development. This advantage over thin-film (MCM-D) and printed wiring board (MCM-L) is reflected in Figure 9-5 for a 20-layer cofired ceramic and a 5× projection for a 100-layer substrate.

9.1.1.3 Best Reliability and Hermeticity

The only packages that achieved the highest reliability have been ceramic packages and, as a consequence, the packages that are in use today for the very stringent reliability requirements are ceramic packages. These uses include most of the defense and aerospace applications and the leading-edge microprocessors (Pentium and Power PC) for commercial applications. This reliability superiority over plastic or printed wiring board is due to three fundamental reasons. First, by their very nature, ceramics are hermetic, as discussed in Chapter 14, "Package Sealing and Encapsulation." They do not absorb and retain moisture nor do they allow permeation of gases. Second, their dimensional stability during and after high-temperature processing is exceptional. Some of the advantages of

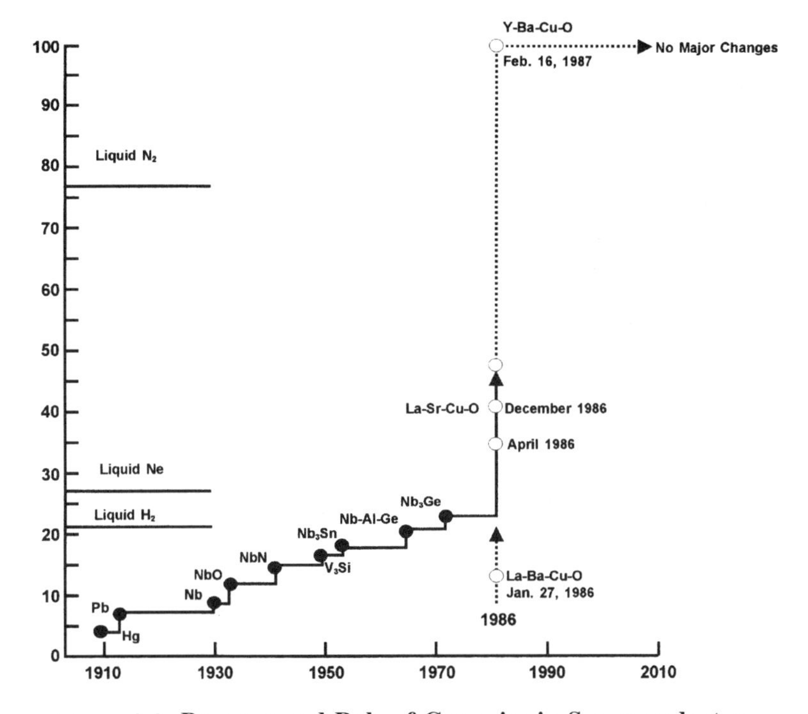

Figure 9-3. Progress and Role of Ceramics in Superconductors

this stability come from the intrinsic low thermal expansion, similar to that of silicon IC devices. Third, the chemical interness of most of the ceramics to water, acids, solvents, and other chemicals is outstanding. The IBM's better than six-sigma achievement for its ceramic multichip module (MCM) reliability, as illustrated in Figure 9-6 and as discussed

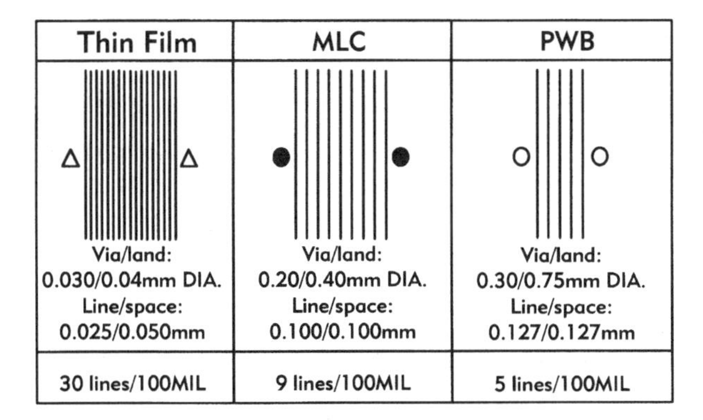

Figure 9-4. Multilayer Ceramic and Printed Wiring Board Compared. (Courtesy of Kyocera Corp.)

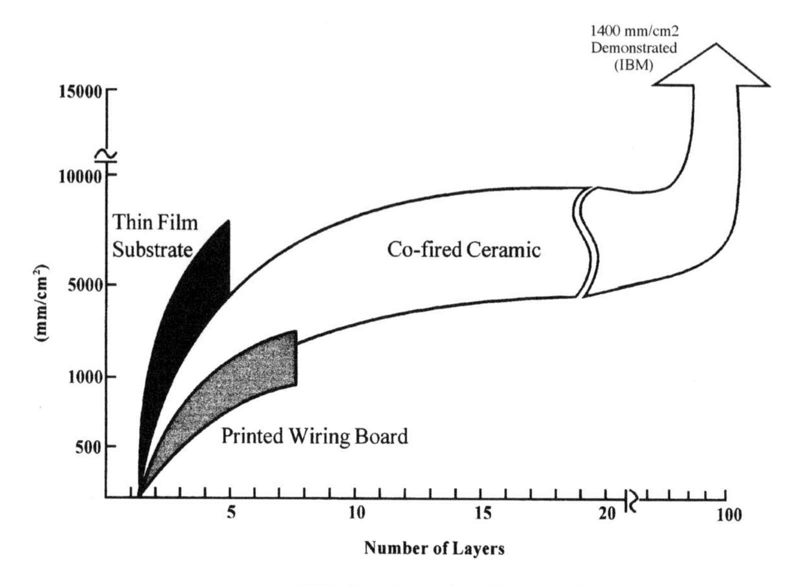

Figure 9-5. Wiring Density Comparison

in Chapter 6, "Manufacturing Considerations for the Packaging Technologist" is a good example of ceramic reliability which has not been achieved by any organic package.

9.1.2 Drawbacks of Ceramics

Ceramics are brittle, making them prone to catastrophic failure and sensitive to stress corrosion unless designed carefully for each application

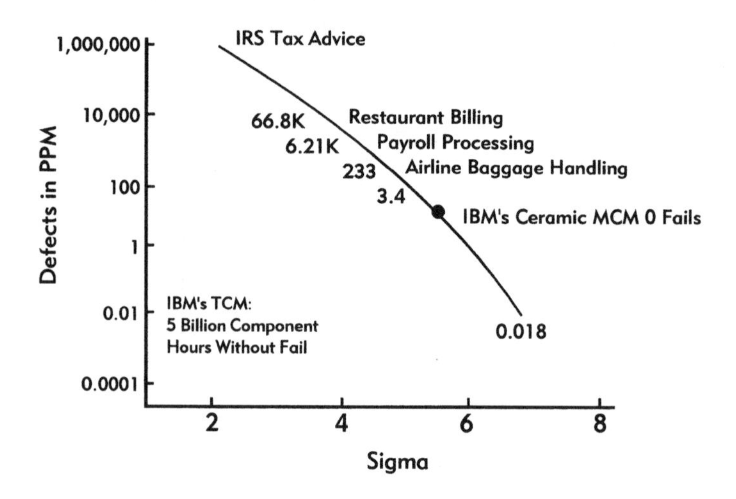

Figure 9-6. Multilayer Ceramic Reliability. (Courtesy of IBM.)

for which they are used. In very-high-performance applications requiring very low dielectric constants and very high packaging densities, ceramics cannot compete with thin-film organic packages. However, ceramics are expected to be used as stable building blocks on which high-performance thin-film wiring can be developed with the use of polymeric materials for high-performance applications.

Ceramic packages offered in a variety of forms, such as dual-in-line packages (DIPs), chip carriers, flat packs, and pin-grid arrays are capable of supporting from 1 chip to as many as 100 or more chips per package, as discussed later in this chapter.

In this chapter, the history and evolution of ceramic packaging are discussed, from simple dual-in-line and solid logic packages to the state-of-the-art ceramic processing, as in fabricating multilayer packages capable of interconnecting in excess of 30 ceramic layers with 30 layers of metal to support devices in excess of 100 chips on a single substrate with all the required power distribution and thermal dissipation to and from each of the devices. Recent developments in the use of ceramics with thermal-expansion coefficients near that of silicon, such as A1N, mullite, glass ± ceramics, are also included. Future technology trends and the scientific areas required to develop the understanding necessary to support the new product developments are indicated.

9.2 EARLY CERAMIC PACKAGING

Early transistors were either solder bonded to glass–epoxy cards or packaged in simple plastic packages (as shown in Fig. 9-7). The need for hermetic sealing led to the development of the outline metal packages,

Figure 9-7. TO Transistor Bonded to Glass–Epoxy

which were sealed by welding nickel lids to gold-plated transistor metal bases.

The initial transistor outline (TO) package was manufactured with as many as 14 leads. In the 1960s the need for many more leads led to the development of rectangular ceramic packages with leads on two opposite sides, known as dual-in-line packages or planar packages. These packages were introduced by Fairchild.

9.2.1 Ceramic Solid Logic Technology

At about this time (1963), IBM introduced for its own use a ceramic package called SLT (Solid Logic Technology) [3], 12 × 12 mm in size, made of 96% dry-pressed alumina on which conductor and resistor pastes were screened and fired at about 800°C. Sixteen connector pins of copper were swaged into the holes in the substrate, and the entire assembly was immersed in a solder bath to coat the conductors and pins, forming electrical contacts between pins and conductor lines. Semiconductor chips were soldered in place and the cap sealed to the module with epoxy (as shown in Fig. 9-8).

9.2.2 Ceramic Dual-in-Line Packages

The availability of dual-in-line packages (DIPs) and the ease with which they can be transported and automatically inserted into the holes of printed-circuit boards and soldered in place made them a industry standard.

Figure 9-8. SLT Package

Ceramic DIPs were fabricated using either ceramic greensheets or dry-pressing processes. Figure 9-9 illustrates the tape processes used in fabricating DIPs, and Figure 9-10 illustrates the same with a dry-pressing process.

With either process, the maximum number of leads provided with DIP packages has been 64. This limit is due to fabrication difficulties, the question of the reliability of solder joints to the board, and excessive lead inductance, among other reasons. The tape process involves forming a slurry of well-dispersed alumina and glass in a suitable organic vehicle, and doctor blading to form tape, which is metallized with tungsten metal paste. The desired number of layers is laminated and sintered in a controlled atmosphere between 1500°C and 1600°C. Exposed tungsten metallizations are nickel and gold plated, and then brazed with a copper–silver alloy.

Pressed ceramic (also known as "Cerdip"), as the name implies, involves molding a mixture of alumina lubricants and binders to form the desired shape, which is then sintered to form a monolithic body. Metal frames are bonded to this body by using low-temperature glasses, as shown in Figure 9-10.

Hermetic sealing posed problems early in the development of DIPs. These problems were traced to the incomplete removal of water prior to sealing. Glasses in the system $PbO–B_2O_3–SiO_2–Al_2O_3–ZnO$, capable of

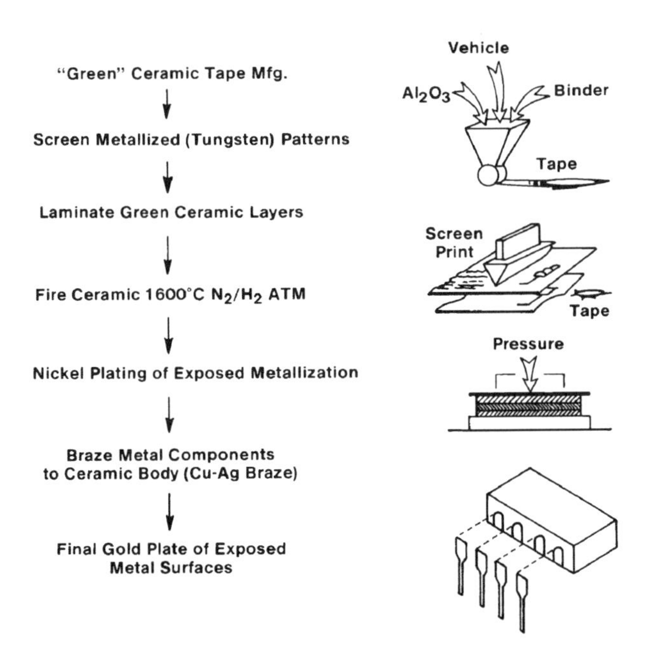

Figure 9-9. DIP—Laminated Ceramic Package. Also called side-brazed package.

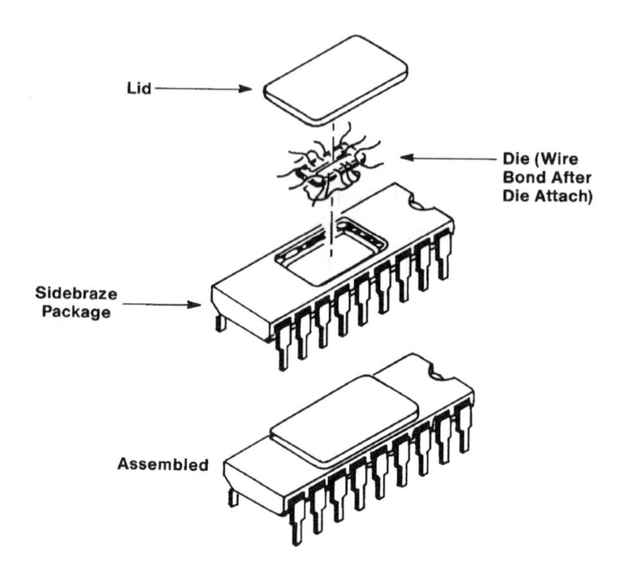

Figure 9-10. Laminated Ceramic Package. Also called a side-brazed ceramic DIP.

meeting all the requirements shown in Table 9-4, have been successfully developed [4] and used to solve the hermeticity problems.

9.2.3 Advanced SLT

Advanced SLT (ASLT) [5] was developed as a follow-up to SLT during the mid-1960s. The basic fabrication technology followed the SLT process closely, and the additional circuit wiring is added by screening conductors on both sides of the substrate. Two such substrates are stacked vertically, as shown in Figure 9-11. Electrical connections between the

Table 9-4. Glass Requirements for Cerdip Sealing

Low-temperature sealing (<450°C)
Good adhesion to both Al_2O_3 ceramic and alloy 42
Thermal expansion matched to ceramic and alloy 42
Electrically insulating
Good chemical resistance
No outgassing of moisture of other contaminants
Low alpha-particle emission for soft error protection
Good fracture toughness
High thermal-shock resistance

Figure 9-11. Stacked Solid Logic Technology

substrates are provided by soldering the pins from the bottom of the top substrate to the top-surface metallization of the bottom substrate.

9.2.4 Monolithic Systems Technology

The monolithic systems technology (MST) [6], consistent with improvements in semiconductor integration at about 25 circuits per 2-mm-size chip, was developed in the late 1960s. This technology is very similar to the SLT and ASLT, but the number of chip connections increased from 3 and 12 with the previous technology to 16 and 18, respectively.

9.2.5 Thick-Film (Crossover) Technology

Parallel to dry-press ceramic technology, thick-film technology began to evolve as a way of depositing thick-film dielectrics and conductors by screen-printing processes. Thick-film technology involving the physics and chemistry of metal and ceramic powders, the organic vehicle systems capable of providing stable thick-film systems, the measurement systems of viscosity, surface tension, and dispersion are discussed in a European handbook [7]. The conductor materials typically used are those that can be fired below 1000°C with good electrical conductivity. These include silver, gold, copper, gold–platinum, silver–palladium, silver–platinum, and palladium–gold. The choice of conductor is usually dependent on one or more of the following considerations for the pastes shown in Table 9-5:

Table 9-5. Primary Components of Some Common Early Resistor Systems

Manufacturer	Series	Components
E.I. du Pont	7800	Palladium oxide/silver/palladium
	8000	
Electroscience Laboratories	6900	Palladium oxide/silver/palladium
	7000	
E.I. du Pont	Birox	Bismuth ruthenate
	1100	
	1400	
EMCA	Firon	Ruthenium oxide/iridium oxide/gold
Plessy Co. (formerly Alloys Unlimited)	A	Ruthenium oxide/silver
	B	
	BB	
Chicago Telephone Supply		Ruthenium oxide
Airco Spear	TGA	Thallium oxide
	TFG	
Electroscience Laboratories	3800	Iridium oxide

1. Resistivity
2. Line definition
3. Compatibility with resistor, dielectric, or other conductor materials
4. Solderability and solder leach resistance
5. Adhesion strength
6. Suitability for ultrasonic wirebonding
7. Suitability for chip bonding
8. Long-term stability (e.g., resistance to migration effects under the influence of voltage gradients)

Glass (frit) additions, generally in the lead–borosilicate family, were included in the paste to provide adhesion to the ceramic on which they were fired. The resistivity that can be achieved with these thick films is in the range of 5–25 $\mu\Omega$ cm.

The dielectrics generally fall into two categories: those involving low-dielectric-constant materials and others involving very-high-dielectric-constant capacitor materials. Each of these have been further classified into those involving crystallizing glasses and those involving mixtures of glasses and ceramics. In addition, thick-film resistors (Table 9-5) are processed in a similar fashion.

9.2.6 Ceramic Packages with Thin Films

With the development of more highly integrated semiconductors, the package-wiring-density requirement became so great that either multiple layers of thick film or a higher density of surface wiring, using thin films, was required. The metallized ceramic shown in Figure 9-12 was fabricated with thin-film wiring on the surface of a dry-pressed alumina substrate [8]. A very thin layer of chromium, followed by a thicker film of copper, and finally a thin layer of chromium were either sputtered or evaporated onto the surface of the ceramic. The first chromium layer was expected to form chromium oxide during substrate processing to form an excellent bond to alumina. The metal conductor patterns are defined by subtractively etching fine lines by conventional lithographic processes. The pinning of the substrate is performed in a way similar to the swaged-pin processes described earlier. The chip connections are performed by the flip-chip solder connections, described in detail in Chapter 8, "Chip-to-Package Interconnections."

9.2.7 Earlier Multilayer Ceramic Technology

The origin of multilayer ceramic (MLC) technology can be traced to RCA in the late 1950s [9], starting with the doctor-blade process for casting greensheets, laminating a number of metallized layers [10], and finally leading to the concept of "via" interconnection from layer to layer [11]. At about this time, the U.S. Army Signal Corps and the Radio

Figure 9-12. Thin Film on Ceramic Package

Corporation of America, in Somerville, New Jersey, started the Micromodule Project [12]. The goal was to find a practical means to increase the value of capacitance available to module designers by a factor of 10. Both fabrication technology and dielectric composition improvements were studied. This project was designed to use 7.6 × 7.6-mm substrates, with three notches on each of the four edges and thicknesses greater than 0.25 mm. The first goal was to develop a laminated capacitor in this form factor. To accomplish this goal, it was necessary to meet the capacitance requirements of up to 3000 pF for a precision capacitor and up to 300,000 pF for a general-purpose capacitor. The equation used to make these calculations was

$$C = 9.6 \frac{KAN}{t} \qquad [9\text{-}1]$$

where

A = the area of the electrode
C = the capacitance (in pF)
K = the dielectric constant
N = the number of layers
t = the thickness of each layer

For a precision capacitor, the dielectric constant was about 30, and for a general-purpose capacitor it was about 3000, which meant that at least 11 layers of 25 μm thickness would have to be achieved in order to meet these specifications. The area of each electrode was about 26 μm^2. Many technologies were tried concurrently until doctor-blading started to show the most promise. Fortuitously, the paint industry was then producing vinyl-based products, and after several months of trials, a plasticized copolymer of polyvinyl chloride acerate binder was discovered. These capacitors had 11 layers, each 25 μm thick, which when sintered met the specifications described. The process steps used are shown in Figure 9-13.

In order to produce a stable slurry with high viscosity and specific gravity that could be used for casting films, the interactions of the ceramic powders, binders, plasticizers, deflocculants, and solvents had to be optimized [13,14]. The early studies correctly pointed out, as we know now, that steric hindrance is the primary mechanism for dispersing ceramic powders in organic systems. Because of the increased interest in tape casting of ceramics for much better dimensional requirements, new research programs have recently been initiated [15] which should improve the understanding of the fundamentals controlling the dispersion behavior of these systems. The casting process was first done with a hand doctor blade, but it was quickly replaced with a continuous caster using a plastic

1. Ball Mill Slip
- **Ceramic Dielectric (Titanates) Powder**
- **Thermo-Plastic Binder and Plasticizer**
- **Deflocculant**
- **Solvents**

2. Doctor Blade a Tape on a Plastic Carrier
3. Dry, Strip, and Cut into Sheets
4. Screen Noble-Metal Paste Electrodes and Dry
5. Laminate Sheets to Design
6. Cut Units to Size
7. Connect Alternate Layers at Edges (Notches)
8. Sinter in Air
9. Tin Terminals
10. Test

Figure 9-13. Early Multilayer Ceramic Flow Sheet for Capacitors

carrier material. The greensheets could be easily stripped from it and inspected.

These sheets were metallized with several of the existing noble metal pastes then commercially available. Palladium was preferred for capacitor applications, because it could be fired in air and withstand the temperatures needed to sinter the titanates. Screening processes provided adequate resolutions for the electrodes.

Assembling the titanate sheets was the most difficult problem to solve. The first attempt was to sequentially cast the ceramic sheets, dry, screen the electrodes, dry and so forth. This method had so many problems with drying, thickness control, yield, and layer-to-layer connections that it led to the development of lamination technology. The criteria for proper greensheet structure and laminating conditions have been described more quantitatively elsewhere [16]. The bond strength between the green ceramic layers is determined by the following relationship:

$$B = K_1 + \mathrm{K}_2 \ln \mathrm{PTt}$$

where

K_1, K_2 = constants for the system used
P = the pressure
T = the temperature
t = the time of hold at the laminating temperature.

All the process steps in fabricating these capacitors are listed in Table 9-6.

About this time in the micromodule program, a new need arose for an alumina hermetic package for a quartz crystal oscillator application that led to the development of alumina multilayer ceramic. This was believed to be the first laminated alumina multilayer ceramic package application. All of the processes described were used, except for the binder. In this application a polyvinyl butyral resin was used for the thicker greensheets. This package used a Mo–Mn paste for its metallization, in order to achieve hermeticity. The iron–nickel–cobalt cap was ultrasonically soldered to a nickel and gold-plated sealing ring on the surface of the three-layer substrate. Because of this breakthrough with the quartz crystal oscillator package, other packages were made for diodes and transistors. Next, the concept of a "via," as illustrated in Figure 9-14, was introduced [11]. This is a hole punched in a green ceramic sheet and subsequently filled with a metal paste in order to make the interconnections from layer to layer within the micromodule substrate. These historical developments are discussed further elsewhere [8]. These basic developments provided the background for the development of the state-of-the-art multilayer ceramic described later in the chapter.

9.2.8 Ceramic Chip Carriers, Flat Packs, and Hybrid Packages

The ceramic dual-in-line package with up to 64 pins on a 2.54-mm pitch became too large for high-performance applications. The ceramic chip carriers, which can be considered as square DIPs with terminals on all four sides at half the pitch of the DIP, have filled an important gap between DIPs and pin-grid arrays. Chip carriers can also be used exten-

Table 9-6. Thick-Film Processes Considered in 1959

Process	Estimated Minimum Thickness (μm)	Concerns
Pressing	125	Too thick
Extruding	75	Control
Spraying	75	Shorts
Doctor-blading	Unknown	Lamination
Screening	15	Shorts
Electrophoresis	2	Shorts
Sputtering or evaporation	0.1	Voltage rating

Source: Data from Ref. 12.

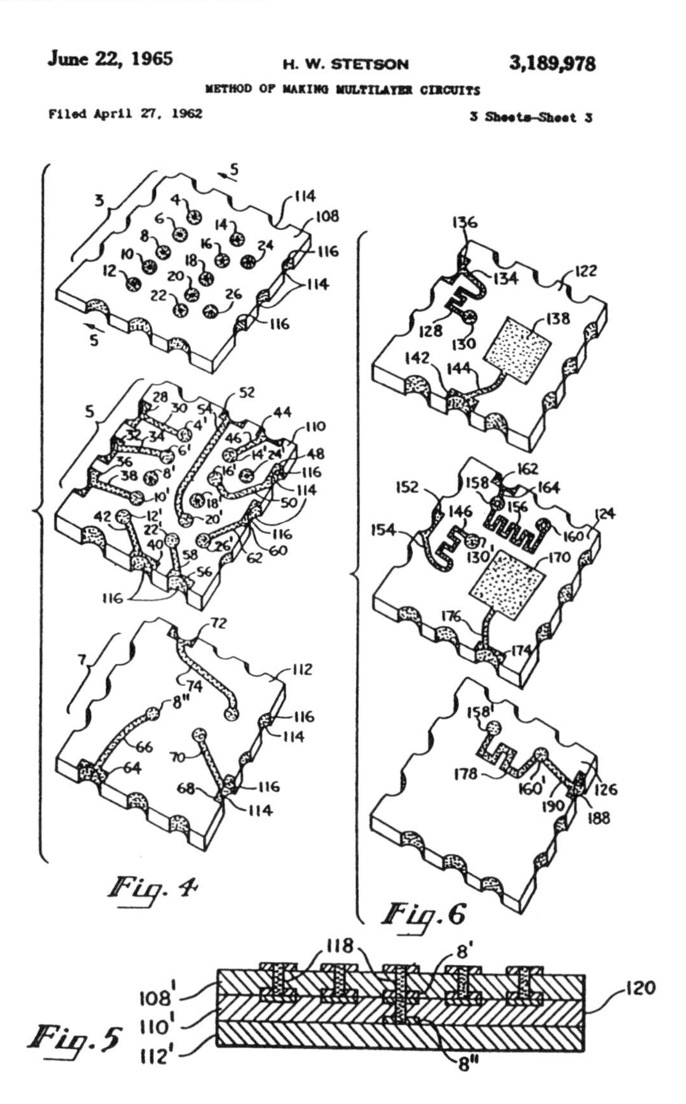

Figure 9-14. Patent Drawings Showing First Ceramic Via. Figure numbers shown in art-work are from patent. (From Ref. 10)

sively in low pin counts to replace DIPs as, for example, in high-density memory packaging. Chip carriers (CC) can be either leaded (LCC) or leadless (LLCC), depending on how it is applied to the second-level card or board technology. An example of a ceramic chip carrier is illustrated in Figure 9-15. These packages are made by using the tape process for multilayer structures and the dry-press process for single-layer DIPs and other plug-in packages (see Fig. 9-16). The materials and processes for

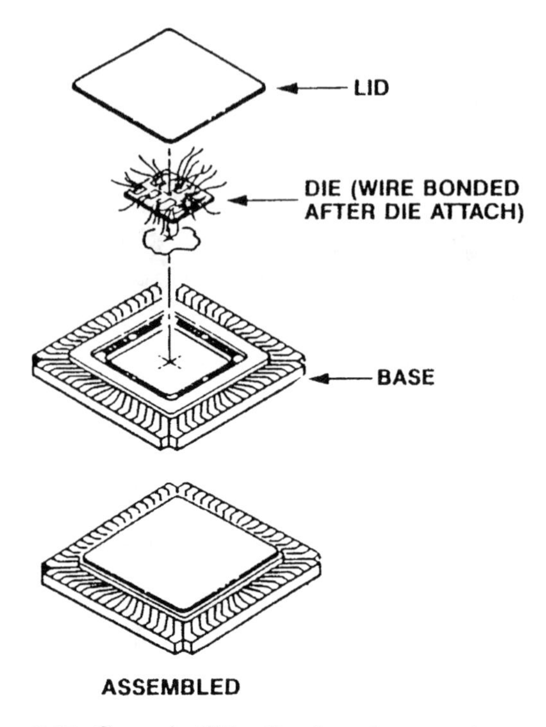

Figure 9-15. Ceramic Chip Carrier. (Courtesy of Intel Corp.)

multilayer substrates are similar to the state-of-the-art ceramic packaging covered in the next section, but the dry-press technology has also been applied to the formation of metallized substrates by first firing in air, followed by screening and sintering the desired metallization. Cavity metallization is accomplished similarly to end up with Cerdip bases. A number of designs (leadless designs A to D and others) of ceramic packages [17], both in pinned and nonpinned versions, have been developed by manufacturers exclusively for their customers' needs. One of these is the ISOPAK pin-grid array developed by General Dynamics [18], where the pins of Kovar are glass-sealed into a chemically milled Kovar pan, which is flush filled with glass and planarized to provide the chip-bonding circuits. The chip itself is mounted on a Kovar platform at the center of the package. So-called multiple-in-line packages (MIP) are rectangular pin-grid arrays, in which the pins are in four rows of 2.5 mm spacing, and the spacing between the first and second, and third and fourth is 7.5 mm. PINBELL, used by AT&T Bell [19], is of this type. In addition to other chip carriers, one variation of leadless chip carrier type, known as single-layer metallized (SLAM), is used primarily by the military market for direct solder attachment to a ceramic substrate [20]. This package has no recess and is sealed by glass as discussed in Chapter 14 "Package Sealing and Encapsulation."

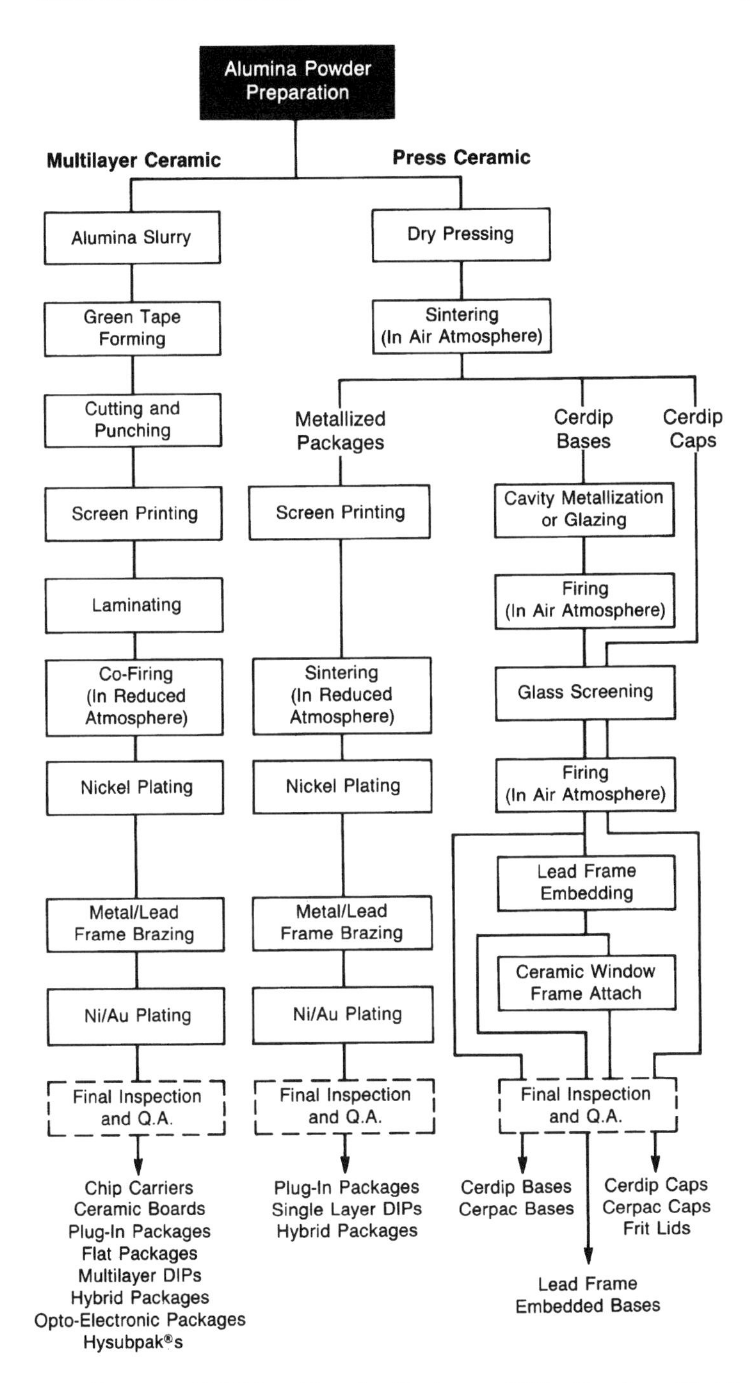

Figure 9-16. Process of Making Alumina Packages (Chip Carriers, Flat Packs, DIPs). (Courtesy of Kyocera Corp.)

Hole-grid array, as opposed to pin-grid array, with holes in ceramic is another variation. The available ceramic chip carriers and DIPs, and their characteristics, are given in Table 9-7, which illustrates the availability of ceramic carriers with I/O spacing down to 0.50 mm (20 mils) and the number of I/Os in excess of 200. The I/O configurations available from Kyocera for these chip carriers are illustrated in Figures 9-17 and 9-18. Figures 9-19 and 9-20 compare the three most common ceramic packages [21] for the area they occupy on second-level organic packages and the signal delay involved in interconnecting 12 silicon chips. For high-performance logic applications, it can be concluded from these figures that the pin-grid array is the best solution, as it uses the least area on the board while providing the fastest signal transmission.

The best package solution may not be entirely technical. In addition to the technical aspects, package decisions are made based on existing assembly methods and the costs of developing and implementing a total package system with minimum risk. Thus, each company must consider its own requirements before selecting a particular package solution.

The use of leadless ceramic chip carriers has fulfilled an important

Table 9-7. Available (1987) Ceramic Chip Carriers, DIPs, PGAs, and Their Leads

Package	Type	I/O Range	I/O Spacing (mm)
BGA	Area array balls	>300	1.00
PGA	Cavity up	64–312	2.54
			1.27 Staggered
			2.54
	Cavity down	64–240	1.27 Staggered
Leadless	JEDEC A	28–156	1.27
	JEDEC B	28–156	1.27
	JEDEC C	16–96	1.00
	Custom 1	200	0.50
	Custom 2	284	0.28
Leaded	Flat package	10–28	1.27
	Quad lead	24–124	1.27
	Quad lead	36–64	1.00
	Quad lead	172–224	0.63
	Quad lead	35–224	0.50
DIP	Standard	8–64	2.54
	Skinny	24–32	2.54
	Shrink	24–64	1.78
	Narrow	24–44	1.27

Leaded Package

CONFIGURATION	PIN/LEAD CENTERLINE SPACING		
	STANDARD	CUSTOM	HDCM
1) SIDE BRAZE	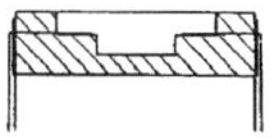 2.54 / .100	1.27 / .050	1.27 / .050
2) TOP BRAZE	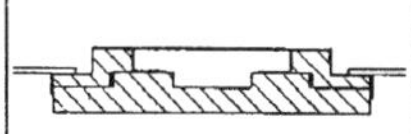 1.27 / .050	0.508 / .020	0.508 / .020
3) BOTTOM BRAZE (Flat leads)	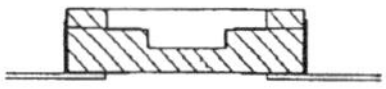 1.27 / .050	0.508 / .020	0.508 / .020
4) BOTTOM BRAZE (L shape leads)	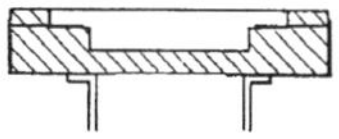 2.54 / .100	1.27 / .050	1.27 / .050
5) PIN GRID ARRAY (Nail head pin)	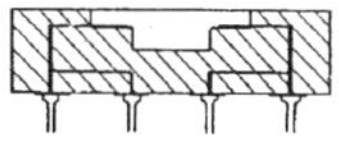 2.54 / .100	1.778 / .070 (Diagonal)	1.778 / .070 (Diagonal)
6) SIP (Nail head pin)	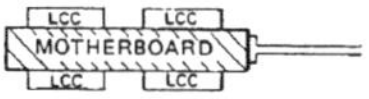 2.54 / .100	1.905 / .075	1.905 / .075
7) SIP (Flat leads)	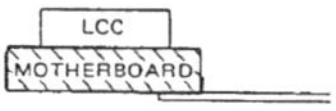 2.54 / .100	1.27 / .050	1.27 / .050

Unit: mm / inch

Figure 9-17. Leaded Package I/O Configurations. (Courtesy of Kyocera Corp.)

need in military applications but posed limitations in commercial applications [22]. Although leadless chip carriers can be fabricated to as low as 0.25 mm (10 mil) lead spacing, surface mounting poses a limit to the size or number of pins a chip carrier can have. This limit has been observed to be about 30–50 mm in size for a standard 96% alumina substrate surface soldered to FR-4 epoxy–glass board [23]. Surface mounting of larger ceramic chip carriers, therefore, requires the development of second-level packages that are closer in thermal expansion than conventional epoxy–glass boards [24], as discussed in Chapter 17, "Printed-Wiring Board Packaging." The surface mount limitations of chip carriers are discussed in Chapter 16, "Package-to-Board Interconnections."

9.3 ALUMINA CERAMIC PACKAGING

The state-of-the-art multilayer ceramic described in this section is excerpted from an earlier publication [25]. The multilayer ceramic packag-

Leadless Package

CONFIGURATION		PIN/LEAD CENTERLINE SPACING		
		STANDARD	CUSTOM	HDCM
1) CHIP CARRIER (Castellations)	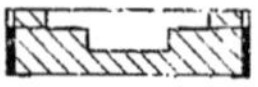	1.27/.050 & 1.016/.040	0.762/.030	.0762/.030
2) CHIP CARRIER (Edge Metallization)	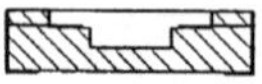	1.27/.050 & 1.016/.040	0.762/.030	0.635/.025
3) CHIP CARRIER (Via Holes)	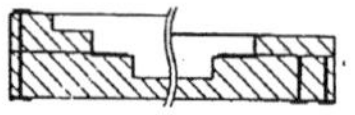	1.27/.050 & 1.016/.040	0.635/.025	0.508/.020
4) PAD GRID ARRAY	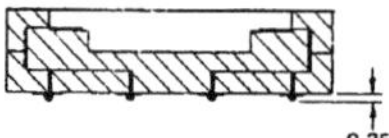 0.254/.010	1.27/.050	0.635/.025	0.508/.020

Unit: mm/inch

Figure 9-18. Leadless Package I/O Configurations. (Courtesy of Kyocera Corp.)

ing capable of providing interconnections for as many as 100 chips currently in a single monolithic alumina–molybdenum substrate is considered the state of the art in ceramic packaging [26]. It should be noted, however, that alumina–tungsten multilayer ceramic, the technology of which is very similar to the technology of alumina–molybdenum technology discussed in this section, is the technology most commonly used in the majority of commercial applications. This technology, which has been in

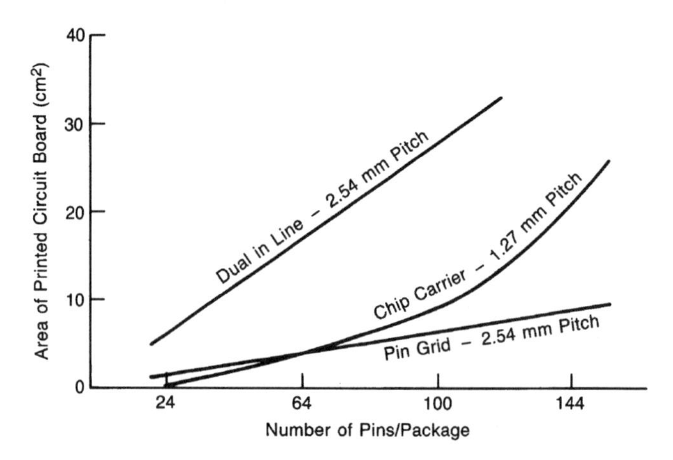

Figure 9-19. Comparison of DIP, Chip Carrier, and PGA for Second-Level Area. (Courtesy of Standard Elektrik.)

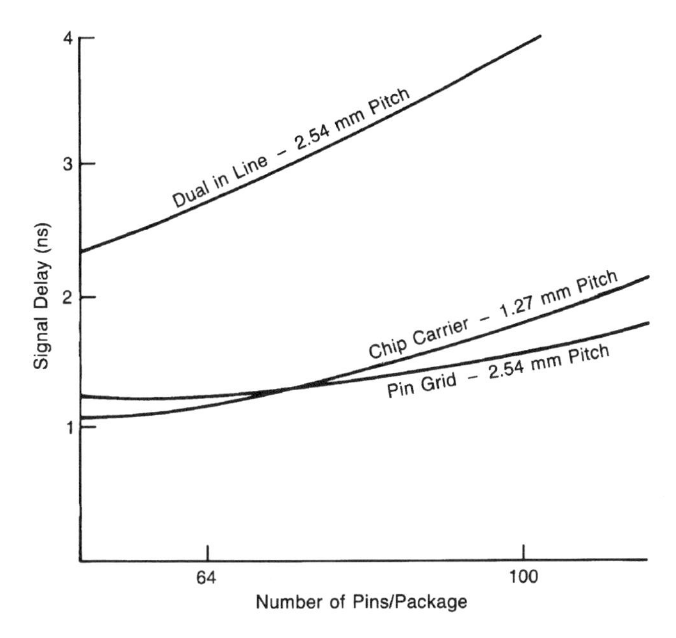

Figure 9-20. Comparison of DIP, Chip Carrier, and PGA for Signal Delay. (Courtesy of Standard Elektrik.)

existence since the early 1980s [24], has been the basis of IBM's mainframe technology, as it provided a revolutionary approach in interconnecting a very large number of chips, providing power to each chip and removal of heat from each chip in a way that was not practiced earlier. A number of Japanese and American manufacturers have developed their own single and multichip, multilayer, ceramic substrate technologies that are variations of the materials, processes, and tools discussed here. Table 9-8, for example, illustrates the characteristics of multilayer ceramic substrates as manufactured by Kyocera. The impact of this technology has been to provide very high performance by minimizing the interchip wiring. It also provides very high reliability due to minimization of the number of interconnections and elimination of one level of packaging. In addition, this ceramic package provides hermeticity for the silicon devices. It is expected that this technology will continue as the basis on which advances will be made in interconnecting more chips with more circuits on each chip, each dissipating more power.

The single-line reference planes have been designed to provide a characteristic impedance of 55 Ω for the selected line geometry by the use of a combination of nominally 0.15-mm- and 0.2-mm-thick ceramic layers with a nominal dielectric constant of 9.4. The bulk resistivity of the sintered molybdenum metallurgy is approximately 10 μΩ-cm.

Table 9-8. Multilayer Substrate Dimensions

	Standard	Custom	High Density
Maximum substrate size (mm)	150 × 150	300 × 300	150 × 150
Tolerances—as fixed			
x-y Dimensions	+1%	±0.5%	±0.5%
Thickness	±10%	±5%	±5%
Flatness (mm/mm)	0.102/25.4	0.076/25.4	0.051/25.4
Number of layers			
Tape layer	7	14	30+
Screened dielectric	4	8	—
Vias diameter (mm)	0.203	0.203	0.102
Pitch	0.762	0.635	0.254
Lines			
Width (mm)	0.254	0.102	0.102
Pitch (mm)	0.508	0.203	0.203

Source: Courtesy of Kyocera Corp.

9.4 STATE-OF-THE-ART ALUMINA PACKAGING AND APPLICATIONS

Table 9-9 illustrates the state-of-the-art evolution in alumina-Mo ceramic packaging during the last 15 years, starting with the original thermal conduction module (TCM) in 1980. The substrate size has grown from 90 mm to 127.5 mm, the number of layers from 33 to 63, the number of vias in the substrate from 350 k to 2 million and the wiring length from 130 to 400 meters. The state-of-the-art is the last one in this table and is described briefly here, but the details can be found elsewhere. [27]

IBM's application of the latest ceramic package for an air-cooled mainframe system is illustrated in Figure 9-21. Assembled, the module measures 166 mm wide by 146 mm deep by 169 mm high; it contains up to 121 logic and array chips with 144 decoupling capacitors, all mounted using controlled collapse chip connection (C4) technology. The alumina thermal-conduction module (TCM) utilizes a new MLC substrate, top-surface thin-film redistribution wiring, and a new air-cooling technology which allows the package to dissipate 600 W and uses 2772 pins to connect with the second-level package. In this module, CMOS and bipolar chip technologies are packaged together on a single TCM for the first time.

The process of fabricating the multilayer ceramic package is outlined in Figure 9-22, and the resulting substrate cross section is illustrated in Figure 9-23.

Table 9-9. Evolution of Alumina Packaging at IBM

	System 3080 TCM (1980)	System 3090 TCM (1985)	System/390 air-cooled TCM (1990)
Size (mm)	90 × 90	110.5 × 117.5	127.5 × 127.5
Layers	33	36–45	63
Via count	350K	470K	2000K
Wiring (m)	130	180	400
Available C4 connections	16K	24K	80.7K
Chip sites	100–133	132	121
I/O pin connections	1800	1800	2772
Terminal metallurgy	Plating	Plating	Plating/thin film
Cooling capacity (W)	300 (water)	520 (water)	600 (air)

Source: From Ref. 27.

9.4.1 Ceramic

Pure alumina, although capable of much higher mechanical strength when combined with glass, presented two basic problems. The thermal-expansion coefficient of pure alumina is $70 \times 10^{-7}/°C$ compared to

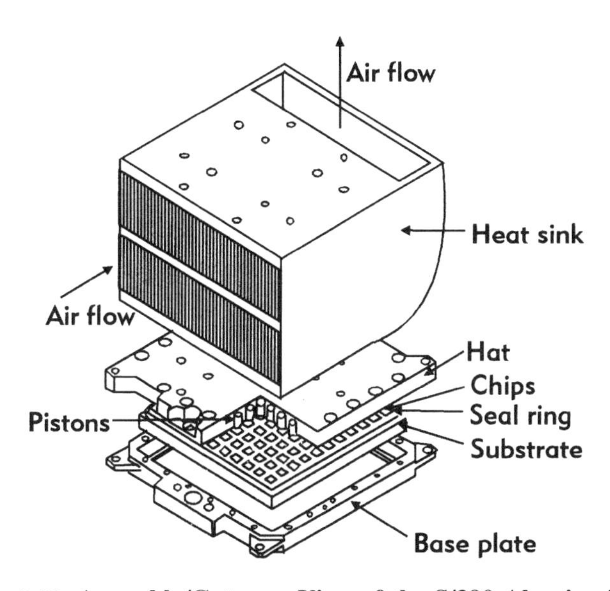

Figure 9-21. Assembly/Cutaway View of the S/390 Alumina TCM

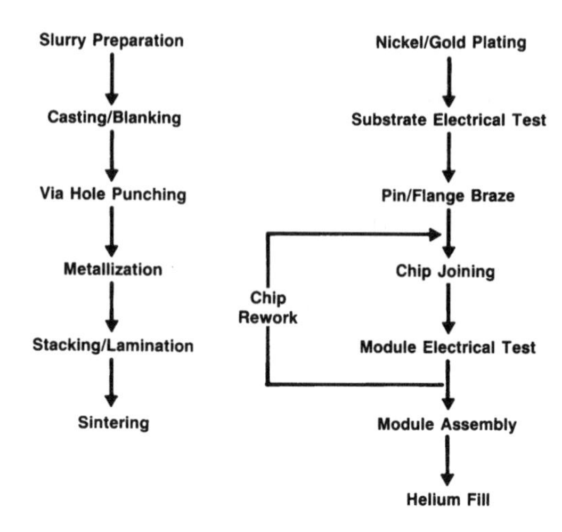

Figure 9-22. Process of Fabricating Multilayer Ceramic

58×10^{-7}/°C for molydenum in the temperature range 20–800°C. This difference resulted in a radial cracking of alumina around each molybdenum via. Additionally, pure alumina required sintering temperatures around 1900°C, which is much too high for the economic operation of manufacturing furnaces. Both these problems were solved by the addition of glass. The properties of alumina used are described in Table 9-10. The

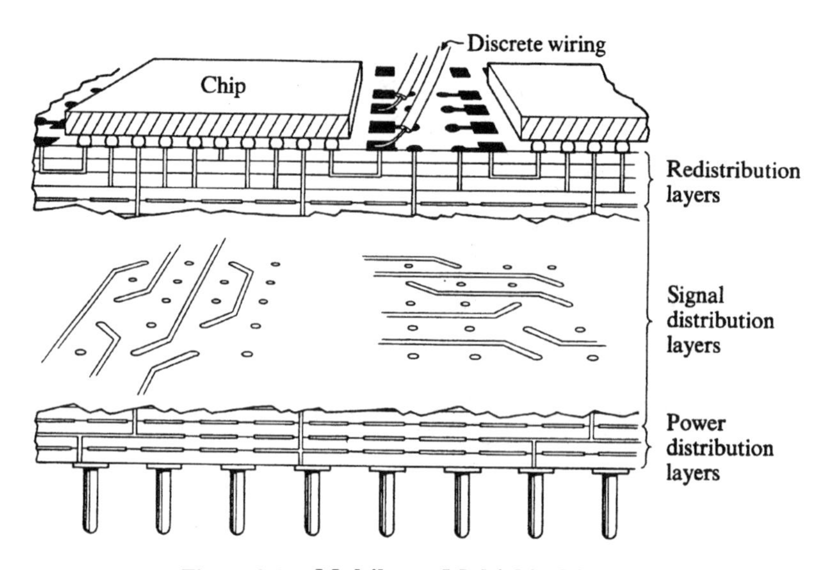

Figure 9-23. Multilayer Multichip Module

Table 9-10. Properties of Alumina Powder for Multilayer Ceramic

Average particle size	3.4–4.0 μm
Surface area	0.75–0.95 m^2/g
Leachable soda	Nominal ± 20 ppm
Bulk density	1.0 g/cm^3

ceramic consists of approximately 90% alumina and 10% glass. A number of glass additives added to alumina powder and the resulting mechanical, thermal, electrical, and dimensional control properties of ceramic led to the choice of calcia–magnesia–alumina silicate glass. The influence of this composition of glass on the nominal shrinkage of the substrate is illustrated in Figure 9-24, further defining the chemical composition of the glass used [28].

9.4.2 Greensheet

The basic building block used in the multilayer ceramic process is the "ceramic greensheet," nominally 0.2 mm or 0.28 mm thick (unfired), which is a mixture of ceramic and glass powder suspended in an organic binder as described below. A key factor in achieving acceptable yields is the formulation of a greensheet that exhibits the necessary strength for handling and processing. In addition, the greensheet must be dimensionally stable to ensure accurate plane-to-plane registration when the layers are

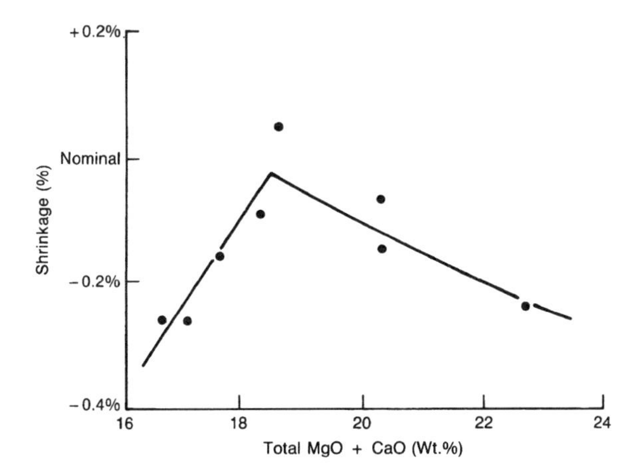

Figure 9-24. Shrinkage of Multilayer Ceramic as Influenced by Glass Composition. (Courtesy of IBM Corp.)

stacked and laminated. Strength and stability have been achieved by the proper selection of binder constituents, a controlled casting process, and the use of a molybdenum paste vehicle that does not interact with the greensheet binder.

A typical greensheet vehicle system [29] consists of binder, solvent, and plasticizer. The purpose of the binder is to "bind" the ceramic particles temporarily in forming greensheets and allowing sheets to be screened with an appropriate paste of the metal powders.

Power is distributed from the substrate pins to the power planes and, in turn, through parallel paths to chip power pads on the surface of the substrate. The maximum voltage drop in the substrate is 16.5 mV, and the power distribution design allows a minimum of 450 off-chip drivers to be switched simultaneously. Solvents play a number of key roles, ranging from deagglomeration of ceramic particles during the ball-milling operation, due to the low viscosity of the grinding liquid the solvent provides, to formation of microporosity as the solvent evaporates from the sheet. The formation of microporosity in the sheet is considered one of the most important features, because it allows the sheet to be compressed around metal lines during lamination. Plasticizers allow the sheet to be "plastic" or flexible due to the lowering of the glass-transition temperature of the binder by the plasticization process.

9.4.3 Binder and Slurry

The most effective binder for providing all the required thermoplastic properties and bond strength between layers is polyvinyl butyral (PVB). Other binders considered and used in certain applications are polyvinyl chloride acetate, polymethyl methacrylate (PMMA), polyisobutylene (PIB), polyalphamethyl styrene (PAMS), nitrocellulose, cellulose acetate, and cellulose acetate butyral. The chemical formula of PVB is shown in Figure 9-25 and the different types of PBVs available, and the solvents and plasticizers used with PVB are listed in Tables 9-11, 9-12, and 9-13 respectively.

The greensheet organic raw materials are combined with the alumina and glass in a ball mill and deagglomerated to achieve a uniform slurry dispersion. A dispersion model, illustrated in Figure 9-26, shows how the greensheet density relates to the breakdown of agglomerates during initial milling. Once the agglomerates are broken, the greensheet density and hence the shrinkage of ceramic remains insensitive to the milling time. The slurry is then fed to a continuous caster and deposited on a constantly moving plastic carrier to form 200-mm-wide ceramic tapes. The caster is fitted with a constant-level slurry reservoir and a doctor-blade assembly to control greensheet thickness. The ceramic webs are passed through a

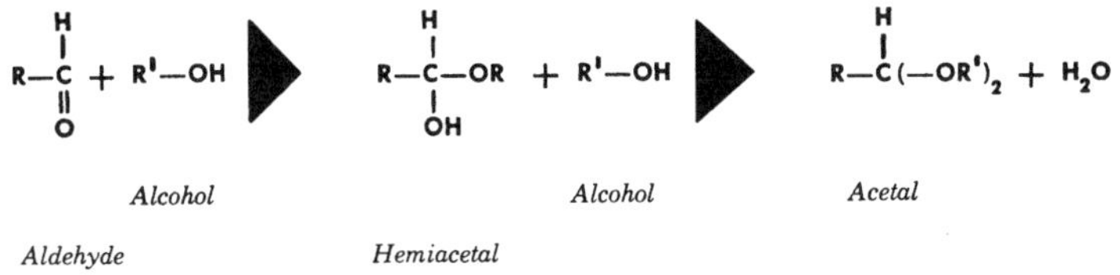

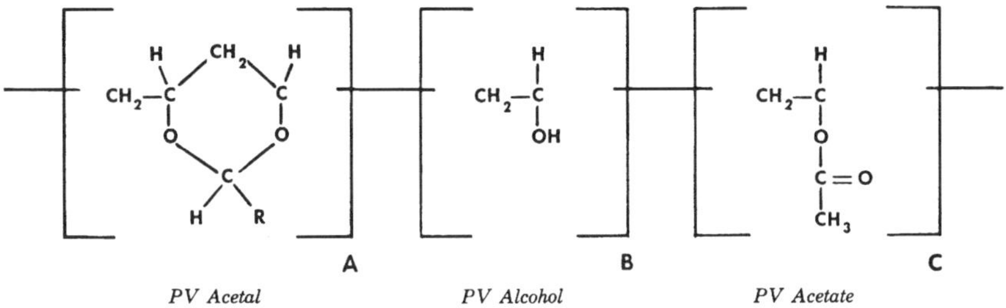

Aldehyde + Alcohol → Hemiacetal + Alcohol → Acetal

$R{-}CH{=}O + R'{-}OH \rightarrow R{-}CH(OH){-}OR + R'{-}OH \rightarrow R{-}CH({-}OR')_2 + H_2O$

A: PV Acetal — B: PV Alcohol — C: PV Acetate

Figure 9-25. Chemical Formula of Polyvinyl Butyral

series of drying ovens with controlled temperature and humidity and are subsequently separated from the plastic carrier and spooled.

Each spool of ceramic tape is fully inspected by a scanning laser and cut into square blocks in preparation for personalization. A representative sample of the blanks is evaluated for density, compressibility, bond and yield strength, and shrinkage.

9.4.4 Via Punching

Computer-controlled step and repeat equipment is used to punch via holes in each greensheet layer. A pallet, used for mounting the greensheets, is an integral part of a precision x-y table that moves the greensheet relative to a stationary die set containing 100 punches at a rate of approximately 10 steps per second. A location hole in each corner of the greensheet is used to position the greensheet on the punch pallet. Up to 36,000 via holes are punched in a single greensheet layer. The punch tool also includes an inspection station that contains a light source and an array of photodiodes to determine the exact location of the hole patterns. If any holes are clogged, the sheet is either repaired or discarded.

An alternative to the via punching process is the photoforming process developed by DuPont and others, as illustrated schematically in Figure 9-27. The photoformed ceramic module (PCM) is formed by the use of photosensitive dielectric and conductive films. These films are subsequently exposed to ultraviolet (UV) light. Linewidths less than 50 μm on 100-μm spacings are being developed with these processes.

Table 9-11. Properties of Polyvinyl Butyral White Free-Flowing Powder Form

Type	Betvar® B-72A	Butvar® B-74	Butvar® B-83	Butvar® B-76	Butvar® B-79	Butvar® B90	Butvar® B98
Volatiles (%)—max.	3.0	3.0	3.0	5.0	5.0	5.0	5.0
Molecular weight (weight average)	180,000–270,000	100,000–150,000	50,000–80,000	45,000–55,000	34,000–38,000	38,000–45,000	30,000–34,000
Hydroxyl content	9.0–17.5	9.0–17.5	17.5–18.0	18.0	—	—	—
expressed as weight % polyvinyl alcohol	21.0	20.0	21.0	13.0	13.0	20.0	20.0
Acetate content expressed as weight % polyvinyl acetate	0–2.5	0–2.5	0–2.5	0–2.5	0–2.5	0–1.0	0–2.5
Butyral content expressed as weight % polyvinyl butyral	80	80	80	88	88	80	80
Specific gravity (23°C/23°C)	1.100	1.100	1.100	1.083	1.083	1.100	1.100
Viscosity (cps)[a]	~1570	~700	~400	~175	~55	~195	~75
Viscosity (cps)[b]	8,000–18,000	4,000–8,000	1,000–4,000	500–1,000	100–200	600–1,200	200–450

[a]Viscosity of a 10% solution in 95% ethanol at 25°C using an Ostwald Viscometer.
[b]Viscosity of a 15% solution in 60 : 40 toluene/ethanol at 25°C using a Brookfield Viscometer.
Note: Butvar® is a trademark of the Monsanto Co.

Table 9-12. Solvents for Polyvinyl Butyral Resins

Solvent	Butvar® B-72A, B-73, B-74	Butvar® B-76, B-79	Butvar® B-90, B-98
Acetic acid (glacial)	S	S	S
Acetone	I	S	SW
n-Butyl alcohol	S	S	S
Butyl acetate	I	PS	I
Carbon tetrachloride	I	PS	I
Cyclohexanone	S	S	S
Diacetone alcohol	PS	S	S
Dioxane (1, 4)	S	S	S
Ethyl alcohol, 95%	S	S	S
Ethyl acetate, 85%	I	S	PS
Ethyl cellosolve	S	S	S
Ethylene chloride	SW	S	PS
Isophoronet	PS	S	S
Isopropyl alcohol, 95%	S	S	S
Isopropyl acetate	I	S	I
Methyl alcohol	S	SW	S
Methyl acetate	I	S	I
Methyl cellosolve	S	S	S
Methyl ethyl ketone	SW	S	PS
Methyl isobutyl ketone	I	S	I
Pentoxol	PS	S	PS
Pentoxone	PS	S	PS
Propylene dichloride	SW	S	PS
Toluene	I	PS	SW
Toluene ethyl alcohol 95% (60 : 40 by weight)	S	S	S

Key: S = soluble; PS = partially soluble; I = insoluble; SW = swells.
Note: Butvar® is a trade mark of the Monsanto Co.

Table 9-13. Plasticizers for Polyvinyl Butyral Resin

Phthalate	Polyester
Phosphate	Ricinoleate
Polyethylene glycol ether	Rosin derivates
Glyceryl mono oleate	Sabacate
Petroleum	Citrate

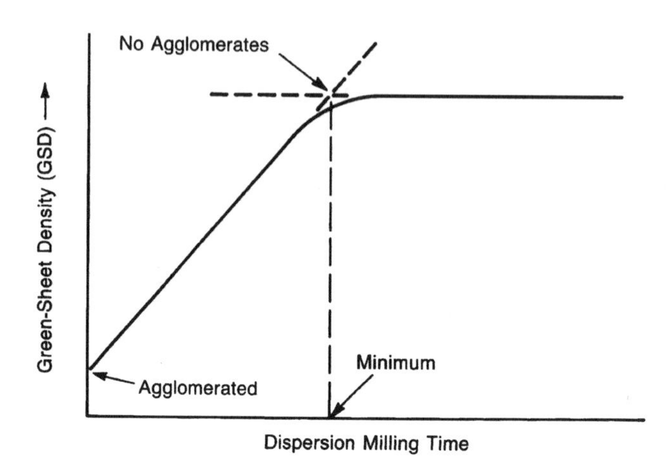

Figure 9-26. Ceramic Dispersion Model

9.4.5 Screening

The screening pastes are tailored for the various patterns shown in Figure 9-28 and consist of molybdenum powder uniformly dispersed in a resin and solvent mixture. Non-IBM multilayer ceramic substrates use tungsten instead of molybdenum, but the paste technology is similar. As with alumina milling for greensheets, the milling process is critical to achieving a proper dispersion of the metal powder in the organic materials used in the paste binder system. Three-roll mills are used for this purpose. The rolls rotate at different speeds to produce a shearing action that breaks up the metal agglomerates and distributes the molybdenum particles uniformly. Metallization of the greensheet is accomplished by extruding the molybdenum paste through a nozzle as the nozzle traverses a metal mask in contact with the greensheet. The vias are filled and the pattern is defined on the surface of the greensheet simultaneously.

Following screening, the metallized layers are dried in a forced-air circulation oven. The drying cycle has been optimized and is carefully controlled to avoid dimensional change and damage to the greensheet, particularly on sheets with dense via arrays.

One of the key advantages of the multilayer ceramic process is the ability to inspect and repair individual metallized layers prior to stacking and lamination. Inspection is accomplished automatically with a system that detects deviations in the screened pattern relative to an optimum configuration.

In order to achieve the required thermal, mechanical, electrical, and dimension control properties of the substrates, three aspects of raw materials technology need to be understood: (1) the physics and chemistry of ceramic and glass powders as well as what can be expected between

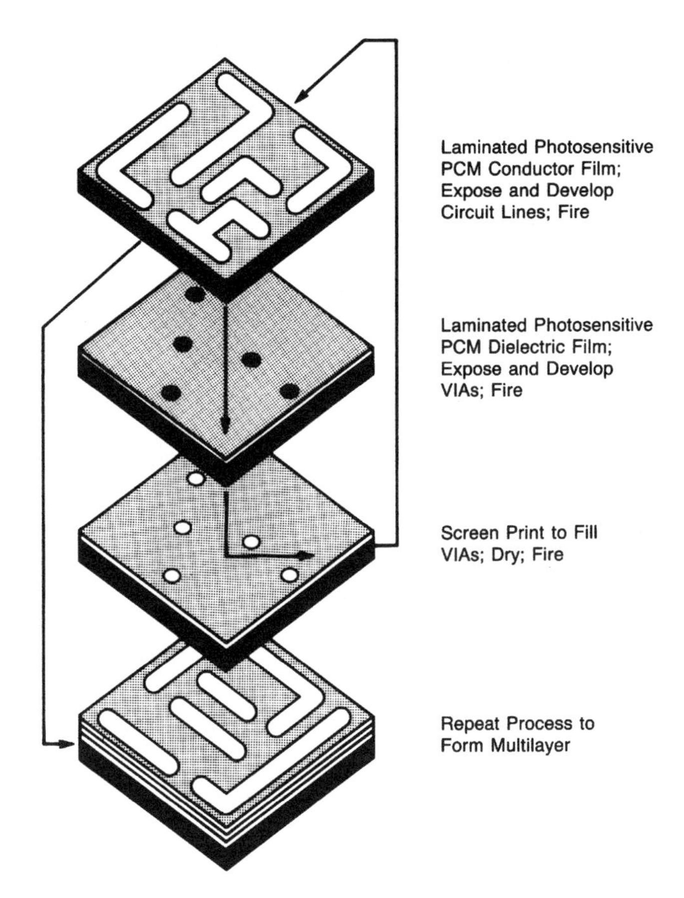

Figure 9-27. Photoformed Ceramic Modules (PCM). (Courtesy of DuPont Corp.)

these two at sintering temperatures; (2) the intrinsic properties of organics such as molecular weight and viscosity as well as the reaction between organics and inorganics to form stable steric forces; (3) properties of slurry and greensheet as affected by the particular slurry, and greensheet tool design. These interactions between organic and inorganic materials as they are processed through the tools are shown in Figure 9-29. A thorough understanding of each of the three, as well as of simultaneous interactions between all three, is necessary to provide the dimensional control required in the green and fired states to align 10,000 vias each of 0.125 mm diameter from layer to layer through 33 layers to form a network of 350,000 interconnections.

The chronological development of multilayer ceramic illustrating the improvements in substrate parameters is shown in Table 9-14 [30]. MLC packaging, as the current dominant multichip technology, owes its

Figure 9-28. Greensheet Screening Patterns

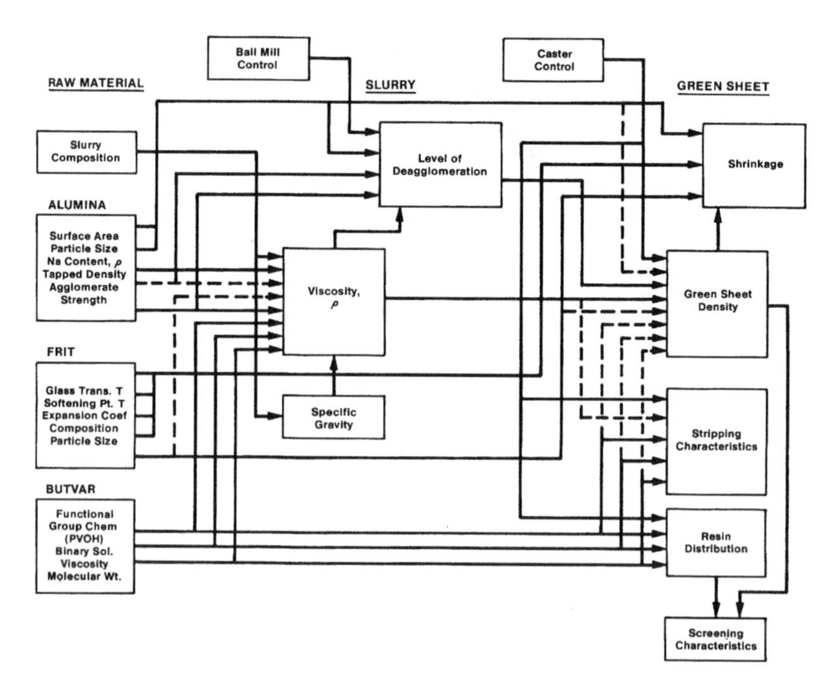

Figure 9-29. Greensheet Materials, Processes, and Tool Interactions. (Butvar® is a trade mark of Monsanto Company. Courtesy of IBM Corp.)

Table 9-14. Laminated Multilayer Ceramic Development

Year Reported	Metallurgy	Min. Via Diam./Min. Spacing (μm)	Min. Line Width/Min. Spacing (μm)	Shrinkage Tolerance	Substrate Layers	Area (mm^2)
1961	Mo, Mo–Mn	125/250	125/250	—	3–24	200
1967	Mo	—	200	—	6	161
1968		200/125	375	—	—	1,290
1969	Pd	125/250	100/200	—	15	323
1971	Noble	250/500	125/250	±0.5%	6	1290
	Mo	250/500	200	—	5–10	323
	W/Mo/Pd/Pt	125/250	100/200	±0.5%	10–20	1,935
1972	Mo–Mn	125/250	200/400	±0.6%	11	323
1973	W	200/400	125/250	—	10	3226
1975	W/Mo/Mo–Mn	125/500	100/275	±0.5%	11	11,613
1977	W/Mo	175/500	100/200	±0.5%	15–20	6,653
1978	W/Mo/Pd/Pt	250	—	±0.1%	—	—
1980	Mo	125/250	—	—	23	2,500
	Mo	100/200	100/250	±0.1%	33	8,100
1983	W	—	—	±0.3%	—	10,000
1992	Mo	75/200	90/225	±0.05	63	16,256

Source: From Ref. 30.

importance to the processes described in this section, leading to the shrinkage control of ±0.1% over as large an area as 127.5 × 127.5 mm with as many as 63 layers. If not for the basic understanding of the materials, slurry, paste, and greensheet processes and the special tools developed to go with each of the processes, multichip ceramic technology would not be the high-performance package it is today. Accordingly, high-performance organic technologies would have been chip-packaging technologies. The figures of merit discussed in Chapter 1, "Microelectronics Packaging Overview," for these two package options brings further evidence to bear on this discussion.

9.4.6 Sintering

The sintering process is the most complex of all the processes used in forming multilayer ceramic substrates. The organic removal of the solvent and plasticizer followed by the pyrolysis of the binder and the oxidation of residual carbonaceous materials in wet H_2 are all complex chemical processes that prevail in excess of 1000°C. The sintering of the substrate begins with the densification of the glass, which leads to the glass–alumina reactions that cause crystallization. The crystals, however, melt at about 1450°C into a fluid glass, the viscosity of which keeps dropping until the maximum temperature is reached. The metal paste similarly goes through organic removal and densification processes. The glass enters the porous tungsten or molybdenum metal vias, thus providing mechanical anchoring as well as any chemical reactions and bonding depending on the oxygen partial pressure of the furnace system. The allowable oxygen partial pressures could be decided from Figure 9-30 for several of the metals and oxides typically used in cofired structures.

The properties of most common metals and ceramics that are typically considered for multilayer cofired ceramic substrates are given in Tables 9-15 and 9-16.

The pyrolytic mechanisms for binder burnout during early stages of firing are shown in Figures 9-31 and 9-32. The thermal decomposition of 23-layer ceramic greensheets, shown in Figure 9-31, were studied by Knudsen Effusion–Mass Spectrometry. This technique is a powerful method for examining decomposition reactions, as it allows one to analyze both equilibrium and activated reactions. The order and activation energy of kinetic processes and heats and entropies of equilibrium reactions can be derived directly from an analysis of the time, temperature, and intensity data provided by the modulated beam mass spectrometric technique.

Figure 9-31 shows the total gas desorption rate as a function of temperature when heated at 4°C/min from 25°C to 800°C. Two organic materials, the binder and the plasticizer, are the principal components that generate gas with this thermal treatment. The plasticizer desorbs without decomposition in a simple first-order process with an activation energy of 90 kJ/mole. The desorption rate peaks at 140°C for a single layer. The relative amounts of gas released for the major components are shown in Figure 9-32. For comparison, the total gas generated as a function of temperature is plotted in the bottom curve of the graph. In the middle two curves, the spectra for water and butyraldehyde (represented by CH_3/CHO^+) show similar behavior, as they are released during the initial breakdown of the side chains of the polymer. The curve for the ion species, $m/z = 44$, is predominantly a fragment from the butyraldehyde molecule. However, this curve also shows a small amount of carbon dioxide being released around 550°C. The sensitivity of the technique is amply demonstrated in the top curve, where the formation of high-molecular-weight aromatic species occurs in the pyrolysis. In this instance, toluene (C_6H_5–CH_3), which is representative of the typical aromatic species desorbing, peaks around 440°C, where the polymer backbone is undergoing decomposition. After this stage, the laminate structure is left with nonvolatile residues of carbon and tars that will require oxidation to carbon dioxide and water for their complete removal.

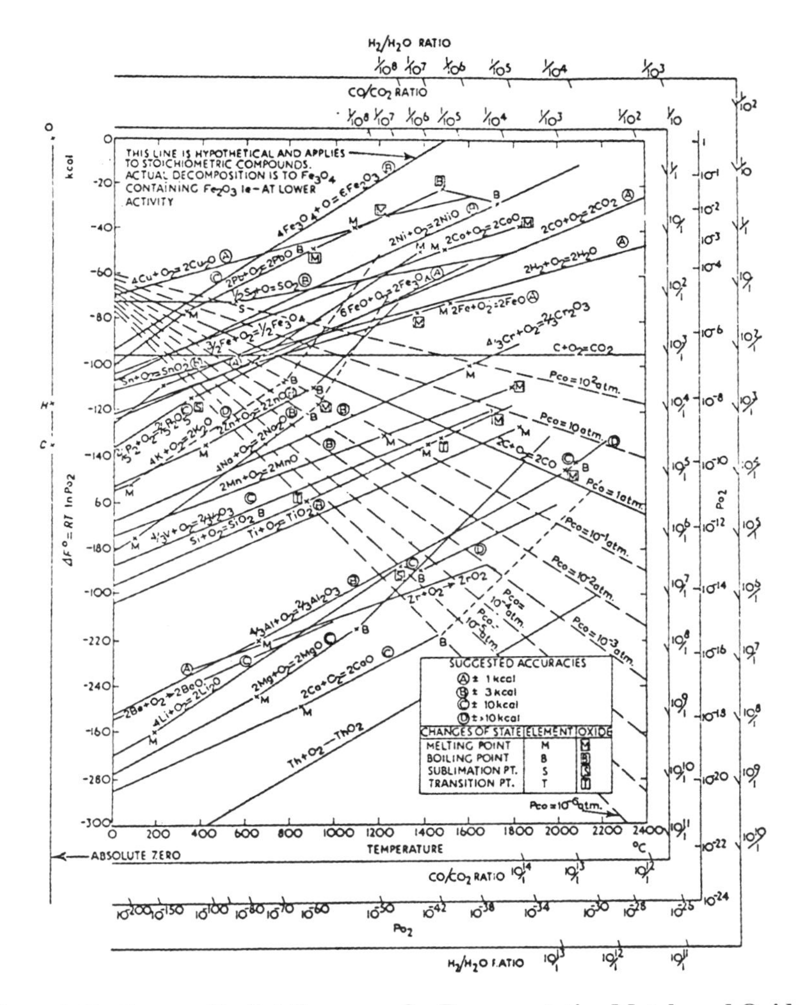

Figure 9-30. Oxygen Partial Pressures for Representative Metals and Oxides. These diagrams are used to determine formation and stability of inorganic oxides as a function of partial pressure of oxygen.

This delay in the gas release would lead to a pressure buildup at the sheet interface and thus be a contributing cause of delamination. This stress at the ground plane is most serious because of the very large amount of metallurgy on this surface, resulting in a weak bond between the layers.

9.4.7 Plating

Nickel is plated on the surface features and is diffusion bonded to the molybdenum base metal to enhance adhesion. Following nickel

Table 9-15. Typical Cofired Metals

	Melting Point (°C)	Electrical Resistivity (10^{-6} Ω m)	Thermal Expansion (10^{-7}/°C)	Thermal Conductivity (W/m K)
Ag	960	1.6	197	418
Au	1063	2.2	142	297
Cu	1083	1.7	170	397
Pd	1552	10.8	110	71
Pt	1774	10.6	90	71
Mo	2625	5.2	50	146
W	3415	5.5	45	201

diffusion, a layer of gold is applied to prevent formation of nickel oxide and to enhance wettability during subsequent soldering and brazing processes. The final plating step is the application of heavy gold on the wiring pads surrounding each chip site to accommodate ultrasonic joining of discrete engineering-change wires on the module surface. After the substrate electric test, the plated pins and flange are simultaneously brazed to the substrate in nitrogen gas atmosphere using a gold–tin alloy. Chips are joined to the substrate using the flip-chip solder-reflow process described in Chapter 8, "Chip-to-Package Interconnections."

Table 9-16(A) Dielectric Constant and Thermal Coefficient of Expansion of Candidate Oxides

	Dielectric Constant	Thermal Expansion Coefficient (10^{-7}/°C)
Al_2O_3	9.6	6.8
AlN		
Mullite ($3Al_2O_3$. $2SiO_2$)	6.6	4.0
BeO	6.8	250
Forsterite (2MgO. SiO_2)	6.2	9.8
Cordierite (2MgO. $2Al_2$ O_3. $5SiO_2$)	4.5–5.3	2.5
Spodumene (Li_2O. Al_2O_3. $4SiO_2$)	6.0	2.0
Steatite (MgO-SiO_2)	5.7	4.2
Quartz, SiO_2	3.9	1.3
Borosilicate Glass	4.0	3.3
Pb-Borosilicate Glass	7.0	7.0
Mg-Aluminosilicate	5.0	3.8

Table 9-16(B). Laminated Ceramics and Their Properties

Alumina (%)	Glass (%)	Dielectric Constant	Thermal Conductivity (W/m K)	Thermal Expansion (10^{-7}/°C)	Bonding Strength (MPa)
96	4	10.2	20.9	71	276
92	8	9.5	16.7	69	315
55	45	7.5	4.2	42	296
Mullite ($3Al_2O_3{\cdot}2SiO_2$)	40	6.2	10	4.5	165
BeO	0	6.8	250	54	231

9.4.8 Chip Joining

The chip joining by solder, tape automated bonding, and wirebonding is covered in great detail in Chapter 8, "Chip-to-Package Interconnections." The chip joining in this section addresses only the joining of the chip to the multilayer multichip substrate. The chip sites, fluxed with a water-white rosin, and the LSI devices, which are terminated in 120 lead–tin solder pads, are positioned semiautomatically on the corresponding substrate. Up to 118 chips are positioned on the substrate in this

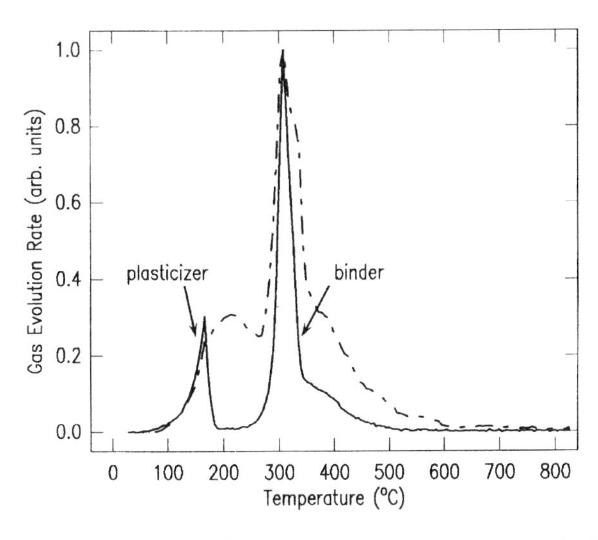

Figure 9-31. Binder Burnout—Processes. The desorption of the plasticizer (Benzoflex) and the decomposition of the binder (Butvar) are plotted for two multilayer ceramic samples heated at 4°/min: single sheet (—) and 23 layers with metallization (—). (From Ref. 29.)

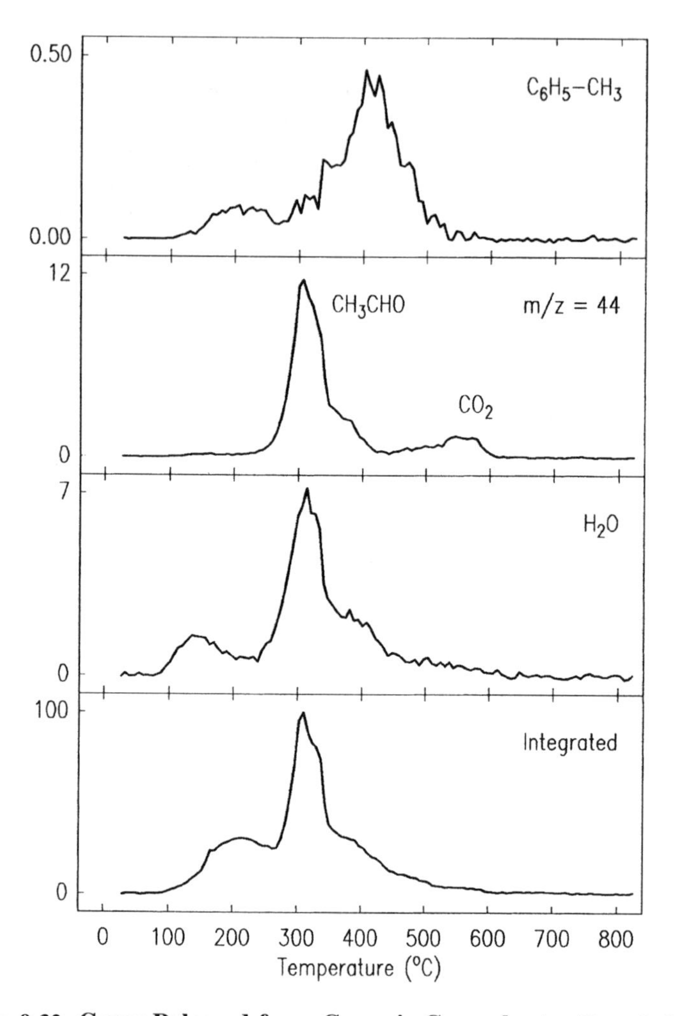

Figure 9-32. Gases Released from Ceramic Greensheets. (From Ref. 29.)

manner and reflowed in a nitrogen environment. Thousands of chip-to-module connections are made simultaneously in this manner. Following chip joining, the assembly is submerged in a nonconducting coolant and tested by contacting the I/O pins and probing the surface pads with power applied to the substrate.

9.5 RECENT DEVELOPMENTS IN CERAMIC PACKAGING

Alumina has been the only workhorse for more than three decades, from the 1960s until the beginning of the 1990s. It has become increasingly clear during the last 5–10 years that ceramic materials with improved properties and lower in cost than alumina are required. These materials fall into two categories: low-temperature ceramics, sometimes referred to as glass–ceramic or glass + ceramic, and aluminum nitride. In limited applications, mullite ceramic has been developed as well. The properties of these substrate materials are compared in Tables 9-17 and 9-18. These materials are being developed very actively so that packages will have improved:

- Signal propagation
- Thermal-expansion matching to silicon, allowing larger chips with more circuits directly solder bonded.
- Electrical conductors for added power distribution
- Thermal conductors for improved power dissipation
- Input/output connection to support more circuits

The recent progress in ceramic substrate materials is discussed in the following sections.

There is a clear trend as indicated below for both ceramic and organic packaging. For most part, the ceramic packages have been pin-grid arrays (PGA) both for pin-through-hole insertion and surface-mount assembly, including butt joining:

$$\text{PGA} \rightarrow \text{SMT} \rightarrow \text{BGA} \rightarrow \text{DCA with SLIM}$$

$$\text{1970s} \rightarrow \text{1980s} \rightarrow \text{1990s} \rightarrow \text{2000}$$

Table 9-17. Recent Ceramic-Package Materials Developments

Package Material	Package Property
Glass ± ceramics	Dielectric constant as low as 5
	Thermal expansion coefficient near that of silicon
	Cofired metallization of high-conductive metal such as gold and copper
	Low thermal conductivity
Mullite	Thermal-expansion coefficient near silicon
	Cofired W metallurgy
	Electrical insulation
	Low dielectric constant (<7)
Aluminum nitride	Thermal-conductivity approaching aluminum
	Thermal expansion near silicon
	Cofired multilayer metallurgy (W)
	Dielectric constant (~9)

Table 9-18. Properties of AlN, Glass ± Ceramics as Compared with 90% Alumina

Substrate Properties	AlN	BeO	Glass ± Ceramics	90% Al_2O_3	Mullite
Thermal conductivity (W/m K)	230	260	5	20	5
Coefficient of thermal expansion (RT ~ 400°C)($\times 10^{-7}$/°C)	43	75	30–42	67	42
Dielectric constant at 1 MHz	8.9	6.7	3.9–7.8	9.4	5.9
Tan δ (1×10^{-4})	2	3	2	2	2
Flexural strength (K_g/cm^2)	3500–4000	2500	1500	3000	2200
Thin-film metals	Ti/Pd/Au NiCr/Pd/Au	—	Cr/Cu, Au	Cr/Cu	Cr/Cu
Thick-film metals	Ag–Pd Cu	— —	Au, Cu, Ag–Pd Cu, Au		Au, Cu Ag–Pd
Cofired metals	W	W	Au, Cu, Ag–Pd	W, Mo	W, Mo
Cofired sheet resistance (mΩ/□)	10	10	3	10	10
Cooling capability					
Air (°C/W)	6	5	60	30	60
Water (°C/W)[a]	<1	<1	<1	<1	<1

[a]External Cooling

It appears that because of thin, light, and portable needs of current and future electronic systems, ball-grid arrays (BGA) and direct chip attach (DCA) of chip to ceramic and organic boards referred to as SLIM, Single Level Integrated Module under development at Georgia Tech, will be the leading-edge technologies.

9.5.1 Aluminum Nitride

Aluminum nitride, a synthetic III–V compound with wurtzite structure, was discovered in the 1860s. Although single-crystal AlN has a theoretical thermal conductivity of 320 W/m K [31], single-crystal or polycrystalline AlN shows a much lower thermal conductivity due to the incorporation of oxygen and metallic impurities into the lattice [32]. This inability to achieve high thermal conductivity in fired parts has historically hindered the use of AlN in commercial applications.

The recent development of effective sintering aids such as CaO and Y_2O_3 has resulted in high-purity and thus higher terminal-conductivity AlN [33]. This, in turn, has led to the commercialization of AlN products for packaging and interconnect applications.

9.5.1.1 AlN Powder and Sintering

Commercial AlN powders are manufactured by two processes: (A) carbothermal reduction of alumina and (B) direct nitridation of aluminum metal:

$$\text{(A) } Al_2O_3 + 3C + N_2 \rightarrow 2AlN + 3CO$$

$$\text{(B) } Al + 0.5N_2 \rightarrow AlN$$

In comparison to the carbothermal reaction, the direct-nitrided reaction produces a broader particle size distribution and a larger average particle size which results in a less sinterable material. Worldwide production of all AlN powders was estimated to be 150 tons in 1991.

Alkali-earth and rare-earth oxides are effective sintering aides for AlN pressureless sintering. Yttrium oxide or calcium oxide are generally used in production. Sintering is done under nitrogen at approximately 1800°C. The sintering aids get oxygen from the surface and/or the lattice of the AlN and form a liquid grain-boundary phase (Y–Al–O or Ca–Al–O) which concentrate at grain-boundary triple points or migrate to the surface of the sintered AlN body. This promotes densification and grain growth and, at the same time, prevents oxygen atoms from diffusing into AlN grains so that high thermal conductivities up to 200 W/m K can be attained [34]. Sintering in a carbon atmosphere is one means of achieving thermal conductivity of > 200W/m K.

The thermal conductivity of AlN is very sensitive to the presence of oxygen and other metallic and nonmetallic impurities such as Mg, Fe, Ca, and Si [32,35]. Oxygen diffusion into the AlN lattice forms nitrogen vacancies, which impede phonon heat transfer. The relationship between thermal conductivity and oxygen content in a fully dense sintered AlN is shown in Figure 9-33.

High-quality AlN ceramics are being produced by conventional ceramic processes such as dry-pressing, tape casting, and injection molding.

9.5.1.2 Low-Temperature Firing

The AlN sintering temperature of > 1800°C prevents the use of traditional alumina furnaces. Recent reports indicate that AlN can be sintered at 1600°C by the addition of fluoride sintering additives [36]. The low-temperature sintering of AlN, if it can be accomplished in alumina furnaces, will lead to cheaper AlN substrates that will expedite its commercialization.

9.5.1.3 AlN Properties

Table 9-18 compared AlN properties with other commercial ceramic substrates and related ancillary materials. The electrical properties of AlN

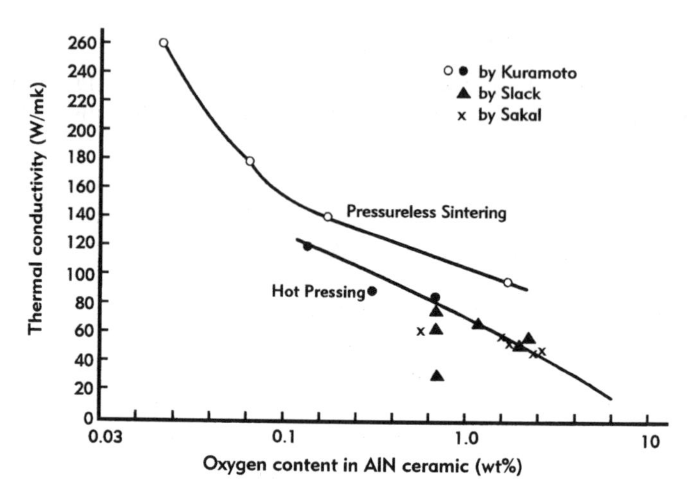

Figure 9-33. Relationship Between Thermal Conductivity and Oxygen in AlN

are comparable to that of alumina and beryllia. AlN shows a better thermal-expansion match to silicon than either alumina or beryllia, as shown in Table 9-18. This results in less stress generation due to thermal mismatch between the silicon chip and the substrate as structures are exposed to temperature excursions. This becomes very important as chips get larger and when silicon-based ICs (2–3.5 cm on a side) are packaged.

The flexural strength of AlN is greater than alumina, whereas the Vickers hardness is half that of alumina. This lower hardness and superior strength allows AlN to be machined into complex shapes. The lower density of AlN results in a 20% weight reduction when compared to equivalent alumina parts.

AlN is resistant to most process chemicals with the following notable exceptions:

- The oxidation characteristics of fired AlN have been examined in detail [37]. The oxidation starts at temperatures in excess of 800°C. Thus, although oxygen must be excluded during AlN firing, air oxidation is not an issue during subsequent fabrication or use.
- Early concerns about AlN centered around its stability in moist environments. The inherent hydrolytic instability of AlN under accelerated aging test conditions is corrected by oxidation of the surface by a short exposure to air at 100°C [38–40] A thin layer of alpha alumina forms, which protects the surface from further oxidation and or hydrolysis. No hydrolysis is observed after 1000 h at 85°C/ 85% **rh** and/or pressure cooker testing (PCT). Loss in thermal conductivity by this process is reported to be 10% or less.

- Strong alkaline solutions attack AlN ceramics [41]; thus basic electroless plating solutions must be avoided. The thin oxide coatings described above also serve to protect the AlN surface from alkaline solutions [40].

9.5.1.4 Metallization of AlN

A summary of thick- and thin-film metallizations is presented in Table 9-19. Thick-film metallizations containing conventional glass frit powders developed for 96% alumina cannot be used on AlN surfaces. Studies have concluded that, during firing, PbO, CuO, and Bi_2O_3 glasses cause oxidation of the AlN surface. This oxidation results in blistering and poor adhesion of the metallized surface [42]. When proper thick-film pastes (Pd/Ag,Au), or thin-film metallizations (Cr/Au, Ti/Pt/Au, Ti/W,A1) were examined, all were found to give acceptable properties[43,44]. Brazing and resistor paste technologies have also been developed [45–47].

9.5.1.5 AlN Applications

Table 9-20 shows AlN's practical application areas. In the field of power devices, bare substrates and thick-film metallized substrates are used in various thyristor and transistor modules. Direct-bond copper (DBC) AlN substrates has overcome early thermal coefficient of expansion (TCE) mismatch problems [48] and are being used in various power applications including IGBT (insulated gate bipolar transistor) modules.

Submount substrates for LD and LED packaging, which are patterned by thin-film Ti/Pt/Au, require mirror surface finish (R_a: 0.5 μm) and high-thermal-conductivity (>200 W/m K) of AlN. Aluminum nitride use in computer packaging has been led by Fujitsu which has used AlN in its

Table 9-19. Thin- and Thick-Film Metallizations for AlN

Process	Metallurgy	Process Temp.
Thick-Film Metallization		
Postfired	Ag–Pd, fritted	920°C in air
	RuO_2, fritted	850°C in air
	Copper	850°C in N_2
	Gold	850°C in air
Molten Metallized	Cu–Ag–Ti–Sn	930°C in H_2
Cofired	Tungsten	1900°C in H_2
Thin-Film Metallization		
Sputtering	NiCr–Pd–Au	100–200°C
Evaporation	Ti–Pd–Au	100–200°C

Table 9-20. AlN Applications in Electronics

	Application	Metallization Example
Power device	Thyristor stack for trains, relay module	Not metallized
	TO-type power module	Mo–TiN, Mo–Mn
	RF package	Mo–Mn
	Heat sink for car ignition	Cofiring W, Mo–TiN
	High-power switching module	Direct bond copper, Ag–Cu–Ti brazed copper
Laser diode	Submount for LD and LED	Ti/Pt/Au/Sn–Pb or Au–Sn
Package	Heat sink for computer packaging	Ti/Ni/Au
	MCM PGA package	Cofiring W, copper polyimide
	QFP	Glass sealed

supercomputer, the VP 2000, as a heat spreader and surface-mount PGA package (17 mm^2 with 462 pins) as shown in Figure 9-34 [49]. Fijitsu has also developed an AlN surface-mount quad flat pack (QFD) for mounting large high-power application-specific integrated circuits (ASICs) (Fig. 9-35). Thermal resistivity measured on a representative QFP is shown in Figure 9-36. This face-down package with a pin-fin can be applied to a chip of about 4 W with air cooling at 1 m/s or 2.5 W without forced air. These packages will be used for the MUSE decoder of high-definition TVs [50].

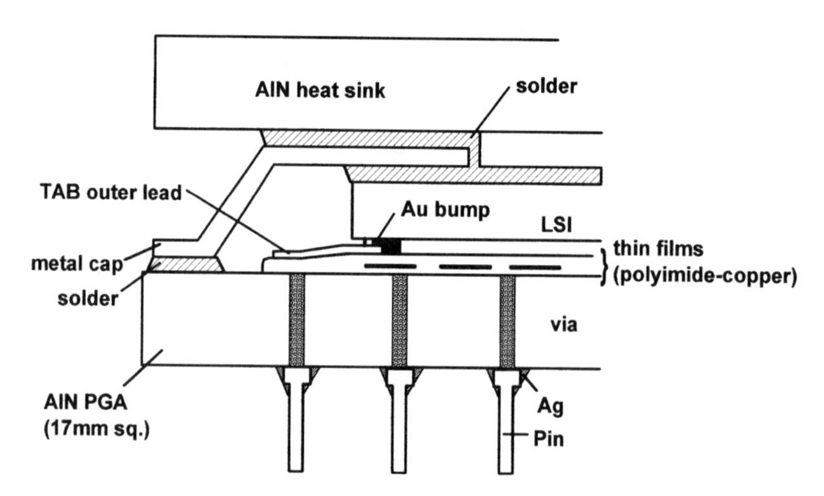

Figure 9-34. Schematic of Fujitsu VP 2000 Module Using Surface-Mount AlN PGA and Heat Sink

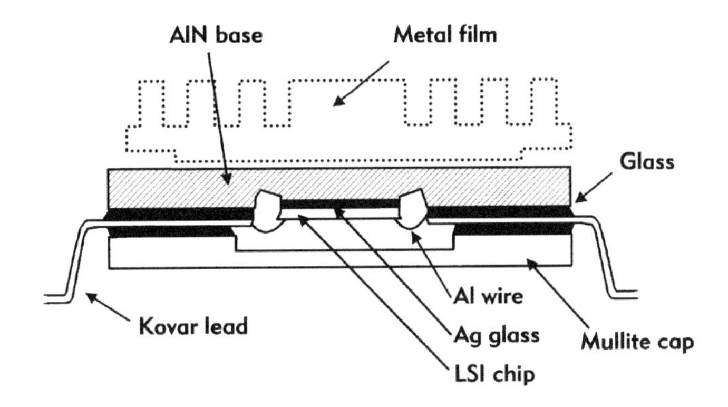

Figure 9-35. Glass-Sealed AlN Quad Flat Package

The Hitachi M-880 processor series adopted a water-cooled thermal solution. The heat generated from the LSI chips goes through the hermetic AlN cap enclosing the chip, the comb-shaped AlN microfins and the large AlN "board" which interfaces the water-cooling jacket as shown in Figure 9-37. The heat density generated on this module (36–41 chips on 106 mm^2) is about 7.5 W/cm^2). In this application, the excellent machinability of AlN is a key factor as well as AlN's thermal conductivity and TCE properties [51].

Multilayer AlN packages with tungsten cofire metallurgy are close to practical application for high-pin-count PGA packages [52,53] The NEC PGA cross section shown in Figure 9-38 is for a 208 I/O AlN package with thermal conductivity of 200 W/m K and tungsten sheet

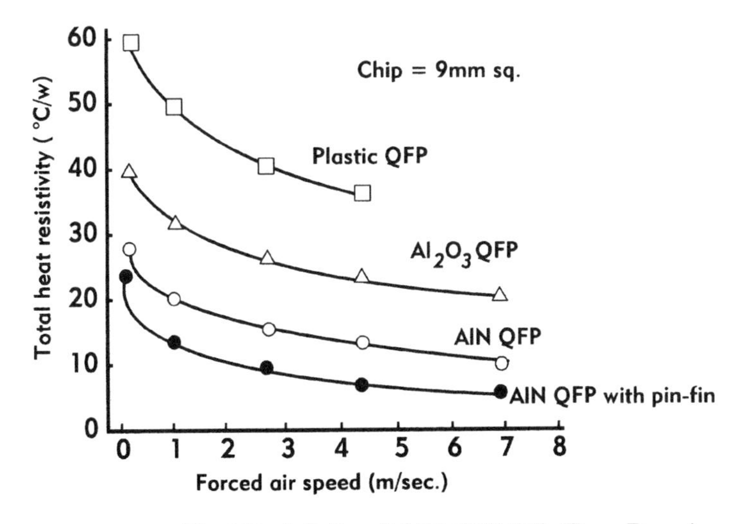

Figure 9-36. Heat Resistivity of AlN QFP 160 (Face Down)

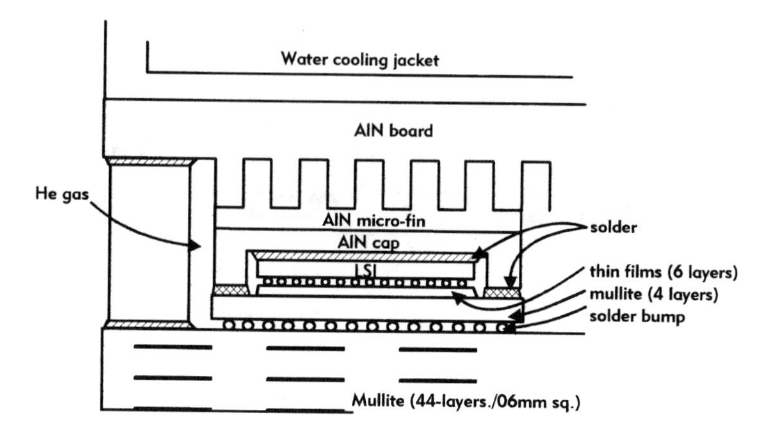

Figure 9-37. Schematic of Hitachi M-880 Processor Module Using AlN Heat Spreaders

resistivity of 7.2 mΩ/□. Compared to an alumina package, this AlN package can handle a 3.7 times higher-powered Si chip (up to 40 W).

Large hermetic enclosures for silicon-on-silicon MCMs [7,25] and MCM cofired substrates (with or without copper–polymer thin-film wiring layers) are also in active development [54,55].

The NOSC (Navan Ocean Systems Center) package (Fig. 9-39), developed by Hughes/Coors/Grace, is a 10-cm^2 hot-pressed package built to hold a large silicon-on-silicon MCM. AlN was chosen because of the need to match TCE with this large silicon area [56].

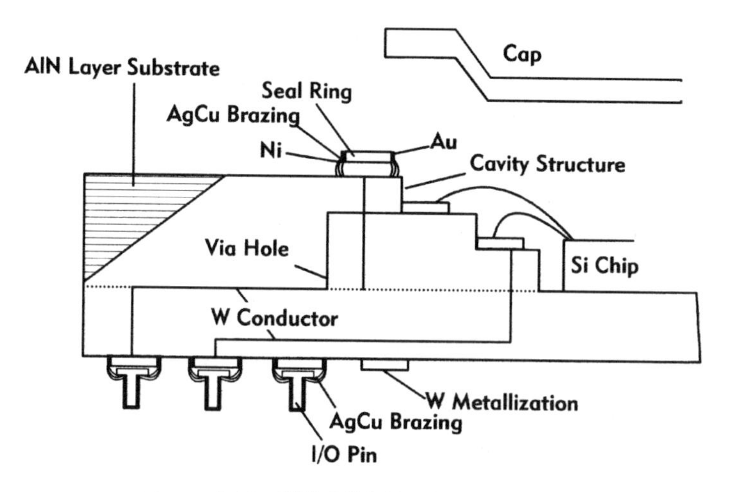

Figure 9-38. AlN PGA Package by NEC

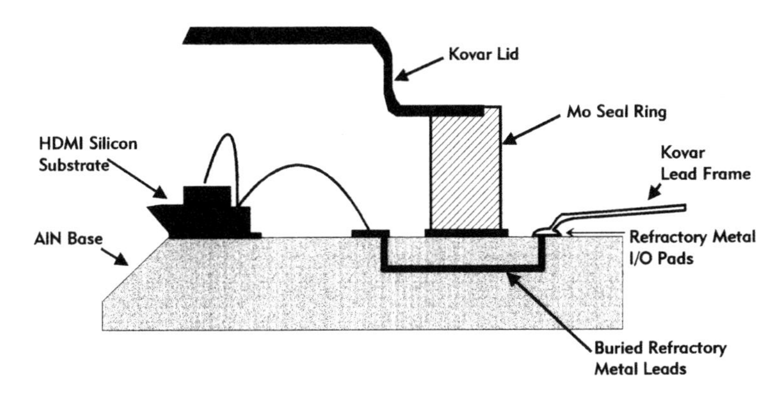

Figure 9-39. Silicon-on-Silicon MCM with AlN

9.5.2 Beryllium Oxide

For years, beryllium oxide has been considered the only ceramic with very-high-thermal conductivity and good dielectric properties for packaging applications. Beryllium oxide packaging, even though not a recent development, is covered here because of its very high thermal conductivity, similar to AlN. The processes described previously in this chapter for alumina using dry-pressing and tape casting processes are applicable to beryllium oxide. Cofired, postfired, and conventional thin-film metallizations have been developed for forming conductors to form ground planes, mounting pads, microstrip circuitry, and other features. Beryllia is used as a heat sink for standard transistor cases, TO-3, TO-8, TO-9, TO-11, and TO-33. It is also used as an electrically insulating heat spreader for high-power transistors covering the frequency range from 60 Hz to 10 GHz. Power hybrids, bipolar chips, LSI devices, and microwave integrated circuits are often housed in beryllia packages. Beryllia is used as a substrate for microwave integrated circuits. The microstrip circuitry is usually applied by means of sputtered or evaporated thin films. Direct bonding of copper using copper–copper oxide eutectic has been reported [57]. Details of these and additional information on other aspects of beryllium oxide can be found elsewhere [58]. The pertinent properties of beryllium oxide are included in Table 9-18. Although the outstanding technical properties of beryllium oxide justify its use, its toxicity—requiring a label, "warning, this product contains BeO"—has limited its use.

9.5.3 Mullite

Conventional mullite has been known for over 30 years. The mullite that is getting the recognition in electronic packaging, however, is mullite with additions to lower the dielectric constant and its sintering temperature.

Table 9-21 lists the hardware technology using mullite that Hitachi developed for its M-880 mainframe computers using small-scale chip carriers called microchip carriers (MCC) with solder ball-grid array connections to the multilayer mullite–glass–ceramic board [59].

To mount such high-speed high-density LSI chips, Hitachi has developed multilayer ceramic boards made of mullite ceramics and tungsten conductors [59]. In contrast to previous mullites, Hitachi mullite has very low dielectric constant, presumably due to SiO_2 addition to conventional mullite ($3Al_2O_3 \times 2SiO_2$).

This mullite ceramic board features the following:

1. Excellent electrical characteristics; specific dielectric constant: 5.9; conductor resistance; 0.6 Ω/cm
2. Thermal expansion coefficient: 3.5×10^{-7}/°C, Very close to that of silicon (3.0×10^{-7}/°C)
3. Flexural strength: 20 kg/mm^2 or more; strong enough to bear MCC soldering, module sealing, and brazing of I/O pins

Table 9-21. M-880 Mainframe Hardware Technology Based on Mullite Ceramic (From Ref. 59)

		M-880	M-680H
High-density module (M-880) or package (M-680H)	Size	106 × 106 × 7 mm	420 × 280 × 3 mm
	Dielectric materials	Mullite ceramic	Polyimide
	Linewidth	100 μm	100 μm
	No. of conductor layers	44	20
	No. of I/O pins	2521	1776
	LSI package	MCC	Flat package
	No. of LSIs	36–41	72
	Cooling	Water	Forced air
	Power density	7.5 W/cm^2	0.4 W/cm^2
	Thermal resistance	≤2°C/W	5°C/W
	Junction temperature	50–60°C	60–80°C
Processor Board	Size	730 × 534 × 7 mm	570 × 420 × 3 mm
	Dielectric materials	MS resin	Poly-imide
	Linewidth/thickness	70 μm/65 μm	100 μm/40 μm
	No. of conductor layers	46	22
	No. of modules	20	21

Source: From Ref. 59.

4. Excellent dimensional stability because of similar thermal-expansion coefficients of the ceramics and the conductor (tungsten)

9.6 LOW-TEMPERATURE CERAMIC OR GLASS + CERAMIC PACKAGING

The limitations of alumina and the advantage of low-temperature ceramic, as indicated in Table 9.22, fall into three categories. The dielectric constant (9.4) of alumina is too high compared to low-temperature ceramics which can be fabricated around 5. The thermal-expansion coefficient of alumina (7×10^{-7}/°C) similarly is too high in those applications in which the silicon chip of TCE, 3×10^{-7}/°C, is bonded directly to the ceramic. This is clearly the case with the flip chip, particularly with next generation of large chips, as indicated in Figure 9-40 for solder-joint fatigue life between the substrate and silicon chip. The resistance of the package conductors has as much, if not more, influence than the dielectric constant on the interconnect delay, particularly in long critical-length applications. Figure 9-36 plots the interconnection delay as a function of linewidth for various line lengths in the critical path of mainframe or workstation applications.

Glass–ceramics, with their dielectric constant around 5, together with their cosinterability with copper or gold, make them potentially one of the best high-performance multilayer ceramic substrates, meeting most of the requirements of an ideal substrate (see Table 9-22). Glass additions to alumina [60–62] to form substrates with a dielectric constant down to 5 and are sinterable with copper or gold have been reported by a number of Japanese and American companies. In this book, "glass + ceramics" refers to those ceramics formed by the addition of glass to the ceramic in forming slurry or greensheet. In contrast, "glass–ceramics" refers to

Table 9-22. Ideal Properties of Ceramic Package

	Ideal Ceramic	Alumina Ceramic
Dielectric constant	<5	9.4
Coefficient of thermal expansion	30×10^{-7}/°C	66×10^{-7}/°C
Package metallization resistivity	2 μΩ cm	5.6 μΩ cm
Metallization layers		
Thick Film	≤75 μm on 500 μm up to 100 layers	Same
Thin Film	≤25 μm on 75 μm, up to 10 layers	Same
Number of contacts to chip	Up to 5000 per chip site	Same
Number of package output pins	>20/cm^2	Same

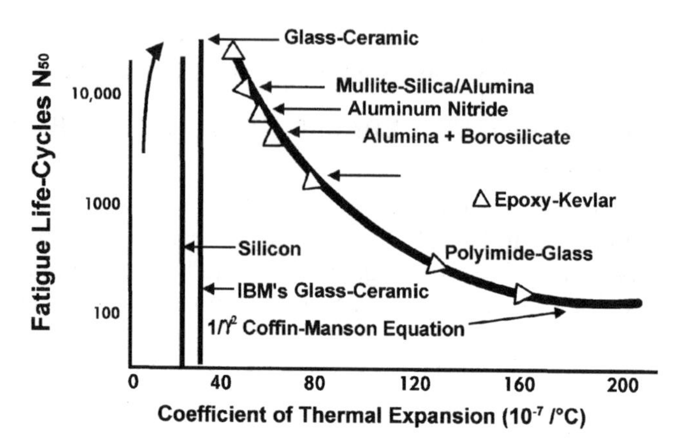

Figure 9-40. Fatigue Life of Flip-Chip Solder Joint as a Function of TCE of Substrate

ceramics formed by controlled crystallization of glass in the slurry or greensheet resulting in ceramics of high crystallinity during firing. "Glass ± ceramics" refers, therefore, to a generic family of both these materials.

Table 9-22, which lists ideal properties of substrate materials, shows that glass–ceramics approach these ideal requirements more than any other group of ceramics. These substrates have low-dielectric-constant, thermal-expansion matching silicon and contain the gold or copper metallization required for high-power distribution. In addition, thin-film metallization and dielectrics could be deposited onto the surface of these substrates, thus making them extendable for high performance. The major disadvantage seems to be their low thermal conductivity. With solder connection (C4) technology, which allows the heat to be removed from the back of the chip, this limitation is largely overcome. This is particularly true for those package applications where the chip is bonded face (active area) down toward the substrate, leaving the back side for air or water cooling. In this case, minimum heat removal takes place through the substrate. Thus, high thermal conductivity of the substrate is desirable but not absolutely required. The impact of poor thermal conductivity in wirebonding applications of these ceramics can be minimized significantly by die bonding to metal or ceramic heat sinks which are placed in the single-chip package cavity and then brazed or glass sealed to the glass ± ceramic package. This subject is further discussed in Section 9.8, "Chip Attachment and Thermal Dissipation of Ceramic Substrates." In this case, minimum heat removal takes place through the substrate, hence, high thermal conductivity of the substrate is desirable but not absolutely required. In addition, for high-performance applications involving bipolar silicon or high performance CMOS, where the integration is expected to be very high, cooling capabilities may be required in excess of 40 W/chip. At these power levels, even the very-high-thermal-conductivity ceramics such as AlN will require thermal enhancements both in face-down and face-up chip-to-substrate interconnections. Low-dielectric-constant glass, glass–ceramic, and various glasses added to various low-dielectric-constant ceramics are listed in Table 9-23.

The concept of glass–ceramic cofired, multilayer substrate containing copper, gold, or silver conductors had its origin in 1976 in IBM. The first multilayer glass–ceramic substrate was reported in 1978 based on vitreous densification of cordierite and spodumene glass powders, followed by crystallization below 1000°C into cordierite and spodumene glass–ceramics [63]. This IBM technology is discussed in great detail later in this chapter. The subject of glass–ceramics is covered in a textbook by McMillan [64]. This process allowed fabrication of high-temperature ceramics using low-temperature vitreous systems, thereby providing opportunities for cofiring in a controlled atmosphere with copper metallization to form multilayer ceramic substrates as reported first in 1980 [65]. Subsequent to this invention, a number of glass ± ceramics, most of them

Table 9-23. Low-Dielectric-Constant Glasses, Ceramics, Glass + Ceramics

Material	Dielectric Constant @ 1 MHz
Borate glass	3.2
Silica glass	3.8
Borosilicate glass	4.0
Lead borosilicate glass	7.0
Magnesium aluminosilicate glass	5.0
Calcium aluminosilicate glass	6.0
Lithium silicate glass	6.5
Cordierite	4.5
Mullite	6.7
Alumina	9.8
Borosilicate + silica	3.9
Borosilicate + alumina	5.6
Borosilicate + mullite	4.5
Borosilicate + cordierite	4.2
Lead borosilicate + alumina	7.8
Calcium aluminosilicate + alumina	7.9
Lithium silicate + silica	5.0
Magnesium aluminosilicate + silica	5.0
Magnesium aluminosilicate + alumina	6.5
Magnesium aluminosilicate + mullite	6.0

Table 9-24. Properties of Glasses, Glass + Ceramics, and Glass–Ceramics

Materials	Dielectric Constant	Coefficient of Thermal Expansion (10^{-7}/°C)	Conductor
Glasses			
B_2O_3–SiO_2–Al_2O_3–Na_2O	4.1	32	Cu
Glass + Ceramics			
$PbO + B_2O_3 + SiO_2 + (Al_2O_3)$	7.5	42	Au, Ag, Ag + Pd
$MgO + Al_2O_3 + SiO_2 + B_2O_3 + (Al_2O_3)$	4.5	30	Au, Ag, Ag + Pd
$B_2O_3 + SiO_2 + (Al_2O_3)$	5.6	45	Au, Ag, Ag + Pd
$2MgO_2Al_2O_3 \cdot 5SiO_2 + (Al_2O_3)$	5.5	30	Au
$CaO + Al_2O_3 + SiO_2 + B_2 + O_3 + (Al_2O_3)$	7.7	55	Ag, Ag–Pd
$Li_2O + SiO_2 + MgO + Al_2O_3 + SiO_2 + (Al_2O_3)$	7.3	59	Au, Ni, Ag–Pd
$Li_2O + Al_2O_3 + SiO_2 + (Al_2O_3)$	7.8	30	Au
Glass–Ceramics			
MgO–Al_2O_3–SiO_2–B_2O_3–P_2O_5	5.0	30	Cu
Li_2O–Al_2O_3–SiO_2–B_2O_3	6.5	25	Cu

as glass additions to alumina, at roughly 50/50 wt.% level were reported. Table 9-24 lists a number of these compositions, the metallization developed to cofire with each, and the properties achieved for each multilayer ceramic substrate. It should be noted, however, that glass–ceramic material technology originated from Corning Glass Works in the late 1950s, and, subsequently, glass–ceramic pastes became available from DuPont and others for sequential, crossover, multilayer applications onto fired alumina substrate.

In those applications that require wirebonding or tape automated bonding (TAB) of chips to these low-dielectric-constant, low-thermal-conductivity ceramics, heat dissipation becomes a problem. A number of compositions containing high-thermal-conductivity additions of beryllia, silicon nitride, and artificial diamonds, shown in Table 9-25, are being explored [66]. In addition to those previously discussed [63], glass–ceramic compositions based on cordierite formation, with a dielectric constant around 5 and thermal expansion near that of silicon [66], are given in Table 9-26.

Copper or gold metallization with glass–ceramic to form a high-conductive (electrical) multilayer ceramic is the most desirable substrate and perhaps the hardest to achieve technically. Copper has the best combination of electrical conductivity, sintering temperature, and wettability to braze and solder. The material is available in abundance, the powder processing is generally known with desired physical and chemical proper-

Table 9-25. High-Thermal-Conductivity Glass + Ceramics

	Various Compositions					
Component	1	2	3	4	5	6
Al_2O_3	50 wt.%	33.3 wt.%	33.3 wt.%	33.3 wt.%	50 wt.%	96 wt.%
B_2O_3–SiO_2	50 wt.%	33.3 wt.%	33.3 wt.%	33.3 wt.%	40 wt.%	—
BeO		33.3 wt.%				
Si_3N_4			33.3 wt.%			
Artificial Diamond				33.3 wt.%	10 wt.%	—
Property						
Thermal conductivity (W/M K)	4	48	32	48	20	20
Dielectric constant	5.6	5.2	5.0	5.5	unknown	9.4

Source: From Ref. 46.

ties, and the paste technology is readily available. Cofired gold metallization is not as desirable, primarily due to the high cost of gold raw material. Thick-film copper metallization into glass–ceramic multilayer ceramic poses chemical and mechanical incompatibility problems, however. The chemical incompatibility arises as follows.

Basically, any conventional atmosphere in which organics could be removed from greensheets during multilayer ceramic fabrication results in oxidation of copper. Conversely, in any reduced atmosphere in which copper could be maintained as pure metallic copper, the greensheet organics remain or "char" the ceramic, thus spoiling its electrical properties. Although it would seem that the standard nitrogen firing of copper thick films should be applicable for greensheet multilayer ceramic laminate, this is not the case for two reasons. Paste organics are chemically different from greensheet organics, and the paste thickness is much thinner than the green laminate thickness (20 μm versus 5000 μm).

Silver, gold, and silver–palladium metallizations requiring no controlled atmospheres for burn-off of binders have been reported [67]. The substrate sintering processes are simpler than those involving copper, but the materials are considered expensive for volume manufacturing. One example of these processes is illustrated in Figure 9-39, wherein Ag–Pd is used in the external layers of substrates which contain silver in the internal layers. Silver–palladium has been reported to provide "hermeticity" to the package.

Both glass + ceramics, by control of volume fractions or the selection

Table 9-26. Selected Glass–Ceramic Compositions Based on Cordierite Formation

Material (wt.%)							Dielectric Constant	Coefficient of Thermal Expansion
SiO_2	Al_2O_3	MgO	B_2O_3	Others		Ceramic Addition	1 MHz	10^{-7}°C
				ZnO	P_2O_5			
56.0	23.5	15.0	1.0	3.5	1.0		5.5	30
				ZnO	P_2O_5			
57.5	25.5	12.0	2.0	2.5	0.5		5.2	25
				TiO_2	P_2O_5	+ Al_2O_3		
52.0	20.0	17.0	2.0	7.0	2.0	20 vol. %	5.5	37
				TiO_2	P_2O_5	+ BN		
52.0	20.0	17.0	2.0	7.0	2.0	20 vol. %	5.0	32
				CaO	ZrO_2			
45.0	32.0	12.0	4.5	4.5	2.0		5.5	22
				CaO	ZrO_2			
50.0	30.0	13.0	3.0	3.0	1.0		5.6	16
				Y_2O_3	ZrO_2			
58.0	22.0	12.0	1.0	6.0	1.0		5.7	35
				Y_2O_3	ZrO_2			
55.0	17.0	21.0	3.0	3 .0	1.0		5.5	30

Source: From Ref. 66.

of ceramics and glass, and glass–ceramics, by selection of glass compositions and heat treatment profiles, could readily be developed with thermal-expansion coefficients near GaAs, the thermal-expansion behavior of which is shown in Figure 9-41. Alumina substrate is considered a good match. However, because it can only be metallized with such metals as molybdenum or tungsten, which have low electrical conductivity, high-performance packaging of GaAs with alumina MLC is considered a major limitation. Low- and medium-performance applications can readily be handled by alumina with C4 bonding or by TAB and wirebonding of chips which pose no thermal mismatch problems.

9.7 STATE OF THE ART IN LOW-TEMPERATURE CERAMIC PACKAGING

9.7.1 IBM's Glass–Ceramic Substrate

IBM's glass–ceramic is the first and true glass–ceramic that starts out with glass and crystallizes to glass–ceramic. IBM developed and

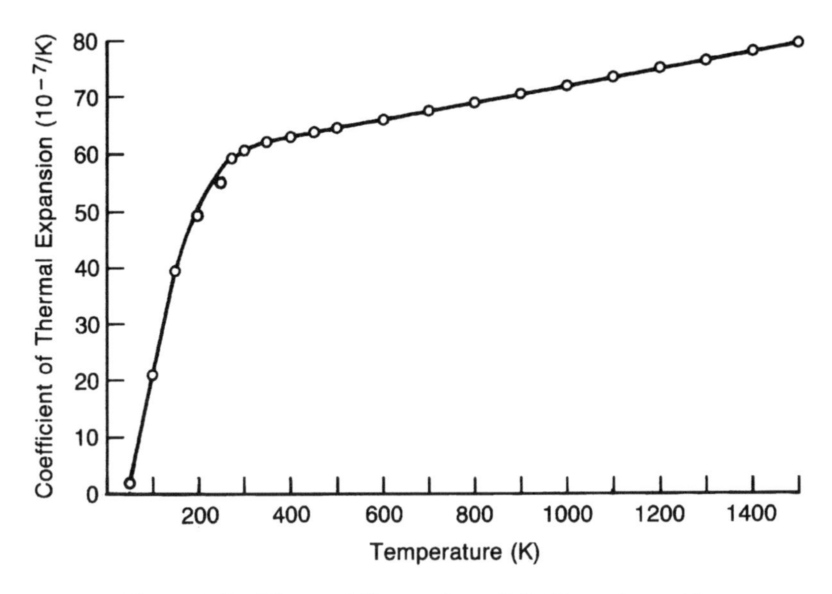

Figure 9-41. Thermal Expansion of Gallium Arsenide

perfected the manufacturing of this technology for all of its top-of-the-line mainframe computers based on a 63-layer substrate. The key technologies in developing the 63-layer glass–ceramic/copper substrate are discussed in the following subsections, excerpted from a recent IBM publication [68]; glass–ceramic material, greensheet, copper powder and paste, personalization, stacking, and lamination, glass–ceramic/copper sintering, and substrate machining and finishing.

9.7.1.1 Glass–Ceramic Material

Table 9-27 lists various potential systems considered for forming glass–ceramic substrates in terms of their thermal and electrical properties as well as sintering temperature required to cofire with highly conductive metal such as copper. None of the systems shown in Table 9-27 met the requirements of a thermal-expansion coefficient near that of silicon (30×10^{-7}/°C), a dielectric constant of 5.0, and sintering compatibility with copper. In addition to these glass–ceramic systems, glass and ceramic systems involving, for example, the addition of a greater amount of glass to alumina than in typical alumina substrates were also considered, such as a low-viscosity lead glass added to alumina ceramic. For the glass-plus-ceramic compositions, binder removal was anticipated to be more difficult because of the need for a low-viscosity glass required to sinter the ceramic high density at about 950°C [69].

The glass developed to form the glass–ceramic (USP: 4,301,324

Table 9-27. Glass–Ceramic Systems Considered

Crystal type	Thermal Expansion	Dielectric Constant	Sintering Temperature
Cordierite $2MgO \cdot 2Al_2O_3 \cdot 5SiO_2$	Too low	Good	Too high
Beta-spodumene $Li_2O \cdot Al_2O_3 \cdot 4SiO_2$	Too low	Too high	Too high
Celsian $BaO \cdot Al_2O_3 \cdot 2SiO_2$	Good	Too high	Too high
Anorthite $CaO \cdot Al_2O_3 \cdot 2SiO_2$	Too high	Too high	Too high
Glass + alumina	High	High	950°C

filed: Feb. 1978) has the following composition—SiO_2: 50–55 wt.%; Al_2O_3: 18–23 wt.%; MgO: 18–25 wt.%; it contains P_2O_5 and B_2O in 0–3 wt.% [70]. Table 9-28 indicates the role of each of the oxides in determining final properties of ceramic.

Crystallizable cordierite glass offers process advantages for multilayer ceramic–Cu substrates. These advantages can be examined with the use of differential thermal analysis (DTA), as shown in Figure 9-42. To form the glass–ceramic substrate with copper, it is important that no sintering occurs below 780°C and that complete crystallization occurs below the melting point of copper (1083°C). In Figure 9-43, both of these conditions are met, allowing adequate binder removal at temperatures below those at which sintering occurs, complete densification prior to crystallization, and crystallization well below 1083°C, the melting point

Table 9-28. Glass Composition (wt.%) to Form IBM's Glass–Ceramic

Component	Wt. % Range	Effect
SiO_2	50–55	Control properties
Al_2O_3	18–23	Control properties
MgO	18–25	Control properties
P_2O_5	0–3	Nucleating agent, sintering aid
B_2O_3	0–3	Delays crystallization, sintering aid

Glass transition temperature: 780°C
Complete densification temperature: 800°C

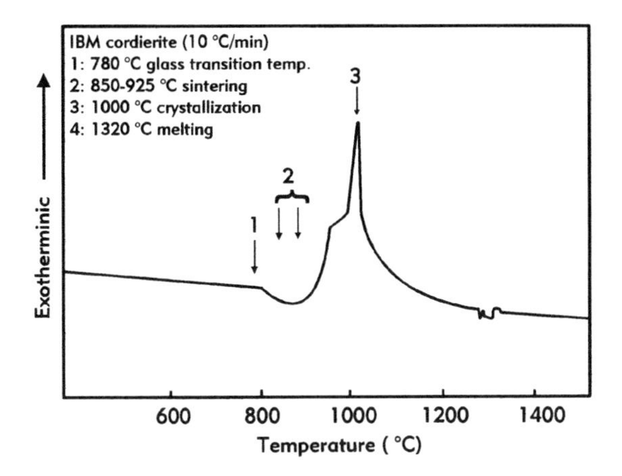

Figure 9-42. Differential Thermal Analysis of Cordierite

of copper. The crystallization peak occurs at approximately 1000°C in the dynamic DTA experiment and at lower temperatures around 950°C for actual substrate formation.

When a stoichiometric cordierite glass is heated in a DTA experiment, the crystallization exotherm initiates at a lower temperature and proceeds at a higher rate. With the shift from stoichiometry and with small amounts of B_2O_3 and P_2O_5, the onset of crystallization is delayed to a higher temperature and the crystallization is reduced. This crystallization temperature increase and crystallization rate decrease help enable the glass to sinter to 100% of theoretical density [71]. Further studies on glass

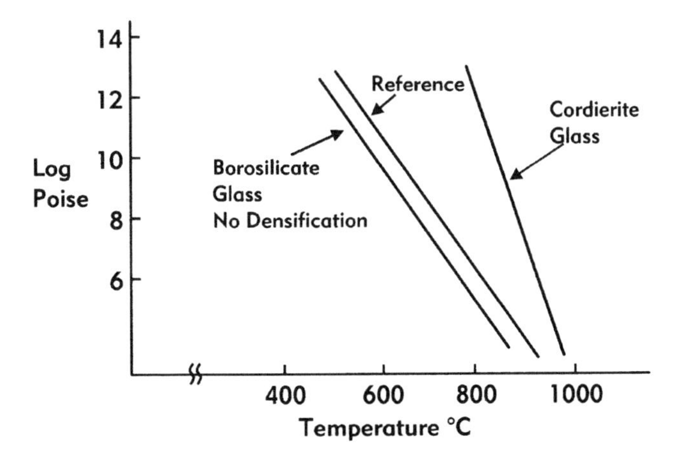

Figure 9-43. Viscosity–Temperature Relation for Borosilicate and Cordierite Glasses

viscosity [72,73], sintering [74], and crystallization [75,76] show the effects of compositional modifications on final density.

The preferred powder particle size is about 3 μm in diameter, and because of surface nucleation, the final grain size is also typically less than 3 μm. The density of this material is more than 99% of theoretical value, and the pores generally measure less than 5 μm. The microstructure and x-ray diffraction data indicate the glass–ceramic to be approximately 95% alpha-cordierite, and the remainder to be enstatite. The properties of this material are shown in Table 9-29.

To meet the stringent requirements of the glass–ceramic substrate, a modification of the binder system and the continuous casting process used previously for the alumina system was developed. It is capable of producing greensheets with the following characteristics:

- Stable dimensional properties after punching and handling
- Microporosity levels which enable sufficient deformability to enclose screened lines and permeability to allow escape of gases during the sintering process
- Wetting by copper paste solvents without dimensional change
- Complete removal of organics in the sintering cycle

Glass powder is formed by grinding the glass in a high-alumina ball mill to an average particle size of about 3.0 μm. The particle size, particle-size distribution, and surface area of the glass prior to incorporation into greensheets are critical control parameters in the fabrication of reproducible greensheets. [77]

The greensheet obtains its stability, porosity, and chemical inertness from a binder system consisting of polyvinylbutyral (PVB), dipropylene-glycol dibenzoate (BPGDB) plasticizer, and a binary solvent of methanol and methylisobutyl ketone [78]. This formulation provides a stable suspension of the glass particles in the slurry with a viscosity suitable for continuous casting of high-quality greensheets having the required me-

Table 9-29. Properties of Cordierite Glass–Ceramic

Property	Value
Dielectric constant	5.0 at 1 MHz
Thermal-expansion coefficient	30×10^{-7}/°C (25–200°C)
Mechanical strength	210 MPa (30 kpsi)
Maximum process temperature	950°C

chanical strength and dimensional stability. The resulting greensheet has about 90% glass powder and 10% binder (PVB/DPGDB) by weight.

The polyvinyl butyral binder and solvent system provides excellent slurry properties. It has a viscosity of about 1500 CP and has a behavior which is pseudosplastic at lower shear rates and Newtonian at higher shear rates typically experienced in continuous casting. These properties permit precise control of the slurry in the caster, resulting in a precise greensheet thickness necessary for the dimensional control of the sintered substrate.

9.7.1.2 Copper Powder and Paste Technology

IBM chose copper as the most desirable metal for cofiring with glass–ceramic because of its high electrical conductivity and low cost as indicated in Table 9-30, compared to other metals. This metal, however, presents major difficulties in forming suitable powders and pastes to fabricate 63-layer glass–ceramic/copper substrates, as discussed previously [69].

A powder suitable for screening 90- and 100-μm vias on 225-m centers as well as fine lines (75 m) must have a small particle size. This requires that the powder be deagglomerated so that it can be screened through a metal mask without clogging the holes in the mask. Unlike many other metal powders, copper is difficult to deagglomerate during paste manufacture. Copper particles tend to deform rather than deagglomerate. When copper thick-film technology originated, no known powder or paste met the requirements of the glass–ceramic/copper substrate described earlier.

As expected, most fine, commercially available copper powders sinter between 600°C and 780°C, whereas the glass begins to densify at

Table 9-30. Package Metallurgy Options

Property/Metal	Mo	Cu	Ag	Ag/Pd	Au
Bulk resistivity (μΩ cm)	5.2	1.72	1.59	7.0	2.35
Thick-film Resistivity (μΩ cm)	10–15	4–6	3.7	2.0	3.75
Sintering temp. (°C)	1500	950	950	850	950
Atmosphere	←Reducing→		←Air→		
TCE X ppm/°C	60	180	170	130	140
Concerns	Sintering atmosphere control		Electromigration		Cost

about 800°C and completes by about 860°C. This mismatch, presented in Figure 9-44, presents major problems in the fabrication of multilayer glass–ceramic substrates, as discussed elsewhere [69].

As stated elsewhere [79], in fabricating the 63-layer glass–ceramic substrate, the copper powder utilized has been suitably modified [80,81] and cosintered with glass.

The interface between the copper and the glass–ceramic was studied by means of a number of material and process options. The bonding methods investigated were mechanical, chemical, and a combination of the two. Compatible glasses and glass–ceramics were used for mechanical interfacial bonding, whereas chemical bonding was achieved by incorporation of suitable oxides in the paste. With both bonding methods, a controlled atmosphere was utilized during the sintering process to enhance the interfacial bonding. The details of these technologies are considered proprietary.

9.7.1.2.1 Copper Thick Film

The IBM molybdenum thick-film (paste) technology used to personalize the alumina TCM as well as thick film used by others are well established as discussed in the alumina section, 9.5. The personalization of the new glass–ceramic/copper package has a higher density with decreased grid spacing as indicated in Table 9-31, requiring significant improvement in the paste technology. For example, via diameters are 90 and 100 μm on 225-μm spacing for glass–ceramic, compared to via diameters of 140 and 150 μm on approximately 300-μm spacing for

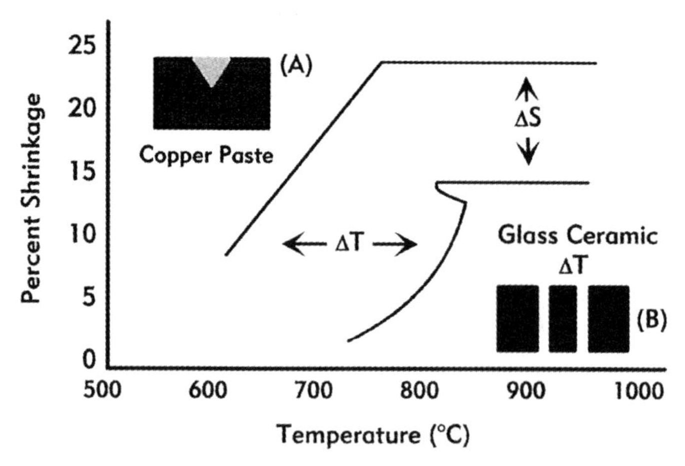

Figure 9-44. Shrinkage Mismatch Between Cu and Glass–Ceramic

Table 9-31. Comparison of Feature Parameters for Alumina and Glass–Ceramic Substrates

Technology	Alumina MLC	Glass–Ceramic MLC
Linewidth (μm)	116	75
Line spacing (μm)	300	225
Wiring length (μm)	130	400
Smallest via diameter (μm)	140	90
Largest via diameter (μm)	150	100
Via spacing (μm)	300	225
Vias per layer	33,000	78,500
Vias per substrate	350,000	2,000,000

original alumina. In addition, more than 78,500 vias are screened simultaneously with lines on each greensheet. These vias extend through the greensheet and are filled simultaneously as the lines are screened.

To form these densely wired structures, paste interaction with the greensheet must be considered. The interaction between the solvent in the paste and the binder of the sheet must be small but not zero. Control of this interaction is essential for dimensional stability of the sheets after screening, thus ensuring proper alignment of the vias in as many as 63 discrete layers during lamination. Solid content, particle-size distribution, rheology stability, and cross-sectional area are essential parameters for reproducible pastes.

A significant challenge in screening was to achieve good leveling of the screened features without spreading of the screened geometry. This is especially important for the glass–ceramic substrate, where the requirement for high electrical conductivity makes it necessary for the paste to have high solid content. The choice of paste solvent (or vehicle system) was critical, as its interaction with greensheets places a limit on the amount of leveling that can occur. A minimal amount of surfactant, combined with an appropriate amount of thixotrope, resulted in a paste that sets up adequately at the edges of the screened features while maintaining sufficient fluidity in the body of the paste to provide excellent leveling, thus producing copper features with a desirable cross-sectional area.

Extensive experimentation was carried out in order to understand the trade-off among the various properties required. Vehicle systems containing various polar and nonpolar solvent such as acetates, oils, esters, and glycols were considered together with polymer systems containing compatible functional groups. Screening process optimization was carried out by means of statistical studies to simultaneously adjust three major parameters as a function of paste composition: line-height variation, line-width variation, and greensheet dimensional control.

In-line controls such as grind gauge for the dispersion of particles, viscosity, and theological measurements were instituted.

The resultant thick film is a balanced blend of pseudoplasticity, thixotropy, and yield strength. A typical thick-film composition includes 80–90 wt.% copper and 20–10 wt.% paste organics and solvents. This thick film is suitable for a wide variety of screened layers (e.g., signal, redistribution, and vias only), which in the past required many different formulations.

9.7.1.3 Personalization, Stacking, and Lamination Technology

After the greensheet is cast, dried, stripped from the carrier and blanked into squares, it is "personalized" in one screening operation. A specific wiring is deposited and a through-via pattern obtained by punching vias through the sheets and filling them. Subsequently, the appropriate individual layers are aligned one on top of another and then laminated together. For the alumina-based thermal-conduction module, the vias were formed by means of mechanical punches, and the screening was done with various formulations of molybdenum metallurgical paste. Although, in principle, the same procedure was used for the glass–ceramic program, several factors presented major challenges with respect to the personalization of the latter system. These included the chemistry of new materials, and, most importantly, the dramatically increased density of features (both vias and lines). Table 9-31 compares some feature parameters for the two systems.

9.7.1.3.1 Punching

The change in materials from alumina to glass–ceramic presented interesting challenges with respect to the personalization of the greensheets. It is well known that the stresses induced in a greensheet during punching cause macroscopic deformations in the sheet [82,83]. All other factors (organic binder, sheet thickness, and punching parameters) remaining constant; this deformation differed significantly in glass–ceramic and alumina greensheets.

The magnitude and causes of distortion influenced by a number of materials and process parameters were investigated. In particular, a comprehensive study of both mechanical and chemical properties of the system was performed.

One mechanical study involved the effect on via location error of the thickness of the greensheet being punched and the difference in diameter between the punch and its associated die bushing, as illustrated in Figure

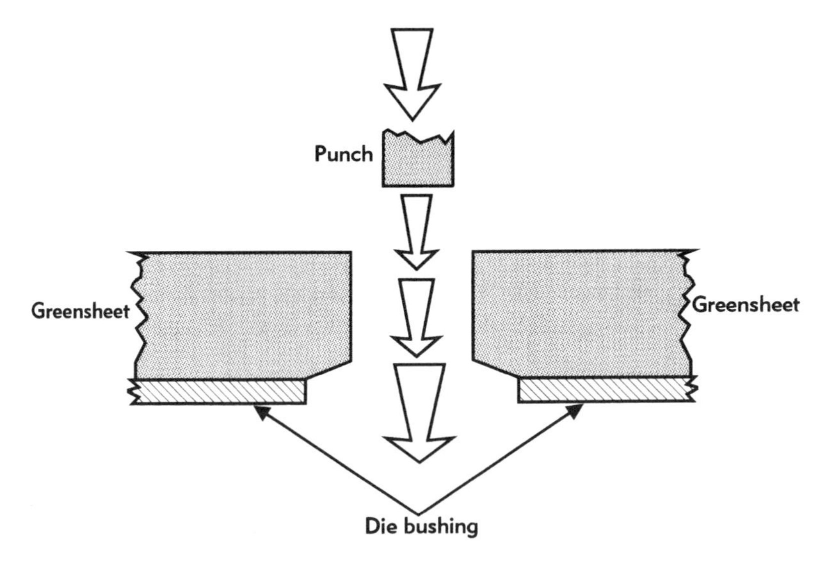

Figure 9-45. Schematic of Greensheet Punching

9-45. Figure 9-46 shows a figure of merit for via location error as a function of one of the controllable parameters, the punch-to-die clearance. It can be seen that if the sheet thickness is kept low enough, the sheet is relatively insensitive to other factors. Beyond a certain critical sheet thickness, relatively small changes in punch-to-die clearance have a large effect on the eventual distortion of the punched sheet. On the other hand,

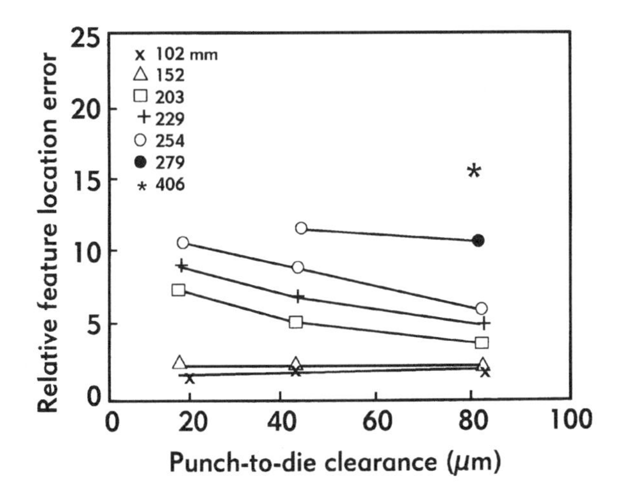

Figure 9-46. Relative Greensheet Feature Location Error as Function of Punch-to-Die Clearance

for any given set of materials and sheet thicknesses, an increase in the clearance between the punch and its associated die bushing results in a decrease in the distortion of the punched sheet.

A fundamental reason for the difference between the alumina and glass–ceramic sheets during punching was found to be the interfacial chemistry between the inorganic components of the greensheets (glass versus alumina) and the organic binder, as discussed in Refs. 84 and 85. This effect was studied utilizing materials with a known degree of surface acidity/basicity (measured by their isoelectric point). There was a direct correlation between feature location accuracy and the known degree of acidity/basicity.

9.7.1.3.2 Screening

A new screening process was developed for the copper paste because it was drastically different in composition and rheology from that used on alumina sheets. In the latter system, the pastes were customized with respect to their solvent and binder materials for individual patterns. For the glass–ceramic system, a single materials set for solvent and binder was used. This required the use of a system which was tailored for limited reaction with the greensheet. This provided sufficient adhesion of the screening paste to the greensheet without causing significant dimensional instability.

The properties of the conductive metal within the paste were radically different. Molybdenum particles have high hardness and a platelet shape, whereas copper particles are malleable and more nearly spherical. The change in density and surface area required extensive cooperation between the screening and the paste formulation development teams in order to fine-tune the rheology of different paste formulations and the screening parameters (pressure, wipe speed, etc.) for the individual layers. Major developments in the fabrication of screening masks were another key ingredient in the final success of the actual screening operation in the manufacture of glass–ceramic substrates.

9.7.1.3.3 Inspection

Automated inspection of screened greensheets ensures that via fill and line dimensions meet specifications. Advances in this technology were required because of the reduction in feature size for glass-ceramic with copper thick film. This operation is critical to achieving high electrical test yields after sintering, because a single defective layer, if not rejected, can cause the loss of the 62 other defect-free layers which combined make up a substrate.

9.7.1.3.4 Stacking and Lamination

During stacking and lamination, the individual sheets must be stacked one on top of another; they must be aligned so that any additional distortion is kept to a minimum during this process. The registration of all the sheets in the stack must be so precise that good electrical contact is made from vias in one layer to vias in the next layer for every one of the nearly two million vias in the laminate. Even after this is achieved, the stack must survive the lamination process, in which the stack is heated (under pressure) to a temperature just above the glass transition temperature of the binder in the sheet. Aside from the tooling challenge of building a press in which the platens have very tight tolerance placed on their flatness, parallelism, and temperature uniformity, a delicate balance had to be struck between conditions of temperature and pressure which would permit enough flow to allow for the consolidation of the individual sheets into a single unit, but were not so severe as to introduce additional distortion due to lateral flow.

9.7.1.4 Glass–Ceramic/Copper Sintering Technology

Sintering a multilayer ceramic substrate requires cofiring a composite matrix of organics, metal powders, and ceramic and glass powders into a complex network of electrically conductive wiring embedded in a dense, insulating ceramic body. The glass–ceramic/copper substrate technology presented the following challenges to co-sintering:

- Low-temperature organics removal (<800°C), without copper oxidation and low-temperature densification (<1000°C)
- Crystallization of glass to form a high-strength ceramic
- Densification of copper powder to high-conductivity copper
- Copper/ceramic interface integrity
- Improved dimensional control to satisfy requirements for via alignment and for thin-film wiring

9.7.1.4.1 Binder Removal and Carbon Oxidation

Various approaches were evaluated to reduce the residual carbon content to the desired low levels. These investigations included polymers that pyrolyze primarily into gaseous species ("unzip"), atmospheric firing cycles for oxidation of carbon and copper, and subsequent reduction of copper oxide to copper and catalysts to aid the binder burn-off and variations in compositions of glass as well as glass added to ceramics. As illustrated in Figure 9-43, the binder burn-off requirement based on viscosity of glass have led to the composition selected. Differences in thermal degradation were noted for the free binder and the binder absorbed on a glassy surface. When the binder was mixed with powdered glass to simulate tape casting, all candidates left excessive residual carbon in the glass. This occurred because the surface of the glass was chemically active and

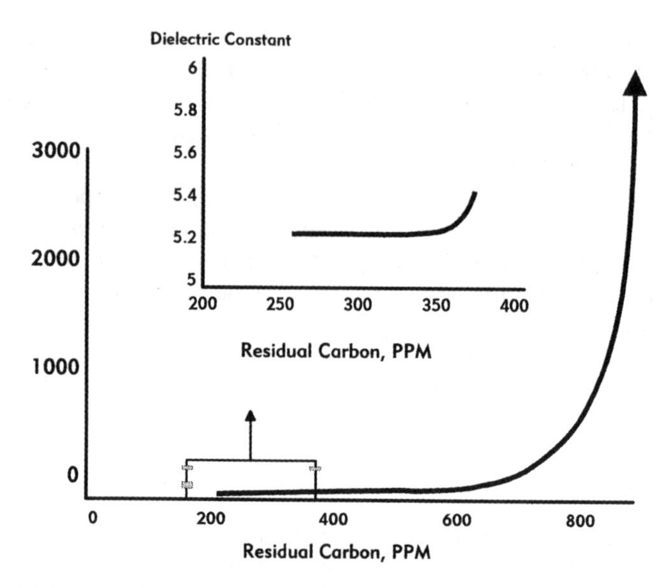

Figure 9-47. Dielectric Constant as a Function of Residual Carbon in Glass–Ceramic. (Courtesy of IBM.)

bonded with the polymer. The result of this residual carbon as affecting the dielectric constant is illustrated in Figure 9-47.

The challenges in cofiring copper paste with glass greensheet to form a 63-layer structure of glass–ceramic/copper substrate can be overcome by the use of steam and hydrogen atmospheres [69]. Figure 9-48 indicates

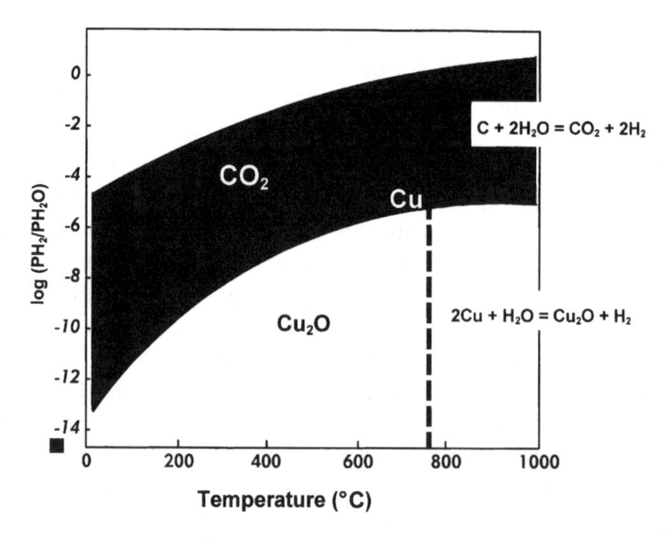

Figure 9-48. Thermodynamics of Cu–O–C System

the P_{H_2}/P_{H_2O} ratio necessary at 800°C to oxidize carbon and reduce copper oxide according to the following chemic reactions:

$$C + H_2O \rightarrow CO_2 + H_2$$
$$Cu_2O + H_2 \rightarrow 2Cu + H_2O$$

Whereas the oxygen partical pressure required to drive the above reactions can be achieved by a number of ambients [86], the only kinetically acceptable process for forming the 63-layer structure in a few hours is the one based on the use of steam.

9.7.1.4.2 Densification

Densification of the glass–ceramic/copper substrate depends on the composition of the glass–ceramic and the particle-size distributions of the glass and metal powers. Greensheet laminate properties such as the ratio of organics to inorganics and laminate density play a role in the control of shrinkage.

The objective of the sintering process is to define a temperature profile, atmospheric conditions, and fixtures to achieve a dense, high-strength ceramic and dense, highly conductive copper wiring. Copper/ceramic interface integrity also depends on materials properties and sintering atmosphere control. The glass–ceramic/copper sintering process development grew out of experience gained in developing the alumina–molybdenum sintering technology and required improvements in tools and processes. Steam furnaces were developed and designed to sinter glass–ceramic/copper substrates, which are believed to be an industry first. The furnaces required significant advances in mass flow control, atmosphere injection systems, and reaction chamber materials. Unique fixtures helped achieve the required mass transport for carbon oxidation and removal which affects materials densification and substrate characteristics. The results of this development effort were illustrated previously in Figure 9-2. This substrate sets new standards for multilayer ceramic packaging, as evidenced by the final properties presented in Table 9-32 [69]. The overall process developed for the glass–ceramic substrate is indicated in Figure 9-49.

9.7.1.4.3 Dimensional Control

Dimensional control during the densification segment of the glass–ceramic/copper sintering is a function of processing before and during sintering [87,88]. Variations in greensheet properties, paste properties, paste loading, and lamination conditions have a significant impact on dimensional control and must be carefully controlled. Improved furnace materials, fixtures, and methods for temperature and atmosphere control

Table 9-32. Sintered Glass–Ceramic/Copper Substrate Properties

Fired ceramic density (g/cm^3) (99.5%)	2.66
Shrinkage tolerance (%)	±0.10
Camber or sintered flatness (μm/cm)	2.3
Ceramic flexural strength (MPa)	210
Copper resistivity (μΩ cm)	3.5
Dielectric constant	5.0
Thermal expansion coefficient (10^{-7}°C)	30
Number of layers	63
Wiring density (cm/cm^2)	800

Source: Courtesy of IBM.

have been developed for sintering glass–ceramic/copper substrates. These advances in sintering technology have resulted in significant improvements beyond state-of-the-art alumina–molybdenum sintering with respect to dimensional control.

With respect to requirements for dimensional control in a sintered substrate, the goal is to have all surface features located exactly within specifications relative to one another and the edges of the substrate. In a multilayer ceramic substrate, deviations from ideal locations are allowed within specified limits. For example, a perfect substrate, where metal features have no deviations from ideal locations, is said to have 0% nominal movement. An MLC substrate tends to deviate from ideal locations by expansion or contraction around the desired value. On the other hand, if

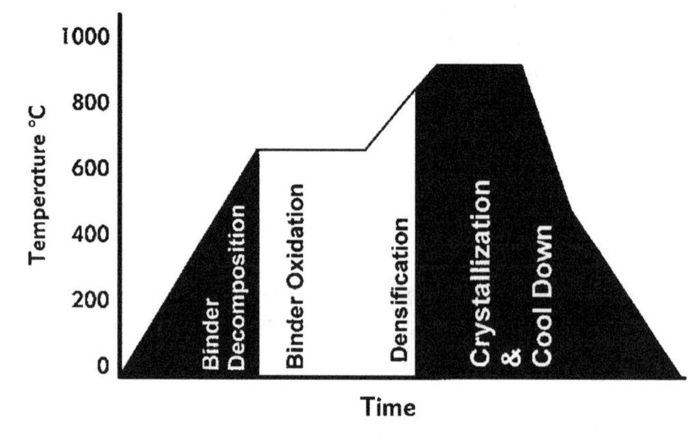

Figure 9-49. Overall Glass-Ceramic/Copper Cosintering Cycle

all the surface features expand or contract linearly, there is zero radial feature displacement or distortion. Radial feature displacement is measured in microns using an optical measurement system.

Figure 9-50 compares the dimensional control in two populations of sintered substrates. The larger area represents alumina–molybdenum substrates, and the smaller area represents glass–ceramic/copper substrates. Perfect dimensional control with ideal linearity (zero distortion) is represented by the two dark lines forming a "V." If, for example, a sintered substrate had a global expansion of +0.10% with ideal linearity, the radial feature displacement would be on top of the right-hand dark line just above the +0.10% mark. Therefore, the goal in dimensional control is twofold: (a) It should be close to zero nominal expansion, and (b) when deviations occur, the expansion should be linear.

As the data in Figure 9-50 illustrate, dimensional control in the glass–ceramic/copper population is significantly better than in the alumina–molybdenum population. Dimensional control in the glass–ceramic/copper population deviates less from the ideal dimensions and is closer to perfect linearity.

9.7.1.5 Substrate Machining and Finishing Technology

As described in the previous sections, the glass–ceramic/copper substrate is a culmination of extensive materials and process development. However, the sintered substrate, advanced as it is at this stage, is not yet complete. The thin-film redistribution wiring, top-surface terminal metallurgy, and back-side I/O pins must still be applied to the top and

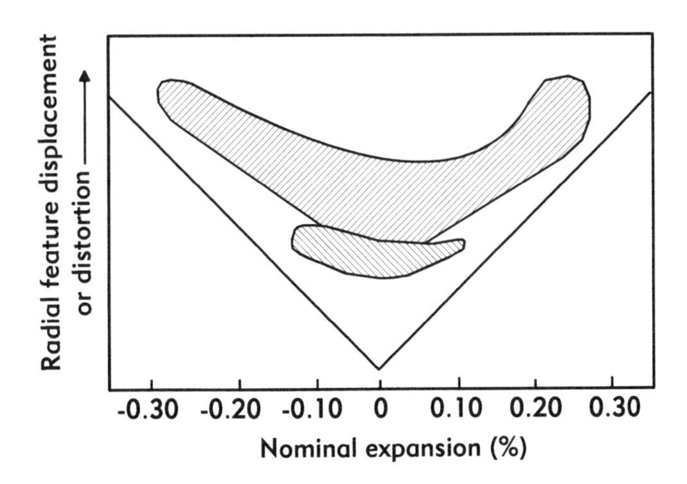

Figure 9-50. Dimensional Control of Alumina TCM Compared to Glass–Ceramic TCM

bottom surfaces and preparation made for mounting and finishing processes prepare the substrate for these final operations.

The application of thin films onto the glass–ceramic substrate generates new requirements not encountered in the fabrication of earlier substrates. Specific requirements pertain to the substrate dimensions, flatness, surface roughness, and the condition of the metal vias exiting the surface. Additionally, a flange must be ground into the periphery of the substrate for module encapsulation. These requirements are met using a series of operations summarized in the process flow given in Table 9.33.

9.7.1.5.1 Postsinter Sizing

Precision diamond-saw sizing is the first machining operation for the sintered substrate. This operation precisely sizes the substrate to 127.5 mm. In this operation, the substrate is cut square, the corners are truncated, and a 45° bevel is machined on all edges. The truncating of the corners and beveling of the edges help to minimize chipping and damage to edges during subsequent operations.

The sizing process uses resin-bonded diamond blades on a double arbor wet saw. A specially formulated coolant is used to minimize the buildup of heat and prevent mechanical damage to the substrate. The saw sizing operation is completely automatic, first cutting the sides of the substrate and then truncating the corners. Precise dimensional control as well as pattern centrality are important for subsequent processing. This is achieved by the use of an integrated optical alignment technique that targets pattern alignment fiducials on the substrate and transfers these alignment data to the saw. After the cutting operations are complete, the newly formed edges are beveled on a deburring machine.

Table 9-33. Details of Substrate Machining and Finishing Processes

Process	Process elements	Results
Sizing	Diamond sawing	Size = 127.5 mm
Planarization	Abrasive slurry on lapping table	Flatness = 2.3 m/m Vias above substrate surface
Flange grind	Abrasive wheel	Surface roughness = 4000 Å Thickness/parallelism = 50 mm
Ceramic polish	Submicron abrasive slurry with polishing pad	Surface roughness = 450 Å Vias coplanar with surface
Metal polish	Submicron abrasive slurry with polishing pad	Reflective vias

9.7.1.5.2 Planarization

The accurate application of thin films to the substrate requires that it be planarized to produce flat surfaces. Further, because the substrate is electrically tested after planarization, the process must leave the copper vias in the surface in order to ensure probe contact during electrical test.

A free-abrasive lapping process is used to planarize the substrate. The process utilizes an abrasive slurry and a lapping table to planarize both the top and bottom surfaces of the substrate. Slurry consistency and its distribution to the part, as well as lapping table speed and the pressure applied to the part, are important in achieving the desired flatness. The machining mechanics of a free-abrasive lapping operation result in the enhanced removal of the ceramic over the more ductile copper [89]. Thus, after planarization, the copper vias protrude above the ceramic surface, as required for electrical testing.

Measurement of the planarized surface is critical to the assessment of process integrity. An automated laser scanning interferometer not only measures the overall flatness of the part but also gives a contour map of the surface, showing the location and the degree of deviation from ideal flatness.

9.7.1.5.3 Seal Flange Grind

Once a substrate has been planarized and electrically tested, those parts passing an electrical test have a seal flange ground into their periphery. The flange is used to mount and seal the top surface of the substrate into the module hardware.

Key elements of the flange are its surface finish and its parallelism. The top surface of the flange must be smooth enough to permit a hermetic seal to the module. The flange surfaces must be parallel, so that an even distribution of pressure to the module seal is obtained without bending moments on the ceramic.

A ground seal flange is not new to ceramic technology; alumina TCM substrates have had this feature for several years [90]. However, machining such a flange into the glass–ceramic/copper substrate required the development of a new process, largely because of the differences in the physical properties between alumina and the crystallized glass–ceramic.

The seal flange is machined into the substrate by means of a cam grinder using a specially shaped, abrasive wheel that grinds both the top and bottom flange surfaces simultaneously. A combination coolant/lubricant is used in conjunction with the process. This coolant, along with the speed of the grinding wheel and the feed rate of the substrate, is critical in preventing damage to the ceramic during processing. Considerable effort has also gone into designing the fixturing used to hold the substrate during grinding. The fixtures must be able to resist the bending and vibrational modes placed on the substrate during grinding while maintaining the dimensional accuracy of the resultant flange. After the flange has been machined, the same abrasive wheel is used to bevel the flange edges to prevent chipping of the ceramic. By means of this process, the flange thickness and parallelism are controlled to within 50 μm, and the surface roughness is maintained at 4500 Å.

9.7.1.5.4 Polishing

The complete surface finishing of a glass–ceramic/copper substrate requires two polishing operations. The first operation, the ceramic polish, produces a smooth surface suitable for the application of thin-film metallization. The second operation, the metal polish, brightens the surface of the copper vias to enable them to be seen by the automatic mapping tools used in thin-film processing. These polishing operations are used on both the top and bottom surfaces of the substrate.

The ceramic polishing operation was developed to provide a smooth surface, maximize surface flatness, and achieve copper vias which are coplanar with the ceramic surface. The ceramic polish uses a submicron abrasive slurry and polishing pad to achieve these goals. Process parameters such as applied pressure, table speed, substrate rotation speed, slurry concentration, and slurry delivery rate are also critical to the quality of the polish. Their relative importance in the process was determined using a robust experimental design methodology [91]. Considerable effort has also gone into developing fixturing for the substrate that ensures a nearly uniform polishing rate across the entire substrate surface. The ceramic polish routinely yields substrates with a surface roughness of 450 Å, minimal change to the surface flatness achieved during the lapping process, and copper vias coplanar with the ceramic surface.

The metal polish operation also uses not only a pad but also a still finer submicron abrasive slurry. The improved reflectivity of the copper achieved by this polish allows for precise optical mapping of via location for subsequent thin-film processing.

9.7.2 Fujitsu's Glass–Ceramic Substrate

Fujitsu developed, manufactured and applied to its mainframe application a very large (245 mm) glass + ceramic substrate with 61 layers of copper wiring [92]. It explored two glass + ceramic systems, one in cordierite + glass and the other in alumina + glass. Alumina–glass, for the reasons discussed below, became a reality.

9.7.2.1 Glass–Cordierite System

Cordierite was chosen as the original ceramic material because of its very low dielectric constant and its low thermal expansion near that of silicon.

Glass powder and ceramic powder were milled with binder and solvent for 20 h. The glass used in this system and in the alumina system is a borosilicate glass. The slurry was cast into 500-μm-thick greensheets using the doctor-blade method. These greensheets were then laminated, giving a thickness of 1 mm, then fired in a N_2 atmosphere at 1000°C for 5 h. The heating rate was 200°C/h.

The glass–cordierite substrate so fabricated exhibited high TCE. The calculated TCE of the system is (2–3) × 10^{-6}/°C, which is slightly lower than that of silicon. The TCE actually measured for that system, however, is 17 × 10^{-6}/°C, or about seven times higher than the calculated value, as shown in Figure 9-51. The formation of cristobalite explains this high TCE.

9.7.2.2 Glass–Alumina System

The same tests as described above were performed on the glass–alumina system. No large difference was observed between the TCE values calculated and those measured for the glass–alumina system. The x-ray diffraction pattern of glass–alumina revealed no formation of cristobalite. The densification mechanism of the alumina–glass system can be explained as liquid-phase sintering. The alumina is uniformly dispersed in the glass matrix. Wettability between the glass and the ceramic powder is essential to densification. A proper amount of glass yields high density with high strength. An optimum firing point for the maximum densification also exists for any given glass-to-ceramic ratio. A combination suitable for co-firing with Cu or Au has an alumina content between 40 and 50

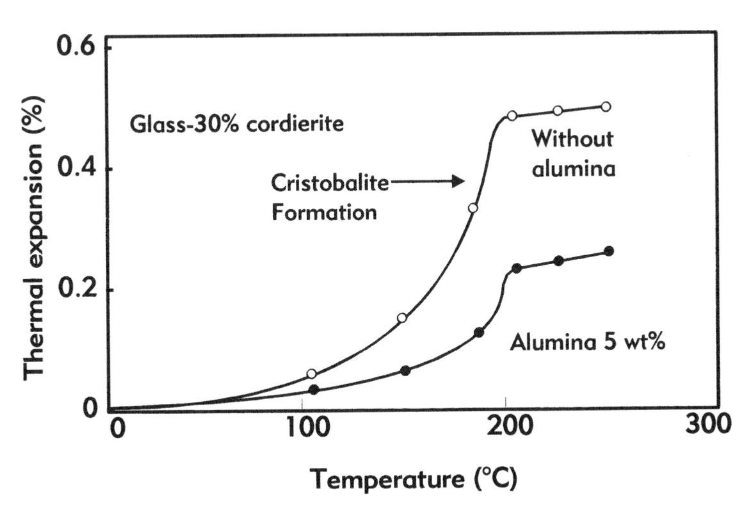

Figure 9-51. Thermal-Expansion Curve of Glass–Cordierite System and Alumina-Added Glass–Cordierite System

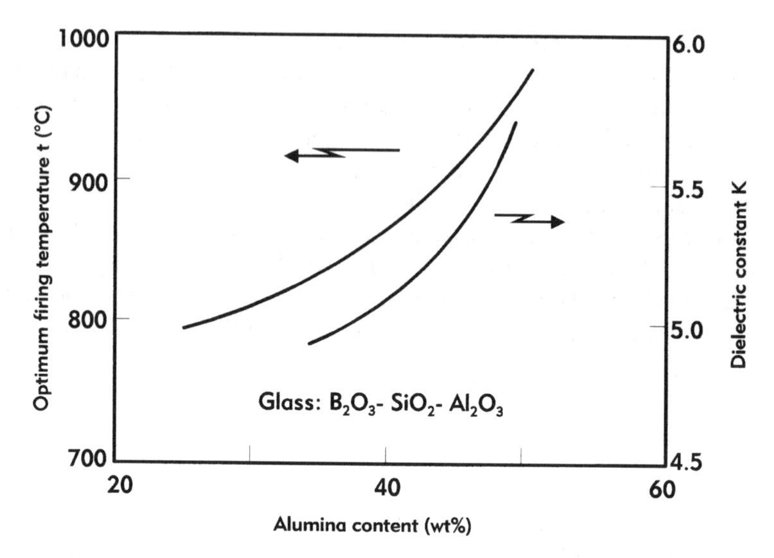

Figure 9-52. Changes in Dielectric Constant and Sintering Temperature of Glass–Alumina System

wt.% (see Fig. 9-52). The powder used for glass–alumina-composite materials is listed in Table 9-34.

To avoid carbonization of the binder, Fujitsu used a newly developed binder system that easily decomposes without oxygen when heated. The binder structure is shown in Figure 9-53. Most binders generally used in ceramic boards show side-chain reaction or random rupture, but the new binder shows depolymerization as illustrated in Figure 9-53 [6]. A 61-layer lamination of greensheets, however, makes the burning out of all binder very difficult, so Fujitsu also developed a new firing process which can decrease the carbon residue in a fired body.

Carbon residue exceeding 100 ppm affected the densification of the fired body as well as the flexural strength. A fired body containing less than 30 ppm of carbon residue approached 100% of theoretical density, but a specimen containing 1000 ppm shows 93% of theoretical density.

Table 9-34. Powder Properties for Glass–Alumina-Composite Material

Powder	Specific gravity (gm/cm^3)	Particle size (mm)	Specific Surface Area (m^2/g)
Alumina	2.5	4.4	2.8
Borosilicate glass	2.2	4.5	4.2

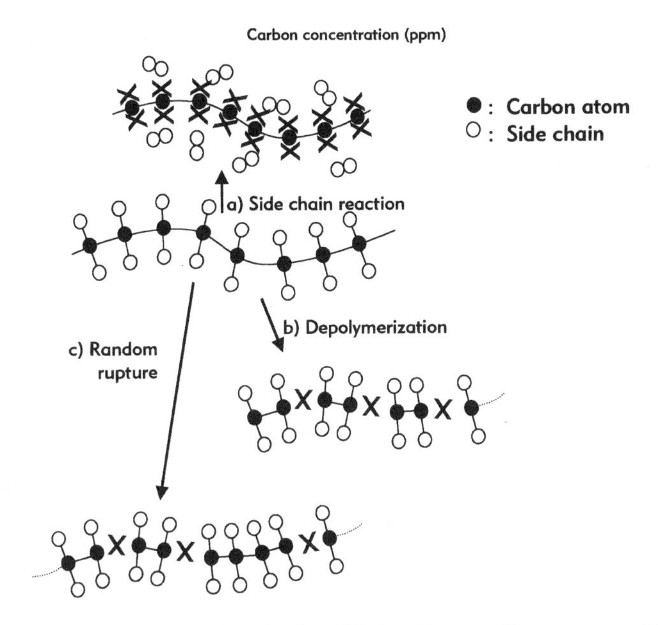

Figure 9-53. Novel Binder for Easier Binder Burn-off. (Courtesy of Fujitsu.)

Carbon also affects the breakdown voltage; a specimen with more than 100 ppm of carbon shows 0 kV/cm voltage breakdown, as illustrated in Figure 9-54. The proposed mechanism involves trapped carbon that is intermixed with the glass matrix, creating short circuits that cause breakdown between the patterns. The multilayer substrates processed, however, in the special fashion do not include more carbon than bodies fired in an air atmosphere.

9.7.2.3 Mainframe Application of Glass–Ceramic

Fujitsu applied the multilayer circuit board for the VP 2000 series supercomputer with total number of 61 layers, including 36 signal layers. About 40,000 signal patterns are formed on 1 layer, and the total wiring length of a signal pattern is about 1 km. The size of the substrate is 24.5 × 24.5 cm and 13 mm thick. The characteristic impedance of the signal pattern is controlled to 65 Ω, and the resistance of the copper pattern is only 100 mΩ/cm. Two signal layers are sandwiched between ground and voltage layers to decrease the cross-talk noise and deviations in characteristic impedance.

The properties of the multilayer glass–ceramic (MLG) substrate are listed in Table 9-35. The MLG made it possible to develop a high-density package that compares favorably with conventional organic printed-circuit boards previously used for large-scale, high-speed computer systems. The

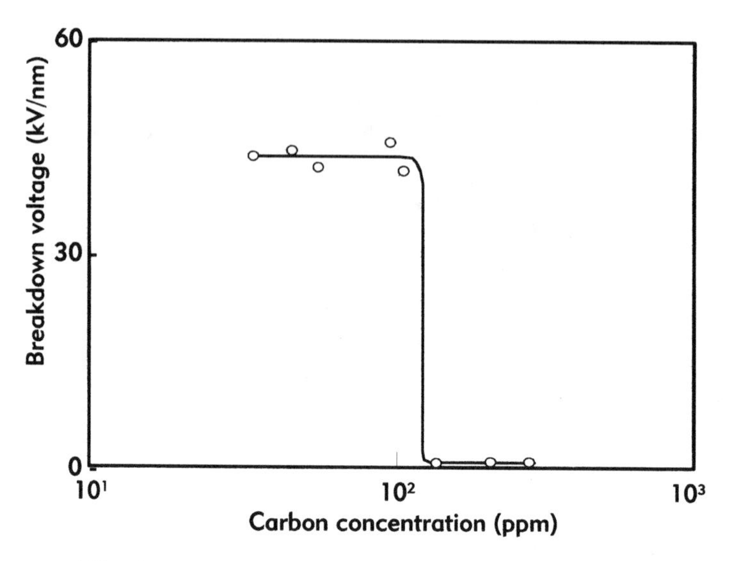

Figure 9-54. The Relation Between Carbon Concentration and Breakdown Voltage

gate density of the MLG used in the VP 2000 series is 10 times greater than that of the Fujitsu M-780 organic printed-circuit board.

9.7.3 NEC's Lead Borosilicate-Alumina Glass + Ceramic

NEC developed its glass–ceramic similar to Fujitsu's except for the nature of glass used [93]. The dielectric material used for the multilayer glass–ceramic (MGC) package is a mixture of alumina and lead borosilicate glass. The composition of the mixture is 55 wt.% alumina and 45

Table 9-35. Properties of 61-Layer VP 2000 Circuit Board

MLG Characteristic	Specifications
Conductor: copper (mΩ/cm)	100 (95 μm width)
Wiring length (m)	1000
Signal layers	36 (40,000 wiring patterns)
Characteristic impedance (Ω)	65
Dielectric constant (at 1 MHz)	5.7
Grid (mm)	0.45

Source: Courtesy of Fujitsu.

wt.% lead borosilicate glass. This material can be sintered at as low as about 900°C. The dielectric constant for this material is 7.8, less than that for alumina [93]. The thermal-expansion coefficient achieved is 42 $\times 10^{-7}$/°C, close to that of silicon chip. When sintered at about 900°C, crystallization takes place in the sintered ceramic body as a result of reaction between glass and ceramic. This reaction leads to high mechanical strength for the multilayer glass–ceramic (MGC) substrate.

Crystallization takes place during sintering around 800°C. If coarse glass particles are used, crystallization within the glass particles takes place, forming crystobalite. However, if fine particles of glass are used, glass reacts with alumina to form anorthite. The resulting substrate has high mechanical strength, a low thermal-expansion coefficient, and a good dielectric constant, as shown in Table 9-36.

This substrate has high flexural strength—more than 300 MPa. Camber is extremely small, even though the substrate size is 225 mm. The dielectric constant is 7.8, and the dissipation factor is less than 0.3% at 1 MHz.

9.7.4 Panasonic's Glass–Ceramic

In contrast to IBM's, Fujitsu's and NEC's glass–ceramics described above, Panasonic developed its own low-temperature cofired technology (LTCC) and began to apply for such consumer products as VCRs at its plant at Saijo. The required low-cost for these type of products necessitated low-cost metallization, which was achieved by simple air firing using Ag–Pd conductors. Figure 9-55 shows the structure, including the embedded capacitors and surface resistors that Panasonic was able to achieve.

Panasonic also designed and developed a low-cost process technol-

Table 9-36. Typical MGC Substrate Properties

Material Characteristic	Property
Flexural strength (MPa)	300
Dielectric constant at 1 MHz	7.8
Dissipation factor at 1 MHz	<0.3%
Thermal-expansion coefficient (°C^{-1})	42×10^{-7}
Thermal conductivity (W/mK)	3.6
Linear shrinkage (%)	13.0
Shrinkage tolerance (%)	±<0.2%

Source: Courtesy of NEC.

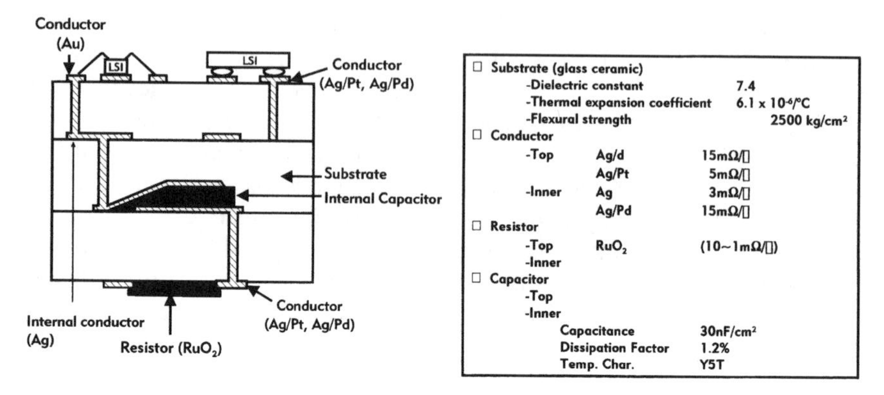

☐ Substrate (glass ceramic)			
	-Dielectric constant		7.4
	-Thermal expansion coefficient		6.1×10^{-6}/°C
	-Flexural strength		2500 kg/cm²
☐ Conductor			
	-Top	Ag/d	15mΩ/□
		Ag/Pt	5mΩ/□
	-Inner	Ag	3mΩ/□
		Ag/Pd	15mΩ/□
☐ Resistor			
	-Top	RuO_2	(10~1mΩ/□)
	-Inner		
☐ Capacitor			
	-Top		
	-Inner		
		Capacitance	30nF/cm²
		Dissipation Factor	1.2%
		Temp. Char.	Y5T

Figure 9-55. Consumer Electronic Ceramic Substrate. (Courtesy of Panasonic.)

ogy for cofiring with copper that is illustrated in Figure 9-56 [94]. This process involves forming thick films with CuO and cofiring in air to initially remove organics from greensheets and pastes and then reducing the oxide in forming gas and finally forming a good bond between the glass that flows from ceramic walls and slightly oxidized copper in the via that is formed in the N_2 atmosphere used in sintering. Panasonic expects to integrate capacitors and resistors into this substrate as well. Figure 9-57 illustrates the glass–ceramic/copper with surface resistors currently practiced by Panasonic.

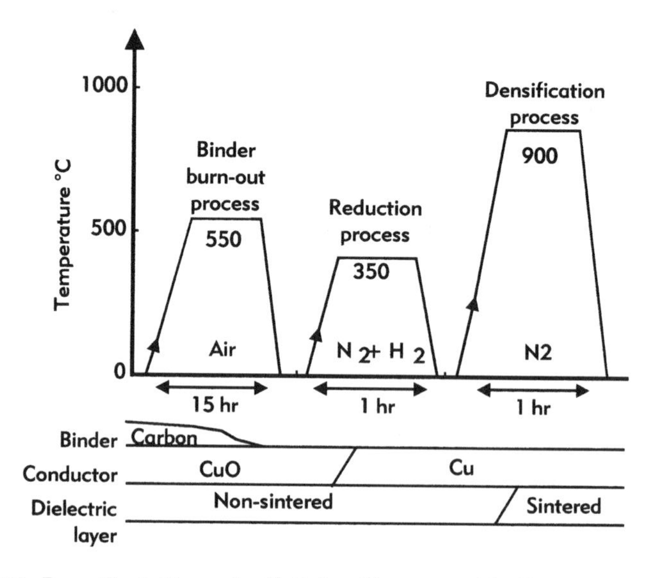

Figure 9-56. Low-Cost Ceramic Cofiring Process with Cu. (Courtesy of Panasonic.)

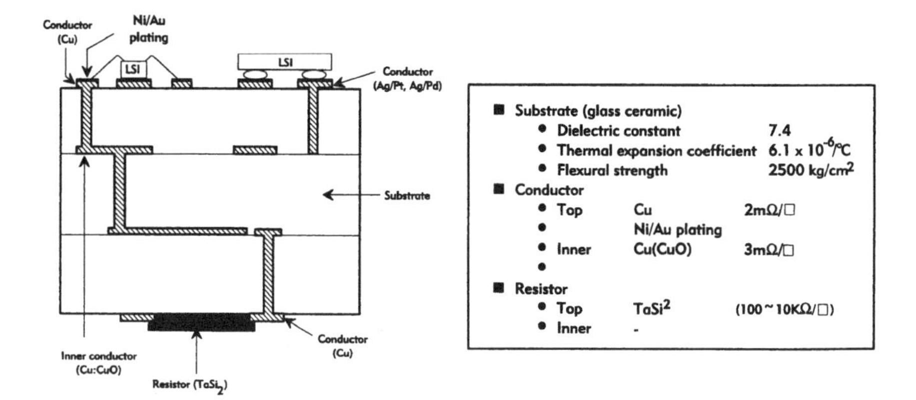

Figure 9-57. Consumer Electronic Ceramic Substrate with Cu. (Courtesy of Panasonic.)

9.7.5 David Sarnoff's Glass–Ceramic on Metal (LTCC-M)

A variation of glass–ceramic that is receiving great attention at David Sarnoff Research Center is that illustrated in Figure 9-58 which is accomplished by lamination of the desired number of layers on to a metal substrate and cofiring the entire substrate [95]. The substrate, obviously, must meet two major requirements. Its thermal-expansion coefficient

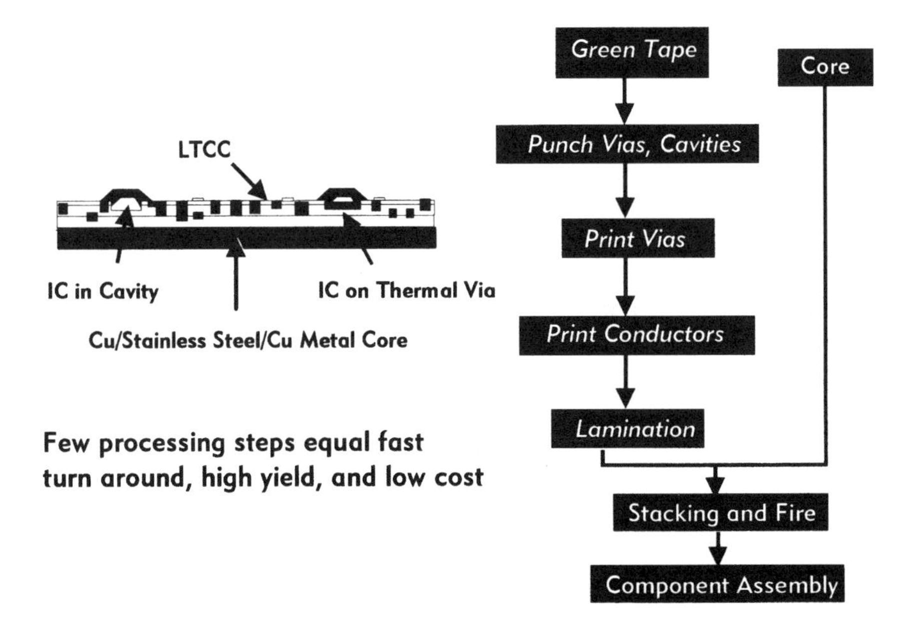

Figure 9-58. Low-Temperature Cofired Ceramic on Metal

should be equal to or slightly higher than the glass–ceramic so as to put the ceramic in compression. It should also form an excellent bond to the metal or alloy. A copper–stainless steel–copper-composite structure is proposed.

The process of fabricating the substrate is illustrated in Figure 9-59 and the advantages of fabricating this way as compared with conventional cofiring of ceramic are listed in Table 9-37. Among the most notable features is the zero shrinkage that is intrinsic to the process because the glass–ceramic is bonded to the metal core all through the process. The thermal conductivity of the substrate is enhanced greatly, and in those applications that require even higher dissipation, the chip can be back-bonded to the metal core and wirebonded to the substrate metal. Silver conductors that provide the very high conductivity and yet provide a fast firing cycle in air are proposed.

9.7.6 Other Ceramic Packages

Silicon nitride, boron nitride, diamond, and mixtures of carbides, nitrides, and oxides with appropriate sintering additives are expected to be explored for ceramic packaging. The addition of such low-dielectric-constant glasses as borosilicates or fused silica (with dielectric constant around 4.0) could be added to alumina, cordierite [93], and other ceramics of low dielectric constant to produce ceramic substrates of dielectric

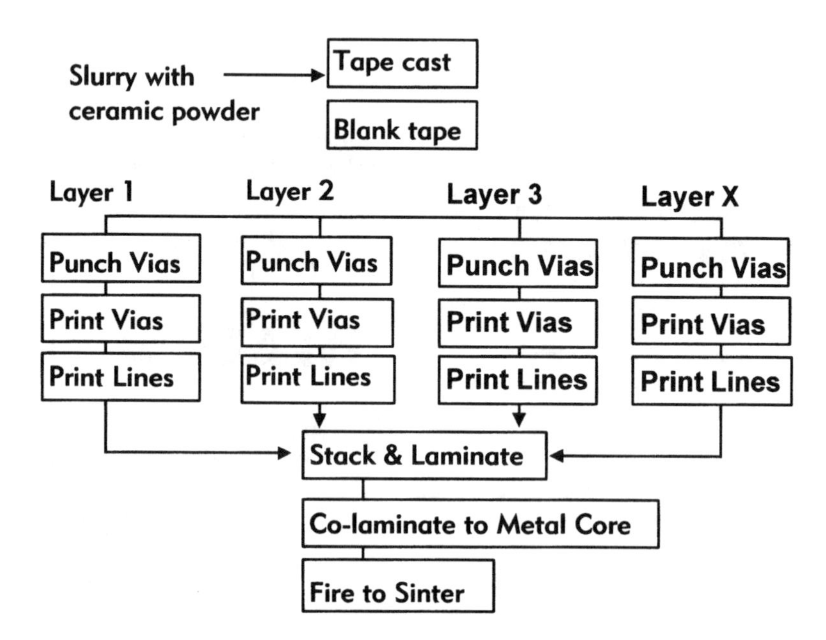

Figure 9-59. LTCC-M Process

Table 9-37. Technology Comparison

	PWB	MCM-L	MCM-D	LTCC	LTCC-M
Wiring density (cm/cm^2)	65	200	800	800	300–600
Thermal conductivity (% of Cu)	0.01	0.01	10	0.1	30
Buried passives	Resistors only Limited range Expensive	Resistors only Limited range Expensive	Resistors only Limited range Expensive	Resistors & capacitors Poor yields	Resistors & capacitors Wide range Good yields
Ruggedness	High	High	Medium	Medium	High
Large-format processing	Yes	Yes	Beginning	No	Yes
Component assembly flexibility	Fair	Fair	Poor	Fair	Good
Flip-chip capability	Good	Good	Good	Good	Good
TCE Match to Si/GaAs	Poor	Poor	Fair	Good	Good
Cost ($\$/in.^2$)	0.95	3–6	10–50	3–5	1–5

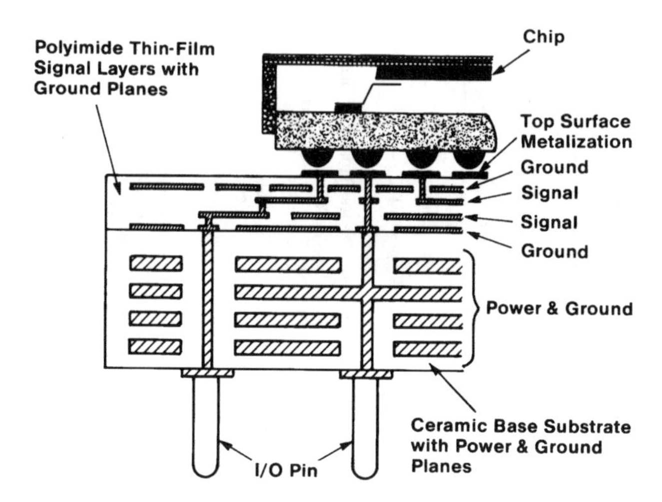

Figure 9-60. Hybrid Thin- and Thick-Film Package (42-Chip Module). (Courtesy of NEC Corp.)

constant around 4.0. Any further improvements can only come from the incorporation of porosity to achieve a dielectric constant around 1 or the use of polymers with a dielectric constant between 2.0 and 4.0. The net result would be composite ceramic substrates fabricated with dielectric constants of less than 4.0 [96]. Because the lowest dielectric constant that can be achieved is always with polymer materials and because most polymer materials have thermal-expansion coefficients much in excess of current and future device materials, hybrid packages of thin films of polymer layers on top of ceramic substrates are expected for very-high-performance single-chip and multichip applications. An example of such a hybrid package, capable of supporting 42 chips [97], by NEC Corporation is shown in Figure 9-60. It is expected that these kinds of packages are the wave of the future, as they use ceramics as building blocks to control thermal and mechanical properties as well as for power distribution, and

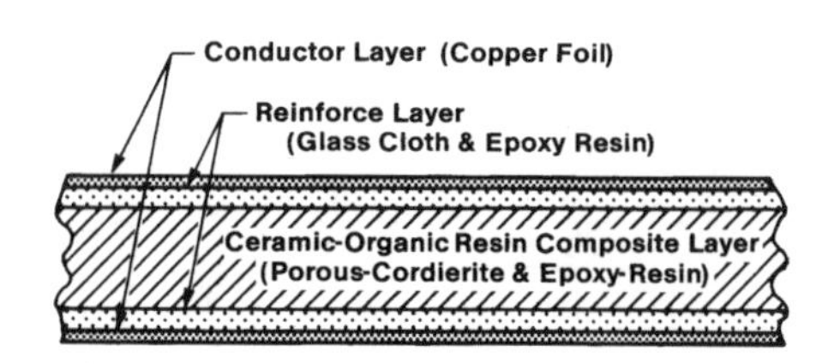

Figure 9-61. Package Structure of Epoxy-Filled Cordierite Ceramic. (Courtesy of Ibiden.)

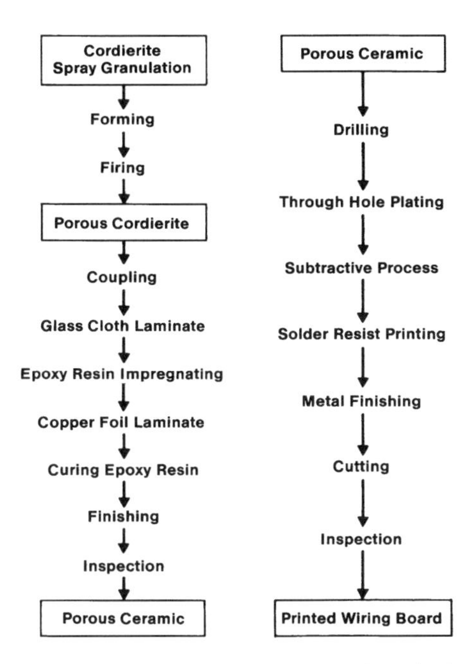

Figure 9-62. Process of Fabricating Epoxy-Filled Cordierite Ceramic Board. (Courtesy of Ibiden.)

polymers for their low dielectric constants as thin-film interchip wiring. In addition, polymer-filled ceramics will be developed to achieve the lowest dielectric constant. An example of this technology involving the fabrication of a porous cordierite-based substrate that is subsequently filled with epoxy, laminated with copper foil, and mechanically drilled to form plated-through via holes has already been reported [98], as shown in Figure 9-61. Such a substrate, fabricated with the processes shown in Figure 9-62 and the properties in Table 9-38, allows ceramics to be fabricated into large-area, low-cost, and high-performance substrates using board technologies. Such substrates could be developed with thermal-expansion coefficients matching silicon or GaAs devices.

9.8 CHIP ATTACHMENT AND THERMAL DISSIPATION OF CERAMIC SUBSTRATES

As discussed in Chapter 8, "Chip-to-Package Interconnections," all three chip-bonding technologies are used with ceramic packages. With solder connection (C4) technology, thermal conductivity of the ceramic is not as important. The thermal limitations of the package, therefore, are the same as for silicon, because the heat is removed from the back of the silicon chip. However, wirebonding is still the predominant technology

Table 9-38. Properties of Epoxy-Filled Ceramic in Comparison with the Other Materials

Property	Cordierite (30 vol.%)		SiC (30 vol.%)		96% Alumina	Epoxy–Glass
Reinforce layer glass cloth thickness	Nothing ×2	0.1 mm ×2	0.2 mm ×2	0.1 mm	—	—
Thickness (mm)	1.0	1.0	1.2	1.0	1.0	1.0
Bending strength (kg/mm^2)	12.0	18.0	19.0	35.2	35.0	45.0
Toughness (kg cm^2/cm)	3.5	11.4	29.2		4.6	54.6
Coefficient of thermal expansion (10^{-7}/°C)	38	38	55	47	70	150
Thermal conductivity (W/m K)	1.3	0.9	0.6	2.8	20.0	0.2
Relative dielectric constant		4.0			9.3	4.8
Density (g/cm^3)	1.9	1.8	1.8	2.5	3.9	1.8
Machinability		2,000	2,000	100	impossible	12,000

Source: From Ref. 98.

in the industry, and it requires that the chip be backbonded to the substrate and wirebonded. In high-power applications, this poses two problems: (1) bonding the die to the ceramic with high-thermal conductive materials and (2) heat removal requirement from the package, sometimes in excess of the capabilities of the alumina or other ceramic used. Bonding of die to the substrate is accomplished by either eutectic alloys (gold–silicon), epoxies, or polyimides filled with precious metals or silver-filled glasses. Eutectic die attachment utilizes gold or silver eutectic composition that forms as a result of a reaction between the gold–silicon preform and the silicon die, as shown in Figure 9-63. The adhesive attachments using organic or inorganic materials filled with such thermally conductive materials as silver are usually cheaper and lower in stress than the eutectics, allowing larger dies to be bonded at lower temperatures. These are further discussed in Chapter 8, "Chip-to-Package Interconnections."

A number of ingenious designs have been reported [99] to handle heat dissipation in applications involving die bonding and wirebonding chips to low thermal-conductive ceramics. These designs, which can be extended to plastic packages and printed-circuit boards, involve bonding of thermal-expansion-matched alloys (Cu-W) or ceramics (AlN, SiC) of very high thermal conductivity to alumina or other ceramics, using copper–silver, gold–tin, or other brazes. This is schematically illustrated in Figure 9-45a for alumina with copper–tungsten alloys and (b) for glass–ceramics, with AlN heat sinks. As shown in Table 7-15 of Chapter 7,

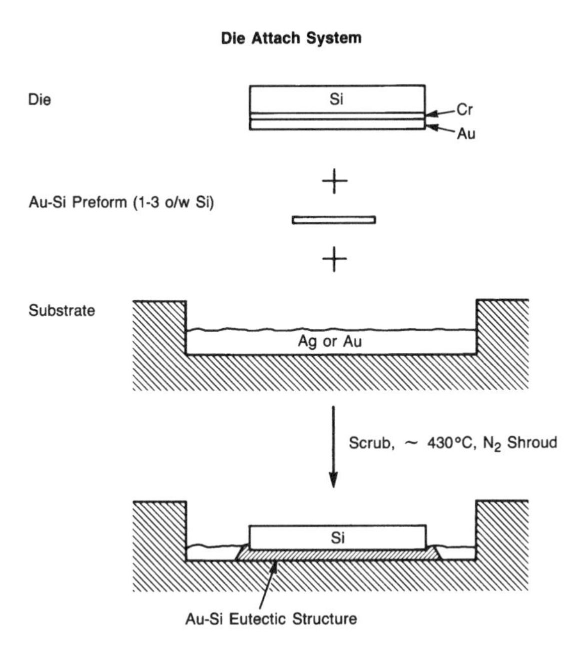

Figure 9-63. Die Attach System

these heat sinks have thermal conductivities in excess of 200 W/m K, providing several watts of capability from each chip. Removal of heat from the chip that is C4-bonded is discussed in Chapter 8, "Chip-to-Package Interconnections." Removal of heat with TAB-bond chips can be accomplished in either of two ways. In those applications involving die bonding, heat removal can be aided with brazed or adhesive-bonded heat sinks, as discussed earlier with wirebonding. In addition, heat sinks can be bonded to chip encapsulation, as discussed in the TAB section of Chapter 8.

9.9 CERAMIC PACKAGE RELIABILITY

Ceramic packages are generally considered "hermetic." However, there have been a number of designs that depended on organic sealants or encapsulants to provide reliability. Reliability of these designs, therefore, is meant to meet the requirements in the absence of hermeticity.

Because ceramic packages consist of ceramics and metals, they are prone to two general types of degradation: those involving metallic corrosion, resulting in open circuit lines, and those involving degradation or cracking of ceramics in the presence of mechanical stresses aided by water vapor. The latter results in stress-corrosion cracking of ceramic. The metal degradation mechanisms are covered in great detail in Chapter

5, "Package Reliability." Stress corrosion of ceramics and glasses is briefly described here.

It is well established that although freshly drawn fibers of glasses and ceramics have tensile strengths around 7×10^3 MPa (one million pounds per square inch) and that their short-term tensile strengths are in the range of 70–350 MPa (10,000 psi for glasses to about 50,000 psi for ceramics), the strengths to which they can be designed are roughly one-third to one-half of these numbers. The degradation mechanism involves growth of Griffith cracks under the influence of residual mismatch stresses and other applied stresses during processing, and are aided by such stress-corrosion atmospheres as water, plating solutions, and so forth. It has been estimated that the residual mismatch stress as a result of cooling from 1600°C alumina–molybdenum multilayer ceramic substrate can be as much as 150 MPa (25,000 psi) at the interface of molybdenum via. Cracks often seen at this interface are directly attributable to the lowering of the strength of the interface as a result of mismatch stress in the presence of water to a level where stress exceeded the strength. More quantitative data on this subject can be found in the literature [100]. Glasses suffer from this type of degradation as well, and the design strengths can often be no more than 35 MPa (5,000 psi) in tension.

9.10 FUTURE CERAMIC PACKAGING

The overall trend in ceramic packaging over the last three decades is summarized in Figure 9-64, indicating how ceramics have evolved from dual-in-line packages to single-chip pin-grid arrays in alumina systems and to low-temperature glass ± ceramic systems in recent years. This figure also shows the trend in mullite and aluminum nitride development. It appears that the trend in future packages, including ceramics, is very clear and is summarized in Table 9-39. The technology trend that can be expected is discussed below for each of the major solutions.

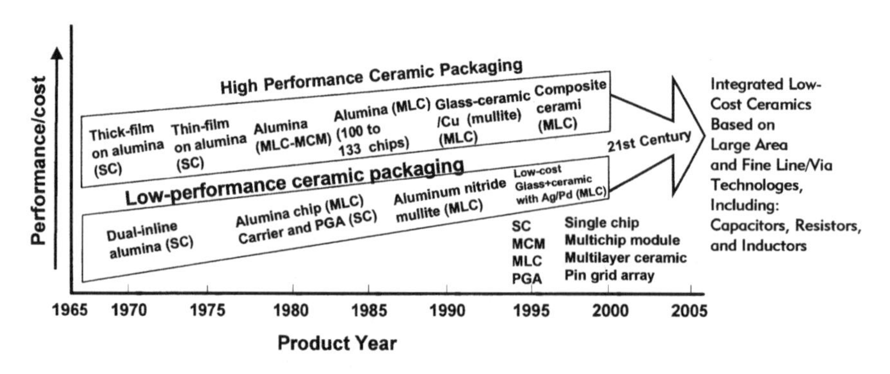

Figure 9-64. Ceramic Packaging Evolution

Table 9-39. Future Requirements of Ceramic Packages

Package Trend	Potential Ceramic Substrate Solution
Thin and light	Fewer layers, thinner layers lower-dielectric-constant materials
Ultralow cost	Low-cost materials and processes
	Large-area substrate fabrication
Compact size	Chip scale packaging
	Ball-grid array, flip-chip connections
	Direct chip attach to ceramic board
Multiple functions	Integrated ceramic substrate with capacitors, resistors, inductors, and other passives.

9.10.1 Future Low-Dielectric-Constant Ceramic Materials

The search for low-dielectric-constant ceramic materials for advanced electronic packages has been reviewed in this section from the viewpoint of technology and manufacturing. Recognizing that speed, size, reliability, and cost are the principal driving forces for the advanced packages, this section then presents a discussion and problems that the presently available low-dielectric constant ceramics offer to meet the above goals. It follows that a material with an ever-decreasing dielectric constant is not enough to increase the packaging speed without a concurrent reduction in circuit dimensions and metallization with the low-resistance metals. In this regard, the lack of a fine-line and via printing technology presents a barrier to the full implementation of the desired circuit feature size that the packaging design rule would allow for low dielectric systems. Similarly, benefits of low-dielectric-constant systems have not been incorporated into the cost equation, and cost remains a complex but highly desirable challenge. This discussion is then followed by a broad overview of the commercial packaging activities, with emphasis given to a currently successful cordierite system with a dielectric constant of about 5. The next breakthrough involves the development of a new, phase-transformation-free, nearly pure SiO_2 dielectric with a dielectric constant of 3.9–4.2, the lowest limit achieved for a single-phase inorganic compound, offering a formidable competition to the high end of a polyimide dielectric with a dielectric constant of 3.5–4.0. Given that this new dielectric has also a coefficient of thermal expansion matched to that of Si, the SiO_2-based ceramic multichip module (MCM-C) can be made as an alternative path to the polyimide-based thin-film multichip module (MCM-D).

9.10.1.1 Borosilicate Glass + Silica

Of particular interest in this category is the dielectric made by reacting a mixture of borosilicate glass + nearly pure SiO_2 filler [101,102] into a single-phase material, with a dielectric constant of 3.9–4.2 and a thermal coefficient of expansion matching that of Si (Table 9-40). Figure 9-65 compares the dielectric constant versus frequency of SiO_2 dielectric as described earlier with the several commercially available glass–ceramics. Dense, multilayered, cofired test vehicles (to characterize both physical and electrical properties as partially indicated in Table 9-41. The unique feature of this dielectric system is that the destructive martensitic phase transformation that is so common to crystalline SiO_2 has been totally prevented by incorporating a *crystal growth inhibitor* (similar to grain growth inhibitor) into the reactive materials systems. As a result, a large quantity of nearly pure SiO_2 can be accommodated in the dielectric composition. The borosilicate glass, to the extent of 30–40 vol.%, has been used to provide a *reactive liquid-phase* sintering [103] of the components at about 950°C in air. The coefficient of thermal expansion was matched to that of Si by alloying with dopants having TCE substantially lower than that of crystalline SiO_2. In other words, a careful design of materials and process has resulted in the prevention of the phase transformation of SiO_2 to cristobalite. The net result is a substrate with a TCE match to Si while allowing sintering to occur below 1000°C by using a glass that reacts with other components to form a single-phase, nearly pure SiO_2 as determined by X-ray diffraction (XRD). By extending the range of doping, TCE can be matched closely to that of GaAs with some sacrifice in dielectric constant from about 4 to about 5.8. Figure 9-66 shows the relation between dielectric constant and TCE versus amount of dopant.

The above progress represents the end of a journey toward achieving a cofireable lowest dielectric constant, single-phase inorganic material with a signal propagation gain of about 50%, and places this dielectric in direct competition with the high end of the polyimide with a dielectric constant of 3.5. In fact, given that polyimides [104], and to a lesser extent BCB [105], are characterized by such unattractive features (Table 9-42) as large TCE, compared to that of Si, combined with poor strength, poor thermal conductivity, lower adhension between metal and polymers, large dissipation factor, significant moisture absorption, and limited number of signal layers (due to bowing of the substrate and sequential processing), makes SiO_2, an attractive alternative for large-scale MCM production, using a well-developed, well-understood multilayer, cofired technology. Investment in plant and equipment is also less for this technology. From the comparison of two systems in Table 9-4 wherein data for the polyimide/Cu system were obtained from AVP technology [106,107] and those for SiO_2/Au were developed at Alcoa, it can hardly be denied that a ceramic multichip module (MCM-C) made from a SiO_2-borosilicate system presents a formidable challenge to deposited multichip module (MCM-D) made from polyimide dielectric, the only limitation being the lack of a fine-line printing technology equivalent to that of photolithography, but this is also being developed as discussed in large-area fine-line processing.

Table 9-40. LTCC Packaging System as Reported by Different Companies

Company	Glass Matrix	Ceramic Filler	Metallization	Dielectric Constant	Dissipation Factor	TCE (ppm/°C)
Alcoa	Borosilicate glass	SiO_2, proprietary dopants	Au, Ag/Pd (internal)	3.9–4.2	<0.003	2.5–3.5
Asahi Glass	BA–Al_2O_3–SiO_2–B_2O_3	Al_2O_3	Au	6.3	0.001	3.8–6.8
Corning	Crystallizable glass	Crystalline cordierite	Au	5.2	—	3.4
DuPont	Alumino borosilicate	Al_2O_3	Ag, Au	7.8	0.002	7.9
DuPont	Crystallizable glass	Crystalline cordierite	Au	4.8	<0.003	4.5
Ferro	Crystallizable glass	Crystalline materials unknown	Ag, Au, Ag/Pd	6.0	0.002	7.0
Fujitsu	Borate glass	Al_2O_3, SiO_2	Cu	4.9	—	4.0
Hitachi	Pb–alumino–borosilicate	Al_2O_3, $CaZrO_3$	Ag/Pd	9–12	0.001–0.003	—
IBM	Crystallizable glass	β-spodumene	Cu	5.0–6.5	0.002	2.0–8.3
IBM	Crystallizable glass	Crystalline cordierite	Cu	5.2–5.7	0.001	2.5–5.5
Kyocera	Pb–borosilicate glass	Al_2O_3, SiO_2	Au	7.9	—	7.9
Kyocera	ZnO–borosilicate	Al_2O_3, SiO_2	Cu	5.0	—	4.4
Matsushita	MgO–CaO–alumino-borosilicate	Al_2O_3	Cu	7.1	0.0025	—
Matsushita	Na_2O–CaO–alumino-borosilicate	Al_2O_3	Cu	7.4	0.002	6.1
Murata	BaO–B_2O_3–Al_2O_3–CaO–SiO_2	—	Cu	6.1	0.0007	8.0
Narumi	CaO–alumino-borosilicate	Al_2O_3	Ag, Au	7.7	0.0003	5.5
NEC	Borosilicate glass	SiO_2, cordierite, 13–49% pores	Au	2.9–4.2	0.002	15–3.2
NEC	Pb–borosilicate	Al_2O_3, SiO_2	Ag/Pd	7.8	0.003	7.9
NGK	ZnO–cordierite glass	Crystalline cordierite	Ag, Au	5.2–5.5	0.001	1.5–1.3
Nihon	—	Al_2O_3	Ag, Au	7.8	—	7.0
Shoel	$BaZr(BO_3)_2$	—	Cu	7.0	0.001	7.7
Taiyo Yuden	Al_2O_3–CaO–SiO_2–ZrO_2–MgO–B_2O_3	—	Cu	6.7	0.001	4.8
Taiyo Yuden	Al_2O_3–SiO_2–ZrO_2–MgO	—	Ag/Pd, Ni	7.3	0.002	5.1
Tektronix	MgO–CaO–silicate	Al_2O_3	Ag, Au	5.8	0.0016	4.6
Toshiba	BaO–SnO_2–TiO_2–B_2O_3	—	Au	7–13	0.0005–0.0008	—
Westinghouse	CaO–B_2O_3 × Al_2O_3	SiO_2	Au	4.6	0.001	9.6

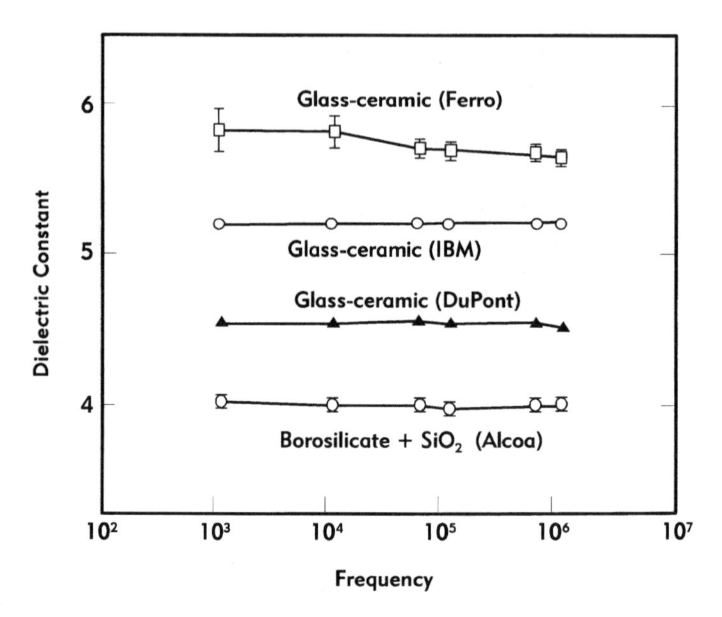

Figure 9-65. Dielectric Constant Versus Frequency for Various Low-Temperature Ceramics

9.10.1.2 Porous Ceramics and Glasses

Moving away from the real world of packaging (i.e., the ability to manufacture repeatedly with high yield), a considerable amount of research work is ongoing to reduce the dielectric constant to values even lower than 4, that of SiO_2. As SiO_2 represents the lowest limit of dielectric constant for a single-phase inorganic material due primarily to its *open* crystal structure, it is only appropriate that the next phase of advancement in the reduction of dielectric constant must come from the creation of void space in the microstructure. (Note that the ultimate dielectric is a vacuum with dielectric constant of 1.) This is equivalent to introducing a second phase in the microstructure with a dielectric constant of 1. Of significant practical value in this regard is the dielectric mixing rules [108] that can be faithfully applied to design the dielectric constant of the composite. In a variety of studies [109–112], the second phase has been chosen to be the pores, deliberately created in the microstructure or hollow microspheres, again deliberately incorporated into the microstructure. It must be acknowledged, however, that the introduction of any open space in the microstructure to reduce the dielectric constant will also degrade both mechanical strength and thermal conductivity and make the material prone to moisture absorption if the pores are open. Compounding these problems are the enormous difficulty of repeatedly manufacturing a body of porous ceramics into a controlled multilayer structure with

Table 9-41. Effects of Matched TCE, Low-*k* and High Conductivity of Au on Package Design and Cost; 4.5 in. (11.43 cm) square module

Parameters	Al_2O_3 Module	SiO_2 Module	Design Advantage of SiO_2 Module	Cost Advantage of SiO_2 Module
Chip size	0.5 cm	1.0 cm	Assumed	Assumed
Chip area	0.25 cm^2	1.0 cm^2	4 times area in larger chip	4 times area in larger chip
Circuits per chip	10^4	4×10^4	4 times # of circuit in larger chip	4 times # of circuit in larger chip
# of chips per module	$20 \times 20 = 400$	$10 \times 10 = 100$	1/4th chips on the SiO_2 module	1/4th cost for chip mounting
Circuit per module	$400 \times 10^4 = 4 \times 10^6$	$100 \times 4 \times 10^4 = 4 \times 10^6$	Same for both modules	Same for both modules
I/O per chip; Rent's rule: I/O = 0.5 (# circuit)$^{0.5}$	50	100	I/O per chip twice in larger chip	I/O per chip twice in larger chip
Total signal I/O per module	$50 \times 400 = 20{,}000$	$100 \times 100 = 10{,}000$	50% less I/O in SiO_2 module	50% less interconnect cost
# of signal planes per module	More	Fewer	Less routing of signal planes in SiO_2 module	Lower routing cost in SiO_2 module
Off-chip delay (microstrip)	86 ps/cm	55 ps/cm	56% gain in speed in SiO_2 module	Higher selling price for SiO_2 module
Wiring density for same Z_0	1 (relative)	1.5	1.5 times closer spacing in SiO_2 module	50% denser lines, fewer planes, lower cost
Wiring length	1 (relative)	0.1–0.5	At least 50% less wire length	At least 50% less gold
Wiring R_S for 25-μm-thick line	$W = 8$ mΩ/□	Au = 1.6 mΩ/□	5 times more line conduction with Au	1/5 less Au for equivalent resistance
Dielectric thickness	1 (relative)	1/2	Thinner SiO_2 module	Less materials consumption; lower cost
Density	3.9 g/cm^3	2.2 g/cm^3	44% less weight for SiO_2 module	Lower shipping cost for SiO_2 module

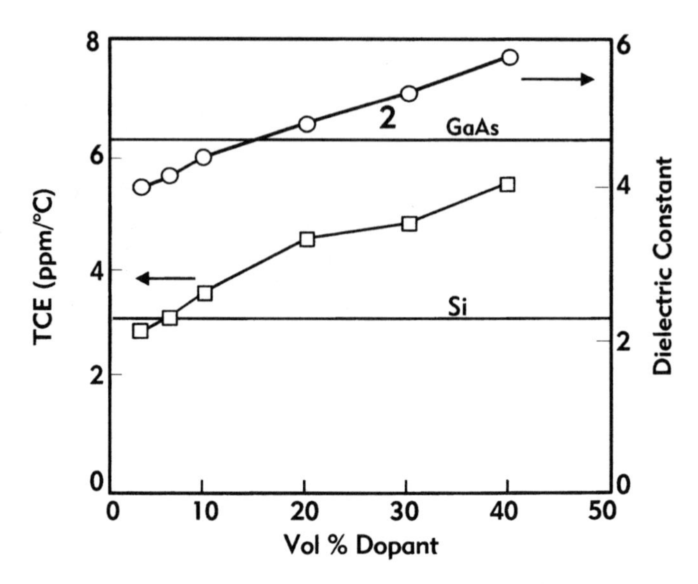

Figure 9-66. TCE and Dielectric Constant as a Function of Dopant in SiO_2-Borosilicate System

reasonable yield, reliability, and throughput. Given that a solution exists for exercising such a high degree of control at the level of manufacturing, convincing the customers to accept the product will be a daunting task.

The most advanced among these approaches that utilize porosity as a second phase is the investigation conducted by NEC on several compositions, as reported [111] in Table 9-40. The amount of porosity introduced in the microstructure varied from 13% to 49%, corresponding to a dielectric constant of 4.2 to 2.9. Predetermined amounts of polystyrene microspheres were added to the glass–ceramic material systems, composed of quartz glass, cordierite, and borosilicate glass, to control the porosity which was generated by the decomposition of the spheres during heat treatment. The merit of the concept was demonstrated by carrying the fabrication process through the entire range of multilayer processing with Au metallization. Different techniques were used by others to introduce porosity as a second phase, although controlling the amount of porosity and keeping the pores isolated were found to be difficult in many instances. Porous SiO_2 films of about 40% porosity and ranging in thickness from 1 to 25 μm with k ~2.2 and tan δ of <0.005 were demonstrated by a sol-gel process [113,114]. Reactively sputtered SiO_2 films [113,115] were prepared with a thickness of 5–10 μm that exhibited k ~3.4 and tan δ of ~0.005. The porous Vycor glass [113] with a pore size of 40Å exhibited dielectric constant values of 2.6–3.3. High-strength SiO_2 microballoons [116] with an average diameter of 65 μm and wall thickness of 1–2 μm

Table 9-42. Comparison of Thin-Film Versus Low-*k* Ceramic Packaging Parameters

Parameters	Thin Film Polymide/Cu	Low k SiO_2/Au
Intrinsic		
Dielectric constant	3.0–3.6 kHz	3.9–4.2 kHz
Dissipation factor	1–2%	0.1%
Volume resistivity	10^{15}–10^{16} cm	>10^{16} cm
Expansion coefficient (Si = 3 ppm/°C)	20–70 ppm/°C	3 ppm/°C
Thermal conductivity	0.17 W/m °C	2.1 W/m °C
Moisture absorption	1–4%	<0.1%
Density	1.42 g/cm^3	2.2 g/cm^3
Strength	5 ksi	15 ksi
Decomposition/melting temperature	400–450°C	1710°C
Conductor resistivity	1.7 Ω cm	2.2 Ω cm
Extrinsic		
Linewidth	10 μm	4–6 mil[a]
Line thickness	2–3 μm	10 μm
Printing technology	Photolithography	Screen printing
Line resistance	8Ω/cm	1.6 Ω/cm
Sheet resistance	8 mΩ/□	1.6 mΩ/□
Speed	1 (relative)	90% of polyimide
Wiring density	1 (relative)	85% of polyimide[b]
Interconnect delay	Low	Low
Process steps	>200	<50
Package support	Requires a substrate	Stand-alone package
Substrate	Si, Al_2O_3, glass–ceramic	Optional: Si, Al_2O_3 metal
TCE match to Si substrate	Poor	Excellent
Flip-chip die attach	Will produce stress	No stress
Equipment/process/compliance cost	High	Low to moderate
Similarity to high-temperature cofire	None	High

[a]Process limited at present, design optimum 25 μm.
[b]Subject to enhancement of printing capability.

were incorporated into calcium aluminate cement to reduce the dielectric constant from a value of about 7 to a value of about 4.7. Further studies have reduced the dielectric constant to 2.7. In a slight variation of this approach [109], quartz bubbles were incorporated into a composite of alumina in glass, resulting in a dielectric constant of about 3.7. In a separate study [117], hollow silica glass microspheres were bonded together with a lead glass frit that also acted as a flux. Samples with 55–68% porosity were fabricated at a temperature of about 550°C. The dielectric constants of these composites ranged from 3.9 to 3.3 and were nearly independent of temperature. Hollow microspheres of mullite [118] with a diameter of 20–

80 mm have recently been used as a second phase in a cordierite glass–ceramic matrix, resulting in a dielectric constant of ~2.5. Contrary to the case of silica microspheres, it is claimed that upon cooling from the sintering temperature, such composites will exhibit higher strength due to mullite microspheres being in tension and matrix glass–ceramic in compression. In search of a still lower dielectric constant, polymer was introduced as yet an additional phase to the pore phase. A SiO_2–polyimide composite with controlled porosity was formed from a spray-dried mixture of colloidal silica and silica fibers that were partially sintered and infiltrated with polyimide [119]. A dielectric constant of ~1.6 and a loss of ~0.005 were reported for this composite. Colloidal SiO_2 film [105] processed by a sol-gel technique also gave a dielectric constant of 1.6. Judging from all these activities, it appears that the days of a perfect material with a dielectric constant of 1 are not too far out in the future! What remains as a challenge in the low-dielectric-constant system is to develop a perfect processing technology to match the perfect material, leading to low-cost. An interesting consequence of such a perfect accomplishment, however, is to put a severe strain on the design rules. Figure 9-67 summarizes the ceramic materials development discussed in this book as a function of dielectric constant.

9.10.2 Low-Cost Based on Large-Area Processing

Referring to Figure 9-68, in which cost of various substrates is plotted against wiring density, it is clear that printed wiring board (PWB),

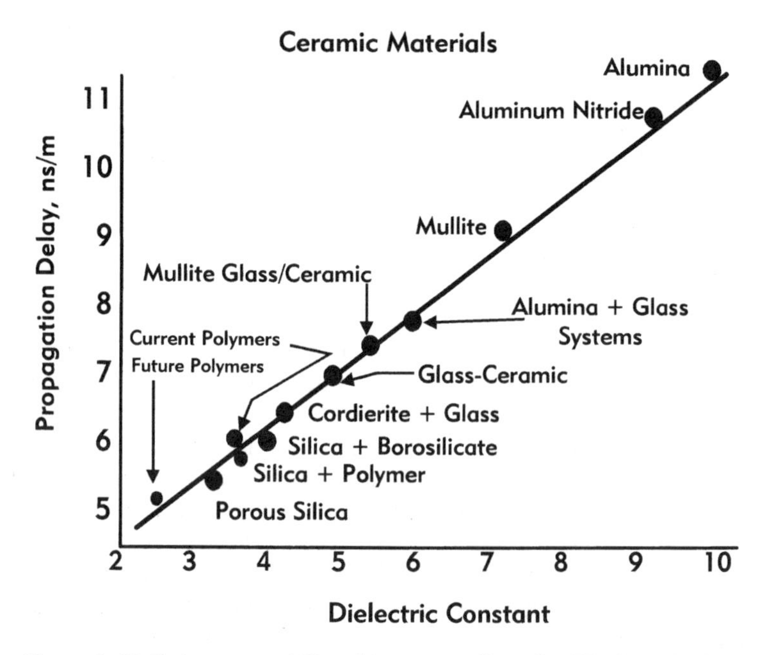

Figure 9-67. Interconnect Density versus Cost for Various MCMs

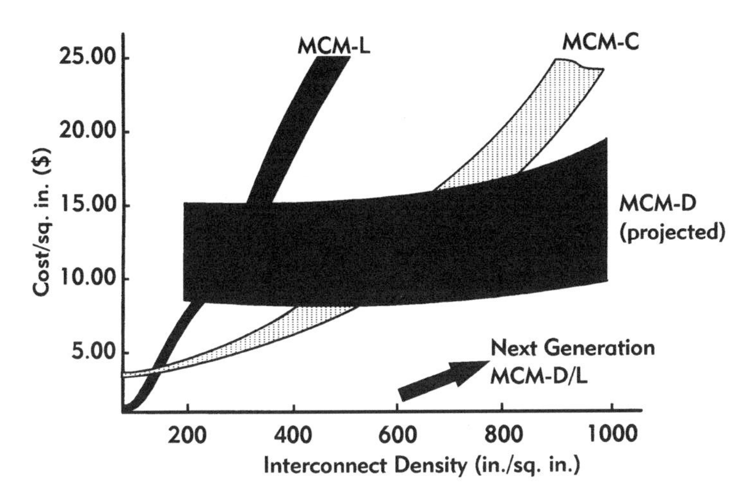

Figure 9-68. Cost of ceramic, Thin-Film, and PWB Technologies as a Function of Wiring Density

even though it is very inexpensive in a few layers and low wiring density, becomes very expensive with high number of layers. This is borne out by at least two examples in which Fujitsu and Hitachi have used a 40-layer board, supposedly at a cost in excess of $100,000 per board [120]. At these densities, cofired alumina ceramic has been established to be very much cheaper, perhaps more like $50/in.^{2} for a 30-layer substrate or one-third of this at $15/in.^{2} for a 10-layer substrate. This realization that wiring density in ceramic is cheaper than PWB has led to the development of very extensive developments by the three mainframe companies: IBM, Fujitsu, and Hitachi. NEC, Semens-Nixdorf, and Digital, however, pursued thin-film multichip technology on either ceramic, metal, or printed wiring board. In Figure 9-68, it is apparent that thin film is a very expensive way of packaging these needs. Nevertheless, because the price of millions of instructions per second (MIPS), is of the order of $25,000 or more, the high cost of packaging has not been a problem in the 1970s and 1980s. The challenge now is quite different. Almost all the electronic needs in computer, telecommunications, consumer, and automotive are now based on CMOS—both for hand-held consumer and desktop high-performance systems—and the cost is number 1 driver of technology. Large-area processing, similar to large-area processing of thin-film packaging discussed in Chapter 12, Thin-Film Packaging," is expected to lower cost as indicated in Figure 9-68.

There are clear indications that a number of larger substrates are being developed throughout the community. IBM has demonstrated a 166-mm substrate including greensheet fabrication, screening, lamination, and sintering. Such a substrate previously shown in Figure 9-2 had 107 layers of glass–ceramic and copper. This substrate was fabricated to zero shrinkage. NEC used a 225-mm alumina substrate with power and ground on which it deposited five to seven layers of thin film wiring using polyimide and gold. This substrate was used in all its mainframe and supercomputers. Similarly, Fujitsu developed and used a 245-mm glass–ceramic/copper substrate for its recent mainframes. It is perhaps conceivable that a 300-mm size substrate from which tens of single-chip and multichip substrates can be fabricated can be thought of to further lower the cost of ceramic packages. It is abundantly clear that large ceramic substrates developments in the future are based on need to lower the cost per unit area of single-chip and multichip substrates as opposed to the use of very large substrates for packaging of leading-edge central processor involving 100 or more ICs on one substrate.

9.10.3 Fine-Line and Via Technologies

Figure 9-69 illustrates the trend in ceramic packaging for linewidth, line pitch, and via dimensions and compares these with thin-film technology based on the polymer–metal structure as discussed in Chapter 12, "Thin-Film Packaging." IBM routinely manufactures its ceramic substrates with 75-μm lines and 90-μm vias by major enhancements in mask fabrication, thick-film paste, and screen-printing technologies.

An alternative to the via punching process is the photoforming process developed by DuPont and others, as illustrated schematically in Figure 9-27. The photoformed ceramic module (PCM) is formed by the use of photosensitive dielectric and conductive films. These films are subsequently exposed to ultraviolet light. Linewidths less than 50 μm on 100-μm spacings are being developed with these processes.

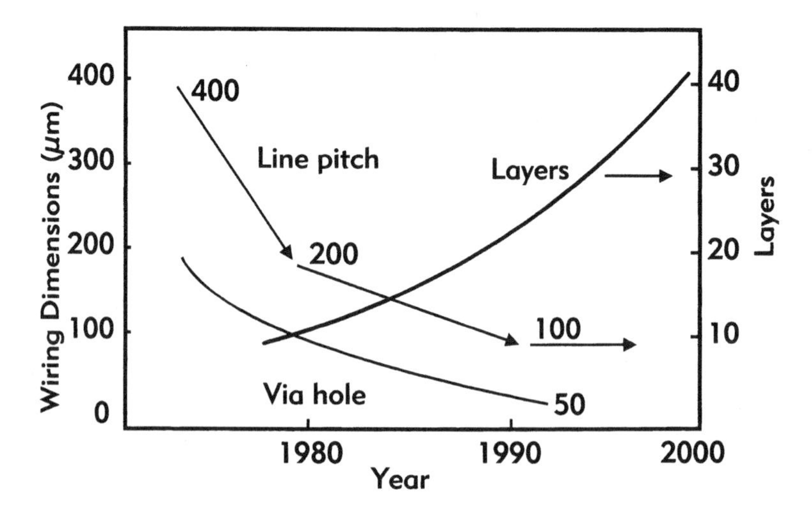

Figure 9-69. Ceramic Packaging Trends

9.10.4 Ceramic Ball-Grid Array and Chip-Scale Packages

Consistent with the need for ultrasmall, thin, light, and low-cost packages, ultimately no bigger than the chip itself, ceramic packaging is being enhanced. Ceramic ball grid arrays are beginning developed at a cost of approximately half of pin grid arrays and size roughly one-fourth.

Figure 9-70 illustrates the cross sections of two ceramic ball-grid arrays, one from IBM [121], called miniball-grid array, and the other from Hitachi, called microchip carrier (MCC) [59], that uses polymer–copper thin film and a $TaSi_2$ resistor on top of the ceramic ball-grid array. IBM's ceramic BGA is a 21-mm-size alumina–Mo substrate with 0.25-mm balls placed on a 0.5-mm pitch, resulting in 1521 I/Os (39 × 39). This BGA also shows, as does the Hitachi, the incorporation of decoupling capacitors. The Hitachi BGA, on the other hand, is a 12 mm with 604 I/Os placed on centers. This subject matter is discussed in greater detail in package-to-board interconnections, along with other nonceramic BGAs and micro-BGAs.

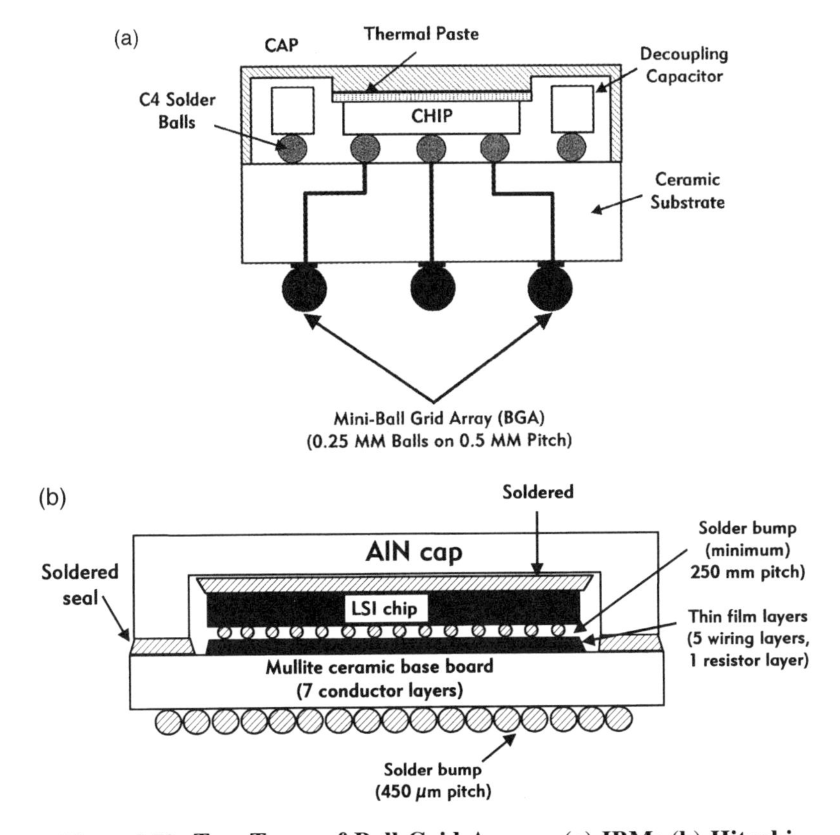

Figure 9-70. Two Types of Ball Grid Arrays: (a) IBM; (b) Hitachi

9.10.5 Integrated Ceramic Packaging with Passives

Discrete ceramic components, as pointed out in Section 9.1, is a large electronic ceramic market estimated at roughly 700 billion units a year. Passive components and their miniaturization are important contributions to the miniaturization of most electronic products. Passives include capacitors, resistors, inductors, transformers, and others. Japanese companies that include Murata, TDK, Kyocera (AVX in the United States), and others lead this market. Some of the ceramic materials and their properties in use today as discrete passives are indicated in Table 9-43. A recent JTEC study [94] revealed the Japanese packaging strategy to be based on the miniaturization of voltage-controlled oscillators (VCO) and capacitors as indicated in Table 9-44, leading to components packaging density that are about 20 currently to almost 50 by year 2000 (Fig. 9-71). It is clear that further enhancements in package miniaturization come from the integration of passives into the ceramic substrate as part of the substrate process.

Potential candidates for integration into ceramic substrates include the following:

- Decoupling capacitors
- Resistors
- Inductors

Table 9-43. High-Dielectric-Constant Ceramics

Item/Materials	SM210 $MgTiO_3$	SZ310 $CaZrO_3$	SH800 Ba Nd_2TiO_{14}	SH110 Ba Nd_2TiO_{14}
Dielectric constant ε_r	21 ± 1	31 ± 1	89 ± 1	110 ± 3
Q value (=1/tan δ)	>8000 (6.0 GHz)	>6000 (4.0 GHz)	>1800 (2.6 GHz)	>1200 (2.3 GHz)
$Q \times F$	>48000	>24000	>4680	>2760
T_ε (temperature coefficient of dielectric constant, ppm/°C)	−25 ± 30	−30 ± 30	−35 ± 30	−70 ± 30
Coefficient of linear thermal expansion (40–400°C), ppm/°C	9.2	8.5	10.1	10.1
Thermal conductivity (cal/cm/s°C)	0.018	0.011	0.005	0.005
Bulk density (g/cm³)	3.7	4.5	5.7	5.7
Flexural strength (kg/cm²)	2000	2000	1800	1800
Water absorption (%)	0	0	0	0
Note	—	Cu thick film is available	—	—

Table 9-44. Mass Production Japanese Strategy for Low-Cost Electronic Products

Technology	Plastic Packages	Discrete Components	PWB	SMT
Today	QFP	1.0 × 0.5 mm size	100-μm lines 250-μm vias 6–8 layers	0.4-mm pitch (QFP)
↓	↓	↓	↓	↓
Tomorrow (Year 2000)	TQFP TAB	0.8 × 0.4 mm built-in capacitors, resistors, inductors	50-μm lines 50-μm vias 100-μm pitch 8 layers	0.15-mm pitch (QFP) 0.5-mm (BGA)

Source: From Ref. 94.

- Optoisolators
- Thin-film batteries
- Driver transistors
- Power transistors
- Power conditioners

9.10.5.1 Requirements for Wireless Communications [122]

Ceramic technologies have become the essence of modern portable radio communicators. The use of low-loss, high-dielectric-constant ceram-

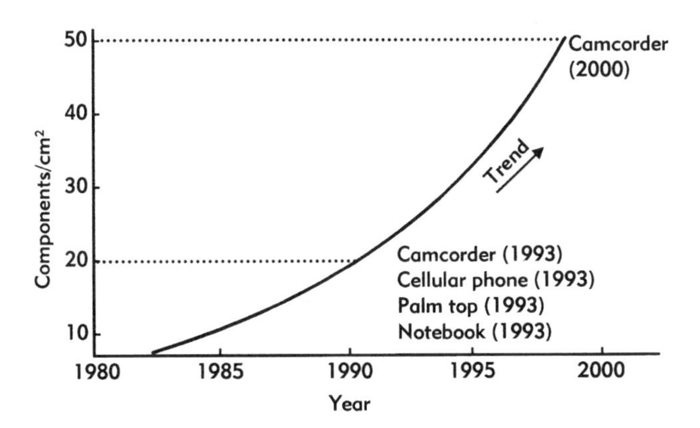

Figure 9-71. Japanese Consumer Product Component Density Trend. (Courtesy of Sony.)

ics for radio frequency (RF) and infrared frequency (IF) filtering, voltage-controlled oscillators, and capacitors has enabled the portable radio to be miniaturized to a practical size while maintaining high performance. In response to the continuous quest for a smaller and thinner profile, ceramic devices are transforming from discrete components into a monolithic lumped circuit component and finally into a totally integrated substrate. Resonators, capacitors, inductors, and resistors are being integrated into a high-density multilayer structure. Compared to the organic circuit board, the multilayer ceramic substrate offers advantages of ease of processing of these integrated passives for lower RF insertion loss, better interference isolation, higher component density, and better temperature properties (Fig. 9-72).

Generally speaking, every passive component behaves like a distributed circuit at RF frequency. The parasitic effect, in many cases, overwhelms the component's original intent. As the frequency increases, the parasitic resistance (skin effect) and the parasitic reactance become so appreciable that an impedance peak can be produced at or near the application frequency, thereby making the electrical properties uncontrollable. This problem is more pronounced at frequencies greater than 500 MHz. Recent ceramic technologies enable the component to be fabricated in chip format and greatly reduce the unwanted lead wire's inductance, distributed parasitic capacitance, and the temperature variation effect.

Using a high-*k* dielectric, one can reduce the coaxial resonator values a thousand times. The coaxial resonator performs better than the equivalent lumped circuit with discrete components. Although higher-dielectric-constant ceramics can logically be used to reduce the size further, the insertion loss and manufacturing problems appear insurmountable.

Higher-dielectric-constant materials have two disadvantages at high frequency. They tend to have higher dielectric loss and temperature sensitivity. To maintain proper impedance, the center conductor becomes too narrow to be formed, due to processing limitations. Narrow center electrodes also lead to Q degradation.

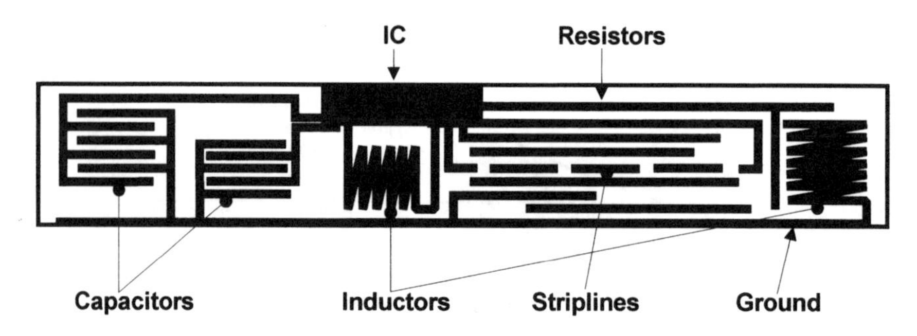

Figure 9-72. Concept of Integrated Ceramic Packaging

9.10.5.2 Integrated Ceramic Size Requirements

It is becoming obvious that only the Personal Computer Memory Card Induced Association (PCMCIA) format or the so-called credit-card-size format provides true portability. The thickness, therefore, needs to be reduced to less than 2 mm. In order to reduce the thickness, the surface-mount approach faces very tough challenges. Instead of mounting components on a printed wiring board, the ceramic components themselves should also function as the board and IC chip carrier (see Fig. 9-72).

All components are embedded into the substrate; recess cavities are provided for mounting active IC chips using the dropped-in, flip-chip technology. Because the resistance of the metal interconnects contribute to RF loss, refractory metals cannot be used. Therefore, the ceramics need to be cofireable with either copper or silver. The sintering temperature should be less than 950°C while maintaining high dielectric and magnetic Q factors. The structure will be like the one in Figure 9-72. All the circuits and components are integrated into three-dimensional structure with minimized thickness. For RF resonators and transmission lines, a virtual space is created using new design technology to provide the required geometry ratio for impedance matching.

9.10.5.3 Some Examples of Passive Integration

Murata [123] has developed LC filters, RF diode switches, hybrid couplers, and delay lines using low-temperature cofireable ceramics with copper electrodes. The low-temperature ceramic is used in the barium–alumina–silica system and the buried capacitor developed to cofire with this ceramic is a calcium zirconate plus glass. The properties of these materials are shown in Table 9-45.

Table 9-45. Ceramic Integrated Capacitors by Murata

	Material Name	
	BAS	CZG
Property		
Basic formulation	BAO, Al_2O_3, SiO_2	$CaZrO_3$, Glass
Dielectric constant (1 MHz)	6.1	25
Q		
(1 MHz)	1400	1700
(5 MHz)	300	700
Temperature coefficient of dielectric constant (ppm/°C)	80 ± 20	0 ± 10
Resistivity (Ω cm)	10^{-12}	10^{-14}
Flexural strength (kg/cm^2)	1600	2150
Thermal Expansion coefficient (× 10^{-6}/°C)	11.6	7.0

9.11 REFERENCES

1. J.R.H. Black. "Technology and Market-Trends in Multilayer Ceramic Devices," in *Advances in Ceramics,* ed. by J. Blum and W. R. Cannon, pp. 3–11, American Ceramics Society, Columbus, OH, 1986.
2. R. R. Tummala. "Multichip Packaging in IBM, Past, Present, and Future," *Proceedings of International Conference on Multichip Modules,* pp. 1–11, 1995.
3. E. M. Davis, W. E. Harding, R. S. Schwartz, and J. J. Corning. "Solid Logic Technology: Versatile High-Performance Microelectronics," *IBM J. Res. Devel.*, 8(2): pp. 102–114, 1964.
4. R. G. Frieser. "A Review of Solder Glasses," *Electrocomponent Sci Technol.* 2(6): pp. 163–199, 1975.
5. R.H.F. Lloyd. "ASLT: An Extension of Hybrid-Miniaturization Techniques," *IBM J. Res. Develop.*, 11(4): pp. 86–92, 1967.
6. P. E. Fox and W. J. Nestork. "Design of Logic-Circuit Technology for IBM System 370 Models 145 and 155," *IBM J. Res. Devel.*, 15(2): pp. 384–390, 1971.
7. P. J. Holmes and R. G. Loasby. *Handbook of Thick Film Technology,* Electrochemical Publications, Ayr, Scotland, 1976.
8. R. W. Gedney. "Trends in Packaging Technology," *16th Annual Proceedings of Reliability Physics,* pp. 127–129, 1978.
9. H. Stetson. "Multilayer Ceramic Technology," *Ceram. Civilization,* 3: pp. 307–322, 1987.
10. W. J. Gyuvk. "Methods of Manufacturing Multilayered Monolithic Ceramic Bodies," U.S. Patent No. 3,192,086, 1965.
11. H. Stetson. "Methods of Making Multilayer Circuits," U.S. Patent No. 3,189,978, 1965.
12. B. Schwartz and H. Stetson. "Ceramics and the Micromodule," *RCA Eng.*, 5(4): p. 56–58. 1960.
13. J. C. Williams. "Doctor Blade Process," in *Treatise on Materials, Science and Technology,* pp. 173–198, Academic Press, New York, 1976.
14. R. E. Mistler, D. J. Shanefield, and R. B. Runk. *Tape Casting of Ceramics in Ceramic Processing Before Firing,* ed. by G. Y. Onoda and L. L. Hench, pp. 411–448, John Wiley and Sons, New York, 1978.
15. M. Sacks. "Milling of Agglomenated Powders," *International Conference on Ultrastructure Processing of Ceramics, Glasses and Composites,* pp. 13–17, 1983.
16. R. A. Gardner and R. W. Nufer. "Properties of Multilayer-Ceramic Green Sheets," *Solid State Technol.* 17(5): pp. 38–43, 1974.
17. J. W. Blade. *VLSI and the Substrate Connection, The Technological Trade-Offs of the Package–Board Interface,* D. Brown Associates, Warrington, PA, 1986.
18. R. B. Lomeson. "High Technology Microcircuit Packaging," *International Electronic Packaging Society Proceedings,* pp. 498–503, 1982.
19. C. Deisch, C. Gopal, and J. Stafford. "Designs of a High Performance DIP-Like Pin Array for Logic Devices," *33rd Electronic Components Conference,* 1983.
20. R. Vernon. "Technology vs. Inflation: The USAF/USN/Ti Low Cost Carrier," *Semiconductor/Microelectronics Symposium,* 1979.
21. H. Reiner. "VLSI Packaging," *Elect. Commun.*, 58: pp. 440–446, 1984.
22. J. W. Balde. "Status and Prospects of Surface Mount Technology," *Solid State Technol.*, 29(6): pp. 99–103, 1986.
23. W. Engelmaier. "Effects of Power Cycling on Leadless Chip Carrier Mounting Reliability and Technology," *Proceedings of International Electronic Packaging Society Conference,* pp. 15–23, 1982.

24. W. Engelmaier. *Surface-Mount Technology,* pp. 87–114, ISHM Press, Silver Spring, MD, 1984.
25. A. J. Blodgett and D. R. Barbour. "Thermal Conduction Module: A High-Performance Multilayer Ceramic Package," *IBM J. Res. Devel.*, 26(1):pp. 30–36, 1982.
26. A. J. Blodgett. "Microelectronic Packaging," *Scientific American,* 249(1): pp. 86–96, 1983.
27. J. U. Knickerbocker, G. B. Leung, W. R. Miller, S. P. Young, S. A. Sands, and R. F. Indyk. "IBM System Air-Cooled Alumina Thermal Conduction Module," *IBM J. Res. Devel.*, 35(5): pp. 330–341, 1991.
28. R. R. Tummala and R. R. Shaw. "Glasses in Microelectronics," *Adv. High-Technol. Ceram.*, 18: pp. 87–102, 1987.
29. L. Anderson, R. W. Nufer, and F. G. Pugliese. "Ceramic Dielectrics," U.S. Patent No. 4104345, 1978.
30. W. S. Young. "Multilayer Ceramic Technology," in *Ceramic Materials for Electronics,* ed. by R. C. Buchanan, pp. 403–424, Marcel Dekker, Inc., New York, 1986.
31. G. A. Slack, R. A. Tanzilli, R. O. Pohl, and J. W. Vanderande. "The Intrinsic Thermal Conductivity of AlN," *J. Phys. Chem. Solids,* 48: p. 641, 1987.
32. N. Kuramoto, H. Taniguchi, and I. Aso. "Sintering and Process of Translucent AlN and Effect of Impurities on Thermal Conductivity of AlN Ceramics," *Yogyo-Kyokai-Shi,* 93: p. 41, 1985.
33. K. Komeya, H. Inoue, and A. Tsuge. "Effects of Various Addatives on the Sintering of AlN," *Yogo Kyokai-Shi,* 89: p. 330, 1981.
34. N. Kuramoto, H. Taniguchi, and I. Aso. "Sintering and Properties of High Purity AlN ceramics," *Adv. Ceram.*, 26: p. 107, 1989.
35. A. K. Knudsen, T. A. Guiton, N. R. Nicholas, K. L. Mills, P. J. Bourns, J. E.Volmering, J. L. Board, D. R. Beaman, D. Susnitzky, P. S. deBranda, and E. Ruh. "The Influence of Imperities on the Optical, Thermal and Electrical Properties of Sintered AlN," *Proceedings International Microelectronics Conference,* p. 270, 1992.
36. N. Ichinose, K. Katakura, and I. Hazeyama. "Electrical and Thermal Properties of AlN Ceramic with Addatives," *Proceedings International Microelectronics Conference,* p. 265, 1992.
37. V. A. Lavrenko and A. F. Alexeev. "Oxidation of Sintered AlN," *Ceram. Int.*, 9: p. 80, 1983.
38. R. Chanchani. "Processability of Thin Film Fine Line Patterns on AlN Substrates for HICs," *IEEE Trans. Components Hybrids Manuf. Technol.*, *CHMT*-11: p. 633, 1990.
39. D. Suryanarayana, L. Matienzo, and D. Spencer. "Behavior of AlN Ceramic Surfaces Under Hydrothermal Oxidation Conditions," *Proceedings Electronic Components Conference,* p. 29, 1989.
40. S. Smith and B. Hazen, "Thin Film Protective Coating for Processing AlN Substrates," *Proc. ISHM,* p. 48, 1991.
41. N. Miyashiro, N. Iwase, A. Tsuje, F. Ueno, M. Nakahashi, and T. Takahashi. "High Thermal Conductivity AlN Ceramic Substrates and Packages," *IEEE Trans. Components Hybrids Manuf. Technol.*, CHMT-13: p. 313, 1990.
42. T. Yamaguchi and M. Kageyama. "Oxidation Behavior of AlN in the Presence of Oxide and Glass for Thick Film Applications," *IEEE Trans. Components Hybrids Manuf. Technol.*, CHMT-12: p. 402, 1989.
43. A. Mohamed, A. Abdo, G. Scarlett, and F. Sherrima. "Effect of Lot Variations on the Manufacturability of Thick and Thin Film AIN Substrates," *Proc. ISHM,* p. 7, 1990.
44. Y. Kurokawa, H. Hamaguchi, Y. Shimada, K. Utsumi, H. Takamizawa, T. Kamata, and S. Noguchi. *Proceedings Electronics Components Conference,* p. 412, 1986.

45. N. Iwase, T. Yasumoto, K. Iyogi, T. Kawakami, Y. Yokono, K. Itoh, and T. Takahashi. "Thin Film and Pin Brazing Technologies for High Performance AlN Packages," *Proc. IEPS,* p. 191, 1990.
46. Y. H. Chiao, A. K. Knudsen, and I. F. Hu. "Interfacial Bonding in Brazed and Cofired Aluminum Nitride," *Proc. ISHM,* p. 460, 1991.
47. Y. Kurihara, S. Takahashi, and K. Yamada. "Thin Film Resistors for AlN Ceramics," *IEEE Trans. Components Hybrid Manuf. Technol.*, CHMT-14: p. 199, 1991.
48. J. B. Blum, "Aluminum Nitride Substrates for Hybrid Microelectronics Applications," *Hybrid Circuit Technol.*, 1989.
49. A. Kaneko, K. Kuwabara, S. Kikuchi, and T. Kano. *Fujitsu Scient. Tech. J.* vol. 27: p. 171–178, 1991.
50. S. Hamano, A. Wanatabe, and H. Takahashi. "High Pin Count AlN Cer-Quad Package," *Electron. Packaging Tech.*, 8: p. 69, 1992.
51. F. Kobayashi, W. Wanatabe, N. Yamamoto, A. Anzai, A. Takahashi, T. Daikoku, and T. Fujita. "Hardware Technology for Hitachi M-880 Processor Group," *Proceedings Electronic Components Conference,* p. 693, 1991.
52. N. Iwase, H. Sawaya, and T. Takahashi. "AlN Multilayer Package Technologies for High Speed ECL and High Pin Count Bi-CMOS Devices," *Proc ISHM,* p. 269, 1992.
53. A. Shibuya, Y. Kurokawa, and Y. Shimada. "High Thermal Conductivity AlN PGA Package," *Proceedings International Microelectronics Conference,* p. 285, 1992.
54. P. E. Garrou, and I. Turlik, "Materials of Construction: Substrate, Dielectric, Metallization," in *Thin Film Multichip Modules,* ed. by G. Messner, I. Turlik, J. Balde, and P. Garrou, ISHM Press, Silver Spring, MD, 1992.
55. Y. Iseki, F. Shimizu, and T. Sudo. "Multichip Module Technology using AlN Substrate for 2-Gbit/s High Speed Switching Module."
56. L. E. Dolhert, J. W. Lau, J. H. Enloe, E. Y. Luh, A. L. Kovacs, and J. Stephan. "Performance and Reliability of Metallized AlN for MCM Applications," *Int. J. Hybrid Microelectron.*, 14: p. 113, 1991.
57. D. E. Neil. *Power Modules Utilizing Beryllium Oxide Direct Bonded to Copper,* Proceedings of International Microelectronic Symposium, pp. 491–495, 1975.
58. A. J. Rothman. "Beryllium Oxide," in *Ceramics for Advanced Technologies,* ed. by J. E. Hove and W. C. Reiley, John Wiley and Sons, New York, 1965.
59. F. Kobayashi, Y. Watanbe, M. Yamamoto, A. Anzai, A. Takahashi, T. Daikoku, and T. Fujita. "Hardware Technology for Hitachi M-880 Processor Group," *Proceedings of 42nd IEEE–ECTC,* pp. 693–703, 1992.
60. Y. Shimade, K. Utsumi, M. Suzuki, H. Takamizowa, M. Nitta, and T. Watari. "Low Firing Temperature Multilayer Glass-Ceramic Substrate," *IEEE Trans. Components, Hybrids Manuf. Technol.*, CHMT-6(4): pp. 382–388, 1983.
61. N. Kamehana, K. Niwa, and K. Murakawa. "Packaging Material for High Speed Computers," *33rd Electronic Components Conference Proceedings,* pp. 388–392, 1983.
62. T. Noro and H. Tozaki. "Constituent Materials for via Conductor in Multilayer Glass–Ceramic Board," Japanese Patent No. 6070799A, 1986.
63. R. R. Tummala, A. H. Kumar, and P. W. McMillan. "Glass Ceramic Structures and Sintered Multilayer Substrates Thereof with Circuit Patterns of Gold, Silver, or Copper," U.S. Patent No. 4,301,324, 1981.
64. P. W. McMillan. *Glass–Ceramics,* Academic Press, New York, 1974.
65. R. R. Tummala, L. W. Herron, and R. Master. "Method of Making Multilayered Glass–Ceramic Structures Having an Internal Distribution of Copper-Based Conductors," U.S. Patent No. 4,234,367, 1980.
66. See for examples Japanese Patients 61-142759 and 60-245154.
67. S. Nichigaki, S. Yano, H. Kawake, J. Fukuta, T. Nonomura, and S. Hebishima. "A New Low-Temperature Multilayer Ceramic Substrate with Gold (Top)-Silver (Internal)-Ti/Mo/Cu (Bottome) Conductor System," *ISHM Proceedings,* 1987.

68. R. R. Tummala, J. U. Knickerbocker et al. "High Performance Glass Ceramic/Copper Multilayer Substrate with Thin Film Redistribution," *IBM J. Res. Devel.*, 36: pp. 889–903, 1992.

69. R. R. Tummala. "Ceramic and Glass Ceramic Packaging in the 1990's," *J. Am. Ceram Soc.*, 74: p. 895, 1991.

70. A. H. Kumar, P. W. McMillan, and R. R. Tummala. "Glass-Ceramic Structures and Sintered Multilayer Substrates Thereof with Circuit Patterns of Gold, Silver, or Copper," U.S. Patent No. 4,301,324, 1981.

71. S. H. Knickerbocker, A. Kumar, and L. W. Herron. "Cordierite Glass-Ceramics for Multi-Layer Ceramic Packaging Applications," *Ceram. Bull.*, in press.

72. E. A. Giess and S. H. Knickerbocker. "Viscosity of $MgO–Al_2O_3–SiO_2–B_2O_3–P_2O_5$ Cordierite-Type Glasses," *J. Mater. Sci. Lett.*, 4: p. 835, 1985.

73. E. A. Giess, J. P. Fletcher, and L. W. Herron. "Isothermal Sintering of Cordierite-Type Glass Powders," *J. Am. Ceram. Soc.*, 67: p. 549, 1984.

74. H. E. Exner and E. A. Giess. "Anisotropic Shrinkage of Cordierite-Type Glass Powder Cylindrical Compacts," *J. Mater. Res.*, 3: p. 122, 1988.

75. K. Watanabe, W. A. Giess, and M. W. Shafer. "The Crystallization Mechanism of High Cordierite Glass," *J. Mater. Sci,* 79: p. 508, 1985.

76. K. Watanabe and E. A. Giess. "Coalescence and Crystallization in Powdered High-Cordierite ($2MgO–2Al_2O_3–5SiO_2$) Glass," *J. Am. Ceram. Sci.*, 68: p. C102, 1985.

77. T. C. Patton. "Film Applicators," in *Paint and Pigment Dispersion,* p. 581, John Wiley and Sons, New York, 1979.

78. R. W. Nuffer, L. E. Anderson, and F. G. Pugliese. "Ceramic Dielectrics," U.S. Patent No. 4,387,131, 1983.

79. R. N. Master, L. W. Herron, and R. R. Tummala. "Cosintering Process for Glass–Ceramic/Copper Multilayer Ceramic Substrate," *IEEE Trans. Components Hybrids Manuf. Technol.* CHMT-14: p. 780, 1991.

80. L. W. Herron, R. N. Master, and R. W. Nufer. "Methods of Controlling the Sintering of Metal Particles," U.S. Patent No. 4,671,928, 1987.

81. L. W. Herron, R. N. Master, and R. W. Nufer. "Methods of Controlling the Sintering of Metal Particles," U.S. Patent No. 4,776,978, 1987.

82. J. R. Piazza and T. G. Steele. "Positional Deviation of Preformed Holes in Substrates," *Bull. Am. Ceram. Soc.*, 51: p. 516, 1972.

83. R. Iwamura, H. Murakami, K. Ichimoto, and M. Takasaki. "Punching of Holes at High Density in Alumina Greensheet," *Proceedings of the American Ceramic Society Meeting,* 1986.

84. F. M. Fowkes and M. A. Mostafa. "Acid-Base Interactions in Polyimide Adsorption," *I&EC Product Res. Develop.*, 17: p. 3, 1978.

85. M. Marmo, H. Jinnai, M. A. Mostafa, F. M. Fowkes, and J. A. Manson. "Acid–Base Interactions in Filler-Matrix Systems," *I&EC Product Res. Devel.*, 15: p. 206, 1978.

86. T. B. Reed. *Free Energy of Formation of Binary Compounds: An Atlas of Charts for High-Temperature Chemical Calculations,* MIT Press, Cambridge, MA, 1971.

87. W. Herron, R. Master, and R. R. Tummala. "Method of Making Multilayered Glass-Ceramic Structures Having an Internal Distribution of Copper-Based Conductors," U.S. Patent No. 4,234,367, 1980.

88. Y. Shimada, Y. Kobayashi, K. Kata, M. Kurano, and H. Takamizawa. "Large Scale Multilayer Glass-Ceramic Substrate for Supercomputer," *IEEE Trans. Components Hybrids Manuf. Technol.*, CHMT 13: p. 751, 1990.

89. J. H. Indge. "Flat Precision Machining Ceramic Materials," *Advanced Ceramics '88,* 1988.

90. M. Bennett. "Grinding Ceramics at IBM," *Inter-Society Symposium for the Machining of Ceramic Materials* (*A. Cer. S., ASME, A. Abrasive Soc.*), 1988.

91. P. L. Flaitz and M. Neisser. "Optimizing a Polish Operation," Proceeding of *91st Annual Meeting of the American Ceramics Society,* p. 92, 1989.
93. Y. Shimada, Y. Shiozawa, M. Suzuki, H. Takamizawa, and Y. Yamashita. "Low Dielectric Constant Multilayer Glass-Ceramic Substrate Ag-Pd Wiring for VSLI Package," *36th Electronic Components Conference Proceedings,* pp. 395–405, 1984.
94. R. R. Tummala and M. Pecht. "Japan's Electronic Packaging Technologies," *Japanese Technology Evaluation Committee (JTEC) Panel Report,* pp. 59–95, 1995.
95. A. H. Kumar. "LTCC Multichip Modules," *Proceedings of International Multichip Module Conference,* 1995.
96. L. E. Gross and T. R. Gurnaraj. "Ultra-Low Dielectric Permittivity Ceramics and Composites for Packaging Applications," *Materials Research Society Symp. Proc.*, 72: pp. 53–65, 1986.
97. T. Watari and H. Murano. "Packaging Technology for the NEC SX Supercomputer," *37th Electronic Components Conference Proceedings,* pp. 192–205, 1985.
98. Y. Iwata, S. Saito, Y. Satoh, and F. Okamura. "Development of Ceramic Composite Porous Ceramic and Resin Composite with Copper Foil," *International Microelectronic Conference in Tokyo,* pp. 65–70, 1986.
99. M. Terasawa, S. Minami, and J. Rubin. "A Comparison of Thin Film, Thick Film, and Co-Fired High Density Ceramic Multilayer," *Int. Hybrid Microelectron.*, 6: pp. 1–11, 1983.
100. J. N. Humenik and J. E. Ritter. "Stress Corrosion of Alumina Substrate," *J. Mater. Sci.*, 14(5): pp. 626–632, 1979.
101. J. H. Jean and T. K. Gupta. "Design of Low Dielectric Materials for Microelectronic Packaging Applications," *Am. Ceram. Soc. Bull.*, 7(9): p. 1326, 1992.
102. J. H. Jean and T. K. Gupta. U.S. Patent Nos. 5,071,793 and 5,079,194, 1992.
103. J. H. Jean and T. K. Gupta. "Isothermal and Nonisothermal Sintering Kinetics of Glass-Filled Ceramics," *J. Mat. Res.*, 7(12): pp. 3342–3347, 1992.
104. G. M. Adema, M. J. Berry, and I. Turlik. "Dielectric Materials for Use in Thin Film Multichip Module," *Electron. Packaging Prodution,* vol. 11, pp. 72–76, 1992.
105. E. W. Rutter, Jr., E. S. Moyer, R. F. Harris, D. C. Frye, V. L. St. Jeor, and F. L. Oaks. "A Photodefinable Benzocyclobutene Resin for Thin Film Microelectronic Applications," *ICMCM Proceedings,* pp. 394–400, 1992.
106. C. J. Bartlett. "Advanced Packaging for VLSI," *Solid-State Technol.*, pp. 119–123, June 1986.
107. C. J. Bartlett, J. M. Segelken, and N. A. Teneketges. "Multichip Packaging Design for VLSI-based systems," *IEEE Trans. Components Hybrids Manuf. Technol.*, CHMT-12: p. 518, 1987.
108. W. D. Kingery, H. K. Bowen, and D. R. Uhlmann. in *Introduction to Ceramics,* chaps. 12, 15, and 18, John Wiley and Sons, New York, 1976.
109. D. W. Kellerman. "The Development and Characterization of a Low Dielectric Constant Thick Film Material," *Proceedings of 37th Electronic Components Conference,* pp. 316–327, 1987.
110. S. J. Stein et al. "Controlled Porosity Dielectrics and Etchable Conductors for High Density Packages," *ISHM 1990 Proceedings,* pp. 725–732, 1990.
111. K. Kata, A. Sasaki, Y. Shimada, and K. Utsumi. "New Fabrication Technology of Low Dielectric Permittivity Multilayer Ceramic Substrate," *ISHM 1990 Proceedings,* pp. 308–315, 1990.
112. J. K. Yamamoto, et al. "Dielectric Properties of Microporous Glass in the Microwave Region," *J. Am Ceram. Soc.* 72(6): pp. 916–21, 1989.
113. L. E. Cross and T. R. Gururaja. "Ultralow Dielectric Permittivity Ceramics and Composites for Packaging Applications," *Mater. Res. Soc. Symp. Proc.*, 72: pp. 53–65, 1986.

114. W. A. Yarbrough, T. R. Gururaja, and L. E. Cross. "Materials for IC Packaging Very Low Permittivity via Colloidal Sol-Gel Processing," *Am. Ceram. Soc. Bull.*, 66(4): pp. 692–698, 1987.
115. A. Das. "Sputter Deposited Silica Films as Substrate for Microelectronic Packaging Applications," Ph.D. thesis, Pennsylvania State University, 1988.
116. P. Silva, L. E. Cross, T. R. Gururaja, and B. E. Scheetz. "Relative Dielectric Permittivity of Calcium Aluminate Cement-Glass Microsphere Composites," *Mater. Lett.* 4(11–12): pp. 475–480, 1986.
117. M. J. Leap. "The Processing and Electrical Properties of Hollow Microsphere Composites," M. S. thesis, Pennsylvania State University, 1989.
118. D. J. Vernetti and D. L. Wilcox, Sr., "Synthesis of Hollow Mullite Microspheres: A New Approach for Low Dielectric Constant Inorganic Packaging Materials," *ICMCM Proceedings,* pp. 300–307, 1992.
119. M. J. Duggan. "Fabrication and Properties of Low Permittivity Silica Fiber Reinforced Silica Substrate," M. S. thesis, Pennsylvania State University, 1991.
120. G. Messner. Private communication.
121. R. N. Master, M. Cole, and G. Martin. "Ceramic Column Grid Array for Flip Chip Applications," *Proceedings of IEEE–ECTC,* pp. 925–929, 1995.
122. Wei-Yean Howng. "Ceramic Technologies for Portable Radio Applications." Private communication.

10

PLASTIC PACKAGING

MICHAEL G. PECHT—*University of Maryland*
LUU T. NGUYEN—*National Semiconductor*

10.1 INTRODUCTION

This chapter presents the information needed to design, manufacture and test microelectronic devices which are encapsulated in "plastic" molding compound. The chapter begins with an historical overview, then discusses advantages and disadvantages of plastic versus ceramic packaging. The next three sections present the materials, manufacturing processes for molding and the handling methods of the finished product. This is followed by a section on test and reliability issues. The chapter concludes with a view of the future of plastic encapsulated microcircuits. The interested reader is referred to the books titled "Plastic Encapsulated Microcircuits" [1], and "Integrated Circuit, Hybrid and Multichip Module Design Guidelines" [2] for more detailed information.

Electronic devices have been packaged in a variety of ways. One of the first methods was a Kovar (high nickel alloy) preformed package. The device was bonded to the bottom, and the top was later secured. Ceramic packages, similar in construction to the Kovar casing, later appeared as a cheaper alternative. The first evidence of plastic encapsulation was seen in the early 1950s, using compression molding of phenolics. The phenolics material was compressed around electrical devices, leaving the electrical connectors protruding. The problem with this method was that the delicate connections were often severed due to the high pressure. By the early 1960s, plastic encapsulation emerged as an inexpensive, simple alternative to ceramics and metal encasing, and during the 1970s, virtually all high-volume Integrated Circuits (ICs) were encapsulated in plastic. For example, one cost comparison of a 14-pin package in 1979 for quantities of 500,000 packages and a gold price of $200 per ounce sets the cost per plastic package at $0.063 and the cost of a ceramic dual-in-line package (DIP) at $0.82.

Although most early devices were compression molded, potting soon emerged as a suitable alternative. It involved positioning of the electrical circuit in a container and pouring the molten encapsulant into the cavity. Figure 10-1 shows a typical transistor encapsulated using the "can and header" method. Transfer molding then gained worldwide acceptance as a economical method best suited for mass production. In transfer molding, the chip is loaded into a mold cavity, constrained, and the encapsulant is transferred from a reservoir into the cavity under pressure. The encapsulant, typically a thermosetting polymer, cross-links and cures in the cavity to form the final assembly.

Figure 10-2 shows one of the first molds used for transfer molding of a full-wave rectifier. In this method, the thin-wirebonds used to form connections from the chip to the outside world were frequently damaged due to the high velocities of the encapsulant. A novel approach to this

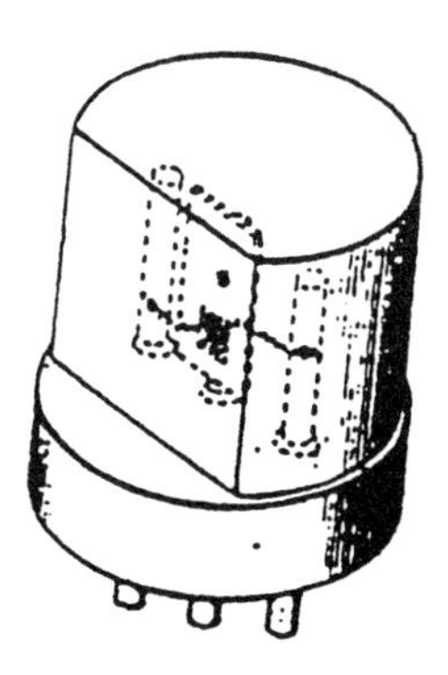

Figure 10-1. "Can and Header" Transistor. (Lanzl, Patent # 3,235,937.)

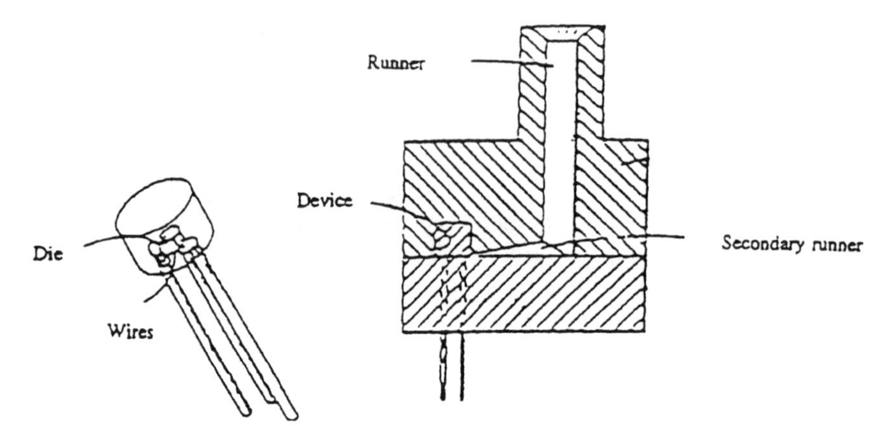

Figure 10-2. Transfer Molding of a Full-wave Rectifier. (Doyle, Patent # 3,367,025.)

problem was to use a bottom-sided gated process in which the molding compound entered the cavity from the opposite side and remote from the bond wires (Fig. 10-3), thereby reducing the chances of breaking the delicate wires.

Although epoxy novolac was the first material used for plastic encapsulation, phenolics and silicones were the dominant plastics of the 1960s. At that time, plastic encapsulated microelectronics (PEMs) were plagued by numerous reliability problems owing largely to the poor quality of the encapsulation system. Over the decade, numerous formulations of epoxies

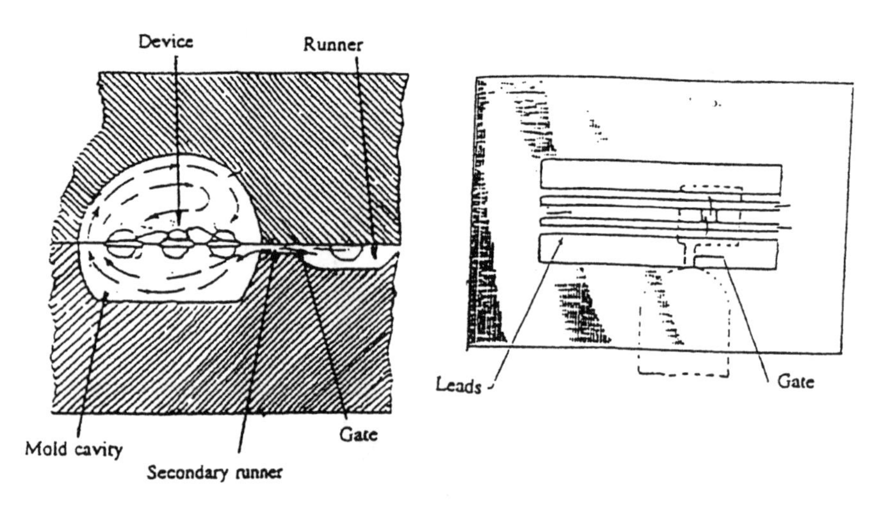

Figure 10-3. Transfer Molding with Bottom-side Gating. (Birchler et al., Patent # 4,043,027)

were developed which had lower curing shrinkages and contamination levels and thus took over as the main encapsulating material in the early 1970s. Although silicones are still sometimes used, a typical encapsulant used today is a complex mixture of cross-linkers, accelerators, flame retardants, fillers, coupling agents, mold release agents, and flexibilizers in an epoxy resin matrix.

Today, a typical plastic package or PEM (Fig. 10-4) consists of a silicon chip, a metal support or leadframe, wires that electrically attach the chip circuits to a leadframe which incorporates external electrical connections, and a plastic epoxy-encapsulating material to protect the chip and the wire interconnects. The leadframe is made of a copper alloy, alloy 42 (42 Ni/58 Fe) or alloy 50 (50 Ni/50 Fe), and is plated with gold and silver or palladium, either completely or in selected areas over nickel or nickel/cobalt. The silicon chip is usually mounted to the leadframe with an organic conductive formulation of epoxy. Wires, generally of gold but also of aluminum or copper, are bonded to the aluminum bonding pads on the chips and to the fingers of the leadframe. The assembly is then typically transfer molded in epoxy. Following the molding operations, the external leads are plated with a lead–tin alloy, cut away from the strip, and formed into desired shapes.

Plastic packages can be premolded or postmolded. In the former, a plastic base is molded; the chip is then placed on it and connected to an I/O fan-out pattern with wire. The die and wirebonds are usually protected by an epoxy-attached lead which forms a cavity. Premolded packages are

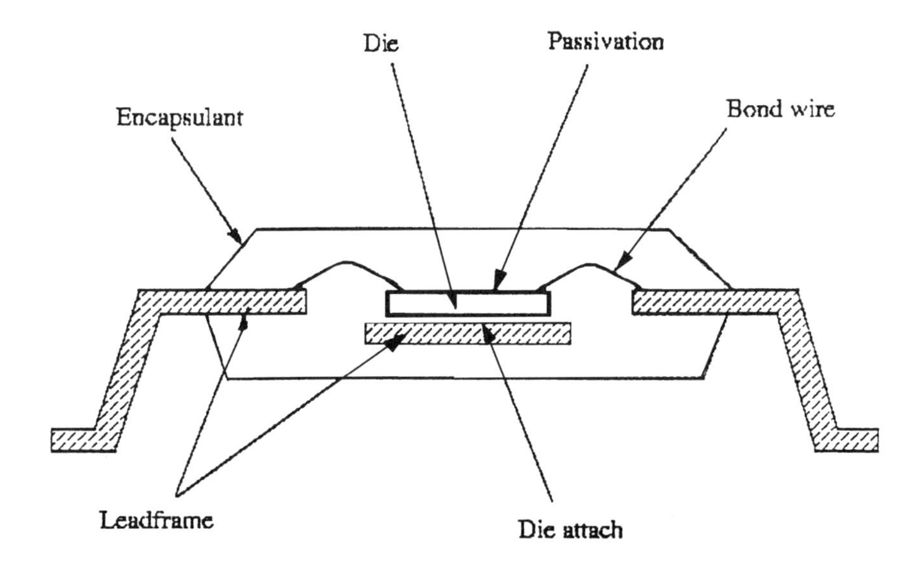

Figure 10-4. Typical Plastic Package Construction (From Ref. 1.)

most often used for high-pin-count devices or pin-grid arrays that are not amenable to flat leadframes and simple fan-out patterns.

In the postmolded type, the die is attached to a leadframe, which is then loaded into a multicavity molding tool and encapsulated in a thermoset molding compound via the transfer-molding process. Postmolded packages are less expensive than premolded ones because there are fewer parts and assembly steps. In the 1990s, about 90% of plastic packages were made using postmolding techniques.

Plastic-encapsulated microelectronics are made in either surface-mount or through-hole configurations. The common families of surface-mounted PEMs are the small-outline package (SOP), the plastic leaded chip carrier (PLCC), and the plastic quad flat pack (PQFP). The common families of through-hole-mounted PEMs are the plastic dual-in-line (PDIP), the single-in-line (SIP), and the plastic pin-grid array (PPGA). These are discussed in overview chapter Sections 1-5 and illustrated in Fig. 1-28.

10.1.1 Through-Hole-Mounted Devices

The common families of through-hole-mounted PEMs are the plastic dual-in-line (PDIP), the single-in-line (SIP), and the plastic pin-grid array (PPGA). The plastic dual-in-line package, the most commonly used PEM in the 1980s, has a rectangular plastic body with two rows of leads, often on 2.54-mm (100-mil) centers on the long sides. The dual-in-line package is conducive to high-volume manufacturing at low cost. Their in-line structure of the leads allows the packages to be shipped in plastic tubes end to end, without contacting the leads. The leads are bent up at a small angle from the package body for through-hole insertion mounting. The package layout allows automated board mounting and offers highly reliable joints.

Single-in-line packages are rectangular in shape with leads on one of the long sides. The leads are typically on 1.27-mm (50-mil) centers at the plastic body interface, formed into two staggered rows spaced 2.54 mm (100 mils) and 1.225 mm (50 mils) apart. These packages offer a high profile but small footprint on the board, maintaining a 2.54-mm (100-mil) hole-mounting standard. They offer all the advantages of dual-in-line packages in ease and low manufacturing cost. Plastic pin-grid arrays are packages with leads located in a grid array under a plastic body. The pins are often located on 2.54-mm (100-mil) centers, although 1.715-mm (70-mil) centers are also available. The pin-grid array package takes advantage of mature through-hole board mounting technology for high-pin-count devices. Plastic pin-grid arrays are thus the highest density through-hole packages and offer the highest available pin counts for PEMs. The grid-array layout of the interconnects allows impedance matching of the package to the chip. Using multilayer printed-board technology, ground and power planes can be incorporated into the package for high-performance devices. Heat sinks can also be incorporated, if needed. Figure 10-5 shows some typical through-hole plastic configurations.

10.1.2 Surface-Mounted Devices

Common families of surface-mounted PEMs are the small-outline packages (SOP), the plastic leaded chip carrier (PLCC), and the plastic quad flat pack (PQFP). These packages are designed for low-profile mounting on printed wiring boards.

Small-outline packages are similar to dual-in-line packages, with leads on two sides of the package body, but the leads are typically formed in a gull-wing configuration.

A variation on the small-outline package is the small-outline J-leaded (SOJ) package, in which the lead is formed in a J-bend configuration and folded under the body. The advantage is an even smaller footprint than that of a gull wing, although solder-joint inspection becomes more difficult.

Plastic leaded chip carrier packages are molded PEMs with leads on all four sides of the plastic body. Typically, from 18 to 124 leads are present on 1.27-mm (50-mil) centers and formed in the J-bend configuration. Because the leads are on all four sides of the plastic body, this package style offers the advantages of dense mounting on the board. Electrically, the interconnect leads of plastic leaded chip carriers are shorter on average and more consistent in length than those of equivalent dual-in-line packages, resulting in a better match of the impedances of the package leads.

Plastic quad flat pack packages (PQFPs) are square or rectangular plastic packages, typically with 40–304 leads distributed on all four sides. Figure 10.5 shows typical surface-mount plastic package configurations.

A new package called Very Small Peripheral Array Package can be considered a type of plastic package because the pins in three rows of periphery are inserted into the preformed plastic as illustrated in Chapter one Fig. 1-54. This simple and elegant package offers the "best of both worlds" in that it is superior to QFP in size and to BGA in pin-to-board reliability. The cost per pin is about the same as QFP. Another major improvement worth noting is its outstanding heat transfer capability that is achieved as a result of the copper heat sink.

10.1.3 Plastic Versus Ceramic

Plastic-encapsulated microelectronics offer many advantages over hermetic packages in the areas of size, weight, performance, cost, reliability, and availability. It is not surprising that plastics account for more than 97% of the worldwide commercial chip-encapsulation market [3].

Molded Plastic Packages

Package Type	Pin Count	Lead Pitch mm	Devices Used for
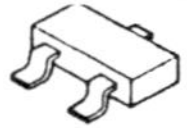 SOT-23, SC-59	3	3.0	Diodes & Transistors
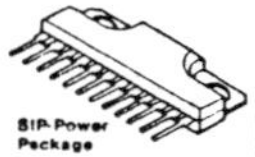 SIP-Power Package	3-12	2.54	Power Transistors, Darlingtons Plus Control Logic
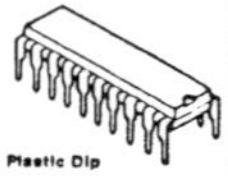 Plastic Dip	8-64	2.54	Linear, Logic, DRAMS, SRAMS, Microprocessors, ROMS, PROMS Gate Arrays
 Zig-Zag In Line Package (ZIP)	16-24	1.3	DRAMS & SRAMS
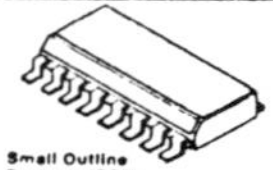 Small Outline Package (SOP)	8-28	1.3	Linear, Logic, SRAMS

Package Type	Pin Count	Lead Pitch mm	Devices Used for
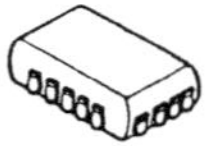 PLCC	18	1.3	64 & 256K DRAM
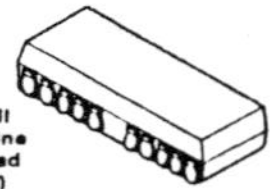 Small Outline J-Lead (SOJ)	20-26	1.3	One & Four Megabit DRAM
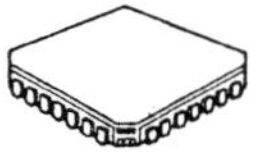 Plastic Leaded Chip Carrier (PLCC)	20-84	1.3	Logic, PROMS, Microprocessors, Gate Arrays, Standard Cell
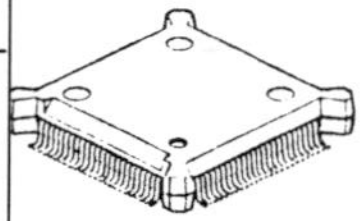 Plastic Quad Flat Pac (PQFP)	88-200	0.64	Gate Array & Standard Cell Logic

Package Type	Pin Count	Lead Pitch mm	Devices used for
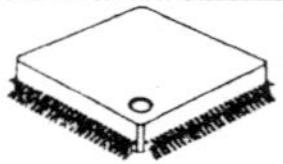 Square Quad Flat Pac (SQFP) 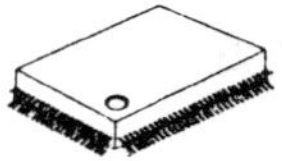 Rectangular Quad Flat Pac (RQFP)	88-200	0.64	Gate Array & Standard Cell
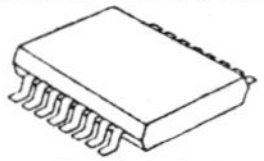 Very Small Outline Package (VSOP)	16-30	0.65	DRAMS
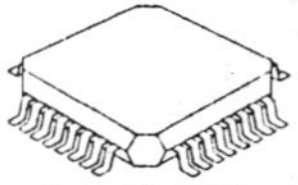 Very Small Quad Flat Pac (VSQF)	32-100	0.5	Gate Array

Figure 10-5. Plastic Package Configurations

10.1.3.1 Size and Weight

Commercial PEMs generally weigh about half as much as ceramic packages. For example, a 14-lead plastic dual-in-line package (DIP) weighs about 1 g, versus 2 g for a 14-lead ceramic DIP. Although there is little difference in size between plastic and ceramic DIPs, smaller configurations, such as small-outline packages (SOPs), and thinner configurations, such as thin small-outline packages (TSOPs), are available only in plastic. The use of SOPs and TSOPs also enables better-performing circuit boards due to higher packing density and consequent reduced component propagation delays. A smaller form factor naturally implies higher board density and more functionality packed into the same precious board real estate. Similarly, a lighter package results in a smaller overall payload for the same board functionality, a concern of critical importance for the avionics industry. The relative size, weight and mounting height of various plastic packages were discussed and illustrated in Chapter 7 ("Microelectronics Packaging—An Overview") Figs. 7-23, 7-24, 7-25 and 7-26.

10.1.3.2 Performance

Plastics have better dielectric properties than ceramics. Although plastic packages, such as DIPs, are not the most effective in propagating high-frequency signals [4], plastic quad flat pack (PQFP), pin-grid arrays, and ball-grid arrays are favored for minimizing propagation delays. For the typical commercial applications, in which frequencies do not exceed 2–3 GHz, plastic packages perform better than their ceramic counterparts in the same form factor. Two main features account for this characteristic: the lower dielectric constant of the epoxy compared with that of the standard cofired ceramic, and the smaller lead inductance of the copper leadframes next to the Kovar leads.

When the application calls for much higher frequencies (up to 20 GHz) however, better and more predictable performance is obtained with ceramic packages. The dielectric constant of typical ceramics stays the same over a wider frequency range than that of plastic molding compounds. This is because the dielectric constant is influenced by the moisture content in the package.

10.1.3.3 Cost

The cost of a complete plastic package is driven by several factors, such as die, encapsulation, production volume, size, cost of assembly and yield, screening, pre-burn-in and its yield, burn-in final testing and its yield, and the mandatory qualification tests. Table 10-1 presents the relative cost for various microcircuits packaging options. Because more than 97%

Table 10-1. Relative Cost of Packaging Options (From Ref. 1.)

Package Type	Relative Cost
Small-outline integrated circuit (SOIC)	0.9
Plastic dual-in-line package (PDIP)	1.0
Plastic leadless chip carrier (PLCC)	1.2
Ceramic dual-in-line package (CERDIP)	4.0
Plastic pin-grid array (PPGA)	10.0
Plastic quad flat pack (PQFP)	50.0
Pin-grid array (PGA)	130.0
Leaded chip carrier (LCC)	150.0

of the integrated-circuit market is plastic packaged, the cost has been lowered by high demand, competition, and high-quality automated volume manufacturing.

Hermetic packages usually have a higher material cost and are fabricated with more labor-intensive manual processes. For example, Thomson-CSF reported a 45% purchase cost reduction for each of 12 printed wiring boards in a manpack transceiver application implemented with plastic rather than ceramic components [5].

Another factor that drives up the cost of hermetically packaged integrated circuits is the rigorous testing and screening required for the low-volume hermetic parts [6]. When both types were screened to customer requirements, ELDEC [7] estimated that purchased components for plastic encapsulation of integrated circuits cost 12% less than their hermetic counterparts, primarily due to the economics of high-volume production.

The cost benefits of PEMs decrease with higher integration levels and pin counts, because of the high price of the die in relation to the total cost of the packaged device. Although these cost benefits may not be realized for complex monolithic very-large-scale integrated circuits, cost advantages may accrue for complex package styles, such as multichip modules, because of the ease of assembly. Indeed, the trend toward future multichip modules in laminates (MCM-L) packaged in form factors such as PQFPs or ball-grid arrays (BGAs) will make plastic packages even more popular. In the MCM-L, several dies and passive components can be combined on a printed-circuit board substrate and integrated into a leadframe to enhance part functionality. This approach shortens the time to market (by using readily available dies to eliminate the need for die integration), increases the process yield (there are no large dies and no mixture of such different technologies as Complimentary Metal Oxide Semiconductor and bipolar), and lowers package cost (compared with a ceramic equivalent). MCM-L products have already been produced for some time and the cost savings achieved will impact the proliferation of plastic packages [8].

10.1.3.4 Reliability

The reliability of plastic-encapsulated microelectronics, which was traditionally a major concern, has increased tremendously since the 1970s, due largely to improved encapsulating materials, die passivation, and manufacturing processes. In particular, modern encapsulating materials now have lower ionic impurities, better adhesion to other encapsulation materials, a higher glass-transition temperature, higher thermal conductivity, and better matching of the coefficients of thermal expansion to the lead-frame.

Advances in passivation include better die adhesion, fewer pinholes or cracks, low ionic impurity, low moisture absorption, and thermal properties well matched to the substrate. The failure rate of plastic packages has decreased from about 100 per million device-hours in 1978 to about 0.05 per million device-hours in 1990 [9]. Figure 10-6 summarizes published improvements in reliability of different PEM technologies since 1976 [7].

Figure 10-7 presents comparative failure-rate data for PEMs and hermetically packaged devices from first-year warranty information on commercial equipment operating primarily in ground-based applications

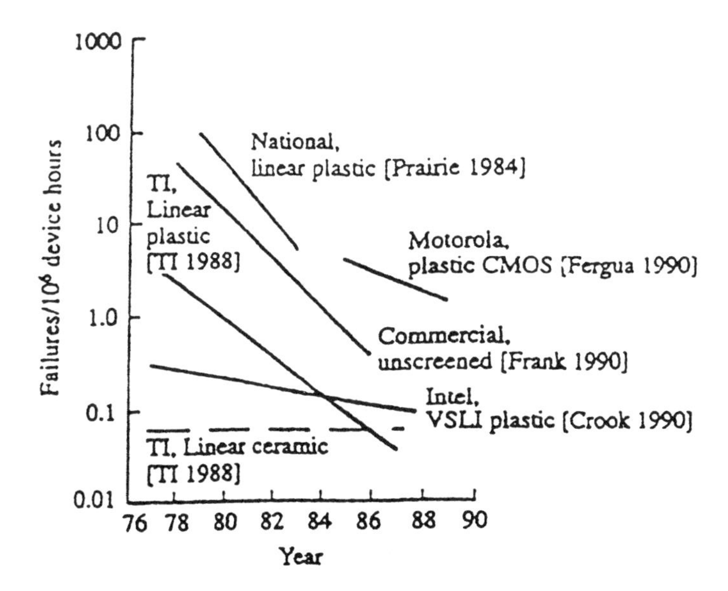

Figure 10-6. Microcircuit Reliability Improvement Trends. (From Ref. 7.)

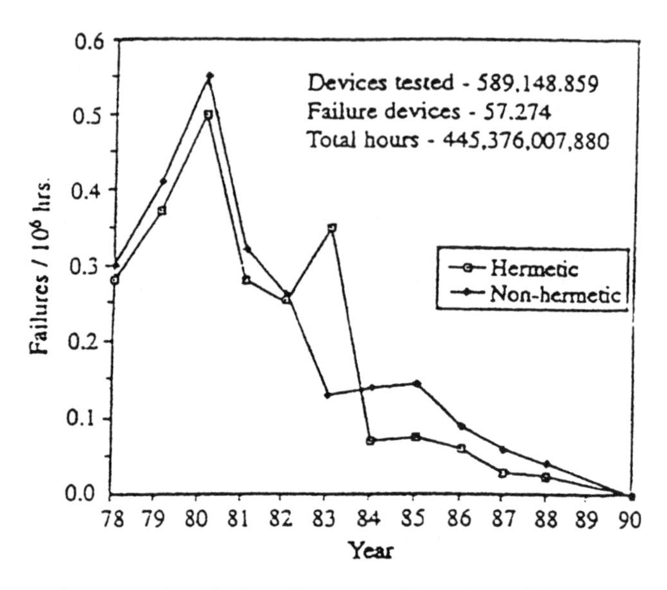

Figure 10-7. Comparative Failure Rate as a Function of Year. (From Ref. 10.)

(office, laboratory, and transportable equipment) from 1978 to 1990; the rates are for the same part (or part function) over time [10]. As shown in Figure 10-8, during this period, both types of packaged devices improved by more than one order of magnitude in early-life failure rate. For PEMs, the early 1990s failure rate was between 0.3 and 3.0 failures per 10^6 device-hours, with less variability between encapsulant materials and their vendors. This very closely correlates with the figures for hermetic parts [11].

A study by Rockwell International, Collins Group [12] compared plastic dual-in-line packages and plastic surface-mounted devices to ceramic dual-in-line packages (CERDIPs) during extended temperature cycling of –40°C to +85°C for 3532h in (4-h cycles). The parts were mounted on circuit boards, biased with 5 V during testing, and subjected to 98% relative humidity. These conditions were designed to stimulate a worst-case avionics environment. The failure rate for the plastic parts was as low as 0.0016% per 1000 h, whereas the failure rate for the hermetic parts was 0.0061% per 1000 h, both at 60% confidence. The large difference in failure rates, shown in Table 10-2, is due to the loss of hermeticity of glass seals. The failure mechanism for both plastic and hermetic parts was metallization damage due to moisture. The sample size of the various lots was 2920 plastic parts and 1200 ceramic parts from 5 different vendors (Texas Instruments, Motorola, National, Signetics, and Fairchild); there were 174.6 million equivalent field device-hours.

Lidback [13], of Motorola's Semiconductor Product Sector, studied

Table 10-2. Thermomechanical Reliability (From Ref. 1.)

Study [Ref.]	Package	Number Failed/ Number Tested	Failure Rate
Griggs	Plastic hermetic	1/2,920	0.0342%/1K h
(Rockwell Collins) [12]		2/1,200	0.1667%/1K h
Villalobos	Plastic hermetic	23/9,177	0.2506%/1K cycles
(Motorola TED)		7/1,844	0.3796%/1K cycles
Lidback	Plastic hermetic	11/133,747	0.083%
(Motorola SPS)		46/46,473	0.099%

133,747 PEMs and 46,473 hermetic packages (obtained from Motorola's Semiconductor Product Sector) that underwent 1000 temperature cycles from –65°C to + 150°C. The failure rate was 0.083% for the PEMs and 0.099% for the hermetic packages.

Villalobos [14] ran 1000 cycles at –65°C to +150°C using a smaller sample size of 9240 plastic packages and 1848 ceramic packages (also obtained from Motorola). The ceramic failure rate was 0.38%, and the plastic failure rate was 0.44%. Both of these studies indicated similar failure mechanisms for plastic and hermetic packages under worst-case conditions.

The automotive industry, which uses some of the commercial industries' most stringent qualification standards, conducts temperature cycling for 1000 cycles, thermal shock (liquid-to-liquid) for 500 cycles, 85°C and 85% rh testing for 1000 h, life testing for 1000 h, high-temperature reverse bias for 1000 h, intermittent operational life testing for 20,000 cycles, and autoclave (live steam) testing for 96 h. The number of rejects allowed for all these tests is zero. Most vendors pass these tests without problems, indicating a broad, industrywide ability to meet or exceed harsh automotive standards [15].

In general, PEMs lend themselves well to automatic assembly techniques, which eliminate manual handling and operator error, resulting in high yields and low assembly costs. On the other hand, even for high-volume production of hermetic packages, automated assembly pick-and-place machines have been known to crack hermetic seals or chip ceramic packages.

10.1.3.5 Availability

Plastic devices are assembled and packaged on continuous production lines, as opposed to the on-demand production of hermetic parts, thus making PEMs much more readily available than hermetic devices.

This is mainly because market forces (cost and volume) encourage most designs to be developed first as plastic encapsulated. Lead times for plastic packages are significantly shorter. The problems associated with restarting a hermetic line are absent in continuous plastic production lines.

Hermetic packages are developed only when there are perceived high-performance requirements and sufficient market interest. Thus, some parts are simply not available from major manufacturers in hermetic form. Furthermore, the U.S. military and government, the major purchasers of hermetic parts, have become a relatively small portion of the total electronics market (~ 1% in 1995), although they accounted for nearly 80% of the total market in the 1960s. With package technology moving to surface mount, development of ceramic packages has lagged further in the microelectronic market, making adaptation of plastic-packaged integrated circuits to government and military applications more critical. With global competition, industrial research in materials and manufacturing processes will continue to focus on PEMs. In 1993, over 97% of all integrated circuits were packaged in plastic. It is estimated by suppliers that, at any given time, 30% more part functions are available in plastic than in ceramic [7].

10.1.4 Summary

Plastic encapsulation has come a long way. It offers attractive benefits over ceramic technology in terms of form factors, weight, performance, cost, and availability. The reliability of plastic packages is no longer a stumbling block to their widespread application.

Plastic-encapsulated microelectronics have generally encountered formidable challenges in gaining acceptance for use in government and military applications. In fact, it was only in the early 1990s that the industry dispelled the notion that hermetic packages were superior in reliability to plastic packages, in spite of the fact that hermetic packages typically had low production and procurement volumes, along with outdated government and defense department standards and handbooks associated with their manufacture and use.

Prior to the 1980s, moisture-induced failure mechanisms, such as corrosion, cracking, and interfacial delamination, were significant [16]. However, improvements in encapsulants, die passivation, metallization technology, and assembly automation have made plastics the encapsulation technology of the future. Currently, the best endorsement for PEMs is from automotive manufacturers, which consume plastic integrated circuits at a rate of over 2.7 million a day [9].

Plastic-encapsulated microcircuits will continue to account for the vast share of the integrated-circuits market in coming years, but hermetic packages, with their special characteristics, will continue to have a unique market in the electronics industry.

10.2 MOLDING COMPOUNDS AND LEAD FRAME MATERIALS

A PEM encapsulant is generally an electrically insulating plastic material formulation that protects an electronic device and die-leadframe assembly from the adverse effects of handling, storage, and operation.

The encapsulant must possess adequate mechanical strength, adhesion to package components, manufacturing and environmental chemical resistance, electrical resistance, matched coefficient of thermal expansion to the materials it interfaces with, and high thermal and moisture resistance in the application temperature range. Thermosetting encapsulation compounds (i.e., molding compounds) based on epoxy resins or, in some niche applications, organosilicone polymers, are the most widely used to encase electronic devices. Polyurethanes, polyimides, and polyesters are used to encase modules and hybrids intended for use under low-temperature and high-humidity conditions. Thermoplastics are rarely used for PEMs, because they require high-temperature and high-pressure processing conditions, are low-purity materials and can result in moisture-induced stresses.

The molding compound is a multicomponent mixture of an encapsulating resin with various types of additives. The principal active and passive (inert) components in a molding compound include curing agents or hardeners, accelerators, inert fillers, coupling agents, flame retardants, stress-relief additives, coloring agents, and mold-release agents. Table 10-3 lists these components and their functions in epoxy molding compounds [17]. Each of these components will be discussed in this section.

10.2.1 Resins

The earliest resin materials used for plastic packaging of microelectronic devices were silicones, phenolics, and bisphenol A epoxies because of their excellent molding characteristics. Silicones were used as molding compounds because of their high-temperature performance and purity, but poor adhesion of silicones with the device and metallization resulted in device failure in salt-spray tests and flux penetrants during soldering. Integrated circuits (ICs) molded with phenolics experienced early corrosion failures caused by the generation of ammonia during postmold curing, the presence of high levels of sodium and chloride ions, and high moisture absorption. Early epoxy materials such as bisphenol A (BPA) offered much better adhesion than silicones but have a low glass-transition temperature in the range of 100–120°C and high concentrations of corrosion agents such as chlorine. As a result, novolac epoxies displaced BPA materials.

Table 10-4 summarizes PEM properties of commonly used polymeric materials. Except for epoxy resins, all others presented have only niche applications in PEM manufacture.

Table 10-3. Components of Epoxy Molding Compounds Used in Electronic Packaging

Component	Concentration (wt. % of Resin)	Major Function	Typical Agents
Epoxy resin	Matrix	Binder	Cresol–novolac
Curing agents (hardeners)	Up to 60	Linear/cross-polymerization	amines, phenols, and acid anhydrides
Accelerators	Very low (<1)	Enhance rate of polymerization	Amines, imidazoles, organophosphines, ureas, Lewis acids, and their organic salts
Inert fillers	68–80	Lower coefficient of thermal expansion, higher TC[a] (w/Al_2O_3), higher E,[b] reduce resin bleed, reduce shrinkage, reduce residual stress	ground fused silica (widely used), alumina
Flame retardants	~10	Retard flammability	Brominated epoxies, antimony trioxide
Mold-release agents	Trace	Aid in release of package from mold	Silicones, hydrocarbon waxes, fluorocarbons, inorganic salts of organic acids
Adhesion parameter	Trace	Enhance adhesion with IC components	Silanes, titanates
Coloring agents	~0.5	Reduce photonic activity; reduce device visibility	Carbon black
Stress-relief additives	Up to 25	Inhibit crack propagation, reduce crack initiation, lower coefficient of thermal expansion	Silicones, acrylonitrile–butadiene rubbers, polybutyl acrylate

[a]Thermal conductivity.
[b]E modulus.
Source: From Ref. 17.

All epoxy resins contain compounds from the epoxide, ethoxylene, or oxirane group, in which an oxygen atom is bonded to two adjacent (end) bonded carbon atoms. Because the resins are cross-linkable (thermosetting), each resin molecule must have two or more epoxy groups. Most commonly, the epoxide group is attached linearly to a chain of $-CH_2-$ and is called a glycidyl group.

This group is attached to the rest of the resin molecule by an oxygen, nitrogen, or carboxyl linkage forming glycidyl ether, amine, or ester. The resin is cured by the reaction of the epoxide group with compounds

Table 10-4. Common Polymeric Encapsulants for Microelectronic Packages

Polymeric Material	Properties	Advantages and Disadvantages
Epoxies	Good chemical and mechanical protection Low moisture absorption Suitable for all thermosetting processing methods Excellent wetting characteristics Ability to cure at atmospheric pressure Excellent adhesion to a wide variety of substrates under many environmental conditions Thermal stability up to 200°C	High stress Moisture sensitivity Short shelf life (can be extended under low-temperature storage conditions)
Silicones	Low stresses Excellent electrical properties Good chemical resistance Low water absorption Thermal stability up to 315°C Good UV resistance	Low tensile tear strength High cost Attacked by halogenated solvents Poor adhesion Long cure time
Polyimides	Good mechanical properties Solvent and chemical resistant Excellent barriers Thermal stability from −190°C to 600°C	High cure temperature Dark color (some) Attacked by alkalis High cost Require surface priming and/or coupling agents to improve adhesion properties Low moisture resistance Low dielectric constant Lower thermal stability Provides improved stress relief
Phenolics	High strength Good moldability and dimensional stability Good adhesion High resistivity Thermal stability up to 260°C Low cost	High shrinkage Poor electrical properties High cure temperature Dark color High ionic concentration
Polyurethanes	Good mechanical properties (toughness, flexibility, resistance to abrasion) Low viscosity Low moisture absorption Ambient curing possible Thermal stability up to about 135°C Low cost	Poor thermal stability Poor weatherability Flammable Dark colors High ionic concentration Inhomogeneities due to poor mixing of two-part system

containing a plurality of reactive hydrogen atoms (curing agents) like primary, secondary, or tertiary amines, carboxylic acid, mercaptan, and phenol; as such, they are coreactants. Catalytic cures are affected by Lewis acids, Lewis bases, and metal salts, resulting in the linear homopolymerization of the epoxide group forming polyether linkages. The viscosity of the base resin for the encapsulant-grade materials is kept low by limiting the average degree of polymerization of the novolac epoxy to values around 5. The different commercial formulations differ primarily in the distribution of the degree of polymerization, with an average molecular weight of about 900. Longer and shorter chain lengths provide looser or tighter cross-link structures, with varying ductility and glass-transition temperatures.

In electrical and electronic applications, three types of epoxy resins are commonly used: the diglycidyl ethers of bisphenol A (DGEBA) or bisphenol F (DGEBF), the phenolic and cresol novolacs, and the cycloaliphatic epoxides. Liquid DGEBAs synthesized from petrochemical derivatives are most common. They are readily adaptable for electrical and electronic device encapsulation. DGEBF is less viscous than DGEBA. The epoxy novolacs, essentially synthesized in the same way as DGEBA, are primarily solids. Because of their relatively superior elevated-temperature performance, they are widely used as molding compounds. The cycloaliphatic epoxides or peracid epoxides, usually cured with dicarboxylic acid anhydrides, offer excellent electrical properties and resistance to environment exposure.

A new class of popcorn-resistant (during reflow soldering in printed wiring assembly) ultralow-stress epoxy encapsulating resins has been synthesized. Ultralow-stress epoxy molding compounds exhibit almost no shrinkage while curing. The absence of stress is achieved by including chemicals that inhibit cross-links between polymer chains in the plastic. Without cross-links, the chains are less likely to pull toward one another as the plastic cures. These new compounds result in PEMs that routinely survive solder immersion shock tests without failure.

10.2.1.1 Fillers and Coupling Agents

Fillers are employed in epoxy resins to modify the properties and characteristics of epoxies. Inert inorganic fillers are added to the molding compound to lower the coefficient of thermal expansion, increase thermal conductivity, raise the elastic modulus, prevent resin bleed at the molding tool parting line, and reduce encapsulant shrinkage during cure (and, thus, reduce residual thermomechanical stress). Microstructurally, particle shape, size, and distribution in the chosen filler dictate the rheology of the molten epoxy molding compound. The advantages and disadvantages of the use of the epoxy are given in Table 10-5. Common fillers and the properties they affect are given in Table 10-6.

Table 10-5. Advantages and Disadvantages of Using Fillers

Advantages	Disadvantages
Reduced shrinkage	Increased weight
Improved toughness	Increased viscosity
Improved abrasion resistance	Increased dielectric constant
Reduced water absorption	
Increased heat-deflection temperature	
Increased thermal conductivity	
Reduced thermal-expansion coefficient	

Historically, the filler with the optimum combination of required properties has been crystalline silica or alpha quartz. A typical crystalline-silica-filled molding compound, loaded to 73% by weight, offers a coefficient of thermal expansion of about 32 ppm/°C and a thermal conductivity of around 15 kW/m °C. Crystalline silica was replaced by ground fused silica, which provided lower density and viscosity. A formulation with similar moldability, as one with 73% crystalline silica, requires 68% fused silica and produces a coefficient of thermal expansion of around 24 ppm/°C and a thermal conductivity about 16 kW/m °C.

The addition of particulate fillers generally reduces strength characteristics such as tensile strength and flexural strength. Fillers do not usually provide any significant enhancement of glass-transition temperature or other measures of heat-distortion temperatures. Fillers also allow the possibility of modifying various thermal characteristics, including the thermal conductivity and thermal expansion coefficient. Thermal conduc-

Table 10-6. Common Fillers and Their Property Modification

Filler	Property
Alumina	Abrasion resistance, electrical resistivity, dimensional stability, toughness, thermal conductivity
Aluminum trioxide	Flame retardation
Beryllium oxide	Thermal conductivity
Calcium Silicate	Tensile strength, flexural strength
Copper	Electrical conductivity, thermal conductivity, tensile strength
Silica	Abrasion resistance, electrical properties, dimensional stability, thermal conductivity, moisture resistance
Silver	Electrical conductivity, thermal conductivity

tivity can be increased by a factor of about 5 by the addition of fillers such as alumina and copper. Generally, an increase in filler concentration increases thermal conductivity. Figure 10-8 shows data on the effect of filler content on thermal conductivity. Thermal conductivity can also be enhanced by the addition of other filler materials such as aluminum nitride, silicon carbide, magnesium oxide, and silicon nitride [18]. However, Proctor and Solc [19] have shown the futility of using filler materials for thermal conductivity more than 100 times of that of the base resin and estimated that a practical limit of thermal conductivity improvement is about 12 times that of the thermoset resin. The influence of varying the thermal conductivity of the epoxy molding compound on the thermal performance of the IC package has also been evaluated [20]. In this case, by varying the filler types (e.g., fused and crystalline silica, aluminum oxide, aluminum and boron nitride, silicon carbide, and diamond), the particle sizes, and the filler distribution, the thermal conductivity of the composites was estimated. The resulting effectiveness in dissipating heat away from the package was computed for three different surface-mount packages, namely SOIC 8-lead, 16-lead wide, and 24-lead wide, The results indicated that an enhanced mold compound can cause a decrease in thermal resistance of the package equivalent to what can be achieved with a molded-in heat spreader. However, when thermally enhanced lead-frames are used, the modified compounds are less effective.

In low-stress epoxy resins, spherical silica particles are usually blended with crushed silica to further lower the coefficient of thermal expansion, by increasing filler loading with a nonlinear increase of viscos-

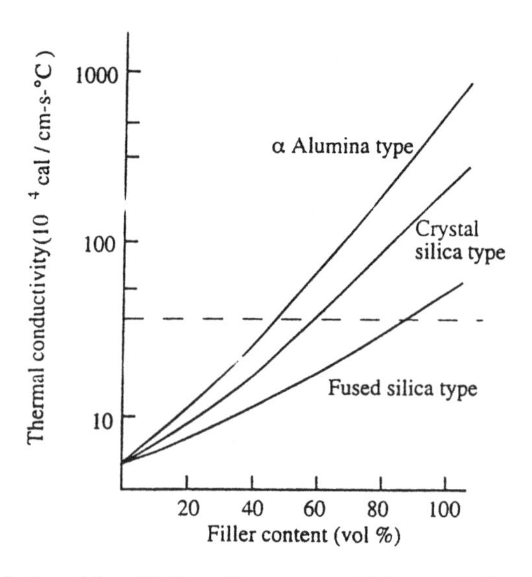

Figure 10-8. Relationship of Filler Content and Thermal Conductivity of the Melting Compound [18]

ity. However, an increase in the melt viscosity of the molding compound can increase void density and increase the difficulty of achieving a uniform encapsulant flow over large areas. Variables such as particle size, particle-size distribution, particle surface chemistry, and particulate volume fraction have been found to be the most important variables necessary for optimum property enhancement. The effect of the filler figure ratio (angular : spherical) as measured by the molding compound spiral flow length for two volume-percentages filler loadings is shown in Figure 10-9. However as Figure 10-10 shows, moisture ingress susceptibility of the molding compound as measured by the "popcorn" effect during 215°C, 90-sec, vapor-phase soldering increases with higher percentages of spherical silica filler.

Rosler [18] evaluated various other fillers for optimizing the coefficient of thermal expansion and the thermal conductivity for specific applications. Figure 10-11 presents the effect of lowering the coefficient of thermal expansion of the molding compound as a function of the crystalline silica, α-alumina, and fused silica volume percentage.

Most polymers undergo shrinkage during the polymerization and cross-linking process. Incorporation of fillers reduces shrinkage by simple bulk replacement of resin with an inert compound which does not participate in the cross-linking process.

Filler addition results in increased viscosity and improved toughness.

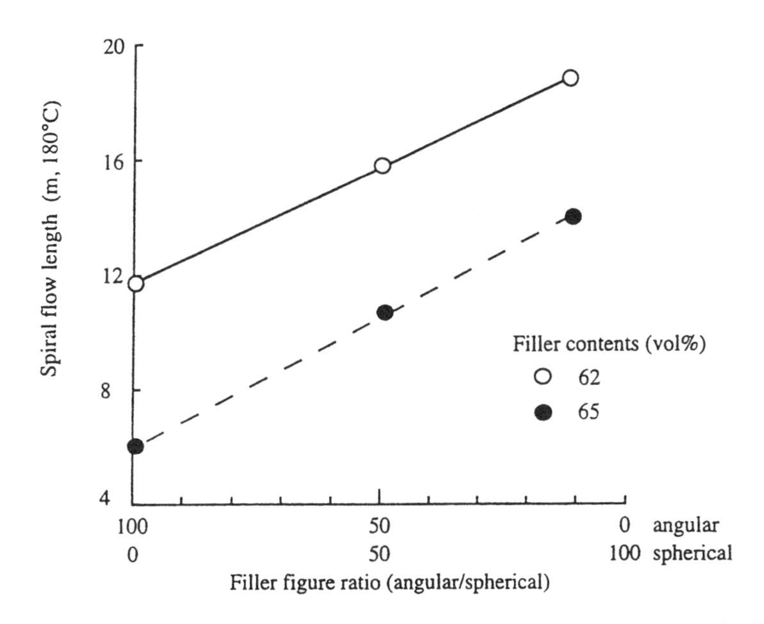

Figure 10-9. The Effect of the Filler Figure Ratio (angular : spherical) as Measured by the Molding Compound Spiral Flow Length for Two Volume-Percentages Filler Loadings (Courtesy of National Semiconductor.)

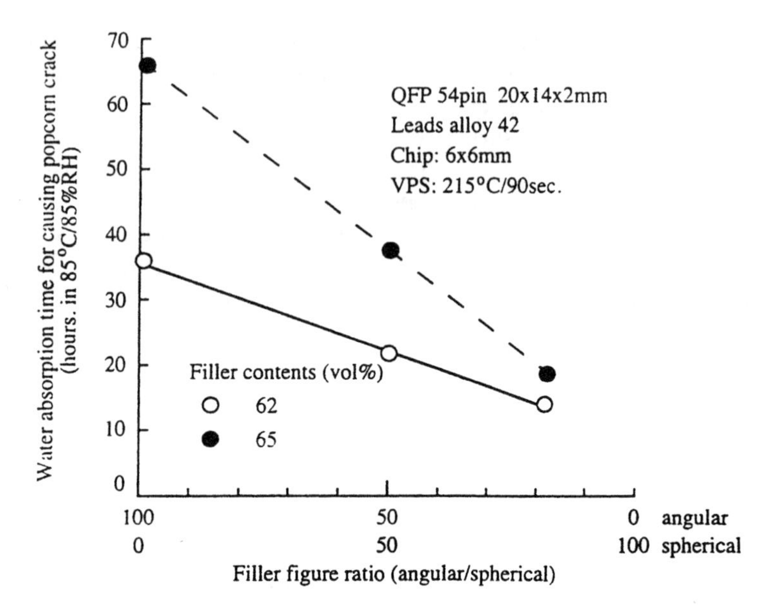

Figure 10-10. Moisture Ingress Susceptibility of the Mold Compound as Measured by the Popcorn Effect

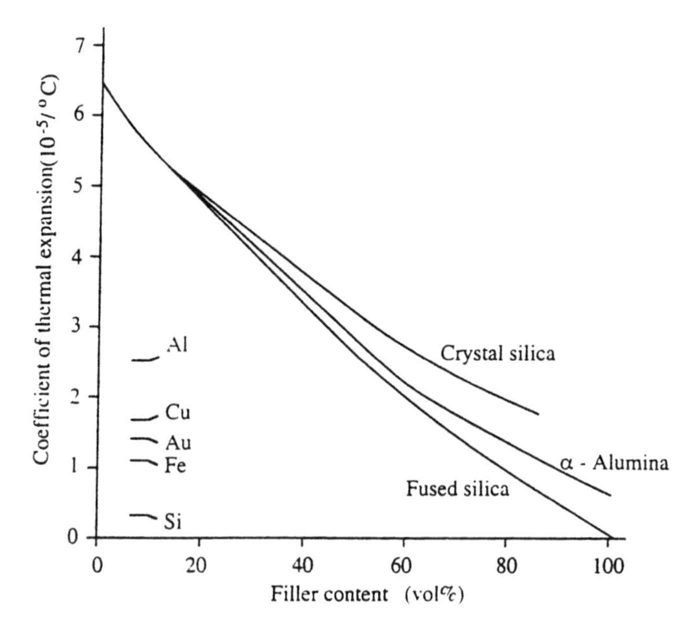

Figure 10-11. Effect of Lowering the Coefficient of Thermal Expansion of the Molding Compound as a Function of the Crystalline Silica, α-alumina, and Fused Silica Volume Percentage

Investigations have shown that the incorporation of particulate fillers such as silica, glass microspheres, and alumina trihydrate can increase the toughness of various epoxy formulations.

A filler can be used to its best advantage if the adhesion between the polymer and itself is good. In particular, the filler particle–polymer interface will not be stress bearing and therefore provides a point of mechanical weakness. Coupling agents are used to increase the adhesion between fillers and polymers by linking them with covalent bonds. Commonly used coupling agents include silanes, titanates, aluminum chelates, and zircoaluminates. Interfacial adhesion enhances mechanical strength and the heat resistance of the encapsulant. Although, coupling agents also improve processability, some adversely affect releasability from the mold if not correctly administered. This adhesion of fillers to the polymer matrix through coupling agents also extends to the device and leadframe, reducing delamination failure.

The main source of radiation in plastic packages comes from fillers. However, the use of synthetic high-purity silica fillers in recent years has brought down the alpha emission rates (AERs) of molding compounds below the detection level of most proportional counters. Ditali and Hasnain [21] studied AER sources in an electronic part and that outside the major alpha-source contamination of the die itself, the leadframe material is more contaminated (per unit surface area) than the plastic encapsulant, and ceramic package materials are 300–400 times more alpha-emitting material contaminated than plastic encapsulants.

An approximate conversion factor of AER to soft-error rate is 0.001 AER $\equiv$ 0.1% Single Event Upset (SEU) per 1000 operating hours for the contaminated material in contact with the die active surface and alpha-particle energies in the range of 4–6 MeV. An AER of about 0.001 can be obtained from 10 ppb of U-238 (4.2 MeV alpha) or 4 ppb of Th-230 (4.7 MeV alpha).

10.2.1.2 Curing Agents and Accelerators

The binder system of a plastic-encapsulant product consists of an epoxy resin, hardener or curing agent, and an accelerating catalyst system. The conversion of epoxies from the liquid (thermoplastic) state to tough, hard thermoset solids is accomplished by the addition of chemically active compounds known as curing agents. Epoxidized phenol novolac and epoxidized cresol novolac are most often used. The specific composition of the resin and the hardener system, therefore, is optimized for the specific application. For example, incorporation of phenol novolacs in the matrix resin can be used to increase cure speed. In general, the glass-transition curing temperature is maximized at an epoxide to hydroxyl ratio of about unity.

Curing agents and accelerators perform the primary function of setting the extent and rate of polymerization of the resin. Consequently, the selection of the agent and cure chemistry are just as important as the choice of resin. Often, the same agent is used for both cure and acceleration.

The cross-linking reactions of epoxy resins are affected by aliphatic or aromatic amines, carboxylic or its derivative acid anhydrides, and complex phenols, whereas homopolymerization is affected by Lewis acids or bases and their organic salts. The most widely used curing agents are amines and acid anhydrides. Although aliphatic amines react rapidly at room temperature, aromatic amine curing agents impart higher thermal stability and improved chemical resistance. Depending on the curing agent used, the reactions proceed through different chemistries. Because this reaction is virtually quantitative, resin and curing agent are mixed in the ratio of one amino hydrogen to one epoxide. In the case of anhydrides and phenols as curatives, cross-linking occurs through pendant hydroxyls. Table 10-7 lists the curing properties of different classes of curatives used with epoxy resins. Because of their excellent moldability, electrical properties, and heat and humidity resistance [22], phenol novolac and cresol novolac hardeners have become the dominant curatives for microelectronic packaging.

To promote cure within a reasonable period of time, it is necessary to use accelerators. Accelerators reduce the in-mold cure time and improve productivity by catalytic activity. Typical accelerators include aliphatic poylamines of tertiary amines, phenols, nonyl phenol, resorcinol, or semi-inorganic-derived accelerators such as triphenyl phosphite and toluene-*p*-sulforic acid.

For transfer molding of microelectronic devices, epoxy resins are B-staged to produce storage-stable thermosetting epoxy slugs known as prepregs. The B-staged resin is formed by reacting terminal epoxides with aminohydrazides at 70–80°C to yield B-staged resins with excellent storage stability.

10.2.1.3 Stress-Release Additives

The toughness and stress-relaxation response of epoxy resins can be enhanced by the addition of flexibilizers and stress-relief agents. Stress-release additives lower the thermomechanical shrinkage that can initiate, as well as propagate, cracks in the molding compound or in the device passivation layer. In terms of molding-compound property modification, stress-release agents lower the elastic modulus, improve toughness and flexibility, and lower the coefficient of thermal expansion.

Inert flexibilizers, like phthalic acid esters or chlorinated biphenyls, remain as a separate phase. Reactive flexibilizers remain in the epoxy

Table 10-7. Pure DGEBA Epoxy Resin Curing Agents Used in Molding Compounds

Curative Type	Concentration Range (Parts/100 Parts of Resin	Typical Cure Time	Typical Cure Temperature (°C)	Typical Postcure Time (h)	Typical Postcure Temperature (°C)	Pot Life at 25°C (h)	Heat Deflection Temperature (°C)
Aliphatic tertiary amines and derivatives (room-temperature cure)	12–50	4–7 days	25	0–1	150–200	0.25–3	55–124
Cycloaliphatic amines (moderate-temperature cure)	4–29	0.5–8 h	60–150	0–3	150	0.25–20	100–160
Aromatic amines (elevated temperature cure)	14–30	4 h	80–200	2	150–200	5–8	145–175
Carboxylic and anhydrides[a] (elevated-temperature cure)	78–134	2–5 h	25–150	3–4	150–200	24–120	74–197
Lewis acids and bases[a] (elevated-temperature cure)	3	7 h	120–200	4	~200	>250	175
Latent curing agents (elevated-temperature cure)	6	1 h	175	1	175	∞	135

[a]Two-step cure.
Source: From Ref. 17.

matrix as a single-phase material and lower the tensile modulus and improve the ductility of the encapsulant. They also reduce the exotherm and, in certain circumstances, reduce shrinkage. Flexibilizers can also improve adhesive joint properties such as lap shear, peel strength, impact strength, and low-temperature crack resistance.

Stress-release additives lower the thermomechanical shrinkage stresses that can initiate as well as propagate cracks in the molding compound or in the device passivation layer. In terms of molding-compound property modification, stress-release agents lower the elastic modulus, improve toughness and flexibility, and lower the coefficient of thermal expansion.

In epoxy molding compounds, the major stress-relief agents used are silicones, acrylonitrile–butadiene rubbers, and polybutyl acrylate (PBA). Silicone elastomers, with their high-purity and high-temperature properties, are the most favored stress-relief agents. Silicone elastomers interface-modified with polymethyl methacrylate (PMMA) possess uniform domain sizes (1–100 μm) and inhibit passivation layer cracking, aluminum line deformation, and cracking.

10.2.1.4 Flame Retardants

Flame retardants in the form of halogens are added to the epoxy resin backbone because epoxy resins are inherently flammable. One of the more important flame retardants is the homologous brominated DGEBA (diglycidyl ethers of bisphenol A). When used with normal nonhalogenated resin, these give self-extinguishing properties, flame retardancy being achieved by the bromine liberated at the decomposition temperature. Approximately 13–15% (wt.) of bromine is required to make an unfilled epoxy flame retardant [17].

Used as a filler, antimony trioxide used as a filler is another commonly used flame retardant. However, the cost is higher. Homologous brominated DGEBA [18] epoxy and heterogenous antimony oxides can also be used together.

Other ways of incorporating flame retardancy in epoxy resins include using a nonreactive phosphorus-containing diluent, chlorinated waxes, aluminum hydrate, and various phosphorus derivatives. These are less popular methods because they can lead to deterioration of physical properties of the system and limit the choice of the curing agent and filler.

Although flame retardants are a source of contamination in encapsulant formulations, the use of nonhydrolyzable compounds and ion getters can control their corrosion-inducing effects. However, studies at Plaskon Electronic Materials [23] show improved high-temperature storage performance of Plaskon 3400 epoxy molding compound when flame retardants are eliminated, as per Figure 10-12.

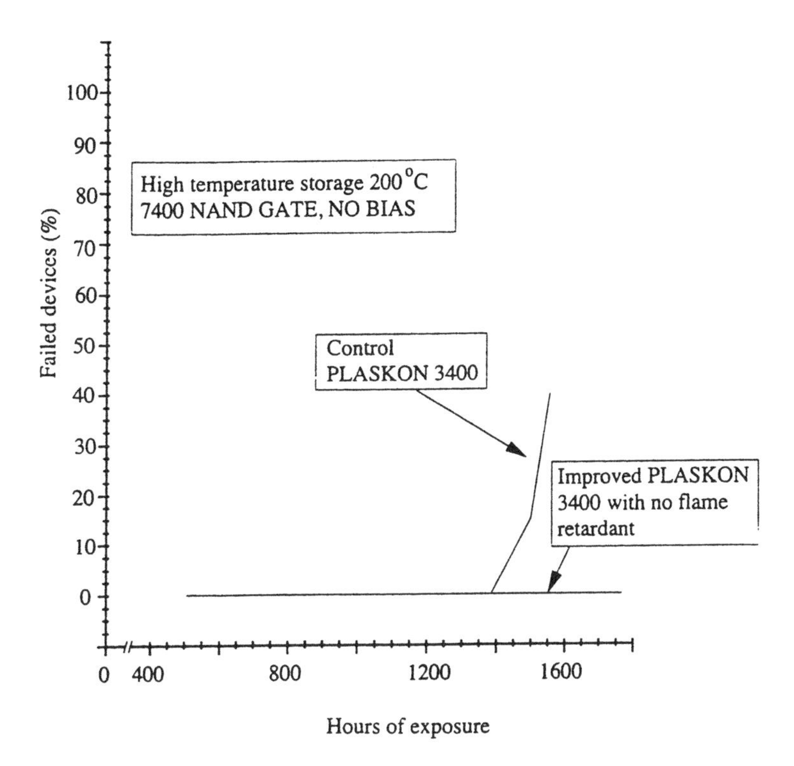

Figure 10-12. Improved High-Temperature Storage Performance of PLASKON 3400 Epoxy Molding Compound When Flame Retardants Are Eliminated. (Courtesy of PLASKON Electronic Materials.)

10.2.1.5 Mold-Release Agents

The excellent adhesion of epoxy resins to all types of surfaces is the major reason for the wide application of release agents in microelectric packaging. But these same properties make release from compounding and molding equipment difficult. Consequently, mold-release agents are needed that do not degrade the epoxy adhesion to the package components. This is achieved by controlling the release-agent activity as a function of temperature.

Mold-release agents are usually in the form of microplates that vary from liquids to pasty solids to finely divided powders. In general, the release agent should be insoluble in the resin mixture, should not melt at the curing temperature, and should be applied in a continuous film. The selection of the release agent is determined by the material of construction of the mold and the type of encapsulation selected.

Different release agents in epoxy molding compounds are used for compounding (~100°C) and molding (~175°C) operations. Room-temperature adhesion of epoxy molding compound to leadframe is provided by the higher temperature (~175°C) active mold-release agent solidification and consequent loss of effectiveness. The release agent used for compounding, if not completely degraded, could become active at near 100°C use temperature and lead to some delamination in the package. Mold-release agents for epoxy molding compounds include silicones, hydrocarbon waxes, inorganic salts of organic acids, and fluorocarbons. Of these, hydrocarbon waxes, such as carnauba wax, are the most common for molding compounds used in microelectronic encapsulation. Silicones and fluorocarbons have poor functional temperature selectivity, whereas organic acid salts can corrode metallic package elements.

10.2.1.6 Ion-Getter Additives

The objective of ion-getter additives is to reduce the conductivity of any accumulated water at the metal–encapsulant interfaces inside the package, and thus to retard any electrolytic corrosive degradation processes. The incorporation of alkali- and halide-ion gettering agents into the epoxy molding compound makes residual Na^+, K^+, Cl^-, and Br^- ions in the epoxy unavailable for dissolution into any diffused, accumulated water in the package. Ion getters in epoxy molding compounds are hydrated metal oxide powders with particles several microns in diameter. These materials react with highly active alkali and halide ions to release OH^- and H^- ions inside the molding compound and form nearly water-insoluble alkali and halide compounds. In about 5 wt.% concentration, ion-getter additives have been reported to reduce aluminum cathodic and chemical corrosion by 1.5–3 times [23]. The main application of such formulated molding compounds is in packages where the encapsulant thickness is around a few mils, as in ultrathin outline packages.

10.2.1.7 Coloring Agents

Coloring agents are normally used to distinguish different device types in packages, reduce photonic activity of the device, and eliminate device visibility through the normally pale yellow epoxy encapsulant. The colors are produced by the addition of thermally stable organic dyes or pigments. Carbon black is used in most plastic-encapsulated silicon integrated circuits, even though it does add slightly to the electrical conductivity of epoxy and can reduce moisture resistance. The concentration of carbon block is usually less than 0.5% to avoid problems related to moisture absorption and impurities.

10.2.2 "Low-Stress" Molding Compounds

On the other hand, reducing the TCE of the encapsulation down to a value matching that of silicon is a difficult and challenging task. Positive and negative TCE values for a number of polymers, metals, and ceramics are listed in Table 10-8. Some general observations can be readily made:

1. TCEs for polymers generally lie in the range from $500\times10^{-7}/°C$ for polyphenylene sulfide to $2,000\times10^{-7}/°C$ for polyethylenes.
2. Metals, on the other hand, except for Invar and SuperInvar (TCE $\simeq 10\times10^{-7}/°C$), display TCEs ranging from $50\times10^{-7}/°C$ for molybdenum and tungsten to close to $240\times10^{-7}/°C$ for tin and aluminum.

Table 10-8. Thermal Expansion of Some Representative Materials at 25°C

Material	Coefficient of Thermal Expansion
POLYMERS	
Polyethylene	1,500–3,000
Polyester	954
Nylon	820
Epoxies (cast rigid)	500
Polyphenylene sulfide	540
Polyphenylene oxide	684
METALS	
Aluminum	235
Boron	48
Copper	170
Gold	141
Invar	2
Molybdenum	48
Nickel	129
Silver	187
Tin	235
CERAMICS	
Al_2O_3	54
MgO	104
Fused silica	9
Quartz	18
TiO_2	75
Si_3N_4	8
Supertemp (Pycobond) graphite pitch material	–45
Corning 9617	—
Nb_2O_5 (sintered)	–20
β-eucryptite ($Li_2O \cdot Al_2O_3 \cdot 2SiO_2$)	–60
$SiO_2 \cdot 12Li_2O \cdot 8Al_2O_3 \cdot 1\ GeO_2$	–28
β · quartz ($64.68SiO_2 \cdot 19.9ZnO \cdot 14.9Al_2O_3$)	–3
$Li_2O \cdot Ta_2O_5$	–110
$Ta_{16}W_{18}O_{94}$	–50

3. Finally, the TCEs of ceramics are much wider, spanning over three orders of magnitude. For instance, Corning 7971 has a TCE of $0.3 \times 10^{-7}/°C$, while magnesium oxide has a TCE of $100 \times 10^{-7}/°C$.

One important conclusion to draw from Table 10-8 is that some materials (e.g., Invar and Corning 7971) can exhibit zero TCE at some temperatures and negative TCE over another, smaller range of temperatures. Studies of fused silica, for instance, indicate that the TCE is negative at low temperatures, becomes positive at higher temperatures, and turns negative again at still higher temperatures. The same behavior is observed with TiO_2-SiO_2 mixtures. At high TiO_2 concentrations, an extended region of low TCE is obtained-around $0.3 \times 10^{-7}/°C$ for Corning 7971[24,25]. Obtaining even lower TCEs in single-phase materials seems to be a very difficult problem since current theories of thermal expansion do not provide any guidance for developing such materials. The TCE of such materials seems to be structure insensitive. Small quantities of second phases and oriented grains do have some effect—typically less than 30%. Porosity, dislocation density, grain-boundary area, and stoichiometry, however, have been found to exert no major influence on the TCE.

It appears that the best approach to achieving low or near-zero TCE materials is through the development of polyphase materials. For instance, composites made of a highly anisotropic second phase inserted into an isotropic matrix can have zero or ultralow TCE for a range of temperatures. Low-expansion glasses, graphite epoxy, or metal composites are good examples. However, the anisotropic thermal expansion rules out their usage in quite a few applications. In PC board applications, for instance, the z axis expansion of Teflon® glass is worse than that of FR-4 epoxy glass laminate, being almost three times that of FR-4 at the latter's glass-transition temperature (120°C). Only particle-reinforced systems can yield a material with isotropic ultralow thermal expansion.

Research into the development of mirror substrates for high-altitude large optics (HALO) and high-energy laser optics (HELO) has pointed out an approach that might be extended to the encapsulation of chips. Requirements of high thermal conductivity, high microyield strength, low density, and distortion of less than 0.1 μm/m in the temperature range of 150°K to 300°K were imposed. Only a metal matrix composite can potentially satisfy such constraints. The two experimental systems selected, $Ni - Nb_2O_5$ and SuperInvar-Nb_2O_5, exhibited TCEs higher than desired for HALO/HELO applications. It was found that a two-phase metal composite will possess ultralow or zero TCE when the matrix has a low positive TCE while the filler should display a large negative TCE [26]. Furthermore, the volume ratio of matrix to dispersed phase requires very close and accurate monitoring. Good interphase bonding has been assumed implicitly.

10.2.3 Molding Formulation Process

Molding compounds are created in automated processing plants that take raw materials and blend them into a molding powder with the properties required for molding ICs. Usually, the formulation process is proprietary. A representative process flow is given in Figure 10-13.

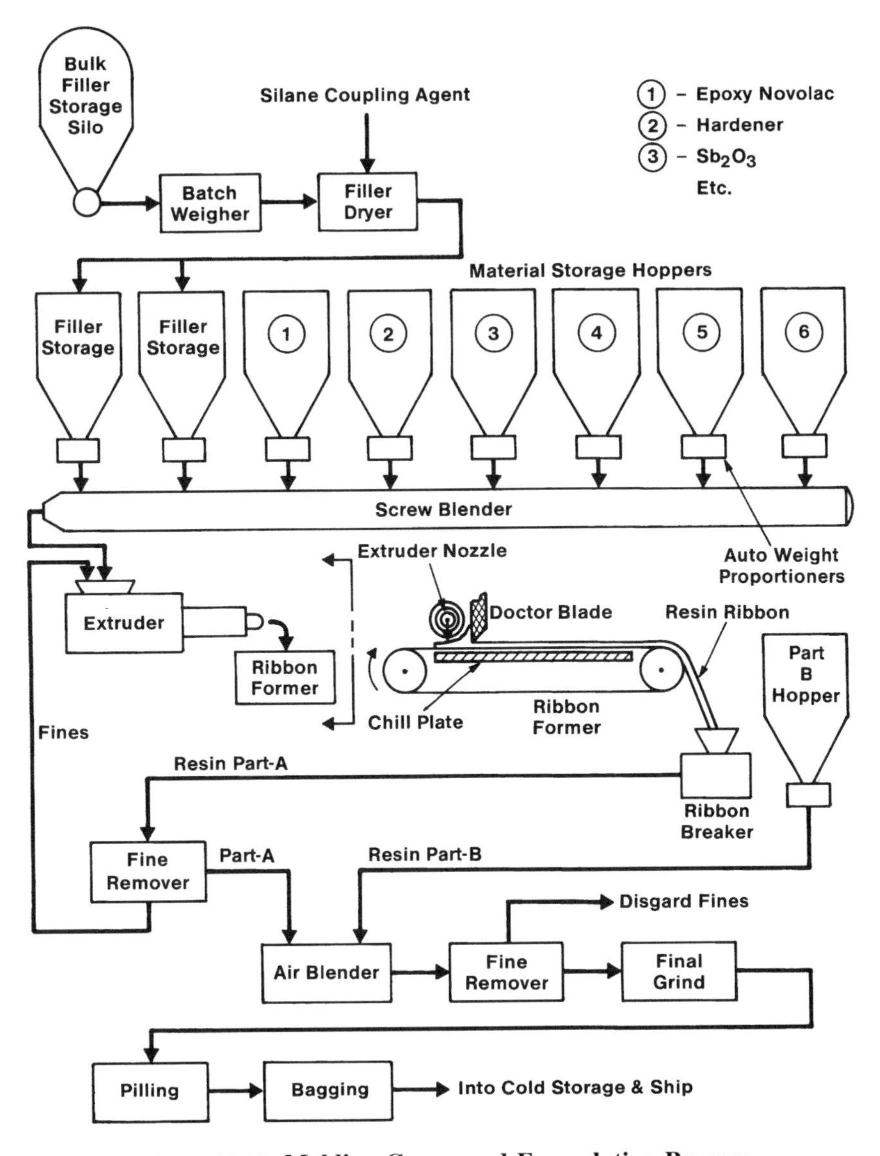

Figure 10-13. Molding Compound Formulation Process

Table 10-9. Properties of Semiconductor Molding Components (Part 1 of 2)

Supplier	Compound No.	Thermal conductivity W/m/°C	Thermal expansion 10^{-7}/°C	Flexural strength GPa	Flexural modulus GPa
Sumitomo	EME-1100-T	0.67	200	0.14	15.66
	EME-1100-H	0.67	200	0.14	15.66
	EME-1100-K	0.67	200	0.14	14.19
	EME-6200	0.67	200	0.13	12.72
	EME-6300	0.42	170	0.13	12.23
	EME-9100-XK	0.67	200	0.14	14.19
	EME-9200	0.67	200	0.13	12.72
	EME-9300	0.42	170	0.13	12.23
	EME-5000LS	0.67	200	0.15	14.68
Plaskon	3100	1.0	270–750	0.14	14.68
	3100LS	0.71	210–600	0.14	14.68
	3200	1.0	270–750	0.15	14.68
	3200LS	0.71	220–600	0.15	14.68
	3300	1.0	270–750	0.15	14.68
	3300LS	0.71	210–600	0.15	15.81
	3300SH	0.71	210–600	0.15	15.81
Hysol	MG15f	0.71	250–700	0.12	13.76
	MG35F	1.46	260–750	0.13	14.44
	MG36f	0.75	170–700	0.13	13.76
Nitto	HC10-2	0.63	200–700	0.15	13.70
	HC20-2	0.92	240–700	0.15	13.70
	HC30-2	1.46	260–700	0.15	14.68
	HC50-2	1.97	250–700	0.15	14.68
	MP119	0.63	180–650	0.13	12.23
	MPX-75	0.71	200–650	0.16	14.19
	MP150	0.71	160–740	0.13	11.74
	MP150-164	0.71	160–740	0.13	11.74

Blending is a batch process and is lot-controlled. Shipping lots range from 2,000 to 18,000 kg. A small sample is generally taken from each lot and sent ahead to the customer for lot-control testing. Resin, hardener, and accelerator are in close proximity after final grinding, and slow curing begins at room temperature. To retard this reaction, the molding compound must be refrigerated until needed at the molding press. A list of IC grade molding compound suppliers is given in Table 10-9.

Table 10-9. Properties of Semiconductor Molding Components (Part 2 of 2)

Extractable Chlorine ppm	Flame Rate UL	T_g °C	Remarks
50	94-VO	155	General purpose molding compound for DIPs with small chips
20	94-VO	155	Same as above, but better humidity resistance
20	94-VO	155	Better temperature-cycle performance
25	94-VO	165	Low stress M. C. for large DIPs & PLCCs
25	94-VO	165	Very low stress compound for SO & large thin packages
20	94-VO	155	High purity, low stress, low alpha particle count for molding DRAMs
25	94-VO	165	Lower stress than 9100-XK low alpha particle count for DRAMs in PLCC & SOJ
25	94-VO	165	Lower stress & low alpha particle count for DRAMs in thin packages
20	94-VO	135	Low-stress molding compound for SOPs that are solder dipped
30	94-VO	150	Standard molding compound for ICs & discretes
30	94-VO	150	Low-stress version of 3100 for resistor networks & MOS devices
20	94-VO	150	General purpose IC molding compound
20	94-VO	155	For large chips in DIPs or PLCCs
20	94-VO	150	For SMT packages & larger DIPs
20	94-VO	155	For gate arrays & microprocessors
20	94-VO	155	For large SMT packages (68-, 84-PIN PLCC) & QFPs
20	94-VO	190	Anhydride cured. Used for high-voltage chips (50-90V)
10	94-VO	160	For high-power discrete
10	94-VO	160	Low-stress compound for large chips in small packages
20	94-VO	162	For small DIP, thin flat packages
20	94-VO	161	28-42-lead chips
20	94-VO	162	Discrete devices high voltage ICs > 30 volts
20	94-VO	170	Power transistors, power packages
10	94-VO	170	Low stress and low ionics for SOIC packages
25	94-VO	151	Good adhesion to leadframe for SMT packages
15	94-VO	150	Very low stress for SOIC & PLCCs
15	94-VO	158	Same as MP150 except low alpha particle count

10.2.4 Lead Frame Design, Materials, and Processes

The leadframe is the backbone of a molded plastic package. Fabricated from a strip of sheet metal by stamping or chemical milling, it serves first as a holding fixture during the assembly process, then, after molding, becomes an integral part of the package. The functional requirements of a leadframe are:

1. A holding fixture that indexes with tool-transfer mechanisms as the package proceeds through various assembly operations.
2. A dam that prevents plastic from rushing out between leads during the molding operation.
3. Chip attach substrate.
4. Support matrix for the plastic.
5. Electrical and thermal conductor from chip to board.

10.2.4.1 Materials

Lead frame material selection depends on factors such as cost, ease of fabrication, and the functional requirements of the frame. There are three categories of materials to choose from when designing leadframes: nickel-iron, clad strip, and copper-based alloys. Many alloys, specifically formulated for plastic IC packages, exist within these categories, as indicated in Table 10-10.

Nickel-Iron Alloys: The most widely used metal for leadframe fabrication is Alloy 42 (42% Nickel-58% Iron). This material was originally formulated as a glass-seal alloy for use as terminal pins on light bulbs and vacuum tubes. The first chip-attach method on hermetic as well as plastic modules was a gold/silicon eutectic alloy bond. Silicon/gold eutectic melts at 370°C, and when solid, has a high elastic modulus. A close match of thermal-expansion coefficient between silicon and substrate is mandatory with this bonding system to avoid chip fracture. Alloy 42, the vacuum tube terminal pin material, was chosen since its expansion coefficient is close to that of silicon (45 vs. 26×10^{-7}/°C). Alloy 42, furthermore, can be heat-treated to obtain optimum tensile strength and ductility for subsequent stamping and lead forming operations. The alloy can also be easily electroplated or solder-dipped directly without a nickel barrier plating.

Low-thermal conductivity is the most important drawback of nickel-iron alloys. Since the leadframe is the main conduit by which heat flows from the chip to the printed circuit board, this can have a profound effect on the package thermal resistance after prolonged device operation. Clad materials and copper-based alloys have been developed to reduce thermal resistance [27].

Clad Materials: A layered composite strip, such as copper-clad stainless steel, was developed to emulate the mechanical properties of Alloy 42 while increasing thermal conductivity. Cladding is accomplished by high-pressure rolling of copper foil onto stainless steel strip and annealing the composite to form a solid-solution weld. Properties are given in Table 10-10 [28]. More detailed discussions on this subject can be found in Chapter 18, "Coated-Metal Packaging."

Copper Alloys: Copper would be an ideal leadframe material from an electrical and thermal conductivity standpoint, but it has properties that must be modified before it can be successfully used in plastic packages [29].

10.2.4.2 Strength

The tensile strength of copper alloys is lower than that of nickel-iron or copper-clad stainless steel. The addition of iron, zirconium, zinc, tin, and phosphorus serves to improve the heat-treating and work-hardening properties of these alloys. The resulting tensile strength and ductility are adequate for leadframe fabrication and function.

Copper-alloy frames can be specified with tempers from one-half hard to spring temper. Half-hard material is less likely to crack when bending, but full-hard or spring temper is preferred for modules that are automatically inserted. Hard copper alloys approach iron-nickel in bending characteristics so that automatic inserters can be set up to run both alloy types without constant adjustments to cut and clinch mechanisms.

10.2.4.3 Thermal Conductivity

The ability of the leadframe to channel heat from chip to board is related to the thermal conductivity of the frame material. The main advantage of copper alloys is their high coefficients of thermal conductivity (see Table 10-10). The largest reduction in package thermal resistance is achieved by the use of copper alloys as shown in Figure 10-14 [30].

10.2.4.4 Thermal Expansion

Copper alloys have high thermal-expansion rates with respect to silicon but nearly match the expansion rate of low-stress molding compounds. This property affects the selection of chip-bonding materials. Silicon/gold eutectic bonding cannot be used with copper frames because its high elastic modulus couples thermally induced bending stress to the silicon, which can cause potential chip fracture. Silver-filled epoxies and polyimide chip-attach adhesives have been developed and are flexible enough to absorb the strain developed between chip and copper.

10.2.4.5 Mechanical Design

Considerations in leadframe mechanical design are:

1. Ease of fabrication.
2. Indexing features for package assembly.
3. Chip-attachment substrate and gold wire span.
4. Coplanarity for wirebonding.

Table 10-10. Leadframe Materials (Part 1 of 2)

	CAR'P		OLIN			HUSSEY		KOBE	
PART NO.	A-42	CDA-151	CDA-194	CDA-195	CDA-155	KPC	KLF-1	KLF-2	KLF-5
TYPICAL ELEMENTS (%)	Fe-58 A-42	Zr-0.1 Cu-bal	Fe-2.3 P-0.03 Zn-0.1 Cu-bal	Fe-1.5 Co-0.8 Sn-0.6 P-0.1 Cu-bal	Ag-0.034 P-0.058 Mg-0.11 Cu-bal	Fe-0.1 P-0.058 Cu-bal	Ni-3.2 Si-0.7 Zn-0.3 Cu-bal	Sn-0.1 Fe-0.1 P-0.03 Cu-bal	Sn-2.0 Fe-0.1 P-0.03 Cu-bal
Melt P (°C)	1425	1000	1009	1090	1002	1083	1090	1083	1068
Spec. Grav.	8.15	8.94	8.8	8.92	8.91	8.9	8.9	8.9	8.9
E (GPa)	144.83	120.37	120.37	118.89	117.43	120.37	124.28	127.22	120.37
Thermal Conductivity (W/m/°C)	15.89	359.8	261.5	196.65	347.27	435.14	219.66	376.56	133.89
Linear Expansion (10^{-7}/°C)	43	177	163	169	177	170	170	170	165
Electrical Conductivity (% IACS)	3.0	90	65	50	86	92	55	82	35
Temper	H	3/4H	H	H		H	H	H	HH
Tensile Strength (GPa)	0.64	0.35	0.41	0.47	0.3 0.3 0.5	0.39	0.54	0.39	0.54
Elongation	10	7 min	3 min	3 min	222	4 min	7 min	4 min	7 min
Vickers HD	210	104	135	147	95 113 137	125	180	140	180

Table 10-10. Lead Frame Materials (Part 2 of 2)

	MIT'B	TAMAGAWA		T.I.	OLIN		PAN
PART NO.	MF-202	TAMAC-2	TAMAC-5	410SS	19750	19700	PMC-1C
TYPICAL ELEMENTS (%)	Sn-2.0 Ni-0.2 -0.15 Cu-bal	Sn-0.15 P-0.006 Cu-bal	Fe-0.75 Sn-1.25 P-0.03 Cu-bal	Cr-12.0 Fe-bal	Fe-0.6 Mg-0.05 P-0.02 Sn-0.23 Cu-bal	Fe-0.6 Mg-0.05 P-0.02 Cu-bal	Ni-1.0 Si-0.2 P-0.03 Cu-bal
Melt P (°C)	1065	1083	1075	—	1085	1086	1090
Spec. Grav.	8.8	8.9	8.8	7.8	8.82	8.84	8.9
E (GPa)	112.54	117.43	117.43	19.96	120.37	118.41	127.22
Thermal Conductivity (W/m/°C)	154.81	376.56	138.07	24.27	261.5	320.4 9	259.41
Linear Expansion (10^{-7}/°C)	169	177	167	110	169	168	169
Electrical Conductivity (% IACS)	30	92	40	30	65	80	60
Temper	H	H	H		S	H	H
Tensile Strength (GPa)	0.59	0.34	0.48	0.62	0.56	0.45	0.55
Elongation	7 min	4 min	4 min	5 min	1.9	1.5	4.8
Vickers HD	185	105	150	220	160	144	160

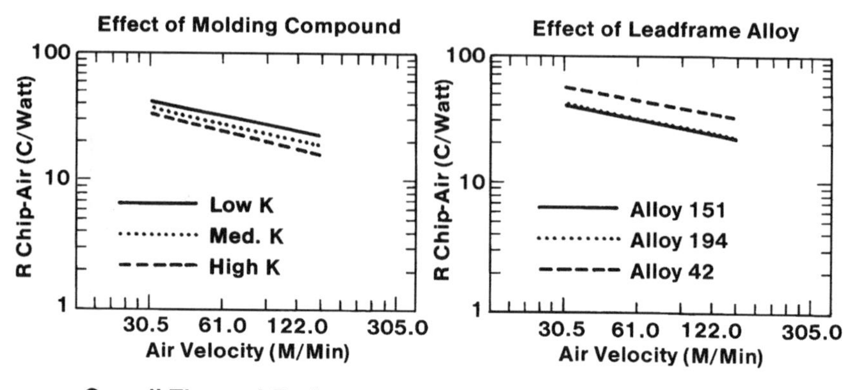

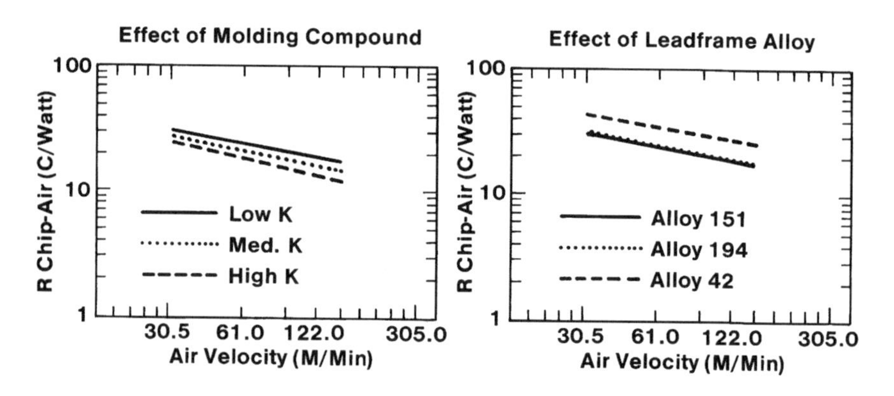

Figure 10-14. Overall Thermal Performance of 84-Lead PLCC-Material Effects.

5. Lead-locking and moisture-inhibiting configuration.
6. Stress relief.
7. Plastic support matrix.
8. Terminal lead and standoff configuration.

Ease of Fabrication: Figure 10-15 illustrates design rules for considerations 1, 3, and 4. Specifying a single temper for strip feedstock avoids proliferation of expensive progressive-stamping dies.

Indexing Features: Design and leadframe strips must consider the package-assembly system. Tooling holes are located along frame-strip-edges. These mate with the transfer-mechanism pins, which are part of the assembly equipment, and include chip bonders, wirebonders, molds,

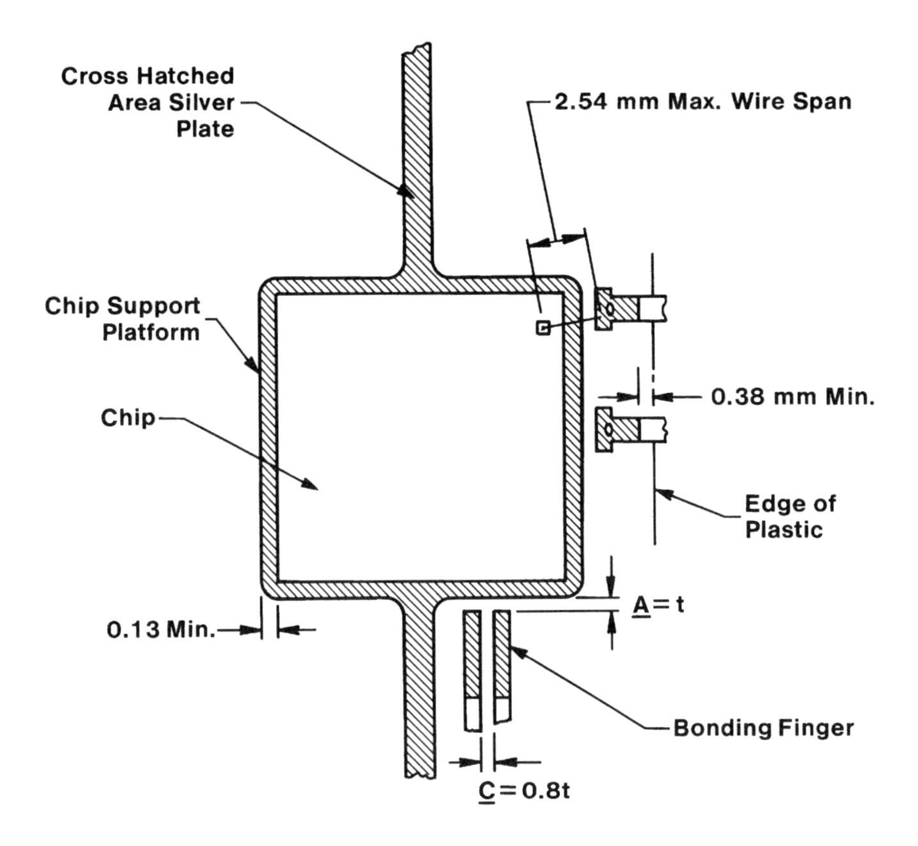

Figure 10-15. Lead Frame Clearances, Wire Span, Plating, Limits, and Coplanarity.

auto-inspection stations, trim and form equipment, and marking machines. In-process storage and transport-magazine-design depend on the frame-strip length.

Lead Locking and Moisture-Inhibiting Configurations: The bond between leadframe and plastic is mechanical. Therefore, any protrusion or hole on the frame that serves to lock the terminal leads within the cured molding compound is desirable. Several mechanical features have been developed to inhibit moisture penetration along the plastic leadframe boundary. These, along with lead-locking configurations, are shown in Figure 10-16.

Stress Relief: Lead frames cause two types of stress that can be

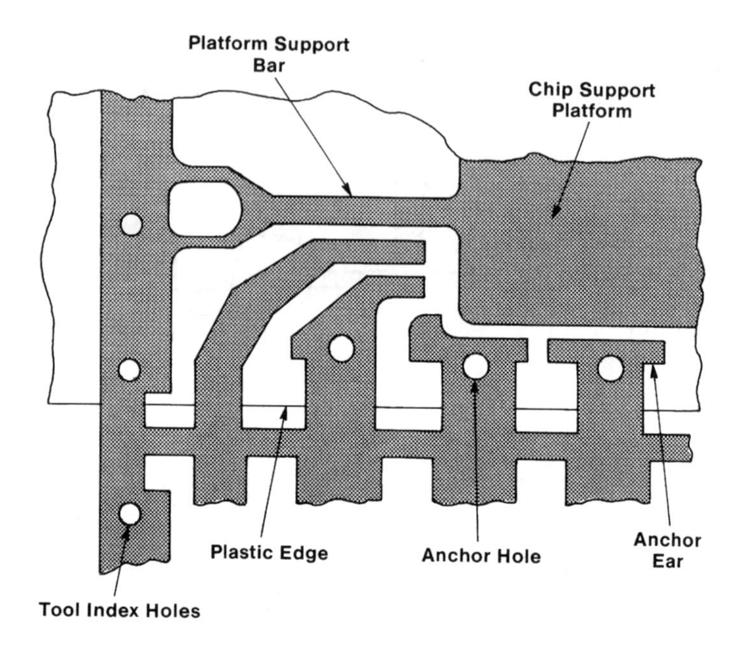

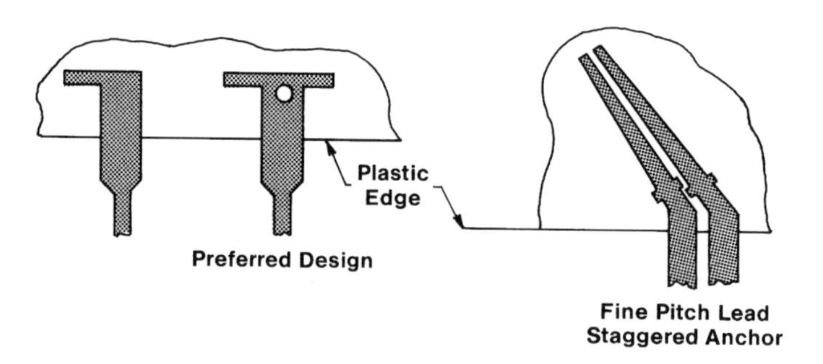

Figure 10-16. Typical Dip Lead Locking Design.

reduced by design. The most severe one is due to the inherently large thermal coefficient of expansion (TCE) mismatch between the plastic and the metal. In the severe cases that may occur during thermal-shock testing, bending of the package can reach sufficient proportions to impair the operation of the device. Such bending stress on the chip surface can be minimized by locating this surface on the neutral bending axis of the package. The chip-attach platform is offset downward to accomplish this effect, as shown in Figure 10-17.

The other stress condition encountered is caused by stress concentra-

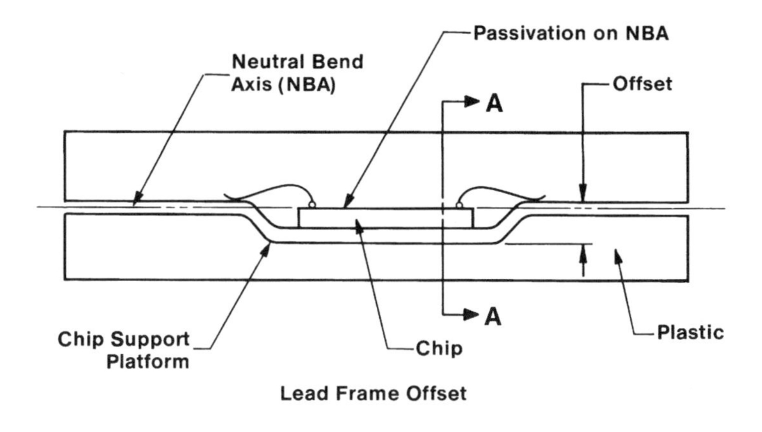

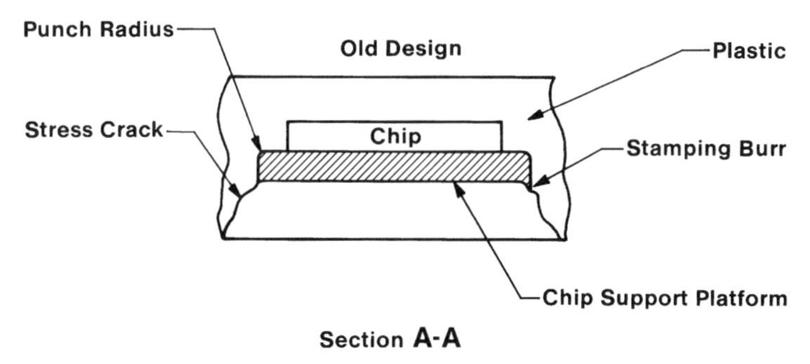

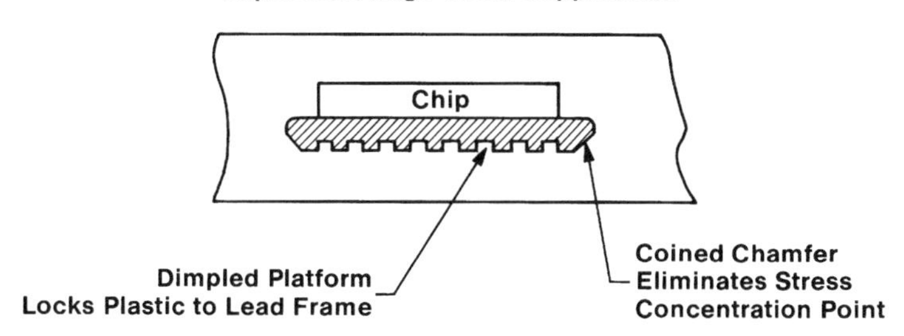

Figure 10-17. Lead Frame Offset—Crack Suppression.

tion points located on the bottom edges of each stamped leadframe, as depicted in Figure 10-17. When a stamping punch drives through the sheet metal to form a radius, it tends to pull material with it by friction. Conversely, on the exit side some metal is extruded, forming a burr. Burrs are sharp and act as stress concentrators. Once cracking occurs, the cracks tend to propagate from the edge burrs on the chip-attach platform toward the bonding figures, forming in the process a water-ingress channel. These internal cracks are insidious since they cannot be detected by conventional x-ray techniques. Burrs are generally removed by adding a coining station to the progressive die set. The burr is in effect forged into an edge chamfer.

Plastic Support Matrix: The leadframe functions as a reinforcement matrix for the molding compound, literally something for the plastic to hang onto while forming the module body. Top and bottom sections of a plastic module hold together because the plastic is contiguous between sections. As the metal area is increased, the force needed to shear top from bottom decreases to a point where fracture occurs along the metal/plastic boundary during lead-bending or soldering operations.

The ability to retard water penetration along the metal/plastic interface is a function of plastic shrinkage around the metal leadframe members. Generally the more plastic coverage, the greater will be the compressive forces. An ideal plastic-to-metal ratio for leadframe design is given by

$$A_m/A_p \leq 1. \qquad [10\text{-}1]$$

where A_m = metal area and A_p = plastic area.

Terminal Leads: There are four lead configurations used on molded plastic packages: integral standoff, gull wing, J-Bend, and butt joint, as illustrated in Figure 10-18. Dual-in-line modules use integral-standoff leads. They are designed to give clearance between the module bottom and the printed circuit board surface for cleaning. Integral standoffs should be designed to provide a clearance between the module and the board of 0.38 mm minimum when placed in a 1.09 mm maximum diameter plated-through-hole. Lead configurations for surface mounting require multiple bends. Therefore, tempers ranging from one quarter to one half hard are chosen to avoid cracking at the bend radii.

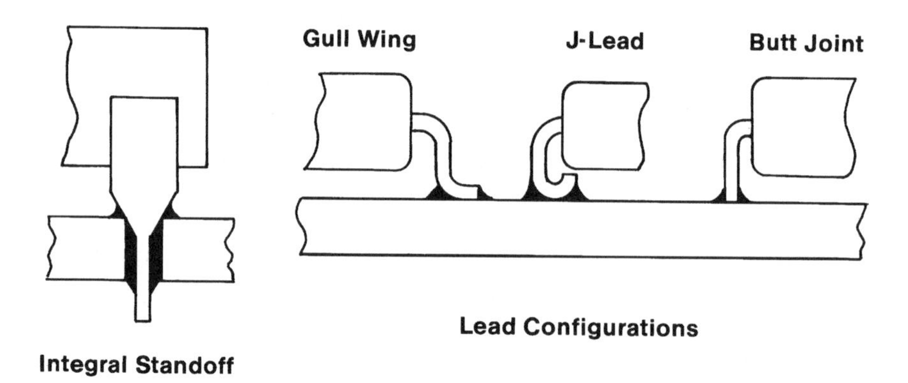

Figure 10-18. Lead Configurations.

10.2.4.6 Lead Frame Fabrication

Lead frames are either chemically milled or mechanically stamped from rolled strip stock. Typical strip thickness is 0.25 mm with thinner material (0.20 mm) used on high lead count packages such as the 84-pin PLCC and quad flat pacs. Chemical milling is a process that uses photolithography and metal-dissolving chemicals to etch a pattern from metal strip. The process begins by punching registration holes in the strip, then coating both sides with a photoresist material. The resist is exposed to ultraviolet (UV) light through a photo mask, developed, and cured. The leadframe is now delineated with cured photoresist and is ready for etching. Chemicals such as ferric chloride or ammonium persulfate are sprayed onto both sides of the strip to etch away exposed metal—leaving the leadframe intact. Chemically milled leadframes cost more than stamped frames, although the corresponding tooling costs are much lower. Furthermore, chemical milling provides a faster turnaround time (typically, one month compared to 6 to 12 months for stamping). This process is ideal for packages under development, short production runs, and frames too complicated for stamping.

Stamped leadframes are fabricated by mechanically removing metal from strip stock with tools called **progressive dies.** The energy required to shear metal is directly proportional to the length of shear. Lead frames have large shear lengths per unit area. Therefore, a large amount of energy is required to stamp a full frame with one press stroke. Progressive dies are usually made of tungsten carbide and are arranged in stations. Each station punches a small area of metal from the strip as it moves through the die set. Dies for complex leadframes can have up to 16 stations. This is shown in Figure 10-19.

Progressive stamping dies are expensive, currently about $140,000 for a 40-pin DIP leadframe, and take 6 to 12 months to design and build. Trouble shooting the die before introducing it to the production line is usually required. Fortunately, stamping rates can be rather impressive, with modern presses able to turn out 900 to 1,000 frames per minute. Stamped frame costs are a factor of 10 lower than etched parts, so tooling costs can be amortized rapidly.

10.2.4.7 Plating

Some leadframes in the late sixties were fully plated with gold or silver. Cost pressures over the years have caused the plating to recede, until only the chip-attach platform and bonding finger ends remain plated with silver. Silver spot plating is done with conventional plating baths. Spot dimensions are controlled by masking and should be specified so the silver limit is 0.38 mm (minimum) from the outer surface of the

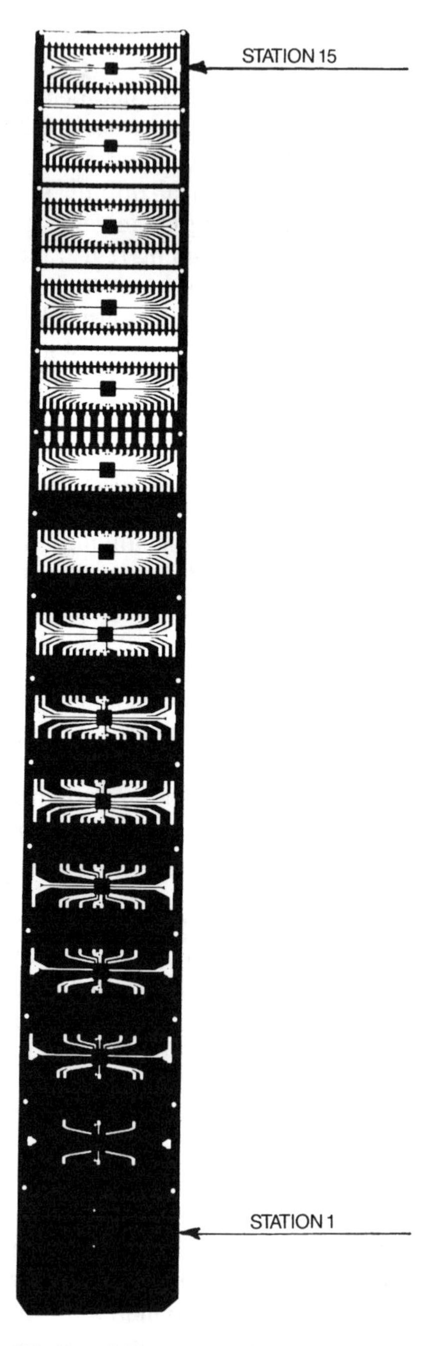

Figure 10-19. 40-Pin Dip Lead Frame, 15-Station Stamping Die Proof. (Courtesy of American Microsystems, Division of Gould Industries.)

module. Silver may be plated directly onto iron-nickel alloy surfaces, but a nickel underplate is recommended for copper alloys.

10.3 CHARACTERIZATION OF MOLDING-COMPOUND PROPERTIES

The molding compound is characterized by a set of properties that determine the suitability of the molding compound for a given application and process. From a manufacturing perspective, viscosity and flow characteristics and curing times and temperatures are important factors in determining which encapsulant materials should be used. From a performance viewpoint, key characteristics include electrical, mechanical, and thermal properties to chemical and humidity resistance, the water solubility of contaminants, and the adhesion to chip–component surfaces. Key properties include the following:

- Mold filling characteristics, resin bleed, hot hardness, and mold staning
- Shear viscosity at low shear rates
- Cure time
- Flow resistance
- Glass-transition temperature
- Thermomechanical properties
- Crack Sensitivity
- Hydrolyzable ionic purity
- Moisture solubility and diffusion rates
- Adhesion strength to leadframe materials
- Moisture sensitivity of viscosity and glass-transition temperature
- Alpha-particle emission (if applicable)
- Popcorn resistance for surface-mount technology

Technical information on the above properties and information on optimum processing conditions for the molding compound is provided by the mature suppliers. Table 10-11 gives an example characterization for a typical molding compound. Properties and characterization methods are given in the following subsections.

10.3.1 Shrinkage Stresses

Shrinkage stresses in a molded plastic package result from polymerization and the disparities in the coefficients of thermal expansion among the various materials that are contained within the package and are in

Table 10-11. Typical Characteristics of Molding-Compound Properties

Property	Value
Coefficient of thermal expansion	16×10^{-6}/°C
Glass-transition temperature	155°C
Flexural modulus	13,700 MPa
Flexural strength	147 MPa
Thermal stress	4.4 MPa
Thermal conductivity	16×10^{-4} cal/cm s°C
Volume resistivity	7×10^{16} Ω cm
Hydrolyzable ionics	<20 ppm
Uranium content	0.4 ppb
Spiral flow length	76 cm (30 in.)
Gel time	23 s
Hot hardness	85 (Shore D)

intimate contact. The following relationship can be used to calculate the magnitude of the thermal shrinkage stress, σ, as the package goes through a thermal excursion from T_1 to T_2:

$$\sigma = c \int_{T_1}^{T_2} \frac{[\alpha_p(T) - \alpha_i]}{1/E_{p(T)} - 1/E_i} dT \qquad [10\text{-}2]$$

where

α = the temperature-dependent coefficient of thermal expansion
E = the modulus of elasticity
c = a design-dependent geometric constant
the subscript p refers to the molding compound and the subscript i refers to either the semiconductor device or leadframe, depending on the desired calculation.

Because $1/E_i << 1/E_p$ and the encapsulant is compliant above the glass-transition temperature, T_g, the above integral can be approximated by

$$\sigma^* = (\alpha_{pg} - \alpha_i)E_{pg}(T_g - T_1) \qquad [10\text{-}3]$$

to provide the stress parameter σ^*. The stress, or the stress parameter, is only a crude approximation of the stress level in the material and does not account for stress concentration points or other geometric and interfacial features that influence delamination, bending, or cracking. The thermomechanical property values needed for calculations are normally provided by suppliers.

The ASTM F-100 test is designed to determine the thermal-stress parameter of Eq. [10-3] for the molding compound. ASTM F-100 is a photoelastic experiment where the molding compound is molded around a glass cylinder. The deformation of the glass at the test temperature is determined by counting the fringe patterns produced by polarized light.

10.3.2 Coefficient of Thermal Expansion and Glass-Transition Temperature

The coefficient of thermal expansion for any material represents a change in dimension per unit change in temperature. The dimension can be either volume, area, or length. However, the rate of expansion varies from material to material and with the temperature. Therefore, the fact that different materials expand differently with the same increase in temperature necessitates that elements attached together have the same or similar coefficients of thermal expansion to avoid the possibility of bending, cracking, or delamination. The glass-transition temperature is an inflection point in the expansion-versus-temperature curve above which the rate of expansion (and therefore the coefficient of thermal expansion) increases significantly.

The coefficient of thermal expansion (TCE) and the glass-transition temperature are two properties measured by thermomechanical analysis (TMA) and reported by most suppliers. The tests are described in ASTM D-696 or SEMI G 13-82 standards. The ASTM D-696 uses the fused quartz dilatometer to measure the TCE. The specimen is placed at the bottom of the outer dilatometer tube with the inner one resting on it. The measuring device, which is firmly attached to the outer tube, is in contact with the top of the inner tube and indicates the variations in length of the specimen with changes in temperature. Temperature changes are brought about by immersing the outer tube in a liquid bath or other controlled-temperature environment maintained at the desired temperature. Typically, a graph of expansion versus temperature is plotted. The coefficient of thermal expansion is the slope of the plot, and the glass-transition temperature is the intersection point between the lower-temperature coefficient of thermal expansion (α_1) and the higher-temperature rubbery region coefficient of thermal expansion (α_2). The glass-transition temperature separates the glassy temperature region from the rubbery temperature region of an amorphous polymer. The glass-transition temperature, being a manifestation of the total viscoelastic response of a polymer material to an applied strain, depends on the rate of strain, the degree of strain, and the heating rate. Because improper molding and postcure conditions can affect both the coefficient of thermal expansion and the glass-transition temperature, most PEM manufacturers remeasure these parameters on a predetermined quality control schedule.

10.3.3 Mechanical Properties

It is generally recognized that lowering the stress parameter leads to improved device reliability and can be accomplished by lowering the modulus, the coefficient of thermal expansion, or the glass-transition temperature. The flexural strength and flexural modulus are derived from the standardized ASTM D-790-71 and ASTM D-732-85 and reported by suppliers. ASTM D-790 suggests two test procedures to determine the flexural strength and flexural modulus. The first method suggested is a three-point loading system utilizing center loading on a simply supported beam. This procedure is designed principally for materials that break at comparatively small deflections. In this procedure, the bar rests on two supports and is loaded by means of a loading nose midway between the supports. The second procedure involves a four-point loading system, utilizing two load points equally spaced from their adjacent support points with a distance of either one-third or one-half of the support span. This test procedure is designed particularly for large deflections during testing. In either of the cases, the specimen is deflected until rupture occurs in the outer fiber. The flexural strength is equal to the maximum stress in the outer fiber at the moment of break. It is calculated using

$$S = \frac{3P_{\text{rupture}}\,l}{2\text{b}^2_{\text{beam}}\,d} \qquad [10\text{-}4]$$

where

S = the flexural strength
P_{rupture} = the load at rupture
l = the support span
b_{beam} = the width of the beam

Flexural modulus is calculated by drawing a tangent to the steepest initial straight-line portion of the load deflection curve and is given by

$$E_B = \frac{l^3 m}{4 b_{\text{beam}} d^3} \qquad [10\text{-}5]$$

where

m = the slope of the tangent to initial straight line portion of the load deflection curve
E_b = the flexural modulus

The tensile modulus, tensile strength, and percent elongation are derived from ASTM D-638 and D2990-77 test methods.

Tensile properties of molding compounds, determined according to ASTM D-638, uses "dog-bone" shaped molded or cut specimens with fixed dimensions and held by two grips at the ends. Care is taken to align the long axis of the specimen and the grips, with an imaginary line joining the points of attachment of the grips to the machine. They are incrementally loaded to obtain stress-strain data at any desired temperature. A typical curve is shown in Figure 10-20. The tensile strength can be calculated by dividing the maximum load (in N) by the original minimum cross-section area of the specimen (in m^2). The percentage elongation is calculated by dividing the extension at break by the original gauge length and this ratio is expressed as a percentage. The modulus of elasticity is obtained by calculating the slope of the initial linear portion of the stress-strain curve. If Poisson's ratio for the material is known or separately determined from tensile strain measurements, the shear modulus of the molding compound can be estimated. It is important to note that the stresses encountered in PEMs are actually a complex mixture of stresses.

The evaluation of the cracking potential of the molding compound is particularly important for devices where a relatively small amount of molding compound surrounds a relatively large die (e.g., memories, SOPs, and ultrathin packages). In the absence of any standard procedure for such evaluation, ASTM D-256A and D-256B Izod impact test procedures

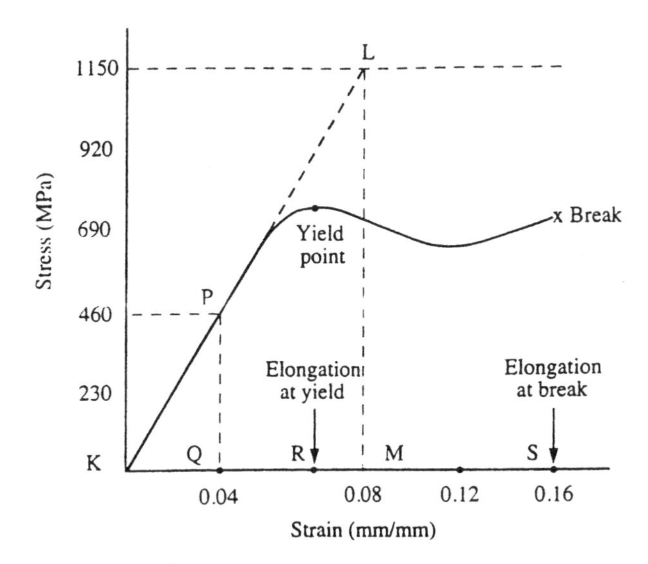

Figure 10-20. A Typical Curve of Incrementally Loaded Specimens from Stress-Strain Data.

are commonly followed. The specimen is held as a vertical cantilever beam in test method ASTM D-256A and is broken by a single swing of the pendulum, with the line of initial contact being at a fixed distance from the specimen clamp and from the centerline of the notch and on the same face as the notch. A variation of this test is the ASTM D-256B, where the specimen is supported as a horizontal simple beam and is broken by the single swing of the pendulum, with the impact line midway between the supports and directly opposite the notch. These are overstress tests for the cracking potential of epoxy molding compounds even under extreme thermomechanical stress conditions and do not test their important region of viscoelasticity. However, they are simulative of trim and form, and handling impact-induced cracking susceptibility.

A fracture test that models the strain history of the package for thermomechanically induced failure is that previously mentioned ASTM D-790-71 three-point flexural bending test used to determine flexural modulus. Here a 0.05-mm (2-mil)-diameter center-notched rectangular specimen is center strained at a rate that simulates the manufacturing cycle [i.e., 20%/min for liquid-to-liquid thermal shock and 0.1%/min for device on–off operation in air (1°C/min)]. The area under the stress-strain curve is proportional to the energy to break at the test temperature. The low-temperature data are usually the discriminating factor between molding compounds because the molded body experiences the greatest stress at lower temperatures, far removed from the molding temperature.

Because the melt viscosity of a molding compound is shear-rate dependent and a typical mold subjects different shear rates at different points of the molding-compound flow channel, the expected shear rates for a specific molding tool need first be calculated under no-slip boundary conditions. The general ranges of experienced shear rates are hundreds of reciprocal seconds in the runner, thousands through the gate, and tens in the cavity. Considerations must also be given to the temperature and time dependence of the shear-rate-dependent molten molding-compound viscosity.

Selection of molding compounds based on shear dependence of viscosity should identify one that has the lowest viscosity at low shear rates and high cavity temperatures for wire sweep and/or paddle shift prone devices and in multicavity molds where complete filling before gelation is a concern [31]. A material whose viscosity is more temperature insensitive performs better in all suboptimal tool designs. The time dependency of molten molding-compound viscosity originates from two opposite phenomena. Although the epoxy curing process will increase the average molecular weight and hence viscosity with time, the decrease of viscosity with increasing molding temperature in the tool will overwhelm it in the early phases. However, ultimately, both the molecular weight and the viscosity go to infinity at gelation. Flow-induced stresses, particularly in distant cavities, could thus be very significant at the latter stages of mold filling. Consequently, molds with longer flow lengths and longer flow times need molding compounds with longer gel times at the 150–160°C mold filling temperature. This requirement is a trade-off for higher productivity.

10.3.4 Spiral Flow Length

The ASTM D-3123 or SEMI G 11-88 test consists of flowing molding compound through a spiral coil of semicircular cross section until the flow ceases. Although it is not a viscometric test, it is a measure of fusion under pressure, melt viscosity, and gelation rate. The spiral flow test is used both to compare different materials and control molding-compound quality. However, it cannot resolve the viscous and kinetic contributions to the flow length. Higher viscosity and longer gel time could compensate each other to provide identical flow lengths. The molding tool used in this test is shown in Figure 10-21.

A "ram-follower" device, specified in SEMI G 11-88, is used to separately record ram displacement versus time and, consequently, separate the viscosity-induced flow time from the gel time in the total spiral length formation time of different molding compounds. The molding tool used in spiral flow test covers the several hundred per second shear-rate range; thus, the test results have no impact on yield and productivity. However, as an indicator of gel time or maximum flow time of thermoset molding compounds, the results of this test can help define mold flow lengths and flow times compatible with the particular molding tool.

10.3.5 Molding Compound Rheology from Molding Trial

This is a test for rheological compatibility of a molding compound with a device to be packaged in a molding tool by a trial molding operation. Rheological incompatibility can cause wire sweep, die paddle shifting, or incomplete filling of the mold cavity resulting in voids. X-ray analysis of the molded packages and cross-sectioning of the molded bodies through the paddle support are the primary methods of evaluation for these trials. Long wirebond spans (> 2.5 mm) and oversized paddle supports are normally used to create worst-case scenarios of wire sweep and paddle shift.

The mold filling characteristics are controlled by the pressure drop through the gates where the molding compound experiences maximum shear stress. Incomplete fill problems originate from molding compounds that have higher viscosities at these high deformation rates in the gates. This flow closely approximates flow through a sudden contraction or a converging channel and has both shear and extensional components.

Statistically confident sampling of all stochastic phenomena, such

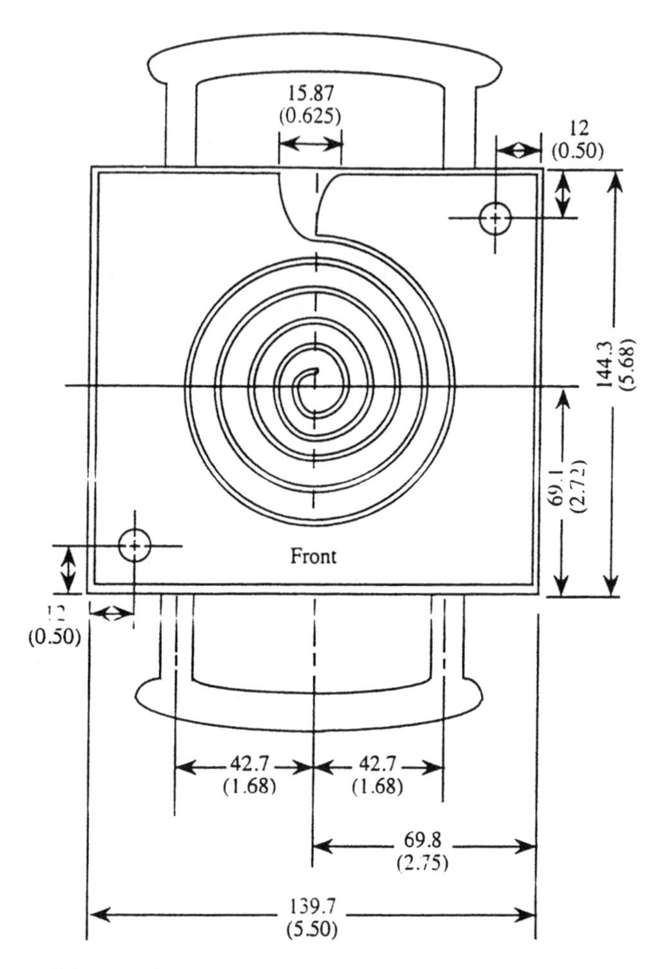

Figure 10-21. Molding Tool Used for the Spiral Flow Length Test from ASTM D-3123

as gate clogging by gel or filler particles, requires a large number of molded packages. The analysis of mold trial results should consider molding-compound density and the consequent package-sectioning analysis for porosity appraisal. Figure 10-22 shows a typical mold map displaying a variety of runners.

Incomplete filling of the mold (because of low packing pressure) in just a few cavities effectively ruins all the devices loaded because it leads to high porosity in the encapsulant that would lead to excessive moisture penetration. It is more common in molding tools with a large number of cavities (> 150) and with packages of large volume such as PQFPs and large chip carriers or with package designs with four-sided leads requiring corner gating.

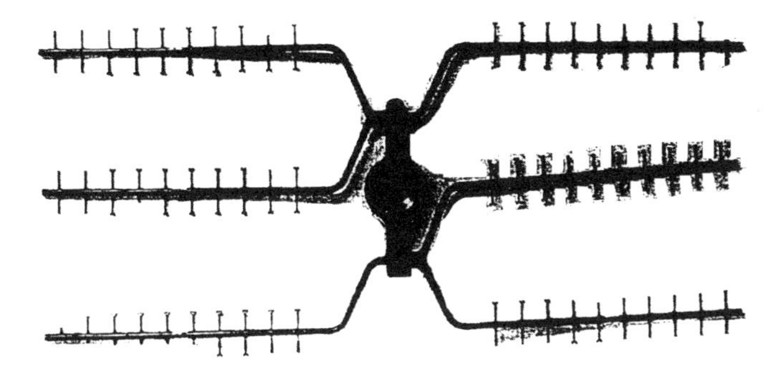

Figure 10-22. Molding Mapping Showing the Runners

Resin bleed and flash are molding problems where molding compound oozes out of the cavity and onto the leadframe at the parting line of the mold. Whereas flash is caused by the escape of the entire molding compound, resin bleed includes only the strained-out resin. Although root causes of these two problems can be traced both to the processing conditions of molding, and mold design and mold defects, resin bleed is considered to be more molding-compound related. Resin bleed occurs more often with formulations containing low-viscosity resin and large filler particles, and with process conditions where excessive packaging pressure is maintained after cavity fills and the use of too low a clamping pressure on the mold halves.

SEMI G 45-88 is a standardized test for assessing a material's potential for resin bleed and flash. It is a transfer-molding experiment that measures the flow of molding compound in a shallow-channel mold (6–75 μm) and simulates flash and bleed in production tools. Propensity of resin bleed and flash from improper molding-compound properties is indicated by long spiral flow lengths obtained in that test.

Moisture has a profound effect of decreasing the viscosity of epoxy molding compound, and the degree of this effect is a function of the additives and curing agents used in different formulations [32]. Compared to dry conditions, this decrease of molten molding-compound viscosity can be 40% or more at ~0.2 wt.% water or higher.

The effect of moisture on the shear-thinning behavior, shown in Figure 10-23 indicates that viscosity is simply lowered with little change in shear-rate dependence and power-law index. Not withstanding the fact that moisture-induced melt viscosity lowering is beneficial in overcoming flow-stress and mold filling problems, excessive moisture content can cause excessive resin bleed and voids. Consequently, moisture uptake sensitivity, determined by the shear-rate-dependent viscosity test, is a necessary factor in the selection of a molding compound.

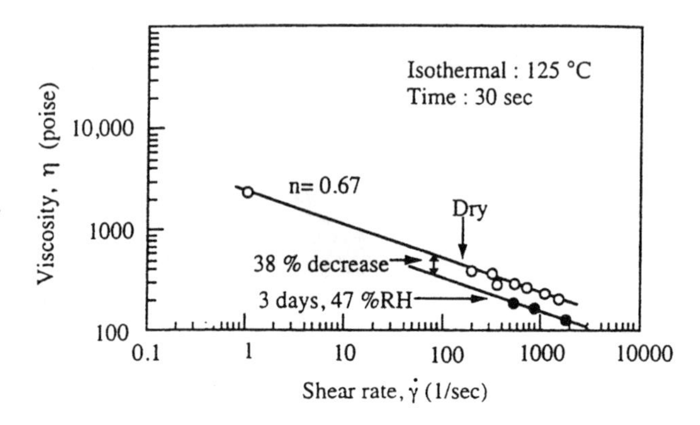

Figure 10-23. Effect of Moisture on the Shear-Thinning Behavior. (From Ref. 33.)

10.3.6 Characterization of Molding-Compound Hardening

Hardening of a molding compound is the progressive thermal polymerization of the matrix resin that leads to gelation in the latter stages of the molding process, an ejectable state of hardness during cure, and complete chemical conversion during postcure. The productivity of plastic package molding depends on the rate of this chemical conversion.

Mold filling can occur in as little as 10 s at 150–160°C (specified by the supplier), and the cure time required before the parts can be ejected from the mold can range from 1 to 4 min. The cure time is about 70% of the molding cycle time. Shorter cure times will generally have shorter flow times into mold cavities before gelation. Multiplunger machines are designed to handle these short flow times and cure times to provide high molding productivity. Most molding tools require molding compounds that flow for 20–30 s and then cure to an ejectable state in less than one additional minute. Standard evaluation tests for these characteristics of molding compounds are discussed below.

10.3.7 Gel Time

The gel time of a thermoset molding compound is usually measured with a gel plate. In gel-time evaluation with a gel plate, a small amount of the molding-compound powder is softened to a thick fluid on a precisely controlled hot plate (usually set at 170°C) and periodically probed to determine gelation. The gel time is a qualitative point in the process where the material cannot be smeared into a thin coating. The SEMI G 11-88 standard recommends using the spiral flow test as a comparative evaluator. Gel times indicate the productivity of a molding compound. Shorter gel times lead to faster polymerizations rates and shorter times for mold cycle, increasing production.

10.3.8 Adhesion

Poor adhesion of a molding compound to the die, die paddle, and leadframe can lead to such failures as "popcorning" during assembly, corrosion, stress concentrations, and subsequent thermomechanical failure, package cracking, chip cracking, and chip metallization deformation. Consequently, adhesion is one of the most important discriminating properties of a molding compound to be chosen for a particular physical and materials design of a package. The theory and practice of adhesion of integrated-circuit molding compounds to package elements have been treated by Kim [33] and Nishimura et al. [34]

A common adhesion test for leadframes to the molding compound is to mold plugs of the molding compound on the properly surface-treated leadframe tab, and then determine adhesion strength in a tensile tester. The molding process used in this test must be identical to that used in production. For maximum simulation of manufacturing conditions, a custom-designed leadframe is used in a production mold to generate the adhesion test specimens. Determination of adhesion strength to other package materials such as silicon die, polyimide, and silicone overcoat is accomplished by molding the compound on the flat surface of the material and measuring the force required to pull them apart. Needless to say, very careful interpretation of the adhesion test data is required to differentiate among most state-of-the-art molding compounds.

10.3.9 Polymerization Rate

The novolac epoxy polymerization reaction includes several competing reactions among three or four reactive species. The chain segments that form are complicated and difficult to predict. Thus, thermal analysis methods, which assume that the fraction of the total heat of reaction liberated is proportional to the fraction of complete chemical conversion, are preferred for these types of highly filled opaque systems. Several different empirical forms have been offered to fit conversion data for the epoxy molding compounds. They do not reflect the molecular dynamics of the reaction but are, instead, phenomenological in that they assess the engineering behavior of the reaction without a theoretical basis for the reaction mechanism or reaction order. Hale et al. [35–37] developed one of the most noteworthy forms:

$$\frac{dX}{dt} = (k_{r1} + k_{r2}X^{m_r})(1 - X)^{n_r} \qquad [10\text{-}6]$$

where the four fitting parameters for the conversion of epoxide groups, X, as a function of reaction time are as follows: m_r, and n_r are the pseudoreaction orders and k_{r1} and k_{r2} are rate constants. For a typical epoxy molding compound, $m_r = 3.33$, $n_r = 7.88$, $k_{r1} = \exp(12.672 - 7560/T)$, and $k_{r2} = \exp(21.835 - 8695/T)$ [36].

Figure 10-24 shows isothermal fractional conversion of epoxide groups with a drop-off in reaction rate near complete conversion. These conversion constants differ from one molding compound to the other and thus form the basis of evaluation of polymerization rate of the compound in question.

Differential scanning calorimetry (DSC) has been used to obtain the degree of polymerization of filled molding compounds [37]. Measuring the heat of reaction versus time during an isothermal cure, it expresses the fractional conversion as a function of time as equal to the fractional total liberation of heat:

$$\frac{\Delta H_{t=t_1}}{\Delta H_{\text{total}}} = \frac{X}{100} \qquad [10\text{-}7]$$

An extensive analysis of the polymerization kinetics is generally not required for material selection. The secondary effects of cure kinetics such as gel time, mechanical properties, and glass-transition temperature are sufficient to effectively compare different molding compounds.

10.3.10 Hardening

Hot hardness or green strength is the stiffness of the material at the end of the cure cycle. A certain degree of this hot hardness is required before the molded strip of parts can be ejected safely from specific molds.

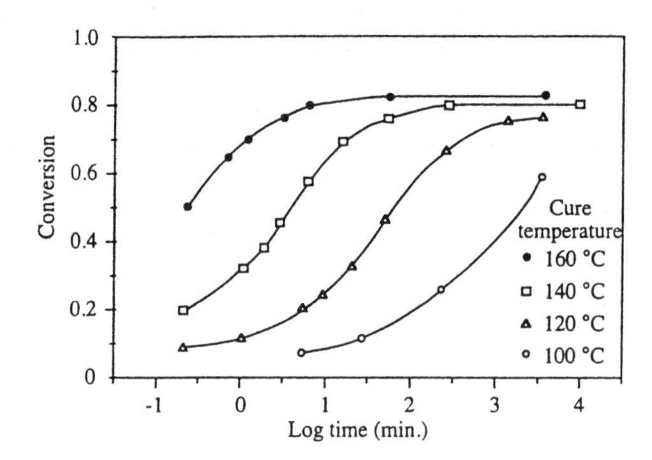

Figure 10-24. A Plot of Conversion Versus Time for an Epoxy Molding Compound During Cure. (From Ref. 35.)

Ejection of the strip from the molding tool is also dependent on the characteristics of the mold. These mold characteristics are the draft angle of the vertical surfaces, the surface finish of the tool, and the number and size of ejector pins. Different molding compounds attain this green strength at different points of cure cycle due to either percentage conversion achieved or low modulus above glass-transition temperature. It is, thus, a productivity issue and can either be determined by molding trial or supplied by the vendor. A hot hardness value of about 80 on the Shore D scale within 10 s of opening the mold is considered acceptable.

10.3.11 Postcure

Most epoxy molding compounds require about 4 h of postcure at 170–175°C for complete cure.

10.3.12 Moisture Effects on the Epoxy Hardening Process

Moisture content of molding compounds has a strong impact on the viscosity, void density, polymerization kinetics, and the properties of the cured material. The lowering of the glass-transition temperature by the use of moist molding compounds indicates that the final inalterable cross-linked network is different under humid conditions. The effect of moisture on the polymerization rate originates from the fact that moisture binding to catalyst sites diminishes its effectiveness. Because of the differences in the formulation of different molding compounds, the effect of moisture content could vary widely and needs to be evaluated from the viewpoints of required encapsulant properties.

10.3.13 Thermal Conductivity

Thermal conductivity is an important property of an encapsulant used for high-heat-dissipating devices or for devices with long duty cycles. Although they are measured with standard thermal-conductance testers, the reported values vary widely, depending on the instrument and the precision used.

10.3.14 Electrical Properties

The state-of-the-art very large-scale integration (VLSI) parts require close control of several electrical properties of the molding compound or the cured encapsulant for superior performance. They include dielectric constant and dissipation factor (ASTM D-150), volume resistivity (ASTM D-257), and dielectric strength (ASTM D-149). ASTM D-257 suggests various electrode systems to determine the volume resistivity by measuring the resistance of the material specimen by a measurement of the voltage or current drop under specified conditions, and the specimen and electrode dimensions. The test specimen may be in the form of flat plates, tapes, or tubes. Figure 10-25 shows the application and electrode arrangement for a flat specimen. The circular geometry shown in the figure is not necessary, although convenient. The actual points of measurements should be uniformly distributed over the area covered by the measuring electrodes. The dimensions of the electrodes, the width of the electrode gap, and the resistance are measured with a suitable device having the required sensitivity and accuracy. The time of electrification is normally 60 s and the applied voltage is 500 ± 5 V. The volume resistivity is given by

$$\rho_v = \frac{A_{elec}}{t} R_v \qquad [10\text{-}8]$$

where

A_{elec} = the effective area of the measuring electrode
R_v = the measured volume resistance
t = the average thickness of the specimen

ASTM D-149 requires that alternating voltage at a commercial power frequency, normally 60 Hz, be applied to a test specimen. The voltage is increased from zero, or from a level well below the breakdown voltage, until dielectric failure of the test specimen occurs. The test voltage is applied using simple test electrodes on opposite faces of specimens. The specimens may be molded, cast, or cut from a flat sheet or plate. Methods of applying voltage include a short-time test, a step-by-step test, and a

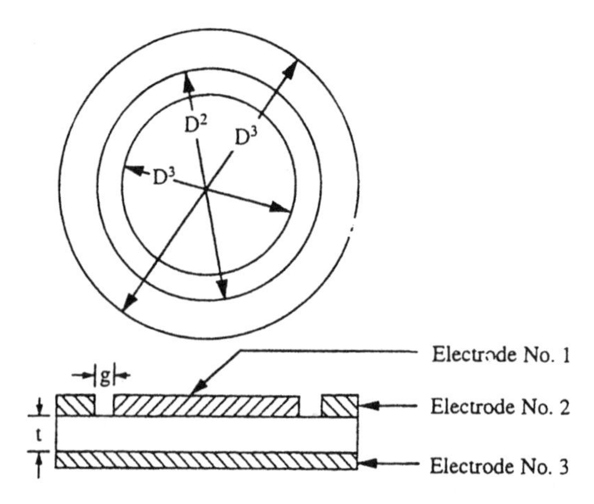

Figure 10-25. Electrode Arrangement for Measurement of Volume Resistance for a Flat Specimen (ASTM D-257 1994)

slow rate-of-rise test. The second and third methods usually give conservative results.

Epoxy composites in dry environments and at room temperatures have similar electrical properties. Deterioration of some materials may occur after being stored in a moist environment at high temperatures.

10.3.15 Chemical Properties

The contamination level of a molding compound ultimately determines the long-term reliability of the parts fabricated with it and used under rugged conditions. The SEMI G 29 standard procedure is used to determine water-soluble ionic level in epoxy molding compounds. A water extract is first tested for electrical conductivity and then quantitatively analyzed by column chromatography. A separate determination of hydrolyzable halides (from the resin, flame retardants, and other impure additives) is particularly crucial in assuring long-term reliability of PEMs. Long-term (48 h), high-pressure, and sometimes hot-water (up to 100°C) extraction from the molding compound and subsequent elemental analysis is needed for such evaluation. Modern molding-compound formulations contain as little as 10 ppm of these corrosion-inducing ionics. Atomic absorption spectroscopy and x-ray fluorescence techniques are used to determine the content of other undesirable contaminants such as sodium, potassium, tin, and iron. Encapsulants used for memory devices, where single-event upsets from alpha-emitting impurities in the filler silica must be minimized, require determination of uranium and thorium content in the molding compound.

10.3.16 Flammability and Oxygen Index

Encapsulation compounds and plastic-encapsulated parts must conform to Underwriters Laboratory flammability ratings (UL 94 V-0, UL 94 V-1, or UL 94 V-2). Molding compounds are evaluated for flammability by UL 94 vertical burn and ASTM D-2836 oxygen index tests. In the UL 94 test, a 127-mm × 12.7-mm (5-in. × ½-in.) cured epoxy test bar of a predetermined thickness is ignited multiple times in a gas flame, and the burn time per ignition, total burn time for 10 ignitions (5 specimens), and extent of burning are recorded for proper UL rating. In the oxygen index test, the minimum volume fraction of oxygen in an oxygen–hydrogen mixture that will sustain burning of a 0.6 × 0.3 × 8-cm molded bar of epoxy molding compound is specified.

10.3.17 Moisture Ingress

All plastic packages prefer low moisture diffusivity for high reliability and minimum soldering heat damage. At room temperature, epoxy molding compounds absorb water to approximately 0.5% by weight on long-term exposure to 100% relative humidity [38]. Moisture absorption rate to saturation is usually determined by weighing molded dry parts exposed to a given relative humidity and temperature for some extended periods [39]. Desorption rates of moisture-saturated parts are similarly determined by drying at an elevated temperature. The selection of an epoxy molding compound from moisture ingress evaluation is a very recent phenomenon and is currently quite subjective even for a specific part and its application environment.

10.4 THE TRANSFER-MOLDING PROCESS

The transfer-molding process is the most popular integrated-circuit encapsulation method for essentially all plastic packages. Transfer molding is a process of forming components in a closed mold from a thermosetting material that is conveyed under pressure, in a hot, plastic state, from an auxiliary chamber, called the transfer pot, through runners and gates into the closed cavity or cavities. Figure 10-26 depicts the transfer molding press.

One of the main limitations of the transfer-molding process is the loss of material. The material left in the pot or well and also in the sprue and runner is completely polymerized and must be discarded. For small parts, this can represent a sizable percentage of the weight of the parts molded. As a result, the trend today's to utilize fully automatic transfer-molding presses which mold a few parts very efficiently. Another limitation of today's transfer-molding process is the need to use standard metal leadframes. Extension of the process to include package designs such as TAB is still not clear.

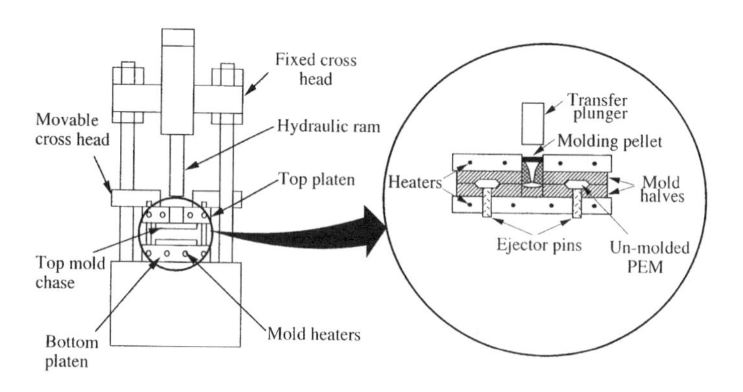

Figure 10-26. Transfer Molding Press.

10.4.1 Molding Equipment

Transfer-molding requires four key pieces of capital equipment: a preheater, a press, the die mold, and a cure oven. The transfer-molding press is normally hydraulically operated. Auxiliary ram-type transfer molds are commonly used in transfer molding. This mold has a built-in transfer pot in which the molding compound is placed. Both the volume and the size of the molding-compound preforms have to be appropriately selected for the press capacity. The mold is then clamped using the clamping pressure. The transfer plunger is activated to apply the transfer pressure to the molding compound. The molding compound is driven through the runners and gates into the cavities.

Since the 1980s, aperture-plate and multiplunger molds have been the dominant approaches to PEM molding. Table 10-12 compares the features of these molding methods. Aperture-plate molds are a patented transfer-molding technology (U.S. Patent No. 4,332,537, 1 June 1982) exclusively developed for PEMs. An aperture-plate mold is constructed by assembling a series of stacked plates. The leadframes form the aperture

Table 10-12. Comparison of Molding Tools

Feature	Aperture-plate molds	Multiplunger molds
Number of cavities	Several thousands possible	From 2 to 100
Flexibility of package types	Extremely flexible	Relatively inflexible
Process setup and control	Moderate	Straight forward for automated tools
Flow-induced stress problems	Minimum	Intermediate
Flash and bleed	Minimum	Average
Molding compound waste (%)	20–40	10–25
Package yield per cycle	High	Very high
Ejection	External ejection	Ejector pins in each cavity
Flow time and cavity material uniformity	Good	Excellent
Packing pressure (MPa)	1.4–2.8	Variable on number of cavities fed from a pot (1.4–4.1)
Temperature profile and stability	Requires care	Excellent
Automation susceptibility	Intermediate	Very high
Capital cost	Low	High (lower for small tools)
Labor cost	High	Low
Maintenance cost	Medium	High, due to automation

plates. The top and bottom of the body are formed by separate plates. The bottom body-forming plate contains the runner system, and the top plate is finished for either laser or ink marking. The gates are positioned between the runners, parallel to the bottom of the body; the aperture-plate cavities can be formed anywhere along this intersection and their width can be any fraction of this length of intersection. This flexibility of gate positioning, along with the much lower pressure drop across the gate in an aperture plate, results in negligible wire sweep and paddle shift during molding. These molds are also highly adaptable for different package types and pinouts.

Multiplunger molds, also called gang-pot molds, have a number of transfer plungers, feeding one to four cavities from each transfer pot. They are highly automated and can be easily set and optimized for a new molding compound.

Figure 10-27 shows a multiplunger mold used for simultaneous encapsulation of DIPs and quad flat packs. Figure 10-28 shows a multiplunger mold with each pot feeding just one cavity.

In general, the mold consists of two halves: the top and bottom. The mating surface of these two halves is called the parting line. Platens are massive blocks of steel used to bolt the two halves to the molding press. Guide pins ensure proper movement of the two halves. The ejector pins aides the ejection of the component after the mold has opened. Gates are located where they can be easily removed and buffed if necessary. Properly designed gates should allow proper flow of material as it enters the mold cavity. Gates should be located at points away from the functioning parts of the molded component. Vents are provided in all transfer molds to facilitate the escape of trapped air. The location of these vents depend on the part design and the locations of pins and inserts. The vent is sufficiently small so that it allows the air to pass through, but a negligible amount of molding compound can pass through it.

10.4.2 Transfer-Molding Process

Figure 10-29 shows the various stages of a typical transfer-molding process. In this process, leadframes are loaded (6–12 in a row) in the bottom half of the mold. For both plate and cavity-chase molds, this is done at a workstation separate from the molding press. A cavity-chase mold uses a loading fixture; most molding operations have automated leadframe loaders. Both the moving platen and the transfer plunger initially close rapidly, but the speed reduces as they close. After the mold is closed and clamping pressure is applied, the preform of molding material (which has usually been preheated to 90–95°C, which is below the transfer temperature) is placed in the pot, and the transfer plunger or ram is activated. Preheating of the molding compound is done by a high-fre-

Figure 10-27. Multiplunger Mold Used for Simultaneous Encapsulation of DIPs and Quad Flat Packs. (Courtesy of National Semiconductor, 1994.)

quency electronic method that works on a principle similar to microwave heating. The transfer plunger then applies the transfer pressure, forcing the molding compound through the runners and gates into the cavities. This pressure is maintained for a certain optimum time, ensuring proper filling of the cavities. The mold then opens, first slowly. This step is

Figure 10-28. Multiplunger Mold with Each Pot Feeding Just One Cavity. (Courtesy of National Semiconductor, 1994.)

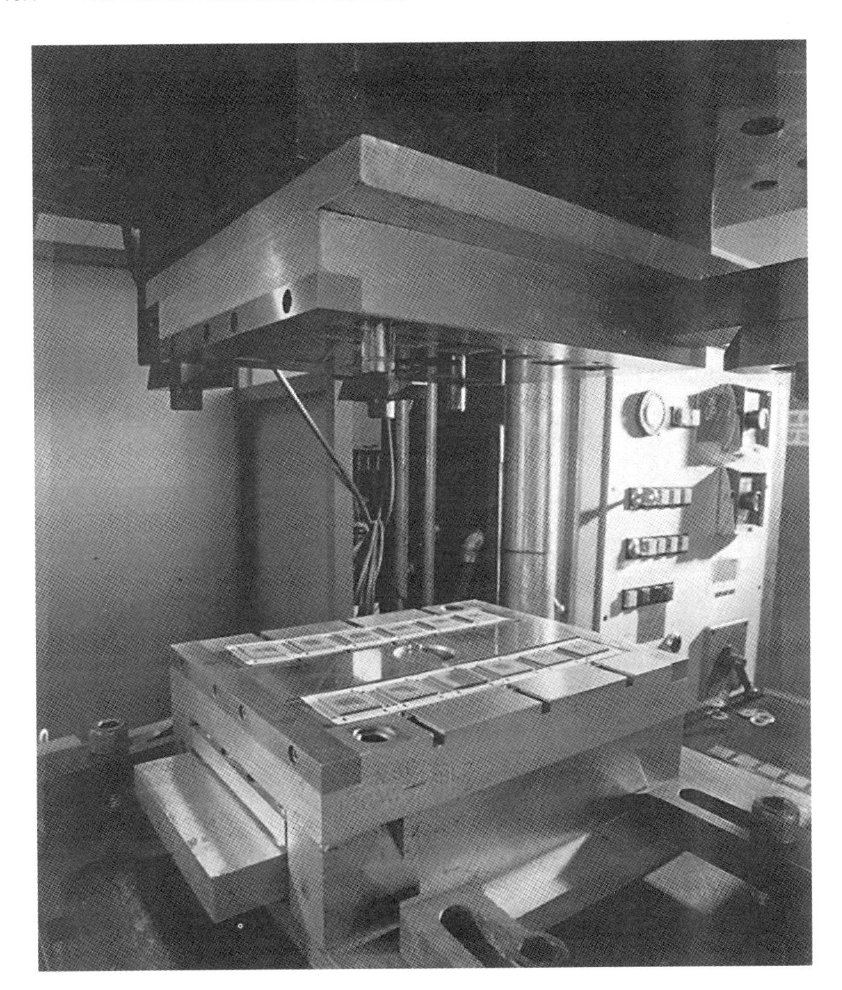

Figure 10-29. Different Parts of a Transfer Mold. (Courtesy of National Semiconductor, 1994.)

known as the slow breakaway. Sometimes it is desirable to have the transfer plunger move forward so that it pushes out the cull, or the material remaining in the pot. Finally the component is ejected using the ejector system in the mold. In an aperture-plate mold, the plates themselves are loaded with the leadframe strips, as they shuttle in and out of the molding press.

The parameters controlled by the press include the temperatures of the pot and the mold, the transfer pressure, the mold-clamping pressure, and the transfer time to fill the mold cavities completely. The mold temperature should be high enough to ensure rapid curing of the part. However, precautions should be taken to control the mold temperature, because too high a mold temperature may result in "precure" or solidification of the molding compound before it reaches the cavities. Electric heating is the most commonly used method for heating the molds. Multiple electric heating cartridges are inserted both in the top and bottom halves of the molds; positioned to supply heat to all the cavities. The applied pressure ensures the flow of material into all parts of the cavity.

Clamping pressure applied on the mold ensures the closure of the mold during polymerization or cure and also against the force of the material entering the cavities. Packing pressure, usually higher than the transfer pressure, is applied once the mold has been filled. While it is being transferred into the mold, the molding compound is reacting. Consequently, the viscosity of the material increases, at first gradually. As the reacting molecules become larger, the viscosity increases more rapidly until gelation occurs, at which point the material is a highly cross-linked network.

The pressure applied by the transfer plunger is critical. It should be sufficient to force the material through the runners and gates into the cavity and hold the material until polymerization. For high throughput, it is desirable to transfer and react the material as rapidly as possible, but decreasing the transfer time requires an increase in the transfer pressure to fill the mold. The high transfer pressures, typically as high as 170 MPa (25 ksi), can cause damage within the package, such as wire-sweep short or, in extreme cases, wirebond lift-off, shear, or fracture. These problems can be avoided by careful design of the package size and the transfer-gate size to minimize shear rate and flow stresses during cavity filling. In general, factors that tend to exacerbate these problems include high wire-loop heights, long wirebonds, bond orientation perpendicular to the advancing polymer flow front, rapid transfer times with corresponding high transfer pressures, high-viscosity molding components, and low-elastic-modulus wire.

An important way to decrease flow-induced stresses and improve molding yield is by velocity reduction. Velocity reduction in epoxy molding compounds reaches its limit before the need for flow-induced stress reduction is satisfied. In constant-pressure programmed transfer-rate control, high flow-induced stresses are created in the cavities nearest to and farthest from the transfer pot, whereas the middle cavities experience the lowest velocities and stresses. Mold mapping of flow-induced defects shows lowest yield in extreme-positioned cavities. A transfer-rate profile that is slow at the start, then higher over the middle cavities, and slow again when the last cavities are filling helps in reducing the flow-induced stresses. Different mold designs and molding compounds require distinctive transfer profiles. Velocity-reducing tool changes are permanent for a particular molding compound and package design. Mold designs aim at balanced mold filling by maintaining nearly uniform pressure fronts through all segments of the mold; in these, channel cross-sectional areas are controlled for volume-flow and pressure-drop uniformities in all cavities. However, in all cases, a velocity surge occurs when the molding machine switches from the transfer pressure to the packing pressure. These transient high velocities, along with the rapid compression of any remaining voids, can cause wire sweep. Using a programmable pressure controller to profile this pressure transition is the best approach to minimizing this problem.

After 1–3 min at the typical molding temperature of 175°C, the polymer is cured in the mold. Following curing, the mold is opened, and ejector pins remove the parts. The molded packages are ready for ejection when the material is resilient and hard enough to withstand the ejection forces without significant permanent deformation. After ejection, the molded leadframe strips are loaded into magazines, which are postcured in a batch (4–16 h at 175°C) to complete the cure of the encapsulant. Postcure normally involves holding the part in an oven at a temperature somewhat lower than the mold temperature but well above the room temperature for several hours. In some instances, postcuring is performed after code marking to eliminate an additional heat-cure cycle. The most important consideration in postmold cure analysis is the development of thermomechanical properties.

10.4.3 Simulation

The transfer-molding process is more complicated and difficult to treat analytically than either thermoplastic extrusion or injection molding, because of the time-dependent behavior of the molding compound, the irregular cross sections of the runners, and the common presence of inserts in the cavities. However, once a good quantitative model of the transfer-molding process has been proven, substantial time and cost can be saved. Long and expensive experimental runs no longer need to be carried out by trial and error to debug a mold or qualify a new compound.

Modeling the dynamics of transfer molding have met with moderate success [31,40]. The approach typically involves formulating a chemorheological description of the epoxy molding compound and coupling it with a network flow model. The gating factor for a successful simulation is a good characterization of the compound to determine its kinetic and rheological behavior. Most often, the curing reaction of the epoxy can be described by an autocalytic expression. The dynamic viscosity, as measured generally through a plate and cone viscometer, is reasonably well covered by the Castro–Macosko model [40]. By incorporating this material information into a flow model that includes the geometrical intricacies of a multicavity mold, an understanding of the filling characteristics of a particular compound can be obtained.

With a rheological model in hand a variety of scenarios can be simulated to optimize processing conditions and mold design [42,43]. For instance, different temperatures, pressures, and packing settings can be estimated to provide optimized filling profiles. To reduce wire sweep, the dominant-yield loss concern for fine-pitch packages, an analytical model of wire deformation can be coupled with rheological and kinetic information to improve process conditions and package layout design rules [42]. Once a mold has been translated into its geometrical equivalent, various combinations of runner sizes, gate locations, gate dimensions, and cavity layout can be evaluated.

10.5 PACKING AND HANDLING

Handling costs for moisture-sensitive plastic-encapsulated microcircuits (PEMs) may constitute as much as 10% of the total cost. Poor handling can lead to loss of product yield during assembly of plastic-encapsulated microcircuits on printed wiring boards. Package cracking during surface mounting is directly associated with improper packing and/or handling of components, but it can be avoided.

Packing and handling involves materials, containers, and precautions that ensure that the quality and reliability of plastic-encapsulated microcircuits are not degraded during shipping, handling, and storage. This necessary step between the manufacture of components and their mounting on printed wiring boards does not add value to the components. However, poor packing and handling can have a dramatically negative impact on board assembly yield, manufacturer's cost, and shipping schedules.

Plastic-encapsulated microcircuits must be adequately protected during shipping, handling, and storage. Precautions taken in packing can provide the requisite protection against high-humidity conditions (which can lead to moisture ingress), against physical damage (which can cause bent or broken leads), and against damage due to electrostatic discharge. Automatic handling of packages is facilitated by providing special features in the packing.

Although all the reasons for special handling procedures are discussed in this chapter, the focus is on issues unique to plastic packages such as dry packing, providing a dry nitrogen environment, and baking before mounting packages on circuit cards.

10.5.1 Considerations in Packing and Handling

Considerations in the packing of plastic-encapsulated microcircuits include prevention of moisture absorption, protection of leads from handling damage, prevention of lead solderability degradation, precautions against electrostatic damage, and packing for automatic handling.

10.5.1.1 Prevention of Moisture

Because plastic-encapsulated microcircuits are nonhermetic, they are susceptible to moisture-induced failures and latent defects. Molding compounds can absorb enough moisture to cause cracking in moisture-sensitive packages when exposed to temperatures typical of furnace-reflow soldering and wave soldering. The phenomenon is known as popcorning because the pop of the packages is audible. High-temperature and temperature ramp rates, prime loads that cause popcorning [41], are experienced by the packages during surface mounting on circuit cards. Of particular concern are solder-reflow processes that heat the package body, such as infrared heating, hot-air furnace, and vapor-phase reflow. Solder-reflow processes that selectively heat the leads and bond pads but not the component body, such as hot-bar reflow, laser soldering, and manual soldering, are not problematic. Wave soldering may be a concern if components dip through the solder wave.

Through-hole mounted plastic-encapsulated microcircuits do not exhibit popcorning because the component body is protected from the solder heat by the board. However, popcorning may occur in through-hole components mounted on double-sided boards if the components come in direct contact with the solder wave.

Possible damage due to popcorning includes delamination of the encapsulant from adjacent metal structures such as the leadframe and die pad, damaged wirebonds, cratering of bond pads, and cracking of the encapsulant. Even if the plastic-encapsulated microcircuit continues to perform its electrical function after popcorning, its reliability is compromised because the cracked and delaminated encapsulant no longer provides protection. Cracks may also let contaminants reach the active circuit via capillary action.

Popcorning in moisture-sensitive plastic packages can be avoided by controlling moisture content with proper packing and handling procedures. The moisture content of a plastic-encapsulated microcircuit is defined as

$$\text{Moisture content (\% wt.)} = \left(\frac{\text{Wet weight} - \text{Dry weight}}{\text{Dry weight}}\right) \times 100\% \qquad [10\text{-}9]$$

Typical moisture absorption curves after a high-temperature bake are shown in Figure 10-30 as a function of temperature and relative humidity. The figure illustrates that temperature is the parameter determining the moisture absorption rate. The saturation limit is a function of both relative humidity and temperature, although relative humidity is the major factor determining the steady-state moisture content of the component. Weight gain by small-outline packages as a function of exposure time in 85°C/

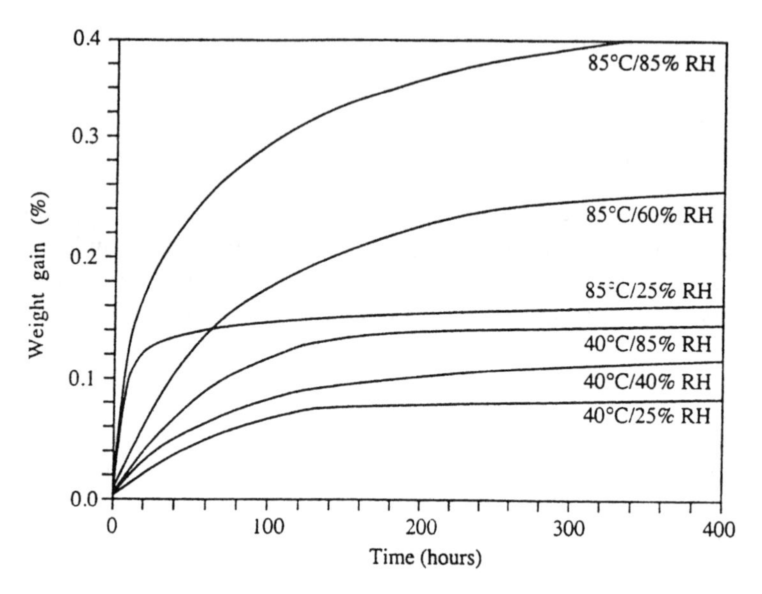

Figure 10-30. Typical Moisture Absorption Curves, After High-Temperature Bake, Shown as a Function of Temperature and Relative Humidity. (From Ref. 42.)

85% **rh** and 85°C/30% **rh** environments is plotted in Figures 10-31a and 10-31b, respectively [43]. The plots show that the mold compound chosen has a significant effect on the amount of moisture absorbed by plastic-encapsulated microcircuits.

For a given package type, there may be safe moisture content which the package may not popcorn. The critical value at which the component exhibits cracking depends on the adhesion and interlocking between the encapsulant and metal surfaces, and on package size, architecture, and lead count. Generally, large and thin plastic packages are more susceptible to popcorning; larger packages generate higher stresses at the corner of the die pad, and thinner packages offer a shorter path for moisture ingress through the molding compound, leading to higher moisture concentration at the interface of encapsulant and metal leadframe.

Dry packing must be used for shipping moisture-sensitive PEMs so they do not absorb more than the safe moisture content. The dry-pack process reduces the saturation moisture content in packages by keeping the relative humidity inside the moisture barrier bag below 20%. Other methods, such as storing in dry nitrogen and baking, may also be employed.

Three methods can be employed to dry components that have absorbed moisture: high-temperature bake, low-temperature bake, and storage in dry nitrogen. Generally, a package with a moisture level at or below 0.05% by weight is considered dry enough for dry packing. Compo-

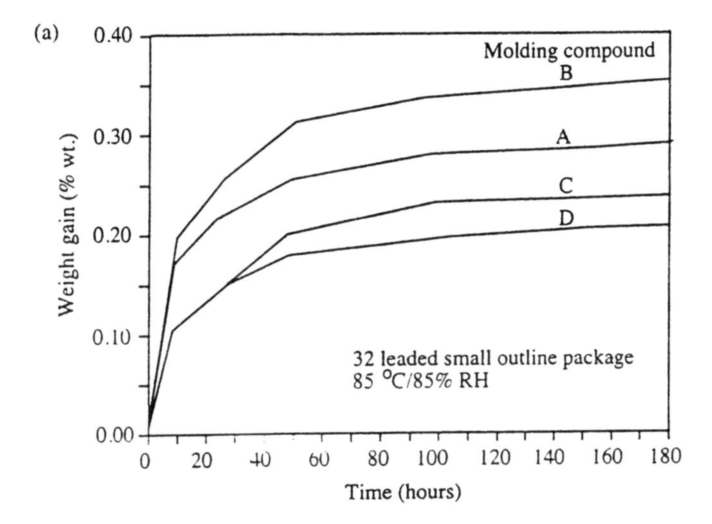

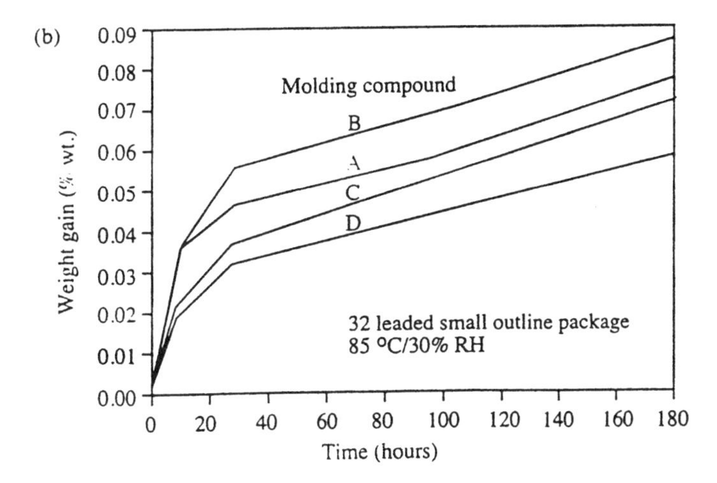

Figure 10-31. Weight Gain by Small-Outline Packages as a Function of Exposure Time in 85°C/85% rh and 85°C/30% rh environments. (From Ref. 43.)

nents are dried to below the safe moisture content to provide the desired time for shipping and handling in a humid environment before the components are mounted on printed wiring boards.

10.5.1.2 Lead Damage and Solderability

Manual handling, together with drying methods, shock, and vibration, can cause damage to leads. Shipping containers are designed for automatic handling and for preventing lead bend, loss of coplanarity, and lead contamination.

Mishandling may occur at various process points, such as trim and form operations, loading of parts into the tester for electrical sorting, or removal of parts after burn-in. Lead deformation can also be induced by such process factors as high molding stresses that result in warpage, or the buckling of thin leadframe strips prior to singulation [44]. Regardless of the cause, damage to fragile leads can be extensive. Six typical defect modes have been characterized: coplanarity, sweep, tweeze, twist, body standoff, and center-to-center spacing [45]. Coplanarity is the vertical deviation of a lead from the seating plane of the device and is the single most important parameter in assembly, packing, and handling.

Packing materials that include shipping trays, reels, tapes, and magazines are made of plastic materials, are often unstable at bake temperature, and outgas products that may condense on package leads, degrading their solderability. Outgassing of hydrocarbons and other organic contaminants from low-temperature stable packing materials is especially of concern during high-temperature bake. The container materials must be stable enough at bake temperature that any outgassing products do not degrade solderability. If solderability is a major concern, then the packages may be baked outside of the shipping containers.

Intermetallic growth is a time- and temperature-dependent diffusion process. Long exposure to high temperatures during baking can form excessive intermetallics and degrade solderability. Often, leads are made of copper, which forms a Cu_6Sn_5 intermetallic with tin (Sn). The growth of Cu_6Sn_5 is one of the causes of lead-finish degradation leading to poor solderability [46]. Limiting the time of exposure, especially at high bake temperatures, is therefore crucial for package lead solderability. If the bake is repeated, the total time of exposure to high temperatures should be limited. Moreover, oxidation of lead surfaces can cause degradation of lead solderability; inert gas flow and a low relative humidity environment is preferred during bake to alleviate this problem.

10.5.1.3 Electrostatic Discharge Protection

Plastic-encapsulated microcircuits may be sensitive to electrostatic discharge (ESD). Electrostatic discharge can either cause overstress failure of the component (which can be exposed by an electrical function test) or induce damage leading to parametric shift and reduced reliability.

Electrostatic discharge occurs most readily in low-humidity environments. The low humidity levels inherent in baking increase the risk of ESD damage. Air ionizers may be used in baking ovens to keep devices from charging, but they cannot dissipate the charge in areas where the ionized air cannot reach because of the flow pattern.

The probability of electrostatic discharge occurring increases with increased handling, which may lead to tribologically produced charges. Charge-dissipative or electrically conductive shipping containers are employed to protect against electrostatic discharge. Charge-dissipative containers are made of electrically conductive materials, such as resin impregnated with conducting graphite fibers. Often, an antistatic coating of carbon is applied on the inner walls of shipping containers; the coating is water soluble and wears out due to friction as components slide around.

Evaluating ESD-preventive characteristics for both graphite-impregnated and antistatic-coated tubes indicates that either type of ESD protection is sufficient, even for low-humidity environments. There is no significant observed difference in component charging for graphite-impregnated tubes stored at ambient humidity and those stored at < 10% **rh.** Although there is a difference in charging of components stored in antistatic-coated tubes in < 10% **rh,** the jeopardy is considered low because the charging is 2.2 picocoulomb (pC)/lead—well below the critical level of 10 pC/ lead [47].

10.5.2 Hierarchy of Packing

Dry packing is required to keep moisture-sensitive plastic surface-mount components dry during shipping and storage. There are generally three levels of packing for plastic-encapsulated microcircuits, as shown in Figure 10-32. The first level involves packing components in suitable shipping containers. The second level involves packing the containers along with desiccant pouches, a humidity-indicator card, and dunnage in a moisture-barrier bag; the third level labels the shipping box with appropriate handling and shelf-life information.

10.5.2.1 First Level Packing

The first level of packing requires placing plastic-encapsulated microcircuits in magazines, trays, or tape reels. ESD-preventive characteristics are built into the containers, either by making the container material electrically conductive or by coating with an antistatic agent.

Magazines are made of plastic or aluminum. Plastic magazines are treated with a water-soluble antistatic agent that wears out as a result of friction caused by package sliding or repetitive use. Magazines come in two forms, one for horizontal stacking of packages and the other for vertical stacking. Most types of plastic packages (e.g., dual-in-line packages, small-outline J-leaded packages, J-Leaded quad flat packs, and plastic-leaded chip carriers) can be stored in magazines. Dual-in-line packages are the only plastic package type shipped exclusively in magazines. Flexible polyvinyl chloride stoppers are inserted into the ends of the magazine to hold the packages in place.

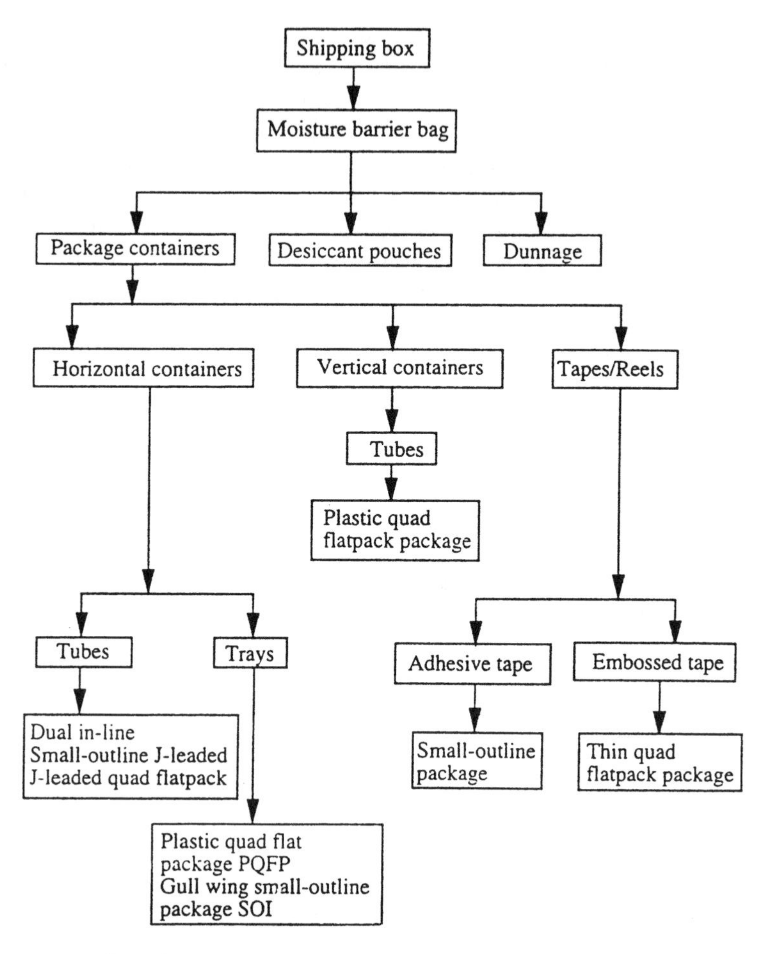

Figure 10-32. Packing Hierarchy for Handling Plastic Encapsulated Microcircuits (From Ref. 1.)

Trays made from injection-molded plastic, such as polyvinyl chloride and polystyrene, are used primarily for storage of quad flat packs and bumped quad flat packs. Antistatic characteristics are provided by impregnating conducting fibers with the injection-molded plastic, or by treating the trays with an antistatic agent, such as carbon. Trays made of polystyrene do not need an antistatic coating because sufficient fiber loading can be added to provide critical conductivity. Polyvinyl chloride trays should be stored between −25°C and +40°C because their shape and color may change above +55°C. As many as 100 plastic packages can be packaged per tray.

Tape reels are used for smaller-size plastic packages, such as small-outline plastic packages and small plastic leaded chip carriers. Adhesive or embossed tapes are used. The adhesive tape holds the plastic-encapsulated microcircuits on carrier tape; the embossed tape has depressed pockets in which the plastic-encapsulated microcircuits reside, and a cover tape ensures that they remain in place. Many of these smaller plastic-encapsulated microcircuits are difficult to handle manually. Tape reels allow packing that is amenable to automatic handling equipment. The carrier tape can have sprocket holes for loading registration on one or both sides, depending on the size of the plastic-encapsulated microcircuits. Tape widths come in a variety of sizes, ranging from 12 to 44 mm. Many more plastic-encapsulated microcircuits can be stored in tape reels than in magazines and trays. Typical tape reels hold 500–3000 PEMs per reel. Nippon Electronic Company has a reel that holds 12,000 small transistors.

10.5.2.2 Second Level Packing

The second level of PEM packing involves enclosing magazines, trays, or tape reels in bags. Moisture-barrier bags are used for dry-packing moisture-sensitive plastic-encapsulated microcircuits. At the time of dry-packing, devices should have a moisture content of less than 0.05% by weight. Not all plastic-encapsulated microcircuits are considered moisture sensitive, so not all are shipped in dry packs. Table 10-13 lists some package types from Texas Instruments, the shipping container used, and whether they are shipped in dry packs.

Table 10-13. Shipping Methods and Quantities

Package[a]	Tube	Tape/Reel	Trays	Dry Pack
56-Pin SSOP	20	500	N/A	No
44-Pin PLCC	27	500	N/A	No
68-Pin PLCC	18/19[b]	250	N/A	Yes
64-Pin SQFP	N/A[c]	N/A	50	Yes
80-Pin SQFP	N/A	N/A	50	Yes
120-Pin SQFP	N/A	N/A	50/84[d]	Yes
80-Pin SQFP	N/A	N/A	50	Yes

[a]SSOP, skinny small-outline package; PLCC, plastic leaded chip carrier; SQFP, small quad flat pack.
[b]Eighteen packages can be packed in a single tube when a pin is used as a tap; 19 packages can be packed when plug is used as a tap.
[c]N/A, not applicable.
[d]Depending on tray size.
Source: From Ref. 48.

Dry-packing uses a moisture-barrier bag (MBB) in shipping containers with dried plastic-encapsulated microcircuits, a desiccant, and a humidity-indicator card. The moisture-barrier bag is sealed and labeled with the seal date.

Moisture-barrier bags are generally made of Tyvek, a multilayered opaque material. MIL-STD-81705B [49] classifies moisture-barrier bag materials into two types. Type I has a water-vapor transmission rate (WVTR) of less than 0.02 g/100 in.^{2} of bag area per 24 h; type II has a WVTR between 0.02 and 0.08 g/100 in.^{2} per 24 h. The bags should meet type I requirements for water-vapor transmission rate, flexibility, electrostatic discharge protection, mechanical strength, and puncture resistance [41]. Most of the air should be removed from the bag by manually compressing the outside of the bag or by using suction equipment. However, the bag should not be completely evacuated, as this would reduce the effectiveness of the desiccant and could also damage the enclosed components.

A desiccant is a moisture-absorbing material that is enclosed in the moisture-barrier bag to absorb moisture from the air inside the bag and maintain its relative humidity below 20%. A number of different materials are used as desiccants. Figure 10-33 shows desiccant moisture absorption rates as a function of desiccant type.

Desiccant material should meet MIL-D-3464 type II [41]. It must be ductless and noncorrosive. The amount of desiccant needed per bag to maintain an interior relative humidity below 20% is a function of bag surface area, bag water-vapor transmission rate, and storage time.

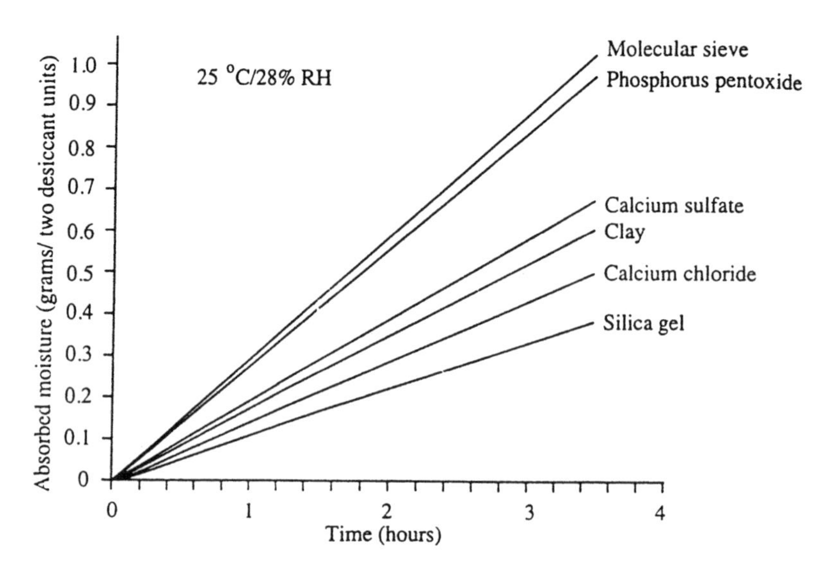

Figure 10-33. Desiccant Moisture Absorption Rates as a Function of Desiccant Type. (From Ref. 42.)

A humidity-indicator card (HIC) is a means of measuring the relative humidity of the air surrounding the card. A card is inserted in the moisture-barrier bag before sealing to help determine the effectiveness of the dry pack. An illustration of a humidity indicator card is given in Figure 10-34. The card has at least three color dots, providing indications from 10% to 30% relative humidity. A pink dot means the card has been exposed to at least the given amount of humidity; a blue dot means it has not been exposed to that much humidity. If the 30% dot starts to change from blue to pink, the relative humidity inside the moisture-barrier bag is higher than 20%, and the desiccant should be replaced. A humidity-indicator card must be dry before it can be packaged in a dry pack; if it is not, it should be baked dry or put in a desiccant bag to dry. The chemical reaction that causes the change in color is reversible, so cards are reusable.

A warning label is mounted on each dry-packed moisture-barrier bag. The label adhesive and markings are water resistant. The labels include a moisture-sensitivity caution symbol, instructional notes, and a bag seal date.

10.5.2.3 Third Level

The third level of PEM packaging is the outer box, often made of corrugated cardboard. Bags and magazines, trays, or tape reels are stacked into the outer carton, which is then taped shut. Bar-code labels, usually placed on the outer box, indicate the presence of plastic-encapsulated microcircuits and the seal date. An example of a desiccant bar-code label is shown in Figure 10-35 [42].

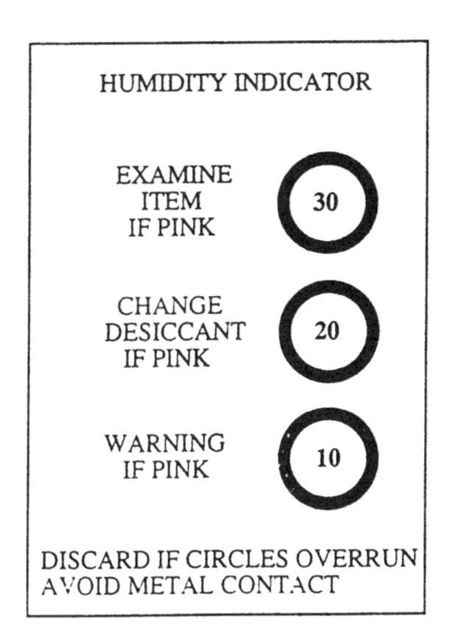

Figure 10-34. An Illustration of a Humidity Indicator Card. (From Ref. 41.)

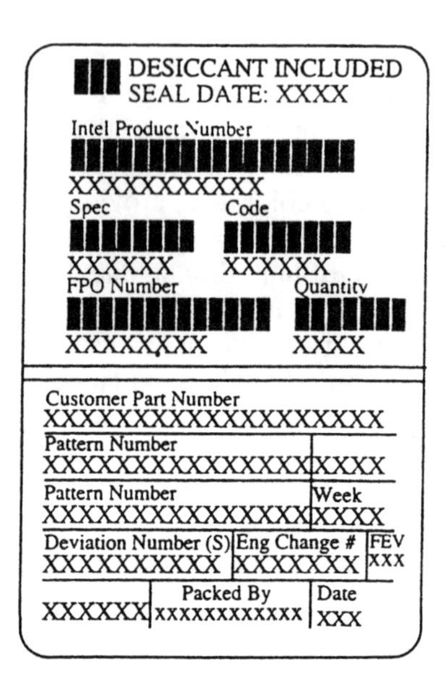

Figure 10-35. An Example of a Desiccant Bar-Code Label. (From Ref. 42.)

10.5.3 Handling of Dry Packages

Dry plastic-encapsulated microcircuits have a finite floor life, after which they absorb moisture above a safe level. Component floor life can be defined as the time period that begins after moisture-sensitive devices are removed from the moisture-barrier bag and the controlled storage environment, and ends when they have absorbed enough moisture to be susceptible to damage during reflow. Floor life characteristics are dependent on many factors. The moisture-barrier bag should not be opened until the components can be board mounted. Before the bag is opened, seal integrity should be verified. Often, the shelf life of dry-packed components is less than 12 months from the seal date (the shelf life is stated on the bag warning label). If leakage is found or the shelf life is exceeded, the plastic-encapsulated microcircuits should be dried before they are used for any reflow soldering.

The length of time moisture-sensitive PEMs may be exposed to the manufacturing floor environment after being removed from the moisture-barrier bag depends on the moisture content of the components and the floor environment temperature and relative humidity. Figure 10-36 shows typical component weight gain versus time for various manufacturing floor environments. For example, plastic leaded chip carriers absorb moisture beyond the safe moisture content of 0.11% by weight in approximately 130 h in a floor environment of 30°C/60% **rh.**

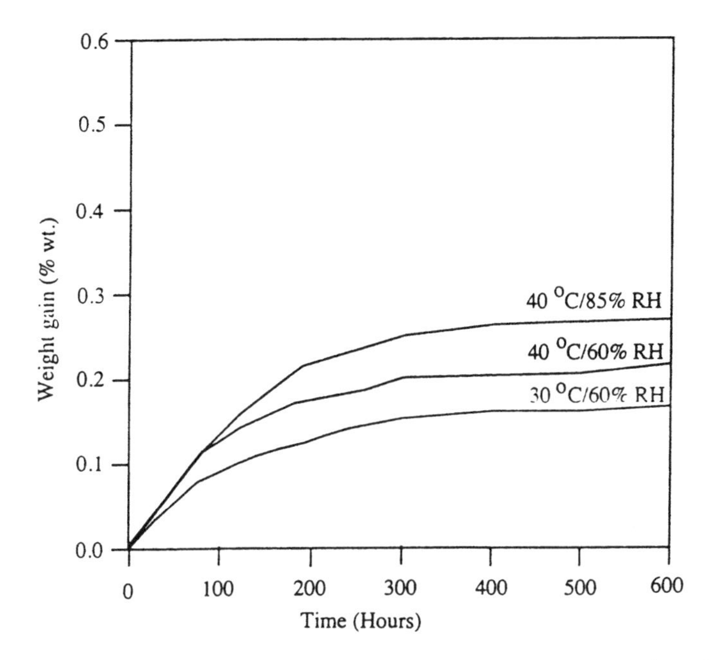

Figure 10-36. A Typical Component Weight Gain Versus Time for Various Manufacturing Floor Environments

10.5.3.1 Classification

Exterior configuration is not the sole indicator of moisture/reflow sensitivity. The same number of days of ambient exposure can result in different amounts of moisture being absorbed by the package, depending on the molding-compound material and package design. The same amount of moisture can cause different amounts of damage in different packages. Packages with the same exterior configuration (lead count, thickness, and so on) may not have the same interior construction, and absorption rate and sensitivity may differ between packages that otherwise look identical. Consequently, the percentage weight gain moisture content is not useful for establishing moisture sensitivity. Different package types, materials, lead counts, and die attach areas will reach different levels of moisture absorption before reflow damage occurs.

Packages are now classified into six levels based on their storage floor life (out of bag) at the board assembly site. Table 10-14 shows the test conditions and the storage floor life of packages at the six levels [41]. The packages are first subjected to the appropriate test conditions based on the level for which they are being tested. They are then subjected to a prescribed reflow simulation and are evaluated for moisture/reflow-induced damage. They are classified into the lowest level at which they

Table 10-14. Classification and Floor Life of Desiccant Packed Components

Level	Storage Floor Life (out of Bag) at Board Assembly Site	Preconditioning moisture[a] °C/%rh	Time (h)
1	Unlimited at ≤ 85% **rh**	85/85	168
2	1 Year at 30°C/60% **rh**	85/60	168
3	1 Week at ≤30°C/60% **rh**	30/60	168 + MET[b]
4	72 h at ≤ 30°C/60% **rh**	30/60	72 + MET
5	As specified on label range: 24–71 h at ≤ 30°C/60% **rh**	30/60	Time on label + MET
6	Mandatory bake; bake before use and once baked must be reflowed within the time limit specified on the label	30/60	Time on label

[a] Preconditioning moisture soak to simulate total exposure time conditions.
[b]MET—Manufacturer's exposure time; that is, the compensation factor that accounts for the time after bake that the component manufacturer requires to process components prior to bag seal. It also includes a default amount of time to account for shipping and handling.
Source: From Ref. 41.

pass the acceptance criteria. The classification technique is discussed in IPC-TM-650 Test Method 2.6.20A: Assessment of Plastic Encapsulated Electronic Components for Susceptibility to Moisture/Reflow Induced Damage.

10.5.3.2 Board-Level Rework of Moisture-Sensitive Components

If packages must be removed from the board, localized heating should be used and the maximum temperature excursion should not exceed 200°C. The component temperature should be measured at the top center of the package body; if the temperature exceeds 200°C, the board may require a bake-out prior to rework. If the part is to be reused, it should be baked dry before remount; new replacements should be kept dry as well. Localized replacement reflow heating avoids resubjecting the entire board to reflow temperature profiles. The impact of nearest-neighbor heating should be evaluated as part of the preconditioning flow for qualification. If the packages are determined to be nonmoisture sensitive after assembly, then controlling the board-level moisture content is not necessary for rework.

If the moisture-barrier bag is opened and closed several times, the cumulative floor exposure time should not exceed the stated floor life; the time may be extended by storing the plastic-encapsulated microcircuits in an environment with humidity less than 20%. The floor life differs from manufacturer to manufacturer. Intel claims that only their plastic leaded chip carriers and plastic quad flat packs with fewer than 100 leads may be in a floor humidity environment of less than 20% for extended periods of time. Hitachi [50] contends that its plastic-packaged devices can be exposed to the floor environment for more than 1 week of cumulative time outside the moisture-barrier bag before reflow soldering.

Once the moisture-barrier bag is opened, component use or handled options may be pursued in the following order of preference:

- The components may be mounted within a specified time before moisture is reabsorbed beyond the safe moisture content level.
- The components may be stored outside the moisture-barrier bag in an environment of less than 20% relative humidity until future use.
- The components may be resealed in the moisture-barrier bag with new fresh desiccant added and resealed and exposure time noted; the total exposure time before mounting must be kept within the allowed time for the exposure conditions.
- The components may be resealed in the moisture-barrier bag using the original desiccant; this method does not allow the floor life of the devices to be extended beyond the time indicated by the seal date.

For lead protection, it is recommended that components not be removed from their original shipping containers until mounting; contact with the leads can cause positional or coplanarity problems. It is also recommended that automatic or semiautomatic handling equipment be used in lieu of manual handling, especially for surface-mounted packages.

Precautions should be taken against electrostatic discharge, especially if the relative humidity is below 20%, as it is during baking; air ionizers can be installed in baking ovens to remove static electricity. Water leakage can wash out the water-soluble antistatic coating in shipping containers or cause the coating to peel off and lose its effectiveness. Storing plastic-encapsulated microcircuits in antistatic-coated containers beyond 6 months should be avoided because the coating may warp over time. Antistatic-coated containers should not be reused.

10.5.4 Environmental Considerations

Protecting delicate leads during handling and transit is the main function of the tubes and trays used in packing. However, this protection comes at a price. Before “green” manufacturing became fashionable, packing paraphernalia was treated as regular, disposable “brown” trash; landfill environmental concerns were never raised. As a result, packing materials are regularly piled and dumped without much consideration of the ecological implications. With sparse land for landfill sites, European countries were the first to raise the possibility of recyclability in electronic manufacturing. Nontoxic, lead-free packaging was one practical outcome of that initial push.

Environmental pressure has forced recyclability to become a prime concern. As the current packing materials are made mostly of thermoplastics, recycling is relatively trouble free. Used polyethylene tubes or polyvinyl chloride trash can be ground, pelletized, and remolded. The end products can be resold as packing grade with the "Made from recyclable materials" label or similar logo. Several industries have already sprung up to address this market niche. Used packing materials are collected from the original equipment manufacturer OEMs or assembly houses, where parts are unloaded from their packing carriers to be mounted. Cleaning, grinding, remelting, and molding the materials can be a profitable enterprise.

10.6 QUALITY AND RELIABILITY

When a product's performance is no longer acceptable, it is said to have failed. Failure occurs when some failure mechanism—a chemical, electrical, physical, mechanical, or thermal process acting at some site in a structure—induces a failure mode, such as an electrical open, a short, or parametric drift. Common failure sites for the various failure mechanisms in plastic-encapsulated microelectronic packages are schematically illustrated in Figure 10-37.

In this section, the emphasis is on understanding the failure mechanisms to avoid failures.

10.6.1 Classification of Failure Mechanisms

Failure mechanisms fall into two broad categories: overstress and wear-out. Overstress failures are often instantaneous and catastrophic. Wear-out failures accumulate damage incrementally over a long period, often leading first to performance degradation and then to device failure. Further classification of failure mechanisms is based on the type of load that triggers the mechanism—mechanical, thermal, electrical, radiation, or chemical.

Mechanical loads include physical shock, vibration (e.g., underneath an automotive engine hood), loads exerted by filler particles on the silicon die (because of encapsulant shrinkage upon curing), and inertial forces (e.g., in the fuse of a cannon shell being fired). Structural and material responses to these loads may include elastic deformation, plastic deformation, buckling, brittle or ductile fracture, interfacial separation, fatigue crack initiation, fatigue crack propagation, creep, and creep rupture.

Thermal loads include exothermic die attach curing, heating prior

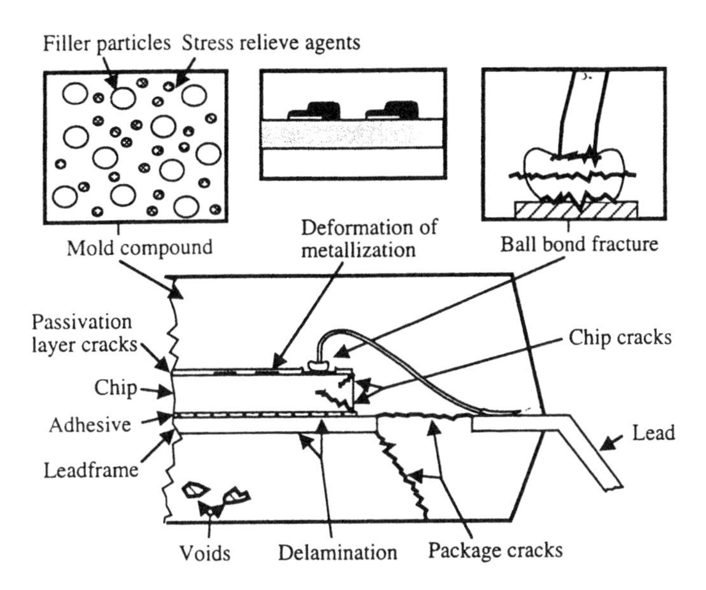

Figure 10-37. Typical Failure Mechanisms, Sites, and Modes in Plastic-Encapsulated Devices (From Ref. 1.)

to wirebonding, encapsulation, postmold exothermic curing, rework on neighboring components, dipping in molten solder, vapor-phase soldering, and reflow soldering. External thermal loads lead to changes in dimensions because of thermal expansion and can change such physical properties as the creep flow rate or cause burning of a flammable material. Mismatches in coefficients of thermal expansion often cause local stresses that can lead to failure of the package structure.

Electrical loads include, for instance, a sudden current surge through the package (e.g., in the ignition system of a car engine during start-up), fluctuation in the line current due to a defective power supply or a sudden jolt of transferred electricity (e.g., from improper grounding procedures), electrostatic discharge, electric overstress, applied voltage, and input current. These external loads may produce dielectric breakdown, surface breakdown of voltage, dissipation of electric power as heat energy, or electromigration. They may also increase electrolytic corrosion, current leakage due to formation of dendrites, and thermally induced degradation.

Radiation loads are principally either alpha particles found in packaging materials from trace radioactive elements, such as uranium and thorium, or cosmic rays that can upset memory devices, degrade performance, and depolymerize the encapsulant. The problem is more acute in memory devices that are highly sensitive to alpha particles, in which even a minor alpha flux through the device can flip the binary state of a memory bit. Due to the stochastic nature of this failure mode, preventive measures usually are built in. A polyimide die coating of proper thickness is a simple measure that can block damaging rays. Compounding ultrapure fillers with low alpha residues in the resin is another way to minimize erratic device behavior due to ionizing charges.

Chemical loads include chemically severe environments that result in corrosion, oxidation, and ionic surface dendritic growth. Moisture in a humid environment can be a major load on a plastic-encapsulated package because of the permeability of moisture through encapsulants. Moisture absorbed by the plastic can leach catalyst residues and polymerization by-products from the encapsulant, then ingress to the die metallization bond pads, semiconductor, and various interfaces and activate failure mechanisms that degrade the package. For instance, reactive flux residues coating the package after assembly can migrate through the encapsulation to reach the die surface.

10.6.2 Analysis of Failures

This section discusses the mechanisms that result in damage or failure of a plastic-encapsulated device.

10.6.2.1 Die Fracture

Surface scratching and cracks in dies may form at the thermal processing, die scribing, and dicing stages. If, after the manufacture of the die, a preexisting initial crack is equal to or greater than the critical crack size (depending on the magnitude of applied stress), the die can catastrophically fracture in a brittle manner. For example, microcracks nucleated at the top surface from surface scratching of the wafer may propagate to an active transistor and cause device failure. Edge cracks developed from die separation damage are most likely to propagate at the corner of the die, due to high longitudinal stresses common in that location when the die–leadframe structure is temperature cycled. Voids in the attachment material may also affect die fracture.

Chiang and Shukla [51] summarized die cracking statistics for samples that had edge or center voids and were subjected to 10 cycles of thermal shock. The devices with center voids showed no cracks, whereas devices with edge voids had nearly a 50% failure rate due to die cracks. Although cracks can originate from edge damage during sawing, fine control of saw speed and saw blade quality has eliminated most of the defects. However, die cracks still originate from flaws on the back side due to backgrinding operations used to manufacture thin packages. Depending on the grit size used, grinding can either reduce or increase the chance of cracking; typically the size of flaws approaches the grit size. Thus, all coarse flaws should be eliminated by either fine grinding or back-side etching.

During die attach, each die is ejected from the backing tape by a transfer plunger. Damage to the die may be introduced at this stage, because the plunger tip and the loading speed govern the size and depth of the indentation.

The thermal stresses due to the TCE mismatches among the various materials in the plastic package can be expressed by

$$\sigma = c_5 \int_{T_1}^{T_2} \left(\frac{\alpha_p - \alpha_i}{1/E_p - 1/E_i} \right) dT \qquad [10\text{-}10]$$

where

c_5 = a geometry-dependent constant
α = the coefficient of thermal expansion
E = the modulus of elasticity
and the subscripts p and i denote the encapsulant material and the material in contact with the encapsulant material (i.e., the chip or leadframe material) [40].

If the thermally induced stress given by Eq. (10-1) is greater than the allowable stress, the package fails. An estimate of allowable stress can be obtained from fracture mechanics principles [52].

10.6.2.2 Loss of Chip Passivation Integrity

The chip passivation layer is usually made of brittle glass films, such as silica (SiO_2), phosphosilicate glass (PSG), or silicon nitride (Si_3N_4). The aluminum bond pad under the passivation film can break from the bonding pressure and cannot sustain the passivation film. Phosphorus, added to the silicon dioxide passivation layer for ion gettering, stress relaxes the layer and avoids cracking. An optimum phosphorus content is 1.8% by weight [53]; a high concentration (> 8%) accelerates corrosion failures of the die due to the formation of corrosive phosphoric acid with ingressed moisture. Corrosion of the die metallization can result in various electrical failures, including device leakage, shorts, or opens. Plasma-enhanced chemical-vapor-deposited (PECVD) silicon nitride has better moisture resistance than phosphorus-doped silicon dioxide.

Invariably, a larger distribution of cracks is found near the edges of the device, rather than in the die center. As mapped out with piezoresistive test chips, high shear stresses are always observed at the edges. The cracks are not easily seen under the optical microscope; however, once exposed to a mild etchant, the affected areas become more visible. The solution diffuses through the microcracks in the passivation, attacking the underlying metal and forming dark areas wherever the metal has been removed.

Cracking of passivation occurs due to shearing stresses imposed by the molding compound [54]. Passivation layer cracking is often associated with ball-bond lift-off and shearing because of the close proximity of these structures to passivation layer damage on the edges of the chip.

Polyimide film can also be used as the passivation layer. The film is relatively soft and sufficiently ductile to withstand bonding pressure and high temperatures; although the film distorts under stress, it does not crack. Inayoshi et al. [55] evaluated passivation cracking and showed that it can be avoided by using a ductile passivation. Polyimide also offers good adhesion with the molding compound, enhancing the interfacial integrity critical to the long-term reliability of the package during thermal excursions [20].

10.6.2.3 Die Metallization Corrosion

Several factors affect corrosion of the aluminum metallization: system acidity, metal composition, encapsulation material, passivation glass, ionic contamination, temperature, relative humidity, applied voltage, and moisture resistance of the package [56].

The most severe condition for the plastic package and a main cause of corrosion is the interaction of moisture with ionic contaminants and ingredients in the molding compound. Moisture can reach the wirebond by diffusion through the encapsulant or through such defects in the package as separation between the leadframe and the encapsulant where the leads enter the package body. Moisture may react with chlorine, in the form of chloride ions present in many manufacturing processes imposed on the package. Bromides and antimony trioxide, both used as flame retardants, can be released from the molding compound when high temperatures occur in the presence of moisture.

Galvanic corrosion requires some voltage potential. The interface of the dissimilar metals gold and aluminum sets up an approximately 3-V potential. In the presence of water and an ionic contaminant, galvanic corrosion can easily take place. Although most of the die metallization is covered by a passivation layer, defect sites can be present in the passivation, allowing corrosion of the metal features. Because there is no passivation in the bond-pad area, this region is also vulnerable to any corrosive electrolyte. Complete coverage of the bond pad by the wire is usually impossible because the bond pad is square and the corresponding ball bond is circular. Unless high bonding force is used to flatten the ball to obtain greater coverage or round bond pads are designed, exposed aluminum will remain a fact of package assembly.

Techniques to passivate bond pads for reliability without hermeticity are expensive and are not typically employed for commercial packages.

Round bond pads have been tried in the past but were abandoned because the process control vision systems of current wirebonders recognize only square features. Good passivation coverage (using phosphorus glass or silicon nitride) is necessary for moisture resistance. Measures to avoid and control corrosion-related failures include reducing ionic impurities in the encapsulant, using impurity-ion catchers or ion scavengers in the encapsulant, increasing the encapsulant-to-leadframe adhesion strength, using fillers in the encapsulant to elongate the path for moisture diffusion, and using a low-water-uptake encapsulant.

10.6.2.4 Metallization Deformation

The aluminum metallization on the chip can be plastically deformed by shrinkage forces imposed by the molding compound. Deformation is highest at the edges of the chip, where the shear stresses are greatest. In this case, the encapsulant is constrained from moving over the chip by its adhesion to the chip surface adjacent to the metal trace; any loss of adhesion between the encapsulant and the chip increases the potential for metallization deformation.

The stress profile on the die can be mapped out with piezoresistive strain gauges embedded in the silicon, which provide a good correlation between measurements and finite element analysis [57]. For a given die and package configuration, the average in-plane shear stress is about one-tenth the magnitude of the corresponding principal stresses. Near the die edges, this scaling factor approaches one-third. The principal stresses in the die center increase with die size, whereas the reverse applies to in-plane shear stresses. Edge stresses are sensitive to die dimensions, increasing with die size.

The amount of free die-paddle space around the die also affects the stress level. Generally, stresses increase as free space increases; thus, tailoring a die pad to closely fit die size optimizes the stress level. Orientation and die aspect ratio also influence edge stresses much more than center stresses. For example, an asymmetrical die with its axis perpendicular to that of the package encounters lower stresses than one aligned parallel to the axis.

The amount of stress imposed on a die depends on the thickness and volume of the surrounding molding compound. Thin packages experience much less stress than thick packages and are the current trend in the packaging industry. The in-plane packaging stresses responsible for metal deformation are caused by lateral compression of the die. A good working knowledge of the die stress profile can help in the design of future products that minimize stress-induced damage.

10.6.2.5 Wire Sweep

Wire sweep usually denotes visible wire deformation, typically a lateral movement in the direction of the compound flow through the cavity.

Under this lateral deformation, ball bonds and sometimes stitch bonds can develop kinks at the attachment point as the wire is pulled off the axis of the bond. The kink is most often noted on the side of the chip nearest the gate through which the molding compound enters the cavity.

Reliability concerns with wire sweep include device shorting and current leakage. Shorting can be from wire to wire, from wire to lead finger, or from wire to die edge. Failure can be immediate or may not show up until the package experiences stress excursions.

Wire sweep can occur from any one of a number of causes: high resin viscosity, high flow velocity, unbalanced flows in the cavities, void transport, late packing, and filler collision [58,59,60].

10.6.2.5.1 Viscosity

During radio-frequency preheating, the pellets of mold compound reach up to 100°C. Compacting the preheated pellets with a plunger in the molding pot causes the compound to heat further through a "fountain flow effect." The material near the wall is hotter than the rest and has a lower viscosity than the colder core. Thus, the compound can be convected to the cavities near the pot while the core is relatively cold. The ensuing high viscosity may cause wire sweep.

10.6.2.5.2 High Flow Velocity

For cavities located far away from the pot, sweep can be induced by a different mechanism. At this stage, sufficient heat has been convected to the compound to lower viscosity to a level within the processing window. The majority of the cavities are already partially filled, resulting in a reduced flow path. As a result, for the same filling pressure, the velocity of the flow front in the end cavities can be quite high, raising the potential for wire sweep.

10.6.2.5.3 Unbalanced Flows in the Cavities

Cavity gates, if improperly designed, may produce flow fronts that are not balanced, above and below the die-paddle assembly, resulting in a "racetrack effect." One front precedes the other and blocks the vent line before complete filling, trapping air in the package. This flow imbalance creates a rolling bending moment that deflects the leadframe and affects the stability of the wirebonds.

10.6.2.5.4 Void Transport

The change in interfacial tension as voids are convected across the die surface by the flow front can move the wirebonds. With a rapid succession of small voids, the wires can oscillate from their equilibrium position; such motion can be captured permanently by the rapidly curing resin. Molding voids tend to be more frequent with conventional molding than with multiplunger (gang-pot) molding, as gang-pot presses provide a shorter flow path for the mold compound and better control of transfer speed and pressure [61].

10.6.2.5.5 Late Packing

The filling profile programmed into standard molding presses usually requires filling 90–95% of the mold first under "velocity control." The profile is then switched to "pressure control," during which constant pressure is maintained to ensure that a certain density is obtained. The switchover must be made while the compound still has a low viscosity; a highly viscous gel could slide across the die and collapse the wirebonds [62].

10.6.2.5.6 Filler Collision

Collision of fillers with fragile wirebonds is another cause of wire sweep. Wirebonds are usually between 23 and 38 μm (0.9 and 1.5 mils), depending on the current-carrying requirements of the device. Fillers, on the other hand, can have multimodal distributions for better packing. Sizes can range from a submicron up to 100 μm. By comparison, the thickness of gates in typical molds is around 200–250 μm (8–10 mils). As the highly filled front flows across the die, collision with any large filler particles can result in plastic deformation of wirebonds.

10.6.2.6 Cratering of a Wire-Bond Pad

Cratering is a failure mechanism that occurs during manufacturing, usually as a result of improper bonder setup. Bonding parameters that affect the quality of a bond include temperature, ultrasonic power, force, and bond time [63].

In ultrasonic bonding, even though the metal flow is equal in all directions, stacking faults in the silicon occur perpendicular to the direction of bonding tool motion, verifying that ultrasonic energy can introduce defects into single-crystal silicon [64,65]. Cratering is common in ultrasonic bonding but not in thermocompression bonding. Similarly, Koyama et al. [66] found that force and temperature cannot cause silicon nodule cratering without ultrasonic energy.

Contamination of the bond pad can cause improper setup of the bonding parameters, including the required bonding energy, because contaminated bond pads require more ultrasonic energy and higher temperature to make strong bonds. Moreover, too high or too low a static bonding force can lead to cratering in wedge bonds [67]. Kale suggests that the optimum bonding force results in more efficient energy transfer, thus lowering the total energy requirement for the bond. Winchell [64] notes the effect of bond-pad thickness on the phenomenon of cratering; the bond pad serves as a cushion protecting the underlying silicon, silicon dioxide, polysilicon, or gallium arsenide from the stresses of the bonding process.

The tendency to crater predominates in thin metallizations. One percent silicon is sometimes added to the aluminum metallization in order to prevent spiking in shallow junctions. Micrometer-size silicon nodules in aluminum bond pads act as stress raisers and crack the underlying glass during thermosonic gold ball bonding. Koch et al. [68], Ching and Schroen [69], and Koyama et al. [66] have generally confirmed the silicon-nodule cratering effect. After bonding is complete, these nodules decrease in the bonding region and damage appears in the insulation. Corrective action includes bonding at higher temperatures (250°C), using lower ultrasonic power, reducing molding stress, and removing a fracture-prone phosphorus glass layer from under the bond pad. To avoid cratering, Ching uses a hard titanium-tungsten underlayer for the pad, a modified bonding profile, and rapid ball touchdown.

10.6.2.7 Loss of Wire Bond

Impurities often cause a loss of surface bondability and premature bond failure during the operational life of the device. Impurities and interdiffusion in wirebonds can cause Kirkendall voids [70]. With gold–aluminum bonds, the interdiffusion front moves through the gold, forming intermetallic alloy phases. Due to the lower solubility of the intermetallic, impurities swept ahead of the front precipitate there and act as sinks for the vacancies produced during the diffusion process.

Plating baths are a major source of such impurities as potassium–gold–cyanide, buffers, lactates, citrates, carbonates, and phosphates. Other impurities, like thallium, lead, or arsenic, are added to the bath to increase plating speed and reduce grain size. Increasing the plating current density increases the codeposited impurity level exponentially. Variations in plating process parameters, such as bath temperature and bath concentration, also cause bondability problems.

10.6.2.8 Wire-Bond Fracture and Lift-Off

Wire can break at the heel of a wedge bond due to the reduced cross section of the wire. Cracks in the heel of a ball bond can arise as a result of excessive flexing of the wire during loop formation, especially when the level of the second bond is significantly lower than the first; repeated flexing and pulling of the wire occurs as the device heats and cools during temperature cycling due to thermal mismatches [71]. Fatigue failure by crack propagation can occur at the heel [72]. A wire can also break at the neck of a wirebond, leading to an electrical open. Thallium, a major source of wirebond neck failures, forms a low-melting eutectic with gold and can be transferred to gold wires from gold-plated leadframes during crescent-bond break-off [73,74]. Thallium diffuses rapidly during bond formation and concentrates over grain boundaries above the neck of the ball, where it forms a eutectic. During plastic encapsulation or temperature cycling, the neck breaks and the device fails [75]. Ball-bond fracture causes bond lift-off. The fracture can be a result of either tensile or shear forces induced by thermal stress or the flow of encapsulant during molding. The latter, also known as wire wash, happens only sporadically; when it does, however, it signals a molding compound that has expired or been improperly conditioned [61]. Generally, bond strength is more a function of temperature cycling than of steady-state temperatures between −55°C and 125°C, although bond strength decreases as a function of temperature above 150°C for gold–aluminum bonds [76] and above 300°C for gold–copper bonds [77].

10.6.2.9 Wire Bond and Bond-Pad Corrosion

Wires inside packages are susceptible to corrosion in the presence of moisture and contaminants. Chlorine residues from the molding compound, introduced by the capillary action of the water along the wire, can concentrate around wirebonds. The aluminum reacts with the dissolved chlorine to produce $AlCl_3$, a process accelerated by the elevated temperatures occurring during burn-in and high-temperature storage. Corrosion of gold wirebonds on aluminum pads results from the liberation of bromine from brominated flame retardants, a process accelerated by high temperatures. Bromide ions react with the aluminum in the gold–aluminum intermetallic phase (Au_4Al), forming $AlBr_3$ and precipitating gold. The aluminum bromide hydrolyzes to Al_2O_3, with the liberated bromide ions providing the driving force until all the intermetallic Au_4Al is consumed.

Bond corrosion may not result directly in failure, but it increases the electrical resistance of the interconnect until the device becomes nonfunctional. Often, the molding compound exerts a compressive force on the die surface and the adjacent wirebonds, and interconnection problems are not revealed until extensive corrosion has occurred. In mild cases, subtle changes take place in parametric functionality; in more severe cases, however, complete metallization attack occurs, resulting in open-circuit failures.

10.6.2.10 Leadframe Corrosion

Although leads generally have a nickel plating under the primary finishes of nickel, tin, and tin–lead to protect the base metal (usually alloy 42 or copper) from corrosion, the assembly process for leadframes can generate high residual stresses and high surface contamination. Any cracks or open voids in this plating can initiate corrosion in the presence of moisture and contaminants. During assembly and handling, the lead fingers are bent; cracks can develop and expose any corrodible surface to the external environment. Stress-corrosion-induced cracking can also occur, especially in alloy 42 leads.

The rate of galvanic corrosion can be high, because the lead finish is often a cathodic metal with respect to the lead base material. Although a leadframe can undergo corrosion anywhere on its external surface, the most sensitive area is at the interface between the molding compound and the leadframe.

10.6.2.11 Lead-Frame Deadhesion and Delamination

Moisture ingress, either through the bulk encapsulant or along the interface between the leadframe and molding compound, can accelerate delamination in plastic packages [78]. Experiments with moisture sensors reveal that when the adhesion between the molding compound and the leadframe is good, the main path of ingress into the package is through the bulk of the encapsulant [58,79]. However, when this adhesion is degraded by improper assembly procedures—for example, oxidation from bonding temperatures, leadframe warpage from insufficient stress relief, or excessive trim and form forces [80]—delamination and microcracks are introduced at the package outline, and water vapor can diffuse along this path.

At each interface, moisture can hydrolyze the epoxy, degrading the interfacial chemical bonds. However, different molding compounds respond differently to moisture exposure. Low-stress epoxy compounds, for instance, with silicone modifiers added for stress reduction tend to be more susceptible to changes induced by moisture than molding compounds without silicone. A low glass-transition temperature also adversely affects moisture absorption.

Surface cleanliness is a crucial requirement for good adhesion. Oxidized surfaces, such as copper-alloy leadframes exposed to high temperatures, often lead to delamination [33,81]. The presence of a nitrogen or forming-gas shroud helps to avoid oxidation and is recommended during high-temperature processes.

Low-affinity surface finishes, such as silver-spot plating, enhance interfacial adhesion. Traditionally, silver plating of the die pad is employed for bias control and to prevent oxidation on leads. Unfortunately, silver plating adheres poorly to molding compounds [31,57]. Newer leadframe designs use spot plating to minimize both the amount of precious metal used and the leadframe coverage, which is susceptible to delamination.

Releasing agents and adhesion promoters in molding compounds have been shown to encourage delamination in plastic packages and must be delicately balanced [78]. Releasing agents facilitate removal of the molded parts from the mold cavities at the risk of greater interfacial delamination. On the other hand, adhesion promoters ensure good interfacial adhesion between the compound and the component, but parts may be hard to pry out of the mold cavity.

10.6.2.12 Encapsulant Rupture

Resistance to crack initiation in a material is strongly dependent on geometry, material properties, and loading. Kitano et al. [82] developed a method for measuring the critical stress intensity factor for an encapsulant ruptured by brittle fracture.

10.6.2.13 Encapsulant Fatigue Fracture

Cyclic stresses in a package occur due to mismatches in thermal expansion caused by temperature changes and temperature gradient. An estimate of package life subjected to thermal cycles is obtained using Paris' law [83] for brittle crack propagation.

The stress intensity factor for a change in temperature is calculated using finite element methods, and the relationship is expressed as a polynomial, $f(a)$. Employing Paris' law, the number of temperature cycles, N_{cycles}, for propagating a crack from length, a_0 to a_c is given by

$$N_{\text{cycles}} = \int_{a_0}^{sa_c} \frac{da}{c_3 f\,(a)^m} \qquad [10\text{-}11]$$

where the material constants c_3 and m have been determined for the encapsulant materials using single-edge-notch specimens [84]. The initial crack length, a_0, is taken to be some small value dependent on manufacturing defects, but the total number of cycles to failure is relatively insensitive to initial crack length, provided the crack is small.

10.6.2.14 Package Cracking ("Popcorning")

Package cracking results from internal stresses generated by the reflow solder temperature profile during the assembly of plastic-encapsulated devices on circuit cards. In brief, during reflow soldering, the assembly temperature is rapidly raised above the glass-transition temperature of the molding compound (approximately 140–160°C) to a temperature of approximately 220°C or more, which is required to melt the solder. At these temperatures, the thermal mismatch between the molding compound and the materials adhering to it, such as the leadframe and silicon die, can be large enough to make the interfaces susceptible to delamination.

Moisture absorbed by the molding compound can vaporize into steam, and the resulting volume change can produce a pressure greater than the adhesion strength of the interface, causing delamination. Delamination may occur between the molding compound and the leadframe, or between the compound and the die surface. In extreme cases, the package may crack, ball bonds may shear from the bond pads, or the silicon may crater beneath the ball bonds. The cracks often propagate from the die paddle to the bottom surface of the package, where they are difficult to see during visual inspection of soldered boards. Occasionally, depending on package dimensions, the cracks may propagate to the top of the package or along the plane of the leads to the package sides, where they become more visible [85]. Proper handling and use after unpacking the assembly from the desiccant will deter this failure mechanism.

10.6.2.15 Electrical Overstress and Electrostatic Discharge

Electrostatic discharge (ESD) is the transfer of electrical charge between two bodies at different potentials, either through direct contact or through an induced electrical field. Electrostatic discharge can occur during manufacturing, handling, or service. It is neither unique nor more prevalent in plastic-encapsulated microcircuits than in hermetic packages. Its failure mechanism can include gate-oxide breakdown and conductive path formation, with or without shorts. The failure site is often at sharp edges, such as corners of bond pads and metallizations, where the electric field strength is the greatest.

Electrostatic charges are generated by frictional sliding contact between dissimilar materials, or by induction charging of a conductor from an external charged insulator. Electrostatic discharge arises primarily from several sources: human contact, leading to static electricity discharge from the human body through the device; discharge of a statically charged device due to sliding against a carrier surface; and discharge of an induced charge in the device conductor from an external, electrostatically charged insulator—for example, a coffee cup carried across a carpet. Thus, the cause is electrical abuse.

Electrical overstress (EOS) is the damage to an electrical circuit due to thermal overstress caused by excessive electrical power dissipation during a transient electrical pulse. Like electrostatic discharge, electrical overstress failures are often caused by misapplication of the device in a circuit, allowing surges in a power line. Failure modes due to electrical overstress include junction spiking, latchup (common in MOS and CMOS devices), melted metallization, metal electromigration, and open bond wires.

Electrical damage from electrostatic discharge and electrical overstress have increased in importance due to the progressive decrease in integrated circuit dimensions. Metal oxide semiconductor (MOS) chips are more susceptible to electrostatic discharge damage because of small geometries and thin gate-oxide dielectrics.

10.6.2.16 Soft Errors

Soft errors are temporary upsets in the state of a memory bit due to electron–hole pairs induced by alpha-particle radiation. Soft errors first showed up in 16K dynamic random access memory devices (DRAMs); the mechanism was discovered by May and Woods [86]. These errors did not show a repeat pattern and were erased with each refresh clock cycle.

The charge difference in memory cells, which distinguishes a “0” state from a “1,” is called the critical charge, which for a 64K DRAM is about $(1–2) \times 10^6$ electrons. An alpha particle produces 1.4×10^6 electrons and an equal number of holes as it penetrates 25 μm (1 mil) into a silicon chip. The holes are electrically rejected by the memory cell, and the electrons collected can change the state of the memory cell until the next refresh cycle restores the proper state. As the dimensions of the memory cells are reduced, the critical charge decreases, compounding the radiation of alpha particles. Like electrostatic discharge, soft errors are not unique to plastic-encapsulated microcircuits and must also be addressed with hermetic packages.

10.6.2.17 Solder-Joint Fatigue

Solder-joint fatigue is the same for PEMs on circuit cards as it is for hermetic-package assemblies. The drivers of solder-joint fatigue are primarily the global mismatch in the TCEs of the component and the circuit card, and the local mismatch in the TCEs of the solder and the lead, and the solder and the circuit card pads. The global mismatch can be further increased under operating conditions when there are temperature differences between the component and the circuit card. It may be advantageous to use a PEM over a hermetic part when the board material coefficient of thermal expansion more closely matches the PEM than the hermetic (metal or ceramic) package; this is the case with most of the common organic board materials, such as FR4, cyanate ester, and bismaleimide triazine. An excellent overview of the solder-joint failure mechanism can be found in Frear et al. [87].

10.6.3 Ranking of Potential Failures

Failure distributions differ from device to device and manufacturer to manufacturer. Table 10-15 presents a summary of failure analysis results in the form of a Pareto ranking of various field failures in very-large-scale integration (VLSI) class devices for demanding commercial

Table 10-15. Pareto Ranking of Failure Mechanisms of 3400 Commercial VLSI-Class Devices from Multiple Sources, Including Manufacturing Fallout, Qualification, Reliability Monitors, and Customer Returns

Failure Mechanisms	% of Failed Integrated Circuits
Electrical overstress and electrostatic discharge	19.9
Unresolved	15.9
Gold ball-bond failure at bond	69.0[a]
Not verified	6.0
Gold ball-bond failure at stitch bond	4.6[a]
Shear stress, chip surface	3.5[a]
Corrosion, chip metallization/assembly	3.2[a]
Dielectric fail, poly–metal, metal–metal	3.0
Oxide defect	2.9
Visible contamination	2.7
Metal short, metal open	2.6[a]
Latch-up	2.4
Misprocessed, wafer fab-related	2.4
Chip damage, cracks/scratches	2.4[a]
Misprogrammed	2.0
Oxide instability	1.9
Design of chip	1.7
Diffusion defect	1.5
Final test escape	1.4
Contact failure	1.2
Bond failure, nongold	1.2[a]
Protective coating defect	0.9
Assembly, other	0.9[a]
Polysilicon/silicide	0.8
External contamination	0.7[a]
Others	5.3

[a]Possible packaging or assembly-related failures.
Source: From Ref. 88.

applications, such as automotive manufacturing and telecommunications [88]. The sources of failures include manufacturing fallout, qualification, reliability monitors, and customer returns. Failures due to electrical overstress and electrical discharge dominate, with electrical overstress due to misapplication and voltage overstress as primary causes. The major causes of failures were poor handling and assembly. Of the remaining failure mechanisms, wirebond failure was significant, and corrosion of die metallization was a relatively small factor.

Vahle and Hanna [89] reported a distribution of early field failures of plastic-encapsulated integrated circuits. Their data indicate that 36.3% of failures are unverified, 22.2% are of undetermined origin, 17% are due to wirebond failure, 8.3% result from electrical overstress, 2.3% are from mask/etch defects, 2.1% are due to electrostatic damage, and 11.9% are attributed other causes.

Another study, by Bloomer [90] presented failures in commercial integrated circuits components (Fig. 10-38). The components analyzed were supplied by a large industrial user from 1984 to 1988. The pie chart includes failures during manufacturing, incoming component testing, final board testing, and field operation. Like other studies, this one found that the major causes of failure are electrical overstress, electrostatic discharge, and wirebond weaknesses.

Finally, Pecht and Ramappan [91] collected electronics system and device failure data and noted some of the reliability problems associated with electronics systems. They also noted that reliability improvements related to inherent component stress considerations are generally ineffective in fault prevention, if the stress levels are within design limits. This contrasts with Arrhenius models, in which increasingly lower stress (steady-state temperature) is believed (but is not physically proven) to provide increasingly improved reliability.

10.6.4 Comparison of Plastic and Ceramic Package Failure Modes

Integrated-circuit packages encapsulated with the first generation of epoxy molding compounds in the 1960s and 1970s exhibited rather poor

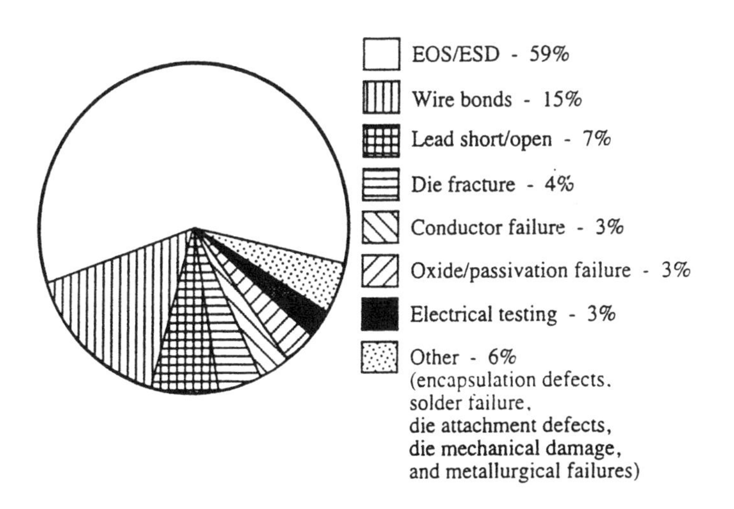

Figure 10-38. Distribution of Failures in Commercial Integrated-Circuit Components (PEMs) from a Large Industrial End User. (From Ref. 90.)

performance when compared with their ceramic counterparts. Corrosion and thermally induced mechanical stress were two common failure modes. Ionic residues like chloride, potassium, and sodium exceeded several hundred parts per million in the early epoxies, compared with the less than 10 ppm obtained with the latest technology. The high level of contaminants, coupled with the poorer quality of the device passivation in the sixties and seventies, accounted for the low mean time to failure of plastic-encapsulated devices. Furthermore, the technology of compound stress modification was not recognized; silicone modifiers and stress absorbers were still in the conceptual stage. Today, with advances in molding compounds and device technology, plastic-encapsulated packages exhibit long-term reliability comparable to and, for many applications, better than hermetic packages [92–95].

Failure modes are similar in plastic and ceramic packages. Table 10-13 shows a Pareto ranking of failures observed in hermetic packages by researchers at Texas Instruments. Comparing the Pareto rankings in Tables 10-15 and 10-16 shows that electrical overstress, electrostatic discharge, corrosion, and thermomechanical stress are reliability concerns for both configurations. However, the relative importance of each mode varies between package types. For example, even though current epoxies are better purified and almost free of ionic residues, external contaminants entrained by absorbed moisture can still reach the die surface in contact with the epoxy. On passivated dies, corrosion often occurs at the exposed bond pads. In hermetic packages, corrosion is often due to organic contaminants from the die attach process mixed with traces of moisture trapped within the hermetic confines of the package.

10.7 FUTURE DEVELOPMENTS IN PLASTIC PACKAGES

This section addresses the current status and trends in encapsulated microcircuit packaging technology. The focus is on trends in encapsulation materials, package design, and manufacturing methods.

10.7.1 Trends in Materials, Design, and Fabrication

Trends in encapsulation materials, package design, and manufacturing methods are presented in this section.

10.7.1.1 Encapsulation Material Trends

Epoxy molding compounds (EMCs) using transfer molding are expected to remain the major force in molded packaging technology. However, it is unclear whether they will be able to meet the yield requirements of high-pin-count complex packages and the heat dissipation needs of future products. The expected drop of supply voltage in future integrated

Table 10-16. Pareto Ranking of Failure Mechanisms of Failed Hermetic Devices

Failure Mechanisms	% of Failed Integrated Circuits
Electrical overstress and electrostatic discharge	16.4
Hermeticity rejects	6.6
Not verified	5.8
Polysilicon/silicide	5.1
Pin rejects	3.8
Visible contamination	3.2
Oxide defect	2.3
Chip damage—cracks/scratches	2.1
Metal open	2.1
Metal short	1.9
Contact failure	1.7
Passivation defect	1.6
Contamination on Porous oxide	1.5
Assembly—other	1.4
Bond failure—nongold	1.3
External contamination	1.0
Dielectric fail, poly–metal	0.8
Waferfab–other	0.8
Misprocessed	0.7
Diffusion defect	0.7
Die attach	0.7
Oxide instability	0.6
Package damage	0.6
Lead adhesion	0.5
Silicon defect	0.4

Source: R. S. Martin, personal communication, 1993.

circuits from 5 to 3.3 to 2.5 V will certainly ease the heat-dissipation demands on PEMs. Among other uncertainties of enhancements in epoxy-molding-compound properties are the coefficient of thermal expansion and viscosity tailoring. Figure 10-39 depicts the current development trend of EMCs, with their specific intended applications.

The following explains some of the terms used in the tables and figures above:

- Antisolder: the properties of molding compounds that make the components robust against potential failures during soldering (e.g., delamination and package cracking).

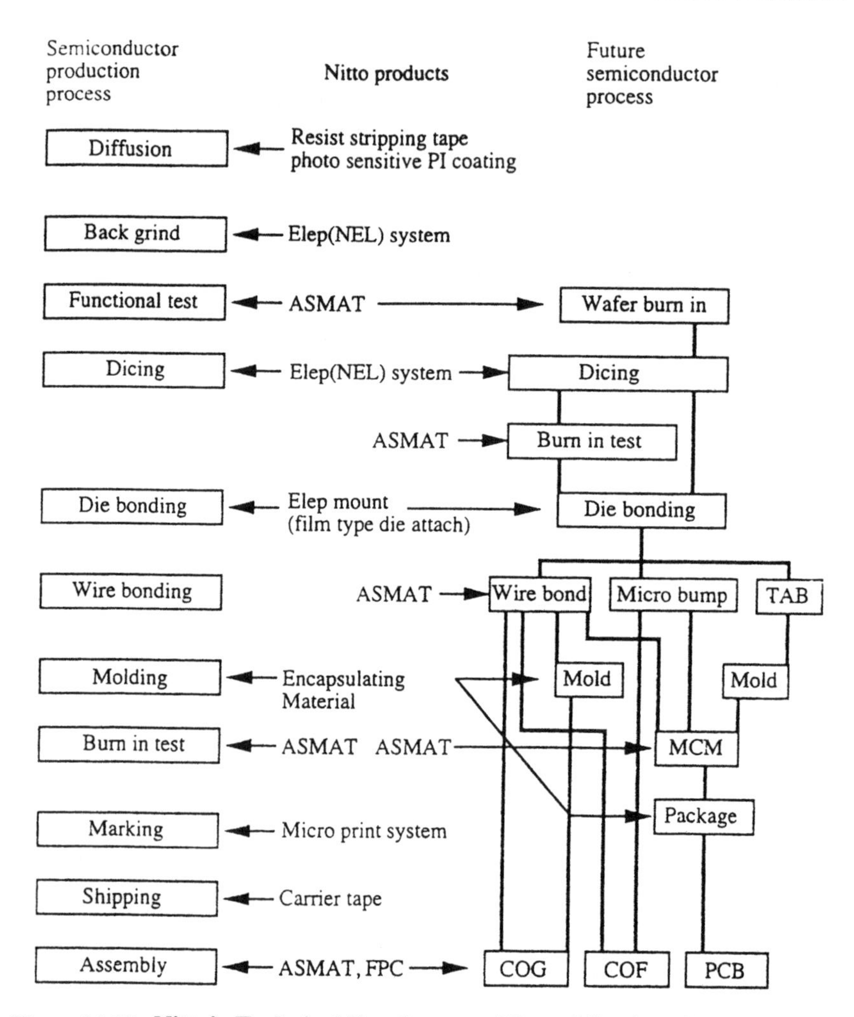

Figure 10-39. Nitto's Technical Development Plan of Semiconductor-Related Materials. (From Ref. 1.)

- Recyclable recovery: development of molding compounds that can be recycled so that the molding compound remaining after the molding process (e.g., material left in the runners) can be remelted by the customer and used for the next molding.
- Bottomless markless: Plastic package markings on the bottom surface are impossible to read unless the package is stripped of the board. Bottomless–markless technology uses a microprint system to eliminate bottom-side markings.
- Single-device burn-in: Burn-in screens for single devices, rather than manufactured lots, is becoming a necessity as customers look for

known good dies (KGD). For example, Nitto Denko is developing application-specific materials (ASMATs) which are unique in that they (1) have polyimide substrates bonded directly to the copper leadframes without any intervening adhesive, (2) can be used for extremely fine-pitch applications, and (3) allow bumps to be made in the leadframe/polyimide substrate to increase the reliability of the die–leadframe bonding during burn-in.

- Dam-barless technology: The leadframe is usually provided with a dam-bar to constrain the molding compound from overflowing during the molding process. Companies such as Nitto Denko are developing thin films which when applied to the leadframe would serve the same purpose while reducing the leadframe's overall material thickness. These new materials are being developed for fine-pitch thin-package applications where the leadframes are thin (and may deform upon removal of the dam-bar).
- Local-stress soft error: The die is susceptible to local stresses due to TCE mismatch effects between the die surface and the molding compound. Because of this, the die functionality can be affected, leading to data-retention errors (soft errors) in memory dies. Presently, companies such as Nitto Denko provide a gel coat on the die surface to act as a buffer and are developing cavity molds which would not require gel coats. Future plans are to develop low-modular materials for molding compounds to provide buffers for local stress in noncavity packages.
- Kneading: a process used for forming a mixture from the elements of the molding compound. During kneading, the molding reaction is allowed to occur to about 10% completion to form a solid continuous mass of the molding compound.
- Pelletizing: The kneaded mold is compressed into tablets to be used during the transfer-molding process.
- Pad tilt: During molding, the die paddle (along with the die) sometimes tilts with the flow of molding compound, leading to the die being exposed outside the package. The goal is to simulate this process to develop molding compounds that are robust against this type of failure.

For postmolded plastic packaging, injection-molded liquid-crystal polymers (LCP) have some promising potential for the future. With a cycle time of only 20–30 s, they can offer several times higher production rates than epoxies. The base resin is very pure, with low viscosity; even with filler particles added for high thermal conductivity, it has a very low coefficient of thermal expansion. Liquid-crystal polymers also have exceptional solvent resistance and are amenable to optical fiber attachment.

Liquid injection and reaction injection molding processes use low-viscosity monomers or oligomers; the former is a one-component system, the latter is two. Chemically, these materials are silicones, epoxies, or epoxy–silicone hybrids. Even with high filler loadings, they have very low viscosities and their properties can be specially formulated.

Polyimides with low coefficients of thermal expansion have been developed with values as low as 4×10^{-7}/°C. They may be especially useful in very-high-precision electronics applications, such as insulation films and plastic-encapsulation underlayers. Most polyimide precursors interact with copper (leadframe), necessitating the use of passivation barrier coatings. However, the polyimide precursor based on ester technology is compatible with copper metallurgy. Its unique isomer ratio enables the development of high solid formulations that allow thick coatings of up to 20 μm.

10.7.1.2 Package-Design Trends

Thin, small-outline packages (TSOPs), with their compact profiles, became a key product of the 1990s. By most estimates, they will also be used increasingly in surface-mount small-outline packages in the future, especially for memories, as demand for space-saving packages grows [96]. Other thin package types include the thin quad flat pack (TQFP), thin sealed small-outline (TSSOP), and thin-body plastic dual-in-line (PDIP) packages. Integrated-circuit packages with width and/or length shrink include the fine-pitch quad flat pack (FPQFP), quarter-size small-outline package (QSSOP), shrink DIP (SDIP), shrink quad flat pack (SQFP), shrink small-outline package (SSOP), and very-small-outline package (VSOP). With these or new shrunken packages, suppliers will continue to provide more functions in less board space, and boards will become smaller. This trend to skinnier housings will challenge die thinning, wafer transportation, chip-pad mounting, leadframe design, lead bonding, board interconnection, molding, and soldering.

JEDEC has standardized a family of 1.4-mm-thick thin quad flat pack (TQFP) packages that range in size from 5 to 28 mm square and 32 to 256 leads. TQFP usage is driven by applications that require small, thin, and lightweight semiconductor chip packaging, including notebook and subnotebook computers, personal digital assistants, portable consumer products, PCMCIA cards, and disk drives. These applications call for chips with larger sizes, greater power dissipation, or higher electrical performance [97]. The trend is toward even larger 1.4-mm-thick TQFP packages than currently covered by JEDEC, such as the 32-mm and 40-mm packages, to handle both higher lead counts and larger chip sizes. However, the molding process gets increasingly complicated when package sizes increase at the same thickness. Olin Interconnect Technologies has developed a high thermal and electrical performance TQFP in which the plastic mold compound is replaced by an anodized aluminum base and lid adhesively sealed to the leadframe [97]. The package uses the same infrared or vapor-phase reflow board mounting profile as a plastic package, weighs the same as a plastic package, and is dimensionally equivalent to a plastic package.

In the area of ultrahigh-heat-dissipation plastic packages, Power Quad 2, a 160- and 208-lead plastic quad flat pack from Amkor Electronics, is a significant development in the 1990s [98]. In this package, thermal management attributes include a large-area solid copper heat sink under the die. For improved electrical and thermal performance, an internal ground plane could be used. With a 8.75 × 8.75-mm (350 × 350-mil) die, $\Theta_{\text{junction-to-case}}$ is less than 0.4°C/W; with a new line of external heat sink and airflow, a junction-to-ambient thermal resistance of 8.0°C/W will result.

Finally, various design tools for both single-chip and multichip packages exist on the market. These tools typically address only some aspects of the IC production flow. For instance, some tools simulate mainly the electrical performance of the packages for various configurations and operations frequencies. Others concentrate only on the thermal management side by portioning the functions on either the die or the module; still others focus only on the mechanical aspect. Such a gap is well recognized by the CAD and modeling tool vendors, and attempts have been made at integrating the disjointed modules into one single package. Understandably, most of the development efforts have been devoted to the high-value-added packages. All the major CAD vendors now profess to have design tools that can partition, route, and simulate the performance of the complex packages, such as multichip modules, in terms of electrical characteristics, thermal dissipation, and mechanical stresses.

Software tools are also available to ensure that reliability, quality, and yield issues are addressed during package design. One such tool for reliability is the Computer Aided Design of Microelectronic Packages (CADMP II) software developed by the CALCE Electronic Package Research Center at the University of Maryland.

10.7.1.3 Package Fabrication Trends

Evolutionary directions in package fabrication include new cleaning methods, improved manufacturing process control, fine-pitch interconnections, and new leadframe fabrication.

Manufacturing technology and the cost of the advanced materials required for high-quality and reliable plastic packaging determine both acquisition and lifetime cost, and drive the technology for the widespread use of PEMs in all markets. Performance criteria naturally adapt to the cost constraints of the market. At near-perfect yield, the present molding process productivity of the most common PLCC packages is about 800 packaged devices per hour. High-yield plastic packages of the future using thinner leadframes, fine-pitch wirebonding, and flip-chip or TAB structures will need low-viscosity molding compounds at a low molding temperature to reduce shear-rate-induced yield losses. Higher production rates will be required to compete with PGAs and premolded packages. Smaller molds with fewer cavities and a total cycle time (including in-mold cure) of 1–2 min in an automated mode will be needed. A high level of automation will result in lower-cost, uniform-quality packages with less damage to fragile high-I/O-count assemblies. A clean manufacturing environment will reduce contamination-related failures.

Automation will play an increasingly important role in PEM manufacture. One approach is partial automation of various labor-intensive aspects of the process, such as preform heating and handling. A robotic arm may place the heated preform into the transfer pot and start the process sequence of transfer, curing, and ejection of the molded leadframes. In this automation approach, a single operator can handle four or five transfer presses. In the total automation approach, the process runs without any operator assistance, although typically 10–20% of an operator's time is spent in moving different cassettes and checking equipment malfunctions. The use of a smaller molding tool with fewer cavities in conjunction with faster-curing molding compounds will further increase productivity. Such a packaging system is usually totally enclosed to maximize process cleanliness and personnel safety. The cassettes of molded leadframes feed an automated trim-and-form press, and then move on to a code-marking station. The partial or total automation of single-pot systems has been superseded by multiplunger technology in multiproduct production environments. In automated multiplunger systems, 6–12 pots feed 6–24 cavities, with 1–4 cavities per transfer pot. Because of very short flow length, they are very effective in minimizing voids and promoting high molding-compound density. State-of-the-art encapsulants—very fast-curing molding compounds with a total cycle time of less than 2 min in the mold—will be used in these manufacturing systems.

The increasing use of analytical techniques and microsensors to fine-tune materials and control process-induced defects falls into the category of enhanced process control and quality assurance. X-ray radiography and C-mode scanning acoustic microscopy/tomography are finding increased use as nondestructive techniques for evaluating plastic delamination; die metal, die attach, and bonding wire deformation; die metal and wirebond voiding; leadframe, die passivation, die attach, wire, wirebond and case brittle fracture; and dendritic growth under bias. Mercury porosimetry has been successfully employed to find the number and size of epoxy and epoxy–metal pores in epoxy-encapsulated packages. Epoxy pores are less than about 0.2 μm in diameter, epoxy–leadframe voids are about 1 μm in diameter, and surface pores range from 5 to 500 μm in diameter. Piezoresistive strain gauges integrated into test chips will continue to be used to directly measure the mechanical stress induced inside a PEM by encapsulation, die bonding, and other factors, either during fabrication or under environmental stress testing. Solid-state moisture microsensors will also be used to measure the moisture content at any specific location inside a PEM.

Challenged by higher integration levels (particularly in ASICs) with tighter bond-pad pitches at the die level, leading-edge pitches will be at 0.060–0.075 mm (2.5–3 mils) by the late 1990s. Beyond a 0.1-mm (4-mil) pitch, bonding with gold wire may be done by wedge bonding. The combination of tight pad pitches and shrink packages demands not only low wire looping (0.09–0.18 mm) but also different loop shapes. This requires modifying the dopants in the wire to offer the right set of mechanical properties. With tight pitches, the physical limits of leadframe technology will also force the placement of lead contacts at ever-increasing distances from the package center to control wire sweep. Low looping also promotes wire-to-die edge shorting when the die has inbound bonding pads to avoid encapsulation stress concentration zones. Edge shorting can be prevented by the use of new wirebonder software, that allows an extra bend in the wire at a predetermined site.

The demands of tightly spaced inner-lead bonding (0.05–0.075 mm) and outer-lead bonding (0.15–0.35 mm), high pin counts, high-end performance-driven applications, and high-volume production could also move bumped tape automated bonding (TAB) to the forefront of the packaging industry [99]; a potential for which Japan already has the capability and infrastructure. Limiting factors for TAB applications have been the expense of tape, the lead time needed to obtain it, and the capital needed for the bonding equipment. Bump fabrication has also been an obstacle. Better tape metallurgy, currently available, now permits burn-in and longer shelf-life before encapsulation. Laser bonding of fine pitch devices is particularly suited to TAB.

Trends in leadframe technology have important implications for the future of molded plastic packaging. Copper alloys are replacing 42 Fe/58 Ni (alloy 42) as the leadframe material for moderate-to-high heat-dissipating devices, such as processors and logics. But their high coefficient of thermal expansion and average mechanical properties could prohibit their use in other applications. It is also becoming increasingly difficult to manufacture thin (≤ 0.15 mm) leadframes with close lead tips for very-high-lead-count packages. Although chemical etching has replaced mechanical punching in these situations, as the leads and lead spacing get smaller, maintaining a lead aspect ratio near unity for the leads by leadframe isotropic etching is becoming extremely difficult. Consistent with the need for a mechanically stable lead geometry, leadframe thickness has decreased from 2.5 mm (100 mils) for low-pin-count devices to under 0.15 mm (6 mils) for high-lead-count (~200) fine-pitch plastic quad flat packs. Several different types of lead tip layouts offer different pad sizes, lead-tip spacings, and wirebond lengths [100]. Thinner leadframes, larger die pads (for short, close-spaced lead tips), and longer wirebond spans pose such molding problems as flow-induced deformation, paddle shift, and wire sweep. There are also problems with handling the delicate leadframe in terms of both inner lead wirebonding yields and outer lead soldering yields. Thus, it appears that leadframe manufacturing limitations could hinder the development of a very fine-pitch and high-lead-count molded plastic packaging.

10.7.2 Trends and Challenges in Circuit Card Assemblies

The trends in printed wiring board (PWB) materials and fabrication technology are likely to include relatively new materials with lower coefficients of thermal expansion, lower dielectric constants, and greater dimensional stability, which can be inexpensively manufactured. In particular, as components become thinner, the TCE is moving to that of the silicon. High-temperature capability will make board distortion a nonissue. Thus, a matched in-plane coefficient of thermal expansion of the PWB will be needed to help reduce solder-joint fatigue.

The use of solder for connecting PEMs to circuit cards also still has reliability challenges. In particular, the assembly process has aggravated popcorn package cracking, particularly in thin and ultrathin package styles. This trend is toward improved molding compounds that resist delamination at the paddle interface, minimize deformation at the reflow solder temperature, and deter cracking when deformed by the vapor expansion.

Because of the ban on chlorofluorocarbons, environmentally acceptable solvent cleansers are being found for assembled-board cleaning operations. In addition to mild detergents used in industrial dishwashers, nonflammable cleaning solvents are being developed for these operations. Molded plastic package material should thus have good solvent resistance.

Although most molding compounds are impervious to these corrosive fluxing agents at the required soldering times and temperatures, the interfacial paths along the leads are particularly vulnerable to ingress by these corrosive materials. Consequently, good sealing and adhesion along these plastic–metal interfaces is needed to prevent flux penetration and package contamination during fluxing. If fluxes are not removed from the assembled board—including the underside—by cleaning operations and effluent testing, temperature cycling eventually opens up more of these interface ingress routes, and premature corrosive failure of PEMs results. The current trend is to use very mild fluxes, such as those made from lemon juice (citric acid), that can be easily washed out during board cleaning.

10.7.3 Standards and Requirements

Methods to ensure high quality and long-term reliability will be established by short-term correlated and tailorable tests. In many cases, qualification-by-design will replace qualification tests. Screening will virtually disappear as a procedure for eliminating defective parts. Military qualification and reliability test standards will be discarded, and critical-

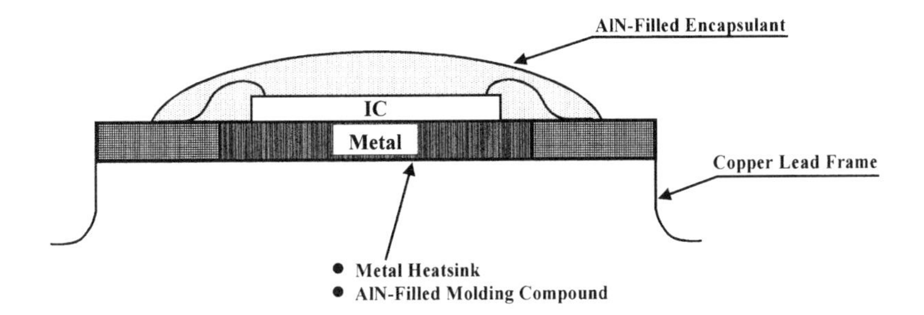

Figure 10-40. Advances in Heat Transfer of Plastic Packages

use environment related-test conditions will be modified to reflect the physics-of-failure processes to be uncovered. Users will probably still demand manufacturer's qualification data for their particular application environments, but they will maintain supplier partnerships with only those vendors that can satisfy parts-reliability criteria from field-failure data, failure analysis, applicable environmental tests, quality assurance techniques, and part-specific design analysis. Qualification requirement matrices will be drawn up to guide suppliers, relating all elements of design, manufacture, and part assembly variables with the required needs.

The future will also have to bring together differences between Japanese and American product standards. For example, the standoff distance between a circuit board and the most popular PEM (PQFP) is 0.25–0.37 mm (10–15 mils) in the U.S.-manufactured product, compared with near zero in its Japanese counterpart. In addition, Japanese lead length requirements on higher pin-count, finer-pitch packages have been shrinking; some lead lengths are as short as 1.0 mm (0.5 mm from the edge of the package to the vertical portion and 0.5 mm for the foot itself). United States manufacturers have reached only 0.8 mm for the lead from the package end to the vertical bend. Differences in lead length standards between U.S. and Japanese manufacturers have had a direct adverse effect on board design, as designers have attempted to accommodate both types of parts. This has led to lower board yield and increased rework. Interestingly, JEDEC standard for package outlines are looser than assembly tolerance requirements, resulting in noninterchangeability of even different U.S. vendor parts made to the same specifications.

10.8 HEAT TRANSFER IN PLASTIC PACKAGES

Plastic packages are notoriously poor for heat-transfer because of their very low thermal conductivity. However, a number of advances are being made to improve the heat-transfer characteristics. Referring to Figure 10-40, these advancements include (1) the use of metal heat sinks such as the VSPA package (see Chapter 7, "Microelectronics Packaging—An

Table 10-17. Chip Bonding Adhesives (Part 1 of 2)

Supplier	Type No.	Resin Type	Filler Type	Components	Electrically Conductive
Epotek	H20E	Epoxy	Silver	2	Yes
	H20E-175	Epoxy	Silver	2	Yes
	H35-175M	Epoxy	Silver	1	Yes
	H41	Epoxy	Silver	1	Yes
	H44	Epoxy	Gold	1	Yes
	H81E	Epoxy	Gold	2	Yes
	P-1011	Polyimide	Silver	1	Yes
Amicon	C-840-4	Epoxy	Silver	1	Yes
	C-868-1	Epoxy	Silver	1	Yes
	C-940-4	Polyimide	Silver	1	Yes
	C-966	Polyimide	Silver	1	Yes
	C-990	Epoxy	Silver	1	Yes
	CT-4042	Epoxy	Silver	2	Yes
	ME-868	Epoxy	Oxide	1	No
Ablestik	71-1	Polyimide	Silver	1	Yes
	71-1LM1	Polyimide		1	Yes
	84-1LM1	Epoxy	Silver	1	Yes
	84-1LM1	Epoxy	Silver	1	Yes
	84-1A	Epoxy	Silver	1	Yes
	84-1LM1S1	Epoxy	Silver	1	Yes
	941-3	Epoxy	Silver	1	Yes

Overview") with copper, copper leadframes and the molding compounds, encapsulants, and die-attach adhesives with aluminum nitride filler. The copper heat sinks raise the thermal conductivity of typical molding compound from 0.5 to as much as 396w/m°C. Aluminum nitride fillers have demonstrated thermal conductivity in excess of 7w/m°C. These developments have enabled the plastic packages to remove as much as 5 watts/chip. These developments are discussed in both Chapter 4 and Chapter 7.

Table 10-17. Chip Bonding Adhesives (Part 2 of 2)

Maximum Cure Temperature °C	Lap Shear Strength kg/mm^2	Remarks
175	1.06	Two-component epoxy adhesive with high electrical & thermal conductivity
180	1.06	Low ionic impurity, dual T_g version of H20E
180	1.27	Similar to H20E except single component, no mixing required
150	2.11	Single-component epoxy adhesive with low cure temp & high lap shear strength
150	2.11	Single-component, gold-filled epoxy adhesive
150	0.91	Two-component gold-filled epoxy adhesive
150	0.50	Low cure temperature polyimide adhesive
200	1.6	Low ionic impurity epoxy adhesives can be pre-cured as B-stage on leadframes
180	1.6	Low ionic impurity epoxy adhesive with low cure temp. can be B-staged
270	1.45	Low ionic impurity polyimide adhesive. Excellent heat resistance to 400°C
180	2.5	Low temp. cure polyimide adhesive with low ionic impurities
275	1.55	Very rapid cure epoxy adhesive, cures on wirebond heater block
150	1.6	Low ionics two-component epoxy adhesives
180	1.6	High thermal conductivity, electrical insulation epoxy adhesive
275	0.70	General purpose polyimide direct-attach adhesive
275	2.04	Very low ionic impurity version of 71-1
150	4.58	Low temp. cure conductive epoxy adhesive with low levels of ionic impurities
150	4.30	High, thermal conductivity version of 84-ILM1
200	4.23	Rapid cure version of 84-1LM1
200	2.11	Low viscosity, low ionic version of 84-1LM1
175	2.11	13-stageable epoxy chip adhesive for preprinting of leadframes

10.9 DIE ADHESIVES

Both silicon/gold eutectic and polymer bonding are used to bond the chip to the leadframe. The operation begins with the loading of the film ring holder, with sawed wafer attached, onto the index stage of the bonding machine. Lead frames are then fed from magazines along a track to a heater block. A small square of silicon/gold alloy (6% *Si,* 94% *Au*) is cut from a feed ribbon and transferred to the chip support platform.

An optical scanner detects a good chip on the sawed wafer causing a probe to push the chip above the wafer plane where a collet picks it up, and transfers it to the heated chip support platform. Chip and eutectic are now scrubbed together by the collet, forming a hard alloy bond. Heater temperature is approximately 420°C. Total cycle time for eutectic bonding is about 6 to 8 seconds.

Polymer bonding is faster than eutectic with a cycle time of 2 seconds. Feed mechanisms for polymer bonding machines are the same as eutectic bonders. However, leadframes are not heated. Silver-filled epoxy or polyimide adhesive paste is transferred to the chip support platform by a print head, and the chip is pressed into the paste immediately after printing. The chip attach adhesive properties are listed in Table 10-17. Chip bonded leadframe strips are loaded into transport magazines. Eutectic bonded frames go directly to wirebonding while magazines containing polymer-bonded frames are routed to ovens for adhesive curing. The curing atmosphere is dry nitrogen, and typical schedules are one hour at 150°C for epoxies, and 30 minutes at 150°C, followed by 30 minutes at 275°C for polyimides.

More recently, improved silicone gels were introduced as an alternative means of fastening the silicon die to the support platform [101]. Pin-grid arrays (PGA) soldered on standard FR-4 boards were used as test vehicles. In this case, the chip was initially bonded with the gel to the epoxy base of the PGA. After wirebonding, the entire PGA cavity was encapsulated with the same gel. Thus, the chip resided in a substantially compliant medium, and could readily withstand thermal and environmental stresses. The higher reliability achieved resulted from using the same gel throughout. Previous attempts at bonding the die with an epoxy-based adhesive, followed by die coating with a gel were unsuccessful for two reasons. First, improper cure conditions can leave low molecular weight amine residues in the epoxy. The unbound amine leached out under environmental testing and reacted adversely with the silicone gel, and, second, although the stiffer epoxy can hold the chip rigidly to the base, thermal mismatch would generate significantly higher stresses ultimately harmful to the integrity of the device.

10.10 REFERENCES

1. M. G. Pecht, L. T. Nguyen, and Edward B. Hakim. *Plastic-Encapsulated Microelectronics Materials, Processes, Quality, Reliability, and Applications.* John Wiley and Sons, New York, 1995.
2. M. G. Pecht. *Integrated Circuit, Hybrid, and Multichip Module Package Design Guidelines: A Focus on Reliability.* John Wiley and Sons, New York, 1994.
3. M. Pecht, R. Agarwal, and D. Quearry. "Plastic Packaged Microcircuits: Quality, Reliability, and Cost Issues," *IEEE Trans. Reliability,* pp. 513–517, 1983.
4. N. Sinnadurai. "Advances in Microelectronics Packaging and Interconnection Technologies—Toward a New Hybrid Microelectronics," *Microelectron. J.,* 16:p. 5, 1985.

5. M. Brizoux, et al. "Plastic-Integrated Circuits for Military Equipment Cost Reduction Challenge and Feasilibity Demonstration," *40th Electronic Components and Technology Conference,* pp. 918–924, 1990.
6. Reliability Analysis Center. A DoD Information Analysis Center Plastic Microcircuit Package—A Technology Review, Reliability Analysis Center, Rome, NY 1992.
7. L. Condra and M. Pecht. "Options for Commercial Microcircuits in Avionic Products," *Defense Electron.* pp. 43–46, 1991.
8. M. Nachnani, L. Nguyen, J. Bayan, and H. Takiar. "A Low-Cost Multichip (MCM-L) Packaging Solution," *Proceedings of International Electron Manufacturing Technology Symposium,* pp. 464–468, 1993.
9. G. F. Watson. "Plastic-Packaged Integrated Circuits in Military Equipment," *IEEE Spectrum,* pp. 46–48, 1991.
10. M. Priore, and J. Farrell. "Plastic Microcircuit Packages: A Technology Review," Report No. CRTA-PEM, Reliability Analysis Center, Rome, NY, 1992.
11. Texas Instruments. *FIFO Surface Mount Package Information,* Texas Instruments, 1992.
12. K. C. Griggs. "Plastic Versus Ceramic Integrated Circuits Reliability Study," *Report No. WP86-2020,* Rockwell International Collins Group, 1986.
13. C. A. Lidback. "Plastic-Encapsulated Products vs. Hermetically Sealed Products," Summary Report, Motorola Inc., Government Electronics Group, January 1987.
14. L. R. Villalobos. "Reliability of Plastic-Integrated Circuits in Military Applications," ESP Report No. PV 620-0530-1, Motorola Government Electronics Group, Tactical Electronics Division, 1989.
15. R. J. Straub. "Automotive Electronic Integrated Circuits Reliability," *Proceedings of Custom Integrated Circuit Conference,* pp. 92–94, 1990.
16. L. W. Condra, G. A. Kromholtz, M. G. Pecht, and E. B. Hakim. "Using Plastic-Encapsulated Microcircuits in High Reliability Applications," *Proceedings Annual Reliability and Maintainability Symposium,* pp. 481–488, 1994.
17. May, C. A., Epoxy Materials. *Electronic Materials Handbook,* 1-Packaging, ASM Intl. pp. 825–837, 1989.
18. R. K. Rosler. "Rigid Epoxies," in *Electronic Materials Handbook, 1 Packaging,* pp. 810–816, ASM International, Materials Park, OH, 1989.
19. P. Proctor and J. Solc. "Improved Thermal Conductivity in Microelectronic Encapsulants," *Proceedings of the 41st IEEE Electronics Components Conference,* pp. 835–842, 1991.
20. A. S. Chen, L. T. Nguyen, and S. A. Gee. "Effects of Material Interactions During Thermal Shock Testing on Integrated Circuits Package Reliability," *Proceedings of the IEEE Electronic Components and Technology Conference,* pp. 693–700, 1993.
21. A. Ditali and Z. Hasnain. Monitoring Alpha Particle Sources During Wafer Processing, *Semiconductor Intl.,* pp. 136–140, June 1993.
22. N. Kinjo, M. Ogata, K. Nish, and A. Kaneda. Epoxy Molding Compounds as Encapsulation Materials for Microelectronics Devices, *Advances in Polymer Science* 88. Springer, Berlin, 1989.
23. B. Bates. "Molding Compounds Technology for Military Applications," *Commercial and Plastic Components in Military Applications Workshop,* 1993.
24. J. B. Austin. "Thermal Expansion of Nonmetallic Crystals," *J. Am. Ceramic Soc.,* 35(10): pp. 243–253, 1952.
25. B. Yates. *Thermal Expansion,* pp. 52–70, Plenum Press, New York, 1972.
26. J. Boustani, "Ultra-Low Expansion Metal Matrix Composition," *M.I.T. S.M. Thesis,* 1981. Cambridge, MA.
27. W. H. Kohn. *Materials and Techniques for Vacuum Devices,* Reinhold Publishing Corp., New York, 1967.

28. F. J. Dance and J. L. Wallace. "Clad Metal Circuit Board Substrates for Direct Mounting of Ceramic Chip Carriers," *Electron. Packag. Prod.*, 22(1): pp. 228–232, 236–237, 1982.
29. J. T. Breedis. "New Copper Alloys for Surface Mount Packaging," *J. of Metals, A. I.M.E.*, p. 48, June 1986.
30. K. E. Manchester and D. W. Bird. "Thermal Resistance, A Reliability Consideration," *I.E.E.E. Trans. on Components, Hybrids, and Manuf. Tech.*, CHMT-3(4): pp. 362–370, 1980.
31. L. T. Nguyen. "Reactive Flow Simulation in Transfer Molding of IC Packages," *Proceedings of the 43rd Electronic Components and Technology Conference,* pp. 375–390, 1993.
32. L. L. Blyler, H. E. Blair, P. Hubbauer, S. Matsuoka, D. S. Pearson, G. W. Poelzing, and R. C. Progelhof. "A New Approach to Capillary Viscometry of Thermoset Transfer Molding Compounds," *Polym. Eng. Sci.,* 26(20): p. 1399, 1986.
33. S. Kim. "The Role of Plastic Package Adhesion in IC Performance," *Proceedings of the 41st Electronic Components and Technology Conference,* pp. 750–758, 1991.
34. A. Nishimura, S. Kawai, and G. Murakami. "Effect of Leadframe Material on Plastic-Encapsulated Integrated Circuits Package Cracking Under Temperature Cycling," *IEEE Trans. Components Hybrids Manuf. Technol.,* CHMT-12: pp. 639–645, 1989.
35. A. Hale, H. E. Bair, and C. W. Macosko. "The Variation of Glass Transition as a Function of the Degree of Cure in an Epoxy–Novolac System," *Proceedings of SPE ANTEC,* 1116, 1987.
36. A. Hale, M. Garcia, C. W. Macosko, and L. T. Manzione. "Spiral Flow Modelling of a Filled Epoxy–Novolac Molding Compound," *Proceedings of SPE ANTEC,* pp. 796–799, 1989.
37. A. Hale. Epoxies Used in the Encapsulation of Integrated Circuits: Rheology, Glass Transition, and Reactive Processing, *Thesis, University of Minnesota, Department of Chemical Engineering,* 1988.
38. R. Gannamani and M. Pecht, "An experimental study of popcorning in plastic-encapsulated microcircuits," *IEEE Trans. Comp. Packaging Mfgrg. Tech. Part A Vol. 19* No. 2 June 1996, pp. 194–201.
39. R. Munamarty, P. McCluskey, M. Pecht, and C. Yip. "Popcorning in Fully Populated and Perimeter Plastic Ball Grid Array Packages," *Soldering and Surface Mount Technology,* No. 22 Feb. 1996, pp. 46–50.
40. L. T. Manzione, J. K. Gillham, and C. A. McPherson. "Rubber Modified Epoxies, Transitions and Morphology," *J. Appl. Polym. Sci.,* 26: p. 889, 1981.
41. P. Yalamanchili, R. Gannamani, R. Munamarty, P. McCluskey, and A. Christou. Optimum Processing Prevents PQFP Popcorning, *Surface Mount Technology,* pp. 39–42, May 1995.
42. Intel Corporation. "Recommended Procedures for Handling of Moisture Sensitive Plastic Packages," in *Intel Corporation Packaging Handbook,* Intel Corp., 1993.
43. S. Altimari, S. Goldwater, P. Boysan, and R. Foehringer. "Role of Design Factors for Improving Moisture Performance of Plastic Packages," *Proceedings of the 42nd Electronic Components and Technology Conference,* pp. 945–950, 1992.
44. L. T. Nguyen, K. L. Chen, and P. Lee. "Leadframe Designs for Minimum Molding-Induced Warpage," *Proceedings of the 44th Electronic Components and Technology Conference,* 1993.
45. F. Linker, B. Levit, and P. Tan. "Ensuring Lead Integrity," *Adv. Packaging,* pp. 20–23, 1993.
46. C. L. Alger, D. E. Pope, P. M. Rehm, and N. Subramaniam. "Solderability Requirements for Plastic Surface Mount Packages," *Proceedings of the 7th IEEE–CHMT, IEMTS,* 1990.

47. E. Pope. "Moisture Barrier Bag Characteristics for PSMC Protection," *Technical Proceedings, SEMICON-East,* pp. 59–69, 1988.
48. Texas Instruments. *Texas Instruments Military Plastic Packaging. Preliminary Handbook,* Texas Instruments, 1992.
49. *MIL-B-81705B. Military specification. Barrier Materials, Flexible, Electrostatic-free, Heat sealable,* U. S. Department of Defense, Washington, DC, 1989.
50. Hitachi. *Surface Mount Package Users Manual* Hitachi, 1991.
51. S. S. Chiang and R. K. Shukla. "Failure Mechanism of Die Cracking Due to Imperfect Die-Attachment," *Proceedings of the IEEE Electronic Components and Technology Conference,* pp. 195–202, 1984.
52. D. Broek, *Elementary Engineering Fracture Mechanics,* 4th ed., Kluwer Academic, Boston, 1991.
53. P. P. Merrett. "Plastic-Encapsulated Device Reliability," in *Plastics for Electronics,* ed. Martin T. Goosey, Elsevier Applied Science Publication, New York, 1985.
54. S. Okikawa, M. Sakimoto, M. Tanaka, T. Sato, T. Toya, and Y. Hava. Stress Analysis of Passivation Film Crack for Plastic Molded LSI Caused by Thermal Stress, *Proceedings International Symposium on Test and Failure Analysis* pp. 275–280, 1983.
55. H. Inayoshi, K. Nishi, S. Okikawa, and Y. Wakashima. "Moisture-Induced Aluminum Corrosion and Stress on the Chip in Plastic-Encapsulated LSIs," *Proceedings of the 17th Annual International Reliability Physics Symposium,* pp. 113–117, 1979.
56. L. J. Gallace, H. J. Khajezadeh, and A. S. Rose. "Accelerated Reliability Evaluation of Trimetal Integrated Circuit Chips in Plastic Packages," *Proceedings of the 14th Annual International Reliability Physics Symposium,* pp. 224–228, 1978.
57. L. T. Nguyen, S. A. Gee, and W. F. Bogert. "Effects of Configuration on Plastic Packages," *J. Electron. Packaging,* 113: pp. 397–404, 1991.
58. L. T. Nguyen. "Moisture Diffusion in Electronic Packages, II: Molded Configurations vs. Face Coatings," *46th SPE ANTEC,* pp. 459–461, 1988.
59. L. T. Nguyen and F. J. Lim. "Wire Sweep during Molding of Integrated Circuits," *IEEE Electronic Components and Technology Conference,* pp. 777–785, 1990.
60. L. T. Nguyen, A. S. Danker, N. Santhiran, and C. R. Shervin. "Flow Modeling of Wire Sweep During Molding of Integrated Circuits," *ASME Winter Annual Meeting,* pp. 27–38, 1992.
61. L. T. Nguyen, R. L. Walberg, C. K. Chua, and A. S. Danker. "Voids in Integrated Circuits Plastic Packages from Molding," *ASME/JSME Conference on Electronic Packaging,* pp. 751–762, 1992.
62. L. T. Nguyen. "Reactive Flow Simulation in Transfer Molding of Integrated Circuits Packages," *IEEE Electronic Components and Technology Conference,* 1993.
63. S. A. Gee, L. T. Nguyen, and V. R. Akylas. "Wire Bonder Characterization Using a *P-N* Junction-Bond Pad Test Structure," *MEPPE FOCUS 91,* pp. 156–170, 1991.
64. V. H. Winchell. "An Evaluation of Silicon Damage Resulting from Ultrasonic Wire Bonding," *Proceeding of the 14th Annual International Reliability Physics Symposium,* pp. 98–107, 1976.
65. V. H. Winchell and H. M. Berg. Enhancing Ultrasonic Bond Development. *IEEE Transactions on Components, Hybrids, and Manufacturing Technology* CHMT-1, pp. 211–219, 1978.
66. H. Koyama, H. Shiozaki, I. Okumura, S. Mizugashira, H. Higuchi, and T. Ajiki. "A Bond Failure Wire Crater in a Surface Mount Device," *Proceedings of the 26th Annual International Reliability Physics Symposium,* pp. 59–63, 1988.
67. V. S. Kale. "Control of Semiconductor Failures Caused by Cratering of Bond Pads," *Proceedings of the International Microelectronics Symposium,* pp. 311–318, 1979.

68. T. Koch, W. Richling, J. Whitlock, and D. Hall. "A Bond Failure Mechanism," *Proceedings of the 24th Annual International Reliability Physics Symposium,* pp. 55–60, 1986.
69. T. B. Ching and W. H. Schroen. Bond Pad Structure Reliability. *Proceedings of the 26th Annual International Reliability Physics Symposium,* pp. 64–70, 1988.
70. C. W. Horsting. Purple Plague and Gold Purity. *Proceedings of the 10th Annual International Reliability Physics Symposium,* pp. 155–158, 1972.
71. G. G. Harman. *Reliability and Yield Problems of Wire Bonding in Microelectronics,* ISHM, 1989.
72. D. O. Harris, R. A. Sire, C. F. Popelar, M. F. Kanninen, D. L. Davidson, L. B. Duncan, Kallis, and J. Hiatt. "Microprobing," *Proceedings of the 18th Annual International Reliability Physics Symposium,* pp. 116–120, 1980.
73. N. C. McDonald and P. W. Palmberg. *Application of Auger Electron Spectroscopy for Semiconductor Technology,* p. 42, IEDM, 1971.
74. N. C. McDonald and G. E. Riach. "Thin Film Analysis for Process Evaluation," *Electron. Packaging Production,* pp. 50–56, 1993.
75. H. K. James. "Resolution of the Gold Wire Grain Growth Failure Mechanism in Plastic-Encapsulated Microelectronic Devices," *IEEE Trans. Components Hybrids Manuf. Technol.,* CHMT-3: pp. 370–374, 1980.
76. J. L. Newsome, R. G. Oswald, and W. R. Rodrigues de Miranda. "Metallurgical Aspects of Aluminum Wire Bonds to Gold Metallization," *14th Annual Proceedings of the IEEE Electronics Components and Technology Conference,* pp. 63–74, 1976.
77. P. M. Hall, N. T. Panousis, and P. R. Manzel. "Strength of Gold Plated Copper Leads on Thin Film Circuits Under Accelerated Aging," *IEEE Trans. Parts, Hybrids, Packaging,* PHP-11(3): pp. 202–205, 1975.
78. S. S. Kim. "Improving Plastic Package Reliability Through Enhanced Mold Compound Adhesion," *IEEE International Reliability Physics Symposium Tutorial,* Topic 2, pp. 2d.1–2d.17, 1992.
79. L. T. Nguyen. "Surface Sensors for Moisture and Stress Studies," in *New Characterization Techniques for Thin Polymer Films,* ed. H-M. Tong and L. T. Nguyen, Wiley, New York, 1990.
80. L. T. Nguyen. "Reliability of Postmolded Integrated Circuits Packages," *SPE RETEC,* pp. 182–204, 1991.
81. O. Yoshioka, N. Okabe, S. Nagayama, R. Yamaguchi, and G. Murakami. "Improvement of Moisture Resistance in Plastic Encapsulants MOS-Integrated Circuits by Surface Finishing Copper Leadframe," *Proceedings of the 39th IEEE Electronic Components and Technology Conference,* pp. 464–471, 1989.
82. M. Kitano, A. Nishimura, and S. Kawai. "A Study of Package Cracking During the Reflow Soldering Process (1st & 2nd Reports, Strength Evaluation of the Plastic by Using Stress Singularity Theory)," *Trans. Japan Soc. Mech. Eng.*, 57 (90): pp. 120–127, 1991.
83. P. C. Paris, M. P. Gomez, and W. E. Anderson. "A Rational Analytical Theory of Fatigue," *Trend Eng.* 13: pp. 9–14. 1961.
84. A. Nishimura, A. Tatemichi, H. Miura, and T. Sakamoto. "Life Estimation for Integrated Circuits Plastic Packages Under Temperature Cycling Based on Fracture Mechanics," *IEEE Trans. Components Hybrids Technol,* CHMT-12(4): pp. 637–642, 1987.
85. S. Ito, A. Kitayama, H. Tabata, and H. Suzuki. "Development of Epoxy Encapsulants for Surface Mounted Devices," *Nitto Technol. Rep.* pp. 78–82, 1987.
86. T. C. May and M. H. Woods. A New Physical Mechanism for Soft Errors in Dynamic Memories. *Proceedings of the 16th Annual International Reliability Physics Symposium,* pp. 33–40, 1978.
87. D. Frear, H. Norgan, S. Burchett, and J. Lau. *The Mechanics of Solder Alloy Interconnects,* Van Nostrand Reinhold, New York, 1994.

88. R. B. Ghate. *Industrial Perspective on Reliability of VLSI Devices,* Texas Instruments, 1992.
89. R. W. Vahle, and R. J. Hanna. Proceedings of the International Congress on Transportation Electronics, Society Automotive Engineering (October 1990) 225.
90. C. Bloomer, R. L. Franz, M. J. Johnson, S. Kent, B. Mepham, S. Smith, R. M. Sonnicksen, and L. S. Walker. "Failure Mechanisms in Through-Hole Packages," in *Electronic Materials Handbook,* 1, *Packaging,* ed. by M. L. Minges, pp. 969–981, ASM International, Materials Park, OH 1989.
91. M. G. Pecht and V. Ramappan. Are Components still the Major Problem: A Review of Electronic System and Device Field Failure Returns. *IEEE Transactions on Components, Hybrids, and Manufacturing Technology,* Vol. 15, No. 6, pp. 1160–1164, Dec. 1992.
92. C. H. Taylor. "Just How Reliable Are Plastic-Encapsulated Semiconductors for Military Applications and How Can the Maximum Reliability Be Obtained?" *Microelectron. Reliability.* 15: pp. 131–134, 1976.
93. C. H. Taylor and B. C. Roberts. Evaluation of a U. K. Specification for the Procurement of Plastic-Encapsulated Semiconductor Devices for Military Use. *Microelectronics and Reliability,* 18 pp. 367–377, 1978.
94. P. V. Robock, and L. T. Nguyen. "Plastic Packaging," in *Microelectronics Packaging Handbook,* ed. R. R. Tummala and E. J. Rymaszewski, Van Nostrand Reinhold, New York, 1989.
95. L. T. Nguyen and J. A. Jackson. "Identifying Guidelines for Military Standardization of Plastic-Encapsulated Integrated Circuits Packages," *Solid-State Technol.,* pp. 39–45, 1993.
96. R. Iscoff. "Thin Outline Packages: Handle With Care!" *Semicond. Int.* pp. 78–82, 1992.
97. P. Hoffman, D. Liang, D. Mahulikar, and A. Parthasarathi. "Development of a High Performance TQFP Package," pp. 57–62, 1994.
98. R. Iscoff. "Amkor Develops Competitor to Multilayer Ceramic, MQUAD Packages," *Semicond. Inter.* p. 34, 1992.
99. R. Iscoff. "Micro SMT Package Avoids Traditional Bonding Methods," *Semicond. Int.* p. 40, 1992.
100. W. E. Jahsman. "Lead Frame and Wire Bond Length Limitations to Bond Densification," *J. Electron. Packaging,* 111: pp. 289–294, 1989.
101. K. Otsuka, Y. Takeo, H. Tachi, H. Ishida, T. Yamada, and S. Kuroda. "High Reliability Mechanism of New Silicone Gel Sealing in Accelerated Environment Text," *Proc. I.E.P.S.,* pp. 720–726, 1986.

ADDITIONAL READINGS

C. Bloomer, R. L. Franz, M. J. Johnson, S. Kent, B. Mepham, S. Smith, R. M. Sonnicksen, and L. S. Walker. "Failure Mechanisms in Through-Hole Packages," in *Electronic Materials Handbook, 1, Packaging,* ed. by M. L. Minges, pp. 969–981, ASM International, Materials Park, OH 1989.

S. Han and K. K. Wang. A Study of the Effects of Fillers on Wire Sweep Related to Semiconductor Chip Encapsulation, *ASME Winter Annual Meeting* pp. 123–130, 1993.

S. Mizugashira, H. Higuchi, and T. Ajiki. "Improvement of Moisture Resistance by Ion-Exchange Process," *IRPS IEEE,* pp. 212–215, 1987.

Nitto Denko Corporation, personal communication, 1993.

G. F. Watson. "Interconnections and Packaging," *IEEE Spectrum,* pp. 69–71, 1992.

SIA 1993

G. Wolfe. "Electronic Packaging Issues in the 1990s," *Electron. Packaging Production,* pp. 76–80, 1990.

BIBLIOGRAPHY

ASTM Annual Book of ASTM Standards, American Society for Testing and Materials, Philadelphia, 1993.

L. W. Condra, G. A. Kromholtz, M. G. Pecht, and E. B. Hakim. "Using Plastic-Encapsulated Microcircuits in High Reliability Applications," *Proceedings Annual Reliability and Maintainability Symposium,* pp. 481–488, 1994.

L. W. Condra, S. O'Rear, T. Freedman, L. Flancia, M. Pecht, and D. Barker. "Comparison of Plastic and Hermetic Microcircuits under Temperature Cycling and Temperature Humidity Bias," *IEEE Transactions on Components, Hybrids, and Manufacturing Technology,* Vol. 15, No. 5, pp. 640–650, Oct. 1992.

L. W. Condra, G. Wenzel, and M. Pecht. "Reliability Evaluation of Simple Logic Microcircuits in Surface Mount Plastic Packages," *ASME Winter Annual Meeting,* New Orleans, Nov. 1993.

A. Gallo, and R. Munamarty. "Popcorning: A Failure Mechanism in Plastic Encapsulated Microcircuits," *IEEE Trans. on Reliability,* Sept. 1995.

R. Gannamani and R. Munamarty. "Techniques to Qualify PEMs against Popcorning," *Electronic Materials and Packaging,* pp. 24–26, Nov. 1995.

R. Gannamani, and M. Pecht. "An Experimental Study on Popcorning in PEMs," *IEEE Trans. on Components, Packaging, and Manufacturing Technology—Part A,* vol. 19, no. 2, pp. 194–201, June 1996.

N. Kelkar, A. Fowler, M. Pecht, and M. Cooper. "Phenomenological Reliability Modeling of Plastic Encapsulated Microcircuits," *International Journal of Microcircuits and Electronic Packaging,* vol. 19, no. 1, March 1996.

R. Munamarty, P. McCluskey, M. Li, P. Yalamanchili, R. Gannamani, and L. Yip. "Delamination and Cracking in PBGAs during IR Reflow Soldering," *BGA Conference,* Berlin Germany, 1995.

L. T. Nguyen. "Wirebond Behavior During Molding of Integrated Circuits," *Polm. Eng. Sci.* 28(4): pp. 926–943, 1988.

L. T. Nguyen and C. A. Kovac. "Moisture Diffusion in Electronic Packages. I. Transport Within Face Coatings," *SAMPE Electronics Materials and Processes Conference,* pp. 574–589, 1987.

L. T. Nguyen. "On Lead Finger Designs in Plastic Packages for Enhanced Pull Strength," *Int. J. Microcircuits Electron. Packaging,* 15(1): pp. 11–33, 1991.

L. T. Nguyen, A. Danker, N. Santhiran, and C. R. Shervin. "Flow Modeling of Wire Sweep During Molding of Integrated Circuits," *ASME Winter Annual Meeting,* pp. 27–38, 1992.

L. T. Nguyen, R. H.Y. Lo, and J. G. Belani. "Molding Compound Trends in a Denser Packaging World, I: Technology Evolution," *IEEE International Electronic Manufacturing Technology Symposium,* 1993.

M. G. Pecht. "A Model for Moisture Induced Corrosion Failures in Microelectronic Packages," *IEEE Transactions on Components, Hybrids, and Manufacturing Technology,* Vol. 13, No. 2, pp. 383–389, June 1990.

M. G. Pecht, and V. Ramappan. "Are Components Still the Major Problem: A Review of Electronic System and Device Field Failure Returns," *IEEE Transactions on Components, Hybrids, and Manufacturing Technology,* Vol. 15, No. 6, pp. 1160–1164, Dec. 1992.

C.G.M. Van Kessel, S. A. Gee, and J. R. Dale. "Evaluating Fracture in Integrated Circuits With Acoustic Emission," *Acoustic Emission Testing,* 5, 2nd ed., vol. 5, pp. 370–388, ed. G. Harman, *American Society for Non-Destructive Testing,* 1987.

P. Yalamanchili, P. Gannamani, R. Munamarty, P. McCluskey, and A. Christou. "Optimum Processing Prevents PQFP Popcorning," *Surface Mount Technology,* pp. 39–42, May 1995.

11

POLYMERS IN PACKAGING

G. CZORNYJ—*IBM*
M. ASANO—*Toray*
R.L. BELIVEAU—*DuPont*
P. GARROU—*Dow*
H. HIRAMOTO—*Toray*
A. IKEDA—*Asahi*
J.A. KREUZ—*DuPont*
O. ROHDE—*Ciba Geigy*

11.1 INTRODUCTION

The technologies in computer electronics are very diverse and span the entire spectrum of engineering and scientific fields. Metals, ceramics, and polymers are at the heart of all leading-edge semiconductors and packages used in the microelectronic Industry [1,2]. Today, dramatic changes are occurring in the applications of semiconductor products which will challenge existing materials and processes in the 21st Century [3]. Key application trends, including scaling down dimensions to 1 gigabit technologies, are expected to affect both package and on-chip interconnection, both requiring major changes in materials [4–10]. The material challenges for semiconductor devices utilizing polymeric materials are primarily for the formation of on-chip interconnection by extremely fine-line lithography [11–15] and for passivation and chip protection [16,17]. The role of packages, beyond simply passive containers in the past, is viewed as setting limits to the ultimate performance of computers. In this role, the packages of the 21st Century need to interconnect, power, cool, and protect devices in such a way that leading-edge performance and reliability result [18,19]. The materials to perform these functions include high-conductivity metals, low-dielectric-constant ceramics, and thermally-stable polymers.

Information processing is performed by the use of semiconductor devices interconnected, powered, and cooled by packages. Both semiconductors and packages depend significantly on materials, particularly thin-film metal, oxides, nitrides, thick-film ceramics, and thin-film polymers. Even though similar polymeric materials are used for semiconductor device on-chip interconnections, this chapter reviews the status and challenges of polymeric materials used for packaging applications, primarily focusing on the use of polymers for interlayer thin-film dielectrics in single and multichip Packages.

11.2 HISTORICAL PERSPECTIVE

Polymers have played and continue to play an important role in the fabrication of thin-film wiring layers on both first- and second-level packages and in semiconductor dielectric applications because of their excellent thermal, mechanical, and solvent resistance properties. Table 11-1 shows the many technology areas where polymers are utilized in semiconductor and packaging applications. The first commercial semiconductor device chip where a polyimide was used as part of a dual dielectric insulating layer in conjunction with quartz was the SAMOS (Silicon Aluminum Metal Oxide Semiconductor) 64Kbit memory chip introduced by IBM in 1978 as shown in Figure 11-1 [20]. The main advantages of

Table 11-1. Microelectronics Applications of Polymers

Organic dielectric insulators	Solder fluxes
Passivation barriers	Sealants
Alpha barriers	Adhesion promoters
Stress buffers	High-temperature adhesives
Lift-off materials	Ceramic binders
Rie barrier materials	Encapsulants
Resists	Heat-transfer materials

Source: Data from Refs. 21–35.

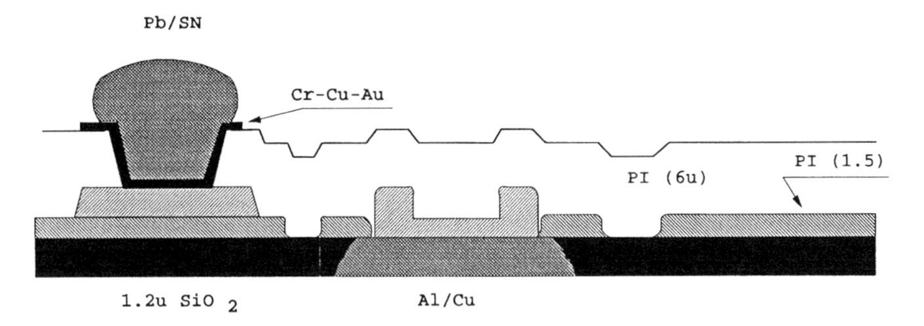

Figure 11-1. Cross Section of a Commercial FET Chip with Polyimide

the polyimide in this application over standard quartz dielectric was a lower dielectric constant of 3.5 versus 4.1 and ease of processing into uniform coatings on silicon wafers. This type of polyimide was utilized subsequently in many electronic applications. These applications include alpha-particle barrier coatings, stress buffer coatings on integrated circuits (ICs), and passivation coatings as well as interlayer dielectrics on ICs and high-density thin-film interconnects for multichip modules (MCM-Ds). NEC is credited with the first use of polymers as dielectrics in 1985 for use in mainframe packaging applications.

IBM announced in 1993 the development of a "thin-film" multichip module for workstations using a silicon substrate. The main attributes of this technology are high performance, high wiring density, reliability without hermeticity, and ability to use semiconductor processes to fabricate thin-film wiring in a polyimide dielectric to achieve low cost. At about the same time, multilayer ceramic modules were fabricated using thin-film polyimide dielectric layers for providing high-density surface wiring typically called "redistribution" in mainframes. The ES9000 mainframe system developed by IBM, for example, utilizes a novel polyimide [formed from a poly(amic ethyl ester) of PMDA–ODA] specially developed to provide ease of processing with excellent mechanical properties [33,34]. A cross section of the high-density thin-film redistribution layers built with this polyimide is shown in Figure 11-2 [35]. In fact, the development of the next generation of higher-density memory and logic chips as well as thin-film package wiring depends to a great extent on the development and availability of improved polymeric dielectric insulators. Today, there are many polymeric systems to choose from, including the newly developed photosensitive polyimides which offer potential reduction in processing steps during thin-film fabrication. In this chapter, these and other

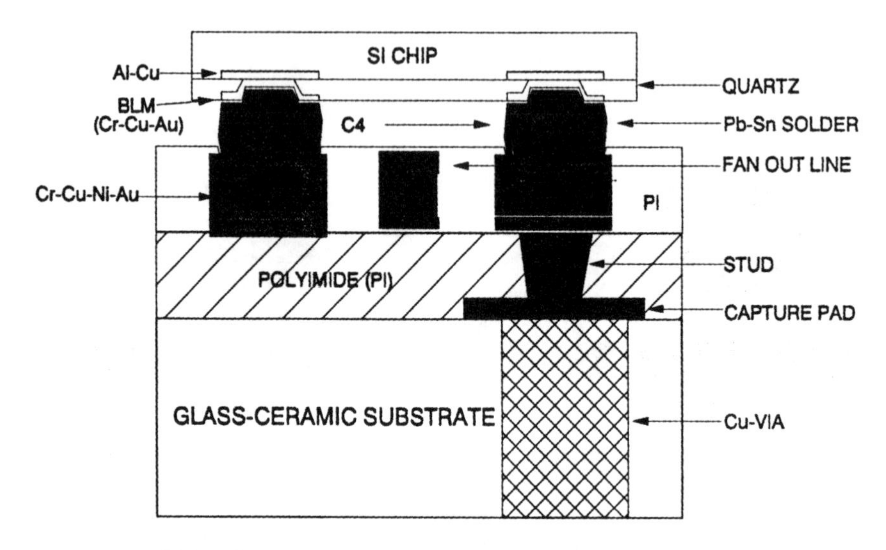

Figure 11-2. Thin-Film Redistribution Showing Glass–Ceramic/Copper Substrate, Polyimide, and Metal Interfaces

materials will be reviewed, properties contrasted, and then summarized for use in selecting the best material for a specific thin-film wiring interconnection, whether it be for packaging or device applications.

11.3 POLYMER THIN-FILM REQUIREMENTS FOR PACKAGING AND SEMICONDUCTORS

To develop high-quality thin films with respect to defects and reliability, one must select materials that match the functional and reliability requirements. The material selection criteria that one needs to consider when selecting high-performance polymers for use in thin-film structure fabrication are summarized in Table 11-2.

The polymeric requirements for high-density interconnects, on-chip and in thin-film packages, have many similarities, as shown in Table 11-2. The main differences that drive these requirements are that semiconductor applications need thinner dielectric films in the range of 1–4 μm and they utilize plasma processing for patterning of finer lines (0.5–3 μm in width). The packaging applications on the other hand utilize thicker films, typically 4–6× thicker, and the dimensions are in the 10–30-μm range.

The packaging applications, to a large extent, will specify the thermal-stability requirement for the polymer whereas the number of polymeric layers determines the requirements for high-end mainframe applica-

Table 11-2. Polymer Requirements for Thin-Film Interconnections

Function	Requirement
Lithographic dimensions	Aspect Ratio = (Height/Width)
Semiconductor device	Aspect ratio of 1–2 in 1–4-μm films
Thin-film package	Aspect ratio of 2–3 in 10–25-μm films
Thermal stability	
Semiconductor device	Withstand metal deposition and anneal temperatures
Thin-film package	Withstand metallization, cure, and chip attach temperatures
Mechanical integrity	Structure and substrate dependent
	Minimize stress to enhance reliability
Process compatibility	No cracking or crazing
	Dimensionally stable
	No delamination of thin films (good adhesion)

tions. In the case of semiconductors, the annealing of the metal interconnects may impose temperatures higher than 400°C, but for shorter time periods, and thus set the thermal-stability requirement for choice of polymer dielectric to this temperature. The mechanical properties and low stress requirements come mainly from the need to build multilayer stacks of films with metals (Cu, Ni, Cr, Ti, Au, W, etc.) on top of low-strength materials like silicon or ceramics. In the case of semiconductors, the total film thickness will be much smaller than in the thin-film packages, so the total stress thickness load imposed on the substrate will be smaller for semiconductor applications. The stress-induced substrate-bow scales up with the number of layers in the interconnect structure and the thickness of each of the layers. Also, semiconductors can benefit more than packages from a lower dielectric constant in reducing signal propagation delays.

11.3.1 Functional Compatibility and Reliability

The robustness of the systems can best be ascertained by subjecting the materials to the processes utilized in the fabrication of the thin-film structures on the substrates. In the fabrication of multilayer thin-film wiring structures such as shown in Figure 11-2, one needs to ascertain the adhesion of polymer to polymer, polymer to the various metals used in the structure fabrication, metal to polymer adhesion, and polymer adhesion to glass–ceramic, ceramic, silicon substrates, PWB, or glass. The various interfaces, described above and previously shown in Figure 11-2, must not only hold together at the time of the initial build but also after all of the stressing to which the structure is subjected [chemical, thermal, mechanical, and temperature and humidity (T & H) stressing]. The polymer dielectric layers must not crack, craze, or delaminate during the thin-film building. They must be compatible with the processing conditions selected for fabricating the multilayer thin-film structure for interconnection of the integrated semiconductor chips or for interconnection on the semiconductor chip.

The key parameters that need to be addressed to assess the reliability are adhesion of the various interfaces, thin-film stresses imparted by local defects or by metal lines or vias used for making the interconnections, mechanical properties before and after thermal cycling, polymer–solvent interactions, compatibility with chip joining and wirebonding schemes, and ease of fabrication. The polymer must be easily patterned to cleanly open up the critical dimensions yielding the proper wall profile angles needed for the subsequent metallization. T & H stressing has the greatest potential deleterious effect for degrading adhesion and via continuity. Once these are enumerated and the correlation of the fundamental physical properties is made to the functional product requirements, one can improve the materials and use the optimized polymer for the thin-film structure or semiconductor dielectric interconnections. In summary, the polymer requirements are thus obtained by matching the functional product requirements to the material physical properties. Table 11-3 summarizes in more detail some typical generic polymer thin-film material requirements such as lithography, thermal stability, mechanical property, and process conditions as discussed above for high end mainframe applications.

An important aspect of these dielectrics that is receiving the greatest attention recently is cost. The cost of typical high temperature polyimide

Table 11-3. Summary of Polymer Requirements

Function	Requirements for Polymer Cured to 400°C
Lithography	Excellent film-forming properties and pin-hole free 1–25 μm; ability to rework film consistently before cure
Thermal stability	Excellent thermal stability at 400°C
Mechanical	Excellent mechanical properties:
	Tensile strength ≅ 150 MPa
	Modulus ≅ 3.0 GPa
	Strain at break > 20%
	Residual stress < 30 MPa
	Self-adhesion (> 50 g/mm)
	Compatible with metals—No reaction
Process	Solvent compatible to survive multiple-layer fabrication
Electrical	Dielectric constant ≤ 3.5, $\tan\delta < 0.03$

is about a third to half of the cost of gold. Because of this and other high-cost processes, tools, and facilities for forming thin-film packages, this technology has not been commercialized beyond mainframes and super-computers. There are several ways to address the cost of polymers for thin-films; (1) by developing more efficient synthetic methods to lower the cost of the polyimides, BCBs, PSPIs, etc. (2) by selection of low-cost polymers such as olefins and epoxies in applications not requiring the attributes (enhanced thermal and mechanical stabilities) exhibited by preceeding list of materials or by (3) the use of large area processing such as extrusion, or meniscus coating of the polymers that don't waste the expensive polymer as in spin coating. These are discussed in more detail in the thin-film chapter.

11.4 POLYMERIC DIELECTRICS

Today, there are many polymeric materials to select from for both semiconductor and packaging applications. Some of the more common and commercialized materials are shown in Table 11-4.

Of these dielectrics, polyimides remain the workhorse interlayer dielectric for many semiconductor and packaging applications and have been used the most since the late 1970s. Enhancements have been made over the last decade to produce lower-stress, lower-dielectric-constant polyimides, including the development of photosensitive polyimides (PSPIs).

Table 11-4. Polymer Dielectrics for Thin-Film Packaging

- Polyimides (PIs) obtained from
  - Polyamic acids
  - Ester precursors
  - Acetylene terminated end groups
  - Low-stress derivatives
- Non-Polyimides
  - Benzocyclobutenes (BCBs)
  - Polyquinolines (PQs)
  - Olefins
- Photosensitive Polymers (negative and positive)
  - Epoxies
  - Preimidized PIs
  - Polyimides with covalently bonded cross-linking groups
  - Polyimides with ionically bonded cross-linking groups

11.5 POLYIMIDES

Polyimides are condensation polymers derived from bifunctional carboxylic acid anhydrides and primary diamines. Aromatic polyimides exhibit outstanding mechanical properties with excellent thermal and oxidative stability. It is well known that the mechanical properties, thermal stability, adhesion, swelling, morphology, dielectric constant, coefficient of thermal expansion, and residual stress are determined to a large extent by the choice of the diamine and dianhydride components used as starting materials. Polyimide materials, which have been synthesized and are commercially available, have been largely made from the monomers shown in Table 11-5, but today new backbone chemistries are being developed to improve material performance and lower the material cost [36,37].

Of the different combinations of monomer components that are possible, there are basically three categories of homopolymer polyimides that one can synthesize from the diamine and dianhydride combinations. These can be categorized from a practical application consideration according to the residual stress levels measured on fully cured films, corresponding to low stress, intermediate stress, and high stress. Typical properties of fully cured films obtained from polyamic acid precursors of BPDA–PDA (low residual stress with low TCE), polyamic acid or ester precursors based on PMDA–ODA chemistry, and acetylene terminated BTDA–type polyimides are summarized in Tables 11–6A and 11-6B. An excellent review of the literature featuring the synthesis and properties of polyimides can be found in reference [37-39]. Comparing commercialized polyimides from Hitachi, DuPont, Ciba Geigy and National Startch and Chemical Co.

Polyimides made from BPDA dianhydride have higher thermal stability than those made from PMDA- or BTDA-based monomers as illustrated in Figure 11-3. In a similar comparison, aromatic diamines such as PDA have greater thermal stability than ODA, MDA, or alkyl-substituted benzidine derivatives. The thermal stability hierarchy described above based on monomer choice of dianhydride and diamine is corraborated by

Table 11-5. Monomers Used for Polyimide Synthesis

Dianhydrides	Diamines
3,3′,4,4′-Biphenyltetracarboxylic acid dianhydride (BPDA)	M- and P-Phenylenediamine (PDA)
Pyromellitic dianhydride (PMDA)	4,4′-Diaminodiphenyl ether (ODA)
Benzophenone dianhydride (BTDA)	4,4′-diaminodiphenylmethane (MDA)

Table 11-6A. Summary of Chemistry Versus Features and Application

Polyimide Chemistry	Features	Applications
PMDA/ ODA	Excellent thermal/mechanical properties	Dielectric, passivation
PMDA/ODA & BTDA/ODA/MPD	Copolymer with ease of processing	Dielectric, passivation
BTDA/ODA/MPD	Improved adhesion, dielectric constant, and planarity	Delectric
BTDA/ODA/MPD esterified	High planarity, high solids, gap filling but reduced thermal properties	Alpha barrier protection
BPDA/PDA	Low TCE, low dielectric constant, low moisture uptake	Dielectric, passivation
6FDA/ODA	Lower dielectric constant, reflows/ cross-links removable	Lift-off layer
Preimidized	Soluble removable coating, long-term room-temperature solution stability	Die coating lift-off layer

the isothermal weight loss shown in Table 11-6B. This thermal stability is seen as lower isothermal weight loss at elevated temperatures above 400°C and, more importantly, as retention of the original mechanical properties after thermal stressing. Table 11-7 summarizes and compares the mechanical, thermal and electrical properties of polyimides derived

Table 11-6B. Typical Commercial Polyimides

	Typical Polyimide Properties		
Property	Low-Stress Polymide	Intermediate-Stress Polyimide	High-Stress Polyimide
Polymer system	BPDA–PDA	PMDA–ODA-based polymers and BTDA-based copolymers	BTDA-based acetylene capped polymers
Residual stress on silicon	6–15 MPa	20–40 MPa	35–60 MPa
Moisture uptake	< 1.0%	2–3%	2–3%
Dielectric constant	3.0	3.2–3.5	3.2–3.5
Isothermal weight loss at 400°C in N_2 or Air	0.03%/h	0.04%/h	0.17%/h
Mechanical properties			
Tensile modulus	9 GPa	2.5 GPa	2.7–3.8 GPa
Tensile strength	390 MPa	150–199 MPa	115–140 MPa
Strain at break	25–34%	90–130%	8–15%

Source: Data from Refs. 22–25.

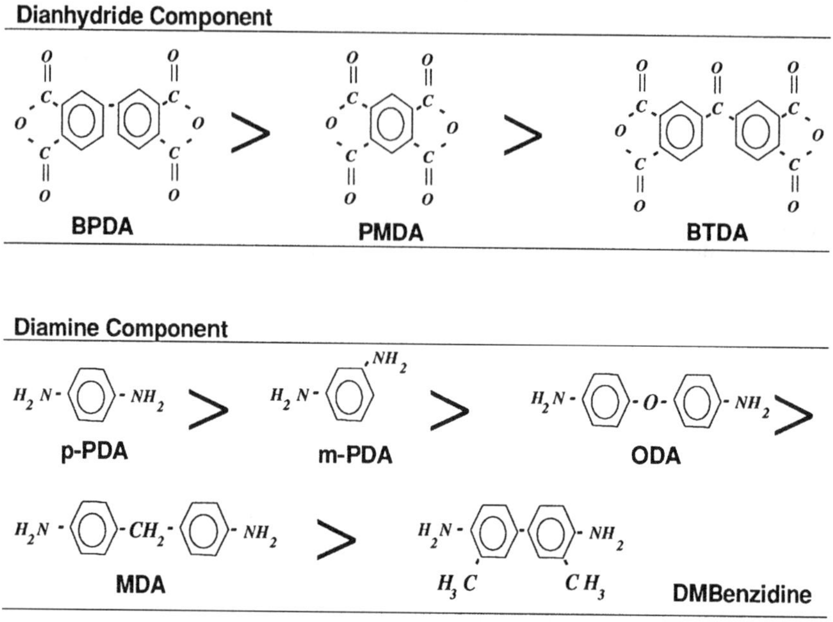

Dianhydride Component

BPDA > PMDA > BTDA

Diamine Component

p-PDA > m-PDA > ODA > MDA > DMBenzidine

Figure 11-3. Effect of Dianhydride and Diamine Components on Thermal Stability of Polyimides (BPDA>PMDA>BTDA and PDA>ODA>MDA) [40,41]

from a variety of chemistries to meet todays demanding electronic semiconductor and packaging applications. During manufacture of thin-film structures, whether for semiconductor chip or package, the removal of solvents from the solvent-based coatings followed by numerous high-temperature heating and cooling cycles generates residual stresses. The residual stresses produced by the difference in the thermal-expansion coefficients of the materials and the moduli of the various material layers can lead to film delamination, cracking of the metal or polyimide, and bowing of thin substrates, resulting in poor photolithographic image transfer. These are the main reasons why low-stress materials are highly desirable in thin-film fabrication [37].

11.5.1 Poly(amic alkyl ester) Development for Thin-Film Application

In 1991, IBM introduced a glass–ceramic substrate with thin-film redistribution as the first-level package utilizing polyimide–copper thin-film redistribution to match the wiring grid of the glass–ceramic substrate to the wiring and interconnection grid of the semiconductor chip as shown in Figure 11-2 [34]. For this application a unique and novel polyimide (PI), formed from a poly(amic ethyl ester) of PMDA–ODA, was developed and called M-PAETE [42].

Table 11-7. Comparison of Polyimide Film Properties Prepared from a Variety Monomer Chemistries Commercialized by DuPont

Cured Film Properties	Units	Homopolymer PMDA/ODA PI-2545	Copolymer PMDA/ODA & BTDA/ODA/MPD PI-2570	BTDA/ODA/MPD PI-2555	BPDA/PDA PI-2610D/11D
Mechanical					
Tensile Strength	kg/mm^2	10.5	10.5	13.5	35
Elongation	%	25–40	40	15	25
Density	g/cm^3	1.42	1.42	1.45	1.46
Modulus	kg/mm^2	140	140	245	845
Moisture Uptake	%	2–3	2–3	2–3	0.5
Stress[1]	MPa	22	15	36	6
Thermal					
Glass Transition Temperature	°C	>400	>400	>320	>400
Melting Point	°C	None	None	None	None
Decomposition Temperature	°C	580	580	550	620
Weight Loss (500°C in air, 2 hrs.)	%	3.6	3.6	2.9	1.0
Coefficient of Thermal Expansion	$(°C)^{-1} \times 10^{-6}$	20	20	40	3
Coefficient of Thermal Conductivity	cal/cm sec °C	37×10^{-5}	37×10^{-5}	35×10^{-5}	—
Specific Heat	cal/g/°C	0.26	0.26	0.26	—
Electrical					
Dielectric Constant (@1kHz, 50% RH)		3.5	3.4	3.3	2.9
Dissipation Factor (@1kHz)		0.002	0.002	0.002	0.002
Dielectric Breakdown Field	V/cm	$> 2 \times 10^6$	$> 2 \times 10^6$	$> 2 \times 10^6$	$> 2 \times 10^6$
Volume Resistivity	Ω-cm	$>10^{16}$	$>10^{16}$	$>10^{16}$	$>10^{16}$
Surface Resistivity	Ω	$>10^{15}$	$>10^{15}$	$>10^{15}$	$>10^{15}$

Note: [1]Stress was measured using a Tropell® Autosort Interferometer. Ref. DuPont product literature.

For this application, the polyimide must withstand the processes enumerated in Ref. 34 and 35: wirebonding, soldering with C4 (controlled collapse chip interconnection), and deletion of signal connection with laser energy. In addition, the polyimide must be compatible with and survive the following processes: laser ablation, oxygen ashing, metallization, planarization process, lift-off stencil fabrication and removal, and metal etching.

11.5.1.1 Background

As previously described in section 11.5 it is well known that the mechanical properties, thermal stability, adhesion, swelling, morphology, dielectric constant, coefficient of thermal expansion, and residual stress are determined to a large extent by the choice of the diamine and dianhydride components used as starting materials. It is now known that the derivative esters enhance the processing of the polyimides and that specific isomers further enhance the solution properties. IBM took advantage of all these characteristics to develop a unique class of PMDA–ODA derivatives and their isomers. These poly(amic alkyl esters) exhibit hydrolytic stability due to the elimination of the "monomer–polymer" equilibrium associated with the poly(amic acids) which are commercially available and used by others. In addition, the processing has been greatly improved because the ester derivatives possess a broad and higher characteristic imidization temperature regime, improved solubility characteristics, and enhanced adhesion and mechanical properties.

11.5.1.2 M–PAETE Synthesis

IBM developed the polyimide chemistry and the scale-up synthesis [42]. The preparation of poly(amic alkyl esters) is shown in Figure 11-4. The diester diacid is obtained from the reaction of the aromatic dianhydride with a suitable alcohol. The meta-isomer is recrystallized from the prepared diester diacid. Then, the diester diacid is converted into a diester diacyl chloride which reacts with an aromatic diamine via low-temperature solution polycondensation in a polar, aprotic solvent like NMP (*N*-methylpyrrolidone) to yield the desired poly(amic alkyl ester) [42,43]. The polymer is then washed, dried, formulated, and filtered to produce a polyimide solution which will meet the thickness requirements for thin-film application. IBM decided to use the predominantly meta-isomer of PMDA-ODA ethyl ester due to its better processing. This meta-isomer of the ethyl ester of PMDA–ODA exhibits excellent coating properties (planarization, low viscosity at high solids), and superior chemical properties (solution stability, nonreactivity to copper metal). After curing, the physical properties are comparable with those of fully cured polyimide derived from PMDA–ODA polyamic acid solution. These properties are summarized in Table 11-8.

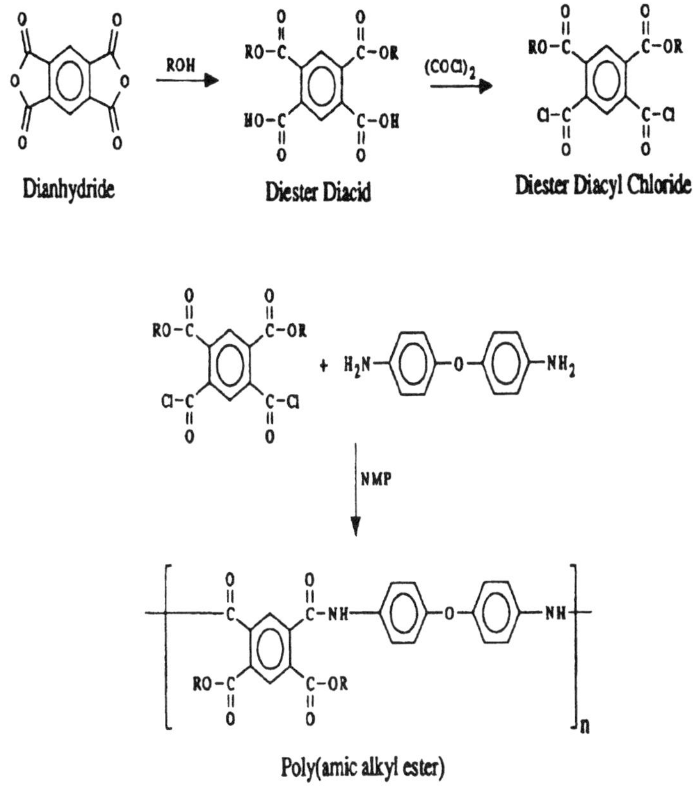

Figure 11-4. Typical Synthesis Scheme for Making Poly(amic alkyl ester) Polymers

11.5.1.3 M–PAETE Solution Properties

In order to meet the film thickness requirements, polymer solutions in the range of 24–31 wt.% in NMP were formulated from the dry powder and filtered to 0.2-μm levels. The intrinsic viscosity of 47–64 ml/g of the polymer corresponds to a molecular weight of 30,000–45,000. To ensure quality, the polymer solutions were analyzed for trace levels of contaminants, especially chloride ions and covalently bound chlorine. The isomer ratio and solvent purity were controlled and monitored to ensure solution stability for thin-film coating of substrates.

11.5.2 Characterization Techniques Used for Evaluation of Thin-Film Polymeric Materials

The experimental techniques developed to fully characterize thin films (10–20 μm thick) of polyimide to ensure manufacturability and reliability of the thin-film redistribution will be reviewed next, illustrating

Table 11-8. Summary of M–PAETE Properties

Meta-poly(amic Ethyl Ester) Precursor Cured to 400°C
Excellent film-forming properties
Planarizes/fills voids
Self-adhesion >100 g/mm
Compatible with Cu—No reaction
Excellent thermal stability at 400°C
Excellent mechanical properties
Tensile strength 160–190 MPa
Modulus 3.0 GPa
Strain at break >70%
Residual stress ~30 MPa
Dielectric constant 3.5
Thermal-expansion coefficient (35–40) × 10^{-6}/° C
Solid powder—can reformulate
Shelf-life stability
Easier to remove alcohol by-product than water

Source: From Ref. 33

the methodology and material characterization used to develop and evaluate the M-PAETE polyimide for the thin-film packaging application. These techniques were developed for the characterization of thin films as defined above and are generally used to characterize polymers for these type of applications [33,38–45]. The same techniques have been applied to characterize polymers for thin film packaging applications.

11.5.2.1 Mechanical Properties

The strength properties of the polyimides were obtained from stress-strain diagrams, which describe the behavior of a homogeneous specimen of uniform cross section subjected to uniaxial tension. The following tensile stress-strain parameters characterize the tensile mechanical properties of polymeric materials: Young's modulus, yielding strength, yielding strain, strain hardening point, stress at break, and strain at break. All these measurements were initially performed on the meta-poly(amic ethyl ester) precursor of PMDA–ODA, as it was being developed, to find the molecular-weight range which produced the best quality films, without crazing or cracking when subjected to processing conditions.

The samples for mechanical testing were prepared by spin-coating the solution of the polymer on silicon wafers, baking for 30 min at 85°C on a hot plate in air, and curing in a convection oven under flowing N_2

Table 11-9. Mechanical Properties of Fully Cured Polyimide

Sample	Cure Schedule	Stress at Break (MPa)	Strain at Break (%)
A	400°C/30 min	156	54
B	400°C/60 min	186	79

with a four-step curing schedule (150°C, 230°C, 300°C, and 400°C). The final cured film thickness was 10 μm.

11.5.2.2 Tensile Characterization

The average Young's modulus on the tested samples was approximately 3 GPa. The average ultimate tensile strength increased from 156 MPa to 186 MPa as the cure time at 400°C was increased from 30 to 60 min. This shows that the best mechanical properties are obtained when the cure cycle is optimized. These results are summarized in Table 11-9.

11.5.2.3 Process Stressing

A summary of the effects of process conditions A to E on the tensile properties, as determined by the changes measured in the strain at break, are shown in Table 11-10. The stressing simulates chip attach, flux cleaning, lift-off, and other process treatment conditions. After thermal aging at 400°C for 16 h, the modulus decreased by 10 MPa units and strain at break decreased by 2 ×. The stress and strain at break decreased slightly upon T & H stressing. The films treated with NMP (85°C) for 60 min and subjected to thermal aging exhibited about a 2 × decrease in the strain at break. The strain at break of the specimen subjected to stressing condition E exhibited about a 4 × reduction in strain at break.

Table 11-10. Effect of Process Conditions on Tensile Properties

Stressing Condition	Strain at Break
A. No stressing (control)	Normalized to 1
B. 16 h at 400°C	Decreased 2×
C. 300 h at T & H	Decreased < ¼×
D. NMP + B	Decreased 2×
E. NaOH + NMP + xylene + B	Decreased 4×

The T & H exposure and 16 h of thermal annealing are accelerated stress conditions used for testing purposes only. Because the tensile mechanical properties and toughness of the virgin polymer are so high, even the changes noted in Table 11-10 and described above provide a large safety margin (robust process) for the thin-film build from a mechanical point of view.

11.5.2.4 Adhesion of Thin-Film Interfaces

Adhesion of the polyimides at the various interfaces shown in Figure 11-2 is important because this is what holds the structure together and provides for the thin-film mechanical integrity.

Basic adhesion is the interfacial bond strength, which depends on the interfacial properties or interactions (electrostatic, chemical, or van der Waals) and mechanical interlocking [46]. As this adhesion cannot be measured mechanically, one must substitute another method (e.g., practical adhesion, which includes the basic adhesion modified by other factors). Peel strength measurements are one method of measuring practical adhesion and were used as a simple guide to assess interface strength.

The adhesion measurements were made at a peel rate of 5 mm/min at a 90° angle as illustrated in Figure 11-5. Peel measurements were made on the virgin films as well as on those that were exposed to various thin-film processing conditions. Table 11-11 shows the adhesion values of the virgin interfaces as measured by peel strength. These values are considered excellent.

11.5.2.4.1 Polyimide-to-Polyimide Adhesion

According to Brown et al. [44], PMDA–ODA adhesion to itself is controlled primarily by the diffusion of the precursor poly(amic acid) into a cured or partially cured polyimide layer. This interdiffusion is controlled by the cure temperatures of the two layers, with good adhesion resulting from a low cure temperature of the first layer. Utilizing this concept, a range of intermediate cures were found between 230°C and 260°C that exhibited excellent adhesion (100–140 g/mm). The meta-poly(amic ethyl ester) precursor is more easily partially cured, because it has a higher onset temperature for imidization and broader temperature regime where imidization takes place [45]. Little or no degradation of self-adhesion is observed after T & H exposure for up to 1000 h as shown in Table 11-12. This is important when one desires to build multiple polyimide layers.

11.5.2.4.2 Polyimide-to-Glass–Ceramic Adhesion

The adhesion of polyimide to glass–ceramic was improved by using surface treatments. After the glass–ceramic was cleaned by oxygen plasma

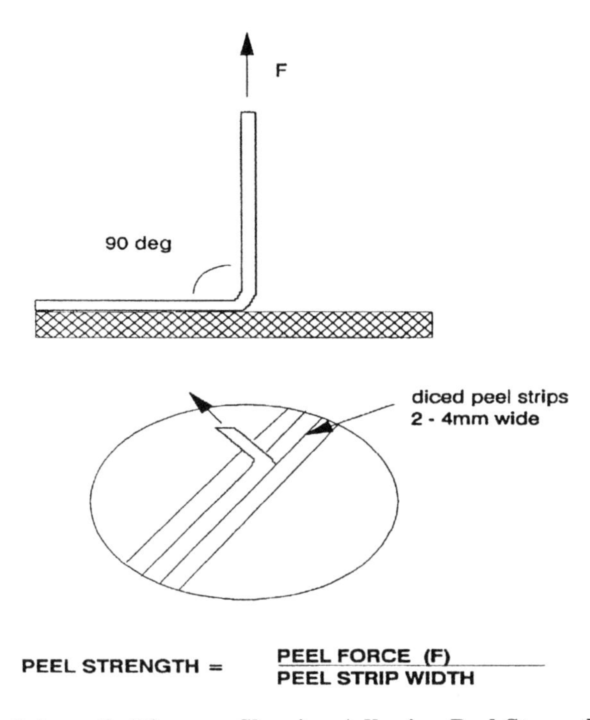

Figure 11-5. Schematic Diagram Showing Adhesion Peel Strength Measurement Being Performed at a 90 Degree Angle from the Test Sample

ashing, an adhesion promoter (silanol amine) was applied prior to overcoating with the polymer film. The adhesion promoter bonds to the oxide surface of the ceramic; this bonding can occur between the oxide and the silanol group, or the amine group can react with the oxide surface. The resulting adhesion is 95–110 g/mm.

11.5.2.4.3 Polyimide-to-Metal Adhesion

In the thin-film structure shown in Figure 11-2, the polyimide is in direct contact with chromium. Polymer solution was spin coated on chromium and then cured to 400°C, resulting in a peel strength of 80–90 g/mm. For polymer-to-copper adhesion, it is well known that poly(amic acid) reacts with Cu, forming a polymer–Cu complex during the spin-coating process. Copper oxide particles are precipitated in the polyimide upon curing of the poly(amic acid) (imidization process) [47-49]. This can cause an increase in the dielectric constant, thus favoring the use of ester-type derivatives of the polymer backbone.

11.5.2.4.4 Metal-to-Polyimide Adhesion

The metal (chromium) to polymer peel strength is 76–90 g/mm. The metal in this case is deposited over fully cured polyimide. The adhesion

Table 11-11. Adhesion Values of Interfaces

Interface	Peel Strength
Polyimide–polyimide	100–140 g/mm
Polyimide–glass–ceramic	95–110 g/mm
Polyimide–Chromium	80–90 g/mm
Chromium–Polyimide	76–90 g/mm

is most affected by coating imperfections which can be reduced by modifying the surface of the polyimide. Because the polyimide has excellent coating properties, T & H stressing has shown the greatest potential effect to reduce adhesion at this interface. Therefore, it is important to bake out any residual water from the films to ensure reliability of this interface.

11.5.2.5 Dynamic Stress Measurement

Residual stress arises from the deposition processes and the mismatch in the thermal coefficient of expansion of the polyimide dielectric layers in the thin-film structure, which consists of metals and glass–ceramic. To assess the development of residual stress in the polyimide, residual stress measurements were dynamically performed in nitrogen ambient during thermal curing of the dried films on wafers through the cure process and subsequent cooling at 1.0°C/min rate. The details of the stress measurement technique, and calculations are reported in Ref. 50.

Films were prepared from the precursor solution on wafers, soft-baked at 80°C for 30 min in a convection oven, and then thermally cured by a four-step cure process: 150°C/30 min, 230°C/30 min, 300°C/30 min, and 400°C/1 h. The thickness of fully cured films was about 10 μm.

Dynamic stress measurements are shown in Figure 11-6 over the temperature range of 25–400°C. Upon heating, thermal imidization starts with loss of solvent, but the residual stress continues to decrease with an

Table 11-12. Self-Adhesion as a Function of T & H Stressing

T & H Stressing Time	Peel Strength
0 h	115–129 g/mm
500 h	114–112 g/mm
1000 h	105–122 g/mm

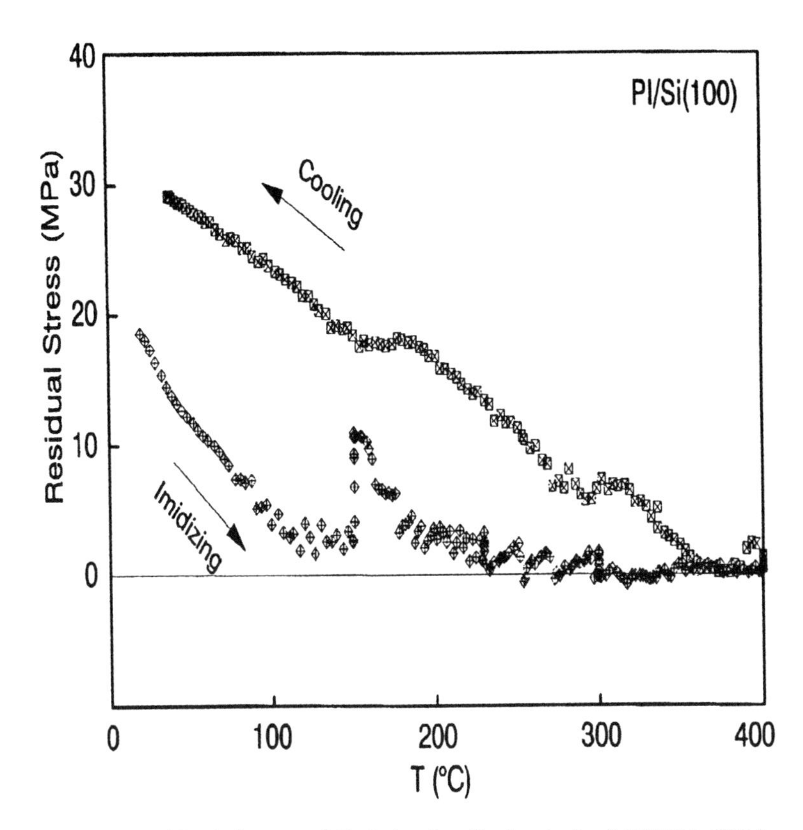

Figure 11-6. Residual Stress of Poly(amic alkyl ester) of PMDA-ODA as a Function of Temperature upon In-situ Heating and During In-situ Cooling of PI Film

abrupt jump in stress to 11 MPa at ~ 150°C. This suggests that the effective softening temperature of this soft-baked film is ~ 150°C. This residual stress then decreases and at ~180°C gradually continues to decrease as more solvent is removed in spite of imidization occurring over this large temperature regime. The residual stress decreases to ~1 MPa at 300°C and with continued heating to 400°C, the stress is unchanged, being < 1 MPa. The stress is very low over a large temperature regime during the curing process. This is a desirable condition for the polyimide in contact with the other materials, comprising the thin-film structure.

11.5.2.6 Dynamic Mechanical Properties

Dynamic mechanical thermal properties (storage modulus E' and loss modulus E'') were measured at a heating rate of 10°C/min and a frequency of 10 Hz in nitrogen ambient over the range of 25–500°C. Figure 11-7 shows the change in the dynamic storage modulus from room

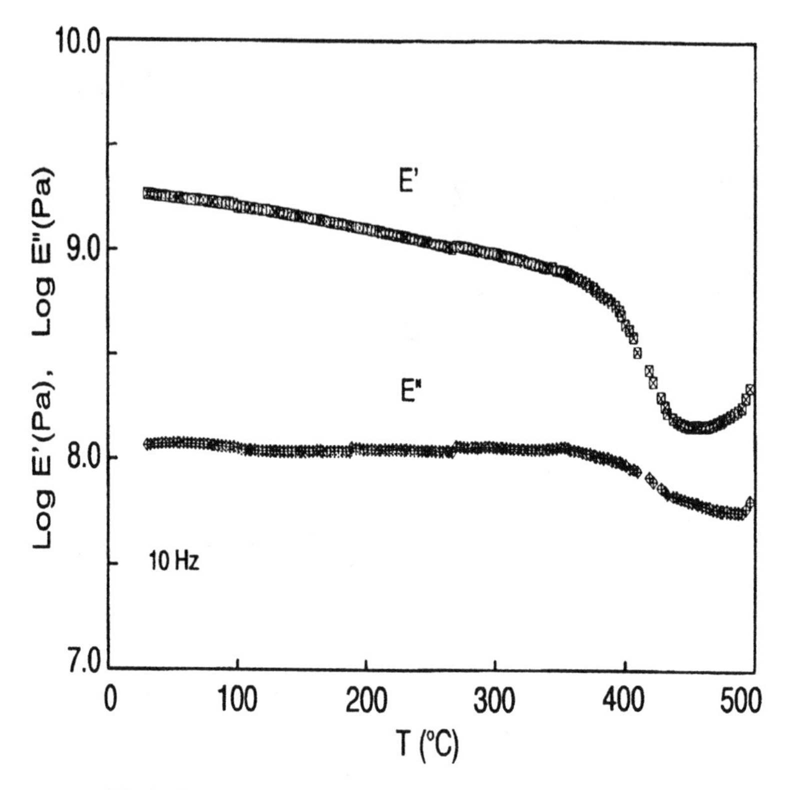

Figure 11-7. Variation of Dynamic Storage Modulus E′ and Loss Modulus E″ as a Function of Temperature from RT to 500°C for a Fully Imidized Film Showing a Tg

temperature to 500°C. The storage modulus gradually decreases from room temperature to ~380°C and shows a broad glass-transition region with a T_g of ~400°C. This high-T_g provides the dimensional stability needed to fabricate the thin-film structure and withstand the chip attach C4 thermal cycles.

M–PAETE was designed to imidize at significantly higher temperatures than its poly(amic acid) counterpart for ease of processing and is devoid of hydrolytic instability. M-PAETE possesses excellent film-forming properties, adhesion, thermal stability, mechanical properties, superior solution properties, and the other processing attributes summarized in Table 11-8.

11.6 COMMERCIAL PREIMIDIZED POLYIMIDE FILMS

Whereas fibers can be considered to be unidirectional by virtue of length, self-supporting films are bidirectional with the added dimension of width. Both structures, of course, have a thickness dimension, and flexible films for use in microelectronics are generally in the range of 25–125 μm. A combination of physical toughness, chemical and thermal resistance, as well as excellent electrical insulating properties make many films such as those from all-aromatic polyimides, attractive as dielectric freestanding film substrates in microelectronic applications.

11.6.1 Desired Properties for Microelectronic Substrates

Advances in microelectronics toward higher circuit density have required the dielectric substrate to be more compatible with active and passive components, so as to achieve functionality during end use. One of the key physical properties is the coefficient of thermal expansion (TCE), which must be close to the value of the component with which it is in contact. Another is the coefficient of hygroscopic expansion (CHE), which must be as low as possible to avoid stresses during relative humidity changes. Also, electrical properties must prevent signal distortion or loss. In general, the dielectric should be thought of as a passive medium to support and protect circuitry, but otherwise it is inert.

Additionally, the dielectric substrate must have properties that allow it to sustain the rigors of processing, as well as to abet processing. Consequently, freedom from melting or severe distortion are required. The material should not shrink during manufacture, otherwise it will induce stresses or misregistration of circuits in multilayers. Physical strengths such as high moduli and high tenacities are important to prevent distortion under tension. Curl and flatness are essential to enable access to equipment without jamming or poor fitting, or to facilitate further processing such as installation of semiconductor chips. Chemical milling, typically with caustic solutions, is an advantage, so as to remove unwanted areas in the final circuit or to provide holes (personality holes or vias) for electrical or mechanical connections.

Many other detailed requirements of dielectric substrates for microelectronics are available, and often requirements vary significantly depending on the specific application. For example, in tape automated bonding (TAB) caustic etchability might be important, but for flexible interconnects, MIT fold endurance might be a demanding need.

Another requirement of the dielectric substrate is that it must adhere well to adjacent materials, whether they be active or passive components, otherwise voids will develop during processing with resulting deterioration in mechanical and electrical integrity. Adhereability is often ignored in stating properties, and although it is a property that will be highly dependent on the application, it is often critical to functionality. In summary, the reader might consult other references for appreciation of needs of dielectric substrates. A number of articles are available that describe these

Table 11-13. Properties of an Ideal Substrate for TAB Applications

Property	Current Requirement	Ideal Requirement
Shrinkage, 200°C (%)	0.01	Nil
TCE, 35–250°C (ppm/°C)	17	As required
Modulus, 23°C (GPa)	5.2	10.0
H_2O absorption % at 90 **rh**	1.5	0.0–0.1
Chemically millable	Yes	Yes
CHE, 30–70 **rh** (ppm/**rh**)	10	Nil
Adhereability (kg/cm)	1.4	2.1

needs. Examples are the account by Arnold et al. [51], and for specific needs of TAB, the comprehensive descriptions by Holzinger [52,53], and Monisinger [54].

Because polyimide films have an excellent balance of thermal durability and physical/dielectrical integrities over wide ranges of temperature and humidity, they have found increasing applications in microelectronics. The properties of these films are undergoing continual improvement because of the ever-increasing demands on them as substrates. A summary of the current property requirement levels for the time frame of the next few years is given in Table 11-13. Along with these immediate goal levels is given an ideal requirement list that might only be attainable with dielectric substrate structures that are considerably altered from what we now know of as linear aromatic polyimide films.

11.6.2 Manufacturing Methods

Commercial preparations of aromatic polyimide films are essentially variations of a two-step sequence expanded from adaptations of laboratory syntheses. They involve (1) preparation of a polyamic acid in a dipolar aprotic solvent (e.g., *N*-methylpyrrolidone) and (2) either chemical or thermal methods of imidization [55]. If the glass transition of aromatic polyimides is under 300°C, it might be possible to alter the film-forming step by direct melt extrusion or even alter both steps by melt polymerization and melt extrusion. However, significant progress toward viable manufacturing processes utilizing melt polymerization/melt extrusion have not been made and have not as yet afforded films for microelectronic packaging.

The chemical method of manufacture involves the addition of an anhydride, usually acetic anhydride, and a tertiary amine, such as pyridine or triethylamine, to the solution of polyamic acid. The film must be formed rather rapidly, because imidization proceeds quickly at room temperature to give an insoluble polyimide/polyamic acid. The polyimide film must then be heated at high temperature under restraint to complete imidization and drying.

The thermal method consists of casting the polyamic acid onto a support and evaporating the solvent with gradual heating. Finally, the partially converted film of polyimide/polyamic acid is separated from the support and heated at high temperature under restraint to complete the curing process of imidization and drying.

Properties of films from the chemical method of manufacture can be different than those from the thermal method of manufacture, despite the fact that the backbone structures are the same. It is outside the context of this handbook to delve into this subject in more detail, except to point out that processing is just as important as polymer backbone in defining properties. Orientation of polymer chains, isotropy, and stress can have profound influences on thermal shrinkage, in-plane TCE, and directional physical strengths [40,41].

11.6.3 Commercial Sources, Types, and Chemical Compositions of Preimidized Polyimide Films

The availability of commercial polyimide films stems from 1965, when DuPont de Nemours & Co. Inc. began large-scale production of Kapton® polyimide film. Since that time, Kanegafuchi Chemical Industry Co., Ltd. and Ube Industries, Ltd. have each commercialized polyimide films in the time frame of the 1980s.

11.6.3.1 Kapton® Polyimide Films

The DuPont products are listed under the trademark Kapton® polyimide film. Detailed property information and availability can be obtained from DuPont High Performance Films and DuPont Japan Ltd.

Films available for microelectronic applications are types H(N), V(N), E(N), and K(N). The optional letter (N) designates the presence (N included) or the absence (N excluded) of an inorganic slip additive in the thousand part per million range to lower the coefficient of friction and to aid winding, such as roll to roll transfer. For most applications, the slip additive is present and the "N" designation is noted. The type H(N) and V(N) films consist of the raw materials, pyromellitic dianhydride (PMDA) and 4,4′-diaminodiphenylether (or 4,4′-oxydianiline; ODA) as illustrated in Figure 11-8. The difference between H(N) and V(N) is that the V(N) has superior dimensional stability [i.e., less shrinkage than H(N) after heating as high as 300°C].

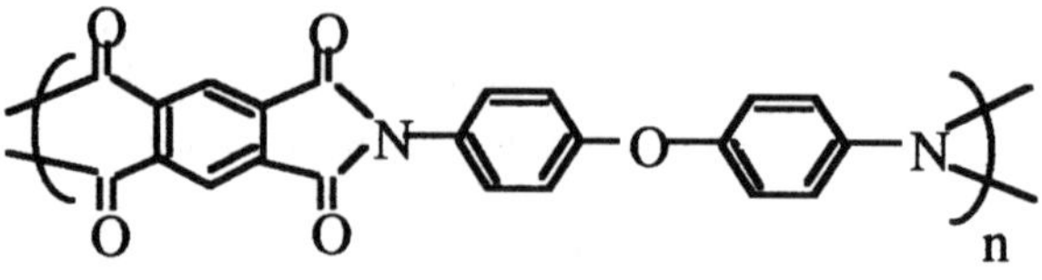

Figure 11-8. PMDA/ODA—H(N); V(N); AV

The type E(N) films also contain PMDA and ODA, but other raw materials are also present. These consist of the dianhydride 3,4,3′,4′-biphenyldianhydride (BPDA) and 1,4-diaminobenzene (or paraphenylene diamine; PPD), as shown in Figure 11-9.

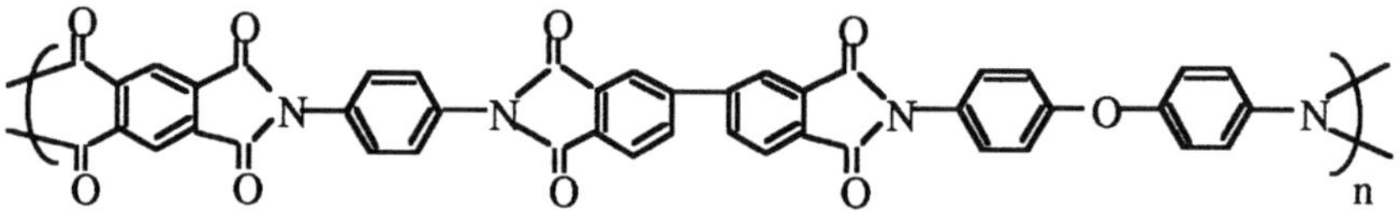

Figure 11-9. PMDA/BPDA/PPD/ODA—E(N) Random distribution of dianhydrides/diamines ratios of dianhydrides and diamines unequal

Type E(N) films are tetrapolyimides. The effects of various monomer mole percentages on properties have been discussed, but precise percentages in the commercial film have not been divulged [56]. The type E(N) films are designed to provide higher modulus, lower TCE, and lower moisture absorption than those of types H(N) or V(N). Advantages of type E(N) films are insignificant distortion of thin films during processing, more compatibility with conductive layers like copper, and less susceptibility to blistering during component attachments. The E(N) films have also been structured so that they will caustic etch.

Type K(N) films consist of terpolyimide that is based on the raw materials of PMDA, ODA, and PPD, but the mole percentages of the diamine components have not been published. The intent of the K(N) product is to provide a film that has a modulus between types H(N) or V(N) and E(N), a TCE that matches copper, and the facility to etch easily in caustic. Moisture absorption is higher than types H(N) V(N), and E(N). A general structural formula for the type K(N) films is given in Figure 11-10.

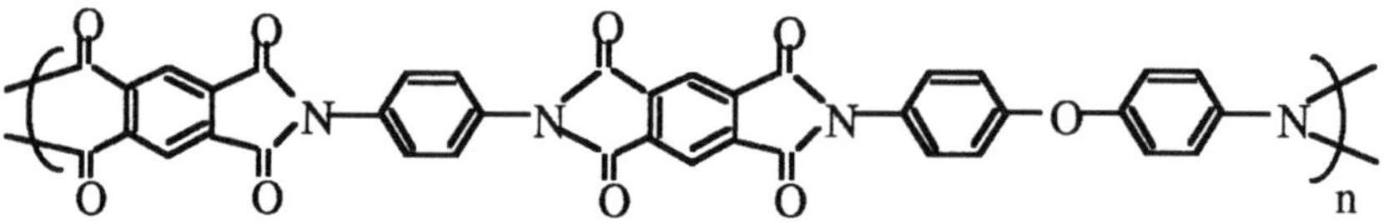

Figure 11-10. PMDA/PPD/ODA—K(N); NP Ratios of diamines unequal

11.6.3.2 Apical® Polyimide Films

Kanegafuchi polyimide films are commercialized under the trademark Apical® polyimide films. These products were originally made in Japan, but in the last few years an additional manufacturing facility has been established in the United States. Detailed product information and availability can be obtained from Allied-Apical Co.

Apical® polyimide films that are available and would have potential use in microelectronics are Types AV and NP. The type AV is generically the same polyimide backbone as is contained in Kapton® types H(N) and V(N), which is shown in Figure 11-8.

The type NP is similar in structure to Kapton® type K(N), which is given by Figure 11-10, but the exact mole percentages of 4,4′-diaminodiphenylether and paraphenylene diamine have not been disclosed. It is speculated that the polyimide film designated by "NP" has an ordered structure that can be considered as a somewhat blocked terpolyimide, but no current documentation is available. The similarity in structure of Apical® type NP as compared to Kapton® type K(N) cause speculation that the properties will be similar and the actual properties, given below (*vida infra*), supports this.

11.6.3.3 Upilex® Polyimide Films

Upilex® polyimide films are made by Ube Industries, Ltd. in Tokyo, Japan and they are marketed in the United States by Uniglobe Kisco, Inc.

The film manufactured by Ube for use in microelectronics is designated as Upilex® type S. It is based on 3,3′,4,4′-biphenyltetracarboxylic dianhydride (BPDA) and paraphenylene diamine (PPD) and the structural formula is given in Figure 11-11. The backbone is very rigid and, consequently, the film therefrom is characterized by a high modulus and low TCE. The film also absorbs very little water, and although it can be hydrolyzed back to the component raw materials with strong caustic at high temperature, it is for all practical purposes nonetchable.

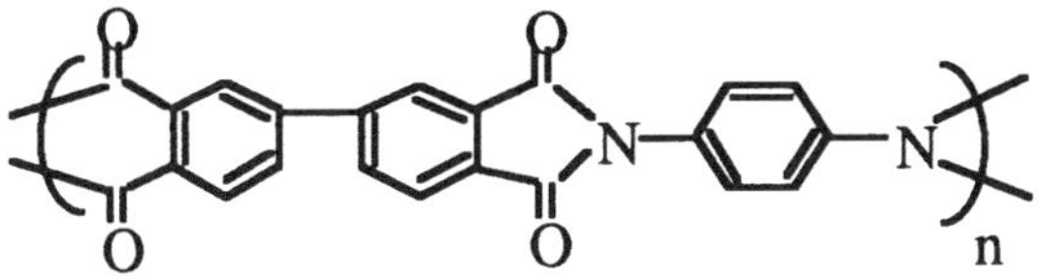

Figure 11-11. BPDA//PPD—S

11.6.3.4 Properties of Commercial Polyimide Films

All of the polyimide properties described in this section are based on films of 25 μm and/or 50 μm thickness [57]. All manufacturers supply their products in thicknesses of 12.5 μm, 25 μm, 50 μm, 75 μm, and 125 μm, and most typical properties are not grossly affected by thickness of the films. A striking exception to this behavior is the dielectric strength, and the values for various thickness are given, if values were reported by the manufacturer. For complete property information over the entire thickness range of films available, the reader is directed to the manufacturer and the likelihood that specific property/gage data are available.

The property data shown in Tables 11-14 through 11-17 are taken from the various manufacturers' property bulletins or from papers by manufacturers, where their products are described [56,57]. No attempt has been made to misrepresent a product by excluding property data. For data not present herein, the reader is advised to contact the specific producer.

Table 11-14. Typical Properties for 25-μm-Thick Films

	Kapton®			Apical®		Upilex®	
Property	H(N) + V(N)	K(N)	E(N)	AV	NP	S	Method
Melting point (°C)	None	None	None	None	None	None	ASTM-E794
Specific heat J/g°K (cal/g/°C)	1.09 (0.261)		1.09	—	—	1.13 (0.27)	DSC[a]
T_g[b] (°C)	410	>400	359	410		>500	DMA[c]
Shrinkage (%)				—	—	—	IPC-TM650; 2.2.4A
30 min/150°C	0.17 H(N)	—	0.05				
2 h/250°C	0.03 V(N)[d]	—	0.1	—	—	0.37	JIS C2318
2 h/400°C	1.25	—	1.1	—	—	0.39	ASTM-D5214
TCE (ppm/°C)							
20–100°C	18	—	12	—	—	8.0	TMA[e]
100–200°C	31	—	14	—	—	10.0	TMA[e]
200–300°C	48	—	16	—	—	16.0	TMA[e]
300–400°C	78	—	—	—	—	24.0	TMA[e]
20–250°C	35	17	14	—	—	14.0	TMA[e]
Flammability	94 V-0	94 V-0	94 V-O				UL-94
Limiting O_2 Index	37	—	48			66	
Smoke gen.	*D* = < 1					0.4	NBS—Kapton JIS-DR01—Uplilex

Note:
Values for coefficients of thermal-expansions (TCEs) are only meaningful after films have been preexposed to as high a temperature as the TCE is desired to measure and under no restraint. The purpose is to remove shrinkage.
[a]DSC = differential scanning calorimetry.
[b]T_g = Glass-transition temperature.
[c]DMA = dynamic mechanical analysis.
[d]V(N) films are thermally stabilized to give low shrinkage.
[e]Thermal mechanical analyzer—exact methods unknown. The Kapton films were initially heat stabilized; it is not known if Upilex films were preheated.

Table 11-15. Mechanical Properties—Typical Values at 23°C–25°C for 25-μm-Thick Films

		Kapton®			Apical®		UPILEX®	
Property	Unit	H (N) + V (N)	K (N)	E (N)	AV	NP	S	Method
Tensile strength	MPa	231	—	306	165	227	392	ASTM-D882
	psi	3.4×10^4	—	4.5×10^4	2.4×10^4	2.3×10^4	5.7×10^4	ASTM-D882
Stress at 5% elong	MPa	90	—	—	—	—	254	ASTM-D882
	psi	1.3×10^4	—	—	—	—	3.7×10^4	ASTM-D882
Elongation	%	72	50	45	60	50	30	ASTM-D882
Tensile modulus	GPa	2.5	3.4	5.1	—	—	8.6	ASTM-D882
	psi	3.7×10^5	5×10^3	7.5×10^5	—	—	1.28×10^6	ASTM-D882
Tear strength								
Graves	N	7.2			—	—	7.9	ASTM-D1004
(Initiation)	lbf	1.6			—	—	1.8	ASTM-D1004
Elmendorf	N	0.07	0.07	0.07	—	—	0.08	ASTM-D1922
(Propagation)	lbf	0.02	0.02	0.02	—	—	0.02	ASTM-D1922
MIT folding endurance	Cycles	2.9×10^5	—	1.5×10^4	—	—	$> 1 \times 10^4$	ASTM-2176
Density	g/cm^3	1.42	1.44	1.46	—	—	1.47	ASTM-D1505
Coeff.—Kinetic Friction (film–film)		0.48	—	—	—	—	0.40	ASTM-D1894

Table 11-16. Electrical Properties—Typical Values at 23–25°C for 25–125-μm-Thick Films

		Kapton®			Apical®		Upilex®[b]	
Property; (Unit) Gauge	Test Cond.	H(N) + V(N)	K(N)	E(N)	AV	NP	S	Method
Dielectric strength (kV/mil)[a]	60 Hz[b]							
25 μm		7.7	—	7.0	6.0	6.0	6.8	ASTM-D149
50 μm		6.1	—	6.5	5.0	5.0	5.1	ASTM-D149
75 μm		5.2	—	—	4.5	4.5	—	ASTM-D149
125 μm		3.9	—	—	3.0	3.0	—	ASTM-D149
Dielectric constant	1 kHz							
25 μm		3.4	3.4	3.2	3.9	3.9	3.5	ASTM-D150
50 μm		3.4	3.4	3.2	3.9	3.9	3.5	ASTM-D150
Dissipation factor	1 kHz							
25 μm		0.0018	0.0018	0.0018	0.0035	0.0035	0.0013	ASTM-D150
50 μm		0.0020	0.0020	0.0020	0.0035	0.0035	0.0013	ASTM-D150
Volume resistivity (Ω/cm)	100 V DC							
25 μm		1.5×10^{17}	10^{17}	10^{17}	10^{12}	10^{12}	10^{17}	ASTM-D257
50 μm		1.5×10^{17}	10^{17}	10^{17}	10^{12}	10^{12}	10^{17}	ASTM-D257

[a] Various thicknesses shown for dielectric strength, because this property is very dependent on thickness.
[b] Data on Upilex S obtained at 50 Hz.

Table 11-17. Chemical Properties of Kapton Type H(N) and V(N); 25-μm

Property	Strength Retained (%)	Elongation Retained (%)	Test Conditions	Test Method
Chemical Resistance				
Isopropyl alcohol	96	94	10 min at 23°C	IP TM-650
Tuolene	99	91		Method 2.2.3B
Methyl ethyl ketone	99	90		
Methylene chloride/ trichlorethylene (1:1)	98	85		
2 *N* Hydrochloric acid	98	89		
2 *N* Sodium hydroxide	82	54		
Fungus resistance	Non-nutrient			IPC TM-650 Method 2.6.1
Moisture absorption	1.8%		50% **rh** at 23°C	ASTM D-570-81 (1988)[a]
	2.8%		Immersion for 24 h at 23°C (73°F)	
Hygroscopic coefficient of expansion	22 ppm/% RH		23°C (73°F), 20–80% **rh**	ASTM D-1434-82 (1988)[a]
Permeability				
Gas	ml/m^2, 24h, MPa	cm^3/(100 in^2 24 h atm)	23°C (73°F), 50% **rh**	ASTM D-1434-82 (1988)[a]
Carbon dioxide	6,840	45		
Oxygen	3,800	25		
Hydrogen	38,000	250		
Nitrogen	910	6		
Helium	63,080	415		ASTM E-96-92
Vapor	g/m^2/24 h	g/(100 in^2 24 h)		
Water	54	3.5		

continued

Table 11-17. (*Continued*)

Property	Strength Retained (%)		Elongation Retained (%)		Test Conditions	Test Method
Resistance to						
10% Sodium hydroxide	80		60	95	Immersion at 25°C for 5 days	ASTM-D882
Glacial acetic acid	100		95	100	Immersion at 110°C for 5 weeks	ASTM-D882
p-Cresol	90		90	95	Immersion at 200°C for 3 weeks	ASTM-D882
Water pH = 1.0	95		85	100	Immersion at 100°C for 2 weeks	ASTM-D882
pH = 4.2	95		85	100	Immersion at 100°C for 2 weeks	ASTM-D882
pH = 8.9	95		85	100	Immersion at 100°C for 2 weeks	ASTM-D882
pH = 10.0	95		85	100	Immersion at 100°C for 4 days	ASTM-D882
Water absorption		1.2%			Immersion in water at 23°C for 24 h	ASTM-D570
		0.9%			Equilibrium at 60% **rh,** 50°C	ASTM-D570
Gas permeability						
Water vapor			1.7 $g/m^2/mil$		At 38°C, **rh** 90% for 24 h	ASTM-E96
Oxygen			0.8 $ml/m^2/mil$			
Carbon dioxide			1.2 $ml/m^2/mil$		At 30°C, 1 atm for 24 h	ASTM-D1434

11.7 HIGH TEMPERATURE NON-POLYIMIDE DIELECTRICS

Materials not based on PI chemistries are summarized below.

11.7.1 Benzocyclobutenes

Benzocyclobutenes are being used commercially as insulators for chip passivation and stress buffering, area array bumping, liquid crystal display and single-chip and multichip thin film packaging applications.

Benzocyclobutenes, **BCBs** [58-60] are a family of thermoset resins commercialized by Dow Chemical under the tradename **Cyclotene**™. The chemical structure of generic BCB (I) is shown below. Altering the bridging "R" group results in a multitude of possible BCB structures. The first BCB to be commercialized is siloxy containing DVS-BCB (II). The resins are supplied partially polymerized, "B-staged", to obtain appropriate viscoelastic handling properties. The Cyclotene™ 3000 series is formulated for dry etchability, the 4000 series is formulated for photo definition (see section 11.12). The 3000 series require no refrigeration. They are stable at room temperature, showing no change in viscosity after months of storage.

The polymerization (curing) of BCB proceeds through a two-step process: a thermally driven ring opening followed by a 4 + 2 Diels Alder reaction producing a structure similar to III. Curing is typically carried out at 210–250°C, several hundred degrees lower than is possible for other thin-film dielectric materials.

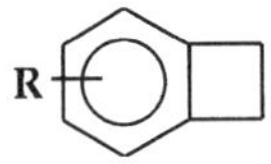

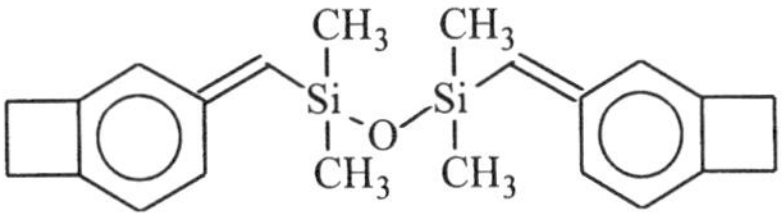

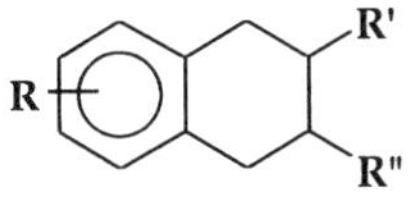

11.7.1.1 Electrical Properties

The dielectric constant (e_r) of Cyclotene 3022 is 2.65 ± 0.05 and is essentially invariant with temperature and frequency [59] as shown in Figure 11-12.

Detailed electrical analyses of Cu/BCB interconnect structures have been published [61-64]. High-frequency transmission line studies at 26 GHz [61] and 50 GHz studies on HFET ansd HBT amplifiers [64] have been reported. Electrical properties are summarized in Table 11-18.

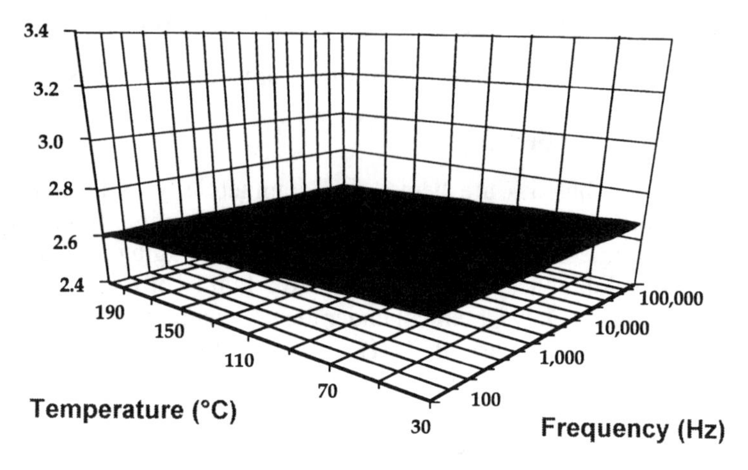

Figure 11-12. Dielectric Constant of Cyclotene™ 3022 [59]

11.7.1.2 Mechanical Properties

The mechanical properties of 25 μm BCB thin films are shown in Table 11-19.

11.7.1.3 Optical Properties

The optical properties of fully cured films of BCB reveal low birefringence and low transmission loss [65]. Loss in passive wave guides, fabricated from BCB is more related to feature roughness due to fabrication technique (etching technology) than to the inherent optical properties of the BCB dielectric. Interfaces created in photodefinable BCB reveal smoother surfaces and thus lower losses than the dry etchable grade. Optical properties are listed in Table 11-20.

11.7.1.4 Water Absorption

The bakeout cycles, normally employed with other dielectric materials having significant moisture absorption in order to avoid blistering and

Table 11-18. Electrical Properties of Cyclotene™ Resins

Dielectric Constant	(1 KHz–1 MHz)	2.65
	(1 GHz–20 GHz)	2.5
Dissipation Factor	(1 KHz–1 mhz)	0.0008
Breakdown Voltage	(V/cm)	3×10^6
Volume Resistivity	(ohm/cm)	1×10^{19}

Table 11-19. Physical and Mechanical Properties of Cyclotene

Cyclotene Series	CTE (ppm)	Tg[a] (°C)	Tensile Modulus (GPa)	Tensile Strength (MPa)	Elongation (%)
3022	52	>350	2.9	87	8

[a]Dependent on % cure.

delamination, are unnecessary with BCB. Water absorption for DVS-BCB is limited to < 0.2% after exposure to high humidity or boiling water [66].

Figure 11-13 shows moisture absorption and desorption rates when humidity is increased to 81% and then decreased to 6% at ambient temperature.

11.7.1.5 Thermal and Thermo-oxidative Stability

BCB's are inherently more susceptible to oxidation than fully aromatic polyimides [58]. To stabilize BCB to thermo-oxidative conditions, Cyclotene resins are formulated with an antioxidant. The thermal stability (determined under nitrogen) and the thermo-oxidative stability (determined in air) have been studied extensively [67]. Oxidation occurs with an activation energy of 25 ± 1.5 Kcal per mole.

11.7.1.6 Planarization

Thin film coatings of BCB exhibit excellent planarization properties, thus eliminating photolithiographic resolution issues caused by the generation of relief when poorly planarizing materials are used to fabricate

Table 11-20. Optical Properties of Cyclotene™ 3022

Optical Property	Cyclotene™ 3022
Birefringence	<0.002
Loss @ 835 nm (dB/cm)	0.08
Refractive Index	
(633 nm)	1.5584
(835 nm)	1.5473
(1320 nm)	1.5398

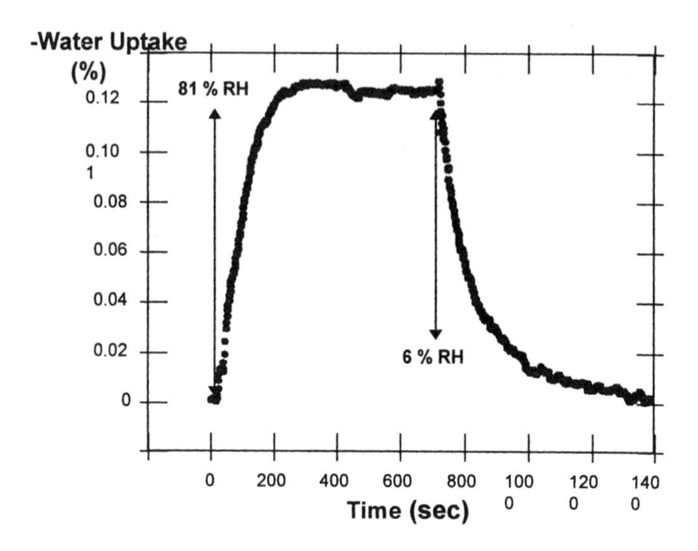

Figure 11-13. Water Absorption / Desorption for Cyclotene™ 3022 [66]

multiple levels of interconnect. The degree of planarization is dependent on the film thickness versus the feature thickness, the number of coatings, the feature width, height and spacing [68]. Typical planarization versus feature width is shown in Figure 11-14.

In conformal via structures, BCB dielectrics not only planarize underlying conductor layers but also planarize the conformal vias, so that the

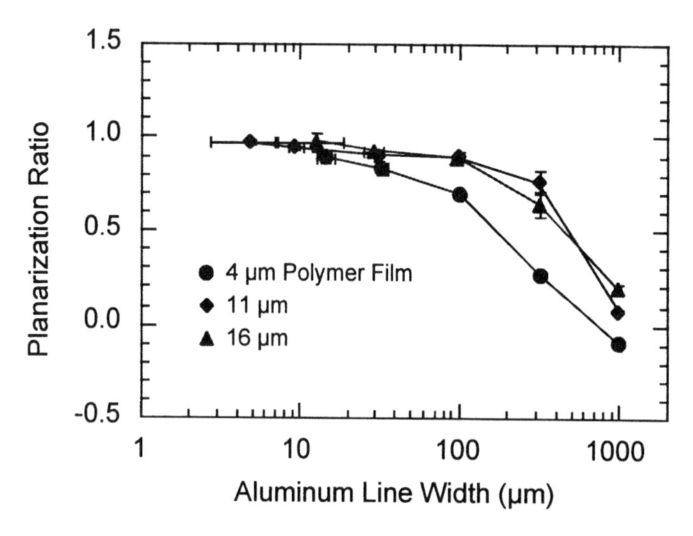

Figure 11-14. Cyclotene™ 3022 Planarization Over 4 μm Thick Isolated Aluminum Lines [68]

next orthogonal layer of conductor can be routed directly over the vias. Under identical conditions Cyclotene 3022 showed 65–70% planarization of via features versus 10–15% for typical polyimide dielectrics [69].

Planarization issues are becoming especially important in the manufacture of Si and GaAs multilevel metal structures [70,71].

11.7.1.7 Adhesion

A siloxy containing adhesion promoter such as [3-Aminopropyltriethoxysilane, APS] or [3-methacryloxypropyl-trimethoxysilane, MOPS] is recommended to enhance the adhesion of BCB to various inorganic interfaces. BCB adhesion to silicon oxide (wafer), alumina, aluminum nitride and to itself has been evaluated by ASTM tape tests, "stud pull" testing after water boil, PCT (pressure cooker testing) and microindentation testing. Such interfaces are reported to show reliable adhesion (zero failures in a cross hatch tape peel test) after PCT. Adhesion of Cyclotene to chromium is reported to be similar to both dry etch and photosensitive PIs [72]. Adhesion to gold or copper is enhanced by a < 500 Å layer of Ti, but not required for copper structures in order to pass reliability tests [64]. Adhesion to copper is also enhanced by texturing the surface with a microetching or black oxide treatment similar to PWB processing [73,74].

11.7.1.8 Thin Film Stress

Tensile stress developed during oven or belt furnace (RTC, "rapid thermal curing"), curing of BCB films is typically measured at 28–32 MPa for oven curing and 32–38 Mpa for belt furnace curing [75,76].

11.7.1.9 Metal Migration

Copper diffusion is not observed when BCB is deposited directly on copper metallization [77,78] even after 1500 hours at 85% RH/85°C in contrast to results with polyamic acid based polyimides [78].

11.7.1.10 Cyclotene 3000 series Processing:

Processing of the dry etch 3000 series will be covered here, processing of the photoimageable 4000 series will be covered in section 11.12.

Deposition: The Cyclotene™ 3000 series have traditionally been applied by spin coating in thicknesses of <1 to 20+ μm. Shrinkage on cure is typically <5%, thus the ultimate thickness needed for each coating is achievable in one spin coat operation by choice of the appropriate resin concentration. This can significantly affect high density substrate fabrication costs [79]. High efficiency coating techniques including spray coating, extrusion coating, and miniscus coating have been successfully evaluated on glass, aluminum and laminate substrates [74,80].

Curing: Extensive kinetic studies have revealed a curing activation energy, Ea, of 36 ± 0.9 Kcal/mol [81]. No volatiles are evolved upon cure thus shrinkage is low (< 5%). Curing is best monitored by FT-IR spectroscopy [82]. Curing must be carried out in an inert (nitrogen) atmosphere (< 100 ppm O_2 recommended). Curing can take place in a typical box oven, on a nitrogen purged hotplate or in a controlled atmosphere belt furnace. Time/temperature conversion curves (Fig. 11-15) which describe the curing process have been developed from the kinetics of the reaction. At vitrification (material solidification) the reaction slows down an order of magnitude but continues in the solid state. Processing to partial (ca. 70–80%) cure is recommended to achieve the best layer to layer adhesion [72,77].

Full cure can be obtained on a time scale of minutes to hours by controlling the curing temperature. In a typical convection oven "soft" curing is achieved in 30 minutes at 210°C and a full "hard" cure is obtained in ca. 1 hour at 250°C. Typical cure schedules with ramp up

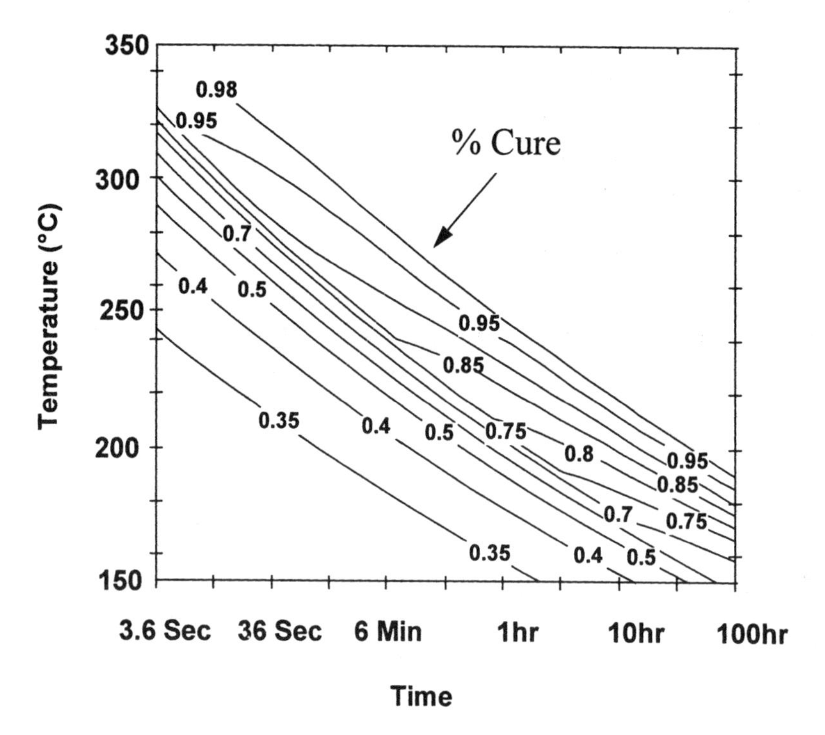

Figure 11-15. Time / Temperature Conversion Derived from Curing Kinetics

and cool down take ca. 4 hours to complete. Rapid thermal curing (RTC), Figure 11-16, in a belt furnace decreases typical cure times from 4–5 hours to <15 minutes and shows no deleterious effect on stress, adhesion or planarization [75]. RTC on the hot plate of a track coater (under nitrogen) has also been described [76].

Patterning: Vias can be formed in films of the BCB dielectric by any of the following techniques:

Photodefinition: Photosensitive BCB formulations are described in section 11.7.12

Dry Etching: Due to the siloxy content of Cyclotene plasma or RIE etching requires fluorinated plasmas such as O_2/CF_4 or O_2/SF_6 [83]. Masking can be accomplished with photoresist soft masks [86] or metal or SiO_2 [83] or Si_3N_4 [85] hard masks. For soft masked vias, via side wall angles are controlled by the degree of baking of the photoresist; for hard masked vias the side wall angles and degree of undercut are controlled by the plasma conditions.

Laser Ablation: The use of scanning eximer laser ablation through a hard metal mask has been successfully used to create vias in DVS-BCB [69,86] with throughputs comparable to RIE based processes. 5 micron vias with aspect ratios of 0.3 have been reportedly generated in BCB

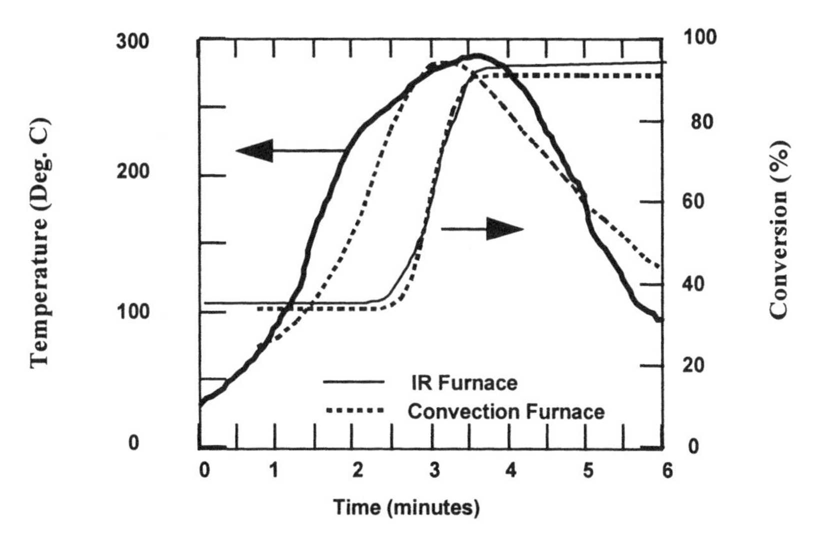

Figure 11-16. Rapid Thermal Curing in Infrared or Convection Belt Furnaces [75]

using this technique [72]. Figure 11-17 shows the ablation rate of Photo-BCB material as a function of laser fluence [87].

The residue (carbonaceous debris) which was left on the surface surrounding the ablated features can be removed using an oxygen/fluorine plasma and/or isopropanol immersion rinsing (with ultrasonication).

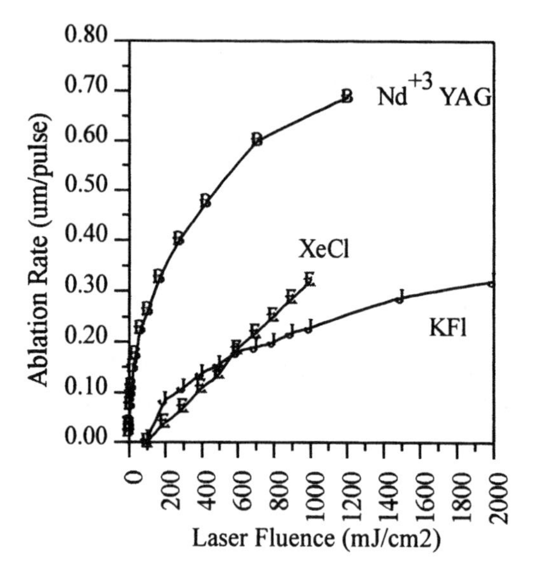

Figure 11-17. Ablation Rate of Photo-BCB vs Laser Fluence [87]

11.7.1.11 Incorporation of Passive Components

Thin film TiN and NiCr resistors and Al/SiON/TiN capacitors have been fabricated as part of thin film BCB structures [61,88,89].

11.7.1.12 Applications

Figure 11-18 shows a Commercial 486 Computer Module System fabricated by Saab/Combitech using Cyclotene 3022. It contains an Intel486™ microprocessor, 1 Mbyte flash PROM, 0.5 Mbyte SRAM and ASIC devices for memory management and I/O handling. It has been qualified by MIL H 38534 [89,90].

11.7.2 Polyquinolines

Polyquinolines (PQs) are linear-chain high-molecular-weight polymers characterized by moderate to high-class transition temperatures, excellent thermal and thermooxidative stability, low dielectric constants, and good planarizing characteristics. They were invented by Stille and

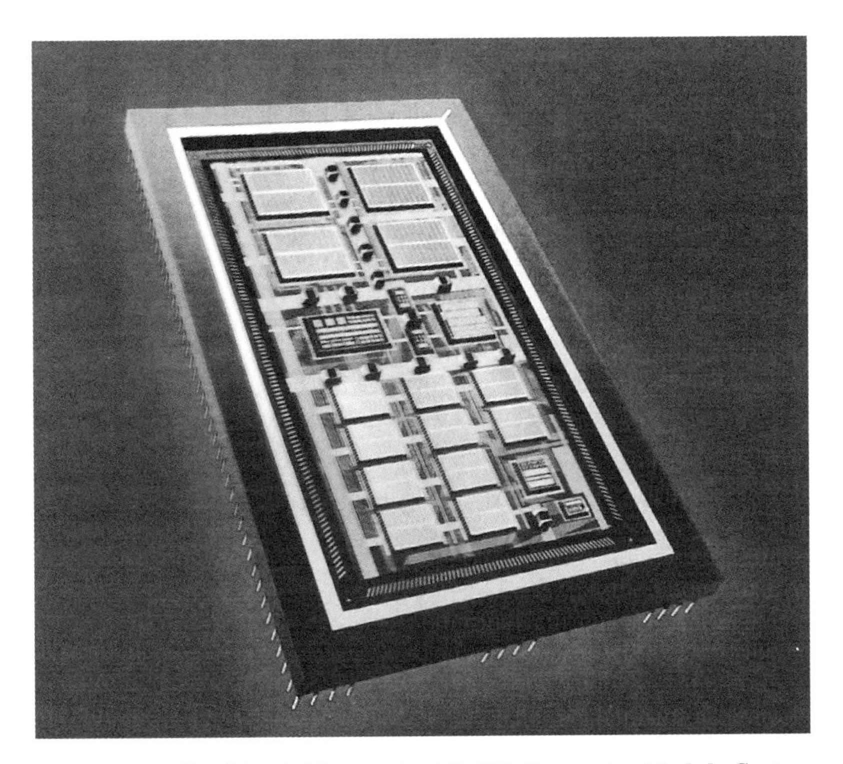

Figure 11-18. Combitech ElectronicsAB 486 Computer Module System

his coworkers [91] who showed that under certain acid-catalyzed conditions and use of model reactions lead them to use difunctional monomers to prepare high molecular weight polyquinolines. The generalized structure of polyquinoline is:

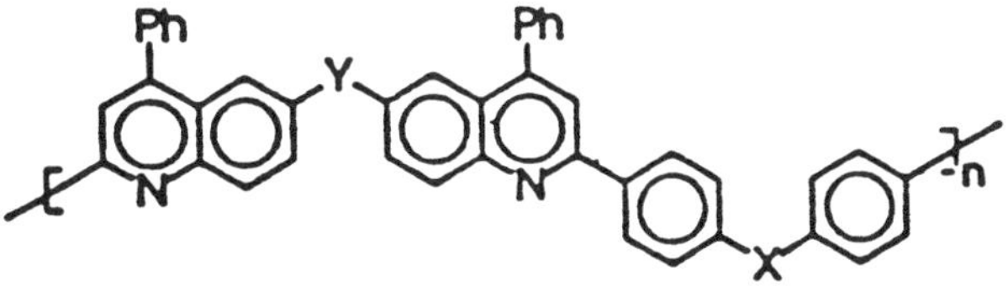

A major distinction in terms of processing thin films of polyimides compared to PQs are that PIs are usually fabricated through a partially cured prepolymer which is subsequently cyclized to the polyimide, soluble polyquinolines are fully cyclized polymers, and therefore require no post-cure.

Polyquinolines, having no carbonyl groups and do not suffer from significant moisture uptake. Furthermore, they are less polar and thus have lower dielectric constants (2.6–2.8) which do not increase appreciably with changes in humidity [92,93].

11.7.3 Olefin Dielectrics

BFG is currently developing a family of dielectric materials marketed as Avatrel™ Dielectric Polymers for use as interlayer dielectric thin films based on cyclic olefins [94]. The generic chemical structure of these polymers is shown below. This polymer family, based principally on polynorbornene, is produced via a transition metal catalyzed polymerization which enables BFG to control the polymerization of bulky, cyclic monomers to form saturated polymers with high T_g. This mechanism allows tailored molecular weights to be designed to specific applications.

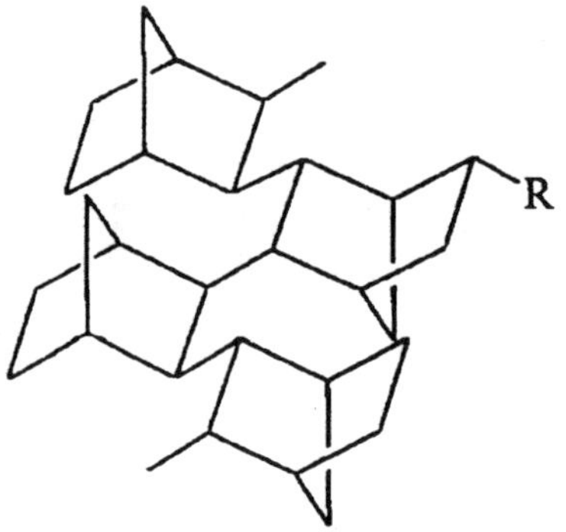

R

The family of polynorbornenes are amorphous and exhibit high T_g (>350 °C) and excellent thermal and electrical stability. These highly transparent polymers are readily soluble in a range of common solvents. Of particular interest are decalin and mesitylene. Polynorbornene polymers have a high glass transition temperature due to the very stiff polymer backbone. The group "R" represents specific functional groups which provide adhesion and sites for crosslinking.

Initial studies of polynorbornene polymers reveal (i) excellent thermal and electrical performance, (ii) adhesion to conductors and substrates without the use of adhesion promoters, (iii) very low moisture content, (iv) low dielectric constant, and (v) simple solution processing. The hydrocarbon nature of polynorbornene yields dielectric thin films with very low moisture absorption (0.1 %) and low dielectric constants (2.6 at 10 kHz). These polymers appear to be quite isotropic from birefringence measurements using a prism coupler technique. Films which are nominally 2 μm thick after being spun onto oxidized silicon wafers exhibit birefringence on the order of 0.0031 ± 0.0001, more than 20 times lower than that reported for a typical polyimide (PMDA-ODA). The refractive index

Table 11-21. Summary of Polynorbornene Properties

Property	Value
Dielectric Constant/Loss	2.6/0.0007
Breakdown Voltage	>29 MV/cm
Moisture Absorption	0.1 wt%
Elongation	10–20%
Tensile Strength	50–60 MPa
Coefficient of Thermal Expansion	60×10^{-6}/°C
Modulus	1.4 GPa
Planarization	70% (2 mil pitch, 3.5 μm height, 6 μm film thickness)
Decomposition Temperature	435°C (5 wt% loss, 10°C/min, inert gas atmosphere)

was measured to be 1.505, and free standing films are clear and colorless. Films are chemically resistant and have revealed no cracking or crazing when exposed to common processing solvents such as: photoresist, aqueous base, PAN etch (phosphoric, acetic, and nitric acid mixture), acetone, and isopropanol. Patterning has been demonstrated using RIE techniques (O_2/CHF_3/Ar plasma) with both soft (photoresist) and hard (silicon dioxide) masks.

Polymer solutions of the polynorbornene are applied by spin coating onto substrates, dried and typically cured above 200°C. These films pass the adhesion Scotch™ tape test on bare copper, gold, silver, aluminum, silicon (and oxides) without any adhesion promoter and in the absence of any metal barrier/adhesion layer (such as chromium, titanium or tantalum). These films remain adherent even after being placed in boiling water for 2 hours. Likewise, metal sputtered on top of the polymer survives these adhesion tests, as do additional layers of the polymer. These films also survive thermal shock tests, 100 cycles from liquid nitrogen to boiling water. Since the polynorbornene polymers are fully polymerized before processing, the base polymer properties are not significantly dependent on the processing conditions and temperatures. Light cross linking is used to enable multi-layer processing. Table 11-21 summarizes the film properties of polynorbornene polymers.

11.8 PHOTOSENSITIVE POLYIMIDES

11.8.1 Introduction

The application of polyimides as a protection layer or an insulation layer has become an important technology in the field of microelectronics packaging because of their excellent thermal, mechanical, and electrical properties [95,96]. Historically, Japanese workers at Hitachi were the first to implement wet-etching techniques capable of patterning fully cured polyimides [97,98,99]. The current trend, as reported in the literature, is to use photosensitive polyimides because of their ability to directly produce patterns in them using standard photolithography techniques. The advantage of using PSPI is the reduction in the number of process steps required to form a patterned polyimide layer for making interconnections as shown in Figure 11-19 and to better control via wall profiles. This development of photosensitive polyimide has contributed to the progress in fabricating high-density devices and packages in the microelectronics field [100-102]. The PSPIs utilized in commercial applications are mostly negative working

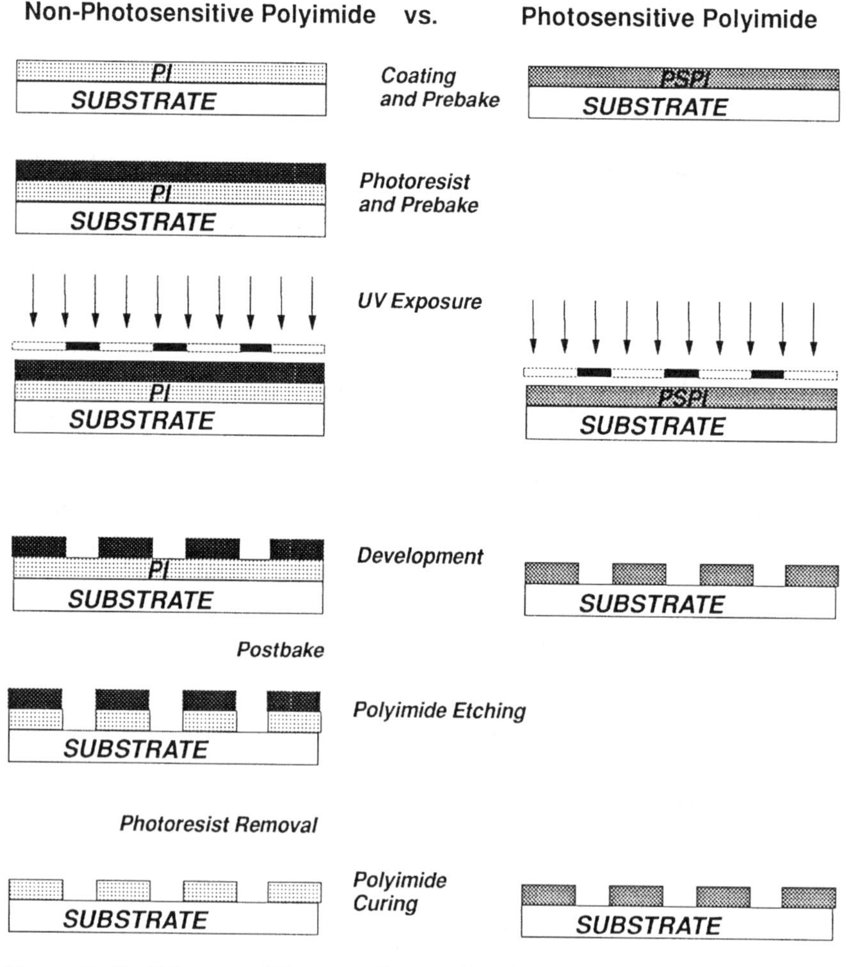

Figure 11-19. Schematic Diagram Comparing Patterning of Non-Photosensitive PI to PSPI

precursor types. Figure 11-20 depicts the current main stream approaches that are being considered for development of photosensitive polyimides. They are classified into three types, covalently bonded ester-type PSPIs, ionic bonded-type PSPI, and fully preimidized soluble polyimides.

Most PSPIs have been made today by using the polyamic acid, which has an ionic bonded amine-linked cross-linking group, or a polyamic ester which has a covalently linked ester cross-linking group to produce a photosensitive precursor as schematically shown in Figure 11-21. The photosensitive polymer precursor is then formulated with a photopackage consisting of initiators, sensitizers, additives, and solvent to produce a PSPI solution. The chemistry, formulation, processing, lithography, physical properties, and applications of the ester-type PSPIs, ionic bonded-type PSPIs and preimidized polymers are described in this section.

11.8.2 Historical Background

The first PSPI was reported by Kerwin et al. of Bell Laboratories in 1971 [103]. It was a mixture of polyamic acid and potassium dichromate, which showed negative working photolithographic properties, but had many practical difficulties, such as very poor shelf life and high metallic contamination.

In 1976 Rubner et al. of Siemens invented the first practical PSPI, to be used for microelectronics applications. This invented technology was based on using a precursor polymer backbone, in which cross-linkable functional groups R* are bonded to carboxylic groups through ester linkages, as shown in Figure 11-22 [104,105]. In this sense, this system is called an "ester-type PSPI." Their efforts were focused on the formulation

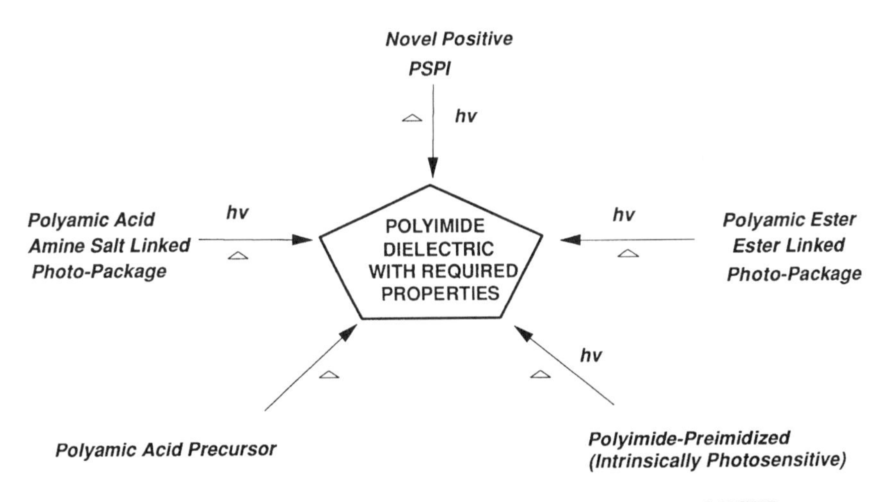

Figure 11-20. Technical Approaches to Development of PSPIs

Photosensitive PI'S

o **PSPI Obtained by Adding Photopackage**

Photopackage----> (X-linker , initiator, sensitizer, additives)

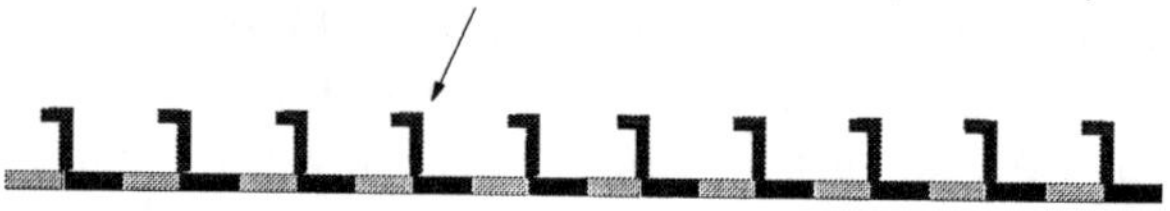

Diamine

Dianhydride

o **Upon Curing Photoactive Group is Released and Volatilized out of PSPI Film**

o **Desire Exposure Wavelength where PI Backbone does not absorb Energy**

- **Since most PI's absorb at i-line / desire g-line exposure for improved sensitivity**

Figure 11-21. Schematic Diagram Showing PSPI Formulation

of a practical composition and a process for polyimide pattern formation, by using methacrylate groups to impart differential solubility in exposed versus unexposed regions of the spun film. Their precursor polymer was based on a pyromellitic dianhydride (PMDA) and oxydianiline (ODA) backbone, which is the same as that of Kapton®, and is illustrated in Figure 11-24. The photosensitivity was improved by adding photoinitiators and co-monomers such as maleimide to the precursor polymer solution [106]. Siemens licenced this fundamental technology to Asahi Chemical in Japan, DuPont in the United States, and Ciba Geigy and Merck in Europe [107].

Each of these chemical companies set out to further develop and improve the licenced technology, particularly, by enhancing the photosensitivity and increasing storage stability of the solution and several other polyimide characteristics in order to produce more advanced PSPIs for commercial use in manufacturing applications [102]. Merrem et al. of Merck focused their efforts on improvement of the photospeed. They used

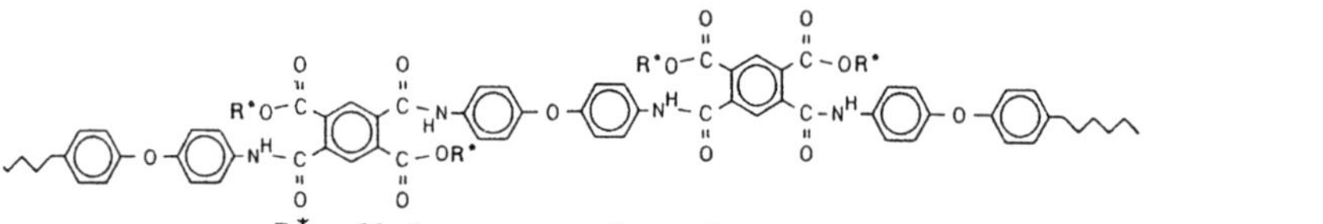

R*: allyl group, methacrylate group, N-methylmaleimide group

Figure 11-22. Schematic Showing Prototype of Rubner's Ester Type PSPI

pentaerythritriallylether as a pendant group [108]. Pottiger et al. of DuPont examined the precursor based on a benzophenone tetracarboxylic dianhydride (BTDA), ODA, and meta-phenylene diamine (m-PDA) backbone. This structure was chosen because of its better solubility than Siemens' original PMDA/ODA polymer and offered a number of advantages over the Siemens' prototype, including better shelf life, higher photosensitivity, and better adhesion characteristics [99,109]. Rohde et al. of Ciba Geigy improved the shortcomings of the prototype such as lower photospeed, excessive volume contraction, and shelf life limitations. By use of novel high-quantum-yield photosensitizer, tailored to the mercury g-line, it was possible to lower exposure times remarkably and to reduce the amount of layer shrinkage during final cure [110]. They reported the achievement of high-resolution patterns even in 70-μm-thick layers after development [111]. Asahi Chemical also focused its endeavors on the improvement of Siemens' products. Main targets were improvements of the purity, photosensitivity, and stability of the polymer solution during storage [112,113].

These licenses have commercialized their improved products, the trade names of which are PIMEL (Asahi Chemical), Probimide (Ciba Geigy), Pyralin PD (DuPont), and Selectilux HTR (Merck), respectively. Toray developed a unique negative working photosensitive polyimide precursor, which is composed of a polyamic acid and tertiary amine having an acryloyl group. This photosensitive PI was given a trade name of "Photoneece."

11.8.3 Thin-Film Processing of PSPIs

11.8.3.1 Polyimide Pattern Formation Process using Ester-type PSPI

An ester-type PSPI is normally supplied in liquid form composed of a precursor polymer and a photopackage including photoinitiators, photosensitizers, adhesion promoters, and stabilizers, as schematically shown in Figure 11-21. Illustrated in Figure 11-23 is the process for polyimide pattern formation using an ester-type PSPI and the corresponding chemical structure in each step. The process steps used for depositing polymer solutions, exposing, developing, and curing are briefly described below:

Coating: Normally the PSPI solution is coated by spin coating onto a substrate such as silicon wafers, glass, or ceramic substrates, followed by partial drying to form a polyimide precursor layer.

Exposure: Then, UV light is exposed to the layer through a photomask. Photochemically generated radical species from the photoinitiator starts polymerizing double bonds of the pendant groups to form a cross-linked structure.

Development: The relief pattern is then obtained by developing the

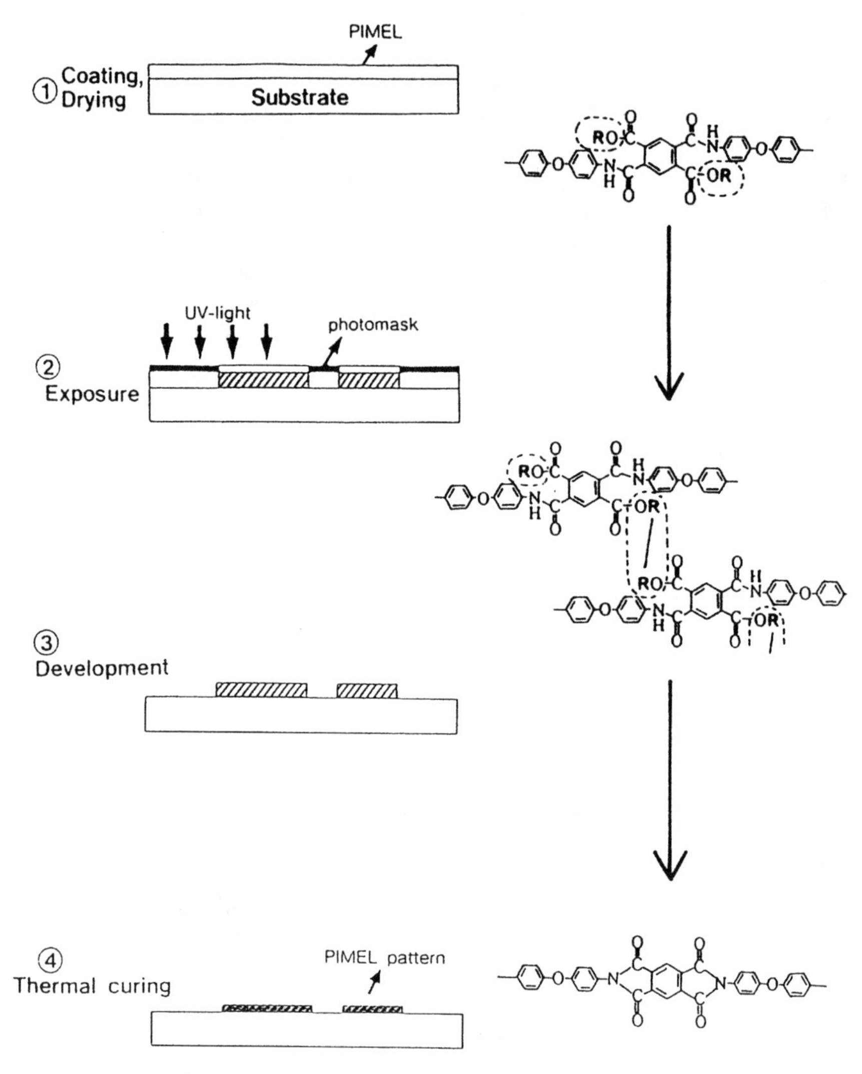

Figure 11-23. The Process for Forming Polyimide Patterns Using PSPI

exposed image with an organic solvent which can dissolve preferentially the unexposed area, leaving the exposed area intact. This process is similar to the one used to pattern negative working photoresist.

Cure: The relief pattern obtained by the development process is heated up to 300–400°C in the cure process, in order to imidize and eliminate the pendant groups from the patterned layer. The solvent is vaporized and photopackage components as well as decomposition by-products are removed from the film.

11.8.3.2 Comparison with Non-PSPI

In the case of the process using non-PSPI [96,97,98], a conventional photoresist process is applied. At first, the non-PSPI solution, which is polyamic acid or ester, is coated on a substrate and then fully or partially cured. Then a photoresist layer is patterned on the polyimide layer to form the etching mask. The polyimide layer is etched with alkaline solution and the overcoated photoresist is removed with chemical stripper and/or dry process, postcure is followed, if necessary. This process with a non-PSPI is considered to have some drawbacks. First, the total process is quite long and, second, the etching step has some problems. Harmful etchants such as hydrazine and ethylene diamine were used in the early days of this process but less toxic (Et4 $N^{+}OH^{-}$ in water) solvents are used. In addition the shape of the designed pattern cannot be reproduced as a polyimide pattern because of sidewall etching due to the isotropicity of wet chemical etching. A partial curing process is often used to avoid usage of hazardous etchants; however, the process margin is narrow because the etching rate depends on the curing condition and on the degree of imidization. Contrary to the non-PSPI process, the process using PSPI is simpler and a fine polyimide pattern with good resolution can be obtained because the PSPI is functionally photolithographic.

Figure 11-24 shows the comparison of the chemical structures of non-PSPI and PSPI, including ionic bonded-type PSPI. Non-PSPI is a

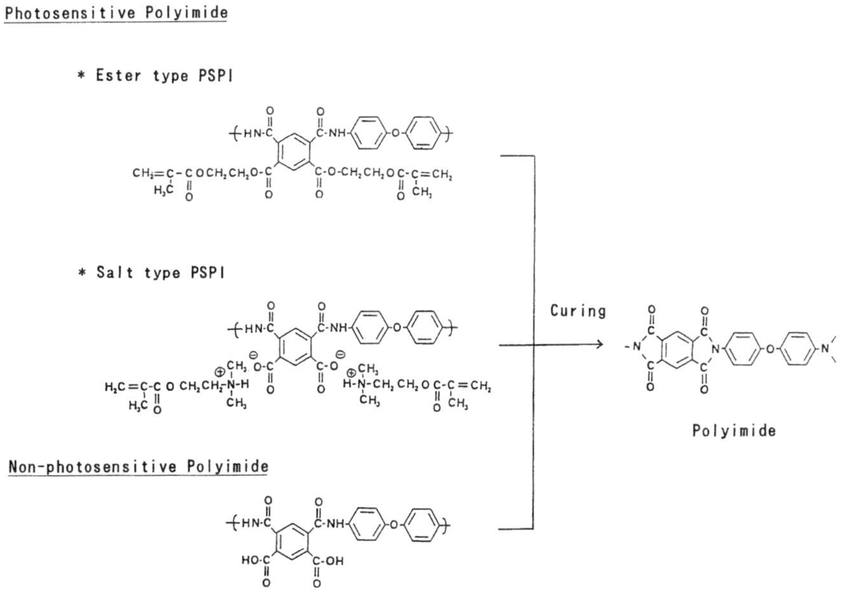

Photosensitive Polyimide

* Ester type PSPI

* Salt type PSPI

Curing

Polyimide

Non-photosensitive Polyimide

Figure 11-24. The Comparison of the Chemical Structure Between PSPI and that of Non-PSPI

polyamic acid itself, which can be thermally converted to a polyimide structure through elimination of water molecules. As mentioned above, there is another type of PSPI called the ionic bonded-type, the structure of which is a mixture of a polyamic acid and a vinyl monomer with an alkyl amino group. It is considered that a photochemically initiated vinyl polymerization occurs to form a poly ion complex comprised of polyamic acid and polyamine in the exposure step. Pattern formation can be achieved by the use of the solubility difference between the exposed area and the unexposed area, because the poly ion complex shows rather poor solubility toward some organic solvents. The reaction in the curing step is an imidization and a decomposition of the polymer of the amine monomer. The ionic bonded-type PSPI may have several practical processing limitations arising from the ionic properties which will be discussed.

11.8.4 Covalently Bonded Ester-Type PSPI

11.8.4.1 Chemistry and Photopackage: Preparation of Precursor Polymers

Rubner prepared originally the precursor polymer using the "acid chloride method" [106]. In this process, diacid is converted to acid dichloride, which is then reacted with diamine to give an ester-type polyimide precursor, as shown in Figure 11-25. Hydrogen chloride and sulfur oxide are involved during the reaction, so the film contains about 10 ppm of the chloride ion, which may subsequently cause corrosion of metallic material in the IC device. Improvements using a "chlorine-free condensation reagent method" have solved this problem.

Siemens and Asahi Chemical independently conducted research on a novel preparation method of ester-type PSPI precursor using carbodiimide as a non-chlorine condensation reagent [101,114]. By the addition of the photoactive alcohol to the aromatic dianhydride, a tetracarboxylic acid diester is produced, which is then polycondensed directly with the diamine in the presence of carbodiimide to yield an ester-type PSPI precursor. Asahi has developed other chlorine-free condensation reagents, of which the phosphine–disulfide system showed the best results [115].

DuPont has reported a unique condensation reaction, in which a part of the aromatic dianhydride is capped with 2-hydroxymethacrylate (HEMA), and aromatic diamine is then added to the reaction mixture to form a capped polyamic acid. It is converted to the corresponding polyiminolactone intermediately by the reaction with trifluoro-acetic anhydride (TFAA), and a treatment of the polyiminolactone with excess HEMA produces the esterified polyimide precursor [109]. This preparation method is referred to as the "iminolacton method" in Figure 11-25.

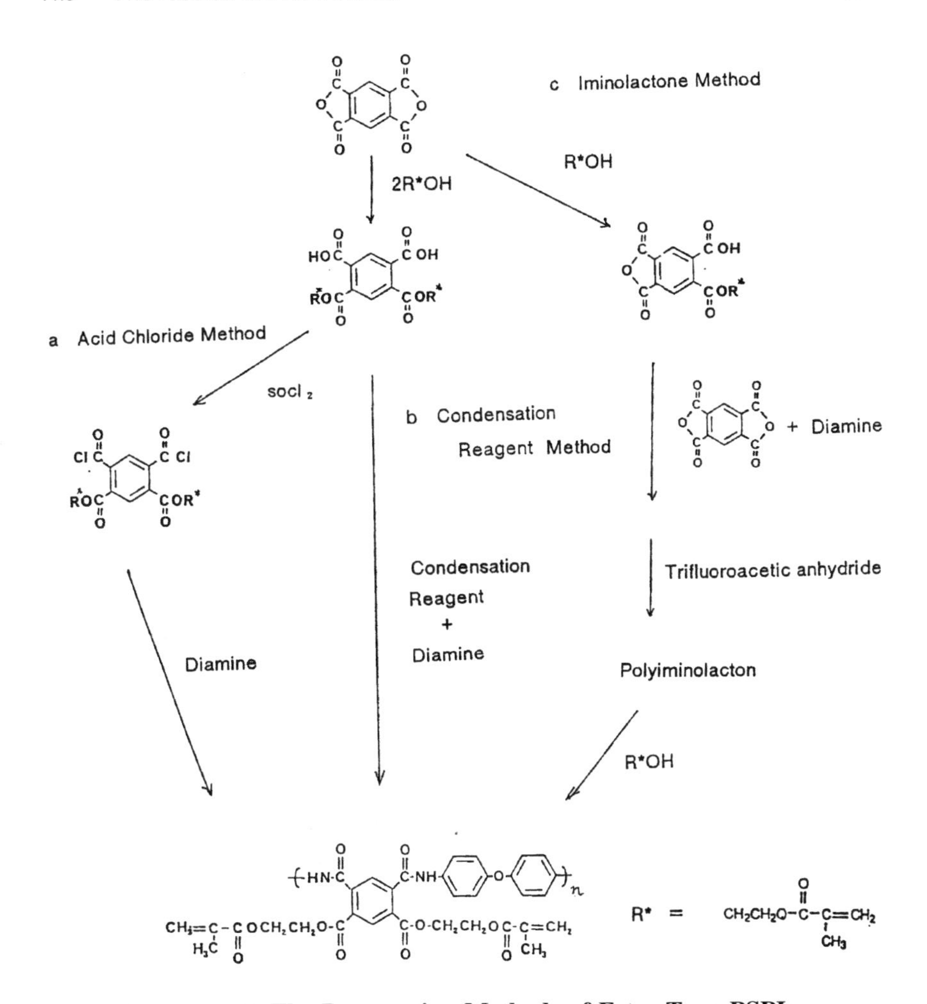

Figure 11-25. The Preparation Methods of Ester Type PSPIs

11.8.4.2 Alternate Approaches to the Preparation of Ester-Type PSPIs

There are some ester-type PSPIs other than Rubner's structure. Minnema of Philips developed a PSPI precursor by esterification of the polyamic acid with isourea, which is the adduct of *N*-hydroxyethyl-*N*-methylmethacrylic amide and *N,N'*-diisopropylcarbodiimide [116]. Rubner and Ahne prepared other ester-type PSPIs by the addition of glycidylmethacrylate to the polyamic acid [117]. Davis of General Electric studied a photosensitive polyimide siloxane. He prepared the precursor polymer by the addition of isocyanatoethyl methacrylate to the polyamic acid containing silicone diamine [118].

11.8.4.3 Processing

11.8.4.3.1 Solution Preparation and Properties

PIMEL has three characteristics which are essential for polyimide coatings used for microelectronic applications: (1) polymer solution purity and stability, (2) photolithographic characteristics capable of producing high photosensitivity and high resolution, and (3) excellent thermal, mechanical, and electrical properties of the polyimide film obtained after full cure. Polymer solutions are prepared by dissolving a precursor polymer in a polar solvent such as *N*-methylpyrollidone (NMP), together with additives (e.g., photopolymerization initiators, sensitizers, adhesion promoters, and stabilizers) and filtering through filters having nominal 1-μm pores. The polymer solution is formulated to different concentrations resulting in viscosities ranging from 17 P, as shown in Table 11-22. These formulations are designed for spin coating on manufacturing tools to produce a range of film thicknesses from 5 to 10 μm when fully cured.

Since ester-type PSPI precursors have very good solubility in polar solvents, it is practical to formulate more concentrated solutions which can be used to spin coat in one-application films thicker than 50 μm. Because the precursor is chemically stable, the viscosity and photosensitivity barely change even after 1 month storage at room temperature, as shown in Figure 11-26, where the photosensitivity is defined as the remaining thickness after development. This excellent shelf-life stability attribute is one of the reasons why ester-type PSPIs have been widely employed in large-scale integration manufacturing production lines.

In contrast, non-PSPIs and ionic bonded-type PSPIs are known to show a decrease in viscosity during storage, which is considered to arise from the autocatalytic decomposition of polyamic acid and often resulting in lowering of the molecular weight of the starting polymer precursor.

Table 11-22. Process Conditions and Liquid Properties of PIMEL™

	Properties	Unit	G -7621A	G-7613N	TL-530
Solution	Viscosity	P	40	75	17
Development	Method		Spray	Spray	Puddle
	Resolution (aspect ratio)		1.5	1.5	2
Curing	Temperature	°C	350	350	450
Film thickness	After prebake	μm	10	20	10
	After cure	μm	5	10	5
Typical values of ionic impurities		ppm	Cl^-<0.5, Na<0.10, K<0.10,Cu<0.10 Fe<0.10		
		ppb	U<0.03		

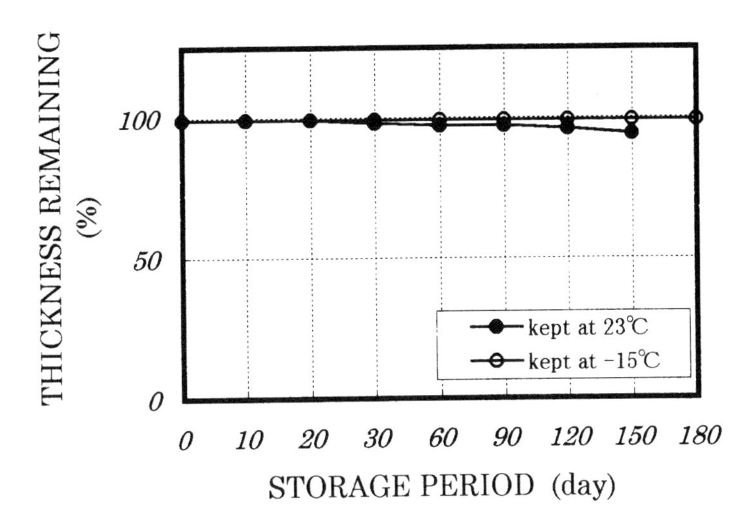

Figure 11-26. Shelf Life of PIMEL

The purity of metal and chloride ions of PIMEL are summarized in Table 11-21 which shows that PIMEL has a sufficient purity for microelectronics applications.

11.8.4.3.2 Photolithographic process

PIMEL is processed in the same manner as photoresists by spin coating on a silicon wafer, ceramic substrate or glass substrate, then the substrate is dried on a hot plate or in a clean oven. Figure 11-27 shows the relation between the rotation speed and the film thickness after prebaking. One advantage of the polyimide coatings over inorganic thin film deposition is a step coverage or planalization capability, which is needed to planarize the rough surface topography of actual devices. Degree of planarization (DOP) is defined by Eq. [11-1].

$$DOP = (1 - D_i D_a) \times 100 \qquad [11\text{-}1]$$

D_i and D_a are expressed in Figure 11-28. Ahne reported that the ester-type PSPI showed better DOP than non-PSPI [107].

PIMEL G-7600 series are designed to yield high photosensitivity to the wide range of wavelengths of light irradiation from the mercury g-line to the i-line, as is shown in the spectral photosensitivity curve of Figure 11-29. Thus, PIMEL enables lithography engineers to achieve high throughput with proximity aligners, g-line steppers, and mirror projection aligners.

The exposure step is composed of two kinds of reactions, which are

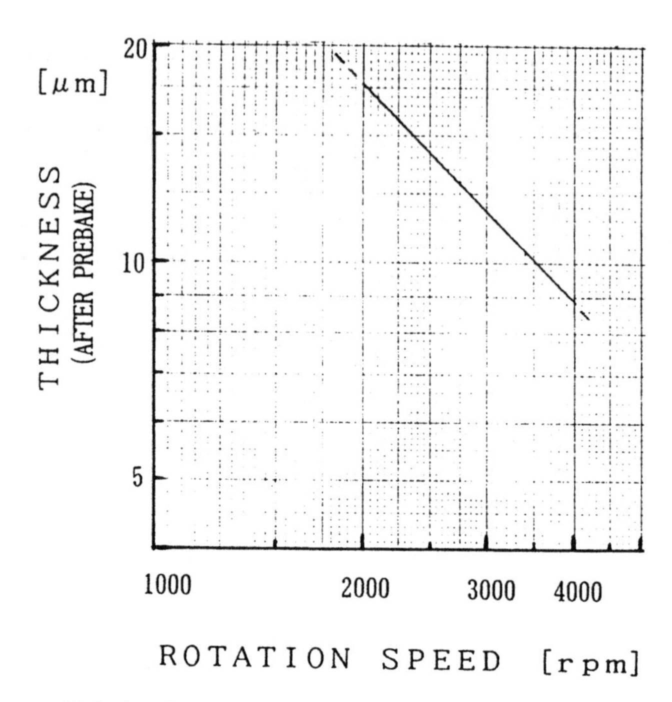

Figure 11-27. Relation Between the Rotation Speed and the Film Thickness after Prebaking

the generation of free radicals by decomposition of photoinitiator and a radical polymerization of vinyl groups. The former reaction is usually accelerated in the presence of a photosensitizer, and the total reaction rate is affected by both oxygen and temperature. Loisel and Abadie studied the photochemical reaction of ester-type PSPIs, DuPont's Pyralin 2732 and 2701, using differential photocalorimetry. The enthalpy of the polymerization, the induction time, the photospeed, and the activation energy were measured under various temperatures and atmospheres [119].

The functional principle for a negative working photoresist to produce patterns is based on the difference in the solubility between unexposed

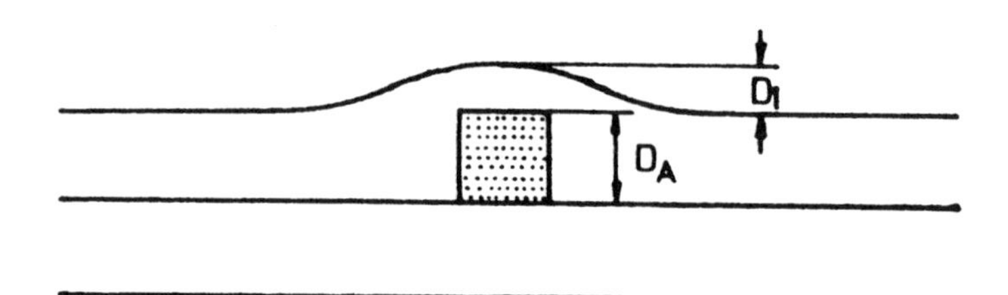

Figure 11-28. Parameters for Degree of Planarization (DOP)

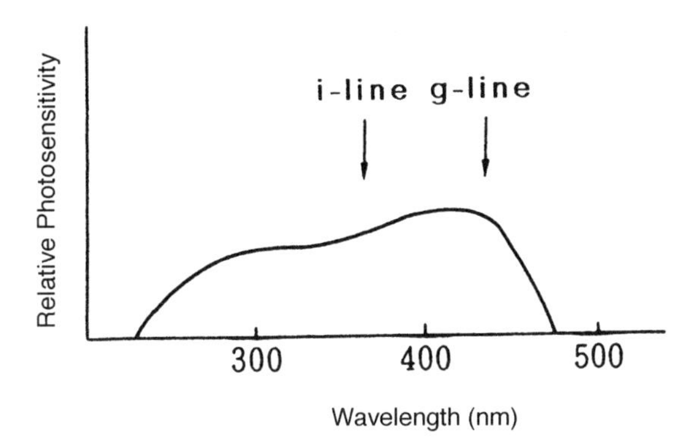

Figure 11-29. The Spectral Photosensitivity Curve of PIMEL G-series

and exposed areas. An ester-type PSPI has good solubility to specific organic solvents, it exhibits high development speed, whereas the cross-linked structure of the exposed portion has good solvent resistance to the developer. Figure 11-30 shows the changes in thickness of exposed and unexposed areas during the development. The wide margin on develop-

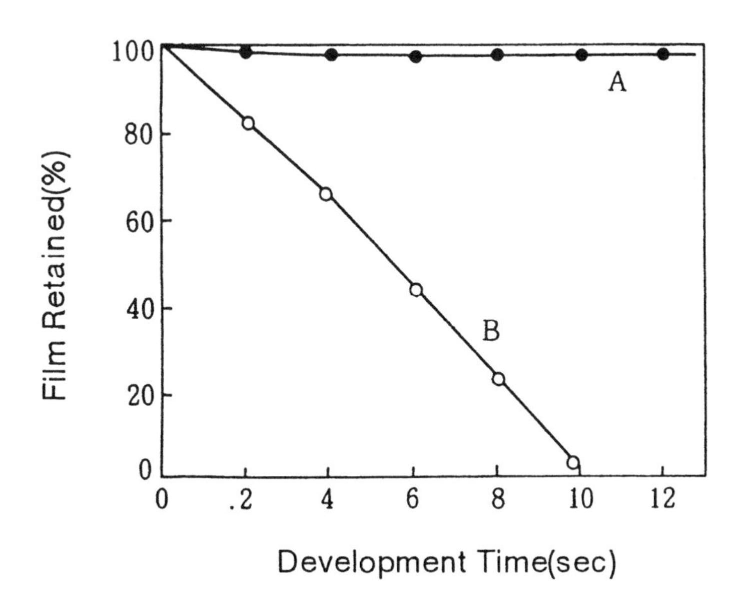

Figure 11-30. Development Characteristics of PIMEL

ment is achieved and a fine-resolution pattern can be obtained due to this large difference in solubility.

Figure 11-30 indicates another feature of ester-type PSPI; that is, the small amount of film thickness loss in the exposed portion during development, which facilitates precise film-thickness control for manufacturing. Figure 11-31 is a typical scanning electron microscopic (SEM) photograph of a highly resolved PSPI pattern. Figure 11-31 is a typical scanning electron microscopic (SEM) photograph of a highly resolved PSPI pattern. The resolved photoimages of PSPIs have an aspect ratio of 1 normally, but, it has been improved up to ~ 2 by modification of the development process.

11.8.4.3.3 Pimel® Curing

The PSPI relief pattern, produced after ultra violet (UV) exposure and development, is converted to the corresponding polyimide upon curing to elevated temperatures, where two kinds of chemical reactions occur. First, a *o*-(alkoxycarbonyl)-substituted benzamide structure is thermally cyclized to form the imide ring, accompanying the elimination of the HEMA monomer. The second reaction is the decomposition and vaporization of cross-linked portion. The imidization reaction of a precursor polymer has been investigated by thermal scanning Fourier transform–infrared (FT-IR) spectrometry. Figure 11-32 shows the plots of the degree of imidization versus temperature of curing, measured at the rate of 5 and 15°C/min. This plot indicates that the imidization reaction occurs between

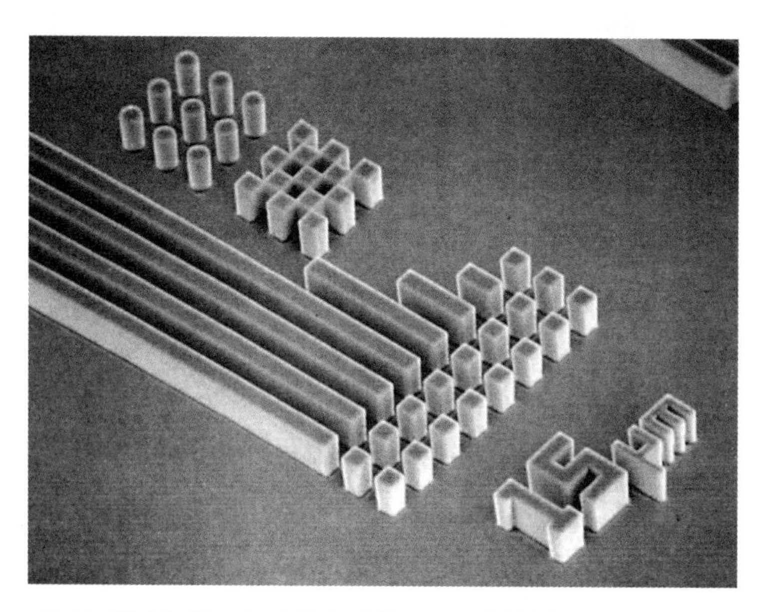

Figure 11-31. Highly Resolved Releaf Pattern of PIMEL after Development

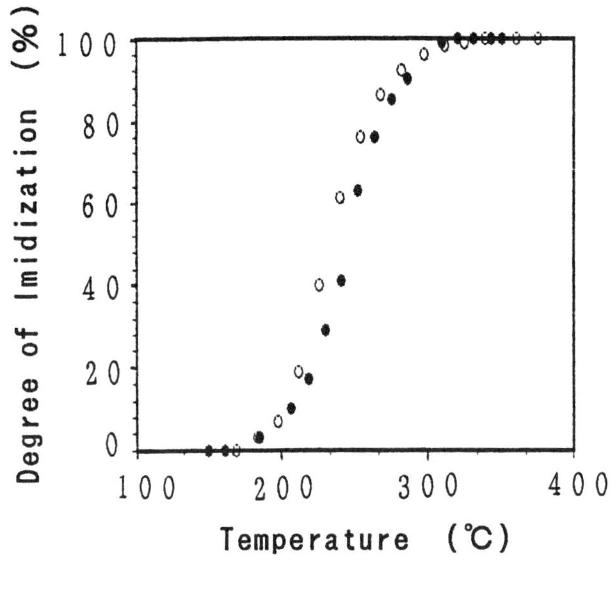

Figure 11-32. Plots of the Degree of Imidization Versus Temperature of Curing

about 200°C and 300°C. This temperature is a bit higher than that of non-PSPI or ionic bonded-type PSPI. [120]

11.8.4.3.4 Pimel Properties

Polyimides are being widely used in microelectronics applications due to their excellent properties as previously discussed. Therefore, PSPIs need to have the same characteristics as those of the corresponding polyimides after they are fully cured. Table 11-23 shows the thermal, mechanical, adhesion, and moisture uptake of cured PIMEL film. Figure 11-33 is a typical thermogravimetric curve of PIMEL G-7600 series film cured at 350°C. The starting onset temperature of the decomposition is nearly at 400°C and the 5% decomposition temperature is at about 550°C, which corroborates the excellent thermal stability of the polyimide layer as being suitable for semiconductor applications. Electrical properties are as good as those of non-PSPIs. PIMEL has a dielectric constant of 3.3 which is suitable as an interlayer dielectric for interconnect structures on multi-chip modules.

Mechanical properties are also very important in these applications. Lead-on-chip (LOC) technology was widely applied in packaging large integrated memory devices such as 16 Mbits DRAMs, because it can

Table 11-23. Properties of Cured PIMEL Film

		Polyimide Types	
		G-7600 Series	TL-530A
Properties		High Elongation Typical dielectric constant	Low TCE Low dielectric constant
Thermal properties			
5% Weight-loss temp.	(°C)	>500	>600
Coefficient of thermal expansion	(ppm/°C)	40–50	10–20
Internal stress	(MPa)	40–50	10–20
Mechanical properties			
Tensile strength	(MPa)	>150	>200
Elongation	(%)	>30	>15
Young's modulus	(GPa)	3.3	6.0
Dielectric constant	(1 kHz)	3.3	3.0
Tan δ	(1 kHz)	0.003	0.002
Adhesion strength after PCT on Si (133°C at 3 atm)			
Tape peel test (h)		>400	>400
Pull strength (MPa)		>70	>70
Moisture Uptake (%) (23°C, 50% **rh** (%), 24 h)		0.8	0.8

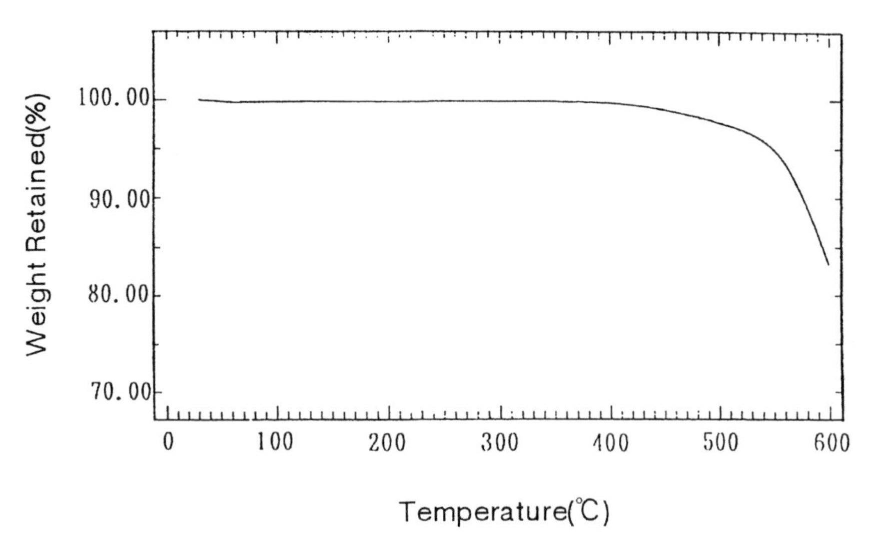

Figure 11-33. Thermogravimetric Curve of PIMEL G-7000 Film Cured at 350°C (heating rate 10°C/min)

dramatically contribute to the reduction of the chip size. However, the buffer-coating layer is exposed to severe mechanical stress during the leadframe mounting process of LOC. It is believed that the mechanical properties of the buffer coat, especially, the elongation at break is an important parameter for LOC packages.

Figure 11-34 is a typical stress-strain (S/S) curve of the cured film of PIMEL G-7000 series, which has been proven to meet the mechanical characteristics for LOC application.

11.8.4.3.5 Functional PSPIs

Polyimides have been developed from a large variety of backbone chemistries to meet the very diversified and demanding recent developments in microelectronics and other fields. PSPIs also have been developed with functional properties other than photosensitivity. Some examples are shown in this chapter.

11.8.4.3.6 Low-Thermal-Expansion PSPIs

The internal stress is the most important factor to consider in multilayer substrates using a Cu wiring layer and a polyimide insulation layer for interconnection. The internal stress is generated in the polyimide layer during cooling after cure. If the internal stress is too high, it will lead to

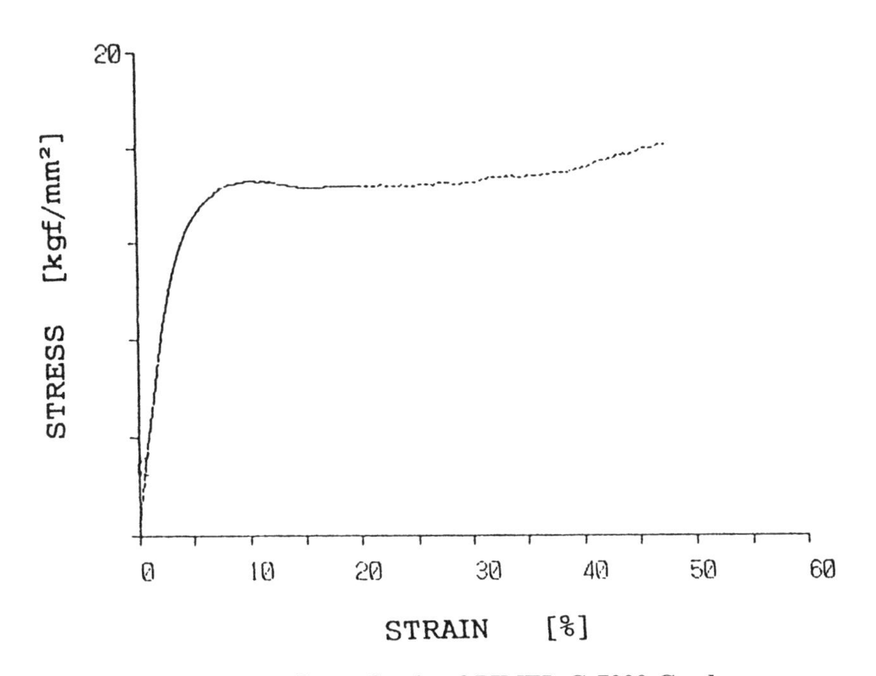

Figure 11-34. Stress-Strain of PIMEL G-7000 Grade

bending and/or cracking of the substrate, and to delamination of the polyimide layer from the lower layer. The internal stress, σ, is sometimes expressed as shown in Equation 11-2:

$$\sigma(t) = K \int_t^{T_g} E(t)[\alpha_1 (t) - \alpha_2 (t)] \, dt. \qquad (11\text{-}2)$$

where

K = constant
T_g = glass transition temperature
E = Young's modulus of film
α_1,α_1 = thermal coefficient of expansion (TCE) of polyimide and substrate, respectively.

It is concluded from Equation (11-2) that one solution to reduce the internal stress is to apply a polyimide with a low TCE, because substrates are generally inorganic materials such as silicon wafers and/or ceramics, which have rather low TCEs. The basic concept for the design of low-thermal-expansion polyimides was proposed by Numata and Kinjo of Hitachi [121]. They concluded that the introduction of rigid components, namely rodlike structure, results in lowering the TCE. Furthermore, the TCE of a polyimide film is influenced by the curing process and conditions. Numata reported that the TCE of the polyimide film cured fixed in a frame was lower than that of the polyimide film without tension, and concluded that it was due to the difference in the degree of orientation in the polyimide molecules [122].

The effect of the molecular weight of PSPI polymers on TCEs have been investigated. Condensation products of pyromellitic acid–1,4-di(2-methacryloyloxyethyl)ester and a mixture of meta/para-phenylenediamine (PDA) [TP], a mixture of ODA and para-PDA [TA], and a mixture of para-PDA and 3,3′-dimethylbenzidine [TT] were prepared.

Figure 11-35 shows the value of tensile strength (Ts) and modulus (E) of polyimide films as a function of a viscosity number (v.n.) of a polyimide precursor, which corresponds to the molecular weight. It was found that the tensile strength is almost independent of the molecular weight of precursors and that the Young's modulus increases with an increase in molecular weight. The TCE of the polyimide is plotted against the viscosity number of the precursor in Figure 11-36 for various thickness of polyimide film. The TCE becomes lower as the molecular weight of the precursor becomes higher and the polyimide film thickness becomes thinner. Figure 11-37 shows the relation between the TCE value and Young's modulus in the case of TP as a polyimide precursor and a film

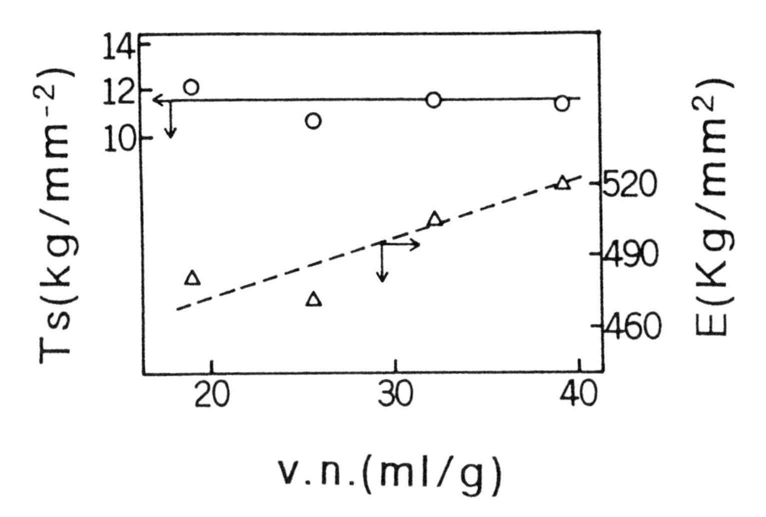

Figure 11-35. Tensile Strength and Young's Modulus of Polyimide Films as a Function of the Molecular Weight of the Polyimide Precursor

thickness of ~ 15 μm. The plots imply that the polyimide film with a low TCE has a higher Young's modulus [120].

A large number of PSPIs with low stress have been developed by many researchers as reported by Numata [121–122]. Asahi Chemical has also developed the TL-500 series, the characteristics of which are summarized in Table 11-22. The TCE is 10–20 ppm/°C, that is about a third of the value of conventional polyimide. Figure 11-38 shows the

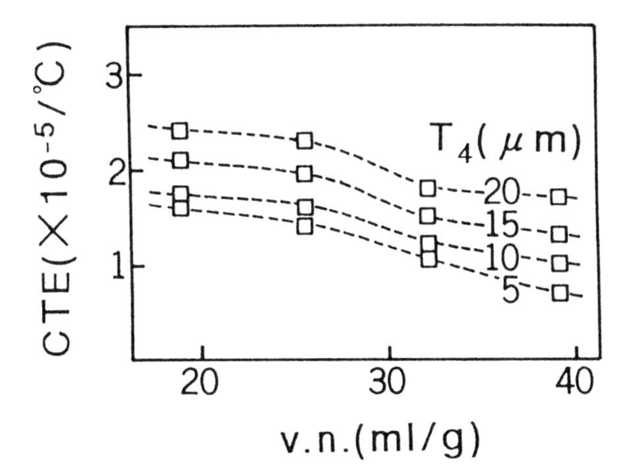

Figure 11-36. Thermal Coefficient of Expansion of Polyimide Films as a Function of the Molecular Weight of the Polymide Precursor in Various Film Thicknesses

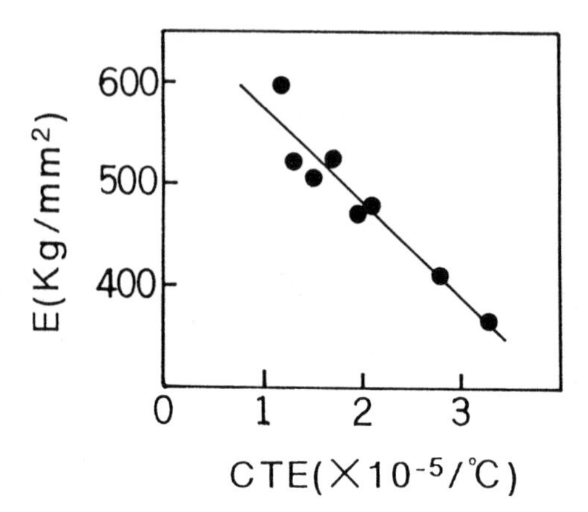

Figure 11-37. Relationship Between Thermal Coefficient of Expansion and Young's Modulus of the Polyimide Films in the Various Sample Preparation Conditions

relation between the internal stress in a cured polyimide layer and film thickness. Curve b shows the internal stress in the five-layered polyimide film with the PIMEL TL-series on a silicon wafer. The stress is dependent on the film thickness; however, the internal stress of a multicoated layer is lower than that of a one-time-coated layer, even if the total thickness

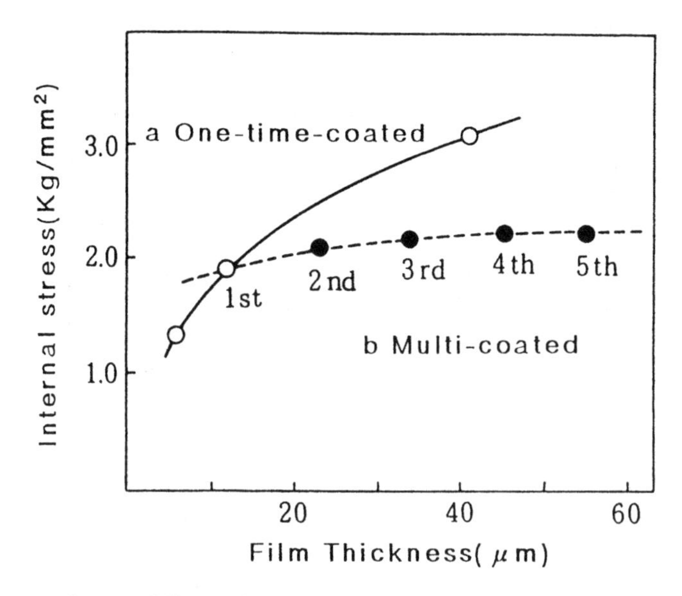

Figure 11-38. Internal Stress in Polyimide Film with PIMEL TL-Series on a Silicon Wafer

is the same, as shown by curve a. These results can be explained by the difference in the degree of the orientation of polyimide molecules found on the top surface versus the bottom side of the film. The thickness dependence of curve b is explained by this anisotropy.

High resolution is one of the outstanding characteristics of the TL grade. The SEM photograph of the PIMEL TL-series after development is shown in Figure 11-39, where the line/space pattern of 15/15 μm is achieved in the film thickness of 35 μm. Another characteristics of the TL grade is its low dielectric constant of 3.0, which is desirable for an interlayer dielectrics application for a multilayer substrate [123].

Nader et al. of DuPont prepared a low-stress ester-type PSPI, the chemistry of which is biphenyltetracarboxylic dianhydride (BPDA) and para-phenylendiamine (PDA). According to them, the prepared spin-coated film showed a TCE of 13 ppm/°C.

Interestingly, the cast film exhibited a TCE of 25 ppm/°C. They speculated that the differences in the TCE values were due to the difference in orientation of the polyimide molecules. The other properties of the cured film that were reported are an elongation of break of 10%, a modulus of 6 GPa, a tensile strength of 192 MPa, and a dielectric constant of 3.5 [124].

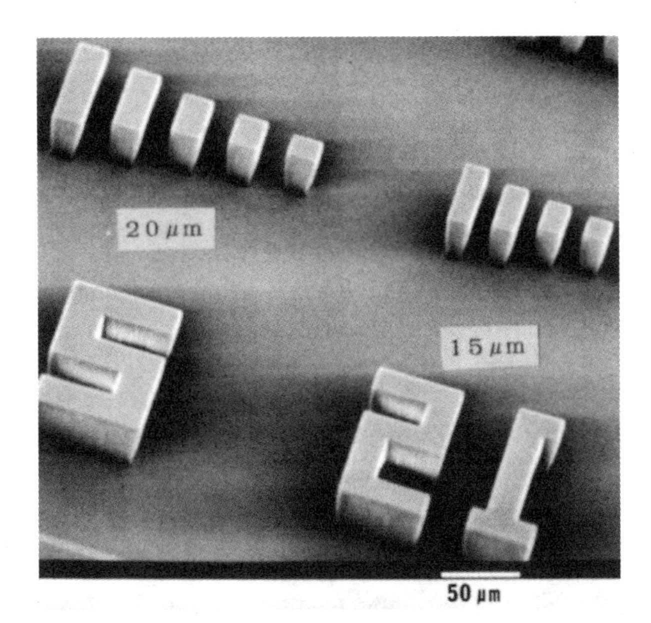

Figure 11-39. SEM Photograph of the Pattern of PIMEL TL-Series after Development

11.8.4.3.7 I-Line photoimageable Ester-Type PSPI

Of the many spectral lines of UV light radiation available, g-line (436 nm) and broad band are utilized for the exposure of conventional PSPIs. Recently, i-line exposure has been replacing g-line exposure in photoresist processes used to fabricate large-scale integration (LSI) devices. Thus, i-line photoimageable PSPI has been desired for exposure of buffer coatings on higher-density devices, since common use of equipment for photoresists and PSPIs is more cost effective in manufacturing. PSPI precursors show a rather high absorption at the i-line, limiting the applicable film thickness. Therefore, a polymer backbone structure with less absorption at the i-line needs to be developed.

Yamaoka et al. proposed an ester-type PSPI containing fluorine. They studied 2,2-bis {4(4-aminophenoxy)phenyl} hexafluoropropane or 2,2-bis(3-amino-4-methylphenyl) hexafluoropropane as diamine component. The i-line transmittance was measured to be more than 60% for a 3-μm-thick film of these PSPI precursor polymers. Conventional ester-type PSPI precursors based on PMDA/ODA chemistry exhibit about 30% i-line transmittance [125]. It was found that there is a substituent effect on the i-line absorption of an ester-type PSPI precursor. The introduction of an electron donating group to an acid group decreases the i-line absorption. On the other hand, a substitution of an electroinductive group for the diamine moiety shows the same effect. Good linearity was found between the absorption per amide unit and a Hammett's value [126]. PSPIs have been developed based on this principle for i-line exposure.

11.8.4.3.8 Base-Catalyzed PSPI

Volksen et al. of IBM reported a very unique PSPI. It is basically composed of ethyl ester-type PSPI precursor and an amine photogenerator. The amine photogenerator is decomposed by exposure to deep UV light to give an amine, which catalyzes the imidization reaction of the precursor polymer. It was also reported that pattern formation was successful due to the large difference in the solubility between the precursor and partially imidized polymer [127].

11.9 IONIC BONDED-TYPE PSPIs

11.9.1 Chemistry and Photopackage

Photosensitive polyimides have continued to gain great attention in the microelectronics industry due to the simple fabrication process based on their direct patternability. Hiramoto et al. developed a unique negative working photosensitive polyimide precursor, which is composed of a poly(amic acid) and tertiary amine having an acryloyl group (Fig. 11-40) [128,129]. Toray's photosensitive polyimide coating, "Photoneece" [130], is developed by this method. The "Photoneece" is used in various applica-

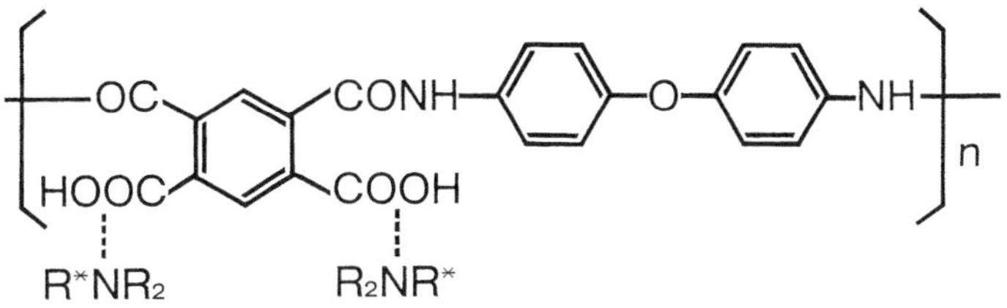

$R^{*}NR_2 : CH_2{=}C(CH_3)COOCH_2CH_2N(CH_3)_2$

Figure 11-40. Ionic Bonded-Type Photosensitive Polyimide Precursors ("Photoneece")—Negative Working

tions, because it has some distinct advantages. The first is that swelling does not occur during a development in spite of it being a negative working type system, producing high resolution. The second is its high photosensitivity, which is comparable to that of a conventional photoresist. The last is that the photosensitive group can be readily eliminated during thermal curing, so as to give a highly reliable polyimide film. These advantages, as will be described later, are closely related to the photosensitive group endowing method.

The mechanism of photosensitivity of the ionic bonded-type photosensitive polyimide precursor was investigated by Tomikawa et al. [131]. The mechanism of the photosensitivity of the system is not photo-cross-linking of the acryloyl group in the tertiary amine as shown in Figure 11-41. The first evidence is that the exposed film can be soluble in NMP completely. The second is that the acryloyl group does not react significantly during photoirradiation as determined by NMR studies. The last is that about half of the tertiary amine (dimethylaminoethyl methacrylate, DMM) remains in the exposed film after chloroform extraction, but all of the remaining DMM in the film is recoverable intact. Since the mechanism for the photosensitivity is not photo-cross-linking of the acryloyl group, this gives "Photoneece" an advantage to yield high resolution patterns and be free from swelling during development in spite of being a negative working system (Fig. 11-42).

In order to improve photosensitivity, the selection of suitable sensitizers is important. Most photoradical initiators do not work effectively except for Michler's ketone and dimethylaminoacetophenone. This result corresponds to the absence of photopolymerization of the methacryloyl group during photoirradiation. Aromatic amines, such as *N*-phenyl-diethanolamine, work as effective sensitizers [132]. In some cases, they work as a cosensitizer, if other photoradical initiators exist in a photosensitive system (Fig. 11-43).

Table 11-24 shows the typical properties of "Photoneece" UR series

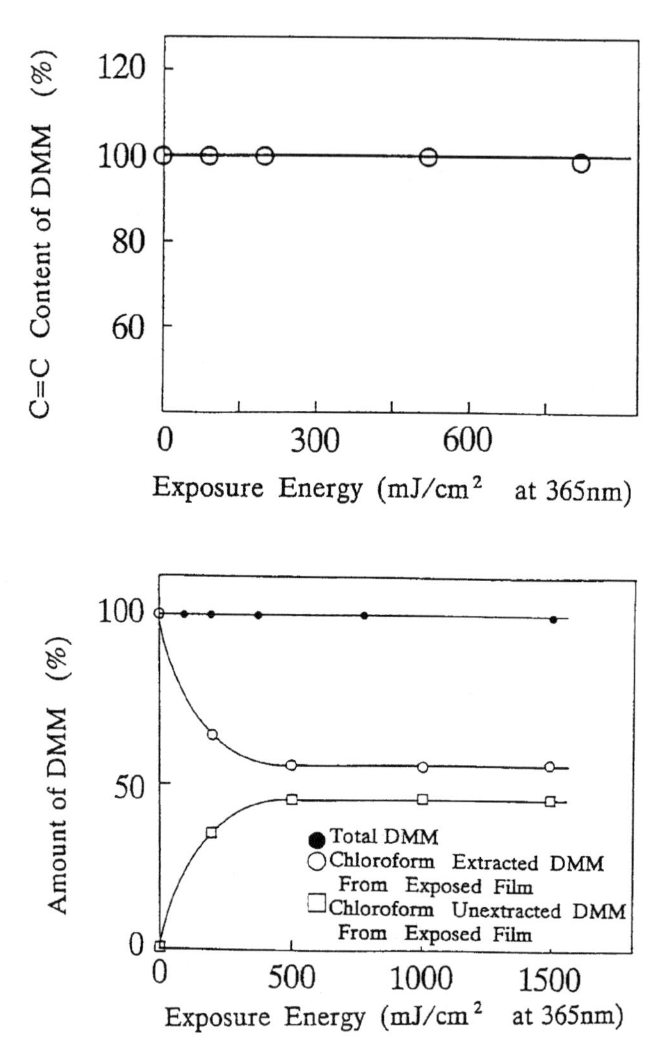

Figure 11-41. Photoreaction of Aminoalkyl Methacrylate (DMM) (From Ref. 128)

produced by Toray. These are composed of different polyimide backbone structures.

11.9.2 Processing "Photoneece" (Exposing, Developing, and Curing)

As previously shown in Figure 11-19, the use of "Photoneece" significantly reduces the number of steps needed for pattern generation compared to the conventional processing using nonphotosensitive poly-

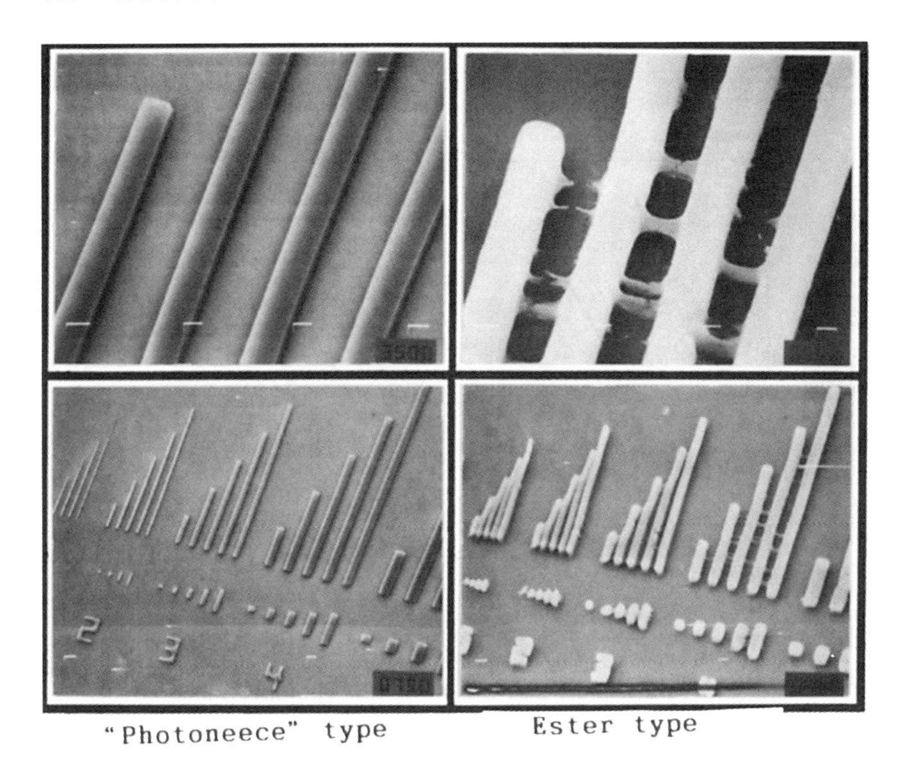

Figure 11-42. Fine Patterns of Photosensitive Polyimides

imide coatings. In general, spin coating is used. Coating thicknesses can vary widely from 1 μm to 40 μm, depending on its applications. As the film thickness is reduced during developing and curing, the final thickness of film will be approximately 50% of the prebaked film thickness.

Prebake: Coated substrates should be prebaked on a hot plate or in a convection oven (in air or nitrogen atmosphere).

Exposure: A contact aligner, mirror projection aligner, or g- or i-line stepper can be used (see Fig. 11-44).

Developing and Rinse: "Photoneece" can be developed by immersion, immersion with ultrasonic assistance or the spray-puddle method, using a "Photoneece" developer (DV series), and rinsed with isopropanol and isobutanol.

Curing: The standard condition for curing "Photoneece" with a convection oven is a three-step curing at 140–180°C and 200–300°C for 30 min each, then 350–400°C for 60 min. It is also

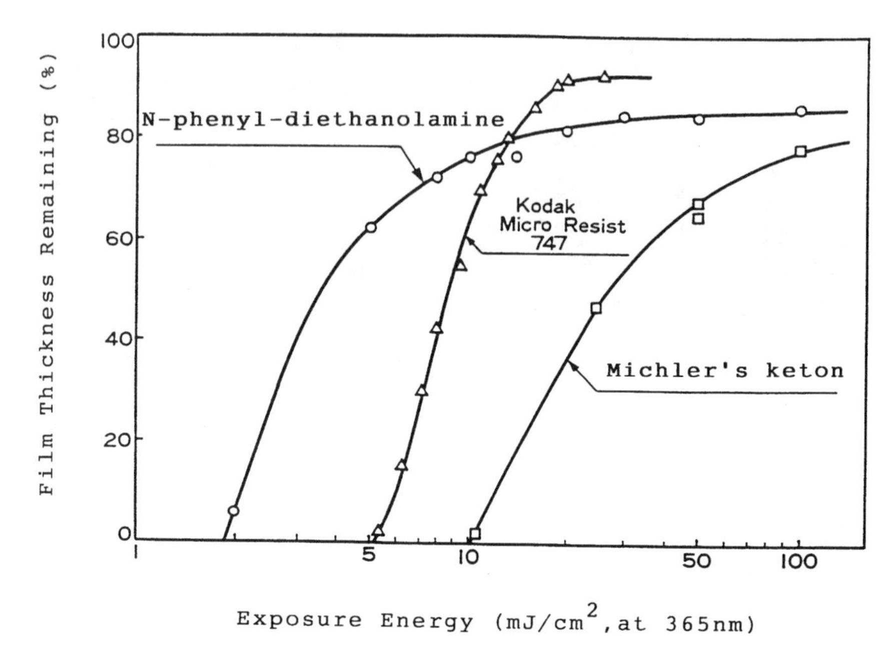

Figure 11-43. Photosensitivity of "Photoneece"

possible to ramp up the temperature from room temperature by 3–5°C/min and hold for 1 h at 350°C.

Removal of "Photoneece" (Rework): Before curing, "Photoneece" film can be striped by using *N*-methyl-2-pyrrolidone or alkaline developers used for positive photoresists. Oxygen plasma etching is also applicable. After curing, hydrazine–hydrate-type etchants or oxygen plasma removal systems should be used.

11.9.3 Lithographic Properties and Characterization

"Photoneece," in spite of being a negative working photosystem, does not swell during development and, therefore, produces very fine patterns. The characteristic curves of "Photoneece" are shown in Fig. 11-45. The film thickness remaining after development is more than 95% at an exposure energy in excess of 200 mJ/cm^2 at 436 nm, in both 20 μm and 40 μm prebaking film thicknesses.

The relationship between the resolution and the film thickness under constant developing conditions is shown in Figure 11-46 [133]. "Photoneece" has a high resolution, with an aspect ratio of more than 2.0. SEM photographs of patterning profiles after developing and curing are shown in Figure 11-47. The patterning profiles of "Photoneece" are not overhanging, even in 40-μm imaged film [133].

Table 11-24. "Photoneece" UR Series Polyimide Properties Measured on Solutions and Cured Films

Solution Properties	UR-3100 Series	UR-3800 Series	UR-4100 Series	UR-5100 Series
Typical grade	UR-3140	UR-3840	UR-4144	UR-5100
Features	Standard	High-sensitivity	Low Young's modulus	High resolution; low thermal expansion
Solid content (%)	17	20	20	22
Viscosity (P/250°C)	50	50	30	100
Metallic ion content (measured by ICP mass spectrometry)	Impurity levels same for all grades Na < 0.1 ppm, K < 0.1 ppm, Fe < 0.1 ppm, U < 0.005 ppb, Th < 0.005 ppb			
Solvent system *N*-methyl-2-pyrrolidone (NMP)	NMP/ methyl cellosolve	NMP/ methyl cellosolve	NMP	NMP
Film Properties				
Mechanical				
Tensile strength (kg/mm^2)	14	14	10	20
Elongation (%)	30	30	10	30
Melting point (°C)	None	None	None	None
Coefficient of thermal expansion (ppm/°C)	40	40	45	45
Flammability	All are self-extinguishing			
Electrical				
Dielectric constant (1 kHz, 25°C)	3.3	3.3	3.3	3.3
Dissipation factor (1 kHz, 25°C)	0.002	0.002	0.002	0.002
Volume resistivity (Ω cm)	$>10^{16}$	$>10^{16}$	$>10^{16}$	$>10^{16}$
Surface resistivity (Ω)	$>10^{16}$	$>10^{16}$	$>10^{16}$	$>10^{16}$
Dielectric strength (kV/mm)	>300	>300	>280	>300

Low-temperature removal of photoreactive groups is another feature of "Photoneece." Most of the photoreactive groups can be removed at a curing stage lower than 300°C. Therefore, the properties of a cured film are primarily determined only by the polyimide backbone structures. This is a very important and desirable characteristic to design into photosensitive polyimides. Figure 11-48 shows the thermogravimetric analysis of an ionic bonded type ("Photoneece") and a covalently bonded ester type [130]. Both have the same polyimide backbone consisting of pyromellitic dianhydride (PMDA) and 4,4′-diaminodiphenylether (DAE). In the case of the ionic bonded type in which dimethylaminoethylmethacrylate (DEM) is

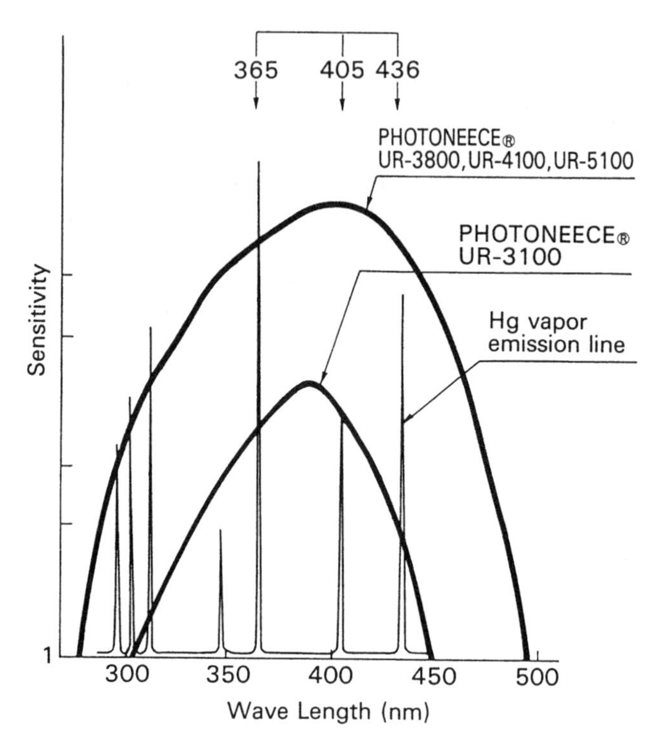

Figure 11-44. Sensitivity Spectrum of "Photoneece"

used, a large weight loss occurs between 100°C and 200°C, followed by small weight loss up to 300°C. A solvent (NMP), water (imidization gives rise to it) and photosensitive group, DEM, are completely removed until 300°C. On the other hand, the weight loss of the ester type starts at higher temperature than that of the ionic bonded type and is still observed up to 500°C. Thermogravimetric curves (TG) shown in Figure 11-49 indicate that the cured films of the ionic bonded type are thermally more stable than those of the ester type and show the same TG curve behavior as exhibited with the nonphotosensitive polyimide of the same structure.

Comparative investigations of the properties of thermally imidized polyimides from different types of photosensitive polyimide precursors were performed by Kojima et al. They studied the thermal, mechanical, and adhesion properties of the three kinds of polyimide precursors [poly(amic acid), ionic bonded-type photosensitive polyimide precursor, ester-type photosensitive polyimide precursor [134]]. As shown in Table 11-25, the weight loss characteristics and mechanical properties of fully cured films prepared from the ionic bonded-type polyimide precursor are comparable to those prepared form poly(amic acid). However, these same properties measured on the ester-type photosensitive polyimide precursor are de-

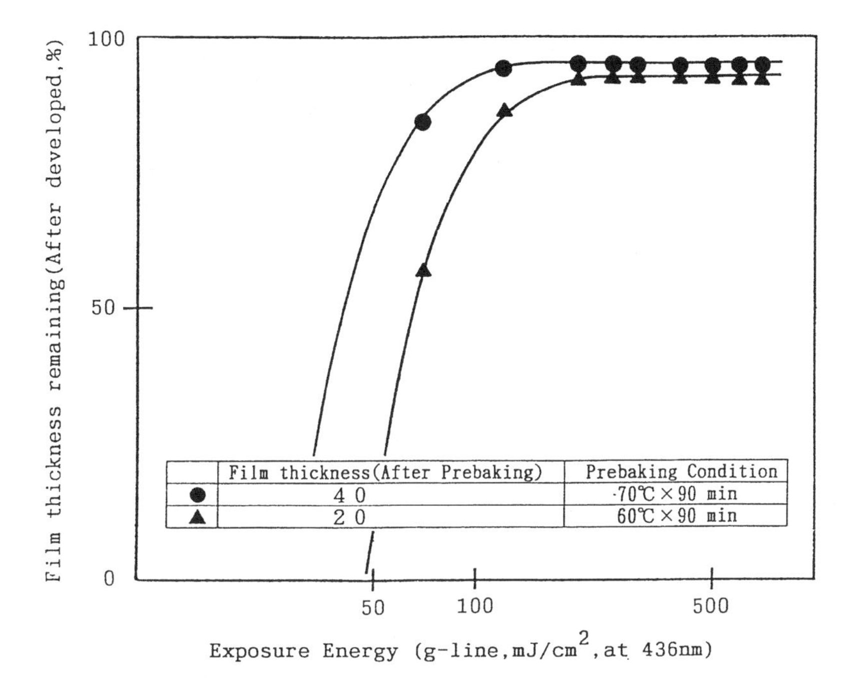

Figure 11-45. Characteristic Curves of "Photoneece"UR-5100 (From Ref. 133)

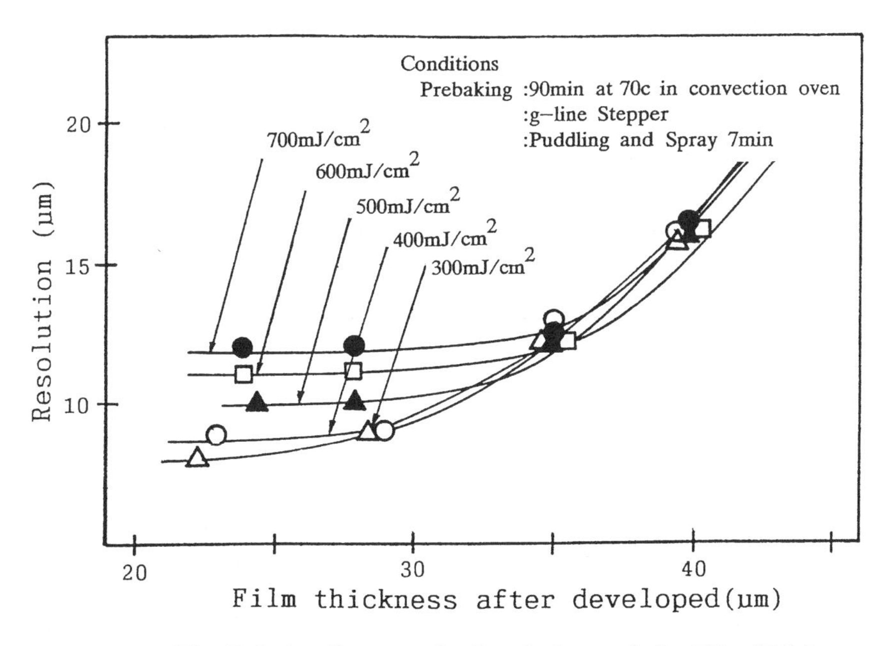

Figure 11-46. The Relation Between the Resolution and the FIlm Thickness of "Photoneece"UR-5100 (From Ref. 130)

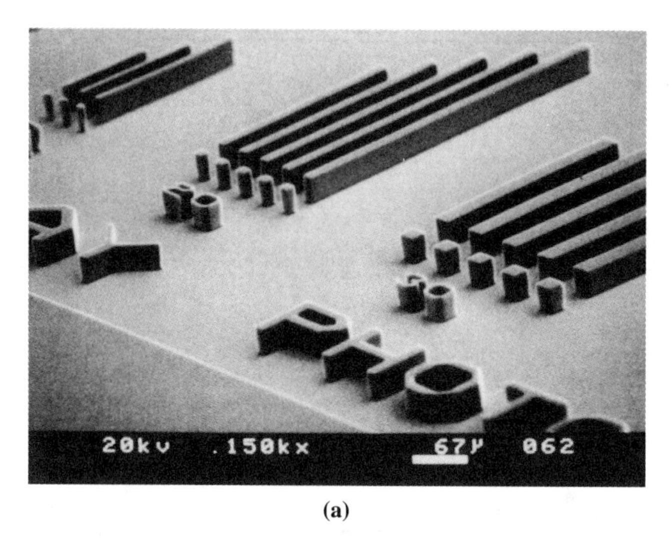

(a)

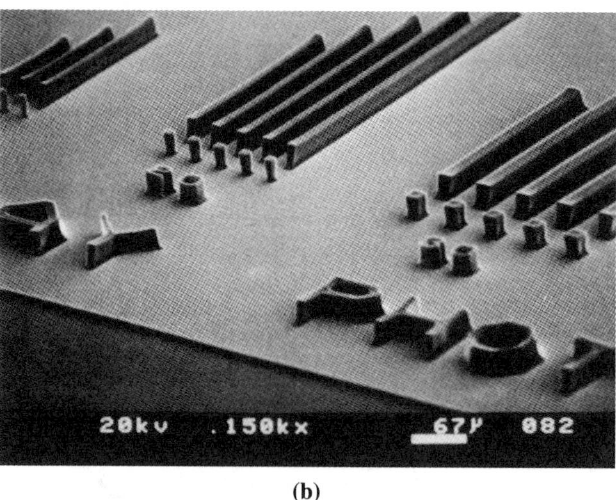

(b)

Figure 11-47. SEM Photographs of Patterning Profiles After Developing and Curing of "Photoneece"UR-5100. (a) Coated film thickness after developing, 40 μm; (b) coated film thickness after curing, 20 μm. (From Ref. 133)

graded when compared to the films prepared from the poly(amic acid), even after imidization at 400°C. This is thought to be due to the film's ability to eliminate the photosensitive groups. The photosensitive group of the ionic bonded type is eliminated faster than that of the ester type. The photosensitive group of the ester type may also not be completely eliminated after being imidized at 400°C because the photosensitive group is attached by a covalent bond, and it reacts with UV exposure.

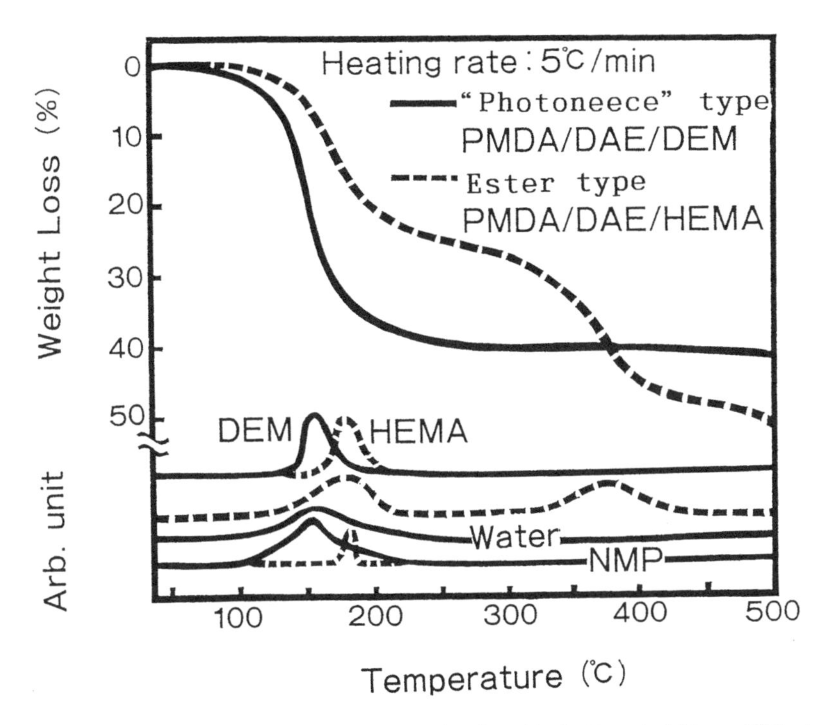

Figure 11-48. Thermogravimetric Analysis of an Ionic Bonded Type ("Photoneece") and a Covalently Bonded Ester Type (From Ref. 130)

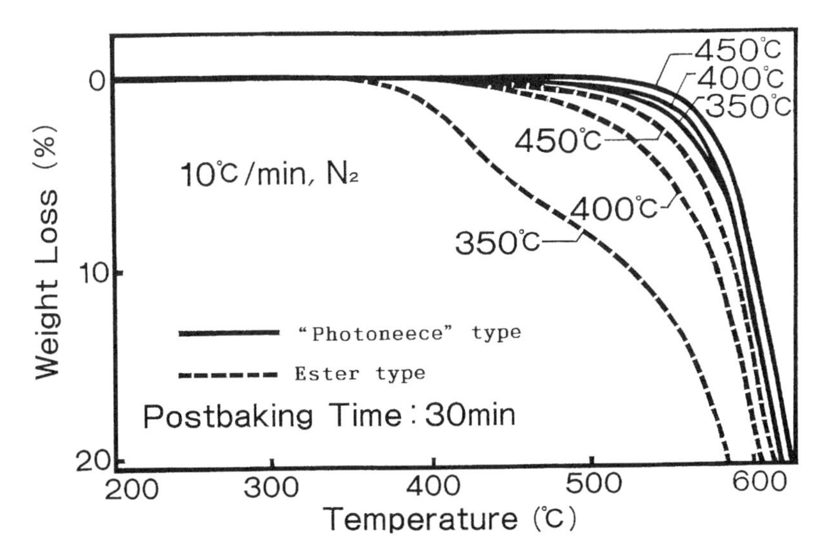

Figure 11-49. Thermogravimetric Curves of Cured Films of an Ionic Bonded Type ("Photoneece") and a Covalently Bonded Ester Type (From Ref. 130)

Table 11-25. Initial Weight-Loss Temperature and Mechanical Properties of Cured Films Formed from Polyimide Precursors

			Mechanical Properties	
Polymer[a]	Curing conditions	Initial Weight-loss Temperature (°C)	Elongation (%)	Tensile Strength (kgf/mm^2)
A	350°C—60 min	450	10	12
	400°C—60 min	450	11	12
B	350°C—60 min	355	<1	<1
	400°C—60 min	385	<1	<1
	430°C—60 min	390	<1	<1
	450°C—60 min	400	<1	<1
C	350°C—60 min	395	10	12
	400°C—60 min	440	10	12

[a]Polymer A = Poly(amic acid); polymer B = ester-type photosensitive polyimide precursor; polymer C = ionic bonded-type photosensitive polyimide precursor. (From Ref. 134).

11.10 PREIMIDIZED PHOTOSENSITIVE POLYIMIDES

11.10.1 Chemical Characterization

PSPIs have been offered by OCG under the trade name Probimide® 400 (Probimide 412 and 414). The descriptions "inherently photosensitive" or "autophotosensitive" have been used synonymously in the literature. The chemical structure of these special PSPIs is

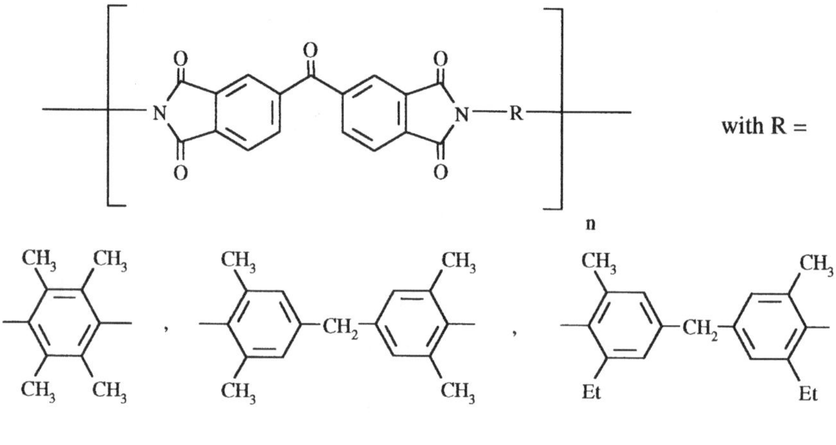

with R =

to produce a variety of PSPI related structures.

They are prepared by standard PI synthesis, first by the corresponding polyamic acid and then by chemical imidization in solution. These solvent soluble polyimides are photosensitive through photocrosslinking by themselves and without the addition of any photosensitizers. The photosystem is an integral part of the PI unit, and the photostructures obtained shrink only slightly upon curing (350–400°C).

The chemistry was first published in 1985 [135], and a review together with new and improved experimental materials was given in 1991/1992 [136]. The polyimides are high molecular weight (M_w by light scattering = 40,000–60,000), and they are offered as ready-to-use solutions with 4-butyrolactone as the main solvent. The selection of monomers is such that an optimum of photospeed, processability, and mechanical properties (T_g, thermal stability, elongation, TCE) is obtained. The key data of these formulations are listed in Table 11-26.

11.10.2 Features

Purity: Due to the simplicity of both the polyimide synthesis and the formulation processes, the high demands with respect to metal-ion contamination can be met, in particular with respect to Na, K, Fe, Cl – (<1 ppm).

Storage Stability: Ready-to-use solutions are storage stable in comparison to a variety of other PSPI systems, with a typical change of viscosity of only about + 2% per year of storage at room temperature.

Spectral Sensitivity/Photospeed: The material is primarily sensitive in the i-line (365 nm) and is compatible with exposure in the contact or proximity mode, with projection tools and with i-line steppers.

Thickness Range: The application range is from 1 to 15 μm (after cure). The best throughput, given the exposure and the developing time, is achieved in the film thickness range of 1–8 μm.

Table 11-26. Probimide 400 Series General-Purpose Formulations

Probimide	% Solids	Solvent	Viscosity (cs)	Filtration (μm)	Thickness (μm)
Probimide® 412	12	GBL	3500	1.0	3.5–12
Probimide® 414	14	GBL	8200	1.0	4–20

Note: Shelf-life stability at room temperature is 18 months for Probimide 400 series.

Resolution/Aspect Ratios: Using contact or i-line stepper exposures, aspect ratios of 1.0 are obtained for films over 5 μm, and 1.5 for films 1–5 μm thick.

Thermal Cross-linking: A thermal cross-linking process (350–400°C, 0.5–2 h typically) renders all photostructures insoluble and inert. It is important for the understanding of these inherently photosensitive polyimides that photo-cross-linking and thermal cross-linking are independent of each other, and obviously follow different pathways. The influence of photo-cross-linking and thermal curing on final material properties has been a subject of intensive studies (see Section 11.10.3 below).

Nonshrinking upon Cure: In contrast to other PSPI systems, inherently photosensitive polyimides of the Probimide 400 type exhibit only minimal shrinkage upon cure (350–400°C). Shrinkage typically is only 7–10%. Thus, photopatterns are preserved and geometrical distortions are minimal. This is illustrated by the SEMs of Figure 11-50, showing 7 μm lines before and after hard bake [137].

Cu Compatibility: Due to their preimidized nature, Probimide 400 polyimides are fully Cu compatible, unlike polyimides derived from poly amic acids which can react with Cu.

Glass-Transition Temperature: The glass-transition temperature of Prohimide 400 is relatively high when compared to other PIs and PSPIs. The exact value depends only slightly on the precise curing conditions (350–400°C) and the method of determination. Thus, DSC gives a value of 369°C, whereas (depending on curing conditions) thermomechanical values from 350°C through 410°C have been reported [137,138].

Reduced Outgassing: Probimide 400 polyimides do not contain thermally instable components/additives like acrylates and photosensitizers that may give rise to outgassing problems.

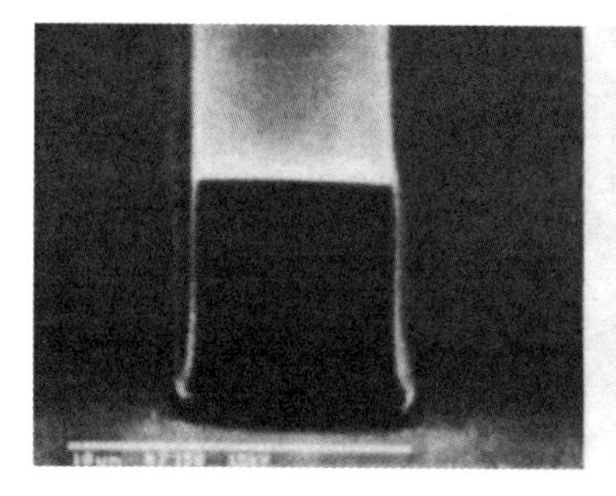
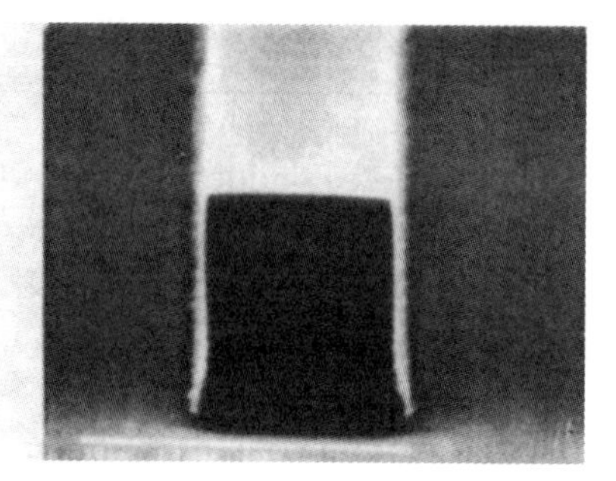

Figure 11-50. SEM Micrographs Showing the Side Wall Profiles of Soft-baked (left) and Hard-Baked (right) 7 μm Lines

11.10.3 Origin of Photo-cross-linking

The photomechanism of Probimide 400 related polyimides has been studied by Lin et al. [139], Scaiano et al. [140,141], Shindo et al. [139], Yamashita et al. [143-145], Horie et al. [146], and Jin et al. [147]. Irradiation experiments from polyimide solutions, as well as from solid films, were reported and the photo-cross-linking efficiency was calculated from Gel Permeation Chromatographic measurements of the irradiated products as well as from gel doses. Also, spectroscopic studies on model compounds were carried out to identify the transient active species:

It is now generally accepted that the photo-cross-linking process of Probimide 400 related polyimides is initiated by H abstraction (photoreduction) of the triplet state of the benzophenone unit according to the following chemistry.

Quantum efficiencies of photo-cross-linking as obtained from irradiation of solid films have been calculated to be only (2.0–3.5) $\times$ 10^{-3}, and less than 50% of all benzophenone units are reactive in the solid. These studies, however, considered only BTDA polyimides with "ortho-mono"-alkylated diamines. Probimide (400-type polyimides are made with ortho-di-alkylated diamines, and are much more photosensitive than polyimides made from (ortho-mono)-alkylated diamines [136,148].

Role of Alkyl Substituents: The two alkyl substituents next to the imide bond have multiple functions:

- Photospeed is improved to a useful range.
- Solubility of the polyimide is greatly improved.
- Glass-transition temperatures are greatly enhanced through locking of the diamine rings in the orthogonal position relative to the tetracarboxylic acid rings.
- Transparency of polyimide is improved such that absorptions at 365 nm (i-line) are ideally suited for lithography in the 5–10-μm thickness range.

Oxidative Stability: Probimide 400 polyimides do not provide thermooxidative stability above about 320°C. Exclusion of 0_2 (<5 ppm), however, allows curing at 350–400°C in nitrogen. It has been shown that thermogravimetric weight loss does not relate to alkyl content when oxygen is excluded [149].

Morphology: Although preimidized polyimides composed of BTDA and ortho-dialkylated diamines may give rise to highly ordered structures which may crystallize from solution, the composition of Prohimide 400 polymers is such that an absolutely amorphous material is obtained. This has been confirmed by Ree et al. [138]. With other experimental products, crystallization has been observed [149].

Processing Conditions: The key processing data of ProbimideX 408, 412, and 414 and the new self-priming developmental product, XB 7003, are listed in the summary in Table 11-27.

Adhesion Promoter: Typically an aminosilane adhesion promoter (QZ 3289/3290) is applied prior to coating. The new developmental product, XB 7003, is fully self-priming and does not need a spun-on adhesion promoter. Its photospeed and resolution capability are the same as with Probimide 414; the processing conditions are practically identical.

Soft Bake: Hot-plate drying is preferred to oven drying, as superior adhesion properties are achieved.

Exposure: The exposure tool must have output at the i-line, 365 nm. i-Line steppers are preferred. The photospeed is at ~ 100 mJ/cm^2 per micron thickness, thus requiring 600 mJ/cm^2 for a 5-μm-thick coating and 1500 mJ/cm^2 for a 15-μm-thick coating (both after hard bake).

Table 11-27. Process Summary for Probimide 400 Series Photosensitive Polyimides

Process and Parameters	Probimide 412	Probimide 414	XB 7003
Adhesion Promoter 1:9 QZ 3289/3290 (rpm/s)	4000/20	4000/20	Self-priming
Spin coating (rpm/s)	2400/25	1500/25	3200/25
Soft bake (°C/min)			
Oven	110/30	110/30	90/30
Hot plate	110/10	110/15	90/10
Exposure energy (i-line stepper) (mJ–cm^2)	600	1500	600
Postexposure bake (°C/min)	110/10	110/10	—
	110/3	110/3	—
Spray Development			
Developer QZ 3501 (rpm/s)	800/100	800/240	800/80
Overlap (rpm/s)	800/10	800/10	800/5
Rinse QZ 3512 (rpm/s)	800/20	800/20	800/5
Spin Dry (N2) (rpm/s)	2000/15	2000/15	2000/10
Hard bake (°C/min)	350/60	350/60	350/60
Final film thickness (μm)	5	15	7

Developing: Developing is by spray-on of developer QZ 3501/rinse QZ 3512 (both xylene free). The development rate is at 16–20 s/μm.

Cure (Hard Bake): Standard hard bake is 350°C, 60 min in high-grade N_2 (< 5 ppm 0_2).

Mechanical Properties: The mechanical properties of Probimide 400 cured films are listed in Table 11-28 [142].

There have been extensive studies to monitor the change/conservation of material properties toward a variety of processing parameters [137,138,145,151]. The most important parameters analyzed and their influence on materials properties are as follows:

Photo-Cross-Linking: Photo-cross-linking with the required exposure doses (< 2000 mJ/cm^2, i-line) has little influence on final mechanical properties. A change of the dynamic storage modu-

Table 11-28. Measured Properties of Probimide 400 Cured Film

Properties	Units	Measured
Physical		
Tensile Strength	GPa	2.9
Modulus	MPa	147
Elongation	%	56
Refractive index (633 nm)		1.66
Density	g/cm^3	1.20
Moisture uptake (50% **rh**)	%	2.0
Stress	MPa	48
Thermal		
Glass-transition temperature	°C	357
Melting point	°C	None
Decomposition temperature	°C	527
Weight loss (400°C in N_2)	%/h	0.07
Thermal coefficient of expansion	ppm/°C	37
Thermal conductivity	cal/cm/s/°C	4.1×10^{-3}
Specific heat	cal/g/°C	0.32
Electrical		
Dielectric constant (1 MHz, **rh** 4%/50%)		2.9/3.7
Dissipation Factor (1 MHz, **rh** 4%/50%)		0.006/0.010
Dielectric breakdown field	V/μm	348
Volume resistivity	Ω-cm	$>6 \times 10^{16}$
Surface resistivity	Ω	$>1 \times 10^{15}$
Adhesion		
90° tape peel test, 72 h in boiling water (ASTM D-3359-83B)		5

lus E' has been seen at the glass-transition region [137], when a hard bake of 350°C, 30 min was applied (lower limit of hard-bake conditions). At higher hard-bake temperatures, however, thermally generated cross-links dominate over photochemically formed cross-links. Also, mechanical stress-strain behavior is primarily dependent on thermal history, not on photochemical history.

Oxygen Content of Curing Atmosphere: Due to thermal oxidation of alkyl side groups, the mechanical properties of Probimide 400 may degrade when cured at 350–400°C in the presence of > 100 ppm 0_2. Mechanical properties are preserved when 0_2 is excluded. 0_2 contents of <10 ppm but preferably <5 ppm are recommended.

Hard Bake—Curing temperature (350°–400°C): Hard-bake temperature is the most important and most thoroughly analyzed processing parameter. If thermal cross-linking is not complete at a hard-bake temperature of 325°C, a minimal hard-bake cycle of 350°C, 0.5 h, preferably 1.0 h, is recommended. On the other hand, a change of mechanical properties is seen at curing at 400°C, 2 h. Thus, the curing window of 350–400°C has been thoroughly analyzed. Below is a list of material properties analyzed that will/will not be affected by the choice of the particular curing temperature within this curing window (1 h cure time):

- Residual stress is 48–52 MPa and independent of cure temperature.
- Elongation at break changes from about 60% (350°C) to about 50% (400°C).
- The E modulus remains constant at 2.8–2.9 GPa.
- T_g remains constant from 350°C to 375°C cure, then rises slightly.
- Solvent swelling (e.g., by *N*-methylpyrrolidone, NMP) decreases sharply as the hard-bake temperature advances through the T_g (365–375°C), and then remains relatively constant.

Hard-bake—curing time 0.5–10 h: Prolonged curing in 0_2-free nitrogen up to 10 h is possible at a cure temperature of 350°C. There are only minor changes in residual stress, elongation at brake, *E* modulus, and T_g. Curing at 400°C for 10 h will, however, noticeably degrade material properties.

Further Experimental Products: Substantial work has been devoted to further improve photospeed, spectral sensitivity, processing conditions, and material properties of Probimide 400 materials. A summary of such work has been given in Ref. 136 extending the working principles of BTDA-bond Probimide 400 to polyimide backbones based on other dianhydrides [152,153].

11.11 POSITIVE PHOTOSENSITIVE POLYIMIDES

Photosensitive polyimides offer cost advantage of eliminating the need for photo-resists, thereby reducing the number of processing steps in the fabrication of microelectronic devices as previously shown in Figure 11-19. Based on photo reactivity, photosensitive polyimides can be classified into two broad categories: (1) negative resist types or (2) positive resist types. Most of the commercialized photosensitive PIs are of the negative resist type and are derived from polyamic acid precursors which have either covalently bonded ester or ionic-bonded photosensitive groups attached to the backbone polymer chain previously shown in Figure 11-21. These types of polymers were discussed extensively in the previous sections. Conversely, positive resist types undergo molecular rearrangement such as chain scission that enhances the solubility of the exposed area. Positive working PSPIs are being developed to overcome the limitations [154-155] of localized swelling often found in negative working systems which may lead to degradation of patterning resolution and use of less toxic organic solvents for fine line/via patterning. The application of positive working PSPIs are limited today due to the narrow film thickness range available but they are deemed preferable for semiconductor and packaging applications [156,157]. They offer better resolution and are developed in aqueous base solutions. In positive PSPIs, adjusting the dissolution rate is critical to avoid film delamination during development.

11.11.1 Positive Working PSPI

There are two approaches to developing positive PSPIs: (I) incorporating a photosensitive group into the backbone or as a side branch of the base polyamic acid or (ii) by blending a photosensitive compound with the polyamic acid precursor, a system analogous to polymer blends.

The development of positive PSPIs by the first approach is more difficult and limited success has been achieved. Preparation of PSPIs with the photosensitive moieties in the main chain are relatively complicated since the introduction of such groups in the main chain may alter the molecular architecture of the basic polyamic backbone. A positive type

PSPI prepared from a maleic anhydride dimer exhibited positive resist like patterning [158]. The cyclobutane ring in the polymer chain undergoes cleavage upon photo-irradiation which imparts differential solubility in the exposed areas to produce a pattern after development leaving the unexposed region unchanged.

Hayase et al. [159] developed a positive working high molecular weight PSPI from polyamic acid esters with phenol moieties which were synthesized from diamines and dicarboxylic acids, polyamic acid and napthoquinone diazide. The phenol moieties in the polyamic acid ester were made through ester linkages. The positive acting PSPI was base developable and the phenol groups were removed during imidization to produce a thermally stable polyimide film after curing to 350°C. A 4 micron line and space pattern was produced by using this PSPI, however as the polyamic acid content was increased to make the resist dissolution rate faster, the resolution degraded to 100 microns in a 3 micron thick film. Therefore controlling the dissolution rate is critical to achieving fine line resist patterns in thicknesses that are required for thin films.

Kubota at Mitsubishi Electric [160] reported a positive working polyimide based on a polyamic acid backbone with o-nitrobenzyl ester side chain. The nitrobenzyl group decomposed into a carboxalic acid upon UV radiation. The exposed part of the film thus became soluble in aqueous alkaline solution.

Takano and Hayase at Toshiba [161,162] showed that a photosensitive PI can be achieved by blending in a photosensitive ester compound such as napthoquinonediazide-sulfonyl ester in the polyamic acid matrix. The mixture reacted as a positive working photoresist and was developed in a aqueous alkaline solution.

Omote and Yamaoka [163] reported a positive working system in a mixture of polyamic acid with a 1,4 dihydropyridine derivative. The solubility in the unexposed area in aqueous alkaline solution was inhibited by the hydrogen bonding, while the exposed area promoted dissolution due to the absence of hydrogen bonding.

Sumitomo Bakelite of Japan [164] recently reported a positive working photosensitive polyimide system consisting of a mixture of polyimide with and o-napthoquinonediazide. These authors also showed that a positive photoresist consisting of a mixture of polybenzoxazole (PBO), o-napthoquinonediazide, and silicone modified polyimide had better adhesion (due to silicone functionality) and higher thermally stability (due to PBO) than the polyimide precursor system.

11.11.2 Properties of Positive Working Photosensitive PI

The properties of photosensitive PI are primarily governed by (i) the photosensitization techniques and (ii) by the modification of the base polyamic acid precursor. Hence, the positive working PI generally exhibit similar properties with those of negative resists since the photosensitization side groups are usually eliminated once the reaction is completed.

Mechanical properties of photosensitive PIs are in general somewhat lower than found in non-photosensitive polyimides with the same backbone [165]. Takechi et al [163] compared properties of (i) a non-photosensitive, (ii) a photosensitive with ester side group, (ii) a photosensitive with ionic side group, and (iv) a positive working polyimide. These polyimides exhibited good mechanical properties and adhesion to a wide variety of substrates except for the ester-type. The ester-type photosensitive PI showed relatively inferior mechanical properties which were attributed to the incomplete removal of the photoreactive group [167]. However, the adhesive and mechanical properties also depends on the constitution of the polymer backbone. Sashida showed that adhesive properties were significantly improved with the incorporation of siloxane groups [168].

11.11.3 Outlook for Positive Photosensitive

The positive resists are expected to become more attractive due to environmental and safety concerns [157,158]. Positive resists tend to reduce the swelling behavior in the exposed area that are predominant in the negative working systems. Positive resists offer better resolution and can be developed in aqueous solution but they have limitations in film thickness and often cause film delamination from the substrate (1-3) during fabrication of thin film structures.

11.12 PHOTOSENSITIVE BENZOCYCLOBUTENE

Photodefinable benzocyclobutene (Photo-BCB), is commercialized by Dow Chemical under the tradename of Cyclotene™ (4000 series). Photo BCB chemistry [169-172] and processing [172] has been reported. The use of Photo-BCB has been reported by several companies, including: NEC [173], Siemens [174], Motorola [175] and MMS [176].

11.12.1 Physical and Chemical Properties

Photodefinable-BCB resins are negative acting materials (unexposed areas are removed during solvent development) which are sensitive to 365 nm radiation (i-line and/or broadband exposure). They are currently available as two formulations which cover film thickness ranges of 3 to 8 μm (Cyclotene 4024) and 7 to 13 μm (Cyclotene 4026). The base resins are supplied as partially polymerized (B-staged) solutions in mesitylene.

The Photo-BCB resins are formulated with thermally sensitive additives to impart photosensitivity. Thus, these photosensitive products re-

Table 11-29. Typical Physical and Chemical Properties of Photo-BCB

Property	Cyclotene-4024	Cyclotene-4046
Solids (% Resin to Solvent)	40	46
Film Thickness Range (mm)	3.5–8.0	7.0–13.0
Viscosity (cSt @ 25°C)	350	1100
Density (gm/cc @ 25°C)	0.97	0.95
Ionics (ppm)	<1.0	<1.0
Extractable Halogens (ppm)	<5.0	<5.0

quire long term storage at −15°C (6 months) and have a pot life of about 1 week at room temperature.

Typical properties of Cyclotene 4000 series products are shown in Table 11.29.

11.12.2 Electrical and Mechanical Properties of Photo BCB

The electrical and mechanical properties of Photosensitive-BCBs are very similar to the dry-etch version Cyclotene®-3022 as shown in Table 11.30 and 11.31.

Table 11-30. Electrical Properties of Photo-BCB

Property	Value
Dielectric Constant (1 KHz-20 Ghz)	2.65
Dissipation Factor (1 KHz-1 MHz)	0.0008
Breakdown Voltage (V/cm)	3×10^6
Volume Resistivity (ohm/cm)	1×10^{19}

Table 11-31. Mechanical Properties of Photo-BCB

Property	Value
CTE (ppm)	52
Elongation (%)	8
Tensile Strength (Mpa)	87
Tg	>350°C
Tensile Modulus (Gpa)	2.9
Poisson's Ratio	0.34
Stress (Mpa)	28–32

11.12.3 Processing Photosensitive-BCB

The processing sequence for Photo-BCB closely resembles that of most common negative acting photoresists. There are seven basic processing steps as shown in Figure 11-51.

11.12.3.1 Surface Preparation

The interface between a deposited dielectric film and other layers (i.e. metal, polymer, and substrate) usually requires some type of surface treatment or preparation to ensure reliable bonding. The adhesion promoter, 3-Aminopropyltriethoxysilane (3-APS), 0.5% in water (or metha-

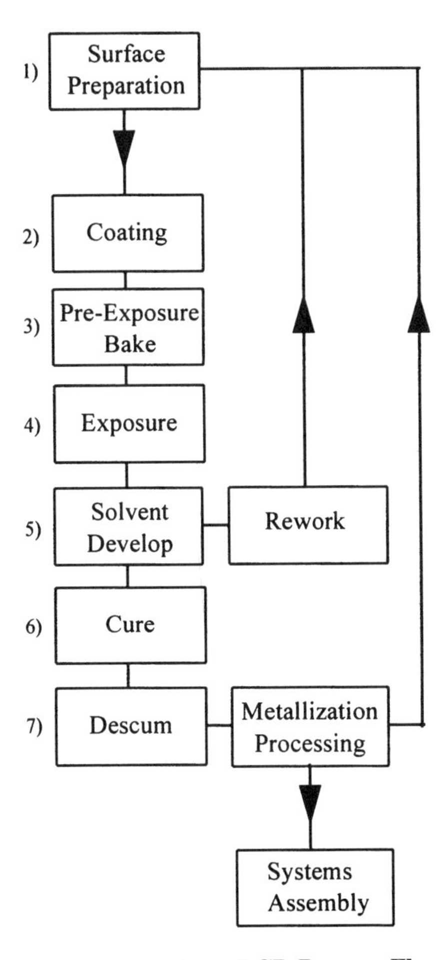

Figure 11-51. Photo-BCB Process Flow

nol), is used to enhance the adhesion of Photo-BCB to copper (power/ ground planes) and silicon-oxide (substrate) [177].

The adhesion between multiple layers of Photo-BCB (polymer-to-polymer adhesion) is enhanced by partially curing the underlying layer, thus leaving unreacted sites for the next layer of Photo-BCB to chemically bond to [178]. This practice produces a monolithic interface between successive Photo-BCB layers. In general, the use of 3-APS is beneficial between all layers where there is a high ratio of metal to polymer, i.e. on top of power and ground planes.

11.12.3.2 Coating

Spin Coating For wafer applications such as IC secondary passivation/buffer coating, wafer redistribution/bumping and MCM-D fabricated on silicon photosensitive-BCB resins can be applied to substrates on a track coater. A 10 sec/500 rpm relaxation spin is used to spread the polymer followed by a 30 sec spin at high speeds (2600 and 2400 rpms) to achieve the final (post cure) polymer thicknesses of 5 μm or 10μm. At the beginning of the spin cycle, a mesitylene backside rinse is employed to clean the substrate backside and to suppress polymer fiber formation in the spinner bowl. A second backside rinse at 800 rpms is used immediately after polymer coating for edge bead removal.

Spin curves for the two Photo-BCB formulations are shown in Figure 11-52. The top curve for each represents the film thickness after deposition and pre-bake (see pre-bake section). The lower curves are after cure (see cure section).

Spray coating, extrusion coating [179] and meniscus coating [180] have been used to deposit photo BCB on 400 mm glass, aluminum and PWB panels. Material utilization for spray cpoating is about 75%, for extrusion and meniscus it is in excess of 90%. Film uniformity across the boards was measured to be in the range of <5% (3 sigma) for meniscus and extrusion coating and <10% for spray coating.

11.12.3.3 Pre-Exposure Bake

After coating the substrates were baked to drive off excess solvent and stabilize the films before going to photo-lithography. This operation is normally carried out in-line on a hot plate station located sequentially after the wafer coating station. Alternatively, a convection oven has been used (off-line) with equally good results. The pre-exposure bake conditions need to be tightly controlled because they effect several subsequent processing operations, such as exposure dose and develop time. Film properties, such as via resolution, film retention, and film quality are also affected. Figure 11-53 shows the effect of pre-exposure bake on via resolution.

The optimum pre-bake conditions for processing are selected based

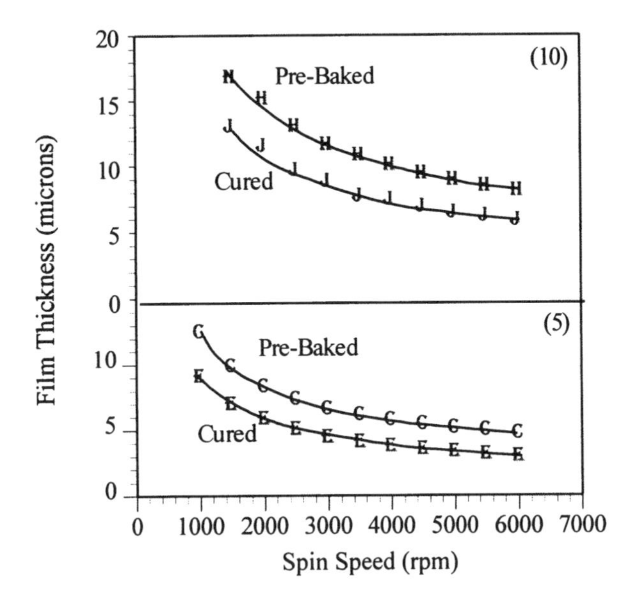

Figure 11-52. Spin Curves for 5 and 10 μm Photo-BCB on 4 inch Silicon Wafers. High Efficiency Coating Techniques

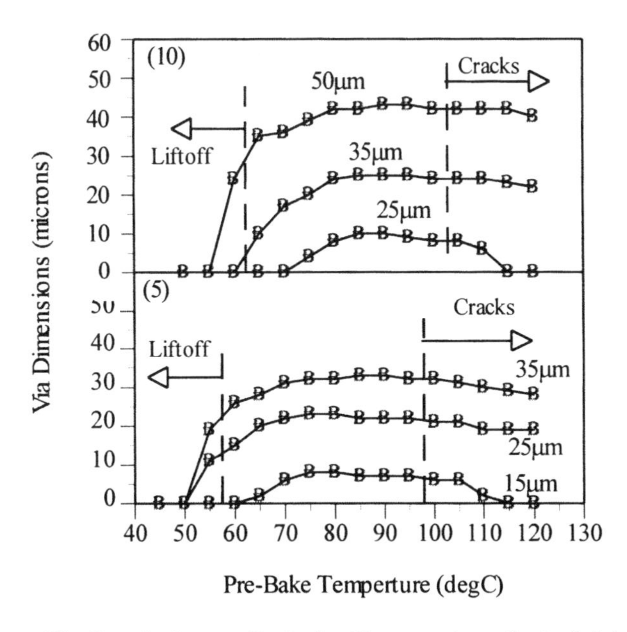

Figure 11-53. Via Resolution vs Prebake Temperature (hot-plate) for 5 μm and 10 μm Photo-BCB

on obtaining maximum via resolution without sacrificing structural properties. At low prebake temperatures the films are washed away (liftoff) during the subsequent solvent development process. At high temperatures the films are susceptible to crack formation during development. The hot plate pre-bake conditions of 75°C for 90 sec (5 μm films) and 90°C for 90 sec (10 μm films) are typical. This gives a robust processing window for each resin.

Oven bakes of 75°C for 20 minutes were found to work well for both formulations (5 μm and 10 μm).

11.12.3.4 Exposure

One of the most important factors in obtaining quality vias with the desired sidewall profile is the exposure process. In general, for photosensitive polyimides, the mask dimensions define the dimensions at the bottom of the vias while the top dimensions change (blow out) during development and cure [181]. For the Photosensitive-BCBs, the mask dimensions define the dimensions at the top of the via; and the bottom via dimensions are determined by the slope of the sidewall and the extent of the "foot" (region of gradually sloping polymer at the bottom of the via). All of the via resolution values reported in this paper refer to the dimensions between the foot which are totally clear of polymer (see Figure 11-54).

Exposures can be performed using: (a) a projection scanner and (b) printers in hard contact, soft contact, and proximity modes. In general, contact and projection techniques produced more vertical side wall profiles than proximity modes because of the reduction in focus effects [182]. The use of contact exposure, however, results in yield loss and an increase in the number of mask cleaning operations required.

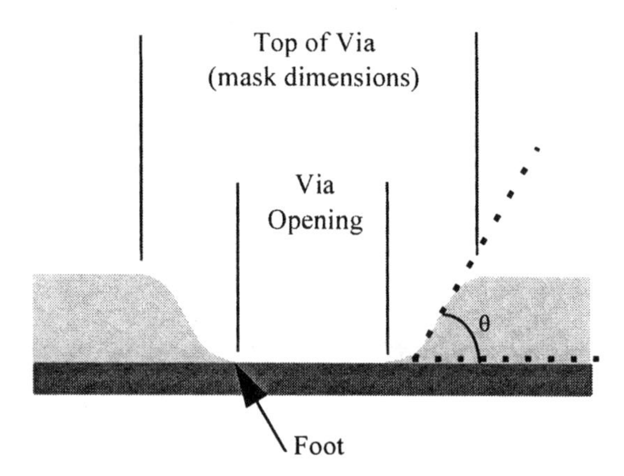

Figure 11-54. Schematic Diagram of a Via in Photo-BCB where θ is the Via Sidewall Angle

Films of the Photo-BCB resins are highly sensitive to i-line radiation, but can be exposed using broad band tools. The absolute dose level required to image Photo-BCB is related to the optical cross-section of the photo-additives contained in the resin, relative to the thickness of the film.

Figure 11-55 is a plot of via resolution versus exposure dose and Figure 11-56 shows via resolution vs exposure gap using a Karl-Suss MA-150 proximity exposure tool. It is clear that resolution increases with a decrease in exposure gap. Figure 11-57 is a plot of via resolution versus exposure dose using a Perkin-Elmer scanning projection printer.

At extremely low doses, Photo-BCB films are insufficiently cross-linked at the bottom of the film, and swelling or liftoff of the films results during the development process. At excessive exposure doses, the vias begin to close up due to light scattering off the underlying surfaces. Exposure doses of 175 to 200 mJ/cm^2 is typical using the 5 μm formulation, and 750 to 900 mJ/cm^2 is typical for the 10 μm formulation.

11.12.3.5 Solvent Develop

The unexposed areas of the film can be developed using a variety of different solvents. Stoddard Solvent had been used to define the vias with good results [169,170]. Isopar-L®/Proglyde-DMM® (isopar/proglyde)

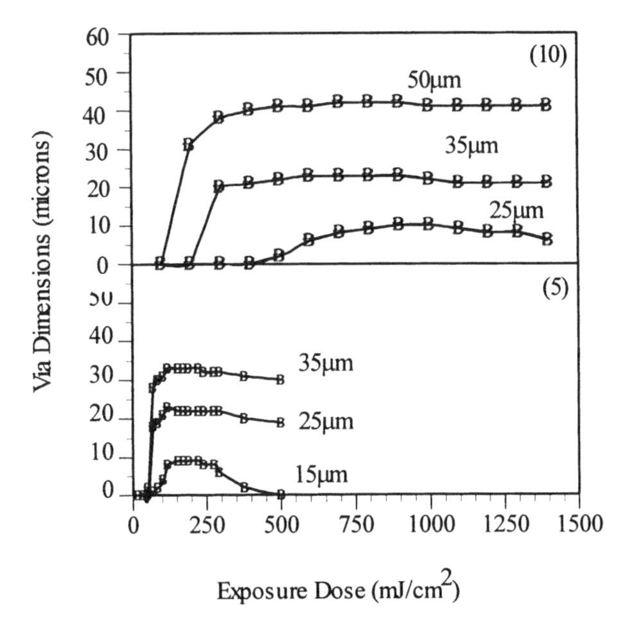

Figure 11-55. Via Resolution vs Exposure Dose for 10 μm and 5 μm Photo-BCB Using a Karl Suss MA-150 (@ 10 μm print gap)

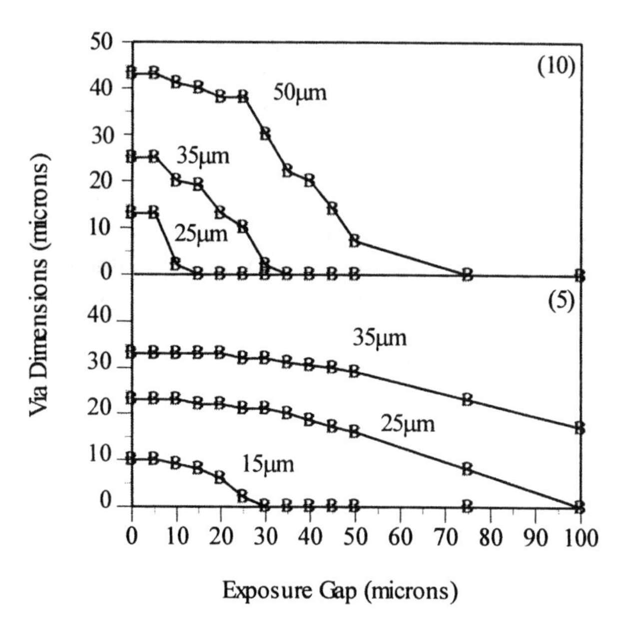

Figure 11-56. Via Resolution vs Exposure Gap for 10 µm and 5 µm Photo-BCB Using a Karl Suss MA-150

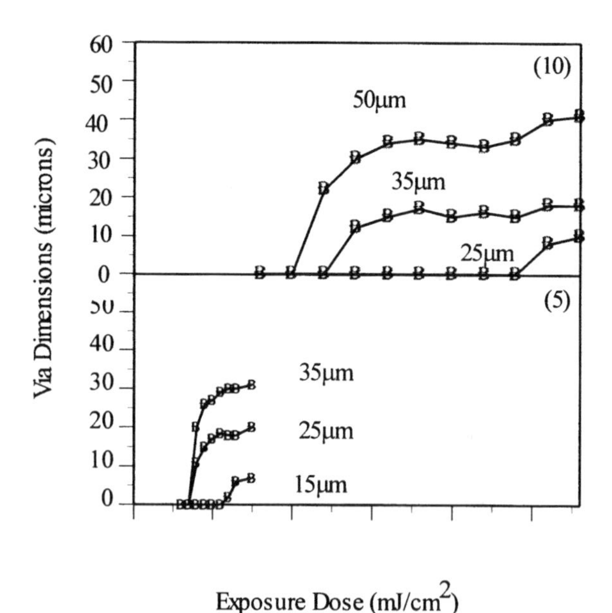

Figure 11-57. Via Resolution vs Exposure Dose for 5 µm and 10 µm Photo-BCB Using a Perkin-Elmer Projection Printer (focus depth = 0, aperture = 4)

(Dow DS2100) and tri-isopropyl benzene (Dow DS 3000) improve the via resolution and exhibit wider processing windows.

In general photo PI looses thickness (about 50+%) due to shrinkage during imidization when water and the cross linking agent are evolved. BCB, however, looses very little thickness during curing (about 5%) but does loose thickness during development due to the disolution of lower molecular weight (Mw), uncrosslinked polymer chains. This loss is dependent on variables such as exposure dose, but is typically 10–20%.

Vias can be opened using "puddle" and "immersion" techniques with isopar/proglyde as the developer. The puddle method consists of placing the exposed substrates onto the stationary chuck of the spin coater and dispensing a layer of solvent (puddle) on the surface. The solvent was left on the film for a predetermined period of time to allow complete dissolution of the non-exposed areas. At the end of the development period the devices are rinsed by spinning, while a stream of the rinse solvent (developing solvent or water based surfactant) is sprayed over the surface.

The key to a successful development is to allow adequate puddle time for completion of the develop process and rinsing briefly (while spinning) to remove the dissolved polymer from the substrate surface. When determining the process windows, the development "end-point" is visually detected as a series of colored interference fringes which disappears when the film is completely dissolved from the non-exposed areas. An additional 10-30 sec of over-development is typically used to ensure complete dissolution (see Figure 11-58).

The develop end-point, is dependent on the film thickness, solvent composition, delay time between prebake and development, and temperature of the solvent. Figure 11-59 illustrates the affect of film thickness on development time. For both formulations, thicker films take longer to develop than thin films.

Figure 11-60 shows the relationship between develop endpoint and pre-bake temperature.

Figure 11-61 illustrates the rate of develop endpoint increase as a function of the delay time between prebake and development steps using the standard 90°C/90 sec (10 μm films) and the 75°C/90 sec (5 μm films) prebakes.

The combination of these phenomena require that either a set protocol be used for developing within a specified time window, or an end-point detection system be used for each wafer (or batch of wafers). In order to cover a 24 hr delay (typical in a manufacturing environment), a puddle develop time of 75 sec is recommended for the 5 μm formulation and 125 sec for the 10 μm formulation. This puddle time should provide an adequate amount of over development and good via resolution while maintaining an acceptable level of film loss. For delays greater than 24 hrs, additional develop time is required.

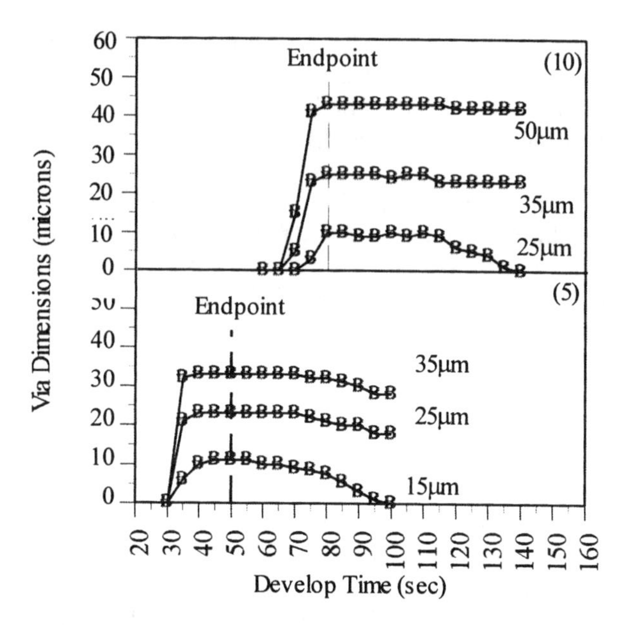

Figure 11-58. Via Resolution vs Develop Time for 10 μm and 5 μm Photo-BCB

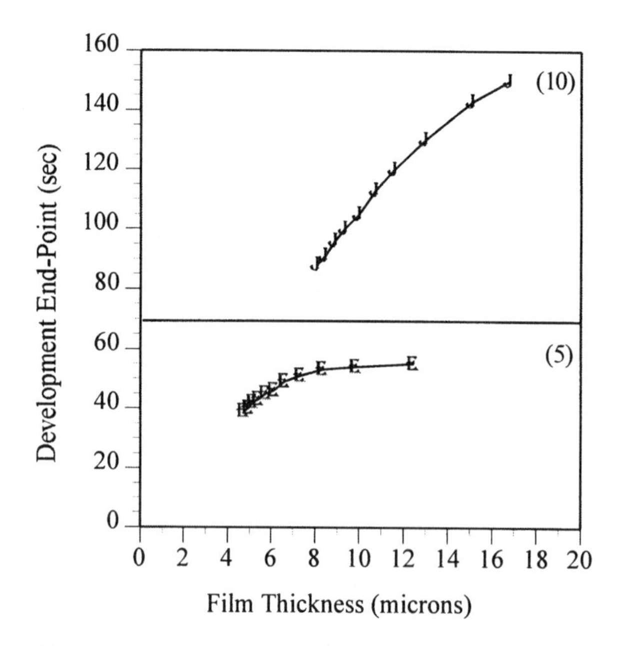

Figure 11-59. Develop End-point vs Film Thickness for 5 μm and 10 μm Photo-BCB

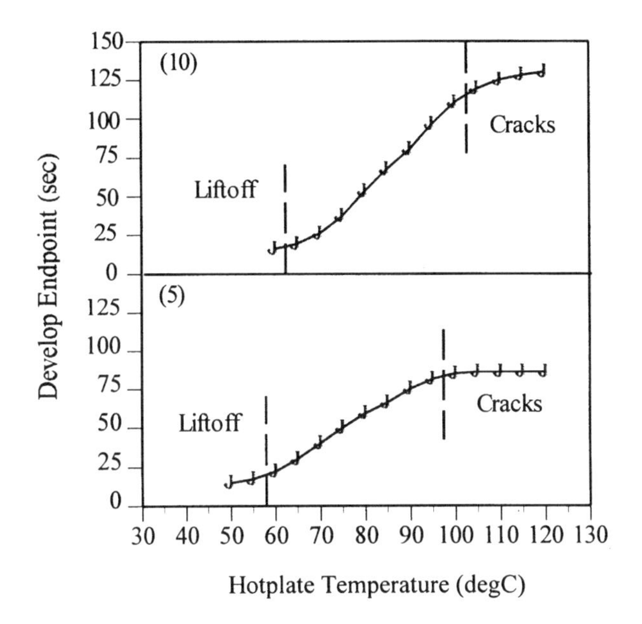

Figure 11-60. Develop End-point vs Hot Plate Temperature for 5 μm and 10 μm Photo-BCB

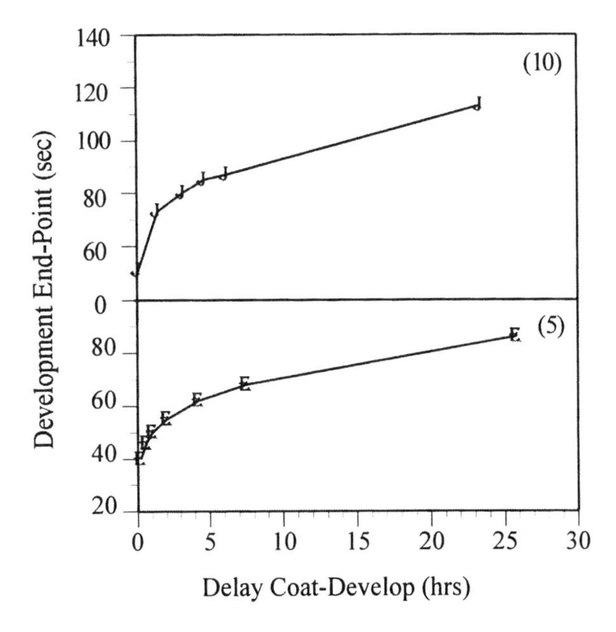

Figure 11-61. Develop End-point vs Delay Time After Coating for 5 and 10 μm Photo-BCB

Figure 11-62 shows the film thickness as a function of over development.

An SEM of a 35 μm via formed by isopar/proglyge development using the puddle technique in a 10 μm layer of photo BCB is shown in Figure 11-63.

Other traditional developing techniques such as tank immersion and spray developing are applicable with triisopropylbenzene (TIB) developer.

11.12.3.6 Rework

Prior to cure, films which are inspected and found to have defects or are misaligned, can be stripped. Parts to be stripped are submerged in a stripper bath solution for about 30 minutes at RT. This process does not effect underlying layers of metal or soft cured polymer.

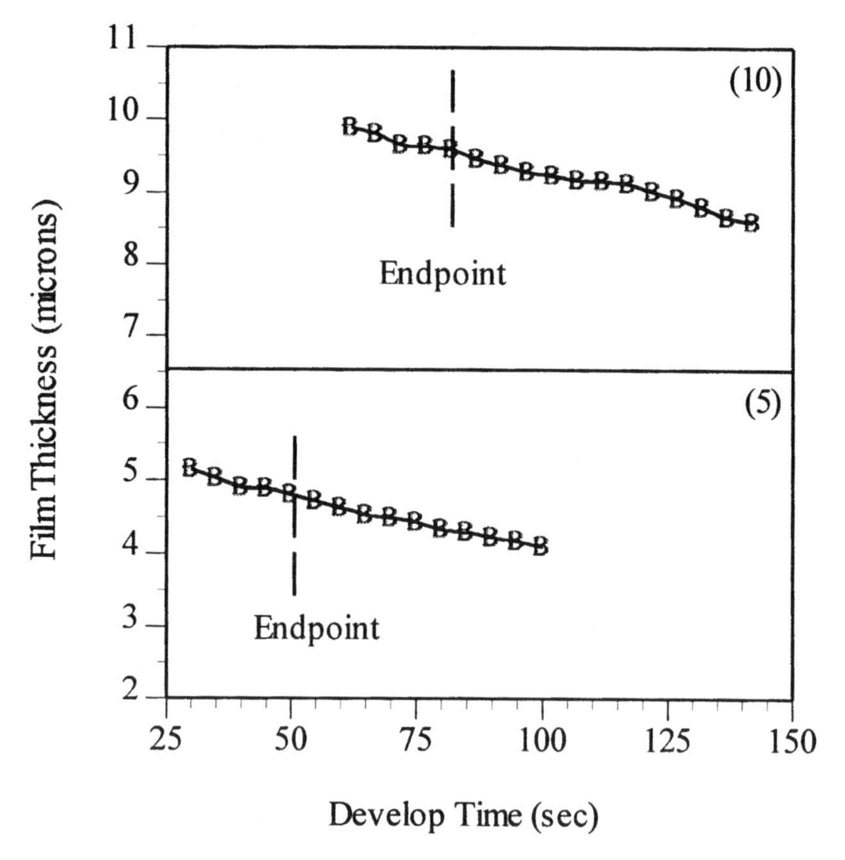

Figure 11-62. Film Thickness as a Function of Over-Development for 5 and 10 μm Photo-BCB

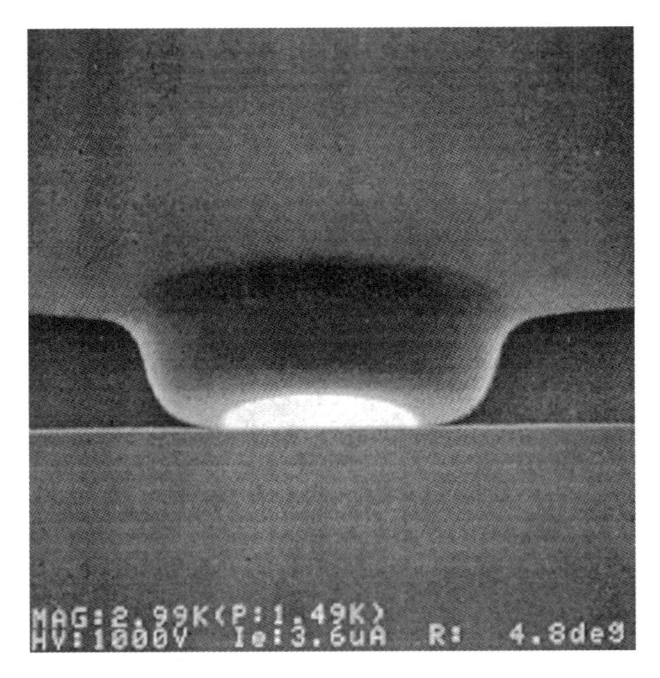

Figure 11-63. SEM Micrographs of 35 mm Vias in 10μm Film of Photo-BCB

11.12.3.7 Cure

At this stage of processing, the polymer film was cured to ensure resistance to subsequent processing operations, e.g. chemical baths, metallizations, and thermal cycling. The cure must be carried out under nitrogen [183]. Experience has shown that adhesion between multiple Photodefinable-BCB layers can be enhanced if each layer of BCB is partially cured to about 75–85% (soft cure). After all processing is complete the films were cured to a 95+% level in a final curing step (hard cure). Alternatively, cure may be performed in a belt oven with equally good results [184].

11.12.3.8 Descum

A descum process is required to clean the vias of debris and/or remove the very thin residual film which is often present (<100 Ångstroms) after the development process. This is typically accomplished with a O_2/ CF_4 mixture.

11.12.3.9 Substrate Dicing

Two different dicing alternatives have been examined for Photosensitive BCB: (1) photo pattern and then solvent develop the scribe streets at each layer during the fabrication process; or (2) laser ablation [185,186] through the full dielectric thickness after all processing is complete. Both alternatives are followed by traditional sawing through the silicon wafer.

11.12.3.10 Applications of Photo BCB

Figure 11-64 shows a 43 sq mm R-3000 workstation fabricated using the Cyclotene™ 4000 series by NEC (174). The chip set includes an R3000 CPU (1 integer Unit, 1 Floating Point Unit) and 1 Mbyte of SRAM memory. The chips have been mounted using NEC flip TAB technology. The conductor is 4 μm × 20 μm plated up copper. No adhesion or barrier layer metallization is present resulting in a conductivity of 1.75 W/cm. BCB layers are 9 μm thick resulting in an overall impedance of 50 W ± 4%. The modules passed MIL 883 long term reliability tests [500 thermal cycles from –45 to 125°C; high temperature storage (125°C) for 3200 hrs; 85% humidity and 85°C for 3200 hrs.] Resistance changes in all cases were <0.2%.

Figure 11-65 shows a commercial MMS Twinstar™ module. The Twinstar contains two Pentium® processor die mounted on a Cu/BCB high density substrate. The substrate is housed in a 347 pin PGA. The Twinstar is used in high end server applications [176].

Figure 11-64. NEC R-3000 Workstation MCM

Figure 11-65. MMS Twinstar Cu/BCB High Density Module

11.13 PHOTOSENSITIVE EPOXY

Epoxies are usually made photodefinable by mixing the base resin with photosensitizing agents. Intrinsically photosensitive epoxies are in the developmental stage and their availability on a commercial scale are limited.

11.13.1 Photodefinable Cyanate-esters and Epoxies

A negative acting acrylate-triazine based photodefinable epoxy was formulated by AT & T for the Polymer Hybrid Integrated Circuit (POLY-CHIC) to provide the dielectric medium between conductors layers [187,188]. The low dielectric constant ($\epsilon \simeq 2.8$ for the base triazine) combined with fine line capability in a controlled impedance environment provide the fabrication capability for MCM applications requiring high frequency, high interconnect density with a large number of I/O's. This polymer is spray coated to produce films 25–50 μm thick, UV patterned and solvent developed. The formulation of this material is described in reference 188. A summary of the polymer properties and their corresponding electrical properties are summarized in Table 11-32.

Table 11-32. Physical and Electrical Properties of Cured Triazine-based Epoxy (PHP-92)

Property	Value	Comments
Moisture Absorption	~2% @ 100% RH $\Delta Z_0/Z_0 < 5\%$	typical of epoxy systems 25% RH-85% RH
Thermal Conductivity	0.2 W/m-K	
TCE	65 ppm/°C	TMA (25-100°C)
T_g	180°C (155-210°C)	DSC & TMA
Thermal	180°C	long term
Stability	300°C	spikes (i.e. soldering)
E_s (Young's Modulus)	2–7 × 10^{10} dynes/cm^2	mechanical tensile tester
Room Temp. Film Stress	2.2–2.9 × 10^8 dynes/cm^2	warpage test with fused quartz disk
Thermal Cycling	>200 cycles	−40°C + 130°C
ε (for resin)	2.8	GHz range: Z_0 from reflection coefficient and propagation
tan δ	<.025	GHz range
Dielectric Strength	35V/μm (900 V/mil)	ASTM D149
Volume Resistivity	>3.7 × 10^{12} Ω-cm	ASTM D257 (instrument-limited)

Researchers at IBM designed several chemically amplified epoxy resists systems by combining commercially available epoxy resins and photoinitiators such as aryldiazonium, diaryliodonium, and triarylsulfonium metal halides [189]. When formulated with the onium salt photoinitiators, epoxy resins undergo cross-linking reactions to produce negative tone resists. Negative tone resists often encounter swelling of the cross-linked matrix during development in organic solvents. Swelling results in distorted images and loss of adhesion to the substrate. Stewart et al [190] formulated an organic solvent developable UV resist capable of providing submicron images with reduced swelling. The reaction involved conversion of optically transparent commercial resins, styrene-allyl alcohol copolymers to glycidyl ethers, thereby providing cationically polymerizable functionalities.

11.13.2 Intrinsically photodefinable epoxy

Ciba offers a higher dielectric constant (4.1 at 1 MHz) photodefinable LMB 7081 epoxy resin with a shelf life of 6 months [191]. In research studies, this material was deposited as a thin dielectric layer (~20 microns) on copper cladded laminates using the following processing scheme: (i) spin coat at 2000 rpm for 35 sec, (ii) tack dry at 60–80° C using a hot plate, (iii) expose at 365–400 nanometer, ~1600 mJ/cm^2, (iv) oven heat at 110°C, 1 hr, (v) develop in γ-butyl lactone, 3–5 minutes, and (vi) a final cure at 150°C for 1 hour. Vias as small as 40 μm in both array and chain configurations were opened without any smearing of epoxy. Preliminary results indicate that the LMB 7081 formulation can be a good candidate material for thin film deposition of MCM-L type application primarily due to its simplified processing and much lower cost compared to polyimides.

11.14 MICROELECTRONIC APPLICATIONS OF PSPIs

Polyimide coatings or films have been widely utilized in the field of microelectronics in both chip and packaging applications due to their excellent thermal and electrical properties. For semiconductor devices, these applications include (a) protective coatings for Si and GaAs devices, (b) insulation dielectric layers, and (c) stress-relief buffer coatings. For thin-film packaging, applications include (a) dielectric insulation layers and (b) corrosion-protective passivating layers to protect metallurgy in-structures fabricated for use in the computer and communications industries. They are also used as protective coatings and dielectric insulation layers in printer thermal head applications and in photolinesensors. Starting in the late 1980s, PSPI coatings have been utilized in these applications because very precise relief patterns could be formed by using conventional photoresist processes. Many researchers also have proposed various kinds of applications of PSPIs in other fields. A sampling of some of the more common applications which use the PSPIs described in chapter are listed below.

11.14.1 Chip Coating Applications

11.14.1.1 Alpha-Particle Protection Layers

In the early 1980s, memory chips had encountered a very serious problem called soft error, which is the disappearance of memory caused by penetration of an alpha particle. The source of the alpha particle was found to be uranium or thorium in the inorganic filler, such as silica or alumina, in encapsulation materials. One of the preventive measures was polyimide coating over the chip as an alpha-particle protection layer. A thick layer of more than 30 μm is necessary to absorb an alpha particle. The first target of Rubner's PSPI was this application. However, alpha-particle shielding is not as critical today because fillers of low uranium content became available. The polyimide coating itself must have on the order of less than 1 parts per trillion of uranium or thorium content [192].

11.14.1.2 Buffer Coating for Stress Relief

The stress generated under plastic-encapslation processing causes a substantial amount of damage to the LSI chip by shearing and moving alumimun wiring and cracking of the inorganic passivation layer. The polyimide chip coating before encapslation can prevent these damages and increase production yields. Therefore, this process is carried out in many LSI production lines, where PSPI is patterned so that the via hole for wiring and the dicing line is opened. Khan et al. studied the effect of buffer coating on the production yields of chips. They applied ester-type PSPI based on BTDA–ODA–MDP to bipolar interface devices containing piezoresistive circuits with plastic encapsulation, which would typically have 10% yield loss due to threshold voltage failure. Accelerated testing with unbiased and biased pressure-cooker storage produced results in favor of the PSPI-coated devices as compared with the uncoated control. [193]

In the past, the use of Probimide 400 as a buffer coat has been delayed since the base of installed i-line steppers required for this technology was limited. Today, i-line steppers are available in many manufacturing facilities which may give new opportunities for Probimide 400 use. Again, for throughput considerations, the thickness range of 2–6 μm is recommended.

11.14.1.3 Thin-Film Interconnect

One main application of "Photoneece" is in semiconductor devices. In this application, the photosensitive polyimide is required to adhere well to silicon, Si_3N_4, and encapsulating resins. "Photoneece" exhibits superior adhesion to these materials, as shown in Table 11-33. It is especially suitable for GaAs substrates which require low curing temperature, below 300°C. Naitoh et al. evaluated "Photoneece" as an insulation layer for five-layered metallization structures having 20K gates CMOS logic LSI [194], where the thickness of the polyimide layer was 1.6 μm and the via hole size was 3 μm.

11.14.2 Packaging Interlayer Applications

11.14.2.1 Interlayer Applications in High-Density Multichip Modules

Polyimide has been utilized as insulation layer in interconnect structures because of the good planarization properties, good thermal performance, and excellent electrical characteristics for the propagation of high-speed signals [195,196].

11.14.2.2 Applications Using Covalently Bonded Ester-Type PSPIs

Nowadays, much attention has been given to PSPI in this field, due to easy patterning of fine via holes for interconnection between signal layers by a very simple process. Ester-type PSPIs have many advantages for this application (e.g., wide process margin, etc.). Chakravorty of Boe-

Table 11-33. Adhesion Properties of "Photoneece" to the Molding Compound[a]

Polyimides	Curing Temperature (°C)	Adhesive Strength (MPa)	
		Initial	After PCT[b]
Poly(amic acid) composed of PMDA and DAE	320	12	0
	350	12	0
	350[c]	15	0.2
	350[d]	24	0.5
	400	18	0
"Photoneece" UR-3140	320	30	18
	350	30	21
	400	39	24
"Photoneece" UR-4144	320	18	12
	350	30	21
	400	30	15

[a]TH7006 produced by Toray.
[b]PCT (120°C, 100% **rh**, 168 h).
[c]O_2 plasma treatment to polyimides.
[d]Aminosilane treatment to polyimides.

ing evaluated Selectilux HTR-3 of EM Industries as a dielectric material in a high-density thin-film Cu/PI interconnect structure on a silicone substrate [197,198]. He investigated the dependence of the PSPI or polyimide profile on exposure dose, linewidth, and bake temperature. He fabricated a high-density multichip memory module containing two layers of orthogonal *X*-*Y* conductor lines with interconnecting vias, with a density of close to 800 lines/cm. The signal layers consisted of 25-μm-wide, 6-μm-thick electroplated copper lines fabricated at a pitch of 80 μm in a polyimide dielectric 6 μm thick. The module consisted of six 64K SRAM chips, two 8-bit buffers, and four discrete ceramic chip capacitors. Transmission characteristics and signal cross-talk between lines were also measured [199-205].

There are many reports in the literature describing the formation of interconnect structures which use ester-type PSPIs. Takahashi and Kolesar of Sacramento Air Logistics Center, applied Selectilux HTR 3-200 to the preparation of a wafer-scale integrated circuit [203]. Myszka et al. of Motorola fabricated multichip module type-D using PSPI, including the ester type. They built up polyimide dielectrics and copper conductor interconnect layers on cofired substrates, and they called this structure a "grafted multilayer" [207].

11.14.2.3 Applications Using Ionic Bonded-Type PSPIs

Another main application field is an assembly package which requires a high-density metallization and low-dielectric-constant insulation layer in order to reduce package size and signal propagation delay [208,209]. A multichip module for a computer is also included in this field [210,211]. Inoue et al. (NEC) reported the multichip module [212] for the NEC SX-3 supercomputer using the "Photoneece" shown in Figure 11-66 having 25-μm-wide, 75-μm center-to-center spacing high-density wiring, polyimide insulating layers 20 μm in thickness and with 60 μm square via holes, four signal layers and 6-ns/cm signal transmissions on a 225-mm square substrate have been achieved by using "Photoneece."

Takasago et al. (Mitsubishi Electric Corp.) has developed a hybrid IC which uses "Photoneece" for the insulation layers [213-15]. They use copper substrates to get good heat dissipation. A multilayer circuit with polyimide insulation layers and copper heat sink for integrated circuits are formed on this substrate. Standard film thickness of "Photoneece" is 10 μm.

11.14.2.4 Application Using Nonshrinking Preimidized PSPI

Due to its dimensional stability upon curing, the most appealing application of Probimide 400 has been in multichip modules. Very substan-

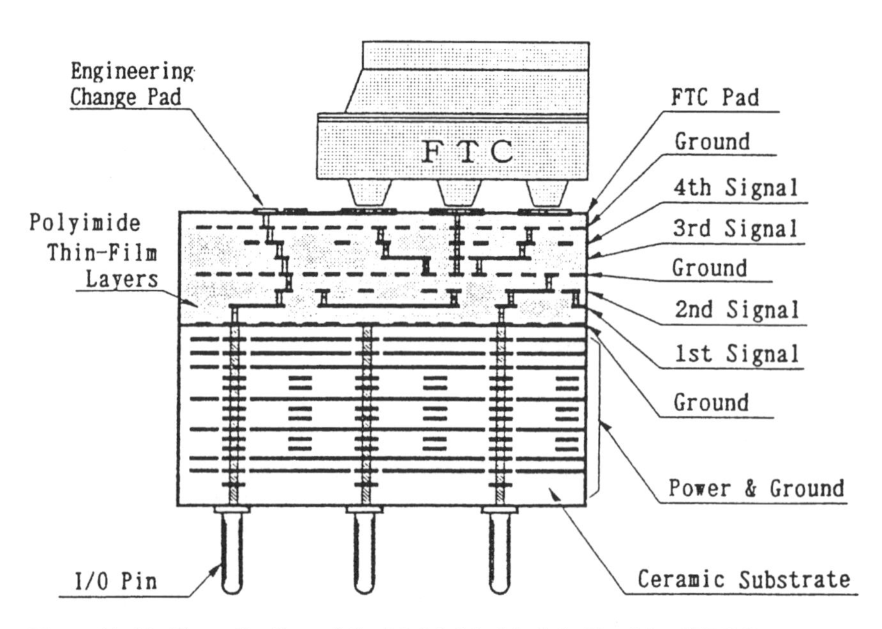

Figure 11-66. Cross Section of the Multichip Module Used for SX-3 Supercomputer Using "Photoneece" (From Ref. 212)

tial work has been done through various groups at Boeing Corp. [197-203]. Thus, in a key paper, a fiber-optic receiver/transmitter set with eight copper/polyimide layers is described, which has copper lines 5 μm thick with a nominal linewidth of 19 μm. When working out this eight-layer MCM process, the Boeing groups could also introduce the following essential processing innovations when working with Probimide 400:

- Probimide 400 has been shown to be Cu compatible, in contrast to other PSPI (ester type).
- Cure time per polyimide layer could be reduced to 5 min/400°C to give enough solvent resistance for the next layer to be applied.
- Solvent diffusion on the other hand can advantageously be exploited for greatly improved planarization.

A cross section of a copper/Probimide 400 interconnect structure with high planarity is shown in Figure 11-67. A top view of improved planarization through solvent diffusion in Probimide 400 is depicted in Figure 11-68.

Figure 11-67. Micrograph (≈ 200×) of a Polished Cross Section of the Optical Transceiver

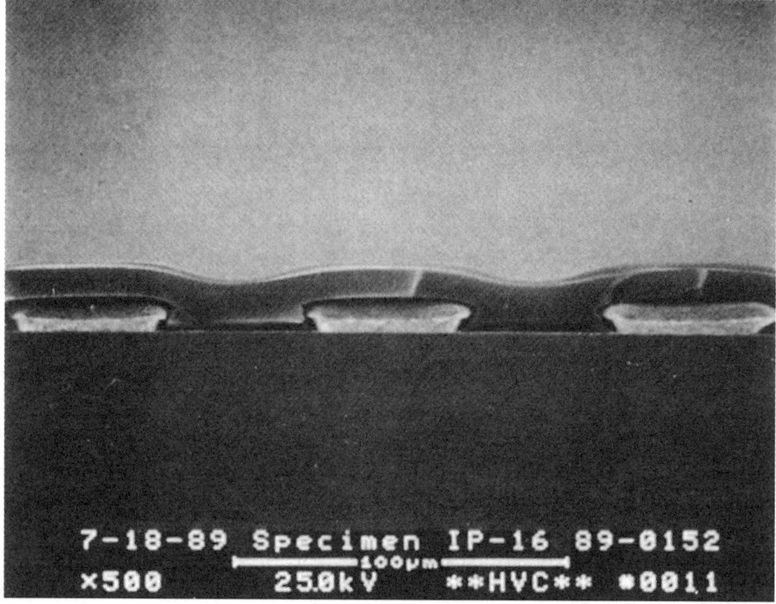

Figure 11-68. SEM Photographs Showing the Superior Planarization Achieved Corresponding to a Denser Conductor Configuration in a Copper/Probimide Interconnect Structure. The fabrication process is outlined in Chakravarty and Tanielian [197]; see Chakravarty *et al.* [198] for other pertinent details.

11.14.2.5 Optical Waveguides

Probimide 400 has been shown to exhibit low optical losses when used as a waveguiding material [198]. Losses of <0.5 dB/cm have been obtained. The generation of embedded channel waveguides through the generation of photolocked regions have been described by Chakravorty [197-200]. Thus, a dopantlike benzyldimethylketal is included in the photo-cross-linkable Probimide 400 matrix and photochemically locked by exposure to form channels of lowered refractive index.

In this context, it appears particularly appealing to combine syntergistically optical and electrical material interconnecting technologies with one polymeric material, as described in Ref. 206.

11.14.2.6 Other Applications

As described in the preceding section, the application of PSPI started in the field of microelectronic packaging. Nowadays, PSPIs are applied in other fields such as optoelectronics [216], display [113], and micromachines [217]. PSPI evaluation in these new fields are mostly carried out by using ester-type PSPI, possibly because ester-type PSPI will meet the diversified requirements of the microelectronics and related industries.

11.15 RESEARCH DEVELOPMENTS/FUTURE POLYMERS

11.15.1 Ultrathin PSPIs (LB Membrane)

A Langmuir–Blodgett (LB) membrane is an unimolecular layer, which recently has attracted a lot of attention from the viewpoint of functional membranes such as photoresist, conductive membrane, sensor, and molecular devices. Uekita-et al. of Kanegafuchi Chemical studied a novel LB membrane of ester-type PSPI and its patterning properties. They prepared a precursor having the heptadecenyl group by condensation reaction of diacid chloride of diheptadecenyl pyromellitate and diamino diphenyl ether. They found that the mixture of the precursor, octadecyl alcohol, and benzoin ethyl ether gave *Y*-type multilayer LB films. The thickness of the monolayer was measured to be 22 Å by ellipsometry. A film consisting of 31 LB layers deposited on a silicon wafer was irradiated by deep UV (254 nm) light. After the exposure, the multilayer was developed in chloroform–dimethylacetoamide and rinsed in ethanol. The photosensitivity, defined as a normalized thickness of 0.5, was 140 mJ/cm^2. Infrared spectra indicated that cross-linking occurred between the double bonds.

After the mixed multilayers (21 layers) were exposed through a photomask, development of the film gave a negative image. A resolution of a 0.5 mm space and a 1.0 mm line was obtained [218].

Lu et al. of Southeast University in China prepared ultrathin conductive films by pyrolysis of PSPI LB films at 1000°C in hydrogen atmosphere. They found that the simple ester-type PSPI based on PMDA/ODA/HEA (hydroxyethyl acrylate) chemistry could form a LB membrane. The conductivity of the pyrolyzed LB films increased with the film thickness and reached the stable value of 180 Scm^{-1}. The pyrolyzed LB films on silicone dioxide substrate were studied by scanning tunneling microscopy [219].

11.16 SUMMARY AND FUTURE CHALLENGES

Most negative PSPIs shrink about 50% upon curing to 400°C, whereas others are preimidized and intrinsically photosensitive. Positive PSPIs are just starting to be developed and have not yet been commercialized. Typical properties of some PSPIs are shown in Table 11-34. Just as in conventional nonphotosensitive polyimides, the push is on to develop low-stress, low-dielectric-constant polymers for the chip and packaging applications. PSPIs are being developed with low-stress and low-dielectric-constant polyimides which can be made into either positive or negative acting photosensitive polyimides.

The challenge the technical community faces is to make the backbone chemistry as transparent as possible to the exposure wavelengths used for patterning fine lines and vias over a variety of substrates. After formulation of the polymer with the photopackage, the base resin properties should remain unchanged, especially after thermal stressing and solvent

Table 11-34. Typical Commercially Available Photosensitive Polyimides

Backbone Chemistry	BTDA–X-DA	BTDA–ODA	BPDA–ODA	BTDA–PMDA–ODA
Lithography photopackage	i-Line	g-Line	i-Line	g-Line
Thermal stability (isothermal weight loss at 400°C)	0.12% h	0.08% h	0.06% h	0.11% h
Residual stress	56 MPa	42 MPa	42 MPa	42 MPa
Mechanical properties (cured films)				
Tensile strength	137 MPa	192 MPa	178 MPa	151 MPa
Modulus	2.8 GPa	3.4 GPa	4.0 GPa	3.5 GPa
Strain at break	59%	69%	40%	44%
Glass transition temperature (T_g)	370°C	352°C	310–450°C	355°C

processing. Finally, the photopackage components should cleanly evolve from the polyimide film before complete imidization, to produce defect-free films. The challenges in developing low-cost PSPIs for both semiconductor and packaging applications with minimum photopackage impact on the base polymer properties and with minimal film shrinkage are formidable.

Acknowledgment: The authors wish to thank Dr. Swapan Bhattacharya of Georgia Tech for sections related to epoxy and positive photopolymer.

11.17 REFERENCES

1. R. R. Tummala. "Electronic Packaging in 1990's, A Perspective from America," *IEEE Trans. Components Hybrids Manuf. Technol.*, CHMT–14: pp. 262–271, 1991.
2. R. R. Tummala. "Ceramic and Glass-Ceramic Packaging in the 1990's," *J. Am. Ceram. Soc.*, 24: pp. 895–908, 1991.
3. Integrated Circuit Eng. Corp. Update & "Worldwide IC Industry Economic Forecast", pp. 2–18, 1991.
4. "Experts: GaAs Applications Booming", *Semicond.* Int., pp. 26–29, October 1991.
5. R. Iscoff. "High Speed GaAs: Still a Niche Technology?" *Semicond. Int.*, pp. 60–66, 1992.
6. "Japan Report," *Semicond. Int.*, p. 35, October 199.
7. G. Larrabee and P. Chatlerjee. "DRAM Manufacturing in the 90s." *Semicond.* Int., pp. 90–92, May 1991.
8. R. J. Kopp. "Navigating to Advanced Wafer Processing," *Semicond. Int.*, pp. 34–41, January 1992.
9. *Electron. Eng. Times*, Issue 687, pp. 1 and 16, April 6, 1992.
10. P. J. Cavill, J. M. Wilkenson, and P. E. Stapleton. "Wafer-Scale Integration," *Microelectron. Manuf. Technol.*, pp. 57–58, May 1991.
11. P. Burggraaf. "Lithography's Leading Edge, Part I: Phase-Shift Technology," *Semicond. Int.*, pp. 42–47, February 1992.
12. I. A. Shoreef, J. R. Maldanado, Y. Vladimirsky, and D. L. Katcoff. "Thermoelastic Behavior of X-ray Litho Masks During Irradiation," *IBM J. of Res. Devel.*, 34: pp. 718–735, 1990.
13. P. Singer. "Trends in CMOS Development," *Semicond. Int.*, pp. 56–60, April 1992.
14. "Wafer Process News," *Semicond. Int.*, p. 32, February 1992.
15. *Proceedings, 8th International IEEE VLSI Multilevel Interconnection Conference*, 1991.
16. M. M. Khan, T. S. Tarter, and H. Fatemi. "Stress Relief in Plastic Encapsulated Integrated Circuit Devices by Die Coating with Photodefinable Polyimide," Proceedings of the 38th Electronics Components Conference, Los Angeles, Ca. May 9–11, 1988.
17. G. Samuelson and S. Lytle. "Reliability of Polyimide in Semiconductor Devices," Symposium on Polymer Materials for Electronic Applications, Coatings, and Plastics Chemical Division, ACS 2nd International Congress, Las Vegas, NV, pp. 1–27, 1980.
18. T. Kwok and P. S. Ho, in *Diffusion Phenomena in Thin Films and Microelectronic Materials*, ed. by D. Gupta and P. S. Ho, pp. 369–425, 1988.
19. M. A. Korhonen, P. Borgesen and Che-Yu Li. "Stress-Induced Voiding and Stress Relaxation in Passivated Metal Lines"; M. Inoue and S. Ogawa. "Japanese Perspec-

tive: Recent Studies on Stress Induced Phenomena in Metallization in Japan"; P. Totta. "Interconnection Materials Reliability U. S. Experience," *First International Workshop on Stress Induced Phenomena*, 1991.

20. R. A. Larson. *IBM J. Res.* Devel., 24(3): p. 268, 1980.
21. H. C. Cook, P. A. Farrar, R. R. Uttecht, and J. P. Wilson. *OBM Tech. Disc. Bull.*, 16: p. 728, 1983.
22. A. W. Lin. Evaluation of Polyimides as Dielectric Materials for Multichip Packages with Multilevel Interconnection Structure, Proceedings 39th Electron. Components Conf. (ECC), pp. 148–154, 1989.
23. B. Schwartz. "Multilayer Ceramics," Mat. Res. Soc. Symp. Proc., Vol. 40, pp. 50–59, 1985.
24. E. S. Anolick, J. R. Lloyd, G. T. Chiu, and V. Korobov, in *Polyimides: Synthesis, Characterization, and Applications*, ed. by K. L. Mittal, vol. 2., pp. 889–904, Plenum Press, New York, 1984.
25. "Overcoats Protect RAMs from Alphas," *Electron. Rev.*, pp. 41–42, September 11, 1980.
26. M. F. Bregman and C. A. Kovac. *Solid State Technol.*, pp. 75–80, 1988.
27. R. J. Jensen, J. P. Cummings, and H. Vora. *IEEE Trans. Components Hybrids Manuf. Technol.*, CHMT-7(4): pp. 384–393. 1984.
28. T. Watari and H. Murano. *IEEE Trans. Components Hybrids Manuf. Technol.*, CHMT-8(4): pp. 462–467, 1985.
29. H. Tsunetsugu, A. Takagi, and K. Moriya. *Int. J. Hybrid Microelectron.* 8: pp. 21–26, 1985.
30. R. J. Jensen. Proc. *ACS Div. Poly. Mater.: Sci. Eng.*, 55: pp. 413–419, 1986.
31. L. B. Rothman. *Solid State Sci. Technol.*, 2216–2220, 1980.
32. K. Sato, S. Harada, A. Saiki, T., Kimura, T. Okubo, and K. Mukai. *IEEE Trans. Parts Hybrid Packaging*, 9PHP-9: p. 176, 1973.
33. G. Czornyj, R. Chen, S. Kim, G. Prada-Silva, M. Ree, H. Souleotis, and D. Yang. *Proc. of the 1991 Spring Meeting of the Materials Research Society*, 1991.
34. T. F. Redmond, C. Prasad, and G. A. Walker. "Polyimide Copper Thin Film Redistribution on Glass Ceramic/Copper Multilevel Substrates, *41st Electronic Components and Technology Conference*, p. 689, 1991.
35. R. R. Tummala, H. R. Potts, and S. Ahmed. "Packaging Technology for IBM's Latest Mainframe Computers (S/390/ES9000), *41st Electronic Components and Technology Conference*, p. 682, 1991.
36. C. Feger, M. M. Khojasteh, and S. E. Molis, Polyimides: Trends in Materials and Applications, Proceedings of 5th International Conf. on Polyimides, Society of Plastics Engineers, Ellenville, N. Y. 1994.
37. D. Wilson, H. D. Stenzenberger and P. H. Hergenrother, Polyimides, Blackie & Son, Ltd., Chapman and Hall, 1990.
38. E. S. Moyer, in Polyimides: Synthesis, Characterization and Adhesion, J. E. McGrath, L. T. Taylor, T. C. Ward, J. P. Wightman, Eds. pp. 89–187, Virginia Tech Center for Adhesive and Sealant, Blacksburg, Virginia, 1992
39. P. Garrou and I. Turlik, Chapter 4, Materials of Construction: Substrate, Dielectric, Metallization. pp. 93–164, Thin Film Multichip Modules, G. Messner, I. Turlik, J. W. Balde and P. E. Garrou, ISHM 1992.
40. S. Numata, N. Kinjo and D. Makino, Chemical Structures and Properties of Low Thermal Expansion Coefficient Polyimides, Polymer Eng. & Sci., Vol. 28, #14, 1988.
41. M. T. Pottiger, J. C., Coburn and J. Edman, The Effect of Orientation on Thermal Expansion Behavior of Polyimide Films, Journal of Polymer Sci.: Part B: Polymer Physics, Vol. 32, pp. 825–837, 1994.

42. A. F. Arnold, Y. Y. Cheng, P. M. Cotts, R. D. Diller, D. C. Hofer, M. Khojasteh, E. Macy, P. R. Shah, and W. Volksen. "Polyimide Coating Compositions Based on Meta-Dialkyldihydrogen Pyromellitate and Aromatic Diamines", *U.S. Patent 4,849,501*, 1989.
43. W. Volksen, D. Y. Yoon, J. L. Hedrick, and D. Hofer. "Chemistry and Characterization of Polyimides Derived from Poly(amic alkyl Esters)", *Materials Science of High Temperature Polymers for Microelectronics*, ed. by D. T. Grubb, I. Mita, and D. Y. Yoon, p. 23, 1991.
44. H. R. Brown, A. C.M. Yang, T. P. Russell, W. Volksen, and E. J. Kramer, *Polymer*, 29: p. 1807, 1988.
45. W. Volksen, D. Y. Yoon, and J. Hedrick. "Polyamic Alkyl Esters: Versatile Polyimide Precursors for Improved Dielectric Coatings," *41st Electronic Components and Technology Conference*, p. 572, 1991.
46. K. L. Mittal. "Adhesion Measurement: Recent Progress, Unsolved Problems, and Prospects," in *Adhesion Measurement of Thin Films, Thick Films, and Bulk Coatings*, ed. by K. L. Mittal, pp. 5–17, American Society for Testing and Materials, Philadelphia, 1978.
47. Y. H. Kim, G. Walker, J. Kim, and J. Park. "Interfaces of Metal to Polyimide," *J. Adhesion Sci. Technol.*, 1(4): p. 331, 1987.
48. Y. H. Kim, J. Kim, G. F. Walker, C. Feger, and S. Kowalczyk. "Interface and Adhesion of Polymers to Metals," *J. Adhesion Sci.Technol.*, 2: p. 95, 1988.
49. S. Kowalczyk, Y. H. Kim, G. F. Walker, and J. Kim, "Polyimide–Copper Interfaces; Growth Sequence Dependence and Solvent Effects," *J Vac. Sci. Technol.*, A6(3): 1989.
50. M. Ree, T. L. Nunes, K.-J. Chen, and G. Czornyj, "Structure and Properties of BPDA–PDA Polyimide from Its Poly(amic acid) Precursor Complexed with an Aminoalkyl Methacrylate," in *Materials Science of High Temperature Polymers for Microelectronics*, ed. by D. T. Grubb, I. Mita, and D. Y. Yoon, p. 211, 1991.
51. F. Arnold, S. Z.D. Cheng, F. W. Harris, and F. Li. *Trends Polymer Sci.* 1(8): p. 243, 1993.
52. S. T. Holzinger. "TAB Types and Material Choices," *EXPO SMT'88 Technical Proceedings*, p. 247, 1988.
53. S. T. Holzinger. "TAB: Mechanical, Chemical, and Thermal Considerations for Three Layer TAB Materials," *EXPO SMT International '89 Technical Proceedings*, p. 149, 1989.
54. M. Monisinger. "Advantages of Polyimide Film in TAB Packaging," *Surface Mount Technol.*, 61, October, 1989.
55. C. E. Sroog, A. L. Endry, S. V. Abramo, C. E. Berr, W. M. Edwards, and K. L. Olivier. *J. Polym. Sci.* A, 3: p. 1373, 1965.
56. J. A. Kreuz, and R. F. Sutton. *Institute of Printed Circuits 33rd Annual Meeting*, 1990.
57. H. Nagano, H. Kawai, and K. Yonezawa. *Symposium on Recent Advances in Polyimides and Other High Performance Polymers*, 1987.
58. R. A. Kirchoff, and K. J. Bruza. "Benzocyclobutenes in Polymer Synthesis," *Prog. Polym. Sci.*, 18: pp. 85–1185, 1993.
59. P. Garrou. "Polymer Dielectrics for Multichip Module Packaging," *Proc. IEEE*, 80: p. 1942, 1992.
60. Burdeaux, D., Townsend, P., Carr., Garrou, P., "Benzocyclobutene Dielectrics for the Fabrication of High Density, Thin Film Multichip Modules", J. Electronic Materials, Vol. 19, 1990, p. 1357.
61. S. F. Gong, J. Strandberg, H. Theide, H. Hentzell, H. Hesselboom, and W. Karner. "Investigation of High Speed Pulse Transmission in MCM-D," *IEEE Trans. Components Hybrids Manuf. Technol.*, CHMT, 16: p. 735, 1993.

62. T. Shimoto, K. Matsui, and K. Utsumi. "Cu/Photosensitive BCB Thin Film Multilayer Technology for High Performance Multichip Modules," *Proceed. Int. Conf. MCM's*, p. 115, 1994.
63. P. Chinoy, and J. Tajadod. "Processing and Microwave Characterization of Multilevel Interconnects Using BCB Dielectric," *IEEE Trans. Components Hybrids Manuf. Technol.*, 16: p. 714, 1993.
64. H. Sakai et al. "A Millimeter-Wave Flip-Chip IC Using Micro-Bumpo Bonding Technology," Proceed. IEEE Int. Solid States Circuit Conf., 1996, p. 408.
65. M. Robertsson, K. Engberg, P. Eriksen, H. Hesselboom, M. Niburg, G. Palmskog. "Optical Interconnects in Packaging for Telecom Applications," Proceed. 10th European Microelectronics Conference, 1995, p. 580.
66. H. Projanto and D. Denton. "Moisture Uptake in BCB Films for Electronic Packaging Applications," *Proc. MRS Symp.*, 203: p. 295, 1991.
67. T. Stokich, D. Burdeaux, C. Mohler, P. Townsend, S. Warrington, J. Tou, B. J. Han, C. Pryde, H. Bair, and G. Johnson. "Thermal and Oxidative Stability of Polymer Thin Films Made from DVS-BCB." T. Stokich, D. Burdeaux, C. Mohler, P. Townsend, M. Dibbs, R. Harris, M. Joseph, C. Fulks, M. McCulloch, and R. Dettman. "Advances in the Thermo-oxidative Stabilization of DVS-BCB Polymer Coatings," *Proc. MRS Sympos.*, 265: p. 275, 1992.
68. T. M. Stokich, C. C. Fulks, M. T. Bernius, D. C. Burdeaux, P. E. Garrou, and R. H. Heistand. "Planarization with Cyclotene 3022 (BCB) Polymer Coatings," *Mater. Res. Soc. Symp.* Proc., 308: p. 517, 1994.
69. T. Tessier. "A Comparison of Common MCM-D Dielectric Material Performance," *Proceedings 6th SAMPE Electronics Conference*, p. 347, 1992.
70. F. A. Sherrima, I. A. Saadat, S. Sekigahama, A. A. Abado, J. O'Brien, and M. Thomas. Manufacturing Studies of BCB as the Interlevel Dielectric Material for Multilevel Interconnect MCM and VLSI Applications," *Proc. ISHM*, p. 596, 1992.
71. S. Bothra, M. Kellam, and P. E. Garrou. "BCB as an Interlevel Dielectric in a Multilevel Metal System," *J. Electron. Mater.*, 23: p. 819, 1994.
72. T. Shimoto, K. Matsui, M. Kimura, and K. Utsumi. "High Density Multilayer Substrate Using BCB Dielectric," *Proceedings International Microelectronics Conference*, p. 325, 1992.
73. T. G. Tessier and E. G. Myszka. "High Performance MCM-LD Substrate Approaches for Cost Sensitive Packaging Applications," *Proceed. Int. Conf. MCM's*, p. 200, 1993.
74. A. J. Strandjord, R. H. Heistand, J. N. Bremmer, P. E. Garrou, and T. G. Tessier. "A Photosensitive BCB on Laminate Technology," *Proc. ECTC*, p. 374, 1994; *IEEE Trans. Components Hybrids Manuf. Technol.* (in press.)
75. P. Garrou, R. Heistand, M. Dibbs, T. Manial, T. Stokich, C. Mohler, P. Townsend, G. Adema, M. Berry, and I. Turlik. "Rapid Thermal Curing of BCB Dielectric," *IEEE Trans. Components Hybrids Manuf. Technol.*, CHMT 16: p. 46, 1993.
76. T. C. Hodge, B. Landmann, S. A. Bidstrupp, and P. A. Kohl. "Rapid Thermal Curing of Polymer Interlayer Dielectrics," *Int. J. Microcircuits Electron. Packaging*, 17: p. 10, 1994.
77. R. Heistand, R. DeVellis, T. Manial, A. Kennedy, T. Stokich, P. Garrou, T. Takahashi, G. Adema, M. Berry, and I. Turlik. "Advances in MCM Fabrication with Benzocyclobutene Dielectric," *Proc. Int. Microelectronics Conf.*, p. 320, 1992; *J. Microcircuits Microelectron. Packaging*, 15: p. 183, 1992.
78. G. Adema, L. Hwang, G. Rinne, and I. Turlik. "Passivation Schemes for Copper/Polyimide Thin Film Interconnections Used in Multichip Modules," *IEEE Trans. Components Hybrids Manuf. Technol.* CHMT 16: p. 53, 1992.

79. R. H. Heistand, D. C. Frye, D. C. Burdeaux, J. N. Carr, and P. E. Garrou. "Economic Evaluation of Deposited Dielectric MCM Manufacturing Costs," *Proc. Int. Conf. MCM*, p. 441, 1993.
80. L. Laursen and P. E. Garrou. "Consortium for Intelligent Large Area Processing," *Proc. Int. MCM Conf.*, 1995.
81. T. M. Stokich, W. M. Lee, and R. A. Peters. "Real Time FT-IR Studies of the Reaction Kinetics for the Polymerization of DVS-BCB Monomer", *Proc. MRS*, p. 103, 1991.
82. C. E. Mohler, A. J. Strandjord, D. J. Castillo, M. R. Stachowiak, R. H. Heistand, P. E. Garrou, and T. G. Tessier. "Micro ATR as a Probe of BCB Layers for MCM-LD Applications," *Proc. MRS Symp*, p. 295, 1993.
83. R. W. Johnson, T. Phillips, W. Weidner, S. Hahn, D. Burdeaux, and P. Townsend. "BCB Interlayer Dielectrics for Thin Film Multichip Modules," *IEEE Trans. Components Hybrids Manuf. Technol.*, CHMT 13: p. 347, 1990.
84. B. Rogers, M. Berry, I. Turlik, P. E. Garrou, and D. Castillo. "Soft Mask for Via Patterning in BCB," *Int. J. Microcircuits Electron. Packaging*, 17: p. 210, 1994.
85. M. Schier. "Reactive Ion Etching of BCB Using a Silicon Nitride Etch Mask," J. Electrochem. Soc., Vol. 142, 1995, p. 3238.
86. T. Tessier and G. Chandler. "Compatibility of MCM-D Dielectrics with Scanning Laser Ablation Via Techniques," *Proc. ECTC*, p. 763, 1992.
87. Strandjord, A., Garrou, P. E., Ida, Y., Cummings, S., Kisting, S., Rogers, B., "MCM-D Fabrication with Photo BCB: Processing, Siolderbumping, Systems Assembly and Test", Proceed ISHM, Los Angeles, 1995, p. 402.
88. W. Radlik, K. Plehnert, R. H. Heistand, C. D. Castillo, and A. Achen. "MCM-D Technology in a Communication Application," *Proc. Int. MCM Conf.*, p. 402, 1994.
89. H. Theide, et al. "High Reliability 4-Layer MCM-D Structure with BCB as Dielectric," *Proc. Int. MCM Conf.*, 1995.
90. J. Carlson. "486 Computer Module System," *Proc. Int. MCM Conf.*, 1995.
91. J. K. Stille, *Macromolecules, 14,* 870, (1981).
92. N. H. Hendricks, "Development of Thermally Stable, Low Dielectric Films for Aerospace Applications", Final Report, NASA Contract No. NAS1-18832, (1989).
93. N. H. Hendricks, M. L. Marrorco, D. M. Stoakley and A. K. St. Clair. "Thermally Stable, Low Dielectric Polyquinolines for Aerospace and Electronics Applications," SAMPE, Vol. 4, Electronics Matls.—Our Future, Edited by R. F. Albred, R. J. Martinez, and K. B. Wirchmann, SAMPE 1990. Society for the Advancement of Matl. and Process Engineering.
94. M. I. Bessonov, M. M. Koton, V. V. Kudryavtsev, and L. A. Laius. *Polyimides: Thermally Stable Polymers,* 2nd edn., Plenum, New York (1987), pp. 1–95.
95. Avatrel Olefinic Dielectric Polymers, BF Goodrich Product Literature.
96. A. Saiki, S. Hirada, T. Okubo, K. Mukai, T. Kimura, J. Electro-chem. Soc: Solid State Sci. Technol., *124*, 1619 (1977).
97. A. Saiki, K. Mukai, S. Harada, Y. Miyadera, Preprints ACS Org. Coat. Plastic Chem., *43*, 459 (1980).
98. Y. Harada, F. Matsumoto, T. Nakakado, J. Electro-chem. Soc.: Solid State Sci. Technol., *130*, 1, 133 (1983).
99. R. D. Rossi. "Polyimides in Microelectronics Package," *Plast. Electron. Packag. Trends Technol.*, pp. 243–251, 1991.
100. R. Rubner, A. Hammerschmidt, R. Leuschner, and H. Ahne. "Photosensitive Organic Dielectrics: Polybenzoxazole versus Polyimides," *Proc. Int. Symp. on Polymer for Microelectronics–Science and Technology—(PME'89),*" pp. 789–810, 1990.

101. H. Ahne, and R. Rubner. "A Simple Way to Producing Industrial Polyimide Patterns," *Siemens Forsh. Entwickl.-Ber.*, 16(3): pp. 112–116, 1987.

102. M. T. Pottiger. "Second Generation Photosensitive Polyimide Systems," *Solid State Technol.*, 32 (12) pp. S1–S4, 1989.

103. R. E. Kerwin and M. R. Goldrick. "Thermally Stable Photoresist Polymer," *Polym. Eng. Sci.*, 11(5): pp. 426–430, 1971.

104. R. Rubner. "Production of Highly Heat-Resistant Film Patterns from Photoreactive Polymer Precursors. Part I. General Principles," *Siemens Forsh. Entwickl.-Ber.*, 5(2): pp. 92–97, 1976.

105. R. Rubner, W. Bart, and G. Bald. "Production of Highly Heat-Resistant Film Patterns from Photoreactive Polymer Precursors. Part 2. Polyimide Film Patterns," *Siemens Forsh. Entwickl.-Ber.*, 5(4): pp. 235–239, 1976.

106. R. Rubner, H. Ahne, E. Kuhn, and G. Kolodziej. "A Photopolymer—The Direct Way to Polyimide Patterns," *Photographic Sci. Eng.*, 23(5): pp. 303–309, 1979.

107. H. Ahne, H Eggers, W. Gross, N. Kokkotakis, and R. Rubner. "New Electronic Application of Polyamic Acid Methacrylate Ester" *Proc. Second Int. Conf. on Polyimide*, pp. 561–574, 1985.

108. H. Merrem, R. Klug, and H. Hartner. "New Developments in Photosensitive Polyimide," *Polyimides Synthesis, Characterization, and Applications*, pp. 919–931, Plenum Press, New York, 1984.

109. M. T. Pottiger, D. L. Golf, and W. J. Lautenberger. "Photodefinable Polyimides: II. The Characterization and Processing of Photosensitive Polyimide Systems," *38th Electron. Components Conf.*, pp. 316–321, 1988.

110. O. Rhode, M. Riediker, and A. Schaffner. "Recent Advances in Photoimagable Polyimides," *SPIE*, 539: pp. 175–180, 1885.

111. O. Rhode, M. Riediker, A. Schaffner, and J. Bateman. "High Resolution, High Photospeed Polyimide for Thick Film Applications," *Solid State Technol.*, (9) pp. 109–112, 1986.

112. A. Ikeda, N. Tsuruta, H. Ai, and T. Isoya. "High Performance Photosensitive Polyimide," *Polyfile*, 27(2): pp. 19–21, 1990.

113. S. Ogitani. "PIMEL: Photosensitive Polyimide Coatings for Electronics," *Proc. Int. Symp. on Polymer for Microelectronics (PME'89)*, pp. 158, 1989.

114. Y. Matsuoka, A. Ikeda, and H. Ai. "Process for Preparing Polyamides," Japanese Patent Kokai, 59-193737, 1984.

115. Y. Matsuoka, A. Ikeda, and H. Ai. "Synthetic Methods of Photosensitive Polyimide Precursor for Electronic Use." *Polymer Preprints*, Japan, 42(3): pp. 741, 1993.

116. L. Minnema and J. M. van der Zande. "Pattern Generation in Polyimide Coatings and its Application in an Electrophoretic Image Display," *Polymer Eng. Sci.*, 28(12): pp. 815–822, 1988.

117. H. Ahne, E. Kuhn. R. Rubner, and E. Schmidt. "Method for the Preparation of Highly Heat-Resistant Relief Structure and the Use Thereof," US Patent 4,311,784, 1982.

118. G. C. Davis. *Photosensitive Polyimide Siloxane*, pp. 259–269, American Chemical Society, Washington, DC, 1984.

119. B. Loisel and M. J.M. Abadie. "Kinetics Studies of Photosensitive Polyimides by Photocalorimetry (DCP)," *Polyimide and Other High-Temperature Polymers.*, (European Technical Symposium on Polyimide and High-Temperature Polymers (2nd)) pp. 471–492, 1991.

120. T. Sakuma, S. Ogitani, and A., Ikeda. "Study on Thermal Curing in Ester Type Photosensitive Polyimide," J. Photopolymer Sci. Technol., 8(2): pp. 277–280, 1995

121. S. Numata and N. Kinjyo. "Chemical Structures and Properties of Low Thermal Expansion Coefficient Polyimides," *Polymer Eng. Sci.*, 28(14): 906–911, 1988.

122. S. Numata, S. Ohhara, K. Fujisawa, J. Imaizumi, and N. Kinjo. "Thermal Expansion Behavior of Various Aromatic Polyimides," *J. Appi. Polymer Sci.*, 31: pp. 101–110, 1986.

123. Y. Matsuoka, K Yokota, S. Ogitani, A. Ikeda, H. Takahashi, and H. Ai. "Ester-Type Photosensitive Polyimide Precursor with Low Thermal Expansion Coefficient," *Polymer Eng. Sci.*, 32(21): pp. 1619–1622, 1992.

124. A. E. Nader, K. Imai, J. D. Craig, C. N. Lazaridis, D. O. Murray, M. T. Pottiger, S. A. Dombchik, and W. J. Lautenberger. "Synthesis and Characterization of a Low Stress Photosensitive Polyimide," *9th Int. Tech. Conf. on Photopolymers*, pp. 333–342, 1991.

125. T. Omote, T. Yamaoka, and K. Koseki. "Preparation and Properties of Soluble and Colorless Fluorine-Containing Photoreactive Polyimide Precursor" *J. Appl. Polym. Sci.*, 38(3): pp. 389–402, 1989.

126. Y. Matsuoka, Y. Kataoka, Y. Tanizaki, and A. Ikeda. "The Design of I-Line Photoimagable Polyimide Precursor," *Polymer Preprints*, Japan, 42(7): pp. 2694–2696, 1993.

127. D. R. McKean, G. M. Wallraff, W. Volksen, N.P. Hacker, M I. Sanchez, and J. W. Labadie. "Base-Catalyzed Photosensitive Polyimide," *Polymer Mater. Sci. Eng.*, 66: pp. 237–238, 1992.

128. H. Hiramoto and M. Eguchi. U. S. Patent 4.24S,743, 19.

129. Y. Yoda and H Hiramoto, *J. Macromol. Sci. Chem.*, A21: p. 1641, 1984.

130. H. Hiramoto. *Mater. Res. Soc. Symp. Proc.*, 167: 87, 1990.

131. M. Tomikawa, M. Asano, G. Ohbayashi, H. Hiramoto, Y. Morishima, and M. Kamschi, *J. Photopolym. Sci. Technol.*, 5: p. 343, 1993.

132. G. Ohbayashi, Umemoto, H. Hiramoto. U. S. Patent 4,547,455, 19.

133. M. Asano, M. Eguchi, K. Kusano, and K. Niwa, *Polym. Adv. Tech.*, 4(4): p. 261, 1993.

134. M. KoJima, H. Sekine, H. Suzuki, H. Satou, and D. Makion, *Proc. 39th Electronic Components Conf.*, p. 920, 1989.

135. J. Pfeifer and O. Rohde. "Polyimides: Synthesis, Characterization, and Applications," *Proc. 2nd Int.* CosXfi Polyimides, p. 130, 1985; *Recent Advances in Polyimide Science and Technologx*, ed. by W. D. Weber and M. R. Gupta, p. 336, Society of Plastics Engineers, Brookfield, CT, 1987.

136. O. Rhode, P. Smolka, P. A. Falcigno, and J. Pfeifer. *Polymer Eng. Sci.* 32 (21): p. 1623, 1992.

137. R. Mo, K. Schlicht, T. Maw, M. Masola, and R. Hopla. in *Proceedings of Advances in Polyimide Science and Technology*, ed. by C. Feger, M. Khojasteh, and M. Htoo, p. 559, Technomic Publishing Company, Lancaster, PA, 1993.

138. M. Ree, K. Chen, and G. Czornyj. *Polymer Eng. Sci.*, 32 (14): p. 924, 1992.

139. A. Lin, V. Sastri, G. Tesoro, and A. Reiser. *Macromolecules*, 21: p. 1165, 1988.

140. J. Scaiano, A. Becknell, and R. Small, *J. Photochem. Photobiol., A: Chem.*, 44: p. 99, 1988.

141. J. Scaiano, J. Netto-Ferreira, A. Becknell, and R. Small. *Polymer Eng. Sci.*, 29(14): p. 942, 1989.

142. Y. Shindo, T. Sujimura, K. Horie, and I. Mita. *Eur. Polym. J.*, 26(6): p. 683, 1990.

143. H. Higuchi, T. Yamashita, K. Horie and I. Mita. *Chem. Mater.* 3: p. 188, 1991.

144. T. Yamashita, K. Horie, and I. Mita. in *Polymers for Microelectronics—Science and Technology*, p. 837, ed. by Y. Tabata, I. Mita, S. Nonogaki, K. Horie, S. Tagawa, Kodansha, 1990.

145. T. Yamashita, T. Kudo, and K. Horie, in *Polymers for Advanced Technologies*, ed. by M. Levin, vol. 4, p. 244, John Wiley and Sons, New York, 1993.

146. K. Horie. *J. Photopolymer Sci. Technol.*, 5(2): p. 315, 1992.

147. Q. Jin, T. Yamashita, and K. Horie, J. Polym. Sci. Part A: Polym. Chem. 32, 503, 1994.

148. O. Rohde. *3rd Annual Int. Conf. Crosslinked Polymers*, p. 197, 1989.
149. H. Lee and Y. Lee. *J. Appl. Polym. Sci.*, 4: p. 2087, 1990.
150. T. Maw, M. Masola, and R. Hopla. *Polymer Mater. Sci. Eng.*, 66: p. 247, 1992.
151. T. Maw and R. Hopla. *Mater. Res. Soc. Symp. Proc.*, 203: p. 71, 1991.
152. W. Chiang and W. Mei. *J. Polymer Sci. A: Polymer Chem.* 31: p. 1195, 1993.
153. W. Chiang and W. Mei. *J. Polymer Sci. A: Polymer Chem.* 50: p. 2191, 1993.
154. J. Sassmannshausen. R. Schultz and E. Bartmann, US Patent, 5,1004,768, April 14, 1992.
155. R. H. Hayase, N. Kihara, N. Oyasato, S. Matake, and M. Oba, SPIE 1446, Vol 8, pp. 439–445, 1991.
156. E. Perfecto, C. Osborn, and D. Berger, ICEMM Proceeding, 1993.
157. M. Asano and H. Hiramoto, in "Photosensitive Polyimides, Fundamental and Applications", Edited by K. Horie and T. Yamashita, Chapter 5, Technomic Publishing Company, Lancaster, PA, 1995.
158. J. Moore and A. Dasheff, Chemistry of Materials, 1, p. 163, 1989.
159. R. Hayase, N. Kihara, N. Oyasato, S. Matake and M. Oba, Polymeric Materials Science and Engineering, Proceedings of the ACS Division of PMSE, Vol. 66, pp. 243–244, 1992.
160. S. Kubota, T. Moriwaki, T. Ando, and A. Fukami, J. Macromol. Sci, Chem., A24, p. 1497, 1987.
161. K. Takano, Y. Mikogami, Y. Nakano, R. Hayase, and S. Hayase, J. Applied Polymer Sci., 46, p. 1137, 1992.
162. S. Hayase, Y. Mikogami, K. Tanako, Y. Nakano, and R. Hayase, Polymer Adv. Tech., 4, p. 308, 1993.
163. T. Omote and T. Yamaoka, Polymer Eng. Sci., 32, p. 1632, 1992.
164. T. Banba, E. Takeuchi, A. Tokoh, and T. Takeda, Proceedings, IEEE 41st Electronic Components and Technology Conference, p. 564, 1991.
165. A. E. Nader, C. N. Lazeridis, D. K. Flattery, and W. J. Lautenberg, MCM Proceedings, Denver, p. 410, 1992.
166. E. Takeuchi, T. Takeda, and T. Hirano Proceedings, IEEE 40th Electronic Components and Technology Conference, p. 818, 1990.
167. M. Kojima, H. Sekine, H. Suzuki, H. Satou, D. Makino, F. Kataoka, J. Tanaka and F. Shoji, Proceedings, IEEE 39th Electronic Components and Technology Conference, p. 920, 1989.
168. H. Sashida, T. Hirano, and A. Tokoh, Proceedings, IEEE 39th Electronic Components and Technology Conference, p. 167, 1989.
169. E. W. Rutter Jr., E. S. Moyer, R. H. Harris, D. C. Frye, V. L. St. Jore, and F. L. Oaks, "A Photodefinable Benzocyclobutene Resin for Thin Film Microelectronic Applications", Proceedings of 1st International Conference on Multichip Modules, Denver, p. 394, 1992.
170. E. S. Moyer, E. W. Rutter Jr., M. T. Bernius, P. H. Townsend, R. F. Harris, H. Pranjoto, and D. D. Denton, "Photodefinable Benzocyclobutene Formulations for Thin Film Microelectronic Applications: Part II", Proceedings IEPS, pp. 37–50, 1993.
171. E. S. Moyer, G. S. Becker, E. W. Rutter, Jr., M. Radler, J. N. Bremmer, M. T. Bernius, D. Castillo, A. J.G. Strandjord, R. Heistand, P. Foster, and R. F. Harris, "Photodefinable Benzocyclobutene Formulations for Thin Film Microelectronic Applications. III. 1 To 20 Micron Patterned Films", MRS Symposium Proceedings, Boston, Vol. 323, pp. 267–276, 1994.
172. Strandjord, A., Garrou, P. E., Ida, Y., Cummings, S., Kisting, S., Rogers, B., "MCM-D Fabrication with Photo BCB: Processing, Siolderbumping, Systems Assembly and Test", Proceed ISHM, Los Angeles, p. 402, 1995.

173. T. Shimoto, K. Matsui, and K. Utsumi, "Cu / Photosensitive-BCB Thin Film Multilayer Technology for High-Performance Multichip Module", Proceedings of the International Conference on Multichip Modules, Denver, p. 115, 1994.

174. W. Radik, K. Plehnert, M. Zellner, A. Achen, R. H. Heistand, D. Castillo, & R. Urscheler, "MCM-D Technology for a Communication Application", Proceedings of the International Conference on Multichip Modules, Denver, p. 402, 1994.

175. Tessier, T., Myszka, E., "Approaches to Cost Reducing MCM-D Substrates", Proceed. ECTC, p. 570, 1993.

176. M. Skinner, P. E. Garrou, D. Castillo, S. Cummings, K. Liu, D. Chazen, R. Reinschmidt, S. Westbrook, C. Ho, B. Rogers, "Twinstar—Dual Pentium Processor Module", Proceed. Int. Conference on MCMs, Denver, p. 75, 1996.

177. P. H. Townsend, D. Schmidt, T. M. Stokich, S. Kisting, D. C. Burdeaux, D. Frye, M. Bernius, M. Lanka, and K. Berry, "Adhesion of CYCLOTENE (BCB) Coatings On Silicon Substrates", MRS Symposium Proceedings, Boston,., Vol. 323, p. 365, 1993.

178. P. Garrou, "Polymer Dielectrics for Multichip Module Packaging", Proceedings of IEEE, Vol. 80, p. 1942, 1992.

179. L. Laursen, P. E. Garrou, "Consortium for Intelligent Large Area Processing", Proceed. Int. MCM Conf., Denver, p. 112, 1995.

180. A. Strandjord, P. E. Garrou, R. Heistand and T. G. Tessier, "MCM-D/L: Large Area Processing Using PS-BCB", IEEE Trans. CPMT, Vol 18(2), 1995.

181. E. D. Perfecto, D. G. Berger, C. T. Osborn, G. White, "Engineering PSPI's for MCM-D Applications", Int. J. Microcircuits & Elect. Pkging.", Vol. 16, p. 319, 1993.

182. E. D. Perfecto, C. Osborn, D. Berger, "Factors That Influence Photosensitive PI Lithographic Performance", Proceed Int Conf MCM's, Denver, p. 40, 1993.

183. D. Burdeaux, P. Townsend, Carr, P. Garrou, "Benzocyclobutene Dielectrics for the Fabrication of High Density, Thin Film Multichip Modules", J. Electronic Materials, Vol. 19, p. 1357, 1990.

184. P. E. Garrou, R. H. Heistand, M. Dibbs, T. A. Manial, C. Mohler, T. Stokich, P. H. Townsend, G. M. Adema, M. J. Berry, and I. Turlik, "Rapid Thermal Curing of BCB Dielectric", Proceedings ECTC, San Diego, p. 770, 1992.

185. T. G. Tessier & G. Chandler, "Compatibility of Common MCM-D Dielectrics with Scanning Laser Ablation Via Generation Processes" IEEE Trans. on CHMT, Vol. 16, p. 39, 1993.

186. T. Shimoto, M. Matsui, M. Kimura, & K. Utsumi, "High Density Multilayer Substrates Using Benzocyclobutene Dielectric", Proceedings 7th IMC, Yokohama, p. 325, 1992.

187. A. V. Shah, E. Sweetman, and C. Hoppes, Proceedings of the National Electronic Packaging and Production Conference, v 2. Publ by Cahner Exposition Group, Des Plaines, Illinois, pp. 850–862, 1991.

188. E. Sweetman, Characteristics and Performance of PH-92: AT&T's Triazine-based Dielectric for POLYHIC MCMs, Microcircuits and Electronic Packaging, Vol. 15, No. 4, pp. 195–203, 1992.

189. H. Ito, and C. Willson, ACS Symposium Series 242. Publ by ACS, Washington, DC, USA, pp. 11–23.

190. K. Stewart, M. Hatzakis, and J. Shaw, Technical Papers, Regional Technical Conference—Society of Plastics Engineers. Publ by Soc of Plastics Engineers, Brookfield Center, CT, pp. 205–213, 1988.

191. Ciba Geigy Product Literature, LMB 7081.

192. H. Ahne, H. Kruger, E. Pammer and R. Rubner. "Polyimide Patterns made Directly from Photopolymers" Polvimides Synthesis, Characterization, and Applications. Plenum Press, New York, pp. 905–918, 1984.

193. M. M. Khan, T. S. Tarter and H. Fatemi. "Stress Relief in Plastic Encapsulated, Integrated Circuit Devices by Die Coating with Photodefinable Polyimide" Electron. Components Conf., 38: 425–431, 1988.
194. N. Naito, N. Kakuda, T., Wada, IERE Electron Device Letters, EDL-6, 589, 1985.
195. H. Ahne, R. Leuschner and R. Rubner. "Recent Advances in Photosensitive Polyimides," *Polymer for Advanced Technologies*, 4, pp. 217–233, 1992.
196. S. Ray, D. Berger, G. Czornyl, A. Kumar and R. Tummala, "Dual-Level Metal (DLM) Method for Fabricating Thin Film Wiring Structure," Proc. Electron. Components Technol. Conf. 43, pp. 538–543, 1993.
197. K. Chakravorty and M. Tanielian, Appl. Phys. Lett. 60 (14), 1670, 1992.
198. K. Chakravorty, in Polymers for Advanced Technologies, 4(4), p. 251, M. Levin ed., John Wiley, 1993.
199. C. Chien, K. Chakravorty, SPIE 1323, 338, 1991.
200. K. Chakravorty, Appl. Optics, 32 (13), 2331, 1993.
201. K. K. Chakravorty, C. P. Chien, J. M. Cech, L. B. Branson, J. M. Atencio, T. M. White, L. S. Lathrop, B. W. Aker, M. H. Tanielian, and P. L. Young. "High Density Interconnection using Photosensitive Polyimide and Electroplated Copper Conductor Line," *Electron. Components Conf.,* pp. 135–142, 1989.
202. K. K. Chakravorty, J. M. Cech, C. P. Chien, L. S. Lathrop, M. H. Tanielian, and P. L. Young. "Photosensitive Polyimide as a Dielectric in High density Thin Film Copper-Polyimide Interconnect Structures," *J. Electrochem. Soc.*, 137(3), pp. 961–966, 1990.
203. G. L. Takahashi and E. S. Kolesar. "Silicon-Hybrid Wafer-Scale Integration Achieved with Multilevel Aluminum Interconnects," *Proc. IEEE National Aerospace Electron. Conf.*, pp. 17–23, 1989.
204. K. Chakravorty, C. Chien, J. Cech, and M. Tanielian. *3rd Int. SAMPE Electron. Conf.*, p. 1213, 1989.
205. J. Cech, A. Burneff, and L. Knapp, *Polymer Eng. Sci.*, 32(21), p. 1646, 1992.
206. D. Smith, D. Scheider, R. Fu, E. Chan, G. LaRue, T. Williams, and M. Tanielian, DoD Fiber-Optics Conference, 1990
207. E. Myszka, G. Demet, H. Fuerhaupter, and T. Tessor. "The Development of a Multilayer Thin Film Copper/Polyimide Process for MCM-D Substrates," *Proc. Int. Conf. on Multichip Modules*," 1992.
208. T. Ohsaki, T. Yasuda, S. Yamaguchi, T. Kon, Proc. IEEE Int. Electron. Manuf. Technol. Symp., 178, 1987.
209. S. Sasaki, T. Kon, T. Ohsaki, Proc 39th Electronic Components Conf., 629, 1989.
210. T. Watari, H. Murano, IEEE Trans. Components, Hybrids, Manuf. Technol., CHMT-8(4), 462, 1985.
211. K. Kimbara. A. Dohya. T. Watari. Mat. Res. Soc. Symp. Proc., 167, 33, 1990.
212. Inoue, K. Seino, K. Tamura, S. Hasegawa, K. Kimbara, J. Photopolym. Sci. Technol., 5, 385, 1992.
213. H. Takasgo, M. Takada, K. Adachi, A. Endo, K. Yamada, T. Makita, E. Gofuku, Y. Ohnishi, Proc. 36th Electronic Components Conf., 481, 1986.
214. M. Takada, T. Makita, E. Gofdku. K. Miyake, A. Endo, K. Adachi, Y. Morihiro, H. Takasago, Proc. Int. Symp. Microelectron, 540, 1988.
215. H. Takasago, K. Adachi. M. Takada, J. Electron. Materials, 8(2), 319, 1989.
216. D. W. Hewak and H. Jerominek. "Channel Optical waveguides in Polyimides for Optical Interconnection by Laser Direct Writing and Contact Printing," SPIE Vol. 1213, Photopolymer Device Physics, Chemistry, and Applications: pp. 86–99, 1990.
217. M. G. Allen. "Polyimide-Based Processes for the Fabrication of Thick Electroplated Microstructures," The 7th International Conference on Solid-State Sensors and Actuators, pp. 60–65, 1993.

218. M. Uekita, H. Awaji, M. Murata, and S. Mizunuma. "Application of Polyimide Langmuir–Blodgett Films to Deep UV Resist," Thin Solid Films, 180: pp. 271–276, 1989.
219. Z. H. Lu, J. Y. Fang, L. Wang, Z. L. Chen, S. J. Xiao and Y. Wei. "Ultrathin Conductive Films Prepared by Pyrolysis of Photosensitive Polyimide Langmuir–Blodgett Films," Solid State Comm., 82(9): pp. 711–713, 1992.

12

THIN-FILM PACKAGING

RAO TUMMALA—*Georgia Tech*
WEIPING LI—*Georgia Tech*
TED TESSIER—*Motorola*
TOM WASSICK—*IBM*

12.1 INTRODUCTION

Thin film refers to a coating layer of thickness typically in the range of from a few (2–3) atomic layers to a few (1–5) microns. However, film thicknesses up to 20–50 μm are frequently regarded as thin-film in electronic packaging. Thin-film packaging is the technology for conductors and dielectrics deposition and patterning in package fabrication, which resembles the technology used for integrated-circuit (IC) chip fabrication. Thin-film packaging is distinguished from thick-film packaging, mainly ceramic and printed wiring board (PWB) packaging, in two aspects: (a) the typical dimensions of conductors and dielectrics are about 2–25 μm versus 100 μm and above in thick-film packaging; (b) the typical methods of thin-film deposition include sputtering, evaporation, chemical vapor deposition (CVD), more recently electro/electroless plating and polymeric solution coating, and other similar methods, which are all typical sequential processes, whereas in thick-film packaging parallel processes are more common, such as lamination and cofiring of ceramic greensheets. Nevertheless, the current trends in electronic packaging are the enhancement of thick-film technology to fabricate finer feature sizes which are traditionally considered only achievable by thin-film technology and the adoption of some of the parallel thick-film processes for thin-film fabrication. Due to these developments, the traditional division between thin- and thick-film packaging technologies is diminishing and the definition of thin-film packaging is becoming less obvious. Still, the mainstream of thin-film packaging can be well identified which allows a fairly extensive compilation on this technology. In this chapter, the need for thin-film packaging, electrical performance considerations, and typical structures in thin-film packages are first elucidated. Subsequently, the materials, processes, and cost analyses of thin-film packaging are discussed in detail. Repair is especially important in thin-film packaging and is dealt with in the context of cost and yield. Reliability of packages has been discussed in Chapter 5, "Package Reliability," and reliability issues which are unique to thin-film packages are identified here. Some major commercial applications of thin-film packaging are reviewed, followed by a summary of the emerging technologies. Integration of various passive components is becoming imperative for thin-film packages to enhance electrical performance and packaging efficiency, which is also included in this chapter along with some applications and future predictions. Finally, the current development and future directions of thin-film packaging technology are summarized and projected.

Packaging must deal with the inexorable trend of semiconductor technology toward higher levels of integration, as pointed out in Chapter 7, "Microelectronics Packaging—An Overview" a trend that is driven by an irresistible economic force and the augmentation of human capabilities by powerful computational facilities [1], and will continue to the limits of physical laws. The level of integration is measured by the number of logic gates on an integrated circuit (IC) chip. High levels of integration have been achieved by miniaturization of the structures on chips. Miniaturization has also conferred high speed on microelectronics; essentially, miniaturization reduces gate capacitance, which can, therefore, be charged more quickly, and gate-to-gate interconnection delay. In addition, speed also depends on device technologies as is illustrated in Figure 7-7 of Chapter 7, along with the other major device attributes. Translating the high speeds available on chips to fast, multichip systems requires minimization of delays attributable to the package. This means placing chips as closely together as possible to reduce chip-to-chip travel times. Each of these chips are expected to have up to 5000 I/O connections as discussed in Table 7-2 of Chapter 7, and all packaging functions must be contained in the smallest possible area. All these requirements as illustrated in Figure 12-1 and previously in Chapter 7, Figure 7-62 [2] point to the need for thin-film packaging. It is seen in Figure 12-1 that, with the chip I/O counts going up to above 600, the pad width and spacing become less that 25 μm and thin-film packaging, characterized by the application of photolithographic technology, is necessary. In Figure 7-62, Chapter 7, the packaging efficiency and wiring dimensions as a function of various thin- and thick-film packages are plotted, illustrating the advantage of thin-film package over other packages with maximized packaging efficiency and minimized size of portable and consumer electronics in the future. For this reason, thin-film packaging is generally considered to be the ultimate in packaging and is usually assumed as **multichip packaging** with minimized interconnection delay between IC chips. Figure 12-2 reaffirms the need for thin-film packaging. With **wiring density** going up to 2000 cm/cm^2 by the end of next decade, thin-film technology would probably be the only choice left [3]. In general, the longest **interconnections** between devices within a package dictate the interconnection delay and form the critical paths. The critical paths restrain the **cycle time** of a **processor** such as for mainframes or supercomputers which, in turn, primarily determines

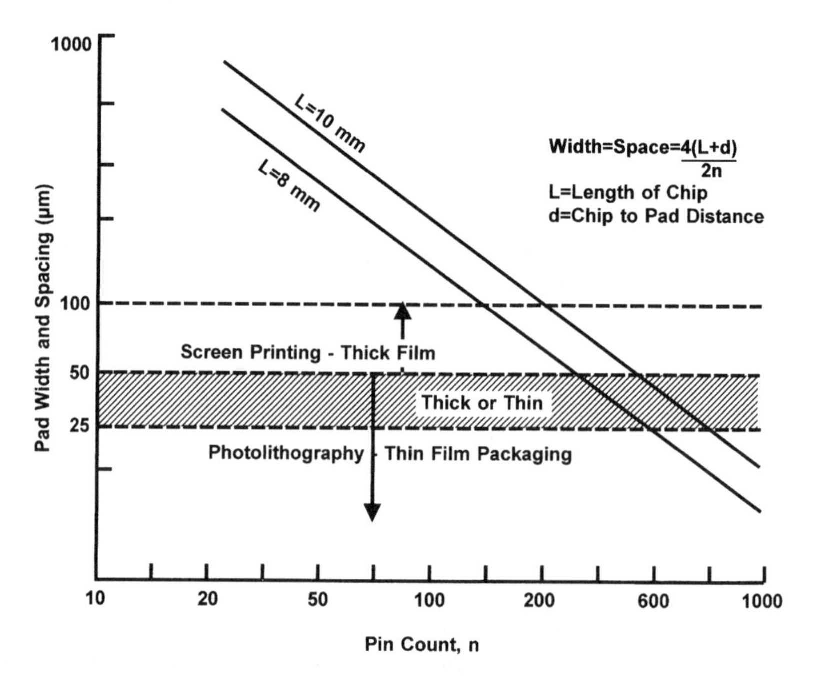

Figure 12-1. Requirements for Thin Film of I/O Pads. (After Ref. 2.)

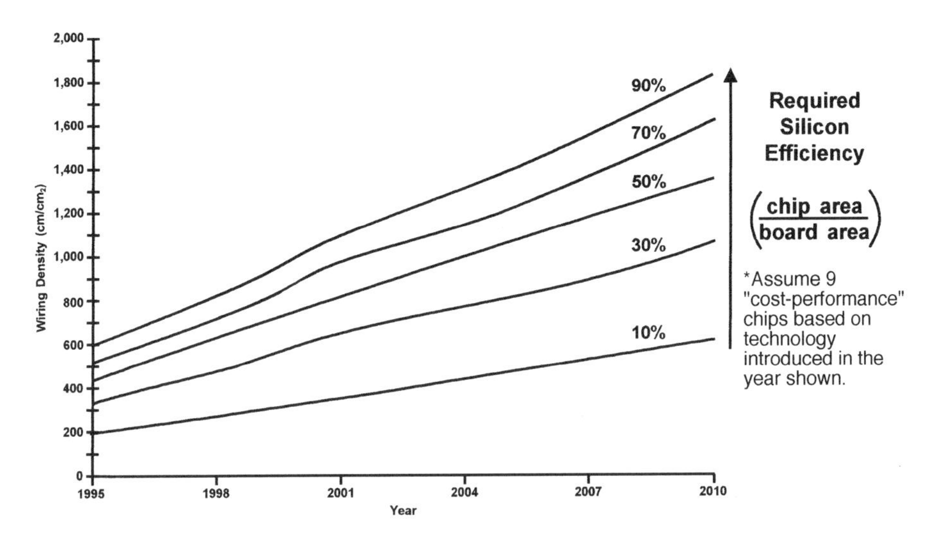

Figure 12-2. Evolution of Wiring Density on Substrate

the processing speed and throughput. A good example of performance enhancement with shorter critical paths is the use of the **multichip modules (MCM)** by IBM in its mainframes since 1980 to reduce the cycle time from about 60 ns to less than 8 ns as illustrated in Figure 12-3. It is noted, however, that ceramic multichip module (MCM-C) packaging rather than thin-film multichip module (MCM-D) packaging was initially employed. Other companies such as Hitachi, Fujitsu, and NEC have basically done the same using somewhat different multichip modules, some with ceramic or **glass-ceramic** (Fujitsu) and some with thin film on ceramic or glass-ceramic substrates (NEC). In the past, high-power **bipolar** devices have been used in all of these ultrahigh-performance systems. However, bipolar devices have continually been and will totally be replaced by **CMOS (complementary metal-oxide-semiconductor)** devices as the performance of CMOS approaches that of bipolar, as indicated in Figure 12-4, yet at much lower power and cost, as indicated in Figure 7-7, Chapter 7.

The limitations on space make packaging of large computers a difficult technology with many facets. A system contains hundreds, and sometimes thousands, of chips. Packaging is also a subject about which it is difficult to generalize, as there is no unanimity among system designers as to the best approach to packaging large, fast machines. There is a much greater variability among the packaging technologies to different systems than among their chip technologies.

The package communicates electrically with the chip by means of I/Os that connect lands on a chip to package wiring. Some of the I/Os are used to supply electrical power to the chip and others are used to

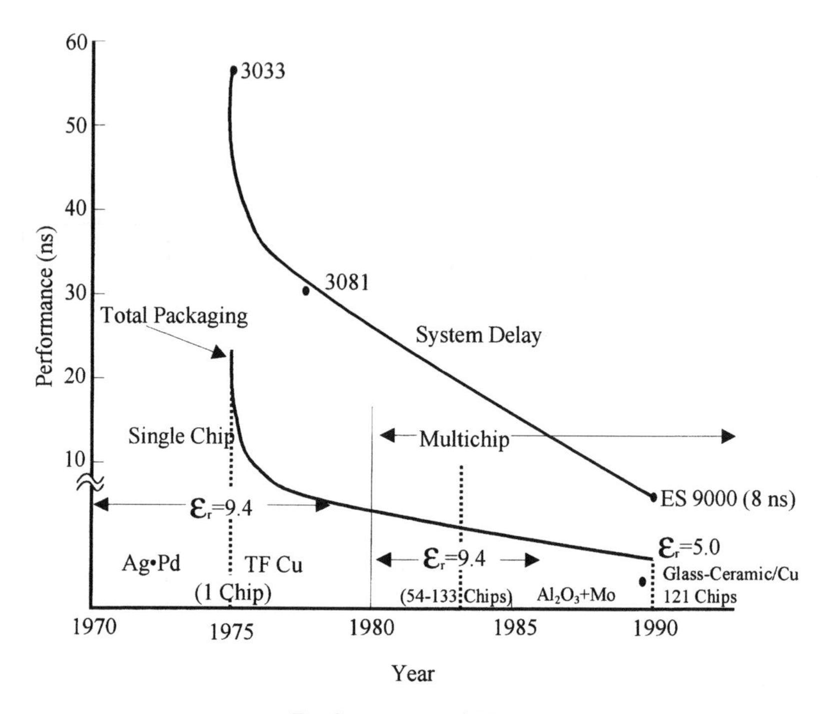

Figure 12-3. Performance of IBM Mainframes

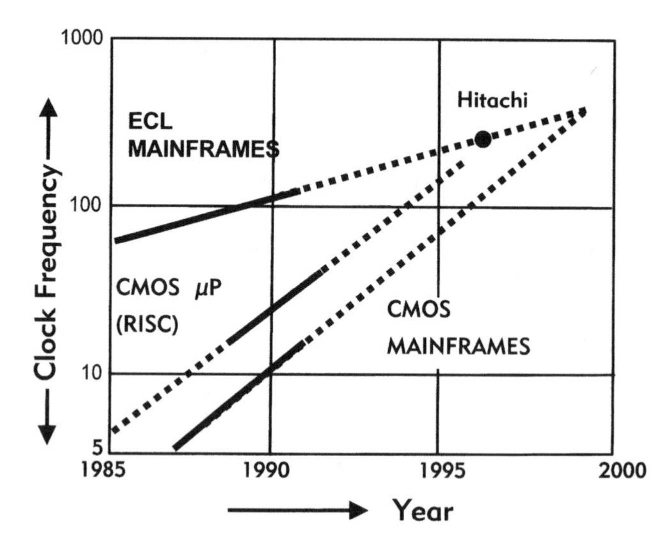

Figure 12-4. Evolution of Bipolar and CMOS- Based Mainframes and RISC Systems

transmit signals. The number of gates on a chip does not directly interact with the package. The number of I/Os on a chip is more relevant to the packaging of the chip than the number of gates it contains.

The **gate arrays** in large computers are part of systems containing a great many chips and hundreds of thousands or millions of logic gates. The results of logic operations performed on a chip are often needed on another chip in a distant part of the system, and many I/Os are necessary to meet the demand for rapid transmission of information throughout the computer. The need for many I/Os in **logic** chips that are part of a large system was recognized long ago [4] and is described by a relation known as Rent's rule, as pointed out in Chapter 7, "Microelectronics Packaging—An Overview". In fact, the rule is not only applicable to chips, but also to multichip modules, boards, or any partition of a large computing system. It has the form

$$P = BN^{s} \qquad [12\text{-}1]$$

Here N is the number of logic gates on a chip or other partition and P is the number of signal connections that must be made to it. B and s are constants which depend on the specific IC type and system design. The general form of this rule is supported by a large number of empirical studies [5–7], as pointed out in other chapters.

It is certainly possible to fabricate a chip with fewer connections than the number given by Equation 12-1, but the basis of Equation 12-1 is that it is then not possible to utilize most of the circuits on the chip as a part of the system. Both I/Os and circuits contribute to the cost of a system, and Equation 12-1 is a compromise aimed at obtaining a certain functional capability at minimal cost. Hardware designers differ as to the optimal way to make the compromises involved. Thus, it is not surprising that different authors disagree as to the best values for the parameters in Equation 12-1. The exponent s is variously given in the range 1/2 to 2/3 and the factor B is between 2 and 4.

It is noteworthy that the number of I/Os grows more slowly than the number of gates on the chip. It appears, as might reasonably be expected, that signal destinations are found on the chip of origin more frequently as the level of integration increases. This fact has also been explicitly noted [8]: the fraction of logic steps in which the result is transmitted to a different chip decreases with increasing level of integration. The reduction of the fraction of off-chip signal paths provides a motivation, in addition to cost reduction, for increasing integration in large machines: the off-chip paths are longer and cause longer delays than paths on a chip; longer series of logic operations on a single chip improves performance. Indeed, this fact is responsible for a significant part of the improvement in computer performance in the last decade.

Interconnections have been moved from packages to chips, where they are shorter, less expensive, and more reliable. Paths that pass from one chip to another at low integration can be completed on a chip at high integration [8,9].

One of the reasons that, in spite of the economic force and the performance advantage pressing toward high levels of integration, the number of gates on chips in large machines lags far behind the maximum capability achieved in memory and microprocessor chips is the large number of I/Os. Higher levels of integration are achieved by miniaturization of features on the chip; the size of the chip are also increasing slowly. The increased number of connections to the chip must be accommodated without a corresponding increase in area allotted to chip mounting, and constant miniaturization of the connections is needed. At present, it is difficult for technology to provide as many as 400 I/Os on a chip, a number that can support less than 5000 logic gates according to Rent's rule. However, the trends in semiconductor technology indicate the need for as many as 1000 I/Os per chip within the next decade. This is only achievable with thin-film packaging technology, as illustrated in Figure 12-1.

The problem of connections between one electronic **package level** and the next does not end at the meeting of the chip and the module. The multichip modules must be mounted on and connected to large boards. The boards contain the wiring that connects the modules to one another. In accord with Rent's rule, very large numbers of connectors are needed on the module. For example, the original IBM Thermal Conduction Module (TCM) had 1800 **pins** to connect to the next higher level of package [10]. Of these, 1200 are allotted to logic signals and 600 to the delivery of power. Another somewhat similarly packaged computer uses 2177 connections to the multichip module [11]. These connections are constructed with a quite different technology than the connections to the chip, as they must be easily separable for purposes of **assembly** and testing, and for replacement in the field to allow **engineering changes (EC)** and repairs. They are much larger in size than the connections between the chip and the module. The area per connection in contemporary large computers is about $3 \times 10^{-3} cm^2$ on chips and 0.06 cm^2 for multichip module to board connectors.

12.1.1 Thin-Film Package Wiring

The large number of connections to the chips and to higher package levels must be joined by a dense network of wires. The IBM Thermal Conduction Module, a ceramic substrate on which 100 chips are mounted in an area of less than 100 cm^2, can contain 288 m of wire. Therefore, each square centimeter of the substrate contains well over 100 cm of wire. The wiring has a pseudorandom character; **wire channels** are provided and wires placed in them to **interconnect** the chips to implement the logical structure of the computer. As the levels of integration increase, more and more logic devices are placed in such multichip modules and increasing wiring densities will be required in the package to support them. A similar trend is evident on chips; chips now contain of the order of 1000 cm of wire per cm^2 of chip area. The high wiring densities on packages are achieved by fabricating the wiring in a few thin-film layers.

The trend to higher wiring density can be expected to continue. In fact, greater wiring density is associated with increasing levels of integration through Equation 12-1. This can be illustrated by a simple model. Let M chips be placed on a multichip module substrate. Then the total number of signal I/Os on the module is MBN^s. If these I/Os are joined in pairs, there are $(1/2)MBN^s$ wires. The wires have an average length that is some fraction f of the linear dimension of the module. Measured in units of the chip-to-chip distance (the chip "pitch"), the side of the module is $M^{1/2}$, and the average wire length can be written as $fM^{1/2}$. Multiplying these terms together, the total amount of wire on a module (in chip pitches) is

$$L_{tot} = \frac{1}{2} f M^{3/2} B N^s \qquad [12\text{-}2]$$

The amount of wire that must be supplied under each chip is

$$L_{tot} / M = \frac{1}{2} f M^{1/2} B N^s \qquad [12\text{-}3]$$

The provision for the large amount of interconnecting wire is one of the factors that limits the space that must be allotted to each chip. Let wires be placed in K layers. Also assume that wire channels have a width W. The channel width W must include space for **vias** that allow one layer to be connected to another, so that W is considerably greater than the actual width of the wire itself. Furthermore, because of the pseudorandom character of the wiring, not all of the channel can be used; the utilization of the available channel space is likely to be only 50% [10]. Additional notation to describe these influences will not be introduced; they are regarded as included implicitly in W and in the average wire length.

If each chip occupies an area A on the module, then the length of wire needed to traverse one chip site is the chip pitch, $A^{1/2}$, and the area that it occupies is $WA^{1/2}$. The area of all of the wiring needed under a chip, now found by using Equation 12-3, must equal the total area available for wiring, KA:

$$KA = \frac{1}{2} f M^{1/2} B N^s \mathrm{W} \mathrm{A}^{1/2} \qquad [12\text{-}4]$$

Solving Equation 12-4 for A yields

$$A = (fB / 2K)^2 MN^{2s} W^2 \qquad [12\text{-}5]$$

The area per chip required for the wiring is inversely proportional to the square of the number of layers. For example, using numbers somewhat representative of the IBM TCM [10], if $f = 1/2$, $K = 16$, $M = 100$, $W = 0.05$ cm, and $N = 500$, and assuming that $B = 2.5$ and $s = 0.6$, A turns out to be 0.7 cm^2. Of course, A must also be greater than the area of the chip plus additional space needed for assembly.

The need for more layers of wiring is also seen by considering the **resistance** of the wire. Let $\rho_\square$ be the **sheet resistance** of the wire. (Again, a factor to account for the fact that the wire is narrower than the channel appears here and is assumed to be absorbed in $\rho_\square$.) Then, because the length of the average wire is $fM^{1/2}A^{1/2}$, its resistance is

$$R = fM^{1/2}A^{1/2}\rho_\square / W \qquad [12\text{-}6]$$

Taking A from Equation 12-5 gives

$$R = f^2 MBN^s \rho_\square / 2K \qquad [12\text{-}7]$$

The resistance is inversely proportional to K. Note also that increasing W does not directly impact R—W must be determined by limitations of the fabrication process and by reduction of the distance between chips, $A^{1/2}$ (Eq.12-5).

Equations such as 12-5 and 12-7 are often more useful when solved for K and used to estimate the number of wire layers needed from other design criteria, for example, the acceptable resistance and module area per chip. Even more layers are needed than suggested by the above analysis of **signal wiring**; conductors to distribute **power and ground planes** to control impedance must be included. The numbers quoted above for the Thermal Conduction Module from IBM are typical of "**thick film**," in which the dimensions of conductors are defined by mechanical masking of the deposition process that is referred to as **paste** screening in Chapter 9, "Ceramic Packaging."

Increasing integration places ever greater demands on the number of layers needed in thick-film technologies. Although **multilayer ceramic** (**MLC**) packaging has managed to cope with a remarkable number of layers, there are inherent disadvantages. Vias that pass through many layers are long and have high inductance. Switching currents through the inductance causes voltage transients that constitute undesirable noise (**ΔI**

noise); the higher the inductance, the greater the noise, as discussed in Chapter 3, "Package Electrical Design."

In addition, a certain degree of compatibility between the spacing of connectors on a chip and the spacing of wires is required. If the spacing of wires is larger than the spacing between pins on a chip, it becomes difficult to fit the wires into the pin pattern. The inevitability of an increasing density of pins has already been mentioned and will increase the difficulty of contacting chips with thick-film techniques.

Thus, for several reasons, the ability of thick-film methods to package high-speed, highly integrated chips is threatened. These limitations of thick-film technology have led to the development of **"thin-film"** processes for packages, in which lines made of high-conductivity metals are defined by lithographic methods akin to those used in IC chip fabrications. Thin-film processes promise to permit the use of much narrower lines on packaging substrates. Tables 12-1 and 12-2 illustrate the differences and the need for thin-film technologies in terms of the above wiring model with representative numbers. Table 12-1 compares the relevant parameters of thick- and thin-film methods. The principal difference in the context of the model is the narrower thin-film channel width. According to Equation 12-7, this translates directly into smaller package areas per chip or fewer layers of wire.

Table 12-2 uses the wiring model to compare the implications of the thick- and thin-film processes for the physical parameters of the package as levels of integration increase. Initially, an area per chip of 1 cm^2 is taken as a goal. The number of wire layers needed to permit the package area in column 3 is calculated and presented in columns 4 and 5 for thick- and thin-film packages, respectively. Thick-film packages have successfully provided 16 layers of signal wire and may be able to cope with two or more times this number. However, the large number of layer-to-layer connections required are certain to eventually limit advances to more layers. If 40 signal layers are regarded as a practical maximum, thick-film packaging fails for levels of integration around 6000 gates per chip. This is a prime motivation for the development of thin-film technology. Other advantages include better **signal transmission** by the nature

Table 12-1. Assumptions Concerning Thick- and Thin-Film Technology

	Thick Film	Thin Film	Advanced Thin Film
Channel width (μm)	300	25	12
Conductor width (μm)	100	12	6
Resistance per unit length (Ω/mm)	0.1	0.3	1.2

Table 12-2. Comparison of Module Wiring with Thick-Film and Thin-Film Techniques

Gates/Chip (N)	Chips/Module (M)	Area/Chip, A (cm^2)	Thick-Film Wire Layers (K)	Thin-Film Wire Layers (K)
1,000	100	1	12	2
2,000	100	1	18	2
4,000	100	1	27	3
6,000	100	1	41	4
16,000	100	1	62	6
32,000	100	1	95	8
32,000	100	2	67	6
32,000	36	1	57	4
32,000*	100	1	—	4
64,000	100	1	144	12
64,000	100	1	101	9
64,000	36	1	86	6
64,000*	100	1	—	6

*Advanced thin film—see Table 12-1.

of low-dielectric polymers and better electrical conductivity provided by thin-film copper wiring. In fact, even in cases in which both techniques might be applicable, the difficulties mentioned above favor thin-film methods.

Part of the need for higher levels of integration has been satisfied by increasing chip size. Chip areas already exceeded 1 cm^2 and would grow larger in the foreseeable future, and an attempt to reflect the possibly larger substrate area simply because of an increase in the area of the chip itself is incorporated in Table 12-2 for levels of integration above 16,000.

12.1.2 Thin-Film Module I/O Connections

Table 12-3 addresses problems that will eventually be encountered by thin-film packages as integration increases. Rent's rule predicts that very large numbers of connections to the next higher package level are needed. However, the area of the substrate does not increase in proportion to the number of connections to the higher level in the scenario of Table 12-2. Thus, the fifth column of Table 12-3 shows that miniaturization of such connections will be required, a difficult task, as these connectors will probably be expected to be mechanically separable in the field. Beyond

Table 12-3. Additional Parameters of Modules with Thin-Film Wiring
M = 100 chips/module, A = 1 cm^2/chip

N	K	N/M	Pins/Module	Area Per Mod. Pin (cm^2)	R (Ω)
1,000	2	1.0×10^5	2,500	0.04	3
4,000	3	4.0×10^5	5,700	0.017	3
16,000	6	1.6×10^6	13,000	0.0076	3
32,000	8	3.2×10^6	20,000	0.0050	3
64,000	12	6.4×10^6	30,000	0.0033	3
32,000*	4	3.2×10^6	20,000	0.0050	12
64,000*	6	6.4×10^6	30,000	0.0033	12

*Advanced thin film— see Table 12-1.

a certain point, Rent's rule may not be applicable to the substrates of Tables 12-2 and 12-3, as they contain over a million gates and can constitute a complete medium-to-large computer.

Although the number of layers is reduced by an order of magnitude as compared to thick-film packaging, a two order-of-magnitude increase in the level of integration uses up this advantage. One can be confident that, as in IC chip technology, package linewidths can be reduced and permit the advantage of fewer layers to be retained for a longer time as is illustrated by the "advanced thin film" in Table 12-2. However, unless unusual **aspect ratios** can be achieved in the wires, increases in the resistance of the long lines on the package will rapidly become unacceptable, as illustrated by the "advanced thin film" in Table 12-3. Another possible direction would be reducing the number of chips on a module, thereby reducing both the number of layers of wire and the length of wires. Examples are included in Table 12-2. However, this would complicate the physical design of the next higher level of package.

A final caveat: Although the characteristics of the chip have been taken as a starting point, the chip and the package are interdependent. The nature of the chips that are actually used will depend on the ability of the package to deal with them, not only for **wireability** but also for **power distribution** and heat dissipation. The latter demands are discussed below.

12.1.3 Power Supply and Heat Removal

The quest for high speed has prevented the power dissipation per circuit in gate arrays for large machines from decreasing very much as the level of integration of the chips has increased. There is a trade-off between power and delay; the highest speed demands that the gates be operated at high power. Many power I/Os are needed to carry the electrical power to the chips. The package must include conductors that distribute the power throughout the system. The power-supply wiring is simpler in the sense that it does not have the pseudorandom character of the signal network. Electrical resistance in the power supply leads must be carefully controlled, however. A module may be powered by 100 A of current, which must reach the chips with less than 5% voltage loss, for example. The number of wiring levels in the module used to deliver current at the supply voltage and return it through ground is not much different from the number needed for signal transmission.

An aspect of the chip that poses a serious problem for the package is the intermittent delivery of large amounts of current to the signal lines by **driver** circuits on the chip [12]. Each such event changes the current drawn from the power supply by the chip by tens of milliamperes. Sometimes many driver circuits switch simultaneously, causing a sudden large change in the current drawn from the power supply, a change of a sizable fraction of an ampere. The possible number of **simultaneous switchings** will apparently increase with the level of integration. The inductance of the power supply leads then produces a significant change of supply voltage at the chip. For example, if a current change of 0.25 A in 250 ps is to produce a voltage transient of less than 0.1 V, the inductance must be less than 0.1 nH. This is approximately the inductance of only 100 μm of wire. The power-supply transient can be controlled with capacitive filters or **decoupling capacitors**, but the package must then provide the capacitor very close to the chip, if not on the chip. One such arrangement of an integrated decoupling capacitor is illustrated elsewhere [13] and discussed in Section 12.9.

The electrical power delivered to the chip is dissipated to heat, and another demand made on the package by high-speed gate array chips is the removal of large amounts of heat. Today, a power of 10–40 W, 10 W in chips being shipped and as much as 40 W in chips being developed, may be produced on a gate-array chip. High **heat fluxes** are carried away from the chip by special cooling structures that transfer the heat to air or another fluid that carries it out of the machine, as discussed in great detail in Chapter 4, "Heat Transfer in Electronic Packages." Chip temperatures must be kept below some maximum value, say about 70°C; only a certain finite temperature difference is available as a driving force for the heat flux. The limited heat-transfer coefficients at interfaces require that sufficient area be provided to allow heat flow from one material to another with an acceptable temperature drop. Apertures through which fluids flow must be large enough to permit the flow to take place at pressures that can be provided by inexpensive and quiet apparatus. All of these restrictions limit the rate of heat removal and constitute another lower bound to the area per chip, particularly with thin-film packaging. Fairly complex structures are needed to achieve even the present rates of heat removal.

It is likely that the power per gate on logic chips will decrease as integration increases. A very rough attempt to extrapolate trends in bipolar chips is presented in Table 12-4 together with the implications for removal of heat from packages. The table clearly illustrates the need for continued increases in cooling capability, as compared with about 5 W/cm^2 in current practice (first row in Table 12-4).

Relief from the heat-transfer limitations may also be found in recent development to build large mainframe computers from CMOS integrated circuits. CMOS circuitry dissipates much less power than high-speed bipolar circuitry. Furthermore, operation at the temperature of liquid nitrogen improves performance by a factor of 2. The electrical resistivity limitation in advanced technology will also be relieved by cryogenic operation and, additionally, motivates interest in it. However, cooling efficiency at these low temperatures will present additional challenges.

12.2 ELECTRICAL PERFORMANCE

The physical structure and the materials used in the construction of a thin-film package are constrained by the electrical properties that are needed to attain the best performance [12,14]. The electrical requirements generally relate to:

Table 12-4. Estimated Power Density of the Thin-Film Modules

Gates (N)	Chip Area, A (cm^2)	Power/Gate (mW)	Module Power Density (W/cm^2)
1,000	1	5	5
4,000	1	5	20
8,000	1	5	40
16,000	1	5	80
32,000	2	5	160
32,000	2	5	80
4,000	1	2.5	10
8,000	1	2.5	20
16,000	1	2.5	40
32,000	1	2.5	80
4,000	1	1	4
8,000	1	1	8
16,000	1	1	16
32,000	1	1	32

- Point-to-point propagation (velocity, attenuation, distortion)
- Noise containment (coupled, ΔI, shielding)
- Impedance (matching terminations, resistance)

Transmission-line characteristics are of special importance in high-speed systems in which short pulse lengths require that high-frequency line impedance and load impedance be matched to prevent reflected pulses from appearing as noise, as discussed below.

Chapter 3, "Package Electrical Design" discusses the factors and trade-offs involved in designing an electronic system. A summary is repeated here in the context of thin-film packaging technology.

The width and thickness of lines determine the DC resistance, whereas the spacing between lines is determined by allowable **cross-talk**. In a single-sided thin- or thick-film structure with no control of characteristic impedance, Z_0, the cross-talk between two adjacent lines is determined by electrostatic and electromagnetic coupling between the two lines. Coupling in this case decreases as the conductor spacing increases. Cross-talk can be reduced by placing a ground plane closer to the line than the line-to-line distance, as in Figure 12-5. Further improvements can be made by having two ground planes, one above and the other below the line, but both closer to the conductor lines than from one line to an adjacent line. The close spacing between signal and ground, however, results in excessive capacitance and affects the characteristic impedance**,** defined as [series inductance (L)/capacitance $(C)]^{1/2}$. Very-high-performance applications require that all lines be terminated in order to keep the signals that are sent down the lines from reflecting at the far end and transmitting backward along the lines. A digital signal can also be partially reflected from a discontinuity in the transmission line. The reflection coefficient, which gives the fraction of the reflected signal, is determined by the characteristic impedance and by the load resistance that terminates the line. If a given transmission line has a characteristic impedance of 100Ω and the load resistance is also 100Ω, the signal is totally absorbed by the load and none of it is reflected back. This is an ideal situation. If the load resistance is 200Ω, however, a third of the signal is reflected and adds to the initial signal on the line. A load resistance of 50Ω also yields a reflection coefficient of one-third, but the signal is subtracted from the initial one. A characteristic impedance range of 50–100Ω is commonly used in designing digital computers. In a **triplate** structure, the linewidth, related to interconnection density, as described previously, determines the thickness of dielectric required between a line and the reference plane for a given characteristic impedance, given a value of **dielectric constant** for the insulating layer. The characteristic impedance Z_0 is related to the inductance and capacitance per unit length (L and C) by

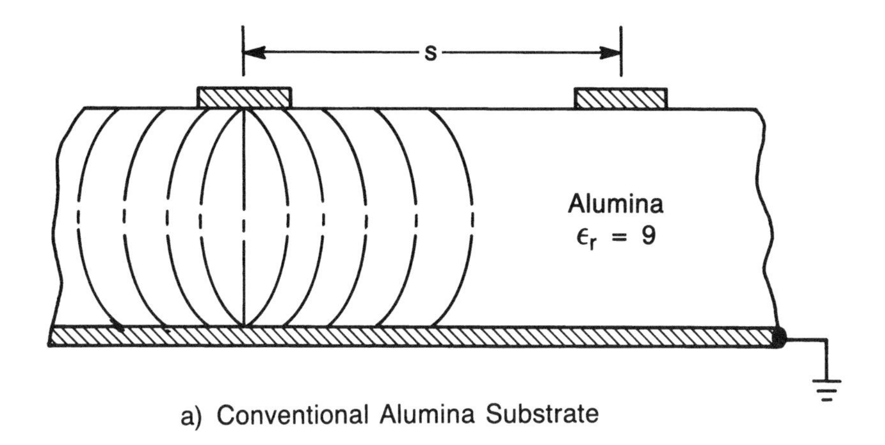

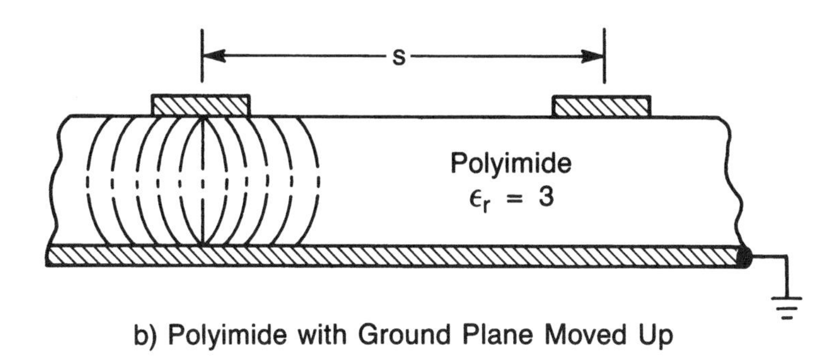

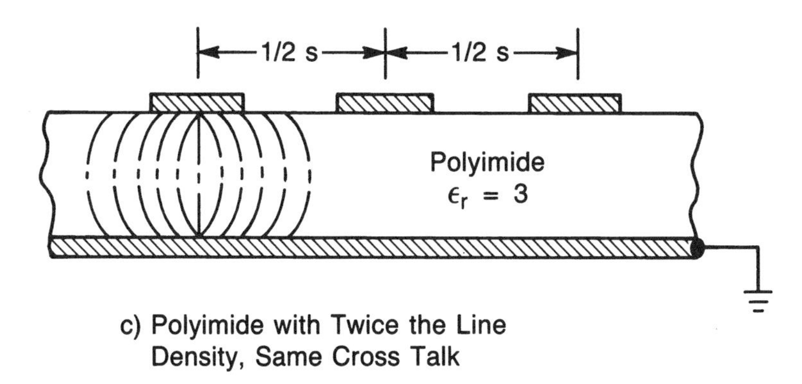

Figure 12-5. Lowering Dielectric Constant Improves Package Density. (a) Alumina substrate; (b) polyimide dielectric; (c) polyimide dielectric with twice the line density, same cross-talk.

$$Z_o = (L / C)^{1/2} \qquad [12\text{-}8]$$

Consequently, the vertical dimensions of a multilayer thin-film (MLTF) structure depend on the electrical parameters desired.

The requirement that **coupled noise** levels be kept low enough to prevent switching a **receiver** on a neighboring line will demand that the line spacing be large enough so the mutual inductance of two neighboring lines will be less than a given value. This requirement will modify the thin-film interconnection structure so that

1. Parallel runs are kept to a certain maximum length
2. A reference plane is used to reduce the coupled energy
3. **Wiring channels** are placed appropriately far apart

Clearly, these approaches are examples of the manner in which electrical requirements modify and restrict the set of thin-film structures that can be used for an application.

Pulse-propagation velocity and dispersion impose materials requirements on the thin-film dielectric. The pulse speed is given by

$$V = c / \varepsilon^{1/2} \qquad [12\text{-}9]$$

where

ε = dielectric constant of the insulating layer
c = speed of light
V = pulse-propagation velocity

For high-speed systems, it is desirable for the thin-film dielectric to have a low value of dielectric constant. A short pulse is composed of a range of high-frequency components (up to 1 GHz and above). Consequently, the dielectric constant of the thin-film dielectric should be relatively constant up to these frequencies, as well as close to unity, to ensure both a minimized pulse broadening and a high velocity of propagation. Additionally, the lowering of the dielectric constant enhances package performance in other ways, as discussed below.

A reduction in dielectric constant in Figure 12-5 can allow a lower thickness of dielectric, for equivalent capacitance, and enable the ground plane to be moved closer to the line. Additional lines can, therefore, be accommodated for the same cross-talk, as illustrated. Thus, the effect of a low dielectric constant, in addition to increased speed of signal, results in improved wiring density of packages, which further results in improved systems performance. The effect of the dielectric constant, therefore, is a synergetic effect of speed improvement from Equation 12-9 and improved packaging density.

Finally, **line resistance** attenuates the signal as it propagates in a driver line (see Chapter 3). A conductor's conductivity, cross section, length, and driver/receiver characteristics must be all coordinated in order for the package interconnections to function properly. Properties of thin films given later can be used in the design of thin-film packages that meet not only geometric interconnection functional needs but also those imposed by the requirements for electrical functionality.

12.3 THIN-FILM VERSUS THICK-FILM PACKAGES

Materials and processes for thin-film technology are distinctly different from those used in other packaging technologies, such as multilayer ceramic or printed wiring boards, as illustrated in Figure 12-6. In the latter technologies, thick-film processes such as doctor-blading and laminating of greensheets and epoxy-glass cloth are the typical processes by which insulating structures are fabricated. These techniques are covered in Chapter 9, "Ceramic Packaging" and Chapter 17, "Printed-Circuit Board Packaging." They are denoted as thick-film approaches because they utilize materials with coarse granularity of about 100 μm (inorganic greensheets and epoxy-glass cloth) and involve processes of low resolution and precision that make them not easily adapted to the fabrication of multilayer structures with feature sizes less than 100 μm. Table 12-5 compares the characteristics of each of these three technologies for wiring dimensions and layers.

Materials utilized for thin-film packaging have granularity generally smaller than 1 μm (photoresist, polymer films, amorphous inorganic structures formed from sputtered materials, reflowed glass, etc.) and the typical feature sizes in thin-film packaging are in the order of 25 μm and below. Consistent with the potential offered by such materials for fine-line patterning, the processes used for thin-film packaging primarily employ **photolithography** for conductor-pattern definition and via generation through the dielectric polymer.

The advantages of thin-film technology over thick-film technologies are better elucidated by specific examples as follows.

12.3.1 Size of Thin-Film and Thick-Film Packages

High-density thin-film packages such as the ones discussed in this chapter are compared in Table 12-6 with thick-film packages involving both sequential buildup of multilayers and cofired multilayer structures discussed in Chapter 9, "Ceramic Packaging." A thorough discussion of various aspects of all these packages is reported elsewhere [15], including:

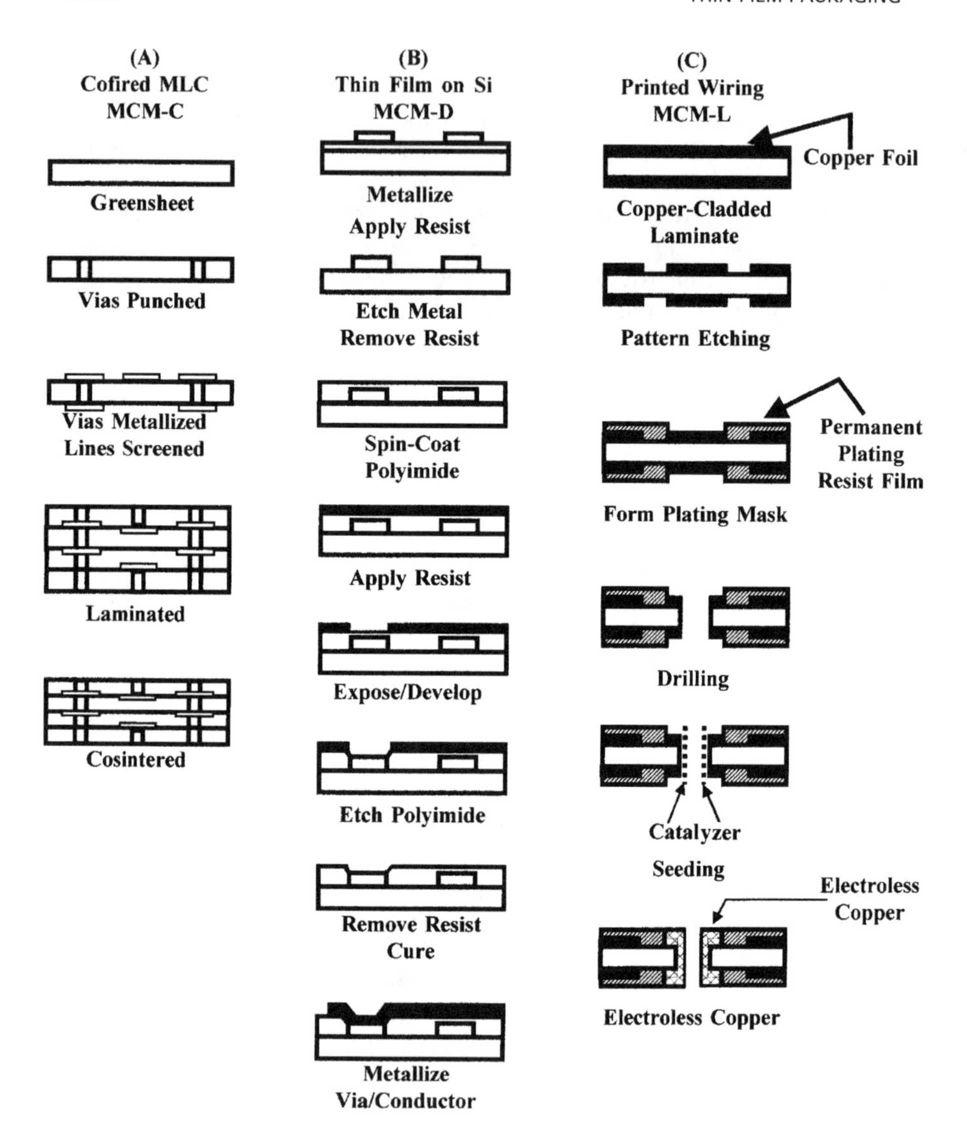

Figure 12-6. Comparison of Thin- and Thick-Film Technologies

- Packaging density
- Package delay
- Package power
- Cost

The analysis, assuming a **very-large-scale-integration (VLSI)** chip of 7.5 mm size, is summarized in Table 12-6. The **printed-wiring board**

Table 12-5. Typical Package Parameters of the Three Multichip Module Types

Characteristic	MCM-L	MCM-C	MCM-D
Description	High-density laminated printed-wiring board	Cofired low-dielectric-constant ceramic substrate	Thin film on silicon
Max. wiring density (cm/cm^2)	300	800	250–750
Min. linewidth (μm)	60–100	75–100	8–25
Line space (μm)	625–2250	125–450	25–75
Max. substrate size (mm)	700	245	50–225
Dielectric constant	3.7–4.5	5–5.9	3.5
Pinout grid (mm)	Array 2.54	Array 2.54 (staggered)	Peripheral 0.63
Max. No. of wiring layers	46	63	4
Via grid (μm)	1250	225–450	25–75
Via diameter (μm)	300–500	100	8–25

in Table 12-6 is assumed to be multilayer epoxy-glass board with four to eight X and Y conductor planes spaced appropriately between ground planes. The conductor widths are assumed to be 125 μm and plated-through-holes (PTH) are on 254-μm centers. Thick-film multilayer ceramic involves sequential deposition and firing of dielectrics and conductors on

Table 12-6. Comparison of Package Figures of Merit—Normalized. After Neugebauer, Ref. [15].

Packaging Approach	Power Delay	Size or Weight	Relative Cost	Packaging Density	Packaging Delay	Overall Figure of Merit
Printed-wiring board	1.00	1.00	1.00	5	7.8	1
Thick-film multilayer ceramic	1.08	0.42	1.02	12	7.8	2
Multilayer ceramic	0.34	0.20	0.65	25	5.6	20
Thin-film multilayer with chips on one side	0.19	0.14	0.60	36	4.8	60
Thin-film multilayer with chips on both sides	0.08	0.07	0.44	72	4.1	400

prefired alumina substrate. The conductor width and dielectric thickness are assumed to be 250 μm and 40 μm, respectively. The multilayer ceramic stands for cofired multilayer alumina and molybdenum substrate described in Chapter 9, "Ceramic Packaging." The thin-film multilayer is the same as the multilevel **polyimide (PI)** or glass thin-film technology discussed in this chapter. Linewidth and dielectric thickness are assumed to be 25 μm and 1 μm, respectively. The thin-film **hybrid** with chips on both sides involves the same multilevel thin-film technology as before, but chips are bonded on both sides with appropriate via connections in between. Referring to Table 12-6, packaging density is calculated by dividing the chip area by the chip footprint. Package-delay calculations are estimated simply from the output capacitance, the on-chip resistance of the output driver, and the buffer delay. **Power-density** calculations are based on power requirements per unit area of the package–thin-film packaging technologies requiring a power supply about 50% more than the multilayer **cofired** ceramic, consistent with packaging-density improvements pointed out earlier. Multilevel thin film in Table 12-6 is clearly shown to provide as much as a sevenfold improvement over the printed wiring board and 50% over the state-of-the-art multilayer ceramic (MLC) substrate. The potential of drastic size reduction of thin-film packages is illustrated in Figure 12-7 for a hypothetical nine-chip multichip package each with 1000 I/Os. In this figure, the printed wiring board is assumed to be on a 50 mil via grid, ceramic on a 3 mil grid, and thin film on a 10 mil grid. This size reduction is the primary motivation for the development of thin

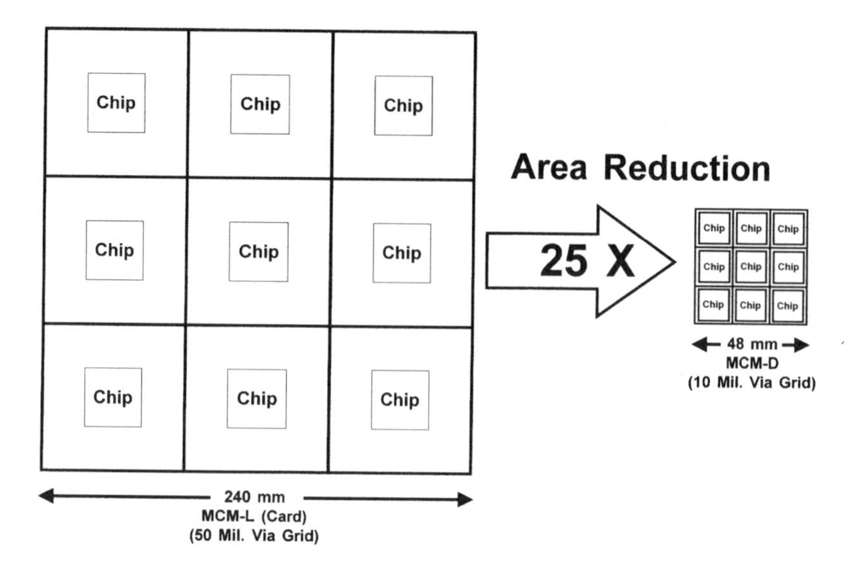

Figure 12-7. Comparison of Sizes of Hypothetical 9-Chip Modules Fabricated by Different Technologies

films on printed wiring boards such as the surface laminar circuitry (SLC) recently developed by IBM Japan as discussed in Section 12.8 [16]. This technology is also referred to in the industry as **MCM-D** on **MCM-L** (printed-wiring board multichip module) or MCM-D/L.

12.3.2 Cost of Thin-Film and Thick-Film Packages

The relative cost of the three packages discussed above is also summarized in Table 12-6, as well as in Figure 12-8 as a function of typical chip I/Os interconnected onto these packages. Cost is by no means a precise or absolute parameter that describes the package; it is a relative measure that changes as a function of time, volume, yield, equipment depreciation, and so on. Nevertheless, it is a good indicator to differentiate various packages as proposed here. Clearly, the printed wiring board technology, based on photolithography patterning and lamination followed by mechanical drilling and plated-through-hole, provides the least wiring but it is also the cheapest technology of the three. It is also obviously the technology that meets most of the low I/O, consumer, and portable product requirements. Two recent breakthroughs, one in direct **flip chip** attach to the printed wiring board by means of underfill encapsulants, as discussed

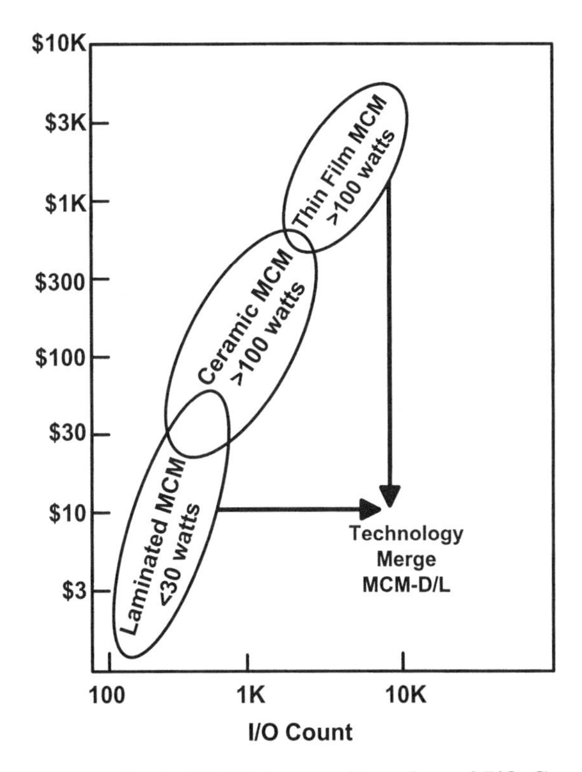

Figure 12-8. Cost of MCMs as a Function of I/O Counts

in Chapter 8, "Chip-to-Package Interconnections," and the other is one that add one or more layers of surface thin-film wiring referred to as surface laminar circuitry (SLC), as discussed later in this chapter as well as in Chapter 17, "Printed-Circuit Board Packaging," provide additional motivation for the use of this technology. These developments are illustrated in Figure 12-9 but are discussed in more detail in the aforementioned chapters. The technology merger in Figure 12-8 allows "best of both world," and thin-film on PWB as the single strategic direction the industry is expected to take.

The cofired-ceramic technology provides improvements over the printed wiring board technology in two areas: finer wiring (75 μm lines and vias) and more layers (up to 100). The cost per I/O of this technology is, however, higher than printed wiring board technology. The thin-film technology discussed in this chapter provides the most wiring, but the cost per I/O is the highest. The significant thin-film costs involve setting up a cleanroom manufacturing facility (class 10–1000). A number of approaches discussed in Section 12.10, however, provide potential solutions to lowering the cost to a level that this technology can be expected to be almost as inexpensive as today's standard printed-wiring board technology.

12.3.3 Single-Layer Thin-Film Packages

Thin-film packages have and will continue to be used both for single-chip and multichip module packages. For **single-chip** packages such as the ones illustrated in Figure 12-10, thin film is used to redistribute or fan-out the chip I/O grid to that of printed wiring board. The chip I/O

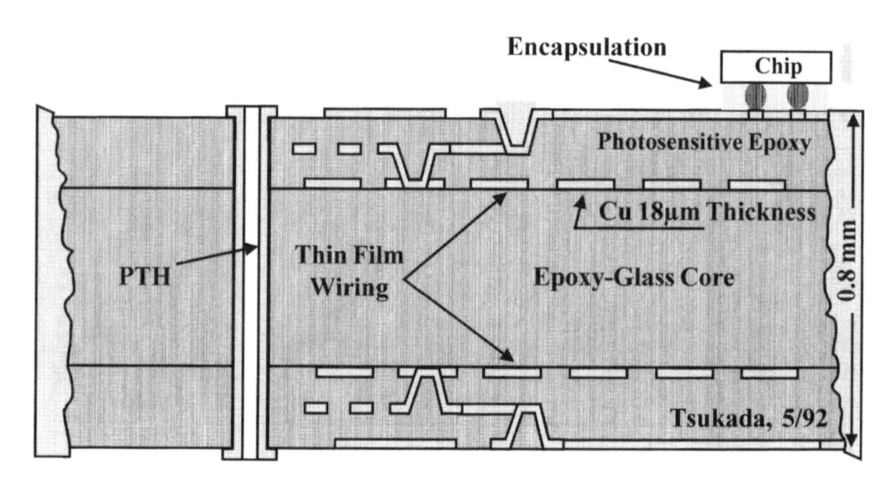

Figure 12-9. Surface Laminar Circuitry by IBM, Japan

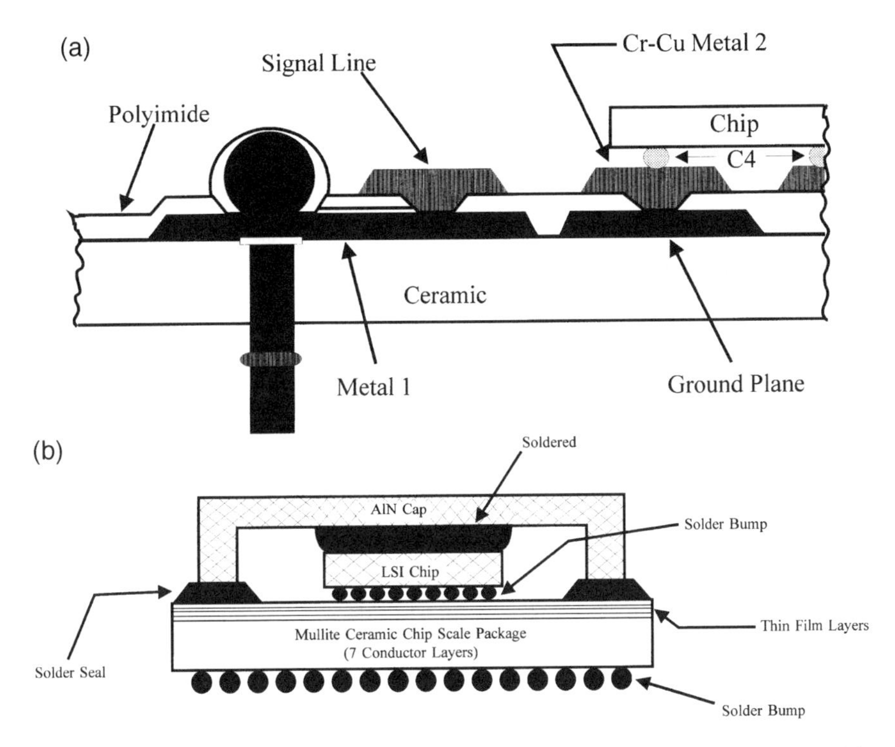

Figure 12-10. First Generation of Single-Chip Thin-Film Packages by (a) IBM, and (b) Hitachi

grid is typically about 100–250 μm and that of the board typically about 0.4–2.5 mm.

The simplest type of thin-film package is one in which the conductor is a single patterned metal thin-film layer on a substrate. The implementation can assume one of two schemes. One of these is a fan-out or **space-transformer** pattern, whereas the other is an **interchip** connection. In both cases, the application does not allow crossing of the point-to-point conduction paths because of the topology of single-layer structure. The space transformer is usually used to connect the signal and power terminations of a chip to the coarser grid of the second-level package. The other type of single layer structure that is used in simple applications predominantly involves parallel transmission of signal and power. Parallel channels, busses, and power supply are examples of such applications.

The substrate for fan-out may be rigid (e.g., ceramic or silicon) or flexible as in **tape automated bonding (TAB)** for chip mounting discussed in Chapter 8, "Chip-to-Package Interconnections." Thin-film processes for forming two-layer and three-layer tapes as described in that chapter employ subtractive or plate-up technologies which are discussed later in this chapter. In the case of parallel signal transmission, fabrication of the conductors on a flexible carrier, such as a polyimide, permits a packaging entity that can be bent during system assembly or repair. This technology is most commonly used in consumer electronics.

Functional requirements determine the materials and structures of a packaging entity. The conductivity of the conductor material, together with its cross-sectional geometry, determines the resistance per unit length. This parameter must be chosen in the case of signal conductors so that the ohmic signal-level drop from driver to receiver is within the system's specifications.

In the case of power-carrying conductors, both **insulation resistance (IR)** drop and conductor heating must be controlled by proper choice of conductor materials and geometry. Coupled noise in simple packages with single-layer thin-film conductors is controlled in two ways, as discussed earlier in this chapter and in Chapter 3, "Packaging Electrical Design." One is to reduce mutual inductance by wider spacing of conductors, whereas the other is to intersperse signal and power lines. This latter technique can also provide control of the high-frequency impedance of signal lines. As in other package geometries, there is a competition between density and functionality.

High wiring density requires close spacing of narrow conductors, whereas power capability, and containment of signal-level drop and coupled noise place an upper bound on wiring density. This competition is usually not an issue in relatively low-performance simple packages, such as single-chip packages, but as performance demands become aggressive (e.g., high I/O count or high-speed chips and systems), multilevel, multichip packaging entities provide extra design degrees of freedom, at a yet higher cost. Furthermore, materials and mechanical effects also place constraints on the construction of multilayer, multichip package. The conductors must adhere to the substrate in the presence of mechanical stress. Such stress may be intrinsic (due to the fabrication processes) or may be externally applied during system construction and use (as can occur in the bending of a flexible substrate) or may be caused by thermal cycling of chips and conductors during power on-off cycles. Severe stress can also develop during system shipment where temperatures can range from as low as −40°C to as high as 120°C, as discussed in Chapter 5, "Package Reliability." More detailed discussion on multilevel, multichip packaging is found in the following sections.

The single-chip packages illustrated in Figure 12-10 and their processes are discussed in greater details in Section 12.7.

12.3.4 Types of Multilayer Thin-Film Structures

Manufacturability, cost, and electrical performance are the three primary factors that are considered when choosing materials, processes, and structures for multilevel thin films (MLTF). From the structure point of view, one needs to choose between planar and nonplanar thin-film processes based on cost/performance trade-offs.

The choice of one type of structure over the other also depends on customer requirements such as wireability, impedance, and global planarity. Therefore, it is desirable to maintain process, materials, and tool commonality between the two types of structure wherever possible. One exception is the mechanical planarization technique that is needed to build planar structures. The criteria that are chosen to compare the processes are (1) packaging density, (2) manufacturability, (3) process and materials maturity, and (4) cost. Multilevel, thin-film interconnect structures can be essentially classified into two types: (1) unfilled via type requiring **staggering of vias** in successive levels and (2) filled via or "**stud**" type which permits stacking of vias in successive levels, as shown in Figure 12-11.

One of the first process considerations in the design of the thin-film package is the choice of the via structure. The conformal via process gives rise to a nonplanar structure, whereas the filled via process gives rise to a planar structure. Planar structures produce the highest wiring density for a given linewidth and pitch. One has to consider the trade-offs between the process complexity and tooling costs, and the electrical advantage of increased wiring density.

12.3.4.1 Planar stacked studs

Planar structures can be formed by first defining the wiring and via pattern in the dielectric and then metallizing the structure. In this approach, **planarization** to remove the excessive metal is done at the wiring level. Two variations exist for this process. In the first variation, the intralevel dielectric (between adjacent conductor lines) and the interlevel dielectric (between two conductor layers, X and Y) are applied separately. After

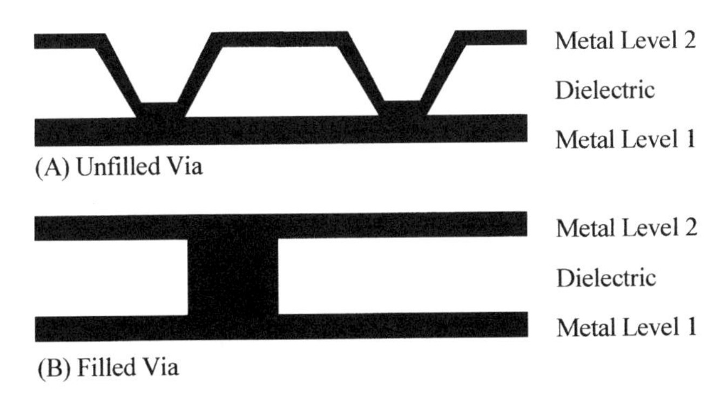

Figure 12-11. Filled and Unfilled Via Interconnect

the application of the intralevel dielectric, the vias are patterned and the polymer is cured. Similarly, after the application of the interlevel dielectric, the conductor features and vias are patterned as shown in Figure 12-12. In the second variation of this approach, the vias can be formed after the conductor lines in the intralevel dielectric are defined. In both cases, the whole dual-level structure is then metallized by electroplating or sputtering. The excessive metal is then removed by mechanical or electrochemical planarization [1]. Because the intralevel dielectric is applied on a planar surface, the thickness is well controlled. On the other hand, the conductor-layer thickness is defined by the planarity of the interlevel dielectric layer and the mechanical planarization process. It is for this reason that the planarization technique chosen should give local planarity, not global flatness.

The alternative is to first define the metal wiring and via features and then applying the dielectric, followed by planarization to remove the excessive dielectric at the stud level. This latter process has been developed and implemented by many companies and institutions [17].

In this process, the wiring level and studs are fabricated by **pattern**

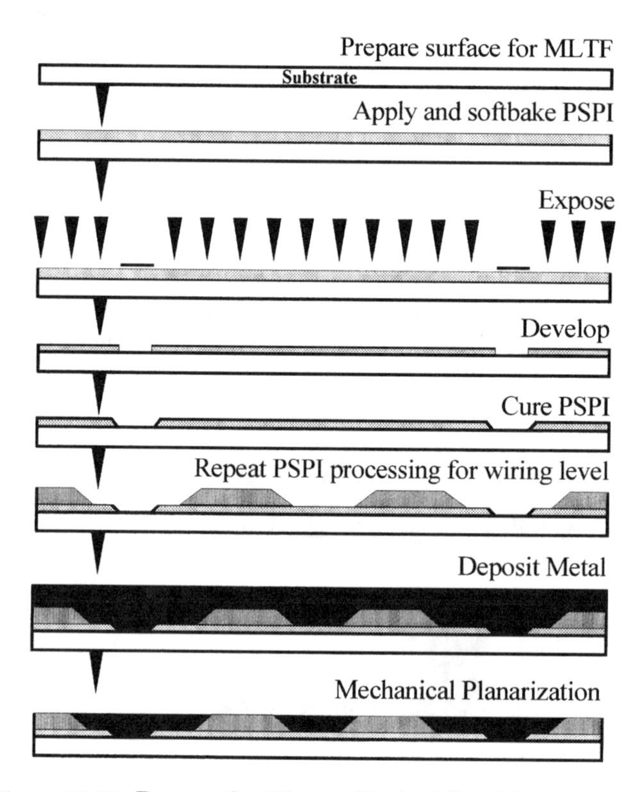

Figure 12-12. Process for Planar Stacked Stud Interconnect

electroplating through a photoresist **stencil** using a common plating seed layer. A polymer dielectric is then applied by spin or spray coating. The excessive polymer is removed by mechanical means. This exposes the tip of the studs for connection to the next level of interconnect and planarizes the entire structure. This process gives the greatest flexibility in the choice of the polymer and the coating technique. The conductor thickness and width is defined by the uniformity of the electroplating and seed-layer etching processes. The dielectric thickness control is much more difficult because it is defined by the mechanical planarization technique.

Figure 12-13 shows the cross sections of (a) a stacked via structure, and (b) the corresponding planar structure.

12.3.4.2 Nonplanar staggered vias

Nonplanar structures are simpler to process and require no mechanical planarization. They also require fewer dielectric application steps and their associated cures. The vias are generated through the dielectric layer, and then the vias and conductor lines are metallized on top of the dielectric layer. Both via formation and conductor line patterning are by lithographic techniques. The connections between the conductors below and above the dielectric layer are made through the via openings in the dielectric. No planarization is produced in the metallization processes of vias and conductor lines. Partial planarization of the topography of the underlying conductor line and via metallization is achieved by the capability of leveling of the dielectric polymer coating. Without further planarization

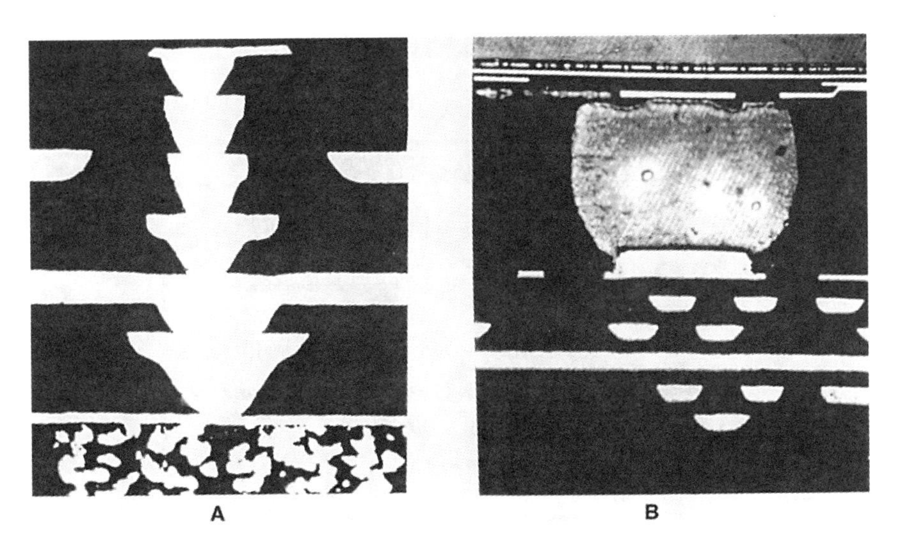

Figure 12-13. (a) Planar Stacked Stud MLTF Structure; (b) Planar MLTF with Flip Chip Connection

steps, a conformal or nonplanar structure resulted and is aggravated with the later built-up of the multilayer thin-film structure. A disadvantage of this process is the use of low leveling or self-planarization of commercial dielectric polyimides, which limits the number of layers. This can be alleviated by incorporating multiple coatings of polyimide which tend to increase the degree of planarization (DOP), or by using dielectric polymers with higher planarization capability.

In the conformal via process shown in Figure 12-14, the via holes are formed by photolithographic methods in a photosensitive polyimide (PSPI) dielectric or laser ablation and metallized conformally by electroplating through a thick positive photoresist mask.

The vias could also be fabricated using reactive ion etching (RIE) or wet etching. Laser ablation and photosensitive polyimide lithography are robust processes and have been successfully implemented. Metallization can be accomplished by subtractive etching, pattern plating, or lift-off. The choice between subtractive etching and pattern electroplating will primarily depend on the ground rules of the package and the tolerances required. The minimum pitch for subtractive etching as a function of

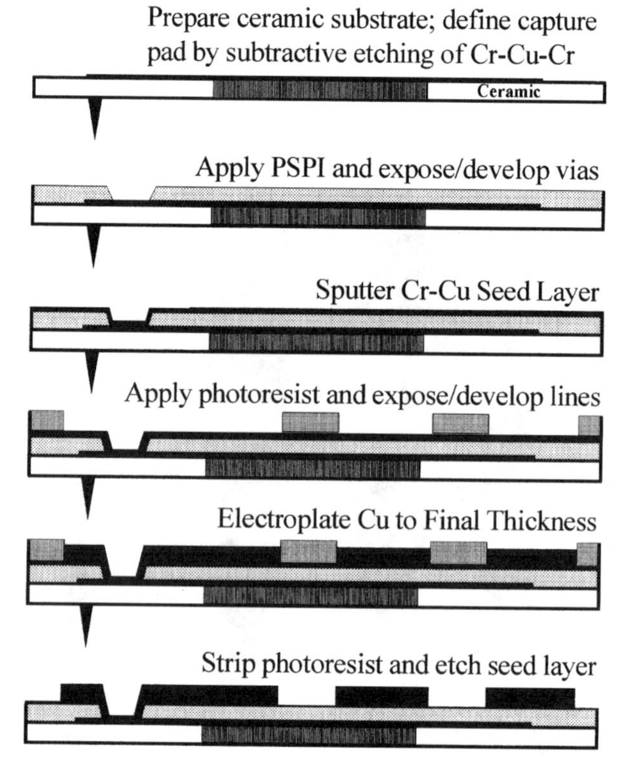

Figure 12-14. Process for Conformal Staggered Via Interconnect

linewidth, undercutting and line thickness is shown in Figure 12-46, and is defined as follows:

$$P = L + M + 2T + U, \qquad [12\text{-}10]$$

where

P = minimum pitch
L = linewidth
M = minimum resolution
T = line thickness
U = undercut

When the ground rules do not allow for subtractive etching to be used as the metallization process, pattern electroplating has to be used.

Figure 12-15 shows the cross section of a nonplanar structure.

12.3.4.3 Electrical Trade-offs

Two fundamental electrical issues need to be addressed when comparing the planar and nonplanar structures. These relate to the differences in the module power distribution due to stacked studs as opposed to staggered vias and electrical variations due to the effect of the dielectric nonplanarity.

A typical triplate wiring structure (two signal levels and two ground/power levels) is shown in Figure 12-16, which consists of two wiring layers sandwiched between two mesh planes. The cross section assumes a planar structure that guarantees a uniform dielectric thickness across the entire module.

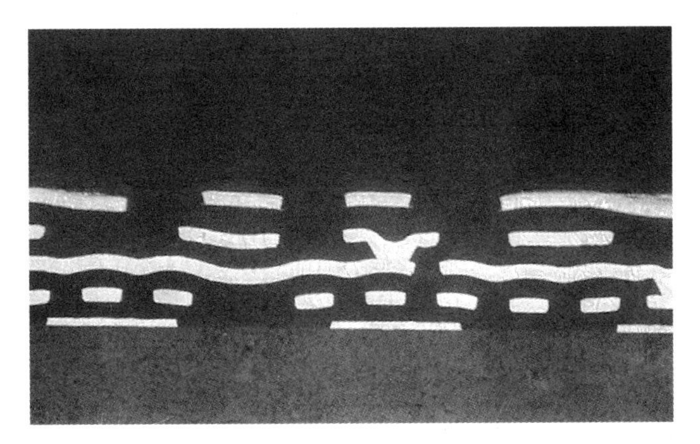

Figure 12-15. Cross Section of One Level of Nonplanar Staggered Via Thin-Film Interconnect

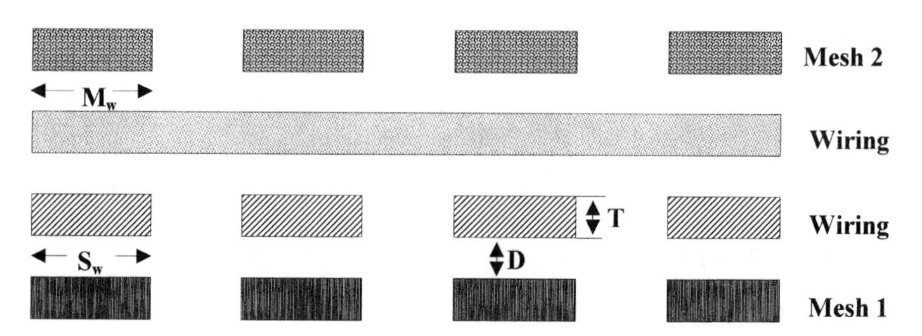

Figure 12-16. Triplate Wiring Structure (Planar Process)

The physical parameters in Figure 12-16 such as the mesh width (M_w), signal linewidth (S_w), dielectric thickness (D), metal thickness (T), and dielectric constant (ε) affect the characteristic impedance Z_0 of the line. The variation of Z_0 as a function of the physical dimensions is shown in Figure 12-17 assuming a planar process. Because the design of the mesh plane plays a critical role in the calculation of Z_0, a parallel mesh is assumed where all lines are completely shielded with an added restriction that $M_w = 2S_w$. The adjacent wiring level adds an additional capacitive loading which has been included in Figure 12-17.

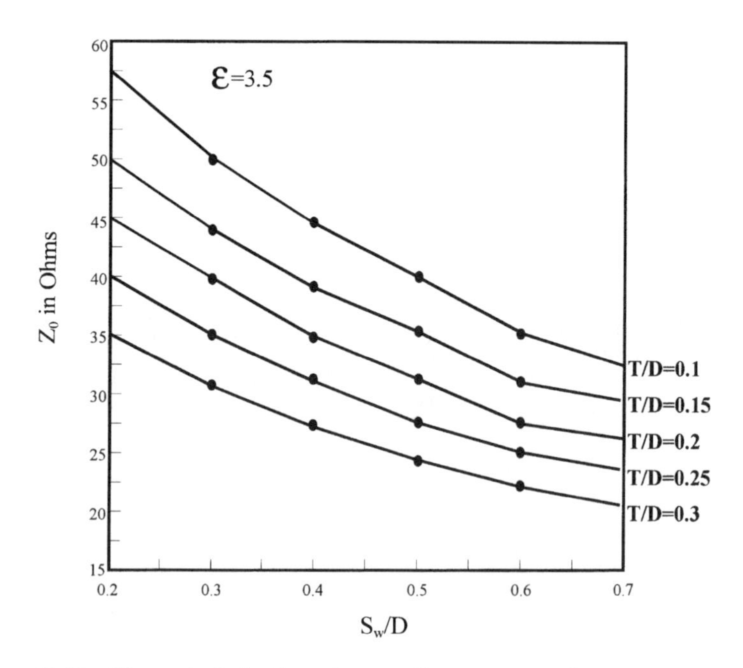

Figure 12-17. Characteristic Impedance, Z_0, versus Thin-Film Dimensions (Planar Process)

The Z_0 variation shown in Figure 12-17 may not be true for a nonplanar process due to the dielectric nonplanarity. In a nonplanar process, the mesh layout has a larger role in deciding the line characteristics than in a planar process. Based on extensive modeling of the nonplanar cross section, the parallel mesh design with all lines completely shielded represents the optimum layout that minimizes the dielectric nonplanarity in critical areas, thus providing a controlled impedance environment for the transmission lines. The cross section of a triplate wiring structure for a nonplanar structure is shown in Figure 12-18.

As seen in Figure 12-18, the two wiring levels do not look alike, thus resulting in different Z_0 for the two levels. The bottom wiring level is similar to that in Figure 12-17 due to the use of a parallel mesh; hence, the variation of Z_0 is the same as in Figure 12-17. However, the top wiring level has variations due to dielectric nonplanarity, which results in a lower Z_0 as compared to the bottom wiring level.

The difference in the power distribution due to the use of stacked studs (planar) as opposed to staggered vias (nonplanar) is best illustrated with an example. Consider a module with a pin-grid array supplying power to the chips through a thin-film triplate wiring structure. Because the vias or studs supply the current to the chip, they decide the module voltage drops.

Assuming $D = 5.5$ μm and $T = 4.5$ μm, the staggered via scheme produces an additional 37.5% voltage drop due to the longer current path as opposed to the stacked via scheme. Hence, it is important that a good power distribution be used in the nonplanar structure with more vias, which decreases the module voltage drop through current sharing.

Although only the Z_0 variation has been considered here, all electrical parameters are affected by the dielectric nonplanarity. Hence, for a good design using the nonplanar structure, it is imperative that the designer be aware of the structural features and the electrical variations so as to create an optimized layout that meets the application requirements.

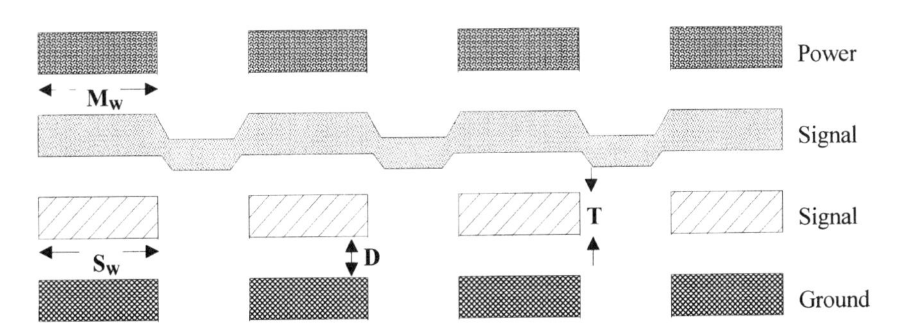

Figure 12-18. Triplate Wiring Structure (Nonplanar Process)

12.4 THIN-FILM MATERIALS AND PROCESSES

Having reviewed the structures to which the functional requirements of systems lead, an overview of materials and processes is in place. Several multilevel thin-film structures [20,21] that are under development or in manufacturing are listed in Table 12-7, along with both the conductor and dielectric materials, as well as their processes. The number of layers ranges from two to six on substrates ranging from silicon to alumina to multilayer ceramic to glass-ceramic to sapphire, and more recently to printed wiring board (PWB). The substrate sizes range up to 100–225 mm at present and are expected to be in excess of 400 mm in the near future. Roll-to-roll continuous fabrication as described in Section 12.8.2 is also being pursued. A list of the properties of common dielectric and conductor materials for thin-film packaging are given in Table 12-8. The set of materials and processes described here to fabricate thin-film packages is reasonably comprehensive. More detailed discussions can be found in many references [2,17,20–42]. Thin-film packaging is one of the fastest evolving packaging technologies today and is expected to continue well into the next century. Consequently, the emergence of novel materials and processes is anticipated.

The general process of fabricating thin-film multichip or single-chip packaging involves five basic steps:

1. Dielectric deposition on substrate
2. Via formation in dielectric
3. Metal deposition
4. Metal patterning
5. Bonding pad or terminal metallurgy deposition

12.4.1 Thin-Film Substrates

The various substrates used for high-density thin-film package manufacturing are as follows:

- Silicon **wafer**
- Dry pressed alumina
- Aluminum nitride
- Cofired alumina
- Cofired low temperature or glass-ceramic
- Polished metals (Al)
- **Printed wiring board (FR4)**

Table 12-7. Some High-Density Multilevel Thin-Film Technologies Reported

Number	1	2	3	4	5	6	7
Metal	Aluminum	Copper	Aluminum	Copper	Copper	Copper	Copper
Thickness (μm)	1	5	1	5	5.5	25	2
Width (μm)	11	25	25	10	8	100	10
Pitch (μm)	22	75	100	25	25	250	20
Resistance (mΩ)	27	1	11	3–4	4	0.1	10
Dielectric	SiO_2	Polyimide	SiO_2/sapphire	Polyimide	Polyimide	Polyimide glass	Polyimide
Dielectric Const.	3.8	3.5	5.5	3.5	3.5	3.8	3.5
Thickness (μm)	1	25	450	8	6	60	10
Deposition	Sputter/CVD	Spray	Sputter/CVD	Spun on	Spun on	Foil	Sputter
Substrate	Silicon	Al_2O_3	Sapphire	Silicon	Alumina	Copper	Silicon
Reference	[20]	[11]	[20]	[20]	[20]	[20]	[21]

Table 12-8. Thin-Film Materials Properties

A. INSULATOR	Dielectric Constant	Coefficient of Thermal Expansion 10^{-7}/°C	Thermal Conductivity W/m K	Method of Deposition
Borosilicate glass	4.1	30	5	Spray
Quartz	3.8	5	7	Sputter
Polyimide	3.5	500	0.2	Spray, spin
Polyimide (PIQ-L 120)	3.5	30	0.2	Spray
Aluminum nitride	8.8	33	230	Sputter
Teflon	2.1	200	0.1	Melt
Lead borosilicate	7.8	35	5	Spray
Boron nitride	5.5	30	250	CVD
Diamond	5.7	23	2000	High pressure
Epoxy	3.2–4.3	150–780	0.18–0.88	Curtain, dry film
Benzocyclobutene (BCB)	2.6	350–660	0.2	Spin, extrusion
B. CONDUCTORS	Resistivity μΩ·cm			
Copper	1.67	170	393	Evap. or sputter
Gold	2.2	142	297	Evap. or sputter
Aluminum	4.3	230	240	Evap. or sputter
Nickel	6.8	133	92	Evap. or sputter

To date, cofired ceramic and silicon wafer are the most common substrates. Extensive research and development efforts are being devoted to the enhancement of wiring density of thin-film structures on printed wiring board substrates typified by the IBM Surface Laminar Circuitry (SLC) technology or MCM-D/L as described elsewhere in this chapter. The inherent low-cost nature of printed wiring board substrates is the biggest driver behind this trend.

Cofired ceramic with wiring inside is referred to as an active substrate and has several advantages over silicon:

1. Wiring channels can be optimized between the ceramic and thin-film layers. The thin-film nets can be used for redistribution and for wiring of cycle-time-determining nets. The rest of the nets, including multidrop nets which run over long distances, can be wired in the ceramic.
2. Use of both thin-film and cofired ceramic wiring reduces thin-film complexity and the number of thin-film layers, thereby increasing thin-film yield.

3. More power planes can be provided when the power consumption and dissipation of the chips are high.
4. Cofired ceramic allows for module level signal and power area array connection. This configuration has lower ΔI noise and inductance compared to power distribution from the side of substrates.

The major advantage of a silicon substrate is the good thermal expansion matching that of IC chips. Higher **thermal conductivity** is also a favorable property. Furthermore, certain active and passive devices can potentially be integrated into the substrate. The major drawback is that no wiring is allowed within the silicon substrate, which is intrinsically a passive substrate.

Surface preparation of cofired ceramic is a very important process step before multilayer thin-film fabrication. The substrate should have a surface roughness of less than 1000 Å and should have a surface flatness of 10 μm or less to be suitable for thin-film photolithography [43]. This is achieved by lapping and polishing the cofired-ceramic substrate. The surface is then cleaned and modified by chemical or physical means prior to thin-film metal or dielectric deposition. The first thin-film layer on cofired ceramic consists of large pads which capture the vias of the ceramic substrate. The use of a capture-pad layer facilitates the matching of a regular photolithographic grid to the irregular ceramic grid. The bottom reference plane is combined with the capture pad level when it is electrically acceptable.

On thin substrates like silicon or cofired ceramic with very few layers, bowing due to the thermal coefficient of expansion (TCE) mismatch between the thin films and the substrate leads to focusing problems during photolithography, whereas this effect is often negligible on thick, rigid substrates. The bowing can be controlled during exposure by the use of a vacuum chuck. One drawback of a cofired ceramic is its low thermal conductivity. In high-power, high-performance systems flip-chip technology is used to connect high-I/O-count chips, with the additional advantage of effective heat removal from the back side of the chip.

12.4.2 Dielectric Materials

The dielectric materials for thin-film structures can be classified as either organic or inorganic. A wide variety of organic materials, all polymers, with adequate electrical, thermal and mechanical properties have been employed as the dielectric layer in thin-film packaging. Polyimides are the most common ones in use today. The inorganic materials typically used are either borosilicate glasses, processed using thick-film techniques, or high-temperature materials such as aluminum nitride (AlN), boron nitride (BN), or quartz (SiO_2) deposited by either sputtering or chemical vapor deposition processes. All these materials with their electrical and thermal properties were listed previously in Table 12-8. The general requirements regardless of which type of dielectric is used for forming multilevel thin-film dielectrics are as follows:

1. Low dielectric constant (less than 4)
2. Thermal-expansion matching to substrate and/or silicon, or good ductility and/or low modulus to minimize stress development
3. Thermal stability to withstand the temperatures in subsequent chip attach and pin-joining high-temperature processes (400°C with 97/3 Pb/Sn, 230°C with 60/40 Pb/Sn)
4. Good adhesion to copper or other conductors
5. Self adhesion
6. Ease of processing, preferably with planar structures
7. Low water absorption

Reliability and cost are other important requirements.

12.4.2.1 Organic Materials and Their Properties

The trend in the industry appears to be for organic dielectrics. Thin-film polymer dielectric layers in conjunction with thin-film copper conductors offer the greatest electrical performance because of the low dielectric constant of polymers and the high electrical conductivity of copper metallization obtained by sputtering or conventional plating technologies. Organic materials considered for this application include fluoropolymers, polyolefins, epoxies, benzocyclobutene (BCB), and derivatives of polyimides in PMDA-ODA, BPDA-PDA, and other systems. All these materials are discussed in great details in Chapter 11, “Polymers in Packaging.” The ideal properties of polymers that can be considered for multilevel thin films are listed below.

- Low intrinsic stress (multilevel structure, reliability)
- High fracture toughness and ductility (processability)
- Excellent self-adhesion and adhesion to metals (Al, Cu, Cr, Ti) and ceramic substrates (alumina, glass-ceramic)
- Nonreactive with conductor metals (especially copper) (reliability)
- Reworkable before full cure (yield enhancement)
- Compatible with other process solvents (processability)
- Dielectric constant of 3.5 or lower (electrical performance)
- Thermally stable to 400°C with 97/3 Pb/Sn or to 230°C with 60/40 Pb/Sn on FR4 (flip-chip and pin-grid array connection)

- Low moisture/solvent uptake (reliability)
- Stable viscosity and long shelf life (manufacturability)
- Formation of pinhole-free films with good planarization ability (yield and manufacturability)
- High solid content and high film retention (processability)
- Via formation by wet etching or laser ablation (processability)

In addition to the above properties, the residual stress developed as a result of mismatch in thermal expansion between the polymer and the substrate to which it is bonded should be considered in selecting the polymer for multilayer thin-film applications. There are three factors that determine this stress:

- Thermal-expansion mismatch
- Stress-relief properties of polymers
- Elastic modulus of polymer

A rule of thumb is to multiply the thermal-expansion mismatch with elastic modulus and arrive at polymer selection based on lowest product of these. PMDA-ODA polyimides, in spite of their very high thermal-expansion coefficient (350×10^{-7}/°C) compared to polyimides based on BPDA-PDA of thermal expansion coefficient around 30×10^{-7}/°C, reflect residual stresses that are only about five times more than the low-expansion material. The majority of these differences can be attributed to the difference in elastic modulus between the two materials.

Referring to Figure 12-19, which shows the thermal expansion of a typical polyimide as a function of temperature up to the chip-joining temperature, it is not surprising that such a material with thermal coefficient of expansion around 450×10^{-7}/°C (room temperature to 400°C) develops so much stress when bonded to silicon, glass-ceramic, or even alumina substrate that either the substrate warps to an unacceptable level or the polyimide film peels off or cracks. Observations supporting these, particularly with thick polyimides, have been reported [44]. A new multilevel thin-film technology based on the use of low-thermal-expansion polyimides has been reported by Hitachi. Polyimides of rodlike backbone structures with thermal expansion coefficients as low as 30×10^{-7}/°C were developed in recent years. Two such polymers coded PIQ-L100 and PIQ-L120 are illustrated for their thermal expansion behavior (Fig. 12-20). Their low thermal expansion is attributed to the restraining of thermal expansion by rodlike crystallized molecules within intermolecular spaces, somewhat similar to the effect of glass fibers in fiber-reinforced plastics. If the lower

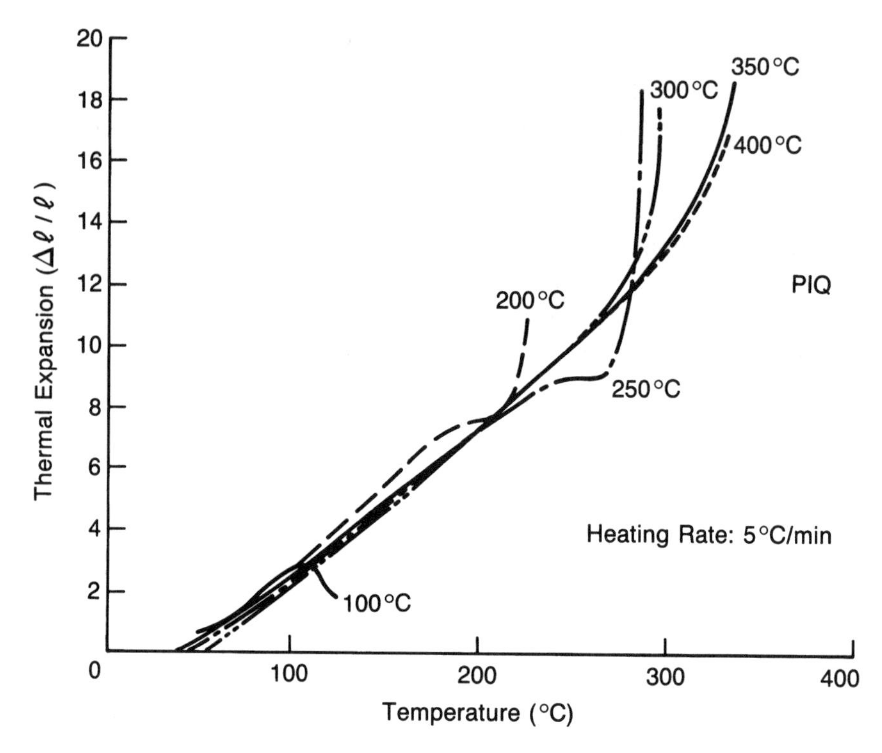

Figure 12-19. Thermal Expansion of Conventional Polyimide. After Ref. 44.

thermal expansion coefficient is attributed to this mechanism, one would expect these low-thermal-expansion films to have a higher modulus in the same direction. Figure 12-21 illustrating the relation between the thermal expansion coefficient and modulus for various polyimide films confirms this proposed mechanism [45].

Deposition of dielectric polymers is usually done by solution-casting, in which a solution of polymer precursor is coated onto the substrate, e.g., by spray or spin coating, and then baked and thermally cured. The two later operations drive off the solvent and cross-link the polymer. Because of the leveling of the polymer precursor solution, the topography of the underlying conductor lines and vias is lessened at the top surface of the cast solution. Despite the fact that the good planarization of the freshly deposited polymer solution is partly lost after baking and curing, partial planarization is still obtained, which is desirable in the construction of a multilayer structure. Such planarization provides a base for the subsequent buildup of multilayer high-density structures without topography problems. Accumulation of topography will be eventually large enough to prevent further buildup of the structure, and a process of planarization step is necessary when more layers are required. Lamina-

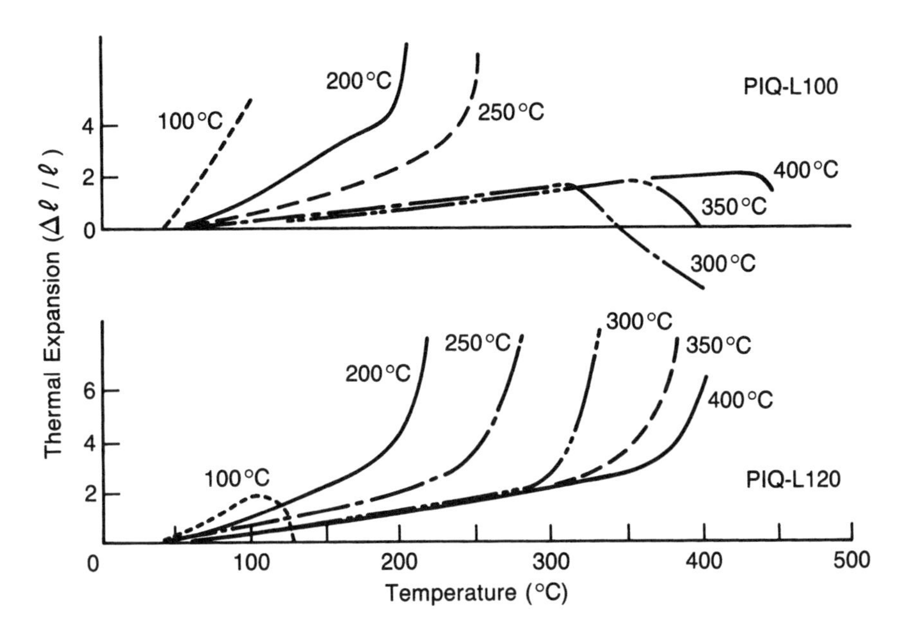

Figure 12-20. Low-Thermal-Expansion Polyimide Films Cured Under Various Conditions. After Ref. 44.

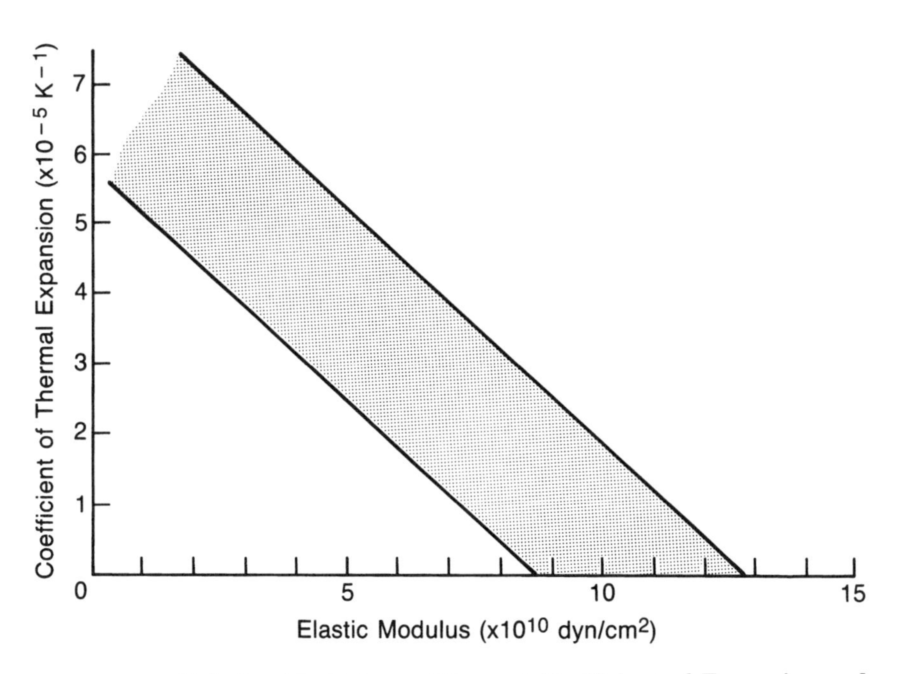

Figure 12-21. Relationship Between Thermal Coefficient of Expansion and Elastic Modulus

tion of the dielectric polymer dry film is another possible approach but is less extendible dimensionally and more prone to defects. Generally, it would not be considered a true thin-film process.

Additional information on the use of polymers for MCM-D is given in Chapter 11, "Polymers in Packaging."

12.4.2.2 Inorganic Materials and Their Properties

Inorganic materials are generally deposited by one of the three following processes:

1. Thick-film processes (spraying, screening, and sedimentation)
2. Sputtering
3. Chemical vapor deposition

Spraying and screening processes are generally limited to about 15–25 μm thickness and are similar to the thick-film processes described in Chapter 9, "Ceramic Packaging." Sedimentation illustrated in Figure 12-22 requires the formation and dispersion of submicron particles in a suitable organic solvent, followed by centrifuging to obtain thin and compact films. Chemical vapor deposition involves formation of volatile compound of the desired metal followed by oxidation of the metal to form the ceramic. Three examples of inorganic dielectrics [46] are illustrated in Table 12-9. The most common of these is SiO_2 + Al, as practiced by nCHIP described in Section 12.7.

One of the unique processes for the formation of inorganic dielectric films involves the development of photosensitive materials that can be mixed with dielectric slurries. A thin-film process based on this unique development, shown in Figure 12-23, involves formation of Au or other conductor lines by one of the three metallization processes described later in this chapter, followed by **screen** printing and drying of a photosensitive inorganic dielectric which is exposed to ultraviolet (UV) light through a mask. The layer is subsequently developed, washing off the undesirable dielectric, and the dielectric is then fired. It is reasonable to think that such a concept could be applied to thick-film process as well, enhancing thick-film technology into the arena of thin-film technology, as discussed in Chapter 9, "Ceramic Packaging."

Planar structures of multiple thin films of inorganic dielectrics could be fabricated this way but involves planarization by chemical, mechanical, or a combination of planarization processes. Such a process, developed in the early seventies [47], involved slurry deposition of borosilicate glass, the thermal expansion of which matches that of alumina substrate, but with the lowest dielectric constant possible (4.2) for that thermal expansion. Such a material is processed in an atmosphere of N_2, followed by

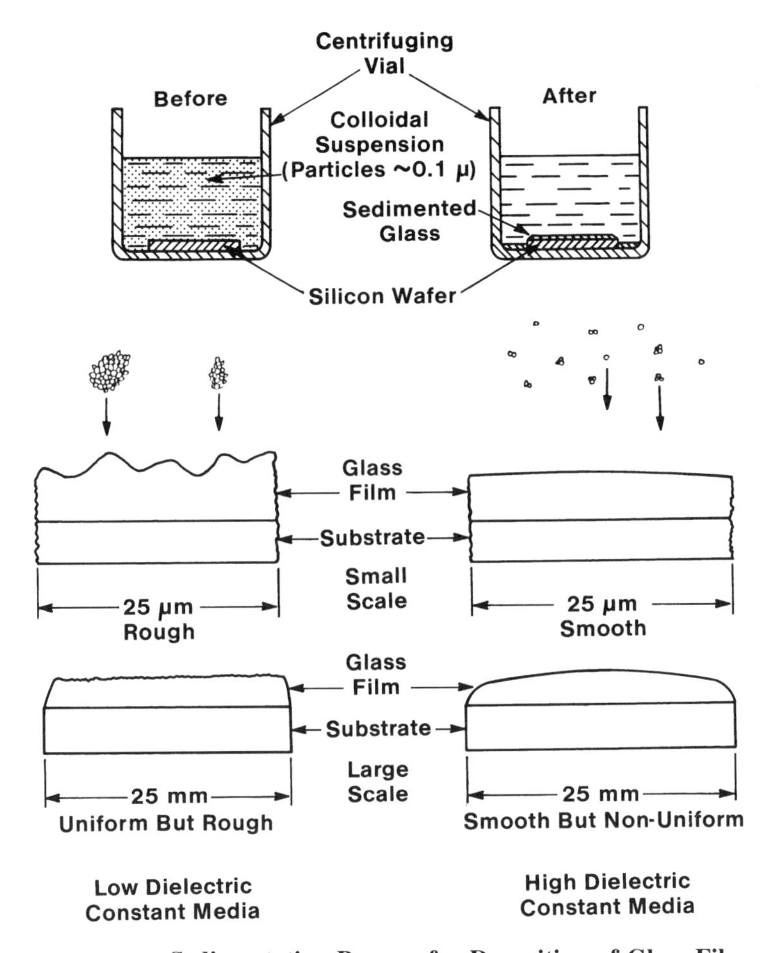

Figure 12-22. Sedimentation Process for Deposition of Glass Films

H_2 up to 50°C above its softening temperature, and then switching gas to N_2 at that temperature [47]. Bubble-free films of 20–50 μm thickness resulted due to out-diffusion of H_2 when its partial pressure became zero in N_2 atmosphere only at temperatures slightly above the softening point of glass, corresponding to viscosity of 10^6 Poises. Such a process, illustrated in Figure 12-24, involved deposition of the capture pad of chromium-copper-chromium, followed by the plate-up of copper stud prior to deposition and firing of glass at about 800°C. The structure was mechanically planarized after every glass firing. A total of five glass and five copper metallization layers have been formed this way. The major disadvantage of this approach using glass, however, is that every time the new glass layer is fired, the previous layers underwent softening resulting in movement of lines and generation of bubbles even at temperatures only slightly above

Table 12-9. Inorganic Thin-Film Dielectrics and Processes

Insulator	AlN + Cu	BN + Au	SiO_2 + Al
No. of layers	5	5	3
Material	AlN	BN	SiO_2
Thickness	0.5 μm/layer	0.2 μm/layer	2 μm/layer
Via hole size	20 μm	20 μm	5–10 μm
Deposition method	Sputter	CVD	PECVD
Conductor			
Material	Cu	Au	Al
Linewidth	20 μm	20 μm	10–20 μm

softening temperature. These problems are largely solved with glass–ceramics available currently. The major advantages, nevertheless, include total thermal compatibility of a materials system from alumina to glass as well as the thermal stability allowing joining processes up to 800°C.

The shortcomings of vitreous glass are in fact solved by the use of crystallized glass in obtaining low-cost multilevel multichip module [48]. The process sequence in building this type of structure is illustrated in Figure 12-25 wherein thin copper conductors are formed by screen printing and firing, followed by nitrogen-fireable crystallizing glass with through holes by screening. The surface smoothness of the glass-ceramic coating is accomplished by a thin coating of vitreous glass. Subsequently, thin Ta_2N/NiCr/Au/ films were deposited, patterned, and the wirebond pads were plated with additional gold.

12.4.3 Thin-Film Processes

A variety of methods have been used or are being developed to fabricate thin-film interconnections. As mentioned at the beginning of this section, the fabrication of thin-film packages involve several basic steps which are summarized again in Figure 12-26. It is noted in this table that patterning is a process step which is required for both dielectric and conductor layers. Multichip module (MCM) fabrication is assumed in this table and most of the following discussions, recognizing the fact that MCM packaging presents the most challenges for thin-film packaging technologies, and is the most important application. Nevertheless, the principles also apply to single-chip packaging. Furthermore, for each of the processing steps listed in Figure 12-26, a variety of methods are

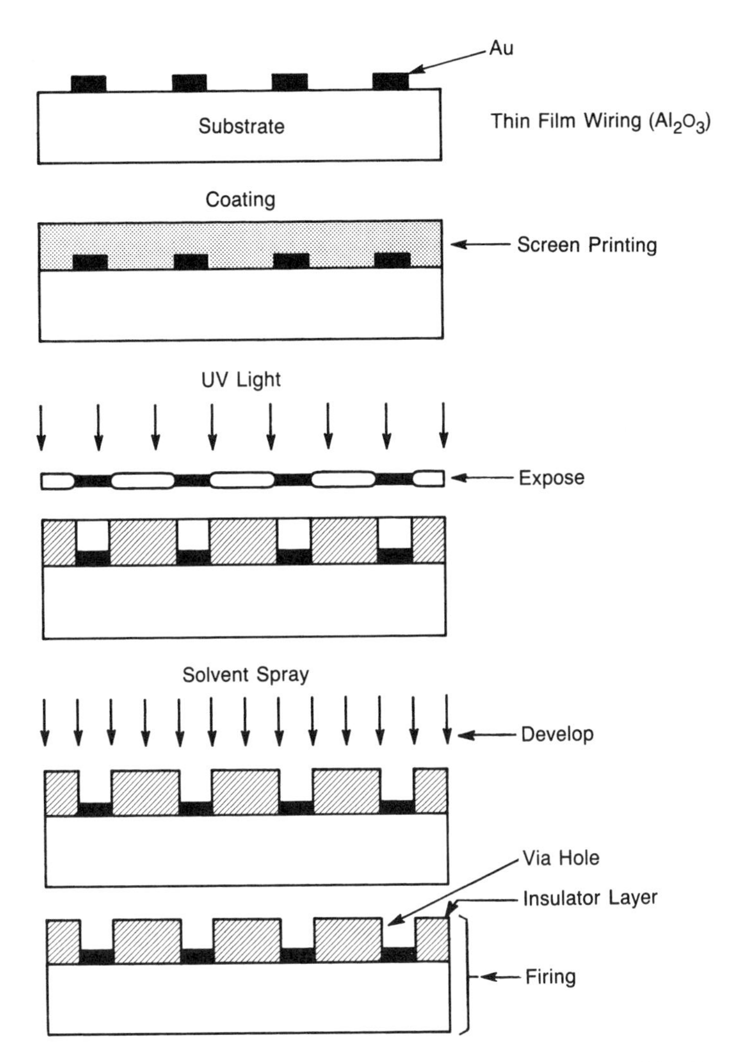

Figure 12-23. Photosensitive Thin-Film Dielectric Process (NEC). Source: MES '87

available and being used, as listed in Figures 12-27 through 12-31. The majority of the techniques involved in thin-film package fabrication as shown in these tables will be discussed in length here. The technological choice is a very complicated issue, and the optimization of fabrication process depends on many factors, such as intended application, design rules, materials selection, manufacturing establishment, and costs, and varies from time to time, and from manufacturer to manufacturer.

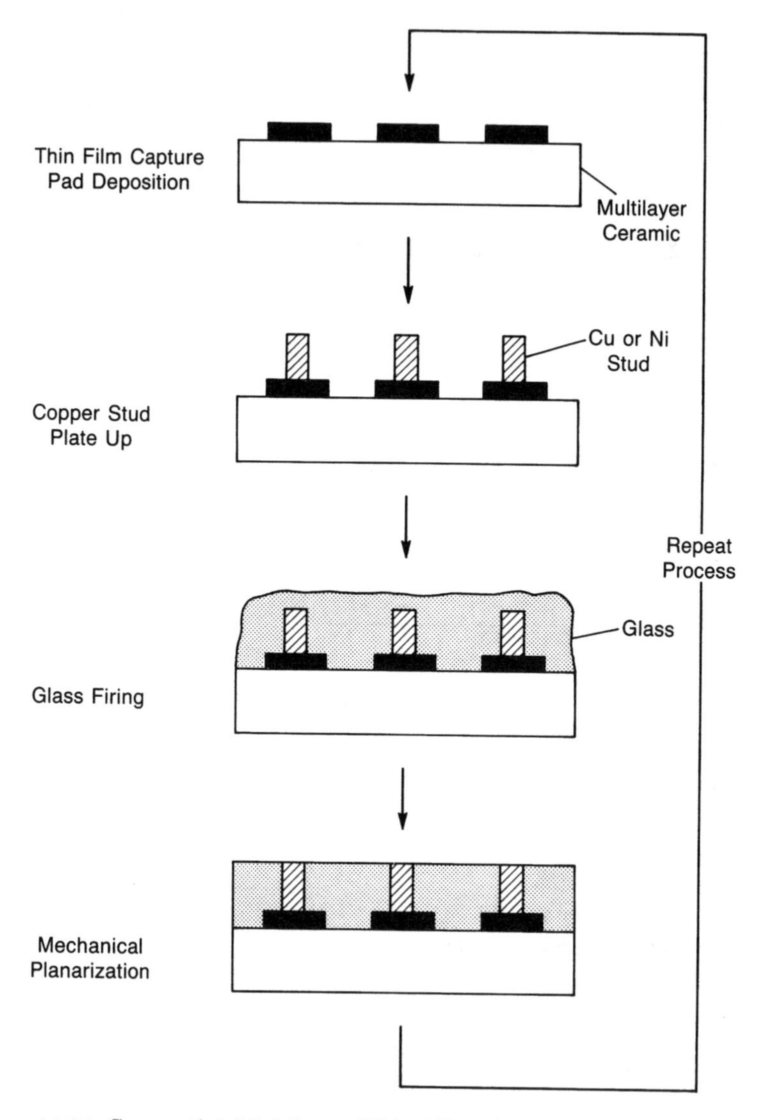

Figure 12-24. Sequential Multilayer Thin-Film Glass/Copper Process. After Ref. 47.

As examples, Table 12-10 compares the most commonly used techniques for the definition of vias, conductors, and bonding pads [49] in terms of process flow. The complexity of thin-film package processing is readily visible. The inherently sequential and *in situ* nature of thin-film package fabrication will become more evident in the discussion that follows.

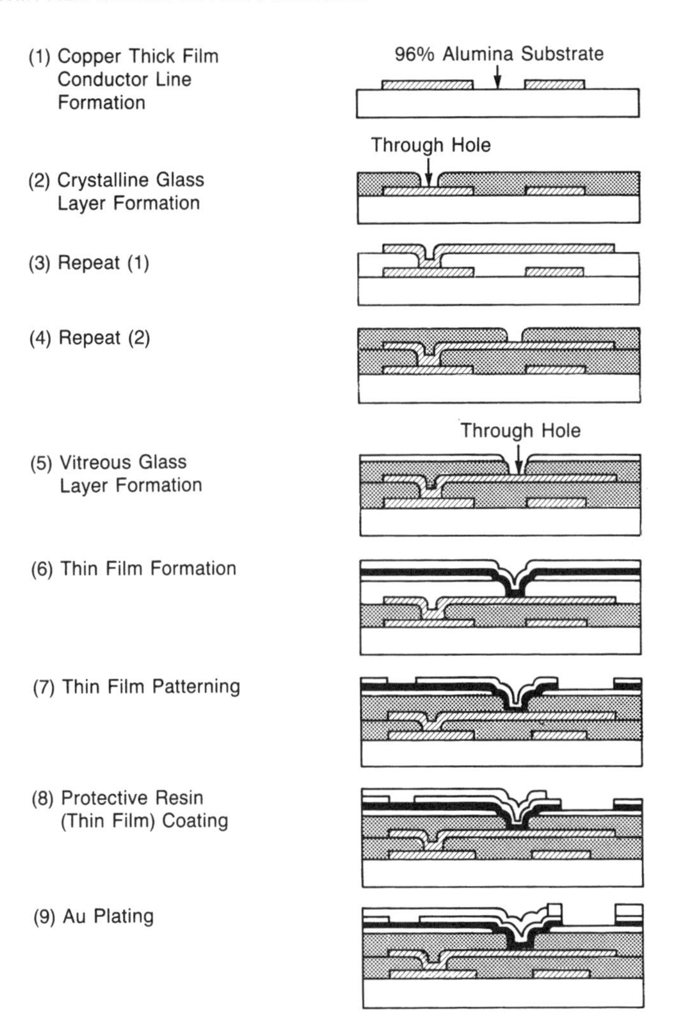

Figure 12-25. Fabrication of Inorganic Thick/Thin-Film Multilayer Substrate. Courtesy of Fujitsu Ltd.

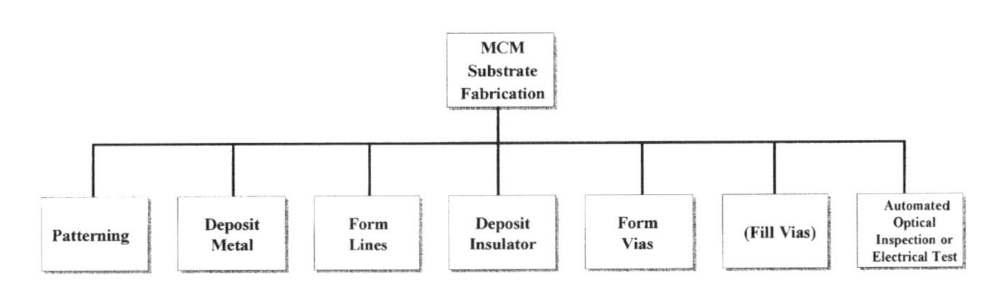

Figure 12-26. Generic MCM Fabrication Processes

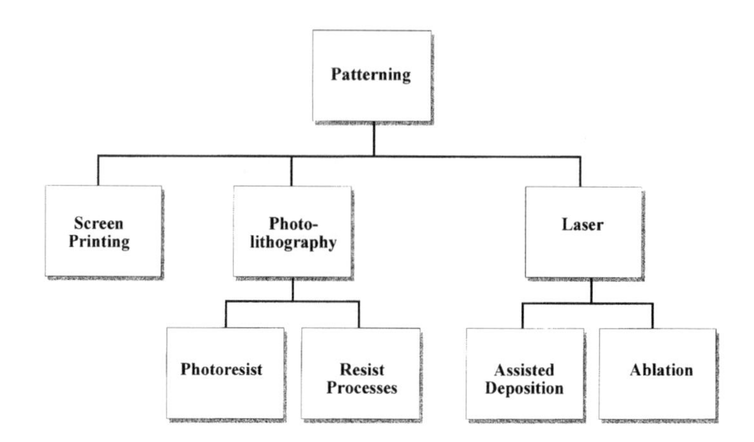

Figure 12-27. Generic Patterning Processes in MCM Fabrication

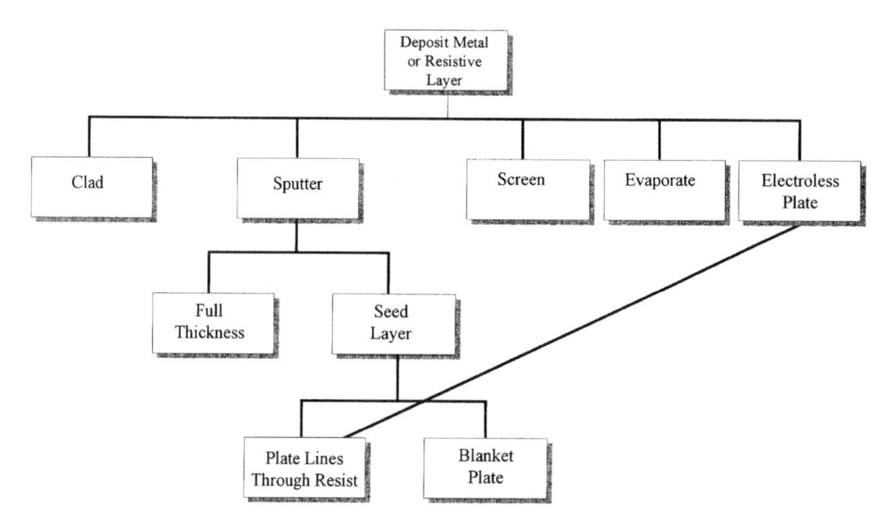

Figure 12-28. Generic Metal- or Resistive-Layer-Deposition Processes in MCM Fabrication

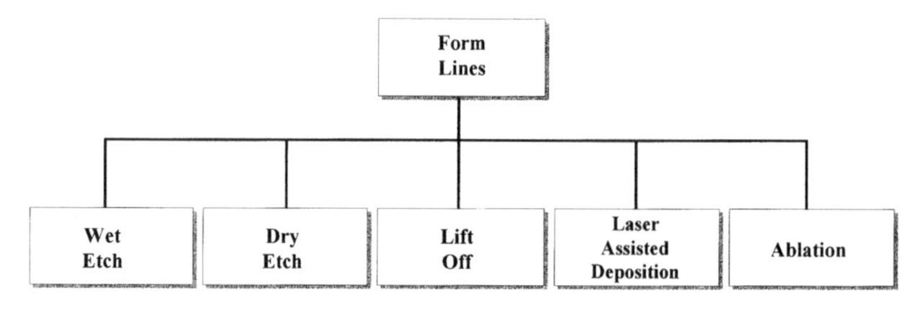

Figure 12-29. Generic Conductor-Line-Formation Processes in MCM Fabrication

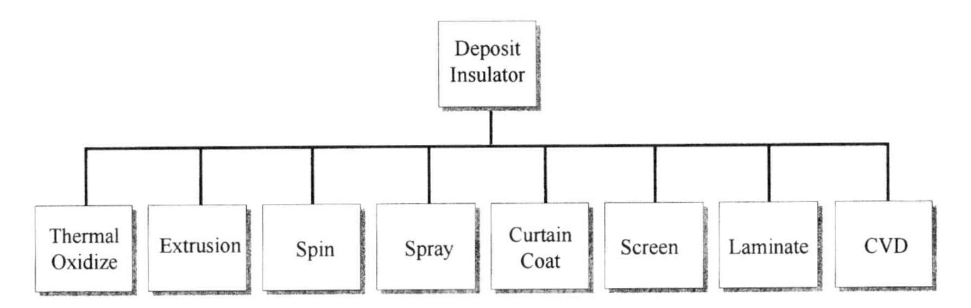

Figure 12-30. Generic Insulator-Deposition Processes in MCM Fabrication

In general, thin-film processes are compared on the basis of

- Number of process steps
- Process complexity
- Process robustness
- Process window
- Defect levels
- Planarization requirements
- Reworkability
- Repairability

12.4.3.1 Polymer Processing

12.4.3.1.1 Deposition of Polymer Thin Films

In general, there are probably more than 20 different ways to deposit thin-film polymers, but this section will only review those that are most prevalent today or expected to be in the future for thin-film packaging, which are summarized below (Table 12-11). Spin coating is currently the standard method for both photoresists and dielectric polymer coating. Spray coating has been borrowed from the painting industry and finds wide application in microelectronic parts coating. Roller coating, meniscus coating, screen printing and extrusion coating are emerging coating meth-

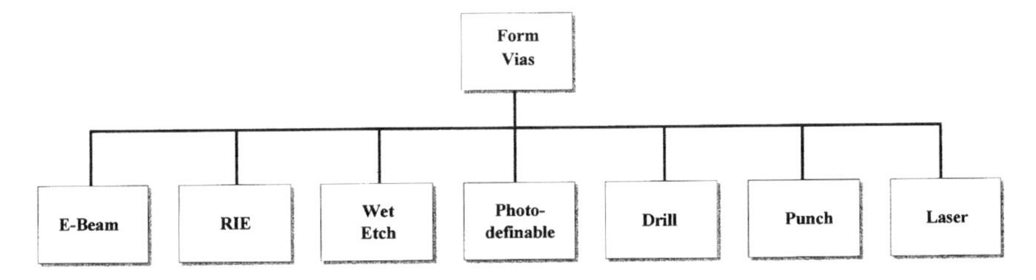

Figure 12-31. Generic Via-Formation Processes in MCM Fabrication

Table 12-10. Comparison of Processes Used for the Definition of Vias, Conductor Lines and Terminal Metals in MCM-D

Via Definition

Laser Ablation	PSPI Lithography	RIE	Wet Etch
Surface preparation Apply adhesion layer Apply PI Cure PI Laser ablate features Plasma clean	Surface preparation Apply adhesion layer Apply PSPI Expose features Develop features Cure PSPI Plasma clean	Surface preparation Apply adhesion layer Apply PI Cure PI Apply photoresist Expose features Develop features RIE Photoresist strip Residue removal	Surface preparation Apply adhesion layer Apply PI Apply photoresist Expose features Develop features Wet etch Photoresist strip Cure PI

Wiring/Terminal Metal Definition

Subtractive Etching (Wiring Only)	Additive Electroplating (Wiring and Terminal Metals)	Lift-Off (Terminal Metals Only)	Evaporation Through Mask (Terminal Metals Only)
Surface preparation Sputter Cr-Cu-Cr or Al Apply photoresist Expose features Develop features Etch features Photoresist strip	Surface preparation Deposit plating seed Apply photoresist Expose features Develop features Plate wiring or terminal metals Photoresist strip Remove seed layer Apply capping layer	Surface preparation Apply adhesion layer Apply soluble polymer Apply RIE barrier Apply photoresist Expose features Develop features RIE Evaporate terminal metals Lift off stencil	Surface preparation Mask attach/align Evaporate terminal metals Mask removal

ods, particularly for large-area-processing-based thin-film package fabrication. Both spin coating and spray coating have been employed for thin-film deposition in MCM-D fabrication, although there is no up-to-date literature on the application of the other coating technologies in this area.

Spin Coating: This method has been used for decades in the semiconductor industry to coat photoresists and dielectric polymers on the tradi-

Table 12-11. Deposition Technologies of Thin-Film Polymers

Coating Method	Film Thickness Range	Advantages	Limitations
Spin	< 15 μm	Excellent thickness control and reproducibility, good uniformity, standard IC fab process	High waste, favors small round wafers
Spray	> 10 μm	High throughput, high material usage, capable of large-area coating	Limited planarization, poor control of thickness and uniformity, poor reproducibility
Roller	—	High throughput, high material usage, large-scale continuous process	Susceptible to contamination and defects, poor quality when film is thin
Meniscus	10–40 μm	High material usage, high throughput, large scale continuous process	Susceptible to contamination and defects, poor quality when film is thin, not suitable for high viscosity solutions
Screen printing	> 5 μm	Good control of thickness, good reproducibility, capable of patterning	Poor uniformity when film is thin, batch process, susceptible to defects
Extrusion	sub-μm to 100 μm	Almost no material waste, good control of film thickness, good uniformity, large-area coating, high throughput	Susceptible to contamination, defects due to slit inhomogeneity

tional round wafers. For dielectric polymer thin-film coating, a liquid solution of a polymer is puddled onto the wafer surface, usually with a volumetric dispenser, and the wafer is then spun to produce a uniform coating. Excess polymer solution is often dispensed onto the substrate to assure good coverage. Film thickness depends on many variables, such as solution viscosity and solid content, the time and angular speed of spinning, and the dispense volume. The most sensitive parameters for controlling film thickness are the spin speed and viscosity. Film thicknesses from 1 to 200 μm are feasible. Thicker films can be obtained with high-viscosity solutions at low spin speed for a short spin time, which adversely affects the resultant film uniformity. Therefore, multiple coatings are preferred for film thicknesses > 15 μm, with the additional advantage of reduced defect rate, notably, pinholes. Spin coating is a simple process and a well-established practice with automated spin coaters widely available. A large data source exists and the spin process has been modeled and remodeled widely in the literature.

Spray Coating: The principle behind spray coating is as following: A nontoxic gas (air or N_2) is forced out of a nozzle to atomize a spray of polymer solution which is swept across the surface of the substrate. Tiny polymer droplets bond to one another as they hit the surface and form a film. In one operation, the nozzle may pass several times over the same spot on the substrate to assure good coverage and uniformity. Process variables include solution viscosity and solid content, solution flow rate, nozzle size and distance from substrate, nozzle pressure, atomization pressure, and sweep speed. It is important to carefully balance between low viscosity, which facilitates the atomization of the polymer solution, and solid content, which determine the flow rate of the solution through the spray nozzle to achieve the desired thickness of polymer on the substrate [50]. However, too low a viscosity in combination with high surface tension may cause pullback of the polymer solution deposited on the substrate. The atomization pressure is also critical; too low results in large droplets coming out of the nozzle which may not flow together to yield an uniform coating, whereas too high causes excessive foaming of the solution and bubbles which may still be present after the drying circle. Spray coating is, in principle, considered a favorable alternative for a thicker film coating (> 15 μm) owing to its capability of large-area coating on a noncircular substrate and high throughput [51]. Uniformity of spray-coated film drops off when film thickness is less than 10 μm, as found for benzocyclobutene (BCB) [28]. It was also found that spray coated BCB exhibited an undesirable orange peel surface finish under certain conditions [52]. Few detailed processes of spray-coating have been released despite some reported applications [33,53]. Thorough characterization of spray-coated thin film and much optimization of the process will be required for spray coating to become a viable coating technique in thin-film packaging.

Roller Coating: The basic idea is to coat one or two rollers with the polymer solution and then roll them across the surface of the substrate. In practice, the substrate is usually dragged underneath the moving roller. One roller will coat one side of the substrate and two rollers can place polymer on both sides of the substrate simultaneously. Film thickness is primarily dependent on the polymer solution properties, such as viscosity, as with spray coating. At this time, an exact thickness range is unavailable. Figure 12-32 depicts this process.

Roller coating is widely used in MCM-L fabrication where one or two wiring layers are placed on each surface of the board. Large-area boards (20 in. by 24 in.) may be roller-coated. Major concerns with roller coating include chronic defect problems and process sensitivity resulting in yield loss.

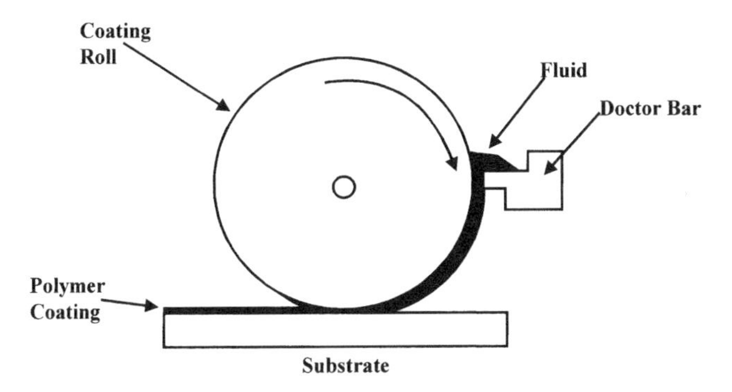

Figure 12-32. A Schematic of Roller Coating

Meniscus Coating: Liquid polymer solution is pumped out of a porous tube over which the substrate slides. Material may be collected under the tube and recirculated into the center of the tube, so it is much more efficient in terms of waste than spin coating. Process parameters include distance from substrate to tube, substrate velocity, surface tension of the coating polymer solution and the substrate surface, viscosity and solvent evaporation rate. Film thickness depends mainly on substrate speed and ranges from 15 to 40 μm. Uniformity across large boards (300 mm by 300 mm) was between 2% and 4% for testing done with BCB at Specialty Coating Systems, Inc. [54]. Figure 12-33 is a schematic of meniscus coating [55].

Screen Printing: This method features deposition of the polymer through a screen. A screen is placed on or very close to the surface of the substrate. Liquid polymer is flowed across the screen and a squeegee is pulled across the screen to force the liquid through the screen and onto the substrate. An important part of the machine is called a floodbar. It

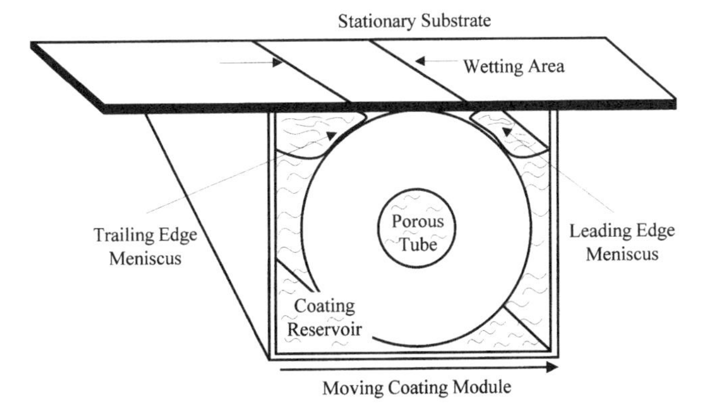

Figure 12-33. A Schematic of Meniscus Coating

pulls the material back onto the screen after the squeegee pushes most of the excess off of the screen and out of range. The major factor determining film thickness and quality is the screen itself. Screen parameters include mesh density and screen emulsion. Film thickness down to 5 µm is possible with variations of ± 1.5 µm under ideal process conditions.

Extrusion Coating: The definition of extrusion is the process of converting a raw material into a product of specific cross section by forcing the material through an orifice or die under controlled conditions. The cross section may vary widely as plastic pipes and spaghetti, which are examples of extruded products. In microelectronic applications, the extruded cross section is a narrow rectangle (i.e., a thin, paperlike film with the cross section consisting of the film thickness and substrate width). Figure 12-34 [56] illustrates the basic idea behind thin-film extrusion. In general, the critical output variable of an extruder being used for polymer deposition is film thickness. It depends on the volumetric flow rate of the polymer into the die, the substrate width and velocity (the substrate moves under the stationary extrusion head), the distance from the head to the substrate, and the shrink factor of the polymer in question.

Film extrusion may be done in two ways: wet or dry. In wet extrusion, the material is softened with or dissolved in solvents. This allows the process to be performed at a low temperature. In dry extrusion, the material is softened and made to flow by the application of temperature and pressure only. As with the other deposition methods previously mentioned, polymers for microelectronic applications are extruded as a solution in organic solvent; correspondingly, wet extrusion would be the appropriate method for thin-film deposition. There are three types of extruders differentiated by the materials feeding method: ram/cylinder, screw, and pump. The earliest machines were all ram-type extruders, but the other versions were

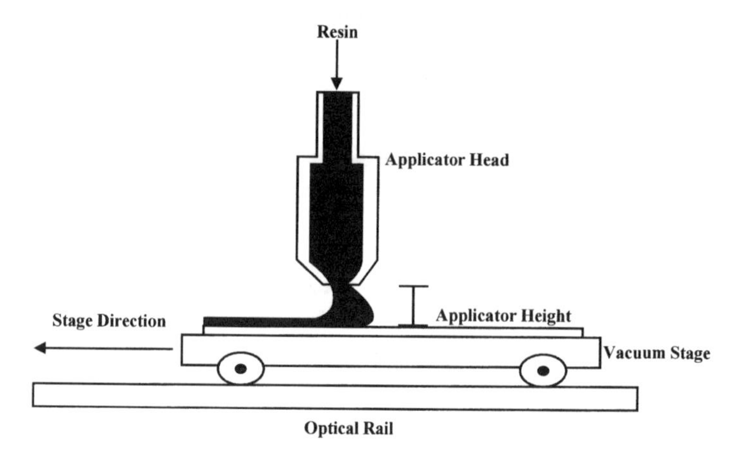

Figure 12-34. A Schematic of Extrusion Coating by FAS Technologies, Inc.

invented to overcome some basic problems associated with ram extrusion. The most fundamental problem with ram-type extruders is that extrusion is not a continuous process. Periodically, the process must be halted so that the cylinder can be refilled. Thus, the screw principle of polymer feeding was adopted and is still the dominant type in the plastic-forming industry today. Screw extruders are not considered suitable for thin-film coating in microelectronics applications, however, due to the propensity of polymer contamination and other inadequacies. A gas-driven diaphragm has been the standard fluid-pumping technology for extruders in the microelectronics industry for more than 30 years. The dispensing rate and volume by this pumping mechanism are very sensitive to the polymer fluid viscosity and the inherent system impedance, and, as a consequence, reproducibility has been proven to be a tough task.

12.4.3.1.2 Via Generation Approaches

Vias in thin-film package substrates are used to provide vertical electrical connections between adjacent layers. Despite the many process options available for via generation, including punching, mechanical drilling, laser ablation, and chemical etching, only four methods as listed in Table 12-10 are suitable for MCM-D substrate fabrication. Small via size and the high requirement on alignment primarily determined the method of choice. The very high routing density achievable with the MCM-D substrate is due as much to its fine-line capabilities as to its ability to very precisely locate small via openings in layers of polymer dielectric. In Figure 12-35, a correlation between MCM-D and MCM-C density is shown for MCM-D linewidth and spacing combinations of 25 on 75 μm and 12 on 25 μm versus the number of cofired MCM-C layers needed to achieve the same interconnect density [27]. Based on these estimates, the use of thin-film wiring can reduce the number of interconnect layers required for a given application by as much as 4 : 1 over that achievable with a cofired-ceramic substrate. Kambe et al. [57] has compared state-of-the-art PWB, high-temperature cofired ceramic, and MCM-D technologies, and has come to the same conclusion (Table 12-12).

Specific via requirements are typically determined by the metallization process that is used in fabrication. For those technologies where vapor-deposited (sputtered or evaporated) metallization schemes are used, the highly directional nature of these metal-deposition techniques dictates the need for sloped via sidewalls in order to ensure adequate coverage. On the other hand, the need for sloped sidewalls can result in significant enlargement in upper via dimensions due to the thick layers (8 to 25 μm) of dielectric commonly used in MCM-D structures resulting in significant loss of routing density. Figure 12-36 shows the difference in via sidewall angles as a result of isotropic and anisotropic via etching. Table 12-13

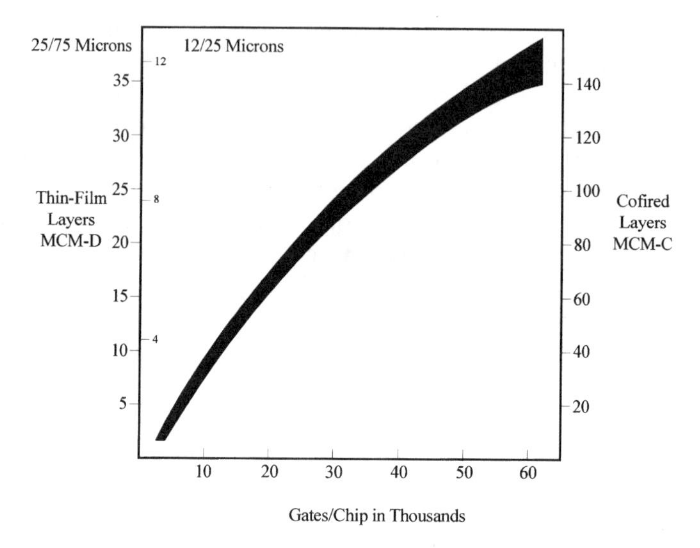

Figure 12-35. Comparison of MCM-D and MCM-C Substrate Density Capabilities

demonstrates the effect of decreasing via sidewall angle and increasing dielectric thickness on the resultant upper via dimension for a nominal 10 μm via opening. In the case of an anisotropic via etch process, the upper via dimension (w_2) would be equal to the lower via dimension (w_1). As the sidewall angle decreases and/or the dielectric thickness increases

Table 12-12. Comparison of Wiring Densities of MCM-D Interconnects Achievable with MCM-C and MCM-L

	Printed-Wiring Board	Alumina Multilayer	Polyimide Multilayer
Linewidth (mm)	0.100	0.100	0.025
Line pitch (mm)	0.250	0.200	0.050
Via size (mm)	0.150 Dia.	0.100 Dia.	0.025 Sq.
Via pitch (mm)	0.300Ⓐ	0.250Ⓑ	0.075Ⓒ
Dielectric constant	4.5Ⓓ	9.5Ⓔ	3.5Ⓕ
Wiring density	1	$\left(\frac{Ⓐ}{Ⓑ}\right)^2 \approx 1.4$	$\left(\frac{Ⓐ}{Ⓒ}\right)^2 \approx 16$
Wiring length	1	$\frac{Ⓑ}{Ⓐ} \approx 0.8$	$\frac{Ⓒ}{Ⓐ} \approx 0.3$
Propagation delay	1	$\frac{Ⓑ}{Ⓐ}\left(\frac{Ⓔ}{Ⓓ}\right)^{1/2} \approx 1.2$	$\frac{Ⓒ}{Ⓐ}\left(\frac{Ⓕ}{Ⓓ}\right)^{1/2} \approx 0.2$

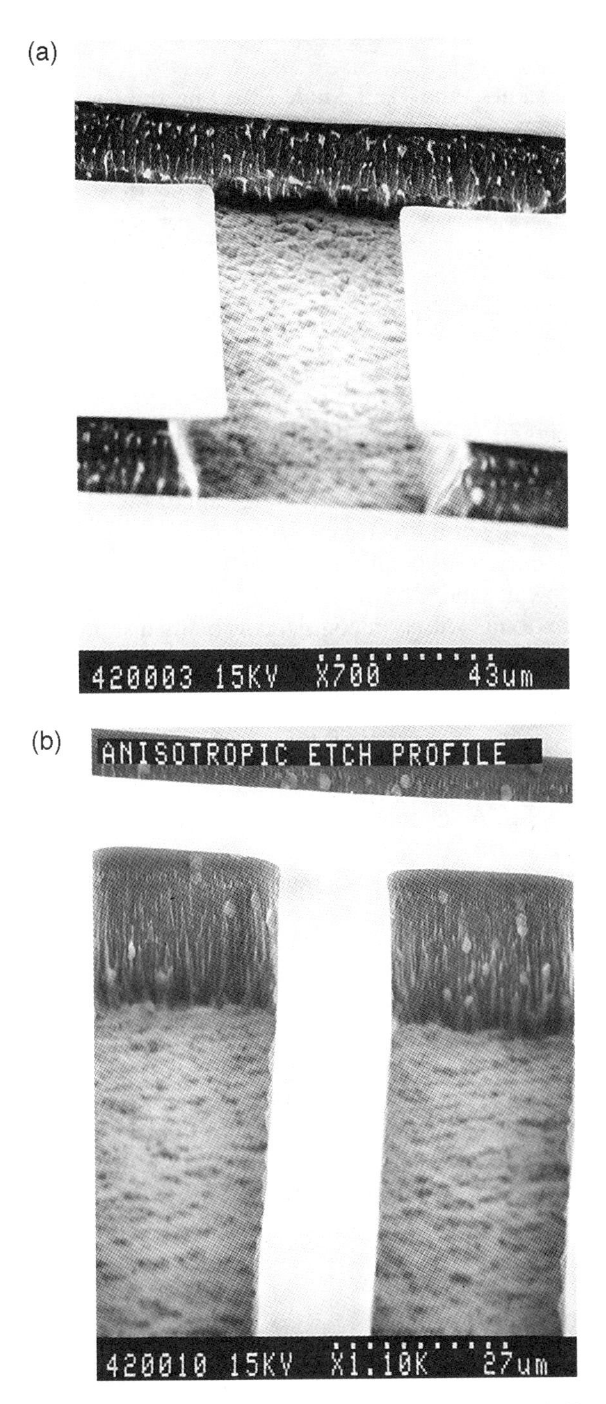

Figure 12-36. (a) Isotropic versus (b) Anisotropic Etch Profiles

Table 12-13. Thickness and Wall Angle Effect on the Upper Via Dimension

Sidewall Angle	Thickness of Dielectric Layer 10 μm	15 μm	20 μm
45°	30.0 μm	45.0 μm	60.0 μm
60°	21.6 μm	27.1 μm	33.1 μm
70°	17.3 μm	20.9 μm	24.6 μm
90°	10.0 μm	10.0 μm	10.0 μm

via enlargements of as much as three to six times the nominal dimension can occur. Most of the commonly used MCM-D via generation processes are typically in the 50° to 80° wall angle range. Metallization processes which are essentially nondirectional like electroless/electroplating of copper and other metals for typical aspect ratios do not impose any constraints on the via sidewall angle.

Via formation is widely recognized as being a significant contributor to the overall MCM-D substrate fabrication costs due to the larger number of processing steps generally required. As a result, a considerable amount of effort has been devoted throughout the industry to the evaluation of alternative via generation processes. As reported in the literature, several of these techniques have been successfully used for via formation in MCM-D substrates. A schematic comparison of the most commonly used alternatives are illustrated in Figure 12-37.

For example, when IBM's metallized-ceramic-polyimide (MCP) process was being defined in the early 1980s, wet etch was the only available low-cost via formation technique. Reactive ion etching (RIE) evolved to maintain compatibility with the semiconductor manufacturing operations which were expected to be required for manufacturing multilayer thin films on a silicon carrier. Laser ablation was developed specifically for large substrates which required a high-yield process. Photosensitive polyimide (PSPI) lithography was developed as a cost reduction to reactive ion etching and wet etching. Based on this experience, the four via formation processes can be compared, considering the following:

- Aspect ratio of the via
- Via sidewall profile
- Availability of tooling

It is clear from Figure 12-37 that projection laser ablation requires the fewest process steps and has only one critical step, making it the

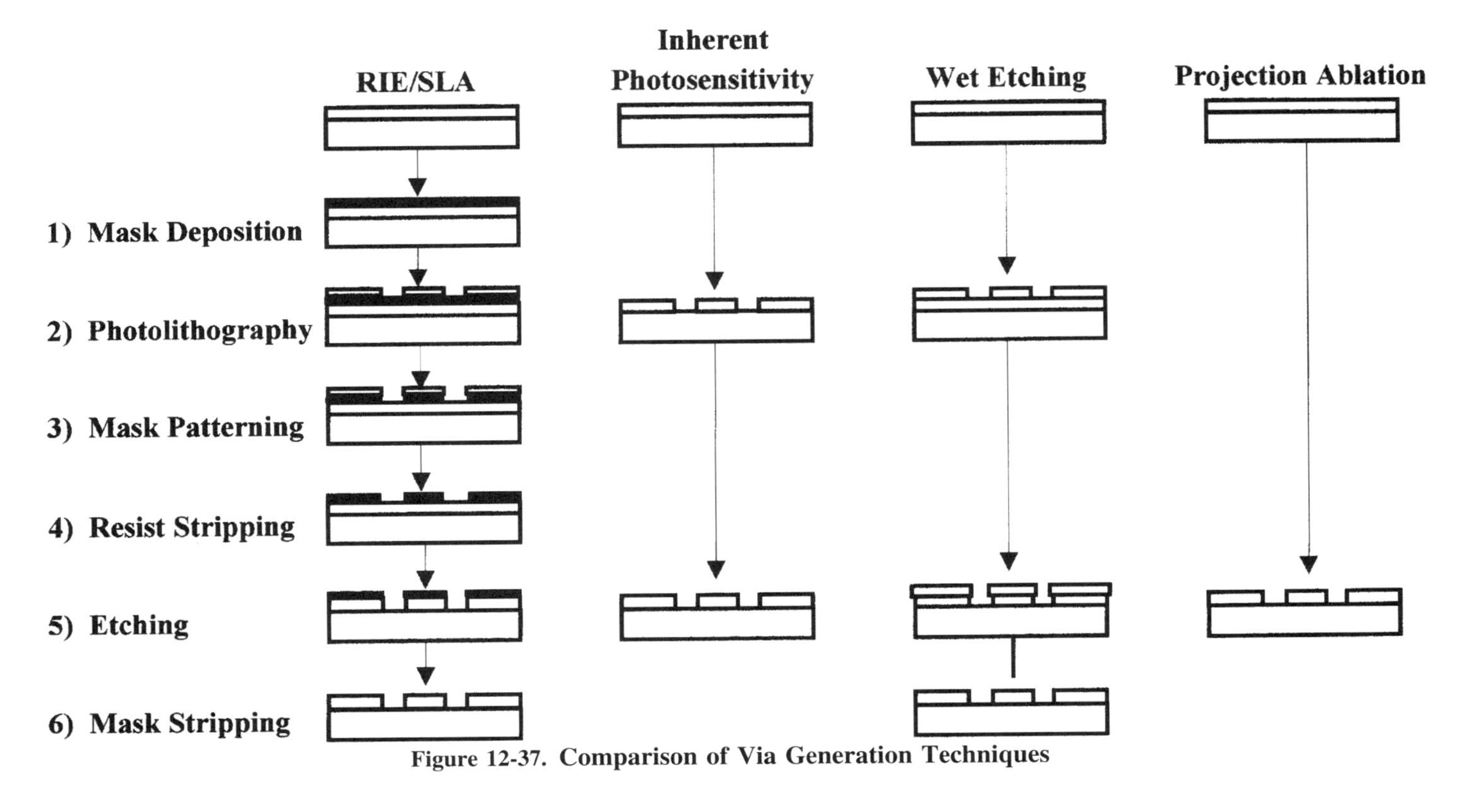

Figure 12-37. Comparison of Via Generation Techniques

process of choice. The main drawback of the projection laser ablation process is that laser tools and dielectric masks are not universally available. Typically, the dielectric mask is also more expensive than the standard chrome masks used for the other three processes. When projection laser ablation tools and mask technology are not available, photosensitive polymer lithography is preferable to wet etch and RIE. However, photosensitive polymer materials are still evolving and are more costly than standard polymers.

Reactive ion etching is not a preferred method for via formation due to the large number of critical process steps. It is a preferred method only when existing semiconductor manufacturing tools need to be used to manufacture MCM-D to reduce the overall cost. The very steep wall profile that can be obtained by the trilevel reactive ion etching process is not required for multilevel thin films. Wet etch, on the other hand, is suitable only when large vias need to be defined in very thick polymers. In most applications for high-performance systems, the dielectric is less than 12 μm thick. For dielectric films less that 12–15 μm thick, photosensitive polymer lithography is preferable to wet etch because it has fewer process steps and critical process steps as well.

Reworkability is a very important factor in the choice of the via formation process. In both laser ablation and reactive ion etching processes, the vias are formed after fully curing the polymer and defects caused during the via formation process cannot be repaired or reworked easily. In the wet etch and photosensitive polymer lithography processes, the vias are formed prior to full cure and hence can be reworked for defects. The large manufacturing database on laser ablation of vias, however, suggests that very few defects are created during the process, and reworkability is not a critical issue for laser ablation.

Laser Ablation: Laser ablation has been widely used to drill through holes in printed wiring board and ceramic substrates. It is also a very competitive candidate technology for via formation in multilevel thin-film packaging as a low-cost, high-throughput, and high-yield technology. Minimized processing steps and almost no limitations on the polymers to be ablated are the most prominent attributes of laser ablation via formation, as can be seen in Table 12-10 and Figure 12-37. Cost modeling as discussed later in this section has indicated that the lowest cost of MCM-D substrates can be achieved by using laser ablation for via formation. Laser ablation is more environmental friendly than photosensitive polymer technology in that no organic chemicals or solutions are employed for via formation. Currently, high cost of ownership, complexity of technology, and equipment reliability concerns are the major impedances for the wide acceptance of this technology. With further improvement in laser technology and lowering of equipment cost, laser ablation would certainly become a more attractive option for via formation.

The basic interaction mechanism between laser light and matter can be of photothermal and/or photochemical nature [58]. Laser-induced material melting and vaporization are the most common photothermal phenomena in laser ablation, whereas **plasma** formation over the incident area of the substrate is a nonthermal process. With the presence of adequate gaseous or liquid substance, more complex interactions among the laser light, the substrate, and the environment may occur, resulting in laser etching [59]. For via formation in polymer thin films, ultraviolet excimer laser at wavelengths of 193 nm (ArF), 248 nm (KrF), 308 nm (XeCl), or 351 nm (XeF) is most commonly used [30,31,59-61]. A short wavelength has the advantage of a low photothermal effect on the substrate but requires much more complex optics and stringent optical material selection. Typically, laser ablation is carried out with a succession of high-power-density pulses. The etching of the surface is a linear function of the number of pulses [62]. Sufficient light absorption of the polymer at the laser wavelength and low thermal conductivity of the polymer are necessary to achieve high ablation efficiency. Either proper selection of the laser radiation source or modification of the polymer can improve light absorption [63]. A threshold of power density is observed below which no significant ablation takes place. This threshold is dependent on the wavelength of the laser and the polymer material being ablated. Direct nonthermal bond breaking is suggested to be an important mechanism for polymer ablation where photon energies in excess of energies for molecular dissociation or band-gap excitation are employed.

Scanning Laser Ablation: Scanning laser ablation (SLA)-based via generation processes have been used successfully in MCM-D [60,64] and high-density printed-wiring-board fabrication [65]. The via generation tool is driven by either a XeCl (308 nm) or KrF (248 nm) excimer laser. Although contact shadow masking approaches have been successfully used for other less demanding laser micromachining applications, the via resolution and placement accuracy achievable is inadequate for most thin-film applications. Consequently, as is commonly the case for RIE via generation processes, an aluminum or copper conformal masking layer has been used to define the via locations. The masked parts are positioned on a computer-controlled precision x-y table and are moved in a serpentine pattern under a stationary homogenized excimer laser beam, thereby exposing the dielectric layer in the open areas of the mask to the ablative power of the laser light. Unlike RIE processing, which is inherently a batch processing technology and performed under vacuum, scanning laser ablation processing is done under ambient conditions. Consequently, one of the major attractions of laser via fabrication approaches is their compatibility with in-line substrate processing.

In establishing a robust, manufacturable SLA via generation capability, a process window must be defined which allows for the ablation of the underlying dielectric layer through the openings in the conformal metal mask at an acceptable rate (>0.25 μm/pulse) while maintaining a fluence well below the mask damage threshold. Soot accumulation from the ablation process around the mask via openings has been found to be an important additional factor in determining the overall robustness of the conformal metal masks. Localized overheating of the masking layer around the via sites can result in excessive soot redeposition. A number of approaches including the use of a He or O_2 process gas has been shown to significantly reduce the amount of soot generated, thereby enhancing the mask stability. Minimum recommended mask thicknesses vary somewhat with the masking metal used; however, thicknesses greater than 3 μm are generally recommended for commonly used dielectric ablation fluences of 250–350 mJ/cm^2. Commercially available SLA tools, with a high-powered (100–150 W) industrial excimer laser, are capable of throughputs that are comparable with an RIE-based via generation process. The need for a much thicker metal mask for SLA via generation than for RIE processing does however adversely impact the processing cost and minimum achievable feature size associated with the deposition and patterning of this thicker masking layer.

Effective soot removal is required to ensure reliable adhesion of the next metal layer. Two main cleaning processes have been used for this application. Aqueous cleaning using a pumice or brush scrubbing process has been used to mechanically remove any residue from the ablation process. Alternatively, contactless, plasma-based, dry soot removal has also been shown to be successful, as described elsewhere [61].

Both RIE- and SLA-based via processes involve a number of preetch and postetch processing steps as demonstrated in Figure 12-37. In order to simplify and thus cost-reduce these via generation processes, efforts have been underway to reduce the number of required processing steps. One such approach has been the use of thick, patterned photoresist layers as erodible RIE or SLA etch masks [66,67]. A successful demonstration of a photoresist-mask-based SLA via generation process used to etch through 10 μm BCB is shown in Figure 12-38 at various stages in the process sequence [67].

Projection Laser Ablation: Projection laser ablation via generation involves the use of excimer laser tools that are very similar in concept and construction to photolithographic steppers used in semiconductor processing. The via sites to be ablated are defined by openings in a discrete mask, in a manner analogous to the exposure of a photoresist layer in a stepper [68]. Changes in tool optics and mask technologies from those used in conventional phototools are required to withstand the high-power pulsed UV laser beam necessary to effect the ablation process. Like the

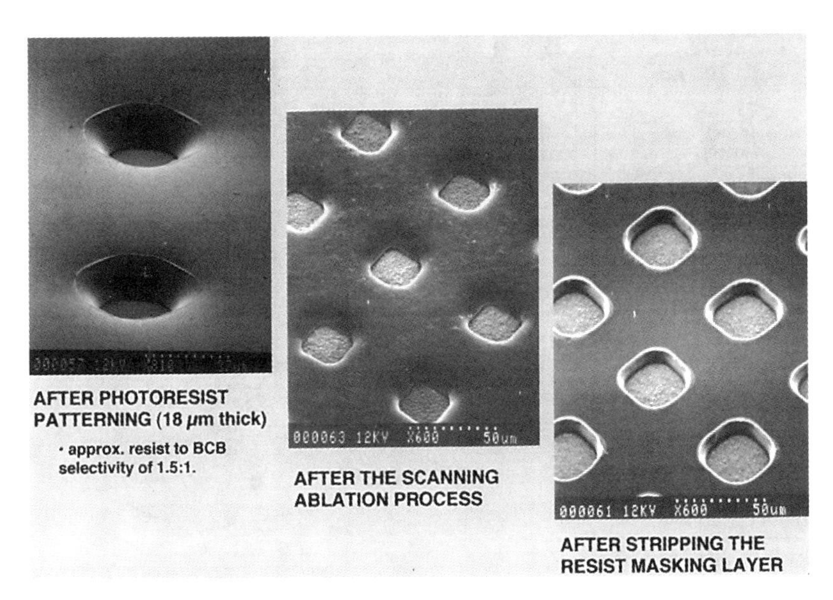

Figure 12-38. Via Generation in BCB Dielectric Using a Photoresist Ablation Mask

SLA equipment described in the previous section, these projection tools are powered by 100–150-W KrF or XeCl excimer lasers. The first commercial application of a projection laser ablation technology was in the manufacture of MCM-C/D substrates for the IBM ES9000 system [61]. A proprietary dielectric on quartz masking technology was developed for this application. Figure 12-39 [61] schematically illustrates the working principle of projection laser ablation tool.

The major process variables influencing the ablation rate as well as the via shape and sidewall angle achievable with a projection ablation process are the fluence, the number of pulses, and the focus of the image of the beam. The importance of the focus on the resultant via sidewall angle was recently reported [61] and is shown in Figure 12-40. A SEM microphotograph of typical via openings generated using a projection ablation process is shown in Figure 12-41. The typical via sidewall angle obtained using a projection ablation process is approximately 65° compared to 75°–85° with a scanning laser ablation process.

Projection laser ablation through a mask using an excimer laser at 308 nm is a very robust process [69], achieving greater than 99% substrate yields on thin-film structures with greater than 100,000 vias per substrate. The wall profiles are very uniform and can be controlled from 30° to 65° by changing the focus conditions [61]. The process is not aspect-ratio limited for packaging applications, and almost any polymer dielectric can be used. The carbonaceous debris formed during laser ablation is removed

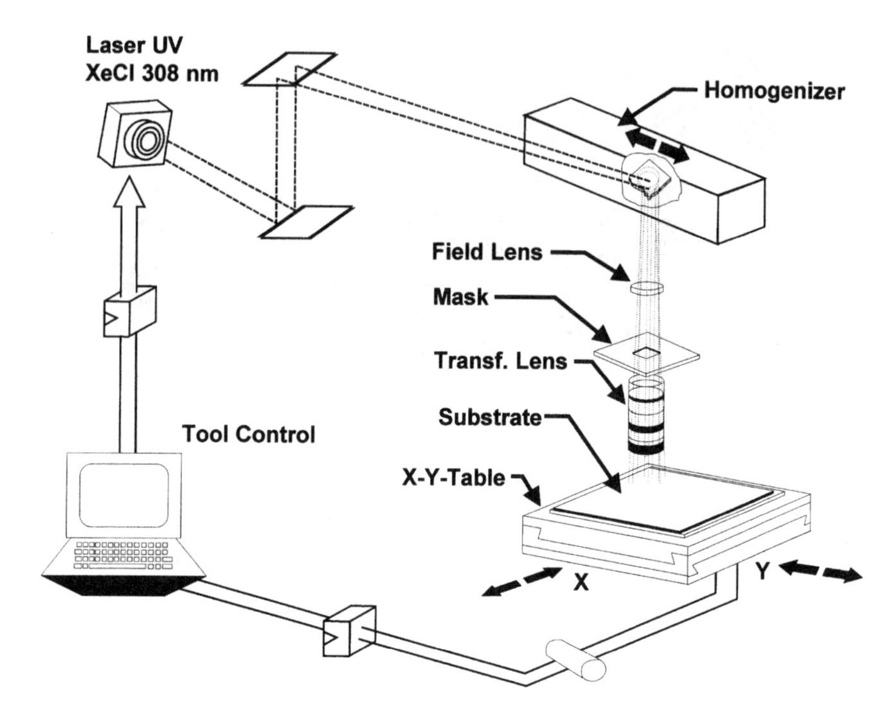

Figure 12-39. A Schematic of Projection Laser Ablation Tool

by plasma cleaning. Defects due to residue are dramatically lower in this process compared to the RIE process [61].

Despite the obvious advantages of projection laser ablation resulting from the elimination of all the preetch and postetch processing steps associated with RIE and SLA processing, until recently, more widespread use of this via generation technology has been hampered by two major factors. First, the high application, capital equipment costs and dedicated functionality of such an ablation tool rendered its use virtually impractical for all but very high-volume MCM-D production as is the case at IBM. Additionally, the unavailability of a commercial source for a proven projection ablation masking technology further discouraged usage of this technology. More recently, efforts initiated at IBM and elsewhere to address these fundamental barriers are expected to broaden the user base of this via generation technology throughout the industry in the future.

Reactive Ion Etching: To date, throughout the industry, reactive ion etching (RIE) has been by far the most widely used via formation technology. Several examples of the successful use of RIE processing for via generation in MCM-D substrate technologies have appeared in the literature [70,71]. For the most part, this preference has been based on

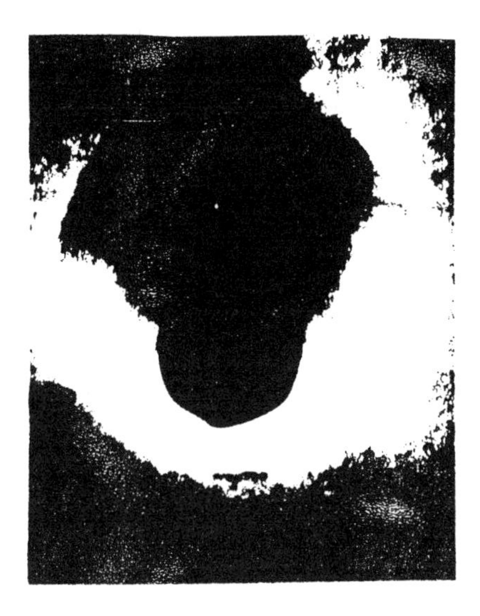

Via Ablated through 10 μm polyimide

Figure 12-40. Effect of Focus on the Resultant Sidewall Angle Achieved by Projection Ablation (Courtesy of Motorola)

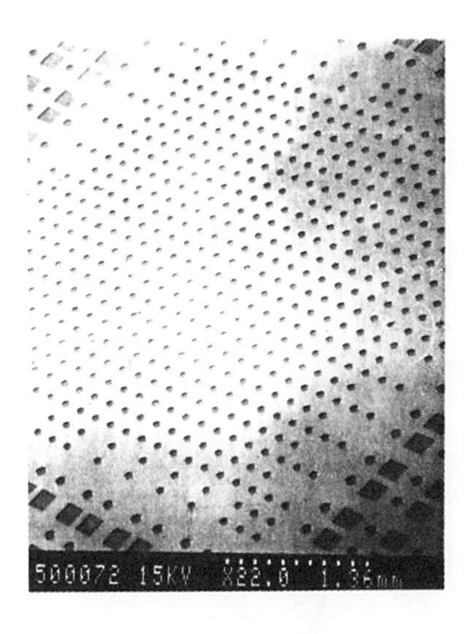

PASSIVATION OPENINGS
ES9000 TCM CHIP SITE

90 μm OPENINGS in POLYIMIDE

Figure 12-41. Projection Ablation of Vias in 12-μm-Thick Polyimide (Courtesy of Motorola)

the evolution of these via etch processes from long-standing semiconductor fabrication techniques and the associated ready availability of suitable parallel-plate RIE tools. An O_2-rich, fluorinated gas (e.g., CF_4, CHF_3, or SF_6) mixture has been commonly used in these processes, thereby providing for etch rates well in excess of 1 μm/min. Gas mixture, power, and pressure are all interactive variables that must be optimized according to the desired via sidewall angle, mask undercutting, and etch rate. Typical via etch results are shown in the SEM microphotographs in Figure 12-42.

Reactive-ion-etching via processing requires a conformal masking layer to define the via sites to be etched. As is seen by the schematic of a typical RIE process flow in Figure 12-37, the bulk of the processing steps in this process are for mask deposition and removal. Because the ultimate resolution capabilities of a RIE-based via generation process depend on the resolution achievable in the patterning of the conformal masking layer, a high etch selectivity to the polymer dielectric layer is desirable so as to minimize the thickness of the masking layer required. Several alternative RIE masking layers have been successfully used and fall into either erodible or nonerodible etch mask categories. Erodible etch masks including SiO_2, Si_3N_4, and photoresist layers are typically etched at an appreciable rate under the same etch conditions used to etch the via openings in the underlying dielectric layer. Because most MCM-

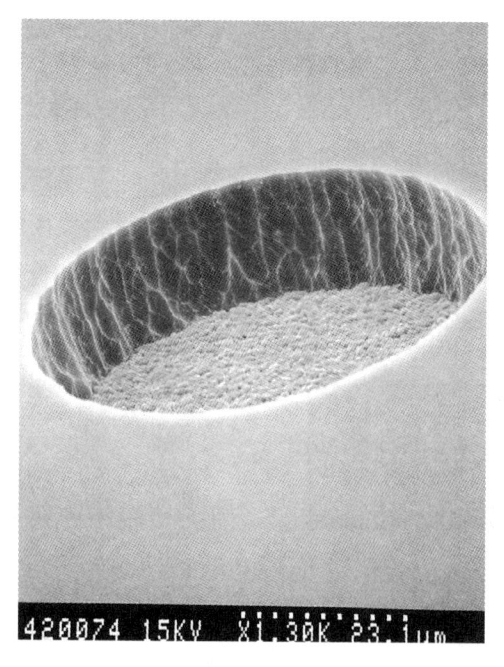

Figure 12-42. SEM Photomicrographs of RIE Generated Via

D technologies involve the use of 8–25-μm-thick dielectric layers, non-erodible, usually metal masking layers have been most commonly used [70,71]. Because these metal masks are not eroded by the dielectric etching process, they can be very thin and are typically in the 3000–5000-Å-thick range. Unlike wet etching, the polymer is reactive-ion-etched after full cure or complete polymerization.

One example of an erodible etch mask system is a simple image transfer through a positive photoresist stencil. To ensure that all the vias are open with no residue and to improve process robustness, the polymer is overetched. To compensate for this overetch, the photoresist is made slightly thicker than the polymer, as the etch rates of polymer and photoresist are nearly the same. The positive photoresist is reflowed prior to reactive ion etching to obtain a tapered wall profile. After etching, organic photoresist strippers are used instead of aqueous strippers to remove the photoresist, because the photoresist is UV hardened during reactive ion etching. Bilevel processes work best for polymer thicknesses below 5 μm.

To increase the range of polymer thickness that can be etched and to improve the wall profile, a trilevel stencil can be used [72]. In this process, a silicon-containing organometallic layer (RIE barrier) is deposited by chemical vapor deposition between the polymer and the photoresist. The image is transferred from the photoresist to the organometallic layer by CF_4 plasma etching. An oxygen plasma, which does not etch the RIE barrier, is subsequently used to transfer the image to the polymer. The RIE barrier is then removed after the vias are defined. The wall angle can be tailored by varying the gaseous composition. Unlike the bilevel process, the photoresist thickness is not critical and is used only to transfer the image to the RIE barrier. The process window is wider for this process and the choice of the polymer dielectric is not restricted.

Backsputtering during reactive ion etching gives rise to redeposition of metallic species in the vias, causing increased defect levels and increased via resistance [73]. The reactive ion etching residue can be cleaned by additional wet or dry processes. The use of low-power reactive ion etching reduces the amount of backsputtering but also decreases the throughput of the process.

Chemical/Wet Processing of Dielectrics: Chemical or wet processing of dielectrics involves the use of aqueous or organic solvents to etch vias in MCM-D dielectric layers. Despite the more limited via design rule capabilities of these "wet" via processing alternatives over those achievable with the dry etch processes described earlier, the process simplicity and inherent lower capital costs of such approaches make them particularly attractive for those more cost-sensitive MCM-D applications. To date, these approaches have fallen into one of two major material/process categories, namely inherently photosensitive dielectrics and wet etchable materials.

Photosensitive Polyimide Lithography: Of the inherently photosensitive dielectric materials available, negative-acting photosensitive polyimides (PSPI) have been by far the most commonly used. The limited resolution and resolvable via aspect ratios achievable with photosensitive polyimides due to the shrinkage (in excess of 50%) that occurs during post-via etch hard baking as well as their generally poor shelf life and stability have somewhat limited their acceptance for MCM fabrication. However, as amply highlighted in Figure 12-37, the process sequence associated with the use of an inherently photosensitive dielectric material is considerably simpler and less capital equipment intensive than the other via generation options. With the recent availability of a wide range of photosensitive polyimides and others, this method of via fabrication has become more attractive. The number of process steps required to form vias is reduced due to the use of a unilevel stencil. The advantages of this process have been well documented [74]. The chemistry of PSPI is complicated because it has to meet all electrical, mechanical and lithographic requirements. The PSPI most commonly used is a negative tone **resist** which cross-links upon exposure to UV radiation and becomes insoluble. The unexposed area is dissolved in an organic-based developer.

Photosensitive polyimides are formulated by adding sensitizers, initiators, and other photoactive components to the base polyimide. Upon curing, these photocomponents and the condensation reaction products are volatilized, leading to a shrinkage of about 50%. This shrinkage can be reduced by using preimidized PSPIs, but these PSPIs have a narrower process window and, in most cases, inferior mechanical, thermal, or solvent-resistance properties. The PSPI lithography process is most cost-effective compared to wet etch and RIE even though the dielectric material is more expensive [75]. The process window is very wide, and PSPI lithography has better resolution and can accommodate higher aspect ratios compared to wet etch. As a typical example, 8-μm vias in 6-μm-thick cured PSPIs have been defined using this process.

Unlike laser ablated vias, the wall profile of vias formed by PSPI lithography is not trapezoidal in shape. The PSPI via wall profile is steeper in the top portion of the via with a very shallow foot. Because the PSPI is cured after the via is defined, a cusp or lip is formed at the top. The top dimension of the via for a particular bottom dimension can be controlled by changing the lithography parameters [76]. Also, the adhesion promoter that is used between dielectric layers is not required because most PSPI materials are formulated with an adhesion promoter, thereby eliminating the need for doing this in a separate step.

Examples of the successful use of photosensitive polyimides in the fabrication of MCM substrates have appeared in the literature [34,77–79] with the bulk of this activity occurring in Japan. Figure 12-43 is a typical via generated in a photosensitive polyimide. In recent years, reports on via formation in a photosensitive benzocyclobutene (BCB) for use in MCM-D substrate fabrication has also appeared in the literature [80].

Wet Etch: Wet etching of conventional nonphotosensitive polyimides have also attracted interest as the industry strived to reduce MCM-D substrate processing complexity and cost. With this type of via generation approach, a relatively thin (1–5 μm thick) patterned photoresist layer serves as the mask template to define the sites of via etching in an underlying dielectric layer, typically when the material is still partially imidized. Because the via photodefinition process involves the use of a positive acting photoresist layer, such a process tends to be significantly less susceptible to particulate-based defects than the negative-acting inherently photosensitive dielectric materials mentioned previously. Preliminary results describing the potential of wet-etch-based via generation processes have been reported in the literature in the past few years [81–83]. The vias are developed by dissolving the exposed areas in an alkaline solution. Process reproducibility is controlled by monitoring the degree of imidization prior to wet etch. For a given bake temperature the degree

Figure 12-43. Vias Generated in a Photosensitive Polyimide

of imidization is also a function of the polyimide thickness [84]. Due to the isotropic nature of the wet etch process and the chemistry involved, the wet etch process gives a very shallow wall profile. Because the final curing of the polyimide is done after via definition, shrinkage of the polyimide will add to the increase in the top via dimension compared to the bottom. An SEM microphotograph of typical vias obtained with this type of process is shown in Figure 12-44.

A 2 : 1 aspect ratio (via bottom dimension : thickness) is preferred for a robust manufacturable process. The minimum feature dimension is recommended to be greater than 15 μm. The wet etch process limits the choice of polyimides for use as the dielectric. The tooling is simple because it is an aqueous-based process. Defect levels are controlled by (1) using a negative tone resist to reduce reaction with the underlying polyimide during stripping of the photoresist after wet etch, (2) implementing a two-pass photolithography process, which involves two coats of resist application to eliminate any pinholes in the photoresist prior to polyimide etching, and (3) introducing a via cleaning process after stripping of the photoresist.

Via Generation Summary: As is clear from this section, a number of via generation alternatives exist for thin-film substrate processing. All

Figure 12-44. Via Generated by Wet Etching in Amoco 4212 Polyimide Film

of these processes have their own advantages and disadvantages. The simple processing sequences such as the wet etching processes tend to provide the lowest-cost solution but at the expense of minimum achievable resolution.

12.4.3.2 Metallization Processes

A number of metallization processes have been used for MCM-D substrate fabrication, with the most common approaches summarized previously in Table 12-10 and illustrated schematically in Figure 12-45. Owing to the origins of this multichip module substrate approach in semiconductor processing, sputter-deposited aluminum metallization has been the most widely used alternative mentioned in the literature. Additionally, because high-resolution subtractive patterning of 2–4 μm-thick sputtered aluminum layers down to feature sizes as small as 8–10 μm is achievable with good reproducibility and minimal lateral etching, several examples of MCM-D substrates based on sputtered aluminum metallization have been reported in the literature [85,86]. As MCM-D technologies have continued to evolve, a general desire to use lower-resistance copper conductors has developed. This general trend has resulted in the need for more complex metallization for the tendency of interaction between bare copper and polyamic acid precursors, resulting in copper diffusion into the dielectric and poor adhesion at the copper/polyimide interface. In order to avoid this undesirable interaction, barrier metals, including Cr, Ti, and Ni, have been used [87–89]. This tends to further complicate subtractive metallization patterning processes where the additional etching of barrier layers is required. Additionally, the sequential etching through upper barrier, core, and lower barrier layers results in conductors with exposed copper on the sidewalls. Recent reports have suggested that the use of polyimide formulations based on polyamic acid ester precursors [90] and DSV-BCB [91] dielectrics are compatible with unprotected copper conductors.

Evaporation/lift-off approaches for MCM-D substrate metallization also owe their origins in semiconductor processing and have been reported in the literature [72,92]. In this additive approach, a thick photoresist layer is deposited and patterned with the desired conductor pattern so as to achieve a retrograde sidewall profile. The desired copper-based conductor with appropriate barrier metals is then sequentially deposited by evaporation onto the substrates. The carefully optimized photoresist profile together with the inherently directional nature of evaporation processes results in a discontinuous metal deposit layer.

Semiadditive copper electroplating metallization processes have also been used in MCM-D processing [77,92,93]. In this approach, a thin seeding layer (3000–5000 Å thick) is deposited by sputtering or electroless

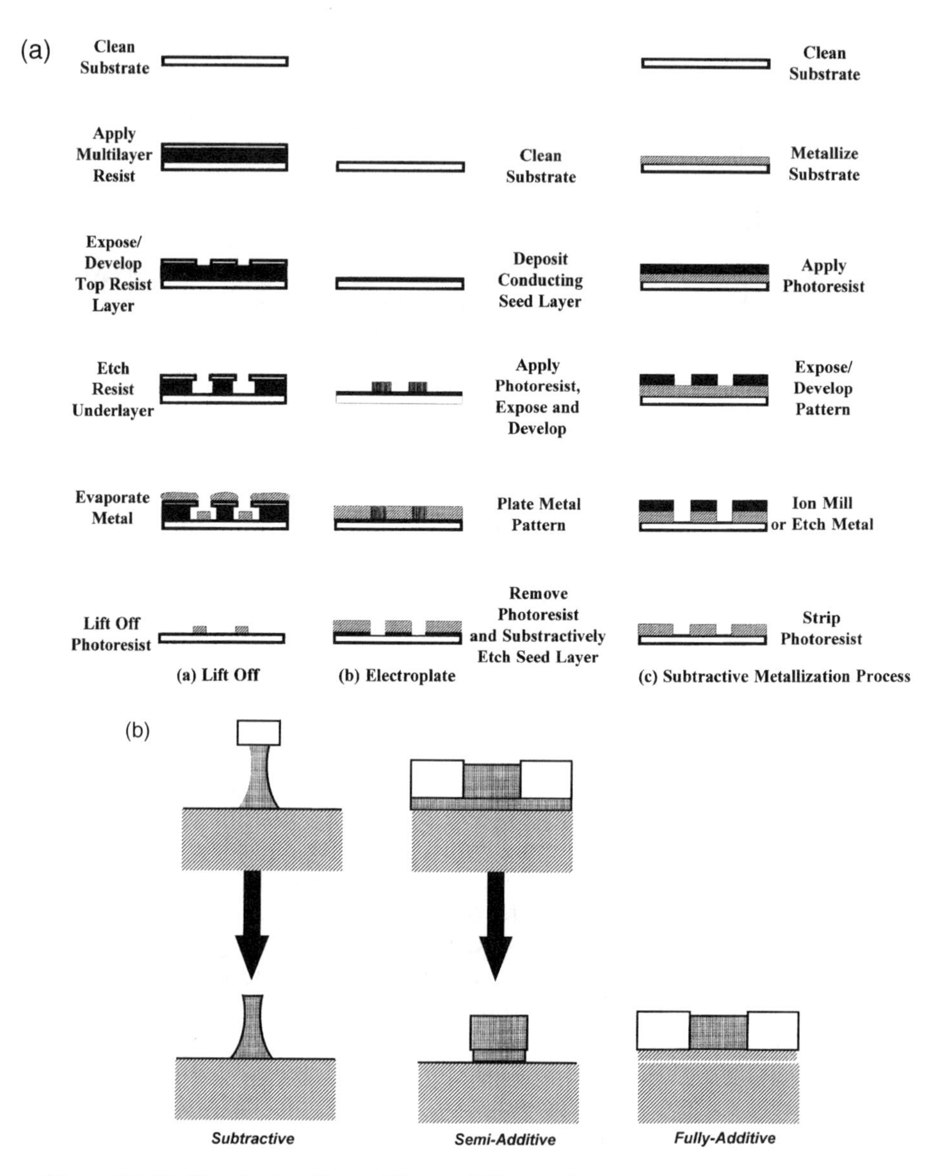

Figure 12-45. Conductor Deposition and Patterning. (a) Process Alternatives; (b) Line Profiles Obtained with Each Processes.

copper seeding. The bulk of the conductor thickness is then built up by electroplating through a patterned photoresist template. Subsequent stripping of the resist layer followed by etching of the thin seeding layer leaves the isolated conductor layers. Due to the significantly lower capital costs associated with electroless/electroplated Cu metallurgy and the compatibility of this process with high-volume, large-format substrate panels as indicated by years of use in the PWB industry, more widespread use of this metallization approach for "low-cost" MCM-D processes is anticipated in the future. IBM's recently announced SLC technology is a case in point [16]. Additional information on the SLC process will be outlined in the following section.

12.4.3.2.1 Wiring-Level Metallization

Aluminum and copper are used as the primary conductor metals for MCM-D. Gold has been used by some MCM-D manufacturers but suffers from high cost and poor adhesion to polyimide. Wiring patterns in aluminum are formed by subtractive etching. Copper conductor lines are defined either by **additive electroplating** or by subtractive etching. Lift-off for wiring-level definition is not favored because of

- Expensive tooling
- Need for RIE barrier and RIE etch stop
- Concern with repeated exposure of the thin-film structure to solvents

The critical elements in the choice of the process are as follows:

- Aspect ratio
- Tolerance requirement
- Conductor metal
- Availability of required tooling

Subtractive Etching: The wiring-level metals are deposited either by sputtering or evaporation. Sputtering is preferred over evaporation due to its higher throughput and its ability to conformally cover the sides of the vias. The line thickness tolerance for this vacuum deposition process is better than 10%. Positive resist lithography is used to define the wiring pattern and the metal is wet-etched in aqueous-based solutions. The linewidth is controlled by the subtractive etching process. Due to the isotropic nature of subtractive etching, the resist mask is sized larger than the desired final metal feature dimensions to compensate for the undercutting during etching (Fig. 12-46). This limits the subtractive etching process to features with low aspect ratios. Subtractive etching of aluminum has less undercut than copper, resulting in a better line profile and linewidth control.

Other factors that control the linewidth uniformity are the following:

- Nonuniformity in etch rate between edge and center of the substrate
- Electrochemical effects
- Topography
- Etchant chemistry

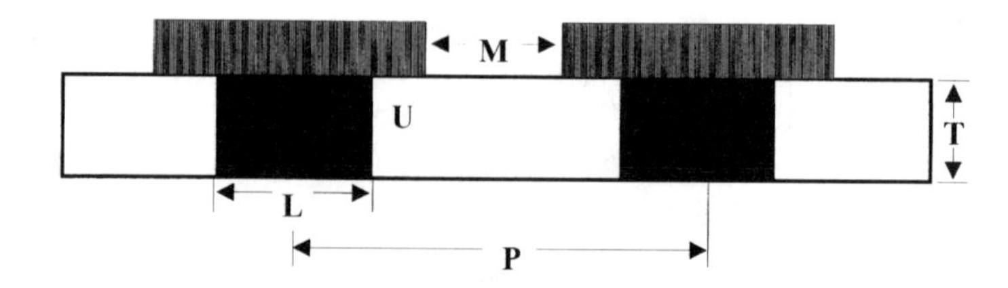

Pitch (P) = Linewidth (L) + Minimum Resolution (M) + 2x Line Thickness (T) + Undercut (U)

Figure 12-46. Undercut of Copper During Subtractive Etching

The nonuniformity in etch rate between the edge and the center of the substrate is minimized by using appropriate spin/spray etch tools. As end-point detection is difficult, differences in the edge-to-center etch rate is compensated by overetching. This factor becomes more important as the size of the substrate increases. Also, as the conductor thickness increases, the linewidth tolerance control becomes more challenging, even if the aspect ratio is favorable.

Multilevel thin-film structures have been fabricated using both copper and aluminum as the conductor metal. To enhance adhesion of copper to polyimide and to prevent its interaction with polyimide, the copper wiring is sandwiched between two thin layers (<300 Å) of Cr. Cr-Cu-Cr is deposited in one sputtering step and the trilevel structure is subtractively etched in two different solutions. In this process, the sidewalls of the wiring channels are not covered with Cr. Because polyimide adheres well to the Cr-coated top surface of the wiring channels, coating of the sidewalls with Cr is not critical. Aluminum metallization does not require the barrier metal because it has good adhesion to polyimide without interaction.

Copper conductors are preferred over aluminum conductors due to its higher electrical conductivity. Aluminum is preferred to reduce cost if a semiconductor manufacturing line is used for manufacturing MCM-D. The primary defect type observed during subtractive etching of copper and aluminum is shorts due to resist flaws and contamination. Overall, subtractive etching is the lowest cost wiring definition process and is used when the aspect ratio allows for it (Table 12-14).

Additive Electroplating: Additive electroplating is feasible for both copper and gold wiring. High-aspect-ratio wiring is possible with additive electroplating, and the linewidth tolerance depends on the photoresist

Table 12-14. Copper Additive Electroplating and Subtractive Etching Metallization Regimes

Thickness	Linewidth/Pitch (μm) 12/25	20/40	20/50	25/75
2 μm	Subetch	Subetch	Subetch	Subetch
3 μm	Electroplating	Subetch	Subetch	Subetch
4 μm	Electroplating	Electroplating	Subetch	Subetch
5 μm	Electroplating	Electroplating	Subetch	Subetch
6 μm	Electroplating	Electroplating	Subetch	Subetch

lithography tolerance. A photoresist developed in IBM gave better than 5% control in linewidth uniformity. The line thickness, on the other hand, is dependent on the uniformity of the electroplating process. Uniformity in pattern density, pattern shape, and feature size across the substrate is critical in controlling the thickness uniformity. By optimizing the electroplating tool and the current density distribution, a thickness uniformity of 10% or better can be achieved, even when the pattern is nonuniform.

Cr-Cu is used as the blanket conductor layer (seed layer) for electroplating. The seed layer is flash etched after electroplating and resist removal. Copper features are covered (capped) with a metal or polymer layer after electroplating; its function is similar to that of Cr in the subtractive etching process. Because the structure is capped after resist removal and etching of the seed layer, the sidewalls of the conductor lines are also capped. The final linewidth uniformity is a function of photoresist lithography, flash etch, and capping uniformity. Line opens are the primary type of defect detected in additive electroplating. Unlike subtractive etching, additive electroplating technology has no restrictions on line thickness or substrate size.

Table 12-14 shows the preferred conductor-line definition process for an assumed undercut of 1 : 1 for the subtractive etch process. In the subtractive etching process, where the metal is sputtered to the final line thickness, the metal thickness on the sidewalls of the vias will be less than on the top surface. This might be a reliability problem for very thin conductor layers on thick polymer dielectric layers. On the other hand, in additive electroplating, only a thin seed layer is deposited by sputtering. The metal thickness is uniform on both the top and the sides of the vias due to the isotropic nature of electroplating. Depending on the reliability requirements and the substrate size, additive electroplating may be preferable to subtractive etching even when the aspect ratio and line pitch are favorable for subtractive etching.

PSPI Lithography/Planarization: Photosensitive polyimide (PSPI) lithography is one of the ways to define the wiring channels in a planar structure [32,94]. As was previously discussed, PSPI lithography involves patterning by photo exposure, development in solvents, and thermal curing. Blanket sputtering or electroplating of Cr-Cu is used to fill the channels defined in PSPI, and mechanical planarization is used to remove the excessive metal. The top layer of the polyimide is also removed during planarization, which puts additional requirements on the mechanical properties of the polyimide. Variation in the density of the wiring pattern leads to variations in the wall angle (profiles) of the PSPI after cure. This is due to a 50% shrinkage of the PSPI thickness on curing, leading to unconstrained shrinkage (Fig. 12-47), especially in regions where the distance between features is smaller than the cured PSPI thickness.

To simplify this complex process, the via and wiring features were metallized during the same metal deposition and planarization step (referred to as Dual-level metal, DLM, in IBM). The thickness tolerance of the wiring channels is controlled by the planarization process and can exceed ±15%. Figure 12-48 is an example of substrates fabricated by this process [17].

12.4.3.2.2 Via Metallization

Several different via metallization processes have been successfully used in the fabrication of MCM-D modules. In most MCM-D technologies described to date, sidewall metallization is carried out simultaneously with

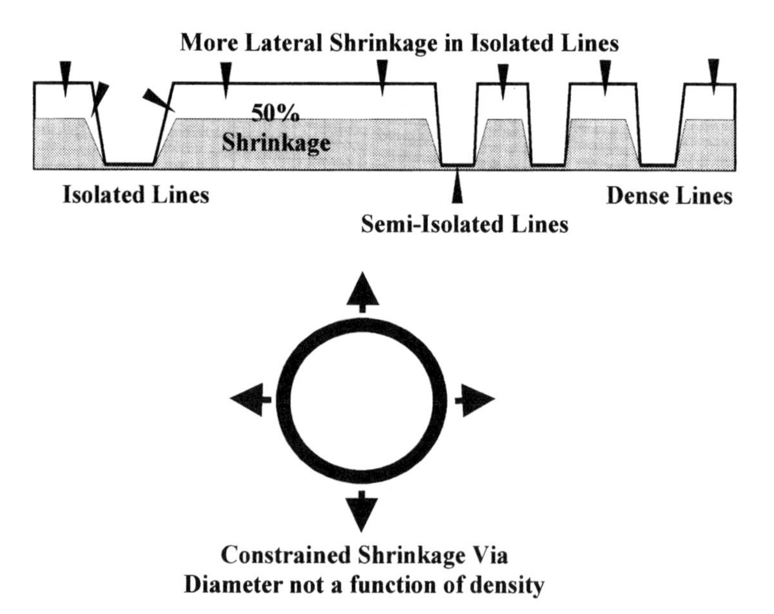

Figure 12-47. Effect of PSPI Shrinkage on Wiring and Via Profile

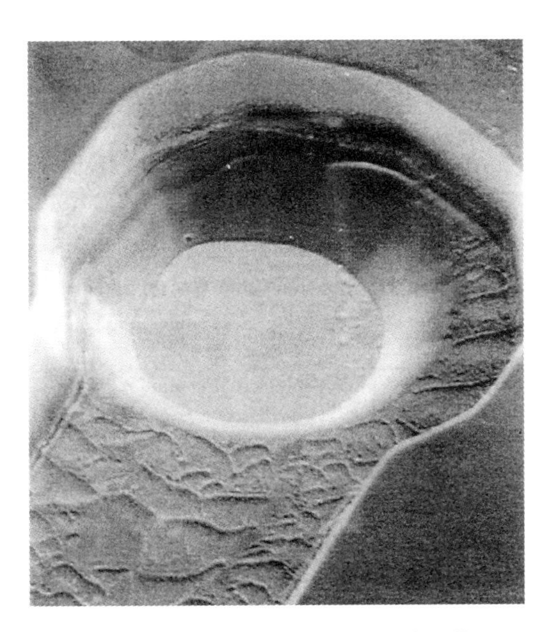

Figure 12-48. Micrographs of Via and Conductor Line Structures Fabricated by Dual-Level Metal (DLM) Process

the deposition of the next metal layer, resulting in electrical connection to the previous metal layer. A cross section of this metallized via structure is shown in Figure 12-15. Because this type of metallization approach generates significant topography in the immediate areas of the vias, stair-stepped via structures are commonly used in these multilayer MCM-D designs. Although this fact significantly affects the maximum interconnect density achievable, most of the MCM-D prototype substrate designs described in the literature could be accommodated with this via structure. Several solid-metal via filling processes have been reported in the literature to satisfy specific, more aggressive MCM design rules where a stacked via configuration is desired/required and the significantly greater number of processing steps involved can be justified. In these higher-density interconnects, substrate planarization is maintained during the whole sequential substrate fabrication process. A schematic comparison of stair-stepped and stacked via arrangements are shown in Figure 12-49. A number of different via filling approaches have been used, including electroless nickel via filling processes [92,93] (Fig. 12-50), via fill by evaporation/lift-off [72], and via post formation with subsequent dielectric deposition and lap back [16,96] (Fig. 12-51).

12.4.3.2.3 Terminal Metals and Their Processing

Terminal metals consist of bonding pads for chip connections and the engineering change features on top of the signal and power layers

(a)

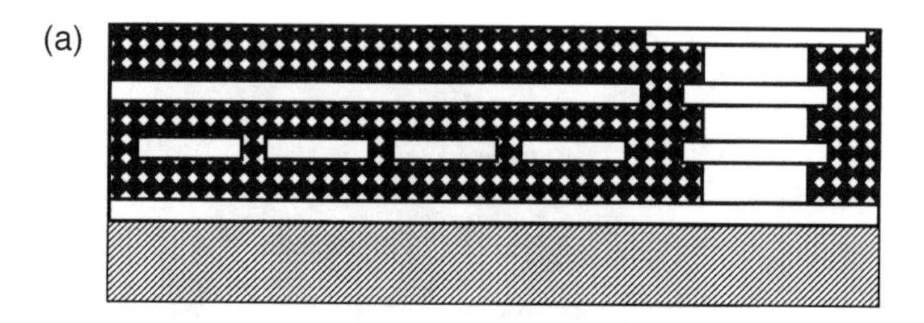

(b)

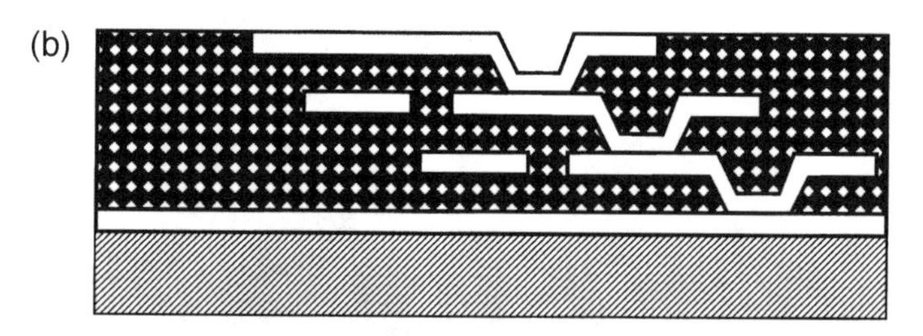

Figure 12-49. A Comparison of (a) Stacked and (b) Conformal/Stair-Stepped Via Arrangements

[97,98]. The terminal-metal process is similar to the wiring-level process, but the range of metals that need to be deposited is more extensive. Bulk and surface composition of the terminal metals is critical due to the sensitivity of the chip connection process to defects and contamination. The terminal-metal stack ranges from Cr-Cu for flip-chip joining with limited rework or reflow requirements to Cr-Cu-Ni-<1500 Å Au for flip-chip connection with multiple reflow and rework requirements. Wirebonding, on the other hand, will need Cr-Cu-Ni->1 μm Au. Nickel serves as a diffusion barrier between gold and copper, thus reducing the formation of Cu-Au intermetallics. Cobalt can also be used as the diffusion barrier. The terminal metals can be defined by the following:

- Evaporation through a metal mask
- Additive electroplating
- Lift-off

If only Cu is needed as the terminal metal, as in the metallized-ceramic-polyimide (MCP) process, subtractive etching is used. The terminal metal process selection depends on

- Tooling availability
- Type of substrate

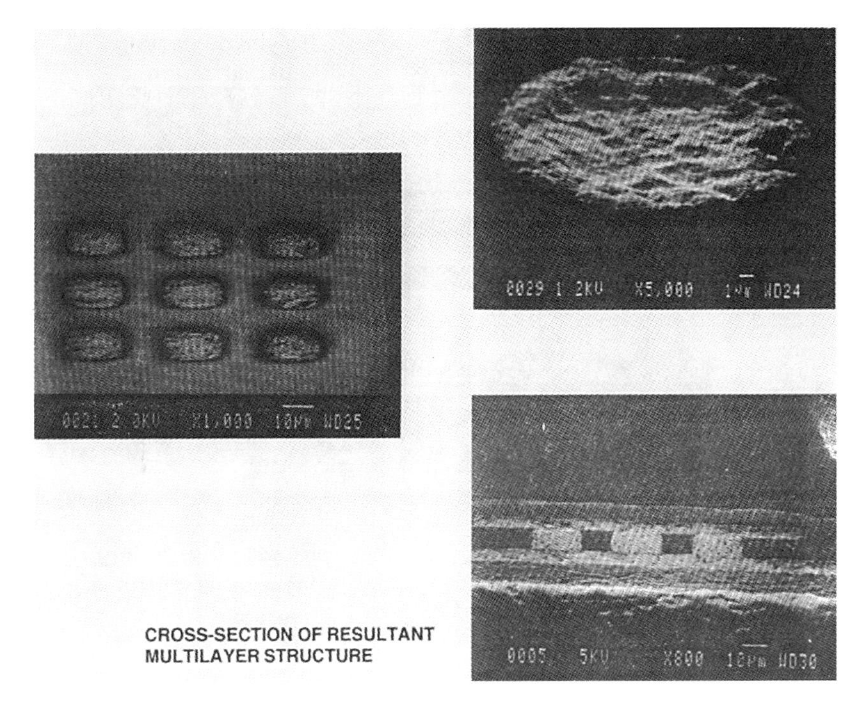

Figure 12-50. SEM Photomicrographs of Vias Filled with an Electroless Ni Process (Courtesy of Motorola)

- Size and shape of top surface features
- Cost.

Table 12-15 compares the process options for terminal-metal fabrication.

Evaporation Through Mask: Evaporation through a metal mask is the simplest method of depositing the terminal metals. Unfortunately, it is not always applicable for multilevel thin-film fabrication. The metal mask has to conform to the substrate for intimate contact between it and the mask. This is possible only when the substrate is flexible, like silicon. The rigidity of cofired-ceramic substrates prohibits the use of evaporation through a mask due to shadowing around the features. This method is suitable for defining bonding pads but not for fine-line wiring which is needed on the top surface for the engineering changes in certain applications.

Additive Electroplating: Additive electroplating was chosen as the terminal-metal process for the multilevel conformal via structures due to its low start-up and operational cost. The process sequence and limitations

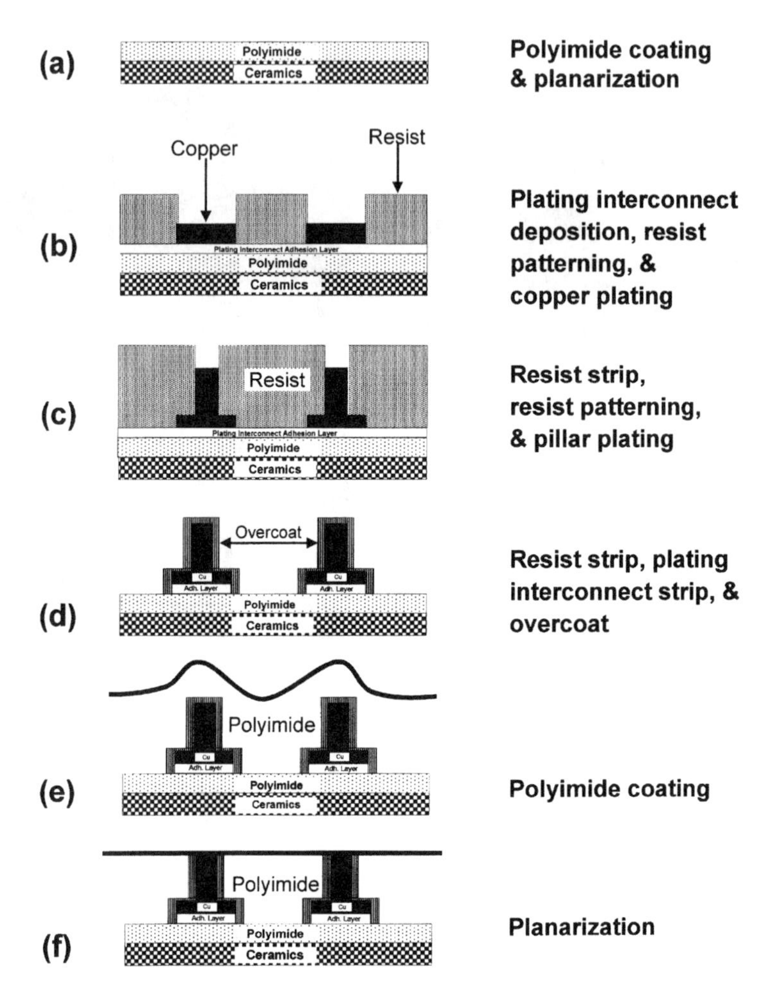

Figure 12-51. MCC Via Post Processing Approaches. (a) Polyimide coating (b) Base metal deposition, resist patterning and plating of copper lines. (c) Resist strip, base metal strip, and nickel overcoat. (c) Polyimide coating and (d) Planarization with mechanical polishing.

are similar to that described in the wiring section except that capping of the features is not required.

Lift-Off: Lift-off technology is practiced for the terminal-metal definition in the IBM ES 9000 processor. The lift-off stencil can either be bilevel or trilevel. Because the terminal metals are evaporated, Cr, Cu, Ni, Co, and Au can be used. The trilevel lift-off stencil is comprised of soluble PI + RIE barrier + photoresist. The features are defined in the photoresist and the image is transferred to the RIE barrier and soluble PI

Table 12-15. Comparison of Terminal-Metal Process Techniques

Metallization Process	Number of Process Steps	Critical Unit Processes	Choice of Metals	Limitations
Evaporation through mask	4	Mask attach/align	Cu, Ni, Co, Au	No wiring features
Subtractive etching	7	Resist lithography	Cu	Only flip chip, no reflows
Additive electroplating	8	Resist lithography, electroplating	Cu, Ni, Co, Au	
Lift-off	12	Resist lithography, RIE residue removal	Cu, Ni, Co, Au	Requires etch stop (3 additional steps)

by reactive ion etching. After evaporation of the terminal metals, the stencil is lifted off in an organic solvent. In this process, aspect ratios of better than 1 : 1 are possible and both engineering change features and bonding pads can be defined.

For the process to be robust, an RIE barrier is needed to protect the underlying wiring and dielectric levels. This is accomplished by the sputtering of blanket Cr-Cu-Cr prior to the stencil build. In the ES 9000 processor process, the Cr-Cu-Cr barrier doubles as a redundant metal layer to improve the yield of the underlying wiring layer. A lithography step is used to cover the terminal metals and the redundant features. The rest of the blanket layer (Cr-Cu-Cr) is subtractively etched. The primary defect type is shorts due to defects during the subtractive etching of the RIE etch barrier.

In summary, evaporation through a metal mask is the terminal-metal process of choice but can be implemented only on silicon substrates. Also, engineering change features cannot be incorporated into the top surface. Additive electroplating is a low-cost process and is the preferred process for cofired-ceramic substrates. The lift-off process requires twice the number of process steps compared to additive electroplating and has the highest number of critical processes. Terminal metals by liftoff is more expensive than either the additive electroplating or the evaporation through mask process.

12.4.3.3 Planarization

Planarization is a very important issue for the multilevel thin-film packaging process regardless which coating technique and via generation and metallization processes are used. It has a great impact on both the package performance and the manufacturing process. Planarization is necessary to maintain properties such as characteristic impedance which is geometry dependent. Variations in impedance may cause signal reflections which lead to logic errors. Geometry also affects line resistance which, in turn, dictates power loss and heat generation. Furthermore, as lines become narrower, the electromigration phenomenon becomes more prominent and thickness variations should be minimized to reduce electromigration-related failure. Photolithographic resolution depends on the power absorption and thickness of the polymer thin film, and variations of film thickness will result in resolution degradation. Another issue is dielectric breakdown. If a dielectric film is too thin at one point, the high electrical field at this point could possibly cause a breakdown, destroying the integration of the package function. Finally, yield and reliability tend to degrade as wiring density becomes high and feature size becomes small. In brief, planar structure is always desirable in terms of electrical performance but, most of the time, would greatly complicate the processing and markedly increase the cost of packaging. Challenges of planarization are growing as aspect ratios increase. More densely packed circuits mean narrower lines. To keep resistance from getting too high, the cross-sectional area is maintained by making conductor lines thick and, as a consequence, the step height is made larger. The goal of planarization is to reduce step height through application of some sort of polymer coating or through mechanical polishing.

The capability of planarizing the surface topography of the underlying substrate is a major processing requirement for polymer thin-film coating. The degree of planarization (DOP) parameter was first introduced by Rothman [99] to characterize this processing property as is shown in Figure 12-52. The degree of planarization is defined as the fractional reduction in surface feature topology realized by the coating; that is,

$$\mathrm{DOP} = \left(1 - \frac{t_{\mathrm{s}}}{t_{\mathrm{Al}}}\right) \times 100 \qquad [12\text{-}11]$$

It is obvious that 100% planarization corresponds to a fully planarized surface, whereas 0% planarization is just the opposite (i.e. a **conformal coating** which serves as a 1 : 1 mapping of the underneath surface topography).

The degree of planarization (DOP) of a polymer coating depends on many factors ranging from material properties of the polymer, to coating technology, to the bake and cure scheme, to the geometry of the underlying features. Solid content and viscosity of the polymer solution are the most important material properties affecting the ability of planarization. High solid content and low viscosity enhance planarization because of

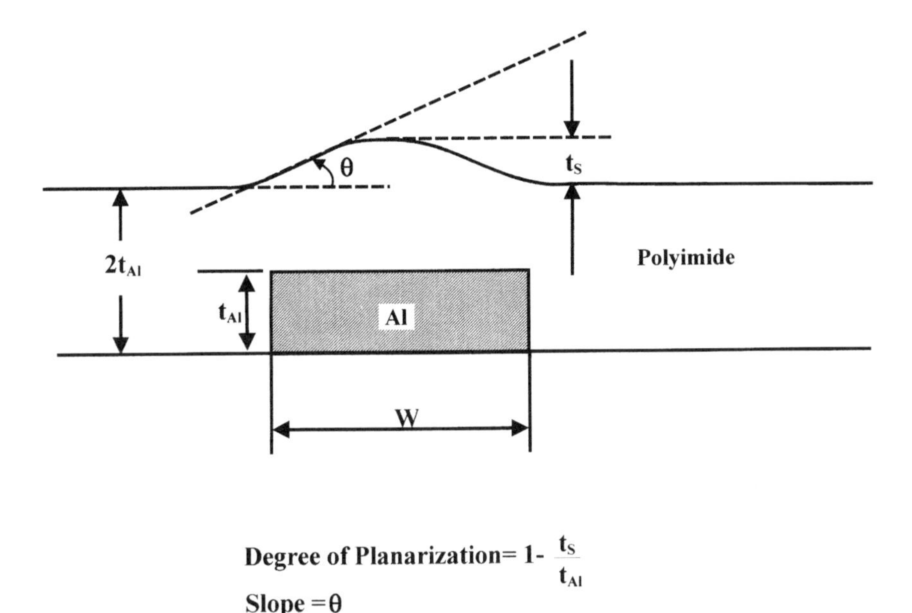

Degree of Planarization= 1- $\frac{t_S}{t_{Al}}$

Slope =θ

Figure 12-52. Degree of Planarization

the low shrinkage and ability to flow [100]. Unfortunately, these two properties often function in contrary to each other (e.g., high solid content increases viscosity). There is very limited information on the effect of coating method on planarization. Studies on the degree of planarization of spin-coated films [101] have shown that high spin speed and solvent evaporation during spinning deteriorate the ability of planarization. Planarization during spinning is dependent on the balance of the capillary, viscous, centrifugal, and gravitational forces. Capillary and gravitational forces tend to planarize the film, whereas centrifugal and viscous forces try to make the film conformal. In addition to these conformal forces, other problems are associated with spinning, particularly as feature sizes and planarization requirements advance. First, film uniformity is not constant across the surface of a spun substrate due to evaporation and turbulence in the air above the substrate. Not only is thickness radially dependent but it varies as the metal line orientation with respect to the radius changes. Material tends to build up where lines are perpendicular to the flow. Compounding this with the effects of shrinkage may make spin coating cease to be a viable method if the feature sizes get too small. The effects of bake, exposure, and curing on planarization are quite complicated and material dependent. There is no generalized processing guidance to be follow. The loss of planarization is a composite of two processes: drying and curing [22]. During baking and curing, the solvent is driven off and the polymer is cross-linked. At some point, there is a transition from liquid to solid, from able to flow to unable to flow. The transition occurs quite quickly and a critical point is associated with the loss of flow [102]. When the polymer is liquid, surface tension tends to keep the top surface flat. After the critical point, the solvent continues to be driven off and the thin-film coating shrinks to a certain percentage of its original height. Because the layer's height varies in the presence or absence of the underlying topography, the total amount of height reduction varies. The effect of shrinkage on planarization is best illustrated in Figure 12-53 by comparing the initial coating surface to the final surface after curing. In principle, a better planarization is obtained when the flow of the coated polymer is facilitated, the time duration of flow is prolonged, and the film shrinkage after the critical point when the solution is unable to flow is minimized. The dependence of degree of planarization of a polymer coating on the surface topography of the underlying substrate fall into three forms. First, it is straightforward to evince that better planarization is obtained over substrates with smaller step heights. Second, planarization is poorer over large features than that over smaller features [101]. In other words, conformal coating is obtained over large features, whereas at least partially planarized coating is obtained over small features. The third substrate topography affecting the planarization is the distribution of the features. Isolated features are easier to planarize than densely populated features.

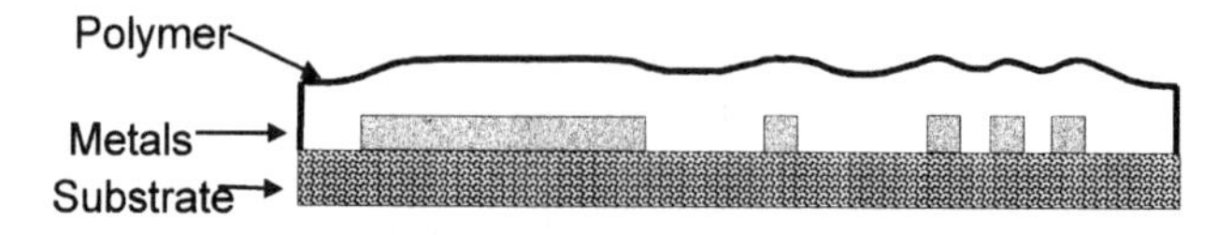

(a) Good planarization of polymer thin film over surface topography after soft bake.

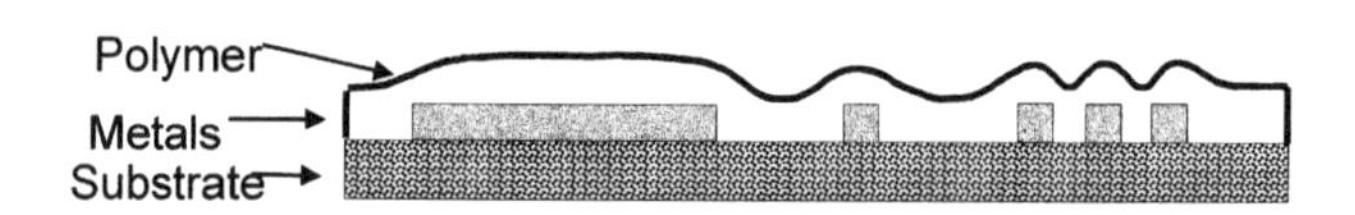

(b) Poor planarization after cure due to film shrinkage.

Figure 12-53. The Effect of Shrinkage on Planarization of Polymer Thin Films

This difference is a result of the fact that there is a large reservation of polymer fluid around the isolated features which flows to or from the feature. Such an ability of flow is necessary for a good planarization, as discussed above.

Mechanical polishing was employed by IBM in the processing of planar multilevel copper-polyimide thin-film substrate for its ES 9000 processors as described elsewhere in this section and Section 12.7 [24,30]. This planar structure has the advantages of superior impedance control and short current path and thereby reduced voltage drop as a result of its stacked stud signal and power via structure. In addition, stacked vias take less realty in the multilevel thin-film structure and enhance wiring density. Two variations of stacked stud process were developed within IBM: the dual-level metal (DLM) process [1] where excessive metal is removed by mechanical polishing for planarization and another process developed at a later time where excessive polyimide is removed by mechanical polishing for planarization. Figure 12-54 [30] is an illustration of the process flow of these two technologies.

There are also other ways to achieve a planarized structure in multilayer interconnection as discussed elsewhere in this chapter. The feasibility of fully planarized structure by fully additive electroless plating of copper was demonstrated by Hitachi [103]. Oki Electric Industry Co., Ltd has developed a plasma etching technology for planarization [104]. In this later process, a sacrifice or erodible photoresist layer was coated on to the polyimide layer to serve as a planarization layer and was then plasma etched away along with the excessive polyimide. The plasma was optimized so that the photoresist and the polyimide are etched at the same rate.

12.4.3.4 Relative Cost Comparison of Thin-Film Processes

The prohibitively high cost of thin-film packaging has been the major impedance for the widespread application of thin-film package in the low-end, cost-sensitive consumer electronics market. It is widely acknowledged that the high cost of thin-film packages must be reduced significantly for this technology to compete effectively in the future market. The goal is to reduce the cost of thin-film MCM substrates from the current level of \$50–\$100/in.^{2} to around \$5/in.^{2}, as is indicated in Table 12-16. Great effort has been devoted and is being proposed to the research and development of inherently low-cost thin-film materials and processes. Cost analysis and modeling are important and indispensable parts of this effort.

A number of cost models have been developed and applied to determine the intrinsically lowest-cost process. The models are typically based on unit operation analysis of the capital, material, utilities, and manpower requirements to deposit dielectric, form vias, metallize vias, and pattern metal conductor wiring. Two rival processing approaches, one based on

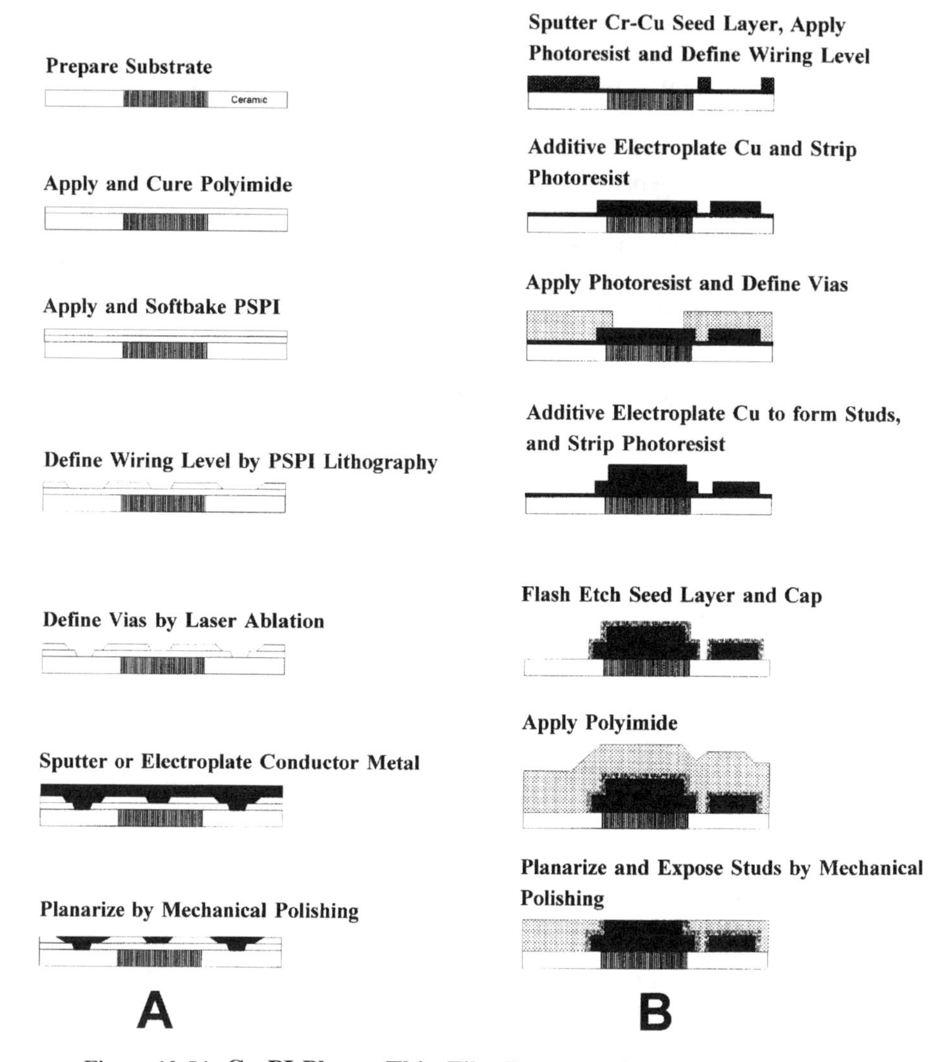

Figure 12-54. Cu-PI Planar Thin-Film Processes for High-End Systems

the 6-in. wafer conventional IC technology and the other based on large-area processing (LAP) technology of flat panel display (FPD), are often compared to demonstrate the impact of large-area processing on the overall cost. Of these cost models, the one by Dow Chemicals [105] assumes a base case of the LAP process with a spin coated dielectric polymer of 12 μm (cured) thickness over a 18 in. × 24 in. glass substrate, laser-etched vias through an erodible photoresist mask, metallization of sputtered bottom and top Cr barrier layers, and sputtered Cu seed followed by electroplated Cu of a total of 4 μm conductor thickness with subtractive

Table 12-16. Comparison of IC and MCM Manufacturing Requirements

	Current IC	Current MCM	High-Volume MCM Goal
Wafer size	6-in.	6-in.	15-in.
Unit size	0.5-in.	1″ × 1″ to 2″ × 2″	3″ × 3″
Number of chips	N/A	4–20 typical	10–50
Linewidth	<1.0 μm	10–20 μm	10–20 μm
Line thinckness	<1.0 μm	5–10 μm	10–20 μm
Photoresist thickness	1 μm	1–5 μm	5–10 μm
Photolithography	Projection/reduction	Proximity/1:1 projection	Projection/enlarge
Die placement	N/A	50 μm typical	<5 μm
Testing (# nodes)	300	2,000–4,000	>10,000
Bonding	Wire	Wire	Area interconnect
Production throughput	Millions/month	500/month	> 50,000/month
Test time (nodes/s)	N/A	1	>100
Lines (in./in.^{2})	20,000 typical	500	2,000
Performance	N/A	80 Mhz	250–400 Mhz
Time to market	50–75 weeks	20 weeks	3 weeks
Yield	1–90%	50%	>98%
Cost/in.^{2}	\$25–5,000	\$50–\$100	\$5

etching for patterning. This base case involved a total of 43 steps, as indicated in Table 12-17. The 6 in. wafer technology and two optimized LAP processes are also listed in this table for the purpose of comparison and process sensitivity analysis of the overall cost.

The model suggests a cost reduction by a factor of more than 5 with the implementation of large-area processing (the base case). A further 50–60% cost reduction could be achieved with the optimized material selection and process enhancement. Such an optimized process is featured by laser ablation for via generation using a holographic mask or, alternatively, a photosensitive dielectric polymer, barrierless Cu layers, and high-usage, high-film retention of dielectric polymer. Figure 12-55 [105] summarizes the cost comparison between the base and optimized processes. As illustrated in Figure 12-55, the development of very low-cost dielectric polymers (about \$50/kg) is believed to be crucial for further cost reduction. Further discussions on the development of low-cost dielectric polymers are found in Section 12.10.2.

Aside from going to large-area fabrication, it was also suggested that the optimization of 6-in. wafer processing technology alone could bring down the cost of MCM-D substrates by as much as 70% [49].

Similar evaluations have also been performed by Amoco, Motorola, IBM, and IBIS associates [67,75,106,107]. All of these studies favored

Table 12-17. Processes in MCM-D Substrate Fabrication

	6-in. Base	LAP Base	Sputtered	Sputtered	Sputtered	Holographic	Base Photo	Optimum
Substrate clean	UV/O_3	UV/O_3	UV/O_3	UV/O_3	UV/O_3	UV/O_3	UV/O_3	UV/O_3
Rinse	SRD	SRD	SRD	SRD	SRD	SRD	SRD	SRD
Dehydration	Oven 1	Oven 1	Oven 1	Oven 1	Oven 1	Oven 1	Oven 1	Oven 1
Sputter Cr	Metallizer	Metallizer	Metallizer	Metallizer		Metallizer	Metallizer	
Sputter Cu	Metallizer	Metallizer 0.1 μ	Metallizer 4μ	Metallizer	Metallizer	Metallizer	Metallizer	
Apply PR	Track	Coater	Coater	Coater	Coater	Coater	Coater	Coater
Bake	Oven 1	Oven 1		Oven 1	Oven 1	Oven 1	Oven 1	Oven 1
Lithography	Lithography	Lithography		Lithography	Lithography	Lithography	Lithography	Lithography
Develop	Wet	Wet		Wet	Wet	Wet	Wet	Wet
Rinse	SRD	SRD		SRD	SRD	SRD	SRD	SRD
Flood expose	Flood	Flood		Flood	Flood	Flood	Flood	Flood
Bake	Oven 1	Oven 1		Oven 1	Oven 1	Oven 1	Oven 1	Oven 1
Rinse	SRD	SRD		SRD	SRD	SRD	SRD	SRD
Plate Cu	Plate	Plate		Plate	Plate	Plate	Plate	Plate
Rinse	SRD	SRD		SRD	SRD	SRD	SRD	SRD
Strip PR	Wet	Wet		Wet	Wet	Wet	Wet	Wet
Rinse	SRD	SRD		SRD	SRD	SRD	SRD	SRD
Etch Cu	Wet	Wet		Wet	Wet	Wet	Wet	Wet
Rinse	SRD	SRD		SRD	SRD	SRD	SRD	SRD
Dehydration				Oven 1	Oven 1			
Sputter Cr			Metallizer	Metallizer				
Apply PR			Coater	Coater				
Bake			Oven 1	Oven 1				
Lithography			Lithography	Lithography				
Electroless Ni	Plate	Plate				Plate	Plate	
Develop			Wet	Wet				
Rinse	SRD	SRD	SRD	SRD		SRD	SRD	
Flood expose			Flood	Flood				

Bake			Oven 1	Oven 1				
Etch Cr	Wet	Wet	Wet	Wet		Wet	Wet	
Rinse	SRD	SRD	SRD	SRD		SRD	SRD	
Etch Cu			Wet					
Rinse			SRD					
Etch Cr			Wet					
Rinse			SRD					
Strip PR			Wet	Wet				
Rinse			SRD	SRD				
Dehydration	Oven 1	Oven 1	Oven 1	Oven 1		Oven 1	Oven 1	
Plasma clean	UV/O_3	UV/O_3	UV/O_3	UV/O_3	UV/O_3	UV/O_3	UV/O_3	UV/O_3
Rinse	SRD	SRD	SRD	SRD	SRD	SRD	SRD	SRD
Dehydration	Oven 1	Oven 1	Oven 1	Oven 1	Oven 1	Oven 1	Oven 1	
Apply dielectric	Track	Coater	Coater	Coater	Coater	Coater	Coater	Coater
Cure dielectric	Oven 2	Oven 2	Oven 2	Oven 2	Oven 2	Oven 2	Oven 2	Oven 2
Apply PR	Track	Coater	Coater	Coater	Coater			
Bake	Oven 1	Oven 1	Oven 1	Oven 1	Oven 1			
Lithography	Lithography	Lithography	Lithography	Lithography	Lithography		Lithography	Lithography
Develop	Wet	Wet	Wet	Wet	Wet		Wet	Wet
Rinse	SRD	SRD	SRD	SRD	SRD		SRD	SRD
Flood expose	Flood	Flood	Flood	Flood	Flood			
Bake	Oven 1	Oven 1	Oven 1	Oven 1	Oven 1			
Rinse	SRD	SRD	SRD	SRD	SRD			
Laser etch	Laser etch	Laser etch	Laser etch	Laser etch	Laser etch	Laser etch		
Strip PR	Wet	Wet	Wet	Wet	Wet	Wet		
Rinse	SRD	SRD	SRD	SRD	SRD	SRD		
Descum	Descum	Descum	Descum	Descum	Descum	Descum	Descum	Descum
Rinse	SRD	SRD	SRD	SRD	SRD	SRD	SRD	SRD
Dehydration	Oven 1	Oven 1	Oven 1	Oven 1	Oven 1	Oven 1	Oven 1	

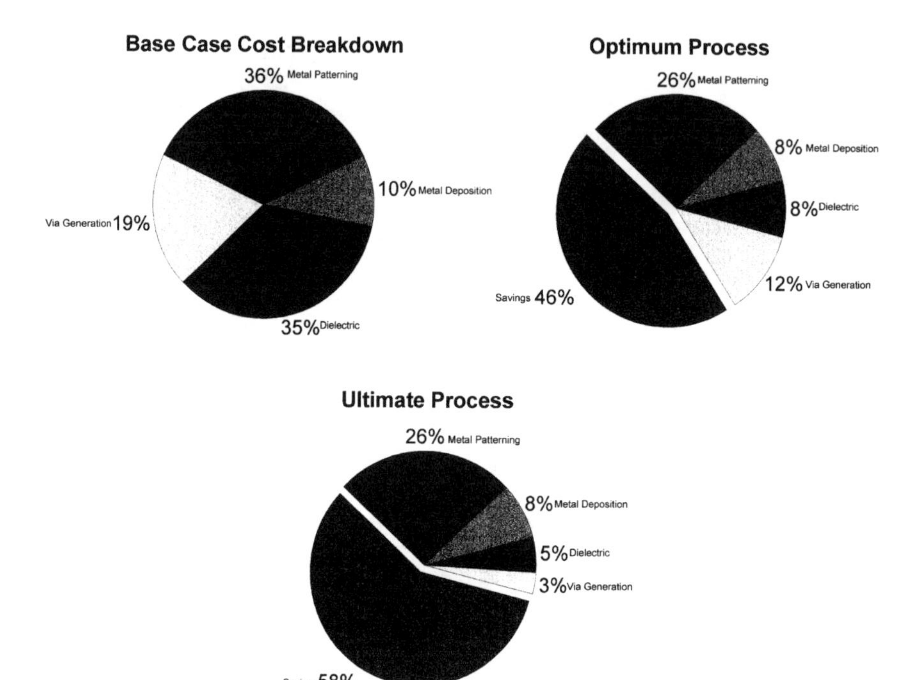

Figure 12-55. Cost Breakdown of the Base Case and the Optimized Cases in Table 12-17

a migration from the conventional IC technology to the LAP process technology of flat panel display. However, the cost issue in thin-film packaging is very complex. Many factors are beyond the capability of these cost models, notably, the availability of equipment, yield associated with a specific process technique, process robustness, and the ground rules, to name a few. Practitioners of thin-film packaging are strongly advised to conduct cost evaluation according to their own established process.

12.5 YIELD/COST CONSIDERATIONS

12.5.1 Repair of Thin-Film Packaging

One of the key challenges in thin-film package fabrication is the ability to create defect-free wiring over the entire surface area of a large substrate. Achieving practical yields with reasonable complexity may require development of novel defect detection and repair techniques for each layer in a thin-film structure. This challenge has been identified [14] as one of the principal obstacles in fabrication of sophisticated multilayer thin-film packaging structures.

The best alternative to repair, however, is eliminating the need for it, and it is more appropriate to focus on controlling the fabrication processes and preventing defects than to expend resources in locating and repairing them. This approach, however, minimizes the realities of the manufacturing environments. Thin-film fabrication processes do not always reach the level of control that eliminates the need for repair, and even in the most well-controlled processes, the impact of defects can be significant.

12.5.2 Leverage of Repair Technologies

There are a number of benefits in having the ability to repair defective conductors or restructure the conductor pattern in a thin-film package. A circuit may need to be rerouted or optimized for design or performance reasons. Yield issues, typically caused by defects occurring during the fabrication process, can be resolved with the addition of new sections of wiring or with the disconnection of bad wiring sections. Typically, thin-film defects are either in the form of unwanted metallization that create shorts or near-shorts or missing metallization that produces opens or near-opens.

Figure 12-56 shows the potential benefit to having repair technologies for yield enhancement. Assuming that the defect distribution is Poisson and that all the defects are repairable, the plot shows the yield gain that

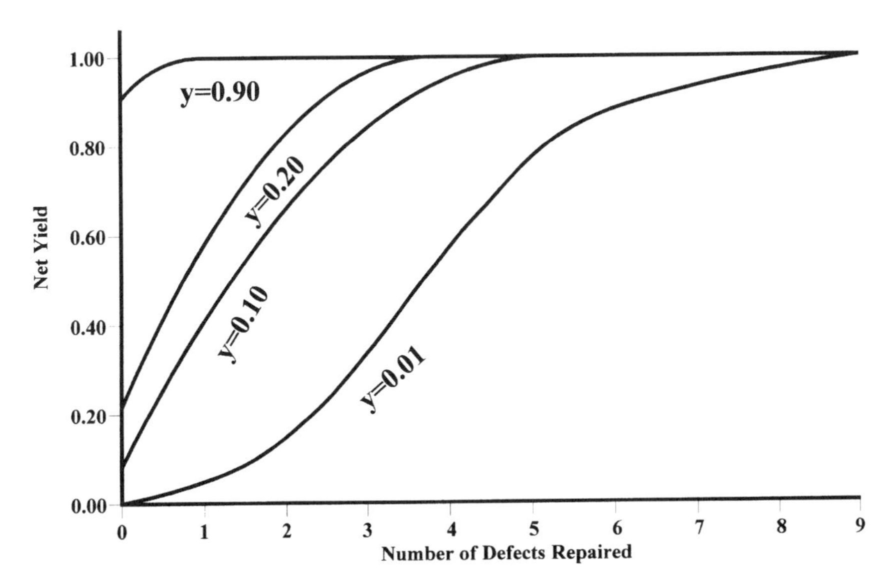

Figure 12-56. Plot Showing Potential Repair Leverage. Note net yield gain by repairing small numbers of opens and/or shorts.

can result by repairing a small number of defects per substrate. The gain is dramatic when the initial yield is low—repairing all parts with less than three defects can raise the effective yield from 10% to almost 80%! As the process matures and the initial yield (those parts with zero defects) increases, the leverage for repair decreases, as shown by the curve showing an initial yield of 90%.

12.5.3 Repair Methodologies

Thin-film packaging structures should be designed and fabricated with repair in mind. For example, in the design of the first-level packages used in the IBM ES 9000 system family, the nonrepairable signal redistribution wiring normally buried in the multilayer ceramic has been replaced with a repairable, top-surface thin-film wiring layer [108].

The choice of thin-film fabrication process can also dramatically affect the nature of defects created, and biasing the process toward a particular defect mode may be desired. For example, an additive process such as lift-off creates many more opens than shorts. By understanding the defect sensitivity of the fabrication technology being considered, it is possible to restrict the defect set to those that are more easily repaired. Because it is generally easier to delete or remove material than to add it, shorts are "more desirable" than opens. With the appropriate repair technologies, any defects created during fabrication can be easily fixed.

Techniques to repair defective conductors are strongly dependent on the size of the features and the materials used. Components built with thin-film technologies, which include almost all semiconductor and high-performance packaging devices and where conductor dimensions are normally less than 25 μm, require a much different set of repair technologies than components built with "thick-film" technologies, where circuit dimensions typically exceed 100 μm.

Repairs of larger-dimension circuit elements, like those used on printed-circuit boards or multilayer ceramic substrates, can use more traditional, mechanically oriented procedures. There are a number of methods available to repair regions of missing or damaged conductors in thick-film circuits [109].

The demands of thin-film dimensions limit the techniques available for repair of defective conductor patterns, and it is common to find lasers and other directed energy sources playing an important role in restructuring and repair of microcircuits. Localized deposition of conductive films and local removal of thin-film material using these sources are now commonplace, especially for repair purposes, and it is these techniques that are the focus of this section.

One obvious drawback of using a surgical approach for defect repair is the necessity of locating and repairing each defect in sequence. One alternative to this serial approach, which relies on redundant or parallel processes, is discussed later. Another approach is found with semicustom MCMs [110], which utilize programmable interconnections. In these schemes, rather than replacing defective nets, defects are routed around as part of the customization, as all wiring resources can be pretested and the defective segments mapped and identified.

12.5.3.1 Defect Detection and Repair

Even in the most well-controlled environments, some defects can be encountered. This is especially true in the earliest stages of hardware build and debug, where rapid turnaround and asset utilization are critical. Besides establishing a strong focus on contamination control to eliminate defects at their source, a system for defect detection and repair should be considered as an effective tool for yield management. A typical defect detection and repair strategy relies on automatic inspection techniques and electrical testing methodologies. Using these approaches, defective nets can be readily identified and the defects pinpointed.

Inspection and test are complementary—focusing on slightly different aspects of product performance. Inspection usually addresses physical variations that could lead to reliability or performance problems and are often associated with a simple visual inspection. As feature ground rules shrink and circuit density increases, machine vision systems play an increasingly important role in the inspection area. Inspections, however, can add significant cost and processing time and, when done manually, are labor intensive as well.

Electrical testing confirms that the assembly is electrically functional, either by simulating the actual operating environment seen by the package, or by checking the performance of each component or circuit path on an individual basis.

Under normal circumstances, neither inspection nor test can identify all the defects that are likely to occur. Many electrical defects are not represented by obvious physical anomalies and so can escape detection. Conversely, many reliability problems do not manifest themselves as immediate electrical problems but are readily apparent when visually inspected. Finding the correct balance between defect detection schemes and product reliability can be a key driver in the success of a manufacturing operation.

IBM [111] has established a representative defect detection and repair system for the thin-film packaging structures used in the System/390 line of computing systems. Using information provided by electrical test and automated inspection tools, defective wiring patterns are quickly identified and the physical coordinate and type of each defect established. This information is transferred electronically to the repair tools, where individual repair operations are performed. The process flow is shown schematically in Figure 12-57.

To take full advantage of the defect detection system, reliable repair techniques for the repair of both opens and shorts are required.

In general, defects created in the fabrication of multilayer thin-film structures can be classified into two major types intralevel and interlevel with both opens and shorts possible. Repair actions are typically established only for intralevel defects, although the repair of interlevel defects is possible in some circumstances [112,113].

12.5.3.2 Repair of Shorts

The use of laser cutting for removal of metal films has become a very important processing operation in the production of electronic

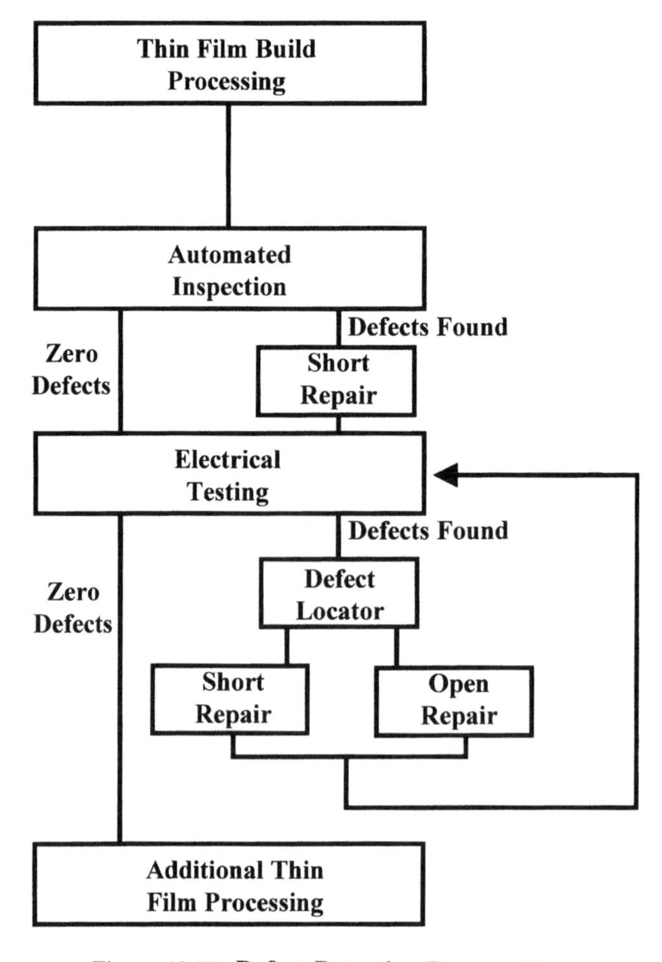

Figure 12-57. Defect Detection Process Flow

components [114], and this technology has been extended to the repair of extraneous metal defects in thin-film circuitry [108,115]. Many conductors used in thin-film packaging possess vaporization temperatures higher than the melting temperatures of the dielectric substrate. This is especially true for polymers, which are finding increasing utilization as the **interlayer dielectric**. Selection of the optimum processing conditions for repair requires an understanding of the practical considerations controlling the laser removal process [116].

To guarantee clean metal removal, it is desirable to use laser spots larger than the features being ablated, and so the dielectric is irradiated by some portion of the incoming beam. As most polymers are only weakly absorbing in the visible and infrared, transmitted laser radiation can create catastrophic damage in thin-film structures when it is absorbed by subsurface structures [111]. Additional damage can be created by direct absorption of the laser radiation by the dielectric or through the transfer of thermal energy from the irradiated region to the underlying substrate. To avoid thermal damage during laser metal removal, short-duration laser pulses such as those obtained from Q-switched solid-state or excimer lasers are commonly used, especially when the dielectric is thermally sensitive, as is the case for most polymers. Longer-duration pulses are usually only used with inorganic dielectrics, such as glasses and ceramics. Besides being stable to higher temperatures, these materials tend to have higher thermal diffusivities than polymers and can more readily dissipate the heat transferred during microsecond or millisecond laser pulses. Other critical laser parameters, such as wavelength and pulse energy, are chosen to minimize thermal or absorption damage to the irradiated areas while providing sufficient intensity to completely sever the conductor. For example, the ultraviolet radiation of excimer lasers is strongly absorbed in most microelectronically useful polymers. This interaction typically results in ablation of the polymer in the irradiated area and can produce loss of dielectric integrity unless carefully controlled.

Figure 12-58 shows optical micrographs of a residual metal defect in a thin-film package repaired using the excimer laser. The micrographs show both the original defect and the repaired site. Note the clean removal of the metal with minimal damage to the underlying polymer dielectric layer.

Typical laser pulse-width conditions used for clean metal removal from thin-film dielectrics are summarized in Table 12-18. Pulse energy and wavelengths are typically application specific.

As dimensions shrink, laser removal techniques become less viable, primarily because the beam focal spot size is limited by the laser wavelength used, achieving a practical limitation of about 0.5 μm. Although lasers remain practical for the dimensions found in today's thin-film packaging structures, the focused ion beam (FIB) has been introduced as an alternative method for material removal. The ion beam removes material

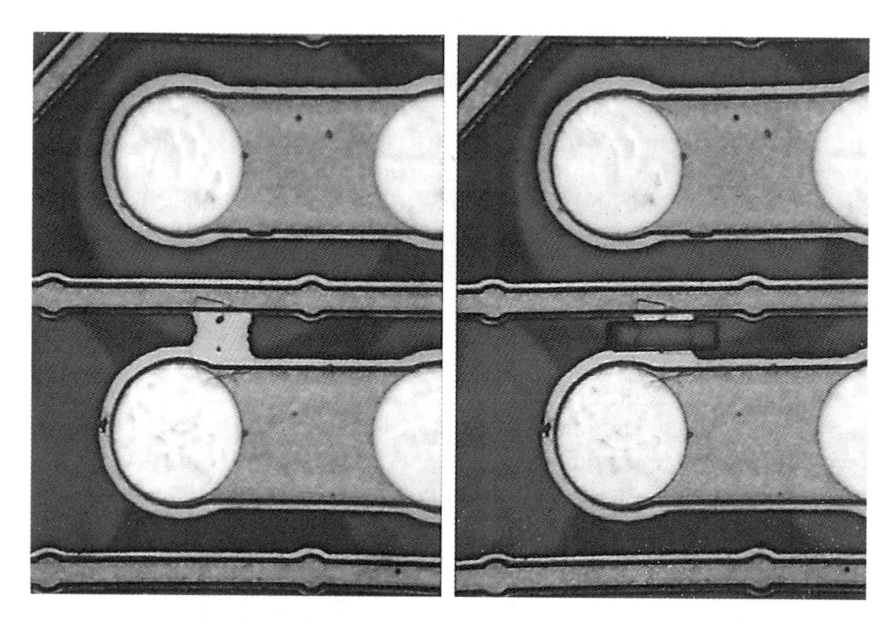

Figure 12-58. Excimer Laser Short Repair on Polyimide Surface

by sputtering, ejecting atoms and molecules from the surface into a vacuum. The FIB eliminates many problems associated with laser removal techniques and is becoming an increasingly common tool, especially for VLSI processing. The FIB has submicron focusing capability, with excellent material selectivity and depth control.

Although process intensive and therefore more costly, traditional lithographic techniques can also be used for material removal. Resist or polymer layers can be used to protect features during a subtractive process, such as wet or reactive ion etching, allowing unwanted material to be removed.

12.5.3.3 Repair of Opens

The method most discussed for repair of thin-film circuitry is laser chemical vapor deposition (LCVD), including deposition from both gas-

Table 12-18. Typical Laser Metal Removal Conditions for Thin Films

Dielectric	Laser Type	Pulse Width
Polymers, ceramics	Excimer, Nd-YAG	20–50 ns
Ceramics/glasses	Nd-YAG	μs–ms

phase and condensed-phase precursors [117–119]. Localized laser deposition may be accomplished photochemically, using the laser to induce photochemical reactions of deposition precursors, or thermally, using a laser or directed energy source to locally heat a substrate surface. In a typical gas-phase process, the substrate is immersed in a vapor environment containing organometallic molecules to be dissociated, and the system is exposed to the laser beam. The beam can activate the molecules in the gas phase or as an adsorbed layer through pyrolysis or photolysis, or, often, by a combination of the two.

Although photochemically stimulated deposition reactions may provide superior resolution, the deposited films are typically contaminated with carbon and have poor electrical properties. This method, therefore, has limited value in direct repair metallization schemes, and the pyrolytic process is almost exclusively used for repair activities.

The pyrolytic process is very analogous to conventional chemical vapor deposition (CVD), where the substrate is locally heated by the laser beam and the reactant atoms are liberated when the parent molecules decompose by collisional excitation with the surface. For sufficient heating to occur, the organometallic molecules must be sufficiently transparent and the substrate or some film over it must be an efficient absorber of the laser radiation. The pyrolytic process is shown schematically in Figure 12-59.

A wide range of precursors are available, permitting the laser deposition of useful microelectronic metals such as aluminum, copper, gold, platinum, tungsten, chromium, titanium, cobalt, nickel, lead, tin, and palladium. Successful film formation requires optimization of the nucleation and growth conditions within the constraints imposed by the thermal environment established by the substrate.

Successful repairs of defective conductors have been made using laser deposition of gold films from dimethyl-Au-trifluroacetylacetonate with an argon-ion laser operating at 514.5 nm on a number of thin-film packaging structures [111].

A typical repair uses parameters similar to those shown in Tables 12-19 and 12-20, depending on the dielectric involved. In Table 12-19, the initial scan conditions are used to induce deposition without damaging the thermally sensitive polymer dielectric. This is followed by additional scans at increasing laser powers to build the deposit to the desired thickness. Actual scanning conditions are determined by the defect type and structure being repaired.

An alternative method for achieving good-quality laser-deposited metal films on thermally sensitive dielectrics involves switching from a continuous-wave (CW) source to a high-repetition-rate pulsed laser. NEC has demonstrated [120] the ability to produce high-quality gold lines from dimethyl-Au-acetylacetonate and its trifluorinated analogue using this ap-

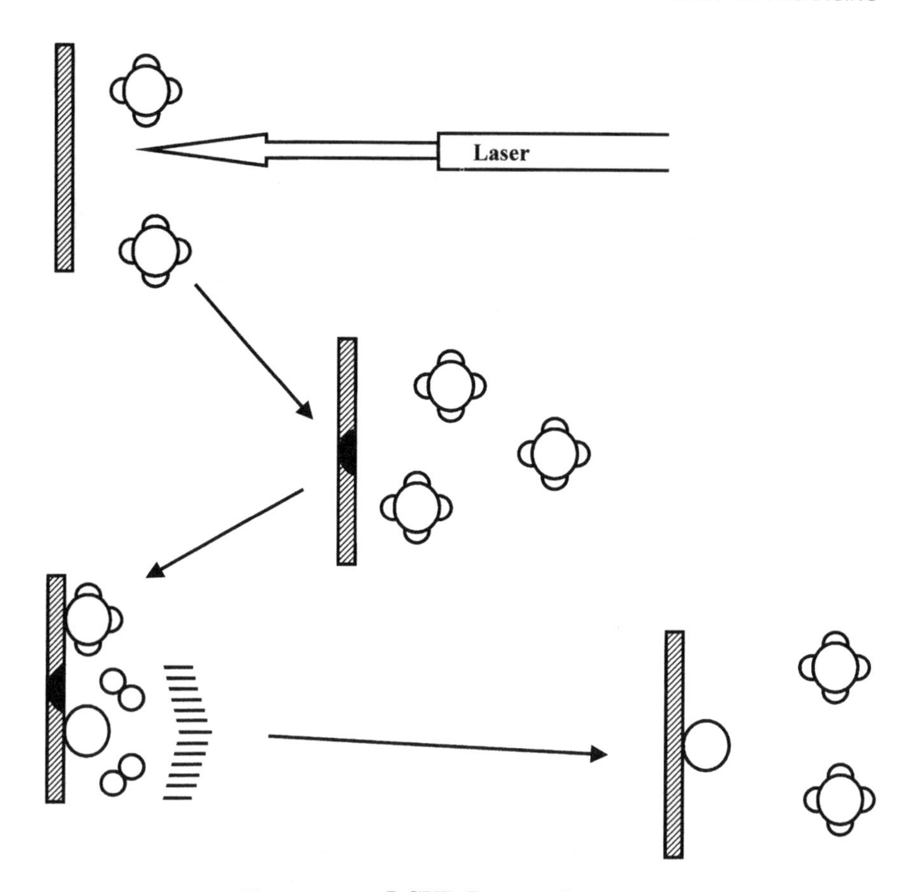

Figure 12-59. LCVD Process Schematic

proach. Using a rectangular-shaped, scanned beam from a high-repetition-rate, frequency-doubled Nd-YAG or copper vapor laser (pulse widths from 20 to 70 ns), chemically pure films up to 10 μm thick and with resistivities as low as 4.0 μΩ·cm have been formed on polyimide dielectrics, with negligible thermal damage to the polymer.

Table 12-19. Representative Laser Writing Conditions for Polymers

Pass	Power (mW)	Scan Rate(μ/s)	No. of Scans
1	5–8	2.0	2
2	40–50	2.0	2
3	150–175	10.0	4

Table 12-20. Representative Laser Writing Conditions for Ceramics

Pass	Power (mW)	Scan Rate (μm/s)	No. of Scans
1	350–1000	10–50	1–4

With inorganic substrates, the risk of thermal damage is minimal and higher-power conditions can be used to create the optimum repair thickness directly. Representative conditions for this type of repair is shown in Table 12-20. The resulting deposit has high chemical purity, low resistivity, and good intimate contact to both the substrate and the existing metal features. The films are typically granular, with some trapped porosity. Although the quality of the deposited film is high, it can be difficult to create a good electrical connection between the newly deposited metal and the existing circuit. For example, the metallurgy in the existing thin-film circuit may consist of multiple layers, where the top layer is a barrier metal that is oxidized, forming a protective, insulating layer. Before a good connection can be made, this top protective layer must be removed. One method of doing this is to use a pulsed laser to partially ablate the metallurgy at the point of interconnection prior to depositing the repair patch [121]. The partial ablation can be readily controlled by adjusting the fluence of the laser beam and by varying the number of pulses.

Figure 12-60 shows how two adjacent thin-film lines can be repaired by this process, using conditions especially developed for a typical thin film on a ceramic structure. Modification of the process conditions allows repairs to be made over a wider range of dielectrics, metallurgies, and line structures. Successful repairs have been made with the LCVD process on alumina, glass-ceramic, and a number of different polymer dielectrics.

Laser writing using solid thin films of metallo-organic precursors are an attractive alternative to gas-phase processes. A suitable precursor for laser writing must be capable of meeting a number of prerequisites. They must be able to form homogeneous films, have high metal content, have sufficient optical absorption, and decompose at relatively low temperatures with clean decomposition and volatilization of by-products.

A metallo-organic complex may meet these requirements, but most materials lack good film-formation characteristics. In those cases, the metal complex is combined with a compatible organic film former, typically a polymer. A wide variety of metals complexes, including Pd, Pt, Ag, Cu, and Au, are available and have been used to form metal films in a variety of circumstances [122].

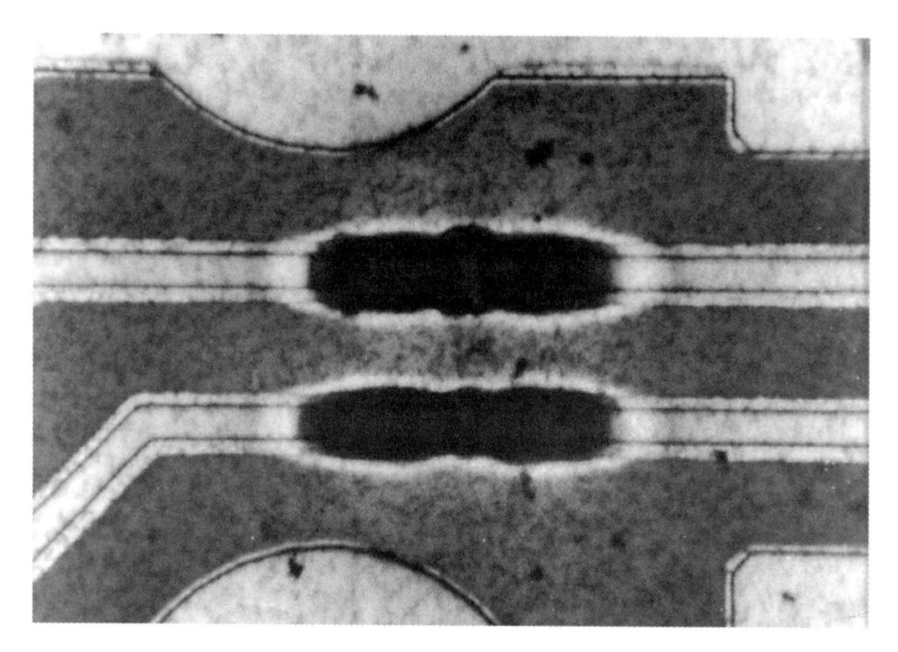

Figure 12-60. LCVD Repair of Open Line Defects

New technology approaches are also being established. Partially open segments of a conductor can be selectively heated by passing a current and inducing the deposition of an electrically conductive material [123]. Repairs made in this way are self-aligning and self-limiting, as the thermal driving force is reduced as the repair is built up. Partridge et al. [124] have expanded this technique for the repair of open conductors by creating a conductive connection across the open before initiating the repair and have designated the technique as "self-induced repair." Although typically done from a plating solution, the use of gas-phase pyrolysis is also possible.

Traditional bonding methods can also be extended to the smaller dimensions found in thin-film packages by utilizing wire or solder technologies, both permitting repair of defects found during the fabrication process and allowing the repair of defects found or created in higher levels of assembly, where the presence of active or passive components may restrict the technology options available [125]. Figure 12-61 shows a repair of a defective thin-film line made using a short segment of a 0.7-mil-diameter wire.

Open repair methods can also take advantage of standard photolithographic processing techniques. The area requiring repair is first covered with a layer of photoresist, followed by exposure of the defective area using an appropriate optical source and development of the resist [126].

Figure 12-61. Wire-Bond Repair of Open Line Defects

In a variation of this scheme, the area can be covered with a suitable polymer film and the opening created using direct UV laser ablation [127]. Metal patches are deposited by conventional evaporative or sputtering techniques. Standard lift-off techniques are then used to remove the undeveloped resist or polymer and the unwanted overlying metallization, leaving the repair metallurgy behind.

12.5.4 Alternative Repair Methodologies

Another alternative to repair is defect tolerance. Instead of repairing defective circuits, redundant or alternative wiring paths can be provided. Although eliminating the need to physically locate and repair the actual defects—the loss of functionality is a sufficient indicator—these methods require the utilization of critical real estate for the spare features.

Processes that reduce defect susceptibility can also be considered defect tolerant. Methods such as coating interlevel dielectrics in multiple coats or providing dual dielectrics are examples of this. Wiring can also be made vertically redundant, in that the effect of a defect in one layer can be eliminated by the presence of a second layer.

12.5.4.1 Redundancy

Redundant wiring structures consisting of two independently processed photolithographic layers have been used to reduce the defect sensitivity associated with fabrication processes. This vertical redundancy can effectively eliminate open defects while increasing the potential level of shorts [108,128]. Redundant or spare features can also provide an escape from the low yields inherent in building larger and more complicated devices. These approaches rely on having excess circuit components present that can be substituted for nonworking components by testing, linking, or opening lines, and retesting to obtain the desired overall functionality.

Although not in wide use for thin-film packaging, this has found wide acceptance in integrated-circuit manufacturing, especially in the fabrication of memory chips. Extra rows or column lines are built, which, after the initial test, can be substituted for nonworking rows or columns. This type of programmable restructuring, which normally involves laser cutting to disconnect circuit elements and laser linking to reconnect elements, can greatly enhance yield and fault tolerance [129].

12.5.4.2 Rerouting

A key feature in high-performance packaging design is the ability to repair the assembly and make engineering changes. Engineering change (EC) is a process in which the chips or modules can be removed from their carrier and replaced individually, or by which wiring can be rerouted. Specific techniques are required for deleting a connection from a signal terminal of a chip to the package wiring network and rerouting the terminal to other networks or chip terminals as desired. Engineering changes are carried out to replace defective signal lines and to make new connections to correct design errors. In manufacturing, such a scheme can be used to enhance yield [130]. Defective nets are isolated by laser delete and rerouted by the addition of top-surface discrete engineering change wires [98].

12.5.5 Performance and Reliability

Repair procedures and methods must be capable of ensuring that a repaired circuit is as reliable as it would be when perfect. Repair integrity is normally assured through the application of functionality and/or reliability testing.

Reliability stress tests are designed to simulate subsequent processing or field use conditions. These can include, but are not limited to the following:

- Thermal cycling test, simulating the high-temperature (360°–380°C) furnace cycles used to join pins and chips, as well as the conditions (0°–100°C) used to simulate machine power on-off cycles.
- Temperature and humidity testing, typically at 85°C, 85% relative humidity (**rh**), with and without electrical bias, to ensure no loss of dielectric integrity due to repair processing.

- **Electromigration testing** under accelerated conditions, ensuring that the new metallurgical structure created during open repair is reliable under long-term current flow conditions.

12.6 RELIABILITY

Reliability of thin-film packages is a major concern. Some polymers like polyimide dielectrics absorb significant amount of water, and unless proper baking precedes the package sealing, corrosion problems discussed in Chapter 5, "Package Reliability," are valid. In addition, the large thermal-expansion mismatches between many polymers (except several low-thermal-expansion polymers like PIQ L100 and L120) and ceramic substrates and the IC chips give rise to huge shear deformations at this interface. Even though short term tests may not expose any real problems, long-term usage at the elevated temperature of package operation in the presence of absorbed moisture may present reliability concerns. It is important, therefore, that the polymer selection include low-moisture absorption. Additionally, direct **passivation** of the polymer or sealing of the package after appropriate outgassing may be necessary. The major concern with moisture being corrosion of the copper conductor, other means such as barrier materials may be used as well. Most of these concerns may simply be due to lack of extensive experience with the use of these materials. Metal-related problems are similar to the reliability concerns of semiconductor thin films and printed-circuit boards, as discussed in Chapter 5, "Package Reliability."

Most of the thin-film structures discussed in this chapter should be submitted to a variety of reliability tests. Table 12-21 is a typical example

Table 12-21. Reliability Tests of Thin-Film Packages

Test	Measurement	Results
1. **High temperature storage** 150° C, 500 h	• Adhesion of metal to polyimide	• Some changes with some metals
2. **Thermal shock** –65°–150° C, 15 cycles	• Adhesion of metal to polyimide	• Some change with some metals
	• Resistance	• <5% change
3. **Temperature, humidity, bias** 85°C, 85% RH, 5V, 1000h	• Resistance	• <5% change
	• Insulation Resistance	• no change
4. **Pressure and temperature (PCT)** 2 Atm., 120° C, 10h	• Adhesion	
	—Metal to polyimide	• No change
	—Polyimide-to-polyimide	• Some change
	—Polyimide-to-ceramic	• Some change

of the reliability tests performed on thin-film packages, which summarized the testing conditions and general observations.

12.7 COMMERCIAL APPLICATIONS OF THIN-FILM PACKAGING

12.7.1 IBM

In this section, multilevel thin-film technologies developed at IBM for use in computer systems, telecommunication products, avionics, and other high-performance applications are discussed. Thin-film wiring has been used at IBM in five different high-performance applications ranging from **chip carriers** for logic and memory devices to multichip modules for large systems. The key elements of multilevel thin-film structures and processes for high-performance multichip modules are (1) cofired-ceramic substrate, (2) conformal via structure, (3) copper conductors for improved conductivity, (4) polyimide dielectric for good mechanical properties, high thermal stability, and low dielectric constant, (5) laser ablation or photosensitive polyimide lithography for via definition, (6) additive electroplating for wiring definition, and (7) additive electroplating for terminal-metal definition. This combination of processes is extendible to finer dimensions. Thin-film wiring on cofired ceramic is capable of flip chip, TAB, and wirebond chip connections, as well as all industry-standard card-level connections.

IBM's technology has moved from **single-chip modules (SCM)**, which needed thin-film wiring for redistribution of chip I/Os, to multichip modules, with multilevel thin-films for signal wiring of highly integrated devices. The first product was a 36-mm-square pin-grid-array package. The next advancement in the technology was two levels of planar thin films for signal redistribution of 121 chips on a 127-mm glass-ceramic substrate for the ES 9000 processor. Planar processes were implemented to satisfy system requirements demanding tighter dimensional control. These devices were primarily bipolar. One plane pair of planar thin films was also implemented on a 166-mm glass-ceramic substrate for redistribution and signal wiring of bipolar devices. For large CMOS devices used in workstations, minicomputers, and servers, a nonplanar thin-film structure on either a silicon wafer or cofired ceramic was practiced. It is interesting to note that the size of the substrate grew from 36 mm square for a single-chip module to 166 mm square for the processor of mainframe computers using bipolar devices, and decreased again in size for computers based on highly integrated CMOS devices.

12.7.1.1 Thin-Film Packaging Evolution

Table 12-22 shows the evolution of multilevel thin-film technology from the small single-chip interconnection to the conformal via interconnection on large substrate.

Table 12-22. Evolution of Multilevel Thin-Film Technology at IBM

Key Features	Metallized Ceramic Polyimide	ES 9000 Processor Substrate	Al-PI Comformal Via Process	Cu-PI Stacked Stud Process	Cu-PI Conformal Via Process
Wiring	Chip I/O redistribution	Chip I/O redistribution	Signal distribution	Signal distribution	Signal distribution
Structure	Nonplanar	Planar and redundant	Nonplanar	Planar	Nonplanar
Size of substrate	36 mm square	127 mm square	125 mm square	166 mm square	127 mm square
Number of levels	2	2	5	7	5–6
Dielectric material	PI	PI	PI	PSPI/PI	PSPI/PI
Via definition	Wet etch	Laser ablation	RIE	Laser	PSPI lithography/ Laser ablation
Conductor metal	Cu	Cu	Al	Cu	Cu
Wiring definition	Subtractive etching	Lift-off	Subtractive etching	PSPI lithography and electroplating	Additive electroplating
Bonding pads	Subtractive etching	Lift-off	Evaporation through mask	Lift-off	Additive electroplating
Wiring pitch	75 μm	55 μm	25 μm	40 μm	>25 μm
Wiring width	25 μm	35 μm	8 μm	15 μm	>10 μm

IBM has developed and implemented multilayer thin films (MLTF) for both high-end and cost-performance systems since the early 1980s. Copper-polyimide and aluminum-polyimide multilayer thin films have been implemented on silicon, alumina, and glass-ceramic substrates. The various MLTF implementations are as follows:

- Two layers of Cu-polyimide interconnection on dry-pressed alumina for single- and dual-chip applications, using wet etch of polyimide and subtractive etch of Cr-Cu-Cr for wiring.
- One to two layers of planar Cu-polyimide interconnection on alumina and glass-ceramic multichip modules for redistribution of chip I/Os to the ceramic vias, using laser ablation for via definition, subtractive etching of Cr-Cu-Cr for wiring definition, and lift-off for bonding-pad definition.
- Four layers of A1-polyimide multichip interconnection on silicon substrates, using reactive ion etching for via definition, subtractive etching of aluminum for wiring definition, and evaporation through a mask for bonding-pad definition.
- Four to five layers of planar Cu-polyimide interconnection on glass-ceramic substrates, using laser ablation for via definition, photosensi-

tive polyimide (PSPI) lithography, and blanket electroplating for wiring definition, and lift-off for bonding-pad definition.

- Four or five layers of nonplanar Cu-polyimide interconnection on alumina, silicon, and glass-ceramic substrates, using laser ablation or photosensitive polyimide lithography for via definition, additive electroplating or subtractive etching of Cu for wiring definition, and additive electroplating for bonding-pad definition.

The evolution of these thin-film technologies, as well as the rationale behind the process choices and their relative merits, are discussed below.

12.7.1.2 Multilayer Thin-Film Implementation

Multilayer thin-film (MLTF) interconnection technology enhances the electrical performance of multichip modules [13,131] providing a variety of features such as

- Low-capacitance redistribution
- High-density wiring
- Ability to support high-I/O-density chips
- Improved propagation delay
- Increased off-chip switching activity

Although thin-film interconnections provide large performance improvements, not all applications require these improvements and hence a custom design is necessary for cost-effectiveness. An important element of this design is the choice between planar and nonplanar thin-film structures [94]. Nonplanar, unfilled via structures require staggering of vias in successive thin-film levels. Planar stud structures permit stacking of vias in successive levels, producing a higher wiring density than nonplanar structures for a given linewidth and pitch. The number of levels of thin films is dependent on the wiring and electrical requirements of the package. Thin films can be used for redistribution of chip I/Os, for a single level of wiring, or for multiple levels of wiring. An additional thin-film layer is utilized to define the bonding pads for chip connections. Depending on the electrical characteristics of the package, a number of thin-film power distribution levels are needed. In this section, the different processes and structures that were developed for different IBM systems over the last decade are described.

12.7.1.3 Single-Layer and Multilayer Single-Chip Packages

Metallized-ceramic-polyimide (MCP) is a pin-grid-array chip carrier. The MCP substrate is a 1.5-mm-thick perforated dry-pressed alumina ceramic wafer with 2.5-mm holes for pins. Two layers of Cr-Cu-Cr wiring isolated by a polyimide layer are used to interconnect the chip I/Os to the module pins. This is shown in Figure 9-8 in Chapter 9, "Ceramic Packaging."

The wiring and the dielectric layers are defined by thin-film technology [132]. The vias in the polyimide dielectric are defined by wet etching. The wiring levels and the bonding pads are defined by subtractive etching of Cr-Cu-Cr. The previous generation of this technology, known as **metallized ceramic (MC)**, had only one level of wiring defined by subtractive etching of Cr-Cu-Cr [133]. Figure 12-62 shows the process flow for MCP.

A **solder dam** (Cr) is used to prevent solder bleed from the bonding pads to the wiring traces during the flip-chip connection process. This is achieved by using an additional lithography step after the wiring traces are defined to selectively remove Cr from the pin and bonding-pad areas, and to leave it on the wiring traces. Copper is used as the bonding-pad metal for flip-chip connection and for the soldering of pins because very few rework and reflow cycles are required during flip-chip connection of

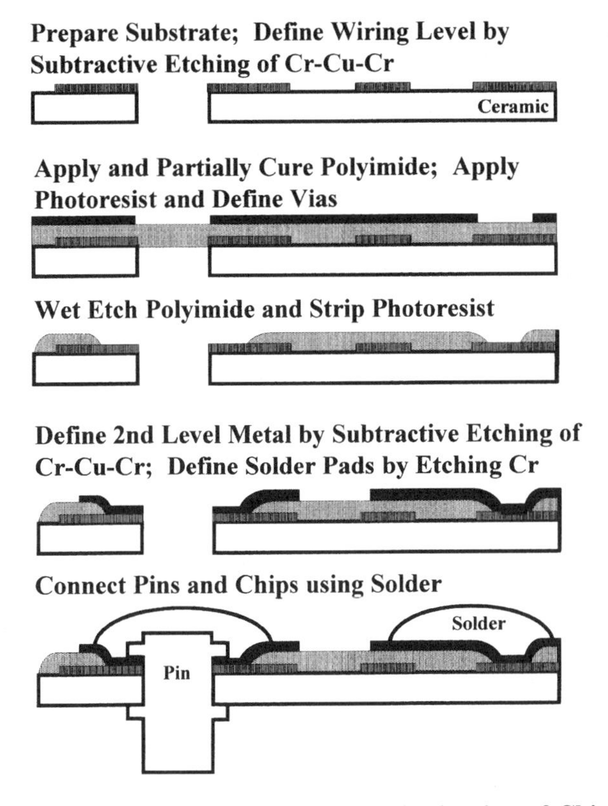

Figure 12-62. MCP Process Flow for Redistribution of Chip I/Os

single-chip modules. The I/O pins are wedged (force fit) and soldered to the wiring layer.

Metallized ceramic (MC) and MCP packages are generally used for packaging of high-performance logic and memory devices. The ability to fabricate two levels of wiring in MCP allows the use of the module as a multichip module for low-I/O chips. Alternatively, the first level of wiring could be used as a ground and/or power plane for high-performance devices.

12.7.1.4 Thin-Film Structure and Process for Mainframe Computers

The ES 9000 processor module was developed for mainframe systems. The ES 9000 processor module was the result of advances in both thin-film and ceramic technologies [108,134]. Some of the advancements included

- 127-mm multilayer glass-ceramic/copper cofired substrates
- Multilayer copper/polyimide thin-film structure
  —Projection laser ablation for via patterning
  —Mechanical polishing for planarization
  —Laser chemical vapor deposition for repairing thin-film opens
  —Laser ablation for repairing thin-film shorts
- Buried conductors for quick engineering changes

A planarized stud process is used to fabricate the high-density thin-film wiring structure. The stud structure is formed by using sputtered Cr-Cu as the conducting metal, and polyimide as the dielectric material. Mechanical polishing is used for planarization. Figure 12-63 shows the process flow used.

The wiring level is defined by subtractive etching of Cr-Cu-Cr and the bonding pads are defined by metal lift-off. Redundancy is incorporated into the wiring level to enhance yield by making use of the reactive ion etching (RIE) etch stop necessary to fabricate the bonding pads by metal lift-off. A polyimide passivation layer is applied on top of the wiring and bonding-pad levels. Openings for flip-chip connection and engineering changes are formed in this passivation level by laser ablation. Figure 12-64 shows the cross section of the thin-film structure.

The design parameters for the thin-film structure in the ES 9000 processor module are 0.065 pF/mm capacitance, 0.18Ω/mm resistance, and 30 mV maximum coupled noise [135].

12.7.1.4.1 Aluminum-Polyimide Multilevel Thin-Film on Silicon

Aluminum-polyimide (Al-PI) multilevel thin-film (MLTF) interconnections on silicon were developed to package high-performance CMOS

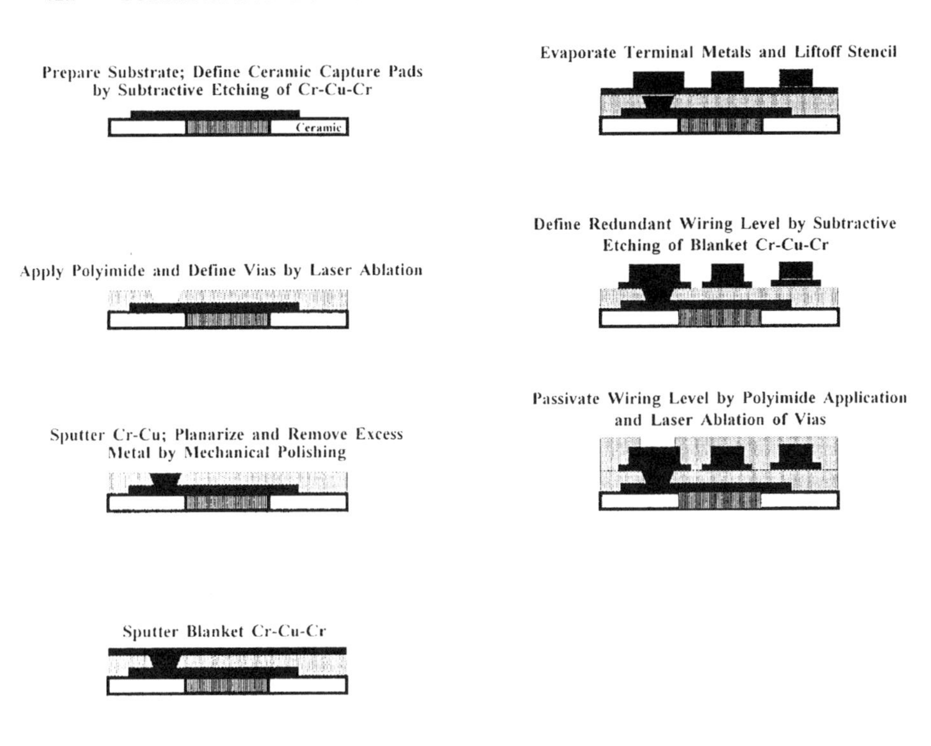

Figure 12-63. Planar Thin-Film Process Flow Used in the Fabrication of the ES 9000

chips in multichip modules [136]. The thin-film structure is defined by a conformal via process, with aluminum as the conductor metal and polyimide as the dielectric. This structure is shown in Figure 12-65.

The polyimide dielectric is patterned by reactive ion etching through a photoresist stencil, and the wiring level is patterned by subtractive etching of aluminum. The process flow is shown in Figure 12-66. Cu-Au flip-chip bonding pads are defined by evaporation through a mask.

The process and the materials were chosen to be compatible with the CMOS manufacturing line, and trade-offs were made between electrical properties and process constraints. Aluminum was used in spite of its high resistivity, and the use of thin dielectric layers led to a conformal dielectric coating and a highly nonplanar structure. The silicon substrate offers the following advantages: (1) It has the same TCE as the VLSI

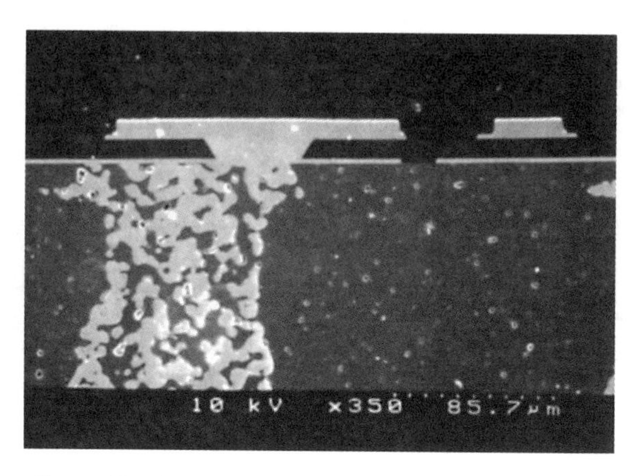

Figure 12-64. Cross Section of Copper/Polyimide Redistribution Layer on Glass-Ceramic Substrate

chips and (2) it is compatible with existing semiconductor tooling and processes. The dimensions of the smallest feature in the substrate were 8 μm wide and 3 μm thick on a 25-μm pitch. The 54-mm square substrate with nine chips contained about 40 m of wiring and more than 15,000 vias, and was fabricated two-up on a 5-in. wafer (two substrates on a wafer).

Figure 12-65. Al-PI MLTF on Silicon Substrate Prior to Dicing

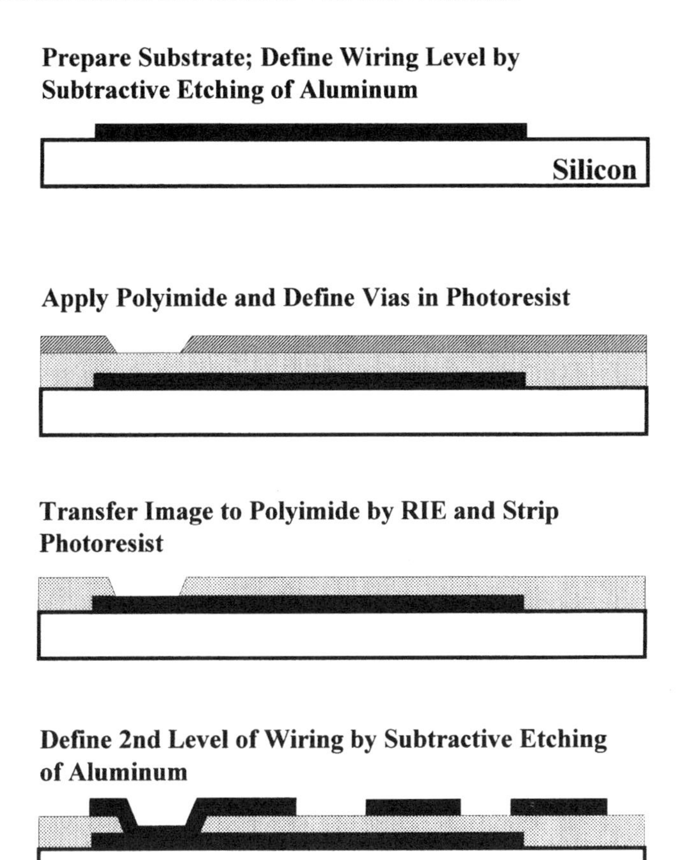

Figure 12-66. A1-PI MLTF Thin-Film Process Flow

12.7.1.4.2 Stacked Stud Copper-Polyimide

The planar stacked stud structure has the advantage of superior electrical characteristics compared to the conformal via structure with staggered vias. The stacked stud structure provides a shorter current path compared to the staggered via structure, thereby reducing the voltage drop in the power via stack.

Two different types of stacked stud processes have been used for high-performance processor applications. The first process is the so-called dual-level metal (DLM) process shown in Figure 12-54.

In this process, a layer of polyimide is first applied and cured, followed by a layer of photosensitive polyimide. The wiring traces are defined in the photosensitive polyimide layer by photolithography and then the vias are patterned in the underlying polyimide layer by projection laser ablation using a dielectric mask [32]. The composite structure is metallized by sputtering or blanket electroplating. Mechanical planarization is used to remove the excessive metal and to define the wiring features. A plane pair (two signal wiring layers and two ground/power layers) is completed by repeating the above process multiple times [25]. Figure 12-13 shows the cross section of the planar stacked stud multilevel thin films built by the DLM process on a glass-ceramic substrate.

The multilayer thin films were built on 166-mm square glass-ceramic substrate with a measured impedance of 39Ω, a line capacitance of 0.17 pF/mm, and a line delay of 6.8 ps/mm [137]. Because the structure is planarized at the wiring level and the polyimide dielectric is applied on a planar surface, precise control of the dielectric thickness is possible.

In the other type of planar thin-film process that was studied, the wiring and the stud layers are formed in separate lithography and plating steps [95]. The electroplated metal features are then overcoated with a single polyimide coating. Planarization of the polyimide coating removes the excessive polyimide over the metal topography and uncovers the tops of the studs. Mechanical polishing is used for planarization. The uniformity of polyimide removal during the planarization process determines the uniformity of the dielectric thickness. Figure 12-67 shows the cross section of one level of planar stacked studs built by this process.

12.7.1.4.3 Conformal Via Copper-Polyimide

Since the introduction of the ES 9000 processor substrate and the development of aluminum-polyimide multilayer thin films, several advancements have been made in the process and structure of multilevel thin films to reduce the cost of the process and tools and to improve the electrical performance of the interconnection [24]. Figure 12-14 shows the process flow for a conformal via process using polyimide as the dielectric and copper as the conductor metal.

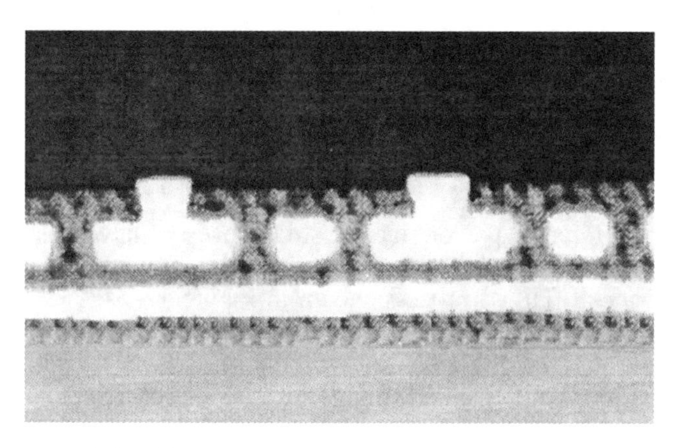

Figure 12-67. Thin-Film Layer Fabricated by Additive Electroplating

Multilevel thin-film structures have been fabricated by combining the photosensitive polyimide or laser ablation via process with the electroplated conductor line process. These two basic processes are repeated multiple times to construct one or more plane pairs of thin-film wiring as shown in Figure 12-15.

For high-performance workstations and minicomputers, the thin-film structure consists of two signal levels, two ground and/or power levels, and one bonding-pad level—a total of five thin-film metal layers, on a cofired-ceramic substrate. It is arranged in an offset **strip line** structure with the bonding pads on the top for interconnection to the semiconductor chips. Before fabricating the signal and power levels, an initial layer of large pads is needed to capture the ceramic vias. This layer is defined by subtractive etching of Cr-Cu-Cr. When it is electrically feasible, the first ground/power level is combined with the capture pad level. The thickness of the conductor lines and the dielectric is tailored to meet the impedance and resistance requirements of the product.

Photosensitive polyimide vias are patterned with a full-field projection printer. The wall profile of the vias is optimized for the metallization process which follows. The degree of planarization (Eq. 12-11) is 35% for a 5-μm-thick PSPI over 5-μm-thick conductor lines on a 25-μm pitch. The topography of the structure provides a processing challenge. By a judicious choice of materials, tools, and process conditions, a highly manufacturable process with very high yields was developed. The conductor lines are patterned in a thick photoresist (developed in house) in conjunction with projection printing. The linewidth control, even over topography, is much better than $\pm$ 10% (3σ) for a linewidth of 10 μm.

The vias for this nonplanar application have also been fabricated by projection laser ablation using a dielectric mask. The laser ablation process and the PSPI lithography process showed equivalent defect levels, and structures fabricated by both processes passed reliability stress tests. Also, yield was very high for both via processes. Figure 12-68 shows the top and bottom surfaces of a 54-mm multichip module.

The plane pair contains nine layers of thin films—five metal levels and four polyimide layers. The thin-film ground rules were as follows:

- Wiring pitch: 25 μm
- Linewidth: 13 μm
- Conductor thickness: 5 μm
- Via diameter: 10 μm
- Dielectric thickness: 5 μm

The substrate size was 54-mm square and was built four at a time on a 127-mm square ceramic. Each of the substrates were diced prior to

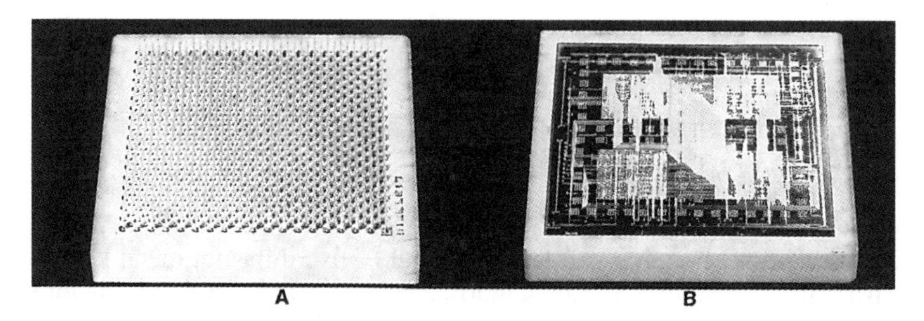

Figure 12-68. Cu-PI Interconnection on PGA Alumina Substrate (Top and Bottom Views)

test. The cycle time for thin-film build, dice, and test was 24 days. The thin-film technology is highly manufacturable and can be implemented on a large substrate to be used as a multichip module. Alternatively, the ceramic carrier can be diced into smaller substrates for single-chip module (SCM) or few-chip module (FCM) implementation. It is built four-up on a 127-mm square cofired alumina substrate using PSPI lithography for vias and additive electroplating for wiring. The top surface has bonding pads for flip-chip connection and the bottom surface has brazed pins for connection to the planar next-level packaging.

Extensive analysis and measurements verified the electrical characteristics of the thin-film package. The signal levels are always arranged in an offset strip-line structure. Various mesh designs were studied to determine their effect on the impedance of the thin-film package. The impedance was measured for a structure whose linewidth, line thickness, line pitch, and dielectric thickness were 13 μm, 5 μm, 25 μm, and 5 μm, respectively. The impedance is a function of the mesh design and the capacitive line loading and was measured between 35 and 45Ω, depending on the design. For a particular mesh design and line loading, the impedance was within 10% of the nominal value. Thin-film structures on a 75-μm pitch and > 12-μm dielectric thickness have also been fabricated for a > 50Ω impedance.

12.7.2 Siemens: Microwiring Thin-Film Technology for Mainframes

Microwiring technology has been developed for multichip module applications used in the Siemens Mainframe H90 or enhanced machines [138].

The multichip module connector has 1850 pins, 1100 of which are signal pins, 847 are power pins, and 2 are for the temperature sensors. Up to 5200 point-to-point connections with a total length of approximately

150 m can be realized as printed wiring on the microwiring substrate. The total wire length per substrate is 480 m, representing a line density of 480 cm/cm^2. The outer dimensions of the module are 115 mm × 115 mm × 26 mm.

12.7.2.1 Microwiring Technology

The 98-mm × 98-mm substrate is approximately 2.1 mm thick and is composed of up to 19 conductive layers. Due to the physical design of the module, the microwiring substrate has to meet certain mechanical specifications with respect to thickness tolerance, flatness, and elasticity, as well as compressive and flexural strength. The substrate consists of a complex of acrylic resin and polyimide layers prepared by a novel relamination process and a fiberglass-reinforced polyimide complex prepared by largely conventional processes [139].

The 17-layer complex comprises one IC connecting layer, two IC redistribution layers, six power layers, and eight signal layers. The reinforcing complex comprises a core with one connector connecting layer and one connector redistribution layer, as well as a wiring layer.

Two adjacent conductive layers are connected by a plated blind via. Layers that are farther apart are electrically connected by a corresponding number of such plated blind vias lined up in the z direction and staggered with respect to one another in the x and y directions because of the laser beam used for drilling the holes. Figure 12-69 shows schematically the module microwiring substrate.

The signal layers contain the printed wiring as well as pads and the via holes spaced 0.5 mm apart. In the channel between two grid lines, there are two copper conductors with a width of 0.08 mm and a thickness of approximately 0.03 mm. The power layers consist of a Cu layer–0.03 mm thick with a contact groove and a clearance. Figure 12-70 shows the design of the signal and power levels.

The electrical parameters of the printed wiring are characteristic impedance Z_0 = 53Ω, signal delay 6.7 ps/mm, and DC resistance 13 Ω·mm. The module heat sink is made of an aluminum-magnesium alloy; it is 9.5 mm thick, has a high-grade surface, and is anodized on all sides for electrical isolation. Its great stiffness guarantees the dimensional stability of the module in the presence of the strong internal forces of 1500 N.

12.7.2.2 Sequential Multilayer Thin-Film Process

Figure 12-71 is a schematic representation of the principal process steps for fabricating a microwiring substrate [140]. A 18-μm-thick Cu foil is laminated with the aid of an adhesive foil onto a previously tested and, if necessary, repaired, defect-free power or signal layer. This is

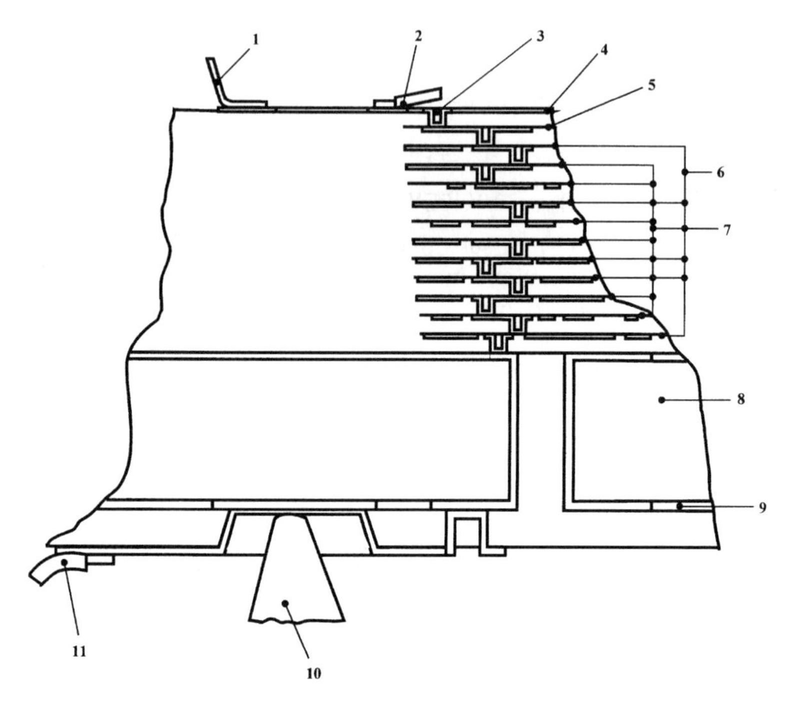

Figure 12-69. Multilayer Structure of a 17-Layer Microwiring Substrate: (1) Spider, (2) engineering change wire, (3) programmable via, (4) pad layer, (5) redistribution layer, (6) power layer, (7) signal layer, (8) reinforcing layer, (9) contact layer, (10) contact spring, (11) repair wire.

followed by the photoimaging and etching of the Cu foil with 80-μm via holes for the electrical interconnection of layers n and n + 1. After stripping the photoresist, a laser beam is used to remove the exposed insulating resin in the via holes down to layer n. A Cu layer is now deposited over the entire surface area, including the blind holes. The top and bottom Cu

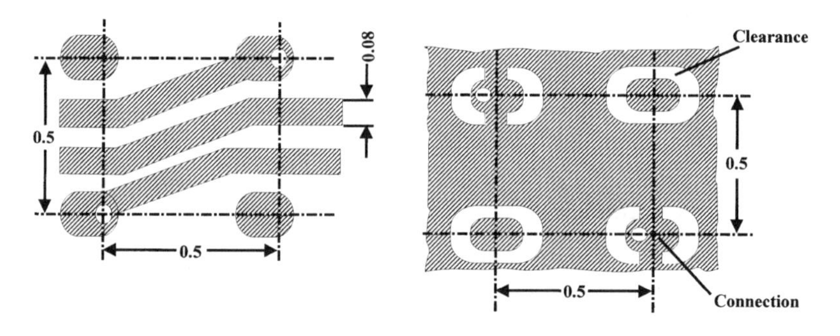

Figure 12-70. Design Layout. (a) Signal wiring; (b) power wiring.

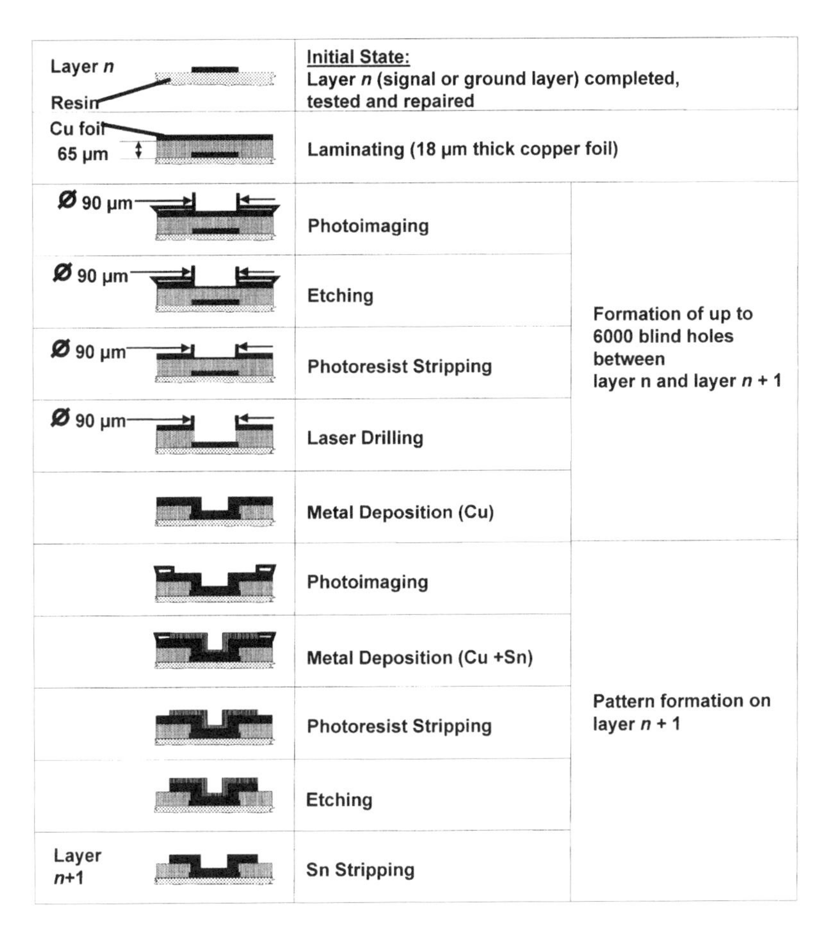

Figure 12-71. Process Steps for Fabrication of Microwiring Substrate

layers are, in this way, electrically interconnected. The interconnecting pattern is generated on the layer n + 1 by photoimaging, electroplating and etching steps. A temporarily deposited thin Sn layer serves as the etch resist. After optical and electrical testing the performance of any repairs that may be found necessary, the described process steps are then repeated for the next layer.

The same sequence of process steps—lamination, laser beam drilling, photoimaging, electroplating, and etching—are likewise used for the fabrication of the chip-mounting layer and the connector layer.

12.7.2.2.1 Reinforcement Substrate

The reinforcement substrate is a 0.6-mm-thick double-sided printed board of polyimide glass material. It has 1848 plated-through-holes of

0.3 mm diameter and nickel/gold plated pads on the connector side. All other layers are added to the reinforcement substrate by a relamination process. The reinforcement substrate has a form of 150 mm × 150 mm.

12.7.2.2.2 Signal, Power, and Redistribution Layers

The microwiring substrates are fabricated as single pieces in a working form of 150 mm × 150 mm. A completed, tested, and repaired layer is laminated with a 18-μm-thick copper foil. The bonding material is a glass-free polyimide layer of 25 μm thickness coated on both sides with an adhesive of 25 μm. Two different bonding materials have been qualified: acrylic adhesive and, more recently, epoxy adhesive.

Lamination is performed using hydraulic presses or, in the future, in an isostatic vacuum gas press which is already qualified. The lamination process and the bonding material meet the following requirements:

1. Void-free filling of the pattern, including the filling of blind holes.
2. The underlying interconnection pattern must not show up in the surface of the applied 18-μm-thick Cu foil.

The desired hole pattern for 80-μm blind holes is transferred by photoimaging from a chromium mask to the Cu layer by a contact exposure process. A liquid aqueous processable positive photoresist is applied by electrostatic spray. For the accurate alignment of mask and substrate, a technique has been developed with which it is possible to align the mask with the inner layer pattern with a tolerance of less than 10 μm.

After developing the photoresist and etching the Cu layer, the hole pattern generated in the Cu foil is checked for completeness and the observance of diameter tolerances by automated optical inspection using CAD data as a reference. Missing or deformed holes can, if necessary, be mechanically reworked.

The copper layer with the etched holes works as a mask during the subsequent laser beam drilling. The intensive UV radiation of a KrF excimer laser decomposes the unprotected resin by photoablation into gases which are exhausted. The laser beam hits an area of 15 mm^2 and travels over the board in a scanning mode.

Laser beam drilling stops at the lower copper layer and assures an absolutely clean base and nearly vertical hole walls.

A Cu-deposition step establishes a reliable electrically conducting connection between the upper and lower layers in the blind holes drilled by the laser beam. The Cu plating of these holes must exhibit high ductility and good adhesion at the base of the hole so that no damage will be caused by stresses that may build up at the elevated temperatures used for fusing and soldering.

Metal deposition starts in a colloidal palladium catalyst with ultrasonic agitation and is followed by an electroless copper bath in which about 0.5 μm of copper is built up. Copper thickness in the blind holes is increased by electroplating to about 10 μm.

For photoimaging of the conductor pattern, a solvent-type photofoil was initially used. Intensive work was done on the replacement of chlorinated hydrocarbon solvents, which has been prohibited by the European Community since 1991, with aqueous type of photofoils and some progress has been achieved.

Pattern plating of copper to a 20-μm thickness in the blind holes and 5 μm of tin follows.

After alkaline etching of the copper and stripping of the tin metal resist, an automatic optical inspection step and an electrical test on continuity and isolation finishes the layer build. For the electrical tests, the inner layers are mechanically contacted with bending needle adapters.

Line discontinuities which are found by optical and/or electrical tests can be repaired by an electrode-gap welding process. Formed elements with all existing line configurations are available. Shorts and insufficient clearances are eliminated by a special machine. An ultrasonic hard-metal stylus removes unwanted copper residues down to a minimum width of 50 μm.

12.7.2.2.3 Chip-Mounting Layer

The process and the base material for the chip-mounting layer are identical with the inner layers. Instead of tin as a metal resist, eutectic tin-lead is electroplated 8 μm thick. This tin-lead deposit is fused in a hot oil bath after etching the chip-mounting layer.

The further development of packaging systems [141] will be decisively influenced by the progress achieved in microwiring substrate technology. Beginning with the present size of the substrates of 100 mm × 100 mm, the substrate area will increase by a factor of 4 or more.

In initial phase, the built-up layers will still be on one side; layers on both sides will be implemented only in a further step. For this, a very intensive development of this technology in relation to the quality of the process and yield is necessary; new materials with a lower dielectric constant have to be evaluated.

To limit the number of layers, the conductor pattern has to become smaller. However, this can be done only to the extent permitted by the requirements for the electrical performance of **ECL (emitter-coupled logic)** systems. For CMOS application the wiring structures will shrink dramatically. Linewidths and via holes in the range of 20 μm are available.

The demand for fine patterns in the device layers will be decisively influenced by the development of the outer-lead pitch of TAB devices. It is expected that in the 1990s this pitch will shrink to 50 μm.

12.7.3 Hitachi: Thin-Film Packaging for Mainframe Computers

Thin-film technology has been extensively applied in the compact chip carrier named MCC (microcarrier for LSI chip) which presents the possibly smallest **hermetic** package and is shown in Figure 12-72 [142]. Hitachi's M880 mainframe introduced a high-performance multichip module [143] which consists of a multilayer ceramic substrate (MLC) and a number of chips, each packaged on a MCC. The MCC has been designed to offer the highest I/O interconnection density and smallest packaging surface area by adopting full-surface flip-chip bonding and built-in termination resistors. It was generally agreed that for TAB or wirebonding where terminals are wired out only from the fringe of the chips, it becomes difficult to connect 400–600 terminals per chip. This is because the pin pitch gets smaller than 100 μm for a chip size of 10 mm × 10 mm, and it can cause interpin electrical interference, problems with reliability, and manufacturability concerns. The design parameters of MCC are summarized in Table 12-23. The MCC is based on a mullite/tungsten cofired substrate of a size of 10 mm × 10 mm to 12 mm × 12 mm, depending on the size of the LSI chip to be mounted. Mullite ceramics were chosen for their low thermal expansion coefficient (3.5×10^{-6}/°C) matching that of silicon (3.0×10^{-6}/°C) and its low dielectric constant (5.9) which gives

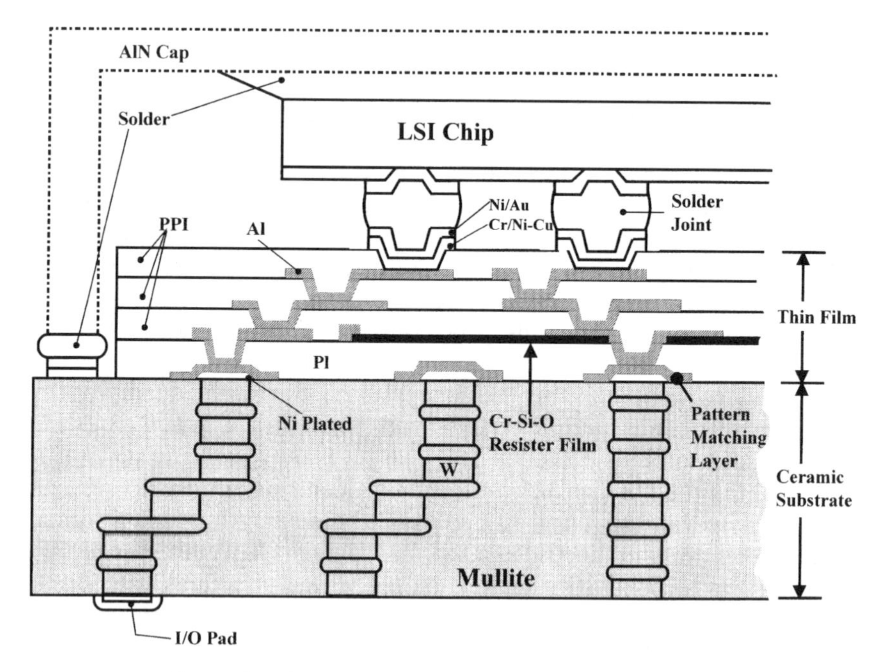

Figure 12-72. Cross-Sectional View of MCC with LSI Chip and A1N Sealing Cap

Table 12-23. Design Parameters of MCC Substrate

	Parameter	M880 10 × 10 – 12 × 12 mm	
Number of pads	Signal:	252	
	Termination Resistor:	99	Total 528
	Power, GND and Monitor:	177	
Pad pitch	MLC Side:	0.450 mm	
BLM pitch	LSI Side:	0.250 mm	
Number of layers	Thick film Conductor:	7	
	Thin Film Conductor:	5	
	Thin Film Resistor:	1	

better electrical performance than conventional alumina. Matching of the thermal expansion coefficients is crucial for highly reliable solder-ball joints.

The thin-film wiring acts as redistribution layers to match the grid pitch of the LSI chip **ball-limiting metallurgy (BLM)** and that of the MLC through-hole. The thin-film conductor is aluminum, and the dielectrics are polyimide (PI) and photosensitive polyimide (PSPI). The MCC has not only thin multilayer circuit patterns but also incorporates thin-film termination resistors in it. The LSI packaging area is considerably reduced owing to these built-in termination resistors compared to that of the conventional modules with discrete resistors.

The fabrication process of the MCC has been well documented [142] and starts with a mullite-tungsten cofired substrate. The substrate has nickel-plated through-hole pads on the top which are arranged in a grid. The ceramic substrates tend to have dimensional inaccuracy caused by process variations, such as firing shrinkage. Hence, a pattern-matching layer is first applied on the ceramic substrate to ensure electrical contacts between the thin-film circuit and the thick-film through-hole pads which may have location variations within the range of tolerance. The pattern-matching layer is shaped into concentric circles around the through-hole pad and is able to absorb locational mismatch between the pads and the vias in the first thin-film layer.

The Al-polyimide combination was chosen as the thin-film multilayer interconnection materials for the reason that plenty of process and reliability data exist and there is a long history of IC fabrication utilizing these materials in Hitachi [144,145]. At first, a wet-etchable polyimide is applied as the first-level dielectric. The polyimide layer is thick enough to effectively planarize the rough topography of the ceramic surface to assure the process latitude for photolithography. The termination resistor material is then deposited as very thin film on the polyimide by RF magnetron sputtering. The Cr-Si-O cermet was chosen because of its suitable sheet resistance when deposited in a processable, submicron thickness range. After fabrication of the thin-film resistors and formation of the first-level vias, the first conductor layer is formed by sputtering A1 followed by photolithographic patterning. The subsequent layers are made by repetition of the above processes, except that a photosensitive polyimide is used instead of the wet-etchable polyimide. Finally, BLM's are formed by sputtering Cr/Ni-Cu and wet etching of the metal layer. The BLM's are finished by nickel and gold chemical plating.

The built-in termination resistor is one of the special attributes of the MCC. One MCC carries about 100 termination resistors. Figure 12-73 is a depiction of the structure of the termination resistors, which are patterned in a doughnutlike shape and located around the first-level vias in the most compact design.

Some other technological breakthroughs for the MCC fabrication process are summarized as follows:

1. Stable and simple via formation through the thick interlevel dielectric polyimide

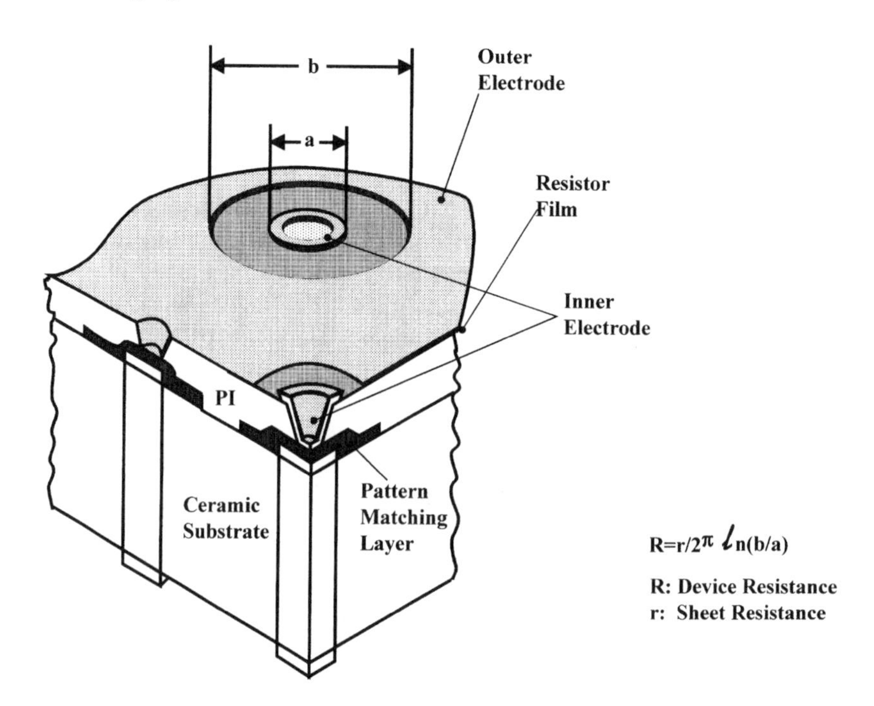

Figure 12-73. Structure of Built-in Termination Resistors

2. Optimum metal composition and its processing to form reliable via contact
3. Homogeneous large-area deposition of the thin resistor film
4. Stable etching processes for metals deposited on the surface of reactive organic materials like polyimides
5. LSIs with over 500 I/O pins in the smallest area

The main purpose of the MCC is to reliably interconnect between the LSI I/Os and the MLC I/Os by matching the I/O pad pitch of the LSI to those of the MLC. However, the MCC offers some additional significant functional merits as follows:

1. Highest chip-level interconnection density achieved at the time of the implementation of this technology
2. Higher packaging density by substituting discrete resistor components with built-in termination resistors
3. Full LSI testability and easy LSI screening
4. Easy replacement and safe handling of individual LSI chips
5. Smallest possible hermetic LSI package

12.7.4 nCHIP: Packaging with Al/SiO_2 on Silicon Substrates

nCHIP's nC1000®/nC2000® Silicon Circuit Board (SICB®) technologies provide high-density interconnect substrates for a wide range of applications. Wiring pitch as low as 25 μm and the ability to route interconnect under the die without interference from thermal vias enables close placement of the ICs. Signal routing is accomplished on two metal layers; the wiring density allows even interconnect-rich devices to be routed in two layers.

nC1000 SiCBs are appropriate for most CMOS/**TTL** (**transistor-transistor logic**) designs. The circuit speeds that can be supported are dependent on the complexity of the design, the size of the SiCB, and the edge speeds of the I/O drivers. For designs incorporating typical CMOS drivers, practical system clock frequencies range from 40 MHz for very large designs to about 100 MHz for smaller designs. It is usually possible to handle a few signals of even higher speed if care is taken in layout. nCHIP's nC2000 SICBs are appropriate for fast BICMOS, ECL or GaAs designs whose performance requirements exceed the nC1000 capabilities.

Two power planes are used for power distribution. One of the planes may be split if an additional power supply is required, Both nC1000 and nC2000 technologies incorporate an integral decoupling capacitor between the two planes, providing all of the decoupling capacitance required for most designs. The integral capacitor is effective to much higher frequencies than surface-mounted chip capacitors because the capacitor is distributed throughout the SICB, the effective inductance between it and any die is only that of the wirebonds to the chip's power supply bond pads (typically 1.2 nH divided by the total number of power supply bond wires).

In addition to providing better power-supply decoupling, the integral capacitor can contribute to a significant reduction in module size. In MCMs built with other technologies, large numbers of chip capacitors can add 10–20% or more to the substrate area required, increasing module size and cost while reducing performance.

The integral decoupling capacitor is one of two major features that distinguish nCHIP's substrate technologies from those used by most other thin-film substrate manufacturers. The other distinguishing feature is the use of SiO_2 as the dielectric between metal layers. The reasons for the choice of SiO_2 as the dielectric are more varied and perhaps less obvious than the reasons for including an integral decoupling capacitor, so they will be discussed at some length.

12.7.4.1 The SiO_2 Dielectric

The cross-section drawing in Figure 12-74 shows the construction of an nC1000 Silicon Circuit Board. All of the intermetal dielectrics are made of SiO_2, except for the thin dielectric forming the decoupling capaci-

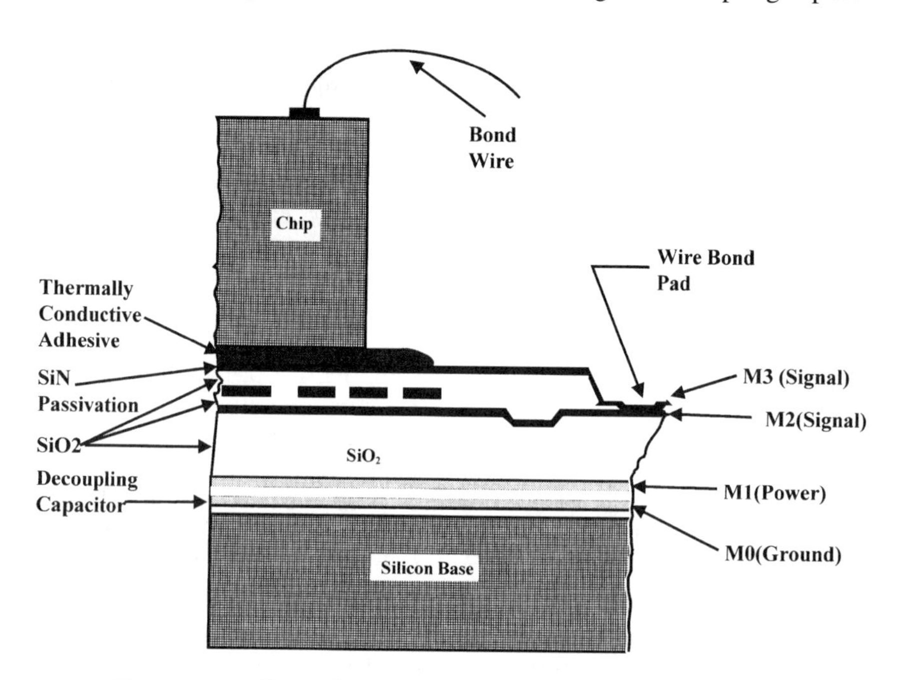

Figure 12-74. Cross Section of nC1000 Silicon Circuit Board

tor. SiO_2 is routinely used as the primary dielectric in integrated circuits (although ICs use much thinner oxides) and is efficiently integrated into the SICB manufacturing process. The selection of the dielectric material is critical to an MCM technology, with issues of cost, electrical and thermal performance, density, manufacturability, and reliability all being important. nCHIP's SiO_2 dielectric was chosen for a combination of properties which make it well suited for MCM applications.

The specific characteristics of SiO_2 which make it attractive can be briefly summarized as follows: high thermal conductivity, low cost, inertness and impermeability, desirable mechanical properties, and good consistency and control.

12.7.4.1.1 Thermal Conductivity

The thermal conductivity of nCHIP's SiO_2 dielectric has been measured (using specially designed thin-film test structures) at room temperature as 1.35 W/m K. This number is approximately an order of magnitude higher than typical polymer dielectric materials (virtually all of which fall in the range of 0.1–0.2 W/m K). For example, the most common form of polyimide, Dupont's Kapton®, has a published room-temperature thermal conductivity of 0.155 W/m K approximately one-ninth that of SiO_2.

This difference in thermal conductivity can have a significant effect on the design of MCMs. It permits modules to be built with no requirement for thermal enhancements such as "thermal vias" or "thermal wells," both of which impair wiring density and hence lead to larger modules. Thin-film terminating resistors, a significant source of local heat (as high as 400 W/cm^2 in typical applications), can also be embedded in the SiO_2 dielectric without problems.

As an example, the thermal resistance for the complete nC2000 series dielectric stack is 0.15 cm^2·°C/V. Thus, a chip dissipating 20 W/cm^2 would incur only a 3°C temperature rise as a result of heat conduction through the dielectric layers. As a result, no thermal vias are necessary and the full area underneath the chip is available for routing. In contrast, an MCM built with a polyimide dielectric of equal thickness would see a 26°C temperature rise, which would normally be unacceptable. Such a chip would have to located in a thermal well (an area where no wiring is permitted underneath the chip), compromising the routability and hence increasing the size of the MCM. Further, some alternative MCM technologies have total dielectric thicknesses substantially greater than that used in the nC2000 architecture; in these technologies, even chips dissipating as little as 2 W/cm^2 may require some use of thermal vias in order to meet a tight thermal budget.

Certain polyimide-based processes use a special form of thermal via, the "solid thermal pillar." Such thermal pillars can be quite compact and highly effective, thereby minimizing the need for added substrate area. The disadvantage of such processes, however, is higher complexity and cost. They are generally believed to be at least twice as expensive to manufacture as the conventional "open" or "staircased" vias, due to the need for planarization or via filling steps at each layer.

12.7.4.1.2 Process Cost

Many factors enter into the cost of an MCM. One key factor is the size of the module. Other things being equal, a denser MCM will allow the fabrication of more substrates per wafer, and with higher substrate yields. As a result, the cost of an MCM increases with area, at a faster than linear relationship (e.g., an area increase of 20% would typically increase cost by more than 20%). In addition, there are other system-related costs (the package for the MCM and the PC board on which it is typically mounted) which also scale in proportion to area.

As a result, to the extent that an MCM technology permits the fabrication of denser modules, it will lead to a lower-cost product, both at the module level and at the system level. Because nCHIP's SiO_2 dielectric does not require the use of thermal vias, it generally allows the design of denser modules.

Another important element of MCM cost is the cost of the materials themselves. A SiO_2 dielectric is economical, representing approximately 10% of the total processed wafer cost (accounting for raw materials, labor, and depreciation). In contrast, the polyimide dielectric in typical multilayer interconnect substrates is estimated as costing 20–30% of the total processed wafer cost, depending on the type of polyimide and other factors.

12.7.4.1.3 Inertness and Impermeability

In some applications, inertness and impermeability can be critical in the choice of dielectrics. The moisture absorption of nCHIP's SiO_2 is too small to be measured by conventional techniques. Moreover, the diffusion constant for moisture, gases, and ionic contaminants is many orders of magnitude less than that of polymers; this is why glass seals are deemed hermetic, whereas polymer seals are not. These attributes are among the reasons why SiO_2 is the preferred interlayer dielectric for most integrated circuits, particularly those intended for nonhermetic packages. (Polyimides are often used as top-level overcoats, principally for alpha-particle protection, but in such cases they are being used as a supplement to SiO_2; the polyimide is seldom used as the interlayer dielectric nor is it the primary passivation.)

Thus far, most thin-film multichip modules have been confined to hermetic packaging schemes, often because of concern about the reliability of the high-density interconnect in the presence of moisture and other contaminants. Polymer-based thin-film substrates can be highly permeable to moisture. For example, a typical polyimide (Kapton H) will exhibit a 30% variation in dielectric constant (from 3.6 to 3.9) as the ambient humidity is varied from 0% RH to 100% RH at room temperature, due entirely to moisture absorption. Besides the variation in electrical properties, there is a reliability concern: one well-known MCM substrate manufacturer has reported that its aluminum-polyimide substrates fail even after relatively brief exposures to the conventional "steam pot" test; they concluded that a copper interconnect is required for nonhermetic applications. Another practitioner with a copper-polyimide substrate system determined that their interconnect required 100% cladding of the copper conductors with nickel in order to reliably pass a steam pot test; this, of course, increases the cost of the process.

The longevity of a thin-film metallization system in an organic dielectric will depend on the precise details of the dielectric chemistry (every polymer dielectric is different), the presence of dissolved gases (O_2, CO_2), and residual ionic contamination present (e.g., from a die attach adhesive). This makes the generation of valid reliability data a complex matter, and the resulting data cannot be generalized to other MCMs employing similar but not identical processes. In contrast, the use of an inert, hermetic SiO_2 dielectric demonstrated high reliability essentially independent of the details of the external environment, and so these results are of general validity.

Severe environmental tests have been performed on substrates using the nCHIP's SiO_2 dielectric. These include the use of a steam pot in which 5% salt (NaCl) has been added as a corrosive agent (50% of the substrate is immersed in the saltwater solution, and the remainder is subject to the salt spray). Such an environment would destroy many MCMs employing polymer dielectrics. However, because of the use of a hermetic dielectric material (SiO_2) as passivation and interlayer dielectric, nCHIP's substrates exhibit no interconnect corrosion, even after 100 h at 127°C.

Finally, the absence of any measurable moisture-absorbing tendency simplifies the manufacturing of SiO_2-based substrates, as there is no need for bakeouts to drive out moisture, nor any requirement for "vent holes" in the metal plates to let moisture escape during the bakeouts.

12.7.4.1.4 Mechanical Properties

SiO_2 films possess mechanical properties which are important in the fabrication of multichip modules. These include a low thermal-expansion coefficient (~0.5 ppm/°C), a high tensile strength (>1 GPa, or 150,000 psi; over six times that of polyimide), a high elastic modulus, and an ability to be deposited under a precisely controlled compressive stress.

One of the difficulties associated with manufacturing MCMs is the large net stresses which can be built up in the substrate, resulting in a tendency for the substrate to bow and perhaps even delaminate. This tendency is quite pronounced when using polymer dielectrics on silicon substrates. Polymer dielectrics inherently have a residual tensile stress, which is the result of their high thermal-expansion coefficient and high cure temperatures. A typical polyimide will have a residual tensile stress of over 40 MPa at room temperature (about 6,000 psi). In addition, metal layers such as copper or aluminum will themselves have residual tensile stresses of around 160 MPa as a result of the thermal annealing that occurs during normal processing at 300–400°C. Thus, both the metal and the polyimide films tend to bow the substrate upward into a concave shape. The bow can be quite large (e.g., 10 mils of warp for a typical process after all the films are deposited). This can pose difficulties for manufacturing, including problems with lithography, vacuum chucking, heat transfer during vacuum processing, and so on. It also leads to difficulties during dicing of the wafer due to the large built-in stresses. For silicon-on-silicon "flip-chip" modules, additional complications can result from the tendency of the substrate to flex during thermal cycling (as both the metal and the polyimide have high expansion coefficients relative to silicon), leading to problems with solder fatigue.

In contrast, nCHIP's SiO_2-based fabrication process produces a stress-balanced substrate. The SiO_2 dielectric layer is deposited in such a way as to be under a moderate residual compressive stress at room temperature (about 50 MPa). This compensates for the tensile stresses in the metals. In fact, because there is on average about three to four times as much dielectric material as there is metal, the net stress in the wafer is very nearly zero (i.e., the final product has essentially no tendency to bow). (nCHIP uses a wafer flatness measurement as one of its in-process SPC steps.) Also, the ratios of thermal expansion, elastic moduli, and thicknesses of metal and SiO_2 are such that the wafer exhibits very little flexing tendency during thermal cycling.

The strength of SiO_2 is an important feature when assembling wire-bonded multichip modules. For example, during **ultrasonic wedge bonding**, local compressive and shear stresses of 200 MPa (30,000 psi) are routinely generated by the bond tool. Such a stress level is essential in order to generate a high-quality weld between the wire and the bond pad. Such stress levels can exceed the yield point of polyimide, causing it to deform during the bonding process. Although the exact mechanisms involving in wirebonding are poorly understood, it is known that wirebond quality can be compromised when bonding onto thin-film metallization atop polyimide dielectrics. Solutions to this difficulty have been found (e.g., by attaching very thick metal pads over the bond-pad areas), but at added cost.

The strength and adhesion of SiO_2 does not degrade with temperature. This is particularly important when performing rework operations on a multichip module (chip removal). A typical chip rework process involves briefly heating the module up to 200°–250°C in order to remove the chip from the substrate. Other investigators have reported an inability to perform such rework procedures without damaging (delaminating) the thin-film interconnect underneath the chip attach site. As a result, they have had to switch to soft, low-T_g adhesives in order to have a viable rework process. Such adhesives introduce compromises such as low strength, excessive moisture absorption, high expansion, and unfavorable creep properties.

12.7.4.1.5 Consistency and Control

SiO_2 is a well-understood material and easy to work with in manufacturing. Compositionally, it is essentially identical to that used in multilevel metal ICs as an interlayer dielectric. Thickness control is typically ±2% across a wafer and wafer to wafer. This leads to better impedance control, which can be important in very high-frequency applications.

In contrast, organic polymer dielectrics are relatively variable materials. Their primary virtue (i.e., that they can readily be chemically engineered to have specific properties for specific applications) makes it more difficult to control a process in high volume. Cure cycles can be complex and must be controlled precisely if the final result is to be consistent. Film thickness can seldom be controlled to better than ±10% through spin or spray coating. Some of the largest electronics manufacturers have decided to manufacture polyimide dielectrics in-house to maximize their control over material consistency.

12.7.4.1.6 SiO_2 Ruggedness

As with all ceramic materials, SiO_2 is a brittle material (i.e., it does not plastically deform prior to fracturing). However, when prepared properly, this fracture strength is very high; nCHIP has determined from extensive flexure tests that its SiO_2 multilayer films are stronger than that of the silicon substrate—that is, their measured flexural strength exceeds 1010 dyn/cm^2 (i.e., 1,000 MPa, or 150,000 psi). This is three times the strength of typical alumina ceramic materials, which are used in most high-reliability single-chip IC packages.

Some people with past experience in semiconductor processing recall issues with SiO_2 cracking at thicknesses beyond 1–2 μm. There is indeed such an issue if the SiO_2 is deposited with conventional techniques that do not control its stress state. Without a controlled compressive stress at deposition, the resultant film will be in a biaxial tensile state and will indeed crack at large thicknesses. nCHIP's deposition process, however, results in a slight compressive stress on the final film, which has no tendency to crack.

12.7.4.2 Key Electrical and Thermal Characteristics

- Signal layer resistance 15 mΩ/square (nC1000), 6 mΩ/square (nC2000)
- Intermetal (SiO_2) dielectric constant ≤3.75
- Power/ground-plane resistance 30 mΩ/square (nC1000), 15 mΩ/square (nC2000)
- Integral decoupling capacitance ≥50 nF/cm^2
- Designed for low-inductance wirebonding—typically 1.2 nH
- SICB thermal resistance (including all standard layers)—0.17 $cm^2 \cdot °C/W$
- Temperature coefficient of resistance (TCR) of aluminum interconnect—0.4%/°C
- Excellent thermal match between silicon base, power/ground/interconnect layers and SiO_2 dielectric. Overall substrate expansion is matched to silicon integrated circuits.

SiO_2 is often listed in references as having a dielectric constant of 3.9. That number is valid at low frequencies for the thermally grown oxides commonly used in integrated circuits. nCHIPs deposited (PECVD) oxide has a somewhat lower dielectric constant, particularly at higher frequencies.

12.7.4.3 nC1000 SICB Feature Sizes

Listed in Table 12-24 are some of the key feature sizes for nC1000 and nC2000 layouts. All are minimum dimensions.

Because of the available routing density of the M2 and M3 signal layers, utilization of the SICB area is generally determined by module assembly requirement rather than by routing constraints.

Table 12-24. SICB Feature Sizes

Feature	nC1000	nC2000
Metal width (M2 and M3)	10 μm	16 μm
Metal space (M2 and M3)	15 μm	20 μm
Metal pitch (M2 and M3)	25 μm	36 μm
Via pitch	35 μm	62 μm
Via diameter	5 μm	10 μm

12.7.4.4 MCM Package Assembly

In the nCHIP assembly process, dice are mechanically and thermally bonded to the SICB using thermally conductive epoxy. Back-side bias, if required, is provided by contact pads underneath the die attachment site. The SICB, in turn, is bonded to the package body or heat spreader with another epoxy layer. These attach methods are proved to have good thermal performance and high reliability.

The module's dice are electrically interconnected through very short (25-mil horizontal length), low-inductance, fine-pitch wirebonds. nCHIP's wirebonding capability employs ultrasonic wedge bonding of 1-mil A1 wire. These can be placed on a 4-mil pitch and have a typical inductance of only 1.2 nH, which is much less than in a conventional IC package because of the very short wire and the matched pad-outs of the chip and SICB.

nCHIP SICBs can be designed for use in a wide variety of packages or carriers. Most common are the well-known pin-grid-array quad flat-pack formats, although unusual configurations can be designed to meet special needs. Because nCHIP SICBs do not contain moisture-absorbing organic materials, they may be used in both hermetic and nonhermetic packaging. For thermal reasons, most packages are of a cavity-down configuration to allow for heat-sink mounting. Low-power or conductively cooled designs, however, can use a cavity-up configuration.

Because MCMs are typically used for high-performance (and therefore high-power) designs, a module's ability to dissipate a large amount of heat is an important issue. nCHIPs SICB technology is well suited to high-power designs, because the SICB materials both tolerate and conduct heat well. Thermally conductive epoxies can be used to attach IC dice to the SICB and to attach the SICB to the package or carrier. The package material(s) may be chosen to meet the module's thermal performance requirements (e.g., copper-tungsten instead of alumina for high-power applications).

12.7.5 Digital: VAX 9000 Multichip Packaging

A high-performance multichip packaging device needs to provide high speed, high density, controlled-impedance interconnections for interchip signal distribution [13,146-148]. It needs to provide a high-current DC power distribution system with small DC voltage drops and also a stable AC power supply system under a large transient, simultaneous current-switching condition. From the packaging point of view, a low-thermal-resistance path is required to cool the high-power chips, and a stable, inert environment ensures long-term reliability in a variety of storage, transport, and use conditions.

As the advanced semiconductor devices used in high-performance computers improve in performance and circuit density, corresponding increases in chip I/O pins, switching speeds, power dissipation, and cooling requirements also result. These requirements then drive toward higher interconnect wiring density with low-defect-tolerant manufacturing, thicker DC power distribution metal layers, and thinner interlayer dielectrics for power decoupling. Furthermore, chip attachment and assembly technologies must be developed to handle the higher chip power density and to maximize the total chip area as a percentage of the package area.

In the design of the VAX 9000 technology [149,150], a wafer fab technology supplemented by a printed wiring board type of technology is chosen to meet the above-mentioned substrate signal and power distribution requirements. The resultant technology is called HDSC (high density signal carrier) which has been shown to be highly manufacturable and extendible as the semiconductor chip technology advances. In chip attachment, packaging and connector approaches, a back-side chip attachment, cooling on a base plate, a two metal layer TAB tape, together with signal and power flex circuits in a mechanical housing form the MCU (multichip unit). This approach facilitates chip and module testing and impingement air cooling and is field replaceable.

In this section, the HDSC and MCU designs are reviewed in some detail and process overviews are described. The technology evolution strategy for future high-end packaging is discussed based on some projected high-performance, high-power semiconductor device characteristics. Several cases of actual multichip packaging designs are given for meeting these chip characteristics with incremental technology improvements for optimized system performance.

12.7.5.1 The HDSC Design and Process

12.7.5.1.1 The HDSC Design

A complete cross section of the HDSC is shown in Figure 12-75 where there is a total of nine Cr/Cu metal layers interfaced with polyimide layers as dielectric isolation. The whole structure is laminated on a Cu/Ni base plate where chip site cutouts are made for semiconductor devices and where cooling fins are mechanically attached for heat removal. The bottom four metal layers of the structure is called the power core for power distribution purposes. Each of the metal layers is 18 μm thick, patterned with drilling pads, breathing holes for moisture evolution during polyimide curing and chip site cutouts, and so on. Between M1 and M2 and also between M3 and M4, a thin polyimide layer of 10 μm is deposited to form two 30-nF capacitors for decoupling. A thicker polyimide layer (25 μm) between M2 and M3 is used to planarize the surface. A top and a bottom polyimide layer are used for electrical isolation and passivation purposes.

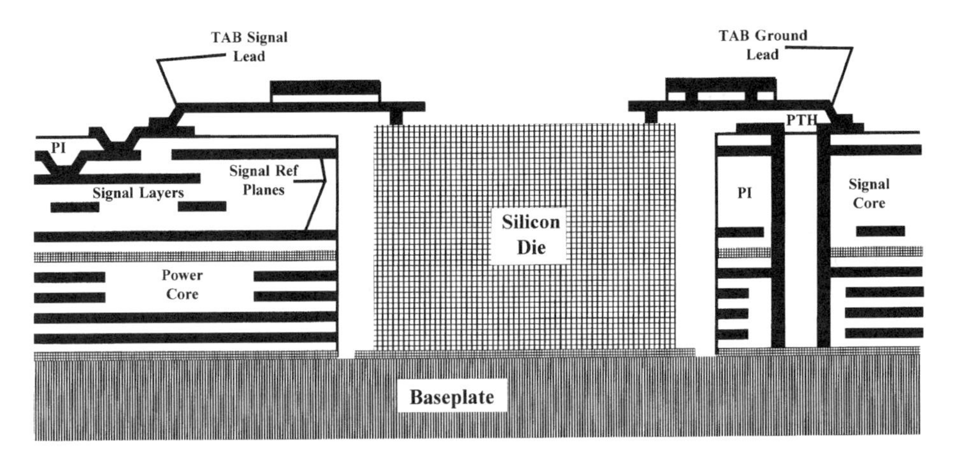

Figure 12-75. Cross Section of High-Density Signal Carrier (HDSC)

The top half of the structure is known as the signal core for signal distribution. It consists of a triplate transmission line with a characteristic impedance of 60 Ω. Each of the *X* and *Y* lines has a resistance of 1.0 Ω/cm. The dimensions are 18 μm wide and 10 μm thick and the wiring tracks are on 75-μm centers. Plated copper posts are used to connect between M2 and M3 and also M3 and M4. Ground and reference metal layers M1 and M4 are 4 μm thick. The total active wiring area on the HDSC is 10 cm × 10.5 cm.

Through-holes are drilled between signal and power cores which are subsequently metallized for electrical connection. The same metallization step is used to pattern the top layer pads for TAB tape attachment and also for signal/power flex circuits. The top metal pads are made of 25-μm-thick copper. The pad pitch is 100 μm for TAB pads and 150 μm for signal flex pads. The interlayer polyimide has a thickness of 25 μm processed with multiple coatings.

12.7.5.1.2 The HDSC Core Processes

Power Core: The power core process starts with an aluminum wafer coated with a 10-μm-thick polyimide layer, a thin Cr/Cu layer is deposited by sputtering which is subsequently plated to 18-μm-thick copper. The top Cr cap layer is then sputtered for adhesion purposes to the subsequent polyimide layer. The metal is then coated with photomasking followed by a wet-etching step to pattern the metal and a dry-etching step by plasma to clear the polyimide surface leakage. A thin layer of polyimide is deposited on top of the metal for isolation. The process then repeats itself three more times to complete the fabrication. In-process tests are performed to monitor the plating and etching steps to stay within specifications.

Signal Core Similar to the power core, a polyimide passivation layer is first deposited on the aluminum wafer. Then the bottom reference plane, a 4-μm-thick Cr/Cu/Cr metal layer, is sputtered on the wafer surface which is masked and etched. The top of the metal is now coated with a 25-μm polyimide layer by multiple spinning and baking, followed by a final cure. Next, a Cr/Cu plating base is sputtered on the polyimide surface. Photoresist and lithography steps are performed to define a stencil of lines and pads. After electrolytic plating in a sulfuric-acid-based solution to plate up the first interconnection copper metal (M2) layer of 10 μm, the photoresist layer is stripped. On top of the lines, a second thicker photoresist layer is coated which is exposed and developed for the post masking. The second plating step is performed to plate up the copper post (M2P) on top of the M2 layer for a thickness of 30 μm. The photoresist layer is then stripped and the plating base is etched.

For dielectric isolation, a polyimide layer (33 μm) is coated on top of the posts. To accomplish planarization, several spinning and baking steps are again taken, which is followed by the final cure. At this point, a photoresist layer is coated on top of the polyimide, patterned, and used as a mask for a plasma etch of the polyimide to form vias. The vias are designed to fall on top of the posts for interlayer contact. The residue photoresist is stripped after plasma etching. The process then repeats for the M3, M3P, and the vias on top of M3P. The 4-μm M4 layer is processed the same way as M1. Finally, a passivation polyimide layer is coated on top of M4 and etched with plasma vias, with the same process as previously described to complete the signal core process. The cross sections of a completed signal core and a power core are shown in Figure 12-76.

12.7.5.1.3 The HDSC Assembly Process

Stainless-steel metal rings are attached on top of the completed signal cores with epoxy adhesive. The front side of the signal core is protected with an organic tape when the A1 wafer is dipped into acid and etched away completely. A copper-polyimide "decal" is now fabricated which is in tension and hence is taut like a drum head. This is because of the thermal-expansion properties of the materials chosen. The decal is aligned on top of a matching power core and laminated onto it with a vacuum press after an epoxy adhesive layer is attached on the top of the power core. Again, a decal is formed as shown in Figure 12-77. Subsequent process steps are similar to those used in manufacturing the advanced printed wiring board.

The decal is drilled with fine (250-μm) holes to connect power pads between the two cores. Sputter and wet chemical steps are taken to seed the top surface and inside the drilled holes. **Dry-film resists** are used to pattern and to form a stencil which is followed by acidic copper plating for the PTH (plated-through-hole) metal and the top metal. After the steps

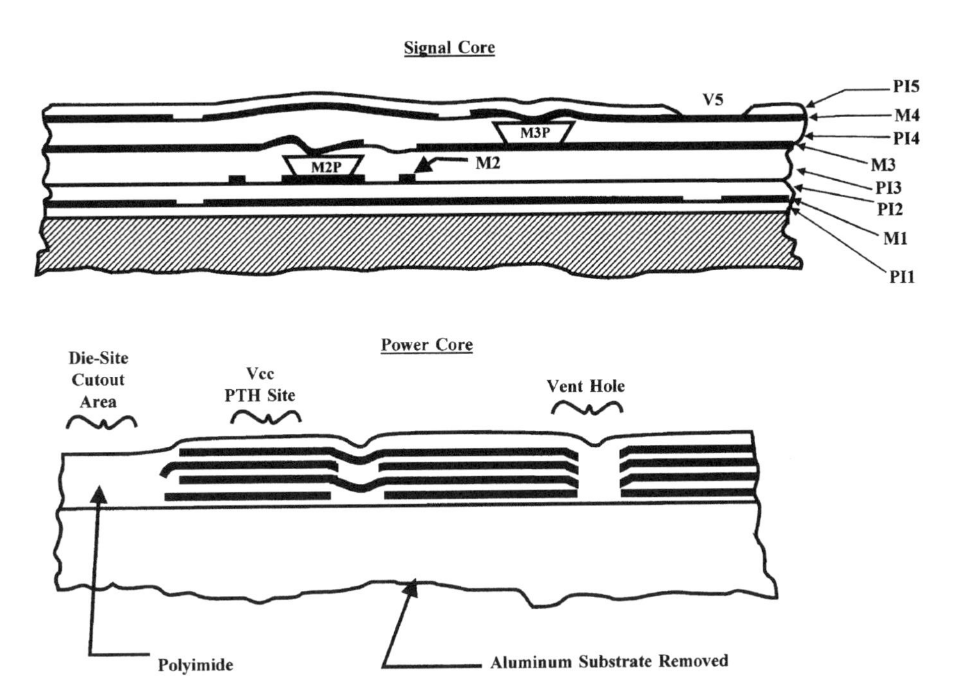

Figure 12-76. Cross Sections of HDSC Cores

of plating, photoresist strip and plating base etch, solder is deposited on top of the metal by screening and reflow. Laser is used to cut through predefined polyimide lines where copper is removed to delineate die site cutout regions. The finished decal is then laminated on a copper-nickel base plate. The extra decal is trimmed away by laser to complete the HDSC process.

12.7.5.2 Evolution of High-Performance Computer Packaging

As the IC technology continues on its path for a narrower linewidth, more on-chip circuit density and speed improvements, the IC chip size, the number of chip I/O pads, pad density, power, cooling requirements, and switching speed will all increase. Furthermore, the use of the 6-in. Cu-polyimide base wafer fab technology to develop the improved HDSC/MCU for meeting the projected IC chip performance requirements will continue. In the technology development planning, packaging of maximal chip density and the number of chips are also assumed; obviously, the actual circuits used on the chip and the number of chips used on the package will have to be dependent on detailed system architectural design and partition, circuit and packaging designs, and other considerations.

A set of projected bipolar ECL chip technology development cases

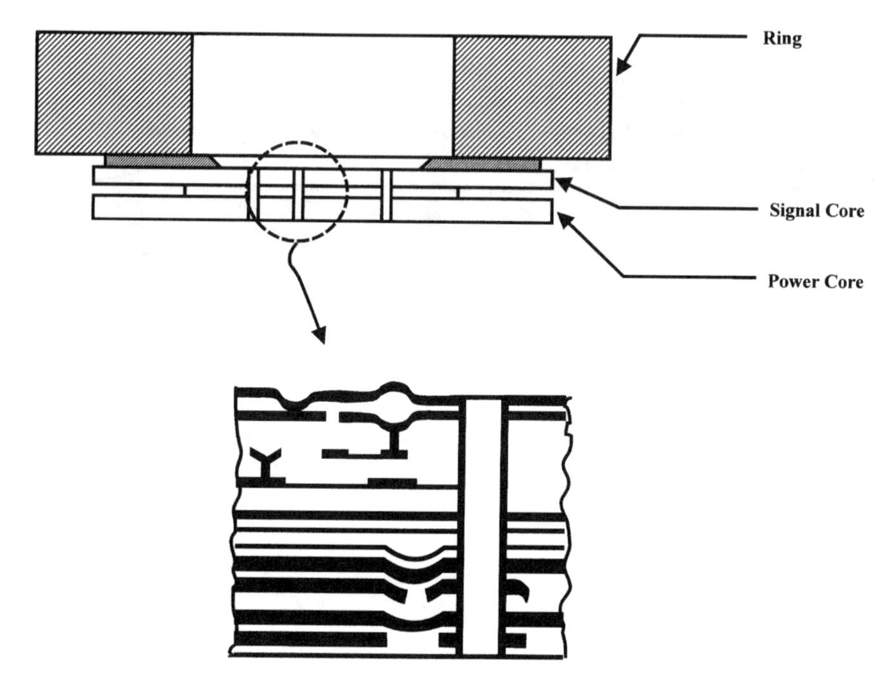

Figure 12-77. Formed Decal for HDSC Assembly

BP1, BP2, and BP3 are listed next to the base case of the Motorola Mosaic 3 technology used in the VAX 9000 MCU as shown in Table 12-25. Some of the key chip technology characteristics are shown together with the corresponding requirements placed on the multichip units which will be referred to as cases A, B, and C. As the linewidth reduces, the circuit speed increases significantly for each step; however, a far larger increase in circuit density is realized such that the number of logic circuits is more than doubled in each case and the on-chip embedded RAM is also provided.

Based on these IC chip requirements, it is clear that every aspect of the multichip interconnect and packaging specifications must be improved (i.e., higher DC and AC power distribution requirements, higher interconnect density, smaller delay, more efficient cooling, etc). From the physical technology of the multichip packaging point of view, this implies thicker DC power core metal, thinner interlayer dielectrics between power core metals, smaller interconnect line pitch for the signal core, and so forth. Some analysis, design, and experimental work were performed to come up with an interconnect and packaging design architecture appropriate for meeting the chip physical and performance requirements.

For MCU power distribution, the VAX 9000 MCU power core process architecture of area plating and wet etch of copper-chromium followed by a thin polyimide coating is believed extendible from the present thickness of 18 μm to at least 25 μm (50% increase), the interlayer

Table 12-25. Chip Technology Development

MCU Technology	MCU	MCU-A	MCU-B	MCU-C
IC Technology	MOSAIC 3	BP1	BP2	BP3
Emitter Size (μm × μm)	1.2 × 3.5	0.5 × 1.5	0.3 × 1.2	0.2 × 1.0
Metal 1 pitch (μm)	4.0	2.8	2.2	1.5
Unloaded gate delay (ps)	80	50	40	30
Chip type	Gate array	Gate array	Standard cell	Standard cell
Max. No. of gates/chip	10K	25K	100K	300K
Embedded RAM bits	None	16K	64K	256K
TAB				
ILB (μm)	100	90	75	Flip chip
OLB μm)	200	140	75	Flip chip
Power/chip (W)	30	45	100	150
Pins/chip				
Signal	256	350	500	940
Power	104	150	200	380
Total	360	500	700	1320
Die size (cm)	0.98	1.18	1.40	1.72

Note:
1. MCU Case B assumes a power TAB to bring power to the center of the chips.
2. MCU Case C assumes flip-chip direct bump to interconnect (no TAB).

dielectric coating from the present thickness of 10 μm to 7 μm on the thicker power core metal layer. Additional DC power distribution improvements can be made by bringing in the DC power from all four sides of the HDSC (the existing design is two sided). The total DC power in-feed can therefore increase by another factor of 2 for the same percentage of DC drop. For AC power distribution, the simultaneous switching requirements on the MCU can improve by 30% by going to a 7-μm polyimide coating. Further improvements on the MCU technology requirements will need a new power core technology architecture; that is, thicker metal layer laminates for DC power and a separate "capacitor core" with thin metal and thin polyimide layer built inside the signal core to provide additional low-inductance capacitance for decoupling.

For a signal interconnect density increase, a straightforward scaling factor of 2 for reducing both the vertical dimensions and horizontal dimensions of the VAX 9000 transmission-line structure can be used. For example, a factor of 2 scaling of the transmission-line dimensions for the existing MCU design will result in a 9-μm × 5-μm line cross section, a dielectric layer thickness of 12.5 μm, and a line pitch of 38 μm, thereby realizing a line density increase of a factor of 2. But a factor of 3 scaling increase will result in a 3× line density increase at an expense of line DC resistance increase by a factor of 9 to 9 Ω/cm, which is too high. However, by allowing the characteristic impedance of the line to drop to 50 Ω from 60 Ω, the use of low-dielectric-constant polyimide (from 3.5 to 2.4) and an increase of the metal line cross-section aspect ratio (thickness : width) from 0.5 to 0.66, the metal linewidth and thickness can both be increased from the scaled-down version and with the corresponding adjustments of dielectric thickness, the line DC resistance can be held to 4 Ω/cm, an acceptable value for these applications.

For die attach and assembly, it is believed that the TAB technology can be extended by reducing both the ILB (inner lead bonder) and **OLB (outer lead bonder)** pitches to 75 μm in one to two steps. By designing the same OLB and ILB pitch, this advanced approach is called an orthogonal TAB in which a fan-out from a tight ILB to a wider OLB is not needed and the TAB lead lengths can significantly be reduced to improve signal integrity. Further significant density increase will require flip-chip bump technology.

Based on the above discussions, the multichip packaging requirements and the trend for the physical features are summarized in Table 12-26.

12.7.5.3 Evolution of High-Performance Packaging Technology

Once the interconnect and packaging materials and process architecture are determined, the physical features options can be developed and

Table 12-26. Evolution of MCU for High Performance Systems

MCU Performance Features	Requirement Trend	Physical Feature Trend
Signal interconnect density	Higher	Smaller wire pitch, more layers
Z_0	60 → 50 Ω	Thinner dielectric, smaller wire pitch
Line resistance	Can increase from 1.0 Ω/cm to 4.0 Ω/cm	Thinner dielectric and metal
DC power	Higher	Thicker metal layer in power-core layer, laminated power core
AC power	Higher transient	Thinner dielectric in power core, separate cap core, orthogonal TAB/wire, flip-chip bumps
Size, number and density of chips, chip I/O number and density	Higher	Orthogonal TAB/wire, flip-chip TAB, flip-chip bumps
Cooling	Higher	Improved die attach on baseplate, top side cooling
Off-chip connector	Higher	Higher density signal flex, pin grid array

evaluated with respect to the performance and density gains on one hand and to the schedule and risk factors on the other hand. In addition, the resultant physical technology features for a given design across different areas (i.e. signal/power core, HDSC assembly and MCU assembly) have to be consistent and compatible to a given chip set; hence, further design trade-offs have to be made. The three MCU designs, A, B, and C for the corresponding bipolar technologies 1, 2, and 3 specified in Table 12-25, are discussed in the following.

12.7.5.3.1 MCU-A

To meet the requirements of bipolar technology of BP1, no change is necessary of the HDSC and MCU materials and process technology architecture from the base case. There is no change of equipment set either. However, only incremental technology design rules are needed to meet the higher-performance requirements, and the technology risks are considered low.

The power core metal thickness is increased to 25 μm per layer and the interlayer polyimide coat is reduced to 7 μm. For the signal core, a 0.7 scaling factor is used for the transmission linewidth, pitch, its thickness and the corresponding dielectric thickness, and the metal post heights. The signal line pitch is therefore 50 μm with a resistance of 2 Ω/cm. The rest of the design and the core processes remain the same. The ILB and OLB pitches become 90 μm and 140 μm, respectively. Signal flex circuits stay the same, power flex circuits are attached on all four sides. BP1 chips in a 3 × 3 array remains the same. The total circuits on the MCU increase by 7.6× with a higher speed, and the power dissipation goes up 50% to 450 W.

12.7.5.3.2 MCU-B

Because of the large increase of the on-chip integration level of the BP2 technology (10× over the base case) and the corresponding increase of chip I/O pad number, switching speeds, and power dissipation, significant process changes are needed for the signal core, as shown in Figure 12-78. A thin polyimide layer of 4 μm is deposited between two power/ground layers to enhance the decoupling capabilities of the signal core. A 2×

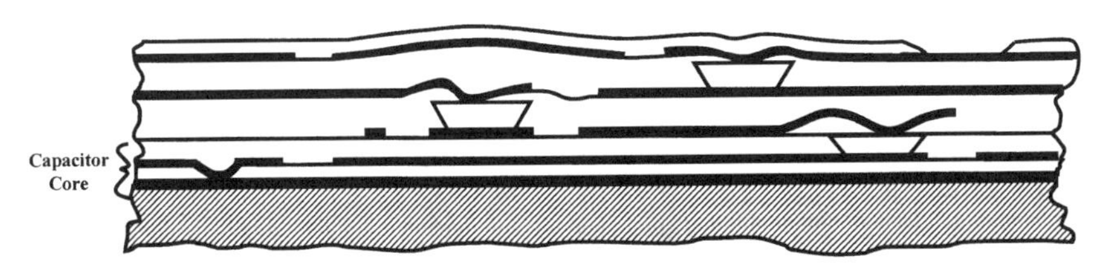

Figure 12-78. MCU-B Signal Core Design

signal line density increase with a line pitch of 36 μm over the base case is needed. The line resistance becomes 3 Ω/cm. The power core stays basically the same as MCU-A, but the power vias and PTHs are increased to account for the lower DC drop and the power supply inductance needed for the maximum 900 W specification for this MCU.

For the chip attach, the TAB ILB and OLB are both 75 μm pitch to bring the total chip I/O to 714. Because of the orthogonal TAB tape design, the leads are much shorter (200 μm) and the tape can be simplified to a single metal layer. For OLB bonding, single-point lead bonding is shown to be more viable compared with gang bonding to reduce bonding defects. In addition, epoxy die bond materials, the bond line, and the pin fin design all need to be improved to account for the 3× power dissipation increase. The base case chip design is still 3 × 3 and several small SRAMS chips can still occupy a chip site.

12.7.5.3.3 MCU-C

As shown in Table 12-25, the BP3 technology further increases the on-chip circuits by a factor of 3 over the previous case of BP2, the chip I/Os increase by a factor of 2 and the power dissipation on chip increases by 50%. Because of the number of chip I/O pads increases, two plane pairs of signal interconnect layers are required, as shown in Figure 12-79. The lower signal plane pair is the same as MCU-B for the long interconnects. For the upper plane pair, the wiring pitch is reduced to 25 μm for high-density localized wiring. The line resistance becomes 4.5 Ω/cm, but the more than 2× wiring density increase over the MCU-B case is expected to support an array of 4 × 4 BP3 chips.

For chip attach, flip-chip solder bumps are used in this case. To support 1320 I/Os on a 1.72-cm chip, 3 rows of perimeter I/O pads with a pad size of 75 μm and a space of 75 μm are needed to meet the requirements.

Two significant requirements of the power core now drives the physical design changes: the total DC power increase to 2000 W and the low thermal-expansion rate of the power core to match that of the silicon to ensure the reliability of the solder bumps. The construction of power core in this case is made of laminates of individual low-thermal-expansion ceramic substrates sheets such as a glass-ceramic or A1N substrate. Each substrate consists of a thin bottom metal layer (4 μm), a thin dielectric film (1.8 μm) for the capacitor, and the thick top metal layer (75 μm) for power distribution. The laminates contains PTH for intersheet connection, bottom metal pads for pin attachment, and top metal pads for signal core decal attachment.

The signal core is attached to the power core by solder/adhesive lamination as shown in Figure 12-79 where the solder bumps on the signal core is reflowed and laminated onto the metal pads on the power core

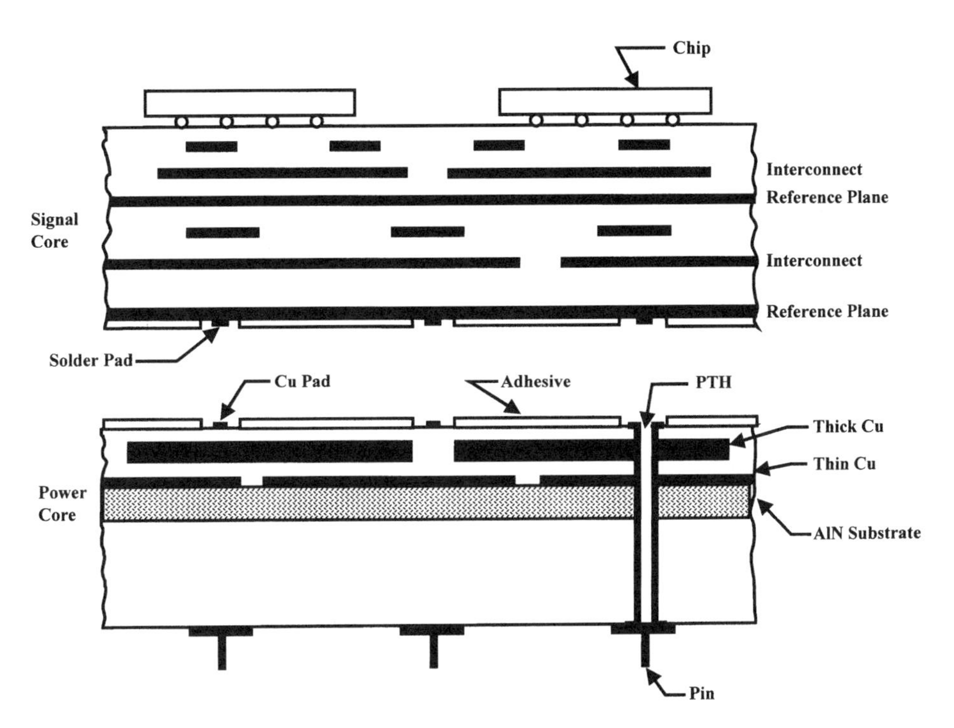

Figure 12-79. Solder/Adhesive Lamination of Signal to Power Core for MCU-C Design

substrate in a heated vacuum press. A high-temperature B-staged thermal setting adhesive starts to activate and cross-link after the solder-reflow temperature to bond the signal core to the power core to complete the lamination process.

The thermal conduction path is now through the back side of the chip and the top side of the package. Well-known approaches that have been reported in the literature such as TCM [151], bellows, and so forth can be used to accomplish the thermal/mechanical requirements.

12.7.5.4 Technology Summary

Under the assumption that the bipolar technology, driven by performance and density considerations, continues to increase in circuit speed, I/O pin number, and power, the challenges for packaging/interconnect designers and materials/process architects are to come up with a multichip module designs that can match these requirements. The challenges are in signal distribution, power distribution, chip attach/rework, thermal management, and environmental protection. Although the focus throughout the section has been on the bipolar circuit technology, similar design concepts could also be applied to other IC technologies such as CMOS and GaAs.

12.7.6 NEC: Thin-Film Packaging for Mainframes and Supercomputers

12.7.6.1 Polyimide-Ceramic Substrate for High-Performance System

As described in Chapter 2, "Packaging Wiring and Terminals," interconnection performance of the wiring substrate is primarily determined by the unit signal delay of the wiring patterns and the packaging density. In order to minimize the unit signal delay, a reduction of electrical capacitance of the signal lines is necessary. The electrical capacitance depends mainly on the substrate technologies, which can be reduced by using a low-dielectric-constant material for the insulator layers and fabrication of fine signal lines. Increased packaging density apparently requires a closer LSI chip mounting and higher cooling rate. When the distance between the LSI chips becomes shorter, the interconnection wiring density must be higher; therefore, either narrower signal lines and pitches with less layers or wider signal lines and pitches with more layers is inevitable. The former demands the implementation of thin-film technology with low-dielectric materials and the latter indicates thick-film technology with inorganic materials for insulator layers such as cofired alumina-ceramic or glass-ceramic. In high-performance multichip modules (MCM), the LSI chip power dissipation is generally high; therefore, a higher thermal-conductive structure is imperative. Such a high-thermal-conductive structure is practicable by separating the heat and wiring paths on the substrate. Usually the heat generated in the LSI chips is removed through a thermal-conductive mechanism attached on to the multichip module like IBM's Thermal Conduction Module [151].

Thin-film wiring technologies are featured by low electric capacitance and fine signal lines of high conductivity as compared to thick-film technology. As a consequence, more LSI chips of high I/O terminals (>400) can be interconnected onto a thin-film substrate with fewer signal layers for faster signal transmission. As discussed previously, polyimide material is a suitable candidate for thin-film dielectric layers due to its excellent electrical, thermal, and mechanical properties. Alumina- or glass-ceramic multilayer base substrates with electrical wiring in them (active substrates) provides more wiring flexibility and the associated performance improvements, in addition to mechanically supporting the LSI chips and cooling mechanism. For these reasons, a package of polyimide thin-film wiring on alumina- or glass-ceramic multilayer base substrates has been developed [11,152]. Figure 12-80 shows a polyimide-ceramic substrate developed by NEC for high-end mainframes and supercomputers [152].

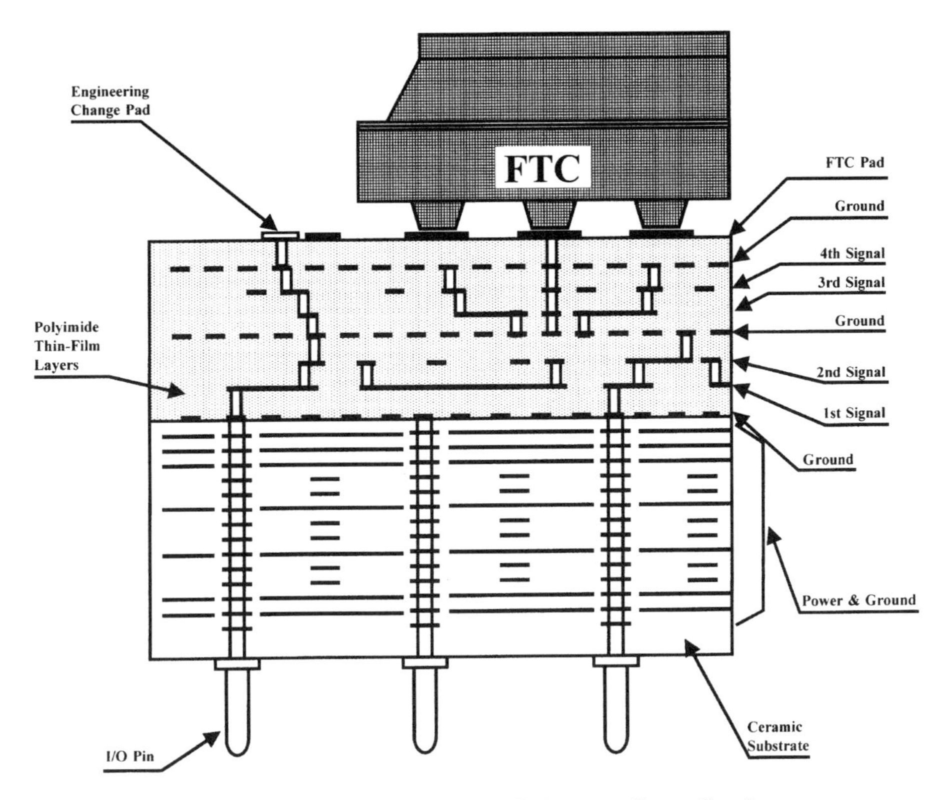

Figure 12-80. NEC Multilayer Substrate Cross Section

The alumina- or glass-ceramic base substrates are fabricated by cofired thick-film technology. Power/ground layers and clock distribution layers are formed inside the base substrate because these wirings are common among the personalized substrates for each system. The NEC thin-film technology is unique for the combination of thick-film inorganic base substrate and organic thin-film signal layers. The major benefits of cofired alumina- and glass-ceramic substrates are good mechanical strength for LSI chip mounting and attachment, compatibility with the cooling mechanism, thermal coefficient of expansion (TCE) matching that of silicon chips, and capability of high I/O terminal number interconnection with high power supply. The advantages of organic thin-film layers are fine signal lines with low electric capacitance for higher system performance. As shown in Figure 12-80, the thin-film multilayer wiring has eight conductive layers, of which four layers are for signal. The signal wiring pitches are 75 μm, wiring width 25 μm, and via holes in the polyimide layer 60 μm square.

The low dielectric constant of polyimide allows high-speed signal transmission at 6 ns/m and high-density interconnections of 53 lines/mm. The size of the substrate is 225 mm × 225 mm square and 5.5 mm thick. Photosensitive polyimide technology is used to fabricate the thin-film multilayer substrate for the simplified patterning process relative to non-photosensitive polyimide [153]. Up to 60,000 via holes of 60-μm square are formed in each polyimide layer of 20 μm thickness using photolithographic processes for mass production.

12.7.6.2 Wafer Scale Substrate for GaAs Packaging

Thin-film substrate is the best choice for attaining low-capacitance signal lines because of its fine-line capability. The currently used alumina-ceramic substrates for thin-film package, however, often have very fine voids on the surface generated by the pulling out of alumina particles during polishing. These voids are undesirable for fine-signal-line patterning. On the other hand, the surface of crystallized silicon or sapphire wafers is smoother and more suitable for fine-line patterning. Apparently, this is the main reason for their wide application in VLSI technology. Initially, these two types of wafers were both employed for the pilot fabrication of a thin-film structure, but it was found that silicon wafers were not good candidates for this purpose. In the process of thin-film layer stacking, silicon wafer of 0.5 mm thickness was broken due to the stress developed within the thin-film structure. However, the sapphire wafer had sufficient strength to withstand the stress produced by a thin-film structure with up to five layers of polyimide without breaking. The sapphire substrate was 10 cm square which was cut off from a 6-in. wafer. Figure 12-81 shows the cross section of the sapphire wafer-based package for supercomputers [154], onto which high-speed bipolar gate arrays (300 gates, 1.5 W max.), GaAs RAMs (T_a = 1.5 ns, 4 W max.), and GaAs gate arrays (2.6K gates, 6.1 W max.) are mounted.

The two layers of thin-film signal lines of 25-μm width and 60-μm center-to-center spacing were formed by sandwiching between ground mesh layers. The conductor was plated copper, and insulator polyimide layers were spin-coated. In this package, TAB-style bipolar gate arrays were directly mounted onto the polyimide thin-film layers face down and leaded chip carrier (LCC)-type GaAs devices were mounted by soldering onto the copper pads on the substrate.

The sapphire wafer is attached to the alumina base substrate by silver epoxy resin. The alumina base substrate has bonding pads on the top periphery, I/O pads on the bottom periphery, and through-hole wirings through the substrate to interconnect the top and bottom pads. The electrical interconnections between the sapphire wafer and alumina substrate are provided by gold ribbons using the gold-to-gold thermal-compression bonding technique. This multichip package was 11 × 11 cm^2 in size with

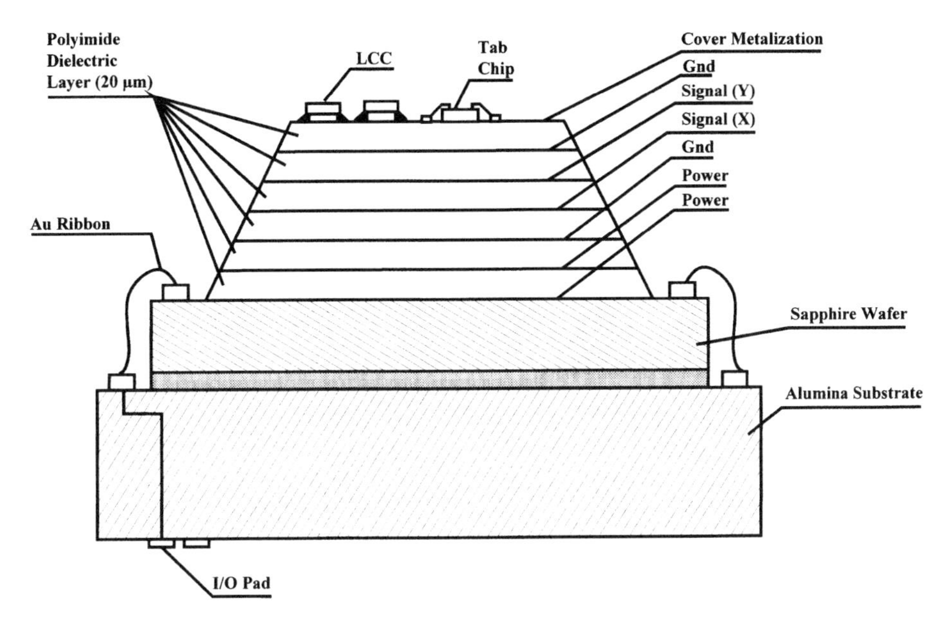

Figure 12-81. Cross Section of Sapphire-Wafer-Based GaAs MCM for NEC Supercomputers

a total of 420 I/Os (228 signals and 192 power and ground) and consumes up to 150 W. To cool this new package, the board with four of these multichip packages mounted on it was immersed into fluorocarbon liquid. By convecting the fluorocarbon liquid at 8 cm/s flow rate, all the semiconductor devices were kept below 55°C of junction temperature.

12.7.7 MicroModule Systems: Thin-Film Packaging Compatible with IC Processing

Thin-film multichip module packages are manufactured using standard automated semiconductor equipment in a Class 100 cleanroom [37]. Multilayer copper-polyimide interconnects were processed on 150-mm aluminum or silicon wafers. The aluminum wafer has the advantages of low cost and good thermal conductivity, and, at the same time, can be employed as the power plane and one electrode of the integrated capacitor, whereas, the silicon wafer provides an excellent thermal-expansion match to that of IC chips, allowing for high-density chip attach. The main features of polyimide processing are spin coating, curing, proximity printing for conformal mask patterning, and plasma etching for signal and thermal via formation through the cured polyimide layers. For copper processing, a sputtered seed layer is followed by electroplating. The metal layer is defect sensitive and, therefore, is patterned by projection lithography to increase the yield. A total of four or more metal layers can be sequentially built up this way. A cost model for the implementation of this technology justified the cost-effectiveness of this technology. At volume production, substrate cost can be as low as $5/in.^{2} and the total cost of a MCM-D, consisting four chips each with a pin count of 100, is around $10–$12. An example of the application of this technology is the fabrication of either 2- or 4-Mbit cache SRAM memories for PCs and workstations where space and cost are the premiums [155].

12.7.8 Other Commercial Schemes

Similar technology capabilities have been reported by other Japanese as well as American companies. Table 12-27 lists an example of thin-

Table 12-27. Thin-Film Technology

Material	Specification	Remarks
Substrate		
Material	Alumina, glass-ceramic	
Size	150 mm maximum	
Conductor Metals		
Conductor metal	Au, Cu, Al	
Middle metal	Pd, Ni, Mo	
Interface layer	Cr, Ti, Mo, Ti/W	
Thickness	5 μm	
Resistance	Electroplating	Cu, Au 5 mΩ/sq.
Linewidth	25 μm minimum	
Line spacing	40 μm minimum	
Adhesion	4–6.5 kg/2 mm^2 (soldered)	Peeled by 6-mm-dia. wire
	≥8 kg/1.5 mm dia (brazed)	45° Pull by 0.4-mm-dia. pin
Resistor		
Material	Ta-N, Ni-Cr-Si	
Sheet Resistance	10–50 Ω/sq.	
TCR	±100 ppm/°C	
Others	Cr-AuTi/W-Au	
Dielectric		
Material	Polyimide	
Thickness	20 μm	
TCE	(200–700) × 10^{-7}/°C	
Dielectric constant	3–4	50 Hz
tan δ	0.2%	50 Hz
Volume resistivity	>10^{16} Ω/cm	20°C
Insulation resistance	>10^6 V/cm	20°C

Source: Courtesy of NTK.

film technology being practiced by NTK on substrates up to 150 mm using multiple levels of polyimide-copper or gold metallization.

Planar polyimide (20–40 μm thick in these examples) is deposited, followed by reactive ion etching of the polyimide to form via holes with a SiO_2 layer as the etching barrier. The holes are subsequently metallized by sputtering a conformal layer of Cu or Au. Patterning of both vias and the next conductor layer is performed with a single photolithography and etching step. Reactive ion etching processes have been successfully used to etch vias as small as 12 μm in diameter in the 20–40-μm-thick polyimide films at an etch rate of 1 μm/min. The sidewall angle is controlled at 65°±2°.

Another example involves the use of photosensitive polyimide and copper plating technology to form very-high-density multilayer interconnections. The outstanding accomplishment by NTT has been the development and application of a selective electroplating and photolithographic process to achieve fine copper conductors (3 μm wide) and very small via holes (15 μm in diameter) through relatively thick photosensitive polyimide [77]. This technology has been claimed to result in low-cost fabrication of high-density multilayer interconnections because of the short process turnaround time.

12.8 EMERGING COMMERCIAL TECHNOLOGIES IN THIN-FILM PACKAGING

As described above, all the multichip module technologies can be generally classified into three generic categories; those based on ceramic (MCM-C), those based on printed wiring board (MCM-L), and those based on thin-film deposition (MCM-D). In practice, however, combinations of these are more common, such as thin films on cofired-ceramic substrates with one layer for redistribution of the chip grid to that of ceramic grid or more layers for interchip connection, as indicated in Figure 12-82. Another combination that is receiving great interest is thin films on PWB by sequential deposition of one or more layers, in particular by additive processes. In addition, a number of companies, IBM and Hitachi in particular, have also practiced the other combination, lamination on ceramic substrate. A number of other alleged low-cost processes have also been reported, mostly based on either large-area or roll-to-roll processes which are capable of high-throughout, high-yield polymer film deposition and processing, as well as metallization. A few of these are described below.

12.8.1 IBM's Surface Laminar Circuitry

The best thin-film package proposed and developed to date is perhaps the surface laminar package developed by IBM Japan in 1986 [16]. In general, it overcomes the limitations of drilled through-holes in printed

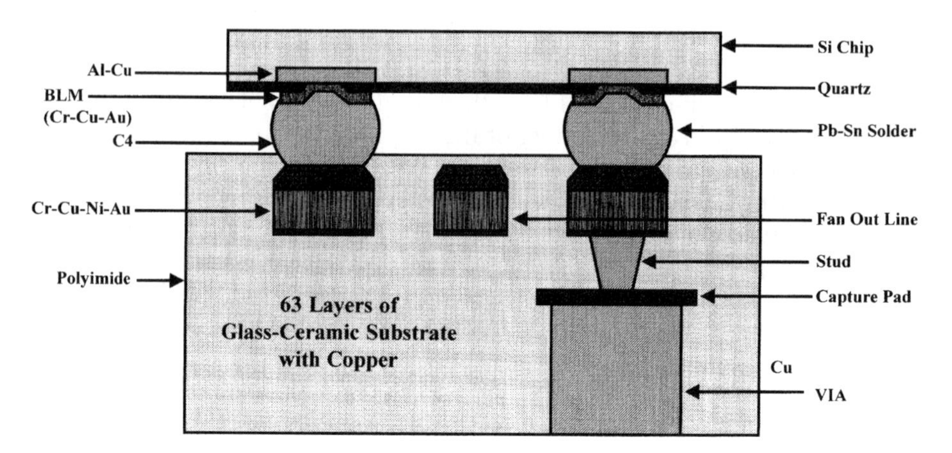

Figure 12-82. Thin-Film Redistribution on Glass-Ceramic Substrate

wiring board technology, enhances wiring density to the level of conventional thin-film photolithography technology and reduces the cost of the total wiring because of large-area processing. The process is relatively simple in that one or two layers of polymer-copper thin films, as illustrated in Figure 12-83, are deposited sequentially on a standard printed wiring board with or without internal wiring, using materials and processes that are very similar to the printed wiring board fabrication. On a printed wiring board about 600 mm × 700 mm in size, a photosensitive epoxy (Probimer 52®) from Ciba Geigy is deposited to about 50 μm thickness followed by exposure and development processes. Copper metallization is subsequently achieved by electroless seeding and **electroplating**. Typical dimensions attainable to date are 100-μm vias and 75-μm linewidths. IBM compared this technology with its other thin-film and ceramic technologies, as discussed in the previous section, for a 200-MHz workstation application and concluded that direct chip attach to this SLC board met all the performance requirements that were previously believed to be met only by ceramic and thin-film technologies.

12.8.2 Sheldahl's Roll-to-Roll Process

Figure 12-84 illustrates the process starting with 200 ft of polyimide film [156]. The typically 50-μm-thick polyimide film has a low modulus with expansion matching that of copper. Via generation is accomplished in 12 in. × 12 in. sections by laser ablation on a 200-μm grid. A 2000-Å thin layer of copper is then sputtered on both sides as well as in the vias, instead of the conventional electroless deposition. Using this as the electrical contact, a 5-μm film of copper is subsequently deposited, and wiring is subtractively etched. In applications where thicker copper is

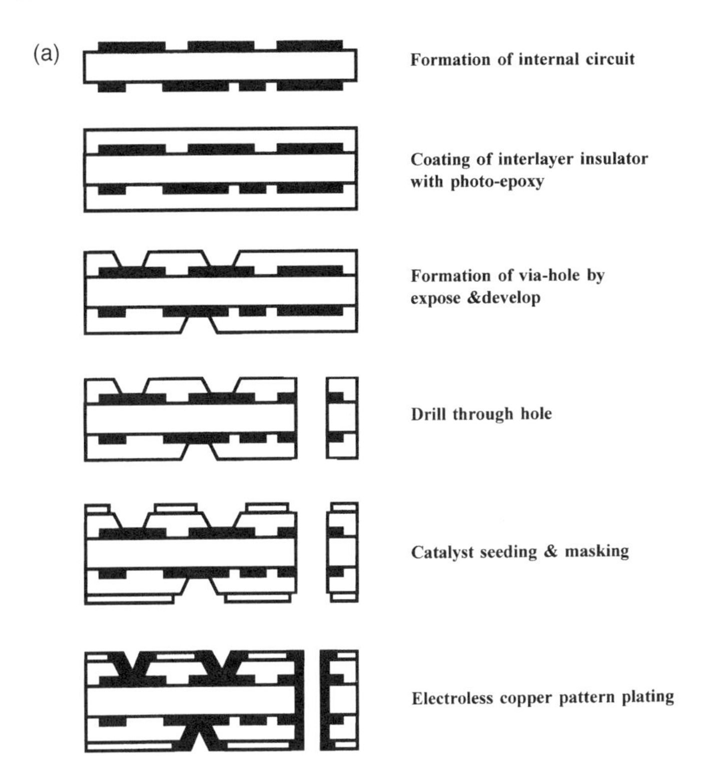

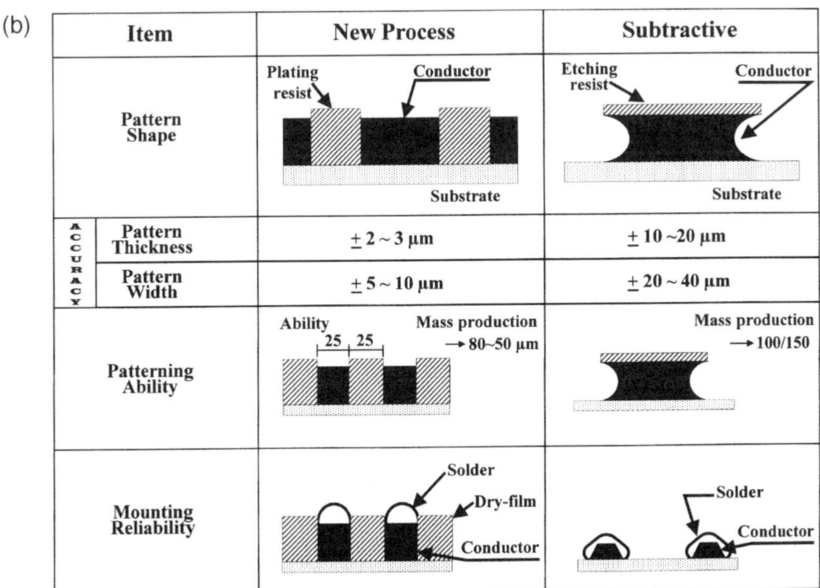

Figure 12-83. Low-Cost Fine-Line Thin-Film Process on PWB

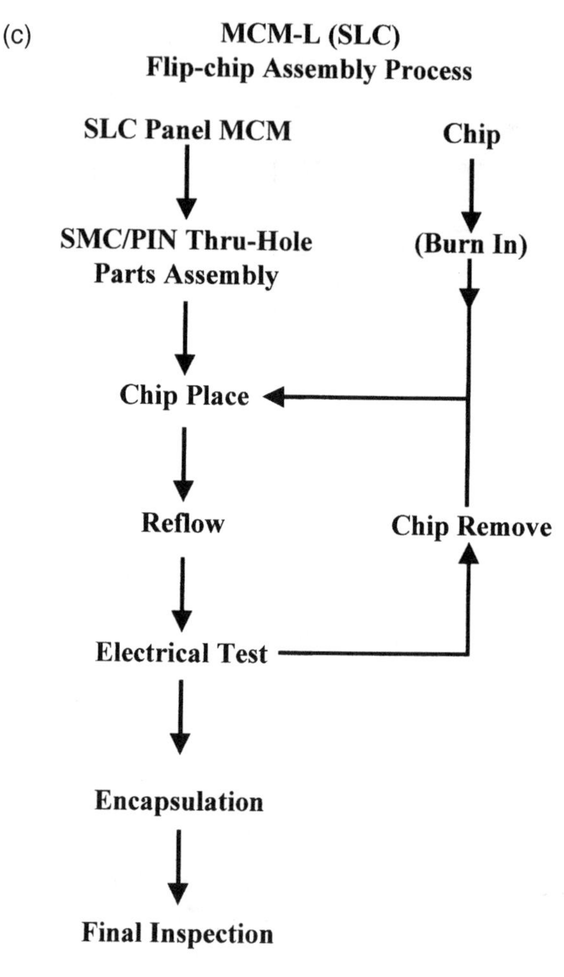

Figure 12-83. (*Continued*)

required, additive plating through photoresist is accomplished. Metallization is critical for stress control within the film. Next, a liquid polyimide is applied to the metallized layer and cured followed by laser ablation to open up opennings in overlay film to expose underlying pads. Electroless Au or Pd is done on electroless Ni to form pads for layer-to-layer connection. A number of these layer pairs are stacked, aligned, and laminated using Z-axis **conductive adhesives**. A typical structure built by this technology is illustrated in Figure 12-85. Typical thin-film dimensions achievable are 25-μm vias, 50-μm lines and spaces, and 125-μm pads.

12.8.3 GE's Conformal Multichip-Flex

Figure 12-86 depicts the overall process of this approach [157]. A typical 1-mil-thick flexible polyimide film with copper patterned traces on both sides and plated vias connecting the bottom and top traces is available from several vendors, such as the one by Sheldahl discussed above. A dielectric layer is deposited on either or both of the two surfaces to protect metallization or serve as a dielectric layer, which is done by the vendors. Chips and components are precisely placed face down onto the bottom side of the panel flex with a layer of polymeric thermoset adhesive. Vias are laser drilled through the flex down to the chip pads to provide chip interconnection. Next, a sputtered Ti barrier/2–6-μm Cu provides both via and upper-level wiring metallization. Thus, a total of three layers of interconnection is fabricated to satisfy most low-cost MCM applications, although more layers can be fabricated onto it. The final operation involves molding the IC and the thin-film package. Typical thin-film dimensions are 2-mil vias and lines spaced 2 mil apart. The low cost of this package is attributed to the large-area as well as low-cost materials and simplified processes. The flexible film provides the additional benefit of conformality.

12.8.4 DYCOstrate Thin-Film Technology

The process developed by Hewlett-Packard GmbH, Germany [158] was claimed to be of low cost enabled by major enhancement of plasma technology that was previously abandoned by IBM due to its uncontrollability and high cost. The concept of DYCOstrate is simple. It starts with a double-copper-cladded 50-μm- thick polyimide foil that is patterned by typical processes of expose, develop, and subtractive etch to expose the polyimide at via sites. The via holes are generated in the polyimide film by dry plasma etching with a shortened processing time of about 20–25 min. The vias are plated for top to bottom connection and lines are patterned and etched afterward. The final step to complete the 2D DYCOstrate fabrication is an electroless Ni/Au. Additional interconnection layers are fabricated by sequentially laminating single-copper-cladded dielectric

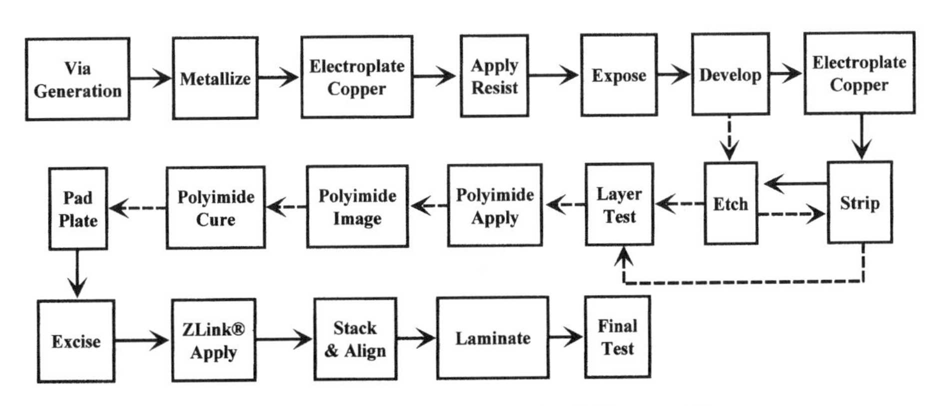

Figure 12-84. Sheldahl Roll-to-Roll Process Flow

material to both sides of the 2D DYCOstrate. Again, the outer copper foil is patterned to expose the dielectric polymer for plasma etching of vias to provide interlayer connections. The sequential buildup process described above can be repeated to obtain substrates with six or more layers. Among many variations of this process, one allows the 2D DYCO-strate to be laminated to a carrier by means of standard B-stage epoxy prepreg. In this case, electrical connection between the 2D DYCOstrate and the FR4 multilayer substrate is accomplished by mechanical drilling and plating. Figure 12-87 illustrates various structures that can be built. The typical wiring dimensions of this substrate are compared in Table 12-28 with standard PWB.

12.8.5 IBM's Integrated Rigid-Flex Thin-Film Package

One of the most sophisticated multilayer thin-film carrier technologies developed to date is a package developed by IBM referred to as high performance carrier (HPC) [159,160]. It preserves the advantages of low cost and volume production characteristic of PWB technology while providing the high wiring densities and high I/O counts of ceramic and thin-film technologies. HPC is also highly integrated in that carrier, flex cable, and connector are integrated into one structure.

The basic building block of HPC technology is a five-layer core consisting of two signal and three power/ground layers as shown in Figure 12-88. The structure has a triplate cross section for reduced coupled noise and EMI protection. The HPC is fabricated using a composite material consisting of amorphous silica filler particles in a polytetrafluoroethylene (PTFE) or Teflon® matrix supplied by Rogers Coporation. Wherever neces-

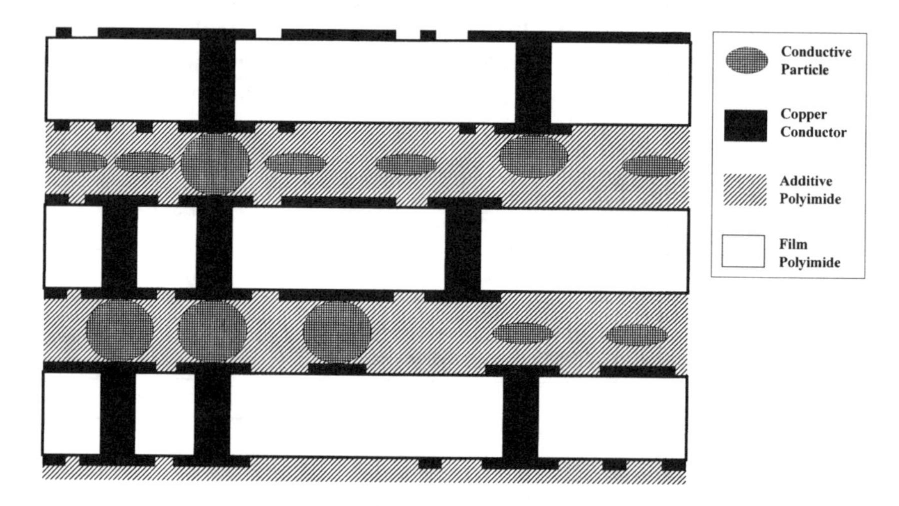

Figure 12-85. Sheldahl Substrate Cross Section

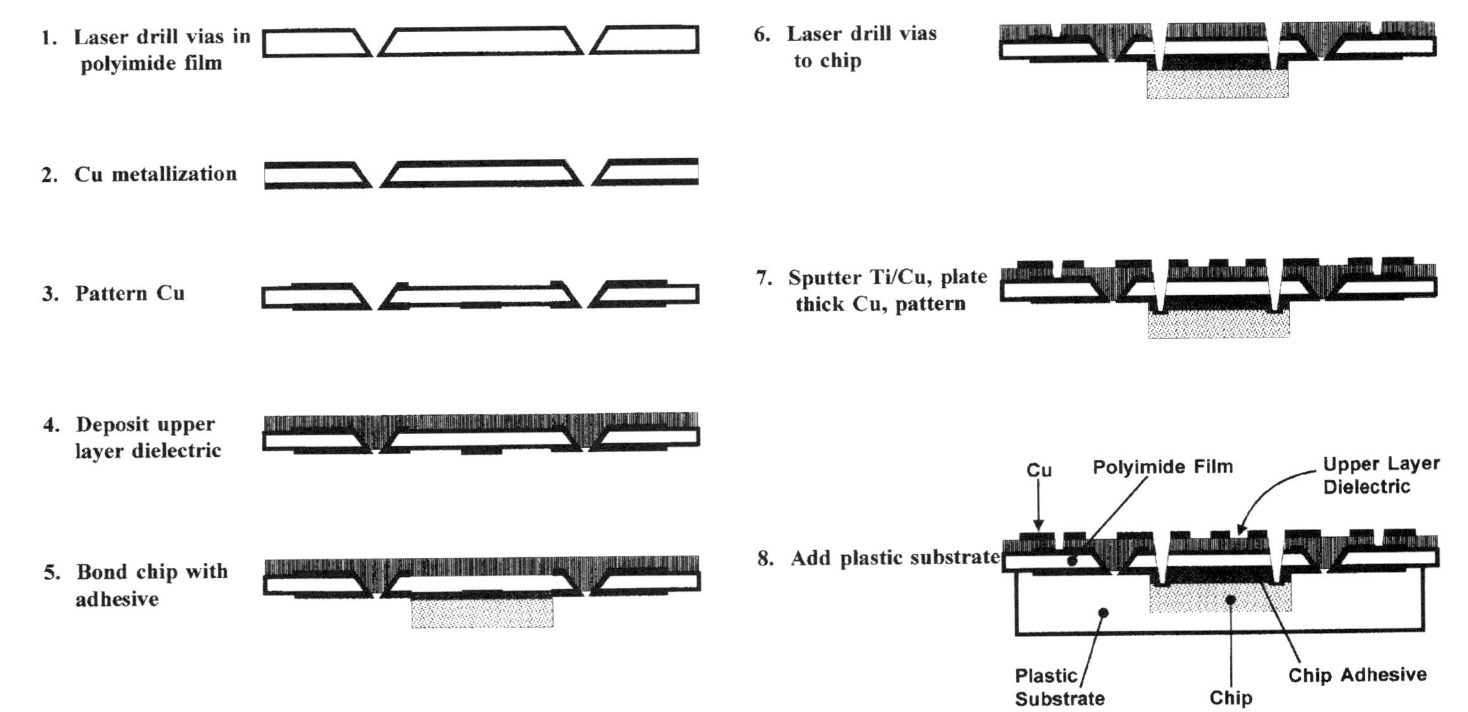

Figure 12-86. GE's Multichip-on-Flex Process Flow

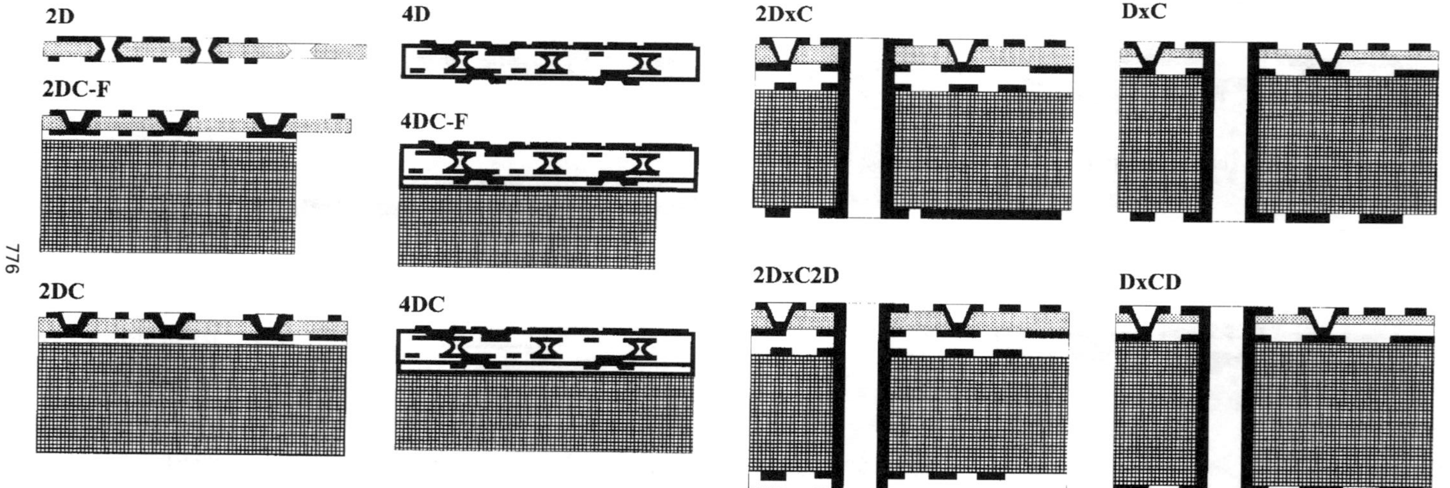

Figure 12-87. Overview of Various DYCOstrate Construction Alternatives

Table 12-28. Comparison of Design Features on Standard PCBs and DYCOstrate

Feature	Standard PWB	DYCOstrate Today	DYCOstrate Tomorrow
Via diameter	225–325 μm	75 μm	30–50 μm
Pad diameter	550–525 μm	200–350 μm	100–150 μm
Conductor width	100–150 μm	75–125 μm	40–60 μm
Space width	100–150 μm	75–125 μm	40–60 μm
Constructions	Not applicable	2D; 2D × C; 2D × C2D	4D, . . ., × D; × D × C × D; +F

sary, a copper-clad **invar** alloy typically 1 mil thick is utilized as an expansion compensator and reference plane to provide thermal-expansion control of the core as is shown in Figure 12-88. The properties of the PTFE and copper-invar-copper compensator used in fabricating the core are listed in Table 12-29. The details of the fabrication are discussed elsewhere [159,160]. The multilayer joining of the cores is accomplished by means of lamination which provides both high-yield metallurgical bonding of via lands on adjacent cores and adhesion of polymer between the lands, as illustrated in Figure 12-89. The key attributes of HPC technology are listed in Table 12-30. The integration of carrier and flex cables into a single structure is illustrated in Figure 12-90.

12.8.6 IBM's Thin-Film Transfer Technology

This technology was developed for the purpose of reducing the cost of MCM-D processing. Its uniqueness lies in that the multilayer polymer-

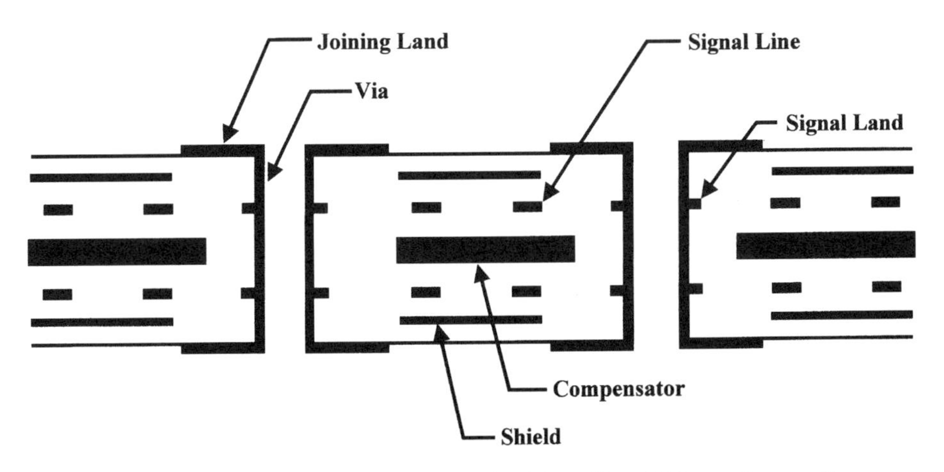

Figure 12-88. High-Performance Carrier Two-Signal/Three-Power Core

Table 12-29. Properties of Dielectric Material and Copper-Invar-Copper Compensator for HPC

	Dielectric	Copper-Clad Invar
ε_r	2.8	
Dissipation	0.0012 at 10 Mhz	
TCE		5.5 ppm/°C (20–120°C) 9.4 ppm/°C (20–380°C)
TCE x,y	25 ppm/°C	
TCE z	40 ppm/°C	
Elastic modulus	1.0 GPa (146 kpsi)	90 GPa (13 Mpsi)
Thickness	0.5–2 mils (12–50 μm)	1 mil (25 μm)
Dimensional stability (Repeatability)		20 ppm
Water absorption	0.15%	
Melt transition temperature	327°C	
Decomposition temperature	400°C	
Elongation	495%	1.4–2.0%
Tensile strength		550 MPa
Mass resistivity		0.60 Ω·gm/m^2

metal thin-film structure is built up on a reusable large-format glass substrate and then transferred onto other product substrates of choice, like laminates, silicon, alumina, cofired ceramic, and so forth. This process allows the tailoring of the thin-film structure before laminating to the product substrates. Cost reduction is achieved from several aspects. First, the thin-film process becomes independent of the choice of product substrate and a single standardized foundry would be able to support the thin-film fabrication for many different product types. As a consequence, the cost of retooling the production line is spared. Second, the reusable large-format glass substrate enables the optimization of processing and the adoption of large-area processing from the flat panel display industry for high-throughput, high-yield and low-cost fabrication. Third, this technology allows the production of low-volume applications at the cost of high-volume manufacturing.

The basic steps and key features of this technology are illustrated as in Figure 12-91 [161] and summarized as follows. A polymer (polyimides as in this technology) is first coated on the glass substrate serving as the release layer. Standard multilayer thin-film interconnection structure is subsequently fabricated. A rigid frame is attached to the surface afterward for the purpose of handling and testing the multilayer thin-film aggregate after release, as well as controlling the distortion of the thin-film pattern when it is separated from the glass substrate. Stresses origi-

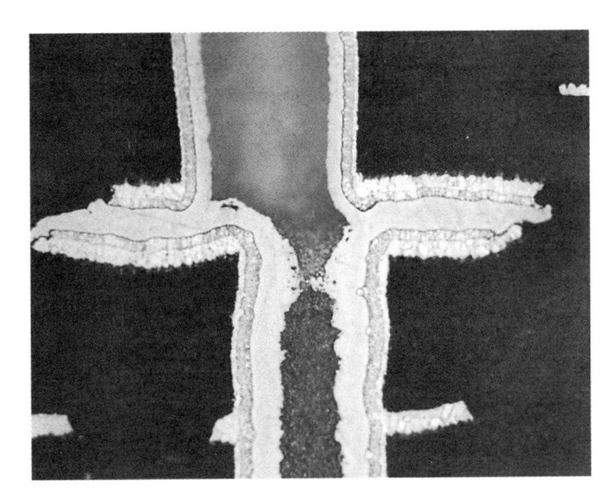

Figure 12-89. High Magnification of Metallized Bond Between Adjacent Core Vias. Courtesy of David Light, IBM.

nated from the processing and material property mismatches can result in severe distortion of the thin-film aggregate after release. With proper frame choice, distortion was controlled well below 0.02%. The separation of the thin-film aggregate from the glass substrate was accomplished by a scanning laser ablation process through the underlying glass substrate. The laser fluence was well controlled so that the polymer release layer is ablated without any damage to the thin-film wiring layers. The glass substrate must be transparent to the laser radiation and the polymer release layer must be thick enough (>10 μm in this case) to avoid any significant

Table 12-30. HPC MCM-L Technology Capabilities

	Typical	Min./Max.
Via pitch, mil (μm)	20 (500)	9 (225)
Via diameter, mil (μm)	4 (100)	3 (75)
Linewidth, mil (μm)	2 (50)	1 (25)
Line spacing, mil (μm)	4 (100)	1.5 (28)
Lines/channel	2	1
Wiring capacity, in./in.^{2}/layer (wires/cm/layer)	100 (40)	100 (40)
Layer count	6–25	36
Cross-talk, %	1.5	0.1
Time delay, ps/cm	55.8	
Z_o, Ω	50 ± 5	50–80

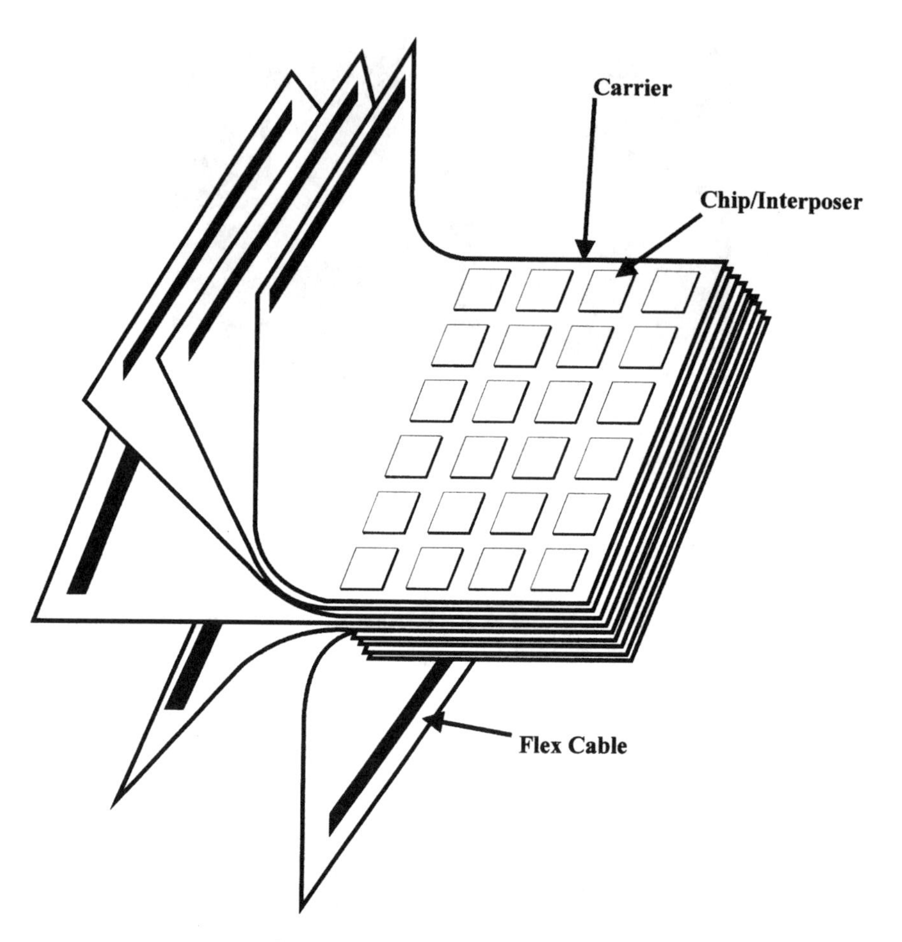

Figure 12-90. Multiple Flex Cables Exiting HPC Carrier

stressing of the thin-film structure induced by the ablation process. Depending on the product substrate onto which the thin-film decal is laminated, the process differs. For passive substrates, the lamination is simply done with polymeric adhesives. For active substrates, surface preparation may be necessary for both the substrate and the thin-film aggregate. Electrical joining can be realized by any of Au/Au thermal-compression, liquid-phase, or transient liquid-phase processes. Uniform application of pressure was found very critical to ensure good electrical joining.

12.9 INTEGRATED PASSIVES IN THIN-FILM PACKAGE

The integration of active circuit functions using silicon integrated circuits is a major success story of the twentieth century. This integration

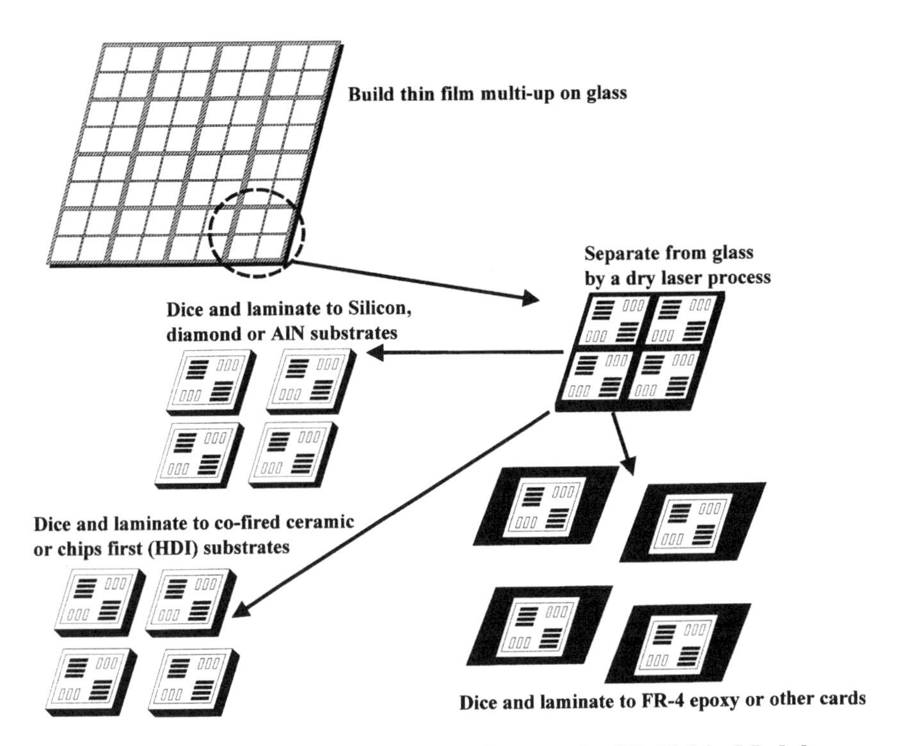

Figure 12-91. Flexible Manufacturing Process for Multichip Modules

has brought forth the development of the computer and telecommunication industries and the information age in which we now live. This success of silicon devices merely further underscores the lack of advancement in integration of passive components, notably resistors, capacitors, and inductors that are the complementary components of active microelectronic circuits.

The role of passive components has changed from a minor portion of electronics system to a major occupant of package realty and important cost driver. The number often exceed by an order of magnitude the number of active ICs in many mixed-signal devices in both consumer and industrial applications. Significant efforts have been and are being made to develop a new class of packages with integrated inductor/capacitor (L/C) and resistor/capacitor (R/C) circuits using advanced materials and processing to provide cost and reliability improvements over conventional nonintegrated components.

The primary advantage of passive integration is the reduced number of electrical contacts or transitions. A traditional assembly has its internal contacts between the components, the transition from the components to the attachment materials, usually solder or epoxy, and finally the transition from the attachment materials to the interconnects on the substrates. By integration, these transitions and the associated electrical performance losses are reduced drastically or eliminated entirely.

A second advantage is increased reliability. Failures occur mainly at mechanical transitions or interfaces between dissimilar materials and geometrical inhomogeneities. Stress concentration often present at these places originated from mismatch of material properties (e.g., thermal coefficient of expansion (TCE) and Young's modulus) and geometry effects. Merely by reducing the number of transitions and simplifying the geometry within the package by means of passive integration, reliability of the package is greatly increased.

An additional advantage of integrated passive components is cost savings. Few additional processing steps are required for passive component integration, whereas a large number of assembly steps are eliminated resulting in cost saving. Less materials are needed for integrated passives than discrete passives resulting in additional cost reduction.

Higher packaging efficiency is another big incentive for passive integration. Component size and component count are the typical drivers for assembly size. By burying the passive components in the substrate, the footprint constraints are removed. The same passive components can be effectively spread "two dimensionally" within the package substrate or the package itself in traditionally unused or waste area.

12.9.1 Current Development and Applications of Integrated Passives

There are various approaches to implement integrated passive components in thin-film packaging. Several materials like tantalum-silicon, TiW, or tantalum nitride have been employed as thin-film resistors available at various resistance values. For integrated capacitors, Si_3N_4, SiO_2, or (Pb, La,Ba, Zr, Ti) oxides with a high permittivity have been used for dielectrics. Nickel-iron permalloy magnetic material combined with copper conductor were used to fabricate integrated inductors. Several investigators have successfully implemented integrated passive components for MCM-D fabrication which are summarized below.

12.9.1.1 Resistors

AT&T Bell Labs [162] utilized existing IC facilities for the development and manufacturing of integrated silicon-on-silicon MCM packages using polyimide/aluminum thin-film processes. They have fabricated Ta-Si thin film resistors by sputter deposition and reactive ion etching (RIE) patterning. Sheet resistance values of 8–20 Ω/square have been achieved.

At W.L. Gore and Associates [163], standard IC equipment was used to fabricate polyimide-Cu on glass-ceramic substrate MCM-D packages with embedded resistors. Better performance and cost savings were claimed for the reduced parasitics, increased routability and component reduction. In a total of eight metal layers, the fourth layer from the bottom was dedicated for thin-film resistors. TiW was chosen as a resistor material for compatibility with the existing thin-film deposition and etching equipment. The resistor was sputtered using a 15/85 wt.% TiW target. The as-deposited sheet resistance was 2.4 Ω/square with a 300-nm film thickness and deposited resistivity of 72 m Ω·cm. Curing processes of the subsequent polyimide layers above the thin-film resistor layer would further increase the sheet resistance to 3.2 /square. The patterning of the thin-film resistors was accomplished by the deposition and patterning of a 500-nm SiO_2 hard mask layer followed by TiW layer etching in hydrogen peroxide. The shape of the thin-film resistors are determined by the 62-Ω termination resistance to the power plane. One potential concern with this technology is the poor thermal conductivity of polyimide which limits the prompt transfer of heat generated by the thin-film resistors.

Sputtered, high-resistance tantalum nitride has been used as a resistor material for thin-film multichip-module application by NTT [34], NTK [57] and Boeing [164]. In NTT's high-speed ATM switching systems, thin layer of tantalum nitride was deposited on a cofired alumina-ceramic substrate and patterned to form resistors. Six layers of Cu-polyimide interconnection were built thereafter. The tantalum nitride thin-film resistors were shown to be very reliable upon current supply and self-heating for 3000 h and humidity and high-temperature tests as well. No significant resistance change was observed. Signal reflection was virtually eliminated with the integration of termination resistors. In the application by Boeing, Ta_2N thin film was deposited directly on the uppermost polyimide layer. The resistor films were patterned using RIE in a SF_6/O_2 mixture. The patterned resistors had a value of 50±1 Ω and a temperature coefficient of resistance (TCR) of ±100 ppm/°C. Experiments were performed to assess the feasibility of laser trimming a resistor located on the polyimide without causing any damage to the underlying interconnects which indicated the possibility of rework on this type of resistors.

The High Density Electronics Center at University of Arkansas has investigated into the feasibility of embedding thin-film resistors, capacitors and inductors in flexible polyimide films for both MCM-D and MCM-L applications [165]. NiCr, TaN, and CrSi were chosen as the resistor materials. Standard fine line lithography was used to fabricate the devices on a 25-μm or 50-μm-thick polyimide film. A wide range of resistance from 50 Ω to 10^6 Ω was obtained. In addition, the same technology was also used to fabricate thin-film capacitors of up to 22-nF capacitance and inductors of up to 137-nH inductance. Ta_xO_y and $BaTiO_x$ were used as the dielectric materials, whereas a Cu line was used to form the spiral inductor.

Deutsche Aerospace (Dasa) has employed NiCr buried layer as the thin-film resistors in their double-face populated MCMs for microwave transmit/receive (TR) modules in radar application [166]. Typical sheet resistance was 35–100 Ω/square. Laser trimming was used for the definition of the resistors.

12.9.1.2 Capacitors

Several inorganic dielectric materials have been studied for integrated thin-film capacitors as listed in Table 12-31 [167]. In the AT&T Bell Labs integrated silicon-on-silicon MCM packages discussed above, single-layer Si_3N_4 or dual-layer SiO_2/Si_3N_4 were the dielectric materials to form the thin-film integrated decoupling capacitors. The additional SiO_2 layer safeguards against power/ground shorts due to defects in the nitride. In principle, the dielectric is sandwiched between the highly doped Si substrate acting as a ground plane and the subsequent aluminum contact. For the formation of floating capacitors, a single Si_3N_4 layer is sandwiched between a Ta-Si bottom electrode and the first-level aluminum contact (Fig. 12-92 [162]). The Si_3N_4 layer was deposited by LPCVD whereas the SiO_2 layer was simply by thermal growth in an oxidation furnace. Capacitance values of 33–40 nF/cm^2 have been obtained. In Fujitsu's process for integrated capacitors [168], $Ba(Zr,Ti)O_3$ (BZT) thin film was deposited on a $Pt/SiO_2/Si$ substrate by multiple cathode RF magnetron sputtering. A dielectric constant value of 146 was achieved at a Zr content of 1.5 mol%. Dimos et al. [169,170] at Sandia National Labs used $(Pb,La)(Zr,Ti)O_3$ (PLTZ) thin film as the dielectric material for integrated capacitors. They have achieved high dielectric constant ($\varepsilon_r \geq 900$), low dielectric loss ($\tan\delta = 0.01$), good leakage resistance ($\rho > 10^{13}$ Ω·cm at 125°C), and good breakdown field strength (E_B ~1 MV/cm). The sol-gel technique was employed for the fabrication of the PZT and PLZT thin films. After spin coating of the solution, the substrate was heated to ~300°C to pyrolyze the organic species in the precursor film followed by firing at 650°C for crystallization. Pt was sputter-deposited to form the

Table 12-31. Dielectric Materials for Thin-Film Capacitors

Material	Dielectric Constant	Band Gap Energy (eV)	Temperature Coefficient of Dielectric Constant (ppm/°C)
SiO_2	4	11.0	
Si_3N_4	9	5.1	
Ta_2O_5	25	4.0	+250
TiO_2	98	3.0	–750

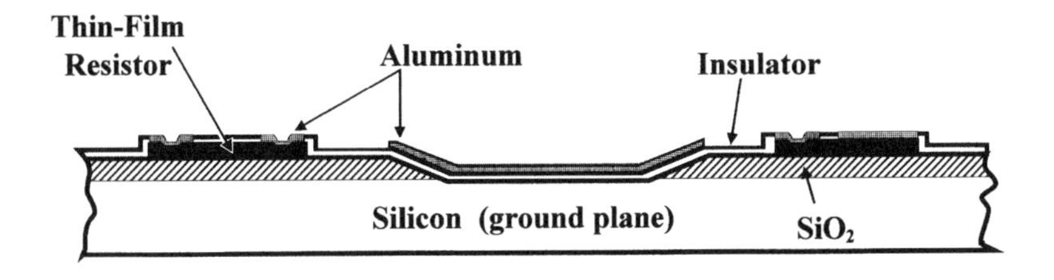

Figure 12-92. Passive Device Structures

bottom and top electrodes of the capacitors. The typical dielectric film thickness was between 90 and 100 nm and thicker films may be obtained by multiple coatings. The heat treatment should be optimized for multilayer films to avoid cracking. nCHIP has adopted another kind of decoupling capacitor technology for its MCM fabrication [171-173]. A 2-μm-thick Al layer was deposited onto a silicon substrate as the ground plane, followed by a 0.15-μm-thick anodized Al_2O_3 layer as the dielectric. Another layer of Al was deposited to 2 μm thickness afterward to form the power plane. Multilayer interconnection was then fabricated using Cu/SiO_2 processing. Figure 12-75 [171] is a schematic showing such a MCM structure. A sheet capacitance of 50 nF/cm^2 was obtained with this integral capacitor. Ground bounce was shown to be greatly reduced by substituting the discrete capacitors with integral decoupling capacitor.

Recently, the Packaging Research Center at Georgia Tech demonstrated a novel technology for the fabrication of thin-film integrated decoupling capacitors [174]. Several photosensitive and nonphotosensitive dielectric polymers (e.g., polyimides) were filled with high-dielectric-constant barium titanate or lead magnesium titanate fine powders to high volume-fraction. These polymer/ceramic composites have the combined advantages of high dielectric constant (up to 65) from the barium titanate and lead magnesium titanate ceramics, and the low processing temperature of polymers. An additional advantage of this type of composites is the capability of patterning by photolithography due to the transparency of the ceramic component in the UV range and wet etching as is with the original polymers. A specific capacitance as high as 22 nF/cm^2 and a loss tangent of <0.032 at 100 KHz were achieved with about 60% ceramic loading as illustrated in Figure 12-93. The dielectric constant of the composite was found to be stable over a wide frequency range. With higher ceramic loadings and thinner films, a higher dielectric constant of the composite and higher specific capacitances are expected to be achieved.

12.9.1.3 Inductors

Two micromachined integrated inductors, bar type and meander type as shown in Figure 12-94, are implemented by Georgia Tech [175] on a silicon wafer using modified, IC-compatible, multilevel metallization techniques. In the case of the bar-type inductor, a 25-μm-thick nickel-iron permalloy magnetic core bar is wrapped with 30-μm-thick multilevel copper conductor lines. For an inductor size of 4 mm × 1 mm × 110 μm thickness with 33 turns of multilevel coils, a specific inductance of approximately 30 nH/mm^2 at 1 MHz is achieved. In the case of the meander-type inductor, the arrangement of conductor wire and magnetic core is reversed (i.e., a magnetic core is wrapped around a conductor wire). This inductor size is 4 mm × 1 mm × 130 μm and consists of 30 turns of a 35-μm-thick nickel-iron permalloy magnetic core around a 10-μm-thick sputtered aluminum conductor lines. A specific inductance of 35 nH/mm^2 is achieved at 1 MHz.

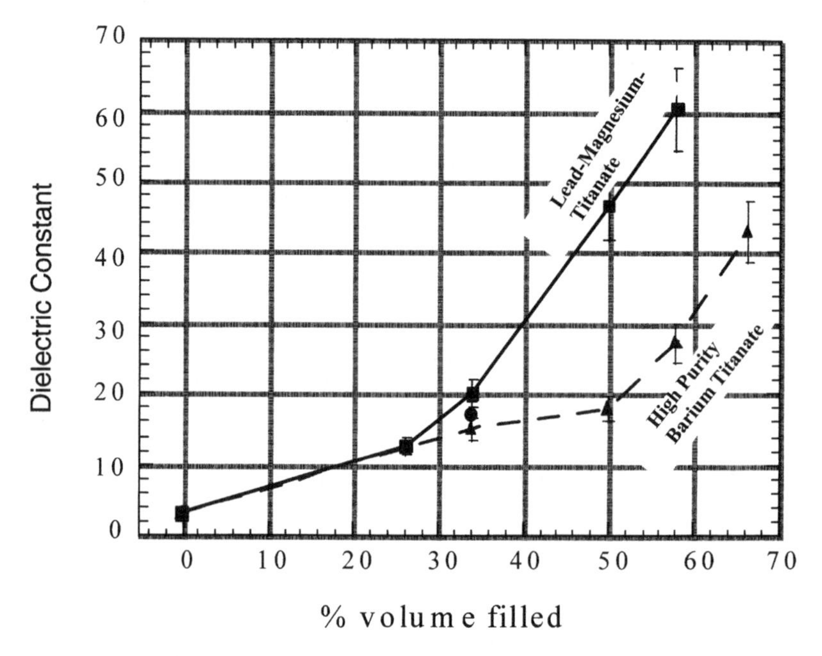

Figure 12-93. Effective Dielectric Constant of Polymer/Ceramic Composite

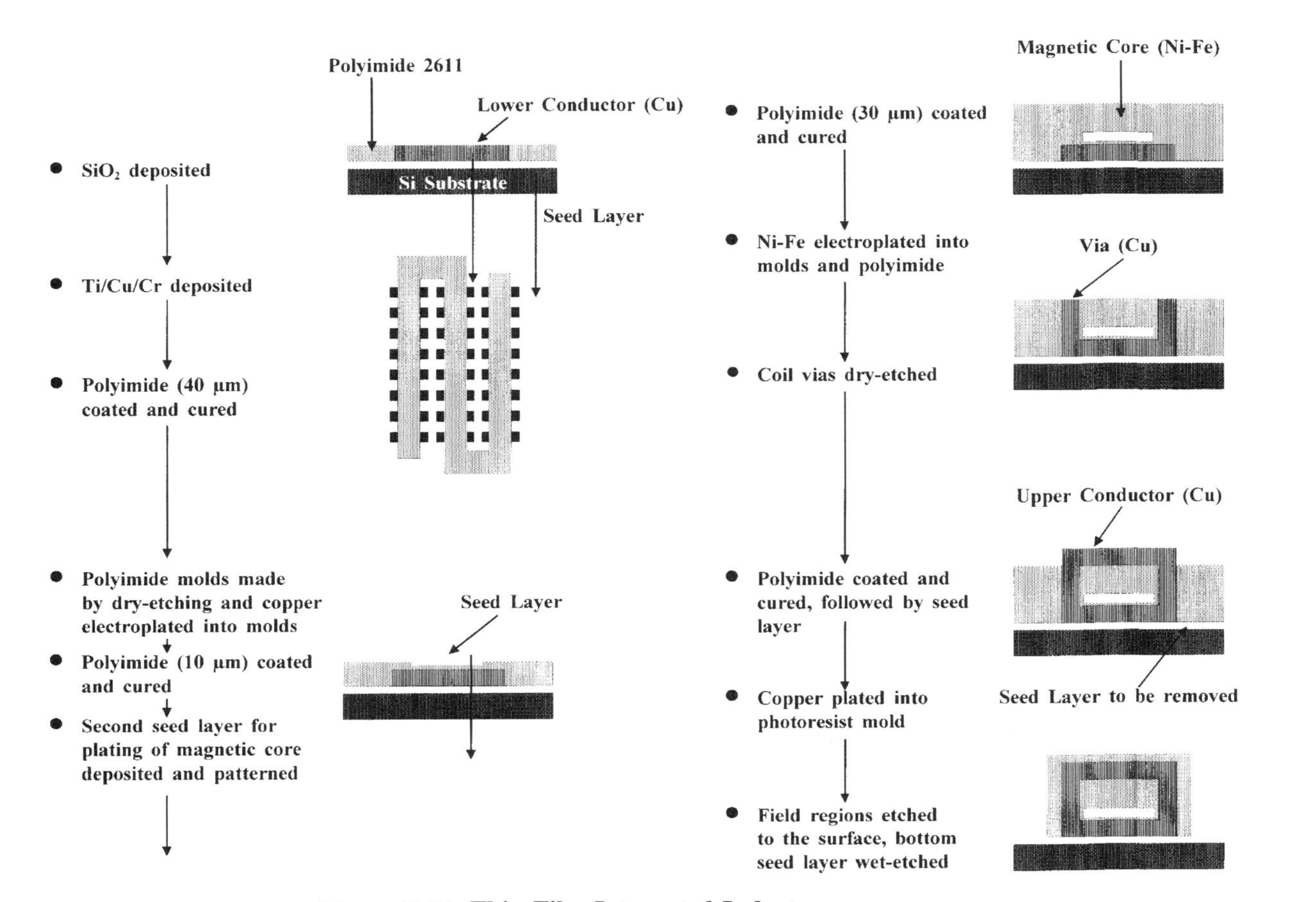

Figure 12-94. Thin-Film Integrated Inductors

12.9.2 Future Directions of Integrated Passives in Thin-Film Package

As discussed above, the integration of passive components has a profound impact on the performance, space, and cost of thin-film packages. A variety of organic and inorganic materials and their composites have been investigated and applied to the fabrication of integrated passives in the form of thin films. Depending on the specific type of passive components (resistors, capacitors, or inductors), the materials employed, and the intended applications, the fabrication processes vary widely. The merits of enhanced performance, specifically, signal integrity and clock speed, and increased packaging efficiency with the implementation of integrated passives are straightforward and technically feasible; it is also the major driver for the current research and development activities. Cost reduction through the integration of passives is less studied and more complicated and controversial. Despite some isolated applications of integrated passives and the current research activities as described in Section 12.9.1 there is still a general lack of experience and knowledge on the technology and the widespread commercial application of integrated passives. Complex materials and processing technology are the key contributors to the cost of integrated passives. As a consequence, the majority of the current applications of integrated passives are mostly performance driven rather than cost driven [42], and integrated passives are mostly found in high-end applications. Significant effort is to be made to lower the cost of integrated passives so that this technology can compete effectively with the vast discrete passive industry in the high-volume, low-end market. This demands major progress in the development of low-cost materials and processes compatible with the thin-film technologies in multichip-module manufacturing.

12.10 FUTURE DIRECTIONS OF THIN-FILM MATERIALS AND PROCESSES

Thin-film or fine-line MCM substrate with wiring density in excess of 400 cm/cm^2 has been and continues to be very expensive, typically \$50–100/in.^{2} for a total of five wiring levels (two for signal, one for power, one for ground, and one for reference plane) as shown in Table 12-32. The technology has not yet taken off in industry, except by IBM, NEC, and Hitachi for their mainframes and supercomputers where performance rather than cost is the priority. The reasons for this high cost are multiple. Materials and processing costs of thin-film substrates are unacceptably high compounded by other obstacles, such as the general lack of industrial infrastructure, lack of known good die (KGD), nonstandard customized design and manufacturing, long turnaround time and small volume. The material and processing costs include expensive dielectric

Table 12-32. Ultralow-Cost Characteristics of Proposed MCM

	Today	PRC Proposed	Approximate Improvement Factor
Size of substrate (in.)	5 × 5	16 × 16	10×
Facilities cost ($/ft^2)	1000 (Class 100)	200 (Class 1000)	5×
Equipment cost ($)	20–150 M	4–20 M	5×
Material cost ($/kg)	1000	100	10×
Material usage (%)	20 (spin)	80 (nonspin)	4×
Raw process time (hr)	200	100	2×
Cost ($/in^2)	30–100	2–5	50×

materials, expensive cleanroom dry processes, expensive tools, and the long processing time to fabricate MCM. Fundamental issues related to MCM substrate materials, processes, and tools to achieve the low-cost objective and meet the industry needs have not been solved yet and, hence, form the basis of future directions.

The future goal of low-cost MCM substrate in the industry is about $2–5/in.^{2} down from the current $50–100/in.^{2} as mentioned above. The strategy to achieve this goal appears to be exploration, development, and prototype of large-area processing: 18 in. × 18 in. (324 in.^{2}) from the current 6-in. round wafer (28 in.^{2}). Figure 12-95 depicts the advantage of large-area processing: a factor of 20× improvement in terms of number of MCMs by replacing the traditional 6-in. round wafer substrate with 18-in. × 18-in. large substrate. Figure 12-96 [176] illustrates the cost reduction both as a function of size (6 in. vs. 24 in.) and production volume of MCMs. Because the cost of manufacturing, in terms of process steps, raw process time, and so forth, is approximately the same for both sizes, the adoption of a large-size substrate should reduce the cost by about 10× from the current $50–100/in^2. Two challenges must be overcome to meet the proposed target of $2–5/in^2. For one, the above 10× improvement assumes the yield of large substrate is the same as small substrate. This is clearly not the case with current fine-line multichip-module technology. In addition, significant cost reduction must be made to bring the cost further down to $2–5/in.^{2}. The science, technology, prototype, and manufacturing leading to the cost goal involves, therefore, low-cost novel materials and chemical processes that are immune to airborne particles, and low-cost novel polymer- and metal-deposition processes. The expected improvements from all of these are listed in Table 12-32.

The low-cost substrate requires both fundamental and practical approaches to materials, processes, and tools. DARPA has taken the initiative

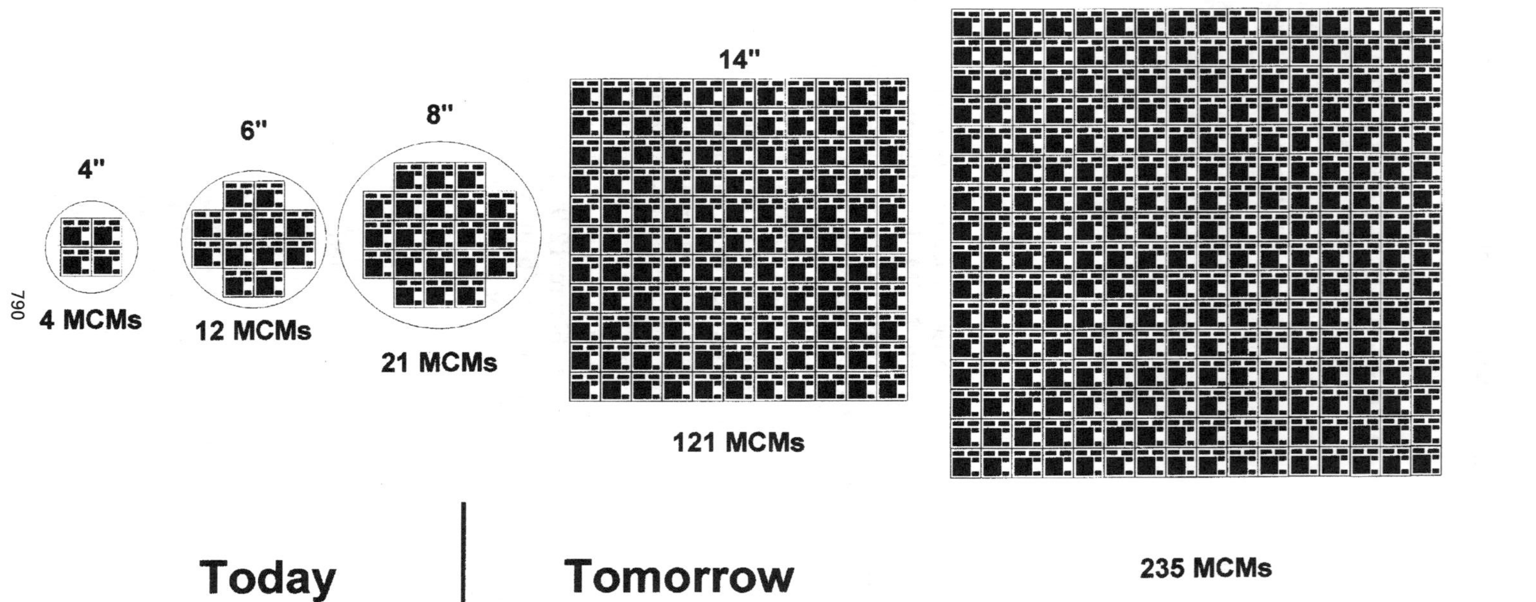

Figure 12-95. The Advantage of Large-Area MCM

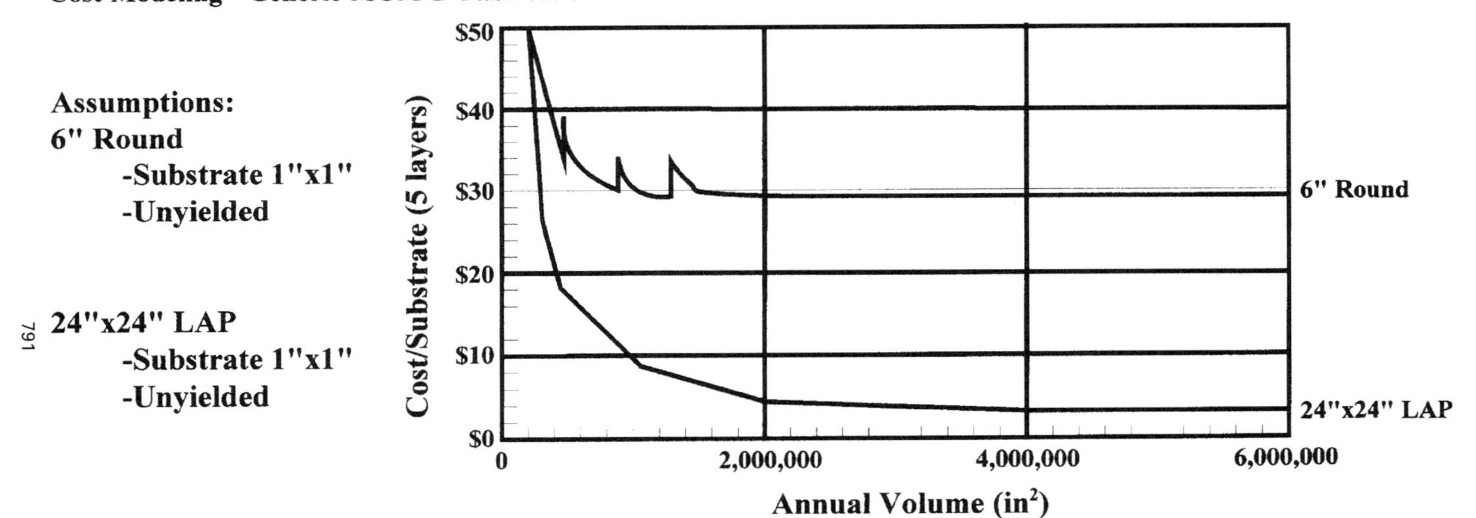

Model Prediction (Generic Model)

-6" round wafer format cost approaches $30/in^2 (1"x1" -5 layer)
-LAP offers the lowest unit cost - approaching $3/in^2 (1"x1" -5 layers)

Figure 12-96. Economics of LAP

to develop large-area tooling in cooperation with Hughes, nChip, Micro-Module Systems, Texas Instrument, and Boeing to alleviate the high cost of production equipment. Georgia Institute of Technology has been awarded a national Engineering Research Center (ERC) by NSF to concentrate on novel materials and processes that are compatible with these tools in complementary mode and to demonstrate the $2–5/in.^{2} goal for MCMs.

Figure 12-97 illustrates one process proposed by Packaging Research Center at Georgia Institute of Technology to achieve low cost. It involves large-area deposition of polymer by extrusion, curtain, or meniscus coating, followed by low-cost via formation in photosensitive polymer by large-area, full-field, and low-cost photolithographic tool. The via and pattern metallization is accomplished by direct electroless and/or electroplating of copper.

There are basically two ways to reduce the cost of polymer deposition. One is to apply conventional $1000/kg polymer onto a large area with high material usage such as by an extrusion coater. The other is to introduce novel dielectric polymers that are about 10–20× lower in cost ($50–100/kg) than those being currently used. When low-cost dielectric polymers are available, the cost of polymer deposition becomes relatively less sensitive to the selection of coating technology, both the existing or modified spinner or, for that matter, extrusion or other large-area coaters can be used. When low-cost dielectric material is not available and cost reduction solely relies on large-area processing, the material cost of a five layer thin-film substrate can make up as much as 70% of the total cost, as illustrated in Figure 12-98 [176]. From this discussion, it is obvious that low-cost materials and large-area processes are vital for future thin-film packaging technology. These are reviewed below.

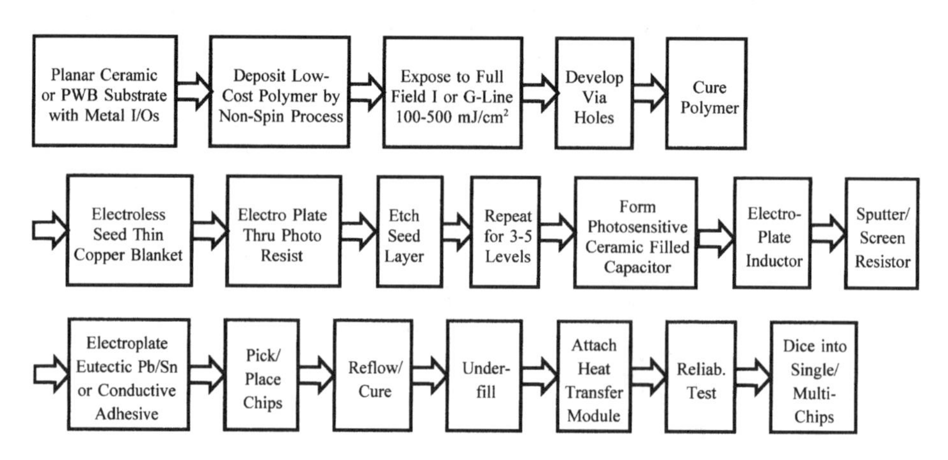

Figure 12-97. PRC Process for Next-Generation Packages

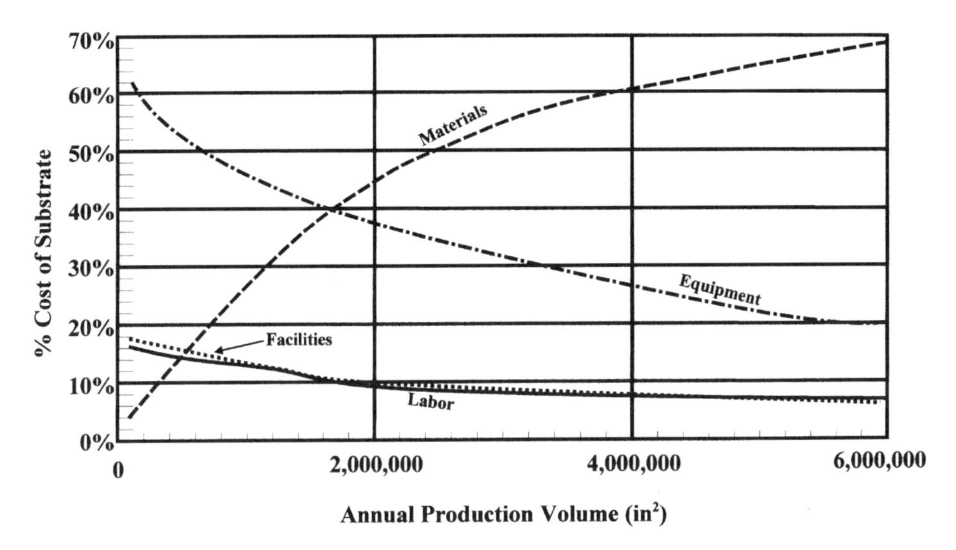

Figure 12-98. MCM-D Substrate Cost Breakdown. As volume increases, the percentage of substrate cost contributed by equipment capital significantly decreases and the percentage contributed by materials increases.

12.10.1 Polymeric Materials

Although both organic and inorganic materials have been used as dielectrics in thin-film packages as discussed in Section 12.4, polymeric materials are the most promising candidates for the future thin-film packages. The versatility of materials and properties, low material cost and process easiness make polymers unparalleled by any other dielectric materials. As in many other fields, polymers are assuming an increasingly important role as dielectrics in thin-film package, despite the unstable (relative to most inorganic materials) and nonhermetic nature of polymers. Excellent electrical and process properties and very low materials cost are the key elements for an idea dielectric polymer, which should form the emphases of future material development.

There are generally three types of dielectric polymers that are in use or development today, as discussed in Chapter 11, "Polymers in Packaging." These are polyimides in a number of chemical families such as PMDA-ODA, BPDA-ODA, and others, Benzocyclobutene (BCB) and epoxy. The important properties of these polymers are summarized in Table 12-33. Polyimides have long been used for thin-film packaging as a dielectric. A variety of derivatives have been developed by a number of companies to fulfill the various requirements as described in Section

Table 12-33. Dielectric Polymers

Polymer Name	Polymer Type	Dielectric Constant	Dissipation Factor	TCE (10^{-6}/°C)	Tg (°C)	Modulus (GPa)	Tensile Strength (MPa)	Percent Elongation	Moisture Absorption (wt.%)
Amoco Ultradel 4212	Fluorinated polyimide	2.9	0.005	50	295	2.7			0.9
Amoco Ultradel 7501	Photosensitive polyimide	2.8	0.004	24	>400	3.4			3.4
Cemota IP 200	Polyphenylquinoxaline	2.7	0.0005	55	365	2.0	117	8-12	0.9
Dow Cyclotene 3022	Benzocyclobutene	2.7	0.0008	52	>350	2.0	85	6	0.2
Du Pont PI-2545	Standard polyimide	3.5	0.002	20	>400	1.4	105	40	2–3
Du Pont PI-2555	Standard polyimide	3.3	0.002	40	>320	2.4	135	15	2–3
Du Pont PI-2610D/2611D	Low stress polyimide	2.9	0.002	3 (x-y)	>400	8.4	350	25	0.5
Du Pont PI-2732/2733	Photosensitive polyimide	2.9		25	>400	6.0	192	8	1.5
Hitachi PIQ-13	Standard polyimide	3.4	0.002	45	>350	3.3	130	20	2.3
Hitachi PIQ-L100	Low stress polyimide	3.2	0.002	3	410	11	380	22	1.3
National Starch EL5010	Preimidized polyimide	3.2	<0.002	34	214	2.8	150	7	1.3
National Starch EL5512	Fluorinated preimidized polyimide	2.8	<0.002	35	225	2.8	150	6	0.8
OCG Probimide 400	Preimidized photosensitive polyimide	3.0	0.003	37	357	2.9	147	56	2.0
OCG Probimide 500	Low-stress polyimide	2.9	0.003	6–7	400	11.6	444	28	0.7
Toray UR-3800	Photosensitive polyimide	3.2	0.002	40	280	3.4	140	11	1.1
Ciba Geigy LMB 7081	Photosensitive epoxy	4.1	0.025		130	3.3			<0.5
Shipley Multiposit XP-9500 CC	Photosensitive epoxy (Novolac)	3.2	0.007	60–70	>180	2.1		6–8	

12.4. Although polyimides are still the major dielectric polymer for today's thin-film package, some inherent concerns with polyimides remain. These include (a) the reactivity with copper necessitating the cladding of copper lines in most applications which complicates the processing and increases the cost, (b) the high material cost, (c) significant moisture uptake causing reliability concern, and (d) the relatively poor adhesion to copper which requires the application of adhesion promoter. Benzocyclobutene (BCB) was developed as an alternative to address the problems associated with polyimides. It has a lower dielectric constant, lower moisture uptake, higher film retention, better planarization, and compatibility with copper metallization and is capable of rapid thermal curing which leads to increased throughput and reduced processing cost. Adhesion promoter is, in general, required. Nevertheless, issues related to materials cost, oxidation resistance, and further process development are to be addressed properly for the wide acceptance of BCB. Epoxies are the prevalent dielectric materials used for printed wiring board fabrication and are known for low-end, low-cost, and high-volume applications. Increased interests have been shown to extend the applications of epoxies into thin-film packages.

Not included in Table 12-33 is a material that has not received much recognition. Polyolefins are long being known for numerous nonelectronic applications and are being developed for electronic applications as dielectrics by B.F. Goodrich. Two key fundamental properties of polyolefins, a very low dielectric constant and a very low moisture absorption, made them very attractive candidates as electrical-grade polymers. Primary results on the polyolefin patented by B.F. Goodrich showed that it can be spin or extrusion coated onto a printed wiring board and cured below 250°C. Its material cost is expected to be about 10–20× lower than that of BCB or polyimide and is hence about the same as epoxy. However, it has superior electrical and thermal properties over all the three aforementioned electronic polymers. The synthesis of polyolefin is by transition-metal catalyzed polymerization of cyclic olefin monomers derived from the Diels-Alders adducts cyclopentadiene. Flexibility in monomer and catalyst selection can lead to polymers with a broad range of properties and structures.

12.10.2 Large-Area Polymer Deposition

As discussed in Section 12.4 the capability of a large-area process (LAP) is essential for future low-cost thin-film packages. A large-area polymer deposition technology with good film thickness uniformity, very low defect rate, high throughput, and low material waste is one of the critical enabling technologies for the implementation of this process. When spin coating is applied to large-format substrate and/or square and rectangular substrates, uniformity becomes a potential concern. There are several fundamental reasons why uniformity is not to be expected from spin coating on large-area, noncircular substrates. In addition, a major drawback with spin coating is that much of the polymer solution (> 85%) is wasted as it is thrown off of the edge of the wafer during the spin process. This increases the fabrication cost of thin-film packaging significantly when expensive dielectric polymer ($1000/kg) is used, as with most of the current applications. It should be noted, however, that flat panel displays of notebook computers are large (6 in. × 12 in.) rectangular substrates and are primarily fabricated by spin coating today. The future of spin coating is unclear despite the general interest in a coating technology capable of large-area, noncircular substrate coating with high polymer utilization at low cost which seems to be heading away from spin coating.

Among the many coating methods, extrusion coating is receiving increased interest and is being investigated as a candidate for potential applications in polymer thin-film coating over a large-format substrate. FAS -Technologies is developing liquid extrusion coating technology and equipment specifically for microelectronic and flat panel display (FPD) applications. A digital electrohydraulic pumping system is used to inject the polymer into the die. It has been shown that this extrusion system is capable of dispensing and coating polymer fluid more precisely than the conventional gas-driven diaphragm system. Materials of higher viscosity (> 400 poise) may be deposited by this extruder, which means that thicker films may be laid down in a single processing step. Uniform thicknesses (< 5% thickness variation) from 0.15 up to 100 μm have been demonstrated in a single pass. Coating speeds of 1–3 in./s can be routinely achieved to ensure a high throughput. Virtually 100% material usage with this extrusion technology would have a significant impact on the cost and environmental issues of thin-film coating technology.

Roller coating, meniscus coating, and screen printing as discussed previously have been used to a much less extent in thin-film packaging and are less investigated. Careful feasibility studies are necessary to further clarify their role in large-area coating of polymer thin films for future thin-film packaging technology.

In comparison, spin coating is the principle method in use today to deposit and planarize polymers because it is easy, cheap, and well established. Nevertheless, the amount of material that is wasted as it gets thrown off of the surface is excessive, on the order of 80%. Spinning by nature favors circular substrates. Because of this, some of the substrate is wasted when the round edges are cut off. Finally, the difficulty of spinning increases with increasing substrate size, and the push is toward larger boards to increase the number of circuits processed at one time. For large-area processing of square or rectangular boards, extrusion appears to be a much better method than spin coating. The material wasted with extrusion is much less than with spin coating and even less than the other methods listed above. The deposition process is very similar to spin coating in that the liquid polymer solution is spread uniformly on the substrate. Thus, aside from the rotary motion, the planarizing mechanisms are the same. Gravitational and capillary forces work together with surface tension to planarize the liquid. In short, the planarizing mechanisms are the same except that the harmful effects (and material waste) of rotational motion are absent. The problem of evaporation due to airflow over the substrate is reduced and shrinkage is the primary factor for planarization. However, extrusion and other candidate coating techniques, including meniscus, roller, and screen printing, are more susceptible to slit inhomogeneities (i.e., misalignment, blockage, and mechanical vibration) [177]. The effects of these factors on the film quality and process robustness must be carefully evaluated and controlled in the development and choice of thin-film coating technology.

In conclusion, coating uniformity and waste reduction seem to indicate a shift away from spin coating in low-cost, high-performance electronic packaging. Extrusion seems to be emerging as a viable and promising alternative. Intelligent use of extrusion and the implementation of feedback control offer the possibility to continue to scale dimensions down and reduce cost while increasing yield on larger substrates.

12.10.3 Via Formation

As is discussed in the previous sections, two rival technologies for via formation in dielectric polymers for high-density interconnections are commonly regarded as low cost: photosensitive polymers and laser ablation. At this time, there is no strong conclusive evidence which favors one technology over the other. Both are employed in actual fabrication of MCM packages. This uncertainty is largely due to the fact that both technologies are under constant evolution which makes any confident prediction and judgment difficult if not impossible. However, a general account of some of the essential attributes of the two technologies can be made and is briefly reviewed in the following. Table 12-34 [30] compares the major attributes of different via formation technologies in polyimides. It is noted that wet etching and reactive ion etching (RIE) are becoming obsolete for both process complexity and cost reasons.

12.10.3.1 Laser Ablation

Both scanning laser ablation and projection laser ablation have been successfully used for via formation in electronic packaging. Scanning laser ablation was first adopted by Siemens in via formation in printed wiring board for MCM-L packaging used for its Siemens 7500 H 90 mainframes [178] Hitachi and Motorola extended this technology later on for via formation in MCM-D substrates [29,60]. A scanning laser

Table 12-34. Comparison of Via Process Techniques

Via Process	No. of Steps	Critical Unit Processes	Choice of Polyimides	Aspect Ratio/ Limitations	Via Wall Profile
Wet etch	9	Polyimide apply and bake, resist lithography, wet etch	Most	2:1, 15 μm min. dia.	Trapezoidal, 50°
RIE	9	Resist lithography, RIE, residue removal	All	2:1, 4μm max. thickness	Trapezoidal, 50°
Laser ablation	6	Laser ablation	All	1:1, 10 μm min. dia., trapezoidal	30°–65°
PSPI	7	PSPI lithography	Few	1:1, 10 μm min. dia.	Barrel, 50°–80°

ablation tool is distinguished from projection laser ablation tool in the utilization of a conformal mask made from patterned metal or photoresist for imaging, and the laser beam is scanned across the substrate to ablate the exposed polymers. Apparently, more processing steps are required for scanning laser ablation relative to projection laser ablation; nevertheless, simpler tool and masking technology are involved. With photoresist as the erodible mask as mentioned in Section 12.4, a significant cost reduction is expected due to the elimination of the metal mask deposition, patterning and stripping steps. More effort is necessary to develop and characterize a robust SLA via generation process using photoresist erodible masks.

Projection laser ablation has been employed by IBM for the fabrication of ES 9000 systems [31]. A via size of 75 μm in polyimide of 18–20 μm thickness was achieved with high yield (99% substrate yield with more than 100,000 vias per substrate) and high throughput (less than 15 min cycle time). The up-to-date smallest achievable via size is 6 μm [31]. Uniform trapezoidal via profile with wall angles ranging from 30° to 65° is readily obtained by altering the focus conditions. Unlike photosensitive polymer processing, laser ablation is not limited by the via aspect ratio and has good tolerance to the nonuniformity of polymer thickness due to the etching-stopping effect of the underlying metal. Key processing parameters include laser fluence, number of pulses, and focus. A plasma cleaning is performed after via formation to remove the carbonaceous debris. The major drawback of this technology is the use of a quartz dielectric mask which is expensive and is not universally available. The issue of reworkability is also a potential concern with laser ablation. It has been shown in Section 12.4 that large-area processing is crucial for cost reduction. This is valid only under the assumption that the costs of large-area tools and masks are not increased significantly. IBM has already demonstrated the capability of laser ablation of a 300 μm square substrate [31]. However, the wide acceptance of this technology in cost-sensitive applications requires further technology simplification and tooling cost reduction.

There are currently many ongoing evolution in laser ablation technology covering all the aspects from radiation source (e.g., TEA CO_2 laser and solid-state laser to lower operation cost), to tool design (e.g., Anvik Corp's scanning system for large-area, high-throughput via drilling) [179], to masking technology [180]. Efforts are also underway to develop a maskless via formation technology by direct laser drilling with a focused fine laser beam. All these are expected to have a great impact on the cost, performance, and reliability of laser ablation technology.

12.10.3.2 Photosensitive Polymers

A variety of photosensitive versions of the conventional dielectric polymers as mentioned in Section 12.10.1 have been formulated by adding sensitizers, initiators, and other photoactive components for the purpose of processing simplification, cost reduction, and improvement of throughput and quality of via formation [30]. The basic requirements on photosensitive polymers are low exposure dose, good resolution, short develop time, and good resistance to solvent attack during develop (e.g., swelling, dissolution, cracking, and crazing), in addition to the common requirements on dielectric polymers as listed in Section 12.4. Mechanical properties of photosensitive polymers are generally inferior to the corresponding nonphotosensitive polymers with the same backbone due to the fact that the photo components cannot be removed completely [181]. Optimization of the overall process of via formation is more a matter of experience and experiment, with many factors ranging from ground rule design to material properties to processing conditions coming into play [76]. Via sidewall profile, aspect ratio and residue at the via bottom are among the most important issues affecting the processibility, performance, yield, and reliability. Vias with aspect ratio of 1 : 1 have been achieved in film thickness less than 8 μm, whereas for thicker films, 1 : 2 is a more realistic figure [76]. Typical vias formed by photosensitive polymer technology has a sidewall profile steeper in the top portion of the via with a very shallow foot [30]. Several factors are the possible causes of residue at the via bottom, including pattern geometry, expose energy, substrate reflectivity, insufficient rinse, and Cu-polymer interaction. However, with good process control and the implementation of postdevelop treatment for residue removal [76,182], residue usually does not pose a severe problem. Cost reduction with the application of photosensitive polymers is realized mainly by the reduced processing steps and low capital costs. However, the process is less controllable in terms of via wall profile and planarization among other processing qualities, which is the future focus of photosensitive polymer material and process development. Relatively high material cost is another shortcoming associated with photosensitive polymers. Additionally, environmental concerns related to the treatment and dispose of organic solutions for photosensitive polymer processing should also be properly addressed.

12.10.4 Planarization

The planarity of a thin-film multilayer substrate has a multiple impact on the electrical performance and reliability of thin-film package and as well as the processability. As discussed in Section 12.4, the requirement on planarization is becoming more compelling and challenging as the wiring density continues to increase. An additional planarization process step (e.g., the mechanical polishing used by IBM) would necessarily accrue to the overall cost of thin-film packaging, while every effort is being made to reduce the cost of thin-film packages to extend the application into the consumer market where low cost could hardly be overstressed. Hence, polymer deposition and/or metallization processes capable of full or partial planarization would avoid the additional planarization steps which had been costly and, meanwhile, technically complex. In the case of planarization by polymer thin-film deposition, both material properties and coating techniques are important as pointed out in Section 12.4. Studies on BCB have shown degree of planarization >85% under controlled baking and curing conditions. In the case of planarization by metallization, the selective deposition of copper allows the filling of vias and trenches through the dielectric polymer layer and planarizes the structure. Both electro- and electroless copper have the capability of selective copper deposition for the purpose of planarization. However, uniformity of electro copper plating degrades when a large substrate is used, and fully electroless copper plating is still under significant investigation and development as briefly discussed in the next section. Therefore, this approach for planarization still does not find wide application.

12.10.5 Electroless Copper Plating

Electroless copper plating has the advantages of simpler equipment and the associated low capital cost compared to electroplating. Among other advantages of electroless plating are the capability of large-area uniform plating over all surfaces—regardless of size and shape, variable plating selectivity, and 100% throwing power. It is most widely used in the printed wiring board (PWB) industry for high-aspect-ratio plated-through-hole (PTH) seeding and/or subsequent buildup of the desired thickness. The cost of current electroless copper plating is, however, much higher compared to electroplating in the PWB industry. Complex bath chemistry, expensive chemicals and catalysts, high requirement of impurity control, and slow plating rate are the major cost drivers, which are compounded by the increasingly environmental and health concerns related to electroless copper plating.

Electroless plating was described as an autocatalytic process of depositing a metal in the absence of an external source of electrical current. Electroless copper solutions generally contain a source of divalent cupric ions, a reducing agent that is capable of reducing the cupric ions to the metal, and a complexing agent to prevent precipitation of the cupric ions and other additives [183]. Such a system is thermodynamically unstable but kinetically inhibited in the absence of catalytic surfaces. Once plating starts, the copper deposition process is generally autocatalytic. Cupric sulfate ($CuSO_4$) is the commonly used source of copper ions, and formaldehyde (HCHO), and, more recently, hypophosphite and dimethylamine borane are employed as the reducing agents. The overall reaction for electroless copper plating with formaldehyde as the reducing agent is

$$Cu^{2+} + 2HCHO + 4OH^- \rightarrow Cu + 2HCOO^- + 2H_2O + H_2 \qquad [12\text{-}12]$$

This reaction is a mixed-potential reaction of the anodic oxidation of formaldehyde and the cathodic reduction of cupric ions at the substrate. In order for the reaction to proceed at an appreciable rate, the pH values of the solution are generally maintained in the range of 11–13 by accelerators, such as NH_3OH, KOH, NaOH, and LiOH [184], to lower the anodic half-cell potential. The accelerators are designated to remove the surfacant and then allow the adsorbed palladium catalyst on the substrate surface to contact with the plating solution. For a bath containing KOH or NaOH, there is a distinct tendency that the deposition rate of electroless copper is reduced with a decrease of the pH value. However, the deposition rate is maintained constant even at low pH in the case of LiOH [185]. Simple cupric salts are almost insoluble at a pH above about 4; the addition of various complexing agents / ligands, such as EDTA (ethylenediamine tetraacetic acid), tartrate, Rochelle salt, and triethanolamine (TEA) is usually required to prevent the precipitation of cupric salt in the alkaline solution. TEA is found to be the most effective ligand for increasing the electroless copper plating rate. The maximum rate is over 20 times faster than the rate observed in the EDTA system [186]. Trace amount of additives are present in the plating baths to provide good-quality deposits [187]. The most commonly used additive is cyanide, which is proven to improve ductility of the deposits. Normally, the reduction of cupric ions initiates at a catalytically active site on the substrate surface. For nonconductive substrates, an additional catalyzation step is required to activate the surface. The most commonly used catalytic solutions are the tin-palladium colloids [188]. The palladium is a catalytically active element, and the tin acts as a protective colloid. When the nonconductive substrates are immersed in the tin-palladium colloid, the Pb and Sn particles are adsorbed on the substrates and become the catalytic sites (Fig. 12-99). Electroless copper deposition then initiates on these Sn-Pd catalytic particles and continues autocatalytically. In the $SnCl_2$-$PdCl_2$ activation treatment, the sites, which are available for copper nucleation on the insulating substrate, depend on the density of small isolated catalytic particles. Basically, the adherence of the catalyst is determined by chemisorption or physisorption to the matrix. A new compound of seeding palladium for electroless copper plating was developed by IBM [189]. The organometallic compound, allylcyclopentadienylpalladium, was found to be easily prepared in high yield due to the unusual volatility at room temperature. Photochemical decomposition of allylcyclopentadienylpalladium results in the deposition of the film that consists of a homogeneous mixture of palladium particles in the insulating matrix. Without further activation, the deposited film is an active catalyst for electroless copper plating to form an uniform plated layer which exhibits good electrical properties. The plating rate of conventional electroless copper baths is generally low (1–5 μm/h) [190]. Reaction kinetics are controlled by the proper selection of additives and by the variation of bath temperatures. A high-build bath

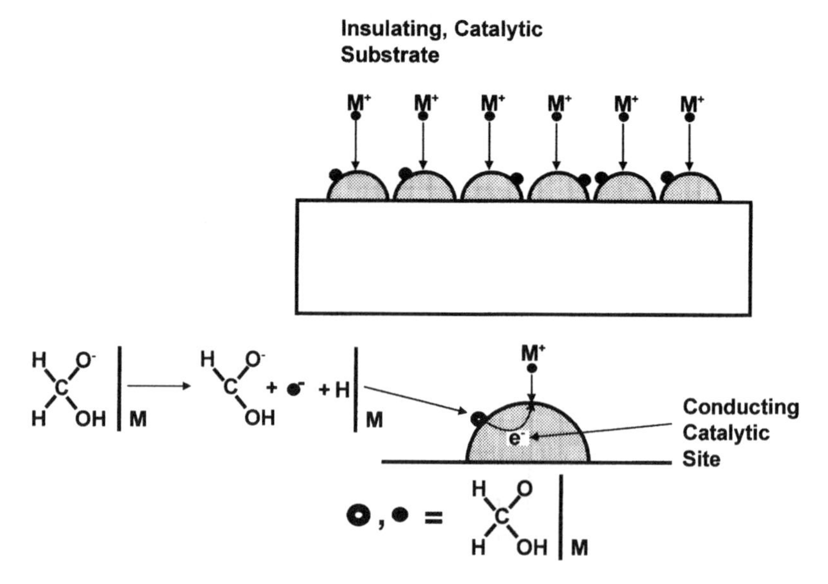

Figure 12-99. Catalyzation of Insulating Substrate in Electroless Copper Plating

with a plating rate up to 23 μm/h has been reported through the joint addition of organics and an increase in the bath temperature [190]. However, bath instability, nodule formation, and inferior deposit morphology are the major problems encountered in the high-speed plating bath. A considerable amount of work is required before optimal balance can be achieved between high plating rate and good-quality deposits. Another new electroless copper plating process for preparing a copper layer with strong adhesion to glass substrate was developed by Yoshiki et al [191]. The new process incorporates a ZnO thin film on the glass substrate, instead of etching and sensitizing the substrate surface as in a conventional electroless plating process. The **peel strength** becomes much greater than that obtained using a conventional plating method. Koyano et al. [192] has proposed that a higher electroless copper plating rate of 7–10 μm/h can be achieved in an alkaline plating solution of a copper-glycerin complex at pH values above 12.2, with formaldehyde as the reducing agent and bipyrizyl as an additive. The stability of the solution is greatly enhanced by controlling the glycerin/Cu molar ratio and by intermittent filtration.

Electroless copper plating at a low pH level is desired to minimize the attack of the caustic bath solution on the dielectric polymers. Common results of caustic solution attack on polymers include swelling, cracking, dissolution, and delamination at interfaces and performance degradation of the polymer dielectric layers. In IBM, a stable bath chemistry was developed which operates at a pH level of <9 [193]. Multidentate nitrogen donor ligands was employed as the cupric-ion complexants in conjunction with triethanolamine (TEA) as the buffer and dimethylamine borane as the reducing agent to enhance the bath stability. High-quality copper deposits was obtained on submicron structures at a deposition rate of 2–3 μm/h.

The operation of an electroless copper plating bath has narrower process window compared to electroplating bath [194]. The range of components and contaminants as well as other processing parameters, such as temperature, must be well controlled to ensure a good deposit quality and prolonged bath life. Impurities, especially, can cause bath decomposition and poor quality of copper deposit and should therefore be minimized. Basic bath controls include replenishing the bath regularly in small amount each time, and checking the bath temperature, formaldehyde, caustic and copper concentrations at certain time intervals and when a problem is suspected. Other important guidelines for maintaining an electroless copper plating bath are redundant air agitation even during shutdown, continuous filtration during plating, use of deionized water and reagent-grade caustic and formaldehyde, and striping of copper plated on the inner surface of the tank and on the racks.

In large-format MCM-D substrate manufacturing, the requirement of uniform metallization on an area of 450 mm × 450 mm becomes critical and is hardly met by electroplating of copper. This provides an unique opportunity for electroless copper. However, technological improvement of electroless plating in many aspects, such as higher plating rate, acid or neutral solution plating to alleviate attack on polymer dielectric and/or photoresist, stable and simple bath chemistry, good process control, improved copper deposit property, and fully additive process must be made before the implementation of this metallization process. In addition, lowered cost of electroless copper plating is essential to survive the consumer market.

12.11 ACKNOWLEDGMENT

The authors wish to acknowlegde the write-ups provided by Dr. Chung Ho of MMS, T. Inoue of Hitachi, T. Watari of NEC, H.R. Krauter of Siemens, and Stan Drobac of nCHIP.

12.12 REFERENCES

1. M. E. Jones, W. C. Holton, and R. Stratton. "Semiconductors: The Key to Computational Plenty," *Proc. IEEE*, 72: pp. 1380–1409, 1982.
2. M. Terasawa, S. Minami, and J. Rubin. "A Comparison of Thin Film, Thick Film, and Co-Fired High Density Ceramic Multilayer with the Combined Technology: T & T HDCM (Thin Film and Thick Film High Density Ceramic Module," *Int. J. Hybrid Microelectron*, 6(1): pp. 607–615, 1983.
3. N. Naclerio. "ARPA Strategy and Programs in Electronic Packaging," *ARPA WWW homepage* (http:// ETO.sysplan.com/ETO/EI-packaging/present), 1996.
4. B. S. Landman and R. L. Russo. "On a Pin vs. Block Relationship for Partitions of Logic Graphs," *IEEE Trans. Computers*, EC-20: pp. 1469–1479, 1971.
5. T. Chiba. "Impact of the LSI on High-Speed Computer Packaging," *IEEE Trans. Computers*, EC-27: pp. 319–325, 1978.
6. C. T. Goddard. "The Role of Hybrids in LSI Systems," *IEEE Trans. Components Hybrids Manuf. Technol.*, CHMT-2: pp. 367–71, 1979.
7. T. S. Steele. "Terminal and Cooling Requirements for VLSI Packages," *IEEE Trans. Components Hybrids Manuf. Technol.*, CHMT-4: pp. 187–191, 1981.
8. D. Balderes and M. L. White. "Package Effects on CPU Performance of Large Commercial Processors," *Proc. 35th ECC*, pp. 351–355, 1985.
9. R. T. Evans. "Interconnection and Packaging of IBM's Large Processors," *Proc. 34th ECC*, pp. 374–378, 1984.
10. A. J. Blodgett. "Microelectronic Packaging," *Scientific American*, 249(1): pp. 86–96, 1983.
11. T. Watari and H. Murano. "Packaging Technology for the NEC SX Supercomputer," *Proc. 35th ECC*, pp. 192–198, 1985.
12. E. E. Davidson. "The Electrical Design Methodology for the Package Used in the IBM 3090 Computer," *IEEE Proc., Wescon/85 Electronic Show and Convention*, p. 1, 1985.
13. C. W. Ho, D. A. Chance, C. H. Bajorek, and R. E. Acosta. "The Thin-Film Module as a High Performance Semiconductor Packaging," *IBM J. Res. Devel.*, 26(3): pp. 286–296, 1982.
14. C. W. Ho. *High-Performance Computer Packaging and the Thin Film Multichip, VLSI Electronics Microstructure Sciences*, pp. 103–143, Academic Press, New York, 1982.

15. C. A. Neugebauer. "Comparison of VLSI Packaging Approaches to Wafer Scale Integration," *Proc. IEEE Int. Conf. Computer Design*, pp. 115–120, 1984.
16. Y. Tsukuda, S. Tsuchida, and Y. Mashimoto. "Surface Laminar Circuitry Packaging," *Proc. 42nd ECTC*, pp. 22–27, 1992.
17. S. Ray, D. Berger, G. Czornyi, A. Kumer, and R. Tummala. "Dual-level Metal (DLM) Method for Fabricating Thin Film Wiring Structures," *Proc. 43rd ECTC*, pp. 538–543, 1993.
18. S. Poon, J. T. Pan, T. C. Wang, and B. Nelson. "High Density Multilevel Copper/Polyimide Interconnects," *NEPCON West*, pp. 426–427, 1989.
19. N. Iwasaki and S. Yamaguchi. "A Pillar-Shaped Via Structure in a Cu-Polyimide Multilayer Substrate," *IEEE Trans. Components Hybrids Manuf. Technol.*, 13(2): pp. 440–443, 1990.
20. J. F. McDonald, J. A. Steakl, C. A. Neugebanner, R. O. Carlson, and A. S. Gergundahl. "Multilevel Interconnections for Wafer Scale Integration," *J. Vacuum Sci. Technol.*, 4(6): pp. 3127–3136, 1986.
21. H. J. Levinstein, C. J. Bartlett, and W. J. Bertram. "Multichip Packaging Technology for VLSI-Based System," *IEEE Computer Soc. Proc.*, March 1987.
22. P. Garrou. "Polymer Dielectrics for Multichip Module Packaging," *Proc. IEEE*, 80(12): pp. 1942–1954, 1992.
23. W. Lautenberger, B. Auman, J. Butera, D. Goff, C. Lazaridis, H. Seiler, and J. Labadie. "Polyimides for Large Area Multichip Modules," *Proc. ICEMCM '95*, pp. 501–506, 1995.
24. E. Perfecto, K. Prasad, P. Wilkens, G. White, R. Tummala, C. Prasad, and T. Redmond. "Multi-level Thin Film Packaging Technology at IBM," *Proc. ICEMM '93*, pp. 474–482, 1993.
25. R. Tummala. "Multichip Packaging in IBM—Past, Present and Future," *Proc. ICEMM '93,* pp. 1–11, 1993.
26. G. Adema, M. Berry, and I. Turlik. "Dielectric Materials for Use in Thin Film Multichip Modules," *Electronic Packaging and Production*, 32(2): pp. 72–76, 1992.
27. R. Tummala, H. Potts, and S. Ahmed. "Packaging Technology for IBM's Latest Mainframe Computers (S/390/ES9000)," *Proc. 41st ECTC*, pp. 682–684, 1991.
28. T. Tessier and P. Garrou. "Overview of MCM Technologies: MCM-D," *Proc. ISHM '92,* pp. 235–247, 1992.
29. K. Miyazaki, A. Takahashi, O. Miura, R. Watanbe, Y. Satsu, T. Miwa, and J. Katagiri. "Fabrication of Thin Film Multilayer Substrate Using Copper Clad Polyimide Sheets," *Proc. 43rd ECTC*, pp. 306–310, 1993.
30. K. Prasad and E. Perfecto. "Multilevel Thin Film Packaging: Applications and Processes for High Performance Systems," *IEEE Trans. Components Packaging Manuf. Technol.—Part B: Advanced Packaging,* 17(1): pp. 38–49, 1994.
31. S. Patel, T. Redmond, C. Tessler, D. Tudryn, and D. Pulaski. "Laser Via Ablation Technology for MCMs Thin Film Packaging—Past, Present, and Future at IBM Microelectronics," *Proc. ISHM '94,* pp. 31–41, 1994.
32. P. B. Chinoy. "Processing and Electrical Characterization of Multilayer Metallization for Microwave Applications," *Proc. ICEMCM '95*, pp. 203–208, 1995.
33. R. J. Jensen. "Polyimides as Interlayer," in *Polymers for High Technology—Electronics and Photonics*, ed. by M. J. Bowden and R. S. Turner, ACS Symposium Series 346, American Chemical Society, Washington DC, 1987.
34. S. Sasaki, T. Kon, and T. Ohsaki. "New Multi-Chip Module Using a Copper Polyimide Multi-Layer Substrate," *Proc. 39th ECC*, pp. 629–635, 1989.
35. T. Shimoto, K. Matsui, and K. Utsumi. "Cu/Photosensitive-BCB Thin-Film Multilayer Technology for High-Performance Multichip Module," *Proc. MCM '94*, pp. 115–120, 1994.

36. D. Darrow and S. Vilmer-Bagen. "Low-Cost Patterned Metallization Technique for High Density Multilayer Interconnect Applications," *Proc. 43rd ECTC*, pp. 544–549, 1993.

37. C. W. Ho and U. A. Deshpande. "A Low Cost Multichip Module-D Technology," *Proc. ICEMM '93*, pp. 483–488, 1993.

38. A. Strandjord, Y. Ida, P. Garrou, B. Rogers, S. Cummings, and S. Kisting. "MCM-D Fabrication with Photosensitive Benzocyclobutene: (Processing, solder Bumping, System Assembly, and Testing)," *Proc. ISHM '95*, pp. 402–417, 1995.

39. T. Dudderar, Y. Degani, J. Spadafora, K. L. Tai, and R. Frye. "AT&T µSurface Mount Assembly: A New Technology for the Large Volume Fabrication of Cost Effective Flip-Chip MCMs," *Proc. MCM '94*, pp. 266–272, 1994.

40. W. Radlik, K. Plehnert, M. Zellner, A. Achen, R. Heistand II, D. Castillo, and R. Urscheler. "MCM-D Technology for a Communication Application," *Proc. MCM '94*, pp. 402–409, 1994.

41. V. Rao, D. Hutchins, J. Reagan, D. Scheid, R. Streif, T. Syrstad, C. Eggerding, and T. Redmond. "Manufacturing of Advanced Cu-PI Multi Chip Modules," *Proc. 43rd ECTC*, pp. 920–934, 1993.

42. M. Chopra and R. Sechler. "Comparison of MCMs for a High Performance Workstation," *Proc. ICEMM '93*, pp. 457–467, 1993.

43. G. B. Leung and S. A. Sands. "A Thin Films on MLC Application," *Proc. 41st ECTC*, pp. 10–13, 1991.

44. S. Numata, S. Oohara, K. Fujisaki, J. Imaizumi, and N. Kinjo. "Low Thermal Expansion Polyimide," *J. Appl. Polymer Sci.*, 31(101): pp. 81–86, 1986.

45. S. Numata, T. Miwa, Y. Misawa, D. Makino, J. Imaizumi, and N. Kinjo. "Low Thermal Expansion Polyimides and Their Applications," *Mater. Res. Soc. Proc.*, p. 1, 1987.

46. See, for example, Japanese Patent 61–111598, 1987.

47. R. Tummala, "Glass Composition for Glass-Metal Package," U. S. Patents 3,640,738, 1971.

48. T. Matsuzaki, K. Sato, T. Suzuki, and M. Terashima. "High Density Multilayer Wiring Substrate Using Copper Thick Film and Thin Film," *IEEE Trans. Components Hybrids Manuf. Technol.*, CHMT-8: pp. 484–491, 1985.

49. R. H. Heistand II, D. C. Frye, D. Burdeaux, J. Carr, and P. Garrou. "Economic Evaluation of Deposited Dielectric MCM Manufacturing Costs," *Proc. 2nd IEPS/ISHM MCM Conf.*, pp. 441–450, 1993.

50. P. Rickerl, J. Stephanie, and P. Jr. Slota. "Evaluation of Photosensitive Polyimides for Packaging Applications", *Proc. 37th ECC*, pp. 220–225, 1987.

51. J. H. Lai, *Polymers for Electronic Applications*, CRC Press, Boca Raton, FL, 1989.

52. M. Berry, T. Tessier, I. Turlik, G. Adema, D. Burdeaux, J. Carr, and P. Garrou "Benzocyclobutene as a Dielectric for Multichip Module Fabrication," *Proc. 40th ECTC*, pp. 746–750, 1990.

53. D. H. Klockow. "Expanding Hybrid Circuits Through Thick and Thin," *AT & T Bell Laboratories Record*, pp. 25–29, March, 1984.

54. A. Strandjord, R. Heistand, J. Bremmer, P. Garrou, and T. Tessier. "A Photosensitive BCB on Laminate Technology (MCM-LD)," *Proc. 44th ECTC*, pp. 374–385, 1994.

55. H. F. Bok. "Process Enhancements in Meniscus Coating for Flat-Panel Displays," *SID 94 Digest*, pp. 940–942, 1994.

56. L. J. Laursen and P. Garrou. "Consortium for Intelligent Large Area Processing (CILAP)—Program Review," *Proc. ICEMCM '95*, pp. 253–258, 1995.

57. R. Kambe, M. Kuroda, R. Imai, and Y. Kimura. "Copper Polyimide Multilayer Substrate for High Speed Signal Transmission," *Proc. 41st ECTC*, pp. 14–16, 1991.

58. E. Fogarassy and S. Lazare. *Laser Ablation of Electronic Materials: Basic Mechanisms and Applications*, pp. 39–53, Elsevier Science Publishers, Amsterdam, 1992.

59. E. Fogarassy and S. Lazare. *Laser Ablation of Electronic Materials: Basic Mechanisms and Applications*, pp. 239–253, Elsevier Science Publishers, Amsterdam, 1992.

60. T. G. Tessier, W. F. Hoffman, and J. W. Stafford. "Via Processing Options for MCM-D Fabrication: Excimer Laser Ablation vs Reactive In Etching," *Proc. 41st ECTC*, pp. 827–834, 1991.

61. T. F. Redmond, J. R. Lankard, J. G. Balz, G. R. Proto, and T. A. Wassick. "The Application of Laser Process Technology to Thin Film Packaging," *Proc. 41st ECTC*, pp. 1066–1071, 1991.

62. R. Srinivasan. "Interactions of Polymer Surfaces with Ultraviolet Laser Pulses," *Photochemistry and Polymeric Systems*, ed. by J. M.Kelly, C. B.McArdle, and M. J.de F.Maunder, pp. 46–53, Royal Society of Chemistry, London, 1992.

63. S. Lazare, H. Hiraoka, A Cros, and R. Gustiniani. "Ultra-Violet Laser Photoablation of Thermostable Polymers: Polyimides, Polyphenylquinoxaline and Teflon AF," in *Polyimides and Other High-Temperature Polymers*, ed. by M. J.M. Abadie and B. Sillion, Elsevier Science Publishers, Amsterdam, pp. 395–406, 1991.

64. T. G. Tessier and G. Chandler. "Compatibility of Common MCM-D Dielectrics with Scanning Laser Ablation Via Processes," *Proc. 42nd ECTC*, pp. 763–765, 1992.

65. F. Bachmann. "Excimer Laser Drill for Multilayer Printed Circuit Boards: From Advanced Development to Factory Floor," *MRS Bull.*, pp. 49–54, 1989.

66. T. Shimoto, K. Matsui, M. Kimura, and K. Utsumi. "High Density Multilayer Substrate Using BCB Dielectric," *Proc. IMC*, pp. 325–326, 1992.

67. T. G. Tessier and E. G. Myszka. "Approaches to Cost Reducing MCM-D Substrate Fabrication," *Proc. 43rd ECTC*, pp. 570–578, 1993.

68. G. E. Wolbold, C. L. Tessler, and D. J. Tudryn. "Characterization, Set-Up and Control of a Manufacturing Laser Ablation Tool and Process," in *Excimer Lasers: Applications, Beam Delivery Systems, and Laser Design*, SPIE Proc. No. 1835, pp. 62–69, SPIE, 1992.

69. J. H. Brannon and J. R. Lankard. "Patterning of Polyimide Films with Ultraviolet Light," US Patent No. 4,508,749, 1985.

70. T. G. Tessier, G. M. Adema, S. M. Bobbio, and I. Turlik. "Low Temperature Etch Masks for High Rate Magnetron RIE of Polyimide Dielectrics in Thin Film Packaging Applications," *Proc. 3rd Int. SAMPE Electronics Conf.*, pp. 85–87, 1989.

71. M. J. Rutter. "Via Formation in Thick Polyimide Layers for Silicon Hybrid Multichip Modules," in *Microelectronic Packaging Technology: Materials and Processes, Proc. 2nd ASM Inter. Electron. Mater. Processing Congress*, ed. by W. T. Shieh, ASM Press, Metals Park, OH, 1989.

72. J. Paraszczak, J. Cataldo, E. Galligan, W. Graham, R. McGouey, S. Nunes, R. Serino, D. Shih, E. Babich, A. Deutsch, and G. Kopcsay. "Fabrication and Performance Studies of Multilayer Polymer/Metal Interconnect Structures for Packaging Applications," *Proc. 41st ECTC*, pp. 362–369, 1991.

73. D. Y. Shih, H. Yeh, C. Narayan, J. Lewis, W. Graham, S. Nunes, J. Paraszczak, R. McGouey, E. Galligan, J. Cataldo, R. Serino, E. Perfecto, C.-A. Chang, A. Deutsch, L. Rothman, and J. Ritsko, "Factors Affecting the Interconnection Resistance & Yield in the Fabrication of Multilayer Polyimide/Metal Thin Film Structure," *Proc. 42nd ECTC*, pp. 1002–1014, 1992.

74. R. R. Tummala and E. J. Rymaszewski. *Microelectronics Packaging Handbook*, von Nostrand Reinhold, New York, 1989.

75. H. J. Neuhaus. "An MCM-D Substrate Fabrication Model: Comparing Dry Etch, Wet Etch and Photosensitive Dielectrics," *Proc. 2nd IEPS/ISHM MCM Conf.*, pp. 46–50, 1993.

76. E. D. Perfecto, C. Osborn, and D. G. Berger. "Factors that Influence Photosensitive Polyimide Lithography Performance," *Proc. 2nd IEPS/ISHM MCM Conf.*, pp. 40–45, 1993.
77. K. Moriya, T. Ohsaki, and K. Katsura. "High Density Multilayer Interconnection with Photosensitive Polyimide Dielectric and Electroplating," *Proc. 34th ECC*, pp. 82–83, 1984.
78. H. Takasago, M. Takada, K. Adachi, K. Yamada, Y. Onishi, and Y. Morihiro. "Fine-Line Multilayer Hybrids with Wet Processed Conductors and Thick Film Resistors," *Proc. 34th ECC*, pp. 324–326, 1984.
79. O. Ahimada, K. Ito, T. Miyagi, S. Kimijima, and T. Sudo. "Electrical Properties of a Multilayer Thin Film Substrate for Multichip Packages," *Proc. Japan IEMT Symp.*, pp. 123–124, 1989.
80. E. W. Rutter, E. S. Moyer, R. F. Harris, D. C. Frye, V. L. St. Jeor, and F. L. Oaks. "A Photodefinable Benzocyclobutene Resin for Thin Film Microelectronics Applications," *1st Int. MCM Conf.*, pp. 394–396, 1992.
81. P. Mukerji, T. G. Tessier, G. Demet, and S. Dasgupta. "MCM Via Generation by Wet Processing," *ISHM Polymer and Ceramic Low K Dielectrics: Advances in Processing and Characterization Workshop*, 1992.
82. H. J. Neuhaus. "A High Resolution, Anisotropic Wet Patterning Processing Technology for MCM Production," *1st Int. MCM Conf.*, pp. 256–263, 1992.
83. T. Rucker, V. Murali, and R. Shukla. "A Wet Etch Polyimide Process for Multichip Modules," *1st Int. MCM Conf.*, pp. 251–254, 1992.
84. C. E. Diener and J. R. Susko. "Etching of Partially Cured Polyimides," *Proc. 1st Technical Conf. Polyimides*, pp. 353–364, 1992.
85. T. Inoue, H. Matsuyama, E. Matsuzaki, Y. Narizuka, M. Tanaka, and T. Takenaka. "Micro Carrier for LSI Chip Used in the HITAC M–880 Processor Group," *Proc. 41st ECTC*, pp. 704–706, 1991.
86. A. Kimura, T. Tsujimura, K. Saitoh, and Y. Kohno. "Fabrication and Characteristics of Silicon Carrier Substrates for Silicon on Silicon Packaging," *1st Int. MCM Conf.*, p. 23, 1992.
87. Y. H. Kim, J. Kim, G. F. Walker, C. Feger, and S. P. Kowalczyk. "Adhesion and Interface Investigation of Polyimide on Metals," *J. Adhesion Sci. Technol.*, 2(2): pp. 95–98, 1988.
88. G. M. Adema, L. T. Hwang, G. A. Rinne, and I. Turlik. "Passivation Schemes for Copper/Polymer Thin Film Interconnections Used in Multichip Modules," *Proc. 42nd ECTC*, pp. 776–782, 1992.
89. G. Czornyj, K. Chen, G. Prada-Silva, A. Arnold, H. Souletis, S. Kim, M. Ree, W. Volksen, D. Dawson, and R. DiPietro. "Polyimide for Thin Film Redistribution on Glass-Ceramic/Copper Multilevel Substrates," *Proc. 42nd ECTC*, pp. 682–687, 1992.
90. S. P. Kowalczyk, Y.-H. Kim, G. F. Walker and J. Kim. "Polyimide on Copper: The Role of Solvent in the Formation of Copper Precipitates," *Appl. Phys. Lett.*, 52(5): pp. 375–376,
91. R. Hiestand, R. DeVellis, P. Garrou, D. Burdeaux, T. Stokich, P. Townsend, T. Mania, and L. Bratton. "Cyclobutene 3022 (BCB) for Non-Hermetic Packaging," *Proc. ISHM '92 Conf.*, pp. 584–590, 1992.
92. T. G. Tessier, I. Turlik, G. M. Adema, D. Sivan, E. K. Yung, and M. J. Berry. "Process Considerations in Fabricating Thin Film Multichip Modules," *Proc. 1989 Int. Electronic Packaging Symp.*, pp. 248–270, 1989.
93. W. J. Bertram. "High Density, Large Scale Interconnection for Improved VLSI System Performance," *Proc. 1987 IEDM*, pp. 100–103, 1987.
94. E. D. Perfecto, K. Prasad, C. Osborn, S. Swaminathan, G. White, and C. Prasad. "Comparison of Planar and Non-Planar Thin Film Processes," *Proc. 43rd ECTC*, (Supplement to the regular proceedings), 1993.

95. J. T. Pan, S. Poon, and B. Nelson. "A Planar Approach to High Density Copper-Polyimide Interconnect Fabrication," *Proc. 8th IEPS*, pp. 174–189, 1988.

96. T. J. Buck. "Substrates for High Density Packaging," *Proc. 1990 NEPCON West*, p. 650, 1990.

97. S. K. Ray, K. Beckham, and R. Mater. "Flip Chip Interconnection Technology for Advanced Thermal Conduction Modules," *Proc. 41st ECTC*, pp. 772–778, 1991.

98. S. K. Ray, K. Seshan, and M. Interrante. "Engineering Change (EC) Technology for Thin Film Metallurgy on Polyimide Films," *Proc. 40th ECTC*, pp. 395–400, 1990.

99. L. B. Rothman. "Properties of Thin Polyimide Films," *J. Electrochem. Soc.: Solid-State Sci. Technol.*, 127(10): pp. 2216–2220, 1980.

100. C. C. Chao and W. V. Wang. "Planarization Enhancement of Polyimides by Dynamic Curing and the Effect of Multiple Coating," *Proc.1st Technical Conf. Polyimides*, pp. 783–793, 1982.

101. L. E. Stillwagon and R. G. Larson. "Topographic Substrate Leveling During Spin Coating," *Proc. Symp. Patterning Sci. Technol.*, 90(1): ed. by R. Gleason, J. Hefferon, and L. White, The Electrochemical Soc., Pennington, NJ, pp. 230–238, 1989.

102. D. R. Day, D. Ridley, J. Mario, and S. D. Senturia. "Polyimide Planarization in Integrated Circuits," *Proc. 1st Technical Conf. Polyimides*, pp. 767–780, 1982.

103. H. Akahoshi, M. Kawamoto, T. Itabashi, O. Miura, A. Takahashi, S. Kobayashi, M. Miyazaki, T. Mutho, M. Wajima, and T. Ishimaru. "Fine Line Circuit Manufacturing Technology with Electroless Copper Plating," *Proc. 44th ECTC*, pp. 367–373, 1994.

104. Y. Kasuya, Y. Takahashi, Y. Uno, Y. Iguchi, and T. Kanamori. "Planarization Process of Copper-Polyimide Thin Film Multilayer Substrate," *Proc. 14th IEEE/CHMT Japan Int. Electronic Manufacturing Technol. Symp.*, pp. 13–17, 1993.

105. D. Frye, M. Skinner, R. Heistand II, P. Garrou, and T. Tessier. "Cost Implications of Large Area MCM Processing", *Proc. MCM '94*, pp. 69–80, 1994.

106. G. White, E. Perfecto, T. DeMercurio, D. McHerron, T. Redmond, and M. Norcott, "Large Format Fabrication—A Practical Approach to Low Cost MCM-D", *Proc. MCM '94*, pp. 86–93, 1994.

107. L. H. Ng, "Economic Impact of Processing Technologies on Thin Film MCMs," *Proc. 42nd ECTC*, pp. 1042–1045, 1992.

108. T. F. Redmond, C. Prasad, and G. A. Walker. "Polyimide-Copper Thin Film Redistribution on Glass Ceramic/Copper Multilevel Substrates," *Proc. 41st ECTC*, pp. 689–693, 1991.

109. A. H. Landzberg. *Microelectronics Manufacturing Diagnostics Handbook*, van Nostrand Reinhold, New York, 1992.

110. G. Messner, I. Turlik, J. W. Balde, and P. E. Garrou. *Thin Film Multichip Modules*, International Society for Hybrid Microelectronics, Reston, VA, 1992.

111. T. A. Wassick. "Repair of Thin Film Wiring with Laser-Assisted Processes," *Proc. 42nd ECTC*, pp. 759–762, 1992.

112. L. Economikos and D. W. Ormond. "Thin Film Repair Process for Interlevel Electrical Connectors," *IBM Tech. Disclosure Bull.*, No. 2, pp. 17–19, 1991.

113. G. P. Flayter, C. H. Perry, and S. K. Ray. "Repair Technique for Inter-Level Short," *IBM Tech. Disclosure Bull.*, 26(1): pp. 242–243, 1983.

114. M. L. Cohen, R. A. Unger, and J. F. Milkovsky. "Laser Machining of Thin Films and Integrated Circuits," *Bell Sys. Tech. J.*, 47: pp. 385–407, 1968.

115. R. L. Waters and G. N. Ravich. "Circuit Surgery Using Xenon and YAG Lasers," *Int. Symp. Testing Failure Analysis*, pp. 86–91, 1982.

116. G. R. Levinson and V. I. Smilga. "Laser Processing of Thin Films (Review)," *Sovi. J. Quantum Electron.*, 6(8): pp. 885–897, 1976.

117. T. H. Baum and P. B. Comita. "Laser-Induced Chemical Vapor Deposition of Metals for Microelectronic Technologies," *Thin Solid Films*, 218(1–2): pp. 80–94, 1992.

118. H. G. Muller, C. T. Galanakis, S. C. Sommerfeldt, T. J. Hirsch, and R. F. Miracky. "Laser Process for Personalization and Repair of Multi-chip Modules," in *Lasers in Microelectronics Manufacturing*, ed. by B. Braren, SPIE Proc. No.1598, pp. 132–140, SPIE, 1991.

119. A. Reisman, D. Temple, and I. Turlik. "Metal-Organic Chemical Vapor Deposition for Repairing Broken Lines in Microelectronic Packages," European Patent Application No. WO 92/08246, 1992.

120. Y. Morishige and S. Kishida. "Highly Conductive Gold Direct Writing by Laser Induced CVD," *Spring Meeting Japan Soc. Appl. Phy.*, Vol. 2, No. 564, 1991.

121. T. H. Baum, P. B. Comita, and R. L. Jackson. "Process for Interconnecting Thin-Film Electrical Circuits," US Patent No. 4,880,959, 1989.

122. M. E. Gross. "Laser Direct-Write Metallization in Thin Metallo-organic Films," *Chemtronics*, 4: pp. 197–201, 1989.

123. R. L. Melcher and N. S. Shiren. "Method for Selective Plating and Etching Using Resistive Heating," *IBM Tech. Disclosure Bull.*, 6(81): pp. 246–247, 1981.

124. J. Partridge, B. Hussey, C. Chen, and A. Gupta. "Repair of Circuits by Laser Seeding and Constriction-Induced Plating," *IEEE Trans. Components Hybrids Manuf. Technol.*, 15(2): pp. 252–257, 1992.

125. E. F. Handford, J. M. Harvilchuck, M. J. Interrante, R. A. Jackson, R. N. Master, S. K. Ray, W. E. Sablinski, and T. A. Wassick. "Method and Structure for Repairing Electrical Lines," US Patent No. 5,153,408, 1992.

126. J. C. Logue, W. J. Kleinfelder, P. Lowy, J. R. Moulic, and W. Wu. "Techniques for Improving Engineering Productivity of VLSI Devices," *IBM J. Res. Devel.*, 25(3): pp. 107–115, 1981.

127. D. L. Klein, P. A. Leary-Renick, and R. Srinivasan. "Ablative Photodecomposition Process for Repair of Line Shorts," *IBM Tech. Disclosure Bull.*, 2(84): pp. 4669–4671, 1984.

128. T. L. Michalka, W. Lukaszek, and J. D. Meindl. "A Redundant Metal-Polyimide Thin Film Interconnection Process for Wafer Scale Dimensions," *IEEE Trans. Semicond. Manuf.*, 3(4): pp. 158–167, 1990.

129. J. I. Raffel. "Laser Linking for Defect Avoidance and Customization," in *Proc. SPIE Lasers in Microlithography*, ed. by D. J. Ehrlich, J. Y. Tsao, and J. S. Batchelder, Vol. 774, pp. 93–100, SPIE, 1987.

130. W. G. Burger and C. W. Welgel. "Multilayer Ceramics Manufacturing," *IBM J. Res. Devel.*, 27(1): pp. 11–19, 1983.

131. M. Swaminathan, E. Perfecto, and K. Prasad. "Thin Film Multichip Module Technology at IBM: Design and Electrical Performance," *IEPS Conf. and Exhibit.*, pp. 780–787, 1993.

132. M. E. Williams. "Production of MCP Chip Carriers," *Proc. 40th ECTC*, pp. 408–411, 1990.

133. D. J. Bendz, R. W. Gedney, and J. Rasile. "Cost/Performance Single-Chip Module," *IBM J. Res. Devel.*, 26(3): pp. 278–285, 1982.

134. R. R. Tummala, J. U. Knickerbocker, S. H. Knickerbocker, L. W. Herron, R. W. Nufer, R. N. Master, M. O. Neisser, B. M. Kellner, C. H. Perry, J. N. Humenik, T. F. Redmond, "High-Performance Glass-Ceramic/Copper Multilayer Substrate with Thin-Film Redistribution.," *IBM J. Res. Devel.*, 36(5): pp. 889–904, 1992.

135. E. E. Davidson, P. H. Harding, G. A. Katopis, M. G. Nealon, and L. L. Wu. "The Design of ES 9000 Module," *Proc. 41st ECTC*, pp. 50–54, 1991.

136. M. G. Bregman, A. Kimura, T. Matsui, H. Nishida, K. Nishiyama, H. Ohkuma, A. Tanaka, C. Kovac, and D. McQueeney. "A Thin Film Multichip Module for Workstation Applications," *Proc. 42nd ECTC*, pp. 968–972, 1992.

137. R. Tummala and B. Clark. "Multichip Packaging Technologies in IBM for Desktop to Mainframe Computers," *Proc. 42nd ECTC*, pp. 1–9, 1992.

138. H. Wessely, W. Turk, K. H. Schmidt, and G. Nagel. "Computer Packaging," *Siemens Res. Devel.*, Rep Bd 17: pp. 234–244, 1988.

139. H. Brosamle, B. Brabetz, V. V. Ehrenstein, and F. Bachmann. "Technology for a Microwiring Substrate," *Siemens Res. Devel.*, Rep Bd 17(5): pp. 249–253, 1988.

140. H. Brosamle. "High Density Multichip Module Based on PWB Technology," *2nd Inter. Symp. on Printed Circuit: Future European Trends and Printed Circuit Technology*, 1991.

141. H. Wessely, O. Fritz, P. Klimke, W. Koschnick, and K. H. Schmidt. "Electronic Packaging in the 1990's: The Perspective from Europe," *IEEE Trans. Components Hybrids Manuf. Technol.*, 14(2): pp. 272–284, 1991.

142. T. Inoue, H. Matsuyama, E. Matsuzaki, Y. Narizuka, M. Ishino, T. Takenaka, and M. Tanaka. "Microcarrier for LSI Chip Used in the Hitachi M–880 Processor Group," *IEEE Trans. Components Hybrids Manuf. Technol.*, 15(1): pp. 7–14, 1992.

143. F. Kobayashi, Y. Watanabe, M. Yamamoto, A. Anzai, A. Takahashi, T. Daikoku, and T. Fujita. "Hardware Technology for Hitachi M–880 Processor Group," *Proc. 41st ECTC*, pp. 693–703, 1991.

144. K. Mukai, A. Saiki, K. Yamanaka, S. Harada, and S. Shoji. "Planar Multi-level Interconnection Technology Employing a Polyimide," *IEEE Trans. Solid-State Circuits*, SC–13: pp. 462–467, 1978.

145. T. Nishida, K. Mukai, T. Inaba, T. Kato, I. Tezuka, and N. Horie. "Moisture Resistance of Polyimide Multilevel Interconnect LSI's," *Proc. IEEE-IRPS*, pp. 148–152, 1985.

146. A. J. Blodgett and D. R. Barbour. "Thermal Conduction Module, a High Performance Multilayer Ceramic Package," *IBM J. Res. Devel.*, 26(1): pp. 30–36, 1982.

147. F. C. Chong, C. W. Ho, K. Liu, and S. Westbrook. "High Density Multichip Memory Package," *WESCON/85 Professional Program Session Record*, Session 7, 1985.

148. H. B. Bakoglu. *Circuits, Interconnections and Packaging for VLSI*, Addison-Wesley, Reading, MA, 1989.

149. P. Dunbeck, R. Dischler, J. McElroy, and F. Swiatowiec. "HDSC and MCU Design and Manufacture," *Digital Tech. J.*, 2(4): pp. 99–108, 1990.

150. D. Marshall and J. McElroy. "VAX 9000 Packaging—The Multi Chip Unit," *35th COMPCON*, pp. 54–57, 1990.

151. P. Hardin, G. Melvin, and M. Nealon. "The ES/9000 Glass Ceramic Thermal Conduction Module Design for Manufacturability," *IEEE/CHMT '91 IEMT Symp.*, pp. 351–355, 1991.

152. A. Dohya, T. Watari, and H. Nishimori. "Packaging Technology for the NEC SX–3/SX-X Supercomputer," *Proc. 40th ECTC*, pp. 525–533, 1990.

153. T. Watari, Private Communication.

154. K. Umezawa, M. Kimura, H. Nishimori, T. Mizuno, K. Kimbara, and R. Nakazaki. "A High-Performance GaAs Multichip Package for Supercomputers," *NEC Res. Devel.*, 33(1): pp. 32–40, 1992.

155. S. Mok. "Volume Implementation of MCM-D Based Cache SRAM Products for Workstation and PC Applications," *Proc. MCM '94*, pp. 320–325, 1994.

156. G. Gengel. "A Process for the Manufacturing of Cost Competitive MCM Substrates", *Proc. MCM '94*, pp. 182–187, 1994.

157. R. Fillion, R. Wojnarowski, B. Gorowitz, W. Daum, and H. Cole. "Conformal Multichip-on-Flex (MCM-F) Technology", *Proc. ICEMCM '95*, pp. 52–58, 1995.

158. M. Moser and T. G. Tessier. "Higher Density PCBs for Enhanced SMT and Bare Chip Assembly Applications," *Proc. ICEMCM '95*, pp. 543–552, 1995.

159. D. N. Light, J. S. Kresge, and C. R. Davis, "Integrated Flex: Rigid-Flex Capability in a High Performance MCM," *Proc. MCM '94*, pp. 430–442, 1994.

160. H. L. Heck, J. T. Kolias, and J. S. Kresge. "High Performance Carrier Technology," *Proc. 1993 IEPS Conf.*, pp. 771–779, 1993.

161. C. Narayan, S. Purushothaman, F. Doany, and A. Deutsch. "Thin Film Transfer Process for MCM Interconnect Wiring," *Proc. MCM '94*, pp. 105–114, 1994.

162. M. Y. Lau, K. L. Tai, R. C. Frye, M. Saito, and D. D. Bacon. "A Versatile, IC Process Compatible MCM-D for High Performance and Low Cost Applications," *Proc. ICEMM '93*, pp. 107–112, 1993.

163. D. Scheid, "Advanced MCM-D with Embedded Resistors," *Proc. MCM '94*, pp. 273–278, 1994.

164. J. M. Cech, A. Burnett, and L. Knapp. "Pre-Imidized Photoimageable Polyimide as a Dielectric for High Density Multichip Modules," *Polymer Sci. Eng.*, 32(21): p. 1647, 1992.

165. T. Lenihan, L. Schaper, Y. Shi, G. Morcan, and J. Parkerson. "Embedded Thin Film Resistors, Capacitors and Inductors in Flexible Polyimide Films," *Proc. 46th ECTC*, pp. 119–124, 1996.

166. M. Oppermann, E. Feurer, and B. Holl. "Development and Realization of a Double-face Populated Multichip Module in Thin Film Technology for High Frequency Application," *Proc. MCM '94*, pp. 279–284, 1994.

167. R. Kambe, R. Imai, T. Takada, M. Arakawa, and M. Kuroda. "MCM Substrate with High Capacitance," *Proc. MCM '94*, pp. 136–141, 1994.

168. M. Tsukada, J. Cross, M. Nishizawa, K. Kurihara, and N. Kamehara. "Preparation and Characterization of Dielectric Thin Films for Decoupling Capacitors," *Proc. ICEMCM '95*, pp. 347–352, 1995.

169. D. Dimos, S. Lockwood, R. Schwartz, and M. Rodgers. "Thin-Film Decoupling Capacitors for Multi-Chip Modules," *Proc. 44th ECTC*, pp. 894–899, 1994.

170. T. Garino, D. Dimos, and S. Lockwood. "Integration of Thin Film Decoupling Capacitors," *Proc. ISHM '94.*, pp. 179–184, 1994.

171. T. Takken and D. Tuckerman. "Integral Decoupling Capacitance Reduces Multichip Module Ground Bounce," *Proc. 43rd ECTC*, pp. 79–84, 1993.

172. D. Tuckerman, D. Benson, H. Moore, J. Horner, and J. Gibbons. "A High-Performance Second-Generation SPARC MCM," *Proc. MCM '94*, pp. 314–319, 1994.

173. L. Matthew, N. Brathwaite, K. Flatow, H. Whittmore, and L. Bauer. "Production Update: Report on Volume MCM Implementation," *Proc. ICEMCM '95,* pp. 229–234, 1995.

174. P. Chahal, R. Tummala, M. Allen, and M. Swaminathan. "A Novel Integrated Decoupling Capacitor for MCM-L Technology," *Proc. 46th ECTC*, pp. 125–132, 1996.

175. C. H. Ahn, Y. J. Kim, and M. G. Allen. "A Fully Integrated Planar Toroidal Inductor with a Micromachined Nickel-Iron Magnetic Bar," *IEEE Trans. Components Packaging Manuf. Technol.*, Part A, 17(3): pp. 463–469, 1994.

176. W. Baker. "Low Cost Thin Film MCMs," *ARPA WWW homepage* (http://eto.sysplan.com/ETO/EI-Packaging/presentation), 1995.

177. M. Parodi, W. Batchelder, J. McKibben, and P. Haaland. "Developments and Trends in Polymer Coating Technologies for Flat-Panel-Display Manufacturing," *SID 94 Digest*, pp. 933–935, 1994.

178. F. Bachmann. "Excimer Lasers in a Fabrication Line for a Highly Integrated Printed Circuit Board," *Chemtronics*, 4(3): pp. 149–152, 1989.

179. K. Jain, J. Hoffman, T. Dunn, W.Folster, and D. Panchal. "Large-Area, High-Throughput, High-Resolution Patterning and Via-Drilling System," *Proc. ICEMCM '95*, pp. 327–335, 1995.

180. R. Patel, W. Advocate, and S. Mukkavilli. "Projection Laser Ablation Mask Alternatives," *Proc. ICEMCM '95*, pp. 320–326, 1995.

181. A. E. Nader. "Photodefinable Polyimides Designed for Use as Multilayer Dielectrics for Multi-Chip Modules", *Proc. MCM '92*, pp. 439–413, 1992.

182. B. D. Kotzias, A. Nader, D. Murray III, and W. Lautenberger. "Application of a Photodefinable Polyimide Process for Manufacturing Multi-Chip Modules," *Proc. ICEMM, '93*, pp. 46–50, 1993.

183. C. H. Ting, M. Paunovic, P. L. Pai, and G. Chiu. "Selective Electroless Metal Deposition for Integrated Circuit Fabrication," *J. Electrochem. Soc.*, 136(2): pp. 462–466, 1989.

184. S. M. Ho, T. H. Wang, H. L. Chen, K. M. Lian, and A. Hung. "Metallization of Polyimide Film by Wet Process," *J. Appl. Polymer Sci.*, 51(8): pp. 1373–1380, 1994.

185. M. Matsuoka, J. Murai, and C. Iwakura. "Kinetics of Electroless Copper Plating and Mechanical Properties of Deposits," *J. Appl. Polymer Sci.*, 139(9): pp. 2466–2470, 1992.

186. K. Kondo, J. Ishikawa, O. Takenada, and T. Matsubara. "Acceleration of Electroless Copper Deposition in the Presence of Excess Triethanolamine," *J. Electrochem. Soc.*, 138(12): pp. 3629–3633, 1991.

187. T. N. Vorob'eva, V. A. Rukhlya, V. V. Sviridov, and E. V. Gert. "Selective Deposition of Copper on a Polyimide Film from a Chemical Metallization Bath," *J. Appl. Chem. USSR*, 59(3): pp. 508–513, 1986.

188. S. Nakahara and Y. Okinaka. "Microstructure and Mechanical Properties of Electroless Copper Deposits," *Annu. Rev. Mater. Sci.*, Vol. 21, pp.93–129, 1991.

189. R. R. Thomas and J. M. Park. "Vapor Phase Deposition of Palladium for Electroless Copper Plating," *J. Electrochem. Soc.*, 136(6): pp. 1661–1666, 1989.

190. E. K. Yung and L. T. Romankiw. "Plating of Copper into Through-Holes and Vias," *J. Electrochem. Soc.*, 136(1): pp. 203–215, 1989.

191. H. Yoshiki, V. Alexandruk, K. Hashimoto, and A. Fujishima. "Electroless Copper Plating Using ZnO Thin Film Coated on a Glass Substrate," *J. Electrochem. Soc.*, 141(5): pp. L56-L58, 1994.

192. H. Koyano, M. Kato, and H. Takenouchi. "Electroless Copper Plating from Copper-Glycerin Complex Solution," *J. Electrochem. Soc.*, 139(11): pp. 3112–3116, 1992.

193. R. Jagannathan and M. Krishnan. "Electroless Plating of Copper at a Low pH Level," *IBM J. Res. Devel.*, 37(2): pp. 117–123, 1993.

194. N. V. Mandich and G. A. Krulik. "Fundamentals of Electroless Copper Bath Operation for Printed Circuit Boards," *Metal Finishing*, 91(1): pp. 33–41, 1993.

13

PACKAGE ELECTRICAL TESTING

MADHAVAN SWAMINATHAN—*Georgia Tech*
ABHIJIT CHATTERJEE—*Georgia Tech*
FRANK CRNIC—*IBM*
BRUCE C. KIM—*Tufts University*
KOPPOLU SASIDHAR—*Georgia Tech*

13.1 INTRODUCTION

Electronic packages provide a means for interconnecting, powering, cooling, and protecting integrated-circuit (IC) chips. This is made possible through interconnections that provide a connection between active devices, such as integrated-circuit chips or to other discrete components, mounted on the package. Because semiconductor chips are expensive, a testing scheme is necessary to ensure the integrity and performance of all the package interconnection paths. This is a two-step process: final testing of the interconnection paths prior to attachment of the active devices (dies or chips) and subsequent testing of the assembled package. In this chapter, an integrated circuit in its barest form (wafer after dicing) is called a die and when it's ready to be packaged is called a chip.

The testing of interconnections prior to die attachment is called "substrate test" and it is used to guarantee a defect-free substrate. This is a screening process to ensure that the interconnections do not contain defects produced during fabrication and ensures that known-good-dies (KGD) (tested dies that are good) are not assembled onto defective substrates, thus minimizing the wastage of expensive dies. Functional testing is a follow-on to substrate testing and represents the testing of the package after die attachment to assure a defect-free system. This ensures that no defects were introduced into the KGD during the assembly operation and that all the terminals of the chips are connected, in the manner desired. Pretested dies, also called KGD, facilitate the functional testing process as opposed to dies that are not pretested, which add additional steps to the testing process.

Both test categories may include repair processes and engineering change (EC) schemes to minimize the wastage of fabricated substrates due to process defects and system design modifications. A typical test flow is shown in Figure 13-1 which consists of a substrate test to guarantee a known-good substrate, assembly of known good die after burn-in to the substrate, and a final functional test prior to shipment. The flow in Figure 13-1 is similar for a wide range of packaging technologies including single

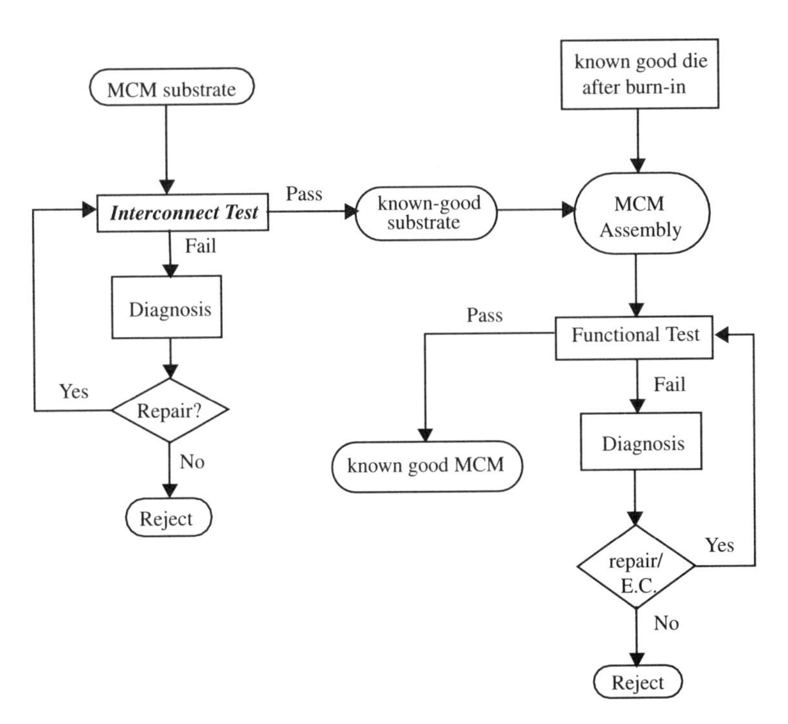

Figure 13-1. MCM Test Flow

chip and multi chip packages. The philosophy and economics of testing have been discussed in detail by Knowles [1] and Davis [2] and serves to provide important feedback to the upstream processes and protects the efficiency of downstream operations. The advantages of electrical testing in packaging are summarized in Table 13-1. Because all packages need to be tested prior to shipment to meet quality, reliability, and functionality goals, electrical testing represents an integral part of manufacturing. Although testing may not add function to the package or product, it adds considerable value and hence is easily justified in a manufacturing environment.

Whereas the electrical testing (substrate and functional) definitely adds value, it could represent a major added cost component in the development and manufacturing of electronic components and systems, with estimates ranging up to 50% of the product cost. This is because manufacturers spend considerable time and effort in assuring the reliability (see Chapter 5, "Package Reliability") and quality of their products, through developmental efforts in design (see Chapter 3, "Package Electrical Design"), test generation, and test application. The testing method or methods selected by the manufacturer may result in expensive test facilities with associated complexity, which are added to the product cost. As a consequence of this, it is widely agreed upon that test issues be considered up front during system design and not as a postdesign effort, so as to select the optimum test method that meets both the cost and the performance goals of the system under development. This is especially true for multichip

Table 13-1. Packaging Operations Flow and Substrate Test Advantages

Operation	Benefits Provided by Substrate Test
Substrate design	Early check of design practices Design changes verified quickly
Substrate materials preparation	Materials problems discovered and diagnosed with less product at risk Shorter cycle of learning for yield improvements
Substrate build	Effective, timely defect diagnostics More accurate build forecasting
Substrate electrical test	
Chip attach (or package to package bonding)	Avoid component loss due to defective substrates
System assembly	Excess capacity not required to process modules with substrate defects
System test/burn-in	Fewer errors, analysis simplified
System use	Product reliability may be improved by substrate specific tests

modules, which are inherently complex and require the use of detailed test procedures to ensure product quality.

This chapter provides details on substrate test and functional test. Both pass/fail testing and performance benchmarking have been explained as part of functional test.

13.2 SUBSTRATE TEST

Consider a multilayered package as discussed in Chapters 7 ("Microelectronics Packaging—An Overview") and 12 ("Thin-Film Packaging"). A simplified cross section of a package has been reproduced in Figure 13-2, consisting of multiple layers of metallization separated by dielectric layers. The multiple metal layers may be necessary due to the density of interconnections required, line pitch allowed by the technology, and electrical performance desired. The metallization layers consist of planes and wiring, the configuration of which is dictated by the electrical specifications. As an example, deposited multi-chip module (MCM-D) substrates may consist of orthogonal (X-Y) wiring layers sandwiched between voltage (VDD) and ground (GND) planes, providing an impedance controlled environment for the interconnections, as shown in Figure 13-2. Other technologies such as ceramic multi-chip module (MCM-C), laminated multi-chip module (MCM-L), MCM-D/C, and so forth may support similar cross sections, with increased layer counts. For example, MCM-C substrates may consist of VDD-X-Y-GND layers, repeated, with 60–120 layers stacked on top of each other. As the layers are fabricated, processes may induce defects on the wiring layers, through the presence or absence of conductive material along the interconnection length or through the

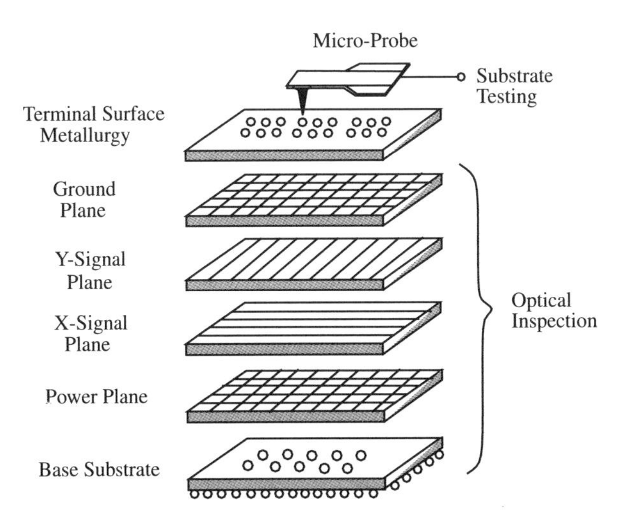

Figure 13-2. Hierarchy of MCM Substrate Test

thinning of dielectric layers. This results in opens on interconnections, shorts between interconnections, via opens or shorts between interconnections and planes, all of which are detrimental to the proper functioning of the product.

In certain other cases, physical imperfections during processing may not render an interconnection functionally open or short but may degrade to an open or shorted condition (near-open and near-short) at a later date. These classes of defects are of particular concern, as they can result in unexpected failures in further processing or during customer usage. Near-opens and near-shorts are classified as latent defects, examples of which are shown in Figure 13-3. The use of very narrow metal lines and polymer insulators (as in thin-film substrates) can increase sensitivity to latent open failures under thermal, mechanical, or bias stress during use. Ionic residues or extraneous metal from photolithographic processing can result in current leakage paths leading to latent shorts.

To ensure the functionality of the product, defects such as opens, shorts, latent opens, and latent shorts have to be detected, diagnosed, and repaired. During processing, defects are detected through visual and optical inspection, where each layer is inspected individually and defective interconnections are either repaired or the substrate is rejected for unrepairable defects. Although every layer is inspected for opens or shorts, temperature and process stressing of subsequent layers may induce defects on interconnections, which therefore need to be inspected prior to die attachment. This obviously is not possible either through visual or optical inspection of the individual layers, because the wiring is embedded in the substrate, as in Figure 13-2. Hence, testing is required through the top and bottom layer pads (terminal metallurgy pads) after all the layers are fabricated. This test is called the substrate test and is used to detect opens, shorts, latent opens, and latent shorts on interconnections buried in the substrate.

Although opens and shorts testing can be used for part sorting, substrates supporting high-performance applications may require parametric measurements to ensure that the interconnections meet a set of design

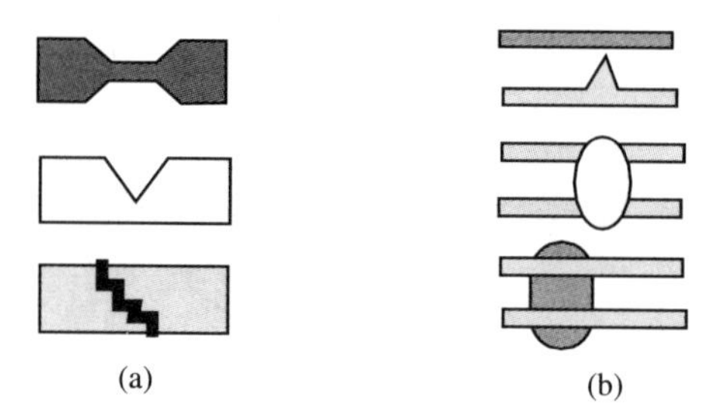

Figure 13-3. Latent Defects due to Near-Opens (a) and Near-Shorts (b)

criteria such as insulation resistance, conductor resistance, and line capacitance. This can be used to achieve less manufacturing variance between substrates to guarantee success for high-speed applications. Parametric measurements to obtain data for process control, in addition to opens and shorts testing, is also part of the substrate test.

Because the ultimate goal of testing is to assure defect-free substrates and lower manufacturing costs, optical inspection can be used to supplement the substrate test. Table 13-2 provides a comparison of optical inspection and substrate test with emphasis on technique, efficiency, diagnosis capability, and application. In manufacturing simpler substrate designs, substrate test is preferred due to its functional and highly efficient defect detection capabilities. However, an effective synergism can be developed between test and inspection for defect detection in substrates. As an example, substrate layers can be optically inspected for defects prior to assembly. Once the substrate is fabricated and tested, defects can be physically analyzed. In this manner, inspection inefficiencies can be understood, corrected and improved. Hence a combination of test and inspection can be a powerful tool for understanding defects and improving the yield of substrates.

Sometimes the material set and smaller feature sizes involved in complex substrates (e.g., thin film) may imply a set of unique test requirements that are not typically practiced in testing substrates. For example, controls introduced or intensified for the thin-film substrate test may include measures to minimize particulate contaminants, metallic debris, organic residue, and tool or handling-related substrate damage. For the

Table 13-2. Comparison of Electrical Test and Optical Inspection for Ceramic Substrate Defect Detection

Test	Inspection
Nature of Technique	
Functional measure of conductor continuity and insulator integrity	Optical observation for presence or absence of conductive material in wiring or insulation channels respectively
Efficiency	
Near 100% defect detection	Variable efficiency
Defect Signature Available for Defect Analysis	
Net identification	Precise x/y location of defects
Application	
Must be fired substrate assembly	Any subassembly with exposed conductors and sufficient conductor/insulator optical contrast

most complex and advanced substrate designs, the electrical test plays a key role in yield management (see Chapter 5, "Package Reliability") for the build process. Tests may be inserted at interim points during substrate build in order to reject defective parts early in the process. The proper selection of an in-process test can be used to significantly reduce the cost of substrates as well as decrease the cycles of yield improvement for new products. Dense substrates used in multi-chip modules (MCMs) necessitate a more sophisticated approach to the substrate test than the simpler single-chip module (SCM) or printed circuit board (PCB) technologies. Hence, although MCMs offer advantages to system designers through tighter integration and offer higher performance, they may require more complex substrate test techniques. Table 13-3 is a summary of the advantages of MCMs and the resulting challenges faced by test engineers.

Numerous techniques are currently being pursued for the substrate test such as capacitance, resistance, combination of resistance/capacitance, electron beam, latent open testing and time domain network analysis (TDNA). In addition, other techniques are at the research and development (R&D) phase. All these methods either require contact probing of the top or bottom surface pads, or use contactless probing as in electron beam testing. The key elements that differentiate test methods are the equipment cost, test time, throughput, and defect resolution capabilities. For example, although capacitance test equipment cost is small, the method allows only for large-defect detection. Hence, capacitance testing has to be augmented by other test techniques (such as resistance) for high defect resolution. This, in turn, may increase equipment cost and impact throughput. However, stand-alone capacitance testing is cheap and easy to use. On the other hand, though stand-alone resistance testing has high defect resolution, it has low throughput, increased test time, and a resulting high test cost. To avoid substrate damage through contact probing, vacuum-based electron beam test systems may be used which are significantly slower and more expensive for volume production testing than the more commonly used

Table 13-3. Advantage of MCM Technology and Implied Test Challenges

MCM Advantage	Test Challenge
Smaller via grid	Smaller probe and tighter positioning accuracy required
More I/O terminals	Premium on test/product contact integrity
Greater circuit density	Higher test speeds required for reasonable throughput
Level of integration	Significant influence on end-product reliability, undetected defects (escapes) more costly
Higher value package	False test errors (overkill) must be minimized

capacitance or resistance testing. The TDNA method [3] may be used for testing substrates that require a high defect resolution, but this requires testing at high frequencies using expensive probes and equipment.

Although TDNA has been proposed as a substrate test, it has been commonly used for high frequency characterization of substrate interconnections. A method invented by IBM called the Electrical Module Test [4] may be used to test substrates to detect latent opens on interconnections, if this is a primary defect causing failures. Because each technique has its own implementation method and diagnostic procedures, test methods need to be compared based on the system application and throughput desired. As wiring densities increase, the applicability and practicality of test systems may vary greatly; hence, the appropriate test method should be selected with care. The following sections discuss the various test methods being used in industry and methods being used in industry and methods still in the R&D phase.

13.3 SUBSTRATE TEST METHODS

13.3.1 Capacitance Testing

Capacitance testing is the most widely practiced test method. It is based on the measurement of the capacitance of a net with respect to a common ground plane, where a net is defined as an interconnection between two or more terminals. It relies on the recognition that a net can be broken down into individual elements, with the total capacitance equaling the sum total of the individual capacitances of the elements. Based on this scheme, the measured net capacitance is smaller than the expected value for an open defect and larger than the expected value for a short defect, with the defects occurring on or between interconnections, respectively. For shorts between interconnections and ground plane, the measured net capacitance is zero. Parametric testing using capacitance measurements may be used to supplement opens/shorts testing to ensure that the nets meet the required specifications for high-speed communication.

The basic concept behind capacitance testing is best illustrated by Marshall et al. [5] as shown in Figure 13-4. Figure 13-4a shows a defect-free net (with two terminals) that has a capacitance C associated with it. The measured capacitance at each end of the net therefore agrees with the expected value C. Figure 13-4b is the same net which is defective with one open defect, resulting in two unconnected elements. The measured capacitance from each end (C1 and C2) for the defective net is therefore less than the expected capacitance C, with the sum total equaling the expected value C. Similarly, Figure 13-4c is the same net with one short defect between two adjacent nets, resulting in a four-terminal net. The measured net capacitance at each terminal is therefore greater than the

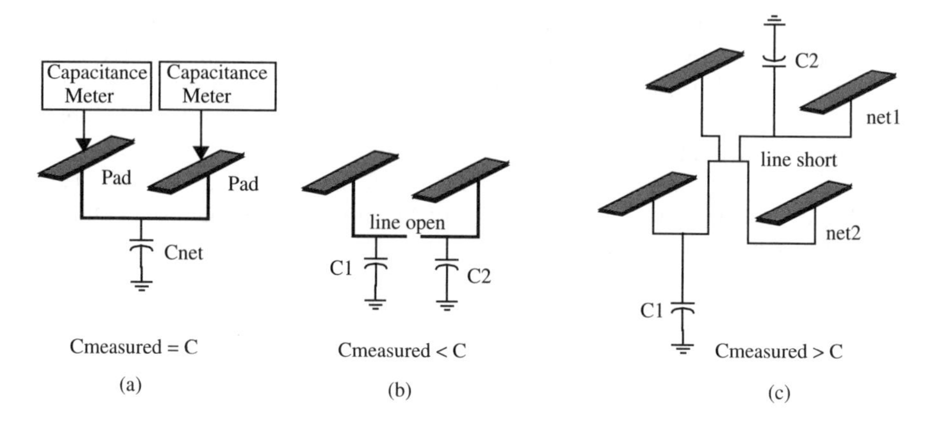

Figure 13-4. Capacitance Measurements of (a) a Defect-Free Net, (b) a Net with an Open Defect, and (c) Two Nets Shorted Together

expected value C and equals the sum total of the individual net capacitances assuming the short circuit has negligible capacitance associated with it. Hence, based on a set of expected capacitances, interconnections can be diagnosed as being defect free, using the above method.

Because probing of the net terminals and ground plane are required, capacitance testing can be performed with a single two-point probe arrangement or with a movable single-point probe and a probe connected to the ground plane. As testing is typically done in the frequency range 1 kHz–10 MHz, close proximity of the two points of the probe is unnecessary. Also, because either the near or far end (not both) of the net requires probing, probe movements are greatly simplified. Marshall et al. [5], Economikos et al. [6], Hamel et al. [7] and Wedwick [8] provide further details on capacitance testing of substrates.

Because capacitance testing is based on expected data which are generated through design and modeling, a careful choice of the minimum and maximum pass/fail limits is key to the successful use of this mode of testing. The pass/fail limits can be explained using Wedwick's [8] defect scenarios shown in Figure 13-5, which are tested using capacitance testing. In the figure, the fail limit is a 100% or greater increase in capacitance at the measurement point for a short, and a 50% or smaller decrease in capacitance for an open. Although a single terminal requires probing at a given time, both ends of the net require probing to completely test that net for opens or shorts. As an example, although the defect is undetectable in Figure 13-5b because the change in capacitance is <100%, probing of terminal B enables short detection due to >100% increase in the measured capacitance, as compared to the expected capacitance of net B. During testing, process variations may introduce an additional 15%

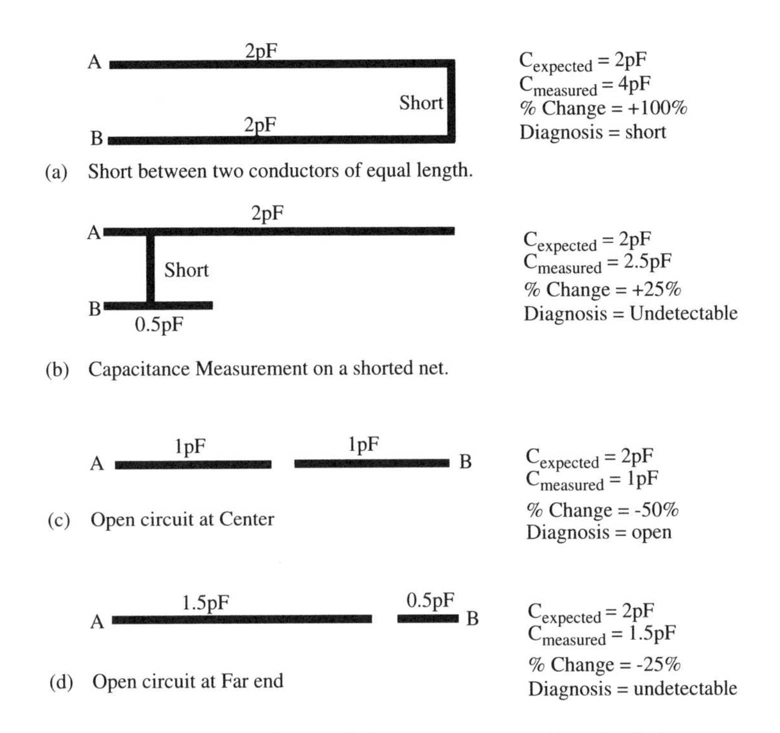

Figure 13-5. Capacitance Measurement at Terminal A

change in expected capacitance values and has to be accounted for during defect detection.

The most difficult defects to detect using capacitance testing is an open circuit that divides the net exactly in two, or a short circuit between two nets of similar lengths. Undetectable multiple defects could also occur as a result of an even number of defects which are equally divided between opens and shorts [5]. However, the occurrence of these defect scenarios is small. Usually, the maximum and minimum fail limits are set no further than ±50% from the expected value to allow for process variations, expected value inaccuracies, and instrumentation tolerances. This margin typically applies for opens testing because an open at the far end causes only a small decrease in measured capacitance at the near end, whereas the measured capacitance at the far end is much larger than the expected value.

13.3.1.1 Test Implementation

The sequence of operations for the successful implementation of capacitance testing of substrates has been discussed by Marshall et al. [5] and shown in Figure 13-6. The data file consists of expected capacitance values of the various nets in the substrate. The data file (also called the

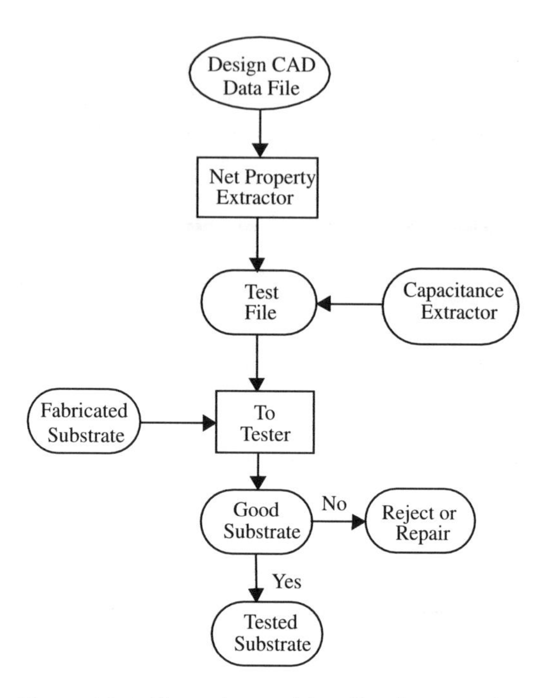

Figure 13-6. Capacitance Test Implementation

test file) can be generated using measured capacitance values from the first few substrates produced. A more popular and robust technique is to use a CAD-based approach that allows for the test file generation before the first substrates are produced. The popularity of the latter technique is because physical design data are used to extract the capacitances and, hence, enables the AC screening of substrates through a correlation of line capacitance to line length. However, the major difficulty in extracting accurate capacitances from the physical layout is the inability to account for complex interactions between multiple conductors and planes. A simplistic approach is generally used for capacitance extraction which is based on the superposition of capacitance contributions from individual elements. As an example, a net may be composed of pads, vias, and lines whose individual capacitances are extracted using numerical simulation tools. The expected capacitance value of the net is the sum total of the individual capacitance contributions. This method obviously does not include the interactions between elements. However, the method is computationally fast, requiring little time for test file generation once the physical layout is complete. Commercial software tools and layout interfaces are available that could be used to automate the process from the physical design to test file generation.

Companies such as Micro Module Systems [5] and IBM [7] have implemented a net property extractor that scans the computer-aided design (CAD) file to extract data such as line lengths with associated geometries (width, thickness, distance from ground plane), via information, pad details, and number of crossovers with other nets. Using two-dimensional (2D) and three-dimensional (3D) parameter extractors, the individual capacitances are extracted and summed to compute the total capacitance of a net. The linear predictor used by Hamel et al. [7] for computing the total net capacitance C_T is based on:

$$C_T = \sum_i C_{Li} \times L_i + \sum_j C_p \times N_j + \sum_k C_v \times N_k + \sum_m C_c \times N_m \quad [13\text{-}1]$$

where C_{Li} is the ith line capacitance per unit length, L_i the ith line length, C_p the pad capacitance, N_j the number of pads, C_v the via capacitance, N_k the number of vias, C_c the cross over capacitance and N_m the number of crossovers. A typical test file generated from the CAD design file is listed in Figure 13-7 providing details on the sample number, net name, net length, expected capacitance, impedance and capacitance per unit length. The test file which is a list of expected capacitance values can then be compared to measured data on a net by net basis to create an error listing file that provides details on the defective nets. A typical measurement system is shown in Figure 13-8 consisting of a moving probe fixture and a capacitance meter operated at 1 MHz. In the figure, the shorting block is used to provide the ground contact for the probe by shorting the VDD and GND I/O's at the bottom surface.

13.3.1.2 Limitations

Capacitance testing, though easy to implement, is limited in its ability to resolve opens and shorts due to the wide test window required (±50%). Since defect detection is subject to process variability and is a function of the frequency used, the test frequency can be varied to improve defect resolution with a trade-off in test time. The typical capacitance test set-up has little capability for providing current or voltage stressing to the product under test. However, capacitance testing is ideally suited for flexible in-line product screening and low volume developmental substrate testing.

13.3.2 Resistance and Continuity Testing

Resistive shorts detection and continuity measurement can be combined to provide good resolution and precise definitions of short and open defects. This method is based on resistance measurements between the extreme ends of the net using two probe heads. If the measured resistance

MAXIMUM	0.1870684257
MINIMUM	0.1328073643
AVERAGE	0.1471803629
STD. DEV	0.006501338716
RANGE	0.0542610614
NO. OBS	1042

SAMPLE NO.	NETNAME	LENGTH	EXPECTED CAP.	Z0	PF/MM
1	1A3AAAB40	253.30	40.40	51.572	0.154
2	1A3AAAB54	220.00	38.30	48.000	0.165
3	1A3AAAA41	193.50	34.50	47.154	0.168
4	1A3AAAB57	188.55	34.00	46.666	0.170
5	1A3AAAA24	182.25	33.20	46.263	0.171
6	1A3AAAA44	178.65	31.90	47.321	0.167
7	1A3AAAB35	171.45	32.40	44.667	0.177
8	1A3AAAB51	169.60	26.80	54.163	0.146
9	1A3AAAB38	166.50	29.30	48.383	0.164
10	10CAAAAC2	163.80	26.60	52.736	0.150
11	1E1AAAA03	162.00	26.50	52.369	0.151
12	1A3AAAA86	160.15	26.90	50.939	0.155
13	1A3AAAA45	159.75	29.90	45.348	0.175
14	1A3AAAE71	158.80	24.70	55.405	0.143
15	1A3AAAE72	158.80	24.90	54.921	0.144
16	1A3AAAF71	158.40	24.70	55.266	0.143
17	1A3AAAF72	158.40	24.90	54.783	0.145
18	1E1AAAB09	158.40	25.60	53.158	0.149
19	1A3AAAE05	157.90	24.60	55.335	0.143
20	1A3AAAB48	156.60	25.40	53.003	0.144

Figure 13-7. Capacitance Test File

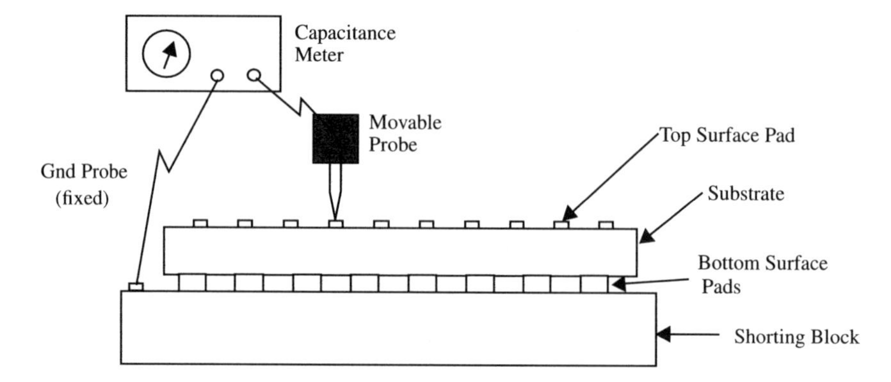

Figure 13-8. Capacitance Measurement System

is within the expected range, the net is assumed to be defect-free. However, any deviation from the expected value represents a defect with the magnitude of the resistance providing details on the defect size and type. The open and short test setup is shown in Figure 13-9. It is based on a four-point probe measurement with probes contacting the terminals of the nets. For the opens test (Figure 13-9a), a known current I is passed through the outer two probes, the potential V is thereby developed and measured across the inner two probes. The ratio of V/I provides the resistance of the net, which can then be compared to expected values. For shorts testing (Figure 13-9c), a potential difference V is applied between two nets, and the current I is measured. The ratio V/I provides details on the dielectric leakage. For good dielectrics, a typical pass limit for dielectric leakage is >100MΩ. The equivalent circuits for opens and shorts testing are shown in Figures 13-9b and 13-9d, respectively.

The resistance test is ideally suited for opens testing due to the increased probings that are required for shorts testing. This is best explained using a substrate containing N nets, each consisting of n terminals that require testing. Terminals represent pads accessible from the top or bottom surface connected to the same net. Assuming two probe heads (each with two probes) are used for resistance measurement, one at each terminal of the net, all the terminals require probing at least once for opens testing, for a total of (n-1)N tests. However, for shorts testing,

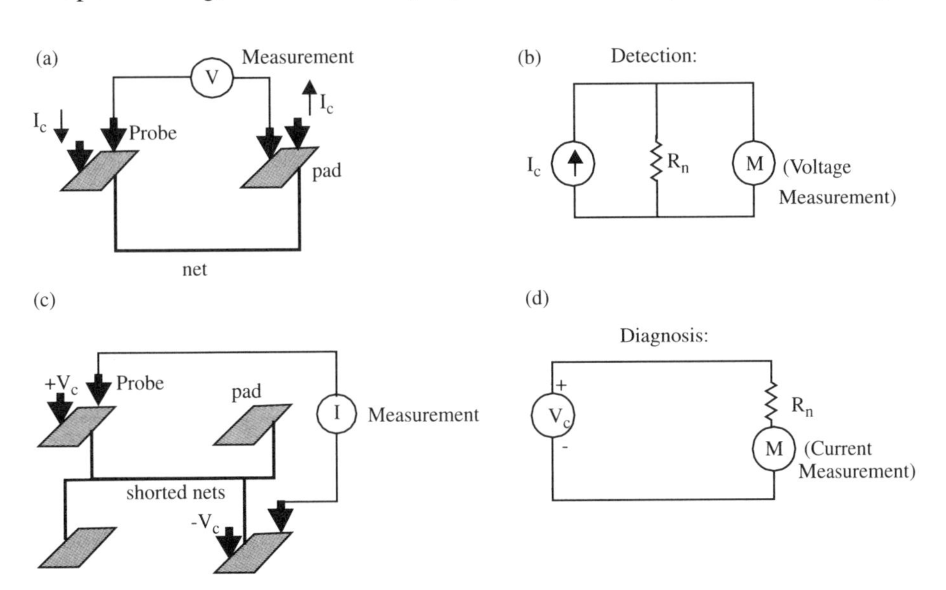

Figure 13-9. Resistance and Continuity Testing. (a) Opens testing; (b) equivalent circuit for opens testing; (c) shorts testing; (d) equivalent circuit for shorts testing. V_c = constant voltage source (current limited); I_c = constant current source (voltage limited); R_n = net under test.

probing on the N-1 remaining nets is required for each net tested, requiring a total of N(N-1)/2 tests assuming no duplication in measurements. Hence the number of tests required is proportional to N for opens testing and proportional to N^2 for shorts testing. For shorts testing, each test requires significant time due to the N(N-1)/2 relative mechanical probe movements, between the two probe heads. Test times can however be reduced by processing the design data from the physical layout, to identify adjacent nets for shorts testing at the expense of increased risk of fault escapes.

Resistance testing's advantage is that it measures opens and shorts directly and can detect low-resistance opens and high-resistance shorts. The large number of probe movements for shorts testing heavily favors the use of a matrix, or cluster probe (also called a "bed of nails" probe) for complex products. This approach employs an array of test probes placed in contact with the substrate and switched, thus greatly minimizing the mechanical movements required of a tester. A cluster probe coupled with mechanical relays or solid-state switching can be used to deliver current and voltage stress to the product in an extremely efficient manner. This method, however, requires a significant outlay for fixturing but provides the best quality and fastest test available. For example, products up to 50 mm square can be tested in less than 20 s, using the cluster probe method. An alternative method is to use "flying probes" for point-to-point measurements, but this method is extremely slow for all but very simple substrates.

Resistance testing is implemented in a manner similar to capacitance testing. A test file is generated from the physical layout of the substrate, providing a listing of the various nets and the associated resistances for continuity testing. The expected resistance values are then matched against measured data. A similar approach is used for shorts testing, where the high resistance between nets is compared to measured data. One source of error in resistance testing is the contact resistance associated with the probes, which may have a significant effect on the measured data, especially for high-density interconnections as in thin film substrates. This error can be minimized by using a four-point probe arrangement (Figure 13-9), as opposed to a two-point probe arrangement connected to a resistance meter.

To eliminate lengthy resistance test time and improve the quality of test, capacitance and resistance testing may be combined. A capacitance tester can be initially used to measure the capacitance of every net in the substrate. The faulty nets can then be verified using resistance testing by applying a stimulus either within the net for an opens test, or between nets for a shorts test, as discussed by Woodard [9].

Resistance testing is ideally suited for thin-film technology, where leakage paths are prone to fail under voltage bias, temperature, or mechanical cycling. The most effective method for detecting a leakage path is through a sensitive resistance test between conductors, as in Figure 13-9, and is practiced by IBM on thin-film substrates. This test operates at a low current to limit the energy injected into the circuit under test and allows defect detection without alteration or arcing. The use of higher test voltages allows for more accurate leakage testing and may permit wetting of poor electrical contacts (contact resistance). High voltages also permit the breakdown of oxides or very thin air gaps that could mask a latent short.

13.3.3 Electron Beam Testing

Noncontact testing has shown promise and offers the advantage of eliminating mechanical probing on the substrate. Electron beam test technology is an attractive alternative for high-throughput, layout-independent, noncontact, nondestructive testing of unpopulated substrates. Because electrons can be positioned at any location on the substrate using computer control, this mode of testing provides high flexibility with respect to layout changes and, hence, can be used to test different products. Electron beams have no mechanical impact and, thus, do not destroy the pad surface or crack fragile substrates. Very fast beam deflection and charge storage allow for high test speed and thus high throughput.

13.3.3.1 Test Implementation

A focused electron beam can be positioned on various locations of the substrate using an electromagnetic deflection system. The beam can be switched on and off by a blanking system in order to form "charge" and "read" pulses. A voltage contrast results as a consequence of switching the beam for a short time duration, which can then be detected and used for discrimination between charged and uncharged pads. Ross et al. [10] describe a means for testing substrates using a voltage-contrast electron beam, as shown in Figure 13-10. Consider a net consisting of three terminals that are accessible from the top surface, as shown in Figure 13-10a. The net is intact between terminals 1 and 2 and contains an open between terminals 2 and 3. In the figure, the net is charged by positioning the electron beam at terminal 1 of the net. The beam is then directed to terminal 2 followed by terminal 3 and pulsed for a short duration at each terminal to read the charge, as shown in Figure 13-10b. The voltage-sensitive detector indicates that terminal 2 is charged and terminal 3 uncharged, indicating an open between terminals 2 and 3. As a final test sequence, all the charge on the substrate is erased by an electron flood gun for subsequent test algorithm implementation, as shown in Figure 13-10c. A similar sequence is used for shorts testing with the difference that charge on the remaining nets are read to detect voltage contrasts as a consequence of charging. To allow for pass/fail limits, charge and read

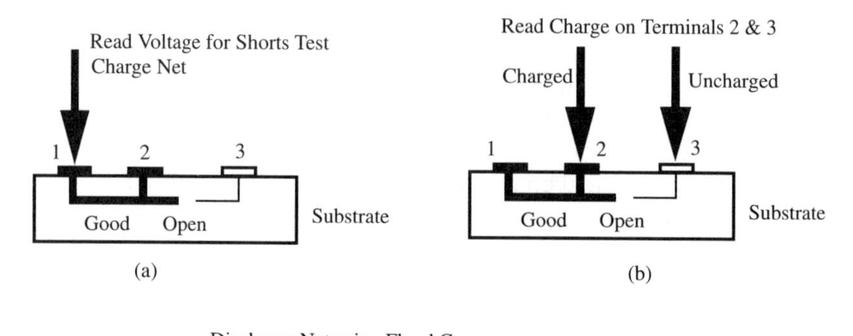

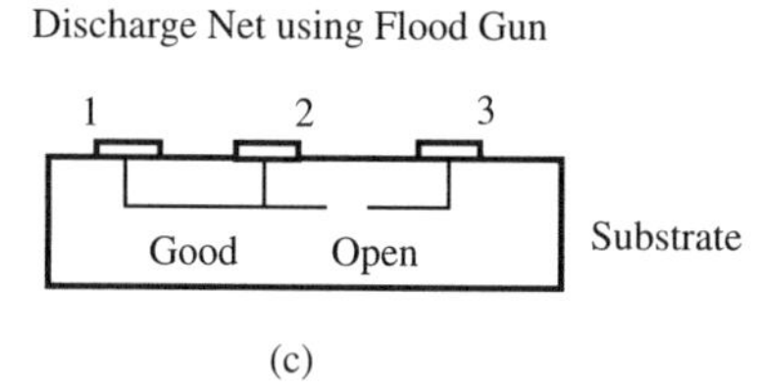

Figure 13-10. Voltage Contrast Electron Beam Test Method. (a) Charge net; (b) charge detected on terminal 2 and none on terminal 3; (c) discharge net for next test.

threshold voltage levels are used in the tester. Nets are charged by using the charging current to bring up their voltage levels to a charge threshold which is above the read threshold. For this purpose, the beam is switched on to form a charge pulse that is typically longer than the read pulse.

Brunner et al. [11] discuss the application of electron beam testing to MCM substrates as in Figure 13-11, which represents a group of substrate nets consisting of multiple terminals which are accessible through pads from the top or bottom surface of the substrate. As an example, net N consists of two terminals while net N+1 consists of three terminals. To test the nets using an electron beam tester, the following sequence is used:

1. One terminal of net N is charged.
2. All other terminals of the same net are read. They are expected to be charged and any voltage contrasts indicating otherwise represents an open.
3. Terminals on net N+1 are next read. They are expected to be uncharged and any voltage contrast suggesting otherwise represents a short to a previously charged net.
4. Net N+1 is next charged.
5. Step 2 is repeated on all the terminals of net N+1 to detect opens.
6. Step 3 is repeated on all the terminals of net N+2.
7. Steps 4, 5 and 6 are repeated on nets N+2,N+3. . . until the test sequence is complete.

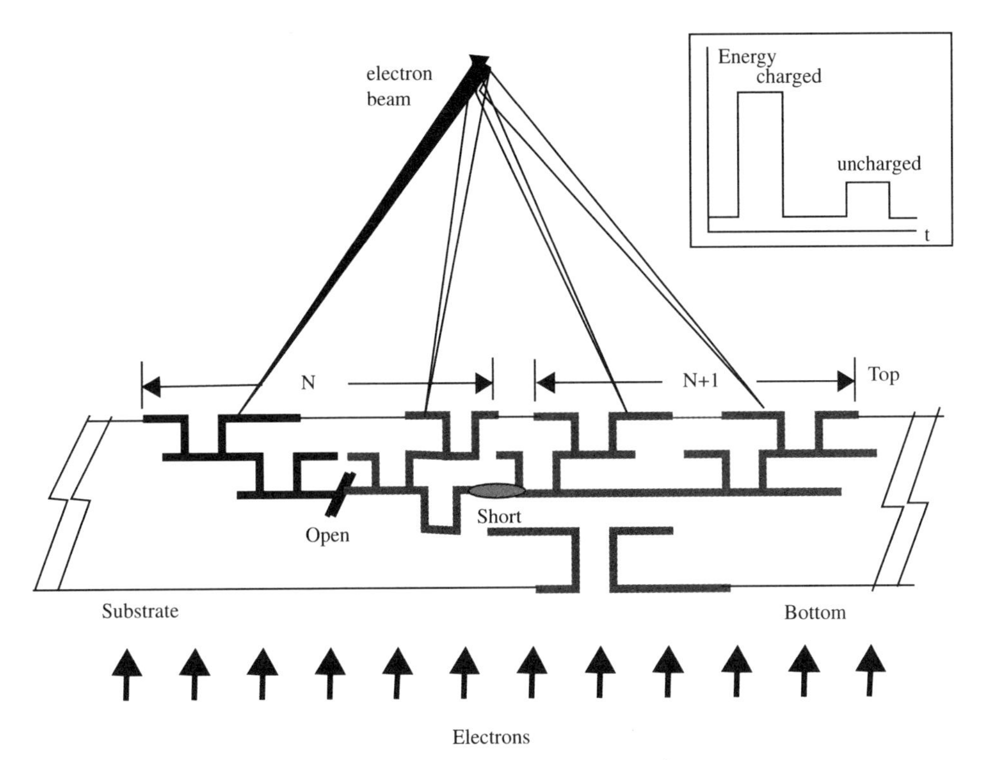

Figure 13-11. Electron Beam Testing

The above test sequence obviously detects and locates opens. However, this sequence is limited for shorts testing, because although it detects shorts, it does not provide information on the nets that are shorted to each other. Short location is done through a separate test cycle. For this purpose, the shorted net is charged and all other nets read with a pulse to detect the nets involved in the short. The second test cycle is short, limited to defective nets and requires little time. The main test sequence requires the access of each terminal twice one each for the detection of opens and shorts. Hence, the test time increases in proportion to the number of terminals and is dependent on the ability of the nets to hold charge during the test duration.

The electron beam can also be used for testing connections to the rear side of the substrate by a separate flood beam which charges all bottom pads simultaneously. Charged pads on the top side belonging to nets connected to bottom pads indicate correct connection. Uncharged pads on the top side belonging to these nets represent opens. Shorts between top to bottom nets show up during the main test sequence from the top side.

13.3.3.2 Limitations

The size of defects detectable using electron beam testing is important in judging the compatibility of the test method with the fabrication process being used. As an example, consider two nets shorted to each other with a resulting shorting resistance R, as shown in Figure 13-12 and discussed in detail by Brunner et al. [11]. The voltage level measured at pad 2 as a result of charging pad 1 is a function of the resistance R and the total capacitance of the shorted net, resulting in a time constant RC, assuming that the charged voltage has reached the appropriate measurable level. Using the electron beam method, the voltage is measurable if the time constant is larger than the time required to read the charge on pad 2. Hence, the resolution for shorts detection depends on the capacitance of the shorted nets, the resistance of the short and the time required for reading the charge. The shorted resistance detectable ranges from less than 1Ω to 100MΩ. Electron beam testing is therefore ideally suited for the detection of dielectric leakage between nets. For opens testing, Figure 13-12 can be modified to include the resistance R as part of a single net. Typical values of open resistances that can be detected are in the range 10–100MΩ. Lower resistances are considered connected; hence, the method is not suited for the detection of low-resistance opens.

For substrate stressing, electron beam testing is plagued by severe current-carrying limitations. The test system is more expensive for volume production testing than the more commonly used resistance or capacitance test systems. Passives integrated into the substrate, such as resistors and capacitors, affect defect detection. As an example, a net terminated by an integrated resistor (of small value) to ground cannot be charged by the electron beam to retain charge for the appropriate time (due to the resulting small time constant) and, hence, shorts between nets cannot be detected, as explained by Brunner et al. [11]. However, opens can be detected on these terminated nets if the defect results in the disconnection of the termination. Integrated capacitors to ground limit the resolution of defects due to their influence on the charge time. Like other test methods,

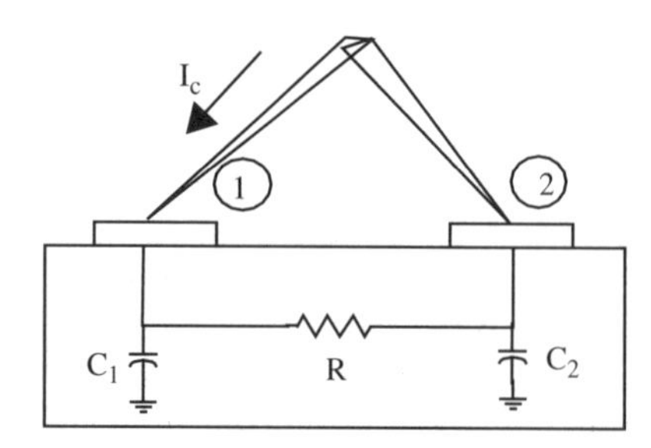

Figure 13-12. Electron Beam Defect Resolution

electron beam testing does not provide details on the precise defect location or the cause of the defect.

13.3.4 Electrical Module Test

As wiring densities increase with a consequent decrease in line width and spacing (as in thin film substrates with line widths of 25μm on a pitch of 75μm), defects such as latent opens have a greater chance to occur. Latent opens are near-opens that transform into complete opens in the field, producing failures. Detection of latent opens such as cracks, notches, via line connections, and so forth are usually based on optical inspection or stress testing. However, optical testing can be applied only to visible areas and stress testing such as thermal cycling or mechanical fatigue tests are time-consuming. For expensive substrates, the lack of a latent open detection technique makes the package vulnerable to failure during subsequent assembly processes. Testing for latent opens as a result of burn-in stressing [12] on thin-film substrates has been addressed by IBM through the formerly proprietary and currently published electrical module test (EMT) or latent opens test [4]. Burn-in stressing involves the application of heat cycles and/or electrical bias to the substrate in order to reduce failure to detect latent opens. Properly chosen burn-in conditions can help to weed out unreliable substrates. However, stress conditions must be carefully developed so that defective circuits can be forced to fail during burn-in without significantly weakening others. Burn-in should also not introduce failure modes inconsistent with field conditions. Burn-in testing is often performed with integrated-circuit chips mounted on the substrate. This, however, could result in expensive perfectly good chips being lost, or requiring rework due to substrate failures. Burn-in or other steps taken at points well removed from substrate build can result in an extended feedback time for latent defect understanding. Once corrective action is implemented, initial parts must be measured at burn-in conditions to verify the solution. Thus, the cycle time to reach a measurement point (from solution to burn-in condition) is a key parameter in determining reliability improvement rates.

The latent open testing adopted by IBM [4] is based on the application of alternating and direct currents through metal interconnections to detect latent electrical opens such as line narrowing, notches, nicks, cracks, weak connections and interface contaminations. Due to the nonlinear relationship between the voltage across and current through the interconnection, distorted signals are generated based on the nature of the defect. The phase of the distorted signal from defective interconnections are compared to the reference phase from defect-free interconnections for fault diagnosis. Application of this technique can improve product reliability and reduce manufacturing cost through early defect detection.

The method is based on non-linear conduction characteristics of metal conductors due to the temperature dependence of the DC resistance and its thermal properties. Halperin et al. [4] explains the method, using Figure 13-13 to show the I-V characteristics of an ideal conductor which is a straight line. As a positively biased sinusoidal current passes through the conductor, the corresponding voltage is also a sinusoid as shown in Figure 13-13. When the conductor has a nonlinear characteristic caused by a resistance change due to an increase in current-induced temperature, a distorted sinusoidal voltage is generated with a corresponding shift in the I-V curve, as shown in Figure 13-13. The amount of distortion depends on the size of the defect. Hence, phase-sensitive detection techniques can be used to compare phases between defective and defect-free interconnections for pass/fail testing. For high defect resolution, the method requires a 90 degree phase shift between good and defective conductors. This can be achieved by selecting an operating frequency range that can generate adequate phase difference and amplitude [4]. The phase detector output [4] as a function of operating frequency is shown in Figure 13-14 for similar-size notch defects on copper thicknesses of 35μm, 3μm and 5μm used in epoxy-glass, polyimide, and alumina substrates respectively, showing the effect of the operating frequency on three different technologies. As an example, an operating frequency of 10 KHz is desired for polyimide substrates, which maximizes the detector output signal.

13.3.4.1 Test Implementation

The low frequency tester implemented by Halperin et al. [4] is ideally suited for printed circuit-board applications and uses a 1KHz drive current.

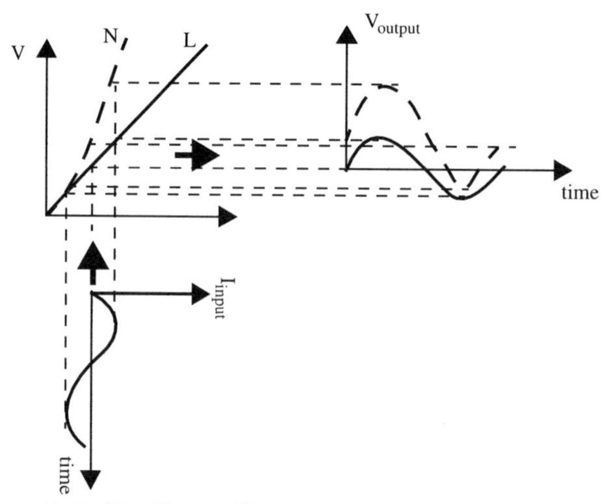

Figure 13-13. Nonlinear Characteristics of Metal Conductors

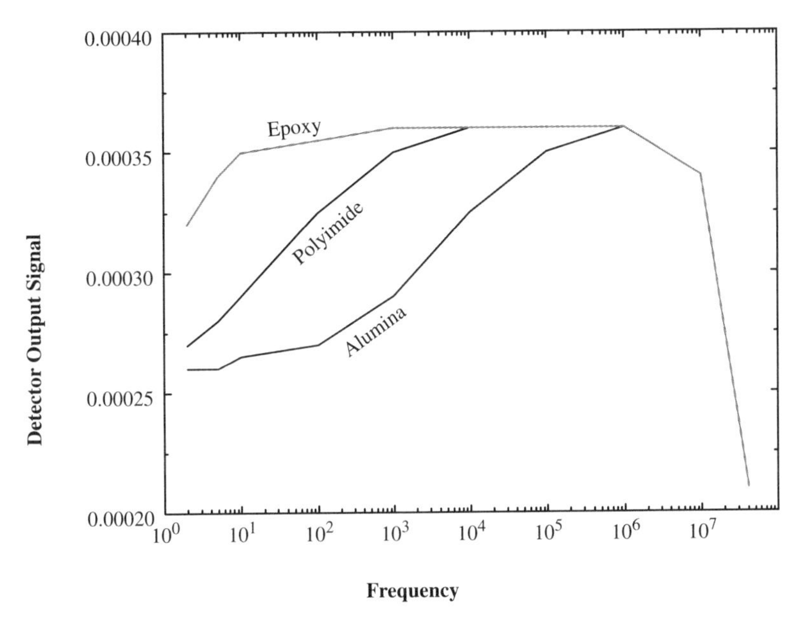

Figure 13-14. Operating Frequency vs. Detector Output for Latent Opens Testing

The method is implemented using two probes, one at each end of the interconnect. DC and superimposed AC current flow through the interconnect via the drive probe and return probe. The sensed voltage passes through receiver circuits which consist of filters for unwanted frequency rejection. Since a sinusoidal input signal oscillating at frequency f is used, the sensed signal consists of the fundamental frequency f and higher frequency harmonic (2f, 3f, 4f.....) components. A linear amplifier amplifies the second harmonic (2f) of the sensed signal which passes through a phase detector. To allow for phase discrimination, the fault signal is phase-sensitive demodulated and converted to a direct current voltage that is amplified and read on a meter. A calibration procedure is initially used to generate the reference phase of good interconnections.

A higher operating frequency (1 MHz) is used for testing thinner conductors as depicted in Figure 13-14 for alumina substrates. The use of the second harmonic for detection in the low frequency regime cannot be applied at higher frequencies due to signal distortion of the AC source produced by commercially available oscillators. This is overcome using the sum of two sinusoids, with the difference frequency ($f_2 - f_1$) used for defect detection.

Using EMT, defects that can be detected are a function of the defect resistance and the total resistance of the line. The low frequency tester is capable of 3 mΩ defect resolution for a 5 Ω line and is capable of testing long defects (125 μm). The high-frequency tester has a 10 mΩ resolution capability for a 20 Ω line resistance and is capable of detecting 20 μm long defects.

13.3.4.2 Limitation

Latent open testing requires probing on both ends of the interconnection and, hence, requires two probe heads, one for the drive signal and the other for the return signal. Contact resistance and I-V non-linearity between the probe and sample must be low to minimize distorted signal generation that can be confused with defect signals. Typically gold plated, palladium alloy or beryllium copper probes are used to minimize contact resistance and I-V non-linearity. EMT is limited due to the need for expensive equipment and is not readily applicable for detecting latent short defects (e.g., line flaring).

13.4 COMPARISON OF TEST METHODS

Table 13-4 provides a qualitative and quantitative comparison among test techniques practiced in the industry. Although TDNA has been used for high-frequency characterization of interconnects and is not a viable test method, it has been included in the table. The number of probe heads depends on whether one end or both ends of the interconnect require probing for a two-terminal net. This is related to probe movements with the complexity arising due to the necessity for the two probe heads to be in synchronization. The complexity also manifests itself through the requirement for expensive test equipment. The test time required for implementing each test method is based on the number of probings required and assumes that the set-up time is similar and small for all the methods. In the table N is the number of nets with n terminals and T_n is the time required per test and assumes simple probes such as flying probes. A test time of 5 tests/s has been assumed in all the cases, although test times of 50–200 tests/s are possible with the current testers. The method used dictates the dwell time on the pads for the probes, assuming similar probes are used (flying probes versus cluster probes).

For example, high-speed capacitance testers with dwell times <10 ms are available today, whereas the dwell times for EMT testers are closer to 50 ms. Test times can be further reduced using the bed-of-nails tester discussed by Crnic [13] and Woodard [9]. A qualitative comparison is provided in Table 13-4 for the total test time assuming high wiring densities, with further details available in Woodard [9]. The opens and shorts resolution in the table provides a measure of the nature of the defects that are detectable by the various test techniques. A small value for opens and large value for shorts represent good resolution. For example,

Table 13-4. Comparison of Test Techniques

Details	Capacitance	Resistance	Capacitance and Resistance	Electron Beam	Latent Open	TDNA
Frequency	1–10MHz	DC	1–10 MHz and DC	—	1 kHz–1 MHz	30–70 GHz
Probe head	1	2	2	None	2	1 / 2
Probe movement	Simple	Complex	Simple	Complex	Complex	Complex
Probe time per test (T_n)	200 ms	200 ms	200 ms	15 ms charge per net (T_n) + 10 ms test	200 ms	Large
Total test time	nNT_n medium	$N(N-1)T_n/2 + (n-1)NT_n$ large	nNT_n medium	$2nNT_n+NT_n$ small	$(n-1)NT_n$ large	Large
Opens resolution	1 MΩ	10 MΩ	1 MΩ	10–100 MΩ	3–10 mΩ	Small
Shorts resolution	1 MΩ	300 MΩ	300 MΩ	1Ω–100 MΩ	Not applicable	Large
Equipment cost	Small	Small	Small	Large	Large	Large

capacitance testing is ideal for opens testing but has poor resolution for shorts testing.

However, electron beam testing is ideally suited for shorts testing. Among the methods shown in Table 13-4, only EMT has the capability of detecting latent opens, with none of the methods providing a capability for detecting latent shorts.

13.5 CONTACTS AND PROBES

The application of the various test methods to products requires the developement of contact techniques using high-density, reliable probes, capable of minimal substrate damage, and incorporation of the probes into testers for automation. For example, ceramic substrate products manufactured by IBM in mid 1990's contain over 7500 usable nets. These designs require probing of over 70,000 top and 3,500 bottom surface features. Due to the presence of interconnections between chips in the substrate and interconnections exiting the substrate through I/O's, the probing scheme must be capable of providing contact to both the top surface pads and bottom-surface I/O's. The top-surface contacts are as small as 90 μm in diameter on centerline spacings of 250 μm. Due to the fine pad dimensions and the number of pads requiring probing, test probes may have to contact many points, simultaneously. Commercially available flying probes (~1996) are capable of contacting ultrafine-pitch geometries down to 50 μm pads on a 50 μm pitch.

IBM has been very successful in applying resistance and continuity test methods through its contact technology for its cofired multilayer ceramic substrates. IBM [17] invented and developed the buckling beam test probe based on the principle of a buckling column. The test probe delivers relatively uniform contact force over a wide range of displacements. This allows for consistent electrical contacts to be made across the surface of the substrate despite Z-dimensional variations. The buckling beam probes used for ceramic substrate testing are round and made from Paliney 7 alloy (includes Ag, Cu, Sn, Au, Pt). The test probe has a built-in offset in the probe guide as shown in Figure 13-15, such that all beams buckle in the same direction under the load. The space transformer design makes replacement of a damaged probe beam possible.

For ceramic-substrate testing, approximately 35 g of force is applied. The contact force applied is a function of the unsupported beam length and the cross section of the beam. In addition to the contact force for testing IBM's ceramic substrates, the probe tips have to contact the terminal pads which are distorted from their design locations due to dimensional changes that occur during the sintering process. This shifts the top-surface features from the nominal grid location. Hence, the pads must be mapped for testing. For current products, IBM has chosen to map the substrate on a

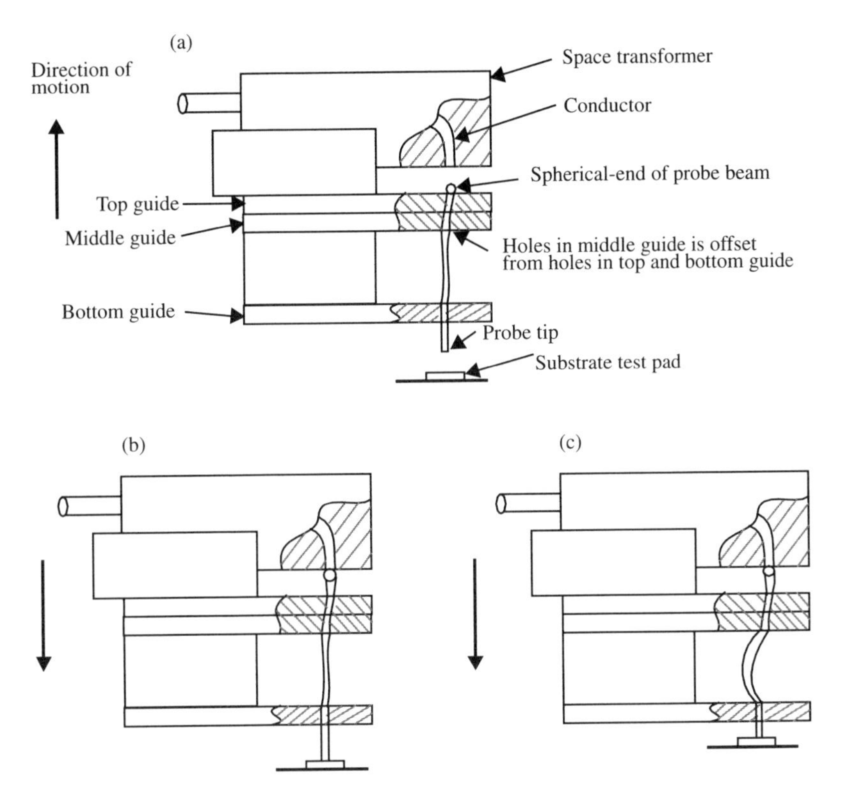

Figure 13-15. Buckling Beam Cross Section. (a) Before contact; (b) at contact; (c) after contact.

separate tool to avoid burdening expensive testers. For future products, alignment requirements have mandated optical mapping and registration capability on each tester. The mapper uses a fixed product stage and a camera mounted on precision X and Y tables. The camera is set up with special lighting to create sufficient contrast between the ceramic and via. Feature recognition is used to define via centers by determining the metal area center of mass. This method is superior to the edge detection method as shown in Figure 13-16, which detects a shifted feature center. The center-of-mass technique is less sensitive to the confusing optical effects of the Cu/glass composite via used by IBM than via perimeter or edge detection. The substrate is held for mapping and tested in a four point locating device. The alignment device allows determination of the substrate's mechanical center. Future testers will incorporate top surface referencing and achieve improved alignment capability with some cost in set-up time. Each mapper includes a benchmark fixture with precision markings for camera pixel site calibration. The benchmark is also used

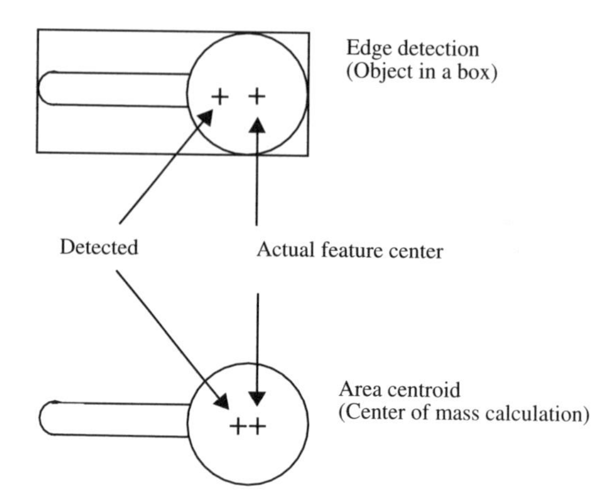

Figure 13-16. Comparison of Feature Recognition by Edge Detection and Area Centroid

as a fixed point for XY distance checking. Glass standards are used to calibrate the mapper and tester. A single feature is mapped with the standard located in the four point fixture and then re-mapped after the standard is rotated 180 degrees. This establishes the machine center. The mapper then collects coordinates for a feature near each of the 100 chip sites on the standard. These coordinates are saved for use as correction factors and must repeat within 3 μm. Buckling beam test probes with over 700 130 μm diameter contact beams in a 100 mm^2 area of a single chipsite are currently in use by IBM. High-density ceramic substrate test probes with over 1000 beams are under development at IBM.

For ceramic substrates, electrical testing is performed prior to I/O pin brazing using an evaporated Cu pad 1.4 mm in diameter for bottom side contact. Contact to the bottom-side I/O pads is made with a "bed of nails" contactor with an array of spring probes as shown in Figure 13-17. The contact pins are a tempered Be-Cu alloy plugged into a Hypertac [18] socket. The 0.9 mm diameter pins have a 0.1 mm contact tip spaced on a 2.50 mm grid, with nearest-neighbor interstices 1.25 mm apart. The I/O contactor has a changeable plate to allow for either 2772 or 3526 I/O's. Testers with "universally switchable" I/O contactors are also under development which does not require a regular I/O grid and can handle over 6000 bottom side features. The I/O contactor provides a contact force of approximately 30 g per pin. The cumulative upward contact force of the bed of nails I/O contactor is balanced by the vacuum hold-down mechanism of the tester. Testers with a hold-down clamping device for very high I/O substrates are also under development.

The test probe used by IBM for testing thin-film substrates is a

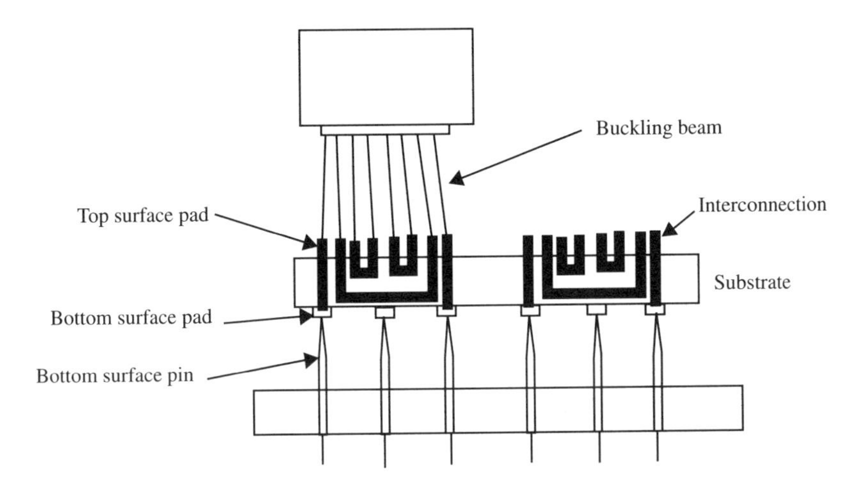

Figure 13-17. Cluster Probe Tester

refinement of the technology used for ceramic substrates. Due to thin-film wiring between contact pads on the top surface (for redistribution, repair and engineering changes), the probe beam is smaller as compared to ceramic substrates. For the thin-film substrates, a test probe of 80 μm in diameter is used. Smaller probe beams are being developed for future products. The thinner probe wire necessitates use of a controlled-direction multiple-buckling beam probe assembly, which was invented and developed by IBM. This assembly, called the sandwich probe, is necessary to achieve sufficient Z dimensional travel in the probe without overstressing the Be-Cu probe beams. A key feature of this device is the probe guide, shown in Figure 13-18. The guide facilitates multiple-buckling action that occurs during probe contact and governs the beam spacing and alignment. A special photolithographic reactive ion etch (RIE) process is used to produce the oblong holes in the guides which control the undulating buckling segments of the sandwich probe. Another unique feature of the sandwich probe is that the probe wire is continuous from the point of contact with the substrate to the connecting plug at the switching matrix of the tester. This feature minimizes resistive loss by reducing contact interfaces and allows for maximum sensitivity test.

To efficiently administer opens and shorts test (OST) for IBM's large thin-film substrates, the full cluster tester was developed. The tester utilizes an array of sandwich probes to contact up to 121 chip sites at a single time. This configuration results in over 26,000 simultaneous top and bottom contacts and a test time of about 80 s. The complete test procedure including load/unload and a re-test after 180 degree substrate rotation takes about 5 min. Approximately 27 miles of probe wire are used to test the substrate.

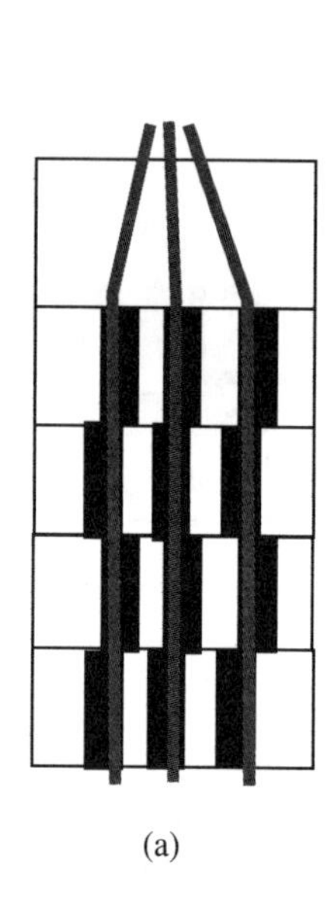

(a)

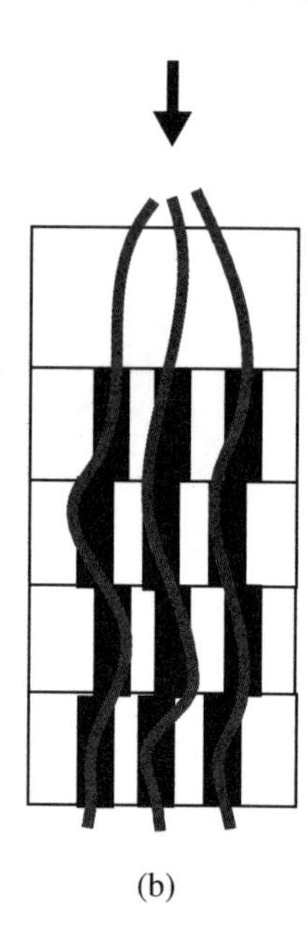

(b)

Figure 13-18. Sandwich Probe Guide. (a) Beam deflection before touching a substrate; (b) beam defection after touching substrate.

The full cluster tester includes two new test features as compared to existing testers. The first of these is a 50 mA current pulse used to wet contacts during opens testing. This pulse is of short duration (up to 0.25 s) so that circuitry is not harmed. The current pulsing feature has been very successful in reducing false open errors. The second is break-down/burn-out detection. This circuitry discerns the transient current spike which occurs at the onset of a dielectric breakdown or arc. The test stimulus is immediately shut down and the incident can be reported, facilitating an investigation. The breakdown detection feature is very useful in high voltage testing of thin-film substrates. Surface residues or contaminants resulting from thin film processing can allow an arc to be struck between metal features. Continued application of the test stimulus can result in burning or "charring" of the polyimide. By interrupting current flow, breakdown detection minimizes the damage that can occur.

A special tester was developed by IBM for latent opens (same as EMT) screening of thin- film interconnections. This tester operates in a serial fashion with two probes placed at the ends of a circuit path for testing and moved to the next circuit when the test is complete. X and Y axis movements are accomplished via commercially available tables. A Z-axis actuator as shown in Figure 13-19 was developed to provide straight line motion with high acceleration and a limited (1mm) stroke.

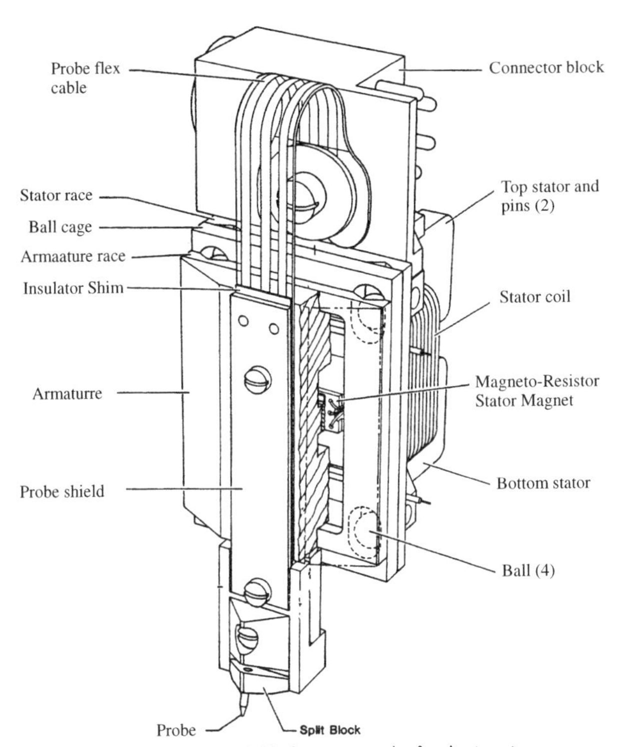

Figure 13-19. Miniature z-Axis Actuator

This device has continuous position control, ability to maintain a steady-state force and controllable impact velocity. The Z actuator operates similar to an electronic stepping motor. The probe is mounted to the armature with a set of balls rolling in V-grooves acting as separators and bearings between the stator and armature. Position sensing is accomplished via an infrared light emitting diode (LED) which differentially illuminates two photo-transistors through an aperture mask which is mounted to the moving armature. With a 2 ms up/down stroke, this device can perform approximately 4 tests/s with the commercial tables, but is designed to operate with the Hummingbird mini-positioner device being developed for use with future IBM testers. The Z actuator mounted on the miniposi-tioner is capable of a 5 mm X-Y probe movement with a total move time of 8.0 ms. Allowing approximately 10 ms for measurement, 40–50 EMT tests/second can be achieved within a 12 mm workplace. A tester utilizing two probe heads and an XYZ positioning system is shown in Figure 13-20. The two point probe tester uses "confidence fixtures," containing examples of conductors whose cross-sections have been reduced. The confidence fixtures are tested before and after each substrate in order to

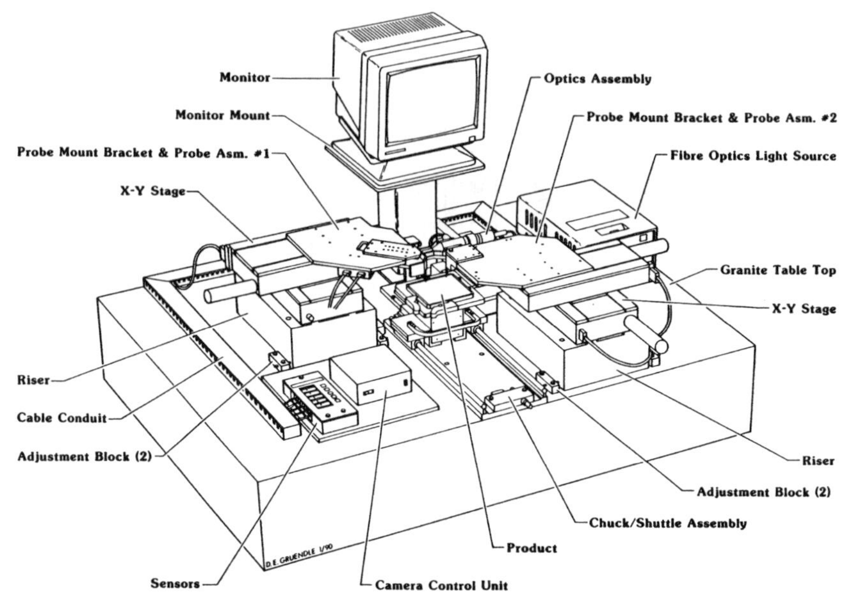

Figure 13-20. IBM Two-Point Probe Tester

guarantee the validity of the test. The two point probe tester uses a top surface reference to register the substrate and can accept maps generated for other IBM substrate testers. Top-to-bottom substrate nets are latent opens tested by evaporating a shorting plate on the top surface and probing from the bottom side. Four point testers with top and bottom probing capability are also in use.

In addition to the requirement for probing fine geometries, the advent of thin-film substrates has lead to new challenges for eliminating the interference between the substrate and the tester. An additional objective is to avoid alteration of either the test output or the substrate due to particulate contaminants, metallic debris, organic residues and tool-handling related substrate damage. Due to the sensitivity of the probes and testers, particulate contamination can affect the test sequence and results. For example, non-conductive debris of almost any size can prevent contact between the test probe and the substrate surface feature. The results of the test could therefore erroneously indicate an open circuit. Similarly, debris from the probe can also be transferred onto contact surfaces and may render bonding or joining to these surfaces problematic. Furthermore, large, hard particulates can result in the mechanical alteration or destruction of substrate features or even the test probes themselves and could create a shorting path if they adhere to the substrate surface. Organic residues such as stains can result in dielectric leakage paths and could also inhibit further processing. Hence particulate control is necessary and is facilitated greatly by the use of a clean room processing area with appropriate garments and procedures. Substrate cleaning is also recommended, both before and after test, to allow for particulates and residues to be removed. To minimize the generation of metallic particles during test, probing of metal surfaces with varying hardness should be avoided. As scratching of pads during probing is sometimes unavoidable, debris and particulate interference should be considered in diagnosing failures that may occur during the test process. Post-test process control inspection of the substrate surface may therefore be necessary and prove to be helpful in the timely detection and correction of substrate damage problems.

One of the most challenging problems related to thin film surface damage is the control of motion in the X/Y, and Z axes, due to fine geometries. Figure 13-21 illustrates the problem, wherein the test probe

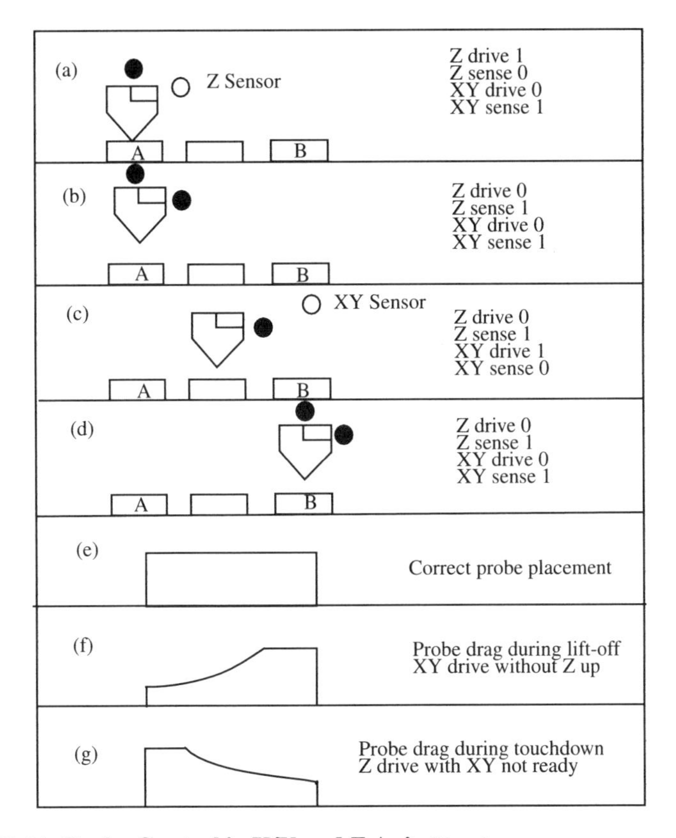

Figure 13-21. Probe Control in X/Y and Z Axis. The above sequences are schematic depictions of test probe or probe cluster movements from one test point to another. Segments (a)–(d) show alignment of the X/Y and Z markers (at right angles on the probe body) with their expected sensor position as indicated by a lighted lamp. Boolean states of sense and drive for the axes involved are shown at right. Segments (e)–(g) are graphs of X/Y and Z positions realized during movement from pad A to pad B.

must be moved from Pad A to Pad B without disturbing the intermediate features or insulation that may lie between the two pads. The intermediate features could represent wiring for redistribution or repair. They are on the same metallization level as the terminal pads. The example shows a single probe and short probe movement but the principle is applicable to clustered probes and long distance probe travel as well. Figures 13-21f and 13-21g show the resulting pad damage due to uncoordinated probe movement in the X,Y and Z directions. The motions described in Figure 13-21 are relative, where either the probe or the substrate may be raised or lowered because both could result in similar damages. To avoid damage, separate sense controls for X/Y, and Z axes, with suitable hardware and software for sense and drive feedback loop, may be necessary, particularly for high-speed test sequences. The feedback loop could, however, have potential downfalls due to false sense or drive signals due to circuit noise, interference between drive and sense signals, consumption of Z dimensional process window by mechanical tolerances (probe to probe, probe plane to product plane skew, table inaccuracy or wear, unexpected product deviation, sensor mounting), oscillation about sense points, and inability to instantly interrupt/reverse drive sequences or timing inconsistencies.

As testing represents a significant portion of the substrate cost, a method of decreasing manufacturing cost is through the use of in-process testing. As shown in Figure 13-22, based on the yield, few substrates require a final test because parts can be rejected early in the manufacturing phase. In-process testing can be achieved either through visual or optical inspection.

13.6 OPENS AND SHORTS TEST IMPLEMENTATION

Test, contact, and probing methods discussed in the earlier sections have been applied here to test substrates. The example used is the ceramic substrate, although the test flow is applicable to other substrates as well.

For complex substrates (such as ceramic), the top-surface features are contacted in two separate tests as shown in Figures 13-23 and 13-24. The two-step method minimizes the complexity of the test probe design as all contact features need not be included in each test head. The first operation is a single-site top-surface probe head making contact with the pads within a chip site which tests circuitry within the chip site and to I/O pins. Two test heads can be operated simultaneously to reduce time, each testing half of the substrate. This test represents an opens/shorts test, also called the single-site common opens/shorts test (COST) which checks voltage planes, redistribution nets, and I/O pins. Neither planes, redistribution, nor I/O nets are repairable. Thus, the relatively inexpensive COST operation can be used to reject substrates with unrepairable defects prior to further processing (e.g., application of thin film to the substrate).

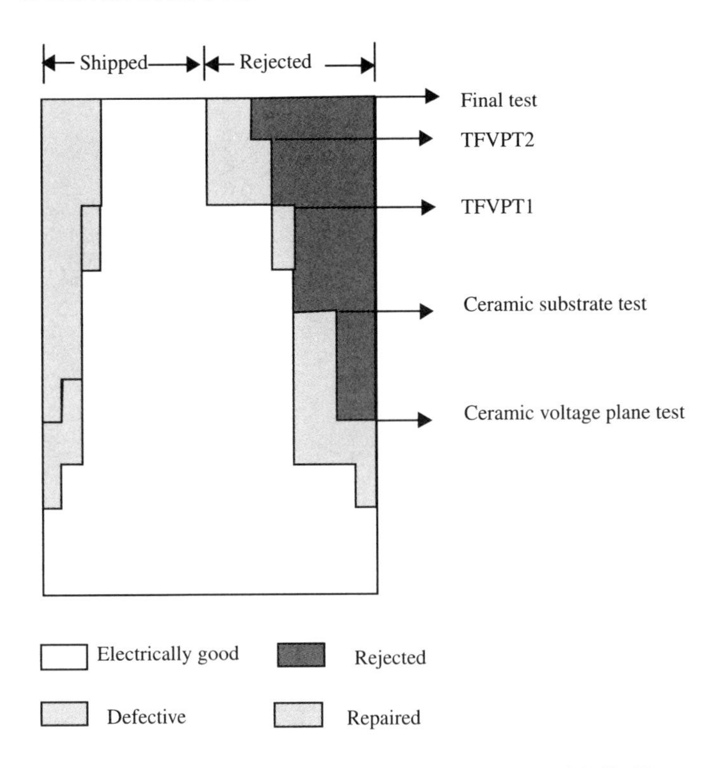

Figure 13-22. Reducing Substrate Manufacturing Cost with In-Process Test. a) Yield advantage from realizing rework/repair opportunities not available later in process; b) labor, material, and capital costs avoided by rejecting defective parts early in process.

The second operation to a complete substrate test is personalized opens/shorts test (POST) which is usually performed with two multi-site test probe heads, as in Figure 13-24. The POST operation consists of one probe head (A head) that is maintained stationary on a chip-site cluster and a second (B) head that is stepped around to all other chip site clusters. The A head then moves to the second chip-site, the B head steps to all but the first two sites, and so on until all chip sites are tested to all others for continuity and shorts. In actual practice, tester movements are optimized to minimize wear of mechanical components. During any test sequence, the probe heads remain stationary on the substrate while the test engine is switched electronically from point to point. The field-effect transistor (FET)-based solid-state switching hardware can develop test speeds of up to 10,000/s with the capability of delivering up to 250 V and 20 mA.

The tester is numerically controlled (NC) using design data. The design data give expected values of opens or shorts to all substrate top and bottom contact features and thus allows complete comparison of

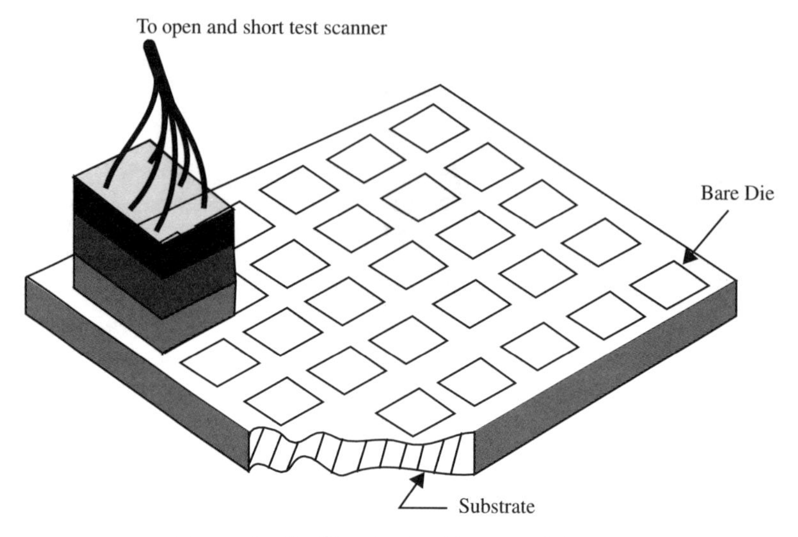

Figure 13-23. Single-site Contact Scheme

product to design expectation. As an example, a 100-chip ceramic substrate fabricated by IBM requires approximately 5 MB of data for testing. NC data can be specified by test engineers and programmers to coincide with the single-chip site contact used for COST or multi-site testing used for POST. NC data can be applied for test through rules-driven tool control software. The software can be adapted for new products of different sizes, shapes, and electrical specifications.

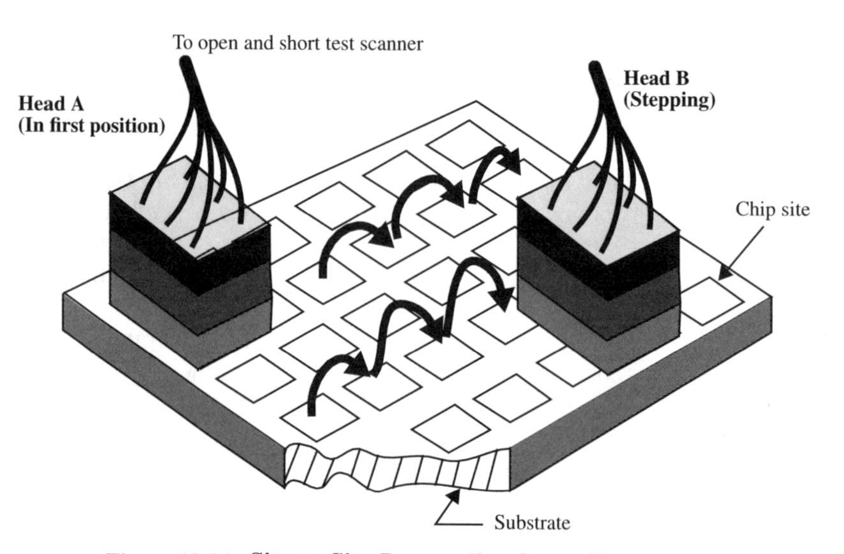

Figure 13-24. Site-to-Site Personality Opens/Shorts Test

Figure 13-25 is a simple example of how opens/shorts testing is performed on substrates. In the figure, both two and three-point nets are shown with all nodes or terminals ordered from top to bottom for shorts testing. In the figure, net 7–8 is shorted to net 10–11 and net 12–13 has an open. Nodes are tested for shorts starting from the bottom and only checked to lower-order points. For example, in Figure 13-25, net 1–2 is tested first for continuity and no shorts testing is done with any other nets. In the next step, net 3–6–9 is tested for opens, and tested for shorts using net 1–2. The sequence is continued until all the nets are tested. As each net node is tested, lower-order nets are electronically coupled, allowing "bulk" examination of the net under test to all lower- order points. If a short is detected, the coupled nets are systematically isolated so that the individual nets can be unambiguously identified as shorted together. Opens in multipoint nets are detected from the ends of the net and isolated to a particular subnet. As an example, in Figure 13-25, the open occurs between points 12 and 13. The open is initially detected between nodes 12 and 14 of the multidrop net, through a continuity test. To obtain the location of the open, new test data is generated on a real time basis and the open is isolated by testing nodes 12 and 13.

To increase test accuracy, error analysis programs may be used for duplicate error checking. For example, the part can be rotated 180 degrees and reprobed for continuity checking. If opens detected on the initial test are not found after rotation, they can be assumed to be false and dropped. This technique has been used very successfully in testing for reducing false open errors, but retesting must be managed carefully to prevent any fault escapes. The rotation test can also be used for adding redundant shorts testing of single-node nets. Reduction in false errors can be enhanced through the use of a verification step following the initial test. This operation is performed on equipment consisting of two top-side and two bottom-side probes or two top-side probes and a bottom-side shorting plate. The probes are fitted with Kelvin tips as shown in Figure 13-26 that allows pad contacts to be verified. Errors called by the prime testers are subject to retest by the verification tool and are declared invalid if not found. Information on verification actions and results are stored to monitor tester efficiency and verification problems. At IBM, aggressive actions to reduce defect counts such as duplicate probing have resulted in very low levels of test escapes (<0.025% for high-volume programs).

For testing, mapping of the substrate is necessary to align the probes, based on coordinates of the terminal pads. For IBM's ceramic substrates (see Chapter 9, "Ceramic Packaging"), this is done through a capacitor pad on a chip site near the center of the substrate. The mapping head then moves one-chip site diagonally toward the upper left corner of the substrate and locates the next capacitor and proceeds until the upper left chip site is reached. The centermost via of that site is then located by the

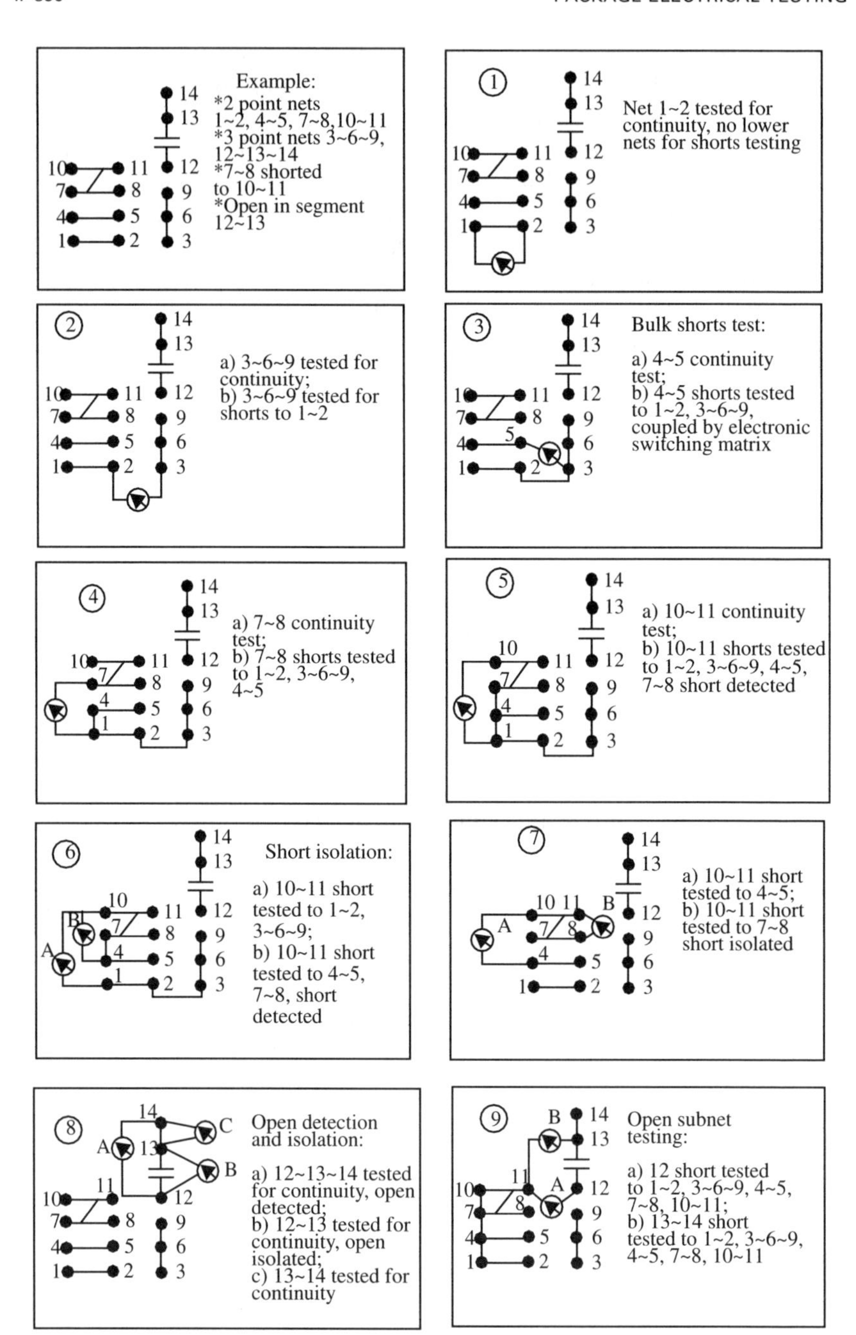

Figure 13-25. Accurate Open/Shorts Testing with Minimal Cycle Time

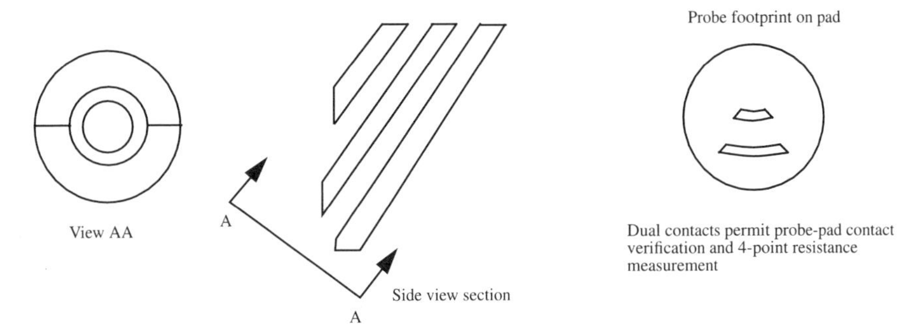

Figure 13-26. Verification Probe

mapper and its coordinates are recorded. The mapper then moves one nominal chip site distance down and continues to map the location of the centermost via of each chip site in a serpentine pattern. The result of the mapping operation is an array of coordinates locating the center of each chip site on the substrate. These coordinates are used by the tester to place the test probes on the substrate terminal pads. The mapping operation is shown in Figure 13-27 where the coordinates of six chip sites in the center row are used to calculate a best-fit overall surface pattern rotation angle. The tester then rotates the substrate by the estimated angle to achieve parallelism between the cluster of probes and the surface metallurgy pattern. An alternate way is to use the glass alignment standard to calibrate the testers, as in Figure 13-28. The standard is located in a four-

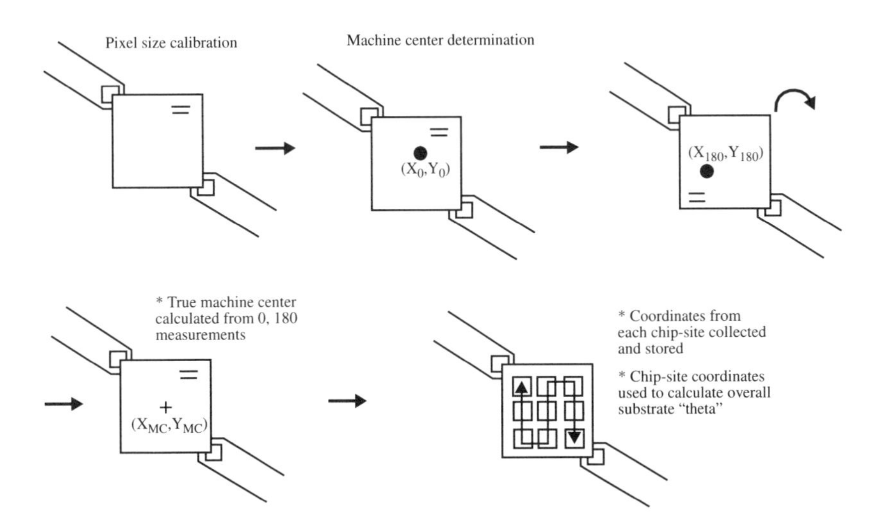

Figure 13-27. Mapper Calibration and Substrate Mapping

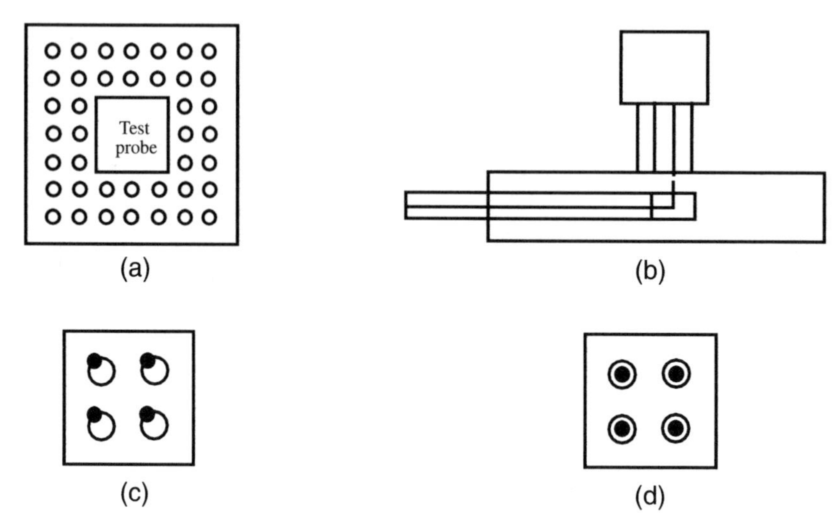

Figure 13-28. Test Probe Alignment Using the Glass Standard. a) Cr-on-glass alignment standard; b) alignment of standard showing optical cavity; c) probe/target alignment before adjustment; d) proper alignment.

point fixture and a fiber-optic bundle is inserted into a tunnel drilled in the body of the standard (Fig. 13-28b). The optics is aimed upward such that the pattern on the standard surface is in view (Fig. 13-28c). The test probe is brought into contact with the glass standard and alignment of probe to pad on the surface is observed. The tester X and Y tables are moved until alignment between the probe and the standard is optimized (Fig. 13-28d). This calibration is performed on a chip site at the center of the standard which can then be repeated at the top and bottom of the standard to check the alignment of the probes.

For opens/shorts testing, test parameters are required to enable pass/fail sorting and for parametric measurements. For IBM's glass-ceramic substrates, to enable shorts testing, a potential of 100 V is applied to the point under test. Up to 1 mA of current is injected and lower-order points tested for leakage currents (as in Fig. 13-25), which allow the detection of resistive shorting paths of up to 10 MΩ. Similarly, for opens testing, a 20-mA constant-current source is used with a voltage of 100 V to detect open circuitry. To enable pass/fail sorting, an open is defined as a product resistance of 35 Ω or greater between points under test. Both opens and shorts testing with the above voltage and current levels stress the circuitry significantly to increase confidence in product reliability. For parametric measurements, a small sample of substrate nets are measured weekly for conformance to limits on resistance and capacitance. Based on the acceptable capacitance for IBM's glass-ceramic nets, capacitance measurements must guarantee an average value between 1.15 and 1.35 pF/cm. Resistance measurements are taken using the four-point Kelvin probes and must demonstrate an average value between 0.15 and 0.30 Ω/cm. Any deviation from these ranges constitutes an unacceptable substrate.

The logistics and control of the substrate-test process is a key factor in the quality and efficiency of the operation. The sequence and completion of operations and flow of data must therefore be carefully managed. The progression of test operations is controlled by a computer-based technology routing or tech route which contains a preset sequence of operations, as in Figure 13-29, controlled by electronic log-on and log-off functions. A job must log on and log off from the first operation before it can be logged on to the second operation, and so on. Several rework options exist for substrate repairs and corrective actions. Product flow through the network operations is computer controlled. Following network completion, the substrate is returned to the primary tech route for continued processing.

Figure 13-30 depicts the flow of data required to successfully perform the desired testing and store test results. Substrate testing requires map data for probe placement and design data for test sequencing. The map data are generated and passed to a storage node. When the substrate part number is released to the manufacturing line for build, numerical control (NC) data are created for build and test. The test data are electronically delivered to a storage node at this time. When the substrate is logged onto a tester, the map data and NC data are automatically retrieved for

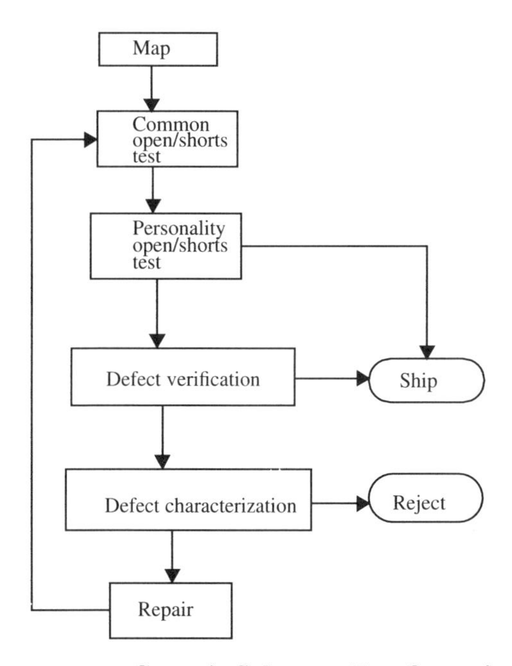

Figure 13-29. Ceramic Substrate Test Operations

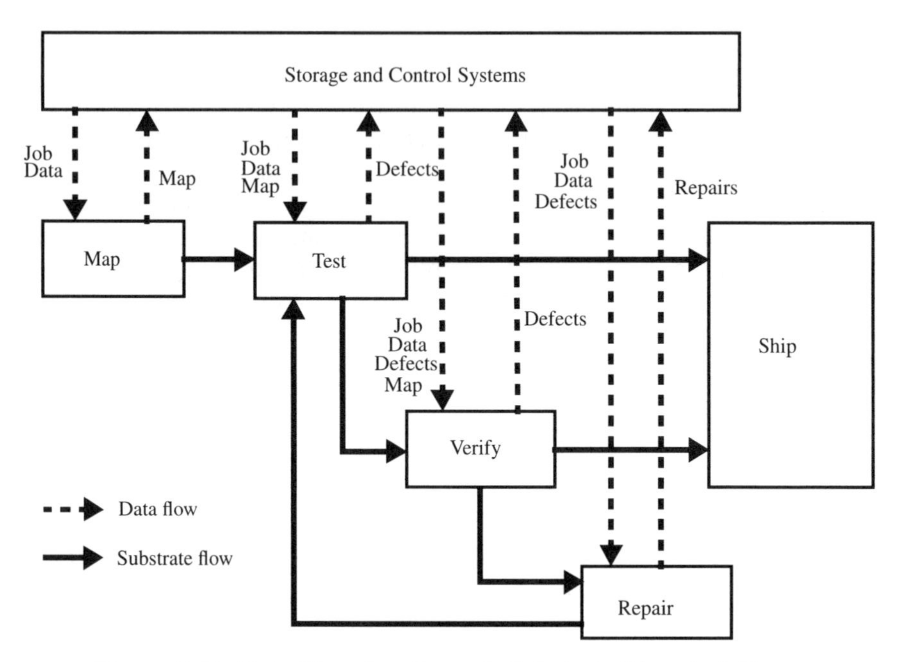

Figure 13-30. Data Flow for Ceramic-Substrate Electrical Test

the particular job and part number. The map data for a particular job number remain in the storage node while the substrate is active in the manufacturing line.

The map data are archived if it is rejected or shipped. Design NC data for frequently used part numbers are stored in the tester. Defect data generated by the tester are sent to a storage node called the serial history file (SHF). Test data for any test pass are stored in SHF. Defect data are passed to the verification tool if errors were detected. Verification results are then recorded in SHF. Completion of all data transmissions are automatically verified. Most substrate signal nets terminate at engineering change (EC) pads and can be repaired if found defective. Defective nets identified during testing can be deleted through a laser cut operation. These nets are eventually rewired on the substrate surface. Net deletion data are stored and used to modify design data for testing such that the deletion is an expected result of the test. All repaired nets are rewired prior to shipment.

13.7 FUTURE ISSUES AND CHALLENGES IN SUBSTRATE TESTING

Small, tightly spaced terminal features and high circuit counts (as in thin-film substrates) place a premium on the test contact technology and on tester speed. Accurate high speed testers are therefore necessary, which require advanced contactor, positioner, and switching technology. To automate advanced testers, carefully conceived tool control software is required for operation. Smaller and more fragile conductors and the increasingly more complex chemical processing required to define them will continue to place significant emphasis on reliability assurance testing. Fine-feature dimensions will create very stringent debris and contamination control requirements for the test processes and equipment. Thus, product damage and test accuracy will remain concerns as substrate integration and its value increases.

Waste or rework of highly functional integrated-circuit chips cannot be tolerated due to the undetected substrate defects. Similarly, capable and valuable substrates cannot be reworked or rejected due to false test errors. High-density substrates with high-reliability specifications require a test scheme capable of detecting latent opens and latent shorts defects. Because manufacturing costs are to be kept low, low-cost test methods which are capable of high defect resolution are necessary. Testing becomes extremely complicated for substrates containing embedded passives such as resistors, inductors, and capacitors. Advancement in tester technology may be necessary because the testing of passives may impose new challenges, such as inaccessibility from the top surface to the pads connected to passives. Control of manufacturing costs for highly integrated substrates can mean additional requirements for in-process testing. Consistent with the need for a new low-cost substrate test technique capable of latent opens and shorts detection, universities such as Georgia Tech, in partnership with industry, are pursuing new methods for substrate testing.

13.8 FUNCTIONAL TESTS

The functional tests follow the substrate test, after the mounting of the semiconductor dies (or chips) onto the MCM substrate, as depicted in the test flow in Figure 13-1. As mentioned earlier, the primary objectives of functional test are as follows:

1. To ensure that no defects were introduced into the known-good-die during the MCM assembly operation in which the bare dies are attached to the tested substrate
2. To ensure that the bare dies are properly bonded to the substrate interconnects and that all the terminals of the attached bare dies are connected to each other in the manner desired
3. To ensure that the MCM performs at speed according to the product specifications (this is accomplished through design for testability (DFT) methodologies, wherein I/O's of the die contained in an MCM are accessed from some MCM I/O's, for test purposes).

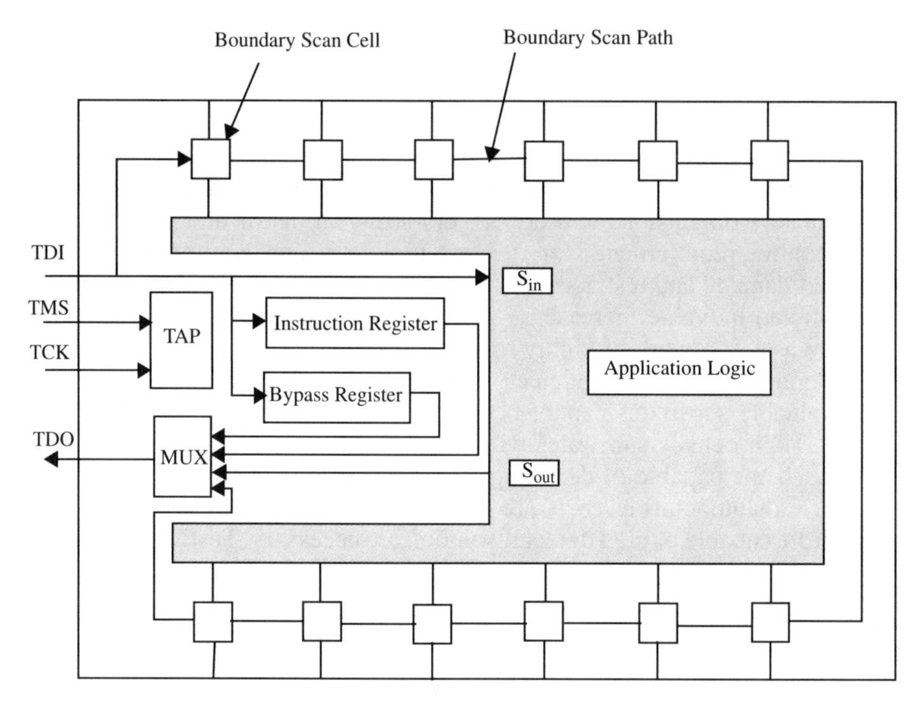

Figure 13-31. Chip Architecture for IEEE 1149.1

This eliminates the need for a bed-of-nails testing procedure to make contact with all the MCM I/O's, as a test access bus may be used to access all the die I/O's, through a test access port. The resulting standard (designed by a consortium of industry partners), called the JTAG 1149.1 standard (JTAG stands for Joint Test Action Group), although initially developed to solve the problem of testing printed-circuit boards, has since evolved and forms the basis for functional test of MCMs. The chip architecture conforming to the above standard is shown in Fig. 13-31. For functional testing, it is important to differentiate MCMs with and without pretested dies (or chips), as the former simplifies the testing process.

Because MCMs contain multiple chips, each one requires a specific set of functional test vectors for excitation, to determine whether defects were introduced into the chip during the MCM assembly process. Because of the large number of test vectors required for testing all the chips, the common practice is to include a built-in self-test (BIST) capability in the chips, which allows for the generation of test stimuli and comparison of the respective responses with reference "signatures". Built in self test is the capability of a circuit (chip, board, MCM or system) to test itself. The BIST procedure may be activated through the JTAG test port using special test commands. The necessary BIST functions may be performed at-speed thereby validating objective (3) of the functional test. If desired, additional functional test vectors may be applied to the MCM through the JTAG interface. Once the individual dies are tested, the remaining task is that of testing the MCM interconnections through the test access bus. With both tests being successful, the test engineer can be assured with high confidence that primary objectives (1) and (2) of the functional test, have been met. In subsequent sections, the boundary scan test standard that provides test access to all the die on an MCM from an external test port is explained, followed by the implementation of BIST through this standard. Some functional test strategies have also been discussed based on the above concepts.

13.8.1 Boundary Scan Test

Boundary scan is a general testability strategy that allows controllability and observability of the pins of a semiconductor IC embedded in a printed-circuit board through a serial test port. The method is equally amenable for accessing I/O's of chips mounted on an MCM. The IEEE/ANSI standard 1149.1-1990 [19] is an industry-wide accepted set of guidelines to implement boundary-scan. The primary components of this standard include the design and use of a test bus which resides on a board, the protocol associated with this bus, elements of a bus master which controls the bus, input/output (I/O) ports that tie a chip to the bus, and some control logic that resides on each chip on the board, to interface the test bus ports to the DFT hardware residing on the application portion of every chip. In addition, the IEEE 1149.1 standard also requires the use of a boundary-scan register to exist on every chip. Boundary scan facilitates isolation and testing of the chips either via the test bus or by built-in-self-test hardware. Testing of interconnections for opens and shorts is also possible through the boundary-scan test hardware.

The boundary-scan bus consists of four ports, a test clock (TCK), a test mode select (TMS), the test data in (TDI) line, and the test data out (TDO) line, as shown in Fig. 13-32. Test instructions and test data

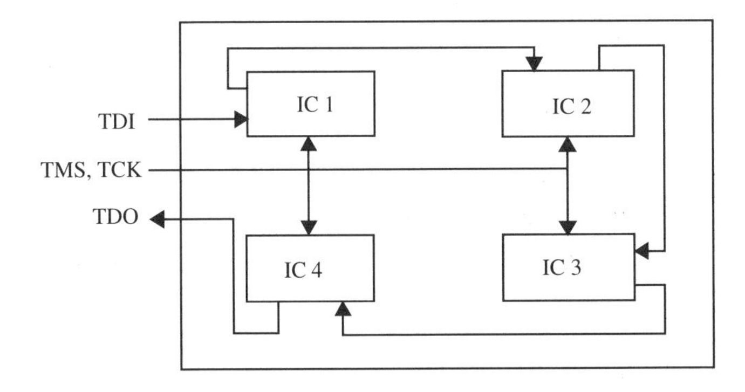

Figure 13-32. Boundary Scan Including an MCM Design

are sent to each chip on the board over the TDI line. Test results and status information are sent from a chip over the TDO line to whatever device is driving the test bus. This information is transmitted serially. The sequence of operations is controlled by a bus master, which can be either automatic test equipment (ATE) or a component that interfaces to a higher level test bus that is part of a hierarchical test and maintenance system. Control of the test-bus circuitry is primarily carried out by the test access port (TAP), which responds to the state transitions on the TMS line.

The test and associated logic operates as follows.

1. An instruction is sent over the TDI line into the instruction register.
2. The selected test circuitry is configured to respond to the instruction. In some cases this may involve sending more data over the TDI line into a register selected by the instruction.
3. The test instruction is executed. Test results are shifted out of selected registers and transmitted over the TDO line to the bus master. New data are shifted into registers using the TDI line while results are shifted out and transmitted over the TDO line.

As an example of boundary-scan test implementation, Figure 13-32 shows a board containing four chips which support IEEE 1149.1. The boundary-scan cells are connected in a single scan path, where the TDO of one chip is tied to the TDI of the next chip in the boundary-scan chain, except for the initial TDI and TDO ports which are tied to distinct terminals of the board. Some of the interconnections between chip pads are also shown in the figure. Using this configuration, various tests may be carried out, such as (1) interconnect test for opens/shorts, (2) snapshot observation of normal system data, and (3) testing of each chip.

The test procedure for multichip modules is similar to that for printed wiring boards. The bare chips on an MCM are analogous to ICs on a printed wiring board from a test perspective. Hence, the boundary-scan test hardware and test procedures are, in general, applicable to MCMs.

There are several issues related with the design and use of the IEEE 1149.1 test bus standard:

(a) The physical structure of the test bus and the rules that govern how the bus is interfaced to the individual chips
(b) The data transfer and control protocol associated with the use of the bus
(c) The on-chip test bus circuitry associated with a chip that allows the chip to communicate with the bus.

The latter includes a set of *boundary-scan registers* and the *test access port (TAP) controller* which is a finite state machine that decodes the state of the bus. Figure 13-31 shows the general form of a chip that supports IEEE 1149.1, which may form part of an MCM or printed-circuit board. The application logic represents the normal chip design prior to the inclusion of logic required to support IEEE 1149.1. This circuitry may include DFT or BIST hardware. If so, the scan paths are connected via the test bus circuitry to the chips's scan-in and scan-out ports. This is illustrated by the connection from the test data input line TDI to S_{in} and S_{out} of the test data output line, TDO. The normal I/O terminals of the application logic are connected through boundary scan cells to the chips I/O pads. The test bus circuitry is referred to as the bus slave and consists of the boundary scan registers, a 1-bit bypass register, an instruction register, several miscellaneous registers, and the TAP controller.

13.8.2 Built-In Self-Test

Built-in self-test (BIST) is the capability of a circuit (chip, board, MCM, or system) to test itself. This section deals primarily with structural off-line BIST. Thus, linear feedback shift registers (LFSRs) are used extensively for BIST purposes. BIST architectures consist of several key elements:

(a) Test-pattern generators (TPG)
(b) Output-response analyzers (ORA)
(c) Circuit under test (CUT)
(d) Distribution system (DIST) for transmitting data from TPGs to CUTs and from CUTs to ORAs
(e) BIST controller for controlling the BIST circuitry and CUT during self-test

Often all or parts of the controller are off-chip. The distribution system consists primarily of direct interconnections (wires), buses, multiplexers, and scan paths. The general form of a centralized BIST architecture is shown in Figure 13-33. Here several CUTs share TPG and ORA circuitry. This leads to reduced overhead but increased test time. During testing, the BIST controller may carry out one or more of the following functions:

1. Single-step the CUTs through some test sequence
2. Inhibit system clocks and control test clocks
3. Communicate with other test controllers, possibly using test buses
4. Control the operation of self-test, including seeding of registers, keeping track of the number of shift commands required in a scan

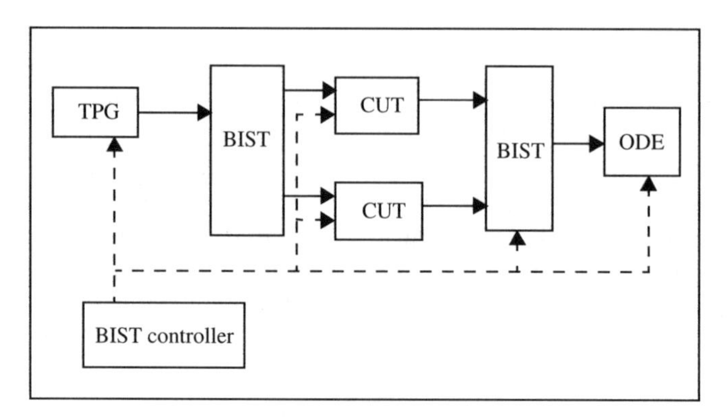

Figure 13-33. Generic Form of BIST Architecture

operation and keeping track of the number of test patterns that have been processed.

Various levels of chip-level BIST have been used during the last decade as explained in Refs. 20–23. The BIST schemes that are needed for MCM test are those that provide very high fault coverage. Most of these schemes run at normal operating speed and typically achieve the required fault coverage.

13.8.3 Functional Testing of MCMs

The testing of MCMs is a difficult task due to both the complexity of MCMs and the fact that no established infrastructure for MCM testing exists [24]. MCM testing combines the complexities of chip and printed-circuit board testing [25–28]. An MCM test which meets product quality requirements has to ensure that all die are properly connected, that they are functionally correct, and that the MCM as a device meets its performance specifications. This requires an interconnect test, a full functional test, and a performance test. Due to very limited accessibility of MCM circuit components (access to other than MCM primary input/output pins may be nonexistent), it is impossible to test the interconnects and the chips with conventional test techniques. Therefore, it is essential to incorporate testability features during functional design of the chips that are to be used in MCMs.

Interconnect test detects mechanical assembly defects and does not identify functional failures. On the other hand, a full-functional at-speed test of the die detects defective dies. The at-speed test is usually not as important with regard to printed-circuit board testing as opposed to testing MCMs, because the packaged chips used in printed circuit boards are fully tested. Due to die-handling and chip-bonding processes during MCM assembly, there is a chance of damaging the dies. Hence, thorough testing is required after mounting the dies on the substrate. In addition, there are thermal stresses which are present while the module is powered. This can also result in damage to the individual dies. Hence, assembled MCMs must be subject to the same kind of rigorous testing as the individual chips.

If failures are detected during testing, an MCM may enter a repair cycle. The repair cycle may sometimes include cap removal if the repair is done after final test and rework. The cost-effectiveness of the repair depends on the MCM technology adopted. In general, the test procedure must include a diagnostic procedure in order to identify the defective element in the MCM.

To enable very high fault coverage, a structured testability approach for testing MCMs is desired and is explained in the next few sections.

13.8.4 A Structured Testability Approach

The testability approach proposed by Zorian [29] is based on implementing chip-level BIST and module-level boundary-scan in the MCM design. Both BIST and boundary-scan, as explained earlier, are design-for-testability techniques; that is, they are meant to improve the controllability and observability of circuit performance to make it easily testable. The use of both techniques necessitates early considerations, because they impact the chip design process.

The most popular testing approach executes BIST through the boundary-scan TAP [30]. The incorporation of BIST and boundary-scan in a chip and techniques to obtain very high fault coverages at the chip and the module level test are covered in the following discussion. This section also addresses the benefits of having BIST and boundary-scan in bare dies and discuss their utilization for bare die and MCM assembly testing. Functional at-speed test of a chip becomes possible due to the incorporation of BIST in its design. In fact, the two main problems of wafer testing, namely obtaining very high fault coverage functional test and running it at system speed for performance test, are greatly simplified due to the use of BIST. The execution of this operation (i.e. running BIST) is autonomous. It only needs the application of boundary-scan standard's RUNBIST instruction [19, 24] through the boundary-scan TAP to start the BIST execution, and upon completion, the BIST response needs to be scanned out for evaluation. This is also done through the boundary-scan TAP [30].

The use of BIST and boundary-scan impacts reliability testing of chips (i.e., dies). During burn-in, BIST can be kept active on the die under test, which needs only the boundary-scan TAP interface. This allows continuous monitoring of the response status for additional in-process information. The above test processes result in known-good-die (KGD) that leverages the assembly of competitively strong MCMs.

13.8.5 Assembled-MCM Test and Diagnosis Procedure

To perform a typical test of MCMs with BIST and boundary-scan the following tests need to be conducted:

(a) Integrity and identity check
(b) Interconnect (substrate) test
(c) Functional at-speed chip test
(d) MCM performance test
(e) MCM parametric test

A brief explanation of each of the operations follows.

13.8.5.1 Integrity and Identity Check

This is to verify the integrity of boundary-scan circuitry prior to its use in subsequent tests. The boundary-scan standard provides a certain methodology for performing this test, yielding relevant diagnostic information regarding the location of the failure if any [11]. Figure 13-32 shows an MCM design composed of four ICs. Each IC contains its corresponding boundary-scan facilities. The scan chain which connects all four ICs is represented by a line starting from TDI (test data input) and ending with TDO (test data output). The remaining boundary scan signals are distributed to all ICs in parallel. Following the integrity test, a test is performed to check the identity of each die. The boundary-scan standard allows permanent storage of chip-level ID codes. Each MCM die has its own ID code, which is read through the boundary-scan TAP and is compared with a reference value. This test detects the existence of an incorrect chip or a chip that is not oriented properly on the substrate. This is especially important, as many of the chips on a MCM may have the same footprint, causing problems with visual inspection methods for reliability.

13.8.5.2 Substrate Test

This test checks whether dies are properly bonded to the substrate interconnections. The existence of a boundary-scan chain in the module, consisting of the boundary-scan registers of each die and the connections between these registers through the substrate, creates a virtual electronic bed of nails built into the module. This bed of nails built into the module is independent of the module's density. Boundary-scan latches provide the stimulus and can monitor logic values across all interconnections. This permits testing for opens or shorts between inputs/outputs of two dies and between inputs/outputs of a die and the inputs/outputs of the MCM. To minimize test application time and optimize detection of interconnect defects, special test algorithms are used. The test processes are automated and also provide diagnostic information identifying the faulty net and the type of fault. This information is very useful for the repair process. Even though the substrate is usually unrepairable, the die to substrate attachment can be reworked or the die can be replaced and rebonded to the substrate.

13.8.5.3 Functional At-Speed Test

It is necessary to run the BIST operations on each die after the latter are bonded to the substrate in order to detect defects that may be introduced into the chip during the die attach process. In order to perform this test, access is needed to the TAP of each die and that is possible through the boundary scan chain of the MCM. Figure 13-34 shows an MCM design with BIST in its individual chips. As an example, the test pattern generator (TPG) and output data evaluator (ODE) of IC1 (integrated circuit chip #1) are illustrated.

The use of BIST also provides diagnostic information. The analysis or automatic comparison of the BIST responses can identify which chip has failed. If some of the chips in an MCM are not designed with BIST capabilities, such as RAM chips, after MCM assembly other measures are taken in order to test them.

13.8.5.4 Performance Test

The MCM performance test is an at-speed test on the assembled MCM. It verifies that the finished module meets all the performance

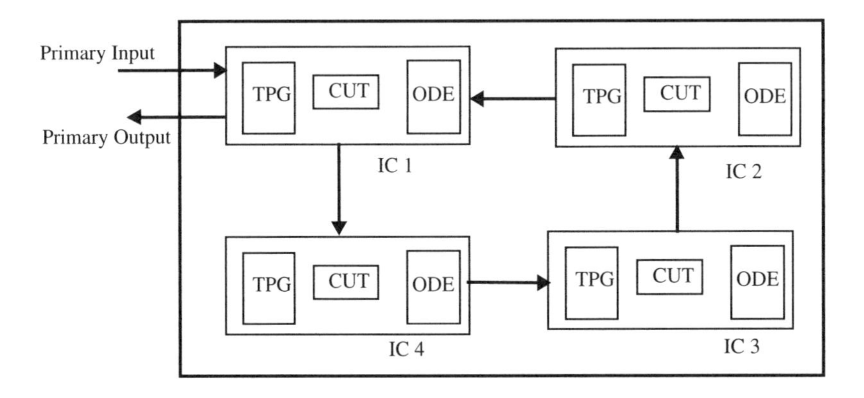

Figure 13-34. An Example of Bare Die BIST in an MCM Design

requirements. Further, it tests propagation delay times, including chip delays and substrate routing delays. This test requires automatic test equipment with high-pin-count and high-speed capabilities. The boundary-scan SAMPLE mode is used to obtain diagnostic information to locate any performance-related problem. In this mode, the state of boundary scan cells are captured at fixed intervals of time (snapshots) and scanned out through the boundary-scan chain.

13.8.5.5 Parametric Test

The boundary-scan chain is used for the input/output parametric test, by allowing proper input/output initialization to predefined values [31].

The above five test operations can be applied during the MCM assembly process, and/or after the assembly completion. It is preferable to have in process testing in order to identify the faulty element(s), and perform repair before the final MCM is capped.

The repair cycle includes faulty element isolation and rework. The diagnostic information, obtained during each one of the five testing operations, is utilized to isolate the faulty element. For instance, the BIST signature identifies the faulty chip, and the Boundary-scan-based interconnect test identifies the failed net. The rework involves the replacement of the faulty element, and this is followed by a retest.

13.9 FUTURE ISSUES AND CHALLENGES IN FUNCTIONAL TESTING

The rapid increase in the complexity of VLSI systems has made the task of test, diagnosis and repair more costly and time consuming. This has necessitated the incorporation of unique design for testability features into system designs. Such features include, for example, hardware for built-in self-test and boundary scan. With the advent of multi-chip modules and other methods of integrating high-density electronic modules, such as 3-D chip and module stacking, the testing problem has been stretched to its limits. Future systems will include ultra-fast processors with gigabytes of static and dynamic memory and multiple co-processors for graphics, video, audio and other utilities. Clearly, methodologies for breaking the testing problem into smaller tractable parts will evolve. Solutions to these smaller problems will then be amalgamated to provide a testability solution for the complete electronic system. We foresee a greater emphasis on intelligent built-in self-test mechanisms in the future. These built-in self-test mechanisms will be tailored to specific classes of hardware and will need to provide high levels of fault coverage (almost 100%) in order for overall system “yields” to be high. Also, these mechanisms will need to accommodate non-conventional fault models such as those for speed-testing of digital circuits. As circuit dimensions become smaller, clock speeds are increased and supply voltages are lowered, issues relating to signal integrity will become an integral part of functional test. Moreover, with increased design of mixed-signal circuitry, standards for boundary scan based test and built-in self-test of mixed-signal systems will evolve. Eventually, schemes for hierarchical boundary-scan based testing of mixed-signal systems will need to be developed and issues such as partitioning of scan chains for ease of testing and reliability of design for testability hardware will draw increased attention.

13.10 RECENT AND FUTURE DEVELOPMENTS

13.10.1 Substrate Testing

To overcome the difficulties and deficiencies with the present techniques, a number of test methodologies are under development at companies (IBM, ISI) and universities (e.g., Georgia Institute of Technology). At IBM and ISI, two new test methods, namely, time domain opens/shorts (TDOS) and high voltage ramp monitor (HVRM) are being used in production applications for advanced ceramic and laminate based MCMs [109]. Both methods suggest a significant improvement in test speed and fault detection capabilities as compared to TDNA and resistance testing. A third test methodology [109], network sensitive pulse generation (NSPG) is under development that is designed to identify opens and shorts up to 10 mΩ resistance, requiring a short dwell time. Trial runs are currently in progress using NSPG on flex and ceramic substrates.

The RF-resonator method being developed at Georgia Institute of Technology is a useful method for latent opens and shorts testing of interconnections. The method is based on the use of a single-probe head in series with a high Q resonator, as discussed by Kim, Swaminathan and Chatterjee [14–16]. The resonator modulates the AC response of the interconnections, producing a change in the frequency response, which changes as a function of the defects. Both the magnitude and phase responses are shown in Fig 13-35a and b, respectively. Interconnect defects change the magnitude response, causing a reduced magnitude as compared to the defect free response, in the vicinity of the resonant frequency. Similarly, the phase response in Fig 13-35b varies with open or short defect, allowing for fault identification. Based on look-up tables created through calibration measurements, the modulated AC response is compared to stored data (for defect-free interconnections), resulting in a differential voltage that provides details on the defects. The look-up table is created through analytical expressions using CAD design data and is similar to the test file generated for capacitance testing. In this method, the magnitude and phase responses (as a function of frequency) are combined to diagnose the defect such as its position, size and so forth. The method is based on the movement of the resonating frequency of the

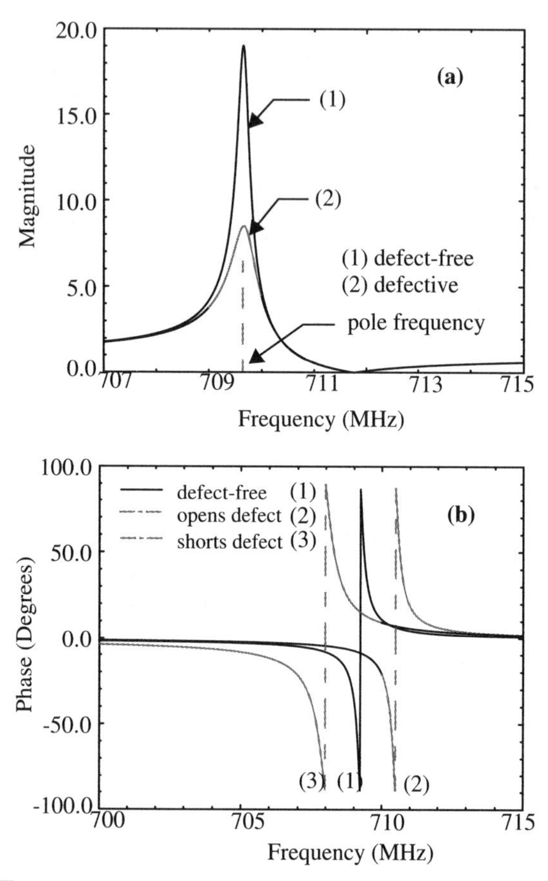

Figure 13-35. Frequency and phase responses of a defect. (a) Frequency response of open (100μm long) for 2 cm long lossy line; (b) phase response of opens (1kΩ) and shorts (4pF) for 2cm long line.

resonator due to the presence of a defect, resulting in a change in the magnitude and phase responses. Hence, the resolution of the method is a function of the quality factor (Q) of the resonator, and for the method to be viable, a high Q resonator is desired.

To further explain the method, consider a 2 cm line fabricated on glass using gold metallization, modulated by using a 711 MHz resonator. Though lower or higher frequency resonators can be used, 711 MHz was chosen since it was commercially available, had an adequate Q (360) and resonated at a frequency which was far removed from the standing wave resonances of interconnections in the range 1-4 cm, which are typical interconnect lengths in MCMs. Table 13-5 shows the movement of the resonating frequency as a function of the position and nature (opens or shorts) of the defect. Assuming a fixed frequency (e.g., a 3-dB point) is

Table 13-5. Resonating Frequency versus Defect for a 2-cm Interconnect

Type of Defect	Frequency
Fault-free	667 MHz
Open (0.5 cm from far end)	681 MHz
Open (1.0 cm from far end)	691 MHz
Open (1.5cm from far end)	701 MHz
Short	639 MHz

used for the measurement, the shift in frequency in Table 13-5 produces a corresponding change in the magnitude and phase responses at the given frequency. This measurement is then used for pass/fail testing of the interconnections. Because passives integrated into the substrate produce shifts in the resonating frequency of the resonator, defects on passives connected to interconnections can be detected by probing from the top-surface metallurgy pads. Due to the use of a single probe head and low test frequency, probe movement, test time, and equipment costs are expected to be low. This method combines the simplicity of capacitance testing with the resolution of latent opens testing.

13.10.2 Functional Testing

In the past, many boundary scan techniques have been proposed for interconnect tests of printed circuit boards and multi-chip modules. A generalized optimal test vector generation approach for interconnect test of boards has earlier been proposed. Further, new techniques for self test of systems and modules have been developed. In the recent past, new deterministic built-in self-test techniques that address specific classes of circuits like RAMs, ROMs, PLAs, adders, multipliers, and so forth, have been developed. Recently, there has been work in automated insertion of built-in self-test hardware into specialized classes of integrated circuit designs. There is ongoing work in the area of improving fault coverage of built-in self-test techniques and in finding optimal combinations of built-in self-test and boundary scan hardware for faster, cheaper and better test of complex systems that have hundreds of ICs.

13.11 REFERENCES

1. T. Knowles. *Automatic Testing—Systems and Applications,* McGraw-Hill, New York, 1976.
2. B. Davis. *The Economics of Automatic Testing,* McGraw-Hill, New York, 1982.
3. A. Deutsch et al. "Defect Detection using 70 GHz TDR," IBM Research Report RC 16699, April 1991.

4. A. Halperin, T. H. DiStefano, and S. Chiang. "Latent Open Testing of Electronic Packaging," *IEEE Multi-Chip Module Conference, MCMC-94,* 1994.
5. J. Marshall et al. "CAD-Based Net Capacitance Testing of Unpopulated MCM Substrate," *IEEE Trans. Components packaging manuf. technol. B,* CPMT-17 (1), 1994.
6. L. Economikos, T. Morrison, and F. Crnic. "Electrical Test of Multichip Substrates," *IEEE Trans. Components packaging manuf. technol. B,* CPMT-17 (1), 1994.
7. H. Hamel, S. Kadakia, and H. Bhatia. "Capacitance Test Technique for the MCM of the 90s," *Proceedings of the International Electronic Packaging Conference,* pp. 855–872, 1993.
8. R. W. Wedwick. "Continuity Testing by Capacitance," *Circuits Manuf.,* pp. 60–61, 1974.
9. O. C. Woodard. "High Density Interconnect Verification of Unpopulated Multichip Modules," *Proceedings of the Eleventh IEEE/CHMT International Electronics Manufacturing Technology (IEMT) Symposium,* pp. 434–439, 1991.
10. A. W. Ross, R. R. Goruganthu, and O. C. Woodard. "High Density Interconnect Verification Using Voltage Contrast Electron Beam," *Proceedings of the Eleventh IEEE/CHMT International Electronics Manufacturing Technology (IEMT) Symposium,* pp. 270–274, 1991.
11. M. Brunner, R. Schmid, R. Schmitt, M. Sturn, and O. Gessner. "Electron-Beam MCM Testing and Probing," *IEEE Trans. Components packaging manuf. technol. B,* CPMT-17 (1), 1994.
12. F. Jensen and N. E. Petersen. *Burn-In: An Engineering Approach and Analysis of Burn-In Procedures,* Wiley, New York, 1982.
13. F. Crnic. "Electrical Test of Multi-Chip Substrates," *Proceedings of International Conference and Exhibition on Multi-chip Modules (ICEMM),* pp. 422–428, 1993.
14. B. Kim, A. Chatterjee, M. Swaminathan, and D. Schimmel. "A Novel Low-Cost Approach to MCM Interconnect Test," *IEEE International Test Conference,* 1995.
15. B. Kim, A. Chatterjee, and M. Swaminathan. "A New Technique for Testing MCM Substrate," *IEEE/ISHM MCM Test II: Advanced Technology Workshop,* 1995.
16. B. Kim, M. Swaminathan, and A. Chatterjee. "A Novel MCM Interconnect Test Technique Based on Resonator Principles and Transmission Line Theory," *Proceedings of IEEE 4th Topical Meeting on Electrical Performance of Electronic Packaging,* 1995.
17. R. Bove, U.S. Patent Number 3,806,801.
18. E. Dalrymple. "Very Low Insertion Force Connectors for High Mating Cycle Applications", *Connection Technol.,* December 1986.
19. *IEEE Standard Test Access Port and Boundary-Scan Architecture,* IEEE Std. 1149.1-1990, IEEE Standards Office, NJ, 1990.
20. V. D. Agarwal, C. J. Lin, P. W. Rutkowski, S. Wu, and Y. Zorian. "Built-in Self-Test for Digital Integrated Circuits," *AT&T Tech J.,* pp. 30–39, 1994.
21. J. A. Jorgenson and R. J. Wagner. "Analyzing the Design-for-Test Techniques in a Multiple Substrate MCM," *Proceedings of 12th IEEE VLSI Test Symposium,* pp. 360–365, 1994.
22. T. Storey. "A Test Methodology for VLSI Chips on Silicon," *Proc. ITC,* pp. 359–368, 1993.
23. M. M. Pradhan, E. O'Brien, S. L. Lam, and J. Beausang. "Circular BIST with Partial Scan," *Proc. IEEE International Test Conf.,* pp. 719–729, 1988.
24. C. M. Maunder and R. E. Tulloss. *The test access port and boundary-scan architecture,* IEEE Computer Society Press, New York, 1990.
25. D. A. Doane and P. D. Franzon. *MultiChip Module Technologies and Alternative The Basics,* pp. 615–658, Van Nostrand Reinhold, New York, 1993.

26. A. Flint. "Testing Multichip Modules," *IEEE Spectrum,* pp. 59–62, 1994.
27. P. R. Mukund and J. F. McDonald. "MCM: The High-Performance Electronic Packaging Technology," *IEEE Computer Mag.* (Special Issue), 1993.
28. Y. Zorian. "A Universal Testability Strategy for Multichip Modules Based on BIST and Boundary-Scan," *Proc. IEEE International Conference on Computer Design,* pp. 59–66, 1992.
29. Y. Zorian. "A structured Approach to Macrocell Testing Using Built-in Self-Test," *Proc. IEEE Custom Integrated Circuits Conf.,* pp. 28.3.1–28.3.4, 1990.
30. C. W. Yau and N. Jarwala. "The Boundary-Scan Master: Target Applications and Functional Requirements," *Proc. IEEE International Test Conference,* pp. 311–315, 1990.
31. K. E. Posse. "A Design-for-Testability Architecture for Multichip Modules," *IEEE International Test Conference,* pp. 113–121, 1991.
32. Y. Zorian. "A distributed BIST control scheme for complex VLSI devices," *Proceedings of 11th IEEE VLSI Test Symposium,* pp. 4-9, 1993.
33. C. J. Lin, Y. Zorian, and S. Bhawmik. "PSBIST: A Partial-Scan Based Built-in Self-Test Scheme," *Proc. IEEE International Test Conference,* pp. 507–516, 1993.
34. M. Nicolaidis. "An Efficient Built-in Self-test Scheme for Functional Test of Embedded RAMs," *Proc. 15th International Symp. Fault-Tolerant Comput.,* pp. 118–123, 1985.
35. Y. Zorian and A. Ivanov. "An Effective BIST Scheme for ROMs," *IEEE Trans. Computing, C- 41(5),* pp. 646–653, 1992.
36. A. J. Van de Goor and Y. Zorian. "An Effective BIST Scheme for Ring-Address FIFOs," *Proc. IEEE International Test Conference,* pp. 96–101, 1994.
37. S.D. Golladay, N.A. Wagner, J.R. Rudert, and R.N. Schmidt. "Electron-Beam Technology for Open/Short Testing of Multi-chip Substrates," *IBM Jo. Res. Devel., 34(2/3):* pp. 250–259, 1990.
38. A. Hopper et al. "A Feasibility Study for the Fabrication of Planar Silicon Multichip Modules Using Electron Beam Lithography for Precise Location and Interconnection of Chips," *IEEE Trans. Components Hybrids Manuf. Technol., CHMT-15(1):* pp. 97–102, 1992.
39. V. Murali, T. Rucker, J. Fu, and R. Shukla. "Yield and Reliability Concerns in Polyimide Based Multi-Chip Modules," *1992 IEEE Multichip Module Conference,* pp. 98–101, 1992.
40. A. Landzberg. *Microelectronics Manufacturing Diagnostics Handbook,* Van Nostrand Reinhold, New York, 1993.
41. S. DeFoster, M. Mancini, and J. Zalesinski. "Automatic Testing of Metallized Ceramic-Polyimide Substrates," *IEEE Trans. Components Hybrids Manuf. Technol.,* CHMT-13(4): 1990.
42. M. Sriram and S.M. Kang. *Physical Design for Multichip Modules,* Kluwer Academic Publishers, Boston, 1994.
43. John H. Lau. *CHIP ON BOARD: Technologies for Multichip Modules,* Van Nostrand Reinhold, New York, 1994.
44. T. Storey et al. "A Test Methodology to Support an ASEM MCM Foundry," *IEEE International Test Conference,* pp. 426–435, 1994.
45. A. Flint. "Test Strategies for a Family of Complex MCMs," *IEEE International Test Conference,* pp. 436–445, 1994.
46. J. M. Jong, B. Janko, and V. Tripathi. "Equivalent Circuit Modeling of Interconnects from Time-Domain Measurements," *IEEE Trans. Components Packaging Manuf. Technol.,* CPMT-16(1): 1993.
47. S. F. Gong et al. "Investigation of High-Speed Pulse Transmission in MCM-D," *IEEE Trans. Components Hybrids Manuf. Technol.,* CHMT-18(1): 1993.

48. G. White et al. "Large Format Fabrication-A Practical Approach to Low Cost MCM-D," *IEEE Trans. Components Packaging Manuf. Technol.,* CPMT-16(1): 1995.

49. C. L. Ratzlaff and L. T. Pillage "RICE: Rapid Interconnect Circuit Evaluation Using AWE," *IEEE Trans. Computer-Aided Design Integrated Circuits Sys.,* CADIC-13(6): 1994.

50. J. Peeters, E. Beyne, and G. Brandli. "A Broad Band Loss Model for MCM Interconnections," *IEEE Multi-Chip Module Conference,* 1993.

51. H. Liao and W. Dai. "Wave Spreading Evaluation of Interconnection Systems," *IEEE Multi-Chip Module Conference,* 1993.

52. L. Prokopchak. "Development of a Solution for Achieving Known-Good-Die," *IEEE International Test Conference,* pp. 15–21, 1994.

53. Y. Eo and W. R. Eisenstadt, "High-Speed VLSI Interconnect Modeling Based on S-Parameter Measurements," *IEEE Trans. Components Packaging Manuf. Technol.,* CPMT-16(5): 1993.

54. X. Zhang and K. K. Mei. "Time-Domain Finite Difference Approach to the Calculation of the Frequency-Dependent Characteristics of Microstrip Discontinuities," *IEEE Trans. Microwave Theory Tech.,* MTT-36(12): pp. 1775–1787, 1988.

55. S. Voranantakul "Crosstalk Analysis for High-Speed Pulse Propagation in Lossy Electrical Interconnections," *IEEE Trans. Components Packaging Manuf. Technol.,* CPMT-16(1): 1993.

56. T. Ueno and Y. Kondoh. "Membrane Probe Technology for MCM Known-Good-Die," *IEEE International Test Conference,* pp. 22–29, 1994.

57. J. R. Brews. "Transmission Line Models for Lossy Waveguide Interconnections in VLSI," *IEEE Trans. Electron Devices,* ED-33(9): pp. 1356–1365, 1986.

58. K. W. Goossen and R. B. Hammond, "Modeling of Picosecond Pulse Propagation in Microstrip Interconnections on Integrated Circuits," *IEEE Trans. Microwave Theory Tech., MTT-37(3):* pp. 469–478, 1989.

59. R. W. Jackson and D. M. Pozar. "Full-Wave Analysis of Microstrip Open-End and Gap Discontinuities," *IEEE Trans. Microwave Theory Tech.,"* MTT-33(10): pp. 1036–1042, 1985.

60. L. Roszel. "MCM Foundry Test Methodology and Implementation," *IEEE International Test Conference,* pp. 369–372, 1993.

61. T. Storey. "A Test Methodology for VLSI Chips on Silicon," *IEEE International Test Conference,* pp. 359–368, 1993.

62. M. Melton and F. Brglez. "Automatic pattern Generation for Diagnosis of Wiring Interconnect Faults," *IEEE International Test Conference,* pp. 389–398, 1992.

63. S. Yao et al. "A Multi-Probe Approach for MCM Substrate Testing," *IEEE Trans. Computer-Aided Design Integrated Circuits Sys.,* CADICS-13(1): pp. 110–121, 1994.

64. F. W. Angelotti "Modeling for Structured System Interconnect Test," *IEEE International Test Conference,* pp. 127–133, 1994.

65. N. Jarwala and C. W. Yau. "A New Framework for Analyzing Test Generation and Diagnosis Algorithms for Wiring Interconnects," *IEEE International Test Conference,* pp. 63–70, 1989.

66. D. Y. Shih et al. "Factors Affecting the Interconnection Resistance and Yield in Multilayer Polyimide/Copper Structures," *IEEE Trans. Components Packaging Manuf. Technol.,* CPMT-16(1): 1993.

67. S. Y. Kim et al. "An Efficient Methodology for Extraction and Simulation of Transmission Lines for Application Specific Electronic Modules," *IEEE Multi-Chip Module Conference,* 1993.

68. W. Blood and W. Yip. "Electrical Analysis of a Thin Film Multichip Module Substrate," *IEEE Multi-Chip Module Conference,* 1992.

69. R. Kambe. "MCM Substrate with High Capacitance," *IEEE Trans. Components Packaging Manuf. Technol.,* CPMT-18(1): 1995.
70. T. S. Chu and T. Itoh. "Generalized Scattering Matrix Method for Analysis of Cascaded and Offset Microstrip Step Discontinuities," *IEEE Trans. Microwave Theory Tech.,* MTT-37(2): pp.280–284, 1986.
71. W. Menzel and I. Wolff. "A Method for Calculating the Frequency-Dependent Properties of Microstrip Discontinuities," *IEEE Trans. Microwave Theory Tech.,* MTT-37(2): pp. 107–112, 1977.
72. A. Gopinath and C. Gupta. "Capacitance Parameters of Discontinuities in Microstrip lines," *IEEE Trans. Microwave Theory Tech.,* MTT-37(10): pp. 831–836, 1978.
73. P. B. Katehi and N. G. Alexopoulos. "Frequency-Dependent Characteristics of Microstrip Discontinuities in Millimeter-Wave Integrated Circuits," *IEEE Trans. Microwave Theory Tech.,* MTT-37(10): pp. 1029–1035, 1985.
74. A. Iqbal, M. Swaminathan, M. Nealon, and A. Omer. "Design Trade-offs Among MCM-C, MCM-D and MCM-D/C Technologies," *IEEE Trans. Components Packaging Manuf. Technol.,* CPMT-17(1): 1994.
75. J. M. Jong, Bozidar Janko, and Vijai Tripathi. "Equivalent Circuit Modeling of Interconnects from Time-Domain Measurements," *IEEE Trans. Components Hybrids Manuf. Technol.,* CHMT-16(1): 1993.
76. N. Nagi, A. Chatterjee, J. A. Abraham. "Fault Simulation of Linear Analog Circuit," *J. Electron. Testing: Theory Applic.,* JETTA-4: pp. 345–360, 1993.
77. N. Nagi, A. Chatterjee, A. Balivada, J. A. Abraham. "Fault-based Automatic Test Generator for Linear Analog Circuits," *IEEE International Conference on Computer Aided Design,* 1993.
78. N. Nagi, A. Chatterjee, and J. A. Abraham. "MIXER: Mixed-Signal Fault Simulator," *Proc. International Conference on Computer Design,* 1993.
79. A. Deutsch, G. Arjavalingam, and G. V. Kopcsay. "Characterization of Resistive Transmission Lines by Short-Pulse Propagation," *IEEE Trans. Microwave Guided Wave Lett.,* MGWL-2(1), 1992.
80. W. H. Kautz. "Testing for Faults in Wiring Networks," *IEEE Trans. Computers,* C-23(4): pp. 358–363, 1974.
81. R. R. Tummala and E. J. Rymaszewski. *Microelectronics Packaging Handbook,* Van Nostrand Reinhold, New York, 1989.
82. A. Mehta et al. "SuperSPARC multichip module," *IEEE Multi-Chip Module Conference,* pp. 19–28, 1993.
83. A. J. Blodgett and D. R. Barbour. "Thermal conduction module: A high Performance Ceramic Package," *IBM J. Res. Devel.,* 26: pp. 30–36, 1982.
84. J. U. Knickerbocker et al. "IBM System/390 Air Cooled Alumina Thermal Conduction Module," *IBM J. Res. Devel.,* 35: pp. 330–341, 1991.
85. Georgia Institute of Technology, "NSF Proposal," Packaging Research Center, Atlanta, GA, 1994.
86. J. Conti. "Case for Moving Probe Testing of Bare Boards," *Surface Mount Technol.,* 5(4): p. 56, 1991.
87. T. Ninomiya and Y. Nakagawa. "Automated Pattern Inspection for Unbaked Layers of Multi-Layer Ceramic Substrates", *IEEE International Workshop on Industrial Applications of Machine Vision and Machine Intelligence,* pp. 346–351, 1987.
88. H. Oka, M. Ando, J. Serizawa, K. Fujihara and H. Tate. "Automatic Optical Pattern Inspection System Using 3D Profile Measurement", *Printed Circuit World Convention,* 1990.
89. K. Smith and G. Rinne. "MCMs: Approaches for Testing and Troubleshooting", *High Speed Digital Symposium and Exhibition,* 1992.

90. R. Allison. "Bare Board Applications for Moving Probe Test Systems", *Proceedings of the Technical Program-National Electronic Packaging and Production Conference,* pp. 792–808, 1992.
91. M. Brunner, D. Winkler and B. Lischke. "Crucial Parameters in Electron Beam Short/Open Testing," in *Microcircuit Engineering 84,* ed. by A. Heuberger and H. Benking, Academic Press, London, pp. 399–410, 1985.
92. D. L. Millard, K. R. Umstadter and R. C. Block. "Noncontact Testing of Circuits Via a Laser-Induced Plasma Electrical Pathway," *IEEE Design and Test of Computers,* 1992.
93. T. Sakata and K. Numata. "Prober for Highly Integrated Multichip Modules", *IEEE/CHMT International Electronics Manufacturing Technology Test Symposium,* pp. 429–433, 1992.
94. P. Slade. "Resistor Trimming", *IEEE Trans. Components Hybrids Manuf. Technol.,* CHMT-14(1): pp. 2–18, 1991.
95. D. Ballew and L. M. Streb. "Board-Level Boundary-Scan Regaining Observability with an Additional IC," *Proceedings IEEE International Test Conference,* pp. 182–189, 1989.
96. J. Bond. "Test dominates MCM assembly," *Test & Meas. World,* pp. 59–64, 1992.
97. M. Brunner, R. Schmid, R. Schmitt, M. Sturm, and O. Gessner. "Electron-beam MCM Testing and Probing," *IEEE Trans. Components Packaging Manuf. Technol.,* CPMT-17(1): pp. 57–62, 1994.
98. R. A. Fillion, R. J. Wojnarowski, and W. Daum. "Bare Chip Test Techniques for Multichip Modules," *Proceedings 40th EIA/IEEE Electronic Components Technology Conference,* p. 554, 1990.
99. A. E. Gattiker, W. Maly, and M. E. Thomas "Are There Any Alternaties to Known-Good-Die," *Proceedings of MCM Conference,* pp. 102-107, 1994.
100. S. C. Hilla, "Boundary-Scan Testing for Multichip Modules," *Proceedings IEEE International Test Conference,* pp. 224–231, 1992.
101. D. C. Keezer. "Bare Die Testing and MCM Probing Techniques," *Proceedings IEEE Multi-Chip Module Conference (MCMC),* pp. 20–23, 1992.
102. M. Lubaszewski, M. Marzouki, and M. H. Touati. "A Pragmatic Test and Diagnosis Methodology for Partially Testable MCMs," *Proceedings IEEE MCM Conference,* pp. 108–113, 1994.
103. R. H. Parker. "Bare Die Test," *Proceedings IEEE Multi-Chip Module Conference (MCMC),* pp. 24–27, 1992.
104. P. Raghavachari. "Circuit Pack BIST for System to Factory—the MCERT Chip," *Proceedings International Test Conference,* p. 641, 1991.
105. R. Roebuck, et al. "Known good die: a practical solution," ICEMM *Proc.,* pp. 177–182, 1993.
106. L. Roszel. "MCM foundary test methodology and implementation," *Proceedings IEEE International Test Conference,* pp. 369–372, 1993.
107. H. N. Scholz, R. E. Tulloss, C. W. Yau, and W. Wach. "ASIC Implementations of Boundary-Scan and BIST," *8th International Custom Microelectronics Conference,* pp. 43.0–43.9, 1988.
108. B. Vasquez, D. Van Overloop, and S. Lindsey. "Known-Good-Die Technologies on the Horizon," *Proceedings of IEEE VLSI Test Symposium,* pp. 356–359, 1994.
109. G. Mezack. "Personal Communication," Integrated Solutions Inc., Tewksbury, MA.

14

PACKAGE SEALING AND ENCAPSULATION

C.P. WONG—*Georgia Tech*
D.B. CLEGG—*Motorola Inc.*
ANANDA H. KUMAR—*IBM*
K. OTSUKA—*Hitachi*
BERHAN OZMAT—*TI*

14.1 INTRODUCTION

The packaging is defined as interconnecting, powering, cooling, and protecting integrated circuits. This chapter addresses the protection of semiconductor devices by means of sealing and encapsulation processes after the device packaging is completed. In addition to providing protection of devices, sealing and encapsulation must also protect the package and its package wiring so as to provide the required reliability of the packaged devices. Figure 14-1 illustrates the basic concepts of sealing and encapsulation.

An advanced ultralarge-scale integration (ULSI) device is a very complex and delicate, three-dimensional structure. It consists of millions of components in a single integrated-circuit (IC) chip. These components

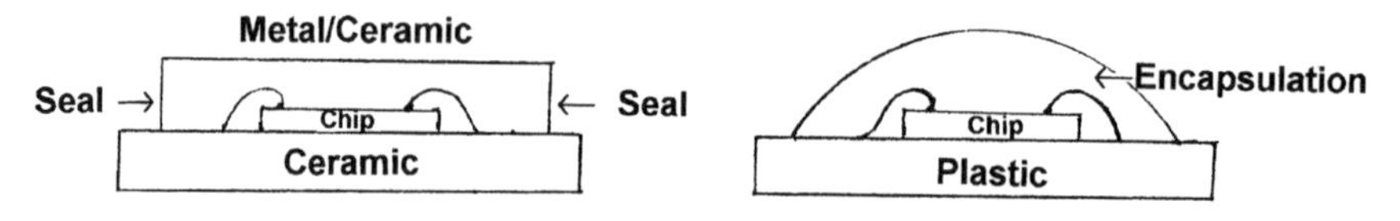

Figure 14-1. Basic Concepts of Sealing and Encapsulation

are densely packaged in a multilayer structure with different metalized conductor lines separated with dielectric insulating layers. The rapid growth of the number of components per chip, the rapid decrease of device dimensions, the steady increase in IC chip size, and the ever-increasing input/output (I/O) interconnects have imposed stringent requirements not only on the IC physical design and fabrication but also on the IC encapsulation and package sealing. The increase of integration in ULSI technology has resulted in the miniaturization of the device size which has reduced the propagation delay due to higher-density packaging and interconnection. As a result, a modern advanced device operates at a faster speed (> 100 MHz), consumes less power, and, consequently, dissipates less heat during operation. In addition, ULSI technology has increased the reliability of the microelectronic devices due to the elimination of the poor interconnections and has decreased the cost per function of the devices which has had a profound impact on the modern electronics industry. However, due to the high-density packaging of these devices, the power consumption per package is also increased, so is the physical dimension of the chip size and multichip modules (MCM). To package, encapsulate, and seal these high-power, large IC devices and the MCM is becoming a challenge [1,2].

The various materials used to fabricate integrated circuits (ICs) and their related interconnect are generally not capable of surviving their end-use operating, or field, environment without some level of additional protection. Metals susceptible to corrosion, dissimilar metals forming galvanic cells, structures that are stress or heat sensitive, and so on, can all lead to performance degradation or failure in the field without proper protection. Field environments can vary greatly for any one device, depending on the product in which it is incorporated or the customer's application of the product. These environmental stresses are not only limited to the thermomechanical stresses associated with device operation but also to the pollutants present in the operating environment.

A particular product may be appropriate for many uses, even a mobile one, so that a single product may see several environments in its lifetime. When factors related to the device are coupled in, such as the cost of manufacture, production volume, expected use environments or markets, expected volume of product going to the various markets, and so forth, packaging issues and decisions become very complex. These complexities require the packaging engineer to make critical decisions on how to most cost-effectively package the device while obtaining acceptable reliability in the field. Packaging selection can be seen as a compromise between cost and performance, where performance includes reliability issues as well as other design considerations. Two earlier chapters of this book have discussed specific packaging methods: Chapter 9, "Ceramic Packaging," usually involving hermetic enclosures; and Chapter 10, "Plastic Packaging." This chapter introduces more general concepts and options for the environmental protection of integrated circuits using inorganic and organic materials, as well as combinations of the two, to obtain acceptable field reliability.

In the early days of single transistors, hermetically sealed metal cases protected sensitive chips from environmental stresses. Moisture-related problems prompted the military and aerospace industries to require hermetic packages in order to achieve high reliability and long life. Hermetic packages utilize a sealed environment that is impervious to gases and moisture to protect the devices. Final sealing is accomplished with caps or lids using glass or metal seals. Moisture content within the package is standardly limited to a volumetric maximum of 5000 parts per million which was chosen to eliminate the possibility of condensation occurring inside the enclosure. For extremely sensitive devices, however, moisture limits are lowered even further, increasing both the time and cost to manufacture and treat these hermetically sealed packages.

Hermetic packaging proved to be such a robust method of obtaining a long field life that little has changed in hermetic packaging technology in recent years. After a slow development and initial reliability problems, hermetically packaged devices gave way to lower-cost nonhermetic plastic-molded packaging, especially in electronics destined for commercial and industrial markets. Today, however, nonhermetic does not imply nonreliable. Plastic packages are not hermetic, yet demonstrate admirable reliability and now account for approximately 90–95% of all device packaging.

Many obstacles had to be overcome with respect to understanding materials and processes associated with plastic packaging before their use was widespread. Organic materials are not hermetic and allow moisture to penetrate and be absorbed. Different organic material families have greatly varying material properties related to their moisture permeation rates and moisture-absorption percentages. In plastic packaging's infancy, corrosion was found to be the primary cause of failure due to poor adhesion and high levels of mobile ionic contaminants in the materials. The presence of moisture along with mobile ions were identified as the major contributors to corrosion, among other failure mechanisms. Improvements in plastic packaging materials and processes have all but removed corrosion as a source of failure and have lead to reliability that approaches that of hermetic packages. Now, stress-related failures from increasing die sizes present the largest obstacle to be solved by the plastic packaging engineer.

These major improvements in plastic packaging were derived from the effective use of large volumes of environmental test data that assisted engineers in understanding factors governing the principal failure modes associated with protection through organic materials. Accelerated environmental stress tests on packaged devices were the primary source of these data. In some cases, combinations of accelerated stress, along with corrosive ambients, were used to predict reliability behavior under severe use conditions and were further extended as a means of assessing the effects of impurities on device lifetimes. Accelerated stress tests such as the 85°C/85% **rh** test (85/85) became useful tools for screening and assessing the coupling between the processes and material systems. In the beginning of plastic packaging, 1000 h of 85/85 was not only difficult to obtain with existing materials and processes but was considered sufficient for many commercial markets. It is now quite common for plastic packages to exceed 10,000 h of 85/85 without failure.

More severe environmental tests have been devised to shorten test and evaluation times, decreasing the cost of the final product. The Highly Accelerated Stress Test (HAST) is the most prominent example and expands upon the 85/85 approach. HAST utilizes a pressurized chamber to allow the temperature to be elevated while maintaining humidity control, to produce the additional activation energy for more accelerated testing. Quite a bit is still unknown with regard to accelerated testing and its correlation to field life. This chapter will consider these issues and their relationship in the environmental protection of integrated circuits.

In both hermetic and nonhermetic packages, it is assumed that one level of protection precedes the sealing or encapsulation, as discussed in this chapter. This first line of defense is what is customarily termed the "passivation layer" and is typically applied at the back end of the semiconductor device circuit fabrication process. The passivation layer covers the entire integrated circuit with the exception of the bond pads, which remain exposed to facilitate signal and power interconnect to the device. The most prevalent circuit passivations consist of chemically vapor-deposited phosphosilicate glass, or silicon nitride, or photodefinable polyimides that are dispensed in spinning process, then exposed, etched, and cured. The passivation layer on the devices, although very important in providing device reliability, is considered a given input to any sealing and encapsulation packaging processes and an art within itself. Therefore, passivations will not be discussed in detail in this chapter.

14.2 HERMETIC VERSUS NONHERMETIC PACKAGING

There is a natural division of definitions between the two most common packaging approaches for integrated circuits: hermetic packaging which presents a sealed, impermeable package to the operating environment, and what we term nonhermetic, which is permeable. The hermetic package usually consists of a cavity package made of ceramic, metal, glass, and so forth, that are, for the most part, resistant to permeation by fluids and gases. On the other hand, nonhermetic packages are generally associated with organic coatings or seals such as epoxies, silicones, polyimides, and so forth, which are not capable of preventing penetration of fluids and gases. There are also new technologies that are designated "hermetic-equivalent" packages that use novel approaches to obtain hermeticity through depositions of metal films to seal the electronics package. Both hermetic and nonhermetic packages are able to house single devices or multichip systems.

Considering the issues, the packaging engineer's objective is to evaluate the criticality of the application, the types of devices and assembly methods that will be contained in the package, and the operational environment in which the package will be used, and to determine which type of packaging approach provides the best trade-off among cost, reliability, and performance. In general, the most critical applications requiring the highest, long-term reliability will utilize hermetic packaging, but today the choice is not always so clear-cut, as there are many factors that must be weighed, including cost, size, and weight. With the great strides that have been made in packaging with polymers in the last 10 years, more and more applications are open to its use. As different as the two packaging approaches are, so are the screening and qualification testing used to evaluate their performance. Hot, wet tests such as 85/85 or HAST do not have much effect on a properly designed and assembled hermetic package. The opposite is true with nonhermetic packages, as organic materials tend to break down, or depolymerize and hydrolyze in these environments. The remainder of this chapter provides an overview of both packaging methods, as well as some of their disadvantages, advantages, and technical details of their design and production-related processes.

14.2.1 Hermetic Packaging

Condensed moisture on a device's surface during operation has historically been shown to lead to the principle causes of failure in the field. The extremely small geometries involved in integrated circuits, differing galvanic potentials between metal structures, and the presence of high electric fields all make the device susceptible to interactions with moisture. One packaging method designed to prevent performance degradation due to moisture's deleterious effects is the hermetic package. Ceramic or metal is used to form an enclosure to isolate the electronic devices from the ambient operating environment. By eliminating condensible moisture from the cavity of the package during the sealing process and preventing the ingress and egress of moisture at the package perimeter during its operating life, excellent long-term reliability can be achieved. The word hermetic is defined as completely sealed by fusion, solder, and so on, so as to keep air or gas from getting in or out; in other words, airtight. In hermetic packaging practice, such seals are nonexistent. Small gas molecules will enter the package over time through diffusion and permeation. Eventually, these gases will reach equilibrium within the cavity of the package. In light of this permeation, long life can still be obtained in the field because of the extremely slow nature of this activity. Accelerated tests and leak testing for screening and qualification are specified in the military standard MIL-STD-883 and represent the foundation and industry-accepted approach to testing for reliability in hermetic packages.

14.2.2 Nonhermetic Packaging

The widespread availability of nonhermetic packaging followed that of hermetic packaging by many years due to problems associated with the initial polymer materials used for encapsulation. Both the traditional plastic-molded package and polymer-encapsulated circuits fell into this category. The inability of these early polymers to sufficiently retard the deleterious effects of moisture once it reached the delicate surfaces of the circuit, interconnect, and substrate led to poor performance both in testing and in the field. Inadequate adhesion, contaminants within the material itself, incompatible thermal expansion, and stress-related problems, in addition to a relatively immature knowledge of filler technology, all combined to prevent plastic packaging's immediate acceptance. With significant efforts in the areas of resins, fillers, material formulations, and process development work, polymer packaging finally began to make its presence felt in the early 1970s, although they had been used in some form as early as the late 1950s. During this time, significant progress was made in improving the quality of the glass passivation layer that is deposited on the active area of the device as a first line of defense against moisture-related problems. The combination of these technological advances provided the essential increase in reliability that was needed for polymer packaging to begin seeing widespread use.

During plastic packaging's infancy period, many failure modes were identified and material concerns addressed that resulted in different polymers and polymer formulations that more closely matched the requirements of the packaging technology and its applications. This has led to recent demonstrations of plastic packages that approach the reliability of hermetic packages in many applications and environments. Certainly, plastic-molded packages are the most dominant packaging method in use today for commercial- and industrial-grade electronics. It is estimated that over 90% of all integrated circuits are marketed in this form. Nonhermetic packaging encompasses not only the aforementioned plastic-molded package but also plastic and ceramic cavity packages that are sealed with a polymer, and the more recent chip-on-board (COB) and multichip module (MCM) approaches to packaging when a hermetic seal is not used. Most of the high-performance MCMs being produced, MCM-C (ceramic) and MCM-D (deposited), are currently packaged in hermetic packages, but the lower-cost MCM-L (laminated) which is a wirebonded or tape automated bonding (TAB) device on a printed wiring board, tends to be encapsulated with organic polymers in an overcoat fashion.

Much remains to be understood in the use of polymers for protection of semiconductor devices. Generalities of the required material properties are known, and how to avoid the failures associated with them, but specific degradation processes and activities at the material–surface interface remain the object of many research dollars. As mentioned earlier, corrosion was the primary cause of failure in early polymer packaging. New formulations with better filler technology resulted in materials that do not impart stress-related failures on devices and their associated interconnect, until the recent emergence of large dies, are attempted. In the newer technologies of COB and MCMs, corrosion is once again a primary cause of failure with the existing encapsulation materials, especially with wirebonded devices as in most plastic-molded packages where the thin aluminum bond pad on the device is exposed to the coating interface.

Adhesion between metallic–organic interfaces is facilitated by a combination of mechanical interlocking and chemical and physical bonding. If the density and strength of the molecular bonds between polymer and substrate is not high enough to prevent an aqueous phase from forming, and the other ingredients are available, corrosion is likely to occur. Different families of materials have different ways of attaining corrosion protection. For example, although typical epoxies do not have the density of bonds at the interface relative to silicones, moisture permeation through epoxies is much, much slower, thereby limiting available moisture at the interface. Although epoxies in general have a much higher strength of adhesion than silicones, they also have a much higher modulus which can lead to degradation of the material interface through mismatches in thermal expansion when thermal excursions are experienced. Corrosion protection and adhesion properties are closely linked, and long-term reliability requires long-term adhesion.

This polymer matrix density perspective carries over into the bulk material as well. Although it is certainly desirable to have a polymer material that is completely avoid of ionic species to begin with, it is also highly desirable for the material to be able to limit the migratory capabilities of ionic species from the ambient environment.

Moisture can also have deleterious effects on the long-term adhesion between organic materials and the substrate being protected. Moisture acts as a debonding agent through one or a combination of the following mechanisms: (1) attack of the metal surface by moisture can form a weak, hydrated oxide surface, (2) moisture-assisted chemical bond breakdown, and (3) moisture-related degradation or depolymerization. The natural protectant of metals, surface oxides, can actually be detrimental to the durability of the polymers adhesion in wet environments. Both metals oxides are relatively polar. Water (H_2O) will preferentially adsorb onto the oxide surface and create a weak boundary layer at the metal–polymer interface that can lead to adhesion problems. In wet environments, long-term corrosion prevention properties of a polymer can be adversely effected by this weakening interface region.

Accelerated testing is usually the means by which nonhermetic packaging is assessed during screening and qualification in the manufacturing process, with hot wet environmental tests the most common. Temperature cycling is the most common thermomechanical environmental test. Temperature cycling does not test the corrosion-resistant properties of the package and polymer system, but it tests the ability of the assembly to endure the stresses imparted by the various materials that make up the device, interconnect, and polymer packaging. An accurate correlation of screening and qualification testing to field life has yet to be accomplished, but individual manufacturers have historical, and most often proprietary, data that prove that the technology achieved acceptable reliability in specific markets and end-use environments. As more and more is understood in relation to the fundamental principles surrounding the use of polymers in packaging, its domination is expected to increase, and more difficult markets, such as the military, will open to it.

14.3 IC FAILURE MECHANISMS

The failure mechanisms associated with moisture-induced failures are discussed in greater detail in Chapter 5, “Package Reliability.” Only a brief description is provided here.

Moisture is the principal cause of nearly all failures in nonhermetic packages. Even in those cases where the presence of a second contaminant is essential to cause the failure, such failures would not normally occur in the absence of moisture. The three common types of problems caused by moisture are the following:

1. Charge Separation in Metal-Oxide Semiconductor (MOS) Devices: Here moisture provides mobility to surface charges on insulators, sometimes leading to the formation of parasitic gates. The failure will occur as soon as the field of the parasitic gate reaches a specific value. This type of failure is not permanent. Gross failures are noted at moisture levels exceeding 2%, whereas below this level, failures depend on the

presence of ionic contaminants such as sodium. [3] Recently, a passivation layer has been added, covering the IC devices and reducing the charge-migration problem.

2. Corrosion of Metallization: A galvanic cell will be formed in the presence of moisture and a galvanic couple. Ionic contaminants can dissolve in condensed moisture, greatly increasing the rate of corrosion. The corrosion of Al–Au and Al–Ag wirebonds in epoxy-bonded microcircuits are two examples of such corrosion [4]. In these cases, small amounts of chlorine were also found on the corroded surfaces. Ultrasonically bonded wires were found to be more susceptible to such corrosion than thermocompression bonded wires, presumably because of the effect of cold-working on the galvanic potential.

An example of direct chemical corrosion is the corrosion of aluminum stripes when moisture and chloride contamination are present, according to the following reactions:

$$6HCl + 2\ Al \rightarrow 2AlCl_3 + 3H_2$$

$$AlCl_3 + 3HOH \rightarrow Al(OH)_3 + 3HCl$$

$$2Al(OH)_3 \rightarrow \text{aging} \rightarrow Al_2O_3 + 3H_2O$$

In the absence of the chloride ion, the aluminum anode stripes readily become passivated with the oxide, but when present, these ions rapidly accumulate at the anode surface, promoting the formation of a nonpassivating oxide [5]. In one instance, it was noted that in the absence of a voltage gradient, failure of aluminum stripes occurred after 2000 h, whereas when a voltage difference of 25V DC between conductors 0.5 mm apart was present, failures took place after only 50 h[6].

Another important cause of aluminum corrosion is its dissolution in orthophosphoric acid formed by the reaction of moisture with excess phosphoros contained in phosphosilicate glass passivating layers on semiconductor devices [7]. The leachability of the phosphoros increases with the P_2O_5 content, the annealing temperature, and time [8]. In plastic packages, the minimum phosphorus level that is compatible with the desired device characteristics has to be used.

3. Electrolytic Conduction: The presence of condensed water and a continuous water path or a very thick absorption layer due to high moisture ambient between two electrically biased stripes of the same or different metals causes an electrolytic current to flow from the anode to the cathode. The anodic metal dissolves in the electrolyte and redeposits at the cathode. The deposited metal forms a projection of the cathode toward the anode, increasing the local field strength at this site, thereby attracting more of the metal to redeposit from the electrolyte. This escalating effect on the field strength gives rise to the characteristic dendritic

(treelike) growth of the metal deposit on the cathode, eventually leading to the disappearance of the anode metal or, more likely, to the formation of a short, causing device failure. The electrolytic corrosion of gold conductor lines in the presence of moisture and chlorine is shown in Figure 14-2 [9]. A definite threshold moisture level of 1.5% was noted for the reaction described, with no failures occurring below this level [9,10]. Other moisture-induced failures are also found to be characterized by minimum threshold levels, some of which are listed in Table 14-1 [11]. Although the mechanisms involved in these electrolytic failures are well understood, the estimation of failure rate for a given field strength and geometry is not trivial.

The failure mechanisms just considered require the presence of condensed water on the device surface at operating temperatures. The temperature at which water vapor of a given partial pressure condenses is the dew point. From the nomograph in Figure 14-3 it will be seen that the moisture content within the package atmosphere must be greater than 6000 parts per million by volume (ppmv) (dew point at 0°C) for any

Figure 14-2. Migrated-Gold Resistive Shorts in Gold Metallized Microcircuit (From Ref. 9, ©IEEE.)

Table 14-1. Device Failure and Water-Vapor Concentration

	Water Vapor Concentration (ppm)[a]	
Failure Mode	Failure-Free Demonstrated Failures	Upper Limits
Nichrome disappearance	5,000–10,000	500
Aluminum disappearance	50,000–250,000	1,000
Gold migration	15,000–150,000	1,000
MOS inversion	5,000–20,000	200

[a]Measured at package ambient temperature of 100°C.
Source: From Ref. 11, ©IEEE.

liquid water to be capable of forming. The actual moisture level that can be tolerated will be lower because of the role that oxide surfaces and cracks and crevices play in causing moisture condensation. For general purposes, the acceptable total water content cited in MIL-STD-883 is 5000 ppmv. The surface resistivity on package surfaces, which is the rate-determining property, is directly proportional to the thickness of the absorbed or condensed water film. This, in turn, is related to the moisture level in the package by the following equation [12]:

$$\text{Number of monolayers of water } L = 3.98 \times 10^{-2}\, CV/S \quad [14\text{-}1]$$

where

C = concentration of water vapor in ppmv
S = internal surface area
V = internal volume in cm^3

For a transistor outline (TO-8) package, the equation calculates that a 6000-ppmv water-vapor level could yield 30 monolayers of water on the package walls if all the water vapor condensed.

Although the connection between the presence of moisture and device failures is beyond any doubt, it has not been possible to establish a direct relationship, other than a statistical one between the level of moisture contamination and device life. However, in a clean surface condition, the number of absorption layers correlates with the leakage current between the electrodes [13]. It is surmised that this may be due to the critical role played by other factors that are not easily characterized, on corrosion rates, such as the presence of ionic contaminants. Statistical models have

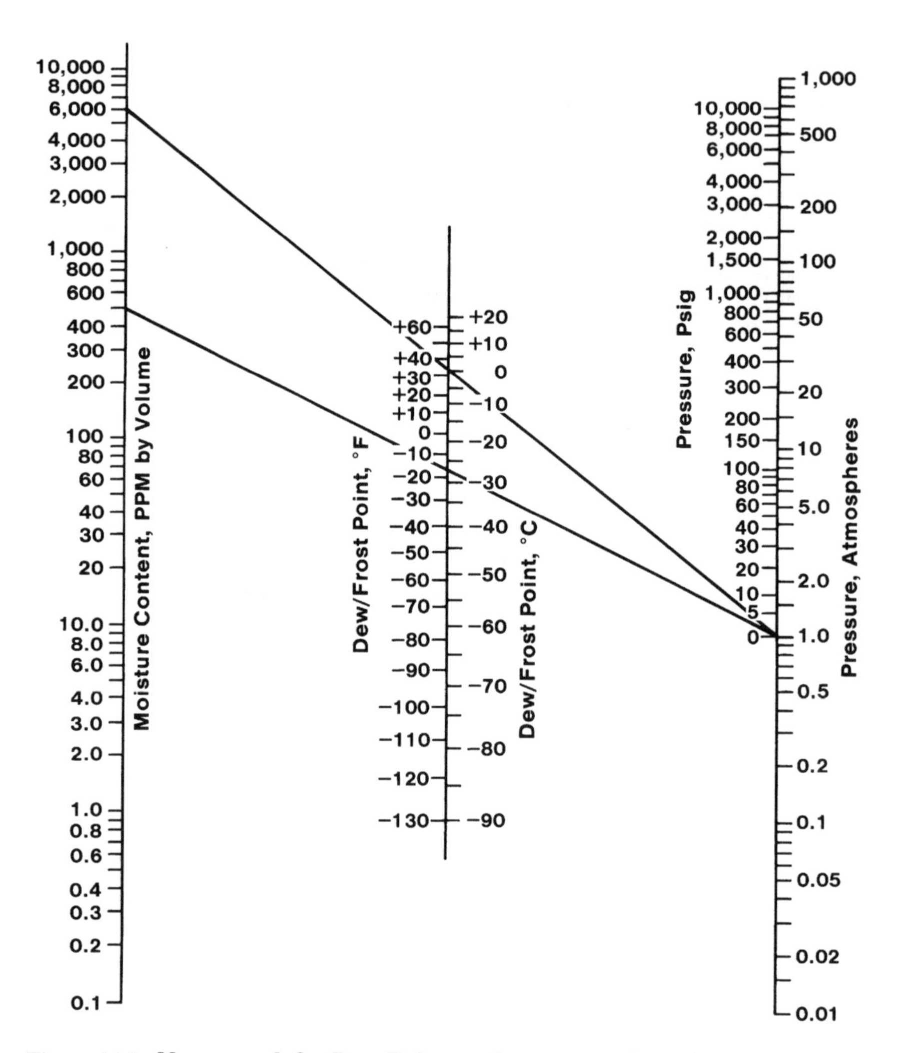

Figure 14-3. Nomograph for Dew Points and ppmv as a Function of Pressure

related failure rate to the temperature and relative humidity in several ways, such as failure rate (1) as proportional to exponential of temperature and relative humidity [14], (2) as proportional to log vapor pressure of water [15], and (3) as proportional to square of the relative humidity [16]. The electrical conductivity on the device surface shows similar relationships to the relative humidity, reflecting the role of ionic mobility in the corrosion reactions.

14.3.1 Sources of Moisture

The three contributors to the moisture within a package are (1) the sealing ambient, (2) the absorbed and dissolved water from the sealing materials, lid, and the substrate, released during the sealing process, and (3) the leakage of external moisture through the seal itself. The sealing ambient has to be very dry, with moisture levels at the sealing zone of the furnaces being kept below 100 ppmv. Hot cap-type sealers use nitrogen gas formed from a liquid-N_2 source which will have moisture levels of 10 ppmv or less.

The problem of moisture evolution from sealing glasses will be discussed later. This is the principal reason for failures in ceramic dual-in-line packages (Cerdips). Even when sealing by soldering and welding techniques, it is important to remove absorbed moisture from all the materials involved by baking the parts to a suitable temperature. Typically, a 200°C, 24-h vacuum baking at a vacuum level of a few torrs should be effective in removing all moisture from packaging materials. Table 14-2 gives the moisture levels associated with several bakeout times and temperatures [3]. Once baked, the package components should be stored under dry conditions such as in N_2 boxes to prevent reabsorption of moisture. It has been noted that the very bakeout operation can often change the moisture-adsorption characteristics of the surfaces for the worse. Thus, although gold-coated surfaces do not absorb water to any significant extent, the bakeout time and temperature could bring out transition-metal atoms from below. The presence of oxygen in the ambient promotes such out-diffusion of transition metals which have a much higher affinity for water molecules. Hence, the control of oxygen levels in bakeout ovens is very important [17].

Another source of moisture within the package is the epoxy die attach material. It has been shown that when silver-epoxy is heated to about 300°C, the unreacted monomers decompose to form water as a reaction product [18]. The moisture levels within brazed packages containing the silver-epoxy die attach compound were found to be over 10,000 ppmv. A preseal bake at 300°C in N_2 significantly reduced the moisture level (2000 ppmv), whereas a change over to weld-sealing, following a

Table 14-2. Moisture Content Versus Bakeout Time and Temperature

		Temperature	
Time (h)	150°C (ppm)	200°C (ppm)	250°C (ppm)
1	15,000	14,000	10,000
10	5,000	1,000	50
100	500	200	10

Source: From Ref. 3, ©IEEE.

preseal bake, lowered the moisture levels within the packages to below 300 ppmv. Significantly, the welded packages when heated to 300°C reached moisture levels of 3000 ppmv. Desiccants are sometimes introduced into the package to absorb the moisture within the package. A Si–Au–Sn alloy, used for die attach, is found to be a powerful desiccant because the silicon in the alloy reacts with water vapor [18].

Once a package is sealed with these precautions to minimize sealed-in-moisture, the only source of moisture left is the leakage through the seal. A hermetic seal implies a seal that will indefinitely prevent the entry of moisture and other contaminants into the sealed cavity. In practice, such seals are nonexistent. Smaller gas molecules will enter the cavity over a period of time, by diffusion or permeation, and ultimately reach equilibrium within the package. Hence, an operational definition of hermeticity is needed. Such a definition is provided for hermetic seals of microelectronic packages in MIL-STD-883, in terms of helium leak rates when the package is tested in a prescribed manner. The acceptable leak rates for military hybrid packages per MIL-STD-883 are given in Table 14-3.

It is suggested that in the absence of an explicit relationship between device life and moisture level, the setting of the these maximum allowable leak rates reflect more the limits of package-sealing and leak-detection methods than any fundamental consideration [20]. For example, for a 1-cm^3 package having a leak rate of 5×10^{-6} atm-cm^3/s, the rate of water buildup will be 10^{-1} ppm/s or 8.6×10^3 ppm per day, indicating that the package will be in equilibrium with the atmosphere within a matter of days. Similar conclusions were arrived at by several investigators [20–22]. Figure 14-4 shows the measured moisture penetration rates for various helium leak rates. A helium leak rate of 10^{-10} atm-cm^3/s is said to assure adequate protection against moisture permeation. It is also noted, however,

Table 14-3. Allowable Hermeticty Leak Rates (Method 1014.4 MIL-STD-883)

Pkg. Volume (cm^3)	(psig)	Exposure Time (h)	Bomb Condition Dwell (h)	Reject Limit (atm·cm^3/s He[a])
<0.40	60±2	2+0.2,–0	1	5×10^{-8}
≥0.40	60±2	2+0.2,–0	1	2×10^{-7}
≥0.40	30±2	4+0.4,–0	1	1×10^{-7}

[a]*Note:* The leak ratios can be expressed in International Units (SI units through the approximate relation 1 atm·cm^3/s = 0.1 Pa m^3/s)

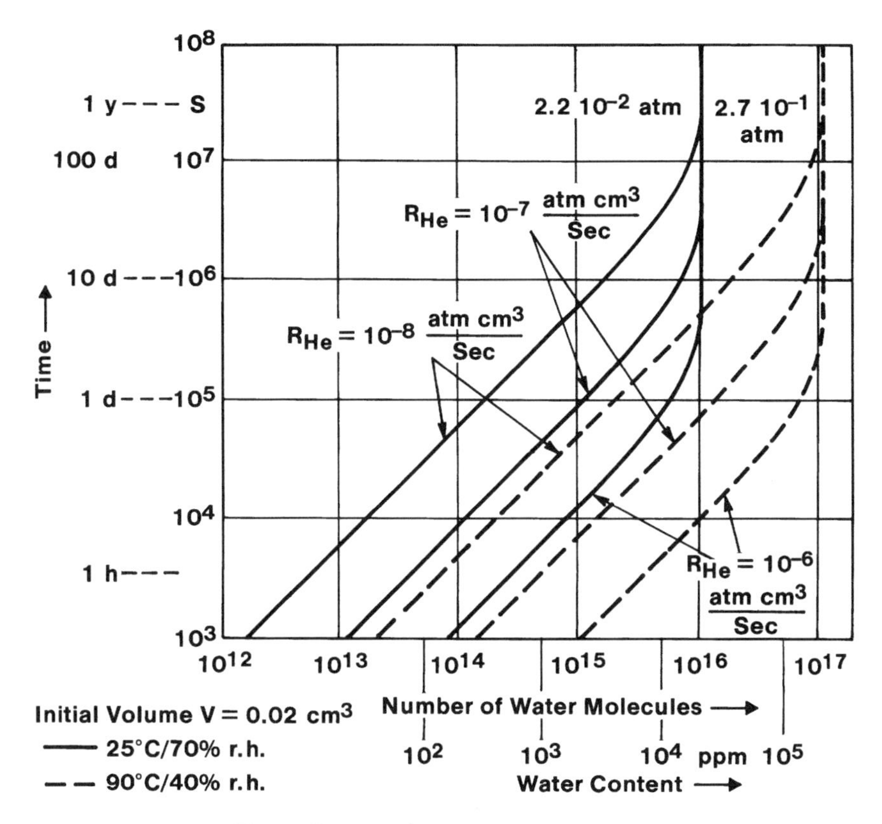

Figure 14-4. Water-Vapor Penetration. The penetration of water vapor into packages for different He leak rates. (From Ref. 20, ©IEEE.)

that the helium rates measured under test conditions could be several orders of magnitude larger than the rate of moisture ingress under use conditions. This is because of the absence of viscous flow contribution to moisture movement in the leak path in the latter case and because of the tendency of moisture to absorb on the walls of the leak path [21]. Actually, most hermetic packages probably have much lower leak rates than the specifications allow.

14.4 PACKAGE SEALING AND ENCAPSULATION

The purpose of sealing and encapsulation is to protect the electronic components from adverse environment and thereby increase their long-term reliability. High-performance and low-cost materials for this purpose are the major driving forces in the 1990s.

14.4.1 Package Sealing and Encapsulations Material Requirements

Electrooxidation (corrosion) and metal migration are attributed to the presence of moisture. Generally speaking, electrocorrosion occurs when enough moisture diffuses through the encapsulant to form a continuous water path on the device in the presence of mobile ions and under an electrical bias. Thus, a major function of an encapsulant is to serve as an effective moisture barrier. Most organic materials are quite permeable to moisture. Figure 14-5 shows the permeability of various materials. Pure crystals and metals are the best materials as moisture barriers. Glass (silicon dioxide) is an excellent moisture barrier but is slightly inferior to pure crystals and metals. Organic polymers, such as fluorocarbons, epoxies, and silicones are a few orders of magnitude more permeable to moisture than glass. Obviously, gases are the most permeable to moisture of all materials shown in Figure 14-5. In general, for each particular material, the moisture diffusion rate is proportional to the water-vapor partial pressure and inversely proportional to the material thickness. This is accurate when moisture diffusion rates are in steady-state permeation. However, moisture transient penetration rates (perhaps more important

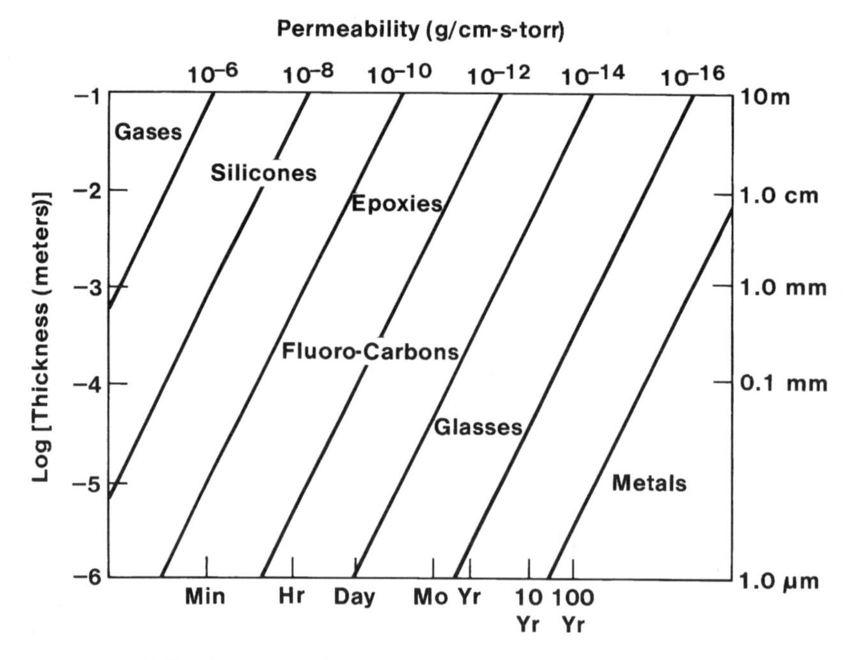

Figure 14-5. Effectiveness of Sealant Materials. The time for moisture to permeate various sealant materials in one defined geometry. (From Ref. 20, ©IEEE.)

because they determine the time it takes for moisture to break through) are inversely proportional to the of material thickness.

Mobile ions, such as sodium or potassium, tend to migrate to the *p-n* junction of the IC device where they acquire an electron, and deposit as the corresponding metal on the *p-n* junction. This consequently destroys the device. Furthermore, mobile ions will also support leakage currents between biased device features, which degrade device performance and ultimately destroy the devices by electrochemical processes such as metal conductor dissolution. For example, chloride and fluoride ions, even in trace amounts (at the ppm level), could cause the dissolution of aluminum metallization of complementary metal-oxide semiconductor (CMOS) devices. Unfortunately, CMOS is likely to be the trend of the very-large-scale integration (VLSI) technology and sodium chloride is a common contaminant. The protection of these devices from the effects of these mobile ions is an absolute requirement. The use of an ultrahigh-purity encapsulant to encapsulate the passivated IC is the answer to some of these mobile ion contaminant problems.

Ultraviolet–visible (UV–VIS) light radiation can cause damage to light-sensitive optoelectronic devices. UV–VIS protection can be achieved by choosing an opaque encapsulant. However, impurities in an encapsulant, such as low levels of uranium in the ceramic or plastic packages can cause appreciable alpha-particle radiation. Cosmic radiation in the atmosphere can also be a source of alpha-particle radiation. The alpha radiation can generate a temporary "soft error" in operating dynamic random access memory (DRAM) devices. This type of alpha-particle radiation has become a major concern, especially in high-density memory devices. Good encapsulants must have alpha-radiation levels less than 0.001 alpha particles/cm^2/h and be opaque in order to protect devices from UV–VIS radiation. Because the alpha particle is a weak radiation, a few micrometers thickness of encapsulant usually will prevent this radiation damage of the DRAM devices.

Hostile environments, such as extreme cycling temperature (values from −65°C to +150°C in MIL-STD-883), high relative humidity (85%–100%), shock and vibration, and high-temperature operating bias are part of the real-life device operation. It is critical for the device to survive these environmental influences. In addition, encapsulants must also have suitable mechanical, electrical, and physical properties (such as minimal stress and acceptable thermal expansion coefficient, etc.) which are compatible with IC devices. In addition to the above low requirements of low moisture permeability, excellent mobile ion barrier, good UV–VIS and alpha-particle protection, and excellent mechanical, electrical, and physical properties, the encapsulant must have a low dielectric constant to reduce the device propagation delay and excellent thermal conductivity to dissipate heat generated by IC devices. Furthermore, the encapsulant must be high-purity material with low-ionic contaminants. Because encapsulation is often the final process step, and some of the devices are expensive, particularly in the high-density multichip modules (MCM), the ability to facilitate repair processes during production and service is desirable. With the proper choice of encapsulant and process, the encapsulation can enhance the reliablity of the fragile IC device, improve its mechanical and physical properties, and improve manufacturing yields.

14.5 TYPES OF HERMETIC PACKAGES

No material is truly hermetic to moisture. The permeability to moisture of glasses, ceramics, and metals, however, is very low and is orders of magnitude lower than for any plastic material. Figure 14-5 illustrates the time scales for moisture penetration for these materials. Although polymeric-sealed devices can be designed to pass even fine leak tests, moisture will move through the seal in hours [23]. Hence, the only true hermetic packages are those made of metals, ceramics, and glasses. There are two basic types of hermetic packages in wide use today:

- Metal packages
- Ceramic packages

The common feature of these packages is the use of a lid or a cap to seal in the semiconductor device mounted on a suitable substrate. The leads entering the package also need to be hermetically sealed. In metal packages, the individual leads are sealed into the metal platform or "header" by separate glass seals. In ceramic packages, the leads are formed by thick-film-screen and fire techniques and are commonly embedded in the ceramic itself. Cerdip packages are unique in that the same sealing member seals in both the cap and the leads. To satisfy the requirement of hermeticity, cap seals in all these packages are formed by fusing a suitable glass or metal. In brazing and soldering, a low-melting braze or solder alloy forms the seal. In welding, the cap itself is partially melted to effect the seal. These different types of seals are described in greater detail later.

14.5.1 Metal Packages

Metal hermetic packages are used extensively in military applications because of their enviable record of reliability under the harshest use conditions. There are four types of metal package: round header (Fig. 14-6a), platform (Fig. 14-6b), flat pack (Fig. 14-6c), and monolithic (Fig. 14-6d). The round header or the T-O packages have been in use since the earliest days of the transistor and are still employed to package discrete

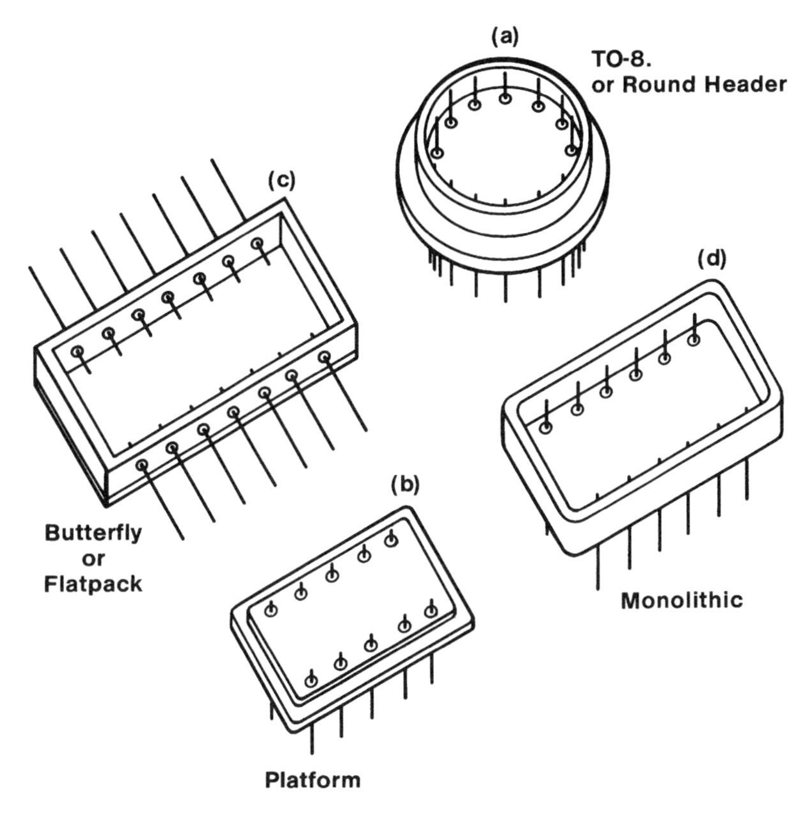

Figure 14-6. Common Types of Metal Package

devices and smaller hybrid circuits. The platform packages provides up to 16 pinouts, whereas the more versatile flat pack is a low-profile package used for LSI and VLSI with up to 200 leads. The monolithic package is a rugged package constructed of a single-piece platform providing up to 88 leads. Nearly 80% of all metal packages are welded, with the remaining being soldered.

14.5.2 Ceramic Packages

A description of several types of ceramic packages is given in Chapter 7 ("Microelectronics Packaging—An Overview"). The brief descriptions of some of these packages here is meant for the purpose of illustrating different methods of package sealing.

14.5.2.1 Ceramic-Dual-in-Line Package

These were once the most widely used hermetic packages because of the ease of their fabrication and the low cost (Fig. 14-7). In these

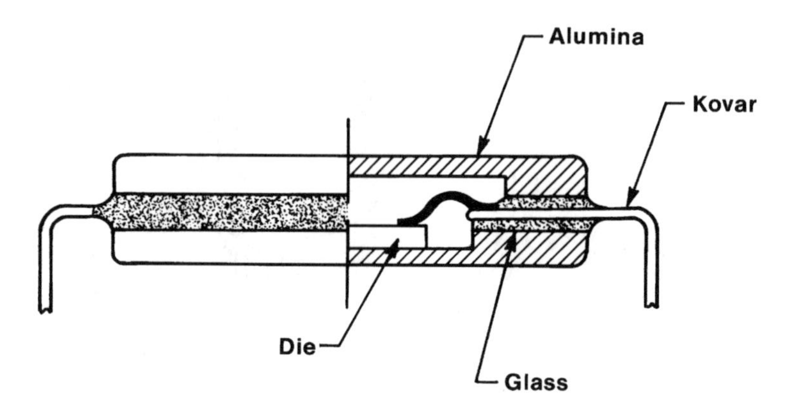

Figure 14-7. Sealing of the Cerdip Package. (From Ref. 24.)

packages, a Kovar® lead frame is first attached to a glazed alumina platform by temporarily softening the glass. After die attach and wirebonding, the glazed ceramic cap is sealed to a subassembly in a conveyor furnace or in the so-called hot cap sealers. The glasses used for sealing Cerdips are high-lead-content vitreous or devitrifying glasses with sealing temperatures of about 400°C.

14.5.2.2 Hard Glass Package

Another type of hermetic package is the so-called hard glass package (Figure 14-8), in which a Kovar lead frame and a Kovar seal ring are first attached to an alumina or beryllia platform using high-melting borosilicate glass at temperatures around 1100°C. After plating the attached lead frame and the seal ring with suitable metals, a metal lid is soldered or brazed

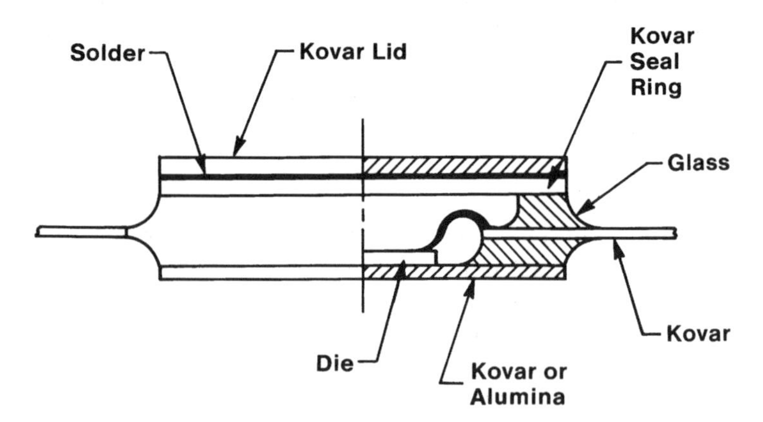

Figure 14-8. Loaded Glass/Hard Glass. Typical hard glass package, which is composed of the glass body through which Kovar leads penetrate. (From Ref. 24.)

to the seal ring. Instead of using a Kovar seal frame, an alumina seal frame is also often used. In this case, a glazed ceramic lid such as those used in Cerdip sealing can be used to seal the package. Figure 14-9 compares the construction of this kind of hard glass package with that of the more economical Cerdip.

14.5.2.3 Other Ceramic Packages

There are several other types of ceramic packages, including the side-brazed dual-in-line package (DIP), bottom- or top-brazed chip carriers, pin-grid arrays, and other multilayer ceramic packages. The DIP, which has been the most widely used package in the last two decades, is now losing ground to the chip carriers and to the pin-grid arrays, including multichip, multilayer substrates. The chip carrier offers higher lead counts than the DIP and is especially designed for economical manufacturing techniques such as tape automated bonding (TAB) and surface-mounting technology (SMT).

When even higher pin counts are required, pin-grid arrays are needed. The largest pin-grid arrays also tend to be multilayer substrates with several layers of cofired circuit lines. In most of these packages, a metal or ceramic cap lid is brazed or soldered to the nickel and gold-plated

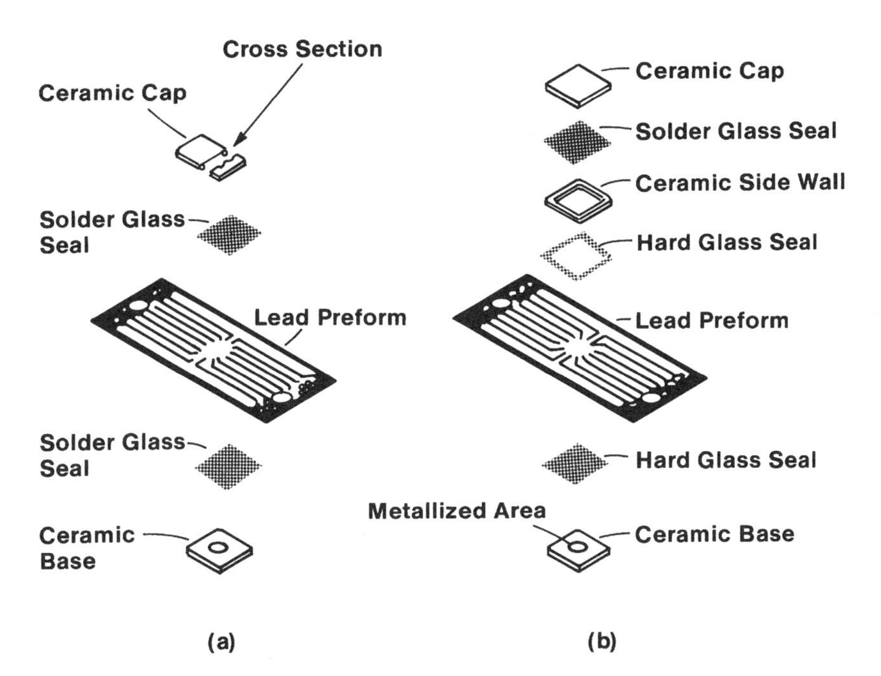

Figure 14-9. Alternative Packages. (a) Single seal; (b) dual seal. (From Ref. 25, reproduced with permission from the Society of Glass Technology, England.)

refractory metal seal band on the substrate. Welding, glass sealing, and metal gasket sealing are other methods used to attach hermetic caps to these substrates. Figure 14-10 illustrates the construction of some ceramic packages.

14.6 TYPES OF HERMETIC SEALS

14.6.1 Fused Metal Seals

Metal hermetic packages with enclosed volumes of a tenth of a milliliter or more are commonly welded, soldered, or brazed. Ceramic packages can also be sealed by glass sealing. To facilitate the sealing of the ceramic substrates by soldering or welding, a metal seal band should be provided on the substrate surface. In hard glass packages described earlier, a seal frame made of Kovar or alloy 42 is first attached to the substrate by using a borosilicate glass. In ceramic packages, the seal band is formed by thick-film, cosintered molybdenum or tungsten metallurgy. The seal frame is then suitably plated, and a metal lid is attached by soldering or welding. The large throughputs, high yields, and reliability associated with the welding technique is spurring a change from glass sealing to welding for ceramic packages [24]. The major considerations in selecting a sealing method are the availability of equipment and the cost of the hybrid circuit. Welding is more economical because of its speed, high process yields (approaching 100%), and reproducibility. Solder or braze sealing is commonly employed when it is required to allow for the ability to remove the cap and reseal the lid. Of these, the most popular method for hermetic sealing is welding.

14.6.1.1 Soldering

The solders used for hermetic sealing are selected based on the required temperature hierarchy for the processes that precede and follow the sealing operation, the desired minimum seal strength, and cost. The lid seal has to remain intact, for example, during the soldering of chip carriers to printed-circuit boards. In this case, the solder for the seal should have a higher melting temperature than the solder for direct mounting. Where cap rework is required for pin-grid array packages, the sealing solder should have a considerably lower melting point than the melting point of the solder or braze used to attach the pins to the substrate. Although straight tin–lead solders are widely used for hermetic sealing, alloying additions such as indium and silver are sometimes added to improve the strength or fatigue resistance. Use of bismuth–tin alloys for sealing has been suggested [26]. It is found that the property of these alloys to slightly expand on solidification (0.0005–0.0007 cm/cm) helps to minimize the shrinkage voids in such seals [27].

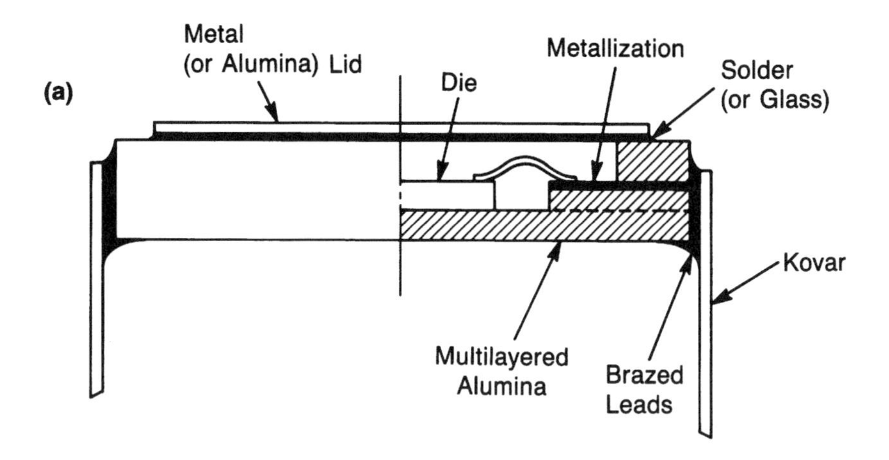

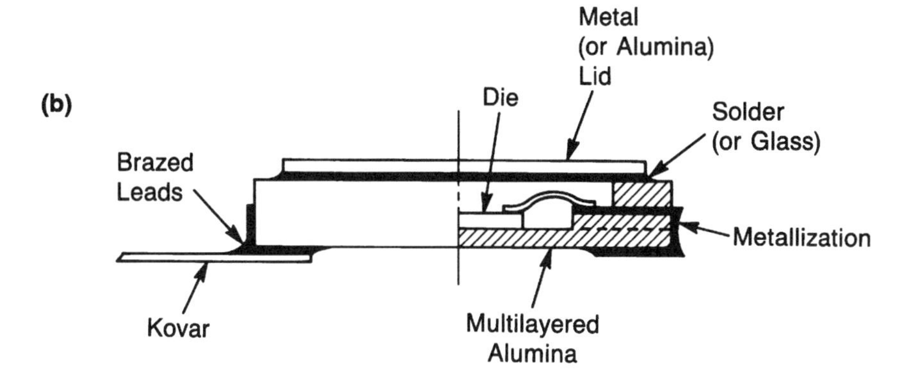

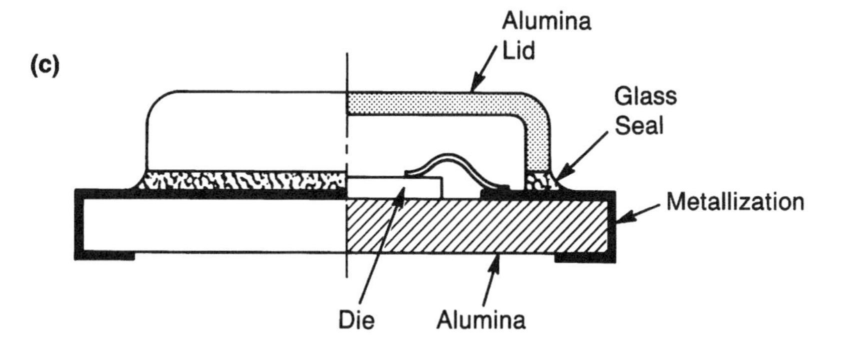

Figure 14-10. Sealing of Other Ceramic Package Types. (a) Side braze ceramic package; (b) chip carrier; (c) chip carrier (SLAM). (From Ref. 24.)

Soldering is less preferable to brazing with a eutectic gold–tin alloy because of its lower strength (less than one-half the strength of gold–tin), greater susceptibility to embrittlement due to intermetallic formation, and the need to employ fluxes during sealing in most cases. The lower solder strength translates into lower fatigue resistance, and, in fact, solder joints in power devices have been found to fail during power cycling [27]. This, however, is not a problem when both the cap and the substrate are made of the same material. The softer the solder, the less likely it is to fatigue.

The brittleness of solder seals has sometimes caused seal failure even during helium-bomb leak testing [27]. These failures were found to be due to the formation of the Au_3Sn_4 intermetallic through its reaction with the plated surfaces of the cap and the substrate and its segregation during solder spreading. It has been observed that at 240°C, a 60 : 40 Sn-Pb alloy can dissolve 10 μm of gold in 1 s [29]. The ductility of this solder with about 8%–10% dissolved gold is practically zero. Although pure gold coatings provide an excellent surface for soldering, the coating thickness has to be kept to a minimum and under close control. Solderability of bright and hard golds are strongly dependent on the alloying additions and organic brighteners used [30]. Some of these additives release gases during soldering, giving rise to gross porosity in the seal. Another reason for solder nonwetting or dewetting is the formation of a gold–tin intermetallic compound ahead of the spreading front. A nickel-alloyed (15%–20% Ni) gold surface, although providing good wettability, avoids the formation of excessive intermetallics because of the slow reaction rate with nickel [31].

The ease with which tin–lead solders oxidize makes it necessary to handle them carefully and store in nitrogen. During sealing, a flux will also be necessary if the solder pastes or preforms are directly used in the operation. Flux entrapment in the die cavity is unavoidable in soldering. Trapped flux can cause corrosion and lead to the deposition of tin on the electrically active areas of the device [37]. Flux usage during solder sealing can be avoided, at some cost, by resorting to the use of previously reflowed solder. Here the solder cream or the preform is first reflowed in nitrogen on to the seal area of the substrate [37] or of the lid [33], using a flux. The flux and, if necessary, the surface of the reflowed solder are cleaned, and a second reflow operation in nitrogen is carried out to smooth and flatten the surface of the reflowed solder. In the actual sealing operation, the lid and the substrate are assembled under a clip or a spring-loaded fixture and reflowed in an inert-gas ambient without using a flux. The absence of the flux increases the criticality of other process parameters such as the metallurgical and surface cleanliness of the sealing surfaces, the spring loading, the furnace profile, and the ambient. The spring load is necessary to force down the cap in order to assist the spreading of the solder. Too high a pressure leads to solder balling and bridging [31].

These tiny solder spheres with oxidized surfaces can cause intermittent shorting inside the package. The furnace ambient in the sealing zone should have no more than 100 ppm oxygen and 50 ppm water vapor.

Solder sealing is commonly performed in a conveyor furnace. A typical furnace cycle consists of a fast preheat period (3–5 min), minimum time (3–5 min) above the liquid's temperature, a peak temperature of 40–80°C above the melting temperature, and a fast cool-down after solidification (Fig. 14-11). The proper amount of solder, conveniently specified as solder thickness, should slightly exceed the total camber tolerances for the substrate and the lid. If thickness is lower, gross leaks can occur; too much solder leads to solder balling.

Among the soldering techniques that avoid the overheating of the device being sealed are seam soldering, which employs the same equipment as seam welding, and platen or hot cap soldering [30]. The latter offers high yields and is generally a low-cost process. In this method, a platen with a controlled heating profile is brought into contact with the lid while the lid, the preform, and the substrate are held in a water-cooled fixture. The operation is performed in a controlled-atmosphere chamber. The chamber pressure is raised slightly, when the solder melts, to counteract the increased pressure within the cavity. The solder fillet, commonly formed in furnace soldering, is not formed in this method of sealing, due to this positive pressure.

Although soldered packages are easily delidded, the rework is seldom

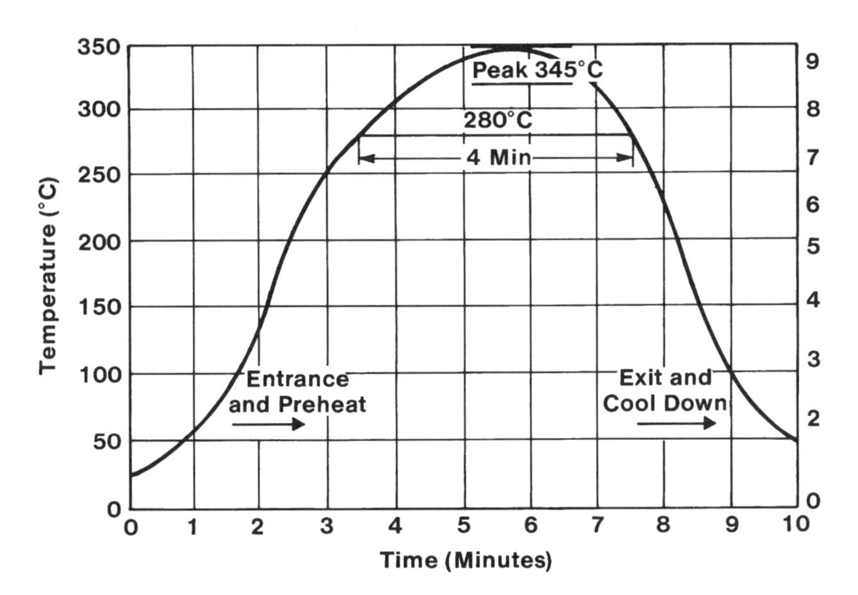

Figure 14-11. Temperatures Versus Time Profile for Reflow Sealing. (From Ref. 33, reprinted with the permission of *Solid State Technology*.)

carried out more than once. This is because of increased gold leaching, dewetting, and the need to use higher and higher temperatures for each successive seal because of solder alloying characteristics [34]. Softer solders allow for more reworks. Gross leakers can often be salvaged by reflowing at a higher temperature.

Solder sealing of large-cavity packages presents special challenges because of the large volume of heated gas inside the package, which tends to form blow holes in the molten solder or draw solder inside as solder balls. One of the largest solder-sealed packages (64 × 64 mm) is the multichip ceramic substrate used in the IBM 4341 System [31].

14.6.1.2 Brazing

Brazing with a eutectic (80 : 20) Au–Sn alloy is used in place of soldering when the need exists for a stronger, more corrosion-resistant seal and where the use of flux has to be avoided. The braze is usually used as a thin, narrow preform tack-welded to a gold-plated Kovar lid. The metal seal band on the substrate is also gold plated for good wettability and corrosion protection. In furnace sealing, the typical reflow time is 2–4 min above the eutectic temperature of 280°C with a peak temperature of about 350°C. Other methods of sealing used for soldering can also be used with advance for braze sealing.

The very steep slope of the liquidus curve in the Au–Sn system on the gold-rich side of the eutectic has given rise to many a hermeticity failure. An increase in just 3–5% in gold content above the eutectic composition raises the liquidus temperature from 280°C to 450°C. This increase can easily and rapidly occur by the dissolution of the gold from the plated surfaces. It is suggested that the problem can be avoided by using a brazing alloy slightly richer in Sn than the eutectic, such as a 78 : 22 alloy [36]. This entails a slightly higher sealing temperature, but in this case, the dissolution of plated gold would actually lower the liquidus and facilitate wetting of the sealing surfaces.

The factors that influence the formation of a reliable seal during brazing are summarized in Table 14-4. These include the flatness of substrate and lid, furnace temperature profile, and atmosphere control [35]. Furnace sealing with AuSn has been reported to provide good yields with combined fine and gross leak losses of 2% or less [33].

14.6.1.3 Welding

The most popular method for sealing high-reliability packages, such as those used in military applications, is welding. One survey indicated that about 80% of military packages are welded [34]. Despite the higher initial cost for equipment, welding is popular because of the high yields and a good history of reliability. In welding, high-current pulses produce

Table 14-4. Factors Affecting Brazing

Parameter	Too Much	Too Little	Impact Magnitude
Metallurgical			
Lid and package gold plating	Gold leaching raises melting temperature	Difficulty wetting	Moderate
Preform volume	Unnecessary wetting of surrounding areas	Difficulty obtaining complete seal	High
Package/lid cleanliness	Cannot be too clean, but too aggressive cleaning methods can produce corrosion	Solder wetting inhibited	Moderate
Package thermal history prior to lidding	Poor wetting caused by base metal diffusion	None	High
Mechanical			
Lid and package flatness	Cannot be too flat	Thicker solder preform and/or use spring clips	High
Package lid and fixture mass	Longer furnace preheat and/or longer furnace dwell	None	Moderate
Lid compliance	Easy deformation from external forces	Difficult to compensate for lack of flatness	Moderate
Preform attach	Tacked or bonded; means easy handling	Chance of preform misalignment	Moderate
Furnace			
Belt speed	Package/lid mass cannot reach melting temperature	Silicon component damage. Excess solder flow.	Moderate
Peak temperature	Solder craws up on lid Silicon component damage	Proper solder flow not achieved	High
Fillet appearance	Hermeticity probable	Hermeticity questionable	Moderate
Atmosphere Moisture and oxygen content	Poor wetting	None	High
Nitrogen flow rate	Wastes nitrogen	Poor wetting	Moderate

Source: From Ref. 35.

local heating between 1000° and 1500°C, fusing the lid or the plating thereon to the package. The local heating prevents damage to internal components.

In parallel seam welding (Fig. 14-12), also called series welding, the package and the lid are passed under a pair of small tapered copper electrode wheels. The transformer produces a series of energy pulses, which are conducted from one electrode across the package lid to the other electrode. The generation of heat at the wheel–lid interface leads to the formation of either a weld or braze type of seal, the nickel or gold plating acting as the brazing medium [37].

In the opposed electrode welding (Fig. 14-13), the package to be sealed is moved under a pair of larger electrode wheels. The power ply, usually a capacitor discharge, produces a series of welding pulses, which pass from the electrode wheel across the lid–package sidewall interface and return via the workbench, that generate heat at the electrode–lid interface and the lid–package sidewall interface.

Welding can accommodate greater deviations from flatness for the package and the lid than soldering or brazing. The higher temperatures at the welding interface can volatilize most contaminants so that cleanliness is not as critical factor as it is in soldering and brazing.

Other less common methods for package welding are electron beam (e-beam) welding and laser welding. Laser welding of large metal packages is attractive to resistance welding because of its high speed, very limited heat input to sensitive areas, ability to handle unconventional seal geometries, and noncontract nature. Continuous-wave (CW) lasers have been found to cause solidification cracking in the weld, whereas the cracking

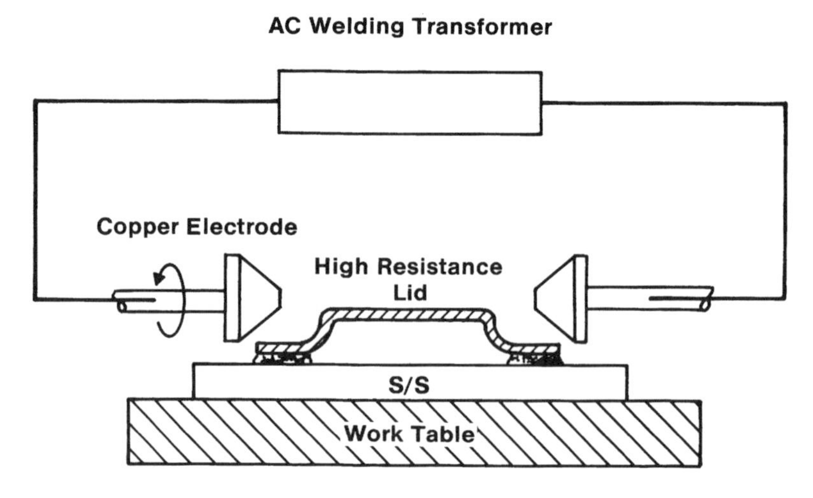

Figure 14-12. Parallel Seam or Welding

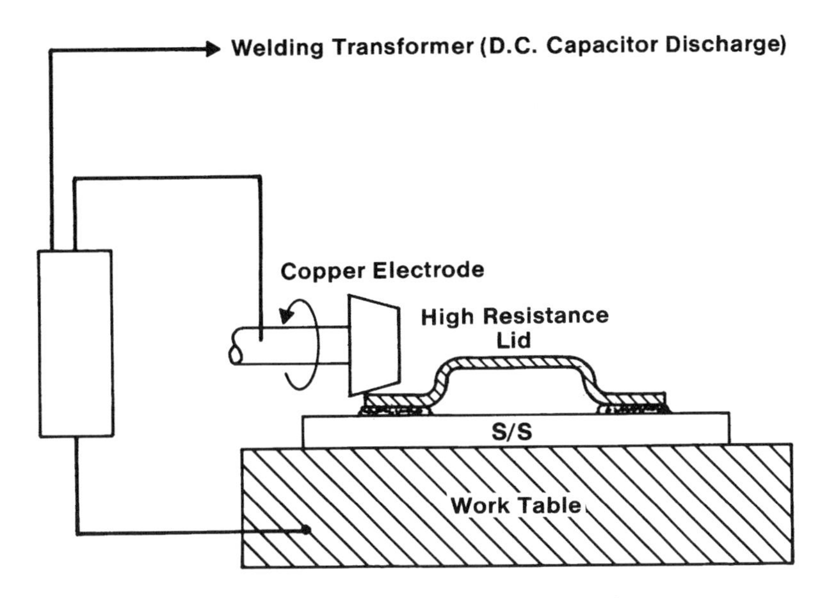

Figure 14-13. **Opposed Electrode Welding**

tendency was significantly lower with pulsed lasers that form a series of overlapping welds. The cracking tendency also depends on the nature of the metal plating in the weld area, with electroless nickel-plated surfaces prone to most cracking than electroplated nickel [37]. Gold plating tended to cause gross leakers and excessive weld splatter according to one report [38], whereas another report found gold plating helped the formation of a sound weld [37]. The availability of reliable, versatile laser equipment should make laser welding more popular in the future. Both CO_2 and Nd–YAG lasers are used in laser welding of hermetic packages.

Delidding of welded packages is most commonly carried out by end milling or by precision sawing [34]. Before resealing by welding, the seal flange has to be reground. Although delidding of a welded package is quite common during development and the prototype-production stage, it is usually not used for production packages because of the prohibition on the use of delidded and resealed packages by military specifications.

14.6.1.4 Glass Sealing

Glasses have been used in semiconductor package sealing since the beginning of the transistor age. The first use of glasses was in device passivation, and, to this day, glasses continue to be used for this important purpose, which serves as the last line of defense for a semiconductor device against external moisture and other contaminants. In this section we only consider the use of glasses in hermetic package sealing.

Glass sealing is encountered in a wide range of package types, from the earliest T-O headers used to package the first transistors to the latest ceramic packages. In the former, glass is used to form hermetic glass-to-metal seals for the package leads fed through holes in a metal plate or header. In the latter, the glass is used to form a sandwich seal between a ceramic cap or lid and the ceramic substrate on which the devices are mounted. In Cerdip packages, the same glass seal performs both these functions.

The versatility of glasses in package-sealing applications resides in their chemical inertness, oxidation resistance, good electrical insulating characteristics, impermeability to moisture and other gases, and the wide choice of thermal characteristics that can be obtained. The principal shortcomings of glasses—their low strength and brittleness—make it imperative to pay particular attention to seal design, choice of sealing glass, and, to some extent, the sealing process used. An excellent review of stresses in different glass-seal configurations has been given by Varsheneya [39] from which the following discussion is drawn.

Glass seals fail for one or more of the following reasons:

1. Poor adhesion between glass and the surfaces to be sealed
2. High stresses in the glass leading to its fracture
3. Fracture due to applied stresses in handling or use
4. Failure in the metal or ceramic parts being sealed

Poor adhesion in glass sealing, although a common problem with glass-to-metal seals, is seldom a problem in glass-to-ceramic seals. There is extensive literature on the mechanisms of glass-to-metal adhesion and the methods for promoting the same for specific metals. For good adhesion, the glass should, in its molten state, exhibit good wetting to the metal surface, which, in turn, promotes a strong chemical bond, mechanical interlocking, or both. Wetting of liquid to a solid surface is governed by the drive to minimize the interfacial tensions in the system. The equilibrium condition for the wetting is given by the classic equation

$$ {}_{s}\gamma_{g} = {}_{s}\gamma_{l} + {}_{l}\gamma_{g} \cos\theta \qquad [14\text{-}2] $$

where

γ = interfacial tensions
θ = contact angle
l, s, g = liquid, solid, and gas, respectively.

The lower the contact angle, the better the wetting. The chemical basis for good wetting is explained by the so-called oxide-saturation theory of Pask and co-workers [40,41].

Stresses in a seal are the result of the differential thermal contraction between the sealing components from the temperature of sealing down to room temperature. This stress can be calculated by

$$\sigma_{\text{ixy}} = -A_{\text{ixy}} \int_{Ta}^{Tr} (\alpha_g - \alpha_m)\, dT \qquad [14\text{-}3]$$

where

A_{ixy} = function of the elastic properties of the glass and metal and the seal configuration
α_g, α_m = thermal contraction coefficients of the glass and the metal, respectively
T_a = set point of the glass
T_r = room temperature.

The set point is generally taken as the temperature 5°C above the strain point ($\log \eta = 14.5$, where η is the viscosity in poises) and 15°C below the annealing point ($\log \eta = 13.0$). It is observed that for vitreous glasses, the set point corresponds to the temperature at which $\log \eta = 13.7$. It is recommended that $\Delta\alpha = (\alpha_g - \alpha_m)$ should be below 500 ppm for good seal integrity.

The most common glass-to-metal seal in microelectronic packages is the so-called bead seal used to seal the feedthrough leads into metal headers. The glass seal also acts to insulate the leads from the header. These seals are of two types—matched or compression seals [42].

In matched seals, the thermal coefficients of expansion of the glass and the metal involved are similar. Sealing is dependent on the formation of a chemical bond between the glass and the metal. To ensure the latter, the metal surface is provided with an adherent oxide layer. The optimum methods for providing such an oxide layer on Kovar alloy, widely used in metal packages, is described in the literature [42–44].

The compression seal does not involve the formation of a chemical bond between the glass and the metal. In this case, the glass is chosen such that its expansion coefficient is lower than that of the metal. When the assembly is sealed at the melting point of the glass and cooled, the metal shrinks more from the sealing temperature and holds the glass tightly in compression. Although this type of seal is generally hermetic and stronger than the matched seal, it is not as thermally reliable as the latter.

The normal glass-to-Kovar seal forms a meniscus on the leads, which signifies good wetting. However, too long a meniscus is known to cause a problem because it can be easily broken off, thus causing hermeticity failures or the exposure of unplated Kovar lead areas, which then readily corrode [24]. A polymeric coating applied to the base of the leads has been shown to be effective in preventing the damage to the meniscus by providing mechanical port to the leads in bending [46].

The resistance of the bead seals to fracture upon sudden temperature variations is tested by a standard thermal-shock test prescribed in the MIL-STD-883. The test involves alternatively immersing the packages in a cold fluid and a hot fluid 15 times, allowing time in each for the package temperature to equilibrate. It has been reported [3] that although the package may be hermetic before and after the thermal shocks, it may, sometimes, lose the hermeticity during the test itself. The reason for this, if true, is obscure. The strength of matched bead seals has been determined to be about 7500 psi [46].

The sandwich seal is the commonly encountered glass-to-ceramic seal in microelectronic packages. These include the low-temperature glass seals in Cerdip packages and the hard glass seals of alumina seal frames to alumina or beryllia substrates with borosilicate glasses.

Most of the soft glass seals are made from lead–zinc–borate glasses. They generally permit sealing below 420°C. These glasses are characterized by a low softening point and low viscosity at the sealing temperature, as well as by a fast-setting viscosity curve. Other required attributes are low water content, good chemical durability, and thermal expansion closely matched to that of the ceramic. Some of these requirements are hard to come by in low-temperature sealing glasses. The thermal expansivity of lead–zinc–borate glasses range between 80×10^{-3}/°C and 100×10^{-7}/°C, compared to a value of 65×10^{-7}/°C for alumina and $(50–55) \times 10^{-7}$/°C for Fe–Co–Ni alloys commonly used in packaging. As such, the stresses in these seals would be extremely high. Hence, these glasses are modified by devitrification and through the addition of certain low-expansivity fillers such as fused silica and beta-eucryptite [25]. Addition of titania to glass composition has been shown to promote the devitrification of the glass as well as the formation of lead titanate crystals having anomalously low expansivity and thereby lowering the overall expansivity of the seal [48]. Typically, the thermal expansivity of the seal glass ranges from 70×10^{-7}/°C to 80×10^{-7}/°C.

It has been noted that in the high PbO region of lead–zinc–borate system, the tendency to devitrify increases rapidly between 5% and 20% ZnO [25]. Devitrification in these glasses is said to occur by homogeneous nucleation despite the availability of enormous surface area for surface nucleation in the powdered glass. Increases in strength and decreases in thermal expansivity generally accompany devitrification, both factors improving the seal integrity.

A major problem in sealing Cerdip packages is the evolution of moisture from the sealing glasses. Water is physically and chemically absorbed on glass surfaces, included in internal pores, and structurally combined with the glass network. The seriousness of this problem is illustrated by the calculation that the evolution of only 0.15 μg of water into the cavity will provide 5000 ppmv of water vapor to 0.04 cm^3 volume of a 16-lead Cerdip and that this amount of moisture is routinely found in package sealing glasses [49]. Figure 14-14 shows the moisture evolution behavior of certain sealing glasses studied by the moisture evolution analysis (MEA) technique [47]. Three distinct regimes of moisture evolution are noted. At around 210°C, the physically absorbed moisture is evolved. The peak around 370°C, near the softening temperature of the glass, is attributed to the evolution of moisture from the internal pores in the sealing glass. At still higher temperatures, in the sealing regime, the chemically combined water is expelled. In an actual sealing operation, some of this moisture gets released into the sealed cavity. The authors were able to show a direct correlation between the moisture measured in the sealing glasses by the MEA technique and the moisture levels actually measured in the hermetic cavities sealed using the same glasses measured by a residual gas analysis (RGA) technique, thereby establishing a monitoring technique for moisture levels in Cerdip. It has been shown that the

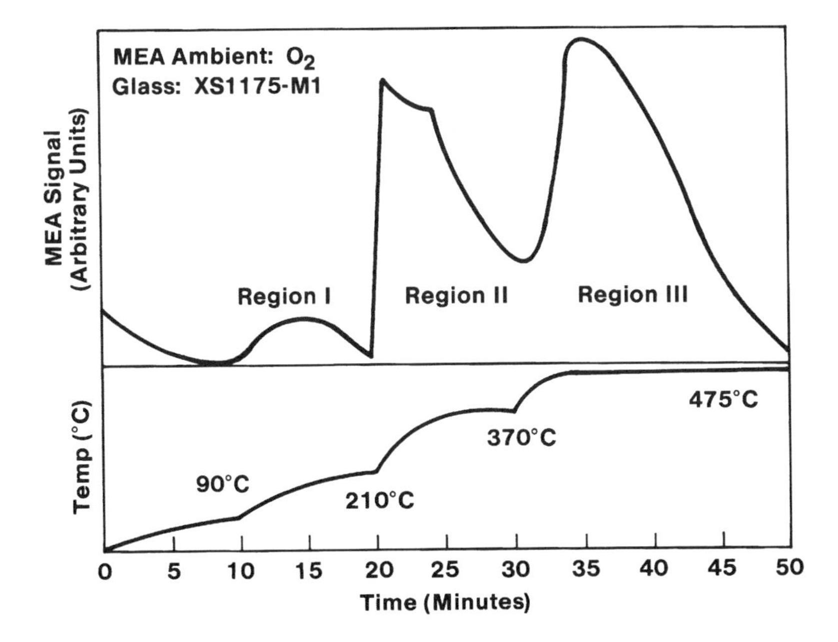

Figure 14-14. Moisture Evolution from Sealing Glasses. (From Ref. 47, reprinted with permission from *Journal of Electronic Materials,* a publication of the Metallurgical Society, Warrendale, PA.)

microporosity in the sealing glasses plays a major role in determining the extent of moisture release into the package even when the glass is used in the preglazed condition [49]. Furthermore, a prebaking in vacuum of the glazed lid at above the softening temperature made a significant difference in the amount of moisture actually entrapped in the package during sealing (4500–6000 ppm for nonprebaked lids to 200–900 ppmv for prebaked lids). Heating to below the softening point during baking was found to be not as effective because the internal porosity still retains the remaining reabsorbed moisture.

Although devitrifying solder glasses offer a better seal integrity, they are also associated with greater degree of moisture within the sealed package compared to packages sealed with vitreous seals. In one study [50], about half of the packages sealed with a devitrifying glass were found to contain over 6000 ppmv water vapor. Of those packages sealed with a nondevitrifying glass containing an aluminum silicate additive, over 90% had enclosed moisture level of less than 500 ppmv, with all of the packages having moisture level below 2000 ppmv. The principal reason for this difference is that crystallized glass has much lower solubility for water than in its glassy state, so that almost all of the chemically combined water gets expelled, some of it into the package cavity, when devitrification occurs during sealing.

14.6.1.5 Glass-Sealing Methods

Furnace sealing is the most common method for glass sealing. The key factors to be controlled are the furnace ambient and the temperature profile. Heating is provided by conventional heaters as well as by infrared (IR) heaters. The lids and the ceramic are usually preglazed with sealing glass A well-controlled glazing process is one that yields good seal dimensions on the ceramic, and good wetting and adhesion to the ceramic and produces little or no prenucleation in the case of devitrifying glasses. Prenucleation causes premature crystallization during the subsequent seal step, thereby reducing the flow, causing insufficient wetting as well as porosity. Differential thermal analysis (DTA) has been found to be a useful tool in determining the occurrence of prenucleation during the glazing process [52]. It is noted that although a 10°C lowering of the crystallization peak temperature from its normal position, as measured by DTA, may be tolerable, a 20°C drop would cause definite yield problems, and a 30–40°C drop could reduce yields by up to 50%. Prenucleation can be pressed by heating rapidly, consistent with the need to remove the organics, during glazing. The same is true during sealing. A heating rate of 75–125°C/min is recommended. For sealing, a 10–20-min soak at the peak temperature is required, after which the package is cooled slowly at about 40°C/min.

Generally, seals are 250–400 μm thick and about 1000 μm wide. A typical furnace profile for glazing and sealing of a semiconductor package sealing glass is shown in Figure 14-15.

In cases where exposure to the relatively high glass-sealing temperatures cannot be tolerated, the hot-cap sealing method has been employed. Here, the lid is heated to a temperature above the melting point of the glass while the substrate is kept at a lower, safer temperature for the device. Even though the substrate is also preheated to avoid a large temperature differential across the interface through sealing glass, the process entails a relatively fast cooling of the seal and, consequently, larger stresses in it. This has sometimes led to seal fracture [51].

A so-called graded viscosity seal is aimed at reducing the danger of moisture entrapment during the sealing process [51] (Fig. 14-16). Here the lead frame is first sealed to the substrate using a devitrified glass. The package is then heated in nitrogen and transferred into a hot-cap sealer in a dry box. The sealing surface of the substrate slopes outward toward its edge so as to favor the outgassing of moisture from vitreous sealing glass away from the package interior.

Another novel sealing method uses a beam of focused infrared light to heat the seal area of a preglazed lid and substrate assembly [51]. The glasses used are specially formulated to absorb infrared radiation. Although the seal area gets heated to about 400°C, very little heating of the interior occurs.

Furnaces and other equipment designed for use in hermetic sealing are described in the literature [51,53,54].

14.6.2 Gasket Seals

One clear trend in packaging for large mainframe systems is to use multilayer substrates capable of accommodating many semiconductor chips. In such packages, the sealing cover is generally a part of the package-cooling system. The cover encloses a large volume and will

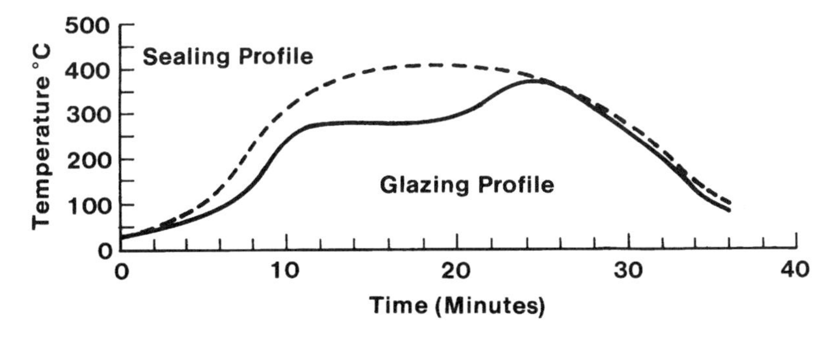

Figure 14-15. A Typical Sealing/Glazing Profile for Cerdips. (From Ref. 51.)

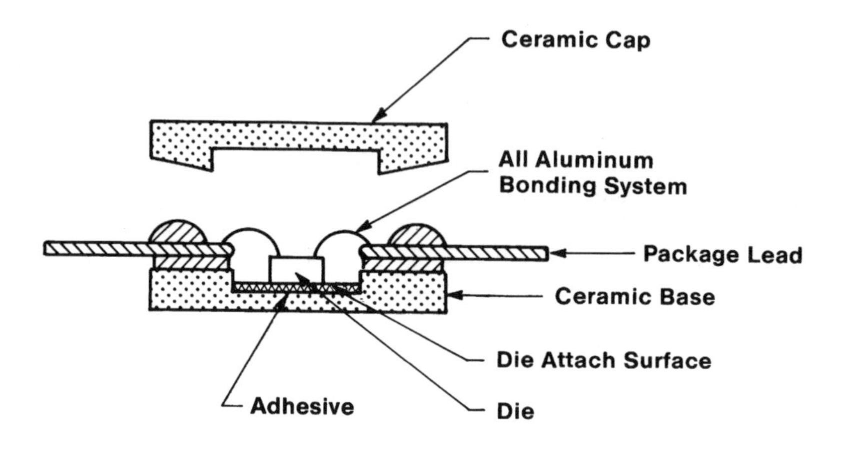

Figure 14-16. Graded Viscosity Seal Package. GVSP features elimination of outgassing within the package. (From Ref. 51.)

have a large seal perimeter. To allow for repair or replacement of the semiconductor chips during assembly, test, or service, the cover should be easily demountable and resealable. The seal should also be capable of accommodating the large difference in the thermal expansivities between the copper or aluminum covers, chosen for their mechanical and thermal properties as well as for the ease of their manufacture, and the ceramic substrate. These considerations clearly rule out fused metal or glass seals. Gasket seals are ideal for this purpose, but traditional rubber gaskets will not provide for a hermetic seal due to the high permeability of these materials to moisture and other gases. However, such gaskets made of a malleable metal are hermetic if sufficiently high pressures are applied and maintained over the seal area. Copper gaskets are commonly used in ultrahigh-vacuum systems, whereas C rings of stainless steel are used in systems operating at high pressures and temperatures. To improve the seal reliability, these rings are covered with a soft metal such as lead or aluminum.

In gasket seals, the possible leak paths are (1) gas permeation through the gasket material itself and (2) the leaks at the contact surfaces. The first consideration eliminates rubber and polymeric materials for gaskets.

Gaskets made of metal are suitable because of the very low permeabilities of metals to gases. The sealing load required for a given maximum acceptable leak rate should be sufficient to elastically and plastically deform the metal to provide a close conformal fit to the microtopography of the surfaces. This load depends on the roughness of the sealing surfaces, the elastic modulii, the yield strength, and the work-hardening characteristics of the gasket material, the friction between the mating surfaces, and the geometry of the gasket. The leak rate of gases through the interfaces of a gasket seal is given by the following equation [55].

$$C = Ar^2 e^{-b\sigma_a/\sigma_y}$$

where

σ_a = the applied average normal sealing stress
σ_y = the yield stress of the gasket metal
A, b = constants dependent on the nature of the gas, the geometry of the gasket, and the temperature
C = the leak conductance
r = mean roughness of the sealing surfaces

It is reported that the leak rate through interfaces of metal gaskets decreases very sharply as the average stress over the seal area reaches a critical value lying between 1.5 and 2.5 times the yield strength of the softer metal [55]. The highest stress ratio was required for radially ground surfaces (1 μm rms), whereas the lowest sealing stress was required for circumferentially machined surfaces ($0.7\sigma_y$, 2.5 μm rms). Among the various cross sections of the gaskets, the circular cross section requires the highest compression ratio, whereas the square cross section require the lowest compression ratio.

The specific elastic and plastic properties of the gasket material must require sealing load levels that are compatible with the strength, toughness, and the static fatigue limit of the ceramic. To keep the sealing stress at safe levels, softer metals like lead or indium could be used as gasket materials. By themselves, these materials will creep under load and relax the sealing stress below that needed to assure effective sealing. The optimum gasket must have both a low-yield stress at the contact surfaces and a large elastic compliance. This unique combination of properties is achieved using a composite gasket. Typical seal loads for metal seal ring with a tubular diameter of 30 μm and a wall thickness of 1.5 μm is about 30 kg/cm [56].

A gasket seal, such as described above, is utilized in a complex, multilayer package used in IBM's 3081 computer systems [57]. The multilayer ceramic package consists of over 30 metallized layers and

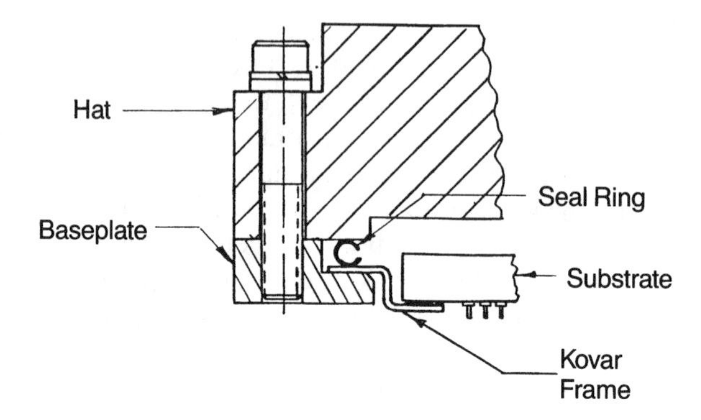

Figure 14-17. Metal C-Ring Seal to Brazed Kovar Frame

carries over 100 high-performance chips. Each chip is contacted by a spring-loaded metal piston embedded in a water-cooled housing or hat that is filled with helium. This arrangement is able to dissipate up to 600 W. In the early versions of this module, the hat assembly was clamped to a Kovar seal frame brazed to the cofired molybdenum seal ring on the bottom surface of the substrate (Fig. 14-17). In a latter version, the hat assembly was directly clamped to the machined ceramic surface of the substrate (Fig. 14-18). In both cases an Inconel C-ring was used.

14.7 TESTING OF HERMETIC PACKAGES

It is implicit that when a package is tested to be hermetic, no other tests are needed to assure its reliability. Hence, 100% leak testing is the principal test for hermetic packages. However, it is not uncommon to subject hermetic packages to temperature–humidity tests as well as to certain thermal and mechanical stress tests aimed at determining seal

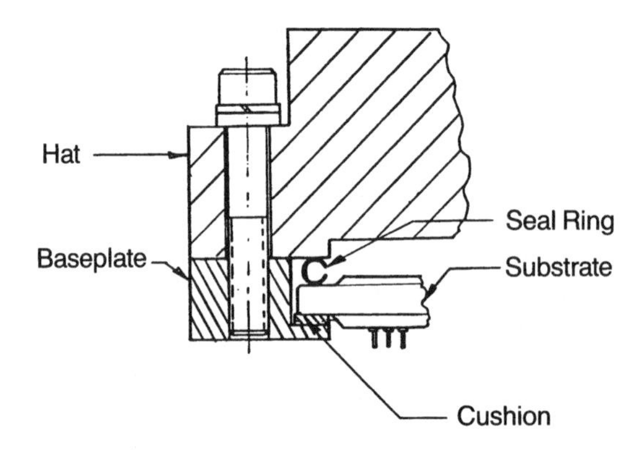

Figure 14-18. Metal C-Ring Seal Directly to Substrate Surface

integrity. It is also often required to determine contaminant levels within the package as part of a quality-control measure or of a failure-analysis effort. Only the methods used for leak testing and for moisture-level measurement within sealed packages are briefly reviewed here. The reader is referred to MIL-STD-883B and other relevant specifications for a more detailed description of the leak-testing methods and of thermal and mechanical characterization of seals.

14.7.1 Leak Testing

The practice of leak testing of hermetic packages has been described in a recent article [58]. The National Bureau of Standards (NBS) Special Publication 400-9, entitled *Hermeticity Testing for Integrated Circuits,* reviews current problems in leak testing [59].

The standard leak rate (L) is defined as that quantity of dry air at 25°C, measured in atmospheric cubic centimeters per second (atm cc/s), flowing through a leak path or paths where the high-side pressure equals 1 atm (760 mm Hg absolute) and the low-side pressure is 1 torr (1 mm Hg absolute). Helium leak rates, R_l, measured under the specified test conditions, converted into the rates that would have been measured under the above standard pressure gradient, are termed the equivalent standard leak rates.

The spectrum of leak rates of interest is broadly divided into two regimes. Leaks in the range 10^{-1}–10^{-4} are termed gross leaks, whereas those in the range 10^{-5}–10^{-12} are termed fine leaks. Leak rates of 10^{-4}–10^{-5} are not easily characterized solely by methods used for either fine or gross leak detection.

14.7.1.1 Gross Leak Testing

Gross leaks are tested by using a fluorocarbon liquid. In the bubble technique, the package to be tested is first subjected to a vacuum-evacuation step (≥5 torrs, 1-h dwell) to facilitate easy entry of the test liquid if a leak exists. Following the evacuation soak, the package is immersed in 3M Electronic Product Division's Fluorocarbon liquid (FC-84) without breaking the vacuum. It is left to soak in it for 30 min. The "filled" packages are then immersed in another fluorocarbon liquid (FC-40) having a higher boiling point than FC-84 and maintained at a temperature above the boiling point of FC-84. Any FC-84 that had leaked into the package during fill phase will now expand in volume and appear as bubbles observable through a magnifying window in the FC-40 liquid container.

An alternative method, called the negative ion detection (NID) method, uses a gas analyzer specially tuned to detect FC-84 molecules from packages filled with the latter in a manner similar to that used in the bubble method. The "filled" packages are heated in a metal chamber maintained at 125°C and attached to the gas analyzer.

14.7.1.2 Fine Leak Testing

The common method of testing for fine leaks uses helium gas as the testing medium. First, the package is subjected to helium gas under pressure in a pressure chamber or "bomb." Typically, the pressure used is 60 psig. and the dwell time is 2 h. After the pressurization step, the packages are introduced, either singly or in small batches, into a vacuum chamber attached to a helium mass spectrometer. If any helium evolution is detected during the evacuation step, the whole batch is rejected.

Another method used for fine leak testing also serves to test for gross leaks. Here, the package is subjected to pressurization in dry nitrogen gas into which is mixed about 1% Kr-85 radioactive tracer gas. After the pressurization, the package enters a scintillation counter, which detects the gamma emission that accompanies the beta decay of Kr-85. Because the gamma rays can easily penetrate the package walls, the scintillation count directly yields a measure of the Kr-85 concentration within the package.

It has been pointed out that the helium leak rates measured in this test could be orders of magnitude larger than helium leak rates that would occur under use conditions—1 atm pressure in air [21]. This difference is due to the predominant contribution of viscous flow of the gas during bombing and subsequent testing in vacuum and to the absence of viscous flow under use conditions. Further, the penetration of moisture into the package could be significantly slower than for He because of the tendency of the former to absorb on the walls of the leak path.

14.7.2 Moisture Measurement and Monitoring

As discussed previously, the actual moisture content within the package is made up of contributions from moisture released from the sealing materials and the package itself, the sealing ambient, and the moisture leaked in through the seal. Thus, it is necessary to measure the actual moisture level within the package on a sampling basis, to make sure that the sealing is being performed under the desired conditions. The moisture content within the package is also of interest for investigating the causes of failure of hermetically sealed devices.

Measurement of moisture level within a package is most commonly carried out by residual gas analyzers (RGA), which use a mass spectrometer to detect and measure water vapor. (See, for example, MIL-STD-883 Method 1018-2.) This technique gives a quantitative measure of the total moisture within the package. The package is broken or punctured in a heated vacuum chamber to release the moisture, which is then led into the mass spectrometer. Because of the extremely small amount of moisture that is being measured, great care is needed in setting up the mass spectrometer, in breaking the package, and in the interpretation of data [3,60]. A very highly sensitive method for measuring moisture content from individual packages using plasma chromatography has also been described [61]. Advantages of the mass spectrometer technique include its widespread use and acceptance, its great sensitivity, its quantitative nature, and the ability to simultaneously measure other volatile constituents of the package besides water. Disadvantages of the RGA method are that it is a costly, time-consuming, and destructive method and requires well-trained personnel to conduct the tests.

A reliable method for determining moisture-penetration rates involves circulating dry He through the package placed in a controlled humidity chamber. The moisture picked up by the circulating gas is continuously monitored using a P_2O_5 cell [20]. As discussed previously, moisture levels in Cerdips can be monitored by determining the moisture content of sealing glasses measured by a moisture evolution analyzer [47].

A nondestructive method for continuously monitoring moisture levels within sealed cavities utilizes several types of in situ monitors. In the commercially available humidity sensors or humistors, chemically altered polymers or oxides are used to absorb the moisture in their ambient, which, in turn, causes changes in their surface or bulk electrical characteristics. These humistors, once calibrated, can be sealed into the package to continuously monitor the moisture level within the package.

One type of moisture sensor that is best suited for use in microelectronic packages consists of a porous aluminum oxide layer sandwiched between two electrodes—an aluminum base and a porous gold electrode. The porous oxide layer is grown by the anodization of aluminum electrode. The thin, evaporated gold electrode is formed on the porous oxide layer and is permeable to moisture. As the moisture content in the package changes, it is reversibly absorbed and desorbed by the porous oxide, thereby modifying the cell's impedance. Because the entire body of the porous aluminum oxide participates in absorbing the moisture, these sensors are said to have "volume effect" and are very sensitive. This type of humistors have a measurement range from 1 to 10,000 ppm moisture level and a sensitivity of about 1 ppm (Fig. 14-19). Earlier versions of this moisture sensor were made on aluminum foil tabs. Newer, more rugged sensors of this type constructed on silicon wafers using thin-film techniques offer improved reliability and enable further miniaturization down to a 15×15 μm size. They are also shown to be stable up to 500°C and exhibit little or no thermal cycling or hysteresis effects. The humistor readings have been shown to correlate well with the industry standard method of mass spectrometric measurements [62,63].

Another type of moisture sensor used in the microelectronics is the

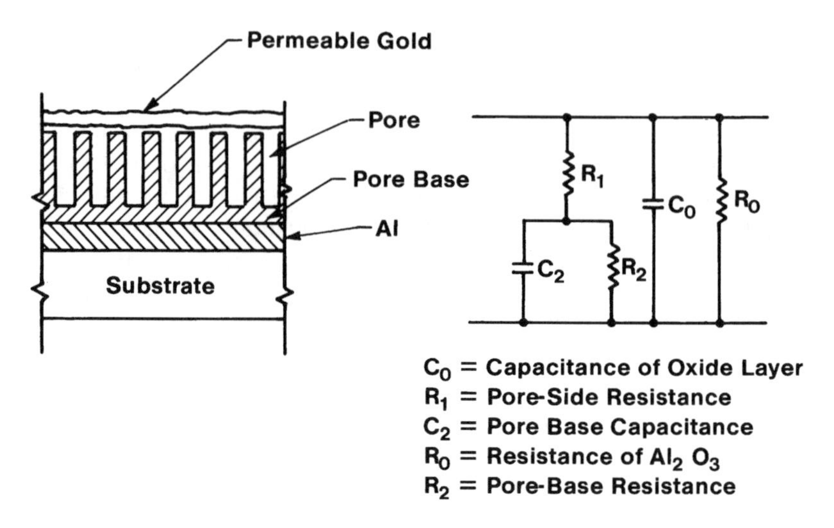

Figure 14-19. Al_2O_3 Moisture Sensor. (a) Cross section; (b) equivalent circuit. (From Ref. 62, ©IEEE.)

dew-point sensor. The operating principle of this sensor is the occurrence of a large decrease in resistance between closely spaced electrodes at the dew point. In principle, the sensor is used to measure the dew point within the package as long as the condensing phase is liquid. The sensor is formed as interdigitated thick- or thin-film electrodes on a chip surface (Fig. 14-20). As the package is slowly cooled, the resistance between the electrodes is monitored. A sudden drop in resistance signals the moisture condensation and, hence, the attainment of dew point. Unlike in the case of the Al_2O_3 sensor, only the surface of the sensor absorbs moisture and causes the measured change in the resistance. For this reason, its detection limit and sensitivity are not as good as for the former.

In both of the above cases, the sensors have to be specially fabricated and enclosed in the package. It has been shown that adjacent metallization tracks on the silicon device itself can be used to measure the dew point [64]. This method makes use of the fact that the stray capacitance between the conductors increases at the moment the chip surface reaches the dew point. In the absence of condensed moisture (Fig. 14-21a), the stray capacitance builds up partly through the chip and partly through the surrounding air. When moisture condenses (Fig. 14-21b), a conducting water film forms on the insulator and the capacitance increases. The actual change that occurs is of the order of a picofarad (pF) and can be measured by a stable capacitance bridge.

The measurements of dew point made using this method have been made shown to agree with those made by the surface-conductivity method. In another version of this type of sensor, the need for cooling the package to the dew point is avoided. Here, the frequency dependence of the

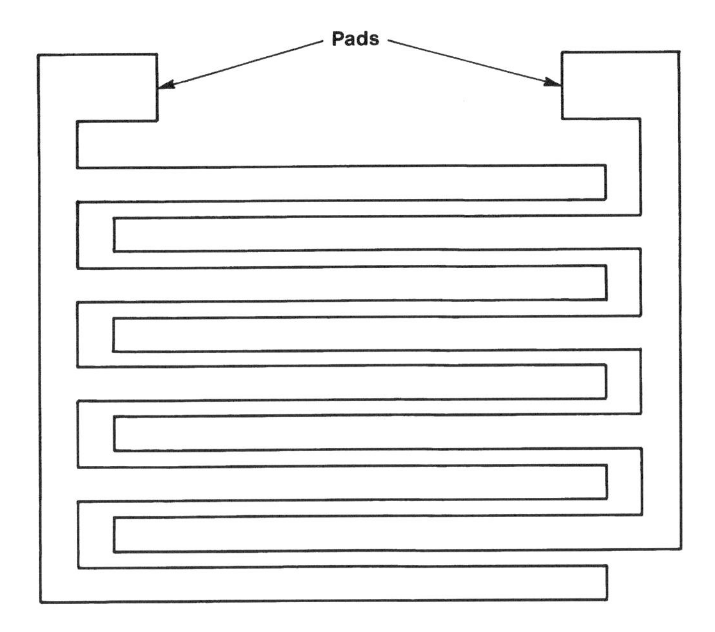

Figure 14-20. Interdigitated Dew-Point Sensor. (From Ref. 62, ©IEEE.)

capacitance between the two metallization tracks is measured at one temperature and is determined over a range of frequencies. For a given moisture level, a sudden change in resistance occurs at a unique frequency. Once calibrated, the sensor can be used to monitor changes in the moisture level within the package [65].

The difficulties involved in leak testing, on the one hand, and the need to determine the actual level of moisture within the package, on the other, have made it desirable to incorporate moisture sensors within the

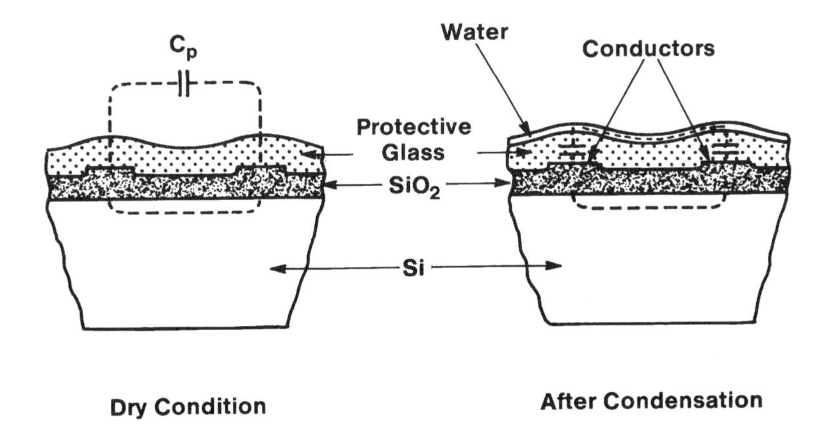

Figure 14-21. IC Chip Cross Section. (From Ref. 64.)

hermetic cavity of the package. The availability of miniature sensors of high reliability and accuracy, such as those described here, makes it attractive to use them routinely, at least in large packages.

14.8 RELIABILITY TESTING

To ensure the long-term reliability of the packaging sealing and encapsulation, reliability testing is an integrated part of the manufacturing process. Effects of moisture, mobile ions, and so forth, are directly affecting the reliability and accelerated testings, such as the conventional 85°C/85% **rh,** and the highly accelerated stress test (HAST) are critical to predict this manufacturing process. These will be discussed as follows.

14.8.1 Correlation of 85°C/85% rh to HAST

Reliability testing results are used to give a degree of confidence of a device's field reliability to the electronics merchant and the customer. The only way this degree of confidence can be given is if there is some correlation between the accelerated test time-to-failure (TTF) date that has been obtained from qualification testing, and the expected field TTF rates. This can be accomplished by formulating reliability models that relate accelerated environmental tests to actual field life. The objective for any reliability model is to be able to estimate device performance accurately outside the range of experimental observations. This can be done in either of two ways: relate the accelerated test TTF results of a specific device to that of the same device at different accelerated test or field conditions; or through comprehensive, fundamental modeling of the physical activities at the molecular level leading to degradation and failure. The latter technique is commonly termed the "physics-of-failure" approach. The former of the two approaches incorporates statistically significant data obtained from well-planned and consistently executed environmental test evaluations to form a probabilistic model allowing extrapolation to other use conditions. Although the physics-of-failure approach is the most desirable because of its fundamental nature, models based on these theories are still at a research level in industry and are just beginning to be investigated in-depth. This section summarizes the evaluation of two existing predictive models for the time-to-failure of nonhermetic packages subjected to constant temperature and relative humidity. The major conclusions of that analysis are summarized here.

Two of the leading candidate models, one proposed by Peck and the other by Denson and Brusius, are essentially empirical in their formulation. To assess the relative consistency of the two models, their functional forms are considered and compared relative to a given set of data. Again, the drive behind these two models and others like them is the need to extrapolate beyond the range of available observations. The main application appears to be for the estimation of acceleration factors from laboratory test conditions, in which temperature and humidity are much higher than operational, or field, conditions. If the acceleration factors are reasonably accurate, tremendous savings in time and testing costs can be realized. If the extrapolation holds into the range of operating conditions, then the model could have great impact for long-term reliability estimations and the cost of reliability testing.

The underlying assumptions in the Peck model are that component TTFs are log-normal random variables and that their variances are equal and constant over a wide range of temperatures (°C) and relative humidities (% **rh**). With these assumptions, an estimated sample median TTF, t_m (°C, % **rh**) was obtained, Peck's model for this sample median is

$$t_m(°\mathrm{C}, \%\mathbf{rh}) = A(\%\mathbf{rh})^{-n} \exp\left[\frac{E_a}{k}(°\mathrm{C} + 273)^{-1}\right] \qquad [14\text{-}4]$$

where

k = Boltzmann's constant, 8.615×10^{-5} eV/K
A = a scaling constant
n, E_a = model parameters.

No doubt, much of the success of the model is due to the fact that a power law for relative humidity and an Arrhenius law for temperature seems quite reasonable.

Because 85°C/85% **rh** (85/85) is an industry standard test condition, Peck defined his observed ratio, R_0, between different tests conditions to be

$$R_0 = t_m(°\mathrm{C}, \%\mathbf{rh})t_m(85{,}85) \qquad [14\text{-}5]$$

R_0 was compared to a calculated ratio, R_c, derived from t_m (°C, % **rh**) in a similar manner to R_0. Specifically, R_c is found to be

$$R_c = \left(\frac{85}{\%\mathbf{rh}}\right)^n \exp\left\{\frac{E_a}{k}[(°\mathrm{C} + 273)^{-1} - 358\ \mathrm{K}^{-1}]\right\} \qquad [14\text{-}6]$$

If the graph of R_0 versus R_c is linear with slope l and intercept 0, then the model is an excellent representation of the experimental data. From a linear least-squares analysis, Peck has estimated the parameters to be $n = 3.0$ and $E_a = 0.90$ eV.

The data used by Peck for his model formulation are shown in Table 14-5. A rather obvious result is that the estimates for the model parameters are dependent on the data. Thus, if the failure mechanisms, materials,

Table 14-5. Published Data of Median TTF for Epoxy-Molded Packages of Unspecified Origin, used by Hallberg and Peck for Their Mathematical Model Formulated to Calculate Accelerated Factors Between Two Accelerated Environmental Tests: 85°C/85% rh and HAST

°C	% rh	Median Life (h)	°C	% rh	Median Life (h)	°C	% rh	Median Life (h)
85	85	3,060	85	85	6,000	150	81	12.4
110	85	500	138	85	359	130	81	85
130	85	90	138	75	450	85	85	4,300
160	85	12	138	50	622	120	85	420
85	85	1,050	127	85	407	130	85	130
111	85	151	85	85	900	140	85	60
120	85	90	110	90	165	85	85	12,000
130	85	50	121	90	104	110	85	2,700
85	85	1,600	131	90	52	127	85	504
111	85	170	85	85	3,360	138	85	300
130	85	66	121	90	239	85	85	3,400
85	85	5,400	131	90	87	110	85	340
110	85	440	85	85	5,300	127	85	130
127	85	160	110	90	700	198	85	54
130	85	220	121	90	240	127	85	140
138	85	70	131	90	85	138	65	270
85	85	1,700	85	85	2,400	127	75	177
130	85	430	85	65	440	85	85	15,000
85	85	7,500	60	85	2,000	145	85	195
130	85	220	111	75	50	85	85	180
85	85	8,000	120	75	30	60	85	1,300
110	85	1,280	130	75	18	110	85	50
130	85	320	111	100	26	125	85	18
85	85	380	120	100	15	85	85	1,000
130	68	66	130	100	8	110	85	180
85	85	245	85	85	210	130	85	50
110	85	57	110	90	30	85	85	6,425
130	85	23	60	90	1,800	110	90	725
85	85	8,300	85	85	15	85	85	210
110	85	2,000	85	20	1,600	60	85	1,300
130	85	300	20	85	7,000	111	85	54
85	85	730	20	58	1,500	121	85	290
110	85	112	85	85	3,920	85	85	5,000
130	85	70	138.5	85	80	110	85	500
85	85	2,500	157.3	85	28	127	85	170
125	85	225	85	85	370	138	85	80
150	85	62	85	70	660	85	85	5,345
85	85	1,440	85	60	1,160	151	85	40
130	81	32	70	81	1,940	133	85	170
150	81	15	55	81	7,390	109	85	1,300
150	50	115	85	85	2,860			
140	65	42	115	81	446			

processes, test methods, and so on differ substantially from those utilized and contained in Peck's data, then a much more extensive set of failure data for these systems must be obtained or the data must be regenerated and models developed for each specific package type. It is expected that the devices and materials and processes are not uniform throughout this data set. As with all statistical models, the more uniform and complete the set of data is, the better the model will be.

Another issue is fundamentally at the heart of the role of acceleration factors. Peck has chosen 85°C/85% **rh** as the reference condition in order to extrapolate to other accelerated test conditions. However, for normal operating conditions, the extrapolation must be from the accelerated reference condition to the decelerated conditions of the operating environment. For the latter, Peck's model does not fit as well, based on statistical criteria [5]. A final issue which may be somewhat theoretical is the choice of a statistical criterion for parameter estimation. Because a linear model is assumed for R_0 versus R_c, the criterion for a best fit could be based on mean square error, best slope, or best intercept. Each yields slightly different results. All things considered, Peck's model has utility for first-order approximations. Nevertheless, caution should be exercised for reliability estimation when accuracy and high levels of confidence are required.

The Denson and Brusius model was developed empirically in a similar fashion to Peck's model. The equation for the median life t_m (°C, % **rh**) for Denson and Brusius model is

$$t_m^{RAC}\ (°\mathrm{C},\ \%\mathbf{rh}) = A\ \exp\left[\frac{N}{\%\mathbf{rh}}\right]\ \exp\left[\frac{E_a}{k(°\mathrm{C} + 273)}\right] \qquad [14\text{-}7]$$

where A is a scaling constant, and N and E_a are model parameters. Note that the temperature dependence in Equation (14-7) is identical to that in Equation [14-4], i.e., both are an Arrhenius law for temperature. However, the relative humidity is considerably different. In Equation (14-7), % **rh** has an exponential dependence, but it is polynomial in Equation 14-4. The analogous equation to Equation 14-6 for the Denson and Brusius model is

$$R_c^{RAC}(°\mathrm{C},\ \%\mathbf{rh}) = \exp\left\{\left(\frac{E_a}{k}\right)\left[(°\mathrm{C} + 273)^{-1} - (358\ \mathrm{K})^{-1}\right]\right\} \qquad [14\text{-}8]$$
$$\exp\left[N\left(\%\mathbf{rh}^{-1} - 85^{-1}\right)\right]$$

and the estimated parameters are $N = 2.96$ and $E_a = 0.30$ eV. These estimates were obtained from data other than that used by Peck and are not readily available for further evaluation.

Figure 14-22 is a graph of the equivalent life relative to 1000 h at

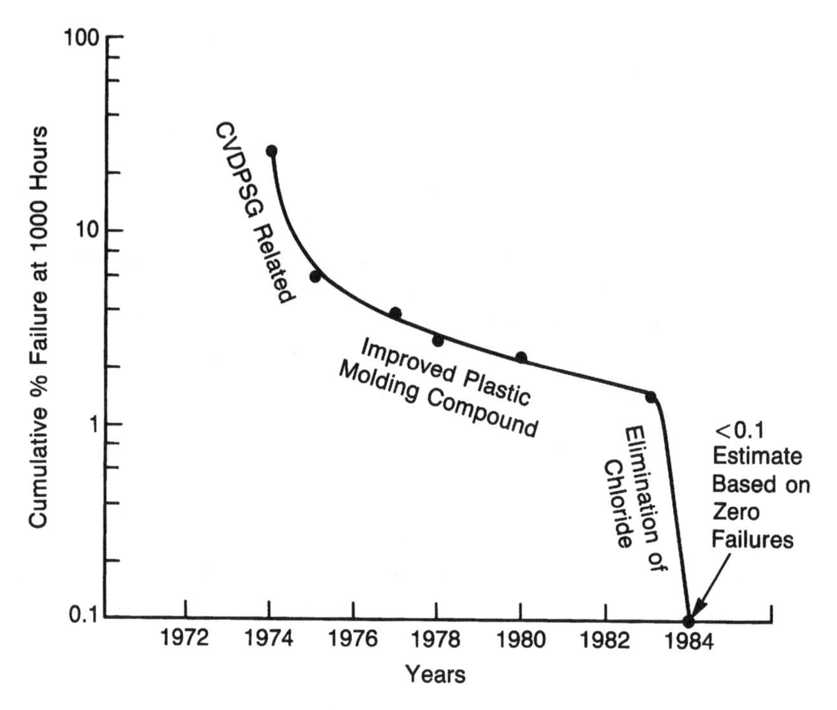

Figure 14-22. Correlation of the Observed Ratio (R_0) of Elevated Temperature–Humidity Test TTFs to 85°C/85% rh TTFs for Plastic Packages Using the Data in Table 14-5 Versus Calculated TTF (R_c) using Hallberg and Peck's Derived Mathematical Model. (From Ref. ??).

85°C/85% **rh** as a function of temperature for fixed relative humidity; that is, a graph of Equations 14-4 and 14-7 for A = 1000 h. The first observation is that there are five curves for Peck's model and only one for the Denson and Brusius model. This is directly attributable to the difference in the functional forms for the relative humidity in the two models. Assume that the range for % **rh** is (50, 100). Thus, 50% **rh** will maximize the effect of relative humidity over the given range, but the contribution of the % **rh** term to Equation 14-7 is only a factor of 1.025, which is graphically indistinguishable. For practical purposes, the Denson and Brusius model is independent of relative humidity; that is, it is simply a function of temperature.

Another striking observation is that the Denson and Brusius model and Peck's model yield similar results for only a very small interval of temperatures around 100°C. In fact, as the temperature drops toward room temperature, the difference in the models is about three orders of magnitude. Obviously, this would yield drastically different predictions in life. This effect is, no doubt, attributable to a value of E_a = 0.30 eV in the Denson and Brusius model versus $E_a = 0.90$ eV in Peck's model. It should be noted that when Peck's data are used to find the parameters in the Denson and Brusius model, the values are $N = 133.24$ and $E_a = 0.86$ eV. These values seem more reasonable. There should be some effect of relative humidity, and because both models have the identical Arrhenius factor, the parameter should be close.

Thus, there are two issues which must be addressed if either of these models are to be used for accurate reliability estimations. One issue is the sensitivity of the model to the data. Would Peck's model be similar to the Denson and Brusius model if the Denson and Brusius data were fit to Peck's model, or is there a major difference between the two? Are the components used in generating Peck's data sufficiently similar to the components and coatings of interest so as to be comparable? How much data are necessary to make a selection for the model? The other issue is the model development. Are there fundamental reasons for selecting one model over the other? What physical principles are applicable for either model? Until some of these questions are resolved, care must be taken when making reliability estimates based on these models.

Both models are primarily empirical and apparently have a great deal of application. However, the lack of sharpness in the estimates of the parameters diminishes their usefulness for estimating failure times beyond the range of observations. Until further information is available, they should be used only for approximate reliability estimations, not for reliability estimations when accuracy and high levels of confidence are required. Data accumulation and probability modeling efforts must continue before we can obtain accurate estimations outside the range of observations, especially from accelerated test conditions back to field conditions. Efforts toward understanding the physics-of-failure approach may provide the fundamental knowledge necessary for the applicable use of reliability models.

14.9 RECENT ADVANCES IN SEALING AND ENCAPSULATION

Although technology advances in hermetic packaging have not been a necessity and, therefore, have been fairly limited in recent years, the technology of using polymers for encapsulating bare devices has progressed from a "black art" to science. The greatest advancements have been made in the materials used for encapsulation. They are cleaner with respect to ionic contamination and possess far better physical and electrical properties. Although plastic-molded packages have been around for some time and their materials are fairly well developed, more demanding and novel packaging approaches such as chip-on-board (COB) and multichip modules (MCMs) require new areas of understanding and entirely different material requirements.

For overcoat applications such as COB and tape automated bonding (TAB), polymer manufacturers are actively pursuing materials that have the specific properties required of these applications. Coating materials for COB and TAB have very different requirements from plastic-molding compounds, not only because of the vastly different processing used for encapsulation but also from the standpoint of physical properties and performance requirements. In general, a partial "wish list" for COB or TAB applications would entail a material that has the following properties:

1. Is easy to dispense with existing robot-syringe equipment
2. Is compatible with the remaining assembly processes
3. Does not require heating of the material during dispensing
4. Possesses an acceptable combination of viscosity and thixotropic properties for the application
5. Has sufficient adhesion to the constituents that are to be protected
6. Has a consistent batch-to-batch ionic content of less than 5 parts per million each of Cl^-, K^+, and Na^+
7. Requires a short duration cure with a maximum temperature of 150°C
8. Has a T_g of 150°C or greater
9. Possesses a residual stress after curing of less than 5 MPa
10. Does not cause stress related failures of the assembly during the temperature cycling required for qualification
11. Provides sufficient mechanical protection to the device and interconnect so additional packaging is not required
12. Is able to stand up to hot-wet-electrically biased environmental tests
13. Maintains acceptable electrical properties in hot and humid environments
14. Exhibits less than 1% linear shrinkage during cure
15. Is nonvolatile and nontoxic
16. Has at least an 8-h pot life and at least a 6-month shelf life
17. Is comparatively low in cost

Granted, many of these properties are application-specific and, therefore, require materials that are also application-specific, but a single-dispense material does not exist today that fits the above criteria for commercial versions of COB, and certainly not for the more advanced MCMs.

With the numerous materials on the market, each application requires different considerations when it comes to decisions on how best to provide environmental protection. Vast ranges of physical properties are available to the packaging engineer. Silicones, in general, possess extremely high coefficients of thermal expansion (TCEs) and an extremely low moduli, whereas epoxies are the opposite in that they have fairly low TCEs and relatively high moduli. For example, a well-known silicone gel on the market has a TCE of well over 325–800 ppm/°C and a modulus of less than 0.001 GPa, and one of the better epoxies for COB has a TCE of less than 22 ppm/°C and a modulus of 10.5 GPa. Also, silicone gels are very clean with respect to Cl^-, K^+, and Na^+. This specific silicone gel is reported to have 2 ppm of each. Epoxies, by the nature of their manufacture, are much higher in ionic content and this epoxy is one of the best with 20 ppm of each. The two materials in this example are at opposite ends of the spectrum with respect to their physical properties, yet both can be used to coat a wirebonded COB device and provide thermomechanical stress resistance and corrosion resistance. The silicone gel needs additional packaging in most cases because it does not provide mechanical protection with its extremely low modulus, and the epoxy will not survive nearly as well as the silicone gel in corrosion tests such as biased 85°C/85% **rh** or biased HAST due to its ionic content and adhesion characteristics [66–76]. Continuing improvements in epoxy resins have resulted in epoxies that have stronger polymer matrices and are better able to withstand the difficult-to-pass hot, wet environments common in accelerated testing. These improved epoxies would appear to be more apt to provide a higher degree of reliability in testing and the field, but they are at a developmental stage today.

New materials have been formulated in an attempt to incorporate the best of both worlds. There are recent resins that combine the physical properties of two materials to produce a composite resin like the silicone-modified epoxies and polyimide siloxanes. If you combine a silicone and an epoxy into one formulation, you obtain a mix of their respective physical properties that may in some cases provide better overall performance. The same is true for the polyimide siloxanes. Material dispense characteristics have been improved to facilitate more consistent dispensing and are formulated with a specific application in mind. TAB is beginning to see more use in commercial electronic assemblies, certainly in consumer-destined electronics coming out of Japan, and normally incorporate an encapsulation coating on the surface of the die and inner lead bond area. TAB encapsulation materials are required to have different properties than those for COB in that smaller particles and higher flow are needed to fully encapsulate around the inner lead bonds. Whereas, for COB, overcoat materials need to have viscosity and thixotropic properties that help hold the desired shape of the coating over the device and interconnect through cure.

Cleaning of the assembly prior to the coating application has been identified as a critical process step in obtaining long-term reliability. As

CFCs are phased out, new cleaning surfactants and processes will be available to the packaging engineer. Cleaning must remove not only ionic species but organic contaminants as well. What removes ionic contaminants, such as a wet cleaning process, does not always remove organic impurities. Surfactant cleaners may be good at eliminating ionic species, but they can leave a residue themselves that may require another cleaning process to remove. When cleaning assemblies such as COB and MCMs, consideration must be given to the effects of the cleaning process on all of the constituents that make up the package. Evaluation methods for determining the effectiveness of a cleaning process on all of the surfaces present is still developing. Some commercial equipment exists that qualitatively measures what contaminants remain that are dissolvable and removable from a device or assembly by gauging the change of resistivity in a rinsing fluid, usually an alcohol. Other means can be used to evaluate cleanliness on specific surfaces by water contact angle measurements. In any case, cleaning is not something to be overlooked or taken lightly.

Silicon dioxides, nitrides, and various polymers can be deposited via chemical vapor deposition. Some research is going on in this area, and a few manufacturers of chemical vapor-deposition equipment are attempting to specialize in producing coatings for assemblies such as COB, as either a first-layer coating to be followed by an overcoat, or as the only coating if housed in a nonhermetic plastic or metal box. The deposition processes, resulting material characteristics, and associated equipment are still in an early stage of development and appear to be immature at present, although not altogether unpromising. Because the deposition requires a vacuum, coating would have to be performed in batches. These coatings may only be feasible for high-end electronics if they are to be cost-effective.

With all of the aforementioned advances in polymer materials technology, a rapid increase in the area of reliability performance has been realized. With this has come the increased difficulty of designing reliability tests that provide the necessary acceleration factors for reasonable test times in the manufacturing environment. Electrically biased 85°C/85% **rh** (85/85) has been the standard test in the plastic-molded package industry for many years. With new molding compounds, 85/85 tests can last several thousand hours before failures begin to appear. These long test times are not conducive to the fast pace of today's electronics design and production floor. The Highly Accelerated Stress Test, better known as HAST, is quickly becoming an accepted alternative to 85/85. HAST utilizes a pressurized chamber to allow the temperature to be elevated while maintaining humidity control, to produce the additional activation energy for more accelerated testing. Electrical bias in 85/85 or HAST provides additional activation energy so that corrosion failures and mechanisms can be identified in a shorter time. Alternative environmental tests such as Battelle's

Flowing Mixed Gas test (FMG), introduce parts-per-million or parts-per-billion quantities of contaminants commonly found in the atmosphere to simulate corrosive processes found in some electronics applications. FMG was designed to test connector technologies and has not yet been used with any regularity in the testing of nonhermetic packages due to the relatively long test times required. Correlation of 85/85, or HAST to field life is still an issue of considerable discussion, as well as the acceleration factors between 85/85 and HAST. A great deal of ongoing work in the correlation and testing area is needed to be able to accurately estimate reliability in the field from models based on these accelerated tests.

Other work has been done where environmental testing is performed in a sequential series of steps. Temperature cycling has historically been used in testing plastic packages as a single-step test to evaluate the thermal-mechanical stresses arising from mismatches of TCEs. The 85/85, HAST, or similar corrosion/polymer degradation tests are also usually run as separate evaluations. Recent studies have been performed using temperature cycling and salt atmosphere as preconditioning tests for 85/85 or HAST. The temperature cycling will induce cracks and delaminations of the polymer matrix and the polymer-to-substrate interface. Next in the sequence, salt atmosphere adds an ionic contaminant (NaCl) to the newly created cracks and voids in the package. Finally, placing the parts in either biased 85/85 or biased HAST to accelerate the corrosion mechanism that is now in place with moisture, NaCl, and electrical potentials. Even with a test sequence this severe, the biased HAST portion of the test sequence has run in excess of 1000 h without significant performance degradation of the wirebonded and coated test chip. With the desire to run ultraclean 85/85 and HAST chambers for the purposes of repeatability and control, among other things, a sequence such as this does have its dangers and unknowns.

On the device side of the test spectrum, the fabrication of test chips that are designed to amplify the failure mechanisms that have been observed by the polymer packaging industry has been greatly advanced in recent years. Sandia National Labs, for one, has taken the sophistication of test chips to new levels. Sandia's family of assembly test chips, designated ATCs, have incorporated assorted detectors and sensors that are of primary interest to those in the polymer encapsulation industry. The ATC family now has five versions in production. The ATC01 is the first-generation test chip designed for studying aluminum corrosion. The ATC01 incorporates eight different triple tracks, three corrosion ladders, and aluminum-sheet resistance, in addition to other corrosion detection structures. The ATC04 is the most recent and advanced test chip in the family, with passivated and unpassivated triple tracks in both single- and double-level metal, some triple tracks have underlying polysilicon heaters for heat dissipation studies, diode thermometers, and 48 addressable piezoresistive stress sensors. Both the ATC01 and ATC04 have features sizes down to 1.25 μm/mm line and space widths. These chips have been made available to industry in an effort to provide an industrywide standard test chip family for comparative studies and data correlation.

The unquestioned superior reliability of hermetic packages will remain a factor in their continued use for military and critical civilian applications despite their higher cost, size, and weight. The predicted increases in device densities will influence failure susceptibilities in several ways. The increased device density will mean increased power densities; hence, generally higher operating temperatures, larger voltage gradients, and less conductor material means there is less material loss required for opens or shorts. Also, the packing of more devices on the same chip, in concert with physically increasing the sizes of chips, greatly increases the probability of chip-level failure. The cumulative effect of these factors serves to increase the susceptibility of failure by an order of magnitude for each doubling in circuit density. Certainly, the group of technologies that form COB and MCM assemblies will continue to evolve, and with it so will polymer encapsulation. These considerations, in combination with the continuing advances in materials and processes for polymer packaging, will constantly redefine the limits of use for nonhermetic versus hermetic packaging.

14.10 FUTURE DEVELOPMENTS

The rapid development of IC technology has created a critical need for packaging sealing and encapsulation using advanced polymeric materials for encapsulants, interlayer dielectrics, and passivation layers. Recent advances in high-performance polymeric materials, such as improved silicone elastomers, ultrasoft silicone gels, low-stress epoxies, low thermal-expansion coefficient and photodefinable polyimides and cyclobutenes have begun to provide polymers to meet reliability requirements of VLSI technology. But to date they have not met all of the requirements, particularly those of defense industry needs. It is becoming increasing clear that this will happen during the next decade or two. The advances in high-density multichip module packaging will undoubtedly require nonhermetic packaging for the ever-increasing large modules. However, the demands for improved properties of materials in the areas of low dielectric constant, high breakdown voltage strength, high sheet resistance, and less dielectric change with humidity will continue to require the development of high-performance polymeric materials. In addition, given the low-cost requirements of all electronic packages, all these requirements need to be accomplished at low cost. Their application for on-chip interconnections, and wafer-scale integration structures with very fast interconnecting networks as well as for high-performance packaging will become apparent. It is a challenge that the collaborative efforts between polymer chemists, material scientists device engineers, and packaging engineers will face in the near future.

14.11 REFERENCES

1. C. P. Wong (ed.). *Polymers for Electronic and Photonic Applications,* Academic Press, San Diego, CA, 1993, and references therein.
2. C. P. Wong. "Overview of IC Device Encapsulants as Device Packaging," *J. Electron. Packaging,* 3: p. 97, 1989.
3. R. W. Thomas. "Moisture, Myths and Microcircuits," *IEEE Trans. Parts, Hybrids Packaging,* PHP-12(3): pp. 167–171, 1976.
4. J. L. Jellison. "Susceptibility of Micro Welds in Hybrids to Corrosion. 'Degradation,' " *13th Annual Proceedings, Reliability Physics Symposium,* pp. 70–79, 1975.
5. H. Koelmans. "Metallization Corrosion in Silicon Devices by Moisture-Induced Electrolysis," *12th Annual Reliability Physics Symposium,* pp. 168–171, 1974.
6. F. R. Neighbor and B. R. White. "Factors Governing Aluminum Interconnection Corrosion in Plastic Encapsulated Microelectronic Devices," *Microelectron. Reliab.,* 16: pp. 161–164, 1977.
7. W. M. Paulson and R. P. Kirk. "The Effects of Phosphorous-Doped Passivating Glass on the Corrosion of Aluminum Metallization," *12th Annual Proceedings, Reliability Physics Symposium,* pp. 172–179, 1974.
8. G. DiGiacomo. "Phosphorous Migration Kinetics from PSG to Glass Passivation Surface," *22nd Annual Proceedings, Reliability Physics Symposium,* pp. 223–228, 1984.
9. A. Shumka and R. R. Piety. "Migrated-Gold Resistive Shorts in Microcircuits," *13th Annual Proceedings, Reliability Physics Symposium,* pp. 93–98, 1975.
10. A. Christou. "Moisture Diffusion Through Hybrid Circuit Encapsulants," *Electron. Packaging Production,* 19(4): pp. 91–99, 1979.
11. R. W. Thomas. "Moisture in Microcircuits," in *Reliability Technology for Cardiac Pacemakers,* ed. by H. A. Schafft, *NBS Special Publication,* 400-9, U.S. GPO, Washington, DC, 1976.
12. M. Byrnes, J. L. Carter, J. Sergent, and D. King. "Considerations in the Hermetic Packaging of Hybrid Microcircuits," *Solid State Technol.,* 27(8): pp. 68–77, 1984.
13. B. Yan, S. I. Meilinle, G. W. Warren, and P. Wymablatt, "Water Absorption and Surface Conductivity Measurement on α-Aluminum Substrates," *IEEE Trans. Components Hybrids Manuf. Technol.,* CHMT-10, No.(2): pp. 247–251, 1987.
14. B. Reich. "Acceleration Factors for Plastic Encapsulated Semiconductor Devices and Their Relation to Field Performance," *Microelectron. Reliab.,* 14(1): pp. 63–66, 1975.
15. F. N. Sinnadurai. "Accelerated Aging of Semiconductor Devices in Environments of High Vapor Pressure of Water," *Microelectron. Reliab.,* 13(1): pp. 23–27, 1974.
16. R. A. Lawson. "Accelerated Testing of Plastic Encapsulated Semiconductor Components," *12th Annual Proceedings, Reliability Physics Symposium,* pp. 243–249, 1974.
17. W. E. Swartz, J. H. Linn, J. M. Ammons, M. Kovac, and K. Wilson. "The Adsorption of Water on Metallic Surfaces," *21st Annual Proceedings, Reliability Physics Symposium,* pp. 52–59, 1983.
18. D. R. Carley, R. W. Nearhoff, and R. Denning. "Moisture Control in Hermetic Leadless Chip Carriers with Silver-Epoxy Die-Attach Adhesive," *RCA Rev.,* 34(2): pp. 278–290, 1984.
19. J. G. Davy. "Calculation of Leak Rates of Hermetic Packages," *IEEE Trans. Parts, Hybrids Packaging,* PHP-11(3): pp. 177–189, 1975.

20. D. Stroehle. "On the Penetration of Water Vapor into Packages with Cavities and on Maximum Allowable Leak Rates," *15th Annual Proceedings, Reliability Physics Symposium,* pp. 101–106, 1977.
21. R. E. Suloff. "A Study of Leak Rate Versus Reliability of Hybrid Packages," *International Microelectronics Symposium,* pp. 121–124, 1978.
22. A. DerMarderosian and V. Ginot. "Water Vapor Penetration into Enclosures with Various Air Leak Rates," *16th Annual Proceedings, Reliability Physics Symposium,* pp. 179–183, 1978.
23. R. K. Traeger. "Hermeticity of Polymetric Lid Sealants," *Proceedings 25th Electronics Components Conference,* pp. 361–367, 1976.
24. D. Nixen. "Package Reliability as Affected by Materials and Processes," *Semicond. Int.*, 5(4): pp. 175–182, 1982.
25. D.W.A. Forbes. "Solder Glass Seals in Semiconductor Packages," *Glass Technol.*, 8(2): pp. 32–42, 1972.
26. H. H. Manko. *Solders and Soldering,* McGraw-Hill, New York, 1964.
27. K. S. Dogra. "A Bismuth-Tin Alloy for Hermetic Seals," *Proceedings IEPS 4th International Conference,* pp. 631–636, 1984.
28. M. D. Zimmer. "Reliability and Thermal Impedance Studies in Soft Soldered Power Transistors," *IEEE Trans. Electron. Devices,* ED-23(8): pp. 843–850, 1976.
29. C. J. Thawaites. "Some Aspects of Soldering Gold Surfaces," *Electroplat. Metal Finishing,* 29(9): pp. 21–26, 1973.
30. G. M. Stoll. "Hermetic Sealing of Thermally Sensitive Circuits with Emphasis on Thermal Conductivity Sealing," *Proceedings International Microelectronics Symposium,* pp. 520–526, 1983.
31. J. M. Morabito. "Recent Advances in Solder Bond Technology," *Thin Solid Films,* 72: pp. 433–442, 1980.
32. J. Brady and M. Courtney. "Hermetic Tin/Lead Solder Sealing for the Air-Cooled IBM 4381 Module," *Paper presented at the IEEE, ICCD Conf. at Portchester, NY,* November 1983.
33. W.M.S. Yang. "Reflow Solder Ceramic Lids for Hermetic Packages," *Solid State Technol.*, 27(12): pp. 137–143, 1984.
34. F. W. Luce. "Delidding and Resealing of Microelectronic Packages," *MICOM Report MM & T Project,* R793438, 1982.
35. G. Bourdelais and E. F. Hill. "Hermeticity Considerations for VISI Packaging," *Proceedings 6th Annual International Electronics Packaging Conference,* pp. 18–24, 1986.
36. D. D. Zimmerman. "A New Gold-Tin Alloy Composition for Hermetic Package: Sealing and Attachment of Hybrid Parts," *Solid State Technol,* 15(1): pp. 44–46, 1972.
37. N. R. Stockham and C. J. Dawes. "Resistance Seam and Laser Welding Hybrid Circuits," *Hybrid Circuits,* 1(9): pp. 509–519, 1983.
38. S. Norrman and P. A. Torstensson. "Hermetic Sealing of Kovar Hybrid Packages by Laser Welding," *Hybrid Circuits,* 1(8): pp. 21–23, 1985.
39. A. K. Varshneya. "Stress in Glass-to-Metal Seals," in *Treatise Materials Science and Technology,* ed. by M. Tomazawa and R. H. Doremus, vol. 22, pp. 241–306, 1982.
40. J. A. Pask and R. M. Fulrath. "Fundamentals of Glass-to-Metal Bonding; VIII. Nature of Wetting and Adherence," *J. Am. Ceram. Soc.*, 45(12): pp. 592–596, 1962.
41. M. P. Borom and J. A. Pask. "Role of Adherance Oxides in the Development of Chemical Bonding at Glass-to-Metal Interfaces," *J. Am. Ceram. Soc.*, 49(1): pp. 1–6, 1966.
42. R. G. Buckley. "Understanding Glass-to-Metal Seals," *Electron. Packaging Production,* 18(10): pp. 74–79, 1978.

43. A. E. Yaniv, D. Katz, J. E. Klein, and J. Sharon. "A New Technique of Surface Preparation for Bonding Glass to Kovar," *Glass Technol.*, 22(5): pp. 231–235, 1981.
44. W. F. Yext, B. J. Shook, W. S. Katzenberger, and R. C. Michalek. "Improving Glass-to-Metal Sealing Through Furnace Atmosphere Composition Control," *IEEE Trans. Components Hybrids Manuf. Technol.* CHMT-6(4): pp. 455–459, 1983.
45. M. B. Miller. "Stress Absorbing Coatings for Glass-to-Metal Seals in Microelectronic Packages," *SAMPE,* pp. 19–21, 1983.
46. K. Kokini and R. W. Perkins. "Estimating the Strength of Annular Glass-to-Metal Seals in Microelectronic Packages: An Experimental Study," *IEEE Trans. Components Hybrids Manuf. Technol.*, CHMT-7(3): pp. 276–279, 1984.
47. R. Shukla and N. Mencinger. "Moisture Evolution Analysis of Sealing Glasses used in Integrated Circuit Packages," *Electron. Mater.*, 14(4): pp. 461–471, 1985.
48. A. H. Kumar and R. R. Tummala. "Titania-Doped Lead-Zinc-Borate," *Bull. Am. Ceram. Soc.*, 57(8): pp. 738–739, 1977.
49. R. W. Vasofsky and R. K. Lowry. "Moisture Evolution from Sealing Glasses Dry CERDIP Packages," *18th Annual Proceedings, Reliability Physics Symposium,* pp. 1–9, 1980.
50. R. K. Lowry, C. J. Van Leeuwen, B. L. Kennimer, and L. A. Miller. "A Reliable, Dry Ceramic Dual-In-Line Package CERDIP," *16th Annual Proceedings, Reliability Physics Symposium,* pp. 207–212, 1978.
51. D. E. Erickson. "Hybrid Circuit Sealing—Problem Prevention Clinic," *Electron. Packaging Production,* 22(11): pp. 133–137, 1982.
52. T. H. Ramsey. "Critical Parameters in Glass Sealed Ceramic Packages," *Solid State Technol.*, 17(9): pp. 51–59, 1974.
53. P. S. Burggraff. "Semiconductor Package Sealing," *Semicond. Int.*, 2(7): pp. 35–49, 1979.
54. W. K. Denby. "Current Trends in Hermetic Package Sealing Equipment," *Microelectron. Manf. Testing,* 4(10): pp. 14–19, 1981.
55. J. Wallach. "Surface Requirements for Seals," *Proceedings International Conference on Surface Technology,* pp. 606–620, 1973.
56. J. G. Davy. "Hermetic Packaging Protects Circuitry from Moisture," *Electron. Packaging Production,* 24(10): pp. 58–62, 1984.
57. A. J. Blodgett and D. R. Barbour. "Thermal Conduction Module: A High-Performance Multilayer Ceramic Package," *IBM J. Res. Devel.*, 26(1): pp. 30–36, 1982.
58. K. A. Jewett. "Practical Methods for Hermetic Seal Testing," in *Electronic Packaging: Materials and Processes,* ed. by J. A. Sartell, pp. 85–88, ASM, Metals Park, OH, 1986.
59. H. A. Schafft. *Hermeticity Testing for Integrated Circuits, NBS Special Publication,* 400-9, p. 43, U.S. GPO, Washington, DC, 1974.
60. R. W. Thomas. "Microcircuit Package Gas Analysis," *14th Annual Proceedings, Reliability Physics Symposium,* pp. 283–294, 1976.
61. T. W. Carr, E. A. Corl, C. L. Lin, and C. G. Majtenji. "Quantitative H_2O Determination in Components Using Plasma Chromatograph-Mass Spectrometer," *16th Annual Proceedings, Reliability Physics Symposium,* pp. 59–63, 1978.
62. M. G. Kovac, D. Chleck, and P. Goodman. "A New Moisture Sensor for In-Site Monitoring of Sealed Packages," *15th Annual Proceedings, Reliability Physics Symposium,* pp. 85–91, 1977.
63. J. Finn and V. Fong. *Recent Advances in A1203 In-Situ Moisture Monitoring Chips for CERDIP Package Application,"* *IEEE,* New York, 1980.
64. N. Bakker. "In-Line Measurement of Moisture in Sealed IC Packages," *Phillips Telecommun. Rev.*, 37(1): pp. 11–19, 1979.

65. R. P. Merret, S. P. Sim, and J. P. Bryant. "A Simple Method of Using the Die of Integrated Circuit to Measure the Relative Humidity Inside Its Encapsulation," *18th Annual Proceeding, Reliability Physics Symposium,* pp. 17–25, 1980.

66. D. Chang, P. Crowford, J. Fulton, R. McBribe, M. Schmidt, R. Sinitiski, and C. P. Wong. "An Overview and Evaluation of Anisotropically Conductive Adhesive Films for Fine Pitch Electronic Assembly," *IEEE Trans. Components Hybrids Manuf. Technol.*, CHMT 16(8): p. 828, 1993.

67. D. Chang, J. Fulton, H. Ling, M. Schmidt, R. Sinitski, and C. P. Wong. "Accelerated Life Test of Z-Axis Conductive Adhesives," *IEEE Trans. Components Hybrids Manuf. Technol.*, CHMP 18(8): p. 836, 1993.

68. C. P. Wong and R. Mcbride. "Robust Titanate-modified Encapsulants for High Voltage Potting Application of Multichip Module/Hybrid IC," *IEEE Trans. Components Hybrids Manuf. Technol.*, CHMT 16(8): p. 868, 1993.

69. C. P. Wong and P. D'Ambra. "Embedment of Electronic Devices," in Kirk–Othmer *Encyclopedia of Chemical Technology,* 4th ed., vol. 9, pp. 377–393, John Wiley & Sons, New York, 1994.

70. C. P. Wong, J. M. Segelken, and C. N. Robinson. "Encapsulation of Chip-on-Board", in *Handbook of Wire Bonding, Tape Automated Bonding, and Flip-Chip on Board for Multichip Module Applications,* ed. by J. Lau, pp. 470–503, Van Nostrand Reinhold, New York, 1994.

71. C. P. Wong. "Recent Advances in Hermetic Equivalent IC Packaging of Microelectronics", *Research Society, Advanced Electroceramics and Packaging Technology, Proceedings of the 1994 International Conference on Electronic Materials,* p. 78, 1994.

72. C. P. Wong and R. McBride. "Preencapsulation Cleaning Method and Control for Microelectronics Packaging," *IEEE Trans. Components Hybrids Manuf. Technol.*, CPMT 17(4): p. 542, 1994.

73. C. P. Wong. "Overview of Microelectronic Packaging: Materials and Processes," *Proceedings of Polymat* '94, p. 369, 1994.

74. C. P. Wong. "Materials for Electronic Packaging," in *Materials for Electronic Packaging,* ed. by D. Chung, p. 270, Butterworth Publisher, Boston, 1995.

75. C. P. Wong. "Recent Advances in Low Cost Plastic Packaging of Flip-chip Multichip Module of Microelectronics," *Proceedings on the 1994 International Conference on Electronic Materials,* p. 73, 1995.

76. C. P. Wong. "Thermal Mechanical Behavior of High Performance Encapsulants in Microelectronic Packaging", *IEEE Trans. Components Packaging Manuf. Technol.*, CPMT-18: p. 270, 1995.

GLOSSARY AND SYMBOLS

The *ISHM Glossary of Hybrid-Circuit Terms (Updated edition—Summer 1987)* by G.S. Szekely provided the foundation for this glossary. Definitions have in some cases been modified to fit this work. Other terms used by the authors and the packaging profession have been added and some terms in the original work have not been deleted. Material chemical symbols are found at the end of this glossary.

A

ACCELERATED STRESS TEST. A test conducted at a stress, e.g., chemical or physical, higher than that encountered in normal operation, for the purpose of producing a measurable effect, such as a fatigue failure, in a shorter time than experienced at operating stresses.

ACCELERATOR. An organic compound which is added to an epoxy resin to shorten the cure time.

ACTIVE COMPONENTS. Electronic components, such as transistors, diodes, electron tubes, thyristors, etc., which can operate on an applied electrical signal so as to change its basic characteristics, i.e., rectification, amplification, switching, etc.

ADDITIVE PLATING. Processing a hybrid circuit substrate by sequentially plating conductive, resistive, and insulative materials, each through a mask, thus defining the areas of traces, pads, and elements.

ADVANCED STATISTICAL ANALYSIS PROGRAM (ASTAP). The "Advanced Statistical Analysis Program" is the IBM circuit analysis simulation program. It performs DC, time domain, and frequency domain simulations. Statistics can be applied to all simulations to predict operating tolerances. Among its many features is a transmission line analysis program. Also see SPICE.

ALLOY. A solid-state solution or compound formation of two or more metals. Alternatively, a combination of metals resulting in a phase or phases containing some of each constituent.

ALPHA PARTICLE. Decay product of some radioactive isotopes. It is a high-energy (mv range) helium nucleus capable of generating electron/hole pairs in microelectronic devices and switching cells, causing soft errors in some devices.

APPLICATION SPECIFIC INTEGRATED CIRCUIT (ASIC). Application Specific Integrated Circuit is an integrated circuit chip with personalization customized for a specific product. Personalization refers to wiring on the integrated circuit chip.

AREA ARRAY TAB. Tape automated bonding where edge-located pads and additional pads on the inner surface area of a chip are addressed in the bonding scheme. This is practiced with extremely complex dice, VLSI etc. Also for use with ICs where peripheral pad pitch cannot be further reduced and all I/Os must be accommodated.

ARRAY. A group of elements (pads, pins) or circuits arranged in rows and columns on one substrate.

ASPECT RATIO. The ratio of the length of hole to the diameter of hole in a board.

ASSEMBLY. A hybrid circuit which includes discrete or integrated components that have been attached to the next level of package, usually a card.

ASSEMBLY/REWORK. Terms denoting joining and replacement processes of microelectronic components. Assembly refers to the initial attachment of device and interconnections to the package. Rework refers to the removal of a device including interconnections, preparation of the joining site for a new device, and rejoining of the new device. Rework is necessary for either repair or engineering change.

B

BACKBONDING. Bonding active chips to the substrate using the back of the chip, leaving the face, with its circuitry face up. The opposite of backbonding is face down bonding.

BACK-END-OF-THE-LINE (BEOL). That portion of the integrated circuit fabrication where the active components (transistors, resistors, etc.) are interconnected with wiring on the wafer. It includes contacts, insulator, metal levels, and bonding sites for chip-to-package connections. Dicing the wafer into individual integrated circuit chips is also a BEOL process. The front-end-of-the-line (FEOL) denotes the first portion of the fabrication where the individual devices (transistors, resistors, etc.) are patterned in the semiconductor.

BACK PANEL. A planar package component holding plugged-in lower-level package components (e.g., cards) as well as discrete wires and cables interconnecting these components.

BACKSIDE METALLURGY (BSM). A metallization pad electrically connected to internal conductors within a multilayered ceramic package, to which pins are brazed.

BALL GRID ARRAY (BGA). A Ball Grid Array is an area array of solder balls joined to a SCM or MCM and used to electrically and physically connect the package to the next level of package, usually a printed circuit board.

BALL LIMITING METALLURGY (BLM). The solder wettable terminal metallurgy which defines the size and area of a soldered connection, such as C4 and a chip. The BLM limits the flow of the solder ball to the desired area, and provides adhesion and contact to the chip wiring.

BANDWIDTH. The maximum pulse rate or frequency that can reliably propagate through a transmission line. For a data bus, bandwidth is commonly used to describe the maximum data rate which is the single line pulse rate multiplied by the number of parallel bus bit lines.

BIFET. The combination of bipolar and FET transistors integrated together on the same piece of silicon for enhanced performance and cost.

BINDER. Materials (organic or inorganic) added to thick-film compositions and to unfired substrate materials to give sufficient strength temporarily for prefire handling.

BIPOLAR TRANSISTOR. Original transistor design in which two semiconductor junctions (regions of opposite polarity doping) are separated by a narrow region, called the base. Minority carriers are injected in the base from the emitter across the base-emitter junction, travel through the base, and are attracted to the collector through the base-collector junction. These transistors consume more power than Field-Effect Transistors (FET), but also achieve higher performance.

BLOCK COPOLYMER. A copolymer compound resulting from the chemical reaction between *n* number of molecules, which are a block of one monomer, and “n” number of molecules, which are a block of another monomer. Example: stearine (rigid) with silicone (elastic).

BOARD. This package element can best be defined as an organic printed-circuit card or board on which smaller cards or modules can be mounted. Its connections to the next higher level involve discrete wire or cables.

BOILING. Phase change and formation of bubbles in a superheated liquid.

BONDABILITY. Those surface characteristics and conditions of cleanliness of a bonding area which must exist in order to provide a capability for successfully bonding an interconnection material by one of several methods, such as ultrasonic or thermocompression wire bonding.

BRAZE. A joint formed between two different materials by formation of liquid at the interface.

BRAZING. Joining of metals by melting a non-ferrous, filler brazing metal, such as eutectic gold-tin alloy, having a melting point lower than that of the base metals. Also known as hard soldering.

BTAB. The acronym for tape automated bonding when the raised bump for each bond site is prepared on the tape material as opposed to the bump being on the chip.

BUMPED TAPE. A tape for the TAB process where the inner-lead bond sites have been formed into raised metal bumps on the tape rather than on the chip. This ensures mechanical and electrical separation between inner lead bonds and the non-pad areas of the chip (die) being bonded.

BURN-IN. The process of electrically stressing a device (usually at an elevated temperature and voltage environment) for an adequate period of time to cause failure of marginal devices.

BURN-OFF. Removal of unwanted materials—typically organics from green-sheets or organic contamination from substrates.

C

CAMBER. A term that describes the amount of overall warpage present in a substrate.

CAPACITANCE. The electrostatic element that stores charge. In packaging systems, it is used in lumped equivalent circuits to represent part of a line discontinuity. It is also used in a distributed system to represent the electrostatic storage property of a transmission line. Because it delivers current in response to a change in voltage, another use is to filter powering systems.

CARD. A printed-circuit panel (usually multilayer) that provides the interconnection and power distribution to the electronics on the panel, and provides interconnect capability to the next level package. It is also known as a daughter board. It plugs into a mother printed-circuit board.

CARD-ON-BOARD. Packaging technology in which multiple printed-circuit panels (cards) are connected to printed-circuit panel (board) at 90° angles.

CELL DESIGN, STANDARD. A semicustom product implemented from a fully diffused or ion implanted semiconductor wafer carrying horizontal rows of primary cells, interlaced with wiring channels (bays). Vertical wiring is supplied by additional processed layers which may use the cell areas or lie in channels on an overhead layer. Channel widths may vary to suit particular chip logic, so that chip sizes are not fixed for all products of a family.

CENTRAL PROCESSOR (CP). Computer processor responsible for fetching, interpreting, and executing program instructions. Also called Processor Unit (PU) and Central Processing Unit (CPU).

CERAMIC. Inorganic, nonmetallic material, such as alumina, beryllia, or glass-ceramic, whose final characteristics are produced by subjection to high temperatures. Often used in forming ceramic-substrates for packaging semiconductor chips.

CERAMIC BALL GRID ARRAY (CBGA). A ceramic package using ball grid array technology. See ball gird array technology.

CERAMIC COLUMN GRID ARRAY (CCGA). A ceramic package using ball grid array technology. See ceramic gird array technology.

CERAMIC DUAL-IN-LINE PACKAGE (DIP). Dual-in-line package in ceramic. See Dual-in-line Package.

CERAMIC QUAD FLAT PACK (CQFP). Quad Flat Pack in ceramic. See Quad Flat Pack.

CERMET. A solid homogeneous material usually consisting of a finely divided admixture of a metal and ceramic in intimate contact. Cermet thin films are normally combinations of dielectric materials and metals.

CHANNELS. Provide communications paths for input to and output from the computer system.

CHARACTERISTIC IMPEDANCE (Z_0). The voltage-to-current ratio of an electric signal propagating through an infinitely long transmission line. If L denotes the inductance per unit length and C denotes the capacitance per unit length, then $Z_0 = (L/C)^{0.5}$.

CHEMICAL VAPOR DEPOSITION (CVD). Depositing circuit elements on a substrate by chemical reduction of vapor of volatile chemical in contact with the substrate.

CHEMORHEOLOGY. The study of the processability or flow (rheology) and the chemistry of the polymer system. Processability parameters include, for instance, heating rates, hold temperatures, injection speeds, and compaction pressures. The chemical aspect, on the other hand, involves the rate of reaction, the mechanisms, the kinetics, and the cessation of the chemical reaction at the end of the polymerization.

CHIP. The uncased and normally leadless form of an electronic component part, either passive or active, discrete or integrated. Also referred to as a die.

CHIP CARRIER. A special type of enclosure or package to house a semiconductor device. It has electrical terminations around its perimeter, or solderpads on its underside, rather than an extended lead frame or plug-in pins.

CHIP DESIGN, DEPOPULATED. A gate array or standard cell array chip in which the wiring capacity (and hence chip area) is deliberately chosen to make automatic wiring possible only for those chips having some amount less than their maximum possible logic cell occupancy. This increases wafer productivity and circuit placement flexibility.

CHIP-ON-BOARD (COB). One of many configurations in which a chip is directly bonded to a circuit board or substrate. These approaches include wirebonding, TAB, or solder interconnections, similar to the C4 structure. In low-end and consumer systems, chip-on-board generally refers to wirebonding of chips directly to board. See also Direct Chip Attach (DCA).

CIRCUIT-BOARD PACKAGING. Packaging of chips by the use of organic printed-circuit boards. See Printed-Circuit Board.

CLADDING. Thin layer of a corrosion-resistant metal coating bonded to a metal core, usually by heating and rolling. Typical examples are steels clad with stainless steel, nickel alloys, or copper alloys. Copper cladding on both sides of invar is accomplished this way.

CLOCK SKEW. A cycle time adder caused by the amount of tolerance associated with the clock signal arrival times at all of the system latch inputs.

CMOS. See Complementary Metal-Oxide Semiconductor.

COATED-METAL CORE SUBSTRATE. A substrate consisting of an organic or inorganic insulation coating bonded to metal. Insulated surface or surfaces are used for circuit deposition.

COEFFICIENT OF THERMAL EXPANSION (CTE). The ratio of the change in dimensions to the change in temperature-per-unit starting length, usually expressed in cm/cm/°C. The acronyms TCE and CTE are synomous.

COFFIN-MANSON EQUATION. A commonly used formula, first proposed by S.S. Manson and L.F. Coffin, relating the fatigue lifetime of a metal to the imposed strain amplitude. Others have extended the formula to incorporate time and temperature dependent phenomena.

COFIRING. Processing thick-film conductors and dielectrics through the firing cycle at the same time to form multilayer structures.

COLORANT. An inorganic or organic compound that is added to a polymeric resin to impart a desired color.

COLUMN GRID ARRAY (CGA). A Column Grid Array is an area array of solder columns joined to an SCM or MCM and used to electrically and physically connect the package to the next level of package, usually a printed circuit board. A Column Grid Array is used when the package performance requires a higher riliability that provided with the similiar Ball Grid Array.

COMPLIANT BOND. A bond which uses an elastically and/or plastically deformable member to impart the required energy to the lead.

COMPLEMENTARY METAL-OXIDE SEMICONDUCTOR (CMOS). This refers to logic in which cascaded field effect transistors (FET) of opposite polarity are used to minimize power consumption.

COMPRESSION SEAL. A seal between an electronic package and its leads. The seal is formed as the heated metal, when cooled, shrinks around the glass insulator, thereby forming a tight compression joint.

CONDUCTION. Thermal transmission of heat energy from a hotter region to a cooler region in a conducting medium.

CONDUCTIVE ADHESIVE. An adhesive material, usually epoxy, that has metal powder added to increase electrical conductivity. Usual conductor added is silver.

CONDUCTIVE EPOXY. An epoxy material (polymer resin) that has been made conductive by the addition of a metal powder, usually gold or silver. Best common conductors are silver, copper, and gold. See also Superconductor, and Conductor Adhesive.

CONDUCTOR, ELECTRICAL. A class of materials that conduct electricity easily. They have very low resistivity which is usually expressed in micro-ohm-cm. The best conductors include silver, copper, gold, and superconducting-ceramics.

CONDUCTOR, THERMAL. A class of materials, such as copper, aluminum, and beryllia, that conduct heat.

CONFORMAL COATING. A thin nonconductive coating, either plastic or inorganic, applied to a circuit for environmental and/or mechanical protection.

CONNECTIONS. The connections belonging to nets interconnecting logic units on a given package level—including connections to terminals on that level—connecting it to the next higher package level.

CONNECTIVITY. See Wiring Density.

CONTACT ANGLE. The angle between the bonding material, usually a liquid-like solder, and the bonding pad. Also called wetting angle.

CONTACT RESISTANCE. Excess electrical resistance in series with the bulk conductor resistance of two contacting electrical conductors arising from the nature of the contact geometry and surface properties of the contacting surfaces.

CONTROLLED COLLAPSE CHIP CONNECTION (C4). A solder joint connecting a substrate and a flip chip, where the surface tension forces of the liquid solder supports the weight of the chip and controls the height (collapse) of the joint.

CONTROLLING COLLAPSE. Controlling the reduction in height of the solder balls in a flip-chip processing operation.

CONVECTION. Transmission of thermal energy from a hotter to a cooler region through a moving medium, such as air or water.

COPLANAR LEADS (FLAT LEADS). Ribbon-type leads extending from the sides of the circuit package, all lying in the same plane for surface mount applications.

COPOLYMER. A compound, resulting from the chemical reaction and polymerization of two chemically different monomers. The resulting larger molecules contain repeating structural units of the original molecules.

CORDIERITE. A crystalline ceramic material of composition $2MgO$-$2Al_2O_3$-$5SiO_2$ that can be crystallized from glass of same composition or sintered from powders.

COUPLED NOISE. (Same as Cross talk) The electromagnetic and electrostatic linkages between two nearby conductors that allow one line to induce a signal on the other. See also Cross Talk.

COUPLER. A chemical agent, frequently an organosilane, used to enhance the bond between a resin and a glass reinforcement.

COUPLING CAPACITOR. A capacitor used to block dc signals, and to pass high-frequency signals between parts of an electronic circuit.

CRAZING. Fine cracks which may extend on or through layers of plastic or glass materials.

CREEP. Nonrecoverable deformation proceeding at relatively low strain rates, less than about 10^{-6}/sec, usually associated with sufficiently high temperature to allow significant rates of diffusion.

CROSSOVER. The transverse crossing of metallization paths without mutual electrical contact. This is achieved by the deposition of an insulating layer between the conducting paths at the area of crossing.

CROSS TALK. Signals from one line leaking into another nearby conductor because of capacitance or inductive coupling or both (i.e., owing to the capacitance of a thick-film crossover.).

CRYSTALLIZATION. Formation of crystalline phase out of amorphous material during high-temperature processing. Undesirable or uncontrollable crystallization is called divitrification.

CUMULATIVE DISTRIBUTION FUNCTION (CDF). Distribution of a parameter as a fraction of the total number of measurements with respect to a statistic, e.g. "probits" or standard deviations relative to a particular statistical distribution, i.e., normal, etc.

CURING AGENT. An inorganic or organic compound which initiates the polymerization of a resin.

CURING CYCLE. For a thermosetting material, commonly a resin compound such as a bonding adhesive, it is the combination of total time-temperature profile to achieve the desired result; for example, the complete irreversible hardening of the material, resulting in a strong bond.

CURRENT CARRYING CAPACITY. The maximum current that can be continuously carried by a circuit without causing objectionable degradation in electrical or mechanical properties.

CURRENT SLEW RATE. The rate of change in current with respect to time (di/dt).

CUSTOM DESIGN. A form of design in which the choice and arrangement of components and wiring on a package may vary arbitrarily within tolerances from a regular array.

CYCLE TIME. Unit of time in which elements of the central processor complete their logical functions. Some elements will require more than one cycle to complete a function. See Cycles per Instruction.

CYCLES PER INSTRUCTION. The number of cycles required to process an instruction.

D

DECOUPLING CAPACITOR. A shunt-placed capacitance that is used to filter transients on a power distribution system.

DELAY EQUATIONS. A set of mathematical terms that are used to predict the propagation times between driving and receiving circuits that are interconnected through signal wires. These equations are usually derived from simulation data using numerical curve fitting techniques.

DELTA-I NOISE (ΔI). See Switching Noise.

DESIGN LIMITS. The fail points that are incorporated into the hardware design rules that drive the computer-aided design system. See noise rules and wiring rules.

DEVITRIFICATION. The undesirable formation of crystals in glass during firing. The desirable process is called crystallization.

DEW POINT. The temperature at which moisture at a given partial pressure becomes saturated, and when cooled below which, it condenses.

DIE. Integrated circuit chip as cut (diced) from finished wafer. See Chip.

DIE BOND. Mechanical attachment of silicon to substrate usually by solder, epoxy, or gold-silicon eutectic, including interface metallurgies on chip and substrate. The die bond is made to the back (inactive) side of the chip with the circuit side (face) up.

DIELECTRIC. Material that does not conduct electricity. Generally used for making capacitors, insulating conductors (as in crossover and multilayered circuits), and for encapsulating circuits.

DIELECTRIC CONSTANT. The term used to describe a material's ability to store charge when used as a capacitor dielectric. It is the ratio of the charge that would be stored with free space to that stored with the material in question as the dielectric.

DIELECTRIC LOSS. The power dissipated by a dielectric as the friction of its molecules opposes the molecular motion produced by an alternative electric field.

DIFFERENTIAL SCANNING CALORIMETRY (DSC). A technique for measuring the physical transitions of a polymer as a function of temperature compared to another material undergoing a similar heating process but not undergoing any transitions or reactions. DSC uses a servo system to supply energy at a varying rate to both sample and reference so as to keep their temperatures equal. A DSC output plots energy supplied vs. average temperature.

DIRECT ACCESS STORAGE DEVICE (DASD). Computer storage hardware subsystem that uses magnetic recording on a rotating disk surface. Access to the information is accomplished with the use of a moveable arm which positions one or more read/write heads along the radius of the disk to the desired track.

DIRECT CHIP ATTACH (DCA). A name applied to any of the chip-to-substrate connections used to eliminate the first level of packaging: see also Chip-on-Board.

DISCRETE COMPONENT. Individual components or elements, such as resistors, capacitors, transistors, diodes, inductors, and others, as self-contained entities.

DISTANCE TO NEUTRAL POINT (DNP). The separation of a joint from the neutral point on a chip. This dimension controls the strain on the joint imposed by expansion mismatch between chip and substrate. The neutral point is usually the geometric center of an array of pads and defines the point at which there is no relative motion of chip and substrate in the *X-Y* plane during thermal cycling.

DOCTOR BLADE. A method of casting slurry into a thin sheet by the use of knife blade placed over moving carrier to control slurry thickness.

DOUBLE-SIDED SUBSTRATE. A substrate carrying active circuitry on both its topside and bottomside, electrically connected by means of metallized through-holes or edge metallization or both.

DRIVER. The off chip circuit that supplies the signal voltage and current to the package lines. Also called an output buffer circuit.

DRY FILM PHOTORESIST. Photoresist material that is processed dry, usually by lamination of prefabricated film.

DRY PRESSING. Pressing and compacting together of dry powdered materials with additives in rigid die molds under heat and pressure to form a solid mass, usually followed by sintering to form shapes.

DUAL-IN-LINE PACKAGE (DIP). A package having two rows of leads extending at right angles from the base and having standard spacings between leads and between rows of leads. DIPs are made of ceramic (Cerdip) and plastic (Pdip).

DYNAMIC FLEX. A form of flexible circuitry developed for applications where continued flexure is necessary. In contrast, static (flex), once installed, remains fixed.

DYNAMIC RANDOM ACCESS MEMORY (DRAM). Electronic information storage that employs transient phenomena, typically charge stored in a leaky capacitor. Refresh cycles are required to restore and thus maintain the information. DRAM is the simplest and least expensive of electronic memories, but it is also the least impressive performer.

E

EFFECTIVE INDUCTANCE (LEFF). A simplified characterization of the goodness of an AC power distribution system. It consists of a single lumped inductance that when multiplied by the total current slew rate predicts the total switching noise across the circuit load.

ELECTRICALLY LONG TRANSMISSION LINE. One in which the delay from the near-end to the far-end is greater than one-half of the near-end signal's transition time. When this occurs, the reflections from the far-end do not distort the near-end signal during its transition time.

ELECTRICALLY SHORT TRANSMISSION LINE. One in which the delay from the near-end to the far-end is less than one-half of the near-end signal's transition time. For this case, reflections from the far-end interfere with the near-end transition waveform causing distortion that usually increases the net's delay.

ELECTROLESS PLATING. Metal deposition, usually in an aqueous medium, which proceeds by an exchange reaction between metal complexes in the solution and the particular metal to be coated; the reaction does not require externally applied electric current.

ELECTROPLATING. Deposition of an adherent metallic coating onto a conductive object placed into an electrolytic bath composed of a solution of the salt of the metal to be plated. Using the terminal as the anode (possibly of the same metal as the one used for plating), a DC current is passed through the solution affecting transfer of metals ions onto the cathodic surface.

ELECTROSTATIC DISCHARGE (ESD). Discharge of static charge on a surface or body through a conductive path to ground. An electronic component or higher-level assembly may suffer damage when it is included in the discharge path.

ELONGATION. The ratio of the increase in wire length at rupture, in a tensile test, to the initial length, expressed in percent.

EMITTER-COUPLED LOGIC (ECL). Emitter-coupled logic is also known as current-switch (logic) circuits. In it, a current source feeds emitters of several transistors. The base of all but one acts as input terminals; the last base is connected to a reference voltage. Very popular circuit for high-performance applications, it is often combined with an emitter-follower output stage to further enhance its performance. It is then called SCEF, for current-switch emitter follower.

ENAMELING. A process that produces pore-free glass dielectric coating over a metal-core substrate.

ENCAPSULATION. Sealing up or covering an element or circuit for mechanical and environmental protection.

END OF LIFE (EOL). The end of the useful operating life of a component or equipment determined by a "wear-out" or life terminating mechanism measured in units of time. EOL is usually specified as an objective in reliability calculations.

ENGINEERING CHANGE (EC). A change in design. An electrical design change is frequently implanted by cutting out or adding an electrical path to the manufactured hardware, e.g., laser deleting a line or adding a wire on a ceramic substrate.

ENTITY. A group of circuits separated from other circuits by a physical package boundary and associated input and output connections.

EUTECTIC. A term applied to the mixture of two or more substances with the the lowest melting point possible between those components.

EXTERNAL RESISTANCE. A term used to represent thermal resistance from a convenient point on the outside surface of an electronic package to an ambient reference point.

F

FAILURE. The temporary or permanent impairment of device function caused by physical, chemical, mechanical, electrical, or electromagnetic disturbance or damage.

FAILURE RATE. The rate at which devices from a given population can be expected (or were found) to fail as a function of time (e.g., %/1000 hr. of operation).

FAST WAVE PROPAGATION. The transmission of energy along a signal line at the speed-of-light (velocity) expected for the dielectric structure. In the case of a low-loss line, the fast wave refers to the portion of the signal that travels at the velocity expected for the dielectric medium.

FATIGUE. Used to describe the failure of any structure caused by repeated application of stress over a period of time.

FERROELECTRIC. A crystalline dielectric that exhibits dielectric hysteresis; an electrostatic analogy to ferromagnetic materials.

FIELD EFFECT TRANSISTOR (FET). A transistor in which a voltage applied to a thin conductor over a thin insulator controls current flow in a semiconductor region (gate) or one polar type. This component originates and terminates in two regions of the opposite polar type located at either end of the gate region.

FIELD REPLACEABLE UNIT (FRU). A component or sub-system of an electronic assembly which may be replaced at the site of installation. A first- or second-level package is commonly an FRU for most computers.

FILLER. A substance, usually ceramic or metal powder, used to modify the properties of fluids or polymers.

FINITE ELEMENT MODELING. A computationally intensive numerical modeling tool in which the body is discretized into small regularly shaped elements.

FIRST-INCIDENT SWITCHING. The case that occurs when all of the receivers on a multi-drop net switch at the first time the signal arrives from the driver. Nets that are not first-incident are referred to as multi-reflection nets.

FLAME RETARDER. An inorganic or organic compound added to a polymer mixture that causes the resulting plastic to self-extinguish after a flame is removed.

FLAT PAC. An integrated circuit package having its leads extending from all four sides and parallel to the base.

FLEXIBLE CIRCUIT CARRIER. Printed circuits employing flexible substrates, processed by patterning copper onto thin flexible Kapton or polyimide films. Originally used only as connector; now employed for multilayers.

FLEXIBLE COATING. A plastic coating that is still flexible after curing.

FLEXURAL STRENGTH. Strength of a material measured by bending, typically used for brittle materials, such as glasses and ceramics; expressed in MPa.

FLIP-CHIP. A leadless, monolithic structure containing circuit elements, which is designed to electrically and mechanically interconnect to the hybrid circuit by means of an appropriate number of bumps, which are covered with a conductive bonding agent, located on its face. Alternatively, bonding of chips with contact pads, face down by solder connection. See also Controlled Collapse Chip Connection (C4).

FLOOR PLANNING. A procedure in physical design which permits approximate shaping and placement of logic and memory circuit groupings on a package before final placement and wiring.

FLOW REGIME, LAMINAR. Flow where fluid layers are undisturbed and smooth.

FLOW REGIME, TURBULENT. Flow where fluid particles are disturbed and fluctuate.

FLUX. In soldering, a material that chemically attacks surface oxides so that molten solder can wet the surface to be soldered, or an inert liquid which excludes oxygen during the soldering process.

FRACTURE TOUGHNESS. A basic property of a material, or of an interface between dissimilar materials, describing its crack resistance in a mechanical or thermomechanical stress field.

FRIT. Glass composition ground up into a powder form and used in thick-film compositions as the portion of the composition that melts upon firing to give adhesion to the substrate and hold the conductive composition together.

FRONT-END-OF-THE-LINE (FEOL). See back-end-of-the-line.

FR-4. Electronic Industries Association's designation for a fireretardant epoxy resin/glass cloth laminate. By common usage, the resin for such a laminate.

G

G-10. It is a grade of epoxy-impregnated glass cloth printed-circuit board material per NEMA (National Electrical Manufacturers Assoc.).

GATE ARRAY. A semicustom product, implemented from a fully diffused or ion-implanted semiconductor wafer carrying a matrix of identical primary cells arranged into columns with routing channels between them in the X and Y directions.

GATE, LOGIC. Usually an electric circuit which combines information of its two inputs to form its output signal in accordance to the logic function it performs.

GATE, SEA OF (see also Gate Array). A form of custom chip layout in which the wiring tracks required for interconnecting a fixed array of logic cells are disposed in rows and columns having widths measured in numbers of wiring tracks per row or column channel which vary to suit the local wiring demand, from point to point and from one logic product to another.

GATE, STRUCTURAL. Term used to designate a hinged frame which contains a number of boards and can swing out for servicing and/or access to the interior of a fixed frame.

GLASS. Inorganic, nonmetallic and amorphous material obtained by melting oxide(s) into glass and retaining the structure by fast cooling.

GLASS-CERAMIC. Inorganic, nonmetallic material obtained by controlled crystallization of glass into nonporous and fine microstructure.

GLASS + CERAMIC. Inorganic, nonmetallic material obtained by admixing crystalline ceramic with glass and sintering composite.

GLASS ± CERAMIC. Refers to family of glass-ceramic and glass + ceramic.

GLASS FABRIC. Cloth woven from glass yarns which are made of filaments.

GLASS TRANSITION TEMPERATURE (T_G). In polymer or glass chemistry, the temperature corresponding to the glass-to-liquid transition, below which the thermal expansion coefficient is low and nearly constant, and above which it is very high.

GLAZED SUBSTRATE. A glass coated ceramic substrate that effects a smooth and nonporous surface.

GLAZING. In the present context, glazing refers to the coating of a smooth, adherent layer of glass, such as a sealing glass, by melting the glass over metal or ceramic surfaces.

GLOB TOP. A glob of encapsulant material surrounding a chip in the chip-on-board assembly process. The attached chip must have already passed pretest and inspection because rework after final curing of the epoxy or silicone globs is virtually impossible.

GLOBAL WIRING. Wiring interconnecting components mounted on a package (as opposed to the wiring inside the components). Also refers to that wiring independent of its detailed allocation to wiring tracks within a channel.

GREEN. A term—unrelated to the actual material color— used in ceramic technology meaning unfired. For example a "green" substrate is one that has been formed, but has not been fired.

GREEN-SHEET. A composite organic-inorganic, flexible sheet ready for metallization, if desired, and lamination to form green substrates which upon removal of organics results in ceramic substrate. Green refers to the unfired state.

GROUND PLANE. A conductive layer on a substrate or buried within a substrate that connects a number of points to one or more grounding electrodes.

GULLWING. A common lead form used to interconnect surface mounted packages to the printed-circuit board. The leads, normally 100 to 250 μm thick, are bent outward, downward, then again outward from the package body, providing feet for solder interconnection, and some degree of mechanical compliance.

H

HARD GLASS. Glasses having a high softening temperature (>700°C), such as the borosilicate glasses used to seal feedthrough leads into metal packages.

HEAT FLUX. The rate of flow of heat energy across or through a surface, measured in watts/cm^2.

HEAT SINK. The supporting member to which electronic components or their substrate or their package bottom are attached. This is usually a heat conductive metal with the ability to rapidly transmit heat from the generating source (component).

HERMETIC. Sealed so that the object is gastight. The test for hermeticity is to fill the object with a test gas—often helium—and observe leak rates when placed in a vacuum. A plastic encapsulation cannot be hermetic as it allows gases to permeate.

HIGH-LEVEL NOISE TOLERANCE (NTH). The receiver noise tolerance that occurs when the input signal is in its UP state.

HOMOGENEOUS MEDIUM. A signal propagation structure in which only a single dielectric is present.

HOT-GAS REFLOW. The technique in which a heated gas, including air, is impinged on a site to be solder-reflowed, usually to form a solder interconnection.

HOT-KNIFE SOLDERING. The technique in which a heated blade (heated electrically or conductively) is used as a heat source for melting solder during package joining. The blade may be used to force mechanical contact throughout the joining process.

HYBRID MODULE. A special carrier of hybrid microcircuits and other components interconnected as a unit, or as a component of an electronic subsystem. The module may be of single construction or made up of submodules, each usually with a compartment to house hermetically packaged hybrids and discrete passive component parts, such as transformers, axle-lead resistors, etc. Nonhermetic hybrid modules generally are parylene coated. In this book, hybrid module is also referred to a module containing a combination of thick and thin films.

I

IMPREGNATION. The process of coating a substrate—say glass cloth—with a resin solution and drying. The dried product is called prepreg.

INDUCTANCE. The electromagnetic element that stores flux lines. In packaging systems, it is used in lumped equivalent circuits to represent part of a line discontinuity. It is also used in a distributed system to represent the electromagnetic storage property of a transmission line. Because it induces an opposing voltage in response to a change in current, it causes package delta I (ΔI) noise.

INERT ATMOSPHERE. A gas atmosphere such as helium or nitrogen that is nonoxidizing or nonreducing to metals.

INFRARED REFLOW (IR). The technique in which primarily long wavelength light is used to heat solder joints to the melting temperature. Normally, a circuit board having prepositioned packages is transported through an IR reflow furnace.

INHOMOGENEOUS MEDIUM. A signal propagation structure in which multiple dielectrics are present.

INJECTION MOLDED. Molding by injecting liquefied plastic into a mold of desired shape.

INJECTION MOLDED CARD (IMC). Card for electronic packages made by injection molding of plastics into a mold cavity of desired shape.

INPUT/OUTPUT TERMINAL (I/O). A chip or package connector (terminal) acting to interconnect the chip to the package or one package level to the physically adjacent level in the hierarchy. Usually refers to the number of contacts necessary to wire to or interconnect an assembly. Pin out, connections, and terminals are other common words to describe the same. Care must be taken to differentiate between the total number of I/Os between levels, signal I/Os, I/Os used to distribute power, and reference I/Os.

INSULATION RESISTANCE (IR). The resistance to current flow when a potential is applied. IR is measured in megohms.

INSULATOR METAL SUBSTRATE TECHNOLOGY (IMST). A substrate, such as one made of porcelainized steel, which is not subject to size limitations and may have superior thermal dissipation characteristics. IMST refers to insulated-metal substrate technology of Sanyo. It is a single-sided aluminum core with epoxy coating and etched copper wiring.

INSULATORS. A class of materials with high resistivity. Materials that do not conduct electricity. Materials with resistivity values of over 10^6 Ω cm are generally classified as insulators.

INTEGRATED CIRCUIT. A microcircuit (monolithic) consisting of interconnected elements inseparably associated and formed in situ on or within a single substrate (usually silicon) to perform an electronic circuit function.

INTERCHIP WIRING. The conducting wiring path connecting circuits on one chip with those on other chips to perform a function.

INTERCONNECTION. The conductive path required to achieve connection from a circuit element to the rest of the circuit.

INTERFACE. The boundary between dissimilar materials, such as between a film and substrate, or between two films.

INTERNAL RESISTANCE. A term used to represent thermal resistance from the junction of a device, inside an electronic package, to a convenient point on the outside surface of the package.

INTERPENETRATING POLYMER NETWORK (IPN). A polymer alloy made up of two or more crosslinked polymers. The networks interact with each other only through permanent physical entanglements rather than through covalent bonding as in a copolymer.

INVAR. A trademark of International Nickel Co., Inc. for a very low thermal expansion alloy of nickel and iron.

ION MIGRATION. The movement of free ions within a material or across the boundary between two materials under the influence of an applied electric field.

ISOPAK. An unique pin-grid array consisting of Kovar pins sealed in glass-to-Klovar plate flush for chip bonding.

J

J-LEAD. An I.C. package terminal lead configuration that resembles in crossection, the letter “J”.

JOSEPHSON SUPERCONDUCTING DEVICE. Superconducting ceramics acting as Josephson devices, at very low temperatures, typically at 4°K.

K

KEEPER BAR, TAB. A strip of dielectric material—such as polyimide—that remains attached to each row of outer-TAB leads following excise operation. It helps to reduce misalignment, non-planarity and damage during outer lead bonding.

KIRKENDAHL VOIDING. Voids induced in a diffusion couple between two metals with different interdiffusion coefficients.

KNOWN GOOD DIE (KGD). IC semiconductor chips that have been tested before being packaged and are known to function as required.

KOVAR. An alloy of iron (53%), Cobalt (17%), and nickel (29%) with thermal expansion matching alumina substrate and certain sealing glasses. Most common lead frame and pin material.

L

LAMINATION. The process of consolidating sheets of prepreg under heat and pressure to form a solid product. Applied also to the consolidation of prepregs and precircuitized subcomposites to form a composite.

LARGE-SCALE INTEGRATION (LSI). Large-scale integration. Term used to designate chips with more than one thousand transistors.

LASER SOLDERING. The technique in which heat to reflow a solder interconnection is provided by a laser, usually a longer wavelength YAG or CO_2 laser. The joints are heated sequentially, and cooled rapidly.

LEAD FRAME. A sheet metal framework on which a chip is attached, wire-bonded, and then molded with plastic.

LEADED-CHIP CARRIER PLASTIC (PLCC). A plastic package containing a chip that has terminal leads emanating from four sides. Each lead has a J-configuration and is designed for surface mounting to a printed-circuit board.

LEADLESS-CHIP CARRIER (LLCC). A surface mounted package having metallized contacts at its periphery (rather than wire leads) which are soldered to metallized contacts on the printed-circuit board or substrate.

LIFT-OFF. Pattering of metal by lift off materials around usually in a solvent.

LINE DISCONTINUITY. A load point, consisting of a lumped equivalent circuit of resistance, capacitance, and inductance anywhere on a transmission line that produces spurious reflections.

LINE LOADING. Externally connected resistance, inductance, and capacitance, or combination of these on a transmission line.

LINE RESISTANCE. Resistance of conductor lines in a package, measured in ohms per unit length or for a given cross section, ohms per square.

LINES PER CHANNEL. The number of conductive lines between through holes in an organic board or ceramic substrate.

LIQUID CRYSTAL DISPLAY (LCD). Display technology based on liquid crystal materials whose light transmission is changeable by the application of an electrical field. LCD devices are used in numeric read outs and for flat screen television receivers.

LOGIC DESIGN. The process of determining the choice and interconnection of logic units (e.g., nands or nors) to accomplish logical functions in an overall digital system.

LOGIC PART. A physical implementation of an interconnected and packaged group of logic circuits used in general more than once in a digital system.

LOGIC PRIMITIVE. A basic logic function embodied as a single unit.

LOGIC SERVICE TERMINAL (LST). A terminal (on a package or package component) carrying logic signals as opposed to one used only for electrical power.

LOSS-LESS TRANSMISSION LINE. A signal path where the total series resistance is less than 10% of the characteristic impedance of the line. For this case, the signal level attenuation is approximately 10% (Vout/Vin) when the line is terminated in it's characteristic impedance.

LOSSY TRANSMISSION LINE. A signal path with total series resistance that exceeds two times the characteristic impedance of the line. When this occurs, the fast-wave portion of the input signal is attenuated by more than "1/e" or 63% (Vout/Vin) when the line is terminated in it's characteristic impedance.

LOW LEVEL NOISE TOLERANCE (NTL). The receiver noise tolerance that occurs when the input signal is in its DOWN state.

LOW-LOSS TRANSMISSION LINE. A signal path where the total series resistance is greater than 0.1 times but less than two times the characteristic impedance of the line. In this range, the signal level attenuation is between 10% and 63% (Vout/Vin) when the line is terminated in it's characteristic impedance.

M

MACRO. A collection of continuous cells defined to be placed as a group within a chip image.

MANHATTAN DISTANCE (LENGTH). Wire length between terminals of a net or connection measured in X or Y directions on a package wiring plane as on city blocks.

MASK. The photographic negative that serves as the master for making thick-film screens and thin-film patterns.

MCM-C. A multi-chip module with a structure that consists of multi-layer co-fired ceramic.

MCM-D. A multi-chip module structure that consists of deposited thin film organic layers.

MCM-L. A multi-chip module structure that consists of laminated organic layers.

MEAN TIME TO FAILURE (MTTF). Applicable to individual parts or devices in reliability technology. It is the arithmetic average of the lengths of time-to-failure registered for parts or devices of the same type, operated as a group under identical conditions.

MEMORY ADDER. An adder, to the basic cycle-per-instruction rate of the Central Processor (CP), due to requirements for data or instructions, not available in the CP when necessary.

METAL MIGRATION. An undesirable phenomenon whereby metal ions, notably silver, are transmitted through another metal or across an insulated surface, in the presence of moisture and an electrical potential.

METALLIZATION. A film pattern (single or multilayer) of conductive material deposited on a substrate to interconnect electronic components.

METALLIZED CERAMIC (MC). Ceramic (fired) substrate metallized with thick and thin films of metals. (In IBM, metallized ceramic refers to thin-film metallization on fired alumina substrate.)

MICRON. An obsolete unit of length equal to a micrometer (μm).

MICROSTRIP LINE. A signal line on the surface of a dielectric with air above it and a reference plane on the opposite side of the dielectric.

MICROSTRUCTURE. Structural features, such as crystal or phase boundaries, or defects or inhomogeneities within a solid, usually resolved at high magnification.

MINER'S RULE. An outgrowth of the cumulative damage concept, this formula predicts the fatigue lifetime of a structural element when the load history encompasses multiple types or various amplitudes of stress.

MODULE. A chip carrier on which the chip terminals are fed out by various means to terminals spaced to suit the spacing and dimensions of wires on the next higher level of package (i.e., card or board). It may also contain wiring planes and power planes interconnecting several of its chips, and thus be used as a card.

MOIRE. A technique to measure in-plane surface displacement, whereby a precision grid of fine lines attached to the surface is deformed relative to a stationary or reference grid. The resulting fringe pattern can be analyzed to provide displacement values throughout the field.

MOLD RELEASE. An organic compound added to a molding compound or powder that migrates to the mold surface to form a waxy layer between the plastic and mold metal and to allow easy removal of the part from the mold.

MONOLITHIC SYSTEMS TECHNOLOGY (MST). Ceramic package manufactured in IBM by screen printing *Ag-Pd* conductors onto alumina substrate swagged with pins and chips bonded with solder connection (C4).

MONTE CARLO ANALYSIS. A statistical analysis method whereby the resultant distribution is generated by using a random number generator to pick a large number of cases from the input distributions.

MULTICHIP MODULE (MCM). A module or package capable of supporting several chips on a single package. Most multichip packages are made of ceramic.

MULTICHIP PACKAGE. An electronic package that carries a number of chips and interconnects them through several layers of conductive patterns. Each one is separated by insulative layer and interconnected via holes.

MULTILAYER CERAMIC (MLC). Ceramic substrate consisting of multiple layers of metals and ceramics interconnected with vias. All with thick film.

MULTILAYER SUBSTRATES. Substrates that have buried conductors so that complex circuitry can be handled, using assembly processes similar to those used in multilayer ceramic capacitors. Fabricated either as a conventional MLC, or a cofired multilayer ceramic (CMC) hybrid structure, in high- and low-temperature versions.

MULTI-REFLECTION SWITCHING. A network that is not first incident. See First-Incident Switching.

N

NEAR INFRARED REFLECTANCE ANALYSIS (NIRA). Infrared spectroscopy that covers the region from 0.75 μm to 2.5 μm. Regular infrared spectrophotometers, on the other hand, use a glowing light source to provide light with wavelengths from 2.5 to about 15 μm.

NET. A group of terminals interconnected to have a common dc electrical potential in a package.

NOISE. In a digital system, noise is any undesirable parasitic effect that causes signal waveform distortion, excessive delay, or false switching. Common types of noise are: reflection, coupled and switching.

NOISE SATURATION. A phenomenon in which switching noise does not increase as the number of simultaneously switching circuits increases beyond an observed number. This effect is caused by a negative feedback between the large noise generated and the current drawn by each of the switching circuits. An increase in circuit delay occurs when noise saturation occurs.

NONLINEAR DIELECTRIC. A capacitor material that has a nonlinear capacitance-to-voltage relationship. .hw Ti-ta-nates (usually barium titanate) ceramic capacitors (Class II) are nonlinear dielectrics. NPO and Class I capacitors are linear by definition.

O

OPTICAL INTERCONNECTS. Composed of the basic optoelectronic devices, these are components and modules used as circuit building blocks. The light-emitting diode (LED) converts electrical energy to light where junction electroluminescence occurs as a result of the application of direct current at low voltage to a suitably doped crystal when forward biased. The light from this source then is detected by the reverse-biased pn-junction photodiode and/or phototransistor. Light, of the proper wavelength, creates a current flow, a photocurrent, in the external circuit proportional to the effective irradiance on the device.

OUTER LEAD BONDING. The process of joining the outer leads of a package, typically TAB (Tape Automated Bonding) to the next level of assembly (usually card or board). The inner leads on the tape are joined to the chip by the process know as inner lead bonding.

OVERCOAT. A thin film of insulating material, either plastic or inorganic (e.g., glass or silicon nitride), applied over integrated circuit elements for the purposes of mechanical protection and prevention of contamination.

OVERFLOW WIRE. See Wiring Overflow.

OVERGLAZE. A glass coating that is grown, or deposited, over another element, normally for physical or electrical protection purposes.

OVERLAY. One material applied over another material.

P

PACKAGE CROSSING. An interconnection which connects a terminal on one package with that on another.

PACKAGE DELAY. The time delays associated with the interconnections between components that complete logical make-up functions. Values depend on materials and distance.

PACKAGING LEVEL. A member of a nested interconnected packaging hierarchy (e.g., chip, chip carrier, card, board in order of low to high level).

PAD-GRID ARRAY PACKAGE (FOR VHSICS). A package embodying a rather novel technology, where solder-contact pads are not just around the package periphery (as with chip carriers) but cover the entire bottom surface in checkerboard fashion.

PASSIVATION. The formation of an insulating layer directly over a circuit or circuit element to protect the surface from contaminants, moisture, or particles.

PASSIVE COMPONENTS (ELEMENTS). Elements or components such as resistors, capacitors, and inductors which do not change their basic character when an electrical signal is applied. In contrast transistors, diodes, and electron tubes are active components.

PASTE. Synonymous with “composition” and “ink” when relating to screenable thick-film materials, usually consisting of metal or ceramic powders dispersed in organic vehicles.

PEEL STRENGTH (PEEL TEST). A measure of adhesion between a conductor and the substrate. The test is performed by pulling or peeling the conductor off the substrate and observing the force required.

PERMEABILITY. The property of a solid plastic material that allows penetration by a liquid or gas.

PHASE DIAGRAM. State of a metal alloy or ceramic over a wide temperature range. The phase diagram is used to identify phases as a function of composition and temperature.

PHOSPHOSILICATE GLASS (PSG). Phosphorus-doped silicon dioxide (also known as P-glass). It is often used as a dielectric material for insulation between conducting layers, for inhibiting the diffusion of sodium impurities, and for planarization since it softens and flows at 1,000 to 1,100°C to create a smooth topography for subsequent metallization.

PHOTOLITHOGRAPHY. The generation of a pattern through a sequence of rubylith, photo-reduction, step-and-repeat, computer-aided design, or finally, stat-of-the-art electron-beam technique. This procedure will generate a product (mask or otherwise) to become the primary tool in transferring an image onto a micro-electronic substrate.

PHYSICAL DESIGN. The process of allocating chip or package components and their interconnections to their appropriate spatial locations or sockets in an overall system arrangement.

PICK-AND-PLACE. The manufacturing process whereby chips are selected and placed on the correct substrate site in preparation for joining (or interconnecting) the chip to the substrate.

PIN. Round, cross-sectional electrical terminal and/or mechanical support. Used in plug-in type packages, either straight or modified as nail-head, upset, pierced, or bent variety. A pin's primary functions are, internally, to support a wirebond or other joint, and, externally, to plug into a second-level package connector.

PIN-GRID ARRAY (PGA). A package or interconnect scheme featuring a multiplicity of plug-in type electrical terminals arranged in a prescribed matrix format or array.

PIN-THROUGH-HOLE (PTH). A term referring to the class of packages or modules that are soldered into plated through holes within the second-level package (printed-circuit board).

PITCH. The center-to-center spacing, between pads, rows of bumps, pins, posts, exit leads, etc. Sometimes also the distance as measured point to corresponding point between two adjacent images in a device matrix, on a semiconductor wafer or its photomask.

PLACEMENT. The manual or automatic placing of chip circuits, chips, chip carriers, and cards in their actual locations on corresponding images at a given package level.

PLANAR MOTOR. Typically a brushless DC servo motor of a flat planar configuration constructed using printed-circuit assembly methods.

PLANAR MOTOR VOICE COIL SERVO. A galvanometer type mechanical positioner similar in design to the conventional audio loudspeaker voice coil assembly. Provides limited excursion capability, but extremely high speed, and with suitable feedback, very high accuracy.

PLASMA. An electrically conductive gas, composed of ionized atoms or molecules, used for dry-etching in the fabrication of devices.

PLASMA ETCHING. The action of an electrically conductive gas, (composed of ionized gas or molecules), to remove unwanted portion of conductive or insulative pattern.

PLASTIC. A polymeric material, either organic (e.g., epoxy, polyimide) or inorganic (e.g., silicone) used for conformal coating, encapsulation, or overcoating.

PLASTIC BALL GRID ARRAY (PBGA). Ball grid array technology on a plastic carrier. See Ball Grid Array.

PLASTIC QUAD FLAT PACK (PQFP). See Quad Flat Pack.

PLATING. A condensation of the word Electroplating or Electroless Plating that describes the coating of a metal on plastic or other surfaces with metal that is electrolytically or chemically deposited from a bath.

POLYIMIDES. A class of resin compounds containing the NH group which are derived from ammonia and are "imidized" from polyamic acid at temperatures high enough to initiate and complete the imide ring closure. Polyimides are useful as organic dielectric interlevel layers in VLSI technologies. They are mostly thermosetting ring-chain polymers, whose useful characteristics include perfect planarity as a (spun) film; high temperature tolerance; excellent weathering and mechanical-wear characteristics; and a low dielectric constant, a decided advantage in reducing propagation delays in multilayer hybrid circuits (faster switching).

PORCELAIN. A mixture of borosilicate glass with minor quantities of zirconia and other ingredients. It is used synonymous with enamel in this book.

PORCELAIN ENAMEL TECHNOLOGY (PET). The technology of coating glass on metal. See Porcelain.

POWER CYCLING. A method of imposing a cyclic stress on an assembly of microelectronic components by applying cyclic power to a heat generating component in the assembly. It is used for accelerated reliability testing of assemblies.

POWER DISTRIBUTION. The network of conductors throughout the package that supplies the operating voltages and currents to the circuits.

PRESSURE CONTACT. Mode of interconnection where the contact points are not fully bonded (as in soldered) but maintain electrical contact by means of a continuously applied force (such as a spring or rubber).

PRINTED-CIRCUIT BOARD (PCB). A composite of organic and inorganic material with external and internal wiring allowing electronic components to be mechanically supported and electrically connected.

PROBABILITY DENSITY FUNCTION (PDF). The normalized frequency of occurrence with respect to a particular statistical distribution, e.g., normal, binomial, Weibull, etc.

PURPLE PLAGUE. One of several gold-aluminum compounds formed when bonding gold to aluminum and activated by re-exposure to moisture. High temperature Purple Plague is purplish in color and is very brittle, potentially leading to time-based failure of the bonds. Its growth is highly enhanced by the presence of silicon to form ternary compounds.

PYROLYZED (BURNED). A material that has gained its final form by the action of heat is said to be pyrolyzed.

Q

QUAD FLAT PAC (QFP). Ceramic or plastic chip carrier with leads projecting down and away from all four sides of a square package.

QUAD IN-LINE PACKAGE (QUIP). A diplike plastic package with leads coming out on 1.27 mm centers. Half of the leads bent close to the body, and the other half projected out for additional 1.27 mm before bent-down.

R

RADIAL-SPREAD COATING. Also known as glob top. A coating process whereby a calibrated amount of resin is dispensed on top of a surface to be encapsulated. The surface can be either a chip or a circuit board. The resin fans out, reacts (by heat input or on contact with air), and forms a solid protective coating.

RADIATION. The combined process of emission, transmission, and absorption of thermal energy between bodies separated by empty space.

REACTION ETCHING. A process wherein a printed pattern is formed by reaction (chemical/plasma ion) removal of the unwanted portion of conductive or insulative pattern.

REACTION INJECTION MOLDING (RIM). A molding process where two (or more) streams of reactants are metered into a small mixing chamber where turbulent mixing breaks up the fluids into finely interspersed striations for faster reaction. The mixture is then delivered to a mold to complete the polymerization.

RECEIVER. The off-chip circuit that accepts the signal voltages and currents from the package lines. Also called an input buffer circuit.

RECEIVER INPUT STABILITY. The assurance that package reactances and high frequency circuit gain and phase shifts do not interact such that excessive ringing or oscillations occur at the receiver's output.

REFLECTION NOISE. Spurious voltage and current wavelets on a transmission lines that are caused by series or shunt networks that disrupt the continuous nature of the characteristic impedance of the line. These wavelets initially travel in a direction that is opposite to the wave that stimulates them. Typical discontinuities that cause reflections are stubs, connectors, vias, missing ground planes and improper terminations.

RELATIVE HUMIDITY (rh). The ratio of partial pressure of water in any gas at a particular temperature to the saturated vapor pressure of water in the same gas at the same temperature, usually expressed in percent.

RELIABILITY. The probability of survival of a component, or assembly, for the expected period of use. Expressed mathematically, R = one minus the probability of failure during the expected life.

REFLOW SOLDERING. A method of soldering involving application of solder prior to the actual joining. To solder, the parts are joined and heated, causing the solder to remelt or reflow.

RENT's RULE. An empirical relation, first recorded by E. Rent of IBM, which states that the number of used input/output terminals on a logic package is proportional to a fractional power of the number of subpackages interconnected in the package.

RESIN. A term used for an organic polymer that when mixed with a curing agent crosslinks to form a thermosetting plastic.

RESIST. A protective coating that will keep another material from attaching or coating something, as in solder resist, plating resist, or photoresist.

RESISTANCE. The property of a conductor that opposes the flow of current by dissipating energy as heat. In packages, it causes voltage and current losses in signal and power distribution systems.

RESISTIVITY (ρ). A proportionality factor that is characteristic of different substances, equal to the resistance that a centimeter cube of the substance offers to the passage of electricity. Expressed $R = \rho L/A$ where R is the resistance of a uniform conductor, L its length, A its cross-sectional area, and ρ its resistivity. Resistivity is usually expressed in ohm-centimeters.

RHEOLOGY. The science dealing with deformation and flow of matter.

ROSIN FLUX. A flux having a rosin base that becomes interactive after being subjected to the soldering temperature.

ROUTING PROGRAM. An automatic program emboding algorithm with prescribed wiring.

RULES DRIVEN DESIGN SYSTEM. The use of sufficiently accurate mathematical expressions that are derived from a limited number of judiciously chosen circuit simulation results using curve fitting techniques to properly design a hardware system in conjunction with a design aids (DA) program. Since these formulas are computationally fast, they can be applied to the design of every data path in the system.

S

SELF-GENERATED NOISE TOLERANCE. A set of differential pulse amplitudes and pulse widths for spurious signals that are generated by the simultaneous switching of internal circuits that can be impressed across a internal circuit's power terminals without falsely setting a downstream latch circuit.

SCREENING. The process whereby the desired film-circuit patterns and configurations are transferred to the surface of the substrate during manufacture by forcing a material through the open areas of the screen using the wiping action of a soft squeegee.

SEALING. Joining the package case header (or chip carrier base or substrate) with its cover or lid into a sealed unit. For hybrids, sealing connotes an important finishing operation in fabricating a hybrid microcircuit, signaling the stage when the assembly, in the form of a populated package, becomes a bona fide hermetic (or nonhermetic) device.

SELF STRETCHING SOLDERING TECHNOLOGY (SST). The acronym for C4 solder joining where two different solder alloys or sized bumps are used so that surface tension forces of the nonfunctional bumps are used to stretch or increase the height of the functional solder joints. Taller connections can withstand higher thermal cycle or power cycle strains.

SHEET RESISTANCE. The electrical resistance of thin sheet of a material with uniform thickness as measured across opposite sides of a unit square pattern. Expressed in ohms per square.

SIGNAL DISTRIBUTION. The network of package conductors that interconnects the drivers and receivers.

SIGNAL WIRING. A conductive path carrying an electric signal.

SILICON EFFICIENCY. The ratio of sum total of area of all silicon chips to the total packaging area—primarily at board level.

SIMULTANEOUSLY SWITCHING DRIVER. A driver circuit that changes state in unison with other drivers on the same chip or nearby chips thereby creating switching noise.

SINGLE-CHIP CARRIER. An electronic package that connects single-chip terminals to second-level package by having a different number of terminations than the chip itself.

SINGLE-CHIP MODULE (SCM). Module or package supporting one chip, as opposed to multichip which supports several.

SINGLE-IN-LINE (SIP). DIP-like package with single line of leads as opposed to two for DIP.

SINGLE-LAYER METALLIZED PACKAGE (SLAM). Ceramic leadless package without cavity, sealed by ceramic or glass to a ceramic cap.

SINTERING. Heating a metal or ceramic powder, thereby causing the particles to bond together to form monolithic body.

SKIN EFFECT. A high frequency effect that causes the resistance of a conductor to increase. This phenomenon occurs because the magnetic fields within the conductor force the current to flow on the outer surface or skin as frequency of the signal increases.

SLOW WAVE PROPAGATION. Energy that travels at less than the expected velocity for a dielectric structure because of series resistance in the signal line or return path.

SLURRY. A thick mixture of liquid and solids. The solids are in suspension in the liquid.

SMALL OUTLINE (SOP). Also called SOIC. Small outline integrated circuit package. It is a rectangular DIP-like package except that it is smaller and leads on 1.27 mm, 1.0 mm, or 0.85 mm spacing. It is meant for surface mounting.

SOFT ERROR. In memory device technology, a memory state error induced by a process which produces no permanent alteration of the physical condition of the device.

SOFT GLASS. Glasses, typically high-lead content glasses, having low softening points that could be used to seal ceramic or metal lids to packages below about 450°C. Also called solder glasses because of their ability to wet most metal surfaces.

SOFTENING POINT. Refers to the temperature at which the log viscosity of glass is 7.6 poises, as defined and measured to ASTM specification.

SOLDER. A low melting-point alloy used in numerous joining applications in microelectronics. The most common solders are lead-tin alloys.

SOLDER DAM. A dielectric composition screened across a conductor to limit molten solder from spreading further onto solderable conductors.

SOLDER GLASSES. Glasses used in package sealing that have a low melting point and tend to wet metal and ceramic surfaces.

SOLDERABILITY. The ability of a conductor to be wetted by solder and to form a strong bond with the solder.

SOLDERING. The process of joining metals by fusion and solidification of an adherent alloy having a melting point below about 300°C.

SOLID LOGIC TECHNOLOGY (SLT). Ceramic package technology practiced by IBM in 1960s by firing *Ag-Pd* conductors on to dry-pressed and fired alumina substrate.

SPACE TRANSFORMER. A package transforming a spatially dense set of chip connections to a less dense set of connection points on package.

SPICE. The "Simulation Program for Integrated Circuit Emphasis" is the industry standard for circuit simulation. It contains many of the features inherent in ASTAP. See ASTAP.

SPUTTER CLEANING. Bombardment of a surface with energetic argon or other noble gas ions to clean the surface of oxide films and residues that could interfere with subsequent electrical or mechanical contact layers. The bombardment knocks off (or sputters) surface atoms to render the surface clean.

SPUTTERING. The removal of atoms from a source by energetic ion bombardment. The ions are supplied by a plasma. The sputtering process is used to deposit films for various thin-film applications.

STATIC FLEX. Flexible wiring circuit carrier, which once installed, remains fixed.

STEINER TREE. See stub.

STENCIL. A planar patterned mask used to transfer images on a surface. Usually metallized patterns on an insulating surface.

STORAGE CONTROL ELEMENT (SCE). Controls the data transfer paths and the interface between the channels, processor storage, and central processor. Also called System Control Element.

STORAGE HIERARCHY. The collection of memory elements (cache, main storage, etc.) and their controls that make up the memory for the processor.

STRESS CORROSION. Refers to the degradation of mechanical properties of brittle materials by crack propagation due to the acceleration of applied stress in the presence of corroding atmospheres such as water.

STRIPLINE. A transmission line that is embedded within a single dielectric medium and sandwiched between two reference planes.

STUB. A short wire which interconnects input at a circuit with the (main) signal line.

STUD. The conductive path that runs vertically from one level of conductors to another in a multilayer substrate.

SUBTRACTIVE PATTERNING. The processing sequence generally followed in producing thin-film networks or circuits. Films are area-deposited (by vacuum evaporation, CVD, or sputtering) and the desired conductive, resistive, etc., pattern is etched into each layer through mask, using appropriate selective etchant fluids.

SUPERCONDUCTOR. Material offering no resistance to the flow of current. In addition to metals, ceramics have been recently discovered to have this property.

SURFACE MOUNT TECHNOLOGY (SMT). A method of assembling hybrid circuits and printed wiring boards, where component parts are mounted onto, rather than into, the printed-wiring board, as in the mounting of components on substrates in hybrid technology.

SURFACE TENSION. An effect of the forces of attraction existing between the molecules of a liquid. It exists only on the boundary surface.

SWITCHING NOISE. An induced voltage on the power distribution system at the circuit terminals caused by the rapidly changing current caused by the simultaneous switching of many drivers.

T

TAPE AUTOMATED BONDING (TAB). The process where silicon chips are joined to patterned metal on polymer tape (e.g., copper on polyimide) using thermocompression bonding, and subsequently attached to a substrate or board by outer lead bonding. Intermediate processing may be carried out in strip form through operations such as testing, encapsulation, burn-in, and excising the individual packages from the tape.

TAPE BALL GRID ARRAY (TBGA). Ball grid array technology on TAB. See TAB.

TEMPERATURE CYCLING. An environmental test where the film circuit is subjected to several temperature changes from a low temperature to a high temperature over a period of time.

TENSILE STRENGTH. The pulling stress which has to be applied to a material to break it, usually measured in MPa.

TERMINAL. A metallic connector or pad to a circuit within a chip or package that permits electrical interconnection to external circuits.

THERMAL COEFFICIENT OF EXPANSION (TCE). The ratio of the change in dimensions to the change in temperature-per-unit starting length, usually expressed in cm/cm/°C. The acronyms TCE and CTE are synomous.

THERMAL CONDUCTION MODULE (TCM). An IBM multichip (100 chips or more) module that is cooled by thermal conduction of pistons in contact with chips.

THERMAL CONDUCTIVITY. The rate with which a material is capable of transferring a given amount of heat through itself.

THERMAL CYCLING. A method to impose a cyclic stress on an assembly of microelectronic components by alternately heating and cooling in an oven. It is used for accelerated reliability testing of assemblies.

THERMAL FATIGUE. Failure of a structural element from repeated temperature excursions, wherein the load develops from thermal expansion mismatch of dissimilar materials.

THERMAL GRADIENT. The plot of temperature variances across the bulk thickness of a material being heated.

THERMAL MISMATCH. Difference in thermal coefficients of expansion of materials which are bonded together.

THERMAL NETWORK. Representation of a thermal space by a collection of conveniently divided smaller parts—each representing the thermal property of its own part and connected to others in a prescribed manner so as not to violate the thermal property of the total system.

THERMAL RESISTANCE (°C/W). The opposition offered by a medium to the passage through it of thermal energy.

THERMOCOMPRESSION BONDING (T/C). A process involving the use of pressure and temperature to join two materials by interdiffusion across the boundary.

THERMOGRAVIMETRIC ANALYSIS (TGA). A technique that measures material weight change as a function of increasing temperature.

THERMOMECHANICAL ANALYSIS (TMA). A technque that measures the linear expansion or other deformations of a material with respect to changes in temperature.

THERMOPLASTIC. A substance that becomes plastic (malleable) on being heated; a plastic material that can be repeatedly melted or softened by heat without change of properties.

THERMOSETTING. The property of some organic materials to irreversibly polymerize and set or harden when heated to some appropriate temperature.

THERMOSONIC BONDING (T/S). A bonding process which uses a combination of thermocompression (TC) bonding and ultrasonic bonding. It is done on what amounts to a gold-wire TC bonder with ultrasonic power applied to the capillary.

THÉVENIN EQUIVALENT. Electrical model describing voltage-current behavior of an electrical network between any two of its nodes. In its simplest form it can be a constant voltage source with a series impedance (resistance in case of direct current) or a constant current source shunted by an impedance. These impedances are often called source impedances.

THICK FILM. A film deposited by screen printing processes and fired at high temperature to fuse into its final form. The basic processes of thick-film technology are screen printing and firing.

THIN FILM. Thin film refers to a coating layer of thickness in the range of from a few (2-3) atomic layers to a few (1-5) microns (micrometers). The important feature distinguishing thin films from thick films, though, is not so much the difference in thickness as the method of deposition which takes place by a variety of techniques such as chemical vapor deposition, evaporation, or sputtering.

THIN-FILM PACKAGING. An electronic package in which the conductors and/or insulators are fabricated using deposition and patterning techniques similar to those used for integrated circuit chips.

THREE-LAYER TAPE. An interconnection medium used in tape automated bonding (TAB), where the tape is comprised of three layers of metallization (usually copper), with polymer and adhesive in between.

THROUGH HOLE. A hole connecting the two surfaces of a printed-circuit structure.

TIME-DOMAIN REFLECTION. A time varying voltage and current disturbance created at a discontinuity on a transmission line that travels in a direction opposite to its stimuli, thereby causing spurious line noises and signal distortions.

TINNED. Literally, coated with tin, but commonly used to indicate coating with solder.

TINNING. To coat metallic surfaces with a thin layer of solder.

TOP SIDE METALLURGY (TSM). An acronym referring to the metallization on the top side of a substrate to which a chip is joined (such as a C4 solder connection).

TOPOGRAPHY. The surface condition of a film; bumps, craters, etc.

TRANSFER MOLDING. An automated type of compression molding in which a preform of plastic (usually an epoxy-based resin) is poured from a pot into a hot mold cavity.

TRANSFER UNITY GAIN POINT. The point on a logic circuit's V_{out} vs. V_{in} transfer curve where the output voltage equals the input voltage. It determines the input signal swing at which noise will propagate and amplify through cascaded logic circuits.

TRANSIENT MISMATCH. Thermal mismatch between elements of a structure which, because of thermal lag, varies with time until reaching a steady-state value.

TRANSISTOR OUTLINE (TO) PACKAGE. An industry standard package designation established by JEDEC of the EIA.

TRANSISTOR-TRANSISTOR LOGIC (TTL). A Nand logic function is implemented by the switching of voltage changes on distinct emitter inputs of bipolar transistors sharing common base and collector voltages.

TRANSMISSION LINE. A conductor that is inductively and capacitively coupled to a nearby return path forming a uniform iternative distributed network with specific properties.

TRANSMITTED NOISE TOLERANCE. A set of pulse amplitudes and pulse widths for spurious signals at the input to a receiver circuit that will not falsely set a downstream latch.

TRI-PLATE LINE. Same as a stripline. See stripline.

TWO-LAYER TAPE. A primary form of tape fabrication for tape automated bonding (TAB), starting with the metallic sputtering and subsequent pattern-plating on Kapton carrier tape. No adhesive is used in bonding copper to Kapton.

U

ULTRA LARGE SCALE INTEGRATION (ULSI). So far, an extreme in the circuit integration, used to indicate presence of one hundred million transistors (or more) on a single semiconductor chip.

ULTRASONIC BONDING. A process involving the use of ultrasonic energy and pressure to join two materials.

V

VACUUM DEPOSITION. Deposition of a metal film onto a substrate in vacuum by metal evaporation techniques.

VAPOR PHASE REFLOW. The technique for solder reflow to form package interconnections. The solder joint is heated by the heat of condensation of an inert vapor. The most common material of choice is a perfluorocarbon.

VERY HIGH SPEED INTEGRATED CHIP (VHSIC). Very high speed integrated circuit, originally referring to 1.0 μm ground rules.

VERY LARGE SCALE INTEGRATION (VLSI). Level of integration with more than approximately ten thousand transistors on a single semiconductor chip. Upper boundary not well defined.

VIA. An opening in the dielectric layer(s) through which a riser passes, or else whose walls are made conductive.

VIA, FIXED. A via built into a package on a predetermined grid, and in general, interconnecting both adjacent and nonadjacent planes.

VIA, PROGRAMMABLE. A via interconnecting adjacent wires on two adjacent wiring planes. Location does not correspond to the same grid locations as fixed vias.

VIA, SEGMENTED. A fixed via interconnection. A predetermined subset of all wiring planes.

VIA, THROUGH. A fixed via passing through all wiring planes.

VISCOSITY. The intrisic property of a fluid that resists internal flow by offering counteracting forces.

VOLTAGE SLEW RATE. The rate of change in voltage with respect to time (dv/dt).

W

WAFER. Commonly, a slice of a semiconductor crystalline ingot used for substrate material when modified by the addition, as applicable, of impurity diffusion (doping), ion implantation, epitaxy, etc., and whose active surface has been processed into arrays of discrete devices or ICs by metallization and passivation.

WAVE SOLDERING. The technique for solder application and reflow in which a jet of liquid solder is directed at the two metallic points to be interconnected. The technique usually involves processing steps to apply flux and remove excess solder.

WEAROUT. The time following the stable failure-rate period during which the expected, or observed, failure rate of an item increases and exceeds a specific value.

WELDING. Joining two metals by applying heat to melt and fuse them with or without a filler metal.

WETTING. The spreading of molten solder or glass on a metallic or nonmetallic surface, with proper application of heat and flux.

WIRE LENGTH, AVERAGE. The average length measured in logic unit pitches, of all connections in a given package level.

WIREABILITY. The capability of a package to permit the interconnection of subpackages mounted on it and terminals attached to it measured as the probability of wiring success. It is near one when sufficient wiring capacity, via availability and terminal access, are present.

WIREBOND. A completed wire connection whose constituents provide electrical continuity between the semiconductor die (pad) and a terminal. These constituents are the fine wire; metal bonding surfaces like die pad and package land; and metallurgical interfaces between wire, and metals on both the chip and substrate.

WIREBONDING. The method used to attach very fine wire to semiconductor components in order to interconnect these components with each other or with package leads.

WIRING (ALSO ROUTING). The manual or automatic prescription of routes (portions of tracks) for wires interconnecting package components or logic cells on chips.

WIRING ASSIGNMENT. The manual or automatic prescription of particular pads, pins, connectors, or terminals to which corresponding wires are to be attached.

WIRING CAPACITY. The total available length of wiring tracks in a package (before any wires are prescribed on the wiring image).

WIRING CHANNEL. A linear region on a package wiring plane containing space for at least one wiring track.

WIRING DEMAND. The product of wiring connection count and average connection length, either locally in a limited region or total over an entire package.

WIRING DENSITY. Total wire length contained within a unit square. Measured in inches per square inch or centimeters per square centimeter.

WIRING OVERFLOW. A wiring connection called for by logic design but not inserted in a proposed package wiring image during prior automatic wiring use of package.

WIRING RULES. A set of electrical constraints that are used in conjunction with a design aids program to control the topological parameters of an interconnection network to assure proper functionality. Wiring rules are usually derived from electrical simulation results.

WIRING TRACK. A linear extent of space in a wiring channel used to contain one (or more if collinear) conducting wires used to interconnect package components.

Y

YELLOW WIRE. Discrete wires that are yellow in color interconnecting terminals on a package. Originally used in reference to all back panel wiring on early electronic assemblies.

Z

ZERO-INSERTION-FORCE CONNECTION (ZIF). A form of connector that allows the connector pins to be brought together under very low force, then wiped and pressed together during cam activation.

SYMBOLS

See Chapter 4, “Heat Transfer in Electronic Packages,” section 4.7, “Nomenclature,” for symbols used in the field of heat transfer.

$AgNO_3$: Silver nitrate
AgO/PdO: Silver oxide/palladium oxide
$AgPd$: Silver-palladium alloy
$A + Pd$: Silver + palladium
$Ag\text{-}Pd$: Silver-palladium
$AgPd\ Au$: Silver-palladium-gold ternary alloy
A/PdO: Silver/palladium oxide
$Al\text{-}Cu$: Aluminum-copper
AlN: Aluminum nitride
Al_2O_3: Aluminum oxide
$3Al_2O_3\ 2SiO_2$: Mullite
$AuAl_2$: Gold-aluminum alloy
Au_5Al_2: Gold-aluminum alloy
$AuGe$: Gold-Germanium alloy
$AuPt$: Gold-platinum alloy
$AuSi$: Gold silicon alloy
$AuSn$: Gold-tin alloy
$AuSn_2$: Gold-tin alloy compound
$AuSn_4$: Gold-tin alloy compound
Au_3Sn_4: Gold-tin alloy compound
BN: Boron nitride
B_2O_3: Boron oxide
$B_2O_3 + SiO_2 + (Al_2O_3)$: Borosilicate glass + alumina
$B_2O_3—SiO_2—Al_2O_3nashNa_2O$: Alumino borosilicate glass
$BaCl_2$: Barium chloride
BeO: Beryllium oxide
$C_6H_5—CH_3$: Toulene
CH_3CHO^+: Butyraldehyde
$C_2H_3O_2$: Acetate
CO_2: Carbon dioxide
C_2O_4: Oxalate ion
$CaCl_2$: Calcium carbide

CaOclon: Calcium oxide
$CaO + Al_2O_3 + SiO_2 + B_2O_3$: calcia-alumino-borosilicate
$Ca(OH)_2$: Calcium hydroxide
Cu—Ag—Ti—Sn: Copper-silver-titanium-tin
$CuAl_2$: Copper-aluminum alloy compound
Cu—Mo(20% *Cu*): Copper-molybdenuim alloy
$CuSO_4$: Copper sulphate
Cu—Sil: Copper-silver alloy
Cu_3Sn: Copper-tin alloy compound
Cu_6Sn_5: Copper-tin alloy compound
Cu—W(20%*Cu*): Copper-tungsten alloy (20% copper)
$Fe(CN)_6$: Ferrous cyanide
FeNi: Iron-nickel alloy
GaAs: Gallium arsenide
GeO_2: Germanium oxide
HCl: Hydrochloric acid
In—Cu—Sil: Indium-copper-silver
KBr: Potassium bronide
KCl: Potassium chloride
$KClO_4$: Potassium chlorate
$K_3Fe(CN)_6$: Potassium ferrocyanide
$K_4Fe(CN)_6$: Potassium ferricyanide
$KHCO_3$: Potassium bicarbonate
KI: Potassium iodide
KIO_4: Potassium iodate
KNO_3: Potassium nitrate
$KReO_4$: Potassium perrhenate
$LaCl_3$: Lanthanum chloride
LiCl: Lithium chloride
$LiClO_4$: Lithium chlorate
$Li_2O + Al_2O_3 + SiO_2Al_2O_3$: Lithium alumino
$Li_2O—Al_2O_3—SiO_2—B_2O_3$: Alumino borosilicate
$Li_2O + SiO_2 + MgO + Al_2O_3 + SiO_2$: Magnesium alumino silicate glass
$MgCl_2$: Magnesium chloride
MgO: Magnesium oxide
$2MgO_2Al_2O_3.5SiO_2$: Cordierite ceramic
$MgO + Al_2O_3 + SiO_2 + B_2O_3 + (Al_2O_3)$: Magnesium alumino borosilicate glass + alumina
$MgO—Al_2O_3—SiO_2—B_2O_3—P_2O_5$: Magnesium alumino borosilicate glass con-
...ining boron oxide and phosphorous oxide
...odium chloride
...dium perchlorate
...lide
...ium acetate

$NaOOCC_2H_5$: Sodium propionate
$NaOOCC_3H_7$: Sodium butyrate
$NaOH$: Sodium hydroxide
Na_2SO_4: Soldium sulphate
Nb_2O_5: Niobium oxide
$Nd—Yag$: Neodium-yag laser
$NiCr$: Nichrome
$NiCr—Pd—Au$: Nichrome-palladium-gold
NH_4: Ammonia radical
NH_4Cl: Ammonium chloride
NH_4ClO_4: Ammonium perchlorate
Ni_3Sn_4: Nickel-tin alloy compound
NO_2: Nitrous oxide
NO_3: Nitrate ion
$PbIn$: Lead indium alloy
$Pb—In$: Lead-indium
50Pb 50In: Lead (50%)-Indium (50%) alloy
PbO: Lead oxide
$PbO + B_2O_3 + SiO_2 + (Al_2O_3)$: Lead borosilicate
$PbO—B_2O_3—SiO_2—Al_2O_3—ZnO$: Zinc lead borosilicate glass
$PbO—ZnO—Al_2O_3—B_3O_3—SIO_2$: Lead-zinc borosilicate-glass (solder glass)
P_2O_5: Phosphorous pentoxide
95 $PB/5$ Sn: Lead (95%-tin (5%) alloy
RuO_2: Ruthenium oxide
S_g: Free sulphur
$Si—Au—Sn$: Silicon-gold-tin
SiC: Silicon carbide
Si_3N_4: Silicon nitride
SiO_2: Silicon dioxide
95Sn 3.5Ag 1.0 Cd, 0.5 Sb: Tin (95%)-Silver (35%)cadmium (1%)-antimony (5%) alloy
42Sn 58Bi: Tin (42%)-bismuth (58%) alloy
50Sn 50In: Tin (50%)-Indium (50%) alloy
40Sn 60Pb: Tin (40%-lead (60%) alloy
60Sn—40Pb: Tin (60%)-lead (40%) alloy
63Sn 37Pb: Tin (63%)-lead (37%) alloy
95Sn 5Sb: Tin (95%)-antimony (5%) alloy
SO_2: Sulphur dioxide
SO_4: Sulfate ion
$SrCl_2$: Strontium chloride
Ta_2N: Tantalum nitride
TiC: Titanium carbide
TiO_2: Titanium dioxide
$Ti—Pd—Au$: Titanium-palladium-gold

$YBa_2Cu_3O_7$: Yttria-baria-copper oxide superconductor
Y_2O_3: Yttrium oxide
$ZnCl_2$: Zinc chloride
ZnO: Zinc oxide
$ZnSO_4$: Zinc sulphate
ZrO_2: Zirconium dioxide

AUTHORS' BIOGRAPHIES

Nanda G. Aakalu—*Advisory Engineer—IBM, Poughkeepsie, NY.* Mr. Aakalu is currently involved with electronic packaging at the card and board level, and with cooling-related development work in mainframe computers. Employed by IBM for the last 19 years, he has been involved with the advanced technology area of direct immersion cooling of VLSI chips, development of thermal compounds for internal thermal enhancement of multichip packages, reliability study of air moving devices, and mechanical analysis of power components. He also directed the development of terminator resistor circuits on porcelain steel substrates. His fields of expertise are heat transfer, stress/strain analysis, and process engineering of thick-film packages. He received his BSME from the University of Mysore, India, and an MSME from Purdue University. He is a member of ISHM and has an ASME publication on condensation heat transfer as well as thirty invention publications in the IBM technical disclosure bulletins related to cooling and packaging of electronics and two patents on cooling and packaging of electronics. (A current biography was not available at the time of publication.)

Vincent W. Antonetti—*Professor and chair; department of mechanical engineering—Manhattan College, Riverdale, NY.* Prior to joining Manhattan College, he was a senior engineer at IBM, where as manager of the IBM Poughkeepsie Thermal Engineering Laboratory, he was responsible for the thermal design of high-end computers. His areas of expertise are heat transfer in electronic equipment and thermal contact resistance. He has a BME from the City College of the City University of New York, a MSME from Columbia University, and a PhD from the University of Waterloo in Canada. He is a fellow of the American Society of Mechanical Engineers, and a licensed professional engineer. He has four IBM Invention Achievement awards, and is the author of 32 publications.

Masaya Asano—*Manager, Electronic & Imaging Materials Research Labs—Toray Industries, Inc., Shiga, Japan.* During his 25 years at Toray, Mr. Asano has been involved with imaging and electronic materials areas including waterless planographic printing plates and polyimides for microelectronics. He received the awards from the Japan Institute of Invention and Innovation for Positive Acting, Waterless Planographic Printing Plate, and from the Society of Polymer Science, Japan, for the development and commercialization of photosensitive polyimide. He received his BS and MS in chemistry form Kyoto University.

Donald E. Barr—*Senior Technical Staff Member and Manager of Site Technical Assurance and Material Engineering—IBM Endicott, NY.* Prior to joining IBM, he was the Technical Director of Research and Development at GAF Photo and Repro Division in Binghamton, New York. He received his Ph.D. from the University of Massachusetts, where he was awarded an NDEA Fellowship, is a member of the American Chemical Society and has publications and patents in the area of electronic applications and fundamental science.

J. Richard Behun, P.E.—*Development Engineering Manager—IBM, Essex Junction, VT.* Mr. Behun is presently charged with definition and implementation of module test handling equipment for memory, logic, and mixed signal products. Prior to this, much of his career was devoted to surface-mount development for multi-layer ceramic products, where he earned the bulk of his 16 inventions including four patents.

Rick was awarded first and second level IBM Invention Achievement awards, an IBM Division award, as well an IBM Achievement Award.

Rick is a member in the American Society of Mechanical Engineers, where he is presently the chairperson of the Design for Manufacturability Committee within the Design Division. Rick is a licensed professional engineer in Vermont and New York. Rick joined IBM in 1980 after earning his B.S. and Masters in Mechanical Engineering at Rensselaer Polytechnic Institute.

Robert L. Beliveau—*Senior Technical Specialist—DuPont High Performance Films, Circleville, Ohio.* During his 22 years with DuPont; Bob has been involved with both materials processing and product development activities in the areas of Flexible Printed Circuits, Tape Automated Bonding and Multi-Chip Modules.

Bob is a member of the IPC, the Society of Vacuum Coaters and the Surface Mount Technology Association.

Bob received his BS in Mechanical Engineering from the University of Massachusetts

Frank J. Bolda—*Advisory Engineer (retired)—IBM.* Frank J. Bolda joined the company in 1956 in the Technical Services Laboratory in Poughkeepsie, where he worked on various plating and heat treating processes, failure analysis, and the development and evaluation of tape slitting devices. From 1960-1965, he worked in the Mechanical Memory Group in the development and packaging of ferrite core memories. In 1966 he became a process engineer in the Research Triangle Park facility in North Carolina supporting the site manufacturing processes. In 1972 he joined the Development Laboratory and was involved in the development of honeycombed cabinetry, keyboard technology, and flexible substrate packaging. More recently he was involved in the development of glass substrate for the scanner products, coated metal substrates applications, and connector development.

He is a co-inventor of six patents and is a member of the American Society for Metals and the Electronic Connector Study Group. He has four publications in the field of coated metal substrates and electronic packaging.

Thomas Caulfield—*Senior Engineer—IBM, East Fishkill, NY.* Dr. Caulfield manages product development, applications, design and analysis for Ceramic

Chip Carriers. Over his 7 year career at IBM he has held various engineering and management positions related to flip-chip and BGA packaging. He holds numerous US patents and is considered a subject-matter expert in electronic packaging. Dr. Caulfield has a Doctorate of Engineering degree from Columbia University (1986) and prior to his work at IBM, he was a Senior Member of the Technical Staff at Philips Laboratories, in Briarcliff Manor, NY.

Abhijit Chatterjee—*Assistant Professor; School of Electrical and Computer Engineering—Georgia Institute of Technology, Atlanta, GA.* Chatterjee received the Ph.D. degree in electrical and computer engineering from the University of Illinois at Urbana-Champaign in 1990 and had ten years experience with General Electric Company before joining Georgia Tech. His research interests are in the fields of mixed-signal MCM testing, fault-tolerant computing, low-power circuit design, computer algorithms and design automation. He is a collaborating partner in NASA's New Millenium Project, is the author of one US patent and has over fifty publications in refereed journals and conferences. He has received two Best Paper awards and has twice been nominated for Best Paper awards. He received the NSF Research Initiation Award in 1993, the NSF CAREER Award in 1995 and is a Senior Member of the IEEE.

William T. Chen—*Senior Technical Staff Member—IBM, Endicott, New York.* He joined IBM in Endicott in 1963. He has worked in various technical and management positions in mathematical sciences, materials laboratory and in the last 17 years in the electronic packaging area. He is a graduate of Queen Mary College, University of London. He received a M.Sc. from Brown University, and a Ph.D. from Cornell University. He is a fellow of the American Society for Mechanical Engineers.

Tsuneyo Chiba—*Senior Chief Researcher, Central Research Laboratory—Hitachi Ltd., Kokubunnji, Tokyo 185, Japan.* He received the B.S. degree in E.E. in 1962 and the Ph.D. degree in I.S. in 1988, both from Kyoto University, Japan.

He joined the CRL, Hitachi Ltd. in 1962. Since then, he has been engaged in the development of many generations of Hitachi's mainframe computers, such as HITAC 5020, 8800, M-180, M-200H/280H, M-680, M-880, and the latest MP5800. He has taken charge of the corporation's large-scale projects to develop LSI-based high- end mainframe computers, leading the development of hardware technologies such as very high-speed ECL gate arrays and their packaging system as well as design technologies. He has also directed and managed the developments of disk files, telecommunication LSIs and network systems, as well as computer systems, serving as department manager in the CRL from 1981 to 1988.

He is currently providing directional guidance in the development of advanced technologies and future computer systems.

Dr. Chiba is a member of the IEEE, the IEICE of Japan, and the IPSJ.

David B. Clegg (A current biography was not available at the time of publication).

Donald S. Cleverley—*Owner and principal consultant—DSC Quality Consultancy, Poughkeepsie, NY.* Mr. Cleverley is a NY Quality Consultant and Certified

ISO-9000 Auditor He assists client companies in ISO-9000 documentation, training, registrar selection, and pre-registration audits. Previous to this, he was an Advisory Engineer with the IBM Microelectronics Division, East Fishkill, NY. He was responsible for Reliability and Quality projections, measurements and improvement actions, including achieving IBM's "six-sigma" quality improvement objectives for semiconductor devices. Previously in IBM, he managed Analog Circuit Test Engineering and introduced the Manufacturing Accelerated Release System (MARS) for new chips. He holds a BSEE degree from Northeastern University and an MSEE degree from Syracuse University. He is a Senior Member of IEEE and an ASQC Certified Quality Engineer and Quality Auditor. He is the author of seventeen papers and conference presentations dealing with reliability and quality improvement and instructs quality courses at the college level.

Marie S. Cole—*Advisory Engineer—IBM, East Fishkill, NY.* Ms. Cole has a B.S.Ch.E. from Rensselaer Polytechnic Institute and an M.S. in Materials Science from Columbia University. She joined IBM in 1984, working until 1988 in Corporate Component Procurement—Assurance and Test qualifying new SMT plastic packages and setting the standards for testing the impacts of SMT assembly on packaging. She earned a Division Award for that work. The author of numerous publications and one patent related to BGA packaging, she has worked on the development of CBGA and CCGA packaging since 1988. Her current assignment is in New Product Applications in the Ceramic Chip Carrier Business Unit, specifically in MLC Packaging Applications.

Frank Crnic—*Brand Acquisition Manager: Mobile Products—IBM, Research Triangle Park, NC.* Frank Crnic joined IBM in 1981 after receiving his Bachelor of Science degree in Materials and Metallurgical Engineering from the University of Michigan. He was involved in a variety of process development efforts in the East Fishkill, NY Semiconductor Laboratory. In March, 1985, he joined the East Fishkill Multi-Layer Ceramic (MLC) Packaging Plant as Manager of Sintering Process Engineering. He became involved in automated optical inspection as Manager of Inspection Systems Engineering in 1987. In 1988 he became Manager of Test Systems Engineering. In this assignment, he was responsible for implementing the Electrical Module Test (EMT) for screening latent opens defects. In May, 1993, Mr. Crnic was assigned to IBMs Kiosk Solutions group in Atlanta, GA as a Project Manager. Since July, 1995, he has been Mobile Computing Brand Acquisition Manager with the IBM Personal Computer Company in Research Triangle Park, NC.

George Czornyj—*Senior Engineer—IBM, Hopewell Junction, NY.* During his 20 years at IBM he has been involved with many technology areas utilizing his expertise in high temperature polymers (polyimides), photosensitive polyimides, plasma polymerization of organo and organometallic films, and adhesion of polymer/metal/ceramic materials to develop materials for thin film packaging applications. He has authored over 40 professional articles and obtained over 20 patents and invention disclosures in the field of thin film packaging, processing and lithography. He has participated on numerous ACS Polymer Division Executive Committees, as a past Chairman of the Speakers Bureau and as a member of the ACS Organic Coatings & Plastics Division, ACS Polymer Division and Materials Research Society. He received the Arthur K. Doolittle Award in 1976 with Dr. B. Wunderlich.

He received his undergraduate schooling in chemistry and physics from Rensselaer Polytechnic Institute, graduate schooling in Polymer Physical Chemistry from Rensselaer Polytechnic Institute and was a Post Doctoral Scientist at IBM Research at San Jose, CA working on Plasma Polymerization.

Evan E. Davidson—*Senior Technical Staff Member—IBM, Poughkeepsie, NY.* Mr Davidson is currently a technology applications engineer for OEM products. Previously, he was the Manager of High Performance Technology for IBM's General Technology Division located in East Fishkill, NY. While in that capacity, he specialized in the electrical package design for the high performance MCMs used in mainframe computers. Before that he was a circuit designer for logic and memory circuits at both IBM and Bell Labs. He received his BEE from Rensselaer Polytechnic Institute and his MSEE form New York University. He is a Senior Member of the IEEE and a member of the Eta Kappa Nu and Tau Beta Pi engineering honor societies. He has written numerous articles and presented many papers on the subject of electrical package design. He holds many patents in the areas of digital circuit design and package design.

Philip E. Garrou—*Chief Scientist—Microelectronics, Dow Chemical, Central Research & New Businesses, Research Triangle Park, NC.* During his 20+ years at Dow, Dr. Garrou has been involved with both electroceramics and polymers for microelectronic packaging and interconnect applications. He has authored 50+ professional articles and co-edited the book Thin Film Multichip Modules.

Dr. Garrou is a senior member of IEEE CPMT where he has served as Chairman of TC-5 (Materials) and was elected to the Board of Governors.

He was elected Technical VP of ISHM and has served as Chairman of the MCM/Advanced Packaging subcommittee of the ISHM National Technical Committee and Chairman of the ISHM Materials Division. He is co-founder of the ISHM/EEE Ojai Workshop on Advanced Materials and was Technical Chair of the 4th International MCM Conference in Denver.

He received his B.S. and Ph.D. in Chemistry from North Carolina State University and Indiana University respectively.

John B. Gillett—*Senior Technical Staff Member (retired)— IBM, Kingston, NY.* At the time of his retirement, Mr. Gillett was the manager of Advanced Systems Packaging working in a wide range of advanced technology activities, from semiconductor devices to total system packaging. Prior to joining IBM, he was with the Automatic Telephone and Electric Company, United Kingdom, designing telephony switching equipment and early electronic computers. His field of expertise is high-performance computer hardware technologies. He received his MSEE in London. He has twelve patents issued in the fields of semiconductor devices and circuits, data storage systems, connectors and power supplies.

Lewis S. Goldmann—*Senior Engineer—IBM, East Fishkill NY.* For the last 15 years, he has been involved in the mechanical design, testing and modeling of microelectronic chips, packages and materials. Previously, he worked in the development of flip chip interconnections and the IBM Thermal Conduction Module. He has published widely in these areas, including many papers on flip chip joining. Mr. Goldmann's formal education includes a pre-engineering liberal arts degree from Queens College, CUNY; a B.S. in Mechanical Engineering from Columbia Univ.; and an M.S.M.E. from MIT. He is a member of ASME and serves as Associate Technical Editor of its Journal of Electronic Packaging.

Dimitry G. Grabbe—*Director, Electronic Interconnection Research—AMP Inc., Harrisburg, PA.* An AMP Fellow since 1989, Dimitry G. Grabbe comes from a background that includes a military-academy education in the former Yugoslavia and machine design education in Germany. He joined AMP in 1973 after having worked for Photocircuits Corporation, the Main Research Corporation (of which he was founder and president) and Rockwell International Autonetics. During his career at AMP, Mr. Grabbe has held the positions of Manager of Advanced Product and Manufacturing Technologies; Director of Applied Technology; Director of Interconnection Research and Applied Technology; and Director of Electronic Interconnection Research. He has won many awards, including the Man of the Year Award from the International Technology Institute, the Leonardo da Vinci Award from the American Society of Mechanical Engineers and the Symposium Honor Award from the International Institute of Connector and Interconnection Technology. He is a member of the International Technology Institute Hall of Fame, and an IEEE Fellow. Mr. Grabbe is AMPs most prolific holder of patents, with a total of some 127 (US and 2 Foreign) of which 116 are with AMP.

Tapan K. Gupta—*Consultant Engineer, Electronic Sensors & Systems Division—Northrop Grumman Corporation, Baltimore, MD.* From 1985 to 1993, he was a Senior Fellow and the Manager of the Electronic Packaging Center at Pittsburgh in charge of packaging development activities for Alcoa Electronic Packaging, Inc., San Diego. From 1967 to 1985, he was an Advisory Scientist at Westinghouse Research Laboratories at Pittsburgh. He earned his MS and Sc.D. from Massachusetts Institute of Technology in Ceramics in 1964 and 1966, respectively. A Fellow of the American Ceramic Society, he has authored or co-authored more than 100 papers, three book chapters, and been issued more than 30 US patents. His present activities lies in the manufacture of advanced microelectronics packaging and other functional devices for digital and microwave applications.

William R. Heller—*Senior Physicist (deceased)—IBM, Poughkeepsie, NY.* Dr. Heller passed away in June of 1994. He was working on this update at the time. His material was passed on to Dr. Rose who incorporated it into this work and completed the revision. Dr. Heller has been missed, but his work continues to show through in this revision. He was a member of the IBM Fellow Department in Poughkeepsie. His expertise was in computer-aided design tools, packaging, and wireability analysis for digital systems. Prior to joining IBM, he was a research associate at the University of Illinois, an assistant professor at Yale University, and worked at the Shell Development Corporation. In 1979, he was a visiting professor of computer science at Caltech. He received a BS in physics from Queens College, City University of New York, an ScM in applied math from Brown University, and a PhD in solid state physics from Washington University. He was a fellow of the American Physical Society, a fellow of the IEEE (Computer Society), and a member of the New York Academy of Sciences. He had approximately forty publications and patents in solid state physics, materials sciences, computer-aided physical design, testing techniques, and wireability analysis.

Hiroo Hiramoto—*Director and General Manager of Electronic and Imaging Materials Research Laboratories—Toray Industries, Inc., Shiga, Japan.* Mr. Hiramoto obtained an B.A.Sc. and M.A.S.c. in Applied Chemistry from the University of Tokyo. He joined Toray Industries, Inc. in 1965. His principal research interests are in electronic materials based on polymers, especially high temperature polymers. He is one of the inventors of ionic-type photo-definable polyimides. In 1991, he received an award from the Society of Polymer Science, Japan for the development and commercialization of photosensitive polyimide.

Robert T. Howard—*Senior Engineer (retired)—IBM, Burlington, VT.* Dr. Howard specialized in Reliability and materials applications and manufacturing procedures for microelectronic packaging. He joined the Federal Systems Division in Huntsville, Alabama where he was lead engineer for establishment of the laboratory for microelectronics packaging. He also pioneered studies for LASER trimming of thick and thin film resistors. In Burlington he was a lead engineer in development of chip joining studies and quality and reliability studies on solders and chemical degradation of packages. Dr. Howard has co-authored several articles on creep and fatigue of solders employed in chip joining.

Prior to joining, IBM, Dr. Howard was Professor of Metallurgy and Materials Engineering at the University of Kansas. Other teaching positions were at Wichita State University and the University of Missouri at Kansas City. He received BS and Sc.D. degrees, both in metallurgy, from the Massachusetts Institute of Technology. He was now consultant in materials, processes and reliability in microelectronics. He was a Registered Professional Engineer in Missouri and Kansas.

Dr. Howard was a senior member of the Institute of Electrical and Electronic Engineers. And he served as Vice-President for Education of the IEEE Components, Packaging, and Manufacturing Technology Society (CPMT). In addition, he has been a member of the Reliability Program Committee of the Electronic Components and Technology Conference (ECTC) for several years, and has chaired many Reliability sessions of ECTC. He was a member of the International Society for Hybrid Microelectronics (ISHM) the American Society for Materials, and the American Welding Society. Dr. Howard has published many papers and articles reporting, research in technical and trade journals.

Akihiko Ikeda—*General Manager, Corporate Research Laboratories, Electronics Materials & Devices Laboratories— Asahi Chemical Industry Co., LTD. Fuji, Japan.* During his 21 Years at Asahi, Dr. Ikeda has been involved with research & development of functional polymers, especially ion exchange resin and photopolymers. He was engaged in research & development of dry film resist for printed circuit board and ester type photosensitive polyimide for microelectronics use. He was co-author of the book "Development in UV curable resin" (in Japanese)

He received his Ph.D. in Engineering from Kyoto University in 1975.

Dexter A. Jeannotte—*Senior Engineer (retired)—IBM, East Fishkill, NY.* Dr. Jeannotte was responsible for materials reliability issues as they relate to lifetime assessment of electronic assemblies, focusing on failure mechanism studies. His concentration had been on solder fatigue and in environmental corrosion studies and electrical contact quality studies. He received a BME degree from Marquette University, an MS in metallurgical engineering from Columbia University, and a PhD in metallurgy from Columbia University. He was chairman of ASTM Committee B04 on Metallic Materials for Thermostats and for Electrical Resistance, Heating and Contacts, and chairs the ASTM B04.04.07 Task Group on an Environmental Testing Standard for Aluminum Electrical Connection Systems. He was vice-chairman of IEEE/CHMT Society TC-1 Committee on Electrical Contacts, and chaired the TC-1 Task Group 4 on Environmental Standards for Electrical Contacts. He was also a member of NACE, ISHM, MRS, IEPS and AAAS. He has coauthored four patents and 15 publications and presentations.

Nan Marie Jokerst—*Associate Professor of Electrical and Computer Engineering—Georgia Institute of Technology, Atlanta, GA.* She received her Ph.D. from the University of Southern California in 1989 in the area of semiconductor nonlinearities. She was a Hewlett Packard Fellow and a Newport Research Award Winner while at USC, and a Summer Intern at the IBM Watson Laboratories in 1982 and 1983. She joined the Electrical Engineering faculty at the Georgia Institute of Technology in 1989, and won a DuPont Young Faculty Award, a National Science Foundation Presidential Young Investigator Award, and three teaching awards. She is co-leader of the Optoelectronics and High Speed Electronics Thrust Area in the Georgia Tech National Science Foundation Engineering Research Center on Packaging. She has published and presented over 70 papers, two book chapters, and has 3 patents. She is on the IEEE Lasers and Electro-Optic Society Board of Governors and the Optical Society of America Engineering Council. Her current research is highly collaborative and interdisciplinary, and focuses on the optimization and integration and packaging of thin film optoelectronic and high speed electronic devices with host substrates such as silicon integrated circuits to form cost effective smart pixel systems. Her professional activities include Chair (1993, 1994) and Programs Chair (1991, 1992: Most Improved Chapter Award) of Atlanta IEEE LEOS, and Secretary and Treasurer of Atlanta IEEE (1995 and 1996). She was elected to the IEEE LEOS Board of Governors in 1995. She has served on the program committee for the IEEE LEOS Annual Meeting (1993, 1994, 1995, 1996); the Topical Meeting on Integrated Optoelectronics (1994); and the Topical Meeting on Smart Pixels (1994). She is a member of the IEEE LEOS Subcommittee on Optical Interconnects and Processing Systems, was Chair of the Optical and Optoelectronic Interconnect, Switching, Processing, and Storage Committee of the Conference on Lasers and Electro-Optics (1995, 1996), and the Program Chair for the LEOS Topical Meeting on

Smart Pixels (1996). She is also a Member of the Engineering Council of the Optical Society of America (1995–1998); Member of the OSA Advisory Board for Optics and Photonics News (1992–1996); Chair (1992) and Member (1991) of the Newport Research Award Committee; Chair (1995) and Member (1996) of the New Focus Award Committee; on the Program Committee of the OSA Optical Computing Conference (1994); and Chair of the Symposium on Smart Pixels, OSA Annual Meeting (1995). She was the Optoelectronics Representative on the National Technical Council of the International Society for Hybrid Microelectronics (ISHM, 1995), was Co-Chair of the Optoelectronics and Sensors Committee of the ISHM Annual Meeting (1994); Conference Chair, ISHM Optoelectronics II (1995); and Co-Chair of the Optoelectronic Materials Committee of the ISHM Materials Packaging Conference (1995). She was also the coordinator of the CO-OP Short Course "Optoelectronics Integrated Onto Silicon VLSI: Devices, Circuits, Systems" (1995).

George A. Katopis—*Senior Engineer—IBM, Poughkeepsie, NY.* Mr. Katopis is the technical leader for High Performance Servers 1st Level Package design. His fields of expertise are electrical package design, electrical noise containment, and signal integrity. He received his MS and MPh in Electrical Engineering and Computer Science from Columbia University in 1972 and 1978 respectively. He has written numerous articles and presented many papers on electrical noise characterization and containment as well as on signal integrity issues. He holds several patents in the area of noise containment and prediction.

Robert W. Keyes—*Research Staff Member Emeritus—IBM T. J. Watson Research Center, Yorktown Heights, NY.* Dr. Keyes has long engaged in research and development activity in modern optical and electronic technologies and their application in digital electronics and electronic devices. He joined the IBM Research Division after several years at the Westinghouse Research Laboratory in Pittsburgh. He received BS, MS, and PhD degrees in physics from the University of Chicago. Dr. Keyes is a member of the National Academy of Engineering and is a Fellow of the Institute of Electrical and Electronic Engineers and of the American Physical Society. He received the IEEE W. R. G. Baker Prize in 1976. He is the author of 140 papers dealing with solid-state electronics and digital information processing, eight issued patents, and a book, Physics of VLSI Systems.

Subash Khadpe—*President—Semiconductor Technology Center, Neffs, PA.* Semiconductor Technology Center is a consulting firm specializing in assembly and packaging. Dr. Khadpe is also the editor and publisher of "Semiconductor Packaging Update", a subscription newsletter. Prior to founding the firm in 1985, he was a member of the technical staff of AT&T Bell Laboratories for seven years and Motorola for four years.

Dr. Khadpe received his M.S.E.E. and Ph.D. degrees from Drexel University in Philadelphia, PA. He also holds B.Sc. (with honors) and M.Sc. degrees from the University of Bombay, India, and the M.Tech. degree from the Indian Institute of Technology, Bombay.

He is a senior member of the IEEE, past associate and guest editor of the

IEEE CHMT Transactions, and past member of the executive board of the International Electronics Packaging Society. He is a member of Tau Beta Pi, Eta Kappa Nu, CPMT, IEPS, ISHM, and SMTA. He has published over 100 papers and articles, and chaired numerous international conferences and workshops.

Bruce C. Kim—*Assistant Professor; Electrical Engineering Department—Tufts University, Medford, MA.* Dr. Kim is currently an Assistant Professor with the Department of Electrical Engineering and Computer Science, Tufts University, Medford, MA. He is a member of International Society for Hybrid Microelectronics (ISHM) and International Electrical and Electronic Engineers (IEEE). He received his Ph.D. in Electrical Engineering from the Georgia Institute of Technology, an MSEE from the University of Arizona, and a BSEE from the University of California at Irvine. He has developed a novel test technique to test Multi-Chip Module substrate interconnections while pursuing his Ph.D. at Georgia Tech. He was previously employed by Georgia Tech Research Institute as a Research Engineer before starting his Ph.D. program at Georgia Tech. He has published 13 conference and journal papers in the areas of testing Multi-Chip Module substrates and mixed-signal circuits. His research interests are in the areas of testing multichip modules and mixed-signal circuits.

Alan (Al) G. Klopfenstein—*President—AGK Enterprises, Hopewell Junction, NY.* Mr. Klopfenstein is involved with the development of strategies for semiconductors and associated packaging for electronic systems. Prior to retiring as a Senior Engineer, he held management, technical, and marketing positions at IBM in semiconductor, packaging, and manufacturing-equipment development. He received a BSME from the University of Connecticut and an MSME from the University of Alabama in Huntsville. He is a member of the National Society of Professional Engineers (NSPE), the International Society for Hybrid Microelectronics, (ISHM), and the International Electronic Packaging Society (IEPS). He is one of the three principal editors of this book.

Nick G. Koopman—*Fluxless Soldering Program Manager— MCNC Center for Microelectronics Systems Technologies, Research Triangle Park, North Carolina.* Dr. Koopman received a BS in Metallurgical Engineering from Lafayette College, and an MS and Ph.D. in Metallurgy and Materials Science from Massachusetts Institute of Technology. He was employed for 23 years with the IBM General Technology Division in East Fishkill, NY, specializing in computer microelectronics interconnection metallurgy development. The last position held was as manager of Bonding and Interconnections where the major responsibility was flip chip development. Since 1960 Nick has been with MCNC developing advanced solder and interconnection technologies. The current assignment involves the development and marketing of the PADS fluxless/no-clean soldering process. He has published extensively in the area of interconnections with 44 papers and holds 61 inventions. He has served as a guest lecturer at several universities and is an associate editor of the International Journal of Microelectronics Packaging.

J. Richard Kraycir—*Board of Directors and Director of Operations—BST Corporation, LaGrange, NY.* Mr. Kraycir has a BS in Engineering, an MS in

Industrial Administration and is a licensed Professional Engineer. Currently, at BST Corporation he specializes in precision plastic injection molding. Formerly, he was Program Manager for IBM Microelectronics in their East Fishkill, NY, MultiLayer Ceramic facility. His group was responsible for designing, tooling, implementing, and qualifying the production facility for the 90 mm-33 layer alumina-based ceramic substrates and later the three-level thin-film wiring program for the glass ceramic package. After product qualification, his group had responsibility for the continuous improvement program required to attain the yield, cost and reliability requirements of the product. He co-chaired the team that accomplished ISO9001 registration of the IBM Microelectronics Multi-Layer Ceramic facility.

John (Jack) A. Kreuz—*DuPont Fellow; High Performance Films,—Dupont, Circleville Research Laboratory, Circleville, Ohio.* Jack has done research on polyimides for 35 years. Most of his efforts have been connected with Kapton® polyimide film, where he has contributed to improvements in monomers, polymer backbones, specifications, adhesion, processing, thermal durability, and hydrolysis resistance, as well as to understandings of imidization/cyclization chemistry. He has been a participative corporate member of the Institute of Printed Circuits (IPC) and of the Materials Research Society (MRS). He is presently a member of the American Chemical Society (ACS). His research has been documented by 14 papers, over 20 US and foreign patents, and numerous seminars.

Jack received his BS from St. Bonaventure University and his Ph.D. in organic chemistry from the University of Notre Dame.

Ananda H. Kumar—*Member of Technical Staff—Applied Materials, Santa Clara, CA.* Ananda has worked in the areas of ceramic and thin film packaging for nearly twenty years, first at IBM Microelectronics for 15 years, and later at the David Sarnoff Research Center, Princeton, NJ. Ananda has written many papers and has earned thirty patents related to many areas of packaging technology. Ananda was the recipient of the top IBM Corporate Technology Award in 1991 for his valuable contributions to the development of glass-ceramic multilayer substrates used in the Company's most advanced computers. He has a Ph.D. degree in Ceramic Engineering from University of Illinois in Urbana.

Nobuyuki Kuramoto—*General Manager—Fujisawa Research Laboratory, Tokuyama Corp., Fujisawa, Japan.* He received the master's degree of science from the Kyushu University, Japan in 1971 and joined Tokuyama Corporation. Since 1977, he has been engaged in the research and development of fine ceramics, especially aluminum nitride. He received the R.M. Fulrath Pacific Award on "Development of Translucent Aluminum Nitride Ceramics" form the American Ceramic Society in 1988, and the Prize of Technologies from the Ceramic Society of Japan in 1996 on "Development of High Purity AlN Powder and It's Related Products". He has 54 Japanese patents, 9 US patents, and 28 from other countries.

Joy Laskar—*Assistant Professor, School of Electrical and Computer Engineering—Georgia Institute of Technology, Atlanta, GA.* Prof. Laskar received the PhD degree in Electrical Engineering from the University of Illinois at Urbana-

Champaign in 1991 and is currently an assistant professor in the School of Electrical and Computer Engineering at Georgia Tech. His research interests include characterization and design techniques with applications to wireless electronics and high speed packages. He has served as a research engineer at IBM's TJ Watson Research Center, Visiting Assistant Professor at the University of Illinois, and Assistant Professor at the University of Hawaii. He serves on the technical program committee for IEEE's MTT symposia and is a co-organizer of the Advanced Heterostructure Workshop. He is a 1995 recipient of the Army Research Office's Young Investigator Award and a 1996 recipient of the National Science Foundation CAREER Award.

Richard F. Levine—*Program Manager, Memory BAT World Wide Manufacturing Program Office—IBM, Fast Fishkill, NY.* Dr. Levine has worked for IBM in various manufacturing and development positions in semiconductor, and packaging. His work has included ceramic packaging sintering and thin films manufacturing engineering, product transfer leadership as well as product engineering. He received his BS and PhD from Rutgers University in Ceramic Engineering and has presented to ISHM and ASME on thin films and product transfer. He holds two patents on semiconductor reliability.

Weiping Li—*Ph.D. student, Packaging Research Center (PRC)—Georgia Institute of Technology, Atlanta, Georgia.* Mr. Li is engaged in via formation processes in photosensitive polymers for the construction of the single level integrated module (SLIM) proposed by Packaging Research Center at Georgia Tech for the next generation electronic package. From 1990 to 1993, he was a visiting scholar in the Institute for Materials Research, GKSS-Research Center, Geesthacht, Germany working on high temperature fracture mechanics of metallic materials. From 1987 to 1990, he was an assistant researcher in the Institute of Materials Science and Engineering, Shanghai Jiao Tong University, China, working on intermetallic compounds. He received both his BS and MS in materials science from Shanghai Jiao Tong University. He is a member of ISHM—the microelectronics society.

Jerry P. Lorenzen—*owner and principal consultant—Lorenzen Consulting, Stone Ridge, NY.* He supports methods to improve quality and reduce cost for all phases of design, development, and production. Dr. Lorenzen consults in the application of Taguchi design of experiments, quality function deployment, statistical process control, basic quality tools, total quality management, and team building. He also teaches classes, seminars, and workshops on a wide range of quality topics for industry, colleges, and government.

Jerry has a Ph.D. in chemistry from Oklahoma State University. He has more than 25 years of academic and industrial quality engineering experience in electronics, chemicals, automotive, ceramics, and data processing. He has written numerous journal articles, is a contributing author to three books, and has two patents. Dr. Lorenzen is an ASQC Certified Quality Engineer.

Wadie F. Mikhail—*Senior Statistician (retired)—IBM, Charlotte, NC.* Dr. Mikhail was involved in the implementation of the design of experiments and data analysis to the manufacturing process with an emphasis on quality control. He developed models for evaluating the reliability and maintenance strategies for the first semiconductors memories with error correction used at IBM. He had worked on models for circuit and product delay, yield analysis, and physical design, where he collaborated in providing methods for estimating wiring space requirements. He was considered an expert in the area of wireability. Dr. Mikhail received his BSc (First Class Honors) and MSc in mathematics from Cairo University, Egypt, and his PhD in statistics from the University of North Carolina. He was a senior member of IEEE, and was the official IBM representative to the American Statistical Association. He has 13 journal and proceedings publications in the areas of physical design, reliability, and statistics, four patents, and 11 publications in the IBM Technical Disclosure Bulletin.

Wataru Nakayama—*Professor; Department of Mechanical & Intelligent Systems Engineering—Tokyo Institute of Technology, Japan.* Wataru Nakayama received a Doctor of Engineering degree from Tokyo Institute of Technology in 1966. From 1970 to 1989 he was with Mechanical Engineering Research Laboratory, Hitachi, Ltd., as a heat transfer specialist. His work experience includes semiconductor packaging and cooling of computers. Since he moved to Tokyo Institute of Technology, he has been teaching thermal management of electronic equipment and modeling of flow and heat transfer processes in industrial equipment. Dr. Nakayama is Past Chairman of the Heat Transfer Society of Japan, a Fellow of ASME, a Senior Member of IEEE, and the recipient of ASME Heat Transfer Memorial Award in 1992.

Luu T. Nguyen—*Engineering Manager, Strategic Planning & Development Group—National Semiconductor, Santa Clara, CA* Dr. Nguyen is with the Package Technology Group working on various aspects of packaging reliability issues and design-for-manufacturability. He is currently in charge of external leveraging with the Government, industry consortia, and research universities. He is also the Focused Technical Advisory Board (FTAB) member of the Sematech Assembly & Packaging Thrust, and the Chair of the Packaging Sciences Technical Advisory Board of the Semiconductor Research Corp. He obtained his Ph.D. in Mechanical Engineering from MIT on a Hertz Foundation Fellowship, and has worked for IBM Research and Philips Research. He has co-edited 2 books on packaging, and has 10 patents, over 15 invention disclosures, and over 100 publications.

Koichi Niwa—*Board Director—Fujitsu Laboratories Ltd., Kanagawa, Japan.* Dr. Niwa received his BS in physics from Chiba University, Japan in 1964. and his Ph.D. in ceramic engineering from Tokyo Institute Technology in 1988.

He has pioneered the development of glass/ceramic composites which combine a low sintering temperature, a low dielectric constant and a low thermal expansion coefficient. He has contributed to the technology of copper co-firing with glass/ceramic sheets. One of the contributions is the organic binder which can be burned out completely in non-oxidizing atmospheres. He discovered a special firing procedure for multilayer ceramic circuit board with copper conductors.

He has more than 50 published papers and more than 10 US patents. He received 1992 Ohkouchi Technical Award for the development of multilayer glass/ceramic circuit board for super-high-density mounting, and Technical Award from Japanese Ceramic Society for the development of the ceramic circuit board for high speed computer, in 1995. He is a member of the Electronics Division in ACerS. He has been the Fellow of ACerS since 1991. He is a Vice President of Ceramic Society of Japan.

Sevgin Oktay—*founder and president—Oktay Enterprises International, Poughkeepsie, NY.* Mr. Oktay is currently involved in projects ranging from forming business partnerships abroad to preparing patent applications for clients in computer technology and semiconductor manufacturing. During his more than thirty year career with IBM, he has held various technical and management positions, published extensively in scientific and professional journals and books in the areas of heat transfer, microelectronics and computer systems technology. He has more than forty inventions of which ten are US Patents. He has conducted workshops and seminars on computer technology world-wide, including Japan, China and Turkey. He is a founding member and an Honorary Member of MIM, the Society of Turkish American Architects, Engineers and Scientists, Inc., and served as its president in 1994. He is a member of ASME, American Society of Mechanical Engineers, where he served as chairperson of various committees and is currently the Secretary and Treasurer in the Executive Committee of the Electrical and Electronic Packaging Division. He is also a member of the Institute of Electrical, Electronic Engineers, and an elected Fellow of ASME. He is listed in Who's Who in Engineering. Mr. Oktay received his BS in Engineering Science from Antioch College, Yellow Springs, Ohio in 1959, and MS and Professional ME in Mechanical Engineering in 1960 and 1963, respectively, from Columbia University, New York, NY.

Kanji Otsuka—*Professor of Meisei University, College of Informatices, Department of Electronics and Computer Science—Tokyo, Japan.* He graduated with a BS in ceramic engineering from Kyoto Institute of Technology in 1958, and received a doctor grade in ceramic material science from Tokyo Institute of Technology in 1987.

He was with Hitachi, Ltd for 34 years, beginning in 1935. He began his career in Ge and Si transistor manufacturing and packaging technology. From period of IC and LSI, he worked in LSI package design and its production issues that included all kind of technology. The job function shifted to the design of system packaging for mainframe computers, especially in high density/high speed packaging and MCMs for coming VLSI era. Final job function was system packaging design in RISC processors for workstation.

He transferred his job from Hitachi to Meisei University in 1993 as the professor in the class of microelectronics. He has taught microelectronics, computer design, electronics circuit and electromagnetic phenomena in bus signal transmission.

He is a board member of the Japan Institute for Interconnection and Packaging Electronic Circuit, and Society for Hybrid Microelectronics, a councilor of The Japan Federation of Engineering Society, a professional member of The Institute of Electronics, Information and Communication Engineers, and a senior member of IEEE engaged in CPMT including TC-6 committee and ECTC program committee.

Burhan Ozmat—*Senior Scientist—Harris Power R&D, Latham, NY.* Since February 1994, Dr. Ozmat has been the leader of the advanced packaging group. His current responsibilities are in advanced packaging materials, processes and cooling technologies for power electronics. He worked for Texas Instruments from February 1987 to 1994. As a Member of Technical Staff, his responsibilities at TI included doing applied research and development for the advanced packaging materials and technologies. MCMs, and multilayer interconnects for microwave and digital applications, light weight constraining core materials for SMT applications, Finite Element Analysis of vibration and thermal fatigue for the solder joints of leaded and leadless ceramic packages, thermal technology develoment for high density high performance systems were among his areas of interest. Previously he worked for IBM in East Fishkill, New York where he was responsible for doing applied research and development on the advanced thermal technologies, packaging materials and bonding for MCMs. He received his Ph.D. in mechanical engineering from Massachusetts Institute of Technology in 1984. His concentration was in the mechanical behavior of materials field. His main area of interest was in the high temperature deformation and failure mechanisms of engineering materials. Burhan has several publications and patents and is a member of ASME, ASM, Sigma Xi and IEPS.

Michael G. Pecht—*Professor and Director; CALCE Electronic Packaging Research Center (EPRC)—University of Maryland, College Park, MD.* The CALCE EPRC is sponsored by over 35 organizations, and conducts reliability research to support the development of competitive electronic products in timely manner. Dr. Pecht has a BS in Acoustics, a MS in Electrical Engineering and a MS and PhD in Engineering Mechanics from the University of Wisconsin. He is a Professional Engineer, an IEEE Fellow, an ASME Fellow and a Westinghouse Fellow. He is the chief editor of the IEEE Transactions on Reliability, an associate editor for the SAE Reliability, Maintainability and Supportability Journal, an associate editor on the International Microelectronics Journal, and on the advisory board of IEEE Spectrum and the Journal of Electronics Manufacturing. He serves on the board of advisors for various companies and consults for the U.S. government, providing expertise in strategic planning in the area of electronics packaging.

Timothy C. Reiley—*Research Staff Member—IBM Almaden Research Center, San Jose, CA.* Dr. Reiley, currently working in the area of micromechanics for data storage, has held several staff and management positions at IBM-Yorktown and IBM-Almaden in the areas of packaging, electrophotographic printing and magnetic storage. Before joining IBM, he worked at the Oak Ridge National Laboratory in experimental solid state physics associated with radiation effects on mechanical behavior. He received a Sc. B. Degree from Brown University in Materials Science Engineering, and M. S. and Ph. D. degrees from Stanford University in Materials Science and Engineering. His current research focus is on the micromechanical miniaturization and integration of disk drive components.

He has co-authored over 50 external publications and has 20 patents awarded or pending.

Ottmar Rohde—*head of the laboratory for research and development of photosensitive polymides (deceased)—Ciba-Geigy, Basel, Switzerland.* Dr. Rohde passed away in December of 1994. He was working on updating the section on preimidized photosensitive polyimides in the polymer chapter at the time. His material was passed on to Dr. Richard Hopla for completion. Dr. Rohde has been missed, but his work continues to show through his contribution to this revision. He received his MS. degree in Chemistry from the University of Oregon and his Ph.D. from the Institute of Polymer Chemistry at the University of Freiburg in Germany. His experience included spin-labeling, photopolymer chemistry, and a visiting professorship in Polymer Chemistry in Taipei, Taiwan. He was a member of the Materiel Research Society, and is the author of several patents and publications related to polyimides and photosensitive polymers.

Kenneth Rose—*Professor; Electrical, Computer, and Systems Engineering Department—Rensselaer Polytechnic Institute, Troy, NY.* He received the BS degree in engineering physics in 1955 and the MS and Ph.D. degrees in electrical engineering in 1957 and 1961, all from the University of Illinois in Urbana-Champaign.

He joined the General Electric Research Laboratory in 1961 and Rensselaer Polytechnic Institute as an Associate Professor in 1965, becoming a Full Professor in 1971. At Rensselaer his research has included the use of superconductors for radiation detection and high performance packaging, the growth and characterization of nitrides and silicon-rich oxides, and the development of CAD tools for VLSI design. A recent interest has been the development of CAD tools for the early estimation of interconnect requirements.

He is a senior member of the IEEE, serving as co-editor of a special issue of the Proceedings of the IEEE on Thick and Thin Films for Electronic Applications in 1971, and a member of the Materials Research Society, serving as a co-chair of MRS Symposia on the role of interfaces in microelectronics processing in 1982 and 1993. He is a recipient of Rensselaer's distinguished faculty award and a founding member of Rensselaer's Center for Integrated Electronics.

Eugene J. Rymaszewski—*Research Professor in Materials Science & Engineering and Associate Director of the Center for Integrated Electronics—Rensselaer Polytechnic Institute, Troy, NY.* His current interests include teaching and research in the high data-rate packaging structures, signal and power distribution systems, thermal management and stress analysis. In 1956 he joined the IBM Research laboratory in Poughkeepsie, NY to work on leading-edge computer technologies, starting with Project STRETCH. Subsequently, as engineer and engineering manager, he had contributed to product development of semiconductor chips, their packaging and design and use of test equipment for many generations of IBM mainframes, notably IBM 7000 series, Systems 360, 370, 308X, 3033, 3090 and, lastly, 390.

From 1950 to 56 he worked in the Microwave Laboratory of C. Lorenz AG, subsidiary of ITT, Pforzheim, Germany on crystal-controlled transmitter section of a microwave TV link and on microwave stages of their 120 telephone channel microwave link. He is a Senior member (now Senior Life Member) of IRE-IEEE since 1957 and a member of Research Society of America-Sigma Xi.

Koppolu Sasidhar—*Georgia Tech, Atlanta, GA.* Mr. Sasidhar received the Btech degree in Computer Science and Engineering from the Indian Institute of Technology, Kharagpur, India, in 1993 and the MS degree in Electrical Engineering from Georgia Institute of Technology, in 1996. Presently, he is working towards his Ph.D. at Georgia Tech. His main research interests include Multi-Chip Module Testing, Parallel and Distributed Algorithms, Graph Theory and Built-In-Self-Test (BIST).

George P. Schmitt—*Senior Engineer (on leave)—IBM Microelectronics Division, Enidcott, NY.* Mr. Schmitt has been involved with the development of materials for advanced second-level packaging. He has worked for a number of corporations in thermosetting polymer systems; the last 23 years have been with IBM. His fields of interest are in the chemorheology of thermosets, chiefly laminating resins, and in photo-patternable coatings. He received a BS in chemistry from Gettysburg College. He is a member of SPE, ACS and Sigma Xi. He has six patents and as many publications, in the area of laminating resins, continuous systems for resin manufacture and photosensitive systems.

Donald P. Seraphim—*IBM Fellow (retired), Consultant—Vestal, NY.* Dr. Seraphim is a consultant in the field of Electronic Packaging and in materials applied science. Previously he headed IBM's Systems Technology Division's Materials Science and Engineering Function which developed advanced flexible and printed circuit board designs including direct-chip attach applications. The function had strong collaborations with IBM Research and Universities. He has been a member of the IBM Corporate Technology Committee reporting to the IBM Chief Scientist. He managed the design and development of IBM's first bipolar integrated circuits and first CMOS circuits. He also managed the development of PCBs and connector packaging for the 4300 and 3091 large systems. He has a masters degree in Applied Science from the University of B.C., Canada and a doctorate in engineering degree from Yale University. He has over 50 published papers and a substantial list of patents. Recently he has been applying his background in packaging to applications in flat panel displays in an entrepreneurial role.

Michael J. Sheaffer—*Director of Technical Support, Packaging Materials Group—Kulicke & Soffa Industries Inc., Willow Grove, PA.* Mr. Sheaffer is responsible for the operation of the K&S worldwide customer support network for packaging materials. This includes regional laboratories and on-site engineers that provide complete solutions to customer wire bonded packages, which includes bonding parameters, wire, tools, and material handling parts. He joined K&S in 1981 and has been actively involved with wirebonder manufacturing and wire bonding processes since 1977. He is the author of several technical papers and articles and holds patents in low looping trajectories and bonding dynamics. He received his MS degrees from Drexel University and Millersville University and his MS degree from Ball State University.

Yuzo Shimada—*Senior Manager, Material Development Center—NEC Corp. Kanagawa, Japan.* Yuzo Shimada received the M.S. degree in chemistry from Kyoto University in 1979. He joined the NEC Corporation in 1979 as inorganic chemist. He engaged in material and process development for high density packaging technology at Material Development Center and Computers Division. Interests of his research and development are in the areas of electrical materials, process technology of multilayer packaging substrate for large scale computer systems and advanced packaging and interconnect technologies for high speed VLSI. Mr. Shimada is a member of the Ceramic Society of Japan and the Institute of Electronic, Information, and Communication Engineers of Japan. He is also a member of the American Ceramic Society.

Robert E. Simons—*Consultant—Electronics Cooling Applications, Poughkeepsie, NY.* Mr. Simons' consulting firm provides thermal analysis, design, and instructional services. Prior to retiring from IBM in 1993, he was a manager in the Poughkeepsie Advanced Thermal Laboratory. He joined IBM in 1966 and throughout his career he participated in the development of new and advanced cooling techniques for computer electronics. As a co-inventor of the cooling scheme for the IBM Thermal Conduction Module (TCM), he received an IBM Outstanding Innovation Award and a Corporate Award. He is also the recipient of ten IBM Invention Achievement awards, holds 18 patents, and has published over 35 papers and book chapters on cooling electronic packages and systems. While at IBM he was elected a member of the IBM Academy of Technology. He has been an active participant and organizer of SEMI-THERM Symposia serving in the capacities of session, program, and general chairman. In recognition of his contributions to the art and science of thermal management for electronic systems, he was awarded the 1995 SEMI-THERM Symposium significant contributor award. He is a past chairman of the ASME Heat Transfer Division K-16 Committee on Heat Transfer in Electronic Equipment and is a member of the IEEE.

Pratap Singh—*President—RAMP Labs, Round Rock, TX.* RAMP Labs is a consulting service specializing in pin-in-hole and surface mount assembly manufacturing processes, packaging reliability and failure analysis.

Prior to this, he worked at IBM Endicott, NY and IBM Austin, TX where he had responsibility in the development of high density muti-layer printed circuit boards, zero insertion-force connectors, BGA sockets and stress-test reliability testing. He has 27 years experience in PCB manufacturing, SMT and PIH assembly processes, statistical process controls and electronics packaging reliability testing.

Pratap has one patent and published 21 invention disclosures in the IBM Technical Disclosure Bulletin. He has authored and co-authored 5 IBM technical reports and 10 technical papers. Pratap was awarded first and second level IBM Invention Awards and two levels of Technical Author Recognition awards.

He has chaired and co-chaired sessions at IEPS in 1982 and at NEPCON in 1993, 1994 and 1995. Pratap was also the member of NEPCON Advisory Council from 1991 to 1995 and is active in Central Texas Electronics Association for the last four years. He is also a member of IEEE since 1970. Pratap received his BS in mechanical engineering from University of Ujjain, India in 1962 and MS in industrial engineering from the University of Iowa, USA in 1969 before joining IBM at Endicott, NY.

Toshio Sudo—*Senior Research Scientist—Toshiba Corporation, Manufacturing Engineering Research Center, Yokoham, Japan.* He is currently developing the board-level electrical design methodology for high-speed digital/analog systems. Mr. Sudo joined the Research and Development Center, Toshiba Corporation, in 1975, where he was engaged in the research of high-speed GaAs packaging and high-density multichip module technology. In 1991, he moved to Semiconductor Device Engineering Laboratory, where he has been engaged in the development of high-performance packages and flip-chip technology for advanced CMOS VLSIs until March 1996. He published several articles for the evaluation of CMOS simultaneous switching noise using a test chip. His research interests include the electrical modeling and characterization of high-speed interconnections and high-performance packages. He received the B.E. and M.E. degrees in electrical engineering from Tohoku University, Sendai, Japan, in 1973 and 1975, respectively. He is a member of IEEE CPMT society, ISHM and IEICE (Institute of Electrical, Information, and Communication Engineering) in Japan.

Madhavan Swaminathan—*Manager, Design and Simulation, Packaging Research Center—Georgia Institute of Technology, Atlanta, GA.* Madhavan Swaminathan received the M.S. and Ph.D. degrees in electrical engineering from Syracuse University in 1989 and 1991, respectively. In 1990, he joined the Advanced Technology Division of the Packaging Laboratory at IBM, E.Fishkill, New York where he was involved with the design, analysis, measurement and characterization of packages for high performance systems. At IBM, he was part of a team that was instrumental in the design, development and prototyping of IBM's low-cost multilayer thin-film technology. He joined the Packaging Research Center at Georgia Tech in 1994 to pursue unique challenges arising in low cost packaging. His research interests are in package design, modeling, measurement and testing. Dr. Swaminathan has 45+ publications in refereed journals and conferences, six inventions, two issued patents and has taught several short courses in packaging.

Masami Terasawa—*General Manager, Technology Strategy Division, Corporate Semiconductor Components Group—Kyocera, Kyoto, Japan.* Dr. Terasawa was born in Gifu, Japan on March 24, 1951. He received the BS, MS, and Ph.D. degrees from Osaka University, Osaka Japan in 1973, 1976, 1979 respectively. In 1979 he joined Kyocera Corporation, Kyoto, Japan and has been engaged in research and development of advanced packages. He developed a Cu/Polyimide package for which he is regarded as a pioneer in the industry. In 1983, he received the best paper award from ISHM, IEPS in the United States. Further he developed various new ceramics—AlN, Mullite and Glass Ceramic.

Theodore (Ted) G. Tessier—*Manager, New Materials and Processes Group, Semiconductor Products Sector—Motorola Inc., Tempe, AZ.* Ted is Manager of the New Materials and Processes Group within Motorola's Semiconductor Products Sector working in the area of chip scale packaging technologies. He has a

Bachelor of Science degree in Organic Chemistry from Laurentian University, Sudbury, Canada and a MSc. degree in Applied Polymer Chemistry from the University of Ottawa. Ted has published more than 45 papers and articles in the area of high density substrate technologies and interconnection.

Paul A. Totta—*IBM Fellow—IBM, East Fishkill, NY.* Mr. Totta is currently an advisor to the General Manager of IBM's packaging business within the Microelectronics Division. Previously, and for many years, he was project manager for the development of chip and package thin film interconnections including IBM's flip chip solder bumps. He is the recipient of ISHM's Outstanding Technical Achievement Award for these activities. In addition he is an ASM Fellow and has over 25 issued US patents in his field.

Rao R. Tummala—*Petit Chair Professor and Director of PRC—Georgia Tech, Atlanta, GA.* Dr. Tummala received the BS degree in Physics, Mathematics and Chemistry from Loyola College, India, the BE degree in Metallurgical Engineering from the Indian Institute of Science, Banglore, India, the MS degree in Metallurgical Engineering from Queen's University, and the Ph.D. degree in Materials Science and Engineering from the University of Illinois. He joined the faculty at Georgia Tech in 1993 as a Pettit Chair Professor in Electronics Packaging and as Georgia State Research Scholar. He is also the Director of the Low-Cost Electronic Packaging Research Center funded by NSF (as one of its Engineering Research Centers), the state of Georgia, and US electronics industry. Prior to joining Georgia Tech, he was an IBM Fellow at the IBM Corporation, where he invented a number of major technologies for IBM's products for displaying, printing, magnetic storage and multichip packaging. He is both a fellow of IEEE and the American Ceramic Society, a member of the National Academy of Engineering, 1996 General Chair of IEEE-ECTC, and 1996 President of ISHM. He was recently named by Industry Week as one of the 50 Stars in the US, for improving US competitiveness.

He is co-editor of widely-used Microelectronics Packaging Handbook. He published 90 technical papers and holds 21 US patents and forty four other inventions. He received a number of awards including: David Sarnoff award, sustained technical achievement award from IEEE, John Wagnon's award from ISHM, Materials Engineering achievements award from ASM-I, distinguished alumni award from University of Illinois, and the Arthur Friedberg Memorial award from American Ceramic Society.

Dr. Tummala's current research interests include packaging materials (metals, ceramics, and polymers) and processes, mechanical properties of materials, thin and thick MCMs, thermal and electrical designs, and integrated passive components.

Elizabeth J. Twyford—*Sr. Member of the Technical Staff—TRW, Redondo Beach, CA.* Dr. Twyford received a BA degree in Philosophy from St. John's College in Annapolis MD, a BS in Electrical Engineering from Florida State University, and a MS and Ph.D. from Georgia Institute of Technology with support from a Kodak Fellowship (1990-92) and an NSF Fellowship (1995). At Georgia Tech, she studied under Dr. Nan Marie Jokerst, focusing on optical interconnects, hybrid integration of III-V semiconductor devices and glass waveguides, and photoelectrochemical etching of diffraction gratings on GaAs/AlGaAs. The Sigma Xi chapter of Georgia Tech awarded her the Doctoral Dissertation Award for 1995-1996. At TRW she is working on fiber-optic analog RF links using linearized optical modulators.

Puligandla Viswanadham—*member of the technology development team in the Circuit Card Assembly unit—Texas Instruments Inc., Lewisville TX.* He is currently involved in the Ball Grid Array and other technology implementation projects. Prior to joining Texas Instruments he worked at the International Business Machines Corporation in Austin TX, Endicott NY, and Rochester MN facilities. He was involved in the process development and qualification of Surface Laminar Circuitry, Assembly and Reliability of fine-pitch quad flat-packs, Thin Small-Outline Packages, and Tape Automated Bonding.

While at IBM Austin he was also site analytical laboratories manager during 1989-1990. As a member of the Materials and Process Engineering group at IBM Rochester Viswanadham was involved in corrosion studies, analytical methods development, plating and contamination control. Prior to joining IBM his research activities included high temperature chemistry and thermodynamics of binary and ternary chalcogenides, atomic absorption, slag-seed equilibria in coal fired magnetohydrodynamics energy generation, and astrophysics.

He has authored or co-authored over 65 technical publications in journals, symposium proceedings, and trade magazines. He has authored or co-authored four book chapters in the areas of microelectronic packaging, tape automated bonding, fine pitch surface mount technology and ball grid array technology. He received the first and second IBM Invention Achievement awards, an IBM Excellence award, and fourth level Technical Author Recognition award. During 1974–78 he was on the faculty of Ohio Dominican College, Columbus, Ohio as Assistant Professor and taught physics and chemistry.

Puligandla Viswanadham has a Ph.D., degree in chemistry from University of Toledo, Ohio, and an M.Sc., degree in chemistry from Saugor University, Saugor, India. He co-authored two patents and 15 invention disclosures.

B.D. Washo—*Member Technical Staff, Technology—AMP Incorporated, Harrisburg, PA.* Formerly he was with the IBM Development Laboratory for 27½ years. His speciality was cable/connector engineering. Dr. Washo works in connector mechanics and teaches "The Electrical Engineering of Signal Connectors", an internal course at AMP Incorporated. He has a BS in Chemical Engineering, Pennsylvania State University, and a Ph.D. in Chemical Engineering/Polymer Science from Rensselaer Polytechnic Institute. He is a member of the American Society of Metals, the Society of Plastics Engineers, and Tau Beta Pi and Sigma Xi honorary scientific societies. He has published 18 papers in the areas of polymer science, electrocoating, plasma polymerized thin films, spin coatings, and the electrical modeling and characterization of cable/connector packages. He has 37 patents publications and holds 5 issued US patents.

Thomas A. Wassick—*Senior Engineer—IBM, East Fishkill, NY.* Mr. Wassick is currently involved in the development of thin film technology for advanced high-performance computer packaging. He is responsible for thin film repair technologies and methodologies and electrical test diagnostics for IBMs thin film packaging programs. His field of expertise is in laser-based processing for deposition and etching of microelectronic materials. He received his BS degree in Biomedical Engineering from Rensselaer Polytechnic Institute and an MS degree in Materials Engineering, also from RPI. He is a member IEEE, ISHM, SPIE, and OSA, He holds a number of patents in the laser processing area, has authored several technical articles involving MCM-D repair technologies and is the co-author of a book chapter on repair and rework of microelectronics.

Toshihiko Watari—*Director and General Manager—PPC Engineering and Production Division, NEC-Niigata, Kashiwazaki City, Niigata, Japan.* Mr. Watari is the director and general manager of PPC Engineering and Production Division, NEC- Niigata. He joined NEC in Tokyo in 1967. Since 1976 to 1994 he was responsible to develop the circuit and packaging technologies for high performance mainframes and Supercomputers. He is a member of IEEE and IEICE Society of Japan. He is a graduate of Hiroshima University, Hiroshima Japan, and he received a BE degree of Electrical Engineering. He has written many articles and presented many papers in packaging technology development. He holds various patents in the areas of computer circuit and packaging design. He is now managing the development and production of Electro-Photography for printers.

Scott Wills—*Assistant Professor; Electrical and Computer Engineering—Georgia Institute Of Technology, Atlanta, GA.* He received his BS in Physics from Georgia Tech in 1983, and his SM, EE, and ScD degrees in Electrical Engineering & Computer Science from MIT in 1985, 1987, and 1990, respectively. His research interests include High Throughput Parallel Architectures, System Integration, Interconnection Networks, and VLSI. He is a member of the IEEE and the Computer Society.

C.P. Wong—*Professor of Materials Science and Engineering and Assistant Director of the Packaging Research Center—Georgia Institute of Technology, Atlanta, GA.* Dr. Wong received the BS. degree in chemistry from Purdue University, and the Ph.D. degree in organic/inorganic chemistry from the Pennsylvania State University. After his doctoral study, he was awarded two years as a postdoctoral scholar with Nobel Laureate Professor Henry Taube at Stanford University.

He joined AT&T Bell Laboratories in January 1977 as a member of the technical staff. He was appointed an AT&T Bell Laboratories Fellow in 1992. In January, 1996 he joined the Georgia Institute of Technology. He is responsible for the Packaging Research Center's Assembly, Reliability, and Thermal Management Areas. His research interests lie in the fields of polymeric materials, high Tc ceramics, materials reaction mechanism, IC encapsulation, in particular, hermetic equivalent plastic packaging, and electronic manufacturing packaging processes. He holds over 38 US patents, numerous international patents, has published over 90 technical papers and 100 key-notes and presentations in the related area.

Dr. Wong is both a Fellow of the IEEE and AT&T Bell Labs. He was president of the IEEE-CPMT Society (1992, 1993). He currently chairs the IEEE Technical Activities Board, Steering Committee on Design and Manufacturing Engineering (1995–).

INDEX

A

B

C

D

F

G

H

I

J

M

N

O

P

Q

R

S

T

U

V

W

X

Y

Z